Setting Yourself up for...
SUCCESS

MAKE THE MOST OF CLASS

1. Attend class. Regular attendance is a key to success.
2. Make sure you have a book and a notebook for homework and notes.
3. Take notes in class. At a minimum, write down what your instructor writes on the board.
4. If you have tried to solve a problem, but need help, ask a question in class when it is appropriate. Many times other students have the same question.
5. Get to know someone in class. Learning is enhanced when you can talk to someone about math.
6. Did you miss something in class? The **Video Lectures on CD or DVD*** offer a short lecture for each section of your text and are taught by math instructors. These videos will help you review material you may have missed in class or found confusing.

HELP IS HERE WHENEVER YOU WANT IT!

1. Find out your instructor's office number and office hours.
2. Ask if there is a tutor room on campus and find out the hours when tutors are available.
3. Before going to office hours or the tutor room, organize your questions. Be sure you try a problem before you ask your instructor or tutor to help you.
4. Don't know why you got the wrong answer? The **Student's Solutions Manual*** contains complete, worked-out solutions (not just answers) to every odd-numbered exercise problem and to ALL the review and test problems in the book.
5. Need some more practice? The **Worksheets for Classroom or Lab Practice*** include extra practice problems for every section of the book.

MASTERING THE EXAM

1. Try the **Pass the Test CD**, found in the front of your book. It contains video clips of an instructor working through the complete solution for all exercises from the chapter tests in your book. **Pass the Test** also includes vocabulary flashcards, tips for time management, and a Spanish glossary.
2. Find a consistent time and place to study every day rather than "pulling an all-nighter" before the exam.
3. Keep all your homework and notes in a notebook. Reading your notes and going through your homework assignments will make reviewing for the exam easier.
4. Find out from your instructor which topics will be covered on the exam.
5. When you think you are ready for the exam, pick a few problems from your assignments *at random*. If you can work them, without any help, you are well on your way to being ready.
6. Carefully show your work whenever possible on your exam.
7. Relax and visualize doing well.

You can SUCCEED in Math!

These items and others can be purchased at www.mypearsonstore.com

Beginning and Intermediate Algebra

with Applications and Visualization

Second Edition

GARY K. ROCKSWOLD

Minnesota State University, Mankato

TERRY A. KRIEGER

Rochester Community and Technical College

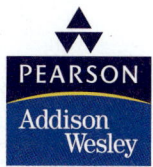

PEARSON

Addison
Wesley

Boston San Francisco New York
London Toronto Sydney Tokyo Singapore Madrid
Mexico City Munich Paris Cape Town Hong Kong Montreal

Editorial Director	Christine Hoag
Acquisitions Editor	Randy Welch
Executive Project Manager	Kari Heen
Senior Project Editor	Lauren Morse
Assistant Editor	Antonio Arvelo
Senior Managing Editor	Karen Wernholm
Senior Production Supervisor	Kathleen A. Manley
Cover Design	Nathaniel Koven and Barbara T. Atkinson
Photo Researcher	Beth Anderson
Supplements Production	Marianne Groth and Kayla Smith-Tarbox
Media Producers	Ceci Fleming and MiMi Yeh
Software Development	Mary Durnwald, TestGen; Tanya Farber, MathXL
Marketing Managers	Michelle Renda and Marlana Voerster
Marketing Assistant	Nathaniel Koven
Senior Author Support/Technology Specialist	Joe Vetere
Senior Prepress Supervisor	Caroline Fell
Rights and Permissions Advisor	Shannon Barbe
Manufacturing Manager	Evelyn Beaton
Senior Media Buyer	Ginny Michaud
Production Coordination	Kathy Diamond
Text Design and Composition	Nesbitt Graphics, Inc.
Illustrations	Techsetters, Inc. and Nesbitt Graphics, Inc.
Cover Photograph	Shutterstock

Photo credits appear on page P-1.

Library of Congress Cataloging-in-Publication Data
Rockswold, Gary K.
  Beginning and intermediate algebra with applications & visualization / Gary K. Rockswold,
    Terry A. Krieger. –2nd ed.
    p. cm.
  Rev. ed. of: Beginning and intermediate algebra with applications and visualization. 2005.
  ISBN 0-321-50005-9 (student)
  1. Algebra-- Textbooks. I. Krieger, Terry A. II. Rockswold, Gary K. Beginning and intermediate algebra
with applications and visualization. III. Title.
QA152.3.R595 2008
512.9–dc22   2007060829

ISBN-13: 978-0-321-50005-2   ISBN-10: 0-321-50005-9

7 8 9 10—QGT—12

In Memory of Susan Briney

CONTENTS

2   LINEAR EQUATIONS AND INEQUALITIES   89

3   GRAPHING EQUATIONS   161

4   SYSTEMS OF LINEAR EQUATIONS IN TWO VARIABLES   251

7 RATIONAL EXPRESSIONS — 433

8 INTRODUCTION TO FUNCTIONS   519

9 SYSTEMS OF LINEAR EQUATIONS   607

10 RADICAL EXPRESSIONS AND FUNCTIONS   649

PREFACE

Beginning and Intermediate Algebra with Applications and Visualization, Second Edition, more than any other conventional textbook, connects the real world to mathematics in ways that are both meaningful and motivational to students. Students using this textbook will have no shortage of realistic and convincing answers to the perennial question, "When will I ever use this?" The early introduction of graphs allows instructors to use graphs and other visualizations to marshal a host of applications of mathematical concepts to galvanize students' attention and comprehension. Real data, graphs, and tables play an important role in the course, giving meaning to the numbers and equations that students encounter. This approach increases students' interest and motivation and, consequently, the likelihood of success.

This textbook is one of three textbooks in a series that includes *Beginning Algebra with Applications and Visualization*, Second Edition, and *Intermediate Algebra with Applications and Visualization*, Third Edition.

APPROACH

We introduce mathematical concepts by moving from the concrete to the abstract, using relevant applications to support students' comprehension of abstract mathematical ideas. We have included a diverse collection of unique, up-to-date applications to buttress students' understanding of difficult concepts. Additionally, we present mathematical concepts using multiple representations, giving students a variety of perspectives for learning the same material—something especially helpful because students have various learning styles.

The primary purpose of the text, of course, is to teach mathematical concepts and skills. Building skills is an objective of every section. Problem-solving skills, which will help students succeed in more advanced mathematics courses, are given a strong emphasis. Throughout the text, problem solving is frequently combined with real-world applications to capture and maintain students' interest while reinforcing problem-solving habits. Standard mathematical definitions, theorems, symbolism, and rigor will prepare students for higher-level courses should they choose to pursue them.

CHANGES INCORPORATED INTO THIS EDITION

This edition contains important changes resulting from numerous comments and suggestions from instructors, students, and reviewers. We have made the text more visual by using more color and adding many photos. The following exciting features have been added to enhance students' learning of mathematical skills and concepts. To emphasize how mathematics can be used in everyday life, **A Look into Math** appears at the start of every section and **Real-World Connections** are highlighted throughout the text. **Now Try Exercises** appear after every example, and **Thinking Generally** concept exercises appear in most exercise sets. There are also new **Cumulative Review** exercise sets for chapters that previously did not have them, and **Chapter Tests** have been updated and are now more comprehensive.

The examples and the exercises have also been updated and expanded throughout the text:

- There are **more than 1000 new exercises**, including additional, optional graphing calculator exercises.
- **Approximately 50 new examples** have been added, many of them multi-part, with an emphasis on solving word problems, factoring, and radicals.
- **Data in applications have been updated** throughout, and many new topics have been added, including applications about iPods, global warming, and video games, giving instructors familiar with the text a fresh, rich set of practice material.

CHAPTER 1: Starting with this chapter, hundreds of applications involving data have been updated throughout the text to make it even more current for students. More examples and exercises have been added that increase student skills and understanding about fractions, absolute values, and powers of positive and negative integers.

CHAPTER 2: Throughout this text there is increased emphasis on having students distinguish between expressions and equations. The number and variety of application problems have been expanded in Section 2.3. An optional early introduction to interval notation now appears in Section 2.5.

CHAPTER 3: Interpretation of graphical and numerical data is expanded throughout this chapter. In Section 3.4, new questions and examples that ask students to interpret slope and intercepts of a graph have been added.

CHAPTER 4: Additional examples and exercises appear throughout this chapter. More emphasis has been given to identifying both the solution to a system of linear equations and the solution to a related application.

CHAPTER 5: New exercises that give students skills to simplify both exponential expressions and products of polynomials now appear.

CHAPTER 6: A new section summarizes factoring techniques and gives students practice factoring a variety of polynomials in the same section. New examples and exercises that require factoring out the greatest common factor first are also included. Additional examples of factoring by grouping and identifying prime trinomials now appear.

CHAPTER 7: A new subsection helps students distinguish between rational expressions and rational equations. Additional explanations and exercises make it easier for students to simplify complex fractions and solve rational equations. More explanation and exercises covering direct, inverse, and joint variation appear.

CHAPTER 8: Arithmetic operations on functions is now introduced in Section 8.1, and is reviewed in later chapters with other types of functions.

CHAPTER 9: Section 9.1 now includes new subsections that discuss systems of equations with no solutions or infinitely many solutions.

CHAPTER 10: There are more explanations, examples, and exercises related to radicals and rational exponents in this chapter.

CHAPTER 11: A new subsection on graphing the equation $y = ax^2 + bx + c$ has been added to this chapter.

CHAPTERS 12-14: Features, such as A Look into Math and Real-World Connections, continue to give meaning to the mathematics that students learn. Additional exercises and examples are provided to enhance a student's opportunity to learn these important concepts.

FEATURES

CHAPTER OPENERS

Each chapter opens with an application that motivates students by giving them insight into the relevance of that chapter's mathematical concepts.

CHAPTER **3**

Graphing Equations

Global warming has been in the news for years. Scientists are documenting that both the arctic and antarctic ice packs are melting. This phenomenon is an indication that global temperatures may be rising. The amount of light from the sun that Earth reflects into space is an important factor. If more light is reflected, then Earth stays cooler. Astronomer Phil Goode from the New Jersey Institute of Technology has developed a simple way to determine how reflective Earth is by measuring the amount of earthshine illuminating the moon. The area of the moon illuminated by earthshine is the darker portion of the moon that is visible but not as visible as the portion illuminated by the sun. Goode found that the percent change in earthshine (from the average) varied with the month. An excellent way to display these data is with the *line graph* shown in the accompanying figure. We discuss line graphs and other types of graphs in this chapter.

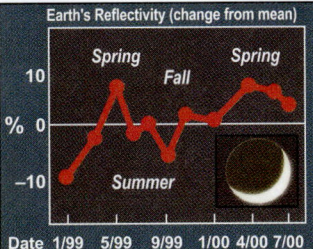

Originality is the essence of true scholarship.

—Nnamdi Azikiwe

Source: Hannah Hoag, *Discover*, November 2002, p. 11. Graphic by Matt Zang. (Reprinted with permission.)

161

MATH IN THE REAL WORLD

NEW!
A LOOK INTO MATH
Nearly every section begins in the context of the practical world that motivates the math topic which follows.

A LOOK INTO MATH ▷   In this section we discuss factoring polynomials having higher degree. Polynomials of degree 2 or higher are often used in applications. For example, the polynomial

$$0.0013x^3 - 0.085x^2 + 1.6x + 12$$

models natural gas consumption in the United States (in trillions of cubic feet) where $x = 0$ corresponds to 1960, $x = 10$ to 1970, and so on until $x = 40$ corresponds to 2000, as shown in Figure 6.8. In Exercises 75–78 we discuss further the consumption of natural gas. (*Source:* Department of Energy.)

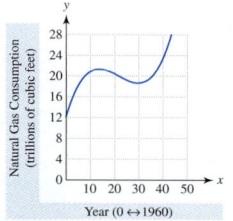

Figure 6.8 Natural Gas Consumption

▶ REAL-WORLD CONNECTION   Suppose that someone leaves a rest stop on an Interstate highway and drives home at a constant speed of 50 miles per hour. The graph in Figure 3.17 reflects the distance of the driver from home at various times. The graph intersects the *y*-axis at 200 miles, which is called the *y-intercept*. In this situation the *y*-intercept represents the initial distance (when $x = 0$) between the driver and home. The graph also intersects the *x*-axis at 4 hours, which is called the *x-intercept*. This intercept represents the elapsed time when the distance of the driver from home is 0 miles.

Figure 3.17

NEW!
REAL-WORLD CONNECTION
Where appropriate, the authors expand on specific math topics and their connections to the everyday world.

GROUP ACTIVITY
WORKING WITH REAL DATA

Directions: Form a group of 2 to 4 people. Select someone to record the group's responses for this activity. All members of the group should work cooperatively to answer the questions. If your instructor asks for the results, each member of the group should be prepared to respond.

Exercises 1–5: In this set of exercises you are to use your knowledge of equations of lines to model the average annual cost of tuition and fees.

1. *Cost of Tuition*   In 2000, the average cost of tuition and fees at *private* four-year colleges was $16,200, and in 2005 it was $20,100. Sketch a line that passes through the points (2000, 16200) and (2005, 20100). (*Source:* The College Board.)

2. *Rate of Change in Tuition*   Calculate the slope of the line in your graph. Interpret this slope as a rate of change.

3. *Modeling Tuition*   Find the slope–intercept form of the line in your sketch. What is the *y*-intercept and does it have meaning in this situation?

4. *Predicting Tuition*   Use your equation to estimate tuition and fees in 2001 and compare it to the known value of $17,300. Estimate tuition and fees in 2010.

5. *Public Tuition*   In 2000, the average cost of tuition and fees at *public* four-year colleges was $3500, and in 2005 it was $5100. Repeat Exercises 1–4 for these data. Note that the known value for 2001 is $3700. (*Source:* The College Board.)

GROUP ACTIVITIES: WORKING WITH REAL DATA
This feature occurs after select sections (1 or 2 per chapter) and provides an opportunity for students to work collaboratively on a problem that involves real-world data. Most activities can be completed with limited use of class time.

PRACTICE

NEW!
EXAMPLES WITH
NOW TRY EXERCISES
Every example directs students to practice exercises patterned after that example for immediate reinforcement of that skill.

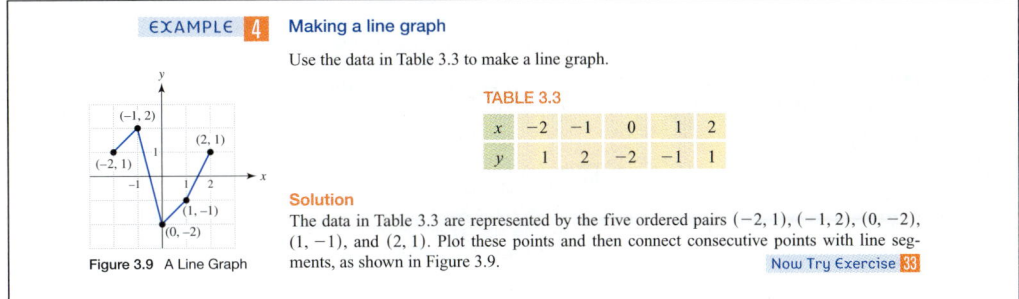

EXAMPLE 4   Making a line graph

Use the data in Table 3.3 to make a line graph.

TABLE 3.3

x	−2	−1	0	1	2
y	1	2	−2	−1	1

Solution
The data in Table 3.3 are represented by the five ordered pairs $(−2, 1)$, $(−1, 2)$, $(0, −2)$, $(1, −1)$, and $(2, 1)$. Plot these points and then connect consecutive points with line segments, as shown in Figure 3.9.

Now Try Exercise 33

Figure 3.9   A Line Graph

TECHNOLOGY NOTE: Intersection of Graphs and Table of Values

A graphing calculator can be used to find the intersection of the two graphs shown in Figure 4.3. It can also be used to create Table 4.1. The accompanying figures illustrate how a calculator can be used to determine that $y_1 = 10x$ equals $y_2 = 40$ when $x = 4$.

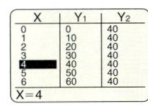

TECHNOLOGY NOTES

Occurring throughout the book, optional Technology Notes offer students guidance, suggestions, and cautions on the use of graphing calculators.

3.3   Exercises

MyMathLab   Math XP PRACTICE   WATCH   DOWNLOAD   READ   REVIEW

CONCEPTS

1. How many points determine a line?

2. The graph of the linear equation $Ax + By = C$ with $A \neq 0$ and $B \neq 0$ has _____ x-intercept(s) and _____ y-intercept(s).

3. The x-coordinate of a point where a graph intersects the x-axis is a(n) _____.

4. To find an x-intercept, let $y =$ _____ and solve for x.

5. The y-coordinate of a point at which a graph intersects the y-axis is a(n) _____.

6. To find a y-intercept, let $x =$ _____ and solve for y.

7. The graph of $y = 3$ has y-intercept _____.

8. The graph of the linear equation $Ax + By = C$ with $A = 0$ and $B = 1$ is a(n) _____ line.

9. A horizontal line with y-intercept b has equation _____.

10. The graph of $x = 3$ has x-intercept _____.

11. The graph of the linear equation $Ax + By = C$ with $A = 1$ and $B = 0$ is a(n) _____ line.

12. A vertical line with x-intercept k has equation _____.

FINDING INTERCEPTS

Exercises 13–20: Identify any x-intercepts and y-intercepts in the graph.

13.

14.

15.

16.

17.

18.

EXERCISES

The exercise sets lead from basic concepts to skill-building, writing, applications, and conceptual mastery.

PRACTICE

GRAPHING CALCULATOR EXERCISES

The icon ▦ is used to denote an exercise that requires students to use a graphing calculator.

---

82. *Dating Artifacts*   Radioactive carbon-14 is found in all living things and is used to date objects containing organic material. Suppose that an object initially contains C grams of carbon-14. After x years it will contain A grams, where

$$A = C(0.99988)^x.$$

▦ **(a)** Let $C = 10$ and graph A over a 20,000-year period. Is this function an example of exponential growth or decay?

(b) How many grams are left after 5700 years? What fraction of the carbon-14 is left?

NEW!

THINKING GENERALLY

Most exercise sets include open-ended conceptual questions encouraging students to synthesize what they've learned.

---

87. **Thinking Generally**   Students sometimes mistakenly apply the "rule" $a^m \cdot b^n \stackrel{?}{=} (ab)^{m+n}$. In general, this equation is *not true*. Find values for a, b, m, and n with $a \neq b$ and $m \neq n$ that will make this equation true.

88. **Thinking Generally**   Students sometimes mistakenly apply the "rule" $(a + b)^n \stackrel{?}{=} a^n + b^n$. In general, this equation is *not true*. Find values for a, b, and n with $a \neq b$ that will make this equation true.

CONTINUOUS REVIEW EXERCISES

After every two sections, ***Checking Basic Concepts*** presents exercises that students can use to review and tie together recently studied concepts.

CHECKING BASIC CONCEPTS
SECTIONS 4.1 AND 4.2

1. Determine graphically the x-value in each equation when $y = 2$.
   (a) $y = 1 - \frac{1}{2}x$ **(b)** $2x - 3y = 6$

2. Determine whether $(-1, 0)$ or $(4, 2)$ is a solution to the system.
$$2x - 5y = -2$$
$$3x + 2y = 16.$$

3. Solve the system of equations graphically. Check your answer.
$$x - y = 1$$
$$2x + y = 5$$

4. Use the method of substitution to solve each system of equations. How many solutions does each have?

   (a) $x + y = -1$
   $y = 2 - x$

   (b) $4x - y = 5$
   $-x + y = -2$

   (c) $x + 2y = 3$
   $-x - 2y = -3$

5. *Room Prices*   A hotel rents single and double rooms for $150 and $200, respectively. The hotel receives $55,000 for renting 300 rooms.
   (a) Let x be the number of single rooms rented and let y be the number of double rooms rented. Write a system of linear equations whose solution gives the values of x and y.
   (b) Use the method of substitution to solve the system. Check your answer.

CONCEPTS

MATH FROM MULTIPLE PERSPECTIVES
Throughout the text, concepts are presented by means of **verbal**, **graphical**, **numerical**, and **symbolic** representations to facilitate student learning.

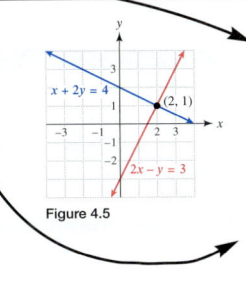

CRITICAL THINKING
One or more Critical Thinking exercises are included in most sections. They pose questions that can be used for classroom discussion or homework assignments.

Solution

Graphically   Begin by writing each equation in slope–intercept form.

$$x + 2y = 4 \quad \text{First equation} \qquad 2x - y = 3 \quad \text{Second equation}$$
$$2y = -x + 4 \quad \text{Subtract } x. \qquad -y = -2x + 3 \quad \text{Subtract } 2x.$$
$$y = -\frac{1}{2}x + 2 \quad \text{Divide by 2.} \qquad y = 2x - 3 \quad \text{Multiply by } -1.$$

The graphs of $y = -\frac{1}{2}x + 2$ and $y = 2x - 3$ are shown in Figure 4.5. Their graphs intersect at the point $(2, 1)$.

Numerically   Table 4.2 shows the equations $y = -\frac{1}{2}x + 2$ and $y = 2x - 3$ evaluated for various values of x. Note that when $x = 2$, both equations have a y-value of 1. Thus $(2, 1)$ is the solution.

Figure 4.5

CRITICAL THINKING
If the graph of a system of linear equations consists of two parallel lines, how many solutions are there?

TABLE 4.2   A Numerical Solution

x	-1	0	1	2	3
$y = -\frac{1}{2}x + 2$	2.5	2	1.5	1	0.5
$y = 2x - 3$	-5	-3	-1	1	3

Now Try Exercise 37

MAKING CONNECTIONS

The Signs in the Binomial Factors

If a trinomial of the form $x^2 + bx + c$ can be factored, the signs of the coefficients in the trinomial can be used to determine the signs in the binomial factors. If c is positive, the binomial factors must have the same signs. If c is negative, the binomial factors must have opposite signs. If b and c represent positive numbers, this can be summarized as follows.

Form of the Trinomial	*Signs in the Binomial Factors*
$x^2 + bx + c$	$(\ + \ )(\ + \ )$
$x^2 - bx + c$	$(\ - \ )(\ - \ )$
$x^2 + bx - c$	$(\ - \ )(\ + \ )$
$x^2 - bx - c$	$(\ - \ )(\ + \ )$

MAKING CONNECTIONS
This feature occurs throughout the text and helps students relate previously learned concepts to new ones, pointing out the interrelatedness of mathematical topics.

PUTTING IT ALL TOGETHER
This helpful feature appears at the end of each section to summarize techniques and reinforce the mathematical concepts presented in the section.

5.2   PUTTING IT ALL TOGETHER

In this section we discussed monomials and polynomials, including how to add, subtract, and evaluate them. The following table summarizes several important concepts related to these topics.

Concept	Explanation	Examples	
Monomial	A number, variable, or product of numbers and variables raised to natural number powers Degree is the sum of the exponents. Coefficient is the number in a monomial.	$4x^2y$ $-6x^2$ $-a^4$ x -8	Degree: 3, coefficient: 4 Degree: 2, coefficient: -6 Degree: 4, coefficient: -1 Degree: 1, coefficient: 1 Degree: 0, coefficient: -8
Polynomial	A sum of one or more monomials	$4x^2 + 8xy^2 + 3y^2$ $-9x^4 + 100$ $-3x^2y^3$	Trinomial Binomial Monomial

CHAPTER MASTERY

CHAPTER SUMMARY

For a quick and thorough review, each chapter is condensed into key terms, topics, and procedures, accompanied by illuminating examples to assist in test preparation.

CHAPTER 3 SUMMARY

SECTION 3.1 ■ INTRODUCTION TO GRAPHING

The Rectangular Coordinate System (xy-plane)

Points   Plotted as (x, y) ordered pairs

Four Quadrants The x- and y-axes divide the xy-plane into quadrants I, II, III, and IV.

NOTE: A point on an axis, such as $(1, 0)$, does not lie in a quadrant.

CHAPTER REVIEW EXERCISES

For extra practice of any topic, students can use these exercises to gain confidence that they've mastered the material.

CHAPTER 3 REVIEW EXERCISES

SECTION 3.1

1. Identify the coordinates of each point in the graph. Identify the quadrant, if any, in which each point lies.

2. Make a scatterplot by plotting the following four points: $(-2, 3), (-1, -1), (0, 3),$ and $(2, -1)$.

Exercises 3 and 4: Use the table of xy-values to make a line graph.

3.

x	-2	-1	0	1	2
y	-3	2	-1	-2	3

4.

x	-10	-5	0	5	10
y	5	-10	10	-5	0

SECTION 3.2

Exercises 5–8: Determine whether the ordered pair is a solution for the given equation.

5. $y = x - 3$ $(6, 3)$

6. $y = 5 - 2x$ $(-2, 1)$

7. $3x - y = 3$ $(-1, 6)$

8. $\frac{1}{2}x + 2y = -8$ $(-4, -3)$

EXTENDED AND DISCOVERY EXERCISES

These capstone projects at the end of every chapter challenge students to synthesize what they've learned and apply it in other college courses.

CHAPTER 3 EXTENDED AND DISCOVERY EXERCISES

Exercises 1 and 2: The table of real data can be modeled by a line. However, the line may not be an exact model, so answers may vary.

1. *Women in Politics* The table lists percentages P of women in state legislatures during year x.

x	1993	1995	1997	1999
P	20.5	20.6	21.6	22.4

x	2001	2003	2005	2007
P	22.4	22.4	22.7	23.5

x	1960	1970	1980	1990	2000
P	179	203	227	249	281

Source: U.S. Census Bureau.

(a) Make a scatterplot of the data.
(b) Find a point–slope form of a line that models these data.
(c) Use your equation to estimate the U.S. population in 2005.

CREATING GEOMETRIC SHAPES

CHAPTER TEST

Students can reduce math anxiety by using these tests as a rehearsal for the real thing.

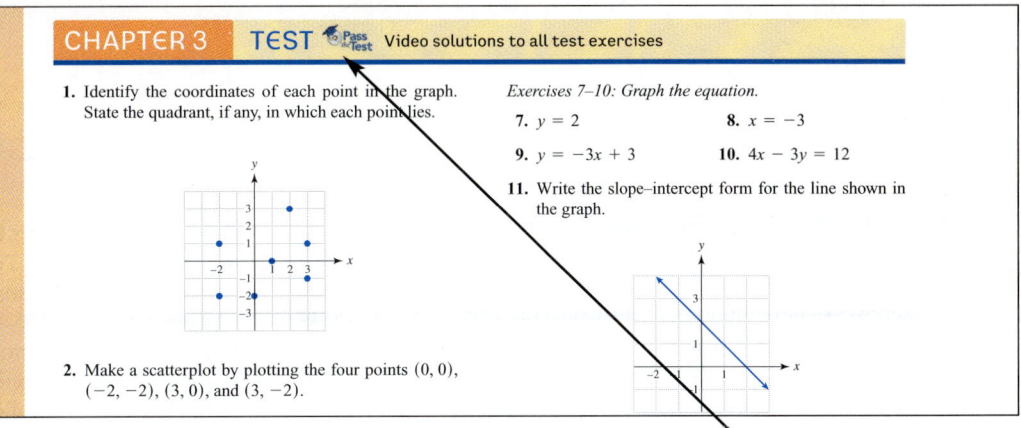

NEW!

PASS THE TEST CD

Included with every new copy of the book, the Pass The Test CD includes the following resources: video clips of an instructor working through the complete solutions for all Chapter Test exercises for each chapter; vocabulary flashcards; tips for time management; and a Spanish glossary.

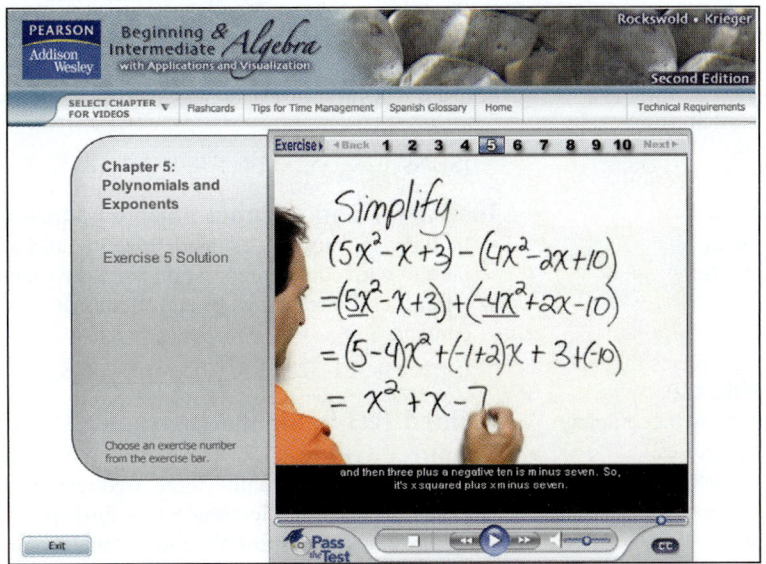

CUMULATIVE REVIEWS

Starting with Chapter 2 and appearing in all subsequent chapters in this edition, these reviews help students see the big picture of math by staying fresh with topics and skills they've already learned.

CHAPTERS 1–3   CUMULATIVE REVIEW EXERCISES

Exercises 1 and 2: Classify the number as prime or composite. If a number is composite, write it as a product of prime numbers.

1. 40

2. 61

Exercises 3 and 4: Translate the phrase into an algebraic expression using the variable n.

3. Ten more than a number

4. A number squared decreased by 2

Exercises 5–8: Evaluate by hand and then simplify to lowest terms.

5. $\frac{4}{3} \cdot \frac{21}{24}$ 6. $\frac{3}{4} \div \frac{9}{8}$

7. $\frac{2}{3} + \frac{4}{3}$ 8. $\frac{7}{10} - \frac{2}{15}$

Exercises 17 and 18: Evaluate the expression by hand.

17. $-12 \div \left(-\frac{2}{3}\right)$ 18. $-\frac{2x}{5y} \div \left(\frac{x}{10y}\right)$

Exercises 19 and 20: Simplify the expression.

19. $3 + 4x - 2 + 3x$ 20. $2(x-1) - (x+2)$

Exercises 21–24: Solve the equation. Check your solution.

21. $x + 5 = 2$ 22. $\frac{1}{3}z = 7$

23. $3t - 5 = 1$ 24. $2(x-3) = -6 - x$

25. Complete the table. Then use the table to solve the equation $6 - 2x = 4$.

x	-2	-1	0	1	2
$6 - 2x$					

STUDENT SUPPLEMENTS

Student's Solutions Manual
- Contains solutions for the odd-numbered section-level exercises (excluding Writing About Mathematics and Group Activity exercises), and solutions to all Concepts exercises, Checking Basic Concepts exercises, Chapter Review Exercises, Chapter Test exercises, and Cumulative Review Exercises.

  ISBNs: 0-321-52359-8, 978-0-321-52359-4

Video Lectures on CD or DVD
- Complete set of digitized videos on CD-ROM or DVD for student use at home or on campus.
- Presents a series of lectures correlated directly to the content of each section of the text.
- Features an engaging team of instructors who present material in a format that stresses student interaction, often using examples and exercises from the text.
- Ideal for distance learning and supplemental instruction.
- Video lectures include optional English captioning.

  CD ISBNs: 0-321-52363-6, 978-0-321-52363-1
  DVD ISBNs: 0-321-53460-3, 978-0-321-53460-6

Math XL Tutorials on CD
- Provides algorithmically generated practice exercises that correlate at the objective level to the content of the text.
- Includes an example and a guided solution to accompany every exercise, and video clips for selected exercises.
- Recognizes student errors and provides appropriate feedback; generates printed summaries of students' progress.

  ISBNs: 0-321-52364-4, 978-0-321-52364-8

Pass the Test: Chapter Test Solutions on CD
- Contains video clips of an instructor working through the complete solution for all Chapter Test exercises.
- Additional resources include tips for time management, vocabulary flashcards, and a Spanish glossary.

  Automatically included in every new copy of the textbook.

Pearson Tutor Center
The Pearson Tutor Center is staffed by qualified mathematics instructors who provide students with tutoring on examples and odd-numbered exercises from the textbook. Tutoring is available via toll-free telephone, toll-free fax, e-mail, and the Internet. White Board technology allows tutors and students to actually see problems worked while they message in real time over the Internet during tutoring sessions.

INSTRUCTOR SUPPLEMENTS

Annotated Instructor's Edition
- Contains Teaching Tips and provides answers to every exercise in the textbook excluding the Writing About Mathematics exercises.
- Answers that do not fit on the same page as the exercises themselves are supplied in the Instructor Answer Appendix at the back of the textbook.

  ISBNs: 0-321-51331-2, 978-0-321-51331-1

Instructor's Solutions Manual
- Provides solutions to all section-level exercises (excluding Writing About Mathematics exercises), and solutions to all Checking Basic Concepts exercises, Chapter Review Exercises, Chapter Test exercises, and Cumulative Review Exercises.

  ISBNs: 0-321-52332-6, 978-0-321-52332-7

Instructor and Adjunct Support Manual
- Includes resources designed to help both new and adjunct faculty with course preparation and classroom management.
- Offers helpful teaching tips and additional exercises for selected content.

  ISBNs: 0-321-52333-4, 978-0-321-52333-4

Printed Test Bank and Instructor's Resource Guide
- The Test Bank contains three free-response test forms per chapter, one multiple-choice test form per chapter, and one free-response and multiple-choice final exam.
- The Resource Guide contains five sets of Cumulative Review Exercises that cover Chapters 1–3, 1–6, 1–9, 1–12, and 1–14, and notes for presenting graphing calculator topics, as well as supplemental activities.

  ISBNs: 0-321-52358-X, 978-0-321-52358-7

TestGen
- Enables instructors to build, edit, print, and administer tests.
- Features a computerized bank of questions developed to cover all text objectives.
- Creates multiple but equivalent versions of the same question or test with the click of a button.
- Instructors can modify questions or add new questions.
- Tests can be printed or administered online.

The software and testbank are available for download from Pearson Education's online catalog.

Worksheets for Classroom or Lab Practice

- Extra practice exercises for every section of the text with ample space for students to show their work.
- These lab- and classroom-friendly workbooks also list the key topics and vocabulary terms for every text section, along with vocabulary practice problems.

ISBNs: 0-321-53967-2, 978-0-321-53967-0

Pearson Adjunct Support Center

The Pearson Adjunct Support Center is staffed by qualified mathematics instructors with more than 50 years of combined experience at both the community college and university level. Assistance is provided for faculty in the following areas:

- Suggested syllabus consultation
- Tips on using materials packed with your book
- Book-specific content assistance
- Teaching suggestions including advice on classroom strategies

MathXL® www.mathxl.com

MathXL is a powerful online homework, tutorial, and assessment system that accompanies Pearson Education textbooks in mathematics and statistics. With MathXL, instructors can create, edit, and assign online homework and tests using algorithmically generated exercises correlated at the objective level to the textbook. They can also create and assign their own online exercises and import Test-Gen tests for added flexibility. All student work is tracked in MathXL's online gradebook. Students can take chapter tests in MathXL and receive personalized study plans based on their test results. The study plan diagnoses weaknesses and links students directly to tutorial exercises for the objectives they need to study and retest. Students can also access supplemental animations and video clips directly from selected exercises. MathXL is available to qualified adopters. For more information, visit our Web site at www.mathxl.com, or contact your sales representative.

MyMathLab® www.mymathlab.com

MyMathLab is a series of text-specific, easily customizable online courses for Pearson Education textbooks in mathematics and statistics. Powered by CourseCompass™ (our online teaching and learning environment) and MathXL® (our online homework, tutorial, and assessment system), MyMathLab gives instructors the tools they need to deliver all or a portion of their course online, whether students are in a lab or working at home or elsewhere. MyMathLab provides a rich and flexible set of course materials, featuring free-response exercises that are algorithmically generated for unlimited practice and mastery. Students can also use online tools, such as video lectures, animations, and a multimedia textbook, to independently improve their understanding and performance. Instructors can use MyMathLab's homework and test managers to select and assign online exercises correlated directly to the textbook, and they can create and assign their own online exercises and import TestGen tests for added flexibility. MyMathLab's online gradebook—designed specifically for mathematics and statistics—automatically tracks students' homework and test results and gives the instructor control over how to calculate final grades. Instructors can also add offline (paper-and-pencil) grades to the gradebook. MyMathLab also includes access to **Pearson's Tutor Center**, which provides students with tutoring via toll-free phone, fax, e-mail, and interactive Web sessions. MyMathLab is available to qualified adopters. For more information, visit our Web site at www.mymathlab.com, or contact your sales representative.

InterAct Math® Tutorial Web site: www.interactmath.com

Get practice and tutorial help online! This interactive tutorial Web site provides algorithmically generated practice exercises that correlate directly to the exercises in the textbook. Students can do an exercise as many times as they like with new values each time for unlimited practice and mastery. Every exercise is accompanied by an interactive guided solution that provides helpful feedback for incorrect answers, and students can also view a worked-out sample problem that steps them through an exercise similar to the one they're working on.

ACKNOWLEDGMENTS

Many individuals contributed to the development of this textbook. We thank the following reviewers, whose comments and suggestions were invaluable in preparing *Beginning and Intermediate Algebra with Applications and Visualization*, Second Edition.

Cristina Berisso, *Shasta College*
Michael Bowen, *Oxnard College*
Barbara Burke, *Hawaii Pacific University*
Linda Chan, *California State Polytechnic University*
Patrick Cross, *University of Oklahoma*
John Driscoll, *Middlesex Community College—Middletown*
Di Dwan, *Yavapai College*
Steve Howard, *Rose State College*
Dr. Philip Kaatz, *Mesalands Community College*
Frederick Lippman, *Shasta College*
Ben Mayo, *Yakima Valley Community College*
Charlotte Newsom, *Tidewater Community College—Virginia Beach Campus*
Caleb Olaleye, *South Suburban College*
Gary Parker, *Blue Mountain Community College*
Amy Petty, *South Suburban College*
Pamelyn Reed, *Cy-Fair Community College*
Matt Shelton, *McLennan Community College*
Eden Thompson, *Utah Valley State College*
Alma Wlazlinski, *McLennan Community College*
Bella Zamansky, *University of Cincinnati*
Loris Zucca, *Kingwood College*

Elina Niemelä, Bob Bohland, David Atwood, Paul Lorczak, Jessica Rockswold, and Namyong Lee deserve special credit for their help with accuracy checking. Without the excellent cooperation from the professional staff at Addison-Wesley Publishing Company, this project would have been impossible. Thanks go to Greg Tobin and Maureen O'Connor for giving their support. Particular recognition is due Randy Welch and Lauren Morse, who gave essential advice and assistance. The outstanding contributions of Kathy Manley, Joe Vetere, Michelle Renda, Marlana Voerster, Nathaniel Koven, Antonio Arvelo, MiMi Yeh, and Ceci Fleming are greatly appreciated. Special thanks go to Kathy Diamond, who was instrumental in the success of this project.

Thanks go to Wendy Rockswold and Carrie Krieger, whose unwavering encouragement and support made this project possible. We also thank the many students and instructors who used the previous edition of this textbook. Their suggestions were insightful and helpful.

Please feel free to send us your comments and questions at either of the following e-mail addresses: gary.rockswold@mnsu.edu or terry.krieger@roch.edu. Your opinions are important to us.

Gary K. Rockswold
Terry A. Krieger

CHAPTER 1

Introduction to Algebra

J ust over a century ago relatively few people were educated past the elementary years. With the industrial revolution came the need for a more educated workforce that possessed the skills necessary to work in a variety of areas, from new factory assembly lines to the rapidly expanding transportation systems. As the skills needed in the workforce increased, the number of people seeking a high school education rose dramatically.

In 1900 only about one in ten workers was in a professional, technical, or managerial occupation. By 1970, two of every ten workers had a job of this type, and today this proportion is nearly one in three.

The study of mathematics is essential for anyone who wants to keep up with the technological changes that are occurring in nearly every occupation. A person who is mathematically competent is more prepared to handle these technological changes as they occur in the workplace with increasing frequency. Mathematics is the *language of technology*.

Students who avoid mathematics classes throughout their education may find that doing so dramatically affects their vocation, income, and lifestyle. As the gap between the wages of the college educated and those with a high school diploma continues to widen, employers continue to seek employees with higher levels of mathematical skill.

It is just a matter of time before the *majority* of the workforce will need the analytic and problem-solving skills that are taught in mathematics classes every day. No matter what career path students choose, a solid background in mathematics will provide opportunities that are not available to those who lack these important skills.

Unless you try to do something beyond what you have already mastered, you will never grow.

—RONALD E. OSBORN

Source: A. Greenspan, "The Economic Importance of Improving Math-Science Education."

1.1 NUMBERS, VARIABLES, AND EXPRESSIONS

Natural Numbers and Whole Numbers ■ Prime Numbers and Composite Numbers ■ Variables, Algebraic Expressions, and Equations ■ Translating Words to Expressions

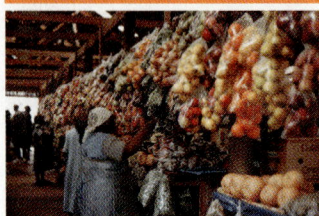

A LOOK INTO MATH ▷ Numbers are an important concept in every society. Without numbers it would be impossible to quantify (count) a group of objects. One number system used in southern Africa consisted of only the numbers from 1 to 20. Numbers larger than 20 were named by counting groups of *twenties*. For example, the number 67 was called *three twenties and seven*. This base-20 number system would not work well in today's technologically advanced world. In this section, we will begin by introducing two sets of numbers that are used extensively in the modern world: natural numbers and whole numbers.

Natural Numbers and Whole Numbers

One important set of numbers found in most societies is the set of **natural numbers**. These numbers comprise the *counting numbers* and may be expressed as follows.

$$1, 2, 3, 4, 5, 6, \ldots$$

Because there are infinitely many natural numbers, three dots are used to show that the list continues in the same pattern without end. A second set of numbers is called the **whole numbers**, and may be expressed as follows.

$$0, 1, 2, 3, 4, 5, \ldots$$

Whole numbers include the natural numbers and the number 0.

▶ **REAL-WORLD CONNECTION**   Natural numbers and whole numbers can be used when data are not broken into fractional parts. For example, Table 1.1 lists the recent and projected U.S. population of persons aged 85 and older during selected years. Note that both natural numbers and whole numbers are appropriate to describe these data because a fraction of a person is not possible.

TABLE 1.1   U.S. Population Aged 85 and Older

Year	1990	2000	2010	2020
Population	3,080,000	4,240,000	6,123,000	7,269,000

Source: U.S. Census Bureau.

CRITICAL THINKING

Give an example from everyday life of natural number or whole number use.

Prime Numbers and Composite Numbers

When two natural numbers are multiplied, the result is another natural number. For example, if the natural numbers 3 and 4 are multiplied, the result is 12, a natural number. We say that the **product** of 3 and 4 is 12 and express this product as $3 \times 4 = 12$. The numbers 3 and 4 are **factors** of 12. Note that other natural numbers also are factors of 12, such as 2 and 6 or 1 and 12.

A natural number greater than 1 that has *only* itself and 1 as natural number factors is a **prime number**. The number 7 is prime because the only natural number factors of 7 are 1 and 7. The following is a partial list of prime numbers. There are infinitely many prime numbers.

$$2, \quad 3, \quad 5, \quad 7, \quad 11, \quad 13, \quad 17, \quad 19, \quad 23, \quad 29$$

A natural number greater than 1 that is not prime is a **composite number**. The natural number 15 is a composite number because $3 \times 5 = 15$, so 15 has factors other than itself and 1. Any composite number can be written as a product of prime numbers. For example, the natural number 120 is composite because 10 and 12 are natural number factors of 120. We can use the tree diagram shown in Figure 1.1(a) to factor 120 into prime numbers. We start by showing the factors 10 and 12 below 120 and then continue downward, factoring **10** and **12**. The bottom of the tree shows

$$120 = 2 \times 5 \times 3 \times 2 \times 2.$$

The tree is not unique because 120 could be factored as 3×40 instead of 10×12, as illustrated in Figure 1.1(b). However, the **prime factorization** is *always* unique.

CRITICAL THINKING

Suppose that you draw a tree diagram for the prime factorization of a prime number. Describe what the tree will look like.

Figure 1.1 Prime Factorization of 120

Because the prime factor tree for a given composite number may not be unique, it is customary to write the prime factors in ascending order. For example,

$$120 = 2 \times 2 \times 2 \times 3 \times 5.$$

MAKING CONNECTIONS

Prime Numbers and the Internet

Have you ever ordered something on the Internet by credit card? To ensure the privacy of your credit card number, security codes are used. Large prime numbers often play an essential role in the development of these codes. One such security code is based on the fact that it is relatively easy to multiply two large prime numbers to generate a very large composite number, but it is enormously difficult to find the original two prime numbers (factors) given the large composite number.

EXAMPLE 1

Classifying numbers as prime or composite

Classify each natural number as prime or composite, if possible. If a number is composite, write it as a product of prime numbers.
(a) 31   **(b)** 1   **(c)** 35   **(d)** 200

Solution
(a) The only factors of 31 are itself and 1, so the natural number 31 is prime.
(b) The number 1 is neither prime nor composite because prime and composite numbers must be greater than 1.
(c) The number 35 is composite because 5 and 7 are factors. It can be written as a product of prime numbers as $35 = 5 \times 7$.

Figure 1.2 Prime Factorization of 200

(d) The number 200 is composite because 10 and 20 are factors. A tree diagram can be used to write 200 as a product of prime numbers, as shown in Figure 1.2. The tree diagram reveals that 200 can be factored as $200 = 2 \times 5 \times 2 \times 2 \times 5$, or it can be factored as $200 = 2 \times 2 \times 2 \times 5 \times 5$.   Now Try Exercises **17, 23, 25**

Variables, Algebraic Expressions, and Equations

There are 12 inches in 1 foot, so 5 feet equal $5 \times 12 = 60$ inches. Similarly, in 3 feet there are $3 \times 12 = 36$ inches. If we need to convert feet to inches frequently, Table 1.2 might help.

TABLE 1.2   Converting Feet to Inches

Feet	1	2	3	4	5	6	7
Inches	12	24	36	48	60	72	84

One difficulty with this table is that a needed value might not occur in it. For example, the table is not helpful in converting 11 feet to inches. We could expand Table 1.2 into Table 1.3 to include $11 \times 12 = 132$. However, expanding the table to accommodate every possible value for feet would be impossible.

TABLE 1.3   Converting Feet to Inches

Feet	1	2	3	4	5	6	7	11
Inches	12	24	36	48	60	72	84	132

Variables are often used in mathematics when tables of numbers are inadequate. A **variable** is a symbol, typically an italic letter such as x, y, z, or F used to represent an unknown quantity. In the preceding example, the number of feet to be converted to inches could be represented by the variable F. The corresponding number of inches could be represented by the variable I. The number of inches in F feet is given by the *algebraic expression* $12 \times F$. That is, to calculate the number of inches in F feet, multiply **F** by **12**. The relationship between feet F and inches I is shown by using the *equation* or *formula*

$$I = \mathbf{12} \times \mathbf{F}.$$

Sometimes a dot (·) is used to indicate multiplication because a multiplication sign ($\times$) can be confused with the variable x. Other times the multiplication sign is omitted altogether. Thus all four formulas

$$I = 12 \times F, \quad I = 12 \cdot F, \quad I = 12F, \quad \text{and} \quad I = 12(F)$$

represent the same relationship between feet and inches. For example, if we wish to find the number of inches in 10 feet, we can replace F in one of these formulas with the number 10. If we let $F = \mathbf{10}$ in the second formula,

$$I = 12 \cdot \mathbf{10} = 120.$$

That is, there are 120 inches in 10 feet.

More formally, an **algebraic expression** consists of numbers, variables, operation symbols, such as $+$, $-$, $\times$, and $\div$, and grouping symbols, such as parentheses. An **equation** is a mathematical statement that two algebraic expressions are equal. Equations *always* contain an equals sign. A **formula** is a special type of equation that expresses a relationship between two or more quantities. The formula $I = 12F$ states that to calculate the number of inches in F feet, multiply F by 12.

EXAMPLE 2   Evaluating algebraic expressions with one variable

Evaluate each algebraic expression for $x = 4$.

(a) $x + 5$ **(b)** $5x$ **(c)** $15 - x$ **(d)** $\dfrac{x}{(x - 2)}$

Solution
(a) Replace x with 4 in the expression $x + 5$ to obtain $4 + 5 = 9$.
(b) The expression $5x$ indicates multiplication of 5 and x. Thus $5x = 5 \cdot 4 = 20$.
(c) $15 - x = 15 - 4 = 11$
(d) Perform all arithmetic operations inside parentheses first.

$$\frac{x}{(x - 2)} = \frac{4}{(4 - 2)} = \frac{4}{2} = 2$$

Now Try Exercises **47, 49, 55**

▶ **REAL-WORLD CONNECTION**   Some algebraic expressions contain more than one variable. For example, if a car travels 120 miles on 6 gallons of gasoline, then the car's *mileage* is $\frac{120}{6} = 20$ miles per gallon. In general, if a car travels M miles on G gallons of gasoline, then its mileage is given by the expression $\frac{M}{G}$. Note that $\frac{M}{G}$ contains two variables, M and G, whereas $12F$ contains only one variable, F.

EXAMPLE 3   Evaluating algebraic expressions with two variables

Evaluate each algebraic expression for $y = 2$ and $z = 8$.

(a) $3yz$ **(b)** $z - y$ **(c)** $\dfrac{z}{y}$

Solution
(a) Replace y with 2 and z with 8 to obtain $3yz = 3 \cdot 2 \cdot 8 = 48$.
(b) $z - y = 8 - 2 = 6$
(c) $\dfrac{z}{y} = \dfrac{8}{2} = 4$

Now Try Exercises **57, 59**

EXAMPLE 4   Evaluating formulas

Find the value of y for $x = 15$ and $z = 10$.

(a) $y = x - 3$ **(b)** $y = \dfrac{x}{5}$ **(c)** $y = 8xz$

Solution
(a) Substitute 15 for x and then evaluate the right side of the formula to find y.

$$\begin{aligned} y &= x - 3 &&\text{Given formula} \\ &= 15 - 3 &&\text{Replace } x \text{ with 15.} \\ &= 12 &&\text{Subtract.} \end{aligned}$$

(b) $y = \dfrac{x}{5} = \dfrac{15}{5} = 3$

(c) $y = 8xz = 8 \cdot 15 \cdot 10 = 1200$

Now Try Exercises **63, 73**

Translating Words to Expressions

Many times in mathematics, algebraic expressions are not given; rather, we must write our own expressions. To accomplish this task, we often translate words to symbols. The symbols $+$, $-$, $\times$, and $\div$ have special mathematical words associated with them. When two numbers are added, the result is called the *sum*. When one number is subtracted from another number, the result is called the *difference*. Similarly, multiplying two numbers results in a *product* and dividing two numbers results in a *quotient*. Table 1.4 illustrates many of the words commonly associated with these operations.

TABLE 1.4   Words Associated with Arithmetic Symbols

Symbol	Associated Words
$+$	add, plus, more, sum, total
$-$	subtract, minus, less, difference, fewer
$\times$	multiply, times, twice, double, triple, product
$\div$	divide, divided by, quotient

EXAMPLE 5   Translating words to expressions

Translate each phrase to an algebraic expression. Specify what each variable represents.
(a) Twice the cost of a CD
(b) Four more than a number
(c) Ten less than the president's age
(d) A number plus 10, all divided by a different number
(e) The product of 6 and a number
(f) The sum of the heights of two children

Solution
(a) If the cost were $15, then twice this cost would be $2 \cdot \mathbf{15} = \$30$. In general, if we let C be the cost of a CD, then twice the cost of a CD would be $2 \cdot \mathbf{C}$, or $2C$.
(b) If the number were 20, then four more than the number would be $\mathbf{20} + 4 = 24$. If we let n represent the number, then four more than the number would be $\mathbf{n} + 4$.
(c) If the president's age were 55, then ten less than the president's age would be $\mathbf{55} - 10 = 45$. If we let A represent the president's age, then ten less would be $\mathbf{A} - 10$.
(d) Let x be the first number and y be the other number. Then the expression can be written as $(x + 10) \div y$. Note that parentheses are used around $x + 10$ because it is "all" divided by y. This expression could also be written as $\frac{x + 10}{y}$.
(e) If n represents the number, then the product of 6 and n is $6 \cdot n$ or $6n$.
(f) Let x be the height of one child and y be the height of the other child. The sum of their heights is $x + y$.
Now Try Exercises 75, 79, 83, 89

▶ **REAL-WORLD CONNECTION**   Cities are made up of large amounts of concrete and asphalt that heat up in the daytime from sunlight but do not cool off completely at night. As a result, urban areas tend to be warmer than surrounding rural areas. This effect is called the *urban heat island* and has been documented in cities throughout the world. The next example discusses the impact of this effect in Phoenix, Arizona.

EXAMPLE **Translating words to a formula**

For each year after 1950, the average nighttime temperature in Phoenix has increased by about 0.1°C.
(a) What was the increase in the nighttime temperature after 20 years, or in 1970?
(b) Write a formula (or equation) that gives the increase T in average nighttime temperature x years after 1950.
(c) Use your formula to estimate the increase in nighttime temperature in 1980.

Solution
(a) The average nighttime temperature has increased by 0.1°C per year, so after **20** years the temperature increase would be $0.1 \times \mathbf{20} = 2.0$°C.
(b) To calculate the nighttime increase in temperature multiply the number of years past 1950 by 0.1. Thus $T = 0.1x$, where x represents the number of years after 1950.
(c) Because $1980 - 1950 = 30$, let $x = \mathbf{30}$ in the formula $T = 0.1x$ to get

$$T = 0.1(\mathbf{30}) = 3.$$

The average nighttime temperature increased by 3°C.   **Now Try Exercise** 95

Another use of translating words to formulas is in finding the areas of various shapes.

EXAMPLE **Finding the area of a rectangle**

The area A of a rectangle equals its length L times its width W, as illustrated in the accompanying figure.
(a) Write a formula that shows the relationship between these three quantities.
(b) Find the area of a standard sheet of paper that is 8.5 inches wide and 11 inches long.

Solution
(a) The word *times* indicates that the length and width should be multiplied. The formula is given by $A = LW$.
(b) $A = 11 \cdot 8.5 = 93.5$ square inches   **Now Try Exercise** 101

1.1 PUTTING IT ALL TOGETHER

The following table summarizes some of the topics discussed in this section.

Concept	Comments	Examples
Natural Numbers	Sometimes referred to as the *counting numbers*	1, 2, 3, 4, 5, . . .
Whole Numbers	Includes the natural numbers and 0	0, 1, 2, 3, 4, . . .
Prime Number	A natural number greater than 1 whose only factors are itself and 1; there are infinitely many prime numbers.	2, 3, 5, 7, 11, 13, 17, and 19 are the primes less than 20

continued on next page

continued from previous page

Concept	Comments	Examples
Composite Number	A natural number greater than 1 that is *not* a prime number; there are infinitely many composite numbers.	4, 9, 25, 39, 62, 76, 87, 91, 100
Variable	Represents an unknown quantity	x, y, z, A, F, and T
Algebraic Expression	May consist of variables, numbers, operation symbols, such as $+, -, \times$, and $\div$, and grouping symbols, such as parentheses	$x + 3, \frac{x}{y}, 2z + 5, 12F,$ $x + y + z,$ $x(y + 5)$
Equation	An equation is a statement that two algebraic expressions are equal. An equation always includes an equals sign.	$2 + 3 = 5, x + 5 = 7,$ $I = 12F,$ $y = 0.1x$
Formula	A formula is a special type of equation that expresses a relationship between two or more quantities.	$I = 12F, y = 0.1x,$ $A = LW,$ $F = 3Y$

1.1 Exercises

MyMathLab

CONCEPTS

1. The natural numbers are also called the _____ numbers.

2. A whole number is either a natural number or the number _____.

3. The factors of a prime number are itself and _____.

4. If a natural number is greater than 1 and not a prime number, then it is called a(n) _____ number.

5. The number 11 is a (prime/composite) number.

6. The number 9 is a (prime/composite) number.

7. Because $3 \times 6 = 18$, the numbers 3 and 6 are _____ of 18.

8. A _____ is a special type of equation that expresses a relationship between two or more quantities.

9. A symbol or letter used to represent an unknown quantity is called a(n) _____.

10. Equations always contain a(n) _____.

11. When one number is added to another number, the result is called the _____.

12. When one number is multiplied by another number, the result is called the _____.

13. The result of dividing one number by another is called the _____.

14. The result of subtracting one number from another is called the _____.

PRIME NUMBERS AND COMPOSITE NUMBERS

Exercises 15–26: Classify the number as prime, composite, or neither. If the number is composite, write it as a product of prime numbers.

15. 4   16. 36

17. 1   18. 0

19. 29   20. 13

21. 92   22. 69

23. 225

24. 900

25. 149

26. 101

Exercises 27–38: *Write the composite number as a product of prime numbers.*

27. 6

28. 8

29. 12

30. 20

31. 32

32. 100

33. 39

34. 51

35. 294

36. 175

37. 300

38. 455

Exercises 39–46: *State whether the given quantity could accurately be described by the whole numbers.*

39. The population of a country

40. The cost of a candy bar in dollars

41. A student's grade point average

42. The Fahrenheit temperature in Antarctica

43. The number of bytes stored on a computer's hard drive

44. The exact winning time in a marathon, measured in hours

45. The number of students in a class

46. The number of bald eagles in the United States

ALGEBRAIC EXPRESSIONS, FORMULAS, AND EQUATIONS

Exercises 47–56: *Evaluate the expression for the given value of x.*

47. $2x$ $x = 5$ **48.** $x + 8$ $x = 7$

49. $8 - x$ $x = 1$ **50.** $7x$ $x = 0$

51. $\dfrac{x}{8}$ $x = 32$ **52.** $\dfrac{5}{x - 3}$ $x = 8$

53. $3(x + 1)$ $x = 5$ **54.** $7(6 - x)$ $x = 3$

55. $\left(\dfrac{x}{2}\right) + 1$ $x = 6$ **56.** $3 - \left(\dfrac{6}{x}\right)$ $x = 2$

Exercises 57–62: *Evaluate the expression for the given values of x and y.*

57. $x + y$ $x = 5, \quad y = 4$

58. $5xy$ $x = 2, \quad y = 3$

59. $4 \cdot \dfrac{x}{y}$ $x = 8, \quad y = 4$

60. $y - x$ $x = 8, \quad y = 11$

61. $y(x - 3)$ $x = 5, \quad y = 7$

62. $(x + y) - 5$ $x = 6, \quad y = 3$

Exercises 63–66: *Find the value of y for the given value of x.*

63. $y = x + 1$ $x = 0$

64. $y = x \cdot x$ $x = 5$

65. $y = 4x$ $x = 7$

66. $y = 2(x - 3)$ $x = 3$

Exercises 67–70: *Find the value of F for the given value of z.*

67. $F = z - 5$ $z = 12$

68. $F = \dfrac{z}{4}$ $z = 40$

69. $F = \dfrac{30}{z}$ $z = 6$

70. $F = z \cdot z \cdot z$ $z = 5$

Exercises 71–74: *Find the value of y for the given values of x and z.*

71. $y = x + z$ $x = 3, \quad z = 15$

72. $y = 3xz$ $x = 2, \quad z = 0$

73. $y = \dfrac{x}{z}$ $x = 9, \quad z = 3$

74. $y = x - z$ $x = 9, \quad z = 1$

TRANSLATING WORDS TO SYMBOLS

Exercises 75–90: *Translate the phrase to an algebraic expression. State what each variable represents.*

75. Three times the cost of a soda

76. Twice the cost of a gallon of gasoline

77. Five more than a number

78. Five less than a number

79. The sum of a number and 5

80. The quotient of two numbers

81. Triple a number

82. A number plus two

83. Two hundred less than the population of a town

84. The total number of dogs and cats in a city

85. A number divided by six

86. A number divided by another number

87. The product of the speed of a car and the time it has been traveling

88. The difference between 220 and a person's heart rate

89. A number plus seven, all divided by a different number

90. One-fourth of a number plus one-tenth of a different number

APPLICATIONS

91. *Yards to Feet*   Make a table of values that converts y yards to F feet. Let $y = 1, 2, 3, \ldots, 7$. Write a formula that converts y yards to F feet.

92. *Gallons to Quarts*   Make a table of values that converts g gallons to Q quarts. Let $g = 1, 2, 3, \ldots, 6$. Write a formula that converts g gallons to Q quarts.

93. *Dollars to Pennies*   Write a formula that converts D dollars to P pennies.

94. *Quarters to Nickels*   Write a formula that converts Q quarters to N nickels.

95. *NASCAR Speeds*   On the fastest speedways, some NASCAR drivers reach average speeds of 3 miles per minute. Write a formula that gives the number of miles M that such a driver would travel in x minutes. How far would this driver travel in 36 minutes?

96. *NASCAR Speeds*   On slower speedways, some NASCAR drivers reach average speeds of 2 miles per minute. Write a formula that gives the number of miles M that such a driver would travel in x minutes. How far would this driver travel in 42 minutes?

97. **Thinking Generally**   If there are 6 blims in every drog, is the formula that relates B blims and D drogs $D = 6B$ or $B = 6D$?

98. *Heart Beat*   The resting heart beat of a person is 70 beats per minute. Write a formula that gives the number of beats B that occur in x minutes. How many beats are there in an hour?

99. *Cost of a CD*   The table lists the cost C of buying x compact discs. Write an equation that relates C and x.

Number of CDs (x)	1	2	3	4
Cost (C)	$12	$24	$36	$48

100. *Gallons of Water*   The table lists the gallons G of water coming from a garden hose after m minutes. Write an equation that relates G and m.

Minutes (m)	1	2	3	4
Gallons (G)	4	8	12	16

101. *Area of a Rectangle*   The area of a rectangle equals its length times its width. Find the area of the rectangle shown in the figure.

9 ft

22 ft

102. *Area of a Square*   A square is a rectangle whose length and width have equal measures. Find the area of a square with length 14 inches.

WRITING ABOUT MATHEMATICS

103. Give an example where the whole numbers are not sufficient to describe a quantity in real life. Explain your reasoning.

104. Explain what a prime number is. How can you determine whether a number is prime?

105. Explain what a composite number is. How can you determine whether a number is composite?

106. When are variables used? Give an example.

1.2 FRACTIONS

Basic Concepts ■ Simplifying Fractions to Lowest Terms ■ Multiplication and Division of Fractions ■ Addition and Subtraction of Fractions ■ Applications

A LOOK INTO MATH ▷ Historically, natural and whole numbers have not been sufficient for most societies. Early on, the concept of splitting a quantity into parts was common, and as a result, fractions were developed. For instance, pouring a quart of milk into four cups gives rise to the concept of a fourth of a quart. In this section we discuss fractions and how to add, subtract, multiply, and divide them.

Basic Concepts

If we divide a circular pie into 6 equal slices, as shown in Figure 1.3, then each piece represents one-sixth of the pie and can be represented by the fraction $\frac{1}{6}$. Five slices of the pie would represent five-sixths of the pie and can be represented by the fraction $\frac{5}{6}$.

The parts of a fraction are named as follows.

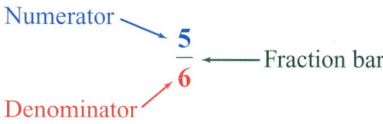

Numerator
$$\frac{5}{6}$$ ← Fraction bar
Denominator

Figure 1.3

In the fraction $\frac{13}{17}$, the numerator is 13 and the denominator is 17. Sometimes we can represent a general fraction by using variables. The fraction $\frac{a}{b}$ can represent any fraction with numerator a and denominator b. However, the value of b cannot equal 0, which is denoted $b \neq 0$. (The symbol $\neq$ means "not equal to.")

NOTE: The fraction bar represents division. For example, the fraction $\frac{1}{2}$ represents the result when 1 is divided by 2, which is 0.5. We discuss this concept further in Section 1.4.

EXAMPLE 1 Identifying numerators and denominators

Give the numerator and denominator of each fraction.

(a) $\dfrac{6}{13}$ **(b)** $\dfrac{ac}{b}$ **(c)** $\dfrac{x-5}{y+z}$

Solution
(a) The numerator is 6, and the denominator is 13.
(b) The numerator is ac, and the denominator is b.
(c) The numerator is $x - 5$, and the denominator is $y + z$.   Now Try Exercise **2**

Simplifying Fractions to Lowest Terms

Consider the amount of pizza shown in each of the three pies in Figure 1.4 on the next page. The first pie was cut into sixths and there are three pieces remaining, the second pie was cut into fourths and there are two pieces remaining, and the third pie was cut into only two pieces with one piece remaining. In all three cases half a pizza remains.

$$\frac{3}{6}\qquad\qquad\frac{2}{4}\qquad\qquad\frac{1}{2}$$

Figure 1.4

Figure 1.4 illustrates that the fractions $\frac{3}{6}$, $\frac{2}{4}$, and $\frac{1}{2}$ are equal. The fraction $\frac{1}{2}$ is in **lowest terms** because its numerator and denominator have no factors in common, whereas the fractions $\frac{3}{6}$ and $\frac{2}{4}$ are not in lowest terms. In the fraction $\frac{3}{6}$, the numerator and denominator have a common factor of 3, and in the fraction $\frac{2}{4}$, the numerator and denominator have a common factor of 2. The fraction $\frac{3}{6}$ can be simplified as follows.

$$\frac{3}{6} = \frac{1\cdot 3}{2\cdot 3}\qquad \text{Factor out 3.}$$

$$= \frac{1}{2}\qquad \frac{a\cdot c}{b\cdot c}=\frac{a}{b}$$

To simplify $\frac{3}{6}$, we used the *basic principle of fractions*: The value of a fraction is unchanged if the numerator and denominator of the fraction are multiplied (or divided) by the same nonzero number. We can also simplify the fraction $\frac{2}{4}$ to $\frac{1}{2}$ by using the basic principle of fractions.

$$\frac{2}{4} = \frac{1\cdot 2}{2\cdot 2} = \frac{1}{2}$$

When simplifying fractions, we usually factor out the *greatest common factor* (GCF) for the numerator and the denominator. For example, to simplify $\frac{27}{36}$ we first find the greatest common factor for 27 and 36. The **greatest common factor** is the largest factor that is common to both 27 and 36. Because $27 = 9 \times 3$ and $36 = 9 \times 4$, their greatest common factor is 9. Note that 3 is a factor of both 27 and 36, but it is not the *greatest* common factor. Greatest common factors can also be found by determining the largest number that divides evenly into each number. Because 9 is the largest number that divides evenly into 27 and 36, it is the greatest common factor. Thus, simplified to lowest terms, $\frac{27}{36} = \frac{3\cdot 9}{4\cdot 9} = \frac{3}{4}$.

SIMPLIFYING FRACTIONS

To simplify a fraction to lowest terms, begin by factoring out the greatest common factor c in the numerator and in the denominator. Then apply the **basic principle of fractions:**

$$\frac{a\cdot c}{b\cdot c}=\frac{a}{b}.$$

NOTE: This principle is true because multiplying a fraction by $\frac{c}{c}$ or 1 does not change the value of the fraction.

EXAMPLE 2   Finding the greatest common factor

Find the greatest common factor (GCF) for each pair of numbers.
(a) 8, 12   **(b)** 24, 60   **(c)** 495, 735

Solution

(a) Because $8 = 4 \cdot 2$ and $12 = 4 \cdot 3$, the number 4 is the largest factor that is common to both 8 and 12. Thus the GCF of 8 and 12 is 4. (Note that 2 is a *common* factor of 8 and 12 because $8 = 2 \cdot 4$ and $12 = 2 \cdot 6$, but 2 is not the *greatest* common factor.)

(b) When we are working with larger numbers, one way to determine the greatest common factor is to find the prime factorization of each number, as in

$$24 = 4 \cdot 6 = \mathbf{2} \cdot \mathbf{2} \cdot 2 \cdot \mathbf{3} \quad \text{and}$$

$$60 = 6 \cdot 10 = 2 \cdot 3 \cdot 2 \cdot 5 = \mathbf{2} \cdot \mathbf{2} \cdot \mathbf{3} \cdot 5.$$

The prime factorizations have two 2s and one 3 in common. Thus the GCF of 24 and 60 is $2 \cdot 2 \cdot 3 = 12$.

(c) Prime factorization trees for 495 and 735 are shown in Figure 1.5. Thus

$$495 = \mathbf{3} \cdot 3 \cdot \mathbf{5} \cdot 11 \quad \text{and}$$

$$735 = \mathbf{3} \cdot \mathbf{5} \cdot 7 \cdot 7.$$

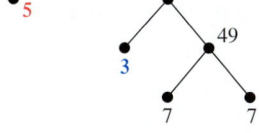

Figure 1.5 Prime Factorizations of 495 and 735

The prime factorizations have one 3 and one 5 in common. Thus the GCF of 495 and 735 is $3 \cdot 5 = 15$.

Now Try Exercises 21, 23

EXAMPLE 3   Simplifying fractions to lowest terms

Simplify each fraction to lowest terms.
(a) $\frac{8}{12}$ **(b)** $\frac{24}{60}$ **(c)** $\frac{42}{105}$

Solution

(a) From Example 2(a), the GCF of 8 and 12 is 4. Thus

$$\frac{8}{12} = \frac{2 \cdot \mathbf{4}}{3 \cdot \mathbf{4}} = \frac{2}{3}.$$

(b) From Example 2(b), the GCF of 24 and 60 is 12. Thus

$$\frac{24}{60} = \frac{2 \cdot \mathbf{12}}{5 \cdot \mathbf{12}} = \frac{2}{5}.$$

(c) The prime factorizations of 42 and 105 are

$$42 = 6 \cdot 7 = 2 \cdot \mathbf{3} \cdot \mathbf{7} \quad \text{and} \quad 105 = 5 \cdot 21 = 5 \cdot \mathbf{3} \cdot \mathbf{7}.$$

The GCF of 42 and 105 is $3 \cdot 7 = 21$. Thus

$$\frac{42}{105} = \frac{2 \cdot \mathbf{21}}{5 \cdot \mathbf{21}} = \frac{2}{5}.$$

Now Try Exercises 31, 39

MAKING CONNECTIONS

Simplifying Fractions in Steps

Sometimes a fraction can be simplified to lowest terms in multiple steps. By using *any* common factor that is not the GCF a new fraction in *lower* terms will result. This new fraction may then be simplified using a common factor of its numerator and denominator. If this process is continued, the result will be the given fraction simplified to lowest terms. The fraction in Example 3(c) could be simplified to lowest terms in two steps.

$$\frac{42}{105} = \frac{14 \cdot 3}{35 \cdot 3} = \frac{14}{35} = \frac{2 \cdot 7}{5 \cdot 7} = \frac{2}{5}$$

Multiplication and Division of Fractions

Figure 1.6

Suppose we cut *half* an apple into *thirds*, as illustrated in Figure 1.6. Then each piece represents one-sixth of the original apple. One-third of one-half is described by the product

$$\frac{1}{3} \cdot \frac{1}{2} = \frac{1}{6}.$$

NOTE: The word "of" in mathematics often indicates multiplication. For example, the phrases "one-fifth of the cookies," "twenty percent of the price," and "half of the money" all suggest that multiplication may be required.

This example demonstrates that the numerator of the product of two fractions is found by multiplying the numerators of the two fractions. Similarly, the denominator of the product of two fractions is found by multiplying the denominators of the two fractions. For example, the product of $\frac{2}{3}$ and $\frac{5}{7}$ is

$$\frac{2}{3} \cdot \frac{5}{7} = \frac{2 \cdot 5}{3 \cdot 7} = \frac{10}{21}.$$

Multiply numerators.

Multiply denominators.

MULTIPLICATION OF FRACTIONS

The product of $\frac{a}{b}$ and $\frac{c}{d}$ is given by

$$\frac{a}{b} \cdot \frac{c}{d} = \frac{ac}{bd},$$

where b and d are not 0.

EXAMPLE 4   **Multiplying fractions**

Multiply each expression and simplify the result when appropriate.

(a) $\frac{4}{5} \cdot \frac{6}{7}$ **(b)** $\frac{8}{9} \cdot \frac{3}{4}$ **(c)** $3 \cdot \frac{5}{9}$ **(d)** $\frac{x}{y} \cdot \frac{z}{3}$

Solution

(a) $\frac{4}{5} \cdot \frac{6}{7} = \frac{4 \cdot 6}{5 \cdot 7} = \frac{24}{35}$

(b) $\frac{8}{9} \cdot \frac{3}{4} = \frac{8 \cdot 3}{9 \cdot 4} = \frac{24}{36}$; the GCF of 24 and 36 is 12, so

$$\frac{24}{36} = \frac{2 \cdot \mathbf{12}}{3 \cdot \mathbf{12}} = \frac{2}{3}.$$

(c) Start by writing 3 as $\frac{3}{1}$.

$$3 \cdot \frac{5}{9} = \frac{3}{1} \cdot \frac{5}{9} = \frac{3 \cdot 5}{1 \cdot 9} = \frac{15}{9}$$

The GCF of 15 and 9 is 3, so

$$\frac{15}{9} = \frac{5 \cdot \mathbf{3}}{3 \cdot \mathbf{3}} = \frac{5}{3}.$$

(d) $\frac{x}{y} \cdot \frac{z}{3} = \frac{x \cdot z}{y \cdot 3} = \frac{xz}{3y}$

When we write the product of a variable and a number, such as $y \cdot 3$, we typically write the number first, followed by the variable. That is, $y \cdot 3 = 3y$.

Now Try Exercises 43, 51, 55

MAKING CONNECTIONS

Multiplying and Simplifying Fractions

When multiplying fractions, sometimes it is possible to change the order of the factors to rewrite the product so that it is easier to simplify. In Example 4(b) the product could be written as

$$\frac{\mathbf{8}}{\mathbf{9}} \cdot \frac{\mathbf{3}}{\mathbf{4}} = \frac{\mathbf{3} \cdot \mathbf{8}}{\mathbf{9} \cdot \mathbf{4}} = \frac{\mathbf{3}}{\mathbf{9}} \cdot \frac{\mathbf{8}}{\mathbf{4}} = \frac{\mathbf{1}}{\mathbf{3}} \cdot \frac{\mathbf{2}}{\mathbf{1}} = \frac{\mathbf{2}}{\mathbf{3}}.$$

Instead of simplifying $\frac{24}{36}$, which contains larger numbers, the fractions of $\frac{3}{9}$ and $\frac{8}{4}$ were simplified first.

EXAMPLE 5   Finding fractional parts

Find each fractional part.
(a) One-fifth of two-thirds   **(b)** Four-fifths of three-sevenths   **(c)** Three-fifths of ten

Solution

(a) The phrase "one-fifth of" indicates multiplication by one-fifth. The fractional part is

$$\frac{1}{5} \cdot \frac{2}{3} = \frac{1 \cdot 2}{5 \cdot 3} = \frac{2}{15}.$$

(b) $\frac{4}{5} \cdot \frac{3}{7} = \frac{4 \cdot 3}{5 \cdot 7} = \frac{12}{35}$

(c) $\frac{3}{5} \cdot 10 = \frac{3}{5} \cdot \frac{10}{1} = \frac{30}{5} = 6$

Now Try Exercise 59

▶ **REAL-WORLD CONNECTION**   In the next application we use fractions.

EXAMPLE 6   **Estimating college completion rates**

About four-fifths of the U.S. population over the age of 25 has a high school diploma. About one-third of those people have gone on to complete 4 or more years of college. What fraction of the U.S. population over the age of 25 has completed 4 or more years of college?

Solution
One-third of four-fifths is

$$\frac{1}{3} \cdot \frac{4}{5} = \frac{1 \cdot 4}{3 \cdot 5} = \frac{4}{15}.$$

About four-fifteenths of the U.S. population over the age of 25 has completed 4 or more years of college.   Now Try Exercise 139

TABLE 1.5   Numbers and Their Reciprocals

Number	Reciprocal
3	$\frac{1}{3}$
$\frac{1}{4}$	4
$\frac{3}{2}$	$\frac{2}{3}$
$\frac{21}{37}$	$\frac{37}{21}$

The **multiplicative inverse** or **reciprocal** of a nonzero number a is $\frac{1}{a}$. Table 1.5 lists several numbers and their reciprocals. Note that the product of a number and its reciprocal is always 1. For example, the reciprocal of 2 is $\frac{1}{2}$, and their product is $2 \cdot \frac{1}{2} = 1$.

Suppose that a group of children wants to buy gum from a gum ball machine that costs a half dollar for each gum ball. If the caregiver for the children has 4 dollars, then the number of gum balls that can be bought equals the number of half dollars that there are in 4 dollars. Thus 8 gum balls can be bought.

This calculation is given by $4 \div \frac{1}{2}$. To divide a number by a fraction, multiply the number by the reciprocal of the fraction. That is, "invert and multiply."

$$4 \div \frac{1}{2} = 4 \cdot \frac{2}{1} \qquad \text{Multiply by the reciprocal of } \frac{1}{2}.$$

$$= \frac{4}{1} \cdot \frac{2}{1} \qquad \text{Write 4 as } \frac{4}{1}.$$

$$= \frac{8}{1} \qquad \text{Multiply the fractions.}$$

$$= 8 \qquad \frac{a}{1} = a \text{ for all values of } a.$$

Justification for multiplying by the reciprocal when dividing two fractions is

$$\frac{a}{b} \div \frac{c}{d} = \frac{\dfrac{a}{b}}{\dfrac{c}{d}} = \frac{\dfrac{a}{b} \cdot \dfrac{d}{c}}{\dfrac{c}{d} \cdot \dfrac{d}{c}} = \frac{\dfrac{a}{b} \cdot \dfrac{d}{c}}{1} = \frac{a}{b} \cdot \frac{d}{c}.$$

These results are summarized as follows.

DIVISION OF FRACTIONS

For real numbers a, b, c, and d, with b, c, and d not equal to 0,

$$\frac{a}{b} \div \frac{c}{d} = \frac{a}{b} \cdot \frac{d}{c}.$$

EXAMPLE 7 **Dividing fractions**

Divide each expression.

(a) $\dfrac{1}{3} \div \dfrac{3}{5}$ **(b)** $\dfrac{4}{5} \div \dfrac{4}{5}$ **(c)** $5 \div \dfrac{10}{3}$ **(d)** $\dfrac{x}{2} \div \dfrac{y}{z}$

Solution

(a) To divide $\frac{1}{3}$ by $\frac{3}{5}$, multiply $\frac{1}{3}$ by $\frac{5}{3}$, which is the reciprocal of $\frac{3}{5}$.

$$\frac{1}{3} \div \frac{3}{5} = \frac{1}{3} \cdot \frac{5}{3} = \frac{1 \cdot 5}{3 \cdot 3} = \frac{5}{9}$$

(b) $\frac{4}{5} \div \frac{4}{5} = \frac{4}{5} \cdot \frac{5}{4} = \frac{4 \cdot 5}{5 \cdot 4} = \frac{20}{20} = 1$. Note that when we divide any nonzero number by itself, the result is 1.

(c) $5 \div \frac{10}{3} = \frac{5}{1} \cdot \frac{3}{10} = \frac{15}{10} = \frac{3}{2}$

(d) $\dfrac{x}{2} \div \dfrac{y}{z} = \dfrac{x}{2} \cdot \dfrac{z}{y} = \dfrac{xz}{2y}$

Now Try Exercises 69, 79, 85

NOTE: In Example 7(c) the answer was left as $\frac{3}{2}$. In arithmetic, $\frac{3}{2}$ is sometimes called an **improper fraction** and is often written as the mixed number $1\frac{1}{2}$. However, in algebra there is nothing "improper" about $\frac{3}{2}$, and fractions are often left as improper fractions.

TECHNOLOGY NOTE

When entering a mixed number into a calculator, it is usually easiest first to convert the mixed number to an improper fraction. For example, enter $2\frac{2}{3}$ as $\frac{8}{3}$. Otherwise, enter $2\frac{2}{3}$ as $2 + \frac{2}{3}$. See the accompanying figure.

```
8/3
              2.666666667
2+2/3
              2.666666667
```

EXAMPLE 8 **Writing a problem**

Describe a problem for which the solution could be found by dividing 5 by $\frac{1}{6}$.

Solution

If five pies are each cut into sixths, how many pieces of pie are there? (Note that one way to answer this question is to determine how many one-sixths there are in 5, which is the equivalent of multiplying 5 by 6.) See Figure 1.7.

Figure 1.7 Five Pies Cut into Sixths

Now Try Exercise 141

Addition and Subtraction of Fractions

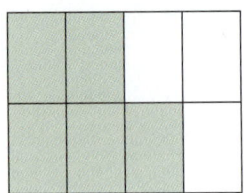

Figure 1.8

FRACTIONS WITH LIKE DENOMINATORS   Suppose that a person cuts a sheet of paper into eighths. If that person picks up two pieces and another person picks up three pieces, then together they have

$$\frac{2}{8} + \frac{3}{8} = \frac{5}{8}$$

of a sheet of paper, as illustrated in Figure 1.8. When the denominator of one fraction is the same as the denominator of a second fraction, the sum of the two fractions can be found by adding their numerators and keeping the common denominator.

Similarly, if someone picks up 5 pieces of paper and gives 2 away, then that person has

$$\frac{5}{8} - \frac{2}{8} = \frac{3}{8}$$

of a sheet of paper. To subtract two fractions with common denominators, subtract their numerators and keep the common denominator.

ADDITION AND SUBTRACTION OF FRACTIONS

To add or subtract fractions with a common denominator d, use the equations

$$\frac{a}{d} + \frac{b}{d} = \frac{a+b}{d} \quad \text{and} \quad \frac{a}{d} - \frac{b}{d} = \frac{a-b}{d},$$

where d is not 0.

EXAMPLE 9   Adding and subtracting fractions with common denominators

Add or subtract as indicated. Simplify your answer to lowest terms when appropriate.
(a) $\frac{5}{13} + \frac{12}{13}$ **(b)** $\frac{11}{8} - \frac{5}{8}$

Solution
(a) Because the fractions have a common denominator, add the numerators and keep the common denominator.

$$\frac{5}{13} + \frac{12}{13} = \frac{5+12}{13} = \frac{17}{13}$$

(b) Because the fractions have a common denominator, subtract the numerators and keep the common denominator.

$$\frac{11}{8} - \frac{5}{8} = \frac{11-5}{8} = \frac{6}{8}$$

The fraction $\frac{6}{8}$ can be simplified to $\frac{3}{4}$.   Now Try Exercise **87**

▶ **REAL-WORLD CONNECTION**   Suppose that one person mows half a large lawn while another person mows a fourth of the lawn. To determine how much they mowed together we need to find the sum $\frac{1}{2} + \frac{1}{4}$. See Figure 1.9(a). Before we can add fractions with unlike denominators, we must write each fraction with a common denominator. The least

$\frac{1}{2} + \frac{1}{4}$

(a)

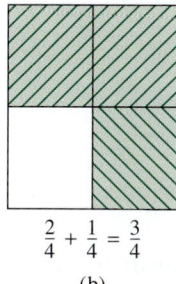

$\frac{2}{4} + \frac{1}{4} = \frac{3}{4}$

(b)

Figure 1.9

common denominator of 2 and 4 is 4. Thus we need to write $\frac{1}{2}$ as $\frac{?}{4}$ by multiplying the numerator and denominator by the *same nonzero number*.

$$\frac{1}{2} = \frac{1}{2} \cdot \frac{\mathbf{2}}{\mathbf{2}} \qquad \text{Multiply by 1.}$$

$$= \frac{2}{4} \qquad \text{Multiply fractions.}$$

Now we can find the needed sum.

$$\frac{\mathbf{1}}{\mathbf{2}} + \frac{1}{4} = \frac{\mathbf{2}}{\mathbf{4}} + \frac{1}{4} = \frac{3}{4}$$

Together the two people mow three-fourths of the lawn, as illustrated in Figure 1.9(b).

FRACTIONS WITH UNLIKE DENOMINATORS   To add or subtract fractions with unlike denominators, we first determine a common denominator. Even though *any* common denominator may be used, it is often best to find the *least* common denominator (LCD). Finding the LCD is described in Example 10. Second, each fraction must be rewritten so that it has the LCD as its denominator. This is accomplished by multiplying each fraction by 1 written in an appropriate form. Example 11 shows how to rewrite each fraction so that its denominator changes to the LCD. Finally, once the rewritten fractions have a common denominator, they may be added or subtracted, as described earlier. This entire process is illustrated in Example 12.

FINDING THE LEAST COMMON DENOMINATOR (LCD)

STEP 1: Find the prime factorization for each denominator.

STEP 2: List each factor that appears in one or more of the factorizations. If a factor is repeated in any of the factorizations, list this factor the maximum number of times that it is repeated.

STEP 3: The product of this list of factors is the LCD.

EXAMPLE 10   Finding the least common denominator

Find the LCD for each set of fractions.

(a) $\frac{5}{6}, \frac{3}{4}$ (b) $\frac{5}{12}, \frac{2}{27}$ (c) $\frac{1}{4}, \frac{2}{5}, \frac{7}{10}$

Solution

(a) **STEP 1:** For the fractions $\frac{5}{6}$ and $\frac{3}{4}$ the prime factorizations of the denominators are

$$6 = 2 \cdot 3 \text{ and } 4 = \mathbf{2} \cdot \mathbf{2}.$$

STEP 2: List the factors: $\mathbf{2}, \mathbf{2}, \mathbf{3}$. Note that, because the factor 2 appears a maximum of two times, it is listed twice.

STEP 3: The LCD is the product of this list, or $2 \cdot 2 \cdot 3 = 12$.

NOTE: Finding an LCD is equivalent to finding the smallest number that each denominator divides into evenly. Both 6 and 4 divide into 12 evenly, and 12 is the smallest such number. Thus 12 is the LCD for $\frac{5}{6}$ and $\frac{3}{4}$.

(b) STEP 1: For $\frac{5}{12}$ and $\frac{2}{27}$ the prime factorizations of the denominators are

$$12 = \mathbf{2 \cdot 2} \cdot 3 \text{ and } 27 = \mathbf{3 \cdot 3 \cdot 3}.$$

STEP 2: List the factors: $\mathbf{2, 2, 3, 3, 3}$.
STEP 3: The LCD is $2 \cdot 2 \cdot 3 \cdot 3 \cdot 3 = 108$.

(c) STEP 1: For $\frac{1}{4}, \frac{2}{5}$, and $\frac{7}{10}$ the prime factorizations of the denominators are

$$4 = \mathbf{2 \cdot 2}, 5 = \mathbf{5}, \text{ and } 10 = 2 \cdot 5.$$

STEP 2: List the factors: $\mathbf{2, 2, 5}$.
STEP 3: The LCD is $2 \cdot 2 \cdot 5 = 20$.

Now Try Exercises 93, 95, 101

EXAMPLE 11 **Rewriting fractions with the LCD**

Rewrite each set of fractions using the LCD.
(a) $\frac{5}{6}, \frac{3}{4}$ **(b)** $\frac{5}{12}, \frac{2}{27}$ **(c)** $\frac{1}{4}, \frac{2}{5}, \frac{7}{10}$

Solution
(a) From Example 10(a) the LCD is 12. To write $\frac{5}{6}$ with a denominator of 12, we multiply the fraction by 1 in the form $\frac{2}{2}$.

$$\frac{5}{6} \cdot \frac{\mathbf{2}}{\mathbf{2}} = \frac{5 \cdot \mathbf{2}}{6 \cdot \mathbf{2}} = \frac{10}{12}$$

To write $\frac{3}{4}$ with a denominator of 12, we multiply the fraction by 1 in the form $\frac{3}{3}$.

$$\frac{3}{4} \cdot \frac{\mathbf{3}}{\mathbf{3}} = \frac{3 \cdot \mathbf{3}}{4 \cdot \mathbf{3}} = \frac{9}{12}$$

Thus $\frac{5}{6}$ can be rewritten as $\frac{10}{12}$ and $\frac{3}{4}$ can be rewritten as $\frac{9}{12}$.

(b) From Example 10(b) the LCD is 108. To write $\frac{5}{12}$ with a denominator of 108, multiply the fraction by $\frac{9}{9}$. To write $\frac{2}{27}$ with a denominator of 108, multiply the fraction by $\frac{4}{4}$.

$$\frac{5}{12} \cdot \frac{9}{9} = \frac{5 \cdot 9}{12 \cdot 9} = \frac{45}{108} \quad \text{and} \quad \frac{2}{27} \cdot \frac{4}{4} = \frac{2 \cdot 4}{27 \cdot 4} = \frac{8}{108}.$$

Thus $\frac{5}{12}$ can be rewritten as $\frac{45}{108}$ and $\frac{2}{27}$ can be rewritten as $\frac{8}{108}$.

(c) From Example 10(c) the LCD is 20.

$$\frac{1}{4} \cdot \frac{5}{5} = \frac{5}{20}, \quad \frac{2}{5} \cdot \frac{4}{4} = \frac{8}{20}, \quad \text{and} \quad \frac{7}{10} \cdot \frac{2}{2} = \frac{14}{20}.$$

With an LCD of 20 the fractions are written $\frac{5}{20}, \frac{8}{20}$, and $\frac{14}{20}$, respectively.

Now Try Exercises 105, 111

EXAMPLE 12 **Adding and subtracting fractions with unlike denominators**

Add or subtract as indicated. Simplify your answer to lowest terms when appropriate.
(a) $\frac{5}{6} + \frac{3}{4}$ **(b)** $\frac{5}{12} - \frac{2}{27}$ **(c)** $\frac{1}{4} + \frac{2}{5} + \frac{7}{10}$

Solution

(a) From Example 10(a) the LCD is 12. Begin by writing each fraction with a denominator of 12 as demonstrated in Example 11(a).

$$\frac{5}{6} + \frac{3}{4} = \frac{5}{6} \cdot \frac{2}{2} + \frac{3}{4} \cdot \frac{3}{3}$$ Change to LCD of 12.

$$= \frac{10}{12} + \frac{9}{12}$$ Multiply the fractions.

$$= \frac{10 + 9}{12}$$ Add the numerators.

$$= \frac{19}{12}$$ Simplify.

(b) Using Example 10(b) and Example 11(b), we perform the following steps.

$$\frac{5}{12} - \frac{2}{27} = \frac{5}{12} \cdot \frac{9}{9} - \frac{2}{27} \cdot \frac{4}{4}$$ Change to LCD of 108.

$$= \frac{45}{108} - \frac{8}{108}$$ Multiply the fractions.

$$= \frac{45 - 8}{108}$$ Subtract the numerators.

$$= \frac{37}{108}$$ Simplify.

(c) Using Example 10(c) and Example 11(c), we perform the following steps.

$$\frac{1}{4} + \frac{2}{5} + \frac{7}{10} = \frac{1}{4} \cdot \frac{5}{5} + \frac{2}{5} \cdot \frac{4}{4} + \frac{7}{10} \cdot \frac{2}{2}$$ Change to LCD of 20.

$$= \frac{5}{20} + \frac{8}{20} + \frac{14}{20}$$ Multiply the fractions.

$$= \frac{5 + 8 + 14}{20}$$ Add the numerators.

$$= \frac{27}{20}$$ Simplify.

Now Try Exercises `113, 123`

Applications

▶ **REAL-WORLD CONNECTION**   Fractions occur in a variety of applications.

EXAMPLE `13`   ### Applying fractions to carpentry

A board measures $35\frac{3}{4}$ inches and needs to be cut into four equal parts, as depicted in Figure 1.10. Find the length of each piece.

Solution

Begin by writing $35\frac{3}{4}$ as the improper fraction $\frac{143}{4}$ ($4 \cdot 35 + 3 = 143$). As the board is to be cut into four equal parts, the length of each piece should be

$$\frac{143}{4} \div 4 = \frac{143}{4} \cdot \frac{1}{4} = \frac{143}{16}, \quad \text{or} \quad 8\frac{15}{16} \text{ inches.}$$

Now Try Exercise `133`

$35\frac{3}{4}$ in.

Figure 1.10

EXAMPLE 14   **Applying fractions to population**

According to a recent study, about $\frac{1}{20}$ of the population in the United States will serve time in prison. Within the part of the population that does serve prison time, $\frac{1}{10}$ will be female. What fraction of the population in the United States is female and will serve time in prison? (**Source:** *Bureau of Justice Statistics.*)

CRITICAL THINKING

Think of a situation in every-day life in which fractions would be needed.

Solution

Because $\frac{1}{10}$ of those persons who will serve prison time will be female, we multiply the fraction of the population that will serve prison time, $\frac{1}{20}$, by $\frac{1}{10}$.

$$\frac{1}{20} \cdot \frac{1}{10} = \frac{1}{200}$$

Thus $\frac{1}{200}$ of the population in the United States is female and will serve time in prison.

Now Try Exercise **140**

1.2   PUTTING IT ALL TOGETHER

The following table summarizes some important concepts related to fractions.

Topic	Comments	Examples
Fraction	The fraction $\frac{a}{b}$ has numerator a and denominator b.	The fraction $\frac{xy}{2}$ has numerator xy and denominator 2.
Greatest Common Factor (GCF)	The GCF of two numbers equals the largest number that divides into both evenly.	The GCF of 12 and 18 is 6 because 6 is the largest number that divides into 12 and 18 evenly.
Simplifying Fractions	Use the principle $$\frac{a \cdot c}{b \cdot c} = \frac{a}{b}$$ to simplify fractions, where c is the GCF of the numerator and denominator.	The GCF of 24 and 32 is 8, so $$\frac{24}{32} = \frac{3 \cdot 8}{4 \cdot 8} = \frac{3}{4}.$$ The GCF of 20 and 8 is 4, so $$\frac{20}{8} = \frac{5 \cdot 4}{2 \cdot 4} = \frac{5}{2}.$$
Multiplicative Inverse or Reciprocal	The reciprocal of $\frac{a}{b}$ is $\frac{b}{a}$, where a and b are not zero.	The reciprocals of 5 and $\frac{3}{4}$ are $\frac{1}{5}$ and $\frac{4}{3}$, respectively. The product of a number and its reciprocal is 1.
Multiplication and Division of Fractions	$$\frac{a}{b} \cdot \frac{c}{d} = \frac{ac}{bd}$$ $$\frac{a}{b} \div \frac{c}{d} = \frac{a}{b} \cdot \frac{d}{c}$$	$$\frac{3}{5} \cdot \frac{4}{9} = \frac{12}{45} = \frac{4}{15}$$ $$\frac{3}{2} \div \frac{6}{5} = \frac{3}{2} \cdot \frac{5}{6} = \frac{15}{12} = \frac{5}{4}$$

Topic	Comments	Examples
Addition and Subtraction of Fractions with Like Denominators	$\dfrac{a}{d} + \dfrac{c}{d} = \dfrac{a+c}{d}$ and $\dfrac{a}{d} - \dfrac{c}{d} = \dfrac{a-c}{d}$	$\dfrac{3}{5} + \dfrac{4}{5} = \dfrac{3+4}{5} = \dfrac{7}{5}$ and $\dfrac{17}{12} - \dfrac{11}{12} = \dfrac{17-11}{12} = \dfrac{6}{12} = \dfrac{1}{2}$
Least Common Denominator (LCD)	The LCD of two fractions equals the smallest number that both denominators divide into evenly.	The LCD of $\dfrac{5}{12}$ and $\dfrac{7}{18}$ is 36 because 36 is the smallest number that both 12 and 18 divide into evenly.
Addition and Subtraction of Fractions with Unlike Denominators	First write each fraction with the least common denominator. Then add or subtract the numerators.	The LCD of $\dfrac{3}{4}$ and $\dfrac{7}{10}$ is 20. $$\dfrac{3}{4} + \dfrac{7}{10} = \dfrac{3}{4} \cdot \dfrac{5}{5} + \dfrac{7}{10} \cdot \dfrac{2}{2}$$ $$= \dfrac{15}{20} + \dfrac{14}{20}$$ $$= \dfrac{29}{20}$$

1.2 Exercises

MyMathLab

PRACTICE   WATCH   DOWNLOAD   READ   REVIEW

CONCEPTS

1. A small pie is cut into 4 pieces. If someone eats 3 of the pieces, what fraction of the pie does the person eat? What fraction of the pie remains?

2. In the fraction $\dfrac{11}{21}$ the numerator is _____ and the denominator is _____.

3. In the fraction $\dfrac{a}{b}$, the variable b cannot equal _____.

4. The fraction $\dfrac{a}{a}$ with $a \neq 0$ equals _____.

5. The numerator of the product of two fractions is found by multiplying the _____ of the two fractions.

6. The denominator of the product of two fractions is found by multiplying the _____ of the two fractions.

7. The fraction $\dfrac{2}{8}$ simplifies to _____.

8. $\dfrac{ac}{bc} =$ _____

9. In the phrase "two-fifths of one-third," the word *of* indicates that we should _____ the fractions $\dfrac{2}{5}$ and $\dfrac{1}{3}$.

10. What fractional part is half of a half?

11. What is the reciprocal of a, provided $a \neq 0$?

12. To divide $\dfrac{3}{4}$ by 5, multiply $\dfrac{3}{4}$ by _____.

13. $\dfrac{a}{b} \cdot \dfrac{c}{d} =$ _____

14. $\dfrac{a}{b} \div \dfrac{c}{d} =$ _____

15. $\dfrac{a}{b} + \dfrac{c}{b} =$ _____

16. $\dfrac{a}{b} - \dfrac{c}{b} =$ _____

17. The greatest common factor of 4 and 6 is _____.

18. The least common denominator for the fractions $\dfrac{1}{4}$ and $\dfrac{5}{6}$ is _____.

19. What is the least common denominator of $\frac{2}{9}$ and 7?

20. To rewrite the fraction $\frac{3}{4}$ with denominator 24, multiply by 1 in the form_____.

LOWEST TERMS

Exercises 21–26: Find the greatest common factor.

21. 4, 12

22. 3, 27

23. 50, 75

24. 45, 105

25. 100, 60, 70

26. 36, 48, 72

Exercises 27–30: Use the basic principle of fractions to simplify the expression.

27. $\frac{3 \cdot 4}{5 \cdot 4}$

28. $\frac{2 \cdot 7}{9 \cdot 7}$

29. $\frac{3 \cdot 8}{8 \cdot 5}$

30. $\frac{7 \cdot 16}{16 \cdot 3}$

Exercises 31–42: Simplify the fraction to lowest terms.

31. $\frac{4}{8}$

32. $\frac{4}{12}$

33. $\frac{5}{15}$

34. $\frac{3}{27}$

35. $\frac{10}{25}$

36. $\frac{5}{20}$

37. $\frac{12}{36}$

38. $\frac{16}{24}$

39. $\frac{12}{30}$

40. $\frac{60}{105}$

41. $\frac{19}{76}$

42. $\frac{17}{51}$

MULTIPLICATION AND DIVISION OF FRACTIONS

Exercises 43–58: Multiply and simplify to lowest terms when appropriate.

43. $\frac{1}{2} \cdot \frac{1}{3}$

44. $\frac{2}{3} \cdot \frac{7}{5}$

45. $\frac{3}{4} \cdot \frac{1}{5}$

46. $\frac{3}{2} \cdot \frac{5}{8}$

47. $\frac{5}{3} \cdot \frac{3}{5}$

48. $\frac{21}{32} \cdot \frac{32}{21}$

49. $\frac{5}{6} \cdot \frac{18}{25}$

50. $\frac{7}{9} \cdot \frac{3}{14}$

51. $4 \cdot \frac{3}{5}$

52. $5 \cdot \frac{7}{10}$

53. $2 \cdot \frac{3}{8}$

54. $10 \cdot \frac{1}{100}$

55. $\frac{x}{y} \cdot \frac{y}{x}$

56. $\frac{x}{y} \cdot \frac{y}{z}$

57. $\frac{a}{b} \cdot \frac{3}{2}$

58. $\frac{5}{8} \cdot \frac{4x}{5y}$

Exercises 59–64: Find the fractional part.

59. One-fourth of three-fourths

60. Three-sevenths of nine-sixteenths

61. Two-thirds of six

62. Three-fourths of seven

63. One-half of two-thirds

64. Five-elevenths of nine-eighths

Exercises 65–68: Give the reciprocal of each number.

65. (a) 5   (b) 7   (c) $\frac{4}{7}$ (d) $\frac{9}{8}$

66. (a) 3   (b) 2   (c) $\frac{6}{5}$ (d) $\frac{3}{8}$

67. (a) $\frac{1}{2}$ (b) $\frac{1}{9}$ (c) $\frac{12}{101}$ (d) $\frac{31}{17}$

68. (a) $\frac{1}{5}$ (b) $\frac{7}{3}$ (c) $\frac{23}{64}$ (d) $\frac{63}{29}$

Exercises 69–86: Divide and simplify to lowest terms when appropriate.

69. $\frac{1}{2} \div \frac{1}{3}$

70. $\frac{3}{4} \div \frac{1}{5}$

71. $\frac{3}{4} \div \frac{1}{2}$

72. $\frac{6}{7} \div \frac{3}{14}$

73. $\frac{4}{3} \div \frac{1}{6}$

74. $\frac{12}{21} \div \frac{4}{7}$

75. $\frac{32}{27} \div \frac{8}{9}$

76. $\frac{8}{15} \div \frac{2}{25}$

77. $\frac{9}{1} \div \frac{8}{7}$

78. $2 \div \frac{3}{4}$

79. $10 \div \frac{5}{6}$

80. $8 \div \frac{4}{3}$

81. $\frac{9}{10} \div 3$

82. $\frac{32}{27} \div 16$

83. $\frac{a}{b} \div \frac{2}{b}$

84. $\frac{3a}{b} \div \frac{3}{c}$

85. $\frac{x}{y} \div \frac{x}{y}$

86. $\frac{x}{3y} \div \frac{x}{3}$

ADDITION AND SUBTRACTION OF FRACTIONS

Exercises 87–92: Add or subtract. Write each answer in lowest terms.

87. (a) $\frac{2}{3} + \frac{1}{3}$ (b) $\frac{2}{3} - \frac{1}{3}$

88. (a) $\frac{5}{12} + \frac{1}{12}$ (b) $\frac{5}{12} - \frac{1}{12}$

89. (a) $\frac{3}{2} + \frac{1}{2}$ (b) $\frac{3}{2} - \frac{1}{2}$

90. (a) $\frac{18}{29} + \frac{7}{29}$ (b) $\frac{18}{29} - \frac{7}{29}$

91. (a) $\frac{5}{33} + \frac{2}{33}$ (b) $\frac{5}{33} - \frac{2}{33}$

92. (a) $\frac{91}{104} + \frac{17}{104}$ (b) $\frac{91}{104} - \frac{17}{104}$

Exercises 93–104: Find the least common denominator.

93. $\frac{1}{5}, \frac{3}{10}$ 　　　　　　**94.** $\frac{5}{12}, \frac{1}{18}$

95. $\frac{4}{9}, \frac{2}{15}$ 　　　　　　**96.** $\frac{1}{11}, \frac{1}{2}$

97. $\frac{2}{5}, \frac{3}{15}$ 　　　　　　**98.** $\frac{8}{21}, \frac{3}{7}$

99. $\frac{1}{6}, \frac{5}{8}$ 　　　　　　**100.** $\frac{1}{9}, \frac{5}{12}$

101. $\frac{1}{2}, \frac{1}{3}, \frac{1}{4}$ 　　　　　**102.** $\frac{2}{5}, \frac{2}{3}, \frac{1}{6}$

103. $\frac{1}{4}, \frac{3}{8}, \frac{1}{12}$ 　　　　**104.** $\frac{2}{15}, \frac{7}{20}, \frac{1}{30}$

Exercises 105–112: Rewrite each set of fractions with the least common denominator.

105. $\frac{1}{2}, \frac{2}{3}$ 　　　　　　**106.** $\frac{3}{4}, \frac{1}{5}$

107. $\frac{7}{9}, \frac{5}{12}$ 　　　　　**108.** $\frac{5}{13}, \frac{1}{2}$

109. $\frac{1}{16}, \frac{7}{12}$ 　　　　　**110.** $\frac{5}{18}, \frac{1}{24}$

111. $\frac{1}{3}, \frac{3}{4}, \frac{5}{6}$ 　　　　**112.** $\frac{4}{15}, \frac{2}{9}, \frac{3}{5}$

Exercises 113–128: Add or subtract. Write your answer in lowest terms.

113. $\frac{1}{2} + \frac{1}{3}$ 　　　　　**114.** $\frac{2}{3} + \frac{1}{4}$

115. $\frac{5}{8} + \frac{3}{16}$ 　　　　**116.** $\frac{1}{9} + \frac{2}{15}$

117. $\frac{1}{2} - \frac{1}{4}$ 　　　　　**118.** $\frac{2}{3} - \frac{2}{9}$

119. $\frac{25}{24} - \frac{7}{8}$ 　　　　**120.** $\frac{4}{5} - \frac{1}{4}$

121. $\frac{11}{14} + \frac{2}{35}$ 　　　　**122.** $\frac{7}{8} + \frac{4}{15}$

123. $\frac{5}{12} - \frac{1}{18}$ 　　　　**124.** $\frac{9}{20} - \frac{7}{30}$

125. $\frac{3}{100} + \frac{1}{300} - \frac{1}{200}$ 　　**126.** $\frac{43}{36} + \frac{4}{9} + \frac{1}{4}$

127. $\frac{7}{8} - \frac{1}{6} + \frac{5}{12}$ 　　　**128.** $\frac{9}{40} - \frac{3}{50} - \frac{1}{100}$

APPLICATIONS

129. *American Flag*　According to Executive Order 10834, the length of an official American flag should be $1\frac{9}{10}$ times the width. If an official flag has a width of $2\frac{1}{2}$ feet, find its length.

130. *American Flag*　The blue rectangle containing the stars on an American flag is called the *union*. On an official American flag, the width of the union should be $\frac{7}{13}$ of the width of the flag. If an official flag has a width of $32\frac{1}{2}$ inches, what is the width of the union?

131. *Accidental Deaths*　For the age group 15 to 24, motor vehicle accidents account for $\frac{31}{42}$ of all accidental deaths, and firearms account for $\frac{31}{1260}$ of all accidental deaths. What fraction of all accidental deaths do vehicle accidents *and* firearms account for? (**Source:** National Safety Council.)

132. *Illicit Drug Use*　For the age group 18 to 25, the fraction of people who used illicit drugs during their lifetime was $\frac{3}{5}$, whereas the fraction who used illicit drugs during the past year was $\frac{7}{20}$. What fraction of this population has used illicit drugs but not during the past year? (**Source:** Department of Health and Human Services.)

133. *Carpentry*　A board measures $64\frac{5}{8}$ inches and needs to be cut in half. Find the length of each half.

134. *Carpentry*　A rope measures $15\frac{1}{2}$ feet and needs to be cut in four equal parts. Find the length of each piece.

135. *Geometry*　Find the area of the triangle shown with base $1\frac{2}{3}$ yards and height $\frac{3}{4}$ yard. (*Hint:* The area of a triangle equals half the product of its base and height.)

136. *Geometry*　Find the area of the rectangle shown.

137. *Distance*　Use the map to find the distance between Smalltown and Bigtown by traveling through Middletown.

138. *Distance*   An athlete jogs $1\frac{3}{8}$ miles, $5\frac{3}{4}$ miles, and $3\frac{5}{8}$ miles. In all, how far does the athlete jog?

139. *Adult Smokers*   About $\frac{1}{5}$ of the adult population in the United States smoked in 2005. Within the part of the population that smoked, $\frac{6}{25}$ of them were aged 25 to 44. What fraction of the adult population in the United States was aged 25 to 44 and smoked? (*Source:* Department of Health and Human Services.)

140. *Smokeless Tobacco Use*   About $\frac{3}{50}$ of the teenage population in the United States used smokeless tobacco in 2004. Within the part of the teenage population that used smokeless tobacco, $\frac{54}{61}$ were males. What fraction of the teenage population was male and used smokeless tobacco? (*Source:* Department of Health and Human Services.)

WRITING ABOUT MATHEMATICS

141. Describe a problem in real life for which the solution could be found by multiplying 30 by $\frac{1}{4}$.

142. Describe a problem in real life for which the solution could be found by dividing $2\frac{1}{2}$ by $\frac{1}{3}$.

CHECKING BASIC CONCEPTS

SECTIONS 1.1 AND 1.2

1. Classify each number as prime, composite, or neither. If a number is composite, write it as a product of primes.
   (a) 19   **(b)** 28   **(c)** 1   **(d)** 180

2. Evaluate $\frac{10}{(x + 2)}$ for $x = 3$.

3. Find y for $x = 5$ if $y = 6x$.

4. Translate the phrase "a number x plus five" into an algebraic expression.

5. Write a formula that converts F feet to I inches.

6. Find the greatest common factor.
   (a) 3, 18   **(b)** 40, 72

7. Simplify each fraction to lowest terms.
   (a) $\frac{25}{35}$ **(b)** $\frac{26}{39}$

8. Give the reciprocal of $\frac{4}{3}$.

9. Evaluate each expression. Write each answer in lowest terms.
   (a) $\frac{2}{3} \cdot \frac{3}{4}$ **(b)** $\frac{5}{6} \div \frac{10}{3}$
   (c) $\frac{3}{10} + \frac{1}{10}$ **(d)** $\frac{3}{4} - \frac{1}{6}$

10. A recipe calls for $1\frac{2}{3}$ cups of flour. How much flour should be used if the recipe is doubled?

1.3 EXPONENTS AND ORDER OF OPERATIONS

Natural Number Exponents ■ Order of Operations ■
Translating Words to Expressions

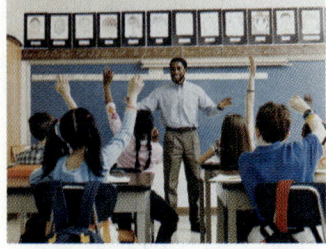

A LOOK INTO MATH ▷   In elementary school you learned addition. Later, you learned that multiplication is a fast way to add. For example, rather than adding

$$3 + 3 + 3 + 3, \quad \longleftarrow 4 \text{ terms}$$

we can multiply $3 \cdot 4$. Similarly, exponents represent a fast way to multiply. Rather than multiplying

$$3 \cdot 3 \cdot 3 \cdot 3, \quad \longleftarrow 4 \text{ factors}$$

we can evaluate the *exponential expression* 3^4. We begin this section by discussing natural numbers as exponents.

Natural Number Exponents

The area of a square equals the length of one of its sides times itself. If the square is 5 inches on a side, then its area is

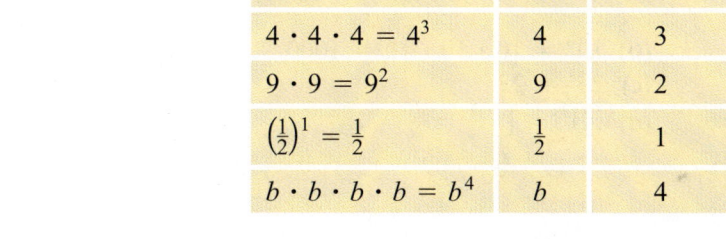

$$5 \cdot 5 = 5^2 = 25 \text{ square inches.}$$

The expression 5^2 is an **exponential expression** with *base* 5 and *exponent* 2. Exponential expressions occur in a variety of applications. For example, suppose an investment doubles 3 times. Then, the calculation

Exponent 3

$$2 \cdot 2 \cdot 2 = 2^3 = 8$$

Factors

shows that the final value is 8 times as large as the original investment. For example, if $10 doubles 3 times, it becomes $20, $40, and finally $80, which is 8 times as large as $10. Table 1.6 contains examples of exponential expressions.

TABLE 1.6

Expression	Base	Exponent
$2 \cdot 2 \cdot 2 \cdot 2 = 2^4$	2	4
$4 \cdot 4 \cdot 4 = 4^3$	4	3
$9 \cdot 9 = 9^2$	9	2
$\left(\frac{1}{2}\right)^1 = \frac{1}{2}$	$\frac{1}{2}$	1
$b \cdot b \cdot b \cdot b = b^4$	b	4

Read 9^2 as "9 squared," 4^3 as "4 cubed," and 2^4 as "2 to the fourth power." The terms *squared* and *cubed* come from geometry. If the length of a side of a square is 3, then its area is

$$3 \cdot 3 = 3^2 = 9$$

Figure 1.11
3 Squared

square units, as illustrated in Figure 1.11. Similarly, if the length of an edge of a cube is 3, then its volume is

$$3 \cdot 3 \cdot 3 = 3^3 = 27$$

cubic units, as shown in Figure 1.12.

NOTE: The expressions 3^2 and 3^3 can also be read as "3 to the second power" and "3 to the third power," respectively. In general, the expression x^n is read as "x to the nth power" or "the nth power of x."

Figure 1.12
3 Cubed

EXPONENTIAL NOTATION

The expression b^n, where n is a natural number, means

$$b^n = \underbrace{b \cdot b \cdot b \cdot \cdots \cdot b}_{n \text{ factors}}.$$

The **base** is b and the **exponent** is n.

EXAMPLE 1 Writing products in exponential notation

Write each product as an exponential expression.
(a) $7 \cdot 7 \cdot 7 \cdot 7$ **(b)** $\frac{1}{4} \cdot \frac{1}{4} \cdot \frac{1}{4}$ **(c)** $x \cdot x \cdot x \cdot x \cdot x$

Solution
(a) Because there are four factors of 7, the exponent is **4** and the base is **7**.
Thus $7 \cdot 7 \cdot 7 \cdot 7 = 7^4$.
(b) Because there are three factors of $\frac{1}{4}$, the exponent is 3 and the base is $\frac{1}{4}$.
Thus $\frac{1}{4} \cdot \frac{1}{4} \cdot \frac{1}{4} = \left(\frac{1}{4}\right)^3$.
(c) Because there are five factors of x, the exponent is 5 and the base is x.
Thus $x \cdot x \cdot x \cdot x \cdot x = x^5$.   Now Try Exercise 17

EXAMPLE 2 Evaluating exponential notation

Evaluate each expression.
(a) 3^4 **(b)** 10^3 **(c)** 23^1 **(d)** $\left(\frac{3}{4}\right)^2$

Solution
(a) The exponential expression 3^4 indicates that 3 is to be multiplied times itself 4 times.

$$3^4 = \underbrace{3 \cdot 3 \cdot 3 \cdot 3}_{\text{4 factors}} = 9 \cdot 9 = 81$$

(b) $10^3 = 10 \cdot 10 \cdot 10 = 1000$
(c) $23^1 = 23$
(d) $\left(\frac{3}{4}\right)^2 = \frac{3}{4} \cdot \frac{3}{4} = \frac{9}{16}$ Now Try Exercise 27

EXAMPLE 3 Writing numbers in exponential notation

Use the given base to write each number as an exponential expression. Check your results with a calculator, if one is available.
(a) 100 (base 10)   **(b)** 16 (base 2)   **(c)** 27 (base 3)

Figure 1.13

Solution
(a) $100 = 10 \cdot 10 = 10^2$
(b) $16 = 4 \cdot 4 = 2 \cdot 2 \cdot 2 \cdot 2 = 2^4$
(c) $27 = 3 \cdot 9 = 3 \cdot 3 \cdot 3 = 3^3$
These values are supported in Figure 1.13, where exponential expressions are evaluated with a calculator by using the "^" key. (Note that some calculators may have a different key for evaluating exponential expressions.)   Now Try Exercises 37, 43

▶ **REAL-WORLD CONNECTION** Computer memory is often measured in bytes, with each *byte* capable of storing one letter of the alphabet. For example, it takes four bytes to store the word "math" in a computer. Bytes of computer memory are often manufactured in amounts equal to powers of 2, as illustrated in the next example.

EXAMPLE 4   **Analyzing computer memory**

In computer technology 1K (kilobyte) of memory equals 2^{10} bytes, and 1MB (megabyte) of memory equals 2^{20} bytes. Determine whether 1K of memory equals one thousand bytes and whether 1MB equals one million bytes.   (***Source:*** D. Horn, *Basic Electronics Theory*.)

Solution
Figure 1.14 shows that $2^{10} = 1024$ and $2^{20} = 1,048,576$. Thus 1K represents slightly more than one thousand bytes, and 1MB represents more than one million bytes.

Now Try Exercise 83

```
2^10
            1024
2^20
         1048576
```

Figure 1.14

CRITICAL THINKING
One gigabyte is often referred to as 1 billion bytes. Use Example 4 to write an expression that gives the number of bytes in a gigabyte. If you have a calculator available, determine whether 1 gigabyte is exactly 1 billion bytes.

Order of Operations

When the expression $10 - 2 \cdot 3$ is evaluated, is the result

$$8 \cdot 3 = 24 \quad \text{or} \quad 10 - 6 = 4?$$

Figure 1.15 shows that a calculator gives a result of 4. The reason is that multiplication is performed before subtraction.

Because arithmetic expressions may contain parentheses, exponents, absolute values, and several operations, it is important to evaluate these expressions consistently. (Absolute value will be discussed in Section 1.4.) To ensure that we all obtain the same result when evaluating an arithmetic expression, the following rules are used.

```
10-2*3
              4
```

Figure 1.15

ORDER OF OPERATIONS

Use the following order of operations. First perform all calculations within parentheses and absolute values, or above and below the fraction bar.
1. Evaluate all exponential expressions.
2. Do all multiplication and division from *left to right*.
3. Do all addition and subtraction from *left to right*.

EXAMPLE 5   **Evaluating arithmetic expressions**

Evaluate each expression by hand.
(a) $10 - 4 - 3$ **(b)** $10 - (4 - 3)$ **(c)** $5 + \frac{12}{3}$ **(d)** $\frac{4}{2 + 6}$

Solution
(a) There are no parentheses, so we evaluate subtraction from *left to right*.

$$10 - 4 - 3 = 6 - 3 = 3$$

(b) Note the similarity between this part and part (a). The difference is the parentheses, so subtraction inside the parentheses must be performed first.

$$10 - (4 - 3) = 10 - 1 = 9$$

(c) In this part we perform division before addition.

$$5 + \frac{12}{3} = 5 + 4 = 9$$

(d) The implication is that both the numerator and the denominator of a fraction have parentheses around them.

$$\frac{4}{2 + 6} = \frac{4}{(2 + 6)} = \frac{4}{8} = \frac{1}{2}$$

Note that the given expression does *not* equal $\frac{4}{2} + 6 = 8$.

Now Try Exercises 51, 53, 65

EXAMPLE 6   Evaluating arithmetic expressions

Evaluate each expression.

(a) $25 - 4 \cdot 6$ **(b)** $6 + 7 \cdot 2 - (4 - 1)$ **(c)** $\dfrac{3 + 3^2}{14 - 2}$ **(d)** $5 \cdot 2^3 - (3 + 2)$

Solution

(a) Multiplication is performed before subtraction, so evaluate the expression as follows.

$$25 - \mathbf{4 \cdot 6} = 25 - \mathbf{24} \qquad \text{Multiply.}$$
$$= 1 \qquad \text{Subtract.}$$

(b) Start by performing the subtraction within the parentheses first and then perform the multiplication. Finally perform the addition and subtraction from left to right.

$$6 + 7 \cdot 2 - (4 - 1) = 6 + 7 \cdot 2 - 3 \qquad \text{Subtract within parentheses.}$$
$$= 6 + 14 - 3 \qquad \text{Multiply.}$$
$$= 20 - 3 \qquad \text{Add.}$$
$$= 17 \qquad \text{Subtract.}$$

(c) First note that parentheses are implied around the numerator and denominator.

$$\frac{3 + 3^2}{14 - 2} = \frac{(3 + 3^2)}{(14 - 2)} \qquad \text{Insert parentheses.}$$
$$= \frac{(3 + 9)}{(14 - 2)} \qquad \text{Do the exponent first.}$$
$$= \frac{12}{12} \qquad \text{Add and subtract.}$$
$$= 1 \qquad \text{Simplify.}$$

(d) Begin by evaluating the expression inside parentheses.

$$5 \cdot 2^3 - (3 + 2) = 5 \cdot 2^3 - 5 \qquad \text{Add within parentheses.}$$
$$= 5 \cdot 8 - 5 \qquad \text{Evaluate the exponent.}$$
$$= 40 - 5 \qquad \text{Multiply.}$$
$$= 35 \qquad \text{Subtract.}$$

Now Try Exercises 67, 69, 71

Translating Words to Expressions

Sometimes before we can solve a problem we must translate words into mathematical expressions. For example, if the gasoline tank in your car holds 20 gallons and there are 11 gallons in it, then "twenty minus eleven," or $20 - 11 = 9$, is the number of gallons of gasoline needed to fill the tank.

EXAMPLE 7   Writing and evaluating expressions

Write each expression and then evaluate it.
(a) Two to the fourth power plus ten
(b) Twenty less five times three
(c) Ten cubed divided by five squared
(d) Sixty divided by the quantity ten minus six

Solution
(a) $2^4 + 10 = 2 \cdot 2 \cdot 2 \cdot 2 + 10 = 16 + 10 = 26$
(b) $20 - 5 \cdot 3 = 20 - 15 = 5$
(c) $\frac{10^3}{5^2} = \frac{1000}{25} = 40$

(d) Here, the word "quantity" indicates that parentheses should be used.

$$60 \div (10 - 6) = \frac{60}{10 - 6} = \frac{60}{4} = 15$$

Now Try Exercise 73

1.3 PUTTING IT ALL TOGETHER

The following table summarizes some topics related to exponential expressions. Remember that exponential expressions are a fast way to multiply.

Topic	Comments	Examples
Exponential Expression	If n is a natural number, then b^n equals $$\underbrace{b \cdot b \cdot b \cdot \,\cdots\, \cdot b}_{n \text{ factors}}$$ and is read "b to the nth power."	$5^1 = 5, 7^2 = 7 \cdot 7 = 49,$ $4^3 = 4 \cdot 4 \cdot 4 = 64,$ and $k^4 = k \cdot k \cdot k \cdot k$
Base and Exponent	The base in b^n is b and the exponent is n.	7^4 has base 7 and exponent 4 and x^3 has base x and exponent 3.

continued on next page

continued from previous page

Without rules for the order of operations, the value of many arithmetic expressions would be ambiguous. For example, without rules of precedence, one person might evaluate $10 - 4 \cdot 2$ to be $6 \cdot 2 = 12$, whereas another person might evaluate it to be $10 - 8 = 2$. However, because multiplication is performed before subtraction, 2 is the correct answer.

To evaluate expressions, use the following order of operations. First perform all calculations within parentheses and absolute values, or above and below the fraction bar.

1. Evaluate all exponential expressions.
2. Do all multiplication and division from *left to right*.
3. Do all addition and subtraction from *left to right*.

1.3 Exercises

CONCEPTS

1. Multiplication is a fast way to _____.

2. Exponents represent a fast way to _____.

3. In the expression 3^6, how many factors of 3 are being multiplied?

4. In the expression 2^5, there are five factors of _____ being multiplied.

5. The expression $a \cdot a \cdot a \cdot a \cdot a \cdot a$ equals _____.

6. In the expression 5^3, the number 5 is called the _____ and the number 3 is called the _____.

7. Use symbols to write "6 squared."

8. Use symbols to write "8 cubed."

9. Write "four to the fifth power" as an exponential expression.

10. Write "the third power of x" as an exponential expression.

11. When evaluating the expression $5 + 6 \cdot 2$, the result is _____ because _____ is performed before _____.

12. When evaluating the expression $10 - 2^3$, the result is _____ because _____ are evaluated before _____ is performed.

13. The expression $10 - 4 - 2$ equals _____ because subtraction is performed from _____ to _____.

14. The expression $16 \div 4 \div 2$ equals _____ because division is performed from _____ to _____.

15. Are the expressions 2^3 and 3^2 equal? Explain.

16. Is 5^2 equal to $5 \cdot 5$ or $5 \cdot 2$?

NATURAL NUMBER EXPONENTS

Exercises 17–26: Write the product as an exponential expression.

17. $2 \cdot 2 \cdot 2 \cdot 2 \cdot 2$ 18. $4 \cdot 4 \cdot 4$

19. $3 \cdot 3 \cdot 3 \cdot 3$ 20. $10 \cdot 10$

21. $\frac{1}{2} \cdot \frac{1}{2} \cdot \frac{1}{2} \cdot \frac{1}{2}$ 22. $\frac{5}{7} \cdot \frac{5}{7} \cdot \frac{5}{7} \cdot \frac{5}{7} \cdot \frac{5}{7}$

23. $a \cdot a \cdot a \cdot a \cdot a$ 24. $b \cdot b \cdot b \cdot b$

25. $(x + 3) \cdot (x + 3)$

26. $(x - 4) \cdot (x - 4) \cdot (x - 4)$

Exercises 27–36: Use multiplication to rewrite the expression, and then evaluate the result.

27. (a) 2^4 (b) 4^2 28. (a) 3^2 (b) 5^3

29. (a) 6^1 (b) 1^6 30. (a) 17^1 (b) 1^{17}

31. (a) 2^5 (b) 10^3 32. (a) 10^5 (b) 3^4

33. (a) $\left(\frac{2}{3}\right)^2$ (b) $\left(\frac{1}{2}\right)^5$ 34. (a) $\left(\frac{1}{10}\right)^3$ (b) $\left(\frac{4}{3}\right)^1$

35. (a) $\left(\frac{2}{5}\right)^3$ (b) $\left(\frac{9}{7}\right)^2$ 36. (a) $\left(\frac{3}{10}\right)^4$ (b) $\left(\frac{3}{4}\right)^4$

Exercises 37–48: (Refer to Example 3.) Use the given base to write the number as an exponential expression. Check your result if you have a calculator available.

37. 8   (base 2)   **38.** 9   (base 3)

39. 25   (base 5)   **40.** 32   (base 2)

41. 49   (base 7)   **42.** 81   (base 3)

43. 1000   (base 10)   **44.** 256   (base 4)

45. $\frac{1}{16}$ $\left(\text{base } \frac{1}{2}\right)$ **46.** $\frac{9}{25}$ $\left(\text{base } \frac{3}{5}\right)$

47. $\frac{32}{243}$ $\left(\text{base } \frac{2}{3}\right)$ **48.** $\frac{216}{343}$ $\left(\text{base } \frac{6}{7}\right)$

ORDER OF OPERATIONS

Exercises 49–72: Evaluate the expression by hand.

49. $5 + 4 \cdot 6$ **50.** $6 \cdot 7 - 8$

51. $6 \div 3 + 2$ **52.** $20 - 10 \div 5$

53. $100 - \frac{50}{5}$ **54.** $\frac{200}{100} + 6$

55. $10 - 6 - 1$ **56.** $30 - 9 - 5$

57. $20 \div 5 \div 2$ **58.** $500 \div 100 \div 5$

59. $3 + 2^4$ **60.** $10 - 3^2 + 1$

61. $4 \cdot 2^3$ **62.** $100 - 2 \cdot 3^3$

63. $(3 + 2)^3$ **64.** $5 \cdot (3 - 2)^8 - 5$

65. $\frac{4 + 8}{1 + 3}$ **66.** $5 - \frac{3 + 1}{3 - 1}$

67. $\frac{2^3}{4 - 2}$ **68.** $\frac{10 - 3^2}{2 \cdot 4^2}$

69. $10^2 - (30 - 2 \cdot 5)$ **70.** $5^2 + 3 \cdot 5 \div 3 - 1$

71. $\left(\frac{1}{2}\right)^4 + \frac{5 + 4}{3}$ **72.** $\left(\frac{7}{9}\right)^2 - \frac{6 - 5}{3}$

TRANSLATING WORDS TO SYMBOLS

Exercises 73–82: Use symbols to write the expression and then evaluate it.

73. Two cubed minus eight

74. Five squared plus nine

75. Thirty less four times three

76. One hundred plus five times six

77. Four squared divided by two cubed

78. Three cubed times two squared

79. Forty divided by ten, plus two

80. Thirty times ten, minus three

81. One hundred times the quantity two plus three

82. Fifty divided by the quantity eight plus two

APPLICATIONS

83. *iPod Memory*   (Refer to Example 4.) Determine the number of bytes on a 512 MB iPod Shuffle.

84. *iPod Memory*   Determine the number of bytes on a 60 GB video iPod. (*Hint:* One gigabyte equals 2^{30} bytes.)

85. *Population by Gender*   One way to measure the gender balance in a given population is to find the number of males for every 100 females in the population. In 1900, the western region of the United States was significantly out of gender balance. In this region, there were 128 males for every 100 females.   (**Source:** U.S. Census Bureau.)
    (a) Find an exponent k so that $2^k = 128$.
    (b) During this time, how many males were there for every 25 females?

86. *Solar Eclipse*   In early December 2048 there will be a total solar eclipse visible in parts of Botswana. Find an exponent k so that $2^k = 2048$. (**Source:** NASA.)

87. *Rule of 72*   Investors sometimes use the *rule of 72* to determine the time required to double an investment. If 72 is divided by the annual interest rate earned on an investment, the result approximates the number of years needed to double the investment. For example, an investment earning 6% annual interest will double in value approximately every $72 \div 6 = 12$ years.
    (a) Approximate the number of years required to double an investment earning 9% annual interest.
    (b) If an investment of $10,000 earns 12% annual interest, approximate the value of the investment after 18 years.

88. *Doubling Effect*   Suppose that a savings account containing $1000 doubles its value every 7 years. How much money will be in the account after 28 years?

WRITING ABOUT MATHEMATICS

89. Explain how exponential expressions are related to multiplication. Give an example.

90. Explain why agreement on the order of operations is necessary.

Directions: Form a group of 2 to 4 people. Select someone to record the group's responses for this activity. All members of the group should work cooperatively to answer the questions. If your instructor asks for your results, each member of the group should be prepared to respond.

Converting Temperatures   To convert Celsius degrees C to Fahrenheit degrees F use the formula $F = 32 + \frac{9}{5}C$. This exercise illustrates the importance of understanding the order of operations.

(a) Complete the following table by evaluating the formula in the two ways shown.

(b) At what Celsius temperature does water freeze? At what Fahrenheit temperature does water freeze?

(c) Which column gives the correct Fahrenheit temperatures? Why?

(d) Explain why having an agreed order for operations in mathematics is necessary.

Celsius	$F = \left(32 + \frac{9}{5}\right)C$	$F = 32 + \left(\frac{9}{5}C\right)$
−40°C		
0°C		
5°C	169°F	41°F
20°C		
30°C		
100°C		

1.4 REAL NUMBERS AND THE NUMBER LINE

Signed Numbers ■ Integers and Rational Numbers ■ Square Roots ■ Real and Irrational Numbers ■ The Number Line ■ Absolute Value ■ Inequality

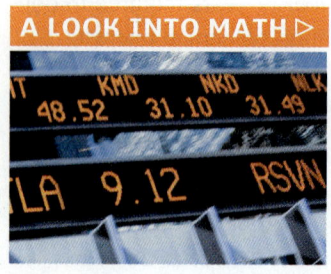

A LOOK INTO MATH ▷

Fractions were invented before decimals and negative numbers. The concept of splitting something into parts was readily understood by most people. However, performing arithmetic with fractions was quite cumbersome, particularly if the fractions had unlike denominators. For example, to find the sum $\frac{1}{2} + \frac{1}{4}$, a common denominator needed to be found. With the invention of decimals, this addition problem became much simpler and could be expressed as $0.50 + 0.25 = 0.75$. One important reason for the invention of decimal numbers is that arithmetic with decimals is much easier to perform than arithmetic with fractions. Even the stock market switched from fractions to decimals in late 2000. In this section we discuss decimal numbers.

Signed Numbers

The idea that numbers could be negative was a difficult concept for many mathematicians. As late as the eighteenth century negative numbers were not readily accepted by everyone. After all, how could a person have −5 oranges?

However, negative numbers make more sense when someone is working with money. If you owe someone 100 dollars, this amount can be thought of as -100, whereas if you have a balance of 100 dollars in your checking account, this amount can be thought of as $+100$. (The positive sign is usually omitted.) The number 0 is neither positive nor negative.

The **opposite**, or **additive inverse**, of a number a is $-a$. For example, the opposite of 25 is -25, the opposite of -5 is $-(-5)$, or 5, and the opposite of 0 is 0. The following double negative rule is helpful in simplifying expressions containing negative signs.

DOUBLE NEGATIVE RULE

Let a be any number. Then $-(-a) = a$.

Thus $-(-8) = 8$ and $-\left(-(-10)\right) = -(10) = -10$.

EXAMPLE 1   Finding opposites (or additive inverses)

Find the opposite of each expression.

(a) 13   **(b)** $-\frac{4}{7}$ **(c)** $5 - \frac{4}{2}$ **(d)** $-(-7)$

Solution

(a) The opposite of 13 is -13.

(b) The opposite of $-\frac{4}{7}$ is $\frac{4}{7}$.

(c) $5 - \frac{4}{2} = 5 - 2 = 3$, so the opposite of $5 - \frac{4}{2}$ is -3.

(d) $-(-7) = 7$, so the opposite of $-(-7)$ is -7.   Now Try Exercises 27, 31

NOTE: To find the opposite of an exponential expression evaluate the exponent first. For example, the opposite of 2^4 is

$$-2^4 = -(2 \cdot 2 \cdot 2 \cdot 2) = -16.$$

EXAMPLE 2   Finding an additive inverse (or opposite)

Find the additive inverse of $-t$, if $t = -\frac{2}{3}$.

Solution

The additive inverse of $-t$ is $t = -\frac{2}{3}$ because $-(-t) = t$ by the double negative rule.

Now Try Exercise 37

Integers and Rational Numbers

In the opening section to this chapter we discussed natural numbers and whole numbers. Because these sets of numbers do not include negative numbers, fractions, or decimals, other sets of numbers are needed. The **integers** include the natural numbers, zero, and the opposites of the natural numbers. The integers are given by the following.

$$\ldots, \quad -3, \quad -2, \quad -1, \quad 0, \quad 1, \quad 2, \quad 3, \quad \ldots$$

A **rational number** is any number that can be expressed as the ratio of two integers, $\frac{p}{q}$, where $q \neq 0$. Rational numbers can be written as fractions, and they include all integers. Rational numbers may be positive, negative, or zero. Some examples of rational numbers are

$$\frac{2}{3}, \quad -\frac{3}{5}, \quad \frac{-7}{2}, \quad 1.2, \quad \text{and} \quad 3.$$

The numbers 1.2 and 3 are both rational numbers because they can be written as $\frac{12}{10}$ and $\frac{3}{1}$.

NOTE: $\frac{-7}{2} = \frac{7}{-2} = -\frac{7}{2}$; the position of the negative sign does not affect the value of the fraction.

The fraction bar can be thought of as a division symbol. As a result, rational numbers have decimal equivalents. For example, $\frac{1}{2}$ is equivalent to $1 \div 2$. The division

$$2)\overline{1.0}^{\,0.5}$$

shows that $\frac{1}{2} = 0.5$. In general, a rational number may be expressed in a decimal form that either *repeats* or *terminates*. The fraction $\frac{1}{3}$ may be expressed as $0.\overline{3}$, a repeating decimal, and the fraction $\frac{1}{4}$ may be expressed as 0.25, a terminating decimal. The overbar indicates that $0.\overline{3} = 0.3333333\ldots$.

▶ **REAL-WORLD CONNECTION**   Integers and rational numbers are used to describe quantities such as change in population from one decade to the next. Table 1.7 lists the actual change in population along with the percent change from 1990 to 2000 for selected U.S. cities. Note that both positive and negative numbers are used to describe these population changes.

TABLE 1.7   Change in Population from 1990 to 2000 for Selected Cities

City	Actual Change	Percent Change
Baltimore	−84,860	−11.5
Cleveland	−27,213	−5.4
Houston	323,078	19.8
Los Angeles	209,263	6.0
New York	685,714	9.4

Source: U.S. Census Bureau.

EXAMPLE 3   Classifying numbers

Classify each number as one or more of the following: natural number, whole number, integer, or rational number.
(a) $\frac{12}{4}$ **(b)** -3 **(c)** 0 **(d)** $-\frac{9}{5}$

Solution
(a) Because $\frac{12}{4} = 3$, the number $\frac{12}{4}$ is a natural number, whole number, integer, and rational number.
(b) The number -3 is an integer and rational number but not a natural number or a whole number.

(c) The number 0 is a whole number, integer, and rational number but not a natural number.

(d) The fraction $-\frac{9}{5}$ is a rational number because it is the ratio of two integers. However, it is not a natural number, a whole number, or an integer.   Now Try Exercises 51, 53, 55

Square Roots

Square roots are frequently used in algebra. The number b is a **square root** of a number a if $b \cdot b = a$. Every *positive* number has one positive square root and one negative square root. For example, the positive square root of 9 is 3 because $3 \cdot 3 = 9$. The negative square root of 9 is -3. (We will show that $-3 \cdot (-3) = 9$ in Section 1.6.) If a is a nonnegative number (a number that is not negative), then the **principal square root** of a, denoted $\sqrt{a}$, is the nonnegative square root of a. For example, $\sqrt{25} = 5$ because $5 \cdot 5 = 25$ and the number 5 is nonnegative. Note that $\sqrt{0} = 0$.

EXAMPLE 4   Calculating principal square roots

Evaluate each square root. Approximate your answer to three decimal places when appropriate.
(a) $\sqrt{36}$ **(b)** $\sqrt{100}$ **(c)** $\sqrt{5}$

```
√(5)
        2.236067977
2.236*2.236
            4.999696
```

Figure 1.16

Solution
(a) $\sqrt{36} = 6$ because $6 \cdot 6 = 36$ and 6 is nonnegative.
(b) $\sqrt{100} = 10$ because $10 \cdot 10 = 100$ and 10 is nonnegative.
(c) $\sqrt{5}$ is a number between 2 and 3 because $2 \cdot 2 = 4$ and $3 \cdot 3 = 9$. We can estimate the value of $\sqrt{5}$ with a calculator. Figure 1.16 reveals that $\sqrt{5}$ *approximately* equals 2.236. However, 2.236 does not exactly equal $\sqrt{5}$ because $2.236 \times 2.236 = 4.999696$, which does not equal 5.   Now Try Exercise 7

CALCULATOR HELP

To evaluate a square root, see Appendix A (page AP-1).

Real and Irrational Numbers

Real numbers can be represented by decimal numbers. That is, if a number can be represented by a decimal number, then it is a real number. Every fraction has a decimal form, so real numbers include rational numbers. However, some real numbers cannot be expressed by fractions. They are called **irrational numbers**. The numbers $\sqrt{2}$, $\sqrt{15}$, and π are examples of irrational numbers. They can be represented by decimals, but not by decimals that either repeat or terminate. Note that for any positive integer a, if $\sqrt{a}$ is not an integer then $\sqrt{a}$ is an irrational number. Examples of real numbers include

$$-17, \quad \frac{4}{5}, \quad -\sqrt{3}, \quad 21\frac{1}{2}, \quad 57.63, \quad \text{and} \quad \sqrt{7}.$$

Any real number may be approximated by a terminating decimal. The symbol $\approx$ represents **approximately equal**. Each of the following real numbers has been approximated to two *decimal places*.

$$\frac{1}{7} \approx 0.14, \quad 2\pi \approx 6.28, \quad \text{and} \quad \sqrt{60} \approx 7.75$$

If you have a calculator, verify these results.

Figure 1.17 shows the relationships of the different sets of numbers. Note that each real number is either a rational number or an irrational number but not both. The natural numbers, whole numbers, and integers are rational numbers.

Real Numbers

Rational Numbers	Irrational Numbers
$-\frac{3}{5}, \frac{1}{3}, \frac{4}{9} = 0.\overline{4}$	$-\sqrt{17}, \pi, \sqrt{6}$

Integers
$..., -2, -1, 0, 1, 2, ...$

Whole Numbers
$0, 1, 2, 3, 4, ...$

Natural Numbers
$1, 2, 3, 4, ...$

Figure 1.17 The Real Numbers

MAKING CONNECTIONS

Rational and Irrational Numbers

Both rational and irrational numbers can be written as decimals. However, rational numbers can be represented by either terminating or repeating decimals. For example, $\frac{1}{2} = 0.5$ is a terminating decimal and $\frac{1}{3} = 0.333...$ is a repeating decimal. Irrational numbers are represented by decimals that neither terminate nor repeat.

EXAMPLE 5   Classifying numbers

Identify the natural numbers, whole numbers, integers, rational numbers, and irrational numbers in the following list.

$$-\sqrt{5}, \quad 9, \quad -3.8, \quad \sqrt{49}, \quad \frac{11}{4}, \quad \text{and} \quad -41$$

Solution

Natural numbers: 9 and $\sqrt{49} = 7$

Whole numbers: 9 and $\sqrt{49} = 7$

Integers: 9, $\sqrt{49} = 7$, and -41

Rational numbers: 9, -3.8, $\sqrt{49} = 7$, $\frac{11}{4}$, and -41

Irrational number: $-\sqrt{5}$

Now Try Exercises 57, 61, 65

Even though a data set may contain only integers, we often need decimals to describe it. One common way to describe data is to find their **average**. To find the average of a set of numbers we add the numbers and then divide by how many numbers there are in the set. For example, the average of 6, 13, and 14 equals 11 because

$$\frac{6 + 13 + 14}{3} = \frac{33}{3} = 11.$$

EXAMPLE 6   Analyzing data

Table 1.8 lists the number of higher education institutions, such as colleges and universities, for various years. Find the average number of institutions during this 4-year period. Is the result a natural number, a rational number, or an irrational number?

TABLE 1.8   Higher Education Institutions

Year	2000	2001	2002	2003
Institutions	4182	4197	4168	4236

Source: National Center for Education Statistics.

CRITICAL THINKING

Think of an example in which the sum of two irrational numbers is a rational number.

Solution

The average number of institutions was

$$\frac{4182 + 4197 + 4168 + 4236}{4} = \frac{16,783}{4} = 4195.75.$$

The average of these four natural numbers is an integer divided by an integer, which is a rational number. However, it is neither a natural number nor an irrational number.

Now Try Exercise 107

The Number Line

Figure 1.18   The Number Line

The real numbers can be represented visually by using a number line, as shown in Figure 1.18. Each real number corresponds to a unique point on the number line. The point associated with the real number 0 is called the **origin**. The positive integers are equally spaced to the right of the origin, and the negative integers are equally spaced to the left of the origin. The number line extends indefinitely both left and right. Other real numbers can also be located on the number line. For example, the number $\frac{1}{2}$ can be identified by placing a dot halfway between the integers 0 and 1. The numbers $-\sqrt{2} \approx -1.41$ and $\frac{5}{4} = 1.25$ can also be placed (approximately) on this number line.

EXAMPLE 7   Plotting numbers on a number line

Plot each real number on a number line.
(a) $-\frac{3}{2}$ **(b)** $\sqrt{3}$ **(c)** π

Figure 1.19   Plotting Real Numbers

Solution

(a) $-\frac{3}{2} = -1.5$. Place a dot halfway between -2 and -1, as shown in Figure 1.19.
(b) A calculator gives $\sqrt{3} \approx 1.73$. Place a dot between 1 and 2 so that it is about three-fourths of the way toward 2, as shown in Figure 1.19.
(c) $\pi \approx 3.14$. Place a dot just past the integer 3, as shown in Figure 1.19.

Now Try Exercises 69, 77

Absolute Value

Figure 1.20

The **absolute value** of a real number equals its distance on the number line from the origin. Because distance is never negative, the absolute value of a real number is *never negative*. The absolute value of a real number a is denoted $|a|$ and is read "the absolute value of a." Figure 1.20 shows that the absolute values of -3 and 3 equal 3 because both have distance 3 from the origin. That is, $|-3| = 3$ and $|3| = 3$.

EXAMPLE 8   **Finding the absolute value of a real number**

Write the expression without the absolute value sign.
(a) $|-7|$ **(b)** $|0|$
(c) $|-a|$, if a is a positive number **(d)** $|a|$, if a is a negative number

Solution
(a) $|-7| = 7$ because the distance between the origin and -7 is 7.
(b) $|0| = 0$ because the distance is 0 between the origin and 0.
(c) If a is positive, then $-a$ is negative. Thus $|-a| = a$.
(d) If a is negative, then $-a$ is positive. Thus $|a| = -a$. For example, if $a = -5$, then $|-5| = -(-5) = 5$.

Now Try Exercises **79, 81, 87**

Our results about absolute value can be summarized as follows.

$$|a| = a, \quad \text{if } a \text{ is positive or } 0.$$
$$|a| = -a, \quad \text{if } a \text{ is negative.}$$

Inequality

Figure 1.21 $-3 < 2$, or $2 > -3$

If a real number a is located to the left of a real number b on the number line, we say that a is **less than** b and write $a < b$. Similarly, if a real number b is located to the right of a real number a, we say that b is **greater than** a and write $b > a$. Thus $-3 < 2$ because -3 is located to the left of 2, and $2 > -3$ because 2 is located to the right of -3. See Figure 1.21. In general, any negative number will always be less than any positive number, and any positive number will always be greater than any negative number.

We say that a is **less than or equal to** b, denoted $a \leq b$, if either $a < b$ or $a = b$ is true. Similarly, a is **greater than or equal to** b, denoted $a \geq b$, if either $a > b$ or $a = b$ is true.

Inequalities are often used to compare the relative sizes of two quantities. This can be illustrated visually as follows.

   <   >

EXAMPLE 9   **Ordering real numbers**

List the following numbers from least to greatest. Then plot these numbers on a number line.

$$-2, \quad -\pi, \quad \sqrt{2}, \quad 0, \quad \text{and} \quad 2.5$$

Solution
First note that $-\pi \approx -3.14 < -2$. The two negative numbers are less than 0, and the two positive numbers are greater than 0. Also, $\sqrt{2} \approx 1.41$, so $\sqrt{2} < 2.5$. Listing the numbers from least to greatest results in

$$-\pi, \quad -2, \quad 0, \quad \sqrt{2}, \quad \text{and} \quad 2.5.$$

Figure 1.22

These numbers are plotted on the number line shown in Figure 1.22. Note that these numbers increase from left to right on the number line.

Now Try Exercise **103**

1.4 PUTTING IT ALL TOGETHER

Because of the variety of information produced and used by our society, different sets of numbers had to be developed. Without numbers, information can only be described qualitatively, not quantitatively. For example, we might say that it rained a lot last night, but we would not be able to give an actual measurement of the rainfall. The following table summarizes several sets of numbers.

Set of Numbers	Comments	Examples
Integers	Include the natural numbers, their opposites, and 0	$\ldots, -2, -1, 0, 1, 2, \ldots$
Rational Numbers	Include integers, all fractions $\frac{p}{q}$, where p and $q \neq 0$ are integers, and all repeating and terminating decimals	$\frac{1}{2}, -3, \frac{128}{6}, -0.335, 0, 0.25 = \frac{1}{4},$ and $0.\overline{3} = \frac{1}{3}$
Irrational Numbers	Any decimal number that neither terminates nor repeats; a real number that is not rational	$\pi, \sqrt{3}$, and $\sqrt{15}$
Real Numbers	Any number that can be expressed in decimal form; include the rational and irrational numbers	$\pi, \sqrt{3}, -\frac{4}{7}, 0, -10, 0.\overline{6} = \frac{2}{3}, 1000,$ and $\sqrt{15}$

A number line can be used to visualize the real number system, as illustrated in the figure. The point associated with the number 0 is called the *origin*.

The absolute value of a number a equals its distance on the number line from the origin. If a number a is located to the left of a number b, then a is less than b (written $a < b$). If a is located to the right of b, then a is greater than b (written $a > b$).

Topic	Notation	Examples	
Absolute Value	If $a \geq 0$, then $$\mid a \mid = a.$$ If $a < 0$, then $$\mid a \mid = -a.$$ $\mid a \mid$ is *never* negative.	$\mid 17 \mid = 17$ $\mid -12 \mid = 12$ $\mid 0 \mid = 0$	
Inequality	Symbols of inequality include: $<, >, \leq, \geq,$ and $\neq$.	$-3 < -2$	less than
		$6 > 4$	greater than
		$-5 \leq -5$	less than or equal
		$18 \geq 0$	greater than or equal
		$7 \neq 8$	not equal

1.4 Exercises

CONCEPTS

1. The opposite of the number b is _____.

2. The additive inverse of -7 is _____.

3. $-(-b) =$ _____.

4. The value of -3^2 is _____.

5. The integers include the natural numbers, zero, and the additive inverses of the _____ numbers.

6. A number that can be written as $\frac{p}{q}$, where p and q are integers with $q \neq 0$, is a(n) _____ number.

7. The equation $4 \cdot 4 = 16$ indicates that $\sqrt{16} =$ ____.

8. The nonnegative square root of a number is called the _____ square root.

9. If a number can be written in decimal form, then it is a(n) _____ number.

10. If a real number is not a rational number, then it is a(n) _____ number.

11. A real number that can be written as a repeating or terminating decimal is a(n) _____ number.

12. A real number that can not be written as a repeating or terminating decimal is a(n) _____ number.

13. Write $0.272727\ldots$ using an overbar.

14. The decimal equivalent for $\frac{1}{4}$ can be found by dividing _____ by _____.

15. Give an example of an irrational number.

16. The symbol $\neq$ is used to indicate that two numbers are _____.

17. The symbol $\approx$ is used to indicate that two numbers are _____.

18. To calculate the average of three numbers, add the three numbers and divide the sum by _____.

19. The origin on the number line corresponds to the number _____.

20. The negative numbers are to the (left/right) of the origin on the number line.

21. The absolute value of a number a gives its distance on the number line from the _____.

22. If $b > 0$, then $|-b| =$ _____.

Exercises 23–26: Insert the symbol $<$, $=$, or $>$ to make the statement true.

23. If a number a is located to the right of a number b on the number line, then a __ b.

24. If $b > 0$ and $a < 0$, then b __ a.

25. If $a \geq b$, then either $a > b$ or a __ b.

26. If $a \geq b$ and $b \geq a$, then a __ b.

SIGNED NUMBERS

Exercises 27–34: Find the opposite of each expression.

27. (a) 9 (b) -9

28. (a) -6 (b) 6

29. (a) $\frac{2}{3}$ (b) $-\frac{2}{3}$

30. (a) $-\left(\frac{-4}{5}\right)$ (b) $-\left(\frac{-4}{-5}\right)$

31. (a) $-(-8)$ (b) $-\left(-(-8)\right)$

32. (a) $-\left(-(-2)\right)$ (b) $-(-2)$

33. (a) a (b) $-a$

34. (a) $-b$ (b) $-(-b)$

35. Find the additive inverse of t, if $-t = 6$.

36. Find the additive inverse of $-t$, if $t = -\frac{4}{5}$.

37. Find the additive inverse of $-b$, if $b = \frac{1}{2}$.

38. Find the additive inverse of b, if $-b = \frac{5}{-6}$.

NUMBERS AND THE NUMBER LINE

Exercises 39–50: Find the decimal equivalent for the rational number.

39. $\frac{1}{4}$ **40.** $\frac{3}{5}$

41. $\frac{7}{8}$ **42.** $\frac{3}{10}$

43. $\frac{3}{2}$ **44.** $\frac{3}{50}$

45. $\frac{1}{20}$ **46.** $\frac{3}{16}$

47. $\frac{2}{3}$ **48.** $\frac{2}{9}$

49. $\frac{7}{9}$ **50.** $\frac{5}{11}$

Exercises 51–56: Classify the number as one or more of the following: natural number, whole number, integer, or rational number.

51. 8 **52.** -8

53. $\frac{16}{4}$ **54.** $\frac{5}{7}$

55. 0 **56.** $-\frac{15}{31}$

Exercises 57–68: Classify the number as one or more of the following: natural number, integer, rational number, or irrational number.

57. -4.5 **58.** π

59. $\frac{3}{7}$ **60.** $\sqrt{25}$

61. $\sqrt{11}$ **62.** $-\sqrt{3}$

63. $\frac{8}{4}$ **64.** -5

65. $\sqrt{49}$ **66.** $3.\overline{3}$

67. $1.\overline{8}$ **68.** $\frac{9}{3}$

Exercises 69–78: Plot each number on a number line.

69. (a) 0 **(b)** -2 **(c)** 3

70. (a) -1 **(b)** -3 **(c)** 4

71. (a) $\frac{1}{2}$ **(b)** $-\frac{1}{2}$ **(c)** 2

72. (a) $-\frac{3}{2}$ **(b)** $\frac{3}{2}$ **(c)** 0

73. (a) 1.3 **(b)** -2.5 **(c)** 0.7

74. (a) 3.2 **(b)** -2.8 **(c)** 0.5

75. (a) -10 **(b)** -20 **(c)** 30

76. (a) 5 **(b)** 10 **(c)** -10

77. (a) π **(b)** $\sqrt{2}$ **(c)** $-\sqrt{3}$

78. (a) $\sqrt{11}$ **(b)** $-\sqrt{5}$ **(c)** $\sqrt{4}$

ABSOLUTE VALUE

Exercises 79–88: Evaluate the expression.

79. $|5.23|$ **80.** $|\pi|$

81. $|-7|$ **82.** $|-\sqrt{2}|$

83. $|2 - 6|$ **84.** $\left|\frac{2}{3} - \frac{1}{3}\right|$

85. $|\pi - 3|$ **86.** $|3 - \pi|$

87. $|b|$, if b is negative **88.** $|-b|$, if b is positive

INEQUALITY

Exercises 89–100: Insert the symbol $>$ or $<$ to make the statement true.

89. $5 \underline{\quad} 7$ **90.** $-5 \underline{\quad} 7$

91. $-5 \underline{\quad} -7$ **92.** $\frac{3}{5} \underline{\quad} \frac{2}{5}$

93. $-\frac{1}{3} \underline{\quad} -\frac{2}{3}$ **94.** $-\frac{1}{10} \underline{\quad} 0$

95. $-1.9 \underline{\quad} -1.3$ **96.** $5.1 \underline{\quad} -6.2$

97. $|-8| \underline{\quad} 3$ **98.** $4 \underline{\quad} |-1|$

99. $|-2| \underline{\quad} |-7|$ **100.** $|-15| \underline{\quad} |32|$

Exercises 101–106: List the given numbers from least to greatest.

101. $-3, 0, 1, -9, -2^3$

102. $4, -2^3, \frac{1}{2}, -\frac{3}{2}, \frac{3}{2}$

103. $-2, \pi, \frac{1}{3}, -\frac{3}{2}, \sqrt{5}$

104. $9, 14, -\frac{1}{12}, -\frac{3}{16}, \sqrt{7}$

105. $-\frac{4}{7}, -\frac{17}{28}, -4^2, \sqrt{7}, \sqrt{2}$

106. $-3.1, 2^3, -3^3, 6, 9.4$

APPLICATIONS

107. *Household Size* The table lists the percentage of U.S. households that were 3-person households during selected years.

Year	1970	1980	1990	2000
Percent	17.2	17.4	17.4	16.5

Source: U.S. Census Bureau.

(a) What was this percentage in 2000?

(b) Mentally estimate the average percentage for these selected years.

(c) Calculate the average percentage. Is your mental estimate in reasonable agreement with your calculated result?

108. *Music Sales* The table lists the percentage of recorded music purchases attributed to children aged 10 to 14 during selected years.

Year	2001	2002	2003	2004
Percent	8.5	8.9	8.6	9.4

Source: Recording Industry Association of America.

(a) What was this percentage in 2001?

(b) Mentally estimate the average percentage for this 4-year period.

(c) Calculate the average percentage. Is your mental estimate in reasonable agreement with your calculated result?

WRITING ABOUT MATHEMATICS

109 What is a rational number? Is every integer a rational number? Why or why not?

110. Explain why $\frac{3}{7} > \frac{1}{3}$. Now explain in general how to determine whether $\frac{a}{b} > \frac{c}{d}$. Assume that a, b, c, and d are natural numbers.

CHECKING BASIC CONCEPTS

SECTIONS 1.3 AND 1.4

1. Write each product as an exponential expression.
(a) $5 \cdot 5 \cdot 5 \cdot 5$　(b) $7 \cdot 7 \cdot 7 \cdot 7 \cdot 7$

2. Evaluate each expression.
(a) 2^3　(b) 10^4　(c) $\left(\frac{2}{3}\right)^3$　(d) -3^4

3. Use the given base to write the number as an exponential expression.
(a) 64 (base 4)　(b) 64 (base 2)

4. Evaluate each expression without a calculator.
(a) $6 + 5 \cdot 4$　(b) $6 + 6 \div 2$
(c) $5 - 2 - 1$　(d) $\frac{6 - 3}{2 + 4}$
(e) $12 \div (6 \div 2)$　(f) $2^3 - 2\left(2 + \frac{4}{2}\right)$

5. Translate the phrase "five cubed divided by three" to an algebraic expression.

6. Find the opposite of each expression.
(a) -17　(b) a

7. Find the decimal equivalent for the rational number.
(a) $\frac{3}{20}$　(b) $\frac{5}{8}$

8. Classify each number as one or more of the following: natural number, integer, rational number, or irrational number.
(a) $\frac{10}{2}$　(b) -5　(c) $\sqrt{5}$　(d) $-\frac{5}{6}$

9. Plot each number on the same number line.
(a) 0　(b) -3　(c) 2
(d) $\frac{3}{4}$　(e) $-\sqrt{2}$

10. Evaluate each expression.
(a) $|-12|$　(b) $|-a|$, if $a > 0$

11. Insert the symbol $>$ or $<$ to make the statement true.
(a) 4 ___ 9　(b) -1.3 ___ -0.5
(c) $|-3|$ ___ $|-5|$

12. List the following numbers from least to greatest.

$$\sqrt{3}, \quad -7, \quad 0, \quad \frac{1}{3}, \quad -1.6, \quad 3^2$$

1.5 ADDITION AND SUBTRACTION OF REAL NUMBERS

Addition of Real Numbers ■ Subtraction of Real Numbers ■ Applications

A LOOK INTO MATH ▷ There are four arithmetic operations: addition, subtraction, multiplication, and division. You probably learned to use arithmetic operations on whole numbers in elementary school. In this section we extend those concepts of addition and subtraction of real numbers to include addition and subtraction of positive *and* negative numbers. Both addition and subtraction require two numbers to calculate an answer. As a result, these operations are called **binary operations**. Multiplication and division are also binary operations. Some operations, such as finding the opposite (negation), are called **unary operations** because they require only one number.

Addition of Real Numbers

In an addition problem the two numbers added are called **addends**, and the answer is called the **sum**. In the addition problem $3 + 5 = 8$, the numbers 3 and 5 are the addends and the number 8 is the sum.

The *opposite* (or *additive inverse*) of a real number a is $-a$. When we add opposites, the result is 0. That is, $a + (-a) = 0$ for every real number a.

EXAMPLE 1   Adding opposites

Find the opposite of each number and calculate the sum of the number and its opposite.
(a) 45   **(b)** $\sqrt{2}$ **(c)** $-\frac{1}{2}$

Solution
(a) The opposite of 45 is -45. Their sum is $45 + (-45) = 0$.
(b) The opposite of $\sqrt{2}$ is $-\sqrt{2}$. Their sum is $\sqrt{2} + (-\sqrt{2}) = 0$.
(c) The opposite of $-\frac{1}{2}$ is $\frac{1}{2}$. Their sum is $-\frac{1}{2} + \frac{1}{2} = 0$.   Now Try Exercise **13**

When adding real numbers, it may be helpful to think of money. A positive number represents income, and a negative number indicates a debt. The sum $9 + (-5) = 4$ would represent being paid \$9 and owing \$5. In this case \$4 would be left over. Similarly, the sum $-3 + (-6) = -9$ would represent debts of \$3 and \$6, resulting in a total debt of \$9. To add two real numbers we can use the following rules.

> **ADDITION OF REAL NUMBERS**
>
> To add two numbers that are either *both positive* or *both negative*, add their absolute values. Their sum has the same sign as the two numbers.
>
> To add two numbers with *opposite signs*, find the absolute value of each number. Subtract the smaller absolute value from the larger. The sum has the same sign as the sign of the number with the larger absolute value. If the two numbers are opposites, their sum is 0.

EXAMPLE 2   Adding real numbers

Evaluate each expression.
(a) $-2 + (-4)$ **(b)** $-\frac{2}{5} + \frac{7}{10}$ **(c)** $6.2 + (-8.5)$

Solution
(a) The numbers are both negative, so we add the absolute values $|-2|$ and $|-4|$ to obtain 6. The signs of the addends are both negative, so the answer is -6. That is, $-2 + (-4) = -6$. If we owe \$2 and then owe \$4, the total amount owed is \$6.
(b) The numbers have opposite signs, so we subtract their absolute values to obtain

$$\frac{7}{10} - \frac{2}{5} = \frac{7}{10} - \frac{4}{10} = \frac{3}{10}.$$

The sum is positive because $\left|\frac{7}{10}\right|$ is greater than $\left|-\frac{2}{5}\right|$. That is, $-\frac{2}{5} + \frac{7}{10} = \frac{3}{10}$. If we spend \$0.40 $\left(-\frac{2}{5} = -0.4\right)$ and receive \$0.70 $\left(\frac{7}{10} = 0.7\right)$, we have \$0.30 $\left(\frac{3}{10} = 0.3\right)$ left.
(c) $6.2 + (-8.5) = -2.3$ because $|-8.5|$ is 2.3 more than $|6.2|$. If we have \$6.20 and we owe \$8.50, we are short \$2.30.   Now Try Exercises **35, 41, 45**

ADDING INTEGERS VISUALLY   One way to add integers visually is to use the symbol ⌢ to represent a positive unit and to use the symbol ⌣ to represent a negative unit. Now adding opposites visually results in "zero," as shown.

For example, to add $-3 + 5$, we draw three negative units and five positive units.

Because the "zeros" add no value, the sum is two positive units, or 2.

EXAMPLE 3   **Adding integers visually**

Perform the addition visually using the symbols ⌢ and ⌣.
(a) $3 + 2$ **(b)** $-6 + 4$ **(c)** $2 + (-3)$ **(d)** $(-5) + (-2)$

Solution
(a) Draw three positive units and then draw two more positive units.

Because no zeros were formed, the sum is five positive units, or 5.
(b) Draw six negative units and then draw four positive units.

Ignoring the zeros that were formed, the sum is two negative units, or -2.
(c)   Draw two positive and three negative units.

The sum is -1.
(d)

The sum is -7.   Now Try Exercises 21, 25

Figure 1.23 $4 + (-3) = 1$

Another way to add integers visually is to use a number line. To add $4 + (-3)$ start at 0 (the origin) and draw an arrow to the right 4 units long from 0 to 4. The number -3 is a negative number, so draw an arrow 3 units long to the left, starting at the tip of the first arrow. See Figure 1.23. The tip of the second arrow is at 1, which equals the sum of 4 and -3.

Figure 1.24 $-2 + (-3) = -5$

To find the sum $-2 + (-3)$, draw an arrow 2 units long to the left, starting at the origin. Then draw an arrow 3 units long to the left, starting at the tip of the first arrow, which is located at -2. See Figure 1.24. Because the tip of the second arrow coincides with -5 on the number line, the sum of -2 and -3 is -5.

Subtraction of Real Numbers

▶ **REAL-WORLD CONNECTION**   The answer to a subtraction problem is the **difference**. Addition and subtraction occur every day at grocery stores, where the costs of various items are added to the total and discounts from coupons are subtracted. When subtracting two real numbers, changing the problem to an addition problem may be helpful.

> ### SUBTRACTION OF REAL NUMBERS
>
> For any real numbers a and b,
>
> $$a - b = a + (-b).$$
>
> To subtract b from a, add a and the opposite of b.

EXAMPLE 4   Subtracting real numbers

Find each difference by hand.
(a) $10 - 20$ **(b)** $-5 - 2$ **(c)** $-2.1 - (-3.2)$ **(d)** $\frac{1}{2} - \left(-\frac{3}{4}\right)$

Solution
(a) $10 - 20 = 10 + (-20) = -10$
(b) $-5 - 2 = -5 + (-2) = -7$
(c) $-2.1 - (-3.2) = -2.1 + 3.2 = 1.1$
(d) $\frac{1}{2} - \left(-\frac{3}{4}\right) = \frac{1}{2} + \frac{3}{4} = \frac{2}{4} + \frac{3}{4} = \frac{5}{4}$ Now Try Exercise **53**

In the next example, we show how to add and subtract groups of numbers.

EXAMPLE 5   Adding and subtracting real numbers

Evaluate each expression.
(a) $5 - 4 - (-6) + 1$ **(b)** $\frac{1}{2} - \frac{3}{4} + \frac{1}{3}$ **(c)** $-6.1 + 5.6 - 10.1$

Solution
(a) Rewrite the expression in terms of addition only, and then find the sum.

$$5 - 4 - (-6) + 1 = 5 + (-4) + 6 + 1$$
$$= 1 + 6 + 1$$
$$= 8$$

(b) Begin by rewriting the fractions with the LCD of 12.

$$\frac{1}{2} - \frac{3}{4} + \frac{1}{3} = \frac{6}{12} - \frac{9}{12} + \frac{4}{12}$$
$$= \frac{6}{12} + \left(-\frac{9}{12}\right) + \frac{4}{12}$$
$$= -\frac{3}{12} + \frac{4}{12}$$
$$= \frac{1}{12}$$

(c) The expression can be evaluated by changing subtraction to addition.

$$-6.1 + 5.6 - 10.1 = -6.1 + 5.6 + (-10.1)$$
$$= -0.5 + (-10.1)$$
$$= -10.6$$

Now Try Exercises 71, 79

Applications

▶ **REAL-WORLD CONNECTION**   In word problems, certain phrases often indicate that we should add two numbers. For example, if a person's age is 10 more than 15, then we add 10 and 15 to obtain 25. The phrase *more than* often indicates addition. Other words that indicate addition are *sum, greater, greater than, plus,* and *increased by.*

Similarly, if the outside temperature is 10°F less than 32°F, then the outside temperature is $32 - 10 = 22$°F. The phrase *less than* often indicates subtraction. Other words that indicate subtraction are *difference, less, minus,* and *decreased by.*

EXAMPLE 6   Calculating temperature differences

The hottest outdoor temperature ever recorded in the shade was 136°F in the Sahara desert, and the coldest outside temperature ever recorded was −129°F in Antarctica. Find the temperature difference between these two temperatures.   (**Source:** *Guinness Book of Records.*)

Solution

The word *difference* indicates subtraction. We must subtract the two temperatures.

$$136 - (-129) = 136 + 129 = 265°\text{F}.$$ Now Try Exercise 97

▶ **REAL-WORLD CONNECTION**   Addition of positive and negative numbers occurs at banks if positive numbers represent deposits and negative numbers represent withdrawals.

EXAMPLE 7   Balancing a checking account

The initial balance in a checking account is $285. Find the final balance if the following represents a list of withdrawals and deposits: −$15, −$20, $500, and −$100.

Solution
Find the sum of the five numbers.

$$285 + (-15) + (-20) + 500 + (-100) = 270 + (-20) + 500 + (-100)$$
$$= 250 + 500 + (-100)$$
$$= 750 + (-100)$$
$$= 650$$

Figure 1.25

The final balance is $650. This result may be supported by evaluating the expression with a calculator, as illustrated in Figure 1.25. Note that the expression has been evaluated two different ways.

Now Try Exercise 93

TECHNOLOGY NOTE: Subtraction and Negation

Calculators typically have *different* keys to represent subtraction and negation. Be sure to use the correct key. Many graphing calculators have the following keys for subtraction and negation.

−		(−)
Subtraction		*Negation*

1.5 PUTTING IT ALL TOGETHER

The following table outlines how to add and subtract real numbers.

Operation	Comments	Examples
Addition	To add real numbers, use a number line or follow the rules found in the box on page 45.	$-2 + 8 = 6$ $0.8 + (-0.3) = 0.5$ $-\frac{1}{7} + \left(-\frac{3}{7}\right) = \frac{-1 + (-3)}{7} = -\frac{4}{7}$ $-4 + 4 = 0$ $-3 + 4 + (-2) = -1$
Subtraction	To subtract real numbers, transform the problem to an addition problem by adding the opposite. $a - b = a + (-b)$	$6 - 8 = 6 + (-8) = -2$ $-3 - 4 = -3 + (-4) = -7$ $-\frac{1}{2} - \left(-\frac{3}{2}\right) = -\frac{1}{2} + \frac{3}{2} = \frac{2}{2} = 1$ $-5 - (-5) = -5 + 5 = 0$ $9.4 - (-1.2) = 9.4 + 1.2 = 10.6$

1.5 Exercises

PRACTICE WATCH DOWNLOAD READ REVIEW

CONCEPTS

1. Addition and subtraction are called binary operations because each requires _____ numbers to calculate an answer.

2. In an addition problem, the two numbers added are called _____.

3. The solution to an addition problem is the _____.

4. When you add opposites, the sum is always _____.

5. If two positive numbers are added, the sum is always a _____ number.

6. If two negative numbers are added, the sum is always a _____ number.

7. If two numbers with opposite signs are added, the sum has the same sign as the number with the larger _____.

8. The solution to a subtraction problem is the _____.

9. When subtracting two real numbers, it may be helpful to change the problem to a(n) _____ problem.

10. To subtract b from a, add the _____ of b to a. That is, $a - b = a +$ _____.

11. The words *sum*, *more*, and *plus* indicate that _____ should be performed.

12. The words *difference*, *less than*, and *minus* indicate that _____ should be performed.

ADDITION AND SUBTRACTION OF REAL NUMBERS

Exercises 13–18: Find the opposite of the number and then calculate the sum of the number and its opposite.

13. 25

14. $-\frac{1}{2}$

15. $-\sqrt{21}$

16. $-\pi$

17. 5.63

18. -5^2

Exercises 19–26: Refer to Example 3. Find the sum visually.

19. $1 + 3$ **20.** $3 + 1$

21. $4 + (-2)$ **22.** $-4 + 6$

23. $-1 + (-2)$ **24.** $-2 + (-2)$

25. $-3 + 7$ **26.** $5 + (-6)$

Exercises 27–34: Use a number line to find the sum.

27. $-1 + 3$ **28.** $3 + (-1)$

29. $4 + (-5)$ **30.** $2 + 6$

31. $-10 + 20$ **32.** $15 + (-5)$

33. $-50 + (-100)$ **34.** $-100 + 100$

Exercises 35–52: Find the sum.

35. $5 + (-4)$ **36.** $-9 + 7$

37. $-1 + (-6)$ **38.** $-10 + (-23)$

39. $\frac{3}{4} + \left(-\frac{1}{2}\right)$ **40.** $-\frac{5}{12} + \left(-\frac{1}{6}\right)$

41. $-\frac{6}{7} + \frac{3}{14}$ **42.** $-\frac{1}{5} + \frac{2}{5}$

43. $-\frac{1}{2} + \left(-\frac{3}{4}\right)$ **44.** $-\frac{2}{9} + \left(-\frac{1}{12}\right)$

45. $0.6 + (-1.7)$ **46.** $4.3 + (-2.4)$

47. $-52 + 86$ **48.** $-103 + (-134)$

49. $8 + (-7) + (-2)$ **50.** $-10 + 20 + (-10)$

51. $\frac{1}{2} + \frac{3}{4} + \left(-\frac{1}{2}\right) + \left(-\frac{3}{4}\right)$

52. $0.1 + (-0.4) + (-1.5) + 0.9$

Exercises 53–68: Find the difference.

53. $5 - 8$ **54.** $3 - 5$

55. $-2 - (-9)$ **56.** $-10 - (-19)$

57. $\frac{1}{3} - \left(-\frac{2}{3}\right)$ **58.** $\frac{3}{4} - \frac{5}{4}$

59. $\frac{6}{7} - \frac{13}{14}$ **60.** $-\frac{5}{6} - \frac{1}{6}$

61. $-\frac{1}{10} - \left(-\frac{3}{5}\right)$ **62.** $-\frac{2}{11} - \left(-\frac{5}{11}\right)$

63. $0.8 - (-2.1)$ **64.** $-9.6 - (-5.7)$

65. $-73 - 91$ **66.** $201 - 502$

67. $-7 - (-6) - 10$ **68.** $-20 - 30 - (-40)$

Exercises 69–82: Evaluate the expression.

69. $10 - 19$ **70.** $5 + (-9)$

71. $19 - (-22) + 1$ **72.** $53 + (-43) - 10$

73. $-3 + 4 - 6$ **74.** $-11 + 8 - 10$

75. $100 - 200 + 100 - (-50)$

76. $-50 - (-40) + (-60) + 80$

77. $1.5 - 2.3 + 9.6$

78. $10.5 - (-5.5) + (-1.5)$

79. $-\frac{1}{2} + \frac{1}{4} - \left(-\frac{3}{4}\right)$ **80.** $\frac{1}{4} - \left(-\frac{2}{5}\right) + \left(-\frac{3}{20}\right)$

81. $|4 - 9| - |1 - 7|$

82. $|-5 - (-3)| - |-6 + 8|$

Exercises 83–92: Write an arithmetic expression for the given phrase and then simplify it.

83. The sum of two and negative five

84. Subtract ten from negative six

85. Negative five increased by seven

86. Negative twenty decreased by eight

87. The additive inverse of the quantity two cubed

88. Five minus the quantity two cubed

89. The difference between negative six and seven (*Hint:* Write the numbers for the subtraction problem in the order given.)

90. The difference between one-half and three-fourths

91. Six plus negative ten minus five

92. Ten minus seven plus negative twenty

APPLICATIONS

93. *Checking Account*   The initial balance in a checking account is $358. Find the final balance resulting from the following withdrawals and deposits: −$45, $37, $120, and −$240.

94. *Savings Account*   A savings account has $1245 in it. Find the final balance resulting from the following withdrawals and deposits: −$189, $975, −$226, and −$876.

95. *Football Stats*   A running back carries the ball five times. Find his total yardage if the carries were 9, −2, −1, 14, and 5 yards.

96. *Football Stats*   A football team starts on the 50 yard line and makes gains of 12, 6, and 22 yards. However, during these three plays the team is assessed penalties of 10 and 15 yards. On what yard line is the team after these plays?

97. *Deepest and Highest*   The deepest point in the ocean is the Mariana Trench, which is 35,839 feet below sea level. The highest point on Earth is Mount Everest, which is 29,029 feet above sea level. What is the difference in height between Mount Everest and the Mariana Trench?   (***Source:*** *The Guinness Book of Records.*)

98. *Greatest Temperature Ranges*   The greatest temperature range on Earth occurs in Siberia, where the temperature can vary between 98°F in the summer and −90°F

in the winter. Find the difference between these two temperatures.   (***Source:*** *The Guinness Book of Records.*)

WRITING ABOUT MATHEMATICS

99. Explain how to add two negative numbers. Give an example.

100. Explain how to subtract a negative number from a positive number. Give an example.

1.6 MULTIPLICATION AND DIVISION OF REAL NUMBERS

Multiplication of Real Numbers ▪ Division of Real Numbers ▪ Applications

A LOOK INTO MATH ▷   In some businesses, such as auto-parts stores, sometimes it is necessary to convert fractions to decimal numbers and decimals to fractions. For example, it might be necessary to convert the size of a $2\frac{3}{8}$-inch bolt to a decimal number before it can be ordered. This conversion makes use of division of real numbers. In this section we discuss both multiplication and division of real numbers.   (***Source:*** NAPA.)

Multiplication of Real Numbers

In a multiplication problem, the numbers multiplied are called the *factors*, and the answer is called the *product*. In the problem $3 \cdot 5 = 15$, the numbers 3 and 5 are factors and 15 is the product. Multiplication is a fast way to perform addition. For example, $5 \cdot 2 = 10$ is equivalent to finding the sum of five 2s, or

$$2 + 2 + 2 + 2 + 2 = 10.$$

Similarly, the product $5 \cdot (-2)$ is equivalent to finding the sum of five −2s, or

$$(-2) + (-2) + (-2) + (-2) + (-2) = -10.$$

Thus $5 \cdot (-2) = -10$. In general, the product of a positive number and a negative number is a negative number.

What sign should the product of two negative numbers have? To answer this question, consider the following patterns.

$$
\begin{array}{ll}
3 \cdot 2 = 6 & 3 \cdot (-2) = -6 \\
2 \cdot 2 = 4 & 2 \cdot (-2) = -4 \\
1 \cdot 2 = 2 & 1 \cdot (-2) = -2 \\
0 \cdot 2 = 0 & 0 \cdot (-2) = 0 \\
-1 \cdot 2 = -2 & -1 \cdot (-2) = ? \\
-2 \cdot 2 = -4 & -2 \cdot (-2) = ?
\end{array}
$$

In the first column, each time the first factor is reduced by 1, the product is found by *subtracting* 2 from the product in the previous row. Note that because multiplication is a fast way to perform addition, the number of 2s being added is one less as we move down the rows in the first column.

In the second column, each time the first factor is reduced by 1, the product is found by *subtracting* -2 from the previous product, which is equivalent to *adding* 2. If the pattern in the second column is to continue, then

$$-1 \cdot (-2) = \mathbf{2}$$
$$-2 \cdot (-2) = \mathbf{4}$$
$$-3 \cdot (-2) = \mathbf{6},$$

and so on. Note that, in general, the product of two negative numbers is a positive number.

SIGNS OF PRODUCTS

The product of two numbers with *like* signs is positive. The product of two numbers with *unlike* signs is negative.

As a result of these two rules, $4 \cdot 5 = 20$, $4 \cdot (-5) = -20$, $-4 \cdot 5 = -20$, and $-4 \cdot (-5) = 20$. Furthermore, multiplying a number by -1 results in the additive inverse (opposite) of the number. For example, $-1 \cdot 4 = -4$ and -4 is the additive inverse of 4. Similarly, $-1 \cdot (-3) = 3$ and 3 is the additive inverse of -3. In general, $-1 \cdot a = -a$.

EXAMPLE 1   Multiplying real numbers

Find each product by hand.
(a) $-9 \cdot 7$ **(b)** $\frac{2}{3} \cdot \frac{5}{9}$ **(c)** $-2.1(-40)$ **(d)** $(-2.5)(4)(-9)(-2)$

Solution
(a) The resulting product is negative because the factors -9 and 7 have unlike signs. Thus $-9 \cdot 7 = -63$.
(b) The product is positive because both factors are positive.

$$\frac{2}{3} \cdot \frac{5}{9} = \frac{2 \cdot 5}{3 \cdot 9} = \frac{10}{27}$$

(c) As both factors are negative, the product is positive. Thus $-2.1(-40) = 84$.
(d) $(-2.5)(4)(-9)(-2) = (-10)(-9)(-2) = (90)(-2) = -180$

Now Try Exercises 19, 29, 35

MAKING CONNECTIONS

Multiplying More Than Two Negative Factors

Because the product of two negative numbers is positive, it is possible to determine the sign of a product by counting the number of negative factors. For example, the product $-3 \cdot 4 \cdot (-5) \cdot 7 \cdot (-6)$ is negative because there are an odd number of negative factors and the product $2 \cdot (-1) \cdot (-3) \cdot (-5) \cdot (-4)$ is positive because there are an even number of negative factors.

The rules for the signs of products allow us to evaluate powers of signed numbers. However, it is important to note that $(-3)^2$ means that -3 should be multiplied with itself two times, but -3^2 means we should negate the product $3 \cdot 3$. The next example illustrates the difference between powers of signed numbers and the opposite of an exponential expression.

EXAMPLE 2   **Evaluating real numbers with exponents**

Evaluate each expression by hand.
(a) $(-4)^2$ (b) -4^2 (c) $(-2)^3$ (d) -2^3

Solution
(a) Because the exponent is outside of parentheses, the base of the exponential expression is -4. The expression is evaluated as
$$(-4)^2 = (-4)(-4) = 16.$$

(b) This is the negation of an exponential expression with base 4. Evaluating the exponent before negating results in $-4^2 = -(4)(4) = -16$.

(c) $(-2)^3 = (-2)(-2)(-2) = -8$

(d) $-2^3 = -(2)(2)(2) = -8$

Now Try Exercises **43, 45**

MAKING CONNECTIONS

Negative Square Roots

Because the product of two negative numbers is positive, $(-3) \cdot (-3) = 9$. That is, -3 is a square root of 9. As discussed in Section 1.4, every positive number has one positive square root and one negative square root. For a positive number a, the positive square root is called the principal square root and is denoted $\sqrt{a}$. The negative square root is denoted $-\sqrt{a}$. For example, $\sqrt{4} = 2$ and $-\sqrt{4} = -2$.

Division of Real Numbers

In the division problem $20 \div 4 = 5$, the number 20 is the **dividend**, 4 is the **divisor**, and 5 is the **quotient**. This division problem can also be written in fraction form as $\frac{20}{4} = 5$. Division of real numbers can be defined in terms of multiplication and reciprocals. The reciprocal, or multiplicative inverse, of a real number a is $\frac{1}{a}$. The number 0 has *no reciprocal*.

TECHNOLOGY NOTE

Try dividing 5 by 0 with a calculator. On most calculators dividing a number by 0 results in an error message. Does it on your calculator?

DIVIDING REAL NUMBERS

For real numbers a and b with $b \neq 0$,
$$\frac{a}{b} = a \cdot \frac{1}{b}.$$
That is, to divide a by b, multiply a by the reciprocal of b.

NOTE: Division by 0 is undefined because 0 has no reciprocal.

EXAMPLE 3   **Dividing real numbers**

Evaluate each expression by hand.

(a) $-16 \div \frac{1}{4}$ (b) $\frac{\frac{3}{5}}{-8}$ (c) $\frac{-8}{-36}$ (d) $9 \div 0$

Solution

(a) $-16 \div \frac{1}{4} = \frac{-16}{1} \cdot \frac{4}{1} = \frac{-64}{1} = -64$

(b) $\dfrac{\frac{3}{5}}{-8} = \frac{3}{5} \div (-8) = \frac{3}{5} \cdot \left(-\frac{1}{8}\right) = -\frac{3}{40}$

(c) $\frac{-8}{-36} = -8 \cdot \left(-\frac{1}{36}\right) = \frac{8}{36} = \frac{2}{9}$

(d) $9 \div 0$ is undefined. The number 0 has no reciprocal. **Now Try Exercises** 51, 59, 71

When determining the sign of a quotient, the following rules may be helpful.

SIGNS OF QUOTIENTS

The quotient of two numbers with *like* signs is positive. The quotient of two numbers with *unlike* signs is negative.

As a result of these two rules, $\frac{20}{4} = 5$, $\frac{-20}{4} = -5$, $\frac{20}{-4} = -5$, and $\frac{-20}{-4} = 5$. You may wonder why a negative number divided by a negative number is a positive number. Remember that division is a fast way to perform subtraction. For example, because

$$20 - 4 - 4 - 4 - 4 - 4 = 0,$$

there are five 4s in 20 and so $\frac{20}{4} = 5$. To check this answer, multiply 5 and 4 to obtain $5 \cdot 4 = 20$. Similarly,

$$-20 - (-4) - (-4) - (-4) - (-4) - (-4) = 0.$$

Thus there are five -4s in -20 and so $\frac{-20}{-4} = 5$. To check this answer, multiply 5 and -4 to obtain $5 \cdot (-4) = -20$.

▶ **REAL-WORLD CONNECTION** In business, employees may need to convert fractions and mixed numbers to decimal numbers. The next example illustrates this process.

EXAMPLE 4 **Converting fractions to decimals**

Convert each measurement to a decimal number.

(a) $2\frac{3}{8}$-inch bolt **(b)** $\frac{15}{16}$-inch diameter **(c)** $1\frac{1}{3}$-cup flour

Solution

(a) Because $\frac{3}{8}$ indicates the division problem $3 \div 8$, begin by dividing 3 by 8.

$$
\begin{array}{r}
0.375 \\
8\overline{)3.000} \\
\underline{24} \\
60 \\
\underline{56} \\
40 \\
\underline{40} \\
0
\end{array}
$$

Thus the mixed number $2\frac{3}{8}$ is equivalent to the decimal number 2.375.

(b) Start by dividing 15 by 16.

$$
\begin{array}{r}
0.9375 \\
16\overline{\smash{)}15.0000} \\
\underline{144} \\
60 \\
\underline{48} \\
120 \\
\underline{112} \\
80 \\
\underline{80} \\
0
\end{array}
$$

Thus the fraction $\frac{15}{16}$ is equivalent to the decimal number 0.9375.

(c) Start by dividing 1 by 3.

$$
\begin{array}{r}
0.333\ldots \\
3\overline{\smash{)}1.000} \\
\underline{9} \\
10 \\
\underline{9} \\
10 \\
\underline{9} \\
1
\end{array}
$$

Thus the mixed number $1\frac{1}{3}$ is equivalent to the decimal number 1.333..., or $1.\overline{3}$.

Now Try Exercises 77, 81, 83

In the next example, numbers expressed as terminating decimals are converted to fractions.

EXAMPLE 5 **Converting decimals to fractions**

Convert each decimal number to a fraction in lowest terms.

(a) 0.06   **(b)** 0.375   **(c)** 0.0025

Solution

(a) The decimal 0.06 equals six hundredths, or $\frac{6}{100}$. Simplifying this fraction gives

$$
\frac{6}{100} = \frac{3 \cdot 2}{50 \cdot 2} = \frac{3}{50}.
$$

(b) The decimal 0.375 equals three hundred seventy-five thousandths, or $\frac{375}{1000}$. Simplifying this fraction gives

$$
\frac{375}{1000} = \frac{3 \cdot 125}{8 \cdot 125} = \frac{3}{8}.
$$

(c) The decimal 0.0025 equals twenty-five ten thousandths, or $\frac{25}{10,000}$. Simplifying this fraction gives

$$
\frac{25}{10,000} = \frac{1 \cdot 25}{400 \cdot 25} = \frac{1}{400}.
$$

Now Try Exercises 89, 93

FRACTIONS AND CALCULATORS (OPTIONAL)   Many calculators have the capability to perform arithmetic on fractions and express the answer as either a decimal or a fraction. The next example illustrates this capability.

EXAMPLE **6**

Performing arithmetic operations with technology

Use a calculator to evaluate each expression. Express your answer as a decimal and as a fraction.

(a) $\frac{1}{3} + \frac{2}{5} - \frac{4}{9}$ **(b)** $\left(\frac{4}{9} \cdot \frac{3}{8}\right) \div \frac{2}{3}$

Solution
(a) Figure 1.26(a) shows that

$$\frac{1}{3} + \frac{2}{5} - \frac{4}{9} = 0.2\overline{8}, \text{ or } \frac{13}{45}.$$

NOTE: Generally it is a good idea to put parentheses around fractions when you are using a calculator.

(b) Figure 1.26(b) shows that $\left(\frac{4}{9} \cdot \frac{3}{8}\right) \div \frac{2}{3} = 0.25$, or $\frac{1}{4}$.

CALCULATOR HELP

To convert answers to fractions, see Appendix A (page AP-2).

```
(1/3)+(2/5)-(4/9
)
           .2888888889
(1/3)+(2/5)-(4/9
)▶Frac
                 13/45
```
(a)

```
((4/9)*(3/8))/(2
/3)
                   .25
((4/9)*(3/8))/(2
/3)▶Frac
                   1/4
```
(b)

Figure 1.26

Now Try Exercise **97**

MAKING CONNECTIONS

The Four Arithmetic Operations

If you know how to add real numbers, then you also know how to subtract real numbers because subtraction is defined in terms of addition. That is,

$$a - b = a + (-b).$$

If you know how to multiply real numbers, then you also know how to divide real numbers because division is defined in terms of multiplication. That is,

$$\frac{a}{b} = a \cdot \frac{1}{b}.$$

Applications

There are many instances when we may need to multiply or divide real numbers. Two examples are provided here.

EXAMPLE 7 Comparing top-grossing movies

Even though *Titanic* (1997) was an extremely successful film that set box office records, it ranks only sixth in top-grossing movies of all time when calculated by using estimated total admissions. The total admissions for *Titanic* are $\frac{10}{11}$ of the total admissions for *The Sound of Music* (1965), which ranks third.   (**Source:** *Exhibitor Relations Co., Inc.*)

(a) If *The Sound of Music* had estimated total admissions of 142 million people, find the estimated total admissions for *Titanic*.
(b) The top-grossing movie of all time is *Gone With the Wind* (1939), which had estimated total admissions of 202 million. How many more people saw *Gone With the Wind* than saw *Titanic*?

Solution

(a) To find the total admissions for *Titanic* we multiply the real numbers 142 and $\frac{10}{11}$ to obtain $142 \cdot \frac{10}{11} = \frac{1420}{11} \approx 129.09$.
Total admissions for *Titanic* were about 129 million people.
(b) The difference is $202 - 129 = 73$ million people.   Now Try Exercise 105

EXAMPLE 8 Analyzing the federal budget

In 2005, the federal government used $\frac{9}{125}$ of its budget to pay interest on loans. Write this fraction as a decimal.   (**Source:** *U.S. Office of Management and Budget.*)

Solution

One method for writing the fraction as a decimal is to divide 9 by 125 using long division. An alternative method is to multiply the fraction by $\frac{8}{8}$ so that the denominator becomes 1000. Then write the numerator in the thousandths place in the decimal.

$$\frac{9}{125} \cdot \frac{8}{8} = \frac{72}{1000} = 0.072$$ Now Try Exercise 107

1.6 PUTTING IT ALL TOGETHER

The following table summarizes some of the information presented in this section.

Topic	Comments	Examples	
Multiplication	The product of two numbers with like signs is positive, and the product of two numbers with unlike signs is negative.	$6 \cdot 7 = 42$	Like signs
		$6 \cdot (-7) = -42$	Unlike signs
		$-6 \cdot 7 = -42$	Unlike signs
		$-6 \cdot (-7) = 42$	Like signs

continued on next page

continued from previous page

Topic	Comments	Examples
Division	For real numbers a and b, with $b \neq 0$, $$\frac{a}{b} = a \cdot \frac{1}{b}.$$ The quotient of two numbers with like signs is positive, and the quotient of two numbers with unlike signs is negative.	$\dfrac{42}{6} = 7$ Like signs $\dfrac{-42}{6} = -7$ Unlike signs $\dfrac{42}{-6} = -7$ Unlike signs $\dfrac{-42}{-6} = 7$ Like signs $\dfrac{\frac{3}{2}}{-6} = \dfrac{3}{2} \cdot \left(-\dfrac{1}{6}\right)$ $\phantom{\dfrac{\frac{3}{2}}{-6}} = -\dfrac{3}{12}$ Unlike signs $\phantom{\dfrac{\frac{3}{2}}{-6}} = -\dfrac{1}{4}$
Converting Fractions to Decimals	The fraction $\frac{a}{b}$ is equivalent to $a \div b$.	The fraction $\frac{2}{9}$ is equivalent to $0.222\ldots$, or $0.\overline{2}$, because the division problem $$\begin{array}{r} 0.222\ldots \\ 9\overline{)2.00000} \\ \underline{18} \\ 20 \\ \underline{18} \\ 20 \\ \underline{18} \\ 2 \end{array}$$ results in a repeating decimal.
Converting Terminating Decimals to Fractions	Write the decimal as a fraction with a denominator equal to a power of 10 and then simplify this fraction.	$0.55 = \dfrac{55}{100}$ $ = \dfrac{11 \cdot 5}{20 \cdot 5}$ $ = \dfrac{11}{20}$

1.6 Exercises

CONCEPTS

1. In a multiplication problem, the numbers multiplied are called _____.

2. The solution to a multiplication problem is the _____.

3. The product of a positive number and a negative number is a _____ number.

4. The product of two negative numbers is _____.

5. The solution to a division problem is the _____.

6. In the problem $15 \div 5$, the number 15 is the _____ and the number 5 is the _____.

7. The reciprocal of a nonzero number a is _____.

8. Division by zero is undefined because zero has no _____.

9. The reciprocal of $-\frac{3}{4}$ is _____.

10. To divide a by b, multiply a by the _____ of b.

11. In general, $-1 \cdot a =$ _____.

12. $(-5)^2 =$ _____

13. $-5^2 =$ _____

14. $\dfrac{a}{b} = a \cdot$ _____

15. A negative number divided by a negative number is a _____ number.

16. A negative number divided by a positive number is a _____ number.

17. Division is a fast way to perform _____.

18. To convert $\frac{5}{8}$ to a decimal, divide _____ by _____.

MULTIPLICATION AND DIVISION OF REAL NUMBERS

Exercises 19–40: Multiply.

19. $-3 \cdot 4$

20. $-5 \cdot 7$

21. $6 \cdot (-3)$

22. $2 \cdot (-1)$

23. $0 \cdot (-2.13)$

24. $-2 \cdot (-7)$

25. $-6 \cdot (-10)$

26. $-3 \cdot (-1.7) \cdot 0$

27. $-\frac{1}{2} \cdot \left(-\frac{2}{4}\right)$

28. $-\frac{3}{4} \cdot \left(-\frac{5}{12}\right)$

29. $-\frac{3}{7} \cdot \frac{7}{3}$

30. $\frac{5}{8} \cdot \left(-\frac{4}{15}\right)$

31. $-10 \cdot (-20)$

32. $1000 \cdot (-70)$

33. $-50 \cdot 100$

34. $-0.5 \cdot (-0.3)$

35. $-2 \cdot 3 \cdot (-4) \cdot 5$

36. $-3 \cdot (-5) \cdot (-2) \cdot 10$

37. $-6 \cdot \frac{1}{6} \cdot \frac{7}{9} \cdot \left(-\frac{9}{7}\right) \cdot \left(-\frac{3}{2}\right)$

38. $-\frac{8}{5} \cdot \frac{1}{8} \cdot \left(-\frac{5}{7}\right) \cdot -7$

39. $(-1) \cdot (-1) \cdot (-1) \cdot (-1)$

40. $5 \cdot (-2) \cdot (-2) \cdot (-2)$

41. **Thinking Generally** Is the product given by the expression $a \cdot (-a) \cdot (-a) \cdot a \cdot (-a)$ positive or negative if $a > 0$?

42. **Thinking Generally** Is the product given by the expression $a \cdot (-a) \cdot (-a) \cdot a \cdot (-a)$ positive or negative if $a < 0$?

Exercises 43–50: Evaluate the expression.

43. $(-1)^3$

44. $(-6)^2$

45. -2^4

46. $-(-4)^2$

47. $-(-2)^3$

48. $3 \cdot (-3)^2$

49. $5 \cdot (-2)^3$

50. -1^4

Exercises 51–76: Divide.

51. $-10 \div 5$

52. $-8 \div 4$

53. $-20 \div (-2)$

54. $-15 \div (-3)$

55. $\frac{-12}{3}$

56. $\frac{-25}{-5}$

57. $\frac{39}{-13}$

58. $-\frac{100}{-20}$

59. $-16 \div \frac{1}{2}$

60. $10 \div \left(-\frac{1}{3}\right)$

61. $0 \div 3$

62. $\frac{0}{-5}$

63. $\frac{0}{-2}$

64. $\frac{-1}{0}$

65. $\frac{1}{2} \div (-11)$

66. $-\frac{3}{4} \div (-6)$

67. $\frac{-\frac{4}{5}}{-3}$

68. $\frac{\frac{7}{8}}{-7}$

69. $\frac{5}{6} \div \left(-\frac{8}{9}\right)$

70. $-\frac{11}{12} \div \left(-\frac{11}{4}\right)$

71. $-\frac{1}{2} \div 0$

72. $-9 \div 0$

73. $-0.5 \div \frac{1}{2}$

74. $-0.25 \div \left(-\frac{3}{4}\right)$

75. $-\frac{2}{3} \div 0.5$

76. $\frac{1}{6} \div 1.5$

CONVERTING BETWEEN FRACTIONS AND DECIMALS

Exercises 77–88: Write the number as a decimal.

77. $\frac{1}{2}$

78. $\frac{3}{4}$

79. $\frac{3}{16}$

80. $\frac{1}{9}$

81. $3\frac{1}{2}$

82. $2\frac{1}{4}$

83. $5\frac{2}{3}$

84. $6\frac{7}{9}$

85. $1\frac{7}{16}$

86. $\frac{11}{16}$

87. $\frac{7}{8}$

88. $6\frac{1}{12}$

Exercises 89–96: Write the decimal number as a fraction in lowest terms.

89. 0.25

90. 0.8

91. 0.16

92. 0.35

93. 0.625

94. 0.0125

95. 0.6875

96. 0.21875

 Exercises 97–104: Use a calculator to evaluate each expression. Express your answer as a decimal and as a fraction.

97. $\left(\frac{1}{3} + \frac{5}{6}\right) \div \frac{1}{2}$

98. $\frac{4}{9} - \frac{1}{6} + \frac{2}{3}$

99. $\frac{4}{5} \div \frac{2}{3} \cdot \frac{7}{4}$

100. $4 - \frac{7}{4} \cdot 2$

101. $\frac{15}{2} - 4 \cdot \frac{7}{3}$

102. $\frac{1}{6} - \frac{3}{5} + \frac{7}{8}$

103. $\frac{17}{40} + 3 \div 8$

104. $\frac{3}{4} \cdot \left(6 + \frac{1}{2}\right)$

APPLICATIONS

105. *Top-Grossing Movies* (Refer to Example 7.) *The Ten Commandments* (1956) is the fifth top-grossing movie of all time. Find the total admissions for *The Ten Commandments* if they were $\frac{13}{20}$ of the total admissions for the top-grossing movie of all time, *Gone With the Wind* (1939), which had total admissions of 202 million. (*Source: Exhibitor Relations Co., Inc.*)

106. *Planet Climate* Saturn has an average surface temperature of $-220°F$. Neptune has an average surface temperature that is $\frac{3}{2}$ times that of Saturn. Find the average surface temperature on Neptune. (*Source: NASA.*)

107. *Uninsured Americans* In 2005, the fraction of Americans who did not have health insurance coverage was $\frac{39}{250}$. Write this fraction as a decimal. (*Source: U.S. Census Bureau.*)

108. *Uninsured Minnesotans* In 2005, the fraction of Minnesotans who did not have health insurance coverage was $\frac{21}{250}$, the lowest in the country. Write this fraction as a decimal. (*Source: U.S. Census Bureau.*)

WRITING ABOUT MATHEMATICS

109. Division is a fast way to subtract. Consider the division problem $\frac{-6}{-2}$, whose quotient represents the number of -2s in -6. Using this idea, explain why the answer is a positive number.

110. Explain how to determine whether the product of three signed numbers is positive or negative.

CHECKING BASIC CONCEPTS

SECTIONS 1.5 AND 1.6

1. Find each sum.
   (a) $-4 + 4$ **(b)** $-10 + (-12) + 3$

2. Evaluate each expression.
   (a) $\frac{2}{3} - \left(-\frac{2}{9}\right)$ **(b)** $-1.2 - 5.1 + 3.1$

3. Write an arithmetic expression for the given phrase and then simplify it.
   (a) The sum of negative one and five
   (b) The difference between four and negative three

4. The hottest temperature ever recorded at International Falls, Minnesota, was 98°F, and the coldest temperature ever recorded was −46°F. What is the difference between these two temperatures?

5. Find each product.
(a) $-5 \cdot (-7)$ (b) $-\frac{1}{2} \cdot \frac{2}{3} \cdot \left(-\frac{4}{5}\right)$

6. Evaluate each expression.
(a) -3^2 (b) $4 \cdot (-2)^3$ (c) $(-5)^2$

7. Evaluate each expression.
(a) $-5 \div \frac{2}{3}$ (b) $-\frac{5}{8} \div \left(-\frac{4}{3}\right)$

8. What is the reciprocal of $-\frac{7}{6}$?

9. Simplify each expression.
(a) $\frac{-10}{2}$ (b) $\frac{10}{-2}$ (c) $-\frac{10}{2}$ (d) $\frac{-10}{-2}$

10. Convert each fraction or mixed number to a decimal number.
(a) $\frac{3}{5}$ (b) $3\frac{7}{8}$

1.7 PROPERTIES OF REAL NUMBERS

Commutative Properties ▪ Associative Properties ▪ Distributive Properties ▪ Identity and Inverse Properties ▪ Mental Calculations

A LOOK INTO MATH ▷
The order in which you perform actions is often important. For example, putting on your socks and then your shoes is not the same as putting on your shoes and then your socks. In mathematical terms, these two actions are not *commutative*. However, the actions of tying your shoes and putting on a sweatshirt probably are *commutative*. In mathematics some operations are commutative and others are not. In this section we discuss several properties of real numbers.

Commutative Properties

The **commutative property for addition** states that two numbers, a and b, can be added in any order and the result will be the same. That is, $a + b = b + a$. For example, if a person buys 4 CDs and then buys 9 CDs or first buys 9 CDs and then buys 4 CDs, the result is the same. Either way the person buys a total of

$$4 + 9 = 9 + 4 = 13 \text{ CDs.}$$

There is also a **commutative property for multiplication**. It states that two numbers, a and b, can be multiplied in any order and the result will be the same. That is, $a \cdot b = b \cdot a$. For example, if one person has 6 boxes each containing 12 greeting cards and another person has 12 boxes each containing 6 greeting cards, then each person has

$$6 \cdot 12 = 12 \cdot 6 = 72 \text{ greeting cards.}$$

We can summarize these results as follows.

COMMUTATIVE PROPERTIES

For any real numbers a and b,

$$a + b = b + a \qquad \text{Addition}$$

and

$$a \cdot b = b \cdot a. \qquad \text{Multiplication}$$

EXAMPLE   Applying the commutative properties

Use the commutative properties to rewrite each expression.
(a) $15 + 100$ **(b)** $a \cdot 8$ **(c)** $x(y + z)$

Solution
(a) By the commutative property for addition $15 + 100$ can be written as $100 + 15$.
(b) By the commutative property for multiplication $a \cdot 8$ can be written as $8 \cdot a$ or $8a$.
(c) We can apply both commutative properties.

$$x(y + z) = (y + z)x \qquad \text{Commutative property for multiplication}$$
$$= (z + y)x \qquad \text{Commutative property for addition}$$

Now Try Exercises 15, 21

While there are commutative properties for addition and multiplication, the operations of subtraction and division are *not* commutative. Changing the order in which numbers are subtracted can affect the result. For example, $5 - 3 = 2$ but $3 - 5 = -2$. Similarly, changing the order in which numbers are divided can affect the result. For example, $4 \div 2 = 2$ but $2 \div 4 = \frac{1}{2}$.

Associative Properties

The commutative properties allow us to interchange the order of two numbers when we add or multiply. The *associative properties* allow us to change how numbers are grouped. For example, if a person earns \$5, \$4, and \$6, then the total amount earned can be calculated either as

$$(5 + 4) + 6 = 9 + 6 = 15 \quad \text{or as}$$
$$5 + (4 + 6) = 5 + 10 = 15.$$

In either case we obtain the same answer, \$15, which is the result of the **associative property for addition**. We did not change the order of the numbers; we only changed how the numbers were grouped. There is also an **associative property for multiplication**, which is illustrated by

$$(5 \cdot 4) \cdot 6 = 20 \cdot 6 = 120 \quad \text{and}$$
$$5 \cdot (4 \cdot 6) = 5 \cdot 24 = 120.$$

We can summarize these results as follows.

ASSOCIATIVE PROPERTIES

For any real numbers a, b, and c,
$$(a + b) + c = a + (b + c) \qquad \text{Addition}$$
and
$$(a \cdot b) \cdot c = a \cdot (b \cdot c). \qquad \text{Multiplication}$$

NOTE: Sometimes we omit the multiplication dot. Thus $a \cdot b = ab$ and $5 \cdot x \cdot y = 5xy$.

EXAMPLE 2   **Applying the associative properties**

Use the associative property to rewrite each expression.
(a) $(5 + 6) + 7$ **(b)** $x(yz)$

Solution
(a) The given expression is equivalent to $5 + (6 + 7)$.
(b) The given expression is equivalent to $(xy)z$.

Now Try Exercises 23, 25

EXAMPLE 3   **Identifying properties of real numbers**

State the property that each equation illustrates.
(a) $5 \cdot (8y) = (5 \cdot 8)y$ **(b)** $3 \cdot 7 = 7 \cdot 3$ **(c)** $x + yz = yz + x$

Solution
(a) This equation illustrates the associative property for multiplication because the grouping of the numbers has been changed.
(b) This equation illustrates the commutative property for multiplication because the order of the numbers 3 and 7 has been changed.
(c) This equation illustrates the commutative property for addition because the order of the terms x and yz has been changed.

Now Try Exercises 55, 57

While there are associative properties for addition and multiplication, the operations of subtraction and division are *not* associative. Changing the grouping in a subtraction problem can affect the result. For example, $(5 - 3) - 2 = 2 - 2 = 0$ but $5 - (3 - 2) = 5 - 1 = 4$. Similarly, changing the grouping of the numbers in a division problem can affect the result. For example, $(24 \div 4) \div 2 = 6 \div 2 = 3$ but $24 \div (4 \div 2) = 24 \div 2 = 12$.

MAKING CONNECTIONS

Commutative and Associative Properties

Both the commutative and associative properties work for addition and multiplication. However, neither property works for subtraction or division.

Distributive Properties

The distributive properties are used frequently in algebra to simplify expressions. An example is

$$4(2 + 3) = 4 \cdot 2 + 4 \cdot 3.$$

The 4 must be multiplied by *both* the 2 and the 3—not just the 2.

The distributive property remains valid when addition is replaced with subtraction. For example,

$$4(2 - 3) = 4 \cdot 2 - 4 \cdot 3$$

is a true statement.

We illustrate a distributive property geometrically in Figure 1.27. Note that the area of one rectangle that is 4 squares by 5 squares is the same as the area of two rectangles: one that is 4 squares by 2 squares and another that is 4 squares by 3 squares. In either case the total area is 20 square units.

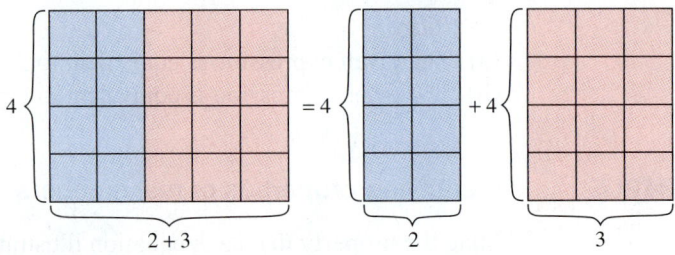

Figure 1.27 $4(2 + 3) = 4 \cdot 2 + 4 \cdot 3$

DISTRIBUTIVE PROPERTIES

For any real numbers a, b, and c,

$$a(b + c) = ab + ac \quad \text{and} \quad a(b - c) = ab - ac.$$

NOTE: Because multiplication is commutative, the distributive properties can also be written as

$$(b + c)a = ba + ca \quad \text{and} \quad (b - c)a = ba - ca.$$

EXAMPLE 4   Applying the distributive properties

Apply a distributive property to each expression.
(a) $3(x + 2)$ **(b)** $-6(a - 2)$ **(c)** $-(x + 7)$ **(d)** $15 - (b + 4)$

Solution
(a) Both the x and the 2 must be multiplied by 3.

$$3(x + 2) = 3 \cdot x + 3 \cdot 2 = 3x + 6$$

(b) $-6(a - 2) = -6 \cdot a - (-6) \cdot 2 = -6a + 12$

(c) $-(x + 7) = (-1)(x + 7)$ In general, $-a = -1 \cdot a$.

$\qquad\qquad\ = (-1) \cdot x + (-1) \cdot 7$ Distributive property

$\qquad\qquad\ = -x - 7$ Multiply.

(d) $15 - (b + 4) = 15 + (-1)(b + 4)$ Change subtraction to addition.

$\qquad\qquad\qquad\ = 15 + (-1) \cdot b + (-1) \cdot 4$ Distributive property

$\qquad\qquad\qquad\ = 15 - b - 4$ Multiply.

$\qquad\qquad\qquad\ = 11 - b$ Subtract.

NOTE: To simplify the expression $15 - (b + 4)$, we subtract *both* the b and the 4. Thus we can quickly simplify the given expression to

$$15 - (b + 4) = 15 - b - 4.$$

Now Try Exercises 33, 39, 43

EXAMPLE 5 **Inserting parentheses using the distributive property**

Use the distributive property to insert parentheses in the expression and then simplify the result.
(a) $5a + 2a$ **(b)** $3x - 7x$ **(c)** $-4y + 5y$

Solution
(a) Because a is a factor in both $5a$ and $2a$, we use the distributive property to write the a outside of parentheses.

$$5a + 2a = (5 + 2)a \qquad \text{Distributive property}$$
$$= 7a \qquad \text{Simplify.}$$

(b) $3x - 7x = (3 - 7)x = -4x$
(c) $-4y + 5y = (-4 + 5)y = 1y = y$

Now Try Exercises 47, 49

EXAMPLE 6 **Identifying properties of real numbers**

State the property or properties illustrated by each equation.
(a) $4(5 - x) = 20 - 4x$ **(b)** $(4 + x) + 5 = x + 9$
(c) $5z + 7z = 12z$ **(d)** $x(y + z) = zx + yx$

Solution
(a) This equation illustrates the distributive property with subtraction.

$$4(5 - x) = 4 \cdot 5 - 4 \cdot x = 20 - 4x$$

(b) This equation illustrates the commutative and associative properties for addition.

$$(4 + x) + 5 = (x + 4) + 5 \qquad \text{Commutative property for addition}$$
$$= x + (4 + 5) \qquad \text{Associative property for addition}$$
$$= x + 9 \qquad \text{Simplify.}$$

(c) This equation illustrates the distributive property with addition.

$$5z + 7z = (5 + 7)z = 12z$$

(d) This equation illustrates a distributive property and commutative properties.

$$x(y + z) = xy + xz \qquad \text{Distributive property}$$
$$= xz + xy \qquad \text{Commutative property for addition}$$
$$= zx + yx \qquad \text{Commutative property for multiplication}$$

Now Try Exercises 59, 61

Identity and Inverse Properties

The **identity property of 0** states that if 0 is added to any real number a, the result is a. The number 0 is called the **additive identity**. Examples include

$$-4 + 0 = -4 \quad \text{and} \quad 0 + 11 = 11.$$

The **identity property of 1** states that if any number a is multiplied by 1, the result is a. The number 1 is called the **multiplicative identity**. Examples include

$$-3 \cdot 1 = -3 \quad \text{and} \quad 1 \cdot 8 = 8.$$

We can summarize these results as follows.

IDENTITY PROPERTIES

For any real number a,

$$a + 0 = 0 + a = a \qquad \text{Additive identity}$$

and

$$a \cdot 1 = 1 \cdot a = a. \qquad \text{Multiplicative identity}$$

The *additive inverse*, or *opposite*, of a number a is $-a$. The number 0 is its own opposite. The opposite of -5 is $-(-5) = 5$ and the opposite of x is $-x$. The sum of a number a and its additive inverse equals the additive identity 0. Thus $-5 + 5 = 0$ and $x + (-x) = 0$.

The *multiplicative inverse*, or *reciprocal*, of a nonzero number a is $\frac{1}{a}$. The number 0 has no multiplicative inverse. The multiplicative inverse of $-\frac{5}{4}$ is $-\frac{4}{5}$. The product of a number and its multiplicative inverse equals the multiplicative identity 1. Thus $-\frac{5}{4} \cdot \left(-\frac{4}{5}\right) = 1$.

INVERSE PROPERTIES

For any real number a,

$$a + (-a) = 0 \quad \text{and} \quad -a + a = 0. \qquad \text{Additive inverse}$$

For any *nonzero* real number a,

$$a \cdot \frac{1}{a} = 1 \quad \text{and} \quad \frac{1}{a} \cdot a = 1. \qquad \text{Multiplicative inverse}$$

EXAMPLE 7 **Identifying identity and inverse properties**

State the property or properties illustrated by each equation.

(a) $0 + xy = xy$

(b) $\frac{36}{30} = \frac{6}{5} \cdot \frac{6}{6} = \frac{6}{5}$

(c) $x + (-x) + 5 = 0 + 5 = 5$

(d) $\frac{1}{9} \cdot 9y = 1 \cdot y = y$

Solution

(a) This equation illustrates use of the identity property for 0.

(b) Because $\frac{6}{6} = 1$, these equations illustrate how a fraction can be simplified by using the identity property for 1.

(c) These equations illustrate use of the additive inverse property and the identity property for 0.

(d) These equations illustrate use of the multiplicative inverse property and the identity property of 1.

Now Try Exercises 69, 75

Mental Calculations

Properties of numbers can be used to simplify calculations. For example, to find the sum

$$4 + 7 + 6 + 3$$

we might apply the commutative and associative properties for addition to obtain

$$(4 + 6) + (7 + 3) = 10 + 10 = 20.$$

Similarly, we might calculate

$$14 + 7 + 16 + 33 \quad \text{as}$$

$$(14 + 16) + (7 + 33) = 30 + 40 = 70.$$

Suppose that we are to add $128 + 19$ mentally. One way is to add 20 to 128 and then subtract 1.

$$
\begin{aligned}
128 + 19 &= 128 + (20 - 1) && 19 = 20 - 1 \\
&= (128 + 20) - 1 && \text{Associative property} \\
&= 148 - 1 && \text{Add.} \\
&= 147. && \text{Subtract.}
\end{aligned}
$$

Similarly, we could find the sum $128 + 98$ by first adding 100 to 128 to obtain 228 and then subtracting 2, resulting in 226.

The distributive property can be helpful when we are multiplying mentally. For example, to estimate the number of people in a marching band with 7 columns and 23 rows we need to find the product $7 \cdot 23$. To evaluate the product mentally, think of 23 as $20 + 3$.

$$
\begin{aligned}
7(20 + 3) &= 7 \cdot 20 + 7 \cdot 3 && \text{Distributive property} \\
&= 140 + 21 && \text{Multiply.} \\
&= 161 && \text{Add.}
\end{aligned}
$$

EXAMPLE 8   **Performing calculations mentally**

Use properties of real numbers to calculate each expression mentally.
(a) $21 + 15 + 9 + 5$ **(b)** $\frac{1}{2} \cdot \frac{2}{3} \cdot 2 \cdot \frac{3}{2}$
(c) $523 + 199$ **(d)** $6 \cdot 55$

Solution
(a) Use the commutative and associative properties to group numbers that sum to a multiple of 10.

$$21 + 15 + 9 + 5 = (21 + 9) + (15 + 5) = 30 + 20 = 50$$

(b) Use the commutative and associative properties to group numbers with their reciprocals.

$$\frac{1}{2} \cdot \frac{2}{3} \cdot 2 \cdot \frac{3}{2} = \left(\frac{1}{2} \cdot 2\right) \cdot \left(\frac{2}{3} \cdot \frac{3}{2}\right) = 1 \cdot 1 = 1$$

(c) Instead of adding 199, add 200 and then subtract 1.

$$523 + 200 - 1 = 723 - 1 = 722$$

(d) Think of 55 as $50 + 5$ and then apply the distributive property.

$$6 \cdot (50 + 5) = 300 + 30 = 330$$

Now Try Exercises 79, 91, 95

CRITICAL THINKING

How could you quickly calculate $5283 - 198$ without a calculator?

The next example illustrates how the commutative and associative properties for multiplication can be used together to simplify a product.

EXAMPLE 9   Finding the volume of a swimming pool

An Olympic swimming pool is 50 meters long, 25 meters wide, and 2 meters deep. The volume V of the pool is found by multiplying the numbers 50, 25, and 2. Use the commutative and associative properties for multiplication to calculate the volume of the pool mentally.

Solution
Because $50 \times 2 = 100$ and multiplication by 100 is relatively easy, it may be convenient to order and group the multiplication as $(50 \times 2) \times 25 = 100 \times 25 = 2500$. Thus the pool contains 2500 cubic meters of water.   Now Try Exercise **107**

1.7   PUTTING IT ALL TOGETHER

The following table summarizes some properties of real numbers.

Property	Definition	Examples
Commutative	For any real numbers a and b, $a + b = b + a$ and $a \cdot b = b \cdot a$.	$4 + 6 = 6 + 4$ $4 \cdot 6 = 6 \cdot 4$
Associative	For any real numbers a, b, and c, $(a + b) + c = a + (b + c)$ and $(a \cdot b) \cdot c = a \cdot (b \cdot c)$.	$(3 + 4) + 5 = 3 + (4 + 5)$ $(3 \cdot 4) \cdot 5 = 3 \cdot (4 \cdot 5)$
Distributive	For any real numbers a, b, and c, $a(b + c) = ab + ac$ and $a(b - c) = ab - ac$.	$5(x + 2) = 5x + 10$ $5(x - 2) = 5x - 10$
Identity (0 and 1)	The identity for addition is 0, and the identity for multiplication is 1. For any real number a, $a + 0 = a$ and $a \cdot 1 = a$.	$5 + 0 = 5$ and $5 \cdot 1 = 5$
Inverse	The additive inverse of a is $-a$, and $a + (-a) = 0$. The multiplicative inverse of a nonzero number a is $\frac{1}{a}$, and $a \cdot \frac{1}{a} = 1$.	$8 + (-8) = 0$ and $\frac{2}{3} \cdot \frac{3}{2} = 1$

1.7 Exercises

CONCEPTS

1. $a + b = b + a$ illustrates the _____ property for _____.

2. $a \cdot b = b \cdot a$ illustrates the _____ property for _____.

3. $(a + b) + c = a + (b + c)$ illustrates the _____ property for _____.

4. $(a \cdot b) \cdot c = a \cdot (b \cdot c)$ illustrates the _____ property for _____.

5. Addition and multiplication are commutative but _____ and _____ are not.

6. Addition and multiplication are associative but _____ and _____ are not.

7. $a(b + c) = ab + ac$ illustrates the _____ property.

8. $a(b - c) = ab - ac$ illustrates the _____ property.

9. $3b + 2b = (3 + 2) \cdot$ _____.

10. $4a - 3a = (4 - 3) \cdot$ _____.

11. $a + 0 = 0 + a = a$ illustrates the _____ property for _____.

12. $a \cdot 1 = 1 \cdot a = a$ illustrates the _____ property for _____.

13. The additive inverse of a is _____.

14. The multiplicative inverse, or reciprocal, of a nonzero number a is _____.

PROPERTIES OF REAL NUMBERS

Exercises 15–22: Use a commutative property to rewrite the expression. Do not simplify.

15. $-6 + 10$

16. $23 + 7$

17. $-5 \cdot 6$

18. $25 \cdot (-46)$

19. $a + 10$

20. $b + c$

21. $b \cdot 7$

22. $a \cdot 23$

Exercises 23–30: Use an associative property to rewrite the expression by changing the parentheses. Do not simplify.

23. $(1 + 2) + 3$

24. $-7 + (5 + 15)$

25. $2 \cdot (3 \cdot 4)$

26. $(9 \cdot (-4)) \cdot 5$

27. $(a + 5) + c$

28. $(10 + b) + a$

29. $(x \cdot 3) \cdot 4$

30. $5 \cdot (x \cdot y)$

31. **Thinking Generally** Use the commutative and associative properties to show that $a + b + c = c + b + a$.

32. **Thinking Generally** Use the commutative and associative properties to show that $a \cdot b \cdot c = c \cdot b \cdot a$.

Exercises 33–44: Use a distributive property to rewrite the expression. Then simplify the expression.

33. $4(3 + 2)$

34. $5(6 - 9)$

35. $a(b - 8)$

36. $3(x + y)$

37. $-1(t + z)$

38. $-1(a + 6)$

39. $-(5 - a)$

40. $12 - (4u - b)$

41. $(a + 5)3$

42. $(x + y)7$

43. $(6 - z)(-3)$

44. $4x - 2(3y - 5)$

45. **Thinking Generally** Use properties of real numbers to show that the distributive property can be extended as follows.
$$a \cdot (b + c + d) = ab + ac + ad.$$

46. **Thinking Generally** Use properties of real numbers to show that the distributive property can be extended as follows.
$$a \cdot (b - c - d) = ab - ac - ad.$$

Exercises 47–54: Use the distributive property to insert parentheses in the expression and then simplify the result.

47. $6x + 5x$

48. $4y - y$

49. $-4b + 3b$

50. $2b + 8b$

51. $3a - a$

52. $-2x + 5x$

53. $13w - 27w$

54. $25a - 21a$

Exercises 55–68: State the property or properties that the equation illustrates.

55. $x \cdot 5 = 5x$

56. $7 + a = a + 7$

57. $(a + 5) + 7 = a + 12$

58. $(9 + a) + 8 = a + 17$

59. $4(5 + x) = 20 + 4x$

60. $3(5 + x) = 3x + 15$

61. $x(3 - y) = 3x - xy$

62. $-(u - v) = -u + v$

63. $6x + 9x = 15x$

64. $9x - 11x = -2x$

65. $3 \cdot (4 \cdot a) = 12a$

66. $(x \cdot 3) \cdot 5 = 15x$

67. $-(t - 7) = -t + 7$

68. $z \cdot 5 - y \cdot 6 = 5z - 6y$

IDENTITY AND INVERSE PROPERTIES

Exercises 69–78: State the property or properties that are illustrated.

69. $0 + x = x$

70. $5x + 0 = 5x$

71. $1 \cdot a = a$

72. $\frac{1}{7} \cdot (7 \cdot a) = a$

73. $\frac{25}{15} = \frac{5}{3} \cdot \frac{5}{5} = \frac{5}{3}$

74. $\frac{50}{40} = \frac{5}{4} \cdot \frac{10}{10} = \frac{5}{4}$

75. $\frac{1}{xy} \cdot xy = 1$

76. $\frac{1}{a + b} \cdot (a + b) = 1$

77. $-xyz + xyz = 0$

78. $\frac{1}{y} + \left(-\frac{1}{y}\right) = 0$

MENTAL CALCULATIONS

Exercises 79–96: Use properties of real numbers to calculate the expression mentally.

79. $4 + 2 + 9 + 8 + 1 + 6$

80. $21 + 32 + 19 + 8$

81. $45 + 43 + 5 + 7$

82. $5 + 7 + 12 + 13 + 8$

83. $129 + 49$

84. $87 + 99$

85. $379 + 98$

86. $4570 + 998$

87. $178 - 99$

88. $500 - 101$

89. $6 \cdot 15$

90. $4 \cdot 56$

91. $8 \cdot 102$

92. $5 \cdot 999$

93. $\frac{1}{2} \cdot \frac{1}{2} \cdot \frac{1}{2} \cdot 2 \cdot 2 \cdot 2$

94. $\frac{1}{2} \cdot \frac{4}{5} \cdot \frac{7}{3} \cdot 2 \cdot \frac{5}{4}$

95. $\frac{7}{6} \cdot \frac{1}{2} \cdot \frac{1}{2} \cdot \frac{1}{2} \cdot \frac{8}{7}$

96. $\frac{4}{11} \cdot \frac{11}{6} \cdot \frac{6}{7} \cdot \frac{7}{4}$

MULTIPLYING AND DIVIDING BY POWERS OF 10 MENTALLY

97. *Multiplying by 10*   Multiplying by 10 is easy in the decimal system. To multiply an integer by 10, append one 0 to the number. For example, $10 \times 23 = 230$. Simplify each expression mentally.
(a) 10×41 **(b)** 10×997
(c) -630×10 **(d)** $-14,000 \times 10$

98. *Multiplying by 10*   To multiply a decimal number by 10, move the decimal point one place to the right. For example, $10 \times 23.45 = 234.5$. Simplify each expression mentally.
(a) 10×101.68 **(b)** $10 \times (-1.235)$
(c) -113.4×10 **(d)** 0.567×10
(e) 10×0.0045 **(f)** -0.05×10

99. *Multiplying by Powers of 10*   To multiply an integer by a power of 10 in the form
$$10^k = \underbrace{100 \ldots 0}_{k \text{ zeros}},$$
append k zeros to the number. Some examples of this are $100 \times 45 = 4500$, $1000 \times 235 = 235,000$, and $10,000 \times 12 = 120,000$. Simplify each expression mentally.
(a) 1000×19 **(b)** $100 \times (-451)$
(c) $10,000 \times 6$ **(d)** $-79 \times 100,000$

100. *Multiplying by Powers of 10*   To multiply a decimal number by a power of 10 in the form
$$10^k = \underbrace{100 \ldots 0}_{k \text{ zeros}},$$
move the decimal point k places to the right. For example, $100 \times 1.234 = 123.4$. Simplify each given expression mentally.
(a) 1000×1.2345 **(b)** $100 \times (-5.1)$
(c) 45.67×1000 **(d)** $0.567 \times 10,000$
(e) 100×0.0005 **(f)** $-0.05 \times 100,000$

101. *Dividing by 10*   To divide a number by 10, move the decimal point one place to the left. For example, $78.9 \div 10 = 7.89$. Simplify each expression mentally.

(a) $12.56 \div 10$ (b) $9.6 \div 10$
(c) $0.987 \div 10$ (d) $-0.056 \div 10$
(e) $1200 \div 10$ (f) $4578 \div 10$

102. *Dividing by Powers of 10* To divide a decimal number by a power of 10 in the form

$$10^k = 1\underbrace{00 \ldots 0}_{k \text{ zeros}},$$

move the decimal point k places to the left. For example, $123.4 \div 100 = 1.234$. Simplify each expression mentally.

(a) $78.89 \div 100$ (b) $0.05 \div 1000$
(c) $5678 \div 10,000$ (d) $-9.8 \div 1000$
(e) $-101 \div 100,000$ (f) $7.8 \div 100$

APPLICATIONS

103. *Earnings* Earning $100 one day and $75 the next day is equivalent to earning $75 the first day and $100 the second day. What property of real numbers does this example illustrate?

104. *Leasing a Car* An advertisement for a lease on a new car states that it costs $2480 down and $201 per month for 20 months. Mentally calculate the cost of the lease. Explain your reasoning.

105. *Gasoline Mileage* A car travels 198 miles on 10 gallons of gasoline. Mentally calculate the number of miles that the car travels on 1 gallon of gasoline.

106. *Gallons of Water* A wading pool is 50 feet long, 20 feet wide, and 1 foot deep.
(a) Mentally determine the number of cubic feet in the pool. (*Hint:* Volume equals length times width times height.)
(b) One cubic foot equals about 7.5 gallons. Mentally calculate the number of gallons of water in the pool.

107. *Dimensions of a Pool* A small pool of water is 13 feet long, 5 feet wide, and 2 feet deep. The volume V of the pool in cubic feet is found by multiplying 13 by 5 by 2.
(a) If you did this calculation mentally would you multiply $(13 \times 5) \times 2$ or $13 \times (5 \times 2)$? Why?
(b) What property allows you to do either calculation and still obtain the correct answer?

108. *Digital Images of Io* Satellites take digital pictures and transmit them back to Earth. The accompanying picture of Jupiter's moon Io is a digital picture, created by using a rectangular pattern of small pixels. This image is 500 pixels wide and 400 pixels high, so the total number of pixels in it is $500 \times 400 = 200,000$ pixels. (*Source:* NASA.)

(a) Find the total number of pixels in an image 400 pixels wide and 500 pixels high.
(b) Suppose that a picture is x pixels wide and y pixels high. What property states that it has the same number of pixels as a picture y pixels wide and x pixels high?

WRITING ABOUT MATHEMATICS

109. To determine the cost of tuition, a student tries to compute $16 \times \$96$ with a calculator and gets $\$15,360$. What would you tell this person?

110. A student performs the following computation by hand.

$$20 - 6 - 2 + 8 \div 4 \div 2 \overset{?}{=} 20 - 4 + 8 \div 2$$
$$\overset{?}{=} 20 - 4 + 4$$
$$\overset{?}{=} 20$$

Find any incorrect steps and explain what is wrong. What is the correct answer?

111. The computation $3 + 10^{20} - 10^{20}$ performed on a calculator gives a result of 0. (Try it.) What is the correct answer? Why is it important to know properties of real numbers even though you have a calculator?

112. Does the "distributive property"

$$a + (b \cdot c) \overset{?}{=} (a + b) \cdot (a + c)$$

hold for all numbers a, b, and c? Explain your reasoning.

Directions: Form a group of 2 to 4 people. Select someone to record the group's responses for this activity. All members of the group should work cooperatively to answer the questions. If your instructor asks for your results, each member of the group should be prepared to respond.

Winning the Lottery   In the multistate lottery game *Powerball,* there are 120,526,770 possible number combinations, only one of which is the grand prize winner. The cost of a single ticket (one number combination) is $1. (**Source:** Powerball.com.)

   Suppose that a *very* wealthy person decides to buy tickets for every possible number combination to be assured of winning a $150 million grand prize.

(a) If this individual could purchase one ticket every second, how many hours would it take to buy all of the tickets? How many years is this?

(b) If there were a way for this individual to buy all possible number combinations quickly, discuss reasons why this strategy would probably lose money.

1.8   SIMPLIFYING AND WRITING ALGEBRAIC EXPRESSIONS

Terms ▪ Combining Like Terms ▪ Simplifying Expressions ▪ Writing Expressions

A LOOK INTO MATH ▷   In arithmetic it is common to simplify expressions such as

$$\frac{1}{2} + \frac{1}{4} + \frac{1}{4}.$$

This expression equals 1, and it is easier to perform calculations with the number 1 than with these fractions. However, simplifying expressions requires mathematical skills. In this section we introduce some of these basic skills.

Terms

One way to simplify expressions is to combine like terms. A **term** is a number, a variable, or a *product* of numbers and variables raised to powers. Examples of terms include

$$4, \quad z, \quad 5x, \quad \frac{2}{5}z, \quad -4xy, \quad -x^2, \quad \text{and} \quad 6x^3y^4.$$

Terms do not contain addition or subtraction signs, but they can contain negative signs. The **coefficient** of a term is the number that appears in the term. If no number appears, the coefficient is either -1 or 1. The coefficient of $\frac{1}{2}xy$ is $\frac{1}{2}$, the coefficient of -15 is -15, and the coefficient of $-y^2$ is -1.

EXAMPLE 1   Identifying terms

Determine whether each expression is a term. If it is a term, identify its coefficient.
(a) 51   **(b)** $5a$ **(c)** $2x + 3y$ **(d)** $-3x^2$

Solution
(a) A number is a term. Its coefficient is 51.
(b) The product of a number and a variable is a term. Its coefficient is 5.

(c) The sum (or difference) of two terms is not a term.

(d) The product of a number and a variable with an exponent is a term. Its coefficient is -3.

Now Try Exercises 11, 15, 21

MAKING CONNECTIONS

Factors and Terms

When variables and numbers are multiplied, they are called *factors*. For example, the expression $4xy$ has factors of 4, x, and y. When variables and numbers are added or subtracted, they are called terms. For example, the expression $x - 5xy + 1$ has terms of x, $-5xy$, and 1.

Combining Like Terms

Suppose that we have two boards with lengths $2x$ and $3x$, where the value of x could be any length such as 2 feet. See Figure 1.28. Because $2x$ and $3x$ are *like terms* we can find the total length of the two boards by applying the distributive property and *adding like terms*.

$$2x + 3x = (2 + 3)x = 5x$$

The combined length of the two boards is $5x$ units.

Figure 1.28 $2x + 3x = 5x$

We can also determine how much longer the second board is than the first board by subtracting terms.

$$3x - 2x = (3 - 2)x = 1x = x$$

Thus the second board is x units longer than the first board.

If two terms contain the same variables raised to the same powers, we call them **like terms**. We can add or subtract like terms but not *unlike* terms. If one board has length $2x$ and the other board has length $3y$, then we cannot determine the total length other than to say that it is $2x + 3y$. See Figure 1.29. The reason is that the lengths of x and y might not be equal. The terms $2x$ and $3y$ are unlike terms and *cannot* be combined.

Figure 1.29 $2x + 3y$

EXAMPLE 2   **Identifying like terms**

Determine whether the terms are like or unlike.
(a) $-4m, 7m$ **(b)** $8x^2, 8y^2$ **(c)** $-x^2, 3x^2$ **(d)** $\frac{1}{2}z, -3z^2$ **(e)** $5, -4n$

Solution
(a) The variable in both terms is m (with power 1), so they are like terms.
(b) The variables are different, so they are unlike terms.

(c) The terms $-x^2$ and $3x^2$ have the same variable raised to the same power (x^2), so they are like terms.

(d) The term $-3z^2$ contains $z^2 = z \cdot z$, whereas the term $\frac{1}{2}z$ contains only z. Thus they are unlike terms.

(e) The term 5 has no variable, whereas the term $-4n$ contains the variable n. They are unlike terms.

Now Try Exercises 23, 29, 33

EXAMPLE 3   **Combining like terms**

Combine terms in each expression, if possible.
(a) $-3x + 5x$ (b) $8y - y$ (c) $-x^2 + 5x^2$ (d) $\frac{1}{2}y - 3y^3$

Solution
(a) Combine terms by applying a distributive property.
$$-3x + 5x = (-3 + 5)x = 2x$$

(b) Note that y can be written as $1y$. They are like terms and can be combined.
$$8y - y = (8 - 1)y = 7y$$

(c) Note that $-x^2$ can be written as $-1x^2$. They are like terms and can be combined.
$$-x^2 + 5x^2 = (-1 + 5)x^2 = 4x^2$$

(d) They are unlike terms, so they cannot be combined.   Now Try Exercises 39, 47, 51

Simplifying Expressions

CRITICAL THINKING

Use rectangles to explain how to add $xy + 2xy$.

The area of a rectangle equals length times width. In Figure 1.30 the area of the first rectangle is $3x$, the area of the second rectangle is $2x$, and the area of the third rectangle is x. The area of the last rectangle equals the total area of the three smaller rectangles. That is,
$$3x + 2x + x = (3 + 2 + 1)x = 6x.$$

The expression $3x + 2x + x$ can be *simplified* to $6x$.

Figure 1.30 $3x + 2x + x = 6x$

EXAMPLE 4   **Simplifying expressions**

Simplify each expression.
(a) $2 + x - 6 + 5x$ (b) $2y - (y + 3)$ (c) $\dfrac{-1.5x}{-1.5}$ (d) $\dfrac{5x}{3x}$

Solution
(a) Combine like terms by applying the properties of real numbers.

$$2 + x - 6 + 5x = 2 + (-6) + x + 5x \qquad \text{Commutative property}$$
$$= 2 + (-6) + (1 + 5)x \qquad \text{Distributive property}$$
$$= -4 + 6x \qquad \text{Add.}$$

(b)

$$2y - 1(y + 3) = 2y + (-1)y + (-1) \cdot 3 \qquad \text{Distributive property}$$
$$= 2y - 1y - 3 \qquad \text{Definition of subtraction}$$
$$= (2 - 1)y - 3 \qquad \text{Distributive property}$$
$$= y - 3 \qquad \text{Subtract.}$$

(c)

$$\frac{-1.5x}{-1.5} = \frac{-1.5}{-1.5} \cdot \frac{x}{1} \qquad \text{Multiplication of fractions}$$
$$= 1 \cdot x \qquad \text{Simplify the fractions.}$$
$$= x \qquad \text{Multiplicative identity}$$

(d)

$$\frac{5x}{3x} = \frac{5}{3} \cdot \frac{x}{x} \qquad \text{Property of fractions}$$
$$= \frac{5}{3} \cdot 1 \qquad \text{Simplify.}$$
$$= \frac{5}{3} \qquad \text{Multiplicative identity}$$

Now Try Exercises 57, 69, 83

NOTE: The expressions in Example 4(c) and 4(d) can be simplified directly by using the basic principle of fractions: $\frac{ac}{bc} = \frac{a}{b}$.

EXAMPLE 5 Simplifying expressions

Simplify each expression.
(a) $7 - 2(5x - 3)$ **(b)** $3x^3 + 2x^3 - x^3$
(c) $3y^2 - z + 4y^2 - 2z$ **(d)** $\dfrac{12x - 8}{4}$

Solution

(a)

$$7 - 2(5x - 3) = 7 + (-2)(5x + (-3)) \qquad \text{Change subtraction to addition.}$$
$$= 7 + (-2)(5x) + (-2)(-3) \qquad \text{Distributive property}$$
$$= 7 - 10x + 6 \qquad \text{Multiply.}$$
$$= 13 - 10x \qquad \text{Combine like terms.}$$

(b)

$$3x^3 + 2x^3 - x^3 = (3 + 2 - 1)x^3 \qquad \text{Distributive property}$$
$$= 4x^3 \qquad \text{Add and subtract.}$$

(c)

$$3y^2 - z + 4y^2 - 2z = 3y^2 + 4y^2 + (-1z) + (-2z) \qquad \text{Commutative property}$$
$$= (3 + 4)y^2 + (-1 + (-2))z \qquad \text{Distributive property}$$
$$= 7y^2 - 3z \qquad \text{Add.}$$

(d)

$$\frac{12x - 8}{4} = \frac{12x}{4} - \frac{8}{4} \qquad \text{Subtraction of fractions}$$
$$= 3x - 2 \qquad \text{Simplify fractions.}$$

Now Try Exercises 69, 71, 81, 89

Writing Expressions

▶ **REAL-WORLD CONNECTION** In real-life situations, we often have to translate words to symbols. For example, to calculate federal and state income tax we might have to multiply taxable income by 0.15 and by 0.05 and then find the sum. If we let x represent taxable income, then the total federal and state income tax is $0.15x + 0.05x$. This expression can be simplified with a distributive property.

$$0.15x + 0.05x = (0.15 + 0.05)x$$
$$= 0.20x$$

Thus the total income tax on a taxable income of $x = \$20,000$ would be

$$0.20(20,000) = \$4000.$$

EXAMPLE 6　**Writing and simplifying an expression**

A sidewalk has a constant width w and comprises several short sections with lengths 12, 6, and 5 feet, as illustrated in Figure 1.31.
(a) Write and simplify an expression that gives the number of square feet of sidewalk.
(b) Find the area of the sidewalk if its width is 3 feet.

Solution
(a) The area of each sidewalk section equals its length times its width w. The total area of the sidewalk is

$$12w + 6w + 5w = (12 + 6 + 5)w = 23w.$$

(b) When $w = 3$, the area is $23w = 23 \cdot 3 = 69$ square feet.　**Now Try Exercise 99**

5 ft

6 ft

w　**12 ft**

Figure 1.31

1.8　PUTTING IT ALL TOGETHER

The following table summarizes some of the concepts presented in this section.

Concept	Comments	Examples
Term	A term is a number, variable, or product of numbers and variables raised to powers.	$12, -10, x, -y$ $-3x, 5z, xy, 6x^2$ $10y^3, \frac{3}{4}x, 25xyz^2$
Coefficient of a Term	The coefficient of a term is the number that appears in the term. If no number appears, then the coefficient is either 1 or -1.	12　Coefficient is 12 x　Coefficient is 1 $-xy$　Coefficient is -1 $-4x$　Coefficient is -4
Like Terms	Like terms have the same variables raised to the same powers.	$5m$ and $-6m$ x^2 and $-74x^2$ $2xy$ and $-xy$
Combining Like Terms	Like terms can be combined by using a distributive property.	$5x + 2x = (5 + 2)x = 7x$ $y - 3y = (1 - 3)y = -2y$ $8x^2 + x^2 = (8 + 1)x^2 = 9x^2$

1.8 Exercises

CONCEPTS

1. A _____ is a number, a variable, or a product of numbers and variables raised to powers.

2. The expression $a + b$ (is/is not) a term, whereas the expression ab (is/is not) a term.

3. The number 7 in the term $7x^2y$ is called the _____ of the term.

4. The coefficient of the term $-xy^2$ is _____.

5. The coefficient of the term 6 is _____.

6. When variables and numbers are multiplied, they are called _____. When they are added or subtracted, they are called _____.

7. If two terms contain the same variables raised to the same powers, we call them (like/unlike) terms.

8. The terms $2x$ and $5x$ are (like/unlike) terms, whereas $9x$ and $9z$ are (like/unlike) terms.

9. We can add or subtract (like/unlike) terms.

10. We can combine like terms in an expression by applying a _____ property.

LIKE TERMS

Exercises 11–22: Determine whether the expression is a term. If the expression is a term, identify its coefficient.

11. 91
12. -12
13. $-6b$
14. $9z$
15. $x + 10$
16. $20 - 2y$
17. x^2
18. $4x^3$
19. $4x - 5$
20. $5z + 6x$
21. $-9xyz$
22. $-a^2b^2$

Exercises 23–36: Determine whether the terms are like or unlike.

23. $6, -8$
24. $2x, 19$
25. $5x, -22x$
26. $19y, -y$
27. $14, 14a$
28. $-33b, -3b$
29. $18x, 18y$
30. $-6a, -6b$
31. $x^2, -15x^2, 6x^2$
32. $xyz, 19xyz, -xyz$
33. $3x^2y, 5xy^2$
34. $12xy^2, 4xy^2, -xy^2$
35. $xy, xz, 2xy$
36. $-8x^2y, 2x^2z, x^2y$

37. **Thinking Generally** Are the terms $4ab$ and $-3ba$ like or unlike?

38. **Thinking Generally** Are the terms $-xyz^2$ and $3yz^2x$ like or unlike?

Exercises 39–56: Combine terms, if possible.

39. $3x + 5x$
40. $6x - 8x$
41. $19y - 5y$
42. $22z + z$
43. $28a + 13a$
44. $41b - 17b$
45. $11z - 11z$
46. $4y + 4y$
47. $5x - 7y$
48. $3y + 3z$
49. $5 + 5y$
50. $x + x^2$
51. $5x^2 - 2x^2$
52. $25z^3 - 10z^3$
53. $8xy - 10xy + xy$
54. $4x^2 + x^2 - 5x^2$
55. $12ab - 7ab - 5ab$
56. $100xy^2 + 25xy^2 - 5xy^2$

SIMPLIFYING AND WRITING EXPRESSIONS

Exercises 57–92: Simplify the expression.

57. $5 + x - 3 + 2x$
58. $x - 5 - 5x + 7$
59. $-\frac{3}{4} + z - 3z + \frac{5}{4}$
60. $\frac{4}{3}z - 100 + 200 - \frac{1}{3}z$
61. $4y - y + 8y$
62. $14z - 15z - z$
63. $-3 + 6z + 2 - 2z$
64. $19a - 12a + 5 - 6$
65. $-2(3z - 6y) - z$
66. $6\left(\frac{1}{2}a - \frac{1}{6}b\right) - 3b$
67. $2 - \frac{3}{4}(4x + 8)$
68. $-5 - (5x - 6)$
69. $-x - (5x + 1)$
70. $2x - 4(x + 2)$

71. $1 - \frac{1}{3}(x + 1)$ **72.** $-3 - 3(4 - x)$

73. $\frac{3}{5}(x + y) - \frac{1}{5}(x - 1)$

74. $-5(a + b) - (a + b)$

75. $0.2x^2 + 0.3x^2 - 0.1x^2$ **76.** $32z^3 - 52z^3 + 20z^3$

77. $2x^2 - 3x + 5x^2 - 4x$ **78.** $\frac{5}{6}y^2 - 4 + \frac{1}{12}y^2 + 3$

79. $a + 3b - a - b$ **80.** $2z^2 - z - z^2 + 3z$

81. $8x^3 + 7 - x^3 - 5$ **82.** $4y - 6z + 2y - 3z$

83. $\dfrac{8x}{8}$ **84.** $\dfrac{-0.1y}{-0.1}$

85. $\dfrac{-3y}{-y}$ **86.** $\dfrac{2x}{7x}$

87. $\dfrac{-108z}{-108}$ **88.** $\dfrac{3xy}{-6xy}$

89. $\dfrac{9x - 6}{3}$ **90.** $\dfrac{18y + 9}{9}$

91. $\dfrac{14z + 21}{7}$ **92.** $\dfrac{15x - 20}{5}$

Exercises 93–98: Translate the phrase into a mathematical expression and then simplify. Use the variable x.

93. The sum of five times a number and six times the same number

94. The sum of a number and three times the same number

95. The sum of a number squared and twice the same number squared

96. One-half of a number minus three-fourths of the same number

97. Six times a number minus four times the same number

98. Two cubed times a number minus three squared times the same number

APPLICATIONS

99. *Street Dimensions*   (Refer to Example 6.) A street has a constant width w and comprises several straight sections having lengths 600, 400, 350, and 220 feet.

(a) Write a simplified expression that gives the square footage of the street.
(b) Find the area of the street if its width is 42 feet.

100. *Sidewalk Dimensions*   A sidewalk has a constant width w and comprises several short sections having lengths 12, 14, and 10 feet.
(a) Write a simplified expression that gives the number of square feet of sidewalk.
(b) Find the area of the sidewalk if its width is 5 feet.

101. *Snowblowers*   Two snowblowers are being used to clear a driveway. The first blower can remove 20 cubic feet per minute, and the second blower can remove 30 cubic feet per minute.
(a) Write a simplified expression that gives the number of cubic feet of snow removed in x minutes.
(b) Find the total number of cubic feet of snow removed in 48 minutes.
(c) How many minutes would it take to remove the snow from a driveway that measures 30 feet by 20 feet, if the snow is 2 feet deep?

102. *Winding Rope*   Two motors are winding up rope. The first motor can wind up rope at 2 feet per second, and the second motor can wind up rope at 5 feet per second.
(a) Write a simplified expression that gives the length of rope wound by both motors in x seconds.
(b) Find the total length of rope wound up in 3 minutes.
(c) How many minutes would it take to wind up 2100 feet of rope by using both motors?

WRITING ABOUT MATHEMATICS

103. The following expression was simplified *incorrectly*.

$$3(x - 5) + 5x \stackrel{?}{=} 3x - 5 + 5x$$
$$\stackrel{?}{=} 3x + 5x - 5$$
$$\stackrel{?}{=} (3 + 5)x - 5$$
$$\stackrel{?}{=} 8x - 5$$

Find the error and explain what went wrong. What should the final answer be?

104. Explain how to add like terms. What property of real numbers is used?

CHECKING BASIC CONCEPTS
SECTIONS 1.7 AND 1.8

1. Use the commutative property to rewrite each expression.
 (a) $y \cdot 18$ **(b)** $10 + x$

2. Use the associative and commutative properties to simplify $5 \cdot (y \cdot 4)$.

3. Simplify each expression.
 (a) $10 - (5 + x)$ **(b)** $5(x - 7)$

4. State the property that the equation $5x + 3x = 8x$ illustrates.

5. Simplify $-4xy + 4xy$.

6. Mentally evaluate each expression.
 (a) $32 + 17 + 8 + 3$ **(b)** $\frac{5}{6} \cdot \frac{7}{8} \cdot \frac{6}{5} \cdot 8$
 (c) $567 - 199$

7. Determine whether the terms are like or unlike.
 (a) $-3xy, -3yz$ **(b)** $4x^2, -2x^2$

8. Combine like terms in each expression.
 (a) $5z + 9z$ **(b)** $5y - 4 - 8y + 7$

9. Simplify each expression.
 (a) $2y - (5y + 3)$
 (b) $-4(x + 3y) + 2(2x - y)$
 (c) $\frac{20x}{20}$ **(d)** $\frac{35x^2}{x^2}$

10. Write "the sum of three times a number and five times the same number" as a mathematical expression with the variable x and then simplify the expression.

CHAPTER 1 SUMMARY
SECTION 1.1 ■ NUMBERS, VARIABLES, AND EXPRESSIONS

Sets of Numbers

Natural Numbers 1, 2, 3, 4, . . .

Whole Numbers 0, 1, 2, 3, . . .

Prime and Composite Numbers

Prime Number Only natural number factors are itself and 1; must be a natural number greater than 1

Examples: 2, 3, 5, 7, 11, 13, and 17

Composite Number A natural number greater than 1 that is not prime; a composite number can be written as a product of two or more prime numbers.

Examples: $24 = 2 \times 2 \times 2 \times 3$ and $18 = 2 \times 3 \times 3$

Important Terms

Variable   A symbol or letter that represents an unknown quantity

                              Examples: $a, b, F, x,$ and y

Algebraic Expression   Consists of numbers, variables, arithmetic symbols, and grouping symbols

                              Examples: $3x + 1$ and $5(x + 2) - y$

Equation   A statement that two algebraic expressions are equal; an equation always contains an equals sign.

                              Examples: $1 + 2 = 3$ and $z - 7 = 8$

SECTION 1.2 ■ FRACTIONS

Parts of a Fraction

Numerator ——— $\dfrac{N}{D}$ ——— Fraction bar

Denominator

Simplifying Fractions $\dfrac{a \cdot c}{b \cdot c} = \dfrac{a}{b}$

Example: $\dfrac{20}{35} = \dfrac{4 \cdot 5}{7 \cdot 5} = \dfrac{4}{7}$

Multiplicative Inverse, or Reciprocal   The reciprocal of a nonzero number a is $\frac{1}{a}$.

Examples: The reciprocal of -5 is $-\frac{1}{5}$, and the reciprocal of $\frac{2x}{y}$ is $\frac{y}{2x}$, provided $x \neq 0$ and $y \neq 0$.

Multiplication and Division

$$\frac{a}{b} \cdot \frac{c}{d} = \frac{ac}{bd} \qquad b \text{ and } d \text{ are nonzero.}$$

$$\frac{a}{b} \div \frac{c}{d} = \frac{a}{b} \cdot \frac{d}{c} \qquad b, c, \text{ and } d \text{ are nonzero.}$$

Examples: $\dfrac{3}{4} \cdot \dfrac{7}{5} = \dfrac{21}{20}$ and $\dfrac{3}{4} \div \dfrac{7}{5} = \dfrac{3}{4} \cdot \dfrac{5}{7} = \dfrac{3 \cdot 5}{4 \cdot 7} = \dfrac{15}{28}$

Addition and Subtraction with a Common Denominator

$$\frac{a}{b} + \frac{c}{b} = \frac{a + c}{b} \qquad \text{and} \qquad \frac{a}{b} - \frac{c}{b} = \frac{a - c}{b}$$

Examples: $\dfrac{3}{5} + \dfrac{1}{5} = \dfrac{3 + 1}{5} = \dfrac{4}{5}$ and $\dfrac{3}{5} - \dfrac{1}{5} = \dfrac{3 - 1}{5} = \dfrac{2}{5}$

Addition and Subtraction with Unlike Denominators   Write the expressions with an LCD. Then add or subtract the numerators, keeping the denominator unchanged.

Examples: $\dfrac{2}{9} + \dfrac{1}{6} = \dfrac{2}{9} \cdot \dfrac{2}{2} + \dfrac{1}{6} \cdot \dfrac{3}{3} = \dfrac{4}{18} + \dfrac{3}{18} = \dfrac{7}{18}$ LCD is 18.

$\dfrac{2}{9} - \dfrac{1}{6} = \dfrac{2}{9} \cdot \dfrac{2}{2} - \dfrac{1}{6} \cdot \dfrac{3}{3} = \dfrac{4}{18} - \dfrac{3}{18} = \dfrac{1}{18}$

SECTION 1.3 ■ EXPONENTS AND ORDER OF OPERATIONS

Exponential Expression

$$\text{Base} \longrightarrow 5^3 \longleftarrow \text{Exponent}$$

Example: $3^4 = 3 \cdot 3 \cdot 3 \cdot 3 = 81$

Order of Operations

Use the following order of operations. First perform all calculations within parentheses and absolute values, or above and below the fraction bar.

1. Evaluate all exponential expressions.
2. Do all multiplication and division from *left to right*.
3. Do all addition and subtraction from *left to right*.

NOTE: Negative signs are evaluated after exponents so $-2^4 = -16$.

Examples:
$$100 - 5^2 \cdot 2 = 100 - 25 \cdot 2 \qquad \text{and} \qquad \frac{4+2}{5-3} \cdot 4 = \frac{6}{2} \cdot 4$$
$$= 100 - 50 \qquad\qquad\qquad\qquad = 3 \cdot 4$$
$$= 50 \qquad\qquad\qquad\qquad\qquad = 12$$

SECTION 1.4 ■ REAL NUMBERS AND THE NUMBER LINE

Opposite or Additive Inverse The opposite of the number a is $-a$.

Examples: The opposite of 5 is -5, and the opposite of -8 is $-(-8) = 8$.

Sets of Numbers

Integers	$\dots, -3, -2, -1, 0, 1, 2, 3, \dots$
Rational Numbers	$\frac{p}{q}$, where p and $q \neq 0$ are integers; rational numbers can be written as decimal numbers that either repeat or terminate.

Examples: $-\frac{3}{4}, 5, -7.6, \frac{6}{3},$ and $\frac{1}{3}$

Real Numbers Numbers that can be written as decimal numbers

Examples: $-\frac{3}{4}, 5, -7.6, \frac{6}{3}, \pi, \sqrt{3},$ and $-\sqrt{7}$

Irrational Numbers Real numbers that are not rational

Examples: $\pi, \sqrt{3},$ and $-\sqrt{7}$

Average To calculate the average of a set of numbers, find the sum of the numbers and then divide the sum by how many numbers there are in the set.

Example: The average of 4, 5, 20, and 11 is

$$\frac{4 + 5 + 20 + 11}{4} = \frac{40}{4} = 10.$$

We divide by 4 because we are finding the average of 4 numbers.

The Number Line

Origin

```
<───┼────┼────┼────┼────┼────┼────┼───>
   -3   -2   -1    0    1    2    3
```

The origin corresponds to the number 0.

Absolute Value   If a is positive or 0, then $|a| = a$, and if a is negative, then $|a| = -a$. The absolute value of a number is *never* negative.

Examples: $|5| = 5$, $|-5| = 5$,   and $|0| = 0$

Inequality   If a is located to the left of b on the number line, then $a < b$. If a is located to the right of b on the number line, then $a > b$.

Examples: $-3 < 6$ and $-1 > -2$

SECTION 1.5 ■ ADDITION AND SUBTRACTION OF REAL NUMBERS

Addition of Opposites

$$a + (-a) = 0$$

Examples: $3 + (-3) = 0$ and $-\frac{5}{7} + \frac{5}{7} = 0$

Addition of Real Numbers

To add two numbers that are either *both positive* or *both negative*, add their absolute values. Their sum has the same sign as the two numbers.

To add two numbers with *opposite signs*, find the absolute value of each number. Subtract the smaller absolute value from the larger. The sum has the same sign as the sign of the number with the larger absolute value. If the two numbers are opposites, their sum is 0.

Examples: $-4 + 5 = 1$, $3 + (-7) = -4$, $-4 + (-2) = -6$,   and $8 + 2 = 10$

Subtraction of Real Numbers   For any real numbers a and b, $a - b = a + (-b)$.

Examples: $5 - 9 = 5 + (-9) = -4$ and $-4 - (-3) = -4 + 3 = -1$

SECTION 1.6 ■ MULTIPLICATION AND DIVISION OF REAL NUMBERS

Important Terms

Factors

Numbers multiplied in a multiplication problem

Example: 5 and 4 are factors of 20 because $5 \cdot 4 = 20$.

Product

The answer to a multiplication problem

Example: The product of 5 and 4 is 20.

Dividend, Divisor, and Quotient

If $\frac{a}{b} = c$, then a is the dividend, b is the divisor, and c is the quotient.

Example: In the division problem $30 \div 5 = \frac{30}{5} = 6$, 30 is the dividend, 5 is the divisor, and 6 is the quotient.

Dividing Real Numbers   For real numbers a and b with $b \neq 0$,

$$\frac{a}{b} = a \cdot \frac{1}{b}.$$

Examples: $\dfrac{8}{\frac{1}{2}} = \dfrac{8}{1} \cdot \dfrac{2}{1} = 16$, $14 \div \dfrac{2}{3} = \dfrac{14}{1} \cdot \dfrac{3}{2} = \dfrac{42}{2} = 21$,   and $5 \div 0$ is undefined.

Signs of Products or Quotients   The product or quotient of two numbers with like signs is positive. The product or quotient of two numbers with unlike signs is negative.

Examples: $-4 \cdot 6 = -24$, $-2 \cdot (-5) = 10$, $\dfrac{-18}{6} = -3$,   and $-4 \div -2 = 2$

Writing Fractions as Decimals   To write the fraction $\dfrac{a}{b}$ as a decimal, divide b into a.

Example: $\dfrac{4}{9} = 0.\overline{4}$ because division of 4 by 9 gives the repeating decimal $0.4444....$

$$
\begin{array}{r}
0.444\dots \\
9\overline{)4.000} \\
\underline{36} \\
40 \\
\underline{36} \\
40 \\
\underline{36} \\
4
\end{array}
$$

SECTION 1.7 ■ PROPERTIES OF REAL NUMBERS

Important Properties

Commutative

$a + b = b + a$ and $a \cdot b = b \cdot a$

Examples: $3 + 4 = 4 + 3$ and $-6 \cdot 3 = 3 \cdot (-6)$

Associative

$(a + b) + c = a + (b + c)$ and $(a \cdot b) \cdot c = a \cdot (b \cdot c)$

Examples: $(2 + 3) + 4 = 2 + (3 + 4)$
$(2 \cdot 3) \cdot 4 = 2 \cdot (3 \cdot 4)$

Distributive

$a(b + c) = ab + ac$ and $a(b - c) = ab - ac$

Examples: $3(x + 5) = 3x + 15$
$4(5 - 2) = 4 \cdot 5 - 4 \cdot 2$

Identity

$a + 0 = 0 + a = a$ Additive identity is 0.
$a \cdot 1 = 1 \cdot a = a$ Multiplicative identity is 1.

Examples: $5 + 0 = 0 + 5 = 5$ and $1 \cdot (-4) = -4 \cdot 1 = -4$

Inverse

$a + (-a) = 0$ and $-a + a = 0$
$a \cdot \dfrac{1}{a} = 1$ and $\dfrac{1}{a} \cdot a = 1$

Examples: $5 + (-5) = 0$ and $\dfrac{1}{2} \cdot 2 = 1$

NOTE: The commutative and associative properties apply to addition and multiplication but *not* to subtraction and division.

SECTION 1.8 ■ SIMPLIFYING AND WRITING ALGEBRAIC EXPRESSIONS

Important Terms

Term	A term is a number, a variable, or a product of numbers and variables raised to powers. **Examples:** 5, $\quad -10x$, $\quad 3xy$, $\quad$ and $\quad x^2$
Coefficient	The number portion of a term **Examples:** The coefficients for the terms $3xy$, $-x^2$, and -7 are 3, -1, and -7, respectively.
Like Terms	Terms containing the same variables raised to the same powers; their coefficients may be different. **Examples:** The following pairs are like terms: $5x$ and $-x$; $6x^2$ and $-2x^2$; $3xy$ and $-\dfrac{1}{2}xy$.

Combining Like Terms   To add or subtract like terms, apply the distributive property.

Examples: $4x + 5x = (4 + 5)x = 9x$ and $5xy - 7xy = (5 - 7)xy = -2xy$

CHAPTER 1   REVIEW EXERCISES

SECTION 1.1

Exercises 1–6: Classify the number as prime, composite, or neither. If the number is composite, write it as a product of prime numbers.

1. 29

2. 27

3. 108

4. 91

5. 0

6. 1

Exercises 7–10: Evaluate the expression for the given values of x and y.

7. $2x - 5$ $x = 4$

8. $7 - \dfrac{10}{x}$ $x = 5$

9. $9x - 2y$ $x = 2$, $y = 3$

10. $\dfrac{2x}{x - y}$ $x = 6$, $y = 4$

Exercises 11–14: Find the value of y for the given values of x and z.

11. $y = x - 5$ $x = 12$

12. $y = xz + 1$ $x = 2$, $z = 3$

13. $y = 4(x - z)$ $x = 7$, $z = 5$

14. $y = \dfrac{x + z}{4}$ $x = 14$, $z = 10$

Exercises 15–20: Translate the phrase into an algebraic expression. State precisely what each variable represents when appropriate.

15. Five times the cost of a CD

16. Five less than a number

17. Three squared increased by five

18. Two cubed divided by the quantity three plus one

19. The product of three and a number

20. The difference between a number and four

SECTION 1.2

21. Use the basic principle of fractions to simplify each expression.
   (a) $\dfrac{5 \cdot 7}{8 \cdot 7}$ **(b)** $\dfrac{3a}{4a}$

22. Simplify each fraction to lowest terms.
   (a) $\dfrac{9}{12}$ **(b)** $\dfrac{36}{60}$

Exercises 23–28: Multiply and then simplify the result to lowest terms when appropriate.

23. $\frac{3}{4} \cdot \frac{5}{6}$ **24.** $\frac{1}{2} \cdot \frac{4}{9}$

25. $\frac{2}{3} \cdot \frac{5}{11} \cdot \frac{9}{10}$ **26.** $\frac{12}{11} \cdot \frac{22}{23} \cdot \frac{1}{2}$

27. $\frac{x}{3} \cdot \frac{6}{x}$ **28.** $\frac{2}{3} \cdot \frac{9x}{4y}$

29. Find the fractional part: one-fifth of three-sevenths.

30. Find the reciprocal of each number.
   (a) 8   **(b)** 1   **(c)** $\frac{5}{19}$ **(d)** $\frac{3}{2}$

Exercises 31–34: Divide and then simplify to lowest terms when appropriate.

31. $\frac{3}{2} \div \frac{1}{6}$ **32.** $\frac{9}{10} \div \frac{7}{5}$

33. $8 \div \frac{2}{3}$ **34.** $\frac{3}{4} \div 6$

Exercises 35 and 36: Find the least common denominator for the fractions.

35. $\frac{1}{8}, \frac{5}{12}$ **36.** $\frac{3}{14}, \frac{1}{21}$

Exercises 37–42: Add or subtract and then simplify to lowest terms when appropriate.

37. $\frac{2}{15} + \frac{3}{15}$ **38.** $\frac{5}{4} - \frac{3}{4}$

39. $\frac{11}{12} - \frac{1}{8}$ **40.** $\frac{6}{11} - \frac{3}{22}$

41. $\frac{2}{3} - \frac{1}{2} + \frac{1}{4}$ **42.** $\frac{1}{6} + \frac{2}{3} - \frac{1}{9}$

SECTION 1.3

Exercises 43–48: Write the expression as an exponential expression.

43. $5 \cdot 5 \cdot 5 \cdot 5 \cdot 5 \cdot 5$ **44.** $\frac{7}{6} \cdot \frac{7}{6} \cdot \frac{7}{6}$

45. $3 \cdot 3 \cdot 3 \cdot 3$ **46.** $x \cdot x \cdot x \cdot x \cdot x$

47. $(x + 1) \cdot (x + 1)$

48. $(a - 5) \cdot (a - 5) \cdot (a - 5)$

49. Use multiplication to rewrite each expression, and then evaluate the result.
   (a) 4^3 **(b)** 7^2 **(c)** 8^1

50. Find a natural number n such that $2^n = 32$.

Exercises 51–62: Evaluate the expression by hand.

51. $7 + 3 \cdot 6$ **52.** $15 - 5 - 3$

53. $24 \div 4 \div 2$ **54.** $30 - 15 \div 3$

55. $18 \div 6 - 2$ **56.** $\frac{18}{4 + 5}$

57. $9 - 3^2$ **58.** $2^3 - 8$

59. $2^4 - 8 + \frac{4}{2}$ **60.** $3^2 - 4(5 - 3)$

61. $7 - \frac{4 + 6}{2 + 3}$ **62.** $3^3 - 2^3$

SECTION 1.4

Exercises 63 and 64: Find the opposite of each expression.

63. **(a)** -8 **(b)** $-(-(-3))$

64. **(a)** $-\left(\frac{-3}{7}\right)$ **(b)** $\frac{-2}{-5}$

Exercises 65 and 66: Find the decimal equivalent for the rational number.

65. **(a)** $\frac{4}{5}$ **(b)** $\frac{3}{20}$

66. **(a)** $\frac{5}{9}$ **(b)** $\frac{7}{11}$

Exercises 67–72: Classify the number as one or more of the following: natural number, whole number, integer, rational number, or irrational number.

67. 0 **68.** $-\frac{5}{6}$

69. -7 **70.** $\sqrt{17}$

71. π **72.** 3.4

73. Plot each number on the same number line.
   (a) 0 **(b)** -2 **(c)** $\frac{5}{4}$

74. Evaluate each expression.
   (a) $|-5|$ **(b)** $|-\pi|$ **(c)** $|\sqrt{2} - 1|$

75. Insert the symbol $>$ or $<$ to make each statement true.
   (a) $-5 \_\_ 4$ **(b)** $-\frac{1}{2} \_\_ -\frac{5}{2}$
   (c) $-3 \_\_ |-9|$ **(d)** $|-8| \_\_ |-1|$

76. List the numbers $\sqrt{3}, -3, 3, -\frac{2}{3},$ and $\pi - 1$ from least to greatest.

SECTIONS 1.5 AND 1.6

Exercises 77 and 78: (Refer to Example 3 in Section 1.5.) Find the sum visually.

77. $-5 + 9$ **78.** $4 + (-7)$

Exercises 79 and 80: Use a number line to find the sum.

79. $-1 + 2$ **80.** $-2 + (-3)$

Exercises 81–92: Evaluate the expression.

81. $5 + (-4)$

82. $-9 - (-7)$

83. $11 \cdot (-4)$

84. $-8 \cdot (-5)$

85. $11 \div (-4)$

86. $-4 \div \frac{4}{7}$

87. $-\frac{5}{9} - \left(-\frac{1}{3}\right)$

88. $-\frac{1}{2} + \left(-\frac{3}{4}\right)$

89. $-\frac{1}{3} \cdot \left(-\frac{6}{7}\right)$

90. $\dfrac{\frac{4}{5}}{-7}$

91. $-\frac{3}{2} \div \left(-\frac{3}{8}\right)$

92. $\frac{3}{8} \div (-0.5)$

Exercises 93 and 94: Write an arithmetic expression for the given phrase and then simplify.

93. Three plus negative five

94. Subtract negative four from two

Exercises 95 and 96: Write the fraction or mixed number as a decimal.

95. $\frac{7}{9}$

96. $2\frac{1}{5}$

Exercises 97 and 98: Write the decimal number as a fraction in lowest terms.

97. 0.6

98. 0.375

SECTION 1.7

Exercises 99–108: State the property that the equation illustrates.

99. $z \cdot 3 = 3z$

100. $6 + (7 + 5x) = (6 + 7) + 5x$

101. $2(5x - 2) = 10x - 4$

102. $5 + x + 3 = 5 + 3 + x$

103. $1 \cdot a = a$

104. $3 \cdot (5x) = (3 \cdot 5)x$

105. $12 - (x + 7) = 12 - x - 7$

106. $a + 0 = a$

107. $-5x + 5x = 0$

108. $-5 \cdot \left(-\frac{1}{5}\right) = 1$

Exercises 109–114: Use properties of real numbers to evaluate the expression mentally.

109. $7 + 9 + 12 + 8 + 1 + 3$

110. $500 - 199$

111. $25 \cdot 99$

112. $4581 + 1999$

113. 54.98×10

114. $4356 \div 100$

SECTION 1.8

Exercises 115–118: Determine whether the expression is a term. If the expression is a term, identify its coefficient.

115. $55x$

116. $-xy$

117. $9xy + 2z$

118. $x - 7$

Exercises 119–130: Simplify the expression.

119. $-10x + 4x$

120. $19z - 4z$

121. $3x^2 + x^2$

122. $7 + 2x - 6 + x$

123. $-\frac{1}{2} + \frac{3}{2}z - z + \frac{5}{2}$

124. $5(x - 3) - (4x + 3)$

125. $4x^2 - 3 + 5x^2 - 3$

126. $3x^2 + 4x^2 - 7x^2$

127. $\dfrac{35a}{7a}$

128. $\dfrac{0.5c}{0.5}$

129. $\dfrac{15y + 10}{5}$

130. $\dfrac{24x - 60}{12}$

APPLICATIONS

131. *Painting a Wall*   Two people are painting a large wall. The first person paints 3 square feet per minute while the second person paints 4 square feet per minute.

  (a) Write a simplified expression that gives the total number of square feet the two people can paint in x minutes.

  (b) Find the number of square feet painted in 1 hour.

  (c) How many minutes would it take for them to paint a wall 8 feet tall and 21 feet wide?

132. *Area of a Triangle*   Find the area of the triangle shown.

133. *Gallons to Pints*   There are 8 pints in 1 gallon. Make a table of values that converts G gallons to P pints. Let $G = 1, 2, 3, \ldots, 6$. Write a formula that converts G gallons to P pints.

134. *Blank CDs*   The table lists the cost C of buying x blank compact discs. Write an equation that relates C and x.

Blank CDs (x)	1	2	3	4
Cost (C)	$0.25	$0.50	$0.75	$1.00

135. *Aging in the United States*   In 2050, about $\frac{1}{5}$ of the population will be aged 65 or over and about $\frac{1}{20}$ of the population will be aged 85 or over. Estimate the fraction of the population that will be between the ages of 65 and 85 in 2050.   (**Source:** U.S. Census Bureau.)

136. *Rule of 72*   (Refer to Exercise 87 in Section 1.3.) If an investment of $25,000 earns 9% annual interest, approximate the value of the investment after 24 years.

137. *Carpentry*   A board measures $5\frac{3}{4}$ feet and needs to be cut in five equal pieces. Find the length of each piece.

138. *Distance*   Over four days an athlete jogs $3\frac{1}{8}$ miles, $4\frac{3}{8}$ miles, $6\frac{1}{4}$ miles, and $1\frac{5}{8}$ miles. How far does the athlete jog in all?

139. *Checking Account*   The initial balance in a checking account is $1652. Find the final balance resulting from the following sequence of withdrawals and deposits: $-$78, $-$91, $256, and $-$638.

140. *Temperature Range*   The highest temperature ever recorded in Amarillo, Texas, was $108°F$ and the coldest was $-16°F$. Find the difference between these two temperatures.   (**Source:** The Weather Almanac.)

141. *Top-Grossing Movies*   *Titanic* (1997) is the sixth top-grossing movie of all time. Find the total admissions for *Titanic* if they were $\frac{16}{25}$ of the total admissions for the top-grossing movie of all time, *Gone With the Wind* (1939), which had total admissions of 202 million.   (**Source:** Exhibitor Relations Co., Inc.)

CHAPTER 1   TEST   🎓Pass the Test   Video solutions to all test exercises

1. Classify the number as prime or composite. If the number is composite, write it as a product of prime numbers.
 (a) 29   **(b)** 56

2. Evaluate the expression $\dfrac{5x}{2x-1}$ for $x = -3$.

3. Translate the phrase "four squared decreased by three" to an algebraic expression. Then find the value of the expression.

4. Simplify $\frac{24}{32}$ to lowest terms.

5. Evaluate each expression. Write your answer in lowest terms.
 (a) $\frac{5}{8} + \frac{1}{8}$ **(b)** $\frac{5}{9} - \frac{3}{15}$ **(c)** $\frac{3}{5} \cdot \frac{10}{21}$
 (d) $6 \div \frac{8}{5}$ **(e)** $\frac{5}{12} + \frac{4}{9}$ **(f)** $\frac{10}{13} \div 5$

6. Write $y \cdot y \cdot y \cdot y$ as an exponential expression.

7. Evaluate each expression.
 (a) $6 + 10 \div 5$ **(b)** $4^3 - (3 - 5 \cdot 2)$
 (c) $-6^2 - 6 + \frac{4}{2}$ **(d)** $11 - \frac{1+3}{6-4}$

8. Classify the number as one or more of the following: natural number, whole number, integer, rational number, or irrational number.
 (a) -1 **(b)** $\sqrt{5}$

9. Plot each number on the same number line.
 (a) -2 **(b)** $\frac{1}{3}$ **(c)** $\sqrt{7}$

10. Insert the symbol $>$ or $<$ to make each statement true.
 (a) $2 \underline{\quad} |-5|$ **(b)** $|-1| \underline{\quad} |0|$

11. Evaluate the expression.
 (a) $-5 \div \frac{5}{6}$ **(b)** $-7 \cdot (-3)$

12. State the property or properties that each equation illustrates.
 (a) $6x - 2x = 4x$ **(b)** $12 \cdot (3x) = 36x$
 (c) $4 + x + 8 = 12 + x$

13. Use properties of real numbers to evaluate $17 \cdot 102$ mentally.

14. Simplify each expression.
 (a) $5 - 5z + 7 + z$ **(b)** $12x - (6 - 3x)$
 (c) $5 - 4(x + 6) + \dfrac{15x}{3}$

15. *Mowing a Lawn*   Two people are mowing a lawn. The first person has a riding mower and can mow $\frac{4}{3}$ acres per hour; the second person has a push mower and can mow $\frac{1}{4}$ acre per hour.

    (a) Write a simplified expression that gives the total number of acres that the two people mow in x hours.

    (b) Find the total acreage that they can mow in an 8-hour work day.

16. A wire $7\frac{4}{5}$ feet long is to be cut in 3 equal parts. How long should each part be?

17. *Cost Equation*   The table lists the cost C of buying x tickets to a hockey game.

Tickets (x)	3	4	5	6
Cost (C)	$39	$52	$65	$78

    (a) Find an equation that relates C and x.

    (b) What is the cost of 17 tickets?

18. The initial balance for a savings account is $892. Find the final balance resulting from withdrawals and deposits of $-\$57$, $\$150$, and $-\$345$.

CHAPTER 1   EXTENDED AND DISCOVERY EXERCISES

1. *Arithmetic Operations*   Insert one of the symbols $+$, $-$, $\times$, or $\div$ in each blank to obtain the given answer. Do not use any parentheses.

$$2 \_ 2 \_ 2 \_ 2 = 0$$
$$3 \_ 3 \_ 3 \_ 3 = 10$$
$$4 \_ 4 \_ 4 \_ 4 = 1$$
$$6 \_ 6 \_ 6 \_ 6 = 36$$
$$7 \_ 7 \_ 7 \_ 7 = 63$$

2. *Magic Squares*   The following square is called a "magic square" because the numbers in each row, column, and diagonal sum to 15.

8	3	4
1	5	9
6	7	2

Complete the following magic square having 4 rows and 4 columns by arranging the numbers 1 through 16 so that each row, column, and diagonal sums to 34. The four corners will also sum to 34.

	2		13
5			
		6	
4			1

Linear Equations and Inequalities

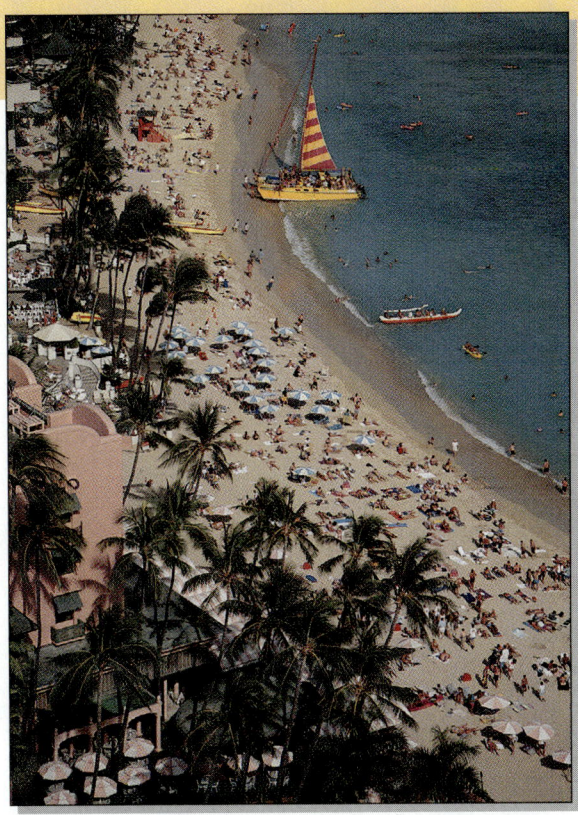

Mathematics is a unique subject because it does not depend on people making experiments, but it is essential for describing, or modeling, events in the real world. For example, sunbathing is a popular pastime, but ultraviolet light from the sun is responsible for both tanning and burning exposed skin. Mathematics lets us use numbers to describe the intensity of ultraviolet light. The table shows the maximum ultraviolet intensity measured in milliwatts per square meter for various latitudes and dates.

Latitude	Mar. 21	June 21	Sept. 21	Dec. 21
0°	325	254	325	272
10°	311	275	280	220
20°	249	292	256	143
30°	179	248	182	80
40°	99	199	127	34
50°	57	143	75	13

If a student from Chicago, located at a latitude of 42°, spends spring break in Hawaii with a latitude of 20°, the sun's ultraviolet rays in Hawaii will be approximately $\frac{249}{99} \approx 2.5$ times as intense as they are in Chicago. Suppose that you travel to the equator for spring break. How many times as intense will the sun be at the equator compared to where you presently live?

> *Education is not the filling of a pail, but the lighting of a fire.*
>
> —WILLIAM BUTLER YEATS

Source: J. Williams, *The USA Today Weather Almanac.*

2.1 INTRODUCTION TO EQUATIONS

Basic Concepts ▪ **The Addition Property of Equality** ▪ **The Multiplication Property of Equality**

A LOOK INTO MATH ▷

A new and exciting technology is the Global Positioning System (GPS), which consists of 24 satellites that travel around Earth in nearly circular orbits. The GPS can be used to determine locations and velocities of cars, airplanes, and hikers with an amazing degree of accuracy. New cars often come equipped with the GPS, and their drivers can determine their cars' locations to within a few feet.   (*Source:* J. Van Sickle, *GPS for Land Surveyors.*)

To create the GPS thousands of equations were solved, and mathematics was essential in finding their solutions. In this section we discuss many of the basic concepts needed to solve equations.

Basic Concepts

▶ **REAL-WORLD CONNECTION**   Suppose that during a storm it rains 2 inches before noon and 1 inch per hour thereafter until 5 P.M. Table 2.1 lists the total rainfall R after various elapsed times x, where $x = 0$ corresponds to noon.

TABLE 2.1   Rainfall *x* Hours Past Noon

Elapsed Time: x (hours)	0	1	2	3	4	5
Total Rainfall: R (inches)	2	3	4	5	6	7

The data suggest that the total rainfall R in inches is **2** more than the elapsed time x. A formula that *models*, or *describes*, the rainfall x hours past noon is given by

$$R = x + 2.$$

For example, **3** hours past noon, or at 3 P.M.,

$$R = 3 + 2 = 5$$

inches of rain have fallen. Even though $x = 4.5$ does not appear in the table, we can calculate the amount of rainfall at 4:30 P.M. with the formula as

$$R = 4.5 + 2 = 6.5 \text{ inches.}$$

The advantage that a formula has over a table of values is that a formula can be used to calculate the rainfall at any time x, not just at the times listed in the table.

At what time have 6 inches of rain fallen? From Table 2.1 the *solution* is 4, or 4 P.M. To find this solution without the table, we can *solve* the equation

$$x + 2 = 6.$$

An equation can be either true or false. For example, the equation $1 + 2 = 3$ is true, whereas the equation $1 + 2 = 4$ is false. When an equation contains a variable, the equation may be true for some values of the variable and false for other values of the variable. Each value of the variable that makes the equation true is called a **solution** to the equation, and the *set of all solutions* is called the **solution set**. *Solving an equation* means finding all of its solutions. Because $4 + 2 = 6$, the solution to the equation

$$x + 2 = 6$$

is 4, and the solution set is {4}. Note that **braces** {} are used to denote a set. Sometimes an equation can have more than one solution. For example, the equation $x^2 = 4$ has two solutions, **−2** and **2**, because

$$(-2)^2 = -2 \cdot (-2) = 4 \quad \text{and} \quad 2^2 = 2 \cdot 2 = 4.$$

For the equation $x^2 = 4$, the solution set is {**−2**, **2**}.

Many times we cannot solve an equation simply by looking at it. In these situations we must use a step-by-step procedure. During each step an equation is transformed into a different but equivalent equation. **Equivalent equations** are equations that have the same solution set. For example, the equations

$$x + 2 = 5 \quad \text{and} \quad x = 3$$

are equivalent equations because the solution set for both equations is {3}.

MAKING CONNECTIONS

Equations and Expressions

Although the words "equation" and "expression" occur frequently in mathematics, they are *not* interchangeable. An equation *always* contains an equals sign but an expression *never* contains an equals sign. We often want to *solve* an equation, whereas an expression can sometimes be *simplified*. Furthermore, the equals sign in an equation separates two expressions. For example, $3x - 5 = x + 1$ is an equation where $3x - 5$ and $x + 1$ are each expressions.

When solving equations, it is often helpful to transform a more complicated equation into an equivalent equation that has an obvious solution, such as $x = 3$. Two important principles used to solve equations are the *addition property of equality* and the *multiplication property of equality*. These properties can be used to transform an equation into an equivalent equation that is easier to solve.

The Addition Property of Equality

When solving an equation, we have to apply the same operation to each side of the equation. For example, one way to solve the equation

$$x + 2 = 5$$

is to add −2 to each side. This step results in isolating the x on one side of the equation.

$x + 2 = 5$	Given equation
$x + 2 + (-2) = 5 + (-2)$	Add −2 to each side.
$x + 0 = 3$	Addition of real numbers
$x = 3$	Additive identity

These four equations are equivalent, but the solution is most apparent in the last equation. To transform one equation to the next, a step-by-step procedure based on mathematical properties is used. The reasons that justify each step are written at the right in blue. When −2 is added to each side of the equation, the addition property of equality is used.

> ### ADDITION PROPERTY OF EQUALITY
>
> If a, b, and c are real numbers, then
>
> $$a = b \quad \text{is equivalent to} \quad a + c = b + c.$$
>
> That is, adding the same number to each side of an equation results in an equivalent equation.

Generally, the addition property of equality is based on the concept that "if equals are added to equals, the results are equal."

NOTE: Because any subtraction problem can be changed to an addition problem, *the addition property of equality also works for subtraction*. That is, if the same number is subtracted from each side of an equation, the result is an equivalent equation.

EXAMPLE **1** **Using the addition property of equality**

Solve each equation.
(a) $x + 10 = 7$ **(b)** $t - 4 = 3$ **(c)** $\frac{1}{2} = -\frac{3}{4} + y$

Solution
(a) When solving an equation, we try to isolate the variable on one side of the equation. If we add **−10** to (or subtract 10 from) each side of the equation, the value of x becomes apparent.

$x + 10 = 7$	Given equation
$x + 10 + (-10) = 7 + (-10)$	Add -10 to each side.
$x + 0 = -3$	Addition of real numbers
$x = -3$	Additive identity

The solution is -3.
(b) To isolate the variable t, add **4** to each side.

$t - 4 = 3$	Given equation
$t - 4 + 4 = 3 + 4$	Add **4** to each side.
$t + 0 = 7$	Addition of real numbers
$t = 7$	Additive identity

The solution is 7.
(c) To isolate the variable y, add $\frac{3}{4}$ to each side.

$\dfrac{1}{2} = -\dfrac{3}{4} + y$	Given equation
$\dfrac{1}{2} + \dfrac{3}{4} = -\dfrac{3}{4} + \dfrac{3}{4} + y$	Add $\frac{3}{4}$ to each side.
$\dfrac{5}{4} = 0 + y$	Addition of real numbers
$\dfrac{5}{4} = y$	Additive identity

The solution is $\frac{5}{4}$.

Now Try Exercises **17, 27**

CHECKING A SOLUTION   To check a solution, substitute it in the given equation to find out if a true statement results. To check the solution for Example 1(c), substitute $\frac{5}{4}$ for y in the given equation. Note that a question mark is placed over the equals sign when a solution is being checked.

$$\frac{1}{2} = -\frac{3}{4} + y \qquad \text{Given equation}$$

$$\frac{1}{2} \stackrel{?}{=} -\frac{3}{4} + \mathbf{\frac{5}{4}} \qquad \text{Replace } y \text{ with } \frac{5}{4}.$$

$$\frac{1}{2} \stackrel{?}{=} \frac{2}{4} \qquad \text{Add fractions.}$$

$$\frac{1}{2} = \frac{1}{2} \qquad \text{The answer checks.}$$

The answer of $\frac{5}{4}$ checks because the left side of the equation equals the right side of the equation.

CRITICAL THINKING

When you are checking a solution, why do you substitute your answer in the *given* equation?

EXAMPLE 2   Solving and checking a solution

Solve the equation $-5 + y = 3$ and then check the solution.

Solution
Isolate y by adding 5 to each side.

$$-5 + y = 3 \qquad \text{Given equation}$$

$$5 + (-5) + y = 5 + 3 \qquad \text{Add 5 to each side.}$$

$$0 + y = 8 \qquad \text{Addition of real numbers}$$

$$y = 8 \qquad \text{Additive identity}$$

The solution is 8. To check this answer substitute **8** for **y** in the given equation.

$$-5 + \mathbf{y} = 3 \qquad \text{Given equation}$$

$$-5 + \mathbf{8} \stackrel{?}{=} 3 \qquad \text{Replace } y \text{ with 8.}$$

$$3 = 3 \qquad \text{The answer checks.}$$

Now Try Exercise **31**

MAKING CONNECTIONS

Equations and Scales

Think of an equation as an old-fashioned scale, where two pans must balance, as illustrated in the accompanying figure. If two weights initially balance the pans, then adding an equal amount of weight to each pan (or subtracting an equal amount from each pan) results in the pans remaining balanced.

The Multiplication Property of Equality

The multiplication property of equality is another important property used to solve equations. We can illustrate this property by considering a formula that converts yards to feet. Because there are 3 feet in 1 yard, the formula $F = 3Y$ computes F, the number of feet in Y yards. For example, if $Y = 5$ yards, then $F = 3 \cdot 5 = 15$ feet.

Now consider the reverse, converting 27 feet to yards. The answer to this conversion corresponds to the solution to

$$27 = 3Y.$$

To find the solution, multiply each side of the equation by the reciprocal of 3, or $\frac{1}{3}$.

$27 = 3Y$	Given equation
$\frac{1}{3} \cdot 27 = \frac{1}{3} \cdot 3 \cdot Y$	Multiply each side by $\frac{1}{3}$.
$9 = 1 \cdot Y$	Multiplication of fractions
$9 = Y$	Multiplicative identity

Thus 27 feet are equivalent to 9 yards.

MULTIPLICATION PROPERTY OF EQUALITY

If a, b, and c are real numbers with $c \neq 0$, then

$$a = b \quad \text{is equivalent to} \quad ac = bc.$$

That is, multiplying each side of an equation by the same nonzero number results in an equivalent equation.

Generally, the multiplication property of equality is based on the concept that "if equals are multiplied by equals, the results are equal."

NOTE: Because any division problem can be changed to a multiplication problem, *the multiplication property of equality also works for division.* That is, if each side of an equation is divided by the same nonzero number, the result is an equivalent equation.

EXAMPLE 3   Using the multiplication property of equality

Solve each equation.
(a) $\frac{1}{3}x = 4$ **(b)** $-4y = 8$ **(c)** $5 = \frac{3}{4}z$

Solution
(a) We start by multiplying each side of the equation by **3**, the reciprocal of $\frac{1}{3}$.

$\frac{1}{3}x = 4$	Given equation
$3 \cdot \frac{1}{3}x = 3 \cdot 4$	Multiply each side by 3.
$1 \cdot x = 12$	Multiplication of real numbers
$x = 12$	Multiplicative identity

The solution is 12.

(b) The coefficient of the y-term is -4, so we can either multiply each side of the equation by $-\frac{1}{4}$ or divide each side by -4. This step will make the coefficient of y equal to 1.

$$-4y = 8 \qquad \text{Given equation}$$

$$\frac{-4y}{-4} = \frac{8}{-4} \qquad \text{Divide each side by } -4.$$

$$y = -2 \qquad \text{Simplify fractions.}$$

The solution is -2.

(c) To change the coefficient of z from $\frac{3}{4}$ to 1, multiply each side of the equation by $\frac{4}{3}$, the reciprocal of $\frac{3}{4}$.

$$5 = \frac{3}{4}z \qquad \text{Given equation}$$

$$\frac{4}{3} \cdot 5 = \frac{4}{3} \cdot \frac{3}{4}z \qquad \text{Multiply each side by } \tfrac{4}{3}.$$

$$\frac{20}{3} = 1 \cdot z \qquad \text{Multiplication of real numbers}$$

$$\frac{20}{3} = z \qquad \text{Multiplicative identity}$$

The solution is $\frac{20}{3}$.

Now Try Exercises **41, 49**

EXAMPLE **4**   Solving and checking a solution

Solve the equation $\frac{3}{4} = -\frac{3}{7}t$ and then check the solution.

Solution

Multiply each side of the equation by $-\frac{7}{3}$, the reciprocal of $-\frac{3}{7}$.

$$\frac{3}{4} = -\frac{3}{7}t \qquad \text{Given equation}$$

$$-\frac{7}{3} \cdot \frac{3}{4} = -\frac{7}{3} \cdot \left(-\frac{3}{7}\right)t \qquad \text{Multiply each side by } -\tfrac{7}{3}.$$

$$-\frac{7}{4} = 1 \cdot t \qquad \text{Multiplication of real numbers}$$

$$-\frac{7}{4} = t \qquad \text{Multiplicative identity}$$

The solution is $-\frac{7}{4}$. To check this answer, substitute $-\frac{7}{4}$ for t in the given equation.

$$\frac{3}{4} = -\frac{3}{7}t \qquad \text{Given equation}$$

$$\frac{3}{4} \overset{?}{=} -\frac{3}{7} \cdot \left(-\frac{7}{4}\right) \qquad \text{Replace } t \text{ with } -\tfrac{7}{4}.$$

$$\frac{3}{4} = \frac{3}{4} \qquad \text{The answer checks.}$$

Now Try Exercise **57**

▶ **REAL-WORLD CONNECTION**   Alaska's major glaciers are melting at an alarming rate. From the mid 1950s to the mid 1990s, these glaciers melted at a rate of 13 cubic miles per year. Recently, Alaska's glaciers have been melting at a rate of 24 cubic miles per year, almost double the earlier rate. Ecologists are concerned that this phenomenon is a sign of global warming.   (**Source:** *USA Today.*)

EXAMPLE 5   **Estimating glacial melting in Alaska**

The glaciers in Alaska are currently melting at a rate of 24 cubic miles per year.
(a) Write a formula that gives the cubic miles of ice I that will melt in x years.
(b) At this rate, determine how long it will take for 300 cubic miles of ice to melt.

Solution
(a) In 1 year **24 · 1** = 24 cubic miles will melt, in 2 years **24 · 2** = 48 cubic miles will melt, and in x years **24 · x** = $24x$ cubic miles will melt. Thus $I = 24x$, where x is in years and I is in cubic miles.
(b) To determine how long it will take for 300 cubic miles to melt let $I = $ **300** in the formula.

$$I = 24x \qquad \text{Formula from part (a)}$$
$$\mathbf{300} = 24x \qquad \text{Let } I = 300.$$
$$\frac{300}{24} = \frac{24x}{24} \qquad \text{Divide each side by 24.}$$
$$12.5 = x \qquad \text{Simplify.}$$

At current rates, it will take 12.5 years for 300 cubic miles of ice to melt.

Now Try Exercise 65

2.1   PUTTING IT ALL TOGETHER

The following table summarizes some of the topics discussed in this section.

Concept	Comments	Examples
Equation	An equation is a mathematical statement that two expressions are equal. An equation can be either true or false.	The equation $2 + 3 = 5$ is true. The equation $1 + 3 = 7$ is false.
Solution	A value for a variable that makes an equation a true statement	The solution to $x + 5 = 20$ is 15, and the solutions to $x^2 = 9$ are -3 and 3.
Solution Set	The set of all solutions to an equation	The solution set to $x + 5 = 20$ is $\{15\}$, and the solution set to $x^2 = 9$ is $\{-3, 3\}$.
Equivalent Equations	Two equations are equivalent if they have the same solution set.	The equations $$2x = 14 \quad \text{and} \quad x = 7$$ are equivalent because the solution set to both equations is $\{7\}$.

Concept	Comments	Examples
Addition Property of Equality	The equations $a = b$ and $a + c = b + c$ are equivalent. This property is used to solve equations.	To solve $x - 3 = 8$ add 3 to each side of the equation. $$x - 3 + 3 = 8 + 3$$ $$x = 11$$ The solution is 11.
Multiplication Property of Equality	The equations $a = b$ and $a \cdot c = b \cdot c$ with $c \neq 0$ are equivalent. This property is used to solve equations.	To solve $\frac{1}{5}x = 10$ multiply each side of the equation by 5. $$5 \cdot \frac{1}{5}x = 5 \cdot 10$$ $$x = 50$$ The solution is 50.
Checking a Solution	Substitute the solution in the given equation and then simplify each side to check it.	To show that 8 is a solution to $$x + 12 = 20$$ substitute 8 for x. $$8 + 12 \stackrel{?}{=} 20$$ $$20 = 20 \quad \text{True}$$

2.1 Exercises

CONCEPTS

1. The equation $1 + 3 = 4$ is _(true/false)_.

2. The equation $2 + 3 = 6$ is _(true/false)_.

3. Each value of a variable that makes an equation true is called a(n) _____.

4. The _____ is the set of all solutions to an equation.

5. To solve an equation, find all _____.

6. _____ equations have the same solution sets.

7. If $a = b$, then $a + c =$ _____.

8. Because any subtraction problem can be changed to an addition problem, the addition property of equality also works for _____.

9. If $a = b$ and $c \neq 0$, then $ac =$ _____.

10. Because any division problem can be changed to a multiplication problem, the multiplication property of equality also works for _____.

11. To solve an equation, transform the equation into a(n) _____ equation that is easier to solve.

12. To check an answer, substitute it in the _____ equation.

THE ADDITION PROPERTY OF EQUALITY

13. To solve $x - 22 = 4$, add _____ to each side.

14. To solve $\frac{5}{6} = \frac{1}{6} + x$, add _____ to each side.

15. To solve $x + 3 = 13$, subtract _____ from each side.

16. To solve $\frac{3}{4} = \frac{1}{4} + x$, subtract _____ from each side.

Exercises 17–34: Solve the equation. Check your answer.

17. $x + 5 = 0$ **18.** $x + 3 = 7$

19. $x - 7 = 1$ **20.** $x - 23 = 0$

21. $a + 41 = 7$ **22.** $a + 30 = 4$

23. $a - 12 = -3$ **24.** $a - 19 = -11$

25. $9 = y - 8$ **26.** $97 = -23 + y$

27. $\frac{1}{2} = z - \frac{3}{2}$ **28.** $\frac{3}{4} + z = -\frac{1}{2}$

29. $t - 0.8 = 4.2$ **30.** $4 = -9 + t$

31. $25 + x = 10$ **32.** $85 = x - 20$

33. $1989 = 26 + y$ **34.** $y - 1.23 = -0.02$

35. Thinking Generally To solve $x - a = b$ for x, add _____ to each side.

36. Thinking Generally To solve $x + a = b$ for x, subtract _____ from each side.

THE MULTIPLICATION PROPERTY OF EQUALITY

37. To solve $5x = 4$, multiply each side by _____.

38. To solve $\frac{4}{3}y = 8$, multiply each side by _____.

39. To solve $6x = 11$, divide each side by _____.

40. To solve $0.2x = 4$, divide each side by _____.

Exercises 41–58: Solve the equation. Check your answer.

41. $5x = 15$ **42.** $-2x = 8$

43. $-7x = 0$ **44.** $25x = 0$

45. $-5a = -35$ **46.** $-4a = -32$

47. $3a = -18$ **48.** $10a = -70$

49. $\frac{1}{2}x = \frac{3}{2}$ **50.** $\frac{3}{4}x = \frac{5}{8}$

51. $\frac{1}{2} = \frac{2}{5}z$ **52.** $-\frac{3}{4} = -\frac{1}{8}z$

53. $25 = 5z$ **54.** $-10 = -4z$

55. $0.5t = 3.5$ **56.** $2.2t = -9.9$

57. $\frac{3}{8} = \frac{1}{4}y$ **58.** $1.2 = 0.3y$

59. Thinking Generally To solve $\frac{1}{a} \cdot x = b$ for x, multiply each side by _____.

60. Thinking Generally To solve $ax = b$, where $a \neq 0$, for x, divide each side by _____.

APPLICATIONS

61. *Rainfall* On a stormy day it rains 3 inches before noon and $\frac{1}{2}$ inch per hour thereafter until 6 P.M.
(a) Make a table that shows the total rainfall R in inches, x hours past noon, ending at 6 P.M.
(b) Write a formula that calculates R.
(c) Use your formula to calculate the total rainfall at 3 P.M. Does the answer agree with the value in your table from part (a)?
(d) How much rain has fallen by 2:15 P.M.?

62. *Cold Weather* A furnace is turned on at midnight when the temperature inside a cabin is $0°$F. The cabin warms at a rate of $10°$F per hour until 7 A.M.
(a) Make a table that shows the cabin temperature T in degrees Fahrenheit, x hours past midnight, ending at 7 A.M.
(b) Write a formula that calculates T.
(c) Use your formula to calculate the temperature at 5 A.M. Does the answer agree with the value in your table from part (a)?
(d) Find the cabin temperature at 2:45 A.M.

63. *Football Field* A football field is 300 feet long.
(a) Write a formula that gives the length L of x football fields in feet.
(b) Use your formula to write an equation whose solution gives the number of football fields in 870 feet.
(c) Solve your equation from part (b).

64. *Acreage* An acre equals 43,560 square feet.
(a) Write a formula that converts A acres to S square feet.
(b) Use your formula to write an equation whose solution gives the number of acres in 871,200 square feet.
(c) Solve your equation from part (b).

65. *Glacial Melting in Alaska* (Refer to Example 5.) The Alaskan glaciers are melting at a rate of 24 cubic miles per year. At this rate, how many years will it take for 420 cubic miles of the glacier to melt? (*Source: USA Today.*)

66. *Arctic Ice Cap*   The Arctic ice cap contains 680,000 cubic miles of ice. If this ice cap melted at a rate of 50 cubic miles per year, how many years would it take for the entire ice cap to melt?   (*Source:* Department of the Interior, Geological Survey.)

67. *Cost of a Car*   When the cost of a car is multiplied by 0.07 the result is $1750. Find the cost of the car.

68. *Raise in Salary*   If an employee's salary is multiplied by 1.06, which corresponds to a 6% raise, the result is $58,300. Find the employee's current salary.

WRITING ABOUT MATHEMATICS

69. A student solves an equation as follows.

$$x + 30 = 64$$
$$x \overset{?}{=} 64 + 30$$
$$x \overset{?}{=} 94$$

Identify the student's mistake. What is the solution?

70. What is a good first step for solving the equation $\frac{a}{b}x = 1$, where a and b are natural numbers? What is the solution? Explain your answers.

2.2 LINEAR EQUATIONS

Basic Concepts ■ Solving Linear Equations ■ Applying the Distributive Property ■ Clearing Fractions and Decimals ■ Equations with No Solutions or Infinitely Many Solutions

A LOOK INTO MATH ▷ Solving equations is an important concern in mathematics. Billions of dollars are spent each year to solve equations that lead to the creation of better products. If our society could not solve equations, we would not have TV sets, DVD players, satellites, fiber optics, CAT scans, computers, or accurate weather forecasts. In this section we discuss linear equations and some of their applications. One of the simplest types of equations, linear equations, can always be solved by hand.

Basic Concepts

Suppose that a bicyclist is 5 miles from home, riding *away* from home at 10 miles per hour, as shown in Figure 2.1. The distance between the bicyclist and home for various elapsed times is shown in Table 2.2.

Figure 2.1 Distance from Home

TABLE 2.2   **Distance from Home**

Elapsed Time (hours)	0	1	2	3
Distance (miles)	5	15	25	35

10 miles   10 miles   10 miles

The bicyclist is moving at a constant speed, so the distance increases by 10 miles every hour. The distance D from home after x hours could be calculated by the formula

$$D = 10x + 5.$$

For example, after **2** hours the distance is

$$D = 10(2) + 5 = 25 \text{ miles}.$$

Table 2.2 verifies that the bicyclist is 25 miles from home after 2 hours. However, the table is less helpful if we want to find the elapsed time when the bicyclist is 18 miles from home. To answer this question, we could begin by substituting **18** for D in the formula to obtain the equation

$$18 = 10x + 5.$$

This equation can be written in a different form by applying the addition property of equality. Subtracting 18 from each side allows us to write an equivalent equation.

$$18 - \mathbf{18} = 10x + 5 - \mathbf{18} \qquad \text{Subtract 18 from each side.}$$
$$0 = 10x - 13 \qquad \text{Simplify.}$$
$$10x - 13 = 0 \qquad \text{Rewrite the equation.}$$

The equation $10x - 13 = 0$ is an example of a *linear equation* and is solved in Example 3(a). Linear equations often model applications in which things either move or change at a constant rate.

LINEAR EQUATION IN ONE VARIABLE

A **linear equation** in one variable is an equation that can be written in the form

$$ax + b = 0,$$

where a and b are constants with $a \neq 0$.

If an equation is linear, writing it in the form $ax + b = 0$ should not require any properties or processes other than the following:

- using the distributive property to clear any parentheses,
- combining like terms,
- applying the addition property of equality.

For example, the equation $18 = 10x + 5$ is linear because applying the addition property of equality results in $10x - 13 = 0$.

Table 2.3 gives examples of linear equations and values for a and b.

TABLE 2.3   Linear Equations

$ax + b = 0$	a	b
$x - 1 = 0$	1	-1
$-5x + 1 = 0$	-5	1
$2.5x = 0$	2.5	0

To determine whether an equation in one variable is linear we must determine whether the equation can be written in the form $ax + b = 0$. An equation *cannot* be written in this form if after clearing parentheses and combining like terms, any of the following statements are true.

1. The variable has an exponent other than 1.
2. The variable appears in a denominator of a fraction.
3. The variable appears under the symbol $\sqrt{}$ or within an absolute value.

EXAMPLE 1   **Determining whether an equation is linear**

Determine whether the equation is linear. If the equation is linear, give values for a and b.

(a) $4x + 5 = 0$ **(b)** $5 = -\frac{3}{4}x$ **(c)** $4x^2 + 6 = 0$ **(d)** $\frac{3}{x} + 5 = 0$

Solution

(a) The equation is linear because it is in the form $ax + b = 0$ with $a = 4$ and $b = 5$.

(b) The equation can be rewritten as follows.

$$5 = -\frac{3}{4}x \qquad \text{Given equation}$$

$$\frac{3}{4}x + 5 = \frac{3}{4}x + \left(-\frac{3}{4}x\right) \qquad \text{Add } \tfrac{3}{4}x \text{ to each side.}$$

$$\frac{3}{4}x + 5 = 0 \qquad \text{Additive inverse.}$$

This equation is linear because it is in the form $ax + b = 0$ with $a = \frac{3}{4}$ and $b = 5$.

NOTE: If 5 had been subtracted from each side, the result would be $0 = -\frac{3}{4}x - 5$, which is an equivalent linear equation with $a = -\frac{3}{4}$ and $b = -5$.

(c) The equation is *not* linear because it cannot be written in the form $ax + b = 0$. The variable has an exponent other than 1.

(d) The equation is *not* linear because it cannot be written in the form $ax + b = 0$. The variable appears in the denominator of a fraction.   Now Try Exercises 13, 19, 21

Solving Linear Equations

Because a linear equation can be written in the form $ax + b = 0$ with $a \neq 0$, it has *exactly one* solution. Showing that this is true is left as an exercise. Solving a linear equation means finding the value of the variable that makes the equation true.

One way to solve a linear equation is to make a table. Making a table is an organized way of checking possible values of the variable to see if any value makes the equation true. For example, if we want to solve the equation $2x - 5 = -7$, we substitute various values for x in the left side of the equation. If one of these values results in -7, the right side of the equation, the value makes the equation true and is the solution. In the next example a table is used to solve this equation.

EXAMPLE 2   **Using a table to solve an equation**

Complete Table 2.4 for the given values of x. Then solve the equation $2x - 5 = -7$.

TABLE 2.4

x	-3	-2	-1	0	1	2	3
$2x - 5$	-11						

Solution

To complete the table, substitute $x = -2, -1, 0, 1, 2,$ and 3 into the expression $2x - 5$. For example, if $x = -2$, then $2x - 5 = 2(-2) - 5 = -9$. The other values shown in Table 2.5 on the next page can be found similarly.

TECHNOLOGY NOTE:

Graphing Calculators
and Tables

Many graphing calculators
have the capability to make
tables. Table 2.5 is shown in
the accompanying figure.

X	Y₁
-3	-11
-2	-9
-1	-7
0	-5
1	-3
2	-1
3	1

Y₁≣2X−5

CALCULATOR HELP

To create a table of values, see
Appendix A (page AP-3).

TABLE 2.5

x	−3	−2	**−1**	0	1	2	3
$2x - 5$	−11	−9	**−7**	−5	−3	−1	1

From the table, $2x - 5$ equals -7 when $x = -1$. Thus the solution to $2x - 5 = -7$ is -1.

Now Try Exercise 27

Although tables can be used to solve some linear equations, the process of creating a table that contains the solution can take a significant amount of time. For some equations, choosing appropriate values to be substituted for the variable can be very difficult. For example, the solution to the equation $9x - 4 = 0$ is $\frac{4}{9}$. However, creating a table that reveals this solution would be quite challenging.

A method for solving equations that is more efficient than making a table is to use the following strategy, which involves the addition and multiplication properties of equality.

STEPS FOR SOLVING A LINEAR EQUATION

STEP 1: Use the distributive property to clear any parentheses on each side of the equation. Combine any like terms on each side.

STEP 2: Use the addition property of equality to get all of the terms containing the variable on one side of the equation and all other terms on the other side of the equation. Combine any like terms on each side.

STEP 3: Use the multiplication property of equality to isolate the variable by multiplying each side of the equation by the reciprocal of the number in front of the variable (or divide each side by that number).

STEP 4: Check the solution by substituting it in the given equation.

When a linear equation does not contain parentheses, we can start with the second step in the strategy shown above. This is the case for the equations in the next example.

EXAMPLE **3**   Solving linear equations

Solve each linear equation. Check the answer for part (b).
(a) $10x - 13 = 0$ **(b)** $\frac{1}{2}x + 3 = 6$ **(c)** $5x + 7 = 2x + 3$

Solution
(a) First, isolate the x-term on the left side of the equation by adding 13 to each side.

$$10x - 13 = 0 \qquad \text{Given equation}$$
$$10x - 13 + 13 = 0 + 13 \qquad \text{Add 13 to each side. (Step 2)}$$
$$10x = 13 \qquad \text{Add the real numbers.}$$

To obtain a coefficient of 1 on the x-term, divide each side by 10.

$$\frac{10x}{10} = \frac{13}{10} \qquad \text{Divide each side by 10. (Step 3)}$$
$$x = \frac{13}{10} \qquad \text{Simplify.}$$

The solution is $\frac{13}{10}$.

NOTE: In the second equation of the solution the addition property is used, and in the fourth equation of the solution the multiplication property is used.

(b) Start by subtracting 3 from each side.

$$\frac{1}{2}x + 3 = 6$$ Given equation

$$\frac{1}{2}x + 3 - 3 = 6 - 3$$ Subtract 3 from each side. (Step 2)

$$\frac{1}{2}x = 3$$ Subtract the real numbers.

$$2 \cdot \frac{1}{2}x = 2 \cdot 3$$ Multiply each side by 2. (Step 3)

$$x = 6$$ Multiply the real numbers.

The solution is 6. To check it substitute **6** for x in the given equation.

$$\frac{1}{2} \cdot 6 + 3 \stackrel{?}{=} 6$$ Replace x with 6. (Step 4)

$$3 + 3 \stackrel{?}{=} 6$$ Multiply.

$$6 = 6$$ The answer checks.

(c) In this equation there are two x-terms. Generally it is a good idea to move all x-terms to one side of the equation and all real numbers to the other side. To accomplish this task, begin by subtracting $2x$ from each side.

$$5x + 7 = 2x + 3$$ Given equation

$$5x - 2x + 7 = 2x - 2x + 3$$ Subtract $2x$ from each side. (Step 2)

$$3x + 7 = 3$$ Combine like terms.

$$3x + 7 - 7 = 3 - 7$$ Subtract 7 from each side. (Step 2)

$$3x = -4$$ Simplify.

$$\frac{3x}{3} = \frac{-4}{3}$$ Divide each side by 3. (Step 3)

$$x = -\frac{4}{3}$$ Simplify the fractions.

The solution is $-\frac{4}{3}$.   Now Try Exercises 33, 39

▶ **REAL-WORLD CONNECTION**   The next example involves a linear equation.

EXAMPLE 4   **Estimating numbers of worldwide Internet users**

The number of Internet users I in millions during year x can be approximated by the formula

$$I = 116x - 231{,}627,$$

where $x \geq 2000$. Estimate the year when there were 950 million Internet users.   (**Source:** Internet World Stats.)

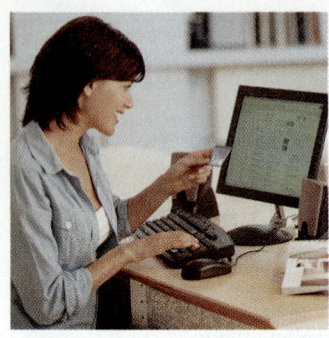

Solution

Let $I = \mathbf{950}$ in the formula $I = 116x - 231{,}627$ and solve for x.

$$950 = 116x - 231{,}627 \qquad \text{Equation to be solved}$$

$$232{,}577 = 116x \qquad \text{Add 231,627 to each side.}$$

$$\frac{232{,}577}{116} = \frac{116x}{116} \qquad \text{Divide each side by 116.}$$

$$\frac{232{,}577}{116} = x \qquad \text{Simplify.}$$

$$x \approx 2005 \qquad \text{Approximate with a calculator.}$$

During 2005 the number of Internet users reached 950 million.   **Now Try Exercise** `75`

Applying the Distributive Property

Sometimes the distributive property is helpful in solving linear equations. The next example demonstrates how to apply the distributive property in such situations. Use of the distributive property was indicated in Step 1 of the strategy for solving linear equations discussed earlier.

EXAMPLE `5`   **Applying the distributive property**

Solve each linear equation. Check the answer for part (a).
(a) $4(x - 3) + x = 0$
(b) $2(3z - 4) + 1 = 3(z + 1)$

Solution

(a) Begin by applying the distributive property.

$$4(x - 3) + x = 0 \qquad \text{Given equation}$$

$$4x - 12 + x = 0 \qquad \text{Distributive property (Step 1)}$$

$$5x - 12 = 0 \qquad \text{Combine like terms.}$$

$$5x - 12 + \mathbf{12} = 0 + \mathbf{12} \qquad \text{Add 12 to each side. (Step 2)}$$

$$5x = 12 \qquad \text{Add the real numbers.}$$

$$\frac{5x}{\mathbf{5}} = \frac{12}{\mathbf{5}} \qquad \text{Divide each side by 5. (Step 3)}$$

$$x = \frac{12}{5} \qquad \text{Simplify.}$$

To check whether $\frac{12}{5}$ is the solution, substitute $\frac{12}{5}$ for x in the given equation.

$$4\left(\frac{\mathbf{12}}{\mathbf{5}} - 3\right) + \frac{\mathbf{12}}{\mathbf{5}} \overset{?}{=} 0 \qquad \text{Replace } x \text{ with } \frac{12}{5}. \text{ (Step 4)}$$

$$4\left(\frac{12}{5} - \frac{15}{5}\right) + \frac{12}{5} \overset{?}{=} 0 \qquad \text{Common denominator}$$

$$4\left(-\frac{3}{5}\right) + \frac{12}{5} \overset{?}{=} 0 \qquad \text{Subtract within parentheses.}$$

$$-\frac{12}{5} + \frac{12}{5} \overset{?}{=} 0 \qquad \text{Multiply.}$$

$$0 = 0 \qquad \text{The answer checks.}$$

(b) Begin by applying the distributive property to each side of the equation. Then move z-terms to the left side and terms containing only real numbers to the right side.

$$2(3z - 4) + 1 = 3(z + 1) \quad \text{Given equation}$$
$$6z - 8 + 1 = 3z + 3 \quad \text{Distributive property (Step 1)}$$
$$6z - 7 = 3z + 3 \quad \text{Add the real numbers.}$$
$$3z - 7 = 3 \quad \text{Subtract } 3z \text{ from each side. (Step 2)}$$
$$3z = 10 \quad \text{Add 7 to each side. (Step 2)}$$
$$\frac{3z}{3} = \frac{10}{3} \quad \text{Divide each side by 3. (Step 3)}$$
$$z = \frac{10}{3} \quad \text{Simplify.} \qquad \text{Now Try Exercises } 43, 49$$

Clearing Fractions and Decimals

Because most people find it easier to do hand calculations without fractions or decimals, clearing an equation of fractions or decimals before solving for the variable is often helpful. To clear fractions or decimals we often multiply each side of the equation by the least common denominator (LCD).

EXAMPLE 6 Clearing fractions from linear equations

Solve each linear equation.
(a) $\frac{1}{7}x - \frac{5}{7} = \frac{3}{7}$ **(b)** $\frac{2}{3}x - \frac{1}{6} = x$

Solution
(a) Multiply each side of the equation by the LCD 7 to clear (remove) fractions from the equation.

$$\frac{1}{7}x - \frac{5}{7} = \frac{3}{7} \quad \text{Given equation}$$
$$7\left(\frac{1}{7}x - \frac{5}{7}\right) = 7 \cdot \frac{3}{7} \quad \text{Multiply each side by 7.}$$
$$x - 5 = 3 \quad \text{Distributive property}$$
$$x = 8 \quad \text{Add 5 to each side.}$$

The solution is 8.
(b) The LCD for 3 and 6 is 6. Multiply each side of the equation by 6.

$$\frac{2}{3}x - \frac{1}{6} = x \quad \text{Given equation}$$
$$6\left(\frac{2}{3}x - \frac{1}{6}\right) = 6 \cdot x \quad \text{Multiply each side by 6.}$$
$$4x - 1 = 6x \quad \text{Distributive property}$$
$$-1 = 2x \quad \text{Subtract } 4x \text{ from each side.}$$
$$-\frac{1}{2} = x \quad \text{Divide each side by 2.}$$

The solution is $-\frac{1}{2}$. 　　　　　Now Try Exercise 55

EXAMPLE 7   **Clearing decimals from a linear equation**

Solve each linear equation.
(a) $0.2x - 0.7 = 0.4$ **(b)** $0.01x - 0.42 = -0.2x$

Solution
(a) The least common denominator for 0.2, 0.7, and 0.4 $\left(\text{or } \frac{2}{10}, \frac{7}{10}, \text{ and } \frac{4}{10}\right)$ is 10. Multiply each side by 10. When multiplying by 10, move the decimal point 1 place to the right.

$$0.2x - 0.7 = 0.4 \qquad \text{Given equation}$$
$$\mathbf{10}(0.2x - 0.7) = \mathbf{10}(0.4) \qquad \text{Multiply each side by 10.}$$
$$2x - 7 = 4 \qquad \text{Distributive property}$$
$$2x = 11 \qquad \text{Add 7 to each side.}$$
$$x = \frac{11}{2} \qquad \text{Divide each side by 2.}$$

The solution is $\frac{11}{2}$, or 5.5.

(b) The least common denominator for 0.01, 0.2, and 0.42 $\left(\text{or } \frac{1}{100}, \frac{2}{10}, \text{ and } \frac{42}{100}\right)$ is 100. Multiply each side by 100. To do this move the decimal point 2 places to the right.

$$0.01x - 0.42 = -0.2x \qquad \text{Given equation}$$
$$\mathbf{100}(0.01x - 0.42) = \mathbf{100}(-0.2x) \qquad \text{Multiply each side by 100.}$$
$$x - 42 = -20x \qquad \text{Distributive property}$$
$$x - 42 \mathbf{+ 20x + 42} = -20x \mathbf{+ 20x + 42} \qquad \text{Add } 20x \text{ and 42.}$$
$$21x = 42 \qquad \text{Combine like terms.}$$
$$x = 2 \qquad \text{Divide each side by 21.}$$

The solution is 2.   Now Try Exercise 53

Equations with No Solutions or Infinitely Many Solutions

Some equations that appear to be linear are not because when they are written in the form $ax + b = 0$ the value of a is 0 and no x-term appears. This type of equation can have no solutions or infinitely many solutions. An example of an equation that has no solutions is

$$x = x + 1$$

because a number x cannot equal itself plus 1. If we attempt to solve this equation by subtracting x from each side, we obtain the equation $1 = 0$, which is *always false*, indicating there are no solutions. This equation has no x-term.

An example of an equation with infinitely many solutions is

$$5x = 2x + 3x,$$

because the equation simplifies to

$$5x = 5x,$$

which is true for any real number x. If $5x$ is subtracted from each side the result is $0 = 0$, which has no x-term and is *always true*.

NOTE: An equation that is always true is called an **identity** and an equation that is always false is called a **contradiction**.

EXAMPLE 8   Determining numbers of solutions

Determine whether the equation has no solutions, one solution, or infinitely many solutions.
(a) $3x = 2(x + 1) + x$ (b) $2x - (x + 1) = x - 1$ (c) $5x = 2(x - 4)$

Solution
(a) Start by applying the distributive property.

$3x = 2(x + 1) + x$	Given equation
$3x = 2x + 2 + x$	Distributive property
$3x = 3x + 2$	Combine like terms.
$0 = 2$	Subtract $3x$ from each side.

Because the equation $0 = 2$ is always false, it is a contradiction and there are no solutions.

(b) Start by applying the distributive property.

$2x - (x + 1) = x - 1$	Given equation
$2x - x - 1 = x - 1$	Distributive property
$x - 1 = x - 1$	Combine like terms.
$x = x$	Add 1 to each side.
$0 = 0$	Subtract x from each side.

Because the equation $0 = 0$ is always true, it is an identity and there are infinitely many solutions. Note that the solution set contains all real numbers.

(c) Start by applying the distributive property.

$5x = 2(x - 4)$	Given equation
$5x = 2x - 8$	Distributive property
$3x = -8$	Subtract $2x$ from each side.
$x = -\dfrac{8}{3}$	Divide each side by 3.

Thus there is one solution.
Now Try Exercises 63, 67, 69

CRITICAL THINKING

What must be true about b and d for the equation

$$bx - 2 = dx + 7$$

to have no solutions? What must be true about b and d for this equation to have exactly one solution?

MAKING CONNECTIONS

Number of Solutions

When solving the general form $ax + b = 0$, the resulting equivalent equation will indicate whether the given equation has no solutions, one solution, or infinitely many solutions.

No Solutions: The result is an equation such as $4 = 0$ or $3 = 2$, which is *always false* for any value of the variable.

One Solution: The result is an equation such as $x = 1$ or $x = -12$, which is true for *only one* value of the variable.

Infinitely Many Solutions: The result is an equation such as $0 = 0$ or $-3 = -3$, which is *always true* for any value of the variable.

2.2 PUTTING IT ALL TOGETHER

In this section a four-step approach to solving linear equations was presented on page 102. Some of the other topics discussed in this section are summarized in the following table.

Concept	Comments	Examples
Linear Equation	Can be written as $$ax + b = 0,$$ where $a \neq 0$; has one solution	The equation $5x - 8 = 0$ is linear, with $a = 5$ and $b = -8$. The equation $2x^2 + 4 = 0$ is not linear.
Solving Linear Equations	Use the addition and multiplication properties of equality to isolate the variable.	$5x - 8 = 0$ $\quad$ Given equation $5x = 8$ $\quad$ Add 8 to each side. $x = \dfrac{8}{5}$ $\quad$ Divide each side by 5.
Equations with No Solutions	Some equations that appear to be linear have no solutions. Solving will result in an equivalent equation that is always false.	The equation $$x = x + 5$$ has no solutions because a number cannot equal itself plus 5.
Equations with Infinitely Many Solutions	Some equations that appear to be linear have infinitely many solutions. Solving will result in an equivalent equation that is always true.	The equation $$2x = x + x$$ has infinitely many solutions because the equation is true for all values of x.

2.2 Exercises

MyMathLab   Math XL PRACTICE   WATCH   DOWNLOAD   READ   REVIEW

CONCEPTS

1. A linear equation can be written in the form _____ with $a \neq 0$.

2. In the linear equation $3x + 2 = 0$, $a =$ _____ and $b =$ _____.

3. How many solutions does a linear equation in one variable have?

4. The equation $x^2 - 1 = 0$ (is/is not) a linear equation.

5. What two properties of equality are frequently used to solve linear equations?

6. What property justifies that $4(x - 3) = 4x - 12$ is true for any value of x?

7. To clear fractions from an equation, multiply each side by the _____.

8. To clear decimals from $0.3x + 1.2 = 0.01$, multiply each side by _____.

9. If solving an equation results in $0 = 4$, how many solutions does it have?

10. If solving an equation results in $0 = 0$, how many solutions does it have?

11. How many solutions does $3x = 2x + x$ have?

12. How many solutions does $x = x + 10$ have?

IDENTIFYING LINEAR EQUATIONS

Exercises 13–26: (Refer to Example 1.) Determine whether the equation is linear. If it is linear, give values for a and b so that the given equation can be written in the form $ax + b = 0$.

13. $3x - 7 = 0$

14. $-2x + 1 = 4$

15. $\frac{1}{2}x = 0$

16. $-\frac{3}{4}x = 0$

17. $4x^2 - 6 = 11$

18. $-2x^2 + x = 4$

19. $\frac{6}{x} - 4 = 2$

20. $2\sqrt{x} - 1 = 0$

21. $1.1x = 0.9$

22. $-5.7x = -3.4$

23. $2(x - 3) = 0$

24. $\frac{1}{2}(x + 4) = 0$

25. $|3x| + 2 = 1$

26. $3x = 4x^3$

SOLVING LINEAR EQUATIONS

Exercises 27–30: Evaluate the expression for each value of x in the table. Then use the table to solve the given equation.

27. $-3x + 7 = 1$

x	0	1	2	3	4
$-3x + 7$	7				

28. $5x - 2 = 3$

x	-1	0	1	2	3
$5x - 2$	-7				

29. $4 - 2x = 6$

x	-2	-1	0	1	2
$4 - 2x$	8				

30. $9 - (x + 3) = 4$

x	-2	-1	0	1	2
$9 - (x + 3)$	8				

Exercises 31–60: Solve the equation and check the solution.

31. $11x = 3$

32. $-5x = 15$

33. $x - 18 = 5$

34. $8 = 5 + 3x$

35. $2x - 1 = 13$

36. $4x + 3 = 39$

37. $5x + 5 = -6$

38. $-7x - 4 = 31$

39. $3z + 2 = z - 5$

40. $z - 5 = 5z - 3$

41. $12y - 6 = 33 - y$

42. $-13y + 2 = 22 - 3y$

43. $4(x - 1) = 5$

44. $-2(2x + 7) = 1$

45. $1 - (3x + 1) = 5 - x$

46. $6 + 2(x - 7) = 10 - 3(x - 3)$

47. $(5t - 6) + 2(t + 1) = 0$

48. $-2(t - 7) - (t + 5) = 5$

49. $3(4z - 1) - 2(z + 2) = 2(z + 1)$

50. $-(z + 4) + (3z + 1) = -2(z + 1)$

51. $7.3x - 1.7 = 5.6$

52. $5.5x + 3x = 51$

53. $-9.5x - 0.05 = 10.5x + 1.05$

54. $0.04x + 0.03 = 0.02x - 0.1$

55. $\frac{1}{2}x - \frac{3}{2} = \frac{5}{2}$

56. $-\frac{1}{4}x + \frac{5}{4} = \frac{3}{4}$

57. $-\frac{3}{8}x + \frac{1}{4} = \frac{1}{2}x + \frac{1}{8}$

58. $\frac{1}{3}x + \frac{1}{4} = \frac{1}{6} - x$

59. $4y - 2(y + 1) = 0$

60. $(15y + 20) - 5y = 5 - 10y$

61. Thinking Generally   A linear equation has exactly one solution. Find the solution to the equation $ax + b = 0$, where $a \neq 0$, by solving for x.

62. Thinking Generally   Solve the linear equation $\frac{1}{a}x - b = 0$ for x.

Exercises 63–72: Determine whether the equation has no solutions, one solution, or infinitely many solutions.

63. $5x = 5x + 1$

64. $2(x - 3) = 2x - 6$

65. $8x = 0$

66. $9x = x + 1$

67. $5(2x + 7) - (10x + 5) = 30$

68. $4(x + 2) - 2(2x + 3) = 10$

69. $4x = 5(x + 3) - x$

70. $x - (3x + 2) = 15 - 2x$

71. $2x - (x + 5) = x - 5$

72. $5x = 15 - 2(x + 7)$

APPLICATIONS

73. *Distance Traveled* A bicyclist is 4 miles from home, riding away from home at 8 miles per hour.
 (a) Make a table that shows the bicyclist's distance D from home after 0, 1, 2, 3, and 4 hours.
 (b) Write a formula that calculates D after x hours.
 (c) Use your formula to determine D when $x = 3$ hours. Does your answer agree with the value found in your table?
 (d) Find x when $D = 22$ miles. Interpret the result.

74. *Distance Traveled* An athlete is 16 miles from home, running *toward* home at 6 miles per hour.
 (a) Write a formula that calculates the distance D that the athlete is from home after x hours.
 (b) Determine D when $x = 1.5$ hours.
 (c) Find x when $D = 5.5$ miles. Interpret the result.

75. *Internet Users* (Refer to Example 4.) The number of Internet users I in millions during year x, where $x \geq 2000$, can be approximated by the formula

$$I = 116x - 231{,}627.$$

Approximate the year in which there were 490 million Internet users for the first time.

76. *HIV Infections* The cumulative number of HIV infections N in thousands for the United States in year x can be approximated by the formula

$$N = 42x - 83{,}197,$$

where $x \geq 2000$. Approximate the year when this number reached 970 thousand. (*Source:* Centers for Disease Control and Prevention.)

77. *State and Federal Inmates* The number N of state and federal inmates in millions during year x, where $x \geq 2002$, can be approximated by the formula $N = 0.03x - 58.62$. Determine the year in which there were 1.5 million inmates. (*Source:* Bureau of Justice.)

78. *Government Costs* From 1960 to 2000 the cost C (in billions of 1992 dollars) to regulate social and economic programs could be approximated by the formula $C = 0.35x - 684$ during year x. Estimate the year in which the cost reached \$6.6 billion. (*Source:* Center for the Study of American Business.)

WRITING ABOUT MATHEMATICS

79. A student says that the equation $4x - 1 = 1 - x$ is not a linear equation because it is not in the form $ax + b = 0$. Is the student correct? Explain.

80. A student solves a linear equation as follows.

$$4(x + 3) = 5 - (x + 3)$$
$$4x + 3 \stackrel{?}{=} 5 - x + 3$$
$$4x + 3 \stackrel{?}{=} 8 - x$$
$$5x \stackrel{?}{=} 5$$
$$x \stackrel{?}{=} 1$$

Identify and explain the errors that the student made. What is the correct answer?

CHECKING BASIC CONCEPTS

SECTIONS 2.1 AND 2.2

1. Determine whether the equation is linear.
 (a) $4x^3 - 2 = 0$ (b) $2(x + 1) = 4$

2. Evaluate $4x - 3$ for each value of x in the table. Then use the table to solve $4x - 3 = 13$.

x	3	3.5	4	4.5	5
$4x - 3$	9				17

3. Solve each equation and check your answer.
 (a) $x - 12 = 6$
 (b) $\frac{3}{4}z = \frac{1}{8}$
 (c) $0.6t + 0.4 = 2$
 (d) $5 - 2(x - 2) = 3(4 - x)$

4. Determine whether each equation has no solutions, one solution, or infinitely many solutions.
 (a) $x - 5 = 6x$
 (b) $-2(x - 5) = 10 - 2x$
 (c) $-(x - 1) = -x - 1$

5. *Distance Traveled* A driver is 300 miles from home and is traveling toward home on a freeway at a constant speed of 75 miles per hour.
 (a) Write a formula to calculate the distance D that the driver is from home after x hours.
 (b) Write an equation whose solution gives the hours needed for the driver to reach home.
 (c) Solve the equation from part (b).

2.3 INTRODUCTION TO PROBLEM SOLVING

Steps for Solving a Problem ■ Percent Problems ■ Distance Problems ■ Other Types of Problems

A LOOK INTO MATH ▷ One important characteristic of human beings is their ability to solve problems. For example, throughout history people have wanted to predict the weather but have found it difficult to do. After all, how could anyone predict the future? Today, with the help of mathematics, science, computers, and years of data collection, meteorologists are providing increasingly accurate weather forecasts. Solving problems requires problem-solving skills, which we introduce in this section.

Steps for Solving a Problem

Word problems are challenging because formulas and equations are not usually given. To solve such problems we need a strategy. The following steps are based on George Polya's (1888–1985) four-step process for problem solving.

STEPS FOR SOLVING A PROBLEM

STEP 1: Read the problem carefully and be sure that you understand it. (You may need to read the problem more than once.). Assign a variable to what you are being asked to find. If necessary, write other quantities in terms of this variable.

STEP 2: Write an equation that relates the quantities described in the problem. You may need to sketch a diagram or refer to known formulas.

STEP 3: Solve the equation. Use the solution to determine the solution(s) to the original problem. Include any necessary units.

STEP 4: Look back and check your solution in the given problem. Does it seem reasonable?

Even if we understand the problem that we are trying to solve, we may not be able to find a solution if we cannot write an appropriate equation. In the next example, we practice the second step in the four-step process by translating sentences into equations.

EXAMPLE **Translating sentences into equations**

Translate the sentence into an equation using the variable x. Then solve the resulting equation.
(a) Three times a number minus 6 is equal to 18.
(b) The sum of half a number and 5 is zero.
(c) Sixteen is 4 less than twice a number.

Solution

(a) The phrase "Three times a number" indicates that we multiply x by 3 to get $3x$. The word "minus" indicates that we then subtract 6 from $3x$ to get $3x - 6$. This expression "equals" 18, so the equation is $3x - 6 = 18$. The solution is 8 as shown here.

$$3x - 6 = 18 \qquad \text{Equation to be solved}$$
$$3x = 24 \qquad \text{Add 6 to each side.}$$
$$\frac{3x}{3} = \frac{24}{3} \qquad \text{Divide each side by 3.}$$
$$x = 8 \qquad \text{Simplify the fractions.}$$

(b) The word "sum" indicates that we add "half a number" and 5 to get $\frac{1}{2}x + 5$. The word "is" implies equality, so the equation is $\frac{1}{2}x + 5 = 0$. The solution is -10 as shown here.

$$\frac{1}{2}x + 5 = 0 \qquad \text{Equation to be solved}$$
$$\frac{1}{2}x = -5 \qquad \text{Subtract 5 from each side.}$$
$$x = -10 \qquad \text{Multiply each side by 2.}$$

(c) To translate "4 less than twice a number" into a mathematical expression, we write $2x - 4$. If this seems backwards, consider how you would calculate "4 less than your age." The equation is $16 = 2x - 4$. The solution is 10 as shown here.

$$16 = 2x - 4 \qquad \text{Equation to be solved}$$
$$20 = 2x \qquad \text{Add 4 to each side.}$$
$$\frac{20}{2} = \frac{2x}{2} \qquad \text{Divide each side by 2.}$$
$$x = 10 \qquad \text{Simplify and rewrite.}$$

Now Try Exercises 11, 13

In the next example we apply the four-step process to a word problem that involves three unknown numbers.

EXAMPLE 2   **Solving a number problem**

The sum of three consecutive natural numbers is 81. Find the three numbers.

Solution

STEP 1: Start by assigning a variable n to an unknown quantity.

n: smallest of the three natural numbers

Next, write the other two natural numbers in terms of n.

$n + 1$: next consecutive natural number

$n + 2$: largest of the three consecutive natural numbers

STEP 2: Write an equation that relates these unknown quantities. The sum of the three consecutive natural numbers is 81, so the needed equation is

$$n + (n + 1) + (n + 2) = 81.$$

STEP 3: Solve the equation in STEP 2.

$n + (n + 1) + (n + 2) = 81$	Equation to be solved
$(n + n + n) + (1 + 2) = 81$	Commutative and associative properties
$3n + 3 = 81$	Combine like terms.
$3n = 78$	Subtract 3 from each side.
$n = 26$	Divide each side by 3.

The smallest of the three numbers is 26, so the three numbers are 26, 27, and 28.

STEP 4: To check this solution we can add the three numbers to find out if their sum is 81.

$$26 + 27 + 28 = 81$$

The solution checks.   Now Try Exercise **19**

▶ **REAL-WORLD CONNECTION**   One way to measure the quality of health care available to babies in various countries is to calculate the infant mortality rate. This rate measures the number of deaths of infants under one year of age per 1000 live births in the same year. A high infant mortality rate may indicate a lack of good quality health care for infants. In the next example, a number problem is solved to find the infant mortality rate in Iceland, which has the lowest rate in the world.

EXAMPLE 3   Solving a number problem

Sierra Leone, in Western Africa, has the highest infant mortality rate in the world at 165. This rate is 15 more than 50 times the rate in Iceland. Find the infant mortality rate in Iceland. (**Source:** World Population Prospectus.)

Solution

If we let x represent the infant mortality rate in Iceland, then "15 more than 50 times the rate in Iceland" can be written $50x + 15$. Because this quantity is equal to 165, the equation that must be solved is $50x + 15 = 165$.

$50x + 15 = 165$	Equation to be solved
$50x = 150$	Subtract 15 from each side.
$x = 3$	Divide each side by 50.

The infant mortality rate in Iceland is 3 deaths per 1000 live births.   Now Try Exercise **31**

Percent Problems

▶ **REAL-WORLD CONNECTION**   Problems involving percentages occur in everyday life. Wage increases, sales tax, and government data all make use of percentages. For example, it is estimated that 10% of the world's population will be older than 65 in 2025. As 1% (one percent) represents one part in one hundred, 10 of every 100 people in 2025 will be older than 65. Percent notation can be changed to either fraction or decimal notation.   (**Source:** World Health Organization.)

PERCENT NOTATION

The expression $x\%$ represents the fraction $\frac{x}{100}$ or the decimal number $x \times 0.01$.

NOTE: To write $x\%$ as a decimal number, move the decimal point in the number x two places to the *left* and then remove the % symbol.

EXAMPLE 4 Converting percent notation

Convert each percentage to fraction and decimal notation.
(a) 23%   **(b)** 5.2%   **(c)** 0.3%

Solution
(a) *Fraction Notation*: To convert to fraction notation divide 23 by 100. Thus $23\% = \frac{23}{100}$.
Decimal Notation: Move the decimal point two places to the left and then remove the % symbol: $23\% = 0.23$.
(b) *Fraction Notation*: $5.2\% = \frac{5.2}{100} = \frac{52}{1000} = \frac{13 \cdot 4}{250 \cdot 4} = \frac{13}{250}$
Decimal Notation: Move the decimal point two places to the left and then remove the % symbol: $5.2\% = 0.052$.
(c) *Fraction Notation*: $0.3\% = \frac{0.3}{100} = \frac{3}{1000}$
Decimal Notation: Move the decimal point two places to the left and then remove the % symbol: $0.3\% = 0.003$.   Now Try Exercise **35**

EXAMPLE 5 Converting to percent notation

Convert each real number to a percentage.
(a) 0.234   **(b)** $\frac{1}{4}$ **(c)** 2.7

Solution
(a) Move the decimal point two places to the *right* and then insert the % symbol to obtain $0.234 = 23.4\%$.
(b) $\frac{1}{4} = 0.25$, so $\frac{1}{4} = 25\%$.
(c) Move the decimal point two places to the right and then insert the % symbol to obtain $2.7 = 270\%$. Note that percentages can be greater than 100%.   Now Try Exercise **43**

PERCENT CHANGE When prices increase (or decrease), the actual amount of the increase (or decrease) is often not as significant as the *percent change* in the price. For example, if the price of a new home increases from \$250,000 to \$251,000, this price increase may not seem as significant as an increase in college tuition from \$4500 to \$5500. Even though both prices increased by \$1000, the percent increase in the price of the home is four-tenths of 1% and the percent increase in the cost of tuition is more than 22%. The increase in tuition is much more dramatic than the increase in the price of the home.

If an amount changes from an **old value** to a **new value**, the **percent change** is given by

$$\frac{\textbf{new value} - \textbf{old value}}{\textbf{old value}} \times 100.$$

The reason for multiplying by 100 is to change the decimal representation to a percentage.

NOTE: A positive percent change corresponds to an increase and a negative percent change corresponds to a decrease.

EXAMPLE 6   Calculating percent increase

From 1995 to 2005 consumer credit increased from $1.1 trillion to $2.5 trillion. Calculate the percent increase in consumer credit from 1995 to 2005.   (**Source:** Federal Reserve Bulletin.)

Solution

The **old value** is **1.1** and the **new value** is **2.5**. The percent increase in consumer credit is

$$\frac{\textbf{new value} - \textbf{old value}}{\textbf{old value}} \times 100 = \frac{2.5 - 1.1}{1.1} \times 100 \approx 127\%.$$

Now Try Exercise 55

As we previously indicated, percentages frequently occur in applications.

EXAMPLE 7   Solving a percent problem

During the 2003–2004 academic year, tuition and fees at public colleges and universities were $4694, on average, and increased by 9.3% during the next year. Find the average cost of tuition and fees during the 2004–2005 academic year. Round your answer to the nearest dollar.

Solution

STEP 1: Let T represent the tuition and fees during 2004–2005.

STEP 2: T equals $4694 plus 9.3% of $4694 and 9.3% = 0.093, so

$$T = 4694 + (0.093)4694.$$

NOTE: The word "of" in percent problems often indicates multiplication.

STEP 3: To find T, evaluate the expression on the right side to get

$$T = 4694 + (0.093)4694 \approx 5131.$$

Tuition and fees were $5131.

STEP 4: To check the answer, calculate the percent increase. Tuition and fees went up by

$$\frac{5131 - 4694}{4694} \times 100 \approx 9.3\%,$$

so the answer checks.

Now Try Exercise 57

> **CRITICAL THINKING**
>
> If your salary increased 200%, by what factor did it increase?

EXAMPLE 8   Solving a percent problem

In 2025, 10% of the world's population, or 800 million people, will be older than 65. Find the estimated population of the world in 2025.

Solution

STEP 1: Let P represent the world's population in millions in 2025.

STEP 2: 10% of P equals 800 million. As 10% = 0.10, this information is described by

$$0.10P = 800.$$

STEP 3: To solve the equation in STEP 2, multiply each side by 10.

$$0.10P = 800 \qquad \text{Equation to be solved}$$
$$P = 8000 \qquad \text{Multiply by 10.}$$

In 2025 the estimated world population is 8000 million, or 8 billion.

STEP 4: To check the answer, determine whether 10% of 8 billion is 800 million.

$$(0.10)8,000,000,000 = 800,000,000$$

The answer checks.   Now Try Exercise **61**

Distance Problems

If a person drives on an interstate highway at 70 miles per hour for 3 hours, then the total distance traveled is $70 \cdot 3 = 210$ miles. In general, $d = rt$, where d is the distance traveled, r is the rate (or speed), and t is time. In this example, the distance is in miles, the time is in hours, and the rate is expressed in miles per hour. In general, the rate in a distance problem is expressed in units of distance per unit of time.

EXAMPLE **9**   Solving a distance problem

A person drives for 2 hours and 30 minutes at a constant speed and travels 180 miles. See Figure 2.2. Find the speed of the car in miles per hour.

180 mi
2 hr 30 min

Figure 2.2

Solution

STEP 1: Let r represent the car's rate, or speed, in miles per hour.

STEP 2: The rate is to be given in miles per hour, so change 2 hours and 30 minutes to 2.5 or $\frac{5}{2}$ hours. Because $d = 180$ and $t = \frac{5}{2}$, the equation $d = rt$ becomes

$$180 = \frac{5}{2} \cdot r.$$

STEP 3: Solve the equation in STEP 2 for r by multiplying each side of the equation by $\frac{2}{5}$, which is the reciprocal of $\frac{5}{2}$.

$$180 = r \cdot \frac{5}{2} \qquad \text{Equation to be solved}$$

$$\frac{2}{5} \cdot 180 = r \cdot \frac{2}{5} \cdot \frac{5}{2} \qquad \text{Multiply each side by } \frac{2}{5}.$$

$$72 = r \qquad \text{Simplify.}$$

The speed of the car is 72 miles per hour.

STEP 4: Because 2 hours and 30 minutes is equivalent to $\frac{5}{2}$ hours, traveling for 2 hours and 30 minutes at a constant rate of 72 miles per hour results in a distance of

$$d = rt = 72 \cdot \frac{5}{2} = 180 \text{ miles.}$$

The answer checks.   Now Try Exercise **71**

EXAMPLE **10**   Solving a distance problem

An athlete jogs at two speeds, covering a distance of 7 miles in $\frac{3}{4}$ hour. If the athlete runs $\frac{1}{4}$ hour at 8 miles per hour, find the second speed.

Solution

STEP 1: Let r represent the second speed of the jogger in miles per hour.

Figure 2.3

STEP 2: The total time spent jogging is $\frac{3}{4}$ hour, so the time spent jogging at the second speed must be $\frac{3}{4} - \frac{1}{4} = \frac{1}{2}$ hour. The total distance of 7 miles is the result of jogging at 8 miles per hour for $\frac{1}{4}$ hour and at r miles per hour for $\frac{1}{2}$ hour. See Figure 2.3. The distance traveled at 8 miles per hour for $\frac{1}{4}$ hour is given by $8 \cdot \frac{1}{4}$ and the distance traveled at r miles per hour for $\frac{1}{2}$ hour is given by $r \cdot \frac{1}{2}$. The sum of these distances must equal 7 miles. Thus

$$8 \cdot \frac{1}{4} + r \cdot \frac{1}{2} = 7.$$

STEP 3: Solve the equation in STEP 2 for r.

$$8 \cdot \frac{1}{4} + r \cdot \frac{1}{2} = 7 \qquad \text{Equation to be solved}$$

$$2 + \frac{r}{2} = 7 \qquad \text{Simplify.}$$

$$\frac{r}{2} = 5 \qquad \text{Subtract 2 from each side.}$$

$$r = 10 \qquad \text{Multiply each side by 2.}$$

The athlete's second speed is 10 miles per hour.

STEP 4: Jogging at a rate of 8 miles per hour for $\frac{1}{4}$ hour results in a distance of $8 \cdot \frac{1}{4} = 2$ miles. Jogging at a rate of 10 miles per hour for $\frac{1}{2}$ hour results in a distance of $10 \cdot \frac{1}{2} = 5$ miles. The total distance is $2 + 5 = 7$ miles and the total time is $\frac{1}{4} + \frac{1}{2} = \frac{3}{4}$ hour, so the answer checks. **Now Try Exercise 75**

Other Types of Problems

Many applied problems involve linear equations. For example, right after people have their wisdom teeth pulled, they may need to rinse their mouth with salt water. In the next example, we use linear equations to determine how much water must be added to dilute a concentrated saline solution.

EXAMPLE 11 Diluting a saline solution

A solution contains 4% salt. How much pure water should be added to 30 ounces of the solution to dilute it to a 1.5% solution?

Solution

STEP 1: Assign a variable x as follows.

$$x\text{: ounces of pure water (0\% salt solution)}$$

$$30\text{: ounces of 4\% salt solution}$$

$$x + 30\text{: ounces of 1.5\% salt solution}$$

In Figure 2.4 on the next page, three beakers illustrate this situation.

Figure 2.4 Mixing a Saline Solution

STEP 2: Note that the amount of salt in the first two beakers must equal the amount of salt in the third beaker. We use Table 2.6 to organize our calculations. The amount of salt in a solution equals the concentration times the solution amount, as shown in the last column of the table.

TABLE 2.6   Mixing a Saline Solution

Solution Type	Concentration (as a decimal)	Solution Amount (ounces)	Salt (ounces)
Pure Water	0% = 0.00	x	$0.00x$
Initial Solution	4% = 0.04	30	$0.04(30)$
Final Solution	1.5% = 0.015	$x + 30$	$0.015(x + 30)$

The amount of salt in the first two beakers is

$$0.00x + 0.04(30) = 0 + 1.2 = 1.2 \text{ ounces.}$$

The amount of salt in the final beaker is

$$0.015(x + 30) \text{ ounces.}$$

Because the amounts of salt in the solutions before and after mixing must be equal, the following equation must hold.

$$0.015(x + 30) = 1.2$$

STEP 3: Solve the equation in STEP 2.

$0.015(x + 30) = 1.2$	Equation to be solved
$0.015x + 0.45 = 1.2$	Distributive property
$0.015x = 0.75$	Subtract 0.45 from each side.
$\dfrac{0.015x}{0.015} = \dfrac{0.75}{0.015}$	Divide each side by 0.015.
$x = 50$	Simplify fractions.

Fifty ounces of water should be added.

STEP 4: Adding 50 ounces of water will yield $50 + 30 = 80$ ounces of water containing $0.04(30) = 1.2$ ounces of salt. The concentration is $\frac{1.2}{80} = 0.015$ or 1.5%, so the answer checks.   Now Try Exercise **79**

▶ **REAL-WORLD CONNECTION** Many times interest rates for student loans vary. In the next example, we present a situation in which a student has to borrow money at two different interest rates.

EXAMPLE 12   Calculating interest on college loans

A student takes out a loan for a limited amount of money at 5% interest and then must pay 7% for any additional money. If the student borrows $2000 more at 7% than at 5%, then the total interest for one year is $440. How much does the student borrow at each rate?

Solution

STEP 1: Assign a variable x as follows.

$$x: \text{loan amount at 5\% interest}$$

$$x + 2000: \text{loan amount at 7\% interest}$$

STEP 2: The amount of interest paid for the 5% loan is 5% of x, or $0.05x$. The amount of interest paid for the 7% loan is 7% of $x + 2000$, or $0.07(x + 2000)$. The total interest equals $440, so we solve the equation

$$0.05x + 0.07(x + 2000) = 440.$$

STEP 3: Solve the equation in STEP 2 for x.

$0.05x + 0.07(x + 2000) = 440$	Equation to be solved
$5x + 7(x + 2000) = 44{,}000$	Multiply by 100 to clear decimals.
$5x + 7x + 14{,}000 = 44{,}000$	Distributive property
$12x + 14{,}000 = 44{,}000$	Combine like terms.
$12x = 30{,}000$	Subtract 14,000 from each side.
$x = 2500$	Divide each side by 12.

The student borrows $2500 at 5% and $2500 + 2000 = \$4500$ at 7%.

STEP 4: The amount of interest on $2500 at 5% is $0.05(2500) = \$125$ and the amount of interest on $4500 at 7% is $0.07(4500) = \$315$. Thus the total interest is given by $125 + 315 = \$440$. Furthermore, the amount borrowed at 7% is $2000 more than the amount borrowed at 5%. The answer checks.   Now Try Exercise 81

2.3 PUTTING IT ALL TOGETHER

In this section on page 111 we presented a four-step approach to problem solving. However, because no approach works in every situation, solving mathematical problems takes time, effort, and creativity.

Concept	Comments	Examples
Percent Notation	$x\%$ represents either the fraction $\frac{x}{100}$ or the decimal $x \times 0.01$.	$17\% = \frac{17}{100} = 0.17$ $1.5\% = \frac{1.5}{100} = \frac{15}{1000} = 0.015$ $234\% = \frac{234}{100} = 2.34$
Converting Fractions to Percent Notation	Divide the numerator by the denominator, multiply by 100, and insert the % symbol.	$\frac{1}{2} = 0.5 = 50\%$ $\frac{2}{3} = 0.\overline{6} \approx 66.7\%$ $\frac{8}{5} = 1.6 = 160\%$

continued on next page

continued from previous page

Concept	Comments	Examples
Percent Change	If a quantity changes from an old value to a new value, then the percent change is $$\frac{(\text{new value}) - (\text{old value})}{(\text{old value})} \times 100.$$	If a price changes from \$1.50 to \$1.35, the percent change is $$\frac{1.35 - 1.50}{1.50} \times 100 = -10\%.$$ The price decreases by 10%.
Distance Problems	Distance d equals rate r times time t, or $$d = r \cdot t.$$	If a car travels 60 mph for 3 hours, then the distance d is $$d = r \cdot t = 60 \cdot 3 = 180 \text{ miles.}$$

2.3 Exercises

CONCEPTS

1. When you are solving a word problem, what is the last step?

2. Given an integer n, what are the next two consecutive integers?

3. The expression $x\%$ equals the fraction _____ .

4. The expression $x\%$ equals the decimal $x \times$ _____ .

5. To write 63.2% as a decimal, move the decimal point 2 places (left/right) and then remove the % symbol.

6. To write 0.349 as a percentage, move the decimal point 2 places (left/right) and then insert the % symbol.

7. If a price changes from P_1 to P_2, then the percent change equals _____ .

8. A positive percent change corresponds to a(n) _____ and a negative percent change corresponds to a(n) _____ .

9. If a car travels at speed r for time t, then distance is given by $d =$ _____ .

10. In general, the rate in a distance problem is expressed in units of _____ per unit of _____ .

NUMBER PROBLEMS

Exercises 11–18: Using the variable x, translate the sentence into an equation. Solve the resulting equation.

11. The sum of 2 and a number is 12.

12. Twice a number plus 7 equals 9.

13. A number divided by 5 equals the number decreased by 24.

14. 25 times a number is 125.

15. If a number is increased by 5 and then divided by 2, the result is 7.

16. A number subtracted from 8 is 5.

17. The quotient of a number and 2 is 17.

18. The product of 5 and a number equals 95.

Exercises 19–28: Find the number or numbers.

19. The sum of three consecutive natural numbers is 96.

20. The sum of three consecutive integers is -123.

21. Three times a number equals 102.

22. A number plus 18 equals twice the number.

23. Five times a number is 24 more than twice the number.

24. Three times a number is 18 less than the number.

25. Six times a number divided by 7 equals 18.

26. Two less than twice a number, divided by 5 equals 4.

27. Four times the sum of a number and 5 equals 64.

28. The opposite of the sum of a number and -5 equals 24.

29. *Finding Age*   In 10 years, a child will be 3 years older than twice her current age. What is the current age of the child?

30. *Weight Loss*   After losing 30 pounds, an individual weighs 110 pounds more than one-third his previous weight. Find the previous weight of the individual.

31. *Hazardous Waste*   In 2005, the number of federal hazardous waste sites in California was 2 less than twice the number of sites in Washington. How many hazardous waste sites were there in Washington if there were 24 such sites in California? (*Source:* Environmental Protection Agency.)

32. *Endangered Species*   There were 77 birds on the endangered species list in 2005. This is 7 more than 5 times the number of reptiles on the list. How many reptiles were on the endangered species list in 2005? (*Source:* U.S. Fish and Wildlife Service.)

33. *Energy Production*   In 1995, U.S. hydroelectric power production hit an all-time high of 311 billion kilowatthours. This is 23 billion kilowatt-hours more than three times the 1950 production level. Find the amount of hydroelectric power produced in 1950. (*Source:* U.S. Department of Energy.)

34. *Open Heart Surgery*   In 2002, open heart surgery was performed 368 thousand times in the United States on persons over the age of 65. This is 22 thousand fewer than 13 times the number of this type of surgery performed on persons under the age of 15. How many times was open heart surgery performed on persons under the age of 15 in 2002? (*Source:* U.S. Department of Health and Human Services.)

PERCENT PROBLEMS

Exercises 35–42: Convert the percentage to fraction and decimal notation.

35. 37%

36. 52%

37. 148%

38. 252%

39. 6.9%

40. 8.1%

41. 0.05%

42. 0.12%

Exercises 43–54: Convert the number to a percentage.

43. 0.45

44. 0.08

45. 1.8

46. 2.97

47. 0.006

48. 0.0001

49. $\frac{2}{5}$

50. $\frac{1}{3}$

51. $\frac{3}{4}$

52. $\frac{7}{20}$

53. $\frac{5}{6}$

54. $\frac{53}{50}$

55. *Voter Turnout*   In the 1980 election for president there were 86.5 million voters, whereas in 2004 there were 122.3 million voters. Find the percent change in the number of voters.

56. *College Degrees*   In 2000, about 457,000 people received a master's degree, and by 2005 this number had increased to 506,000. Find the percent change in the number of master's degrees received over this time period.   (*Source:* The College Board.)

57. *Wages*   A part-time instructor is receiving $950 per credit taught. If the instructor receives a 4% increase, how much will the new per credit compensation be?

58. *Tuition Increase*   Tuition is currently $125 per credit. There are plans to raise tuition by 8% for next year. What will the new tuition be per credit?

59. *Income Tax Returns*   In 2004, the Internal Revenue Service processed 131 million individual income tax returns, up 0.46% from the preceding year. How many individual income tax returns were processed in 2003? (*Source:* Internal Revenue Service.)

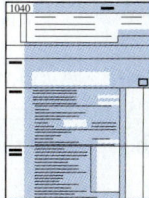

60. *Medicare Enrollment*   In 2004, 41.2 million people were enrolled in Medicare, an increase of 5.1% from 2000. How many people were enrolled in Medicare in 2000?   (*Source:* Health Care Financing Administration.)

61. *AIDS Deaths*   There were 18,017 AIDS deaths in 2003, or 34.5% of the 1995 AIDS deaths. Determine the number of AIDS deaths in 1995.   (*Source:* Centers for Disease Control and Prevention.)

62. *Rural Forestland*   There are about 13.6 million acres of rural forestland in Wisconsin. This is about 38% of the state's total area. Approximate the total area of Wisconsin in millions of acres.   (*Source:* U.S. Department of Agriculture.)

63. Thinking Generally   Calculate the percent change if the price of an item increases from $1.20 to $1.50. Now calculate the percent change if the price of the item decreases from $1.50 to $1.20.

64. Thinking Generally   Refer to the previous exercise. Suppose an amount increases from A_1 to A_2 and the percent change is calculated to be 30%. If that amount now decreases from A_2 back to A_1, will the percent change be -30%?

DISTANCE PROBLEMS

Exercises 65–70: Use the formula $d = rt$ to find the value of the missing variable.

65. $r = 4$ mph, $t = 2$ hours

66. $r = 70$ mph, $t = 2.5$ hours

67. $d = 1000$ feet, $t = 50$ seconds

68. $d = 1250$ miles, $t = 5$ days

69. $d = 200$ miles, $r = 40$ mph

70. $d = 1700$ feet, $r = 10$ feet per second

71. *Driving a Car*   A person drives a car at a constant speed for 4 hours and 15 minutes, traveling 255 miles. Find the speed of the car in miles per hour.

255 mi
4 hr 15 min

72. *Flying an Airplane*   A pilot flies a plane at a constant speed for 5 hours and 30 minutes, traveling 715 miles. Find the speed of the plane in miles per hour.

73. *Jogging Speeds*   One runner passes another runner traveling in the same direction on a hiking trail. The faster runner is jogging 2 miles per hour faster than the slower runner. Determine how long it will be before the faster runner is $\frac{3}{4}$ mile ahead of the slower runner.

74. *Distance Running*   An athlete runs 8 miles, first at a slower speed and then at a faster speed. The total time spent running is 1 hour. If the athlete runs $\frac{1}{3}$ hour at 6 miles per hour, find the second speed.

75. *Jogging Speeds*   At first an athlete jogs at 5 miles per hour and then jogs at 8 miles per hour, traveling 7 miles in 1.1 hours. How long does the athlete jog at each speed? (*Hint:* Let t represent the amount of time the athlete jogs at 5 mph. Then $1.1 - t$ represents the amount of time the athlete jogs at 8 mph.)

t hr at $1.1 - t$ hr
5 mph   at 8 mph

7 mi

76. *Distance and Time*   A bus is 160 miles east of the North Dakota–Montana border and is traveling west at 70 miles per hour. How long will it take for the bus to be 295 miles west of the border?

77. *Distance and Time*   A plane is 300 miles west of Chicago, Illinois, and is flying west at 500 miles per hour. How long will it take for the plane to be 2175 miles west of Chicago?

500 mph ——   ← 300 mi —●
Chicago

2175 mi

78. *Finding Speeds*   Two cars pass on a straight highway while traveling in opposite directions. One car is traveling 6 miles per hour faster than the other car. After 1.5 hours the two cars are 171 miles apart. Find the speed of each car.

OTHER TYPES OF PROBLEMS

79. *Saline Solution*   (Refer to Example 11.) A solution contains 3% salt. How much water should be added to 20 ounces of this solution to make a 1.2% solution?

80. *Acid Solution*   A solution contains 15% hydrochloric acid. How much water should be added to 50 milliliters of this solution to dilute it to a 2% solution?

81. *College Loans*   (Refer to Example 12.) A student takes out two loans, one at 5% interest and the other at 6% interest. The 5% loan is $1000 more than the 6% loan, and the total interest for 1 year is $215. How much is each loan?

82. *Bank Loans* Two bank loans, one for $5000 and the other for $3000, cost a total of $550 in interest for one year. The $5000 loan has an interest rate 3% lower than the interest rate for the $3000 loan. Find the interest rate for each loan.

83. *Mixing Antifreeze* How many gallons of 70% antifreeze should be mixed with 10 gallons of 30% antifreeze to obtain a 45% antifreeze mixture?

84. *Mixing Antifreeze* How many gallons of 65% antifreeze and how many gallons of 20% antifreeze should be mixed to obtain 50 gallons of a 56% mixture of antifreeze? (*Hint:* Let x represent the number of gallons of 65% antifreeze. Then $50 - x$ represents the amount of 20% antifreeze.)

85. *Hydrocortisone Cream* A pharmacist needs to make a 1% hydrocortisone cream. How many grams of 2.5% hydrocortisone cream should be added to 15 grams of cream base (0% hydrocortisone) to make the 1% cream?

86. *Credit Card Debt* A person carries a balance on two credit cards, one with a monthly interest rate of 1.5% and the other with a monthly rate of 1.75%. The balance on the 1.5% card is $600 less than the balance on the 1.75% card. If the total interest for the month is $49.50, what is the balance on each card?

WRITING ABOUT MATHEMATICS

87. State the four steps for solving a word problem.

88. The cost of living has increased about 600% during the past 50 years. Does this percent change correspond to a cost of living increase of 6 times? Explain.

Directions: Form a group of 2 to 4 people. Select someone to record the group's responses for this activity. All members of the group should work cooperatively to answer the questions. If your instructor asks for your results, each member of the group should be prepared to respond.

Exercises 1–5: In this set of exercises you are to use your mathematical problem-solving skills to find the thickness of a piece of aluminum foil without measuring it directly.

1. *Area of a Rectangle* The area of a rectangle equals length times width. Find the area of a rectangle with length 12 centimeters and width 11 centimeters.

2. *Volume of a Box* The volume of a box equals length times width times height. Find the volume of the box shown, which is 12 centimeters long, 11 centimeters wide, and 5 centimeters high.

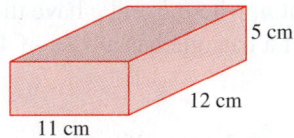

5 cm

12 cm

11 cm

3. *Height of a Box* Suppose that a box has a volume of 100 cubic centimeters and that the area of the bottom of the box is 50 square centimeters. Find the height of the box.

4. *Volume of Aluminum Foil* One cubic centimeter of aluminum weighs 2.7 grams. If a piece of aluminum foil weighs 5.4 grams, find the volume of the aluminum foil.

5. *Thickness of Aluminum Foil* A rectangular sheet of aluminum foil is 50 centimeters long and 20 centimeters wide, and weighs 5.4 grams. Find the thickness of the aluminum foil in centimeters.

2.4 FORMULAS

Basic Concepts ■ Formulas from Geometry ■ Solving for a Variable ■
Other Formulas

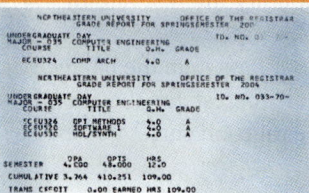

Have you ever wondered how the registrar's office calculates your grade point average (GPA)? A formula is used that involves the number of credits earned at each possible grade. Once a formula has been derived, it can be used over and over by any number of people. This fact makes formulas a convenient and easy way to solve certain types of recurring problems. In this section we discuss several types of formulas.

Basic Concepts

A *formula* is an equation that can be used to calculate an unknown quantity by using known values of other quantities. Formulas also establish a relationship between different quantities. For example, to calculate the area A of a rectangular room with length L and width W, the formula

$$A = LW$$

can be used. If a dormitory room is 16 feet by 10 feet, then its area is

$$A = 16 \cdot 10 = 160 \text{ square feet.}$$

The formula $A = LW$ calculates area by using the quantities of length and width.

Another useful formula is $M = \frac{D}{G}$, which can be used to calculate a car's gas mileage M after it has traveled D miles on G gallons of gasoline.

EXAMPLE 1   Calculating mileage of a trip

A tourist starts a trip with a full tank of gas and an odometer that reads 45,682 miles. At the end of the trip, it takes 9.7 gallons of gas to fill the tank, and the odometer reads 45,903 miles. Find the gas mileage for the car.

Solution
The distance traveled is $D = 45,903 - 45,682 = \textbf{221}$ miles and the number of gallons used is $G = \textbf{9.7}$. Thus

$$M = \frac{D}{G} = \frac{\textbf{221}}{\textbf{9.7}} \approx 22.8 \text{ miles per gallon.}$$ Now Try Exercise **81**

Formulas from Geometry

Formulas from geometry are frequently used in various fields, including surveying and construction. In this subsection we discuss several important formulas from geometry.

TRIANGLES   If a triangle has base b and height h, where b and h have the *same* units, then its area is $A = \frac{1}{2}bh$, as shown in Figure 2.5(a). If a triangle has a base of **4** feet and a height of 18 inches, then its area is

$$A = \frac{1}{2}bh = \frac{1}{2}(\textbf{4})(\textbf{1.5}) = 3 \text{ square feet,}$$

as illustrated in Figure 2.5(b). Note that 18 inches needs to be converted to **1.5** feet so that all the units of length are the same. If the base had been converted to 48 inches, then the area would have been

$$A = \frac{1}{2}bh = \frac{1}{2}(48)(18) = 432 \text{ square } inches.$$

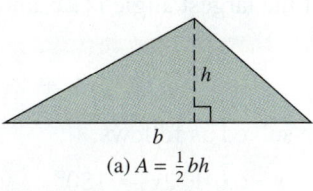

(a) $A = \frac{1}{2}bh$ (b)

Figure 2.5 Triangles

EXAMPLE 2   **Calculating area of a region**

A residential lot is shown in Figure 2.6. It comprises a rectangular region and an adjacent triangular region.
(a) Find the area of this lot.
(b) An acre contains 43,560 square feet. How many acres are there in this lot?

Figure 2.6

Solution
(a) The rectangular portion of the lot has length **405** feet and width **235** feet. Its area A_R is

$$A_R = LW = \mathbf{405} \cdot \mathbf{235} = \mathbf{95{,}175} \text{ square feet.}$$

The triangular region has base **124** feet and height **235** feet. Its area A_T is

$$A_T = \frac{1}{2}bh = \frac{1}{2} \cdot \mathbf{124} \cdot \mathbf{235} = \mathbf{14{,}570} \text{ square feet.}$$

The total area A of the lot equals the sum of A_R and A_T.

$$A = \mathbf{95{,}175} + \mathbf{14{,}570} = 109{,}745 \text{ square feet}$$

(b) Each acre equals 43,560 square feet, so divide 109,745 by 43,560 to calculate the number of acres.

$$\frac{109{,}745}{43{,}560} \approx 2.5 \text{ acres}$$

Now Try Exercise 29

Figure 2.7

ANGLES   Angles are often measured in degrees. A **degree** (°) is $\frac{1}{360}$ of a revolution, so there are 360° in one complete revolution. In any triangle, the sum of the measures of the angles equals 180°. In Figure 2.7, triangle ABC has angles with measures x, y, and z. Therefore

$$x + y + z = 180°.$$

EXAMPLE 3   Finding angles in a triangle

In a triangle the two smaller angles are equal in measure and are half the measure of the largest angle. Find the measure of each angle.

Solution

Let x represent the measure of each of the two smaller angles, as illustrated in Figure 2.8. Then the measure of the largest angle is $2x$, and the sum of the measures of the three angles is given by

$$x + x + 2x = 180°.$$

Figure 2.8

This equation can be solved as follows.

$$x + x + 2x = 180° \qquad \text{Equation to be solved}$$

$$4x = 180° \qquad \text{Combine like terms.}$$

$$\frac{4x}{4} = \frac{180°}{4} \qquad \text{Divide each side by 4.}$$

$$x = 45° \qquad \text{Divide the real numbers.}$$

The measure of the largest angle is $2x = 2 \cdot 45° = 90°$. Thus the measures of the three angles are $45°$, $45°$, and $90°$.   **Now Try Exercise 35**

BOXES   The box in Figure 2.9 has length L, width W, and height H. Its volume V is given by

$$V = LWH.$$

The surface of the box comprises six rectangular regions: top and bottom, front and back, and left and right sides. The total surface area S of the box is given by

$$S = LW + LW + WH + WH + LH + LH.$$

(top + bottom + front + back + left side + right side)

Figure 2.9
Rectangular Box

When we combine like terms, this expression simplifies to

$$S = 2LW + 2WH + 2LH.$$

EXAMPLE 4   Finding the volume and surface area of a box

Find the volume and surface area of the box shown in Figure 2.10.

Solution

Figure 2.10 shows that the box has length $L = \mathbf{10}$ inches, width $W = \mathbf{8}$ inches, and height $H = \mathbf{5}$ inches. The volume of the box is

$$V = LWH = \mathbf{10} \cdot \mathbf{8} \cdot \mathbf{5} = 400 \text{ cubic inches.}$$

5 in.

10 in.

8 in.

Figure 2.10

The surface area of the box is

$$S = 2LW + 2WH + 2LH$$

$$= 2(\mathbf{10})(\mathbf{8}) + 2(\mathbf{8})(\mathbf{5}) + 2(\mathbf{10})(\mathbf{5})$$

$$= 160 + 80 + 100$$

$$= 340 \text{ square inches.} \qquad \text{Now Try Exercise 43}$$

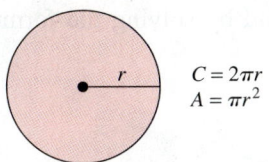

$C = 2\pi r$
$A = \pi r^2$

Figure 2.11 Circle

CIRCLES The *radius* of a circle is the distance from its center to the perimeter of the circle. The perimeter of a circle is called its **circumference** C and is given by $C = 2\pi r$, where r is the radius of the circle. The area A of a circle is given by $A = \pi r^2$. See Figure 2.11. The distance across a circle, through its center, is called the *diameter*. Note that a circle's radius is half of its diameter. (Recall that $\pi \approx 3.14$.)

EXAMPLE 5 Finding the circumference and area of a circle

A circle has a diameter of 25 inches. Find its circumference and area.

Solution
The radius is half the diameter, or **12.5** inches.

$$\text{Circumference: } C = 2\pi r = 2\pi(\mathbf{12.5}) = 25\pi \approx 78.5 \text{ inches.}$$
$$\text{Area: } A = \pi r^2 = \pi(\mathbf{12.5})^2 = 156.25\pi \approx 491 \text{ square inches.}$$

Now Try Exercise **39**

Figure 2.12
Cylinder

CRITICAL THINKING
Write a formula for the circumference and area of a circle having diameter *d*.

CYLINDERS A soup can is usually made in the shape of a cylinder. The volume of a cylinder having radius r and height h is $V = \pi r^2 h$. See Figure 2.12.

EXAMPLE 6 Calculating the volume of a soda can

A cylindrical soda can has a radius of $1\frac{1}{4}$ inches and a height of $4\frac{3}{8}$ inches.
(a) Find the volume of the can.
(b) If 1 cubic inch equals 0.554 fluid ounces, find the number of fluid ounces in the can.

Solution
(a) Changing mixed numbers to improper fractions gives the radius as $r = \frac{5}{4}$ inches and the height as $h = \frac{35}{8}$ inches.

$$V = \pi r^2 h \qquad \text{Volume of the soda can}$$
$$= \pi\left(\frac{5}{4}\right)^2\left(\frac{35}{8}\right) \qquad \text{Substitute.}$$
$$= \pi\left(\frac{875}{128}\right) \qquad \text{Multiply the fractions.}$$
$$\approx 21.48 \text{ cubic inches} \qquad \text{Approximate.}$$

(b) To calculate the number of fluid ounces in 21.48 cubic inches, multiply by 0.554.

$$21.48(0.554) \approx 11.9 \text{ fluid ounces}$$

Note that a typical aluminum soda can holds 12 fluid ounces.   Now Try Exercise **51**

Solving for a Variable

A formula establishes a relationship between two or more variables (or quantities). Sometimes a formula is not solved for the needed variable. For example, if the area A and the width W

of a rectangular region are given, then its length L can be found by solving the formula $A = LW$ for L.

$$A = LW \qquad \text{Area formula}$$

$$\frac{A}{W} = \frac{LW}{W} \qquad \text{Divide each side by } W.$$

$$\frac{A}{W} = L \qquad \text{Simplify the fraction.}$$

$$L = \frac{A}{W} \qquad \text{Rewrite the equation.}$$

If the area A of a rectangle is 400 square inches and its width W is 16 inches, then the rectangle's length L is

$$L = \frac{A}{W} = \frac{400}{16} = 25 \text{ inches.}$$

Once the area formula has been solved for L, the resulting formula can be used to find the length of *any* rectangle whose area and width are known.

▶ **REAL-WORLD CONNECTION**   The use of formulas is common in almost any field. For example, there is a useful formula for calculating how far away a bolt of lightning is. It is based on the fact that sometimes we see a flash of lighting before we hear the thunder. This phenomenon happens because light travels much faster than sound. The farther away the lightning, the greater is the delay between our seeing the flash and hearing the thunder. In general, if the delay is x seconds, then the lightning is $D = \frac{x}{5}$ miles away. For instance, if there is a 10-second delay, then the lightning is $D = \frac{10}{5} = 2$ miles away.

EXAMPLE 7   **Calculating the delay between lightning and thunder**

Doppler radar shows an electrical storm 3.5 miles away. If you see lightning from this storm, how long will it be before you hear the thunder?

Solution
Solve the formula $D = \frac{x}{5}$ for x.

$$D = \frac{x}{5} \qquad \text{Given formula}$$

$$5D = x \qquad \text{Multiply each side by 5.}$$

If $D = \mathbf{3.5}$ miles, then the delay is

$$x = 5D = 5(\mathbf{3.5}) = 17.5 \text{ seconds.} \qquad \boxed{\text{Now Try Exercise } 83}$$

In the next example we apply the area formula for a trapezoid.

EXAMPLE 8   **Finding the base of a trapezoid**

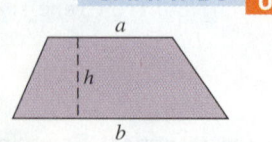

Figure 2.13 Trapezoid

The area of the trapezoid shown in Figure 2.13 is given by

$$A = \frac{1}{2}(a + b)h,$$

where a and b are the bases of the trapezoid and h is the height.

(a) Solve the formula for b.

(b) A trapezoid has area $A = 36$ square inches, height $h = 4$ inches, and base $a = 8$ inches. Find b.

Solution

(a) To clear the equation of the fraction, multiply each side by 2.

$$A = \frac{1}{2}(a + b)h \qquad \text{Area formula}$$

$$2A = (a + b)h \qquad \text{Multiply each side by 2.}$$

$$\frac{2A}{h} = a + b \qquad \text{Divide each side by } h.$$

$$\frac{2A}{h} - a = b \qquad \text{Subtract } a \text{ from each side.}$$

$$b = \frac{2A}{h} - a \qquad \text{Rewrite the formula.}$$

(b) Let $A = 36$, $h = 4$, and $a = 8$ in $b = \frac{2A}{h} - a$. Then

$$b = \frac{2(36)}{4} - 8 = 18 - 8 = 10 \text{ inches.} \qquad \boxed{\text{Now Try Exercise } 57}$$

EXAMPLE 9   Solving for a variable

Solve each equation for the indicated variable.

(a) $c = \frac{a + b}{2}$ for b **(b)** $ab - bc = ac$ for c

Solution

(a) To clear the equation of the fraction, multiply each side by 2.

$$c = \frac{a + b}{2} \qquad \text{Given formula}$$

$$2c = a + b \qquad \text{Multiply each side by 2.}$$

$$2c - a = b \qquad \text{Subtract } a.$$

The formula solved for b is $b = 2c - a$.

(b) In this equation c appears in two terms. We will combine the terms containing c by using the distributive property. Begin by moving the term on the left side containing c to the right side of the equation.

$$ab - bc = ac \qquad \text{Given formula}$$

$$ab - bc + \mathbf{bc} = ac + \mathbf{bc} \qquad \text{Add } bc \text{ to each side.}$$

$$ab = (a + b)c \qquad \text{Combine terms; distributive property.}$$

$$\frac{ab}{a + b} = \frac{(a + b)c}{(a + b)} \qquad \text{Divide each side by } (a + b).$$

$$\frac{ab}{a + b} = c \qquad \text{Simplify the fraction.}$$

CRITICAL THINKING

Are the formulas $c = \frac{1}{a - b}$ and $c = \frac{-1}{b - a}$ equivalent? Why?

The formula solved for c is $c = \frac{ab}{a + b}$. $\boxed{\text{Now Try Exercise } 61}$

Other Formulas

To calculate a student's GPA, the number of credits earned with a grade of A, B, C, D, and F must be known. If a, b, c, d, and f represent these credit counts respectively, then

$$GPA = \frac{4a + 3b + 2c + d}{a + b + c + d + f}.$$

This formula is based on the assumption that a 4.0 GPA is an A, a 3.0 GPA is a B, and so on.

EXAMPLE 10 Calculating a student's GPA

A student has earned 16 credits of A, 32 credits of B, 12 credits of C, 2 credits of D, and 5 credits of F. Calculate the student's GPA to the nearest hundredth.

Solution
Let $a = 16$, $b = 32$, $c = 12$, $d = 2$, and $f = 5$. Then

$$GPA = \frac{4 \cdot 16 + 3 \cdot 32 + 2 \cdot 12 + 2}{16 + 32 + 12 + 2 + 5} = \frac{186}{67} \approx 2.78.$$

The student's GPA is 2.78.   **Now Try Exercise 69**

EXAMPLE 11 Converting temperature scales

In the United States, temperature is measured with either the Fahrenheit or the Celsius temperature scales. To convert Fahrenheit degrees F to Celsius degrees C, the formula $C = \frac{5}{9}(F - 32)$ can be used.
(a) Solve the formula for F to find a formula that converts Celsius degrees to Fahrenheit degrees.
(b) If the outside temperature is 20°C, find the equivalent Fahrenheit temperature.

Solution
(a) The reciprocal of $\frac{5}{9}$ is $\frac{9}{5}$, so multiply each side by $\frac{9}{5}$.

$$C = \frac{5}{9}(F - 32) \qquad \text{Given equation}$$

$$\frac{9}{5}C = \frac{9}{5} \cdot \frac{5}{9}(F - 32) \qquad \text{Multiply each side by } \tfrac{9}{5}.$$

$$\frac{9}{5}C = F - 32 \qquad \text{Multiplicative inverses}$$

$$\frac{9}{5}C + 32 = F \qquad \text{Add 32 to each side.}$$

The required formula is $F = \frac{9}{5}C + 32$.

(b) If $C = 20°$C, then $F = \frac{9}{5}(20) + 32 = 36 + 32 = 68°$F.   **Now Try Exercise 73**

2.4 PUTTING IT ALL TOGETHER

In this section we discussed formulas and how to solve for a variable. The following table summarizes some of these formulas.

Concept	Formula	Examples
Area of a Rectangle	$A = LW$, where L is the length and W is the width.	If $L = 10$ feet and $W = 5$ feet, then the area is $$A = 10 \cdot 5 = 50 \text{ square feet.}$$
Gas Mileage	$M = \frac{D}{G}$, where D is the distance and G is the gasoline used.	If a car travels 100 miles on 5 gallons of gasoline, then its mileage is $M = \frac{100}{5} = 20$ miles per gallon.
Area of a Triangle	$A = \frac{1}{2}bh$, where b is the base and h is the height.	If $b = 5$ inches and $h = 6$ inches, then the area is $$A = \frac{1}{2}(5)(6) = 15 \text{ square inches.}$$
Angle Measure in a Triangle	$x + y + z = 180°$, where x, y, and z are the angle measures.	If $x = 40°$ and $y = 60°$, then $z = 80°$ because $$40° + 60° + 80° = 180°.$$
Volume and Surface Area of a Box	If a box has length L, width W, and height H, then its volume is $$V = LWH$$ and its surface area is $$S = 2LW + 2WH + 2LH.$$	If a box has dimensions $L = 4$ feet, $W = 3$ feet, and $H = 2$ feet, then $$V = 4(3)(2) = 24 \text{ cubic feet}$$ and $$S = 2(4)(3) + 2(3)(2) + 2(4)(2)$$ $$= 52 \text{ square feet.}$$

continued on next page

continued from previous page

Concept	Formula	Examples
Circumference and Area of a Circle	If a circle has radius r, then its circumference is $$C = 2\pi r$$ and its area is $$A = \pi r^2.$$	If $r = 6$ inches, $C = 2\pi(6) = 12\pi \approx 37.7$ inches and $A = \pi(6)^2 = 36\pi \approx 113.1$ square inches. 6 in.
Volume of a Cylinder	$V = \pi r^2 h$, where r is the radius and h is the height.	If $r = 5$ inches and $h = 20$ inches, then the volume is $$V = \pi(5^2)(20) = 500\pi \text{ cubic inches.}$$ 5 in. 20 in.
Distance from Lightning	$D = \frac{x}{5}$, where x is the delay in seconds between seeing the flash and hearing the thunder. D is in miles.	If $x = 15$ seconds, then the lightning is $D = \frac{15}{5} = 3$ miles away.
Area of a Trapezoid	$A = \frac{1}{2}(a + b)h$, where a and b are the bases and h is the height.	If $a = 4$, $b = 6$, and $h = 3$, then the area is $$A = \frac{1}{2}(4 + 6)(3) = 15 \text{ square units.}$$ 4 3 6
Calculating Grade Point Average (GPA)	GPA is calculated by $$\frac{4a + 3b + 2c + d}{a + b + c + d + f},$$ where a, b, c, d, and f represent the credits earned with grades of A, B, C, D, and F, respectively.	10 credits of A, 8 credits of B, 6 credits of C, 12 credits of D, and 8 credits of F results in a GPA of $$\frac{4(10) + 3(8) + 2(6) + 12}{10 + 8 + 6 + 12 + 8} = 2.0.$$
Converting Between Fahrenheit and Celsius Degrees	$C = \dfrac{5}{9}(F - 32)$ $F = \dfrac{9}{5}C + 32$	212°F is equivalent to $$C = \frac{5}{9}(212 - 32) = 100°\text{C.}$$ 100°C is equivalent to $$F = \frac{9}{5}(100) + 32 = 212°\text{F.}$$

2.4 Exercises

CONCEPTS

1. A(n) _____ can be used to calculate one quantity by using known values of other quantities.

2. The area A of a rectangle with length L and width W is $A =$ _____.

3. If a car that travels D miles uses G gallons of gasoline, the mileage is $M =$ _____.

4. The area A of a triangle with base b and height h is $A =$ _____.

5. One degree equals _____ of a revolution.

6. There are _____ degrees in one revolution.

7. The sum of the measures of the angles in a triangle equals _____ degrees.

8. The volume V of a box with length L, width W, and height H is $V =$ _____.

9. The surface area S of a box with length L, width W, and height H is $S =$ _____.

10. The circumference C of a circle with radius r is $C =$ _____.

11. The area A of a circle with radius r is $A =$ _____.

12. The volume V of a cylinder with radius r and height h is $V =$ _____.

13. If thunder is heard x seconds after lightning is seen, the distance in miles to the storm is $D =$ _____.

14. The area A of a trapezoid with height h and bases a and b is $A =$ _____.

FORMULAS FROM GEOMETRY

Exercises 15–22: Find the area of the region shown.

15.

6 ft
3 ft

16.

4 yd
2.5 yd

17.

3 in.
6 in.

18.

1 mi
3 mi

19.

4 ft

20.

6 in.

21.

6 ft
2 ft
5 ft

22.

3 mi
2 mi
4 mi

23. Find the area of a rectangle having a 7-inch width and a 13-inch length.

24. Find the area of a rectangle having a 5-foot width and a 7-yard length.

25. Find the area of a triangle having a 12-inch base and a 6-inch height.

26. Find the area of a triangle having a 9-foot base and a 72-inch height.

27. Find the circumference of a circle having an 8-inch diameter.

28. Find the area of a circle having a 9-foot radius.

29. *Area of a Lot* Find the area of the lot shown, which consists of a square and a triangle.

73 ft
52 ft
52 ft

30. *Area of a Lot*   Find the area of the lot shown, which consists of a rectangle and two triangles.

Exercises 31 and 32: *Angle Measure*   *Find the measure of the third angle in the triangle.*

31.   **32.**

33. A triangle contains two angles having measures of 23° and 76°. Find the measure of the third angle.

34. The measures of the angles in an *equilateral triangle* are equal. Find their measure.

35. The measures of the angles in a triangle are x, $2x$, and $3x$. Find the value of x.

36. The measures of the angles in a triangle are $3x$, $4x$, and $11x$. Find the value of x.

37. In a triangle the two smaller angles are equal in measure and are each one third of the measure of the larger angle. Find the measure of each angle.

38. In a triangle the two larger angles differ by 10°. The smaller angle is 50° less than the largest angle. Find the measure of each angle.

39. The diameter of a circle is 12 inches. Find its circumference and area.

40. The radius of a circle is $\frac{5}{4}$ feet. Find its circumference and area.

41. The circumference of a circle is 2π inches. Find its radius and area.

42. The circumference of a circle is 13π feet. Find its radius and area.

Exercises 43–46: *A box with a top has length L, width W, and height H. Find the volume and surface area of the box.*

43. $L = 22$ inches, $W = 12$ inches, $H = 10$ inches

44. $L = 5$ feet, $W = 3$ feet, $H = 6$ feet

45. $L = \frac{2}{3}$ yard, $W = \frac{2}{3}$ foot, $H = \frac{3}{2}$ feet

46. $L = 1.2$ meters, $W = 0.8$ meter, $H = 0.6$ meter

Exercises 47–50: *Use the formula $V = \pi r^2 h$ to find the volume of a cylindrical container for the given r and h. Leave your answer in terms of π.*

47. $r = 2$ inches, $h = 5$ inches

48. $r = \frac{1}{2}$ inch, $h = \frac{3}{2}$ inches

49. $r = 5$ inches, $h = 2$ feet

50. $r = 2.5$ feet, $h = 1.5$ yards

51. *Volume of a Barrel*   (Refer to Example 6.) A cylindrical barrel has a diameter of $1\frac{3}{4}$ feet and a height of 3 feet. Find the volume of the barrel.

52. *Volume of a Can*   (Refer to Example 6.)
   (a) Find the volume of a can with a radius of $\frac{3}{4}$ inch and a height of $2\frac{1}{2}$ inches.
   (b) Find the number of fluid ounces in the can if one cubic inch equals 0.554 fluid ounces.

SOLVING FOR A VARIABLE

Exercises 53–66: Solve the formula for the given variable.

53. $A = LW$ for W **54.** $A = \frac{1}{2}bh$ for b

55. $V = \pi r^2 h$ for h **56.** $V = \frac{1}{3}\pi r^2 h$ for h

57. $A = \frac{1}{2}(a + b)h$ for a **58.** $C = 2\pi r$ for r

59. $V = LWH$ for W **60.** $P = 2x + 2y$ for y

61. $s = \dfrac{a + b + c}{2}$ for b **62.** $t = \dfrac{x - y}{3}$ for x

63. $\dfrac{a}{b} - \dfrac{c}{b} = 1$ for b **64.** $\dfrac{x}{y} + \dfrac{z}{y} = 5$ for z

65. $ab = cd + ad$ for a

66. $S = 2LW + 2LH + 2WH$ for W

67. *Perimeter of a Rectangle*   The perimeter of a rectangle equals the sum of the lengths of its four sides. If the width of a rectangle is 5 inches and its perimeter is 40 inches, find the length of the rectangle.

68. *Perimeter of a Triangle*   Two sides of a triangle have lengths of 5 feet and 7 feet. If the triangle's perimeter is 21 feet, what is the length of the third side?

OTHER FORMULAS AND APPLICATIONS

Exercises 69–72: (Refer to Example 10.) Let a represent the number of credits with a grade of A, b the number of credits with a grade of B, and so on. Calculate the corresponding grade point average (GPA). Round your answer to the nearest hundredth.

69. $a = 30, b = 45, c = 12, d = 4, f = 4$

70. $a = 70, b = 35, c = 5, d = 0, f = 0$

71. $a = 0, b = 60, c = 80, d = 10, f = 6$

72. $a = 3, b = 5, c = 8, d = 0, f = 22$

Exercises 73–76: (Refer to Example 11.) Convert the Celsius temperature to an equivalent Fahrenheit temperature.

73. $25°C$ **74.** $100°C$

75. $-40°C$ **76.** $0°C$

Exercises 77–80: (Refer to Example 11.) Convert the Fahrenheit temperature to an equivalent Celsius temperature.

77. $23°F$ **78.** $98.6°F$

79. $-4°F$ **80.** $-31°F$

81. *Gas Mileage* A truck driver leaves a gas station with a full tank of gas and the odometer showing 87,625 miles. At the next gas stop, it takes 38 gallons to fill the tank and the odometer reads 88,043 miles. Find the gas mileage for the truck.

82. *Gas Mileage* A car that gets 34 miles per gallon is driven 578 miles. How many gallons of gasoline are used on this trip?

83. *Lightning* (Refer to Example 7.) The time delay between a flash of lightning and the sound of thunder is 12 seconds. How far away is the lightning?

84. *Lightning* (Refer to Example 7.) Doppler radar shows an electrical storm 2.5 miles away. If you see the lightning from this storm, how long will it be before you hear the thunder?

WRITING ABOUT MATHEMATICS

85. A student solves the formula $A = \frac{1}{2}bh$ for h and obtains the formula $h = \frac{1}{2}bA$. Explain the error that the student is making. What is the correct answer?

86. Give an example of a formula that you have used and explain how you used it.

CHECKING BASIC CONCEPTS
SECTIONS 2.3 AND 2.4

1. Translate the sentence into an equation containing the variable x. Then solve the resulting equation.
  (a) The product of 3 and a number is 36.
  (b) A number subtracted from 35 is 43.

2. The sum of three consecutive integers is -93. Find the three integers.

3. Convert 9.5% to a decimal.

4. Convert $\frac{5}{4}$ to a percentage.

5. *Serious Crime* In 2003, New York City experienced 2652 serious crimes per 100,000 people. This figure represented a 25% decrease from the number in 2000. Find the rate for serious crimes per 100,000 people in 2000.

6. *Driving a Car* How many hours does it take the driver of a car to travel 390 miles at 60 miles per hour?

7. *College Loans* A student takes out two loans, one at 6% and the other at 7%. The 6% loan is $2000 more than the 7% loan, and the total interest for one year is $510. Find the amount of each loan.

8. *Gas Mileage* A car that gets 28 miles per gallon is driven 504 miles. How many gallons of gasoline are used on this trip?

9. *Height of a Triangle* The area of a triangle having a base of 6 inches is 36 square inches. Find the height of the triangle.

continued on next page

continued from previous page

10. *Area and Circumference*   Find the area and circumference of the circle shown.

3 ft

11. *Angles*   Find the value of x in the triangle shown.

2x
3x
x

12. *Solving a Formula*   Solve $A = \pi r^2 + \pi r l$ for l.

2.5   LINEAR INEQUALITIES

Solutions and Number Line Graphs ▪ The Addition Property of Inequalities ▪ The Multiplication Property of Inequalities ▪ Applications

A LOOK INTO MATH ▷

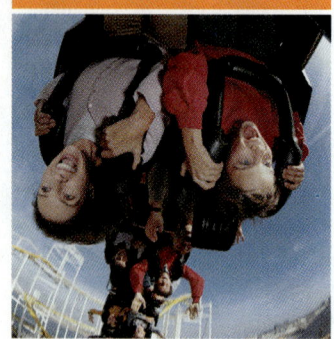

At an amusement park, a particular ride might be restricted to people at least 48 inches tall. A child who is x inches tall may go on the ride if $x \geq 48$ but may not go on the ride if $x < 48$. A height of 48 inches represents the boundary between being allowed on the ride and being denied access to the ride. A posted height restriction, or *boundary*, allows parents to easily determine if their child may go on the ride.

Solving linear inequalities is closely related to solving linear equations because equality is the boundary between *greater than* and *less than*. In this section we discuss techniques used to solve linear inequalities.

Solutions and Number Line Graphs

A **linear inequality** results whenever the equals sign in a linear equation is replaced with any one of the symbols $<, \leq, >,$ or $\geq$. Examples of linear equations include

$$x = 5, \quad 2x + 1 = 0, \quad 1 - x = 6, \quad \text{and} \quad 5x + 1 = 3 - 2x.$$

Therefore examples of linear inequalities include

$$x > 5, \quad 2x + 1 < 0, \quad 1 - x \geq 6, \quad \text{and} \quad 5x + 1 \leq 3 - 2x.$$

A **solution** to an inequality is a value of the variable that makes the statement true. The set of all solutions is called the **solution set.** Two inequalities are *equivalent* if they have the same solution set. Inequalities frequently have infinitely many solutions. For example, the solution set for the inequality $x > 5$ includes all real numbers greater than 5.

A number line can be used to graph the solution set for an inequality. The graph of all real numbers satisfying $x < 2$ is shown in Figure 2.14(a), and the graph of all real numbers satisfying $x \leq 2$ is shown in Figure 2.14(b). (The symbol $\leq$ is read "less than or equal to." Similarly, the symbol $\geq$ is read "greater than or equal to.") A parenthesis ")" is used to show that 2 is not included in Figure 2.14(a), and a bracket "]" is used to show that 2 is included in Figure 2.14(b).

(a) $x < 2$ (b) $x \leq 2$

Figure 2.14

EXAMPLE 1   Graphing inequalities on a number line

Use a number line to graph the solution set to each inequality.
(a) $x > 0$ **(b)** $x \geq 0$ **(c)** $x \leq -1$ **(d)** $x < 3$

Solution
(a) First locate $x = 0$ (or the origin) on a number line. Numbers greater than 0 are located to the right of the origin, so shade the number line to the right of the origin. Because $x > 0$, the number 0 is not included, so place a parenthesis "(" at 0, as shown in Figure 2.15(a).
(b) Figure 2.15(b) is similar to the graph in part (a) except that a bracket "[" is placed at the origin because 0 is included in the solution set.
(c) First locate $x = -1$ on the number line. Numbers less than -1 are located to the left of -1. Because -1 is included, a bracket "]" is placed at -1, as shown in Figure 2.15(c).
(d) Real numbers less than 3 are graphed in Figure 2.15(d).

(a) $x > 0$ (b) $x \geq 0$

(c) $x \leq -1$ (d) $x < 3$

Figure 2.15

Now Try Exercises **13, 17**

INTERVAL NOTATION (OPTIONAL)   Each number line graph in Figure 2.15 represents an interval of real numbers that corresponds to the solution set to an inequality. These solution sets can also be represented in a convenient notation called **interval notation**. Rather than draw the entire number line, we can use brackets or parentheses to indicate the interval of values that represent the solution set. For example, the number line shown in Figure 2.15(a) can be represented by the interval $(0, \infty)$, and the number line shown in Figure 2.15(b) can be represented by the interval $[0, \infty)$. The symbol ∞ refers to infinity and is used to indicate that the values increase without bound. Similarly, $-\infty$ can be used when the values decrease without bound. The number lines shown in Figures 2.15(c) and (d) can be represented by $(-\infty, -1]$ and $(-\infty, 3)$, respectively.

EXAMPLE 2   Writing solution sets in interval notation

Write the solution set to each inequality in interval notation.
(a) $x > 4$ **(b)** $y \leq -3$ **(c)** $z \geq -1$

Solution
(a) Real numbers greater than 4 are represented by the interval $(4, \infty)$.
(b) Real numbers less than or equal to -3 are represented by the interval $(-\infty, -3]$.
(c) The solution set is represented by the interval $[-1, \infty)$.

Now Try Exercise **27**

CHECKING SOLUTIONS   We can check possible solutions to an inequality in the same way that we checked possible solutions to an equation. For example, to check whether 5 is a solution to $2x + 3 = 13$, we substitute $x = 5$ in the equation.

$$2(\mathbf{5}) + 3 \stackrel{?}{=} 13 \qquad \text{Replace } x \text{ with 5.}$$
$$13 = 13 \qquad \text{A true statement}$$

Thus 5 is a solution to this equation. Similarly, to check whether 7 is a solution to $2x + 3 > 13$, we substitute $x = 7$ in the inequality.

$$2(7) + 3 \overset{?}{>} 13 \qquad \text{Replace } x \text{ with 7.}$$
$$17 > 13 \qquad \text{A true statement}$$

Thus 7 is a solution to the inequality.

.EXAMPLE 3   Checking possible solutions

Determine whether the given value of x is a solution to the inequality.
(a) $3x - 4 < 10$, $x = 6$ **(b)** $4 - 2x \le 8$, $x = -2$

Solution
(a) Substitute **6** for x and simplify.

$$3(6) - 4 \overset{?}{<} 10 \qquad \text{Replace } x \text{ with 6.}$$
$$14 \overset{?}{<} 10 \qquad \text{A false statement}$$

Thus 6 is *not* a solution to the inequality.
(b) Substitute **−2** for x and simplify.

$$4 - 2(-2) \overset{?}{\le} 8 \qquad \text{Replace } x \text{ with } -2.$$
$$8 \le 8 \qquad \text{A true statement}$$

Thus −2 is a solution to the inequality.   Now Try Exercise 35

Just as solving a linear equation means finding the value of the variable that makes the equation true, solving a linear inequality means finding the *values* of the variable that make the inequality true.

Making a table is an organized way of checking possible values of the variable to see if any value makes an inequality true. For example, if we want to find some of the solutions to the inequality $2x - 6 < 0$, we substitute various values for x in the left side of the inequality. If a value gives a result that is less than 0, the value makes the inequality true and is a solution. In the next example, we use a table to find solutions to an equation and related inequalities.

EXAMPLE 4   Finding solutions to equations and inequalities

In Table 2.7 the expression $2x - 6$ has been evaluated for several values of x. Use the table to determine any solutions to each equation or inequality.
(a) $2x - 6 = 0$ **(b)** $2x - 6 > 0$ **(c)** $2x - 6 \ge 0$ **(d)** $2x - 6 < 0$

TABLE 2.7

x	0	1	2	3	4	5	6
$2x - 6$	−6	−4	−2	0	2	4	6

Solution
(a) From Table 2.7, $2x - 6$ equals **0** when $x = $ **3**.
(b) The values of x in the table that make the expression $2x - 6$ greater than 0 are 4, 5, and 6. These values are all greater than 3, which is the solution found in part (a). It follows that $2x - 6 > 0$ when $x > 3$.
(c) The values of x in the table that make the expression $2x - 6$ greater than or equal to 0 are 3, 4, 5, and 6. It follows that $2x - 6 \ge 0$ when $x \ge 3$.
(d) The expression $2x - 6$ is less than 0 when $x < 3$.   Now Try Exercise 43

The Addition Property of Inequalities

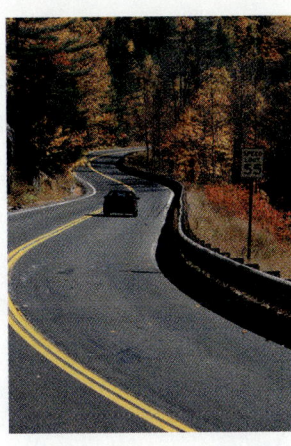

▶ **REAL-WORLD CONNECTION** Suppose that the speed limit on a country road is 55 miles per hour, which is 25 miles per hour faster than the speed limit in town. If x represents lawful speeds in town, then x satisfies the inequality

$$x + 25 \leq 55.$$

To solve this inequality we **add −25** to each side of the inequality.

$$x + 25 + (-25) \leq 55 + (-25) \qquad \text{Add −25 to each side.}$$
$$x \leq 30 \qquad \text{Add the real numbers.}$$

Thus drivers are obeying the speed limit in town when they travel at 30 miles per hour or less. To solve this inequality the addition property of inequalities was used.

ADDITION PROPERTY OF INEQUALITIES

Let a, b, and c be expressions that represent real numbers. The inequalities

$$a < b \quad \text{and} \quad a + c < b + c$$

are equivalent. That is, the same number may be added to (or subtracted from) each side of an inequality. Similar properties exist for $>$, $\leq$, and $\geq$.

To solve some inequalities, we apply the addition property of inequalities to obtain a simpler, equivalent inequality.

EXAMPLE 5   Applying the addition property of inequalities

Solve each inequality. Then graph the solution set.
(a) $x - 1 > 4$ **(b)** $3 + 2x \leq 5 + x$

Solution
(a) Begin by adding 1 to each side of the inequality.

$$x - 1 > 4 \qquad \text{Given inequality}$$
$$x - 1 + 1 > 4 + 1 \qquad \text{Add 1 to each side.}$$
$$x > 5 \qquad \text{Add the real numbers.}$$

The solution set is given by $x > 5$ and is graphed as follows.

```
 ←——+———+———+———+——(———+———→
    −2   0   2   4   6   8
```

(b) Begin by subtracting x from (or adding $-x$ to) each side of the inequality.

$$3 + 2x \leq 5 + x \qquad \text{Given inequality}$$
$$3 + 2x - x \leq 5 + x - x \qquad \text{Subtract } x \text{ from each side.}$$
$$3 + x \leq 5 \qquad \text{Combine like terms.}$$
$$3 + x - 3 \leq 5 - 3 \qquad \text{Subtract 3 from each side.}$$
$$x \leq 2 \qquad \text{Subtract the real numbers.}$$

The solution set is given by $x \leq 2$ and is graphed as follows.

```
 ←——+———+———+———+———]———+——→
   −2  −1   0   1   2   3
```

Now Try Exercise 51

MAKING CONNECTIONS

The addition property of inequalities can be illustrated by an old-fashioned pan balance. In the figure on the left, the weight on the right pan is heavier than the weight on the left because the right pan rests lower than the left pan. If we add the same amount of weight to (or subtract the same amount of weight from) both pans, the right pan still weighs more than the left pan by the same amount, as illustrated in the figure on the right.

In the next example we apply the addition property of inequalities and graph the solution set.

EXAMPLE 6 **Applying the addition property of inequalities**

Solve $5 + \frac{1}{2}x \leq 3 - \frac{1}{2}x$. Then graph the solution set.

Solution
Begin by subtracting 5 from each side of the inequality.

$$5 + \frac{1}{2}x \leq 3 - \frac{1}{2}x \qquad \text{Given inequality}$$

$$\frac{1}{2}x \leq -\frac{1}{2}x - 2 \qquad \text{Subtract 5 from each side.}$$

$$\frac{1}{2}x + \frac{1}{2}x \leq -\frac{1}{2}x + \frac{1}{2}x - 2 \qquad \text{Add } \tfrac{1}{2}x \text{ to each side.}$$

$$x \leq -2 \qquad \text{Combine like terms.}$$

The solution set is given by $x \leq -2$ and is graphed as follows.

Now Try Exercise **57**

The Multiplication Property of Inequalities

The multiplication property of inequalities differs from the multiplication property of *equality*. When we multiply each side of an inequality by the same nonzero number, we may need to reverse the inequality symbol to make sure that the resulting inequality remains true. Table 2.8 shows various results that occur when each side of a true inequality is multiplied by the same nonzero number.

TABLE 2.8 Determining When the Inequality Symbol Should Be Reversed

True Statement	Multiply Each Side By	Resulting Inequality	Is the Result True or False?	Reverse the Inequality Symbol
$-3 < 5$	4	$-12 \overset{?}{<} 20$	True	Not needed
$-4 \leq 2$	-5	$20 \overset{?}{\leq} -10$	False	$20 \geq -10$
$7 > -1$	-2	$-14 \overset{?}{>} 2$	False	$-14 < 2$
$0 \geq -6$	8	$0 \overset{?}{\geq} -48$	True	Not needed
$-2 > -5$	3	$-6 \overset{?}{>} -15$	True	Not needed
$4 < 9$	-11	$-44 \overset{?}{<} -99$	False	$-44 > -99$

Table 2.8 indicates that the inequality symbol must be reversed when each side of the given inequality is multiplied by a negative number. This result is summarized in the following.

MULTIPLICATION PROPERTY OF INEQUALITIES

Let a, b, and c be expressions that represent real numbers with $c \neq 0$.

1. If $c > 0$, then the inequalities $a < b$ and $ac < bc$ are equivalent. That is, each side of an inequality may be multiplied (or divided) by the same positive number.
2. If $c < 0$, then the inequalities $a < b$ and $ac > bc$ are equivalent. That is, each side of an inequality may be multiplied (or divided) by the same negative number, provided the inequality symbol is reversed.

Note that similar properties exist for $\leq$ and $\geq$.

NOTE: It is important to reverse the inequality symbol when either multiplying or *dividing* by a negative number.

EXAMPLE 7 Applying the multiplication property of inequalities

Solve each inequality. Then graph the solution set.
(a) $3x < 18$ **(b)** $-7 \leq -\frac{1}{2}x$

Solution
(a) To solve for x, divide each side by 3.

$$3x < 18 \qquad \text{Given inequality}$$

$$\frac{3x}{3} < \frac{18}{3} \qquad \text{Divide each side by 3.}$$

$$x < 6 \qquad \text{Simplify fractions.}$$

The solution set is given by $x < 6$ and is graphed as follows.

(b) To isolate x in $-7 \leq -\frac{1}{2}x$, multiply each side by -2 and reverse the inequality symbol.

$$-7 \leq -\frac{1}{2}x \qquad \text{Given inequality}$$

$$-2(-7) \geq -2\left(-\frac{1}{2}\right)x \qquad \text{Multiply by } -2; \text{ reverse the inequality.}$$

$$14 \geq 1 \cdot x \qquad \text{Multiply the real numbers.}$$

$$x \leq 14 \qquad \text{Rewrite the inequality.}$$

The solution set is given by $x \leq 14$ and is graphed as follows.

Now Try Exercise 61

SET-BUILDER NOTATION   Because $x \leq 14$ is an inequality with infinitely many solutions and is not itself a set of solutions, a mathematical notation called **set-builder notation** has been devised for writing the solutions to an inequality as a set. For example, the solution set consisting of "all real numbers x such that x is less than or equal to 14" can be written $\{x \mid x \leq 14\}$. The vertical line segment "$\mid$" is read "such that."

In the next example, the solution sets are expressed in set-builder notation. However, this notation is not widely used throughout this text.

EXAMPLE 8   Applying both properties of inequalities

Solve each inequality. Write the solution set in set-builder notation.
(a) $4x - 7 \geq -6$ **(b)** $-8 + 4x \leq 5x + 3$ **(c)** $0.4(2x - 5) < 1.1x + 2$

Solution
(a) Start by adding 7 to each side.

$$4x - 7 \geq -6 \qquad \text{Given inequality}$$

$$4x \geq 1 \qquad \text{Add 7 to each side.}$$

$$x \geq \frac{1}{4} \qquad \text{Divide each side by 4.}$$

In set-builder notation, the solution set is $\left\{x \mid x \geq \frac{1}{4}\right\}$.
(b) Begin by adding 8 to each side.

$$-8 + 4x \leq 5x + 3 \qquad \text{Given inequality}$$

$$4x \leq 5x + 11 \qquad \text{Add 8 to each side.}$$

$$4x - 5x \leq 5x + 11 - 5x \qquad \text{Subtract } 5x \text{ from each side.}$$

$$-x \leq 11 \qquad \text{Combine like terms.}$$

$$-1 \cdot (-x) \geq -1 \cdot 11 \qquad \text{Multiply by } -1; \text{ reverse the inequality.}$$

$$x \geq -11 \qquad \text{Simplify.}$$

The solution set is $\{x \mid x \geq -11\}$.

(c) Begin by multiplying each side by 10 to eliminate decimals.

$$0.4(2x - 5) < 1.1x + 2 \qquad \text{Given inequality}$$
$$4(2x - 5) < 11x + 20 \qquad \text{Multiply each side by 10.}$$
$$8x - 20 < 11x + 20 \qquad \text{Distributive property}$$
$$8x < 11x + 40 \qquad \text{Add 20 to each side.}$$
$$-3x < 40 \qquad \text{Subtract } 11x \text{ from each side.}$$
$$x > -\frac{40}{3} \qquad \text{Divide by } -3; \text{ reverse the inequality.}$$

The solution set is $\left\{ x \mid x > -\frac{40}{3} \right\}$.

Now Try Exercises 71, 81, 91

CRITICAL THINKING

Solve $-5 - 3x > -2x + 7$ without having to reverse the inequality symbol.

Applications

To solve applications involving inequalities, we often have to translate words to mathematical statements. For example, the phrase "at least" can be translated to "greater than or equal to" and the phrase "at most" can be translated to "less than or equal to."

EXAMPLE 9 Translating words to inequalities

Translate each phrase to an inequality. Let the variable be x.
(a) A number that is more than 30
(b) An age that is at least 18
(c) A grade point average that is at most 3.25

Solution
(a) The inequality $x > 30$ represents a number x that is more than 30.
(b) The inequality $x \geq 18$ represents an age x that is at least 18.
(c) The inequality $x \leq 3.25$ represents a grade point average x that is at most 3.25.

Now Try Exercises 101, 107

▶ **REAL-WORLD CONNECTION** In the lower atmosphere the air temperature generally becomes colder as the altitude increases. One mile above Earth's surface the temperature is about 19°F colder than the ground-level temperature. As the air cools, there is an increased chance of clouds forming. In the next example we estimate the altitudes where clouds may form. (*Source:* A. Miller and R. Anthes, *Meteorology.*)

EXAMPLE 10 Finding the altitude of clouds

If the ground temperature is 79°F, then the temperature T above Earth's surface is given by the formula $T = 79 - 19x$, where x is the altitude in miles. Suppose that clouds form only where the temperature is 3°F or colder. Determine the heights at which clouds may form.

Solution
Clouds may form at altitudes at which the temperature T is less than or equal to 3°F. Thus we must solve the inequality $79 - 19x \leq 3$.

$$79 - 19x \leq 3 \qquad \text{Inequality to be solved}$$
$$-19x \leq -76 \qquad \text{Subtract 79 from each side.}$$
$$\frac{-19x}{-19} \geq \frac{-76}{-19} \qquad \text{Divide by } -19; \text{ reverse the inequality.}$$
$$x \geq 4 \qquad \text{Simplify the fraction.}$$

Clouds may form at 4 miles or higher.

Now Try Exercise 123

EXAMPLE **11** **Modeling AIDS Research**

AIDS research funding F in billions of dollars increased from 1995 to 2005 and can be modeled by the formula $F = 0.153x - 303.85$, where x is the year. Estimate when the funding F was greater than or equal to $2.15 billion. (**Source:** National Institute of Health.)

Solution
We must solve the inequality $F \geq 2.15$.

$$0.153x - 303.85 \geq 2.15 \qquad \text{Inequality to be solved}$$

$$0.153x \geq 306 \qquad \text{Add 303.85 to each side.}$$

$$x \geq \frac{306}{0.153} \qquad \text{Divide each side by 0.153.}$$

Because $\frac{306}{0.153} = 2000$, funding for AIDS was greater than or equal to $2.15 billion during 2000 and after. Now Try Exercise **125**

EXAMPLE **12** **Calculating revenue, cost, and profit**

For a computer company, the cost to produce one laptop computer is $1320 plus a one-time fixed cost of $200,000 for research and development. The revenue received from selling one laptop computer is $1850.
(a) Write a formula that gives the cost C of producing x laptop computers.
(b) Write a formula that gives the revenue R from selling x laptop computers.
(c) Profit equals revenue minus cost. Write a formula that calculates the profit P from selling x laptop computers.
(d) How many computers need to be sold to yield a positive profit?

Solution
(a) The cost of producing the first laptop is

$$1320 \times 1 + 200{,}000 = \$201{,}320.$$

The cost of producing two laptops is

$$1320 \times 2 + 200{,}000 = \$202{,}640.$$

And, in general, the cost of producing x laptops is

$$1320 \times x + 200{,}000 = 1320x + 200{,}000.$$

Thus $C = 1320x + 200{,}000$.
(b) Because the company receives $1850 for each laptop, the revenue for x laptops is given by $R = 1850x$.
(c) Profit equals revenue minus cost, so

$$P = R - C$$

$$= 1850x - (1320x + 200{,}000)$$

$$= 530x - 200{,}000.$$

Thus $P = 530x - 200{,}000$.

(d) To determine how many laptops need to be sold to yield a positive profit, we must solve the inequality $P > 0$.

$$530x - 200{,}000 > 0 \qquad \text{Inequality to be solved}$$

$$530x > 200{,}000 \qquad \text{Add 200,000 to each side.}$$

$$x > \frac{200{,}000}{530} \qquad \text{Divide each side by 530.}$$

Because $\frac{200{,}000}{530} \approx 377.4$, the company must sell at least 378 laptops. Note that the company cannot sell a fraction of a laptop.

Now Try Exercise 119

2.5 PUTTING IT ALL TOGETHER

In this section we discussed linear inequalities and how to solve them. A linear inequality has infinitely many solutions and can be solved by using the addition and multiplication properties of inequalities. When multiplying or dividing an inequality by a negative number, we must reverse the inequality symbol. The following table summarizes some of the concepts presented in this section.

Concept	Comments	Examples
Linear Inequality	If the equals sign in a linear equation is replaced with $<$, $>$, $\le$, or $\ge$, a linear inequality results.	*Linear Equation* $\quad$ *Linear Inequality* $4x - 1 = 0 \qquad 4x - 1 > 0$ $2 - x = 3x \qquad 2 - x \le 3x$ $4(x + 3) = 1 - x \quad 4(x + 3) < 1 - x$ $-6x + 3 = 5 \qquad -6x + 3 \ge 5$
Solution to an Inequality	A value for a variable that makes the inequality a true statement	5 is a solution to $2x > 5$ because $2(5) > 5$ is a true statement.
Set-Builder Notation	A notation that can be used to identify the solution set to an inequality	The solution set for $x - 2 < 5$ can be written as $\{x \mid x < 7\}$ and is read "the set of real numbers x such that x is less than 7."
Solution Set to an Inequality	The set of all solutions to an inequality	The solution set to $x + 1 > 5$ is given by $x > 4$ and can be written in set-builder notation as $\{x \mid x > 4\}$.
Number Line Graphs	The solutions to an inequality can be graphed on a number line.	$x < 2$ is graphed as follows. $x \ge -1$ is graphed as follows.

continued on next page

continued from previous page

Concept	Comments	Examples
Addition Property of Inequalities	$a < b$ is equivalent to $a + c < b + c,$ where a, b, and c represent real number expressions.	$x - 5 \geq 6$ Given inequality $x \geq 11$ Add 5. $3x > 5 + 2x$ Given inequality $x > 5$ Subtract $2x$.
Multiplication Property of Inequalities	$a < b$ is equivalent to $ac < bc$ when $c > 0$, and is equivalent to $ac > bc$ when $c < 0$.	$\dfrac{1}{2}x \geq 6$ Given inequality $x \geq 12$ Multiply by 2. $-3x > 5$ Given inequality $x < -\dfrac{5}{3}$ Divide by -3; reverse the inequality symbol.

2.5 Exercises

MyMathLab

  PRACTICE   WATCH   DOWNLOAD   READ   REVIEW

CONCEPTS

1. A linear inequality results whenever the _____ in a linear equation is replaced by any one of the symbols _____, _____, _____, or _____.

2. Equality is the boundary between _____ and _____.

3. A(n) _____ is a value of the variable that makes an inequality statement true.

4. Two linear inequalities are _____ if they have the same solution set.

5. When a linear equation is solved, the solution set contains <u>(one/infinitely many)</u> solution(s).

6. When a linear inequality is solved, the solution set contains <u>(one/infinitely many)</u> solution(s).

7. The solution set to a linear inequality can be graphed by using a _____.

8. The value of 5 <u>(is/is not)</u> a solution to the inequality $3x < 10$.

9. The addition property of inequalities states that if $a > b$, then $a + c$ _____ $b + c$.

10. The multiplication property of inequalities states that if $a < b$ and $c > 0$, then ac _____ bc.

11. The multiplication property of inequalities states that if $a < b$ and $c < 0$, then ac _____ bc.

12. Are $-4x < 8$ and $x < -2$ equivalent inequalities? Explain.

SOLUTIONS AND NUMBER LINE GRAPHS

Exercises 13–20: Use a number line to graph the solution set to the inequality.

13. $x < 0$

14. $x > -2$

15. $x > 1$

16. $x < -\frac{5}{2}$

17. $x \leq 1.5$

18. $x \geq -3$

19. $z \geq -2$

20. $z \leq -\pi$

Exercises 21–26: Express the set of real numbers graphed on the number line with an inequality.

21.

22.

23.

24.

25.

26.

Exercises 27–32: Write the solution set to the inequality in interval notation.

27. $x \geq 6$

28. $x < 3$

29. $y > -2$

30. $y \geq 1$

31. $z \leq 7$

32. $z < -5$

Exercises 33–42: Determine whether the given value of the variable is a solution to the inequality.

33. $x + 5 > 5$ $x = 4$

34. $x - 7 < 0$ $x = 6$

35. $5x \geq 25$ $x = 5$

36. $-3x \leq -8$ $x = -2$

37. $4y - 3 \leq 5$ $y = -3$

38. $3y + 5 \geq -8$ $y = -3$

39. $5(z + 1) < 3z - 7$ $z = -7$

40. $-(z + 7) > 3(6 - z)$ $z = 2$

41. $\frac{3}{2}t - \frac{1}{2} \geq 1 - t$ $t = -2$

42. $2t - 3 > 5t - (2t + 1)$ $t = 5$

TABLES AND LINEAR INEQUALITIES

Exercises 43–46: Use the table to solve the inequality.

43. $3x + 6 > 0$

x	-4	-3	-2	-1	0
$3x + 6$	-6	-3	0	3	6

44. $6 - 3x \leq 0$

x	1	2	3	4	5
$6 - 3x$	3	0	-3	-6	-9

45. $-2x + 7 > 5$

x	-1	0	1	2	3
$-2x + 7$	9	7	5	3	1

46. $5(x - 3) \leq 4$

x	3.2	3.4	3.6	3.8	4
$5(x - 3)$	1	2	3	4	5

Exercises 47–50: Complete the table. Then use the table to solve the inequality.

47. $-2x + 6 \leq 0$

x	1	2	3	4	5
$-2x + 6$	4				-4

48. $3x - 1 < 8$

x	0	1	2	3	4
$3x - 1$	-1				

49. $5 - x > x + 7$

x	-3	-2	-1	0	1
$5 - x$	8				4
$x + 7$	4				8

50. $2(3 - x) \geq -3(x - 2)$

x	-2	-1	0	1	2
$2(3 - x)$					
$-3(x - 2)$					

SOLVING LINEAR INEQUALITIES

Exercises 51–58: Use the addition property of inequalities to solve the inequality. Then graph the solution set.

51. $x - 3 > 0$

52. $x + 6 < 3$

53. $3 - y \leq 5$

54. $8 - y \geq 10$

55. $12 < 4 + z$

56. $2z \leq z + 17$

57. $5 - 2t \geq 10 - t$

58. $-2t > -3t + 1$

Exercises 59–66: Use the multiplication property of inequalities to solve the inequality. Then graph the solution set.

59. $2x < 10$

60. $3x > 9$

61. $-\frac{1}{2}t \geq 1$

62. $-5t \leq -6$

63. $\frac{3}{4} > -5y$

64. $10 \geq -\frac{1}{7}y$

65. $-\frac{2}{3} \leq \frac{1}{7}z$

66. $-\frac{3}{10}z < 11$

Exercises 67–72: Solve the linear inequality and write the solution in set-builder notation.

67. $x + 6 > 7$

68. $x + 4 < 1$

69. $-3x \leq 21$

70. $4x \geq -20$

71. $2x - 3 < 9$

72. $-5x + 4 < 44$

Exercises 73–100: Solve the linear inequality.

73. $3x + 1 < 22$

74. $4 + 5x \leq 9$

75. $5 - \frac{3}{4}x \geq 6$

76. $10 - \frac{2}{5}x > 0$

77. $45 > 6 - 2x$

78. $69 \geq 3 - 11x$

79. $5x - 2 \leq 3x + 1$

80. $12x + 1 < 25 - 3x$

81. $-x + 24 < x + 23$

82. $6 - 4x \leq x + 1$

83. $-(x + 1) \geq 3(x - 2)$

84. $5(x + 2) > -2(x - 3)$

85. $3(2x + 1) > -(5 - 3x)$

86. $4x \geq -3(7 - 2x) + 1$

87. $-(7x + 5) + 1 \geq 3x - 1$

88. $3(2 - x) - 5 > -4(5 - x)$

89. $1.6x + 0.4 \leq 0.4x$

90. $-5.1x + 1.1 < 0.1 - 0.1x$

91. $0.8x - 0.5 < x + 1 - 0.5x$

92. $0.1(x + 1) - 0.1 \leq 0.2x - 0.5$

93. $-\frac{1}{2}\left(\frac{2}{3}x + 4\right) \geq x$

94. $-5x > \frac{4}{5}\left(\frac{10}{3}x + 10\right)$

95. $\frac{3}{7}x + \frac{2}{7} > -\frac{1}{7}x - \frac{5}{14}$

96. $\frac{5}{6} - \frac{1}{3}x \geq -\frac{1}{3}\left(\frac{5}{6}x - 1\right)$

97. $\frac{x}{3} + \frac{5x}{6} \leq \frac{2}{3}$

98. $\frac{3x}{4} - \frac{x}{2} < 1$

99. $\frac{6x}{7} < \frac{1}{3}x + 1$

100. $\frac{5x}{8} - \frac{3x}{4} \leq 8$

TRANSLATING PHRASES TO INEQUALITIES

Exercises 101–108: Translate each phrase to an inequality. Let x be the variable.

101. A speed that is greater than 60 miles per hour

102. A speed that is at most 60 miles per hour

103. An age that is at least 21 years old

104. An age that is less than 21 years old

105. A salary that is more than $40,000

106. A salary that is less than or equal to $40,000

107. A speed that does not exceed 70 miles per hour

108. A speed that is not less than 70 miles per hour

APPLICATIONS

109. *Geometry*   Find values for x so that the perimeter of the rectangle is less than 50 feet.

110. *Geometry*   A rectangle is twice as long as it is wide. If the rectangle is to have a perimeter of at least 36 inches, what values for the width are possible?

111. *Geometry*   A triangle with height 12 inches is to have area less than 120 square inches. What must be true about the base of the triangle?

112. *Geometry*   A trapezoid with height 6 inches is to have area not more than 120 square inches. What must be true about the sum of the bases of the trapezoid? (*Hint:* $A = \frac{1}{2}h(a + b)$, where a and b are the bases of the trapezoid.)

113. *Grade Average*   A student scores 74 out of 100 on a test. If the maximum score on the next test is also 100 points, what score does the student need to maintain at least an average of 80?

114. *Grade Average*   A student scores 65 and 82 on two different 100-point tests. If the maximum score on the next test is also 100 points, what score does the student need to maintain at least an average of 70?

115. *Parking Rates*   Parking in a student lot costs $2 for the first half hour and $1.25 for each hour thereafter. A partial hour is charged the same as a full hour. What is the longest time that a student can park in this lot for $8?

116. *Parking Rates* Parking in a student lot costs $2.50 for the first hour and $1 for each hour thereafter. A nearby lot costs $1.25 for each hour. In both lots a partial hour is charged as a full hour. In which lot can a student park the longest for $5? For $11?

117. *Car Rental* A rental car costs $25 per day plus $0.20 per mile. If someone has $200 to spend and needs to drive the car 90 miles each day, for how many days can that person rent the car? Assume that the car cannot be rented for part of a day.

118. *Car Rental* One car rental agency charges $20 per day plus $0.25 per mile. A different agency charges $37 per day with unlimited mileage. For what mileages is the second rental agency a better deal?

119. *Revenue and Cost* (Refer to Example 12.) The cost to produce one compact disc is $1.50 plus a one-time fixed cost of $2000. The revenue received from selling one compact disc is $12.
- **(a)** Write a formula that gives the cost C of producing x compact discs. Be sure to include the fixed cost.
- **(b)** Write a formula that gives the revenue R from selling x compact discs.
- **(c)** Profit equals revenue minus cost. Write a formula that calculates the profit P from selling x compact discs.
- **(d)** What numbers of compact discs need to be sold to yield a positive profit?

120. *Revenue and Cost* The cost to produce one laptop computer is $890 plus a one-time fixed cost of $100,000 for research and development. The revenue received from selling one laptop computer is $1520.
- **(a)** Write a formula that gives the cost C of producing x laptop computers.
- **(b)** Write a formula that gives the revenue R from selling x laptop computers.
- **(c)** Profit equals revenue minus cost. Write a formula that calculates the profit P from selling x laptop computers.
- **(d)** How many computers need to be sold to yield a positive profit?

121. *Distance and Time* Two athletes are jogging in the same direction along an exercise path. After x minutes the first athlete's distance in miles from a parking lot is given by $\frac{1}{6}x$ and the second athlete's distance is given by $\frac{1}{8}x + 2$.
- **(a)** When are the athletes the same distance from the parking lot?
- **(b)** When is the first athlete farther from the parking lot than the second?

122. *Sales of CDs and Cassettes* From 1987 to 2004, the percentage of U.S. music sales in CD format can be modeled by

$$C = 4.64(x - 1987) + 11.5,$$

and the percentage of sales in cassette tape format can be modeled by

$$T = -3.58(x - 1987) + 62.5.$$

In both expressions x represents the year.   (*Source:* Recording Industry Association of America.)
- **(a)** Estimate the years when cassette tape sales were greater than or equal to CD sales.
- **(b)** What happened to sales of CDs and cassette tapes after 1995?

123. *Altitude and Temperature* (Refer to Example 10.) If the temperature on the ground is 90°F, then the air temperature x miles high is given by $T = 90 - 19x$. Determine the altitudes at which the air temperature is less than 4.5°F.   (*Source:* A. Miller.)

124. *Altitude and Dew Point* If the dew point on the ground is 65°F, then the dew point x miles high is given by $D = 65 - 5.8x$. Determine the altitudes at which the dew point is greater than 36°F.   (*Source:* A. Miller.)

125. *Life Expectancy* Because of medical advances and improved health care, people in the U.S. can expect to live longer. From 1980 to 2005, the number of years Y that a 65-year-old man could expect to live beyond age 65 (years remaining), can be approximated by

$$Y = 0.11(x - 1980) + 14.1,$$

where x represents the year. Determine the years when a 65-year-old man could expect to live an additional 16.3 years or more.   (*Source:* Department of Health and Human Services.)

126. *Size and Weight of a Fish*   If the length of a bass is between 20 and 25 inches, its weight W in pounds can be estimated by the formula $W = 0.96x - 14.4$, where x is the length of the fish.   (*Source:* Minnesota Department of Natural Resources.)

(a) What length of bass is likely to weigh 7.2 pounds?

(b) What lengths of bass are likely to weigh less than 7.2 pounds?

WRITING ABOUT MATHEMATICS

127. Explain each of the terms and give an example.
(a) Linear equation
(b) Linear inequality

128. Suppose that a student says that a linear equation and a linear inequality can be solved the same way. How would you respond?

CHECKING BASIC CONCEPTS
SECTION 2.5

1. Use a number line to graph the solution set to the inequality $x + 1 \geq -1$.

2. Express the set of real numbers graphed on the number line by using an inequality.

3. Determine whether -3 is a solution to the inequality $4x - 5 \leq -15$.

4. Complete the table. Then use the table to solve the inequality $5 - 2x \leq 7$.

x	-2	-1	0	1	2
$5 - 2x$					1

5. Solve each inequality.
(a) $x + 5 > 8$
(b) $-\frac{5}{7}x \leq 25$
(c) $3x \geq -2(1 - 2x) + 3$

6. Translate the phrase "a price that is not more than \$12" to an inequality using the variable x.

7. *Geometry*   The length of a rectangle is 5 inches longer than twice its width. If the perimeter of the rectangle is more than 88 inches, find possible widths for the rectangle.

CHAPTER 2   SUMMARY
SECTION 2.1 ■ INTRODUCTION TO EQUATIONS

Equations   Every equation contains an equals sign. An equation can either be true or false.

Important Terms

Solution	A value for a variable that makes the equation true
Solution Set	The set of all solutions
Equivalent Equations	Equations that have the same solution set
Checking a Solution	Substituting the solution in the given equation to verify that the equation is true

Example: 3 is the solution to $4x - 2 = 10$ because $4(3) - 2 = 10$ is a true statement.

Properties of Equality

Addition Property

$a = b$ is equivalent to $a + c = b + c$.

Example: $x - 3 = 0$ and $x - 3 + 3 = 0 + 3$ are equivalent equations.

Multiplication Property

$a = b$ is equivalent to $ac = bc$, provided $c \neq 0$.

Example: $2x = 5$ and $\frac{1}{2} \cdot 2x = \frac{1}{2} \cdot 5$ are equivalent equations.

SECTION 2.2 ■ LINEAR EQUATIONS

Linear Equation   Can be written in the form $ax + b = 0$, where $a \neq 0$

Examples: $3x - 5 = 0$ is linear, whereas $5x^2 + 2x = 0$ is *not* linear.

Solving Linear Equations   The following steps can be used as a guide for solving linear equations.

STEP 1: Use the distributive property to clear any parentheses on each side of the equation. Combine any like terms on each side.

STEP 2: Use the addition property of equality to get all of the terms containing the variable on one side of the equation and all other terms on the other side of the equation. Combine any like terms on each side.

STEP 3: Use the multiplication property of equality to isolate the variable by multiplying each side of the equation by the reciprocal of the number in front of the variable (or divide each side by that number).

STEP 4: Check the solution by substituting it in the given equation.

Distributive Properties $a(b + c) = ab + ac$ or $a(b - c) = ab - ac$

Examples: $5(2x + 3) = 10x + 15$ and $5(2x - 3) = 10x - 15$

Clearing Fractions and Decimals   When fractions or decimals appear in an equation, multiplying each side by the least common denominator can be helpful.

Examples: Multiply each side of $\frac{1}{3}x - \frac{1}{6} = \frac{2}{3}$ by 6 to obtain $2x - 1 = 4$.

Multiply each side of $0.04x + 0.1 = 0.07$ by 100 to obtain $4x + 10 = 7$.

Number of Solutions   Equations that can be written in the form $ax + b = 0$, where a and b are *any* real number, can have no solutions, one solution, or infinitely many solutions.

Examples: $x + 3 = x$ is equivalent to $3 = 0$.   (No solutions)

$2y + 1 = 9$ is equivalent to $y = 4$.   (One solution)

$z + z = 2z$ is equivalent to $0 = 0$.   (Infinitely many solutions)

SECTION 2.3 ■ INTRODUCTION TO PROBLEM SOLVING

Steps for Solving a Problem   The following steps can be used as a guide for solving word problems.

STEP 1: Read the problem carefully and be sure that you understand it. (You may need to read the problem more than once.) Assign a variable to what you are being asked to find. If necessary, write other quantities in terms of this variable.

STEP 2: Write an equation that relates the quantities described in the problem. You may need to sketch a diagram or refer to known formulas.

STEP 3: Solve the equation. Use the solution to determine the solution(s) to the given problem. Include any necessary units.

STEP 4: Look back and check your solution in the given problem. Does it seem reasonable?

Percent Problems

The expression x%

Represents the fraction $\frac{x}{100}$ or the decimal number given by $x \times 0.01$.

Examples: $45\% = \dfrac{45}{100} = 0.45$

$7.1\% = \dfrac{7.1}{100} = \dfrac{71}{1000} = 0.071$

Percent Change

If a quantity changes from an old value to a new value, then the percent change equals $\dfrac{(\text{new value}) - (\text{old value})}{(\text{old value})} \times 100$.

Example: If a price increases from \$2 to \$3, then the percent change equals $\dfrac{3-2}{2} \times 100 = 50\%$.

Distance Problems   If an object travels at speed (rate) r for time t, then the distance d traveled is calculated by $d = rt$.

Example: A car moving at 65 mph for 2 hours travels

$$d = rt = 65 \cdot 2 = 130 \text{ miles.}$$

SECTION 2.4 ■ FORMULAS

Formula   A formula is an equation that can be used to calculate a quantity by using known values of other quantities.

Example: The formula $M = \frac{D}{G}$ can be used to calculate the gas mileage M obtained by a car traveling D miles on G gallons of gasoline.

Formulas from Geometry

Area of a Rectangle

$A = LW$, where L is the length and W is the width.

Area of a Triangle

$A = \frac{1}{2}bh$, where b is the base and h is the height.

Degree Measure

There are 360° in one complete revolution.

Angle Measure

The sum of the angles in a triangle equals 180°.

Volume of a Box	$V = LWH$, where L is the length, W is the width, and H is the height.
Surface Area of a Box	$S = 2LW + 2WH + 2LH$, where L is the length, W is the width, and H is the height.
Circumference	$C = 2\pi r$, where r is the radius.
Area of a Circle	$A = \pi r^2$, where r is the radius.
Volume of a Cylinder	$V = \pi r^2 h$, where r is the radius and h is the height.
Area of a Trapezoid	$A = \frac{1}{2}(a + b)h$, where h is the height and a and b are the bases of the trapezoid.

Other Formulas

Distance from Lightning	$D = \frac{x}{5}$, where x is the number of seconds between seeing the flash and hearing the thunder.
GPA	$GPA = \frac{4a + 3b + 2c + d}{a + b + c + d + f}$, where a represents the number of A credits earned, b the number of B credits earned, and so on.
Temperature Scales	$F = \frac{9}{5}C + 32$ and $C = \frac{5}{9}(F - 32)$, where F is the Fahrenheit temperature and C is the Celsius temperature.

See Putting It All Together in Section 2.4 for examples.

SECTION 2.5 ■ LINEAR INEQUALITIES

Linear Inequality   When the equals sign in a linear equation is replaced with any one of the symbols $<$, $\leq$, $>$, or $\geq$, a linear inequality results.

Examples: $x > 0$, $6 - \frac{2}{3}x \leq 7$,   and $4(x - 1) < 3x - 1$

Number Line Graphs   A number line can be used to graph the solution set to a linear inequality.

Example: The graph of $x \leq 1$ is shown in the figure.

Properties of Inequality

Addition Property $a < b$ is equivalent to $a + c < b + c$.

Example: $x - 3 < 0$ and $x - 3 + 3 < 0 + 3$ are equivalent inequalities.

Multiplication Property   When $c > 0$, $a < b$ is equivalent to $ac < bc$.
When $c < 0$, $a < b$ is equivalent to $ac > bc$.

Examples: $2x < 6$ is equivalent to $2x\left(\frac{1}{2}\right) < 6\left(\frac{1}{2}\right)$ or $x < 3$.
$-2x < 6$ is equivalent to $-2x\left(-\frac{1}{2}\right) > 6\left(-\frac{1}{2}\right)$ or $x > -3$.

CHAPTER 2   REVIEW EXERCISES

SECTION 2.1

Exercises 1–8: Solve the equation. Check your solution.

1. $x + 9 = 3$ **2.** $x - 4 = -2$

3. $x - \frac{3}{4} = \frac{3}{2}$ **4.** $x + 0.5 = 0$

5. $4x = 12$ **6.** $3x = -7$

7. $-0.5x = 1.25$ **8.** $-\frac{1}{3}x = \frac{7}{6}$

SECTION 2.2

Exercises 9–12: Decide whether the equation is linear. If the equation is linear, give values for a and b so that it can be written in the form ax + b = 0.

9. $5x - 3 = 0$ **10.** $-4x + 3 = 2$

11. $\frac{1}{x} + 3 = 0$ **12.** $\frac{3}{8}x^2 - x = \frac{1}{4}$

Exercises 13–22: Solve the equation. Check the solution.

13. $4x - 5 = 3$ **14.** $7 - \frac{1}{2}x = -4$

15. $5(x - 3) = 12$ **16.** $3 + x = 2x - 4$

17. $2(x - 1) = 4(x + 3)$

18. $1 - (x - 3) = 6 + 2x$

19. $3.4x - 4 = 5 - 0.6x$

20. $-\frac{1}{3}(3 - 6x) = -(x + 2) + 1$

21. $\frac{2}{3}x - \frac{1}{6} = \frac{5}{12}$

22. $2y - 3(2 - y) = 5 + y$

Exercises 23–26: Determine whether the equation has no solutions, one solution, or infinitely many solutions.

23. $4(3x - 2) = 2(6x + 5)$

24. $5(3x - 1) = 15x - 5$

25. $8x = 5x + 3x$

26. $9x - 2 = 8x - 2$

Exercises 27 and 28: Complete the table. Then use the table to solve the given equation.

27. $-2x + 3 = 0$

x	0.5	1.0	1.5	2.0	2.5
$-2x + 3$	2				

28. $-(x + 1) + 3 = 2$

x	-2	-1	0	1	2
$-(x + 1) + 3$					0

SECTION 2.3

Exercises 29–32: Using the variable x, translate the sentence into an equation. Solve the resulting equation.

29. The product of a number and 6 is 72.

30. The sum of a number and 18 is -23.

31. Twice a number minus 5 equals the number plus 4.

32. The sum of a number and 4 equals the product of the number and 3.

Exercises 33 and 34: Find the number or numbers.

33. The sum of four consecutive natural numbers is 70.

34. The sum of three consecutive integers is -153.

Exercises 35–38: Convert the percentage to fraction and decimal notation.

35. 85%   **36.** 5.6%

37. 0.03%   **38.** 342%

Exercises 39–42: Convert the number to a percentage.

39. 0.89   **40.** 0.005

41. 2.3   **42.** 1

Exercises 43–46: Use the formula d = rt to find the value of the missing variable.

43. $r = 8$ miles per hour, $t = 3$ hours

44. $r = 70$ feet per second, $t = 55$ seconds

45. $d = 500$ yards, $t = 20$ seconds

46. $d = 125$ miles, $r = 15$ miles per hour

SECTION 2.4

Exercises 47 and 48: Find the area of the region shown.

47.

48.

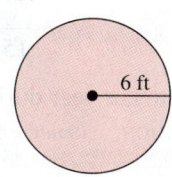

49. Find the area of a rectangle having a 24-inch width and a 3-foot length.

50. Find the area of a triangle having a 13-inch base and a 7-inch height.

51. Find the circumference of a circle having an 18-foot diameter.

52. Find the area of a circle having a 5-inch radius.

53. Find the measure of the third angle in the triangle.

54. The angles in a triangle have measures x, $3x$, and $4x$. Find the value of x.

55. If a cylinder has radius 5 inches and height 25 inches, find its volume. (*Hint:* $V = \pi r^2 h$.)

56. Find the area of a trapezoid with height 5 feet and bases 3 feet and 18 inches. (*Hint:* $A = \frac{1}{2}(a + b)h$.)

Exercises 57 and 58: Find the area of the figure shown.

57.

58.

Exercises 59–64: Solve the formula for the specified variable.

59. $a = x + y$ for x

60. $P = 2x + 2y$ for x

61. $z = 2xy$ for y

62. $S = \dfrac{a + b + c}{3}$ for b

63. $T = \dfrac{a}{3} + \dfrac{b}{4}$ for b

64. $cd = ab + bc$ for c

Exercises 65 and 66: Let a represent the number of credits with a grade of A, b the number of credits with a grade of B, and so on. Calculate the grade point average (GPA). Round your answer to the nearest hundredth.

65. $a = 20$, $b = 25$, $c = 12$, $d = 4$, $f = 4$

66. $a = 64$, $b = 32$, $c = 20$, $d = 10$, $f = 3$

67. Convert 15°C to an equivalent Fahrenheit temperature.

68. Convert 113°F to an equivalent Celsius temperature.

SECTION 2.5

Exercises 69–72: Use a number line to graph the solution set to the inequality.

69. $x < 2$

70. $x > -1$

71. $y \geq -\frac{3}{2}$

72. $y \leq 2.5$

Exercises 73 and 74: Express the set of real numbers graphed on the number line with an inequality.

73.

74.

Exercises 75–78: Determine whether the given value of x is a solution to the inequality.

75. $2x + 1 \leq 5$ $\qquad x = -3$

76. $5 - \frac{1}{2}x > -1$ $\qquad x = 4$

77. $1 - (x + 3) \geq x$ $\qquad x = -2$

78. $4(x + 1) < -(5 - x)$ $\qquad x = -1$

Exercises 79 and 80: Complete the table and then use the table to solve the inequality.

79. $5 - x > 3$

x	0	1	2	3	4
$5 - x$	5				

80. $2x - 5 \le 0$

x	1	1.5	2	2.5	3
$2x - 5$	-3				

Exercises 81–86: Solve the inequality.

81. $x - 3 > 0$

82. $-2x \le 10$

83. $5 - 2x \ge 7$

84. $3(x - 1) < 20$

85. $5x \le 3 - (4x + 2)$

86. $3x - 2(4 - x) \ge x + 1$

Exercises 87–90: Translate the phrase to an inequality. Let x be the variable.

87. A speed that is less than 50 miles per hour

88. A salary that is at most $45,000

89. An age that is at least 16 years old

90. A year before 1995

APPLICATIONS

91. *Rainfall* On a stormy day 2 inches of rain fall before noon and $\frac{3}{4}$ inch per hour fall thereafter until 5 P.M.
  (a) Make a table that shows the total rainfall at each hour starting at noon and ending at 5 P.M.
  (b) Write a formula that calculates the rainfall R in inches, x hours past noon.
  (c) Use your formula to calculate the total rainfall at 5 P.M. Does your answer agree with the value in your table from part (a)?
  (d) How much rain had fallen at 3:45 P.M.?

92. *Cost of a Laptop* A 5% sales tax on a laptop computer amounted to $106.25. Find the cost of the laptop.

93. *Distance Traveled* At noon a bicyclist is 50 miles from home, riding toward home at 10 miles per hour.
  (a) Make a table that shows the bicyclist's distance D from home after 1, 2, 3, 4, and 5 hours.
  (b) Write a formula that calculates the distance D from home after x hours.
  (c) Use your formula to determine D when $x = 3$ hours. Does your answer agree with the value shown in your table?
  (d) For what times was the bicyclist at least 20 miles from home? Assume that $0 \le x \le 5$.

94. *High School Graduates* The number of high school graduates N in millions during year x, $x \ge 1995$, can be approximated by the formula

$$N = \frac{1}{15}x - 130.4.$$

Estimate the year during which this number reached 2.8 million. (*Source:* Department of Education.)

95. *Hazardous Waste* In 2005 the number of federal hazardous waste sites in Wisconsin was 12 less than three times the number of sites in Texas. How many hazardous waste sites were there in Texas if there were no federal hazardous waste sites in Wisconsin? (*Source:* Environmental Protection Agency.)

96. *Master's Degree* In 1971, about 230,500 people received a master's degree, and in 1997, about 419,400 did. Find the percent change in the number of master's degrees received between 1971 and 1997.

97. *Car Speeds* One car passes another car on a freeway. The faster car is traveling 12 miles per hour faster than the slower car. Determine how long it will be before the faster car is 2 miles ahead of the slower car.

98. *Gas Mileage* An SUV that gets 18 miles per gallon is driven 504 miles. How many gallons of gasoline are used on this trip?

99. *Saline Solution* A saline solution contains 3% salt. How much water should be added to 100 milliliters of this solution to dilute it to a 2% solution?

100. *Investment Money* A student invests two sums of money, $500 and $800, at different interest rates, receiving a total of $55 in interest after one year. The $500 investment receives an interest rate 2% lower than the interest rate for the $800 investment. Find the interest rate for each investment.

101. *Dimensions of a Rectangle* The width of a rectangle is 10 inches less than its length. If the perimeter is 112 inches, find the dimensions of the rectangle.

102. *Lightning* The time delay between a flash of lightning and the sound of thunder is 9 seconds. How far away is the lightning?

103. *Geometry* A triangle with height 8 inches is to have an area that is not more than 100 square inches. What lengths are possible for the base of the triangle?

104. *Grade Average* A student scores 75 and 91 on two different tests of 100 points. If the maximum score on the next test is also 100 points, what score does the student need to maintain an average of at least 80?

105. *Parking Rates* Parking in a lot costs $2.25 for the first hour and $1.25 for each hour thereafter. A partial hour is charged the same as a full hour. What is the longest time that someone can park for $9?

106. *Profit* The cost to produce one DVD player is $85 plus a one-time fixed cost of $150,000. The revenue received from selling onc DVD player is $225.

(a) Write a formula that gives the cost C of producing x DVD players.

(b) Write a formula that gives the revenue R from selling x DVD players.

(c) Profit equals revenue minus cost. Write a formula that calculates the profit P from selling x DVD players.

(d) What numbers of DVD players sold will result in a loss? (*Hint:* A loss corresponds to a negative profit.)

CHAPTER 2 TEST · Pass the Test · Video solutions to all test exercises

Exercises 1–4: Solve the equation. Check your solution.

1. $9 = 3 - x$

2. $4x - 3 = 7$

3. $4x - (2 - x) = -3(2x + 6)$

4. $\frac{1}{12}x - \frac{2}{3} = \frac{1}{2}\left(\frac{3}{4} - \frac{1}{3}x\right)$

5. Determine the number of solutions to the equation

$$6(2x - 1) = -4(3 - 3x).$$

6. Complete the table. Then use the table to solve the equation $6 - 2x = 0$.

x	0	1	2	3	4
$6 - 2x$	6				

Exercises 7 and 8: Translate the sentence into an equation, using the variable x. Then solve the resulting equation.

7. The sum of a number and -7 is 6.

8. Twice a number plus 6 equals the number minus 7.

9. The sum of three consecutive natural numbers is 336. Find the three numbers.

10. Convert 3.2% to fraction and decimal notation.

11. Convert 0.345 to a percentage.

12. Find 7.5% of $500.

13. *Speed of Sound* Sound can travel one mile (5280 feet) in 5 seconds. Find the speed of sound in feet per second.

14. *Area* Find the area of the triangle shown.

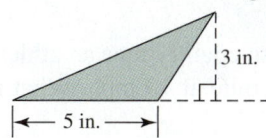

15. Find the circumference and area of a circle with a 30-inch diameter.

16. The measures of the angles in a triangle are x, $2x$, and $3x$. Find the value of x.

Exercises 17 and 18: Solve the formula for x.

17. $z = y - 3xy$

18. $R = \frac{x}{4} + \frac{y}{5}$

Exercises 19 and 20: Solve the inequality.

19. $-3x + 9 \geq x - 15$

20. $3(6 - 5x) < 20 - x$

21. *Snowfall* Suppose 5 inches of snow fall before noon and 2 inches per hour fall thereafter until 10 P.M.

(a) Write a formula that calculates the snowfall S in inches, x hours past noon.

(b) Use your formula to calculate the total snowfall at 8 P.M.

(c) How much snow had fallen at 6:15 P.M.?

22. **Mixing Acid**   A solution is 45% hydrochloric acid. How much water should be added to 1000 milliliters of this solution to dilute it to a 15% solution?

23. **Medical Malpractice**   Malpractice insurance premiums at Thomas Jefferson University Hospital in Philadelphia increased from $8 million in 1998 to $32 million in 2003. Calculate the percent change in malpractice insurance premiums over this 5-year period.   (**Source:** The New York Times.)

CHAPTER 2   EXTENDED AND DISCOVERY EXERCISES

Exercises 1–4: Average Speed   *If someone travels a distance d in time t, then the person's average speed is $\frac{d}{t}$. Use this fact to solve the problem.*

1. A driver travels at 50 mph for the first hour and then travels at 70 mph for the second hour. What is the average speed of the car?

2. A bicyclist rides 1 mile uphill at 5 mph and then rides 1 mile downhill at 10 mph. Find the average speed of the bicyclist. Does your answer agree with what you expected?

3. At a 3-mile cross-country race an athlete runs 2 miles at 8 mph and 1 mile at 10 mph. What is the athlete's average speed?

4. A pilot flies an airplane between two cities and travels half the distance at 200 mph and the other half at 100 mph. Find the average speed of the airplane.

5. **A Puzzle About Coins**   Suppose that seven coins look exactly alike but that one coin weighs less than any of the other six coins. If you have only a balance with two pans, devise a plan to find the lighter coin. What is the minimum number of weighings necessary? Explain your answer.

6. **Global Warming**   If the global climate were to warm significantly as a result of the greenhouse effect or other climatic change, the Arctic ice cap would start to melt. This ice cap contains the equivalent of some 680,000 cubic miles of water. More than 200 million people live on land that is less than 3 feet above sea level. In the United States several large cities have low average elevations. Three examples are Boston (14 feet), New Orleans (4 feet), and San Diego (13 feet). In this exercise you are to estimate the rise in sea level if the Arctic ice cap were to melt and to determine whether this event would have a significant impact on people living in coastal areas.

    (a) The surface area of a sphere is given by the formula $4\pi r^2$, where r is its radius. Although the shape of Earth is not exactly spherical, it has an average radius of 3960 miles. Estimate the surface area of Earth.

    (b) Oceans cover approximately 71% of the total surface area of Earth. How many square miles of Earth's surface are covered by oceans?

    (c) Approximate the potential rise in sea level by dividing the total volume of the water from the ice cap by the surface area of the oceans. Convert your answer from miles to feet.

    (d) Discuss the implications of your calculation. How would cities such as Boston, New Orleans, and San Diego be affected?

    (e) The Antarctic ice cap contains some 6,300,000 cubic miles of water. Estimate how much the sea level would rise if this ice cap melted.   (**Source:** Department of the Interior, Geological Survey.)

Exercises 1 and 2: Classify the number as prime or composite. If the number is composite, write it as a product of prime numbers.

1. 45   **2.** 37

Exercises 3 and 4: Multiply or divide and then simplify to lowest terms when appropriate.

3. $\frac{4}{3} \cdot \frac{3}{8}$ **4.** $\frac{2}{3} \div 6$

Exercises 5 and 6: Add or subtract and then simplify to lowest terms when appropriate.

5. $\frac{11}{12} - \frac{3}{8}$ **6.** $\frac{2}{3} + \frac{1}{5}$

Exercises 7 and 8: Classify the number as one or more of the following: natural number, whole number, integer, rational number, or irrational number.

7. -1 **8.** $\sqrt{3}$

Exercises 9–14: Evaluate the expression by hand.

9. $15 - 4 \cdot 3$ **10.** $30 \div 6 \cdot 2$

11. $23 - 4^2 \div 2$ **12.** $11 - \frac{3+1}{6-4}$

13. $-14 - (-7)$ **14.** $-\frac{2}{3} \cdot \left(-\frac{9}{14}\right)$

Exercises 15 and 16: Simplify the expression.

15. $5x^3 - x^3$ **16.** $4 + 2x - 1 + 3x$

Exercises 17–20: Solve the equation.

17. $x - 3 = 11$ **18.** $4x - 6 = -22$

19. $5(6y + 2) = 25$

20. $11 - (y + 2) = 3y + 5$

Exercises 21 and 22: Determine whether the equation has no solutions, one solution, or infinitely many solutions.

21. $6x + 2 = 2(3x + 1)$

22. $2(3x - 4) = 6(x - 1)$

23. Find three consecutive integers whose sum is 90.

24. Convert 4.7% to decimal notation.

25. Convert 0.17 to a percentage.

26. Find the speed of a car that travels 325 miles in 5 hours.

27. The angles in a triangle have measures $2x$, $3x$, and $4x$. Find the value of x.

28. Find the area of a circle having a 10-inch diameter.

Exercises 29 and 30: Solve the formula for x.

29. $a = 3xy - 4$ **30.** $A = \dfrac{x + y + z}{3}$

Exercises 31 and 32: Solve the inequality and graph the solution set.

31. $7 - 3x > 4$ **32.** $6x \le 5 - (x - 9)$

33. *Yards to Inches* There are 36 inches in 1 yard. Write a formula that converts Y yards to I inches.

34. *Checking Account* The initial balance in a checking account is $468. Find the final balance resulting from the following sequence of withdrawals and deposits: $-\$14$, $\$200$, $-\$73$, $-\$21$, and $\$58$.

35. *Acid Solution* How much of a 4% acid solution should be added to 150 milliliters of a 10% acid solution to dilute it to a 6% acid solution?

36. *Bank Loans* An individual has two low-interest loans, one at 4% interest and the other at 6% interest. The amount borrowed at 6% is $250 more than the amount borrowed at 4%. If the total interest for one year is $165, how much money is borrowed at each rate?

Graphing Equations

Global warming has been in the news for years. Scientists are documenting that both the arctic and antarctic ice packs are melting. This phenomenon is an indication that global temperatures may be rising. The amount of light from the sun that Earth reflects into space is an important factor. If more light is reflected, then Earth stays cooler. Astronomer Phil Goode from the New Jersey Institute of Technology has developed a simple way to determine how reflective Earth is by measuring the amount of earthshine illuminating the moon. The area of the moon illuminated by earthshine is the darker portion of the moon that is visible but not as visible as the portion illuminated by the sun. Goode found that the percent change in earthshine (from the average) varied with the month. An excellent way to display these data is with the *line graph* shown in the accompanying figure. We discuss line graphs and other types of graphs in this chapter.

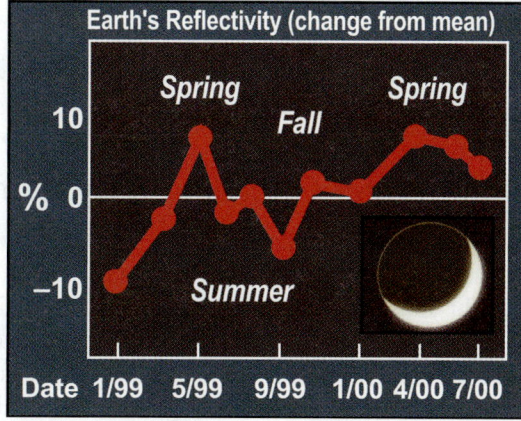

Originality is the essence of true scholarship.

—Nnamdi Azikiwe

Source: Hannah Hoag, *Discover*, November 2002, p. 11. Graphic by Matt Zang. (Reprinted with permission.)

3.1   INTRODUCTION TO GRAPHING

The Rectangular Coordinate System ▪ Scatterplots and Line Graphs

A LOOK INTO MATH ▷

Table 3.1 lists the per capita (per person) personal income for the United States for selected years. These income amounts have *not* been adjusted for inflation. When a table contains only four data values, as in Table 3.1, we can easily see that income increased significantly from 1990 to 2005. If a table listed income for every year from 1990 to 2005, seeing trends in the data would be more difficult. And if a table contained 1000 data values, determining trends in the data would be extremely difficult. In mathematical problems, there are frequently infinitely many data points!

Rather than always using tables to display data, presenting data on a graph is often more useful. For example, the data in Table 3.1 are graphed in Figure 3.1. This line graph is more visual than the table and shows the trend at a glance.

TABLE 3.1

U.S. Per Capita Income

Year	Amount
1990	$18,667
1995	$23,562
2000	$29,469
2005	$34,586

Source: Bureau of Economic Analysis.

Figure 3.1 Per Capita Income

NOTE: In Figure 3.1 the symbol –∕∕– on the *x*-axis indicates a break in the *x*-axis data values. The years before 1990 have been skipped.

The Rectangular Coordinate System

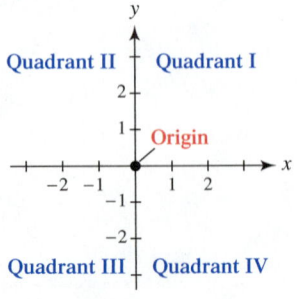

Figure 3.2 The *xy*-plane

One common way to graph data is to use the **rectangular coordinate system**, or *xy*-**plane**. In the *xy*-plane the horizontal axis is the *x*-**axis**, and the vertical axis is the *y*-**axis**. The axes can be thought of as intersecting number lines. The point of intersection is called the **origin** and is associated with zero on each axis. Negative values are located left of the origin on the *x*-axis and below the origin on the *y*-axis. Similarly, positive values are located right of the origin on the *x*-axis and above the origin on the *y*-axis. The axes divide the *xy*-plane into four regions called **quadrants**, which are numbered I, II, III, and IV counterclockwise, as shown in Figure 3.2.

Before we can plot data, we must first understand the concept of an **ordered pair** (x, y). In Table 3.1 we can let *x*-values correspond to the year and *y*-values correspond to the per capita income. Then the fact that the per capita income in **1990** was $**18,667** can be summarized by the ordered pair (**1990**, **18667**). Similarly, the ordered pair (2000, 29469) indicates that the per capita income was $29,469 in 2000.

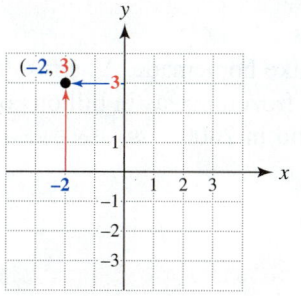

Figure 3.3 Plotting a Point

Order is important in an ordered pair. The ordered pairs given by $(1950, 2025)$ and $(2025, 1950)$ are different. The first ordered pair indicates that the per capita income in 1950 was \$2025, whereas the second ordered pair indicates that the per capita income in 2025 will be \$1950.

To plot the ordered pair $(-2, 3)$ in the *xy*-plane, begin at the origin and move left to locate $x = -2$ on the *x*-axis. Then move upward until a height of $y = 3$ is reached. Thus the point $(-2, 3)$ is located 2 units left of the origin and 3 units above the origin. In Figure 3.3, the point $(\mathbf{-2, 3})$ is plotted in quadrant II.

NOTE: A point that lies on an axis is not located in a quadrant.

EXAMPLE 1 **Plotting points**

Plot the following ordered pairs on the same *xy*-plane. State the quadrant in which each point is located, if possible.

(a) $(3, 2)$ **(b)** $(-2, -3)$ **(c)** $(-3, 0)$

Solution

(a) The point $(3, 2)$ is located in quadrant I, 3 units to the right of the origin and 2 units above the origin. See Figure 3.4.

(b) The point $(-2, -3)$ is located in quadrant III, 2 units to the left of the origin and 3 units below the origin. See Figure 3.4.

(c) The point $(-3, 0)$ is not in any quadrant because it is located 3 units left of the origin on the *x*-axis. See Figure 3.4.

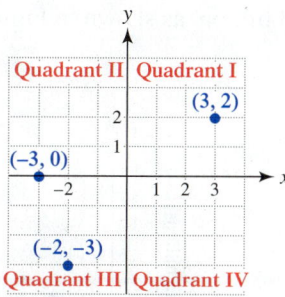

Figure 3.4

Now Try Exercises 11, 15

Choosing appropriate scales for the axes is important when plotting points and making graphs. This can be accomplished by looking at the *coordinates* of the ordered pairs to be plotted. When plotting in the *xy*-plane, the first value in an ordered pair is called the **x-coordinate** and the second value is called the **y-coordinate**. In Example 1, the *x*-coordinates of the three points are 3, -2, and -3 and the *y*-coordinates are 2, -3, and 0. Because there are both positive and negative values in these lists and no coordinate is more than 4 units from the origin, the scale shown in Figure 3.4 is appropriate.

In Figure 3.5 of the next example, the *increment* on the *x*-axis is 2 because each step from one vertical grid line to the next represents a change of 2 years. Similarly, the increment on the *y*-axis is 0.5 because each step from one horizontal grid line to the next represents a \$0.5 billion change in sales. This example demonstrates that the scale and increment on one axis are not always the same as those on the other axis.

EXAMPLE 2   Reading a graph

Frozen pizza makers have improved their pizzas to taste more like homemade. As a result, their sales have increased. Figure 3.5 shows U.S. retail sales of frozen pizzas in billions of dollars. Use the graph to estimate frozen pizza sales in 1994 and in 2000.   (**Source:** Business Trend Analyst.)

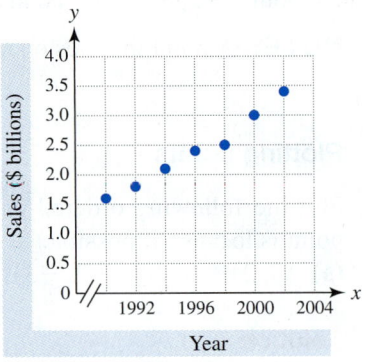

Figure 3.5 Frozen Pizza Sales

Solution

To estimate frozen pizza sales in 1994, first locate 1994 on the *x*-axis. Then move upward to the data point and approximate its *y*-coordinate. Figure 3.6(a) shows that there were about $2.1 billion in frozen pizza sales in 1994. Similarly, frozen pizza sales in 2000 were about $3 billion, as shown in Figure 3.6(b).

(a)

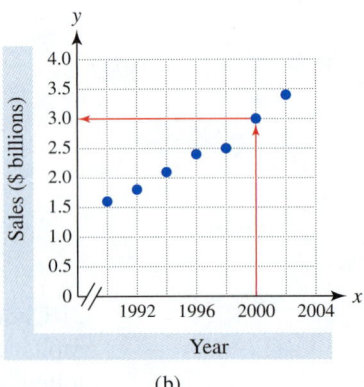

(b)

Figure 3.6

Now Try Exercise **39**

Scatterplots and Line Graphs

If distinct points are plotted in the *xy*-plane, then the resulting graph is called a **scatterplot**. Figure 3.5 is an example of a scatterplot that shows frozen pizza sales. A different scatterplot is shown in Figure 3.7, in which the points $(1, 3)$, $(2, 2)$, $(3, 1)$, $(4, 4)$, and $(5, 1)$ are plotted.

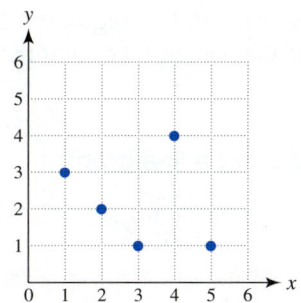

Figure 3.7 A Scatterplot

The next example illustrates how to make a scatterplot by using real data.

EXAMPLE **3**   **Making a scatterplot of gasoline prices**

Table 3.2 lists the average price of a gallon of gasoline for selected years. Make a scatterplot of the data. These prices have *not* been adjusted for inflation.

TABLE 3.2 Average Price of Gasoline

Year	1950	1960	1970	1980	1990	2000
Cost (per gal)	27¢	31¢	36¢	119¢	115¢	156¢

Source: Department of Energy.

Solution

The data point (**1950**, **27**) can be used to indicate that the average cost of a gallon of gasoline in **1950** was **27**¢. Plot the six data points (1950, 27), (1960, 31), (1970, 36), (1980, 119), (1990, 115), and (2000, 156) in the *xy*-plane. The *x*-values vary from 1950 to 2000, so label the *x*-axis from 1950 to 2000 every 10 years. The *y*-values vary from 27 to 156, so label the *y*-axis from 0 to 175 every 25¢. Note that the *x*- and *y*-scales must be large enough to accommodate every data point. Figure 3.8 shows the scatterplot.

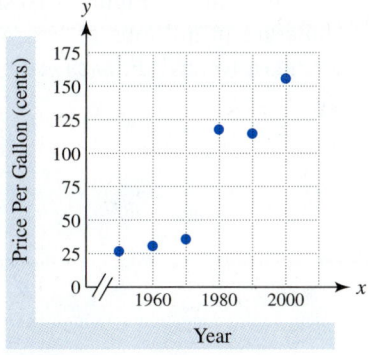

Figure 3.8 Price of Gasoline   Now Try Exercise **23**

Sometimes it is helpful to connect consecutive data points in a scatterplot with line segments. This type of graph visually emphasizes changes in the data and is called a **line graph**. When making a line graph, be sure to plot all of the given data points *before* connecting the points with line segments. The points should be connected consecutively, from left to right, on the scatterplot even if the data are given "out of order" in a table.

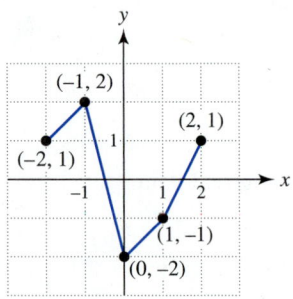

Figure 3.9 A Line Graph

EXAMPLE 4 **Making a line graph**

Use the data in Table 3.3 to make a line graph.

TABLE 3.3

x	-2	-1	0	1	2
y	1	2	-2	-1	1

Solution
The data in Table 3.3 are represented by the five ordered pairs $(-2, 1)$, $(-1, 2)$, $(0, -2)$, $(1, -1)$, and $(2, 1)$. Plot these points and then connect consecutive points with line segments, as shown in Figure 3.9.

Now Try Exercise 33

TECHNOLOGY NOTE: Scatterplots and Line Graphs

Graphing calculators are capable of creating both line graphs and scatterplots. The line graph in Figure 3.9 is shown to the left and the corresponding scatterplot is shown to the right.

CALCULATOR HELP

To make a scatterplot or a line graph, see Appendix A (page AP-4).

EXAMPLE 5 **Analyzing a line graph**

The line graph in Figure 3.10 shows the per capita energy consumption in the United States. Units are in millions of Btu, where 1 Btu equals the amount of heat necessary to raise 1 pound of water 1°F.   (*Source:* Department of Energy.)

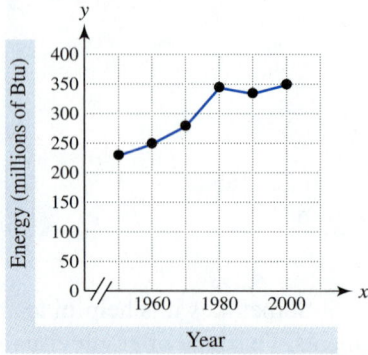

Figure 3.10 U.S. Energy Consumption

(a) Did energy consumption ever decrease during this time period? Explain.
(b) Estimate the energy consumption in 1960 and in 2000.
(c) Estimate the percent change in energy consumption from 1960 to 2000.

Solution

(a) Yes, energy consumption decreased slightly between 1980 and 1990.
(b) From the graph, per capita energy consumption in 1960 was about 250 million Btu. In 2000 it was about 350 million Btu.

NOTE: When different people read a graph, values they obtain may vary slightly.

(c) The percent change from 1960 to 2000 was

$$\frac{350 - 250}{250} \times 100 = 40,$$

so the increase was 40%. (To review the percent change formula, refer to Section 2.3.)

Now Try Exercise 47

CRITICAL THINKING

When analyzing data, do you prefer a table of values, a scatterplot, or a line graph? Explain your answer.

3.1 PUTTING IT ALL TOGETHER

The rectangular coordinate plane, or xy-plane, can be used to plot ordered pairs in the form (x, y). Application data can frequently be represented by ordered pairs. For example, the ordered pair $(12, 45)$ might indicate that the average high temperature in the twelfth month (December) is $45°$F.

Concept	Explanation	Examples
Ordered Pair	Has the form (x, y), where the order of x and y is important	$(1, 2)$, $(-2, 3)$, $(2, 1)$ and $(-4, -2)$ are distinct ordered pairs.
Rectangular Coordinate System, or xy-plane	Consists of a horizontal x-axis and a vertical y-axis Can be used to graph ordered pairs	The points $(1, 2)$, $(0, -2)$, and $(-1, 1)$ are plotted in the graph.

continued on next page

continued from previous page

Concept	Explanation	Examples
Scatterplot	Individual points that are plotted in the xy-plane	
Line Graph	Similar to a scatterplot except that line segments are drawn between consecutive data points	

3.1 Exercises

CONCEPTS

1. Another name for the rectangular coordinate system is the _____.

2. The intersection point of the x-axis and y-axis is the _____.

3. The origin corresponds to the point _____ in the xy-plane.

4. How many quadrants are there in the xy-plane?

5. The point $(-2, 3)$ is located in quadrant _____.

6. If both x and y are negative, then the point (x, y) is located in quadrant _____.

7. A point that lies on one of the _____ is not located in a quadrant.

8. In the xy-plane, the first value in an ordered pair is called the _____ -coordinate and the second value is called the _____ -coordinate.

9. If distinct points are plotted in the xy-plane, the resulting graph is called a _____.

10. If the consecutive points in a scatterplot are connected with line segments, the resulting graph is called a _____ graph.

CARTESIAN COORDINATE PLANE

Exercises 11–14: Identify the coordinates of each point in the graph.

11.

12.

13.

14.

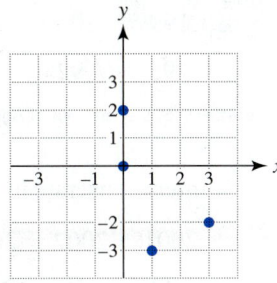

Exercises 15–20: Identify the quadrant, if any, in which each point is located.

15. (a) $(1, 4)$ **(b)** $(-1, -4)$

16. (a) $(-2, 3)$ **(b)** $(2, -3)$

17. (a) $(7, 0)$ **(b)** $(0.1, 7)$

18. (a) $(100, -3)$ **(b)** $(-100, -3)$

19. (a) $\left(-\frac{1}{2}, \frac{3}{4}\right)$ **(b)** $\left(\frac{3}{4}, -\frac{1}{2}\right)$

20. (a) $(1.2, 0)$ **(b)** $(0, -1.2)$

21. Thinking Generally Which of the four quadrants contain points whose *x*- and *y*-coordinates have the same sign?

22. Thinking Generally Which of the four quadrants contain points whose *x*- and *y*-coordinates have different signs?

Exercises 23–32: Make a scatterplot by plotting the given points. Be sure to label each axis.

23. $(0, 0), (1, 2), (-3, 2), (-1, -2)$

24. $(0, -3), (-2, 1), (2, 2), (-4, -4)$

25. $(-1, 0), (4, -3), (0, -1), (3, 4)$

26. $(1, 1), (-2, 2), (-3, -3), (4, -4)$

27. $(2, 4), (-4, 4), (0, -4), (-6, 2)$

28. $(4, 8), (8, 4), (-8, -4), (-4, 0)$

29. $(5, 0), (5, -5), (-10, -20), (10, -10)$

30. $(10, 30), (-20, 10), (40, 0), (-30, -10)$

31. $(0, 0.1), (0.2, -0.3), (-0.1, 0.4)$

32. $(1.5, 2.5), (-1, -1.5), (-2.5, 0)$

Exercises 33–38: Use the table to make a line graph.

33.

x	-2	-1	0	1	2
y	2	1	0	-1	-2

34.

x	-4	-2	0	2	4
y	4	-2	3	-1	2

35.

x	-10	-5	0	5	10
y	20	-10	10	0	-20

36.

x	1	2	3	4	5
y	2	3	1	5	4

37.

x	-5	5	-10	10	0
y	10	20	30	20	40

38.

x	3	-2	2	1	-3
y	4	3	3	-2	-3

Exercises 39 and 40: Identify the coordinates of each point in the graph. Then explain what the coordinates of the first point indicate.

39. Cigarette consumption in the United States (**Source:** U.S. Department of Health and Human Services.)

40. Number of unhealthy days per year in Pittsburgh from 2000 to 2004, based on air quality (**Source:** Environmental Protection Agency.)

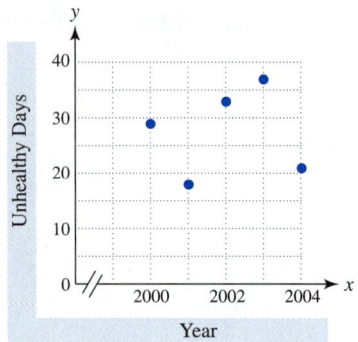

GRAPHING REAL DATA

Exercises 41–46: **Graphing Real Data**   *The table contains real data.*

   (a) *Make a line graph of the data. Be sure to label the axes.*

   (b) *Comment on any trends in the data.*

41. Federal income tax receipts I in billions during year t

t	1970	1980	1990	2000
I	90	244	467	1003

Source: Office of Management and Budget.

42. Percent P of total music sales that were digital downloads during year t

t	2001	2002	2003	2004	2005
P	0.2%	0.5%	1.3%	2.9%	5.7%

Source: Recording Industry Association of America.

43. Welfare beneficiaries B in millions during year t

t	1970	1980	1990	2000
B	7	11	12	6

Source: Administration for Children and Families.

44. U.S. cotton production C in millions of bales during year t

t	2002	2003	2004	2005
C	17	18	23	24

Source: U.S. Department of Agriculture.

45. Number of farms in Iowa F in thousands during year x

x	1999	2001	2003	2005
F	95	92	90	89

Source: U.S. Department of Agriculture.

46. U.S. Internet users y in millions during year x

x	2001	2002	2003	2004	2005
y	143	158	162	189	201

Source: Department of Commerce.

47. The line graph shows the U.S. infant mortality rate for selected years.

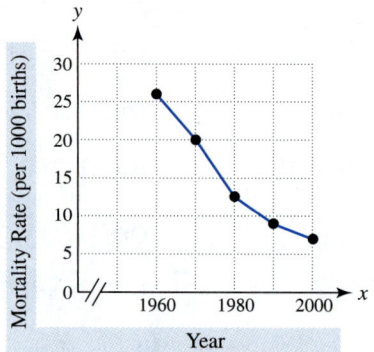

   (a) Comment on any trends in the data.

   (b) Estimate the infant mortality rate in 1990.

   (c) Estimate the percent change in the infant mortality rate from 1960 to 2000.

48. The line graph shows the population in millions of the midwestern states for selected years.

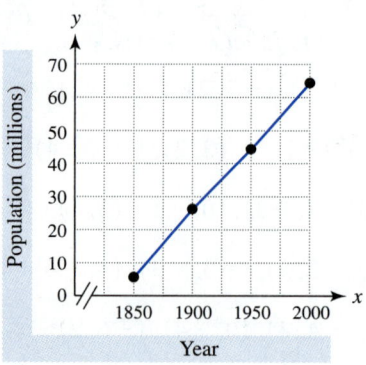

   (a) Comment on any trends in the data.

   (b) Estimate this population in 1900.

   (c) Estimate the percent change in population from 1850 to 2000.

49. Explain how to identify the quadrant that a point lies in if it has coordinates (x, y).

50. Explain the difference between a scatterplot and a line graph. Give an example of each.

3.2 LINEAR EQUATIONS IN TWO VARIABLES

Basic Concepts ▪ Tables of Solutions ▪ Graphing Linear Equations in Two Variables

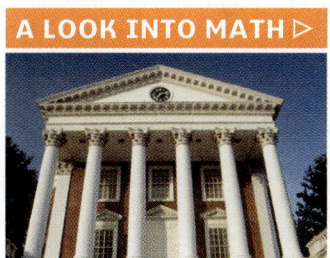

A LOOK INTO MATH ▷ Figure 3.11 shows a scatterplot of average college tuition and fees at public colleges and universities, together with a line that models the data. If we could find an equation for this line, then we could use it to estimate tuition and fees for years without data points. In this section we discuss linear equations, whose graphs are lines. Linear equations and lines are often used to approximate data.

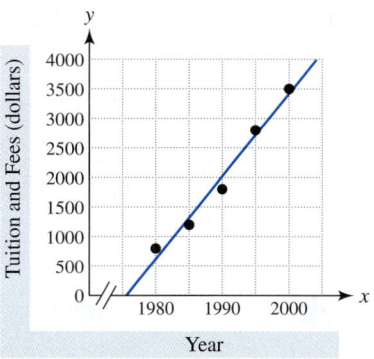

Figure 3.11   Public Institutions

Basic Concepts

Equations can have any number of variables. In Chapter 2 we solved equations having one variable. The following are examples of equations with two variables.

$$y = 2x, \quad 3x + 2y = 4, \quad z = t^2, \quad \text{and} \quad a - b = 1$$

A solution to an equation with one variable is one number that makes the statement true. For example, the solution to $x - 1 = 3$ is **4** because $4 - 1 = 3$ is a true statement. A solution to an equation with two variables consists of two numbers, one for each variable, which can be expressed as an ordered pair. For example, one solution to the equation $y = 5x$ is given by $x = 1$ and $y = 5$ because $5 = 5(1)$ is a true statement. This solution can be expressed as the ordered pair $(1, 5)$. Another solution to $y = 5x$ is $(2, 10)$, because $10 = 5(2)$ is also a true statement. In fact the equation $y = 5x$ has infinitely many solutions because, for every value of x that we choose, there is a corresponding value of y that makes the equation true.

EXAMPLE 1 **Testing solutions to equations**

Determine whether the given ordered pair is a solution to the given equation.

(a) $y = x + 3$, $(1, 4)$ **(b)** $2x - y = 5$, $\left(\frac{1}{2}, -4\right)$ **(c)** $-4x + 5y = 20$, $(-5, 1)$

Solution

(a) Let $x = 1$ and $y = 4$ in the given equation.

$$y = x + 3 \qquad \text{Given equation}$$
$$4 \overset{?}{=} 1 + 3 \qquad \text{Substitute.}$$
$$4 = 4 \qquad \text{A true statement}$$

The ordered pair $(1, 4)$ is a solution.

(b) Let $x = \frac{1}{2}$ and $y = -4$ in the given equation.

$$2x - y = 5 \qquad \text{Given equation}$$
$$2\left(\frac{1}{2}\right) - (-4) \overset{?}{=} 5 \qquad \text{Substitute.}$$
$$1 + 4 \overset{?}{=} 5 \qquad \text{Simplify the left side.}$$
$$5 = 5 \qquad \text{A true statement}$$

The ordered pair $\left(\frac{1}{2}, -4\right)$ is a solution.

(c) Let $x = -5$ and $y = 1$ in the given equation.

$$-4x + 5y = 20 \qquad \text{Given equation}$$
$$-4(-5) + 5(1) \overset{?}{=} 20 \qquad \text{Substitute.}$$
$$20 + 5 \overset{?}{=} 20 \qquad \text{Simplify the left side.}$$
$$25 \overset{?}{=} 20 \qquad \text{A } \textit{false} \text{ statement}$$

The ordered pair $(-5, 1)$ is *not* a solution. **Now Try Exercises** 11, 15

Tables of Solutions

A table can be used to list solutions to an equation. For example, Table 3.4 lists solutions to $x + y = 5$, where the sum of each xy-pair equals 5.

TABLE 3.4 $x + y = 5$

x	-2	-1	0	1	2
y	7	6	5	4	3

Most equations in two variables have infinitely many solutions, so it is impossible to list all solutions in a table. However, when you are graphing an equation, having a table that lists a few solutions to the equation is often helpful. The next two examples demonstrate how to complete a table for a given equation.

EXAMPLE 2 **Completing a table of solutions**

Complete the table for the equation $y = 2x - 3$.

x	-4	-2	0	2
y				

Solution

Start by determining the corresponding y-value for each x-value in the table. For example, when $x = -2$, the equation $y = 2x - 3$ implies that $y = 2(-2) - 3 = -4 - 3 = -7$. Filling in the y-values results in Table 3.5.

TABLE 3.5 $y = 2x - 3$

x	-4	-2	0	2
y	-11	-7	-3	1

Now Try Exercise **19**

EXAMPLE 3   Making a table of solutions

Use $y = 0, 5, 10,$ and 15 to make a table of solutions to $5x + 2y = 10$.

Solution

Begin by listing the required y-values in the table. Next determine the corresponding x-values for each y-value by using the equation $5x + 2y = 10$.

TABLE 3.6 $5x + 2y = 10$

x	2	0	-2	-4
y	0	5	10	15

When $y = 0$,
$$5x + 2(0) = 10$$
$$5x + 0 = 10$$
$$5x = 10$$
$$x = 2$$

When $y = 5$,
$$5x + 2(5) = 10$$
$$5x + 10 = 10$$
$$5x = 0$$
$$x = 0$$

When $y = 10$,
$$5x + 2(10) = 10$$
$$5x + 20 = 10$$
$$5x = -10$$
$$x = -2$$

When $y = 15$,
$$5x + 2(15) = 10$$
$$5x + 30 = 10$$
$$5x = -20$$
$$x = -4$$

Filling in the x-values results in Table 3.6.

Now Try Exercise **23**

▶ **REAL-WORLD CONNECTION**   Formulas can sometimes be difficult for people to understand. As a result, newspapers, magazines, and books often list numbers in a table rather than presenting a formula for the reader to use. The next example illustrates a situation in which a table might be preferable to a formula.

EXAMPLE 4   Calculating appropriate lengths of crutches

People who sustain leg injuries often require crutches. An appropriate crutch length L in inches for an injured person who is t inches tall is estimated by $L = 0.72t + 2$.
(*Source: Journal of the American Physical Therapy Association*.)

(a) Complete the table. Round values to the nearest inch.

t	60	65	70	75	80
L					

(b) Use the table to determine the appropriate crutch length for a person 5 feet 10 inches tall.

Solution

(a) For the formula $L = 0.72t + 2$, if $t = \mathbf{60}$, then $L = 0.72(\mathbf{60}) + 2 = 43.2 + 2 = 45.2$, or about 45 inches. If $t = \mathbf{65}$, then $L = 0.72(\mathbf{65}) + 2 = 46.8 + 2 = 48.8 \approx 49$. Other values in Table 3.7 are found similarly. (You may want to use a calculator to help complete the table.)

TABLE 3.7   Crutch Lengths

t	60	65	70	75	80
L	45	49	52	56	60

(b) A person who is 5 feet 10 inches tall is $5 \cdot 12 + 10 = 70$ inches tall. Table 3.7 reveals that a person 70 inches tall needs crutches that are about 52 inches long.

Now Try Exercise 69

Graphing Linear Equations in Two Variables

In the preceding section we showed that a line graph visually displays data. Many times graphs are used in mathematics to make concepts easier to understand.

EXAMPLE 5   Graphing an equation with two variables

Make a table of values for the equation $y = 2x$, and then use the table to graph this equation.

Solution

Start by selecting a few convenient values for x, such as $x = -1, 0, 1$, and 2. Then complete the table by doubling each x-value to obtain the corresponding y-value.

NOTE: A table of values can be either horizontal or vertical. Table 3.8 is presented in a vertical format.

TABLE 3.8 $y = 2x$

x	y
-1	-2
0	0
1	2
2	4

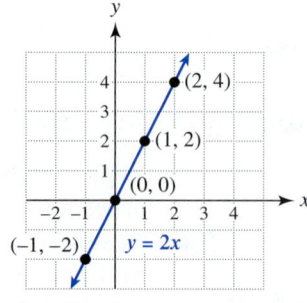

Figure 3.12

To graph the equation $y = 2x$, start by plotting the points $(-1, -2)$, $(0, 0)$, $(1, 2)$, and $(2, 4)$ given in Table 3.8, as shown in Figure 3.12. Note that all four points appear to lie on the same (straight) line. Because there are infinitely many points that satisfy the equation, we can draw a line through these points.

Now Try Exercise 35

An equation whose graph is a single, straight line is called a *linear equation in two variables*. Linear equations in two variables can be written in the following **standard form**.

LINEAR EQUATION IN TWO VARIABLES

A **linear equation in two variables** can be written as

$$Ax + By = C,$$

where A, B, and C are fixed numbers (constants) and A and B are not both equal to 0. The graph of a linear equation in two variables is a line.

NOTE: In mathematics a line is always *straight*.

In Example 5, the equation $y = 2x$ is a linear equation because it can be written in standard form by adding $-2x$ to each side.

$y = 2x$	Given equation
$-2x + y = -2x + 2x$	Add $-2x$ to each side.
$-2x + y = 0$	Simplify.

The equation $-2x + y = 0$ is a linear equation in two variables because it is in the form $Ax + By = C$ with $A = -2$, $B = 1$, and $C = 0$. Hence, the graph of $y = 2x$ is a line.

EXAMPLE 6   Graphing linear equations

Graph each linear equation.
(a) $y = \frac{1}{2}x - 1$ **(b)** $x + y = 4$

Solution

(a) Because $y = \frac{1}{2}x - 1$ can be written in standard form as $-\frac{1}{2}x + y = -1$, it is a linear equation in two variables and its graph is a line. Two points determine a line. However, it is a good idea to plot three points to be sure that the line is graphed correctly. Start by choosing three values for x and then calculate the corresponding y-values, as shown in Table 3.9. In Figure 3.13, the points $(-2, -2)$, $(0, -1)$, and $(2, 0)$ are plotted and the line passing through these points is drawn.

TABLE 3.9 $y = \frac{1}{2}x - 1$

x	y
-2	-2
0	-1
2	0

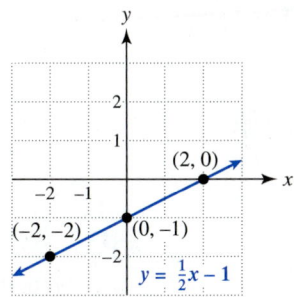

Figure 3.13

(b) The equation $x + y = 4$ is a linear equation in two variables, so its graph is a line. If an ordered pair (x, y) is a solution to the given equation, then the sum of x and y is 4. Table 3.10 shows three examples. In Figure 3.14, the points $(0, 4)$, $(2, 2)$, and $(4, 0)$ are plotted with the line passing through each one.

TABLE 3.10 $x + y = 4$

x	y
0	4
2	2
4	0

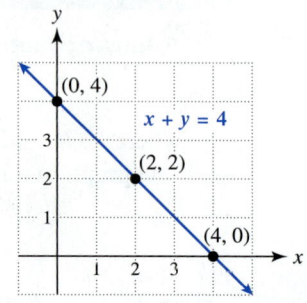

Figure 3.14

Now Try Exercises 39, 49

TECHNOLOGY NOTE: Graphing Equations

CALCULATOR HELP

To graph an equation, see Appendix A (page AP-5).

Graphing calculators can be used to graph equations. Before graphing the equation $x + y = 4$, solve the equation for y to obtain $y = 4 - x$. A calculator graph of Figure 3.14 is shown, except that the three points have not been plotted.

When a linear equation in two variables is given in standard form, it is sometimes difficult to create a table of solutions. Solving such equations for y often makes it easier to select x-values for the table that will make the y-values simpler to calculate. This is demonstrated in the next example.

EXAMPLE 7   Solving for y and then graphing

Graph each linear equation by solving for y first.
(a) $4x - 3y = 12$ **(b)** $-2x + 4y = 8$

Solution
(a) First solve the given equation for y.

$$4x - 3y = 12 \qquad \text{Given equation}$$

$$-3y = -4x + 12 \qquad \text{Subtract } 4x \text{ from each side.}$$

$$\frac{-3y}{-3} = \frac{-4x + 12}{-3} \qquad \text{Divide each side by } -3.$$

$$\frac{-3y}{-3} = \frac{-4x}{-3} + \frac{12}{-3} \qquad \text{Property of fractions, } \frac{a + b}{c} = \frac{a}{c} + \frac{b}{c}$$

$$y = \frac{4}{3}x - 4 \qquad \text{Simplify fractions.}$$

Note that dividing each side of an equation by -3 is equivalent to dividing each term in the equation by -3.

Select multiples of 3 (the denominator of $\frac{4}{3}$) as x-values for the table of solutions. For example, if $x = \mathbf{6}$ is chosen, $y = \frac{4}{3}(\mathbf{6}) - 4 = \frac{24}{3} - 4 = 8 - 4 = 4$. Table 3.11 lists the solutions $(0, -4)$, $(3, 0)$, and $(6, 4)$, which are plotted in Figure 3.15 with the line passing through each one.

TABLE 3.11 $y = \frac{4}{3}x - 4$

x	y
0	-4
3	0
6	4

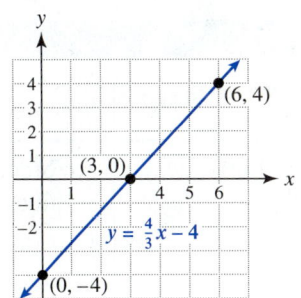

Figure 3.15

(b) First solve the given equation for y.

$$-2x + 4y = 8 \qquad \text{Given equation}$$
$$4y = 2x + 8 \qquad \text{Add } 2x \text{ to each side.}$$
$$\frac{4y}{4} = \frac{2x}{4} + \frac{8}{4} \qquad \text{Divide each side by 4.}$$
$$y = \frac{1}{2}x + 2 \qquad \text{Simplify fractions.}$$

Table 3.12 lists the solutions $(-2, 1)$, $(0, 2)$, and $(2, 3)$, which are plotted in Figure 3.16 with the line passing through each one. Note that the selected x-values are multiples of 2.

TABLE 3.12 $y = \frac{1}{2}x + 2$

x	y
-2	1
0	2
2	3

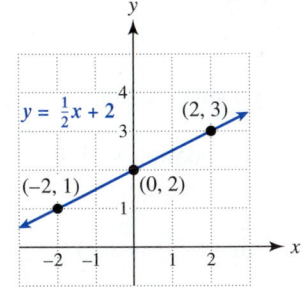

Figure 3.16   **Now Try Exercise** `57`

MAKING CONNECTIONS

Graphs and Solution Sets

A graph visually depicts the set of solutions to an equation. Each point on the graph represents one solution to the equation.

NOTE: In a linear equation, each x-value determines a unique y-value. Because there are infinitely many x-values, there are infinitely many points located on the graph of a linear equation. Thus the graph of a linear equation is a *continuous* line with no breaks.

3.2   PUTTING IT ALL TOGETHER

The following table summarizes several important topics covered in this section.

Concept	Explanation	Examples
Equation in Two Variables	An equation that has two variables	$y = 4x + 5$, $4x - 5y = 20$, and $u - v = 100$
Solution to an Equation in Two Variables	An ordered pair (x, y) whose x- and y-values satisfy the equation Equations in two variables often have infinitely many solutions.	$(1, 3)$ is a solution to $3x + y = 6$ because $3(1) + 3 = 6$ is a true statement. The equation $y = 2x$ has infinitely many solutions, such as $(1, 2)$, $(2, 4)$, $(3, 6)$, and so on.
Linear Equation in Two Variables	Can be written as $$Ax + By = C,$$ where A, B, and C are fixed numbers and A and B are not both equal to 0	$3x + 4y = 5$ $y = 4 - 3x$ (or $3x + y = 4$) $x = 2y + 1$ (or $x - 2y = 1$) The graph of each equation is a line.

3.2   Exercises

MyMathLab   Math XL PRACTICE   WATCH   DOWNLOAD   READ   REVIEW

CONCEPTS

1. $4x + 6y = 24$ is an equation in _____ variables.

2. A solution to an equation with two variables consists of two numbers expressed as a(n) _____.

3. The ordered pair $(1, 3)$ is a(n) _____ to the equation $y = 3x$.

4. A(n) _____ equation in two variables can be written in the form $Ax + By = C$.

5. The equation $4x - 3y = 10$ (is/is not) a linear equation in two variables.

6. A linear equation in two variables has infinitely many _____.

7. An equation's _____ visually depicts its solution set.

8. The graph of a linear equation in two variables is a(n) _____.

SOLUTIONS TO EQUATIONS

Exercises 9–18: Determine whether the ordered pair is a solution to the given equation.

9. $y = x + 1$, $(5, 6)$

10. $y = 4 - x$, $(6, 2)$

11. $y = 4x + 7$, $(2, 13)$

12. $y = -3x + 2$, $(-2, 8)$

13. $4x - y = -13$, $(-2, 3)$

14. $3y + 2x = 0$, $(-2, 3)$

15. $y - 6x = -1$, $\left(\frac{1}{2}, 2\right)$

16. $\frac{1}{2}x + \frac{3}{2}y = 0$, $\left(-\frac{3}{2}, \frac{1}{2}\right)$

17. $0.31x - 0.42y = -9$, $(100, 100)$

18. $0.5x - 0.6y = 4$, $(20, 10)$

Exercises 19–24: Complete the table for the given equation.

19. $y = 4x$

x	-2	-1	0	1	2
y	-8				

20. $y = \frac{1}{2}x - 1$

x	0	1	2	3	4
y	−1				

21. $y = x + 4$

x	−8				
y	−4	0	4	8	12

22. $2x - y = 1$

x	−1				
y	−3	−1	0	1	3

23. $3y + 2x = 6$

x					
y	−2	0	2	4	8

24. $3x - 5y = 30$

x	−5	0	5	10	15
y					

Exercises 25–32: Use the given values of the variable to make a table of solutions for the equation.

25. $y = 3x$ $x = -3, 0, 3, 6$

26. $y = 1 - 2x$ $x = 0, 1, 2, 3$

27. $y = \dfrac{x + 4}{2}$ $x = -8, -4, 0, 4$

28. $y = \dfrac{x}{3} - 1$ $x = 0, 2, 4, 6$

29. $x + y = 6$ $y = -2, 0, 2, 4$

30. $2x - 3y = 9$ $y = -3, 0, 1, 2$

31. $y - 4x = 0$ $y = -2, -1, 0, 1$

32. $-4x = 6y - 4$ $y = -1, 0, 1, 2$

33. **Thinking Generally** If a student wishes to avoid fractional y-values when making a table of solutions for $y = \frac{3}{5}x - 7$, what must be true about any selected integer x-values?

34. **Thinking Generally** If a student wishes to avoid fractional y-values when making a table of solutions for $y = \frac{a}{b}x$, where a and b are natural numbers, what must be true about any selected integer x-values?

Exercises 35–40: Make a table of solutions for the equation, and then use the table to graph the equation.

35. $y = -2x$ **36.** $y = 2x - 1$

37. $x = 3 - y$ **38.** $x = y + 1$

39. $x + 2y = 4$ **40.** $2x - y = 1$

GRAPHING EQUATIONS

Exercises 41–56: Graph the equation.

41. $y = x$ **42.** $y = \frac{1}{2}x$

43. $y = \frac{1}{3}x$ **44.** $y = -2x$

45. $y = x + 3$ **46.** $y = x - 2$

47. $y = x - 4$ **48.** $y = x + 2$

49. $y = 2x + 1$ **50.** $y = \frac{1}{2}x - 1$

51. $y = 4 - 2x$ **52.** $y = 2 - 3x$

53. $y = 7 + x$ **54.** $y = 2 + 2x$

55. $y = -\frac{1}{2}x + \frac{1}{2}$ **56.** $y = -\frac{3}{4}x + 2$

Exercises 57–68: Graph the linear equation by solving for y first.

57. $2x + 3y = 6$ **58.** $3x + 2y = 6$

59. $x + 4y = 4$ **60.** $4x + y = -4$

61. $-x + 2y = 8$ **62.** $-2x + 6y = 12$

63. $y - 2x = 7$ **64.** $3y - x = 2$

65. $5x - 4y = 20$ **66.** $4x - 5y = -20$

67. $3x + 5y = -9$ **68.** $5x - 3y = 10$

APPLICATIONS

69. *U.S. Population* For the years 2010 to 2050, the projected percentage P of the U.S. population that will be over the age of 65 during year t is estimated by $P = 0.178t - 344.6$. (*Source:* U.S. Census Bureau.)

(a) Complete the table. Round values to the nearest tenth.

t	2010	2020	2030	2040	2050
P					

(b) Use the table to find the year when the percentage of the population over the age of 65 is expected to reach 16.7%.

70. *U.S. Population*   For the years 2010 to 2050, the projected percentage P of the U.S. population that will be 18 to 24 years old during year t is estimated by $P = -0.025t + 60.35$.   (*Source:* U.S. Census Bureau.)
(a) Complete the table. Round values to the nearest tenth.

t	2010	2020	2030	2040	2050
P					

(b) Use the table to find the year when the percentage of the population that is 18 to 24 years old is expected to be 9.4%.

71. *Solid Waste in the Past*   In 1960 the amount A of garbage in pounds produced after t days by the average American is given by $A = 2.7t$.   (*Source:* Environmental Protection Agency.)
(a) Graph the equation for $t \geq 0$.
(b) How many days did it take for the average American in 1960 to produce 100 pounds of garbage?

72. *Solid Waste Today*   Today the amount A of garbage in pounds produced after t days by the average American is given by $A = 4.5t$.   (*Source:* Environmental Protection Agency.)
(a) Graph the equation for $t \geq 0$.
(b) How many days does it take for the average American to produce 100 pounds of garbage today?

73. *Compact Disc Sales*   From 1985 to 2000 the percent P of total music sales with a compact disc format is modeled by $P = 5.9t - 11,709$, where t is the year. (*Source:* Recording Industry Association of America.)

(a) Evaluate P for $t = 1985$ and for $t = 2000$.
(b) Use your results from part (a) to graph the equation from 1985 to 2000.
(c) In what year was $P = 79.2\%$?

74. *HIV Infections in the United States*   The cumulative number of HIV infections I in thousands during year t is modeled by the equation $I = 42t - 83,197$ where $t \geq 2000$.   (*Source:* Department of Health and Human Services.)
(a) Evaluate I for $t = 2000$ and for $t = 2005$.
(b) Use your results from part (a) to graph the equation from 2000 to 2005.
(c) In what year was $I = 971$?

WRITING ABOUT MATHEMATICS

75. The number of welfare beneficiaries B in millions during year t is shown in the table. Discuss whether a linear equation might work to approximate these data from 1970 to 2000.

t	1970	1980	1990	2000
B	7	11	12	6

Source: Administration for Children and Families.

76. The Asian-American population P in millions during year t is shown in the table. Discuss whether a linear equation might model these data from 2000 to 2004.

t	2000	2002	2004
P	11.2	12.0	12.8

Source: U.S. Census Bureau.

CHECKING BASIC CONCEPTS
SECTIONS 3.1 AND 3.2

1. Identify the coordinates of the four points in the graph. State the quadrant, if any, in which each point lies.

2. Make a scatterplot of the five points $(-2, -2)$, $(-1, -3)$, $(0, 0)$, $(1, 2)$, and $(2, 3)$.

3. *U.S. Population*   The table gives the percentage P of the U.S. population that was over the age of 85 during year t. Make a line graph of the data in the table, and then comment on any trends.   (*Source:* U.S. Census Bureau.)

t	1970	1980	1990	2000
P	0.7	1.0	1.2	1.5

4. Determine whether $(-2, -3)$ is a solution to the equation $-2x - y = 7$.

5. Complete the table for the equation $y = -2x + 1$.

x	-2	-1	0	1	2
y	5				

6. Graph the equation.
(a) $y = \frac{1}{2}x$ (b) $4x + 6y = 12$

7. *Total Federal Receipts*   The total amount of money A in \$ trillions collected by the federal government in year t from 1995 to 2001 can be approximated by $A = 0.115t - 228$. (*Source: Internal Revenue Service.*)
(a) Find A when $t = 1995$ and $t = 2001$. Interpret each result.
(b) Use your results from part (a) to graph the equation from 1995 to 2001. Be sure to label the axes.
(c) In what year were the total receipts equal to \$1.54 trillion?

3.3 MORE GRAPHING OF LINES

Finding Intercepts ▪ Horizontal Lines ▪ Vertical Lines

The graph of a linear equation is a line, which can be used to model data. Two points, such as those associated with the *x*- and *y*-intercepts, can be used to sketch such a line. We begin this section by discussing intercepts and their significance in applications and then consider horizontal and vertical lines.

Finding Intercepts

▶ **REAL-WORLD CONNECTION** Suppose that someone leaves a rest stop on an Interstate highway and drives home at a constant speed of 50 miles per hour. The graph in Figure 3.17 reflects the distance of the driver from home at various times. The graph intersects the *y*-axis at 200 miles, which is called the *y-intercept*. In this situation the *y*-intercept represents the initial distance (when $x = 0$) between the driver and home. The graph also intersects the *x*-axis at 4 hours, which is called the *x-intercept*. This intercept represents the elapsed time when the distance of the driver from home is 0 miles.

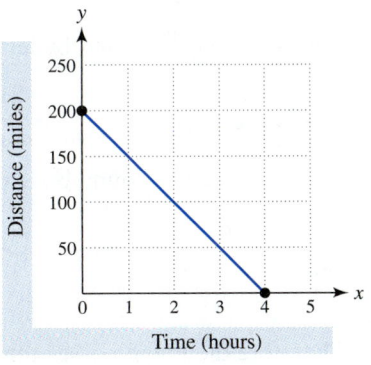

Figure 3.17

FINDING x- AND y-INTERCEPTS

The x-coordinate of a point where a graph intersects the x-axis is an **x-intercept**. To find an x-intercept, let $y = 0$ in the equation and solve for x.

The y-coordinate of a point where a graph intersects the y-axis is a **y-intercept**. To find a y-intercept, let $x = 0$ in the equation and solve for y.

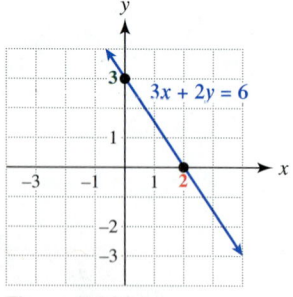

Figure 3.18

The graph of the linear equation $3x + 2y = 6$ is shown in Figure 3.18. The x-intercept is 2 and the y-intercept is 3. These two intercepts can be found without a graph. To find the x-intercept let $y = 0$ in the equation and solve for x.

$$3x + 2(0) = 6 \qquad \text{Let } y = 0.$$
$$3x = 6 \qquad \text{Simplify.}$$
$$x = 2 \qquad \text{Divide each side by 3.}$$

To find the y-intercept let $x = 0$ in the equation and solve for y.

$$3(0) + 2y = 6 \qquad \text{Let } x = 0.$$
$$2y = 6 \qquad \text{Simplify.}$$
$$y = 3 \qquad \text{Divide each side by 2.}$$

Note that the x-intercept **2** corresponds to the point $(2, 0)$ on the graph and the y-intercept **3** corresponds to the point $(0, 3)$ on the graph.

CRITICAL THINKING

If a line has no x-intercept, what can you say about the line?

If a line has no y-intercept, what can you say about the line?

EXAMPLE 1   Using intercepts to graph a line

Use intercepts to graph $2x - 6y = 12$.

Solution

The x-intercept is found by letting $y = 0$.

$$2x - 6(0) = 12 \qquad \text{Let } y = 0.$$
$$x = 6 \qquad \text{Solve for } x.$$

The y-intercept is found by letting $x = 0$.

$$2(0) - 6y = 12 \qquad \text{Let } x = 0.$$
$$y = -2 \qquad \text{Solve for } y.$$

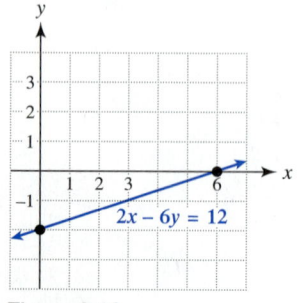

Figure 3.19

Therefore the graph passes through the points $(6, 0)$ and $(0, -2)$, as shown in Figure 3.19.

Now Try Exercise 25

In the next example a table of solutions is used to determine the x- and y-intercepts.

EXAMPLE 2   Using a table to find intercepts

Complete the table. Then determine the x-intercept and y-intercept for the graph of the equation $x - y = -1$.

x	-2	-1	0	1	2
y					

Solution

Substitute -2 for x in $x - y = -1$ to find the corresponding y-value.

$$-2 - y = -1 \qquad \text{Let } x = -2.$$
$$-y = 1 \qquad \text{Add 2 to each side.}$$
$$y = -1 \qquad \text{Multiply each side by } -1.$$

The other y-values can be found similarly. See Table 3.13.

The x-intercept corresponds to a point on the graph whose y-coordinate is 0. Table 3.13 reveals that the y-coordinate is 0 when $x = -1$. Thus the x-intercept is -1. Similarly, the y-intercept corresponds to a point on the graph whose x-coordinate is 0. Table 3.13 reveals that the x-coordinate is 0 when $y = 1$. Thus the y-intercept is 1.

TABLE 3.13 $x - y = -1$

x	-2	-1	0	1	2
y	-1	0	1	2	3

Now Try Exercise **21**

EXAMPLE 3   Modeling the velocity of a toy rocket

A toy rocket is shot vertically into the air. Its velocity v in feet per second after t seconds is given by $v = 160 - 32t$. Assume that $t \geq 0$ and $t \leq 5$.

(a) Graph the equation by finding the intercepts. Let t correspond to the horizontal axis (x-axis) and v correspond to the vertical axis (y-axis).

(b) Interpret each intercept.

Solution

(a) To find the t-intercept let $v = 0$.

$$0 = 160 - 32t \qquad \text{Let } v = 0.$$
$$32t = 160 \qquad \text{Add } 32t \text{ to each side.}$$
$$t = 5 \qquad \text{Divide each side by 32.}$$

To find the v-intercept let $t = 0$.

$$v = 160 - 32(0) \qquad \text{Let } t = 0.$$
$$v = 160 \qquad \text{Simplify.}$$

Therefore the graph passes through (**5**, **0**) and (**0**, **160**), as shown in Figure 3.20.

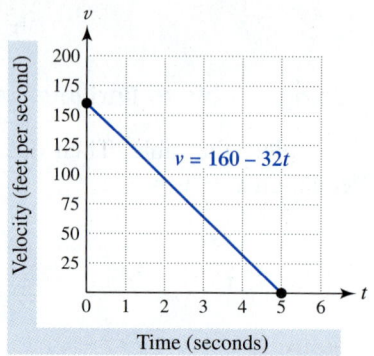

Figure 3.20

(b) The *t*-intercept indicates that the rocket had a velocity of 0 feet per second after 5 seconds. The *v*-intercept indicates that the rocket's initial velocity was 160 feet per second.

Now Try Exercise 85

Horizontal Lines

▶ **REAL-WORLD CONNECTION**   Suppose that someone drives a car on a freeway at a constant speed of 70 miles per hour. Table 3.14 shows the speed *y* after *x* hours.

TABLE 3.14   Speed of a Car

x	1	2	3	4	5
y	70	70	70	70	70

We can make a scatterplot of the data by plotting the five points $(1, \mathbf{70})$, $(2, \mathbf{70})$, $(3, \mathbf{70})$, $(4, \mathbf{70})$, and $(5, \mathbf{70})$, as shown in Figure 3.21(a). The speed is always 70 miles per hour and the graph of the car's speed is a horizontal line, as shown in Figure 3.21(b). The equation of this line is $y = \mathbf{70}$ with *y*-intercept 70. There are no *x*-intercepts.

(a)

(b)

Figure 3.21

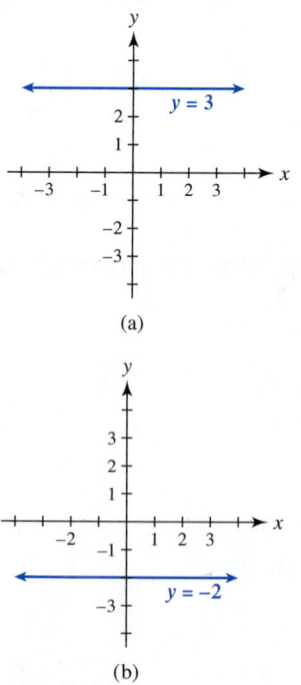

(a)

(b)

Figure 3.22

In general, the equation of a horizontal line is $y = b$, where *b* is a constant that corresponds to the *y*-intercept. Examples of horizontal lines are shown in Figure 3.22. Note that every point on the graph of $y = \mathbf{3}$ in Figure 3.22(a) has a *y*-coordinate of **3**, and that every point on the graph of $y = -\mathbf{2}$ in Figure 3.22(b) has a *y*-coordinate of $-\mathbf{2}$.

HORIZONTAL LINE

The equation of a horizontal line with y-intercept b is $y = b$.

The equation $y = b$ is an example of a linear equation in the form $Ax + By = C$ with $A = 0$, $B = 1$, and $C = b$. (Note that in general B and b do not represent the same number.)

EXAMPLE 4   **Graphing a horizontal line**

Graph the equation $y = -1$ and identify its y-intercept.

Solution

The graph of $y = -1$ is a horizontal line passing through the point $(0, -1)$, as shown in Figure 3.23. Its y-intercept is -1.

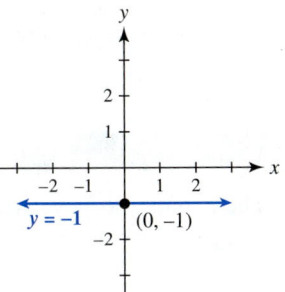

Figure 3.23

Now Try Exercise 47(a)

Vertical Lines

TABLE 3.15
Falling Object

x (seconds)	y (feet)
1	16
1	15
1	18
1	17

▶ **REAL-WORLD CONNECTION**   A scientist is experimentally determining the distance y that an object falls in 1 second. Because it is an experiment, the distances vary slightly on each trial. Table 3.15 shows the results.

We can make a scatterplot of the data by plotting the points $(1, 16)$, $(1, 15)$, $(1, 18)$, and $(1, 17)$, as shown in Figure 3.24(a). In each case the time is always 1 second and each point lies on the graph of a vertical line, as shown in Figure 3.24(b). This vertical line has the equation $x = 1$ because each point on the line has an x-coordinate of 1 and there are no restrictions on the y-coordinate. This line has x-intercept 1 but no y-intercept.

(a)

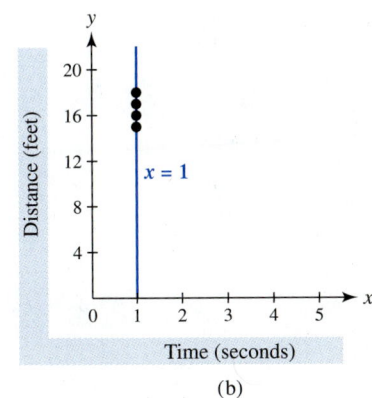

(b)

Figure 3.24

In general, the graph of a vertical line is $x = k$, where k is a constant that corresponds to the x-intercept. Examples of vertical lines are shown in Figure 3.25. Note that every point on the graph of $x = 3$ in Figure 3.25(a) has an x-coordinate of **3** and that every point on the graph of $x = -2$ shown in Figure 3.25(b) has an x-coordinate of -2.

CALCULATOR HELP

To graph a vertical line see Appendix A (page AP-6).

 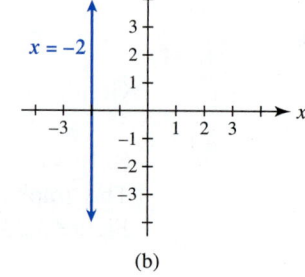

(a)   (b)

Figure 3.25

VERTICAL LINE

The equation of a vertical line with x-intercept k is $x = k$.

The equation $x = k$ is an example of a linear equation in the form $Ax + By = C$ with $A = 1, B = 0$, and $C = k$.

EXAMPLE 5   Graphing a vertical line

Graph the equation $x = -3$, and identify its x-intercept.

Solution

The graph of $x = -3$ is a vertical line passing through the point $(-3, 0)$, as shown in Figure 3.26. Its x-intercept is -3.

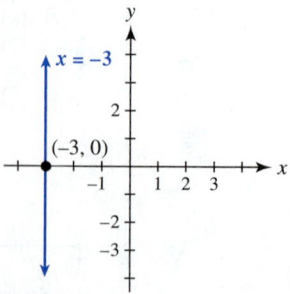

Figure 3.26

Now Try Exercise 47(b)

EXAMPLE 6 Writing equations of horizontal and vertical lines

Write the equation of the line shown in each graph.

(a)

(b)

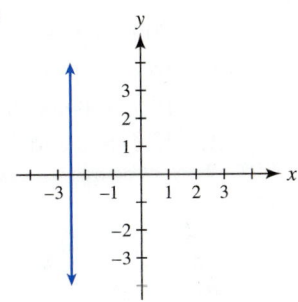

Solution

(a) The graph is a horizontal line with y-intercept 2. Its equation is $y = 2$.

(b) The graph is a vertical line with x-intercept -2.5. Its equation is $x = -2.5$.

Now Try Exercises 55, 57

EXAMPLE 7 Writing equations of horizontal and vertical lines

Find an equation for a line satisfying the given conditions.

(a) Vertical, passing through $(2, -3)$

(b) Horizontal, passing through $(3, 1)$

(c) Perpendicular to $x = 3$, passing through $(-1, 2)$

Solution

(a) A vertical line passing through $(2, -3)$ has x-intercept 2, as shown in Figure 3.27(a). The equation of a vertical line with x-intercept 2 is $x = 2$.

(b) A horizontal line passing through $(3, 1)$ has y-intercept 1, as shown in Figure 3.27(b). The equation of a horizontal line with y-intercept 1 is $y = 1$.

(c) Because the line $x = 3$ is vertical, a line that is perpendicular to this line is horizontal, as shown in Figure 3.27(c). The equation of a horizontal line with y-intercept 2 is $y = 2$.

(a)

(b)

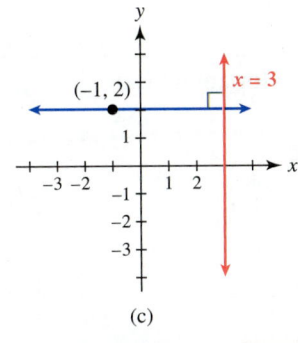

(c)

Figure 3.27 Horizontal and Vertical Lines

Now Try Exercises 73, 75, 77

MAKING CONNECTIONS

Lines and Linear Equations

The equation of any line can be written in the standard form $Ax + By = C$.

1. If $A = 0$ and $B \neq 0$, the line is horizontal.
2. If $A \neq 0$ and $B = 0$, the line is vertical.
3. If $A \neq 0$ and $B \neq 0$, the line is neither horizontal nor vertical.

3.3   PUTTING IT ALL TOGETHER

The following table summarizes several important topics presented in this section.

Concept	Explanation	Examples
x- and y-Intercepts	The x-coordinate of a point at which a graph intersects the x-axis is called an x-intercept. The y-coordinate of a point at which a graph intersects the y-axis is called a y-intercept.	 x-intercept, -3 y-intercept, 2
Finding Intercepts	To find x-intercepts let $y = 0$ in the equation and solve for x. To find y-intercepts let $x = 0$ in the equation and solve for y.	Let $4x - 5y = 20$. x-intercept: $4x - 5(0) = 20$ $x = 5$ y-intercept: $4(0) - 5y = 20$ $y = -4$ The x-intercept is 5, and the y-intercept is -4.
Horizontal Line	Has equation $y = b$, where b is a constant Has y-intercept b and no x-intercept if $b \neq 0$	 y-intercept, b

Concept	Explanation	Examples
Vertical Line	Has equation $x = k$, where k is a constant Has x-intercept k and no y-intercept if $k \neq 0$	 x-intercept, k

3.3 Exercises

CONCEPTS

1. How many points determine a line?

2. The graph of the linear equation $Ax + By = C$ with $A \neq 0$ and $B \neq 0$ has _____ x-intercept(s) and _____ y-intercept(s).

3. The x-coordinate of a point where a graph intersects the x-axis is a(n) _____.

4. To find an x-intercept, let $y =$ _____ and solve for x.

5. The y-coordinate of a point at which a graph intersects the y-axis is a(n) _____.

6. To find a y-intercept, let $x =$ _____ and solve for y.

7. The graph of $y = 3$ has y-intercept _____.

8. The graph of the linear equation $Ax + By = C$ with $A = 0$ and $B = 1$ is a(n) _____ line.

9. A horizontal line with y-intercept b has equation _____.

10. The graph of $x = 3$ has x-intercept _____.

11. The graph of the linear equation $Ax + By = C$ with $A = 1$ and $B = 0$ is a(n) _____ line.

12. A vertical line with x-intercept k has equation _____.

FINDING INTERCEPTS

Exercises 13–20: Identify any x-intercepts and y-intercepts in the graph.

13.

14.

15.

16.

17.

18.

19.

20.

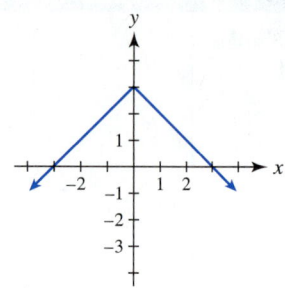

Exercises 21–24: Complete the table. Then determine the x-intercept and the y-intercept for the graph of the equation.

21. $y = x + 2$

x	-2	-1	0	1	2
y					

22. $y = 2x - 4$

x	-2	-1	0	1	2
y					

23. $-x + y = -2$

x	-4	-2	0	2	4
y					

24. $x + y = 1$

x	-2	-1	0	1	2
y					

Exercises 25–44: Find any intercepts. Then graph the linear equation.

25. $-2x + 3y = -6$ **26.** $4x + 3y = 12$

27. $3x - 5y = 15$ **28.** $2x - 4y = -4$

29. $x - 3y = 6$ **30.** $5x + y = -5$

31. $6x - y = -6$ **32.** $5x + 7y = -35$

33. $3x + 7y = 21$ **34.** $-3x + 8y = 24$

35. $40y - 30x = -120$ **36.** $10y - 20x = 40$

37. $\frac{1}{2}x - y = 2$ **38.** $x - \frac{1}{2}y = 4$

39. $-\frac{x}{4} + \frac{y}{3} = 1$ **40.** $\frac{x}{3} - \frac{y}{4} = 1$

41. $\frac{x}{3} + \frac{y}{2} = 1$ **42.** $\frac{x}{5} - \frac{y}{4} = 1$

43. $0.6y - 1.5x = 3$ **44.** $0.5y - 0.4x = 2$

45. **Thinking Generally** Find any intercepts for the graph of $Ax + By = C$.

46. **Thinking Generally** Find any intercepts for the graph of $\frac{x}{A} + \frac{y}{B} = 1$.

HORIZONTAL AND VERTICAL LINES

Exercises 47–54: Graph each equation.

47. **(a)** $y = 2$ **(b)** $x = 2$

48. **(a)** $y = -2$ **(b)** $x = -2$

49. **(a)** $y = -4$ **(b)** $x = -4$

50. **(a)** $y = 0$ **(b)** $x = 0$

51. **(a)** $y = -1$ **(b)** $x = -1$

52. **(a)** $y = -\frac{1}{2}$ **(b)** $x = -\frac{1}{2}$

53. **(a)** $y = \frac{3}{2}$ **(b)** $x = \frac{3}{2}$

54. **(a)** $y = -1.5$ **(b)** $x = -1.5$

Exercises 55–62: Write an equation for the line shown in the graph.

55.

56.

57.

58.

59.

60.

61.

62.

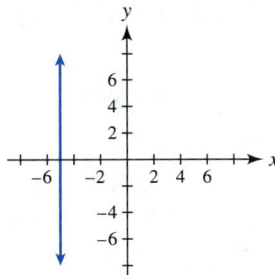

Exercises 63–66: Write an equation for the line that passes through the points shown in the table.

63.

x	-2	-1	0	1	2
y	1	1	1	1	1

64.

x	0	1	2	3	4
y	-10	-10	-10	-10	-10

65.

x	-6	-6	-6	-6	-6
y	5	4	3	2	1

66.

x	20	20	20	20	20
y	-2	-1	0	1	2

Exercises 67–72: Write the equations of a horizontal line and a vertical line that pass through the given point. (Hint: Make a sketch.)

67. $(1, 2)$

68. $(-3, 4)$

69. $(20, -45)$

70. $(-5, 12)$

71. $(0, 5)$

72. $(-3, 0)$

Exercises 73–80: Find an equation for a line satisfying the following conditions.

73. Vertical, passing through $(-1, 6)$

74. Vertical, passing through $(2, -7)$

75. Horizontal, passing through $\left(\frac{3}{4}, -\frac{5}{6}\right)$

76. Horizontal, passing through $(5.1, 6.2)$

77. Perpendicular to $y = \frac{1}{2}$, passing through $(4, -9)$

78. Perpendicular to $x = 2$, passing through $(3, 4)$

79. Parallel to $x = 4$, passing through $\left(-\frac{2}{3}, \frac{1}{2}\right)$

80. Parallel to $y = -2.1$, passing through $(7.6, 3.5)$

81. **Thinking Generally** Write the equation of the x-axis. (*Hint:* The x-axis is a horizontal line.)

82. **Thinking Generally** Write the equation of the y-axis. (*Hint:* The y-axis is a vertical line.)

APPLICATIONS

Exercises 83 and 84: **Distance** *The distance of a driver from home is illustrated in the graph.*

   (a) Find the intercepts.

   (b) Interpret each intercept.

83.

84.

Exercises 85 and 86: **Modeling a Toy Rocket**   *(Refer to Example 3.) The velocity v of a toy rocket in feet per second after t seconds of flight is given, where t ≥ 0.*

(a) *Find the intercepts and then graph the equation.*

(b) *Interpret each intercept.*

85. $v = 128 - 32t$ **86.** $v = 96 - 32t$

Exercises 87 and 88: **Water in a Pool**   *The amount of water in a swimming pool is depicted in the graph.*

(a) *Find the intercepts.*

(b) *Interpret each intercept.*

87.

88.

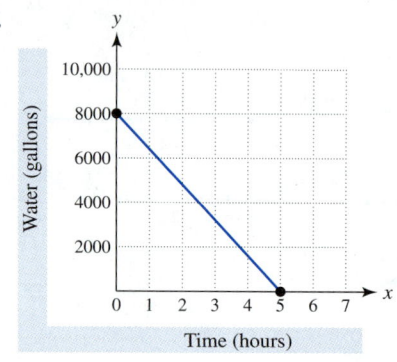

WRITING ABOUT MATHEMATICS

89. Given an equation, explain how to find an x-intercept and a y-intercept.

90. The form $\frac{x}{a} + \frac{y}{b} = 1$ is called the **intercept form** of a linear equation. Explain how you can use this equation to find the intercepts. (*Hint:* Graph $\frac{x}{2} + \frac{y}{3} = 1$ and find its intercepts.)

GROUP ACTIVITY
WORKING WITH REAL DATA

Directions: Form a group of 2 to 4 people. Select someone to record the group's responses for this activity. All members of the group should work cooperatively to answer the questions. If your instructor asks for the results, each member of the group should be prepared to respond.

1. *Radio Stations*   The approximate number of radio stations on the air for selected years from 1950 to 2000 is shown in the table.

x (year)	1950	1960	1970
y (stations)	2800	4100	6800

x (year)	1980	1990	2000
y (stations)	8600	10,800	12,600

Source: M. Street Corporation, 2001.

Make a line graph of the data. Be sure to label both axes.

2. *Estimation*   Discuss ways to estimate the number of radio stations on the air in 1975. Compare your estimates with the actual value of 7700 stations. Repeat this estimate for 1985 and compare it to the actual value of 10,400. Discuss your results.

3. *Modeling Equation*   Substitute each x-value from the table into the equation $y = 196x - 379,400$ and determine the corresponding y-value. Do these y-values give reasonable approximations to the y-values in the table? Explain your answer.

4. *Making Estimates*   Use $y = 196x - 379,400$ to estimate the number of radio stations on the air in 1975 and 1985. Compare the results to your answer in Exercise 2.

3.4 SLOPE AND RATES OF CHANGE

Finding Slopes of Lines ■ Slope as a Rate of Change

Figure 3.28 shows some graphs of lines, where the horizontal axis represents time.

Which graph might represent the distance traveled by you if you are walking?

Which graph might represent the temperature in your freezer?

Which graph might represent the amount of gasoline in your car's tank while you are driving?

To be able to answer these questions, you probably used the concept of slope. In mathematics, slope is a real number that measures the "tilt" or "angle" of a line. In this section we discuss slope and how it is used in applications.

(a)

(b)

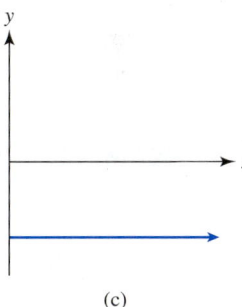
(c)

Figure 3.28

Finding Slopes of Lines

▶ **REAL-WORLD CONNECTION**   The graph shown in Figure 3.29 illustrates the cost of parking for x hours. The graph tilts upward from left to right, which indicates that the cost increases as the number of hours increases. Note that, for each hour of parking, the cost increases by \$2. The graph *rises* 2 units for every unit of *run*, and the ratio $\frac{\text{rise}}{\text{run}}$ equals the *slope* of the line. The slope m of this line is 2, which indicates that the cost of parking is \$2 per hour.

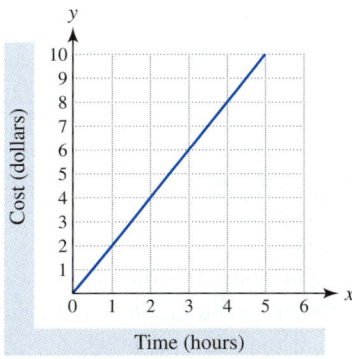

Figure 3.29 Cost of Parking

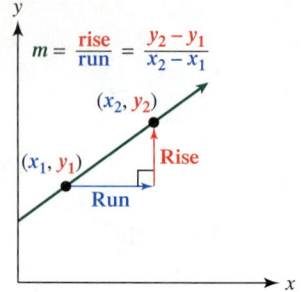

Figure 3.30 Slope m of a Line

A more general case of the slope of a line is shown in Figure 3.30 where a line passes through the points (x_1, y_1) and (x_2, y_2). The **rise**, or *change in y*, is $y_2 - y_1$, and the **run**, or *change in x*, is $x_2 - x_1$. Slope m is given by $m = \frac{y_2 - y_1}{x_2 - x_1}$.

NOTE: The symbol x_1 has a **subscript** of 1 and is read "x sub one" or "x one". Thus x_1 and x_2 are used to denote two different x-values. Similar comments apply to y_1 and y_2.

SLOPE

The **slope** m of the line passing through the points (x_1, y_1) and (x_2, y_2) is

$$m = \frac{\text{rise}}{\text{run}} = \frac{y_2 - y_1}{x_2 - x_1},$$

where $x_1 \neq x_2$. That is, slope equals rise over run.

NOTE: If $x_1 = x_2$, the slope is undefined.

EXAMPLE 1   Calculating the slope of a line

Use the two points labeled in Figure 3.31 to find the slope of the line. What are the rise and run between these two points? Interpret the slope in terms of rise and run.

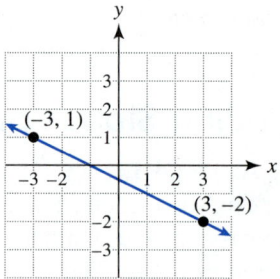

Figure 3.31

Solution
The line passes through the points $(-3, 1)$ and $(3, -2)$, so let $(x_1, y_1) = (-3, 1)$ and $(x_2, y_2) = (3, -2)$. The slope is

$$m = \frac{y_2 - y_1}{x_2 - x_1} = \frac{-2 - 1}{3 - (-3)} = \frac{-3}{6} = -\frac{1}{2}.$$

Starting at the point $(-3, 1)$, count 3 units downward and then 6 units to the right to return to the graph at the point $(3, -2)$. Thus the "rise" is -3 units and the run is 6 units. See Figure 3.32(a). The ratio $\frac{\text{rise}}{\text{run}}$ is $\frac{-3}{6}$, or $-\frac{1}{2}$. Figure 3.32(b) shows an alternate way of finding this slope. Starting at the point $(3, -2)$, count 3 units upward and then 6 units to the left to return to the graph at the point $(-3, 1)$. Here, the rise is 3 units and the run is -6 units so that the ratio $\frac{\text{rise}}{\text{run}}$ is $\frac{3}{-6}$, or $-\frac{1}{2}$. In either case, the slope is $-\frac{1}{2}$, indicating that the graph falls 1 unit for every 2 units of run (to the right).

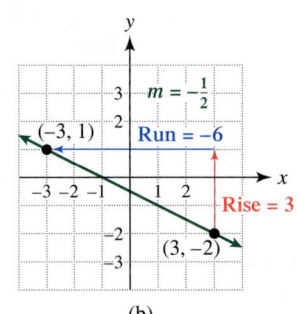

(a)   (b)

Figure 3.32

Now Try Exercise 25

NOTE: In Example 1 the same slope would result if we let $(x_1, y_1) = (3, -2)$ and $(x_2, y_2) = (-3, 1)$. In this case the calculation would be

$$m = \frac{y_2 - y_1}{x_2 - x_1} = \frac{1 - (-2)}{-3 - 3} = \frac{3}{-6} = -\frac{1}{2}.$$

Two points determine a line and its slope, as demonstrated in the next example.

EXAMPLE 2 Calculating the slope of a line

Calculate the slope of the line passing through each pair of points. Graph the line.
(a) $(-2, -2), (0, 2)$ **(b)** $(-2, 3), (2, 1)$ **(c)** $(-1, 3), (2, 3)$ **(d)** $(-3, 3), (-3, -2)$

Solution

(a) $m = \frac{y_2 - y_1}{x_2 - x_1} = \frac{2 - (-2)}{0 - (-2)} = \frac{4}{2} = 2$. This slope indicates that the line rises 2 units for every 1 unit of horizontal run, as shown in Figure 3.33(a).

(b) $m = \frac{y_2 - y_1}{x_2 - x_1} = \frac{1 - 3}{2 - (-2)} = \frac{-2}{4} = -\frac{1}{2}$. This slope indicates that the line falls 1 unit for every 2 units of horizontal run, as shown in Figure 3.33(b).

(c) $m = \frac{y_2 - y_1}{x_2 - x_1} = \frac{3 - 3}{2 - (-1)} = \frac{0}{3} = 0$. The line is horizontal, as shown in Figure 3.33(c).

(d) Because $x_1 = x_2 = -3$, the slope formula does not apply. If we try to use it, we obtain $m = \frac{y_2 - y_1}{x_2 - x_1} = \frac{-2 - 3}{-3 - (-3)} = \frac{-5}{0}$, which is an undefined expression. The line has undefined slope and is vertical, as shown in Figure 3.33(d).

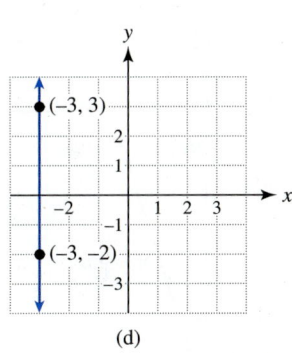

(a)   (b)   (c)   (d)

Figure 3.33

Now Try Exercises 37, 39, 41

If a line has **positive slope**, it *rises from left to right*, as shown in Figure 3.34(a). If a line has **negative slope**, it *falls from left to right*, as shown in Figure 3.34(b). Slope 0 indicates that a line is horizontal, which is shown in Figure 3.34(c). Any two points on a vertical line have the same *x*-coordinate so the run always equals 0. Thus the slope is undefined for a vertical line, which is shown in Figure 3.34(d).

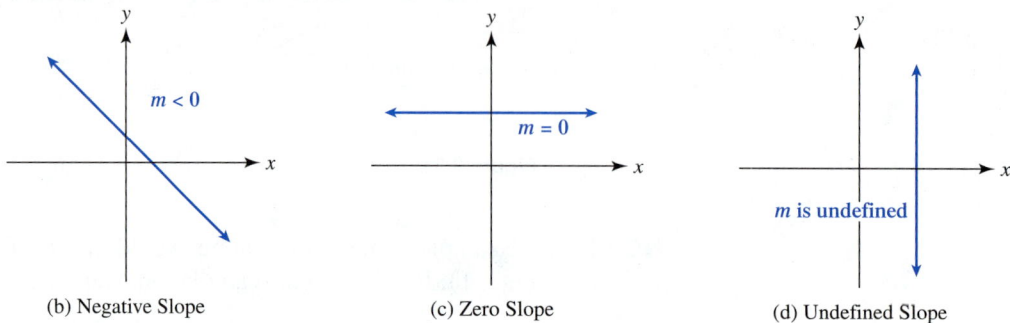

(a) Positive Slope   (b) Negative Slope   (c) Zero Slope   (d) Undefined Slope

Figure 3.34

SLOPE OF A LINE

1.   A line that rises *from left to right* has positive slope.
2.   A line that falls *from left to right* has negative slope.
3.   A horizontal line has slope 0.
4.   A vertical line has undefined slope.

EXAMPLE 3   **Finding slope from a graph**

Find the slope of each line.

(a)

(b)

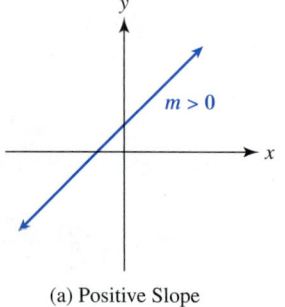

Figure 3.35

Solution

(a) The graph rises 3 units for each unit of run. For example, the graph passes through $(1, 1)$ and $(2, 4)$, so the line rises $4 - 1 = 3$ units with a $2 - 1 = 1$ unit run, as shown in Figure 3.35. Therefore the slope is

$$m = \frac{\text{rise}}{\text{run}} = \frac{3}{1} = 3.$$

(b) The line is horizontal, so the slope is 0.

Now Try Exercises   15, 19

A point and a slope also determine a line, as illustrated in the next example.

EXAMPLE 4   Sketching a line with a given slope

Sketch a line passing through the point $(1, 4)$ and having slope $-\frac{2}{3}$.

Solution
Start by plotting the point $(1, 4)$. A slope of $-\frac{2}{3}$ indicates that the y-values *decrease* 2 units each time the x-values increase by 3 units. That is, the line *falls* 2 units for every 3-unit increase in the run. Because the line passes through $(\mathbf{1}, \mathbf{4})$, a **2**-unit decrease in y and a **3**-unit increase in x results in the line passing through the point $(\mathbf{1} + \mathbf{3}, \mathbf{4} - \mathbf{2})$ or $(4, 2)$. See Figure 3.36.

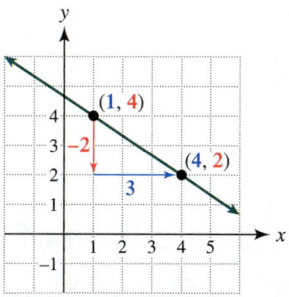

Figure 3.36 Slope $m = -\frac{2}{3}$

Now Try Exercise 55

EXAMPLE 5   Sketching a line with a given y-intercept

Sketch a line with slope -2 and y-intercept 3.

Solution
For the y-intercept of 3, plot the point $(0, 3)$. The slope is -2, so the y-values decrease 2 units for each unit increase in x. Increasing the x-value in the point $(\mathbf{0}, \mathbf{3})$ by **1** and decreasing the y-value by **2** results in the point $(\mathbf{0} + \mathbf{1}, \mathbf{3} - \mathbf{2})$ or $(1, 1)$. Plot $(0, 3)$ and $(1, 1)$ and then sketch the line, as shown in Figure 3.37.

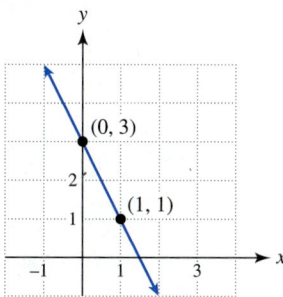

Figure 3.37 Slope: -2; y-intercept: 3

Now Try Exercise 51

If we know the slope of a line and a point on the line, we can complete a table of points, as demonstrated in the next example.

EXAMPLE 6   Completing a table of values

A line has slope 2 and passes through the first point listed in the table. Complete the table so that each point lies on the line.

x	-2	-1	0	1
y	1			

Solution

Slope 2 indicates that $\frac{\text{rise}}{\text{run}} = 2$. Because consecutive x-values in the table increase by 1 unit, the run from one point in the table to the next is 1 unit. Substituting **1** for the **run** in the slope equation results in $\frac{\text{rise}}{1} = 2$. Thus the **rise** is **2** and consecutive y-values shown in Table 3.16 increase by 2 units.

TABLE 3.16

x	-2	-1	0	1
y	1	3	5	7

Now Try Exercise 63

Slope as a Rate of Change

▶ **REAL-WORLD CONNECTION**   When lines are used to model physical quantities in applications, their slopes provide important information. Slope measures the **rate of change** in a quantity. We illustrated this concept in the next four examples.

EXAMPLE 7   Interpreting slope

The distance y in miles that an athlete training for a marathon is from home after x hours is shown in Figure 3.38.
(a) Find the y-intercept. What does the y-intercept represent?
(b) The graph passes through the point $(1, 10)$. Discuss the meaning of this point.
(c) Find the slope of this line. Interpret the slope as a rate of change.

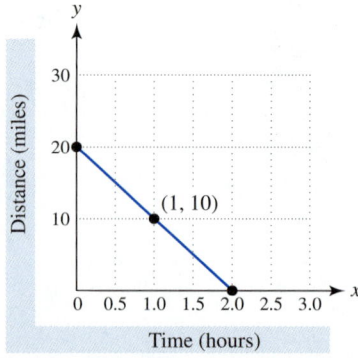

Figure 3.38   Distance from Home

Solution

(a) On the graph the y-intercept is 20, so the athlete is initially 20 miles from home.
(b) The point $(1, 10)$ means that after 1 hour the athlete is 10 miles from home.

(c) The line passes through the points $(0, 20)$ and $(1, 10)$. Its slope is

$$m = \frac{10 - 20}{1 - 0} = -10.$$

Slope -10 means that the athlete is running *toward* home at 10 miles per hour. A *negative* slope indicates that the distance between the runner and home is decreasing.

Now Try Exercise 93

EXAMPLE 8   **Interpreting slope**

When a company manufactures 2000 MP3 players, its profit is $10,000, and when it manufactures 4500 MP3 players, its profit is $35,000.

(a) Find the slope of the line passing through $(2000, 10000)$ and $(4500, 35000)$.

(b) Interpret the slope as a rate of change.

Solution

(a) $m = \frac{35{,}000 - 10{,}000}{4500 - 2000} = \frac{25{,}000}{2500} = 10$

(b) Profit increases, *on average*, by $10 for each additional MP3 player made.

Now Try Exercise 99

EXAMPLE 9   **Analyzing tetanus cases**

Table 3.17 lists numbers of reported cases of tetanus in the United States for selected years.

TABLE 3.17

Year	1950	1960	1970	1980	1990	2000
Cases of Tetanus	486	368	148	95	64	45

Source: Department of Health and Human Services.

(a) Make a line graph of the data.

(b) Find the slope of each line segment.

(c) Interpret each slope as a rate of change.

Solution

(a) A line graph connecting the points $(1950, 486)$, $(1960, 368)$, $(1970, 148)$, $(1980, 95)$, $(1990, 64)$, and $(2000, 45)$ is shown in Figure 3.39.

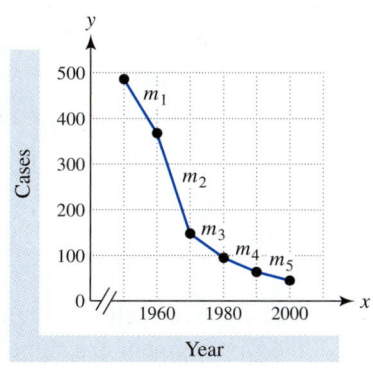

Figure 3.39 Tetanus Cases

(b) The slope of each line segment may be calculated as follows.

$$m_1 = \frac{368 - 486}{1960 - 1950} = -11.8, \qquad m_2 = \frac{148 - 368}{1970 - 1960} = -22.0,$$

$$m_3 = \frac{95 - 148}{1980 - 1970} = -5.3, \qquad m_4 = \frac{64 - 95}{1990 - 1980} = -3.1,$$

$$m_5 = \frac{45 - 64}{2000 - 1990} = -1.9$$

(c) Slope $m_1 = -11.8$ indicates that, *on average*, the number of tetanus cases *decreased* by 11.8 cases per year between 1950 and 1960. The other four slopes can be interpreted similarly.

NOTE: The number of tetanus cases did not decrease by *exactly* 11.8 cases per year between 1950 and 1960. Why? However, the yearly *average* decrease was 11.8 cases.

Now Try Exercise 97

EXAMPLE 10   Sketching a model

During a storm, rain falls at the rate of 2 inches per hour from 1 A.M. to 3 A.M., 1 inch per hour from 3 A.M. to 4 A.M., and $\frac{1}{2}$ inch per hour from 4 A.M. to 6 A.M.
(a) Sketch a graph that shows the total accumulation of rainfall from 1 A.M. to 6 A.M.
(b) What does the slope of each line segment represent?

CRITICAL THINKING

An athlete runs 10 miles per hour for 30 minutes away from home and then jogs back home at 5 miles per hour. Sketch a graph that shows the distance between the athlete and home. What does the slope of each line segment represent?

Solution

(a) At 1 A.M. the accumulated rainfall is 0, so place a point at $(1, 0)$. Rain falls at a constant rate of 2 inches per hour for the next 2 hours, so at 3 A.M. the total rainfall is 4 inches. Place a point at $(3, 4)$. Because the rainfall is constant, sketch a line segment from $(1, 0)$ to $(3, 4)$, as shown in Figure 3.40. Similarly, during the next hour 1 inch of rain falls, so draw a line segment from $(3, 4)$ to $(4, 5)$. Finally, 1 inch of rain falls from 4 A.M. to 6 A.M., so draw a line segment from $(4, 5)$ to $(6, 6)$.

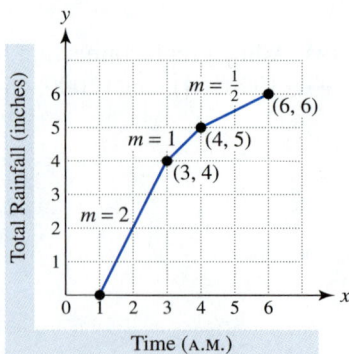

Figure 3.40 Total Rainfall

(b) The slope of each line segment represents the rate at which rain is falling. For example, the first segment has slope 2 because rain falls at a rate of 2 inches per hour during that period of time.

Now Try Exercise 89

3.4 PUTTING IT ALL TOGETHER

The "tilt" of a line is called the slope and equals rise over run. A positive slope indicates that the line *rises* from left to right, whereas a negative slope indicates that the line *falls* from left to right. A horizontal line has slope 0, and a vertical line has undefined slope. When a quantity can be approximated with a line, its slope indicates a rate of change in the quantity. The following table summarizes some basic concepts about slope.

Concept	Comments	Example
Rise, Run, and Slope	Rise is a vertical change in a line, and run is a horizontal change in a line. The ratio $\frac{\text{rise}}{\text{run}}$ is the slope m when run is nonzero.	$$m = \frac{\text{rise}}{\text{run}} = \frac{1}{2}$$
Calculating Slope	For any two points (x_1, y_1) and (x_2, y_2) slope m is $$m = \frac{y_2 - y_1}{x_2 - x_1},$$ where $x_1 \neq x_2$.	The slope of the line passing through $(-2, 3)$ and $(1, 5)$ is $$m = \frac{5 - 3}{1 - (-2)} = \frac{2}{3}.$$ The line rises 2 units for every 3 units of run along the x-axis.
Slope as a Rate of Change	Slope indicates how fast the graph of a line is changing.	The graph shows that water is being pumped from a swimming pool. Slope $m = -66\frac{2}{3}$ indicates that water is *leaving* the pool at the rate of $66\frac{2}{3}$ gallons per hour.

3.4 Exercises

CONCEPTS

1. Run is the change in the (horizontal/vertical) distance along a line.

2. Rise is the change in the (horizontal/vertical) distance along a line.

3. Slope m of a line is _____ over _____.

4. Slope 0 indicates that a line is _____.

5. Undefined slope indicates that a line is _____.

6. If a line passes through (x_1, y_1) and (x_2, y_2), then $m = $ _____.

7. A line that rises from left to right has _____ slope.

8. A line that falls from left to right has _____ slope.

Exercises 9–14: State whether the slope of the line is positive, negative, zero, or undefined.

9.

10.

11.

12.

13.

14.

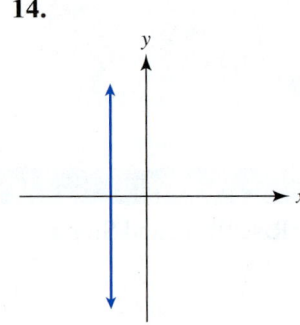

FINDING SLOPES OF LINES

Exercises 15–26: If possible, find the slope of the line. Interpret the slope in terms of rise and run.

15.

16.

17.

18.

19.

20.

21.

22.

23.

24.

25.

26.

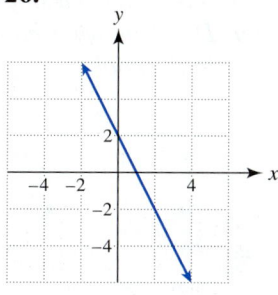

Exercises 27–34: If possible, find the slope of the line passing through the two points. Graph the line.

27. $(1, 2), (2, 4)$

28. $(-4, 7), (7, -4)$

29. $(2, 1), (2, 4)$

30. $(-1, 3), (-1, -1)$

31. $(1, 3), (-2, 5)$

32. $(0, -4), (-4, 6)$

33. $(2, -1), (-2, -1)$

34. $(-2, 3), (1, 3)$

Exercises 35–50: If possible, find the slope of the line passing through the two points.

35. $(4, -2), (-3, -7)$

36. $(15, -3), (20, 9)$

37. $(-3, 4), (4, -2)$

38. $(1, -3), (3, -5)$

39. $(-3, 5), (2, 5)$

40. $(-3, 3), (-5, 3)$

41. $(-1, 6), (-1, -4)$

42. $\left(\frac{1}{2}, -\frac{2}{7}\right), \left(\frac{1}{2}, \frac{13}{17}\right)$

43. $(1980, 5), (2000, 18)$

44. $(1989, 10), (1999, 16)$

45. $(1950, 6.1), (2000, 10.6)$

46. $(1900, 10), (1950, 35)$

47. $\left(\frac{1}{3}, -\frac{2}{7}\right), \left(-\frac{2}{3}, \frac{3}{7}\right)$

48. $(-1.3, 5.6), (-2.6, -2.5)$

49. $(12, -34), (14, 64)$

50. $(-25, 105), (60, 55)$

Exercises 51–58: (Refer to Example 4.) Sketch a line passing through the point and having slope m.

51. $(0, 2), m = -1$

52. $(0, -1), m = 2$

53. $(1, 1), m = 3$

54. $(1, -1), m = -2$

55. $(-2, 3), m = -\frac{1}{2}$

56. $(-1, -2), m = \frac{3}{4}$

57. $(-3, 1), m = \frac{1}{2}$

58. $(-2, 2), m = -3$

Exercises 59–62: The table lists points located on a line. Find the slope, x-intercept, and y-intercept of the line.

59.

x	0	1	2	3
y	-2	0	2	4

60.

x	-1	0	1	2
y	0	5	10	15

61.

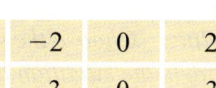

x	-2	-1	0	1
y	9	0	-9	-18

62.

x	-4	-2	0	2
y	6	3	0	-3

Exercises 63–68: (Refer to Example 6.) A line has the given slope m and passes through the first point listed in the table. Complete the table so that each point in the table lies on the line.

63. $m = 2$

x	0	1	2	3
y	−4			

64. $m = -\frac{1}{2}$

x	0	1	2	3
y	2			

65. $m = -3$

x	1	2	3	4
y	4			

66. $m = -1$

x	−1	0	1	2
y	10			

67. $m = \frac{3}{2}$

x	−4	−2	0	2
y	0			

68. $m = 3$

x	−2	0	2	4
y	−4			

Exercises 69–76: Do the following.

   (a) Graph the equation.
   (b) Find the slope of the line.

69. $y = 2x - 1$ **70.** $y = x - 2$

71. $3x + y = 2$ **72.** $-\frac{1}{2}x + y = -1$

73. $-x + 3y = 0$ **74.** $2x + y = 0$

75. $y = 2$ **76.** $y = -3$

SLOPE AS A RATE OF CHANGE

*Exercises 77–80: **Modeling** Choose the graph (a.–d.) in the next column that models the situation best.*

77. Cost of buying x gum balls at a price of 25¢ each

78. Total number of movies in VHS format (videotape) purchased during the past 5 years

79. Average cost of a new car over the past 30 years

80. Height of the Empire State Building after x people have entered it

a.

b.

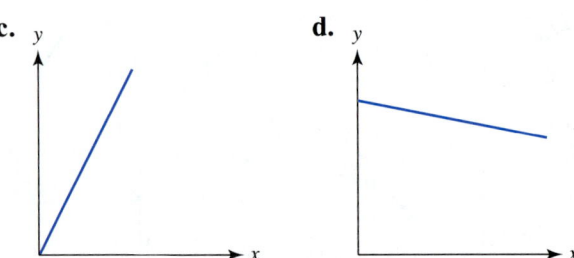

c.

d.

*Exercises 81 and 82: **Modeling** The line graph represents the gallons of water in a small swimming pool after x hours. Assume that there is a pump that can either add water to or remove water from the pool.*

   (a) Estimate the slope of each line segment.
   (b) Interpret each slope as a rate of change.
   (c) Describe what happened to the amount of water in the pool.

81.

82.

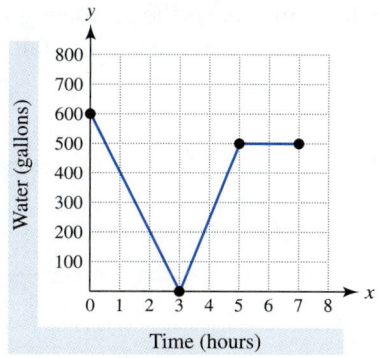

Exercises 83 and 84: *Modeling*   An individual is driving a car along a straight road. The graph shows the distance that the driver is from home after x hours.

(a) Find the slope of each line segment in the graph.
(b) Interpret each slope as a rate of change.
(c) Describe both the motion and location of the car.

83.

84.

85. Thinking Generally   If a line is used to model physical data where the *x*-axis is labeled "Time (minutes)" and the *y*-axis is labeled "Volume (cubic meters)," what are the units for the rate of change represented by the slope of the line?

86. Thinking Generally   If a line is used to model physical data where the *x*-axis is labeled "Cookies" and the *y*-axis is labeled "Chocolate Chips," what are the units for the rate of change represented by the slope of the line?

Exercises 87–90: *Sketching a Model*   Sketch a graph that models the given situation.

87. The distance that a boat is from a harbor if the boat is initially 6 miles from the harbor and arrives at the harbor after sailing at a constant speed for 3 hours

88. The distance that a person is from home if the person starts at home, walks away from home at 4 miles per hour for 90 minutes, and then walks back home at 3 miles per hour

89. The distance that an athlete is from home if the athlete jogs for 1 hour to a park that is 7 miles away, stays for 30 minutes, and then jogs home at the same pace

90. The amount of oil in a 55-gallon drum that is initially full, then is drained at a rate of 5 gallons per minute for 4 minutes, is left for 6 minutes, and then is emptied at a rate of 7 gallons per minute

APPLICATIONS

91. Older Mothers   The number of children born to mothers 40 years old or older in 2000 was 70 thousand and in 2004 it was 110 thousand.   (**Source:** National Center for Health Statistics.)
(a) Calculate the slope of the line passing through (2000, 70) and (2004, 110).
(b) Interpret the slope as a rate of change.

92. Profit from Laptops   When a company manufactures 500 laptop computers, its profit is $100,000, and when it manufactures 1500 laptop computers, its profit is $400,000.
(a) Find the slope of the line passing through the points (500, 100000) and (1500, 400000).
(b) Interpret the slope as a rate of change.

93. *Revenue*   The graph shows revenue received from selling x screwdrivers.

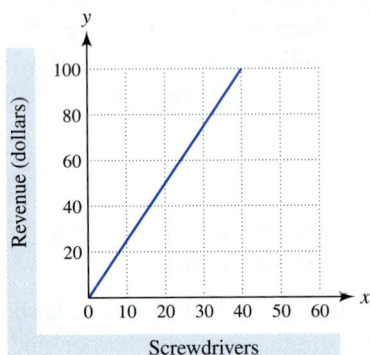

(a) Find the slope of the line shown.
(b) Interpret the slope as a rate of change.

94. *Electricity*   The graph shows how voltage is related to amperage in an electrical circuit. The slope corresponds to the resistance in ohms. Find the resistance in this electrical circuit.

95. *Two-Cycle Engines*   Two-cycle engines used in jet skis, snowmobiles, chain saws, and outboard motors require a mixture of gas and oil to run properly. For certain engines the gallons of oil O that should be added to G gallons of gasoline is given by $O = \frac{1}{50}G$.
(*Source:* Johnson Outboard Motor Company.)
(a) How many gallons of oil should be added to 50 gallons of gasoline and to 100 gallons of gasoline?
(b) Find the slope of the graph of the equation.
(c) Interpret the slope as a rate of change.

96. *Plant Growth*   The graph shows the seasonal growth of a type of grass for various amounts of rainfall.

(a) Find the slope of the line shown.
(b) Interpret the slope as a rate of change.

97. *Walking for Charities*   The table lists the amount of money M in dollars raised for walking various distances in miles for a charity.

x	0	5	10	15
M	0	100	250	450

(a) Make a line graph of the data.
(b) Calculate the slope of each line segment.
(c) Interpret each slope as a rate of change.

98. *Insect Population*   The table lists the number N of black flies in thousands per acre after x weeks.

x	0	2	4	6
N	3	4	10	18

(a) Make a line graph of the data.
(b) Calculate the slope of each line segment.
(c) Interpret each slope as a rate of change.

99. *Median Household Income*   In 1980, median family income was about $18,000, and in 2000 it was about $42,000.   (*Source:* Department of the Treasury.)
(a) Find the slope of the line passing through the points (1980, 18000) and (2000, 42000).
(b) Interpret the slope as a rate of change.
(c) If this trend continues, estimate the median family income in 2005.

100. *Minimum Wage* In 1990, the minimum wage was about \$3.50 per hour, and in 2005 it was \$5.15. (*Source:* Department of Labor.)

(a) Find the slope of the line passing through the points (1990, 3.5) and (2005, 5.15).

(b) Interpret the slope as a rate of change.

(c) If this trend continues, estimate the minimum wage in 2015.

101. *Rate of Change* Suppose that $y = -2x + 10$ is graphed in the first quadrant of the xy-plane where the x-axis is labeled "Time (minutes)" and the y-axis is labeled "Distance (feet)." If this graph represents the distance y that an ant is from a stone after x minutes, answer each of the following.

(a) Is the ant moving toward or away from the stone?

(b) Initially, how far from the stone is the ant?

(c) At what rate is the ant moving?

(d) What is the value of x (time) when the ant reaches the stone?

102. *Rate of Change* Suppose that $y = 15x + 8$ is graphed in the first quadrant of the xy-plane where

the x-axis is labeled "Time (minutes)" and the y-axis is labeled "Distance (feet)." If this graph represents the distance y that a frog is from a tree after x minutes, answer each of the following.

(a) Is the frog moving toward or away from the tree?

(b) Initially, how far from the tree is the frog?

(c) At what rate is the frog moving?

(d) What is the value of x when the frog is 53 feet from the tree?

WRITING ABOUT MATHEMATICS

103. If you are given two points and the slope formula $m = \frac{y_2 - y_1}{x_2 - x_1}$, does it matter which point is (x_1, y_1) and which point is (x_2, y_2)? Explain.

104. Suppose that a line approximates the distance y in miles that a person drives in x hours. What does the slope of the line represent? Give an example.

105. Describe the information that the slope m gives about a line. Be as complete as possible.

106. Could one line have two slopes? Explain.

CHECKING BASIC CONCEPTS
SECTIONS 3.3 AND 3.4

1. Identify the x- and y-intercepts in the graph.

2. Complete the table for the equation $2x - y = 2$. Then determine the x- and y-intercepts.

x	-2	-1	0	1	2
y	-6				

3. Find any intercepts for the graphs of the equations and then graph each linear equation.

(a) $x - 2y = 6$ (b) $y = 2$ (c) $x = -1$

4. Write the equations of a horizontal line and a vertical line that pass through the point $(-2, 4)$.

5. If possible, find the slope of the line passing through each pair of points.

(a) $(-2, 3), (2, 6)$ (b) $(-5, 3), (0, 3)$

(c) $(1, 5), (1, 8)$

6. Find the slope of the line shown.

7. Sketch a line passing through the point $(-3, 1)$ and having slope 2.

continued on next page

continued from previous page

8. *Modeling* The line graph shows the depth of water in a small pond before and after a rain storm.
   (a) Estimate the slope of each line segment.
   (b) Interpret each slope as a rate of change.
   (c) Describe what happened to the amount of water in the pond.

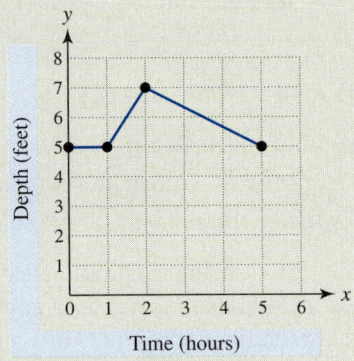

9. *Minnesota County Population* Dodge County has an area of 450 square miles with a population of 18,100. Meeker County has an area of 600 square miles with a population of 22,300. (*Source:* U.S. Census Bureau.)
   (a) Find the slope of the line passing through the points (450, 18100) and (600, 22300).
   (b) Interpret the slope as a rate of change.
   (c) Is this rate of change a valid predictor of the population of a county with an area of 700 square miles? Explain your reasoning.

3.5 SLOPE–INTERCEPT FORM

Finding Slope–Intercept Form ▪ Parallel and Perpendicular Lines

A LOOK INTO MATH ▷   For any two points in the xy-plane, we can draw a unique line passing through them, as illustrated in Figure 3.41(a). Another way we can determine a unique line is to know the y-intercept and the slope. For example, if a line has y-intercept 2 and slope $m = 1$, then the resulting line is shown in Figure 3.41(b). In this section we discuss how to find the equation of a nonvertical line given its slope and y-intercept.

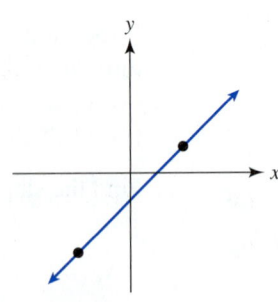

(a) Two points determine a line.   (b) One point and a slope determine a line.

Figure 3.41

Finding Slope–Intercept Form

The graph of $y = 2x + 3$ passes through $(0, 3)$ and $(1, 5)$, as shown in Figure 3.42. The slope of this line is

$$m = \frac{5 - 3}{1 - 0} = 2.$$

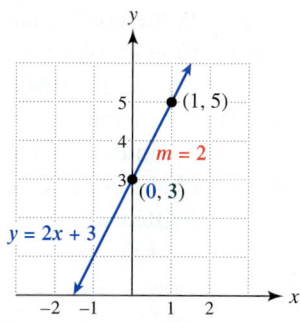

Figure 3.42 Slope 2, y-intercept 3

If $x = \mathbf{0}$ in $y = 2x + 3$, then $y = 2(\mathbf{0}) + 3 = \mathbf{3}$. Thus the graph of $y = \mathbf{2}x + \mathbf{3}$ has slope **2** and y-intercept **3**. In general, the graph of $y = mx + b$ has slope m and y-intercept b. The form $y = mx + b$ is called the *slope–intercept form*.

> ### SLOPE–INTERCEPT FORM
>
> The line with slope m and y-intercept b is given by
> $$y = mx + b,$$
> the **slope–intercept form** of a line.

The graph of the equation $y = 4x - 3$ has slope 4 and y-intercept -3, and the graph of the equation $y = -\frac{1}{2}x + 8$ has slope $-\frac{1}{2}$ and y-intercept 8.

EXAMPLE 1   Using a graph to write the slope–intercept form

For each graph write the slope–intercept form of the line.

(a)

(b)
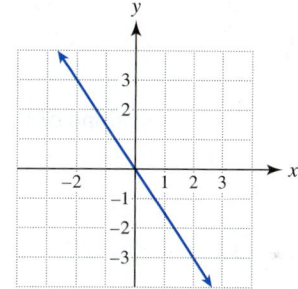

Solution

(a) The graph intersects the y-axis at -2, so the y-intercept is $-\mathbf{2}$. Because the graph rises **1** unit for each **2**-unit increase in x, the slope is $\frac{1}{2}$. The slope–intercept form of the line is $y = \frac{1}{2}x - \mathbf{2}$.

(b) The graph intersects the y-axis at 0, so the y-intercept is 0. Because the graph falls 3 units for each 2-unit increase in x, the slope is $-\frac{3}{2}$. The slope–intercept form of the line is $y = -\frac{3}{2}x + 0$, or $y = -\frac{3}{2}x$.

Now Try Exercise **17**

EXAMPLE 2   Sketching a line

Sketch a line with slope $-\frac{1}{2}$ and y-intercept 1. Write its slope–intercept form.

Solution

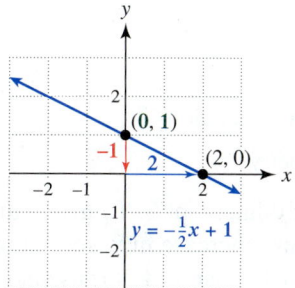

Figure 3.43

For the y-intercept of 1 plot the point $(\mathbf{0}, \mathbf{1})$. Slope $-\frac{1}{2}$ indicates that the graph falls **1** unit for each **2**-unit increase in x. Thus the line passes through the point $(\mathbf{0 + 2}, \mathbf{1 - 1})$, or $(\mathbf{2}, \mathbf{0})$, as shown in Figure 3.43. The slope–intercept form of this line is $y = -\frac{1}{2}x + 1$.

Now Try Exercise **27**

When a linear equation is given in slope–intercept form, $y = mx + b$, the coefficient of the x-term gives the slope and the constant term gives the y-intercept for the graph of the line. For example, the graph of $y = -2x + 5$ has slope -2 and y-intercept 5. However, when a linear equation is *not* given in slope–intercept form, the coefficient of the x-term may not represent the slope and the constant term may not represent the y-intercept. For example, the graph of $2x + 3y = 12$ does *not* have slope 2 and it does *not* have y-intercept 12. To find the correct slope and y-intercept, the equation can first be written in slope–intercept form. This is demonstrated in the next example.

EXAMPLE 3   Writing an equation in slope–intercept form

Write each equation in slope–intercept form. Then give the slope and y-intercept of the line.
(a) $2x + 3y = 12$ **(b)** $x = 2y + 4$

Solution

(a) To write the equation in slope–intercept form, solve for y.

$$2x + 3y = 12 \qquad \text{Given equation}$$
$$3y = -2x + 12 \qquad \text{Subtract } 2x \text{ from each side.}$$
$$y = -\frac{2}{3}x + 4 \qquad \text{Divide each side by 3.}$$

The slope of the line is $-\frac{2}{3}$, and the y-intercept is **4**.

(b) This equation is *not* in slope–intercept form because it is solved for x, not y.

$$x = 2y + 4 \qquad \text{Given equation}$$
$$x - 4 = 2y \qquad \text{Subtract 4 from each side.}$$
$$\frac{1}{2}x - 2 = y \qquad \text{Divide each side by 2.}$$
$$y = \frac{1}{2}x - 2 \qquad \text{Rewrite the equation.}$$

The slope of the line is $\frac{1}{2}$, and the y-intercept is -2.   **Now Try Exercise 43**

EXAMPLE 4   Graphing an equation in slope–intercept form

Write the equation $y = 3 - 2x$ in slope–intercept form and then graph it.

Solution

First write the given equation in slope–intercept form.

$$y = 3 - 2x \qquad \text{Given equation}$$
$$y = 3 + (-2x) \qquad \text{Change subtraction to addition.}$$
$$y = -2x + 3 \qquad \text{Commutative property}$$

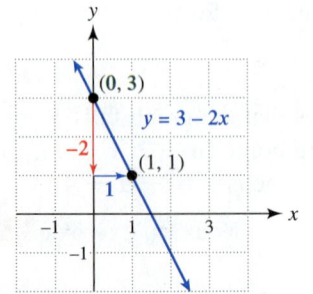

Figure 3.44

The slope–intercept form is $y = -2x + 3$, with slope -2 and y-intercept 3. To graph this equation plot the point $(\mathbf{0, 3})$. The line falls **2** units for each **1** unit increase in x, so plot the point $(\mathbf{0 + 1, 3 - 2}) = (1, 1)$. Sketch a line passing through $(0, 3)$ and $(1, 1)$, as shown in Figure 3.44.   **Now Try Exercise 59**

EXAMPLE 5   Modeling cell phone costs

Roaming with a cell phone costs $5 for the initial connection and $0.50 per minute.
(a) If someone talks for 23 minutes, what is the charge?
(b) Write the slope–intercept form that gives the cost of talking for x minutes.
(c) If the charge is $8.50, how long did the person talk?

Solution
(a) The charge for 23 minutes at $0.50 per minute plus $5 would be

$$0.50 \times 23 + 5 = \$16.50.$$

(b) The rate of increase is $0.50 per minute with an initial cost of $5. Let $y = 0.5x + 5$, where the slope or rate of change is 0.5 and the y-intercept is 5.

(c) To determine how long a person can talk for $8.50, we can solve the following equation.

$0.5x + 5 = 8.5$	Equation to solve
$0.5x = 3.5$	Subtract 5 from each side.
$x = \dfrac{3.5}{0.5}$	Divide each side by 0.5.
$x = 7$	Simplify.

The person talked for 7 minutes. Note that this solution is based on the assumption that the phone company did not round up a fraction of a minute.   Now Try Exercise 77

Parallel and Perpendicular Lines

Slope is an important concept for determining whether two lines are parallel. If two lines have the same slope, they are parallel. For example, the lines $y = 2x$ and $y = 2x - 1$ are parallel because they both have slope 2, as shown in Figure 3.45.

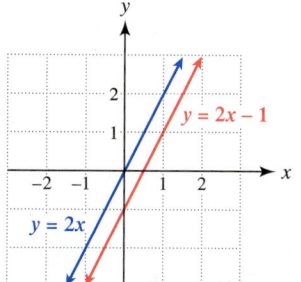

$y = 2x - 1$
$y = 2x$

Figure 3.45

> **PARALLEL LINES**
>
> Two lines with the same slope are parallel.
>
> Two nonvertical parallel lines have the same slope.

NOTE: Two vertical lines are parallel and the slope of each is undefined.

EXAMPLE 6   Finding parallel lines

Find the slope–intercept form of a line parallel to $y = -2x + 3$ and passing through the point $(-2, 3)$. Sketch a graph of each line.

Solution
Because the line $y = -2x + 3$ has slope -2, any parallel line also has slope -2 with slope–intercept form $y = -2x + b$ for some b. The value of b can be found by substituting the point $(-2, 3)$ in the slope–intercept form.

$y = -2x + b$	Slope–intercept form
$3 = -2(-2) + b$	Let $x = -2$ and $y = 3$.
$3 = 4 + b$	Multiply.
$-1 = b$	Subtract 4 from each side.

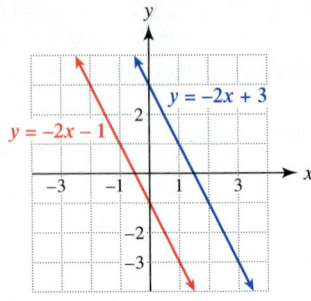

Figure 3.46

The y-intercept is -1, and so the slope–intercept form is $y = -2x - 1$. The graphs of the equations $y = -2x + 3$ and $y = -2x - 1$ are shown in Figure 3.46. Note that they are parallel lines, both with slope -2 but with different y-intercepts, 3 and -1.

Now Try Exercise 67

The lines shown in Figure 3.47 are perpendicular because they intersect at a 90° angle. Rather than measure the angle between two intersecting lines, we can determine whether two lines are perpendicular from their slopes. The slopes of perpendicular lines satisfy the following properties.

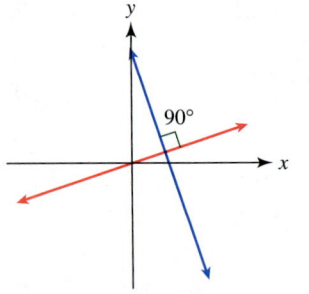

Figure 3.47 Perpendicular Lines

PERPENDICULAR LINES

If two perpendicular lines have nonzero slopes m_1 and m_2, then $m_1 \cdot m_2 = -1$.

If two lines have slopes m_1 and m_2 such that $m_1 \cdot m_2 = -1$, then they are perpendicular lines.

NOTE: A vertical line and a horizontal line are perpendicular.

Table 3.18 shows examples of slopes m_1 and m_2 that result in perpendicular lines. Note that $m_1 \cdot m_2 = -1$ and that $m_2 = -\frac{1}{m_1}$. That is, the product of the two slopes is -1, and the two slopes are negative reciprocals of each other.

TABLE 3.18 Slopes of Perpendicular Lines

m_1	1	$-\frac{1}{2}$	-4	$\frac{2}{3}$	$\frac{3}{4}$
m_2	-1	2	$\frac{1}{4}$	$-\frac{3}{2}$	$-\frac{4}{3}$

We can use these concepts to find equations of perpendicular lines, as illustrated in the next two examples.

EXAMPLE 7 Finding perpendicular lines

Find the slope–intercept form of a line passing through the origin that is perpendicular to each line.

(a) $y = 3x$ **(b)** $y = -\frac{2}{5}x + 5$ **(c)** $-3x + 4y = 24$

Solution

(a) If a line passes through the origin, then its y-intercept is 0 with slope–intercept form $y = mx$. The given line $y = 3x$ has slope $m_1 = 3$, so a line perpendicular to it has slope

$$m_2 = -\frac{1}{m_1} = -\frac{1}{3}.$$

The required slope–intercept form is $y = -\frac{1}{3}x$.

(b) The given line $y = -\frac{2}{5}x + 5$ has slope $m_1 = -\frac{2}{5}$, so a line perpendicular to it has slope $m_2 = \frac{5}{2}$. The required slope–intercept form is $y = \frac{5}{2}x$.

(c) To determine the slope of the given line, first write the equation in slope–intercept form.

$$-3x + 4y = 24 \qquad \text{Given equation}$$

$$4y = 3x + 24 \qquad \text{Add } 3x \text{ to each side.}$$

$$y = \frac{3}{4}x + 6 \qquad \text{Divide each side by 4.}$$

The slope of the given line is $m_1 = \frac{3}{4}$, so a line perpendicular to it has slope $m_2 = -\frac{4}{3}$. The required slope–intercept form is $y = -\frac{4}{3}x$.

Now Try Exercise **71**

EXAMPLE 8 Finding a perpendicular line

Find the slope–intercept form of the line perpendicular to $y = -\frac{1}{2}x + 1$ and passing through the point $(1, -1)$. Sketch each line in the same xy-plane.

Solution
The line $y = -\frac{1}{2}x + 1$ has slope $m_1 = -\frac{1}{2}$. Any line perpendicular to it has slope $m_2 = 2$ with slope–intercept form $y = 2x + b$ for some b. The value of b can be found by substituting the point $(1, -1)$ in the slope–intercept form.

$$y = 2x + b \qquad \text{Slope–intercept form}$$

$$-1 = 2(1) + b \qquad \text{Let } x = 1 \text{ and } y = -1.$$

$$-1 = 2 + b \qquad \text{Multiply.}$$

$$-3 = b \qquad \text{Subtract 2 from each side.}$$

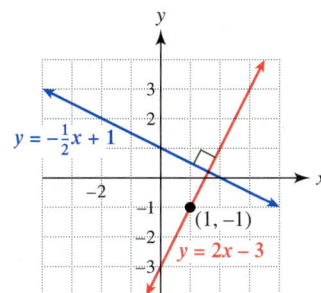

Figure 3.48

The slope–intercept form is $y = 2x - 3$. The graphs of $y = -\frac{1}{2}x + 1$ and $y = 2x - 3$ are shown in Figure 3.48. Note that the point $(1, -1)$ lies on the graph of $y = 2x - 3$.

Now Try Exercise **73**

3.5 PUTTING IT ALL TOGETHER

The following table shows important forms of an equation of a line.

Concept	Comments	Example
Slope–Intercept Form $y = mx + b$	A unique equation for a line, determined by the slope m and the y-intercept b	An equation of the line with slope $m = 3$ and y-intercept $b = -5$ is $y = 3x - 5$.
Parallel Lines	$y = m_1x + b_1$ and $y = m_2x + b_2$, where $m_1 = m_2$ Nonvertical parallel lines have the same slope. Two vertical lines are parallel.	The lines $y = 2x - 1$ and $y = 2x + 2$ are parallel because they both have slope 2.

continued on next page

continued from previous page

Concept	Comments	Example
Perpendicular Lines	$y = m_1x + b_1$ and $y = m_2x + b_2$, where $m_1m_2 = -1$ Perpendicular lines which are neither vertical nor horizontal have slopes whose product equals -1. A vertical line and a horizontal line are perpendicular.	The lines $y = 3x - 1$ and $y = -\frac{1}{3}x + 2$ are perpendicular because $$m_1m_2 = 3\left(-\frac{1}{3}\right) = -1.$$

3.5 Exercises

CONCEPTS

1. The slope–intercept form of a line is _____.

2. In the slope–intercept form of a line, m represents the _____ of the line.

3. In the slope–intercept form of a line, b represents the _____ of the line.

4. If $m = 0$ in the slope–intercept form of a line, then its graph is a _____ line.

5. If $b = 0$ in the slope–intercept form of a line, then its graph passes through the _____.

6. If a line passes through $(0, 5)$ with slope 2, then its slope–intercept form is _____.

7. Two lines with the same slope are _____.

8. Two nonvertical parallel lines have the same _____.

9. If m_1 and m_2 are the slopes of two lines where $m_1 \cdot m_2 = -1$, the lines are _____.

10. If two perpendicular lines have nonzero slopes, the slopes are negative _____.

Exercises 11–16: Match the description with its graph (a.–f.).

11. A line with positive slope and negative y-intercept

12. A line with positive slope and positive y-intercept

13. A line with negative slope and y-intercept 0

14. A line with negative slope and nonzero y-intercept

15. A line with no x-intercept

16. A line with no y-intercept

a.   **b.**

c.   **d.**

e.

f.

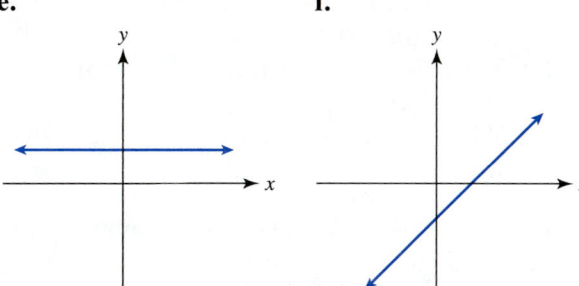

SLOPE–INTERCEPT FORM

Exercises 17–26: Write the slope–intercept form for the line shown.

17.

18.

19.

20.

21.

22.

23.

24.

25.

26.

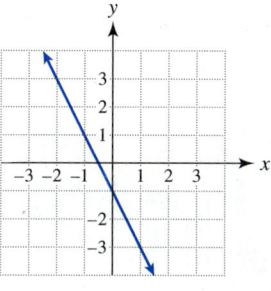

Exercises 27–36: Sketch a line with the given slope m and y-intercept b.   Write its slope–intercept form.

27. $m = 1, b = 2$

28. $m = -1, b = 3$

29. $m = 2, b = -1$

30. $m = -3, b = 2$

31. $m = -\frac{1}{2}, b = -2$

32. $m = -\frac{2}{3}, b = 0$

33. $m = \frac{1}{3}, b = 0$

34. $m = 3, b = -3$

35. $m = -2, b = 1$

36. $m = \frac{1}{2}, b = -1$

Exercises 37–48: Do the following.

   (a) Write the equation in slope–intercept form.

   (b) Give the slope and y-intercept of the line.

37. $x + y = 4$

38. $x - y = 6$

39. $2x + y = 4$

40. $-4x + y = 8$

41. $x - 2y = -4$

42. $x + 3y = -9$

43. $2x - 3y = 6$

44. $4x + 5y = 20$

45. $x = 4y - 6$

46. $x = -3y + 2$

47. $\frac{1}{2}x + \frac{3}{2}y = 1$

48. $-\frac{3}{4}x + \frac{1}{2}y = \frac{1}{2}$

Exercises 49–60: Graph the equation.

49. $y = -3x + 2$

50. $y = \frac{1}{2}x - 1$

51. $y = \frac{1}{3}x$ **52.** $y = -2x$

53. $y = 2$ **54.** $y = -3$

55. $y = -x + 3$ **56.** $y = \frac{2}{3}x - 2$

57. $y = -\frac{1}{2}x + 1$ **58.** $y = -2x + 2$

59. $y = 2 - x$ **60.** $y = 2 - 3x$

Exercises 61–64: The table shows points that all lie on the same line. Find the slope–intercept form for the line.

61.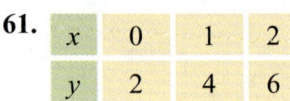

x	0	1	2
y	2	4	6

62.

x	−1	0	1
y	4	8	12

63.

x	−2	0	2
y	−4	−2	0

64.

x	0	2	4
y	6	3	0

PARALLEL AND PERPENDICULAR LINES

Exercises 65–74: Find the slope–intercept form of the line satisfying the given conditions.

65. Slope $\frac{4}{7}$, y-intercept 3

66. Slope $-\frac{1}{2}$, y-intercept -7

67. Parallel to $y = 3x + 1$, passing through $(0, 0)$

68. Parallel to $y = -2x$, passing through $(0, 1)$

69. Parallel to $2x + 4y = 5$, passing through $(1, 2)$

70. Parallel to $-x - 3y = 9$, passing through $(-3, 1)$

71. Perpendicular to $y = -\frac{1}{2}x - 3$, passing through $(0, 0)$

72. Perpendicular to $y = \frac{3}{4}x - \frac{1}{2}$, passing through $(3, -2)$

73. Perpendicular to $x = -\frac{1}{3}y$, passing through $(-1, 0)$

74. Perpendicular to $6x - 3y = 18$, passing through $(4, -3)$

APPLICATIONS

75. *Rental Cars*   Driving a rental car x miles costs $y = 0.25x + 25$ dollars.
  (a) How much would it cost to rent the car but not drive it?

(b) How much does it cost to drive the car 1 *additional* mile?

(c) What is the y-intercept of $y = 0.25x + 25$? What does it represent?

(d) What is the slope of $y = 0.25x + 25$? What does it represent?

76. *Calculating Rainfall*   The total rainfall y in inches that fell x hours past noon is given by $y = \frac{1}{2}x + 3$.
  (a) How much rainfall was there at noon?
  (b) At what rate was rain falling in the afternoon?
  (c) What is the y-intercept of $y = \frac{1}{2}x + 3$? What does it represent?
  (d) What is the slope of $y = \frac{1}{2}x + 3$? What does it represent?

77. *Long-Distance Phone Service*   Long-distance phone service costs $3.95 per month plus $0.07 per minute. (Assume that a partial minute is not rounded up.)
  (a) During July, a person talks a total of 50 minutes. What is the charge?
  (b) Write an equation in slope–intercept form that gives the monthly cost C of talking long distance for x minutes.
  (c) If the long-distance charge for one month is $8.64, how much time did the person spend talking on the phone?

78. *Electrical Rates*   Electrical service costs $8 per month plus $0.10 per kilowatt-hour of electricity used. (Assume that a partial kilowatt-hour is not rounded up.)
  (a) If the resident of an apartment uses 650 kilowatt-hours in 1 month, what is the charge?
  (b) Write an equation in slope–intercept form that gives the cost C of using x kilowatt-hours in 1 month.
  (c) If the monthly electrical bill for the apartment's resident is $43, how many kilowatt-hours were used?

79. *Cost of Driving*   The cost of driving a car includes both fixed costs and mileage costs. Assume that it costs $164.30 per month for insurance and car payments and $0.35 per mile for gasoline, oil, and routine maintenance.
  (a) Find values for m and b so that $y = mx + b$ models the monthly cost of driving the car x miles.
  (b) What does the value of b represent?

80. *Antarctic Ozone Layer*   The ozone layer occurs in Earth's atmosphere between altitudes of 12 and 18 miles and is an important filter of ultraviolet light from the sun. The thickness of the ozone layer is frequently measured in Dobson units. An average value is 300 Dobson units. In 1991, the reported minimum in the antarctic *ozone hole* was about 110 Dobson units.

(*Source:* R. Huffman, *Atmospheric Ultraviolet Remote Sensing.*)

(a) The equation $T = 0.01D$ describes the thickness T in millimeters of an ozone layer that is D Dobson units. How many millimeters thick was the ozone layer over the antarctic in 1991?

(b) What is the average thickness of the ozone layer in millimeters?

WRITING ABOUT MATHEMATICS

81. Explain how the values of m and b can be used to graph the equation $y = mx + b$.

82. Explain how to find the value of b in the equation $y = 2x + b$ if the point $(3, 4)$ lies on the line.

GROUP ACTIVITY
WORKING WITH REAL DATA

Directions: Form a group of 2 to 4 people. Select someone to record the group's responses for this activity. All members of the group should work cooperatively to answer the questions. If your instructor asks for the results, each member of the group should be prepared to respond.

Exercises 1–5: In this set of exercises you are to use your knowledge of equations of lines to model the average annual cost of tuition and fees.

1. *Cost of Tuition*   In 2000, the average cost of tuition and fees at *private* four-year colleges was $16,200, and in 2005 it was $20,100. Sketch a line that passes through the points (2000, 16200) and (2005, 20100).
(*Source:* The College Board.)

2. *Rate of Change in Tuition*   Calculate the slope of the line in your graph. Interpret this slope as a rate of change.

3. *Modeling Tuition*   Find the slope–intercept form of the line in your sketch. What is the y-intercept and does it have meaning in this situation?

4. *Predicting Tuition*   Use your equation to estimate tuition and fees in 2001 and compare it to the known value of $17,300. Estimate tuition and fees in 2010.

5. *Public Tuition*   In 2000, the average cost of tuition and fees at *public* four-year colleges was $3500, and in 2005 it was $5100. Repeat Exercises 1–4 for these data. Note that the known value for 2001 is $3700.
(*Source:* The College Board.)

3.6 POINT–SLOPE FORM

Derivation of Point–Slope Form ■ Finding Point–Slope Form ■ Applications

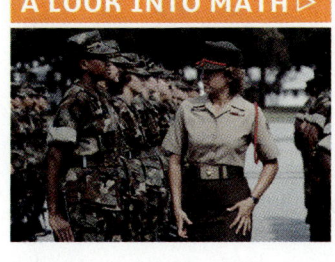

A LOOK INTO MATH ▷ In 1995, there were 690 female officers in the Marine Corps, and by 2003 this number had increased to about 1090. This growth is illustrated in Figure 3.49 on the next page, where the line passes through the points (1995, 690) and (2003, 1090). Because two points determine a unique line, we can find the equation of this line and use it to *estimate* the number of female officers in other years. In this section we discuss how to find this equation by using the *point–slope form*, rather than the slope–intercept form of the equation of a line.

(*Source:* Department of Defense.)

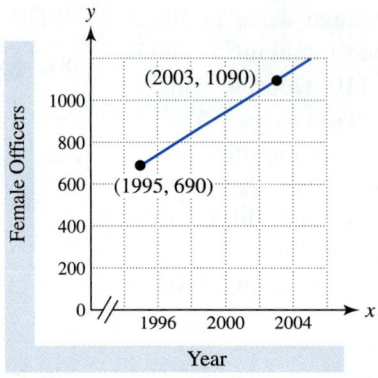

Figure 3.49 Female Officers in the Marine Corps

Derivation of Point–Slope Form

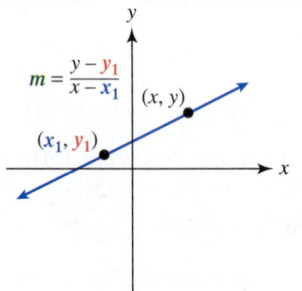

Figure 3.50 Slope of a Line

If we know the slope and y-intercept of a line, we can write its slope–intercept form, $y = mx + b$, which is an example of an **equation of a line**. The point–slope form is a different type of equation of a line.

Suppose that a (nonvertical) line with slope m passes through the point (x_1, y_1). If (x, y) is a different point on this line, then $m = \dfrac{y - y_1}{x - x_1}$. See Figure 3.50. Using this slope formula, we can find the point–slope form.

$$m = \frac{y - y_1}{x - x_1} \qquad \text{Slope formula}$$

$$m \cdot (x - x_1) = \frac{y - y_1}{x - x_1} \cdot (x - x_1) \qquad \text{Multiply each side by } (x - x_1).$$

$$m(x - x_1) = y - y_1 \qquad \text{Simplify.}$$

$$y - y_1 = m(x - x_1) \qquad \text{Rewrite the equation.}$$

$$y = m(x - x_1) + y_1 \qquad \text{Add } y_1 \text{ to each side.}$$

The equation $y - y_1 = m(x - x_1)$ is traditionally called the *point–slope form*. An equivalent form that is helpful when graphing is $y = m(x - x_1) + y_1$. Both equations are in *point–slope form*.

POINT–SLOPE FORM

The line with slope m passing through the point (x_1, y_1) is given by

$$y - y_1 = m(x - x_1), \quad \text{or equivalently,}$$

$$y = m(x - x_1) + y_1,$$

the **point–slope form** of a line.

Finding Point–Slope Form

In the next example we find a point–slope form for a line. Note that *any* point that lies on the line can be used in its point–slope form.

EXAMPLE 1 Finding a point–slope form

Use the labeled point in Figure 3.51 to write a point–slope form for the line and then simplify it to the slope–intercept form.

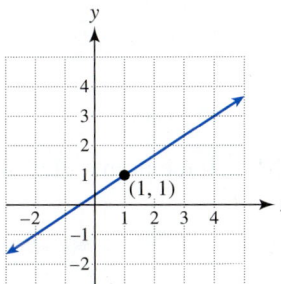

Figure 3.51

Solution

The graph rises 2 units for each 3 units of horizontal run, so the slope is $\frac{2}{3}$. Let $m = \frac{2}{3}$ and $(x_1, y_1) = (1, 1)$ in the point–slope form.

$$y - y_1 = m(x - x_1) \qquad \text{Point–slope form}$$

$$y - 1 = \frac{2}{3}(x - 1) \qquad \text{Let } m = \frac{2}{3}, x_1 = 1, \text{ and } y_1 = 1.$$

A point–slope form for the line is $y - 1 = \frac{2}{3}(x - 1)$.

To simplify to the slope–intercept form, apply the distributive property.

$$y - 1 = \frac{2}{3}x - \frac{2}{3} \qquad \text{Distributive property}$$

$$y = \frac{2}{3}x + \frac{1}{3} \qquad \text{Add 1, or } \frac{3}{3}, \text{ to each side.}$$

The slope–intercept form is $y = \frac{2}{3}x + \frac{1}{3}$.

Now Try Exercise 15

EXAMPLE 2 Finding a point–slope form

Find a point–slope form for a line passing through the point $(-2, 3)$ with slope $-\frac{1}{2}$. Does the point $(2, 1)$ lie on this line?

Solution

Let $m = -\frac{1}{2}$ and $(x_1, y_1) = (-2, 3)$ in the point–slope form.

$$y - y_1 = m(x - x_1) \qquad \text{Point–slope form}$$

$$y - 3 = -\frac{1}{2}\left(x - (-2)\right) \qquad x_1 = -2, y_1 = 3, \text{ and } m = -\frac{1}{2}$$

$$y - 3 = -\frac{1}{2}(x + 2) \qquad \text{Simplify.}$$

To determine whether $(2, 1)$ lies on the line, substitute $x = 2$ and $y = 1$ in the equation.

$$1 - 3 \overset{?}{=} -\frac{1}{2}(2 + 2) \qquad \text{Let } x = 2 \text{ and } y = 1.$$

$$-2 \overset{?}{=} -\frac{1}{2}(4) \qquad \text{Simplify.}$$

$$-2 = -2 \qquad \text{A true statement}$$

The point $(2, 1)$ lies on the line because it satisfies the point–slope form.

Now Try Exercises 9, 19

In the next example we use the point–slope form to find the equation of a line passing through two points.

EXAMPLE 3   Finding an equation of a line

Use the point–slope form to find an equation of the line passing through the points $(1, -4)$ and $(-2, 5)$.

Solution

Before we can apply the point–slope form, we must find the slope.

$$m = \frac{y_2 - y_1}{x_2 - x_1} \qquad \text{Slope formula}$$

$$= \frac{5 - (-4)}{-2 - 1} \qquad x_1 = 1, y_1 = -4, x_2 = -2, \text{ and } y_2 = 5$$

$$= -3 \qquad \text{Simplify.}$$

We can let either the point $(1, -4)$ or the point $(-2, 5)$ be (x_1, y_1) in the point–slope form. If $(x_1, y_1) = (\mathbf{1}, -\mathbf{4})$, then the equation of the line becomes the following.

$$y - y_1 = m(x - x_1) \qquad \text{Point–slope form}$$

$$y - (-\mathbf{4}) = -\mathbf{3}(x - \mathbf{1}) \qquad x_1 = 1, y_1 = -4, \text{ and } m = -3$$

$$y + 4 = -3(x - 1) \qquad \text{Simplify.}$$

If we let $(x_1, y_1) = (-2, 5)$, the point–slope form becomes

$$y - 5 = -3(x + 2). \qquad \text{Now Try Exercise } 25$$

Although the two point–slope forms in Example 3 might appear to be different, they actually are equivalent because they simplify to the same slope–intercept form.

$y + 4 = -3(x - 1)$	$y - 5 = -3(x + 2)$	Point–slope forms
$y = -3(x - 1) - 4$	$y = -3(x + 2) + 5$	Addition property
$y = -3x + 3 - 4$	$y = -3x - 6 + 5$	Distributive property
$y = -3x - 1$	$y = -3x - 1$	Identical slope–intercept forms

MAKING CONNECTIONS

Slope–Intercept and Point–Slope Forms

The slope–intercept form, $y = mx + b$, is unique because any nonvertical line has one slope m and one y-intercept b. The point–slope form, $y - y_1 = m(x - x_1)$, is *not* unique because (x_1, y_1) can be any point that lies on the line. However, any point–slope form can be simplified to a unique slope–intercept form.

EXAMPLE 4   Finding equations of lines

Find the slope–intercept form for the line that satisfies the conditions.
(a) Slope $\frac{1}{2}$, passing through $(-2, 4)$
(b) x-intercept -3, y-intercept 2
(c) Perpendicular to $y = -\frac{2}{3}x$, passing through $\left(\frac{2}{3}, 3\right)$

Solution
(a) Substitute $m = \frac{1}{2}$, $x_1 = -2$, and $y_1 = 4$ in the point–slope form.

$$y - y_1 = m(x - x_1) \qquad \text{Point–slope form}$$

$$y - 4 = \frac{1}{2}(x + 2) \qquad \text{Substitute and simplify.}$$

$$y - 4 = \frac{1}{2}x + 1 \qquad \text{Distributive property}$$

$$y = \frac{1}{2}x + 5 \qquad \text{Add 4 to each side.}$$

(b) The line passes through the points $(-3, 0)$ and $(0, 2)$. The slope of the line is

$$m = \frac{2 - 0}{0 - (-3)} = \frac{2}{3}.$$

Because the line has slope $\frac{2}{3}$ and y-intercept 2, the slope–intercept form is

$$y = \frac{2}{3}x + 2.$$

(c) The slope of the given line is $m_1 = -\frac{2}{3}$, so the slope of a line perpendicular to it is $\frac{3}{2}$. Let $m = \frac{3}{2}$, $x_1 = \frac{2}{3}$, and $y_1 = 3$ in the point–slope form.

$$y - y_1 = m(x - x_1) \qquad \text{Point–slope form}$$

$$y - 3 = \frac{3}{2}\left(x - \frac{2}{3}\right) \qquad \text{Substitute.}$$

$$y - 3 = \frac{3}{2}x - 1 \qquad \text{Distributive property}$$

$$y = \frac{3}{2}x + 2 \qquad \text{Add 3 to each side.}$$

Now Try Exercises 45, 49, 53

In the next example the point–slope form is used to find the slope–intercept form of a line that passes through several points given in a table.

EXAMPLE 5   Using a table to find slope–intercept form

The points in the table lie on a line. Find the slope–intercept form of the line.

x	2	4	6	8
y	2	1	0	-1

Solution
The y-values in the table decrease one unit for every two-unit increase in the x-values so the line has a "rise" of -1 when the run is 2. The slope is $m = \frac{\text{rise}}{\text{run}} = \frac{-1}{2} = -\frac{1}{2}$. Because the

x	y
2	2
4	1
6	0
8	−1

y-intercept is not given in the table, the value of b in the equation $y = mx + b$ is not immediately available. However, *any* point from the table, which is repeated in vertical form in the margin, can be used to obtain a point–slope form of the line, which can then be simplified to slope–intercept form. Letting $(x_1, y_1) = (2, 2)$ and $m = -\frac{1}{2}$ in the point–slope form yields the following result.

$$y - y_1 = m(x - x_1) \qquad \text{Point–slope form}$$

$$y - 2 = -\frac{1}{2}(x - 2) \qquad \text{Substitute.}$$

$$y - 2 = -\frac{1}{2}x + 1 \qquad \text{Distributive property}$$

$$y = -\frac{1}{2}x + 3 \qquad \text{Add 2 to each side.}$$

This result can be checked by substituting each x-value from the table in the equation. For example, when $x = 4$, the corresponding y-value is $y = -\frac{1}{2}(4) + 3 = -2 + 3 = 1$. This agrees with the table.

Now Try Exercise 55

NOTE: The equation in Example 5 can be obtained without using the point–slope form. The value of b can be found by letting $(x, y) = (2, 2)$ and $m = -\frac{1}{2}$ in the slope–intercept form.

$$y = mx + b \qquad \text{Slope–intercept form}$$

$$2 = -\frac{1}{2}(2) + b \qquad \text{Substitute.}$$

$$2 = -1 + b \qquad \text{Multiply.}$$

$$3 = b \qquad \text{Add 1 to each side.}$$

Because $m = -\frac{1}{2}$ and $b = 3$, the slope–intercept form is $y = -\frac{1}{2}x + 3$.

Applications

In the next example we find the equation of the line that models the Marine Corps data presented at the beginning of this section.

EXAMPLE 6 Modeling numbers of female officers

In 1995, there were 690 female officers in the Marine Corps, and by 2003 this number had increased to about 1090. Refer to Figure 3.49 at the beginning of this section.
(a) Use the point (1995, 690) to find a point–slope form of the line shown in Figure 3.49.
(b) Interpret the slope as a rate of change.
(c) Use Figure 3.49 to estimate the number of female officers in 2000. Then use your equation from part (a) to approximate this number. How do your answers compare?

Solution
(a) The slope of the line passing through (1995, 690) and (2003, 1090) is

$$m = \frac{1090 - 690}{2003 - 1995} = 50.$$

If we let $x_1 = 1995$ and $y_1 = 690$, then the point–slope form becomes

$$y - 690 = 50(x - 1995) \quad \text{or} \quad y = 50(x - 1995) + 690.$$

(b) Slope $m = 50$ indicates that the number of female officers increased, *on average,* by about 50 officers per year.

(c) From Figure 3.49, it appears that the number of female officers in 2000 was about 950. To estimate this value let $x = \mathbf{2000}$ in the equation found in part (a).

$$y = 50(\mathbf{2000} - 1995) + 690 = 940$$

Although the graphical estimate and calculated answers are not exactly equal, they are approximately equal. Estimations made from a graph usually are not exact.

Now Try Exercise 67

In the next example we review several concepts of lines.

EXAMPLE 7   Modeling water in a pool

A small swimming pool is being emptied by a pump that removes water at a constant rate. After 1 hour the pool contains 5000 gallons, and after 3 hours it contains 3000 gallons.
(a) How fast is the pump removing water?
(b) Find the slope–intercept form of a line that models the amount of water in the pool. Interpret the slope.
(c) Find the y-intercept and the x-intercept. Interpret each.
(d) Sketch a graph of the amount of water in the pool during the first 6 hours.
(e) The point (2, 4000) lies on the graph. Explain its meaning.

Solution

CRITICAL THINKING

Suppose that a line models the amount of water in a swimming pool. What does a positive slope indicate? What does a negative slope indicate?

(a) The pump removes $5000 - 3000 = 2000$ gallons of water in 2 hours, or 1000 gallons per hour.

(b) The line passes through the points (1, 5000) and (3, 3000), so the slope is

$$m = \frac{3000 - 5000}{3 - 1} = -1000.$$

One way to find the slope–intercept form is to use the point–slope form.

$$y - y_1 = m(x - x_1) \qquad \text{Point–slope form}$$
$$y - 5000 = -1000(x - 1) \qquad m = -1000, x_1 = 1, \text{ and } y_1 = 5000$$
$$y - 5000 = -1000x + 1000 \qquad \text{Distributive property}$$
$$y = -1000x + 6000 \qquad \text{Add 5000 to each side.}$$

Slope -1000 means that the pump is *removing* 1000 gallons per hour.

(c) The y-intercept is 6000 and indicates that the pool initially contained 6000 gallons. To find the x-intercept let $y = \mathbf{0}$ in the slope–intercept form.

$$\mathbf{0} = -1000x + 6000 \qquad \text{Let } y = 0.$$
$$1000x = 6000 \qquad \text{Add } 1000x \text{ to each side.}$$
$$x = \frac{6000}{1000} \qquad \text{Divide by 1000.}$$
$$x = \mathbf{6} \qquad \text{Simplify.}$$

An x-intercept of 6 indicates that the pool is empty after 6 hours.

(d) The x-intercept is **6**, and the y-intercept is **6000**. Sketch a line passing through (**6**, 0) and (0, **6000**), as shown in Figure 3.52.

(e) The point (2, 4000) indicates that after 2 hours the pool contains 4000 gallons of water.

Now Try Exercise 63

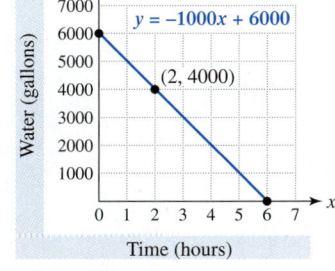

Figure 3.52 Water in a Pool

3.6 PUTTING IT ALL TOGETHER

The following table summarizes the point–slope form of a line. Note that the slope–intercept form is unique for a given line, whereas the point–slope form depends on the point used for (x_1, y_1).

Concept	Comments	Example
Point–Slope Form $$y - y_1 = m(x - x_1) \quad \text{or}$$ $$y = m(x - x_1) + y_1$$	Used to find an equation of a line, given two points or one point and the slope Can always be simplified to slope–intercept form	For two points $(1, 2)$ and $(3, 5)$, first compute $m = \frac{5-2}{3-1} = \frac{3}{2}$. An equation of this line is $$y - 2 = \frac{3}{2}(x - 1) \quad \text{or}$$ $$y = \frac{3}{2}(x - 1) + 2.$$

3.6 Exercises

CONCEPTS

1. How many lines are determined by two distinct points?

2. How many lines are determined by a point and a slope?

3. Give the slope–intercept form of a line.

4. Give the point–slope form of a line.

5. If the point–slope form is used to write an equation of a line that passes through $(1, 3)$, then $x_1 =$ _____ and $y_1 =$ _____.

6. To write a point–slope equation in slope–intercept form, use the _____ property to clear the parentheses.

7. Is the slope–intercept form of a line unique? Explain.

8. Is the point–slope form of a line unique? Explain.

POINT–SLOPE FORM

Exercises 9–14: Determine whether the given point lies on the line.

9. $(-3, 3)$ $y - 1 = -\frac{2}{3}x$

10. $(4, 0)$ $y + 1 = \frac{1}{4}x$

11. $(1, 4)$ $y - 3 = -(x - 1)$

12. $(3, -11)$ $y + 1 = -2(x + 3)$

13. $(0, 4)$ $y = \frac{1}{2}(x + 4) + 2$

14. $(2, -8)$ $y = 3(x - 5) - 1$

Exercises 15–18: Use the labeled point to write a point–slope form for the line.

15.

16.

17.

18.

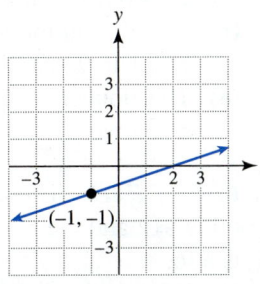

Exercises 19–30: Find a point–slope form for the line that satisfies the stated conditions. When two points are given, use the first point in the point–slope form.

19. Slope 4, passing through $(-3, 1)$

20. Slope -3, passing through $(1, -2)$

21. Slope $\frac{1}{2}$, passing through $(-5, -3)$

22. Slope $-\frac{2}{3}$, passing through $(-3, 6)$

23. Slope 1.5, passing through $(2000, 30)$

24. Slope -10, passing through $(2004, 100)$

25. Passing through $(2, 4)$ and $(-1, -3)$

26. Passing through $(3, -1)$ and $(1, -4)$

27. Passing through $(5, 0)$ and $(0, -3)$

28. Passing through $(-2, 0)$ and $(0, -1)$

29. Passing through $(1990, 15)$ and $(2000, 65)$

30. Passing through $(1999, 5)$ and $(2004, 30)$

Exercises 31–42: Write the point–slope form in slope–intercept form.

31. $y - 4 = 3(x - 2)$

32. $y - 3 = -2(x + 1)$

33. $y + 2 = \frac{1}{3}(x + 6)$

34. $y - 1 = \frac{2}{5}(x + 10)$

35. $y - \frac{3}{4} = \frac{2}{3}(x - 1)$

36. $y + \frac{2}{3} = -\frac{1}{6}(x - 2)$

37. $y = -2(x - 2) + 5$

38. $y = 4(x + 3) - 7$

39. $y = \frac{3}{5}(x - 5) + 1$

40. $y = -\frac{1}{2}(x + 4) - 6$

41. $y = -16(x + 1.5) + 5$

42. $y = -15(x - 1) + 100$

43. Thinking Generally   Find the y-intercept of the line given by $y - y_1 = m(x - x_1)$.

44. Thinking Generally   Find the x-intercept of the line given by $y - y_1 = m(x - x_1)$.

Exercises 45–54: Find the slope–intercept form for the line satisfying the conditions.

45. Slope -2, passing through $(4, -3)$

46. Slope $\frac{1}{5}$, passing through $(-2, 5)$

47. Passing through $(3, -2)$ and $(2, -1)$

48. Passing through $(8, 3)$ and $(-7, 3)$

49. x-intercept 3, y-intercept $\frac{1}{3}$

50. x-intercept 2, y-intercept -3

51. Parallel to $y = 2x - 1$, passing through $(2, -3)$

52. Parallel to $y = -\frac{3}{2}x$, passing through $(0, 20)$

53. Perpendicular to $y = -\frac{1}{2}x + 3$, passing through the point $(6, -3)$

54. Perpendicular to $y = \frac{3}{5}(x + 1) + 3$, passing through the point $(1, -2)$

Exercises 55–58: The points in the table lie on a line. Find the slope–intercept form of the line.

55.

x	1	2	3	4
y	-3	-5	-7	-9

56.

x	2	3	4	5
y	5	8	11	14

57.

x	-1	1	3	5
y	-3	-2	-1	0

58.

x	-1	5	11	17
y	1	-3	-7	-11

GRAPHICAL INTERPRETATION

59. *Distance and Speed*   A person is driving a car along a straight road. The graph shows the distance y in miles that the driver is from home after x hours.
   (a) Is the person traveling toward or away from home?
   (b) The graph passes through (1, 250) and (4, 100). Discuss the meaning of these points.
   (c) How fast is the driver traveling?
   (d) Find the slope–intercept form of the line. Interpret the slope as a rate of change.

60. *Distance and Speed*   A person rides a bicycle at 10 miles per hour, first away from home for 1 hour and then toward home for 1 hour. Sketch a graph that shows the distance d between the bicyclist and home after x hours.

61. *Water and Flow*   The graph shows the amount of water y in a 500-gallon tank after x minutes have elapsed.
   (a) Is water entering or leaving the tank? How much water is in the tank after 4 minutes?
   (b) Find the y-intercept. Explain its meaning.
   (c) Find the slope–intercept form of the line. Interpret the slope as a rate of change.
   (d) After how many minutes will the tank be full?

62. *Water and Flow*   A hose is used to fill a 100-gallon barrel. If the hose delivers 5 gallons of water per minute, sketch a graph of the amount A of water in the barrel during the first 20 minutes.

63. *Change in Temperature*   The outside temperature was 40°F at 1 A.M. and 15°F at 6 A.M. Assume that the temperature changed at a constant rate.
   (a) At what rate did the temperature change?
   (b) Find the slope–intercept form of a line that models the temperature T at x A.M. Interpret the slope as a rate of change.
   (c) Assuming that your equation is valid for times after 6 A.M., find and interpret the x-intercept.
   (d) Sketch a graph that shows the temperature from 1 A.M. to 9 A.M.
   (e) The point (4, 25) lies on the graph. Explain its meaning.

64. *Cost of Fuel*   The cost of buying 5 gallons of fuel oil is $12 and the cost of buying 15 gallons of fuel oil is $36.
   (a) What is the cost of a gallon of fuel oil?
   (b) Find the slope–intercept form of a line that models the cost of buying x gallons of fuel oil. Interpret the slope as a rate of change.
   (c) Find and interpret the x-intercept.
   (d) Sketch a graph that shows the cost of buying 20 gallons or less of fuel oil.
   (e) The point (11, 26.40) lies on the graph. Explain its meaning.

APPLICATIONS

65. *Inmates*   From 1995 to 2005, the number of state and federal inmates y in millions can be modeled by $y = 0.032(x - 1995) + 1.13$, where x is the year. (*Source:* Bureau of Justice.)
   (a) Find the number of inmates in 1995 and 2005.
   (b) What is the slope of the graph of y? Interpret the slope as a rate of change.

66. *Municipal Waste*   From 1960 to 2000, municipal solid waste y in millions of tons can be modeled by $y = 3.6(x - 1960) + 87.8$, where x is the year. (*Source:* Environmental Protection Agency.)
   (a) Find the tons of waste in 1999.
   (b) What is the slope of the graph of y? Interpret the slope as a rate of change.

67. *Tuition and Fees*   The graph models average tuition and fees at public four-year colleges from 2000 to 2006. (**Source:** The College Board.)
  (a) The line passes through the points (2002, 4081) and (2004, 5133). Explain the meaning of each point.
  (b) Find a point–slope form for this line. Interpret the slope as a rate of change.
  (c) Use the graph to estimate tuition and fees in 2005. Then use your equation from part (b) to estimate tuition and fees in 2005.

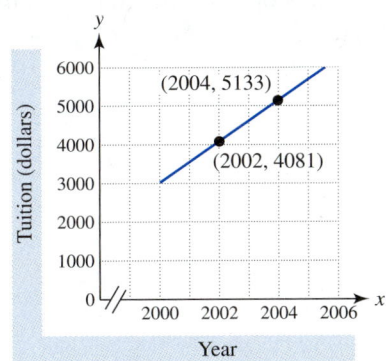

Year

68. *Western Population*   In 1950 the western region of the United States had a population of 20.2 million, and in 1960 this population was 28.4 million. (**Source:** U.S. Census Bureau.)
  (a) Find the slope–intercept form of a line passing through (1950, 20.2) and (1960, 28.4).

(b) Use your equation from part (a) to estimate the western population in 1967.

69. *HIV Infection Rates*   In 2004, there were an estimated 39 million HIV infections worldwide with an annual infection rate of 4.9 million. (**Source:** U.S. Centers for Disease Control and Prevention.)
  (a) Find the slope–intercept form of a line that approximates the cumulative number of HIV infections in millions during year x, where $x \geq 2004$.
  (b) Estimate the number of HIV infections in 2010.

70. *Farm Pollution*   In 1978, the number of farm pollution incidents reported in England and Wales was about 1500, and in 1988 it was about 4000. (**Source:** C. Mason, *Biology of Freshwater Pollution*.)
  (a) Find a slope–intercept form of a line passing through (1978, 1500) and (1988, 4000).
  (b) Use your equation from part (a) to estimate the number of incidents in 1986.

WRITING ABOUT MATHEMATICS

71. Explain how to find the equation of a line passing through two points with coordinates (x_1, y_1) and (x_2, y_2). Give an example.

72. Explain how slope is related to rate of change. Give an example.

CHECKING BASIC CONCEPTS
SECTIONS 3.5 AND 3.6

1. Write the slope–intercept form for the line shown in the graph.

2. Write the equation $4x - 5y = 20$ in slope–intercept form. Give the slope and y-intercept.

3. Graph $y = \frac{1}{2}x - 3$.

4. Write the slope–intercept form of a line that satisfies the following.
  (a) Slope 3, passing through $(0, -2)$
  (b) Perpendicular to $y = \frac{2}{3}x$, passing through the point $(-2, 3)$
  (c) Passing through $(1, -4)$ and $(-2, 3)$

5. Write a point–slope form for a line with slope -2, passing through $(-1, 3)$.

6. Write the equation $y + 3 = -2(x - 2)$ in slope–intercept form.

continued on next page

continued from previous page

7. Find the slope–intercept form of the line passing through the points in the table.

x	-3	-1	1	3
y	-3	1	5	9

8. *Distance and Speed*   A bicyclist is riding at a constant speed and is 36 miles from home at 1 P.M. Two hours later the bicyclist is 12 miles from home.
   (a) Find the slope–intercept form of a line passing through (1, 36) and (3, 12).

(b) How fast is the bicyclist traveling?
(c) When will the bicyclist arrive home?
(d) How far was the bicyclist from home at noon?

9. *Snowfall*   The total amount of snowfall S in inches t hours past noon is given by $S = 2t + 5$.
   (a) How many inches of snow fell by noon?
   (b) At what rate did snow fall in the afternoon?
   (c) What is the S-intercept for the graph of this equation? What does it represent?
   (d) What is the slope for the graph of this equation? What does it represent?

3.7   INTRODUCTION TO MODELING

Basic Concepts ■ Modeling Linear Data

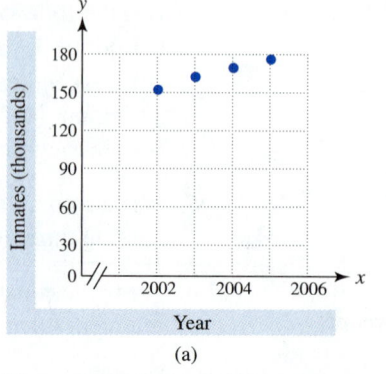

For centuries people have tried to understand the world around them by creating models. For example, a weather forecast is based on a model. Mathematics is used to create these weather models, which often contain thousands of equations.

   A model is an *abstraction* of something that people observed. Not only should a good model describe *known* data, but it should also be able to predict *future* data. In this section we discuss linear models, which are used to describe data that have a constant rate of change.

Basic Concepts

▶ **REAL-WORLD CONNECTION**   Figure 3.53(a) shows a scatterplot of the number of inmates in the federal prison system from 2002 to 2005. The four points in the graph appear to be "nearly" collinear. That is, they appear almost to lie on the same line. Using mathematical modeling, we can find an equation for such a line. Once we have found it, we can use it to make estimates about the federal inmate population. An example of such a line is shown in Figure 3.53(b). (See Exercise 51 at the end of this section.)

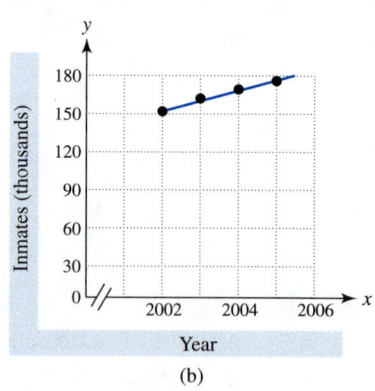

Figure 3.53   Federal Prison Population

Generally, mathematical models are not exact representations of data. Data that might appear to be linear may not be. Although the line in Figure 3.53(b) appears to touch every point, it does not pass through all four points exactly. Figure 3.54(a) shows data modeled *exactly* by a line, whereas Figure 3.54(b) shows data modeled *approximately* by a line. In applications a model is more apt to be approximate than exact.

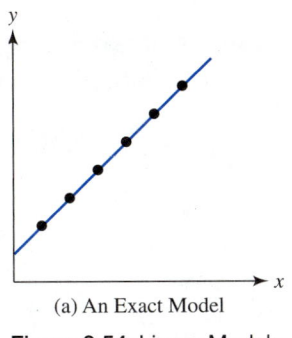

(a) An Exact Model   (b) An Approximate Model

Figure 3.54 Linear Models

EXAMPLE 1

Determining whether a model is exact

A person can vote in the United States at age 18 or over. Table 3.19 shows the voting-age population P in millions for selected years x. Does the equation $P = 3.25x - 6297$ model the data exactly? Explain.

TABLE 3.19

Voting-Age Population

x	2000	2002	2004
P	203	211	216

Source: U.S. Census Bureau.

Solution

To determine whether the equation models the data exactly, let $x = 2000$, 2002, and 2004 in the given equation.

$$x = \mathbf{2000}: \quad P = 3.25(\mathbf{2000}) - 6297 = 203$$

$$x = \mathbf{2002}: \quad P = 3.25(\mathbf{2002}) - 6297 = 209.5$$

$$x = \mathbf{2004}: \quad P = 3.25(\mathbf{2004}) - 6297 = 216$$

The model is *not exact* because it does not predict a voting-age population of 211 million in 2002.   Now Try Exercise 15

Modeling Linear Data

A line can model linear data. In the next example we use a line to model gas mileage.

ЄXAMPLЄ **2**   Determining gas mileage

TABLE 3.20

x	2	4	6	8
y	30	60	90	120

Table 3.20 shows the number of miles y traveled by an SUV on x gallons of gasoline.
(a) Plot the data in the xy-plane. Be sure to label each axis.
(b) Sketch a line that models the data. (You may want to use a ruler.)
(c) Find the equation of the line and interpret the slope of the line.
(d) How far could this SUV travel on 11 gallons of gasoline?

Solution
(a) Plot the points (2, 30), (4, 60), (6, 90), and (8, 120), as shown Figure 3.55(a).

(a)   (b)

Figure 3.55 Gas Mileage of an SUV

(b) Sketch a line similar to the one shown in Figure 3.55(b). This particular line passes through each data point.

(c) First find the slope m of the line by choosing two points that the line passes through, such as (**2**, **30**) and (8, 120).

$$m = \frac{120 - 30}{8 - 2} = \frac{90}{6} = 15$$

Now find the equation of the line passing through (2, 30) with slope 15.

$y - y_1 = m(x - x_1)$	Point–slope form
$y - \mathbf{30} = 15(x - \mathbf{2})$	$x_1 = 2, y_1 = 30$, and $m = 15$
$y - 30 = 15x - 30$	Distributive property
$y = 15x$	Add 30 to each side.

The data are modeled by the equation $y = 15x$. Slope 15 indicates that the mileage of this SUV is 15 miles per gallon.

(d) On 11 gallons of gasoline the SUV could go $y = 15(11) = 165$ miles.

Now Try Єxercise **53**

ЄXAMPLЄ **3**   Modeling linear data

Table 3.21 contains ordered pairs that can be modeled approximately by a line.
(a) Plot the data. Could a line pass through all five points?
(b) Sketch a line that models the data and then determine its equation.

TABLE 3.21

x	1	2	3	4	5
y	3	5	6	10	11

(a)

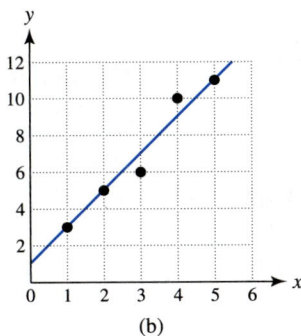

(b)

Figure 3.56

Solution

(a) Plot the ordered pairs $(1, 3)$, $(2, 5)$, $(3, 6)$, $(4, 10)$, and $(5, 11)$, as shown in Figure 3.56(a). The points are not collinear, so it is impossible to sketch a line that passes through *all* five points.

(b) One possibility for a line is shown in Figure 3.56(b). This line passes through three of the five points, and is above one point and below another point. To determine the equation of this line, pick two points that the line passes through. For example, the points $(1, 3)$ and $(5, 11)$ lie on the line. The slope of this line is

$$m = \frac{11 - 3}{5 - 1} = \frac{8}{4} = 2.$$

The equation of the line passing through $(1, 3)$ with slope 2 can be found as follows.

$y - y_1 = m(x - x_1)$	Point–slope form
$y - 3 = 2(x - 1)$	$x_1 = 1, y_1 = 3,$ and $m = 2$
$y - 3 = 2x - 2$	Distributive property
$y = 2x + 1$	Add 3 to each side.

The equation of this line is $y = 2x + 1$.   Now Try Exercise **33**

NOTE: The equation found in Example 3 represents one possible linear model. Other linear models are possible.

▶ **REAL-WORLD CONNECTION** When a quantity increases at a constant rate, it can be modeled with the linear equation $y = mx + b$. This concept is illustrated in the next example.

EXAMPLE 4   Modeling worldwide HIV/AIDS cases in children

At the beginning of 2004, a total of 2.5 million children (under age 15) were living with HIV/AIDS. The rate of new infections was 0.7 million per year.

(a) Write a linear equation $C = mx + b$ that models the total number of children C in millions that were living with HIV/AIDS, x years after January 1, 2004.

(b) Estimate C at the beginning of 2010.

Solution

(a) In the equation $C = mx + b$, the rate of change in HIV infections corresponds to the slope m, and the initial number of cases at the beginning of 2004 corresponds to b. Therefore the equation $C = 0.7x + 2.5$ models the data.

(b) The beginning of 2010 is 6 years after January 1, 2004, so let $x = 6$.

$$C = 0.7(6) + 2.5 = 6.7 \text{ million}$$

NOTE: There were a total of 2.5 million children living with HIV/AIDS at the beginning of 2004 with an infection rate of 0.7 million per year. After 6 years there would be an additional $0.7(6) = 4.2$ million infected children, raising the total number to $2.5 + 4.2 = 6.7$ million.   Now Try Exercise **49**

EXAMPLE 5   **Modeling with linear equations**

Find a linear equation in the form $y = mx + b$ that models the quantity y after x days.
(a) A quantity y is initially 500 and increases at a rate of 6 per day.
(b) A quantity y is initially 1800 and decreases at a rate of 25 per day.
(c) A quantity y is initially 10,000 and remains constant.

Solution

(a) In the equation $y = mx + b$, the y-intercept b represents the initial amount and the slope m represents the rate of change. Therefore $y = 6x + 500$.
(b) The quantity y is decreasing at the rate of 25 per day with an initial amount of 1800, so $y = -25x + 1800$.
(c) The quantity is constant, so $m = 0$. The equation is $y = 10,000$.

Now Try Exercises 27, 29, 31

> **MODELING WITH A LINEAR EQUATION**
>
> To model a quantity y that has a constant rate of change, use the equation
> $$y = mx + b,$$
> where $m = $ (constant rate of change) and $b = $ (initial amount).

3.7 PUTTING IT ALL TOGETHER

The following table summarizes important concepts related to linear modeling.

Concept	Comments	Example
Linear Model	Used to model a quantity that has a constant rate of change	If a total of 2 inches of rain falls before noon, and if rain falls at the rate of $\frac{1}{2}$ inch per hour, then $y = \frac{1}{2}x + 2$ models the total rainfall x hours past noon.
Modeling Linear Data with a Line	1. Plot the data. 2. Sketch a line that either passes through or nearly through the points. 3. Pick two points on the line and find the equation of the line.	To model $(0, 4)$, $(1, 3)$, and $(2, 2)$, plot the points and sketch a line as shown in the accompanying figure. Many times one line cannot pass through all the points. The equation of the line is $y = -x + 4$.

3.7 Exercises

MyMathLab Math XL PRACTICE WATCH DOWNLOAD READ REVIEW

CONCEPTS

1. A model is a(n) _____ of something that people observe.

2. A good model should be able to _____ future data.

3. Linear data are modeled by a(n) _____ equation.

4. If a line passes through all the data points, it is a(n) _____ model.

5. If a line passes near, but not through each data point, it is a(n) _____ model.

6. Linear models are used to describe data that have a(n) _____ rate of change.

7. If a quantity is modeled by the equation $y = mx + b$, then m represents the _____.

8. If a quantity is modeled by the equation $y = mx + b$, then b represents the _____.

a.

b.

c.

d.

e.

f.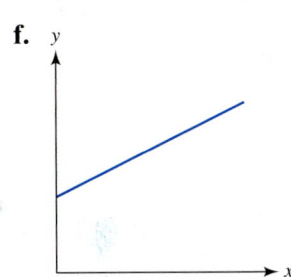

*Exercises 9–14: **Modeling** Match the situation to the graph (a.–f. in the right-hand column) that models it best.*

9. College tuition from 1980 to 2000

10. Yearly average temperature in degrees Celsius at the North Pole

11. Profit from selling boxes of candy if it costs $200 to make the candy

12. Height of Mount Hood in Oregon

13. Total amount of water delivered by a garden hose if it flows at a constant rate

14. Sales of 8-track music tapes from 1970 to 1980 (*Hint:* 8-track tapes are obsolete.)

MODELING LINEAR DATA

Exercises 15–20: State whether the ordered pairs in the table are modeled exactly by the linear equation.

15. $y = 2x + 2$

x	0	1	2
y	2	4	6

16. $y = -2x + 5$

x	0	1	2
y	5	3	0

17. $y = -4x$

x	-1	0	1
y	4	0	-8

18. $y = 5 - x$

x	-2	1	4
y	7	4	1

19. $y = 1.4x - 4$

x	0	5	10
y	-4	3	9

20. $y = -\frac{4}{3}x - \frac{13}{3}$

x	-7	-4	-1
y	5	1	-3

Exercises 21–26: State whether the linear model in the graph is exact or approximate. Then find the equation of the line.

21.

22.

23.

24.

25.

26.

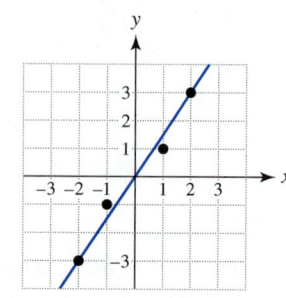

Exercises 27–32: Find an equation $y = mx + b$ that models the quantity y after x units of time.

27. A quantity y is initially 40 and increases at a rate of 5 per minute.

28. A quantity y is initially -60 and increases at a rate of 1.7 per minute.

29. A quantity y is initially -5 and decreases at a rate of 20 per day.

30. A quantity y is initially 5000 and decreases at a rate of 35 per day.

31. A quantity y is initially 8 and remains constant.

32. A quantity y is initially -45 and remains constant.

Exercises 33–40: **Modeling Data**   *For the ordered pairs in the table, do the following.*

(a) Plot the data.   Could a line pass through all five points?

(b) Sketch a line that models the data.

(c) Determine an equation of the line. For data that are not exactly linear, answers may vary.

33.

x	0	1	2	3	4
y	4	2	0	-2	-4

34.

x	0	1	2	2	3
y	7	6	5	4	3

35.

x	-2	-1	0	1	2
y	4	1	0	-1	-4

36.

x	-2	-1	0	1	2
y	7	5	1	-3	-5

37.

x	-6	-4	-2	0	2
y	1	0	-1	-2	-3

38.

x	-4	-2	0	2	4
y	1	2	3	4	5

39.

x	-6	-3	0	3	6
y	-3	-2	-0.5	0	0.5

40.

x	-3	-2	-1	0	1
y	1	2	3	4	5

APPLICATIONS

Exercises 41–48: Write an equation that models the described quantity. Specify what each variable represents.

41. A barrel contains 200 gallons of water and is being filled at a rate of 5 gallons per minute.

42. A barrel contains 40 gallons of gasoline and is being drained at a rate of 3 gallons per minute.

43. An athlete has run 5 miles and is jogging at 6 miles per hour.

44. A new car has 141 miles on its odometer and is traveling at 70 miles per hour.

45. A worker has already earned $200 and is being paid $8 per hour.

46. A gambler has lost $500 and is losing money at a rate of $150 per hour.

47. A carpenter has already shingled 5 garage roofs and is shingling roofs at a rate of 1 per day.

48. A hard drive has been spinning for 2 minutes and is spinning at 7200 revolutions per minute.

49. *Kilimanjaro Glacier* Mount Kilimanjaro is located in Tanzania, Africa, and has an elevation of 19,340 feet. In 1912, the glacier on this peak covered 5 acres. By 2002 this glacier had melted to only 1 acre. (*Source:* NBC News.)
  (a) Assume that this glacier melted at a constant rate each year. Find this *yearly* rate.
  (b) Use your answer in part (a) to write a linear equation that gives the acreage A of this glacier t years past 1912.

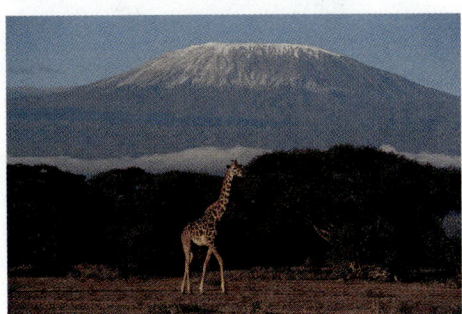

50. *World Population* In 1987, the world's population reached 5 billion people, and by 2012 the world's population is expected to reach 7 billion people.
(*Source:* U.S. Census Bureau.)
  (a) Find the average yearly increase in the world's population from 1987 to 2012.
  (b) Write a linear equation that estimates the world's population P in billions x years after 1987.

51. *Prison Population* The data points in Figure 3.53 are (2002, 152), (2003, 162), (2004, 169), and (2005, 176) where the y-coordinates are in thousands.
  (a) Use the first and last data points to determine a line that models the data. Write the equation in slope–intercept form.
  (b) Use the line to estimate the population in 2009.

52. *Niagara Falls* The average flow of water over Niagara Falls is 212,000 cubic feet of water per second.
  (a) Write an equation that gives the number of cubic feet of water F that flow over the falls in x seconds.

(b) How many cubic feet of water flow over Niagara Falls in 1 minute?

53. *Gas Mileage* (Refer to Example 2.) The table shows the number of miles y traveled by a car on x gallons of gasoline.

x (gallons)	3	6	9	12
y (miles)	60	120	180	240

  (a) Plot the data in the xy-plane. Be sure to label each axis.
  (b) Sketch a line that models these data. (You may want to use a ruler.)
  (c) Calculate the slope of the line. Interpret the slope.
  (d) Find an equation of the line.
  (e) How far could this car travel on 7 gallons of gasoline?

54. *Air Temperature* Generally, the air temperature becomes colder as the altitude above the ground increases. The table lists typical air temperatures x miles high when the ground temperature is $80°$F.

x (miles)	0	1	2	3
y ($°$F)	80	61	42	23

  (a) Plot the data in the xy-plane. Be sure to label each axis.
  (b) Sketch a line that models these data. (You may want to use a ruler.)
  (c) Calculate the slope of the line. Interpret the slope.
  (d) Find the slope–intercept form of the line.
  (e) Estimate the air temperature 5 miles high.

WRITING ABOUT MATHEMATICS

55. In Example 2 the gas mileage of an SUV is modeled with a linear equation. Explain why it is reasonable to use a linear equation to model this situation. (*Hint:* Compare the amounts of gasoline that the SUV uses traveling 30 miles and 60 miles.)

56. Explain the steps for finding the equation of a line that models a table of data points.

CHECKING BASIC CONCEPTS
SECTION 3.7

1. State whether the ordered pairs shown in the table are modeled exactly by $y = -5x + 10$.

x	-2	-1	0	1
y	20	15	10	5

2. State whether the linear model shown in the graph is exact or approximate. Then find the equation of the line.

3. Find an equation, $y = mx + b$, that models the quantity y after x units of time.

(a) A quantity y is initially 50 pounds and increases at a rate of 10 pounds per day.

(b) A quantity y is initially 200°F and decreases at a rate of 2°F per minute.

4. The table contains ordered pairs.
   (a) Plot the data. Could a line pass through all four points?
   (b) Sketch a line that models the data.
   (c) Determine the equation of the line.

x	-2	0	2	4
y	2	1	0	-1

5. *Global Warming*   Since 1945 the average annual recorded temperature on the Antarctic Peninsula has increased by 0.075°F per year.
   (a) Write an equation that models the average temperature *increase T*, x years after 1945.
   (b) Use your equation to calculate the temperature increase between 1945 and 2005.

CHAPTER 3   SUMMARY
SECTION 3.1 ■ INTRODUCTION TO GRAPHING

The Rectangular Coordinate System (*xy*-plane)

Points   Plotted as (x, y) ordered pairs

Four Quadrants   The x- and y-axes divide the xy-plane into quadrants I, II, III, and IV.

NOTE: A point on an axis, such as $(1, 0)$, does not lie in a quadrant.

xy-plane

Scatterplot

Line Graph

SECTION 3.2 ■ LINEAR EQUATIONS IN TWO VARIABLES

Equations in Two Variables   An equation with two variables and possibly some constants

Examples: $y = 3x + 7$ and $x + y = 100$

Solution to an Equation in Two Variables   The solution to an equation in two variables is an ordered pair that makes the equation a true statement.

Example: $(1, 2)$ is a solution to $2x + y = 4$ because $2(1) + 2 = 4$ is true.

Graphing a Linear Equation in Two Variables

Linear Equation $y = mx + b$ or $Ax + By = C$

Graphing   Plot at least three points and connect them with a line.

Example: $y = 2x - 1$

x	y
0	−1
1	1
2	3

SECTION 3.3 ■ MORE GRAPHING OF LINES

Intercepts

x-Intercept   The x-coordinate of a point at which a graph intersects the x-axis; to find an x-intercept, let $y = 0$ in the equation and solve for x.

y-Intercept   The y-coordinate of a point at which a graph intersects the y-axis; to find a y-intercept, let $x = 0$ in the equation and solve for y.

Example: $x + 3y = 3$

x-intercept: Solve $x + 3(0) = 3$ to find the x-intercept of 3.

y-intercept: Solve $0 + 3y = 3$ to find the y-intercept of 1.

x-intercept: 3, y-intercept: 1

Horizontal and Vertical Lines

The equation of a horizontal line with y-intercept b is $y = b$.

The equation of a vertical line with x-intercept k is $x = k$.

Example: The horizontal line $y = -1$ has y-intercept -1 and no x-intercepts. The vertical line $x = 2$ has x-intercept 2 and no y-intercepts.

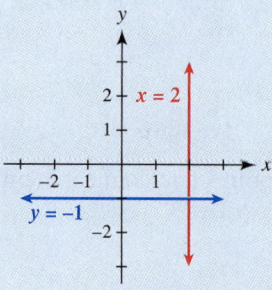

SECTION 3.4 ■ SLOPE AND RATES OF CHANGE

Slope   The ratio $\frac{\text{rise}}{\text{run}}$, or $\frac{\text{change in } y}{\text{change in } x}$, is the slope m of a line when run (change in x) is nonzero. A positive slope indicates that a line rises from left to right, and a negative slope indicates that a line falls from left to right.

Example: Slope $\frac{2}{3}$ indicates that a line rises 2 units for every 3 units of run along the x-axis. The line shown in the graph has slope $\frac{2}{3}$.

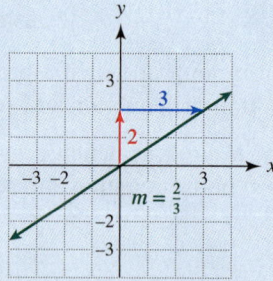

Calculating Slope   A line passing through (x_1, y_1) and (x_2, y_2) has slope

$$m = \frac{y_2 - y_1}{x_2 - x_1}, \text{ where } x_1 \neq x_2.$$

Example: The line through $(-2, 2)$ and $(3, 4)$ has slope

$$m = \frac{4 - 2}{3 - (-2)} = \frac{2}{5}.$$

Horizontal Line   Has slope 0

Vertical Line   Has undefined slope

Slope as a Rate of Change   Slope indicates how fast a line is rising or falling. In applications, slope can indicate how fast a quantity is changing.

Example: The line shown in the graph has slope -2 and depicts an initial outside temperature of $6°F$. Slope -2 indicates that the temperature is *decreasing* at a rate of $2°F$ per hour.

SECTION 3.5 ■ SLOPE–INTERCEPT FORM

Slope–Intercept Form

$$y = mx + b \qquad \text{Slope } m, \text{ } y\text{-intercept } b$$

Example: $y = -\frac{1}{2}x + 2$ has slope $-\frac{1}{2}$ and y-intercept 2, as shown in the graph.

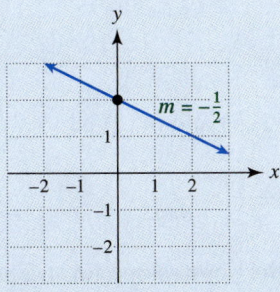

Parallel Lines   Lines with the same slope are parallel; nonvertical parallel lines have the same slope. Two vertical lines are parallel.

Example: The equations $y = -2x + 1$ and $y = -2x$ determine parallel lines because $m_1 = m_2 = -2$. See the graph on the next page.

Perpendicular Lines

If two perpendicular lines have nonzero slopes m_1 and m_2, then $m_1 \cdot m_2 = -1$.

If two lines have nonzero slopes satisfying $m_1 \cdot m_2 = -1$, then they are perpendicular.

A vertical line and a horizontal line are perpendicular.

Example: The equations $y = -\frac{1}{2}x$ and $y = 2x - 2$ determine perpendicular lines because $m_1 \cdot m_2 = -\frac{1}{2} \cdot 2 = -1$. See the graph on the next page.

Parallel Lines

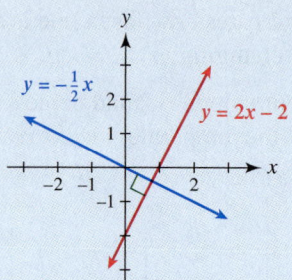

Perpendicular Lines

SECTION 3.6 ▪ POINT–SLOPE FORM

Point–Slope Form   An equation of the line passing through (x_1, y_1) with slope m is

$$y - y_1 = m(x - x_1) \quad \text{or} \quad y = m(x - x_1) + y_1.$$

Example: If $m = -2$ and $(x_1, y_1) = (-2, 3)$, then the point–slope form is either

$$y - 3 = -2(x + 2) \quad \text{or} \quad y = -2(x + 2) + 3.$$

Example: The slope of the line passing through $(-2, 5)$ and $(4, 2)$ is

$$m = \frac{2 - 5}{4 - (-2)} = \frac{-3}{6} = -\frac{1}{2}.$$

Either $(-2, 5)$ or $(4, 2)$ may be used in the point–slope form. The point $(4, 2)$ results in the equation

$$y - 2 = -\frac{1}{2}(x - 4) \quad \text{or} \quad y = -\frac{1}{2}(x - 4) + 2.$$

SECTION 3.7 ▪ INTRODUCTION TO MODELING

Mathematical Modeling   Mathematics is used to describe or approximate the behavior of real-life phenomena.

Exact Model   The equation describes the data precisely without error.

Example: $y = 3x$ models the data in the table exactly.

x	0	1	2	3
y	0	3	6	9

Approximate Model   The equation describes the data approximately. An approximate model occurs most often in applications.

Example: The line in the graph models the data approximately.

Modeling with a Linear Equation   To model a quantity y that has a constant rate of change, use the equation $y = mx + b$, where

$$m = \text{(constant rate of change)} \quad \text{and} \quad b = \text{(initial amount)}.$$

Example: If the temperature is initially $100°F$ and cools at $5°F$ per hour, then

$$T = -5x + 100$$

models the temperature T after x hours.

CHAPTER 3   REVIEW EXERCISES

SECTION 3.1

1. Identify the coordinates of each point in the graph. Identify the quadrant, if any, in which each point lies.

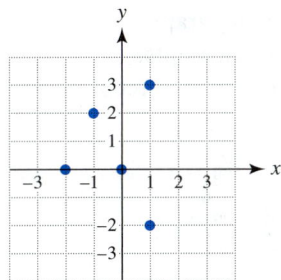

2. Make a scatterplot by plotting the following four points: $(-2, 3)$, $(-1, -1)$, $(0, 3)$, and $(2, -1)$.

Exercises 3 and 4: Use the table of xy-values to make a line graph.

3.

x	-2	-1	0	1	2
y	-3	2	-1	-2	3

4.

x	-10	-5	0	5	10
y	5	-10	10	-5	0

SECTION 3.2

Exercises 5–8: Determine whether the ordered pair is a solution for the given equation.

5. $y = x - 3$ $(6, 3)$

6. $y = 5 - 2x$ $(-2, 1)$

7. $3x - y = 3$ $(-1, 6)$

8. $\frac{1}{2}x + 2y = -8$ $(-4, -3)$

Exercises 9 and 10: Complete the table for the given equation.

9. $y = -3x$

x	−2	−1	0	1	2
y					

10. $2x + y = 5$

x					
y	−3	−1	0	1	3

Exercises 11–14: Use the given values of the variable to find solutions for the equation. Put them in a table.

11. $y = 3x + 2$ $x = -2, 0, 2, 4$

12. $y = 7 - x$ $x = 1, 2, 3, 4$

13. $y - 2x = 0$ $y = -1, 0, 1, 2$

14. $2y + x = 1$ $y = 1, 2, 3, 4$

Exercises 15–22: Graph the equation.

15. $y = 2x$

16. $y = x + 1$

17. $y = \frac{1}{2}x - 1$

18. $y = -3x + 2$

19. $x + y = 2$

20. $3x - 2y = 6$

21. $-4x + y = 8$

22. $2x + 3y = 12$

SECTION 3.3

Exercises 23 and 24: Identify the x- and y-intercepts.

23.

24.

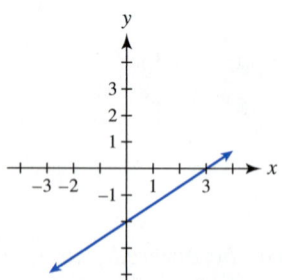

Exercises 25 and 26: Complete the table. Then determine the x- and y-intercepts for the graph of the equation.

25. $y = 2 - x$

x	−2	−1	0	1	2
y					

26. $x - 2y = 4$

x	−4	−2	0	2	4
y					

Exercises 27–30: Find any intercepts for the graph of the equation and then graph the linear equation.

27. $2x - 3y = 6$

28. $5x - y = 5$

29. $0.1x - 0.2y = 0.4$

30. $\dfrac{x}{2} + \dfrac{y}{3} = 1$

31. Graph each equation.
    (a) $y = 1$ **(b)** $x = -3$

32. Write an equation for each line shown in the graph.

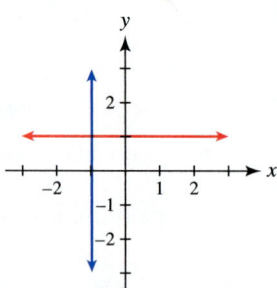

Exercises 33 and 34: Write an equation for the line that passes through the points shown in the table.

33.

x	−2	−1	0	1	2
y	1	1	1	1	1

34.

x	3	3	3	3	3
y	−2	−1	0	1	2

35. Write the equations of a horizontal line and a vertical line that pass through the point $(-2, 3)$.

36. *Distance* The distance a driver is from home is illustrated in the graph.
    (a) Find the intercepts.
    (b) Interpret each intercept.

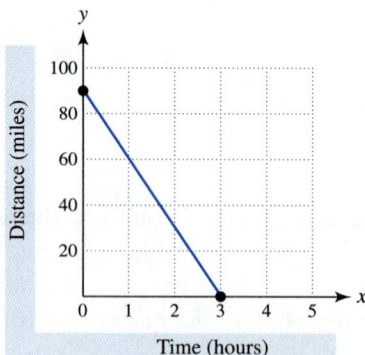

SECTION 3.4

Exercises 37–40: Find the slope, if possible, of the line passing through the two points.

37. $(2, 3), (4, 7)$ **38.** $(-3, 1), (2, -1)$

39. $(2, 1), (5, 1)$ **40.** $(-5, 6), (-5, 10)$

Exercises 41 and 42: Find the slope of the line shown in the graph.

41.   **42.**

          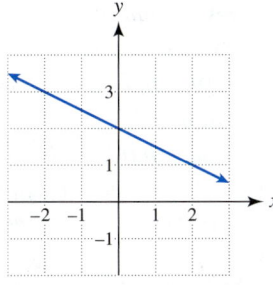

Exercises 43–46: Do the following.

   (a) *Graph the linear equation.*

   (b) *What are the slope and y-intercept of the line?*

43. $y = -2x$ **44.** $y = x - 1$

45. $x + 2y = 4$ **46.** $2x - 3y = -6$

Exercises 47–50: Sketch a line passing through the given point and having slope m.

47. $(0, -3), m = 2$ **48.** $(0, 1), m = -\frac{1}{2}$

49. $(-1, 1), m = -\frac{2}{3}$ **50.** $(2, 2), m = 1$

51. The table lists points located on a line. Find the slope and intercepts of the line.

x	-1	0	1	2
y	-4	-2	0	2

52. A line with slope $\frac{1}{2}$ passes through the first point shown in the table. Complete the table so that each point in the table lies on the line.

x	0	1	2	3
y	1			

SECTION 3.5

Exercises 53 and 54: Write the slope–intercept form for the line shown in the graph.

53.   **54.**

          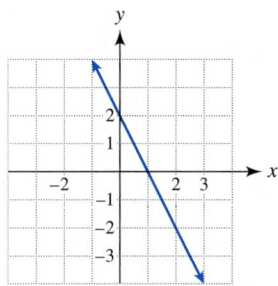

Exercises 55 and 56: Sketch a line with slope m and y-intercept b. Write its slope–intercept form.

55. $m = 2, b = -2$ **56.** $m = -\frac{3}{4}, b = 3$

Exercises 57–60: Do the following.

   (a) *Write the equation in slope–intercept form.*

   (b) *Give the slope and y-intercept of the line.*

57. $x + y = 3$ **58.** $-3x + 2y = -6$

59. $20x - 10y = 200$ **60.** $5x - 6y = 30$

Exercises 61–68: Graph the equation.

61. $y = \frac{1}{2}x + 1$ **62.** $y = 3x - 2$

63. $y = -\frac{1}{3}x$ **64.** $y = 3x$

65. $y = 2$ **66.** $y = -1$

67. $y = 4 - x$ **68.** $y = 2 - \frac{2}{3}x$

Exercises 69 and 70: All the points shown in the table lie on the same line. Find the slope–intercept form for the line.

69.

x	0	1	2
y	-5	0	5

70.

x	-1	0	1
y	2	0	-2

Exercises 71–74: Find the slope–intercept form for the line satisfying the given conditions.

71. Slope $-\frac{5}{6}$, y-intercept 2

72. Parallel to $y = -2x + 1$, passing through $(1, -5)$

73. Perpendicular to $y = -\frac{3}{2}x$, passing through $(3, 0)$

74. Perpendicular to $y = 5x - 3$, passing through the point $(0, -2)$

SECTION 3.6

Exercises 75 and 76: Determine whether the given point lies on the line.

75. $(-3, 1)$ $y - 1 = 2(x + 3)$

76. $(3, -8)$ $y = -3(x - 1) + 2$

Exercises 77–84: Find the slope–intercept form for the line that satisfies the conditions given.

77. Slope 5, passing through $(1, 2)$

78. Slope 20, passing through $(3, -5)$

79. Passing through $(-2, 1)$ and $(1, -1)$

80. Passing through $(20, -30)$ and $(40, 30)$

81. x-intercept 3, y-intercept -4

82. x-intercept $\frac{1}{2}$, y-intercept -1

83. Parallel to $y = 2x$, passing through $(5, 7)$

84. Perpendicular to $y - 4 = \frac{3}{2}(x + 1)$, passing through $(-1, 0)$

Exercises 85–88: Write the given point–slope form in slope–intercept form.

85. $y - 2 = 3(x + 1)$ **86.** $y - 9 = \frac{1}{3}(x - 6)$

87. $y = 2(x + 3) + 5$ **88.** $y = -\frac{1}{4}(x - 8) + 1$

SECTION 3.7

89. State whether the ordered pairs shown in the table are modeled exactly by $y = -x + 4$.

x	0	1	2	2
y	4	3	2	1

90. State whether the linear model shown in the graph is exact or approximate. Then find the equation of the line.

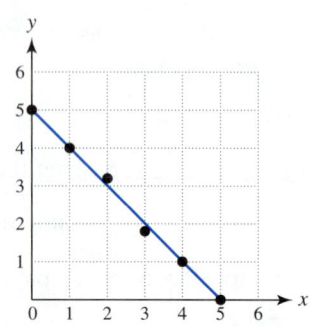

Exercises 91–94: Find an equation $y = mx + b$ that models y after x units of time.

91. y is initially 40 pounds and decreases at a rate of 2 pounds per minute.

92. y is initially 200 gallons and increases at 20 gallons per hour.

93. y is initially 50 and remains constant.

94. y is initially 20 feet *below* sea level and rises at 5 feet per second.

Exercises 95 and 96: For the ordered pairs in the table, do the following.

    (a) *Plot the data. Could a line pass through all five points?*

    (b) *Sketch a line that models the data.*

    (c) *Determine an equation of the line. For data that are not exactly linear, answers may vary.*

95.

x	0	1	2	3	4
y	10	6	2	-2	-6

96.

x	-4	-2	0	2	4
y	1	2.1	3	3.9	5

APPLICATIONS

97. *Graphing Real Data*   The table contains real data on divorces D in millions during year t.
    (a) Make a line graph of the data. Label the axes.
    (b) Comment on any trends in the data.

t	1960	1970	1980	1990	2000
D	0.4	0.7	1.2	1.2	1.2

Source: National Center for Health Statistics.

98. *Water Usage*   The average American uses 100 gallons of water each day.
    (a) Write an equation that gives the gallons G of water that a person uses in t days.
    (b) Graph the equation for $t \geq 0$.
    (c) How many days does it take for the average American to use 5000 gallons of water?

99. *Modeling a Toy Rocket*   The velocity v of a toy rocket in feet per second after t seconds of flight is given by $v = 160 - 32t$, where $t \geq 0$.
    (a) Graph the equation.
    (b) Interpret each intercept.

100. *Modeling*   The accompanying line graph represents the insect population on 1 acre of land after *x* weeks. During this time a farmer sprayed pesticides on the land.

(a) Estimate the slope of each line segment.

(b) Interpret each slope as a rate of change.

(c) Describe what happened to the insect population.

Time (weeks)

101. *Sketching a Graph*   An athlete jogs 4 miles away from home at a constant rate of 8 miles per hour and then turns around and jogs back home at the same speed. Sketch a graph that shows the distance that the athlete is from home. Be sure to label each axis.

102. *Nursing Homes*   In 1985, there were 19,100 nursing homes, and in 2005, there were 17,000.   (*Source:* National Center for Health Statistics.)

(a) Calculate the slope of the line passing through (1985, 19100) and (2005, 17000).

(b) Interpret the slope as a rate of change.

103. *School Enrollment*   In 1995, school enrollment at all levels was 65 million students, and in 2005, it was 71 million.   (*Source:* National Center for Education Statistics.)

(a) Find the slope of the line passing through the points (1995, 65) and (2005, 71).

(b) Interpret the slope as a rate of change.

(c) If this trend continues, estimate the school enrollment in 2030.

104. *Rental Cars*   The cost *C* in dollars for driving a rental car *x* miles is $C = 0.2x + 35$.

(a) How much would it cost to rent the car but not drive it?

(b) How much does it cost to drive the car one *additional* mile?

(c) What is the *C*-intercept of the graph of $C = 0.2x + 35$? What does it represent?

(d) What is the slope of the graph of $C = 0.2x + 35$? What does it represent?

105. *Distance and Speed*   A person is driving a car along a straight road. The accompanying graph shows the distance *y* in miles that the driver is from home after *x* hours.

(a) Is the person traveling toward or away from home? Why?

(b) The graph passes through (1, 200) and (3, 100). Discuss the meaning of these points.

(c) Find the slope–intercept form of the line. Interpret the slope as a rate of change.

(d) Use the graph to estimate the distance from home after 2 hours. Then check your answer by using your equation from part (c).

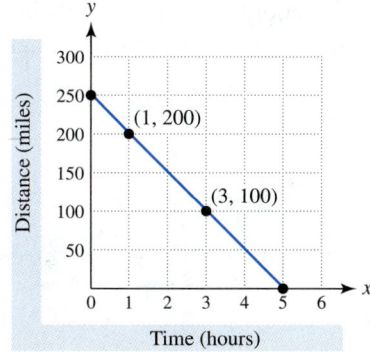

Time (hours)

106. *Arctic Glaciers*   Owing to global warming the arctic ice cap has been breaking up. As a result there has been an increase in the number of icebergs floating in the Arctic Ocean. The table gives an iceberg count *I* for various years *t*.   (*Source:* NBC News.)

t	1970	1980	2000
I	400	600	1000

Source: National Broadcasting Company.

(a) Make a scatterplot of the data.

(b) Find the slope–intercept form of a line that models the number of icebergs *I* in year *t*. Interpret the slope of this line.

(c) Is the line you found in part (b) an exact model for the data in the table?

(d) If current trends continue, what will be the iceberg count in 2005?

107. *Deaths from Pneumonia*   From 1998 to 2002, deaths D in thousands from pneumonia can be modeled by $D = -6.05(x - 1998) + 90.1$, where x is the year.   (*Source:* National Center for Health Statistics.)

   (a) Find the number of deaths from pneumonia in 1998 and 2002.

   (b) What is the slope of the graph of D? Interpret the slope as a rate of change.

108. *Gas Mileage*   The table shows the number of miles y traveled by a car on x gallons of gasoline.

x	2	4	8	10
y	40	79	161	200

   (a) Plot the data in the xy-plane. Be sure to label each axis.

   (b) Sketch a line that models these data. (You may want to use a ruler.)

   (c) Calculate the slope of the line. Interpret the slope of this line.

   (d) Find the equation of the line. Is your line an *exact* model?

   (e) How far could this car travel on 9 gallons of gasoline?

CHAPTER 3   TEST   Pass the Test   Video solutions to all test exercises

1. Identify the coordinates of each point in the graph. State the quadrant, if any, in which each point lies.

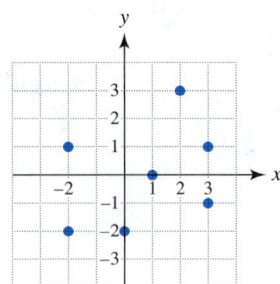

2. Make a scatterplot by plotting the four points $(0, 0)$, $(-2, -2)$, $(3, 0)$, and $(3, -2)$.

3. Complete the table for the equation $y = 2x - 4$. Then determine the x- and y-intercepts for the graph of the equation.

x	-2	-1	0	1	2
y					

4. Determine whether the ordered pair $(1, -3)$ is a solution for the equation $2x - y = 5$.

5. Sketch a line passing through the point $(2, 1)$ and having slope $-\frac{1}{2}$.

6. Find the x- and y-intercepts for the graph of the equation $5x - 3y = 15$.

Exercises 7–10: Graph the equation.

7. $y = 2$

8. $x = -3$

9. $y = -3x + 3$

10. $4x - 3y = 12$

11. Write the slope–intercept form for the line shown in the graph.

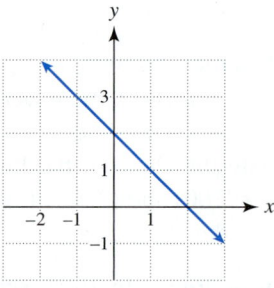

12. Write the equation $-4x + 2y = 1$ in slope–intercept form. Give the slope and the y-intercept.

13. Write the equation of a horizontal line and a vertical line passing through the point $(1, -5)$.

14. Find the slope of a line passing through the points $(-4, 3)$ and $(5, 1)$.

Exercises 15–18: Find the slope–intercept form for the line that satisfies the given conditions.

15. Slope $-\frac{4}{3}$, y-intercept -5

16. Parallel to $y = 3x - 1$, passing through $(2, -5)$

17. Perpendicular to $y = \frac{1}{3}x$, passing through $(1, 2)$

18. Passing through $(-4, 2)$ and $(2, -1)$

19. Write $y - 3 = \frac{1}{2}(x + 4)$ in slope–intercept form.

20. State whether the linear model shown is exact or approximate. Then find the equation of the line.

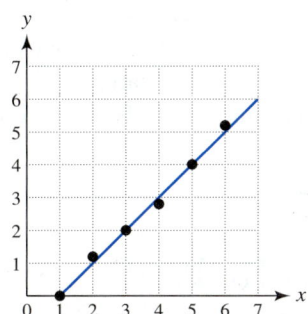

21. *Sketching a Graph*   A cyclist rides a bicycle at 10 miles per hour for 2 hours and then at 8 miles per hour for 1 hour. Sketch a graph that shows the total distance d traveled after x hours. Be sure to label each axis.

22. *Modeling*   The line graph represents the total fish population P in a small lake after x years. One winter the lake almost froze solid.
   (a) Estimate the slope of each line segment.
   (b) Interpret each slope as a rate of change.
   (c) Describe what happened to the fish population.

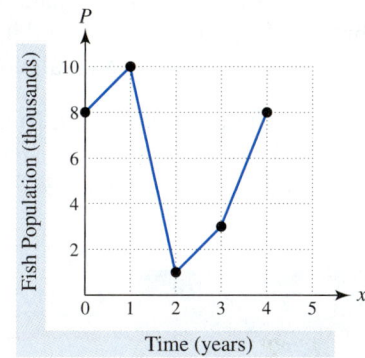

23. *Modeling Insects*   Write an equation in slope–intercept form that models the number of insects N after x days if there are initially 2000 insects and they increase at a rate of 100 insects per day.

CHAPTER 3   EXTENDED AND DISCOVERY EXERCISES

Exercises 1 and 2: The table of real data can be modeled by a line. However, the line may not be an exact model, so answers may vary.

1. *Women in Politics*   The table lists percentages P of women in state legislatures during year x.

x	1993	1995	1997	1999
P	20.5	20.6	21.6	22.4

x	2001	2003	2005	2007
P	22.4	22.4	22.7	23.5

Source: National Women's Political Caucus.

   (a) Make a scatterplot of the data.
   (b) Find a point–slope form of a line that models these data.
   (c) Use your equation to estimate the percentage of women in state legislatures in 2010.

2. *U.S. Population*   The population P of the United States in millions is shown in the table during year x.

x	1960	1970	1980	1990	2000
P	179	203	227	249	281

Source: U.S. Census Bureau.

   (a) Make a scatterplot of the data.
   (b) Find a point–slope form of a line that models these data.
   (c) Use your equation to estimate the U.S. population in 2005.

CREATING GEOMETRIC SHAPES

Exercises 3–6: Many geometric shapes can be created with intersecting lines. For example, a triangle is formed when three lines intersect at three distinct points called vertices. Find equations of lines that satisfy the characteristics described. Sketching a graph may be helpful.

3. *Triangle*   A triangle has vertices $(0, 0)$, $(2, 3)$, and $(3, 6)$.
   (a) Find slope–intercept forms for three lines that pass through each pair of points.

(b) Graph the three lines. Is a triangle formed by the line segments connecting the three points?

4. *Parallelogram*   A parallelogram has four sides, with opposite sides parallel. Three sides of a parallelogram are given by the equations $y = 2x + 2$, $y = 2x - 1$, and $y = -x - 2$.
  (a) If the fourth side of the parallelogram passes through the point $(-2, 3)$, find its equation.
  (b) Graph all four lines. Is a parallelogram formed?

5. *Rectangle*   The two vertices $(0, 0)$ and $(4, 2)$ determine one side of a rectangle. The side parallel to this side passes through the point $(0, 3)$.

(a) Find slope–intercept forms for the four lines that determine the sides of the rectangle.
(b) Graph all four lines. Is a rectangle formed? (*Hint:* If you use a graphing calculator be sure to set a square window.)

6. *Square*   Three vertices of a square are $(1, 2)$, $(4, 2)$, and $(4, 5)$.
  (a) Find the fourth vertex.
  (b) Find equations of lines that correspond to the four sides of the square.
  (c) Graph all four lines. Is a square formed?

CHAPTERS 1–3   CUMULATIVE REVIEW EXERCISES

Exercises 1 and 2: Classify the number as prime or composite. If a number is composite, write it as a product of prime numbers.

1. 40   **2.** 61

Exercises 3 and 4: Translate the phrase into an algebraic expression using the variable n.

3. Ten more than a number

4. A number squared decreased by 2

Exercises 5–8: Evaluate by hand and then simplify to lowest terms.

5. $\frac{4}{3} \cdot \frac{21}{24}$ **6.** $\frac{3}{4} \div \frac{9}{8}$

7. $\frac{2}{3} + \frac{4}{3}$ **8.** $\frac{7}{10} - \frac{2}{15}$

Exercises 9–12: Evaluate the expression by hand.

9. $20 - 2 \cdot 3$ **10.** $14 - 5 - 2$

11. $\frac{1 + 4}{1 + 2}$ **12.** -3^2

Exercises 13 and 14: Classify the number as one or more of the following: natural number, whole number, integer, rational number, or irrational number.

13. $-\frac{4}{5}$ **14.** $\sqrt{3}$

15. Plot each number on the same number line.
  (a) 0   **(b)** -1 **(c)** $\frac{3}{2}$

16. Evaluate $|3 - 5|$.

Exercises 17 and 18: Evaluate the expression by hand.

17. $-12 \div \left(-\frac{2}{3}\right)$ **18.** $-\frac{2x}{5y} \div \left(\frac{x}{10y}\right)$

Exercises 19 and 20: Simplify the expression.

19. $3 + 4x - 2 + 3x$ **20.** $2(x - 1) - (x + 2)$

Exercises 21–24: Solve the equation. Check your solution.

21. $x + 5 = 2$ **22.** $\frac{1}{3}z = 7$

23. $3t - 5 = 1$ **24.** $2(x - 3) = -6 - x$

25. Complete the table. Then use the table to solve the equation $6 - 2x = 4$.

x	-2	-1	0	1	2
$6 - 2x$					

26. Translate the sentence "Twice a number increased by 2 equals the number decreased by 5" to an equation, using the variable n. Then solve the equation.

27. The sum of four consecutive integers is -98. Find the integers.

28. If $r = 10$ mph and $d = 80$ miles, use the formula $d = rt$ to find t.

Exercises 29 and 30: Find the area of the figure shown.

29.

40 ft
18 ft

30.

5 ft
8 ft

Exercises 31 and 32: Solve the formula for the specified variable.

31. $A = \frac{1}{2}bh$ for b

32. $P = 2W + 2L$ for L

33. Use an inequality to express the set of real numbers graphed.

34. Complete the table. Then use the table to solve the linear inequality $2x - 3 \le 1$.

x	0	1	2	3	4
$2x - 3$					

Exercises 35 and 36: Solve the inequality. Write the solution set in set-builder notation.

35. $3 - 6x < 3$

36. $2x \le 1 - (2x - 1)$

37. Identify the coordinates of each point in the graph. State the quadrant, if any, in which each point lies.

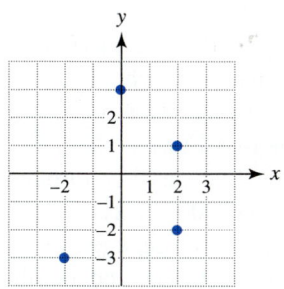

38. Make a scatterplot and a line graph with the points $(-2, 1)$, $(-1, 3)$, $(0, 0)$, $(1, -2)$, and $(2, 3)$.

39. Determine whether $(5, -1)$ is a solution to the equation $x - 3y = 8$.

40. Complete the table for the equation $x + 2y = 4$.

x	-2	-1	0	1	2
y					

Exercises 41–46: Graph the equation.

41. $y = -\frac{1}{2}x + 1$

42. $x + y = 3$

43. $2x + 3y = -6$

44. $5x - 2y = 10$

45. $x = 3$

46. $y = -\frac{3}{2}$

Exercises 47 and 48: Identify the x- and y-intercepts. Then write the slope–intercept form of the line.

47.   **48.**

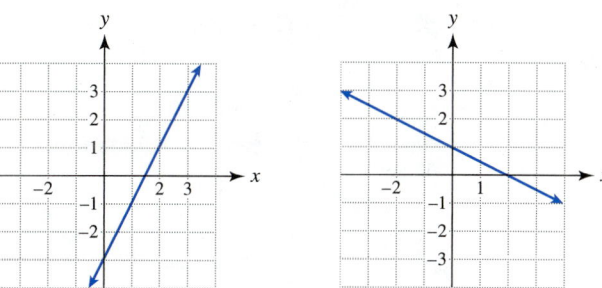

49. Determine the intercepts for the graph of the equation $-4x + 5y = 40$.

50. Write an equation for each line shown in the graph.

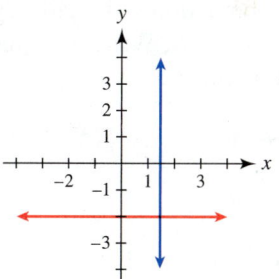

51. Sketch a line with slope $m = -2$, passing through the point $(-1, 2)$.

52. Sketch a line with undefined slope, passing through the point $(2, -1)$.

53. The table lists points located on a line. Write the slope–intercept form of the line.

x	-1	0	1	2
y	-6	-3	0	3

54. Write the equation $3x - 5y = 15$ in slope–intercept form. Graph the equation.

Exercises 55–58: Find the slope–intercept form for the line satisfying the given conditions.

55. Parallel to $y = -\frac{1}{3}x - 5$, passing through $(-3, 8)$

56. Perpendicular to $3x - 2y = 6$, passing through the point $(0, -3)$

57. Passing through $(-1, 3)$ and $(2, -3)$

58. x-intercept -2, y-intercept $\frac{1}{2}$

59. State whether the linear model shown in the graph is exact or approximate. Then find the equation of the line.

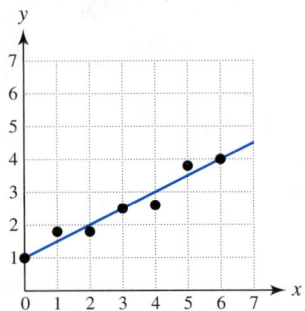

60. Let the initial value of y be 100 pounds, increasing at 5 pounds per hour. Find an equation in slope–intercept form that models y after x hours.

61. An insect population is initially 20,000 and increasing at 5000 per day. Write an equation that gives the number of insects I after x days.

62.

x	0	1	2	3
y	5	3	1	-1

(a) Plot the data in the table. Could a line pass through all four points?
(b) Sketch a line that models the data.
(c) Determine the equation of the line.

APPLICATIONS

63. *Pizzas*   The table lists the cost C of buying x large pepperoni pizzas. Write an equation that relates C and x.

x	2	3	4
C	\$16	\$24	\$32

64. *U.S. Postal Service*   In 2000, about $\frac{11}{20}$ of the mail consisted of first-class mail and periodicals. For every periodical there were 9 pieces of first-class mail.

Estimate the fraction of the mail that was first-class mail.   (***Source:*** U.S. Postal Service, *2000 Annual Report.*)

65. *U.S. Per Capita Income*   The per capita income I in dollars from 1970 to 2000 can be modeled by

$$I = 807.4x - 1{,}587{,}300,$$

where x is the year.   (***Source:*** U.S. Census Bureau.)
(a) What has been the yearly average increase in per capita income?
(b) Estimate the year when this income reached \$19,400.

66. *Income Tax Returns*   In 2000, about 127 million income tax returns were filed. By 2004, this number had increased to about 131 million. Find the percent change in the number of income tax returns filed between 2000 and 2004.   (***Source:*** Internal Revenue Service.)

67. *Investment Money*   A student invests two sums of money at 3% and 4% interest, receiving a total of \$110 in interest after 1 year. Twice as much money is invested at 4% than at 3%. Find the amount invested at each interest rate.

68. *School Lunch Program*   In 1970, 22.4 million students participated in the national school lunch program. By 2000, this number increased to 27.2 million students.   (***Source:*** Department of Agriculture.)
(a) Find the slope of the line passing through the points (1970, 22.4) and (2000, 27.2).
(b) Interpret the slope as a rate of change.
(c) If this trend continues, estimate the number of participants in 2005.

69. *Rental Cars*   The cost C in dollars for driving a rental car x miles is $C = 0.3x + 25$.
(a) How much does it cost to drive the car 200 miles?
(b) How much does it cost to rent the car but not drive it?
(c) How much does it cost to drive the car 1 *additional* mile?

CHAPTER 4

Systems of Linear Equations in Two Variables

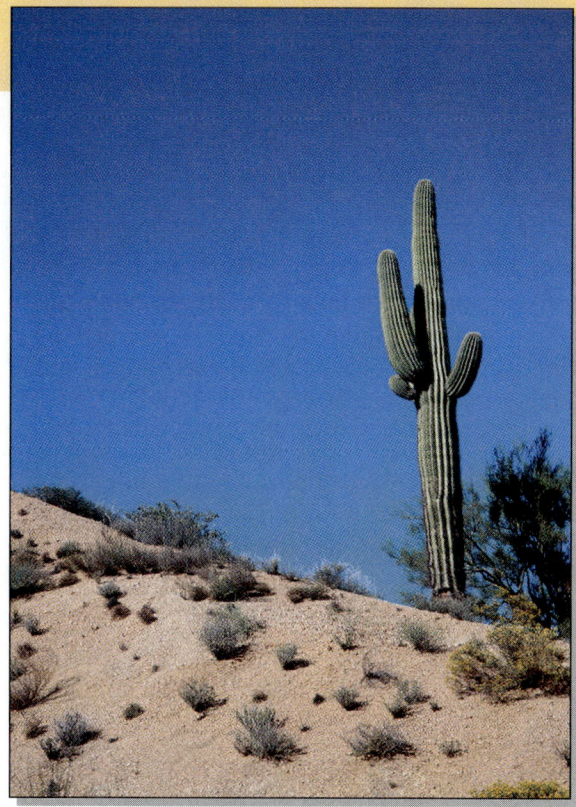

4.1 **Solving Systems of Linear Equations Graphically and Numerically**

4.2 **Solving Systems of Linear Equations by Substitution**

4.3 **Solving Systems of Linear Equations by Elimination**

4.4 **Systems of Linear Inequalities**

Two factors that have a critical effect on plant growth are temperature and precipitation. If a region has too little precipitation, it will be a desert. Forests often grow in regions where trees can exist at relatively low temperatures and there is sufficient rainfall. At other levels of temperature and precipitation, grasslands may prevail. The accompanying figure illustrates the relationships among forests, grasslands, and deserts suggested by average annual temperature T in degrees Fahrenheit and precipitation P in inches. For example, Casa Grande, Arizona, has an average annual temperature of about 70°F and average annual precipitation of about 8 inches. What does the graph predict about the plant growth near Casa Grande?

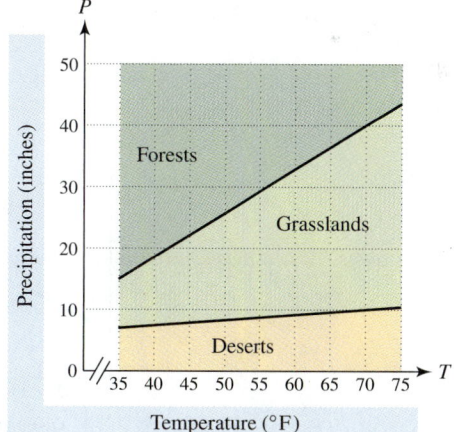

This graph illustrates systems of linear inequalities, which we discuss in this chapter. Mathematics is essential to creating these types of graphs and solving application problems.

We can do anything we want to do if we stick to it long enough.

—HELEN KELLER

Source: A. Miller and J. Thompson, *Elements of Meteorology.*

4.1 SOLVING SYSTEMS OF LINEAR EQUATIONS GRAPHICALLY AND NUMERICALLY

Basic Concepts ■ Solutions to Systems of Equations

A LOOK INTO MATH ▷

Many equations in everyday life involve more than one variable. To calculate the area of a rectangle we need to know its width and its length. To calculate wind chill we need to know the outside temperature and the wind velocity. In this section we consider linear equations in two variables. The concepts discussed in this section are used in many applications.

Basic Concepts

In Chapter 3 we showed that the graph of $y = mx + b$ is a line with slope m and y-intercept b, as illustrated in Figure 4.1. Each point on this line represents a solution to the equation $y = mx + b$. Because there are infinitely many points on a line, there are infinitely many solutions to this equation. However, many applications require that we find one particular solution to a linear equation. One way to find such a solution is to graph a second line in the same xy-plane and determine the point of intersection (if one exists).

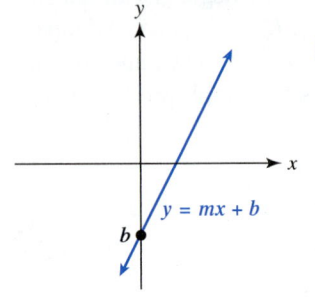

Figure 4.1

▶ **REAL-WORLD CONNECTION** Consider the following application of a line. If renting a moving truck for one day costs $25 plus $0.50 per mile driven, then the equation $C = 0.5x + 25$ represents the cost C in dollars of driving the rental truck x miles. The graph of this line is shown in Figure 4.2(a) for $x \geq 0$.

(a) Renting a Truck

(b) A $75 Rental Cost

Figure 4.2

Suppose that we want to determine the number of miles that the truck is driven when the rental cost is $75. One way to solve this problem *graphically* is to graph both $C = 0.5x + 25$ and $C = 75$ in the same coordinate plane, as shown in Figure 4.2(b). The lines intersect at the point (**100**, **75**), which is a solution to $C = 0.5x + 25$ and to $C = 75$. That is, if the rental cost is **$75**, then the mileage must be **100** miles. This graphical technique of solving two equations is sometimes called the **intersection-of-graphs method**. To find a solution with this method, we locate a point where two graphs intersect.

EXAMPLE 1   ### Solving an equation graphically

The equation $P = 10x$ calculates an employee's pay for working x hours at $10 per hour. Use the intersection-of-graphs method to find the number of hours that the employee worked if the amount paid is $40.

Solution
Begin by graphing the equations $P = 10x$ and $P = 40$, as illustrated in Figure 4.3.

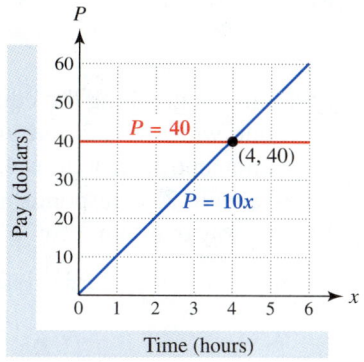

Figure 4.3 Wages

The graphs intersect at the point $(\mathbf{4}, \mathbf{40})$, which indicates that the employee must work **4** hours to earn **$40**.   Now Try Exercise 59(a), (b)

Equations can be solved in more than one way. In Example 1 we determined graphically that $P = 10x$ is equal to 40 when $x = 4$. Because we want to know the value of x that makes $P = 40$, we could solve this problem *symbolically* by substituting 40 for P in the equation $P = 10x$. The resulting equation, $40 = 10x$, can be solved using methods discussed in Chapter 2. By dividing each side by 10 and simplifying, we obtain $x = 4$. We could also solve this problem by making a table of values, as illustrated by Table 4.1. Note that, when $x = \mathbf{4}$, $P = \mathbf{\$40}$. A table of values provides a *numerical solution*.

TABLE 4.1   Wages Earned at $10 per hour

x (hours)	0	1	2	3	4	5	6
P (pay)	$0	$10	$20	$30	$40	$50	$60

NOTE: Although graphical, symbolic, and numerical methods are different, all methods should give the same solution. Exceptions might occur, for example, when reading a graph precisely is difficult, or when a needed value does not appear in a table.

TECHNOLOGY NOTE: Intersection of Graphs and Table of Values

A graphing calculator can be used to find the intersection of the two graphs shown in Figure 4.3. It can also be used to create Table 4.1. The accompanying figures illustrate how a calculator can be used to determine that $y_1 = 10x$ equals $y_2 = 40$ when $x = 4$.

CALCULATOR HELP

To find a point of intersection see Appendix A (page AP-7).
To make a table, see Appendix A (page AP-3).

X	Y₁	Y₂
0	0	40
1	10	40
2	20	40
3	30	40
4	40	40
5	50	40
6	60	40

X=4

For the remainder of this section we focus on solving two equations by graphing and by making a table of values. Methods for solving two equations symbolically will be discussed later in the chapter. The next example illustrates a graphical approach.

EXAMPLE 2   Solving an equation graphically

Use a graph to find the x-value when $y = 3$.
(a) $y = 2x - 1$ (b) $-3x + 2y = 12$

(a)

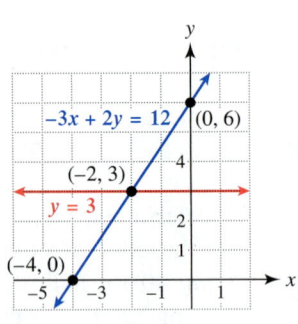
(b)

Figure 4.4

Solution
(a) Begin by graphing the equations $y = 2x - 1$ and $y = 3$. The graph of $y = 2x - 1$ is a line with slope 2 and y-intercept -1. The graph of $y = 3$ is a horizontal line with y-intercept 3. In Figure 4.4(a) their graphs intersect at the point $(2, 3)$. Therefore an x-value of **2** corresponds to a y-value of **3**.
(b) One way to graph $-3x + 2y = 12$ is to write this equation in slope–intercept form.

$$-3x + 2y = 12 \qquad \text{Given equation}$$
$$2y = 3x + 12 \qquad \text{Add } 3x \text{ to each side.}$$
$$y = \frac{3}{2}x + 6 \qquad \text{Divide each side by 2.}$$

The line has slope $\frac{3}{2}$ and y-intercept 6. Its graph and the graph of $y = 3$ are shown in Figure 4.4(b). Their graphs intersect at $(-2, 3)$. Therefore an x-value of -2 corresponds to a y-value of **3**.

Now Try Exercises 13, 15

MAKING CONNECTIONS

Different Ways to Graph the Same Line

In Example 2(b) the line $-3x + 2y = 12$ was graphed by finding its slope–intercept form. A second way to graph this line is to find the x- and y-intercepts of the line. If $y = 0$, then $x = -4$ makes this equation true, and if $x = 0$, then $y = 6$ makes this equation true. Thus the x-intercept is -4, and the y-intercept is 6. Note that the (slanted) line in Figure 4.4(b) passes through the points $(-4, 0)$ and $(0, 6)$, which could be used to graph the line.

Solutions to Systems of Equations

In Example 2(b) we determined the x-value when $y = 3$ in the equation $-3x + 2y = 12$. This problem can be thought of as solving the following *system of two equations in two variables*.

$$-3x + 2y = 12$$
$$y = 3$$

The solution is the ordered pair $(-2, 3)$, which indicates that when $x = -2$ and $y = 3$, each equation is a true statement.

$$-3(-2) + 2(3) = 12 \qquad \text{A true statement}$$
$$3 = 3 \qquad \text{A true statement}$$

Suppose that the sum of two numbers is 10 and that their difference is 4. If we let x and y represent the two numbers, then the equations

$$x + y = 10 \qquad \text{Sum is 10.}$$
$$x - y = 4 \qquad \text{Difference is 4.}$$

describe this situation. Each equation is a linear equation in two variables, so we call these equations a **system of linear equations in two variables**. Its graph typically consists of two lines. A **solution to a system** of two equations is an ordered pair (x, y) that makes *both* equations true. This ordered pair gives the coordinates of a point where the two lines intersect.

NOTE: When two lines do not coincide (their graphs are two distinct lines), there can be no more than one intersection point. If such an intersection point exists, the ordered pair corresponding to it represents the only solution to the system of linear equations. In this case, we say the ordered pair is *the* solution to the system of equations.

EXAMPLE **3**   **Testing for solutions**

Determine whether (4, 6) or (7, 3) is a solution to

$$x + y = 10$$
$$x - y = 4.$$

Solution   To determine whether (4, 6) is a solution, substitute $x = 4$ and $y = 6$ in each equation. It must make *both* equations true.

$x + y = 10$	$x - y = 4$	Given equations
$4 + 6 \stackrel{?}{=} 10$	$4 - 6 \stackrel{?}{=} 4$	Let $x = 4, y = 6$.
$10 = 10$ (True)	$-2 = 4$ (False)	Second equation is false.

Because (4, 6) does not satisfy *both* equations, it is not a solution for the system of equations. Next let $x = 7$ and $y = 3$ to determine whether (7, 3) is a solution.

$x + y = 10$	$x - y = 4$	Given equations
$7 + 3 \stackrel{?}{=} 10$	$7 - 3 \stackrel{?}{=} 4$	Let $x = 7, y = 3$.
$10 = 10$ (True)	$4 = 4$ (True)	Both are true.

Because (7, 3) makes *both* equations true, it is a solution for the system of equations.

Now Try Exercise **17**

In the next example we find the solution to a system of linear equations graphically and numerically.

EXAMPLE **4**   **Solving a system graphically and numerically**

Solve the system of linear equations

$$x + 2y = 4$$
$$2x - y = 3$$

with a graph and with a table of values.

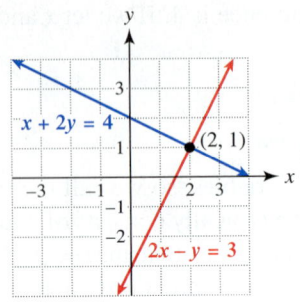

Figure 4.5

Solution

Graphically   Begin by writing each equation in slope–intercept form.

$x + 2y = 4$	First equation	$2x - y = 3$	Second equation
$2y = -x + 4$	Subtract x.	$-y = -2x + 3$	Subtract $2x$.
$y = -\dfrac{1}{2}x + 2$	Divide by 2.	$y = 2x - 3$	Multiply by -1.

The graphs of $y = -\frac{1}{2}x + 2$ and $y = 2x - 3$ are shown in Figure 4.5. Their graphs intersect at the point $(2, 1)$.

Numerically   Table 4.2 shows the equations $y = -\frac{1}{2}x + 2$ and $y = 2x - 3$ evaluated for various values of x. Note that when $x = 2$, both equations have a y-value of **1**. Thus $(\mathbf{2}, \mathbf{1})$ is the solution.

CRITICAL THINKING

If the graph of a system of linear equations consists of two parallel lines, how many solutions are there?

TABLE 4.2   A Numerical Solution

x	-1	0	1	2	3
$y = -\frac{1}{2}x + 2$	2.5	2	1.5	1	0.5
$y = 2x - 3$	-5	-3	-1	1	3

Now Try Exercise 37

EXAMPLE 5   Solving a system graphically

Solve the system of equations graphically.

$$y = 2x$$
$$2x + y = 4$$

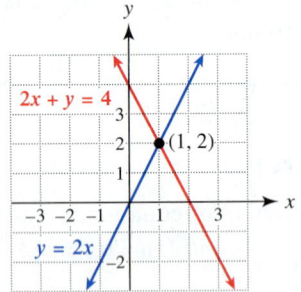

Figure 4.6

Solution

The equation $y = 2x$ can be written in slope–intercept form as $y = 2x + 0$. Its graph is a line passing through the origin with slope 2, as shown in Figure 4.6. The equation $2x + y = 4$ can be graphed by finding the intercepts. If we let $x = 0$, the resulting equation is $y = 4$. If we let $y = 0$, the resulting equation is $2x = 4$ which is equivalent to $x = 2$. Thus the y-intercept is 4 and the x-intercept is 2. This line is also graphed in Figure 4.6. Because the intersection point is $(1, 2)$, the solution to the system of equations is the ordered pair $(1, 2)$.

Now Try Exercise 51

In the next example, we use a four-step process to solve an application involving a system of linear equations. These steps are based on the four-step process discussed in Section 2.3, with Step 3 split into two parts to emphasize the importance of using the solution to the system of equations to determine the solution to the given problem.

EXAMPLE 6   Traveling to watch sports

In 2002, about 50 million Americans traveled to watch either football or basketball. About 10 million more people traveled to watch football than basketball. How many Americans traveled to watch each sport?   (***Source:*** *Sports Travel Magazine.*)

Solution

STEP 1: *Identify each variable.*

x: millions of Americans who traveled to watch football
y: millions of Americans who traveled to watch basketball

STEP 2: *Write a system of equations.* The total number of Americans who watched either sport is 50 million, so we know that $x + y = 50$. Because 10 million more people watched football than basketball, we also know that $x - y = 10$. Thus a system of equations representing this situation is

$$x + y = 50$$
$$x - y = 10.$$

STEP 3A: *Solve the system of equations.* To solve this system graphically, write each equation in slope–intercept form.

$$y = -x + 50$$
$$y = \ \ \ x - 10$$

Their graphs intersect at the point (30, 20), as shown in Figure 4.7.

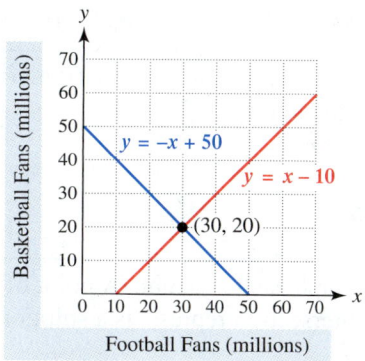

Figure 4.7

STEP 3B: *Determine the solution to the problem.* The point (30, 20) corresponds to $x = 30$ and $y = 20$. Thus about 30 million Americans traveled to watch football and 20 million Americans traveled to watch basketball.

STEP 4: *Check your solution.* Note that $30 + 20 = 50$ million Americans traveled to watch either football or basketball and that $30 - 20 = 10$ million more Americans watched football than basketball.   **Now Try Exercise** `61`

TECHNOLOGY NOTE: Checking Solutions

CALCULATOR HELP

To find a point of intersection, see Appendix A (page AP-7).

The solution to the system in Example 6 can be checked with a graphing calculator, as shown in the accompanying figure where the graphs of $y_1 = -x + 50$ and $y_2 = x - 10$ intersect at the point (30, 20). However, a graphing calculator cannot read Example 6 and write down the system of equations. A human mind is needed for these tasks.

4.1   PUTTING IT ALL TOGETHER

Some types of situations are described by using two equations in two variables. For example, suppose that a person walks a total of 8 miles in 2 days by walking 2 miles farther on the first day than on the second day. If x represents the number of miles walked on the first day and if y represents the number of miles walked on the second day, then the system of linear equations

$$x + y = 8$$
$$x - y = 2$$

describes this situation. The methods discussed in this section reveal that the solution to this system is (5, 3), which indicates that the person walked 5 miles on the first day and 3 miles on the second day. The following table summarizes some important concepts of systems of equations.

Concept	Explanation	Example
System of Linear Equations in Two Variables	Can be written as $$Ax + By = C$$ $$Dx + Ey = F$$	$$x + y = 8$$ $$x - y = 2$$
Solution to a System of Equations	An ordered pair (x, y) that satisfies *both* equations	The solution to the preceding system is (5, 3) because, when $x = 5$ and $y = 3$ are substituted, both equations are true. $$5 + 3 \stackrel{?}{=} 8 \qquad \text{True}$$ $$5 - 3 \stackrel{?}{=} 2 \qquad \text{True}$$
Graphical Solution to a System of Equations	Graph each equation. A point of intersection represents a solution.	The graphs of $y = -x + 8$ and $y = x - 2$ intersect at (5, 3).
Numerical Solution to a System of Equations	Make a table for each equation. A solution occurs when one x-value gives the same y-values in both equations.	Make a table for $y = -x + 8$ and $y = x - 2$. When $x = 5$, $y = 3$ in both equations, so (**5, 3**) is the solution.

	x	4	5	6
$y = -x + 8$		4	3	2
$y = x - 2$		2	3	4

4.1 Exercises

CONCEPTS

1. A solution to a system of two equations in two variables is a(n) _____ pair.

2. A graphical technique for solving a system of two equations in two variables is the _____ method.

3. When two lines are not identical, there can be no more than _____ intersection point(s).

4. The graphs of the equations shown intersect at $(11, 9)$.
$$x + y = 20$$
$$x - y = 2$$
What is the solution to the system?

5. Is $(0, 0)$ a solution to the given system?
$$2x + 3y = 8$$
$$5x - 4y = -3$$

6. To find a numerical solution to a system, start by creating a _____ of values for the equations.

7. If a graphical method and a numerical method (table of values) are used to solve the same system of equations, then the two solutions should be (the same/different).

8. One way to graph a line is to write its equation in slope–intercept form. A second method is to find the x- and y-_____.

SOLVING SYSTEMS OF EQUATIONS

Exercises 9–16: Determine graphically the x-value when $y = 2$ in the given equation.

9. $y = 2x$

10. $y = \frac{1}{3}x$

11. $y = 4 - x$

12. $y = -2 - x$

13. $y = -\frac{1}{2}x + 1$

14. $y = 3x - 1$

15. $2x + y = 6$

16. $-3x + 4y = 11$

Exercises 17–22: Determine which ordered pair is a solution to the system of equations.

17. $(0, 0), (1, 1)$
$$x + y = 2$$
$$x - y = 0$$

18. $(-1, 2), (1, -2)$
$$2x + y = 0$$
$$x - 2y = -5$$

19. $(-1, -1), (2, -3)$
$$2x + 3y = -5$$
$$4x - 5y = 23$$

20. $(2, -1), (-2, -2)$
$$-x + 4y = -6$$
$$6x - 7y = 19$$

21. $(2, 0), (-1, -3)$
$$-5x + 5y = -10$$
$$4x + 9y = 8$$

22. $\left(\frac{1}{2}, \frac{3}{2}\right), \left(\frac{3}{4}, \frac{5}{4}\right)$
$$x + y = 2$$
$$3x - y = 0$$

Exercises 23–28: The graphs of two equations are shown. Use the intersection-of-graphs method to identify the solution to both equations. Then check your answer.

23.

24.

25.

26.

27.

28.

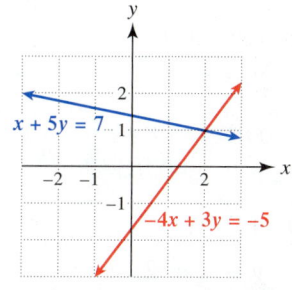

Exercises 29–32: A table for two equations is given. Identify the solution to both equations.

29.

x	1	2	3	4
$y = 2x$	2	4	6	8
$y = 4$	4	4	4	4

30.

x	2	3	4	5
$y = 6 - x$	4	3	2	1
$y = x - 2$	0	1	2	3

31.

x	1	2	3	4
$y = 4 - x$	3	2	1	0
$y = x - 2$	-1	0	1	2

32.

x	1	2	3	4
$y = 6 - 3x$	3	0	-3	-6
$y = 2 - x$	1	0	-1	-2

Exercises 33 and 34: Complete the table for each equation. Then identify the solution to both equations.

33.

x	0	1	2	3
$y = x + 2$				
$y = 4 - x$				

34.

x	-5	-4	-3	-2
$y = 2x + 1$				
$y = x - 1$				

Exercises 35–40: Use the specified method to solve the system of equations.
  (a) Graphically
  (b) Numerically (table of values)

35. $y = 2x + 3$
    $y = 1$

36. $y = 2 - x$
    $y = 0$

37. $y = 4 - x$
    $y = x - 2$

38. $y = 2x$
    $y = -\frac{1}{2}x$

39. $y = 3x$
    $y = x + 2$

40. $y = 2x - 3$
    $y = -x + 3$

Exercises 41–52: Solve the system of equations graphically.

41. $x + y = -3$
    $x - y = \ \ 1$

42. $x - y = 3$
    $2x + y = 3$

43. $2x - y = 3$
    $3x + y = 2$

44. $x + 2y = 6$
    $-x + 3y = 4$

45. $-4x + 2y = \ \ \ 0$
    $x - \ \ y = -1$

46. $4x - y = 2$
    $y = 2x$

47. $2x - \ \ y = 4$
    $x + 2y = 7$

48. $x - y = 2$
    $\frac{1}{2}x + y - 4$

49. $x = -y + 4$
    $x = 3y$

50. $x = 2y$
    $y = -\frac{1}{2}x$

51. $x + y = 3$
    $x = \frac{1}{2}y$

52. $2x - 4y = \ \ \ 8$
    $\frac{1}{2}x + \ \ y = -4$

*Exercises 53–58: **Number Problems**   For each problem, complete each of the following.*
  (a) Write a system of equations for the problem.
  (b) Find the unknown numbers by solving the system of equations graphically.

53. The sum of two numbers is 4 and their difference is 0.

54. The sum of two numbers is -5 and their difference is 1.

55. The sum of twice a number and another number is 7. Their difference is 2.

56. Three times a number subtracted from another number results in 1. Their sum is 5.

57. One number is triple another number. Their difference is 4.

58. Half of a number added to another number equals 5. Their difference is 1.

APPLICATIONS

59. ***Renting a Truck***   A rental truck costs $50 plus $0.50 per mile.
    (a) Write an equation that gives the cost of driving the truck x miles.
    (b) Use the intersection-of-graphs method to determine the number of miles that the truck is driven if the rental cost is $80.
    (c) Solve part (b) numerically with a table of values.

60. *Renting a Car*   A rental car costs $25 plus $0.25 per mile.
  (a) Write an equation that gives the cost of driving the car x miles.
  (b) Use the intersection-of-graphs method to determine the number of miles that the car is driven if the rental cost is $100.
  (c) Solve part (b) numerically with a table of values.

61. *Recorded Music*   In 2004, rock and hip hop music accounted for 36% of all music sales. Rock music sales were 2 times greater than hip hop music sales.
  (**Source:** Recording Industry Association of America.)
  (a) Let x be the percentage of sales due to rock music and let y be the percentage of music sales due to hip hop music. Write a system of two equations that describes the given information.
  (b) Solve your system graphically.

62. *Sales of Radios*   During 2000 and 2001, about 37 thousand radios were sold, excluding car radios. The 2000 sales exceeded the 2001 sales by 3 thousand radios.   (**Source:** M. Street Corporation.)
  (a) Let y be the radio sales in thousands during 2000 and x be the radio sales in thousands during 2001.

Write a system of two equations that describes the given information.
  (b) Solve your system graphically.

63. *Dimensions of a Rectangle*   A rectangle is 4 inches longer than it is wide. Its perimeter is 28 inches.
  (a) Write a system of two equations in two variables that describes this information. Be sure to specify what each variable means.
  (b) Solve your system graphically. Interpret your results.

64. *Dimensions of a Triangle*   An isosceles triangle has a perimeter of 17 inches with its two shorter sides equal in length. The longest side measures 2 inches more than either of the shorter sides.
  (a) Write a system of two equations in two variables that describes this information. Be sure to specify what each variable means.
  (b) Solve your system graphically. Explain what your results mean.

WRITING ABOUT MATHEMATICS

65. Use the intersection-of-graphs method to help explain why you typically expect a linear system in two variables to have one solution.

66. Could a system of two linear equations in two variables have exactly two solutions? Explain your reasoning.

67. Give one disadvantage of using a table to solve a system of equations. Explain your answer.

68. Do the equations $y = 2x + 1$ and $y = 2x - 1$ have a common solution? Explain your answer.

4.2 SOLVING SYSTEMS OF LINEAR EQUATIONS BY SUBSTITUTION

The Method of Substitution ▪ Types of Systems of Linear Equations ▪ Applications

A LOOK INTO MATH ▷   In Section 4.1 we solved systems of linear equations by using graphs and tables. A disadvantage of a graph is that reading the graph precisely can be difficult. A disadvantage of using a table is that locating the solution can be difficult when it is either a fraction or a large number. In this section we introduce the method of substitution, in which we solve systems of equations using only symbols. The advantage of this method is that the *exact* solution can always be found (provided it exists).

The Method of Substitution

If you and a friend earned $120 together, then there is no way to determine how much each one earned. You might have earned $70 while your friend earned $50, or vice versa. If x represents how much your friend earned and y represents how much you earned, then the equation $x + y = 120$ describes this situation. However, if we know that you earned twice as much as your friend, then we can include a second equation, $y = 2x$. The amount that each of you earned can now be determined by *substituting* $2x$ for y in the first equation.

$x + y = 120$	First equation
$x + 2x = 120$	Substitute $2x$ for y.
$3x = 120$	Combine like terms.
$x = 40$	Divide each side by 3.

Thus your friend earned $40, and you earned twice as much, or $80.

This technique of substituting an expression for a variable and solving the resulting equation is called the **method of substitution**.

EXAMPLE 1   **Using the method of substitution**

Solve each system of equations.
(a) $2x + y = 10$ **(b)** $-2x + 3y = -8$
$\qquad\quad y = 3x$ $\qquad\qquad\quad x = 3y + 1$

Solution
(a) From the second equation, substitute $3x$ for y in the first equation.

$2x + y = 10$	First equation
$2x + 3x = 10$	Substitute $3x$ for y.
$5x = 10$	Combine like terms.
$x = 2$	Divide each side by 5.

The solution to this system is an *ordered pair*, so we must also find y. Because $y = 3x$ and $x = 2$, it follows that $y = 3(2) = 6$. The solution is $(2, 6)$. (Check it.)

(b) The second equation, $x = 3y + 1$, is solved for x. Substitute $(3y + 1)$ for x in the first equation. Be sure to include parentheses around the expression $3y + 1$ since this entire expression is to be multiplied by -2.

$-2x + 3y = -8$	First equation
$-2(3y + 1) + 3y = -8$	Substitute $(3y + 1)$ for x.
$-6y - 2 + 3y = -8$	Distributive property
$-3y - 2 = -8$	Combine like terms.
$-3y = -6$	Add 2 to each side.
$y = 2$	Divide each side by -3.

To find x, substitute 2 for y in $x = 3y + 1$ to obtain $x = 3(2) + 1 = 7$. The solution is $(7, 2)$. (Check it.) Now Try Exercises 11, 19

NOTE: When an expression contains two or more terms, it is usually best to place parentheses around it when substituting it for a single variable in an equation. In Example 1(b), the distributive property would not have been applied correctly without the parentheses.

Sometimes it is necessary to solve for a variable before substitution can be used, as demonstrated in the next example.

EXAMPLE 2 Using the method of substitution

Solve each system of equations.

(a) $\begin{aligned} x + y &= 8 \\ 2x - 3y &= 6 \end{aligned}$ (b) $\begin{aligned} 3a - 2b &= 2 \\ a + 4b &= 3 \end{aligned}$

Solution

(a) Neither equation is solved for a variable, but we can easily solve the first equation for y.

$$x + y = 8 \qquad \text{First equation}$$
$$y = 8 - x \qquad \text{Subtract } x \text{ from each side.}$$

Now we can substitute $(8 - x)$ for y in the second equation.

$$2x - 3y = 6 \qquad \text{Second equation}$$
$$2x - 3(8 - x) = 6 \qquad \text{Substitute } (8 - x) \text{ for } y.$$
$$2x - 24 + 3x = 6 \qquad \text{Distributive property}$$
$$5x = 30 \qquad \text{Combine terms; add 24.}$$
$$x = 6 \qquad \text{Divide each side by 5.}$$

Because $y = 8 - x$ and $x = 6$, $y = 8 - 6 = 2$. The solution is $(6, 2)$.

(b) Although we could solve either equation for either variable, solving the second equation for a is easiest because the coefficient of a is 1.

$$a + 4b = 3 \qquad \text{Second equation}$$
$$a = 3 - 4b \qquad \text{Subtract } 4b \text{ from each side.}$$

Now substitute $(3 - 4b)$ for a in the first equation.

$$3a - 2b = 2 \qquad \text{First equation}$$
$$3(3 - 4b) - 2b = 2 \qquad \text{Substitute } (3 - 4b) \text{ for } a.$$
$$9 - 12b - 2b = 2 \qquad \text{Distributive property}$$
$$-14b = -7 \qquad \text{Combine terms; subtract 9.}$$
$$b = \frac{1}{2} \qquad \text{Divide each side by } -14.$$

To find a, substitute $b = \frac{1}{2}$ in $a = 3 - 4b$ to obtain $a = 1$. The solution is $\left(1, \frac{1}{2}\right)$.

Now Try Exercises 23, 29

NOTE: When a system of equations contains variables other than x and y, we will list them alphabetically in an ordered pair.

Types of Systems of Linear Equations

A system of linear equations typically has exactly one solution. However, in the next example we solve two systems of equations that do not have exactly one solution.

EXAMPLE 3   **Solving other types of systems**

If possible, use substitution to solve the system of equations. Then use graphing to help explain the result.

(a) $3x + y = 4$ **(b)** $x + y = 2$
$\quad\ \ 6x + 2y = 2$ $\quad\ \ 2x + 2y = 4$

Solution

(a) Solve the first equation for y to obtain $y = 4 - 3x$. Next substitute $(4 - 3x)$ for y in the second equation.

$$6x + 2y = 2 \qquad \text{Second equation}$$
$$6x + 2(\mathbf{4 - 3x}) = 2 \qquad \text{Substitute } (4 - 3x) \text{ for } y.$$
$$6x + 8 - 6x = 2 \qquad \text{Distributive property}$$
$$8 = 2 \ (\text{False}) \qquad \text{Combine terms.}$$

The equation $8 = 2$ is *always false*, which indicates that there are *no solutions*. One way to graph each equation is to write the equations in slope–intercept form.

$3x + y = 4$	First equation	$6x + 2y = 2$	Second equation
$y = -3x + 4$	Subtract $3x$.	$y = -3x + 1$	Subtract $6x$, divide by 2.

The graphs of these equations are parallel lines with slope -3, as shown in Figure 4.8(a). Because the lines *do not intersect* there are *no solutions* to the system of equations.

(b) Solve the first equation for y to obtain $y = 2 - x$. Now substitute $(2 - x)$ for y in the second equation.

$$2x + 2y = 4 \qquad \text{Second equation}$$
$$2x + 2(\mathbf{2 - x}) = 4 \qquad \text{Substitute } (2 - x) \text{ for } y.$$
$$2x + 4 - 2x = 4 \qquad \text{Distributive property}$$
$$4 = 4 \ (\text{True}) \qquad \text{Combine terms.}$$

The equation $4 = 4$ is *always true*, which means that there are *infinitely many solutions*. One way to graph these equations is to write them in slope–intercept form first.

$x + y = 2$	First equation	$2x + 2y = 4$	Second equation
$y = -x + 2$	Subtract x.	$y = -x + 2$	Subtract $2x$, divide by 2.

Because the equations have the same slope–intercept form, their graphs are identical, resulting in a single line, as shown in Figure 4.8(b). *Every point on this line represents a solution* to the system of equations, so there are infinitely many solutions.

Now Try Exercises 53, 55

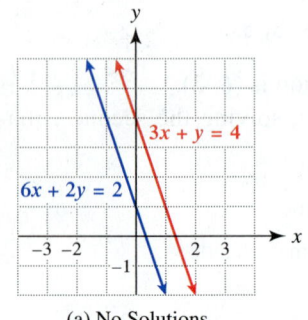

(a) No Solutions

(b) Infinitely Many Solutions

Figure 4.8

TYPES OF EQUATIONS AND NUMBER OF SOLUTIONS

A system of linear equations can have no solutions, one solution, or infinitely many solutions. It cannot have any other number of solutions. If a system has

1. no solutions, it is an **inconsistent system**. Graphing the equations results in parallel lines.
2. one solution, it is a **consistent system**, and its equations are **independent equations**. Graphing the equations results in two lines that intersect at one point.
3. infinitely many solutions, it is a **consistent system**, and its equations are **dependent equations**. Graphing the equations results in identical lines.

EXAMPLE 4 Identifying types of equations

Graphs of two equations are shown. State the number of solutions to each system of equations. Then state whether the system is consistent or inconsistent. If it is consistent, state whether the equations are dependent or independent.

(a) **(b)** **(c)**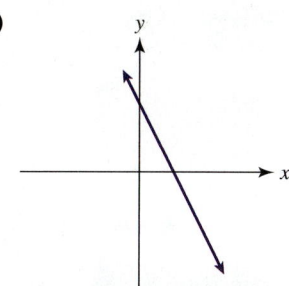

Solution
(a) The lines are parallel, so there are no solutions. The system is inconsistent.
(b) The lines intersect at one point, so there is one solution. The system is consistent, and the equations are independent.
(c) There is only one line, which indicates that the graphs are identical, or coincide, so there are infinitely many solutions. The system is consistent and the equations are dependent.

Now Try Exercises 39, 41, 43

Applications

In the next two examples we use the method of substitution to solve applications.

EXAMPLE 5 Determining pizza sales

In 2002, combined sales of frozen and ready-to-eat pizza reached $28.9 billion. Ready-to-eat pizza sales were 7.5 times frozen pizza sales. Find the amount of sales for each type of pizza.   (*Source: Business Trend Analyst.*)

Solution
STEP 1: *Identify each variable.* Clearly identify what each variable represents.

x: sales of ready-to-eat pizza, in billions of dollars
y: sales of frozen pizza, in billions of dollars

STEP 2: *Write a system of equations.*

$$x + y = 28.9 \qquad \text{Sales total \$28.9 billion.}$$

$$x = 7.5y \qquad \begin{array}{l} \text{Ready-to-eat pizza sales } x \text{ are} \\ \text{7.5 times frozen pizza sales } y. \end{array}$$

STEP 3A: *Solve the system of linear equations.* Substitute $7.5y$ for x in the first equation.

$x + y = 28.9$	First equation
$7.5y + y = 28.9$	Substitute $7.5y$ for x.
$8.5y = 28.9$	Combine like terms.
$y = \dfrac{28.9}{8.5}$	Divide each side by 8.5.
$y = 3.4$	Simplify.

Because $x = 7.5y$, it follows that $x = 7.5(3.4) = 25.5$.

STEP 3B: *Determine the solution to the problem.* The solution $y = 3.4$ and $x = 25.5$ indicates that frozen pizza sales were \$3.4 billion and ready-to-eat pizza sales were \$25.5 billion in 2002.

STEP 4: *Check the solution.* The sum of these sales was $25.5 + 3.4 = \$28.9$ billion, and ready-to-eat pizza sales were $\frac{25.5}{3.4} = 7.5$ times frozen pizza sales. The answer checks.

Now Try Exercise **87**

Determining airplane speed and wind speed

An airplane flies 2400 miles into (or against) the wind in 8 hours. The return trip takes 6 hours. Find the speed of the airplane with no wind and the speed of the wind.

Solution

STEP 1: *Identify each variable.*

x: the speed of the airplane without wind
y: the speed of the wind

STEP 2: *Write a system of equations.* The speed of the airplane against the wind is $\frac{2400}{8} = 300$ miles per hour, because it traveled 2400 miles in 8 hours. The wind slowed the plane, so $x - y = 300$. Similarly, the airplane flew $\frac{2400}{6} = 400$ miles per hour with the wind because it traveled 2400 miles in 6 hours. The wind made the plane fly faster, so $x + y = 400$.

$x - y = 300$	Speed against the wind
$x + y = 400$	Speed with the wind

STEP 3A: *Solve the system of linear equations.* Solve the first equation for x to obtain $x = y + 300$. Substitute $(y + 300)$ for x in the second equation.

$x + y = 400$	Second equation
$(y + 300) + y = 400$	Substitute $(y + 300)$ for x.
$2y = 100$	Combine like terms; subtract 300.
$y = 50$	Divide each side by 2.

Because $x = y + 300$, it follows that $x = 50 + 300 = 350$.

Now Try Exercise 83

CRITICAL THINKING

A boat travels 10 miles per hour upstream and 16 miles per hour downstream. How fast is the current?

STEP 3B: *Determine the solution to the problem.* The solution $x = 350$ and $y = 50$ indicates that the airplane can fly 350 miles per hour with no wind, and the wind speed is 50 miles per hour.

STEP 4: *Check the solution.* The plane flies $350 - 50 = 300$ miles per hour into the wind, taking $\frac{2400}{300} = 8$ hours. The plane flies $350 + 50 = 400$ miles per hour with the wind, taking $\frac{2400}{400} = 6$ hours. The answers check.

4.2   PUTTING IT ALL TOGETHER

A system of linear equations can be solved symbolically by using the method of substitution. This method always provides the exact solution (provided one exists). A system of linear equations can have no solutions, one solution, or infinitely many solutions. These concepts are summarized in the following table.

Concept	Explanation	Example
Method of Substitution	Can be used to solve a system of equations **STEP 1:** Solve one equation for one variable. **STEP 2:** Substitute the result in the other equation and solve. **STEP 3:** Use the solution for the first variable to find the other variable. **STEP 4:** Check the solution.	$x + y = 5$ Sum is 5. $x - y = 1$ Difference is 1. **STEP 1:** Solve for x in the second equation. $x = y + 1$ **STEP 2:** Substitute $(y + 1)$ in the first equation for x. $(y + 1) + y = 5$ $2y = 4$ $y = 2$ **STEP 3:** $x = 2 + 1 = 3$ **STEP 4:** $3 + 2 = 5$ $3 - 2 = 1$ (3, 2) checks.
Inconsistent System	System of linear equations with no solutions. Graphs result in parallel lines.	 No Solutions

continued on next page

continued from previous page

Concept	Explanation	Example
Consistent System with Independent Equations	System of linear equations with 1 solution Graphs result in intersecting lines.	 One Solution
Consistent System with Dependent Equations	System of linear equations with infinitely many solutions Graphs result in identical lines, or lines that coincide.	 Infinitely Many Solutions

4.2 Exercises

CONCEPTS

1. The technique of substituting an expression for a variable in an equation is called the method of _____.

2. A system of linear equations can have _____, _____, or _____ solutions.

3. One advantage of solving a linear system using the method of substitution rather than graphical or numerical methods is that the _____ solution can always be found (provided it exists).

4. When substituting an expression that contains two or more terms for a single variable in an equation, it is usually best to place _____ around it.

5. Suppose that the method of substitution results in the equation $1 = 1$. What does this indicate about the number of solutions to the system of equations?

6. Suppose that the method of substitution results in the equation $0 = 1$. What does this indicate about the number of solutions to the system of equations?

7. If a system of linear equations has at least one solution, then it is a(n) (consistent/inconsistent) system.

8. If a system of linear equations has no solutions, then it is a(n) (consistent/inconsistent) system.

9. If a system of linear equations has exactly one solution, then the equations are (dependent/independent).

10. If a system of linear equations has infinitely many solutions, then the equations are (<u>dependent/independent</u>).

SOLVING SYSTEMS OF EQUATIONS

Exercises 11–38: Use the method of substitution to solve the system of linear equations.

11. $x + y = 9$
     $y = 2x$

12. $x + y = -12$
     $y = -3x$

13. $x + 2y = 4$
     $x = 2y$

14. $-x + 3y = -12$
     $x = 5y$

15. $2x + y = -2$
     $y = x + 1$

16. $-3x + y = -10$
     $y = x - 2$

17. $x + 3y = 3$
     $x = y + 3$

18. $x - 2y = -5$
     $x = 4 - y$

19. $3x + 2y = \frac{3}{2}$
     $y = 2x - 1$

20. $-3x + 5y = 4$
     $y = 2 - 3x$

21. $2x - 3y = -12$
     $x = 2 - \frac{1}{2}y$

22. $\frac{3}{4}x + \frac{1}{4}y = -\frac{7}{4}$
     $x = 1 - 2y$

23. $2x - 3y = -4$
     $3x - y = 1$

24. $\frac{1}{2}x - y = -1$
     $2x - \frac{1}{2}y = \frac{13}{2}$

25. $x - 5y = 26$
     $2x + 6y = -12$

26. $4x - 3y = -4$
     $x + 7y = -63$

27. $\frac{1}{2}y - z = 5$
     $y - 3z = 13$

28. $3y - 7z = -2$
     $5y - z = 2$

29. $10r - 20t = 20$
     $r + 60t = -29$

30. $-r + 10t = 22$
     $-10r + 5t = 30$

31. $3x + 2y = 9$
     $2x - 3y = -7$

32. $5x - 2y = -5$
     $2x - 5y = 19$

33. $2a - 3b = 6$
     $-5a + 4b = -8$

34. $-5a + 7b = -1$
     $3a + 2b = 13$

35. $-\frac{1}{2}x + 3y = 5$
     $2x - \frac{1}{2}y = 3$

36. $3x - \frac{1}{2}y = 2$
     $-\frac{1}{2}x + 5y = \frac{19}{2}$

37. $3a + 5b = 16$
     $-8a + 2b = 34$

38. $5a - 10b = 20$
     $10a + 5b = 15$

Exercises 39–44: The graphs of two equations are shown. State the number of solutions to each system of equations. Then state whether the system is consistent or inconsistent. If it is consistent, state whether the equations are dependent or independent.

39.   **40.**

41.   **42.**

43.   **44.**

 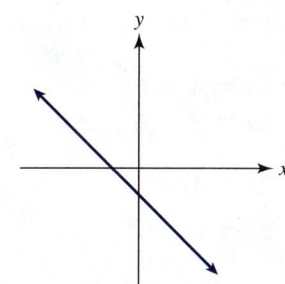

Exercises 45–52: Use the method of substitution and the intersection-of-graphs method to solve the system of linear equations. Then state whether the system is consistent or inconsistent. If it is consistent, state whether the equations are dependent or independent.

45. $x + y = 4$
     $x + y = 2$

46. $x + y = 10$
     $x - y = 2$

47. $2x - y = 3$
     $x - 2y = 0$

48. $-2x + y = 0$
     $y = 2x + 1$

49. $x - y = 1$
$2x - 2y = 2$

50. $x + 2y = 4$
$-x - 2y = -4$

51. $x - 2y = 4$
$x = 2y$

52. $-3x + y = 4$
$y = 3x + 4$

Exercises 53–70: Use the method of substitution to solve the system of linear equations. These systems may have no solutions, one solution, or infinitely many solutions.

53. $x + y = 9$
$x + y = 7$

54. $x - y = 8$
$x - y = 4$

55. $x - y = 4$
$2x - 2y = 8$

56. $2x + y = 5$
$4x + 2y = 10$

57. $x + y = 4$
$x - y = 2$

58. $x - y = 3$
$2x - y = 7$

59. $x - y = 7$
$-x + y = -7$

60. $2u - v = 6$
$-4u + 2v = -12$

61. $u - 2v = 5$
$2u - 4v = -2$

62. $3r + 3t = 9$
$2r + 2t = 4$

63. $2r + 3t = 1$
$r - 3t = -5$

64. $5x - y = -1$
$2x - 7y = 7$

65. $y = 5x$
$y = -3x$

66. $a = b + 1$
$a = b - 1$

67. $5a = 4 - b$
$5a = 3 - b$

68. $3y = x$
$3y = 2x$

69. $2x + 4y = 0$
$3x + 6y = 5$

70. $-5x + 10y = 3$
$\frac{1}{2}x - y = 1$

71. Thinking Generally If the method of substitution results in an equation that is always true, what can be said about the graphs of the two equations?

72. Thinking Generally If the method of substitution results in an equation that is always false, what can be said about the graphs of the two equations?

APPLICATIONS

73. *Rectangle* A rectangular garden is 10 feet longer than it is wide. Its perimeter is 72 feet.

(a) Let W be the width of the garden and L be the length. Write a system of linear equations whose solution gives the width and length of the garden.

(b) Use the method of substitution to solve the system. Check your answer.

74. *Isosceles Triangle* The measures of the two smaller angles in a triangle are equal and their sum equals the largest angle.

(a) Let x be the measure of each of the two smaller angles and y be the measure of the largest angle. Write a system of linear equations whose solution gives the measures of these angles.

(b) Use the method of substitution to solve the system. Check your answer.

75. *Complementary Angles* The smaller of two complementary angles is half the measure of the larger angle.

(a) Let x be the measure of the smaller angle and y be the measure of the larger angle. Write a system of linear equations whose solution gives the measures of these angles.

(b) Use the method of substitution to solve your system.

(c) Use graphing to solve the system.

76. *Supplementary Angles* The smaller of two supplementary angles is one-fourth the measure of the larger angle.

(a) Let x be the measure of the smaller angle and y be the measure of the larger angle. Write a system of linear equations whose solution gives the measures of these angles.

(b) Use the method of substitution to solve the system.

77. *Average Room Prices* In 2002, the average room price for a hotel chain was $1.68 less than in 2001. The 2002 room price was 98% of the 2001 room price. (*Source:* Smith Travel Research.)

(a) Let x be the average room price in 2001 and y be this price in 2002. Write a system of linear equations whose solution gives the average room prices for each year.

(b) Use the method of substitution to solve the system.

78. *Ticket Prices*   Two hundred tickets were sold for a baseball game, which amounted to $840. Student tickets cost $3, and adult tickets cost $5.

   (a) Let x be the number of student tickets sold and y be the number of adult tickets sold. Write a system of linear equations whose solution gives the number of each type of ticket sold.

   (b) Use the method of substitution to solve the system.

79. *NBA Basketball Court*   An official NBA basketball court is 44 feet longer than it is wide. If its perimeter is 288 feet, find its dimensions.

80. *Football Field*   A U.S. football field is 139.5 feet longer than it is wide. If its perimeter is 921 feet, find its dimensions.

81. *Number Problem*   The sum of two numbers is 70. The larger number is two more than three times the smaller number. Find the two numbers.

82. *Number Problem*   The difference of two numbers is 12. The larger number is one less than twice the smaller number. Find the two numbers.

83. *Speed Problem*   A tugboat goes 120 miles upstream in 15 hours. The return trip downstream takes 10 hours. Find the speed of the tugboat without a current and the speed of the current.

84. *Speed Problem*   An airplane flies 1200 miles into the wind in 3 hours. The return trip takes 2 hours. Find the speed of the airplane without a wind and the speed of the wind.

85. *Mixture Problem*   A chemist has 20% and 50% solutions of acid available. How many liters of each solution should be mixed to obtain 10 liters of a 40% acid solution?

86. *Mixture Problem*   A mechanic needs a radiator to have a 40% antifreeze solution. The radiator currently is filled with 4 gallons of a 25% antifreeze solution. How much of the antifreeze mixture should be drained from the car if the mechanic replaces it with pure antifreeze?

87. *Great Lakes*   Together, Lake Superior and Lake Michigan cover 54 thousand square miles. Lake Superior is approximately 10 thousand square miles larger than Lake Michigan. Find the size of each lake. (*Source:* National Oceanic and Atmospheric Administration.)

88. *Longest Rivers*   The two longest rivers in the world are the Nile and the Amazon. Together, they are 8145 miles with the Amazon being 145 miles shorter than the Nile. Find the length of each river.   (*Source:* National Oceanic and Atmospheric Administration.)

WRITING ABOUT MATHEMATICS

89. State one advantage that the method of substitution has over the intersection-of-graphs method. Explain your answer.

90. When applying the method of substitution, how do you know that there are no solutions?

91. When applying the method of substitution, how do you know that there are infinitely many solutions?

92. When applying the intersection-of-graphs method, how do you know that there are no solutions?

CHECKING BASIC CONCEPTS
SECTIONS 4.1 AND 4.2

1. Determine graphically the x-value in each equation when $y = 2$.
   (a) $y = 1 - \frac{1}{2}x$ (b) $2x - 3y = 6$

2. Determine whether $(-1, 0)$ or $(4, 2)$ is a solution to the system

   $$2x - 5y = -2$$
   $$3x + 2y = 16.$$

3. Solve the system of equations graphically. Check your answer.

   $$x - y = 1$$
   $$2x + y = 5$$

4. Use the method of substitution to solve each system of equations. How many solutions does each have?

   (a) $x + y = -1$
   $\quad\ \ y = 2 - x$
   (b) $4x - y = \ \ \ 5$
   $\ -x + y = -2$
   (c) $\ \ \ x + 2y = \ \ \ 3$
   $\ -x - 2y = -3$

5. *Room Prices*   A hotel rents single and double rooms for \$150 and \$200, respectively. The hotel receives \$55,000 for renting 300 rooms.
   (a) Let x be the number of single rooms rented and let y be the number of double rooms rented. Write a system of linear equations whose solution gives the values of x and y.
   (b) Use the method of substitution to solve the system. Check your answer.

4.3 SOLVING SYSTEMS OF LINEAR EQUATIONS BY ELIMINATION

The Elimination Method ▪ Recognizing Other Types of Systems ▪ Applications

A LOOK INTO MATH ▷ Two methods for solving systems of linear equations are the intersection-of-graphs method and the substitution method. The intersection-of-graphs method is a graphical method because it involves the use of graphs to find the solution. The substitution method is a symbolic method because it involves the use of only symbols and algebra to find the solution. In this section we introduce a second symbolic method called the *elimination method*. This method is very efficient for solving some types of systems of linear equations.

The Elimination Method

The elimination method is based on the addition property of equality. Simply put, it is based on the concept that "if equals are added to equals, the results are equal." That is, if

$$a = b \quad \text{and} \quad c = d,$$

then

$$a + c = b + d.$$

For example, if the sum of two numbers is 20 and their difference is 4, then the system of equations

$$x + y = 20$$
$$x - y = \ \ \ 4$$

describes these two numbers. By the addition property of equality, the sum of the left sides of these equations equals the sum of their right sides.

$$(x + y) + (x - y) = 20 + 4$$ Add left sides of these equations.
                                   Add right sides of these equations.
$$2x = 24$$ Combine terms.

Note that the y-variable is eliminated by adding the left sides. The resulting equation, $2x = 24$, simplifies to $x = 12$. Thus the value of x in the solution is 12. The value of y in the solution can be found by substituting 12 for x in either of the given equations. Substituting 12 for x in the first equation, $x + y = 20$, results in $12 + y = 20$ or $y = 8$. The solution to the system of equations is $x = 12$, $y = 8$, which can be written as the ordered pair (12, 8).

To organize the elimination method better, we can carry out the addition vertically.

$$x + y = 20$$
$$\underline{x - y = 4}$$
$$2x + 0y = 24$$ Add left sides and right sides.
$$2x = 24$$ Simplify.
$$x = 12$$ Divide each side by 2.

Once the value of one variable is known, in this case $x = 12$, don't forget to find the value of the other variable by substituting this known value in either of the given equations. By substituting 12 for x in the second equation, we obtain $12 - y = 4$ or $y = 8$. The solution is the ordered pair (12, 8).

EXAMPLE 1   **Applying the elimination method**

Solve each system of equations. Check each solution.
(a) $2x + y = 1$ (b) $-2a + b = -3$
    $3x - y = 9$ $2a + 3b = 7$

Solution
(a) Adding these two equations eliminates the y-variable.

$$2x + y = 1$$ First equation
$$\underline{3x - y = 9}$$ Second equation
$$5x = 10, \quad \text{or} \quad x = 2$$ Add and solve for x.

To find y, substitute 2 for x in either of the *given* equations.

$$2(2) + y = 1$$ Let $x = 2$ in first equation.
$$y = -3$$ Subtract 4 from each side.

The solution is the *ordered pair* $(2, -3)$, which can be checked by substituting 2 for x and -3 for y in the given equations.

$$2x + y = 1 \qquad\qquad 3x - y = 9$$ Given equations
$$2(2) + (-3) \stackrel{?}{=} 1 \qquad 3(2) - (-3) \stackrel{?}{=} 9$$ Let $x = 2$ and $y = -3$.
$$4 - 3 \stackrel{?}{=} 1 \qquad\qquad 6 + 3 \stackrel{?}{=} 9$$ Simplify.
$$1 = 1 \qquad\qquad\qquad 9 = 9$$ The solution checks.

(b) Adding these two equations eliminates the *a*-variable.

$$-2a + b = -3 \qquad \text{First equation}$$
$$2a + 3b = 7 \qquad \text{Second equation}$$
$$\overline{4b = 4,} \quad \text{or} \quad b = 1 \qquad \text{Add and solve for } b.$$

To find *a*, substitute 1 for *b* in either of the *given* equations.

$$2a + 3(1) = 7 \qquad \text{Let } b = 1 \text{ in second equation.}$$
$$2a = 4 \qquad \text{Subtract 3 from each side.}$$
$$a = 2 \qquad \text{Divide each side by 2.}$$

The solution is the *ordered pair* (**2**, **1**), which can be checked by substituting 2 for *a* and 1 for *b* in the given equations.

$$-2a + b = -3 \qquad 2a + 3b = 7 \qquad \text{Given equations}$$
$$-2(2) + 1 \stackrel{?}{=} -3 \qquad 2(2) + 3(1) \stackrel{?}{=} 7 \qquad \text{Let } a = 2 \text{ and } b = 1.$$
$$-4 + 1 \stackrel{?}{=} -3 \qquad 4 + 3 \stackrel{?}{=} 7 \qquad \text{Simplify.}$$
$$-3 = -3 \qquad 7 = 7 \qquad \text{The solution checks.}$$

Now Try Exercises 17, 19

Adding two equations does not always eliminate a variable. For example, adding the following equations eliminates neither variable.

$$3x - 2y = 11 \qquad \text{First equation}$$
$$4x + y = 11 \qquad \text{Second equation}$$
$$\overline{7x - y = 22} \qquad \text{Add the equations.}$$

However, by the multiplication property of equality, we can multiply the second equation by 2. Then adding the equations eliminates the *y*-variable.

$$3x - 2y = 11 \qquad \text{First equation}$$
$$8x + 2y = 22 \qquad \text{Multiply the second equation by 2.}$$
$$\overline{11x = 33,} \quad \text{or} \quad x = 3 \qquad \text{Add and solve for } x.$$

In the next example, we use the multiplication property of equality.

EXAMPLE **2** Multiplying before applying elimination

Solve each system of equations.

(a) $5x - y = -11$ **(b)** $3x + 2y = 1$
       $2x + 3y = -1$ $2x - 3y = 5$

Solution

(a) We multiply the first equation by 3 and then add to eliminate the *y*-variable.

$$15x - 3y = -33 \qquad \text{Multiply first equation by 3.}$$
$$2x + 3y = -1 \qquad \text{Second equation}$$
$$\overline{17x = -34,} \quad \text{or} \quad x = -2 \qquad \text{Add and solve for } x.$$

We can find y by substituting -2 for x in the second equation.

$$2(-2) + 3y = -1 \qquad \text{Let } x = -2 \text{ in second equation.}$$
$$3y = 3 \qquad \text{Add 4 to each side.}$$
$$y = 1 \qquad \text{Divide each side by 3.}$$

The solution is $(-2, 1)$.

(b) For this system we must apply the multiplication property to both equations. If we multiply the first equation by 3 and the second equation by 2, then the coefficients of the y-variables will be opposites. Adding eliminates the y-variable.

$$9x + 6y = 3 \qquad \text{Multiply the first equation by 3.}$$
$$\underline{4x - 6y = 10} \qquad \text{Multiply the second equation by 2.}$$
$$13x = 13, \quad \text{or} \quad x = 1 \qquad \text{Add and solve for } x.$$

To find y, substitute 1 for x in the first *given* equation.

$$3(1) + 2y = 1 \qquad \text{Let } x = 1 \text{ in first equation.}$$
$$2y = -2 \qquad \text{Subtract 3 from each side.}$$
$$y = -1 \qquad \text{Divide each side by 2.}$$

The solution is $(1, -1)$.

Now Try Exercises 29, 33

In part (b) of the previous example the y-variable was eliminated. However, the system of equations could have been solved by first eliminating the x-variable. In practice, it is possible to eliminate *either* variable from a system of linear equations. It is often best to choose the variable that requires the least amount of computation to complete the elimination. In the next example, we solve a system of equations twice—first by using the multiplication property of equality to eliminate the x-variable and then using it to eliminate the y-variable.

EXAMPLE 3   **Multiplying before applying elimination**

Solve the system of equations two times, first by eliminating x and then by eliminating y.

$$2y = -6 - 5x$$
$$2x = -5y + 6$$

Solution

It is best to write each equation in the standard form: $Ax + By = C$.

$$5x + 2y = -6 \qquad \text{First equation in standard form}$$
$$2x + 5y = 6 \qquad \text{Second equation in standard form}$$

Eliminate x   If we multiply the first equation in standard form by -2 and the second equation in standard form by 5, then we can eliminate the x-variable by adding.

$$-10x - 4y = 12 \qquad \text{Multiply by } -2.$$
$$\underline{10x + 25y = 30} \qquad \text{Multiply by 5.}$$
$$21y = 42, \quad \text{or} \quad y = 2 \qquad \text{Add and solve for } y.$$

To find x, substitute **2** for **y** in the first given equation, $2y = -6 - 5x$.

$$2(2) = -6 - 5x \qquad \text{Let } y = 2 \text{ in first equation.}$$
$$10 = -5x \qquad \text{Add 6 to each side.}$$
$$-2 = x \qquad \text{Divide by } -5.$$

The solution is $(-2, 2)$.

Eliminate y   If we multiply the first equation in standard form by -5 and the second equation in standard form by 2, then we can eliminate the y-variable by adding.

$$-25x - 10y = 30 \qquad \text{Multiply by } -5.$$
$$\underline{4x + 10y = 12} \qquad \text{Multiply by 2.}$$
$$-21x \qquad = 42, \quad \text{or} \quad x = -2 \qquad \text{Add and solve for } x.$$

To find y, substitute -2 for x in the second given equation, $2x = -5y + 6$.

$$2(-2) = -5y + 6 \qquad \text{Let } x = -2 \text{ in second equation.}$$
$$-10 = -5y \qquad \text{Subtract 6 from each side.}$$
$$2 = y \qquad \text{Divide by } -5.$$

Now Try Exercise **31**

The solution is $(-2, 2)$.

In the next example we use three different methods to solve a system of equations.

EXAMPLE 4   Solving a system with different methods

Solve the system of equations symbolically, graphically, and numerically.

$$x + y = 2$$
$$x - 3y = 6$$

Solution

Symbolic Solution   Both the method of substitution and the elimination method are symbolic methods. The elimination method is used here. We can solve the system by multiplying the second equation by -1 and adding to eliminate the x-variable.

$$x + y = 2 \qquad \text{First equation}$$
$$\underline{-x + 3y = -6} \qquad \text{Multiply by } -1.$$
$$4y = -4, \quad \text{or} \quad y = -1 \qquad \text{Add and solve for } y.$$

We can find x by substituting -1 for y in the first equation, $x + y = 2$.

$$x + (-1) = 2 \qquad \text{Let } y = -1 \text{ in first equation.}$$
$$x = 3 \qquad \text{Add 1 to each side.}$$

The solution is $(3, -1)$.

Graphical Solution   For a graphical solution, we solve each equation for y to obtain the slope–intercept form.

$x + y = 2$	First equation	$x - 3y = 6$	Second equation
$y = -x + 2$	Subtract x.	$-3y = -x + 6$	Subtract x.
		$y = \dfrac{1}{3}x - 2$	Divide by -3.

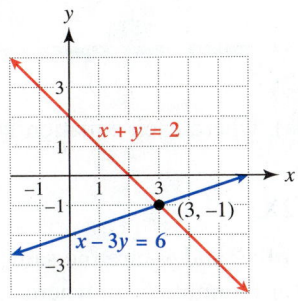

Figure 4.9

The graphs of $y = -x + 2$ and $y = \frac{1}{3}x - 2$ are shown in Figure 4.9. They intersect at $(3, -1)$. (These graphs could also be obtained by finding the x- and y-intercepts for each equation.)

Numerical Solution   A numerical solution consists of a table of values, as shown in Table 4.3. Note that when $x = 3$, both y-values equal -1. Therefore the solution is $(3, -1)$.

TABLE 4.3

	x	0	1	2	**3**	4
$y = -x + 2$		2	1	0	-1	-2
$y = \frac{1}{3}x - 2$		-2	$-\frac{5}{3}$	$-\frac{4}{3}$	-1	$-\frac{2}{3}$

Now Try Exercise **43**

Recognizing Other Types of Systems

In Section 4.2 we discussed how a system of linear equations can have no solutions, one solution, or infinitely many solutions. Elimination can also be used on systems that have no solutions or infinitely many solutions.

EXAMPLE **5**   **Solving other types of systems**

Solve each system of equations by using the elimination method. Then graph the system.
(a) $\quad x - 2y = \quad 4$ **(b)** $3x + 3y = 6$
$\quad -2x + 4y = -8$ $\quad x + \ \ y = 1$

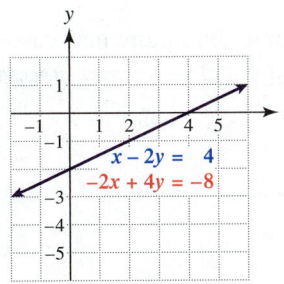

Figure 4.10

Solution

(a) We multiply the first equation by 2 and then add, which eliminates both variables.

$$
\begin{array}{ll}
2x - 4y = \ \ 8 & \text{Multiply first equation by 2.} \\
\underline{-2x + 4y = -8} & \text{Second equation} \\
0 = \ \ 0 \quad \text{(True)} & \text{Add.}
\end{array}
$$

The equation $0 = 0$ is *always true*, which indicates that the system has *infinitely many solutions*. A graph of the two equations is shown in Figure 4.10. The two lines are identical so there actually is only one line, and *every point on this line represents a solution*. For example, $(0, -2)$ and $(4, 0)$ lie on the line and are both solutions.

(b) We multiply the second equation by -3 and add, eliminating both variables.

$$
\begin{array}{ll}
3x + 3y = \ \ 6 & \text{First equation} \\
\underline{-3x - 3y = -3} & \text{Multiply second equation by } -3. \\
0 = \ \ 3 \quad \text{(False)} & \text{Add.}
\end{array}
$$

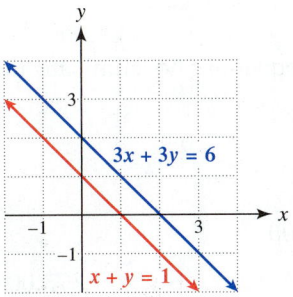

Figure 4.11

The equation $0 = 3$ is *always false*, which indicates that the system has *no solutions*. A graph of the two equations is shown in Figure 4.11. Note that the two lines are parallel and thus do not intersect.

Now Try Exercises **53, 57**

MAKING CONNECTIONS

Numerical and Graphical Solutions

The left-hand figure shows a numerical solution for Example 5(a), where each equation is solved for y to obtain

$$Y_1 = (4 - X)/(-2) \quad \text{and} \quad Y_2 = (-8 + 2X)/4.$$

Note that $y_1 = y_2$ for each value of x, indicating that the graphs of y_1 and y_2 are the same line. Similarly, the right-hand figure shows a numerical solution for Example 5(b), where

$$Y_1 = (6 - 3X)/3 \quad \text{and} \quad Y_2 = 1 - X.$$

Note that $y_1 \neq y_2$ and the difference $y_1 - y_2$ is 1 for every value of x, indicating that the graphs of y_1 and y_2 are parallel lines that do not intersect.

CALCULATOR HELP

To make a table, see Appendix A (page AP-3).

X	Y1	Y2
0	-2	-2
1	-1.5	-1.5
2	-1	-1
3	-.5	-.5
4	0	0
5	.5	.5
6	1	1
Y1∎(4−X)/(-2)		

X	Y1	Y2
-1	3	2
0	2	1
1	1	0
2	0	-1
3	-1	-2
4	-2	-3
5	-3	-4
Y1∎(6−3X)/3		

Applications

▶ **REAL-WORLD CONNECTION**   In the next two examples we use elimination to solve applications relating to new cancer cases and burning calories during exercise.

EXAMPLE 6   Determining new cancer cases

In 2005, there were 1,373,000 new cancer cases. Men accounted for 47,000 more new cases than women. How many new cases of cancer were there for each gender?   (**Source:** American Cancer Society.)

Solution

STEP 1: *Identify each variable.*

x: new cancer cases for men in 2005
y: new cancer cases for women in 2005

STEP 2: *Write a system of equations.*

$$x + y = 1{,}373{,}000$$
$$x - y = 47{,}000$$

STEP 3A: *Solve the system of linear equations.* Add the two equations to eliminate the y-variable.

$$
\begin{aligned}
x + y &= 1{,}373{,}000 \\
\underline{x - y} &= \underline{47{,}000} \\
2x &= 1{,}420{,}000, \quad \text{or} \quad x = 710{,}000
\end{aligned}
$$

Substituting 710,000 for x in the first equation results in $710{,}000 + y = 1{,}373{,}000$ or $y = 663{,}000$. The solution is $x = 710{,}000$ and $y = 663{,}000$.

STEP 3B: *Determine the solution to the problem.* There were 710,000 new cases of cancer for men and 663,000 new cases for women in 2005.

STEP 4: *Check the solution.* The total number of cases was

$$710{,}000 + 663{,}000 = 1{,}373{,}000.$$

The number of new cases for men exceeded the number of new cases for women by

$$710{,}000 - 663{,}000 = 47{,}000.$$

The answer checks.   Now Try Exercise 59

EXAMPLE 7   Burning calories during exercise

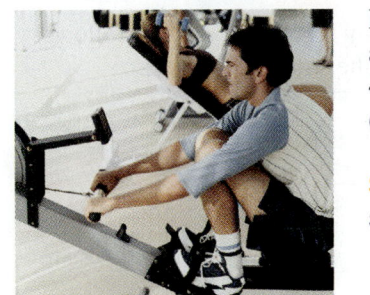

During strenuous exercise, an athlete can burn 10 calories per minute on a rowing machine and 11.5 calories per minute on a stair climber. If an athlete burns 433 calories in a 40-minute workout, how many minutes does the athlete spend on each machine? (*Source: Runner's World.*)

Solution

STEP 1: *Identify each variable.*

x: number of minutes on a rowing machine
y: number of minutes on a stair climber

STEP 2: *Write a system of equations.* The total workout takes 40 minutes, so $x + y = 40$. The athlete burns $10x$ calories on the rowing machine and $11.5y$ calories on the stair climber. Because the total number of calories equals 433, it follows that $10x + 11.5y = 433$.

$$
\begin{aligned}
x + \quad\; y &= \;\;40 \qquad &&\text{Workout is 40 minutes.}\\
10x + 11.5y &= 433 \qquad &&\text{Total calories is 433.}
\end{aligned}
$$

STEP 3A: *Solve the system of linear equations.* Multiply the first equation by -10 and add the two equations.

$$
\begin{aligned}
-10x - \quad 10y &= -400 \qquad &&\text{Multiply by } -10.\\
\underline{10x + 11.5y} &= \quad \underline{433} \qquad &&\text{Second equation}\\
1.5y &= \quad 33, \quad \text{or} \quad y = \frac{33}{1.5} = 22 \qquad &&\text{Add and solve for } y.
\end{aligned}
$$

Because $x + y = 40$ and $y = 22$, it follows that $x = 18$.

STEP 3B: *Determine the solution to the problem.* The athlete spends 18 minutes on the rowing machine and 22 minutes on the stair climber.

STEP 4: *Check your answer.* Because $18 + 22 = 40$, the athlete works out for 40 minutes. Also,

$$10(18) + 11.5(22) = 433,$$

so the athlete burns 433 calories. The answer checks.   Now Try Exercise 61

4.3 PUTTING IT ALL TOGETHER

The method of substitution and the elimination method are two symbolic techniques that can be used to solve systems of linear equations. The elimination method makes use of the addition and multiplication properties of equality and is based on the idea that "if equals are added to equals, the results are equal." The following table highlights the important aspects of elimination.

Concept	Explanation	Example
Elimination Method	If $a = b$ and $c = d$, then $$a + c = b + d.$$ May be used to solve systems of equations	$x + y = 5$ $x - y = -1$ $2x = 4$, or $x = 2$ Add. Because $x + y = 5$ and $x = 2$, it follows that $y = 3$. The solution is $(2, 3)$.
Other Types of Systems	Elimination can be used to recognize systems having **1.** no solutions **2.** infinitely many solutions.	**1.** $-x - y = -4$ $x + y = 2$ $0 = -2$ Add. Because $0 = -2$ is always false, there are no solutions. **2.** $x + y = 4$ $2x + 2y = 8$ Multiply the first equation by -2. $-2x - 2y = -8$ $2x + 2y = 8$ $0 = 0$ Add. Because $0 = 0$ is always true, there are infinitely many solutions.

4.3 Exercises

CONCEPTS

1. Name two symbolic methods for solving a system of linear equations.

2. The elimination method is based on the _____ property of equality.

3. The addition property of equality states that if $a = b$ and $c = d$, then $a + c$ _____ $b + d$.

4. The multiplication property of equality states that if $a = b$, then ca _____ cb.

5. When you are using elimination to solve

$$2x + y = 6$$
$$x - y = 2,$$

what is a good first step?

6. When you are using elimination to solve

$$x + 2y = 8$$
$$3x - 5y = 2,$$

what is a good first step?

7. Suppose that the elimination method results in the equation $1 = 1$. What does this indicate about the number of solutions to the system of equations?

8. Suppose that the elimination method results in the equation $0 = 1$. What does this indicate about the number of solutions to the system of equations?

USING ELIMINATION

Exercises 9–16: If possible, use the given graph to solve the system of equations. Then use the elimination method to verify your answer.

9. $x - y = 0$
$\ \ \ \ x + y = 2$

10. $\ \ x + y = 6$
$\ \ \ \ \ \ 2x - y = 3$

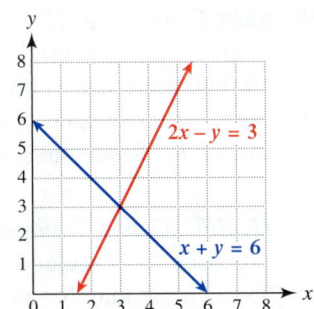

11. $2x + 3y = -1$
$\ \ \ \ \ 2x - 3y = -7$

12. $-2x + y = -3$
$\ \ \ \ \ \ \ 4x - 3y = \ \ 7$

13. $x + y = \ \ \ 3$
$\ \ \ \ x + y = -1$

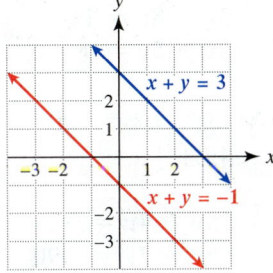

14. $\ \ \ 2x - y = \ \ \ 4$
$\ \ \ \ \ -2x + y = -4$

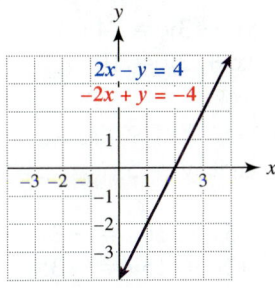

15. $2x + 2y = 6$
$\ \ \ \ x + \ \ y = 3$

16. $-x + 3y = 4$
$\ \ \ \ \ \ x - 3y = 3$

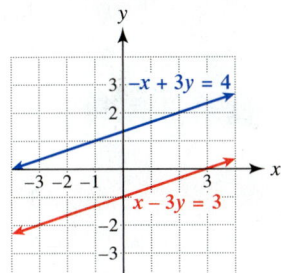

Exercises 17–38: Use the elimination method to solve the system of equations.

17. $x + y = 7$
$\ \ \ \ x - y = 5$

18. $x - y = 8$
$\ \ \ \ x + y = 4$

19. $-x + y = 5$
$\ \ \ \ \ \ x + y = 3$

20. $\ \ \ x - y = 10$
$\ \ \ \ \ -x - y = 20$

21. $2x + y = 8$
$\ \ \ \ 3x - y = 2$

22. $-x + 2y = 3$
$\ \ \ \ \ \ x + 6y = 5$

23. $-2x + \ \ y = -3$
$\ \ \ \ \ \ 2x - 4y = \ \ \ 0$

24. $2x + 6y = -5$
$\ \ \ \ 7x - 6y = -4$

25. $a + 6b = \ \ \ 2$
$\ \ \ \ a + 3b = -1$

26. $5a - 6b = -2$
$\ \ \ \ 5a + 5b = \ \ \ 9$

27. $3r - t = 7$
$\ \ \ \ 2r - t = 2$

28. $-r + 2t = 0$
$\ \ \ \ \ \ 3r + 2t = 8$

29. $3u + 2v = -16$
$\ \ \ \ 2u + \ \ v = \ \ -9$

30. $5u - \ \ v = \ \ \ \ 0$
$\ \ \ \ 3u + 3v = -18$

31. $5x - 7y = 5$
$-2x + 2y = -2$

32. $2x + 7y = 6$
$4x - 3y = -22$

33. $5x - 3y = 4$
$3x + 2y = 10$

34. $-3x - 8y = 1$
$2x + 5y = 0$

35. $\frac{1}{2}x - y = 3$
$\frac{3}{2}x + y = 5$

36. $x - \frac{1}{4}y = 4$
$-4x + \frac{1}{4}y = -9$

37. $-5x - 10y = -22$
$10x + 15y = 35$

38. $-15x + 4y = -20$
$5x + 7y = 90$

Exercises 39–42: A table of values is given for two linear equations. Use the table to solve this system.

39.

x	0	1	2	3	4
$y = -x + 5$	5	4	3	2	1
$y = 2x - 4$	-4	-2	0	2	4

40.

x	-3	-2	-1	0	1
$y = x + 1$	-2	-1	0	1	2
$y = -x - 3$	0	-1	-2	-3	-4

41.

x	-2	-1	0	1	2
$y = 3x + 1$	-5	-2	1	4	7
$y = -x + 1$	3	2	1	0	-1

42.

x	-2	-1	0	1	2
$y = 2x$	-4	-2	0	2	4
$y = -x$	2	1	0	-1	-2

USING MORE THAN ONE METHOD

Exercises 43–48: Solve the system of equations
(a) symbolically,
(b) graphically, and
(c) numerically.

43. $2x + y = 5$
$x - y = 1$

44. $-x + y = 2$
$3x + y = -2$

45. $2x + y = 5$
$x + y = 1$

46. $-x + y = 2$
$3x - y = -2$

47. $6x + 3y = 6$
$-2x + 2y = -2$

48. $-x + 2y = 5$
$2x + 2y = 8$

ELIMINATION AND OTHER TYPES OF SYSTEMS

Exercises 49–58: Use elimination to determine whether the system of equations has no solutions, one solution, or infinitely many solutions. Then graph the system.

49. $2x - 2y = 4$
$-x + y = -2$

50. $-2x + y = 4$
$4x - 2y = -8$

51. $x - y = 0$
$x + y = 0$

52. $x - y = 2$
$x + y = 2$

53. $x - y = 4$
$x - y = 1$

54. $-2x + 3y = 5$
$4x - 6y = 10$

55. $x - y = 5$
$2x - y = 4$

56. $6x + 9y = 18$
$4x + 6y = 12$

57. $4x - 8y = 24$
$6x - 12y = 36$

58. $x - 3y = 2$
$-x + 3y = 4$

APPLICATIONS

59. *Skin Cancer* In 2005, there were 66,000 new cases of skin cancer in the United States. Men represented 10,000 more cases than women. How many new cases of skin cancer were there for men and for women? (*Source:* American Cancer Society.)

60. *Health Care Expenses* In 2002, out-of-pocket expenses for an elderly person in poor health were $3353 more than for an elderly person in good health. Combined out-of-pocket expenses for one person in poor health and one person in good health were $6213. Find the out-of-pocket expenses for each type of person. (*Source:* "Trends in Medicare-Choice Benefits and Premiums, 1999–2002." *Commonwealth Fund.*)

61. *Burning Calories* During strenuous exercise an athlete can burn 9 calories per minute on a stationary bicycle and 11.5 calories per minute on a stair climber. In a 30-minute workout an athlete burns 300 calories. How many minutes does the athlete spend on each type of exercise equipment? (*Source:* Runner's World.)

62. *Distance Running* An athlete runs at 9 mph and then at 12 mph, covering 10 miles in 1 hour. How long does the athlete run at each speed?

63. *River Current* A riverboat takes 8 hours to travel 64 miles downstream and 16 hours for the return trip. What is the speed of the current and the speed of the riverboat in still water?

64. *Airplane Speed* An airplane travels 3000 miles with the wind in 5 hours and takes 6 hours for the return trip into the wind. What is the speed of the wind and the speed of the airplane without any wind?

65. *Investments* A total of $5000 is invested at 3% and 5% annual interest. After 1 year the total interest equals $210. How much money is invested at each interest rate?

66. *Mixing Antifreeze* A car radiator holds 2 gallons of fluid and initially is empty. If a mixture of water and antifreeze contains 70% antifreeze and another mixture contains 15% antifreeze, how much of each should be combined to fill the radiator with a 50% antifreeze mixture?

67. *Number Problem* The sum of two integers is -17, and their difference is -69. Find the two integers.

68. *Supplementary Angles* The measures of two supplementary angles differ by 74°. Find the two angles.

69. *Picture Dimensions* The figure at the top of the next column shows a red graph that gives possible dimensions for a rectangular picture frame with perimeter 120 inches. The blue graph shows possible dimensions for a rectangular frame whose length L is twice its width W.

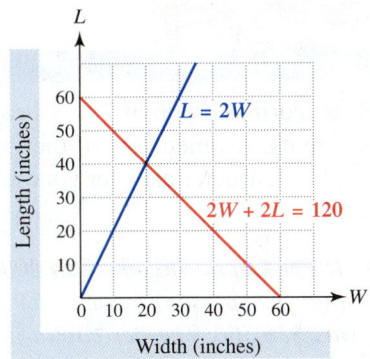

(a) Use the figure to determine the dimensions of a frame with a perimeter of 120 inches and a length that is twice the width.

(b) Solve this problem symbolically.

70. *Sales of CDs and Tapes* A company sells compact discs d and cassette tapes t. The figure shows a red graph of $d + t = 2000$. The blue graph shows a revenue of $15,000 received from selling d compact discs at $12 each and t tapes at $6 each.

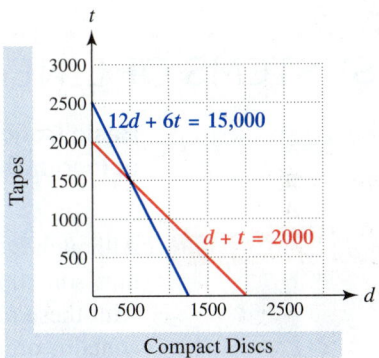

(a) If the total number of discs and tapes sold is 2000, determine how many of each were sold to obtain a revenue of $15,000.

(b) Solve this problem symbolically.

WRITING ABOUT MATHEMATICS

71. Suppose that a system of linear equations is solved symbolically, numerically, and graphically. How do the solutions from each method compare? Explain your answer.

72. When you are solving a system of linear equations by elimination, how can you recognize that the system has no solutions?

GROUP ACTIVITY
WORKING WITH REAL DATA

Directions: Form a group of 2 to 4 people. Select someone to record the group's responses for this activity. All members of the group should work cooperatively to answer the questions. If your instructor asks for your results, each member of the group should be prepared to respond.

*Exercises 1–4: **Per Capita Income***   *In 2004, the average of the per capita (per person) incomes for Massachusetts and Maine was $36,500. The per capita income in Massachusetts exceeded the per capita income in Maine by $11,000.*

1. Set up a system of equations whose solution gives the per capita income in each state. Identify what each variable represents.

2. Use substitution to solve this system. Interpret the result.

3. Use elimination to solve this system.

4. Solve this system graphically. Do all your answers agree?

4.4 SYSTEMS OF LINEAR INEQUALITIES

Basic Concepts ■ Solutions to One Inequality ■ Solutions to Systems of Inequalities ■ Applications

A LOOK INTO MATH ▷

Although there is no *ideal* weight for a person, government agencies and insurance companies sometimes recommend a *range* of weights for various heights. *Inequalities* are used with these recommendations. One example is shown in Figure 4.12, where the blue region contains ordered pairs (w, h) that give recommended weight–height combinations. Describing this region mathematically requires an understanding of systems of linear inequalities, which we discuss in this section.   (**Source:** Department of Agriculture.)

Figure 4.12

Basic Concepts

▶ **REAL-WORLD CONNECTION** Suppose that a college student works both at the library and at a department store. The library pays $10 per hour, and the department store pays $8 per hour. The equation $A = 10L + 8D$ calculates the amount of money earned from working L hours at the library and D hours at the department store. If the cost of one college credit is $80, then solutions to the equation

$$10L + 8D = 80$$

are ordered pairs (L, D) that result in the student earning enough to pay for one credit. Its graph is the line shown in Figure 4.13(a). The point $(4, 5)$ lies on this line, which indicates that, if the student works 4 hours at the library and 5 hours at the department store, then the pay is $80.

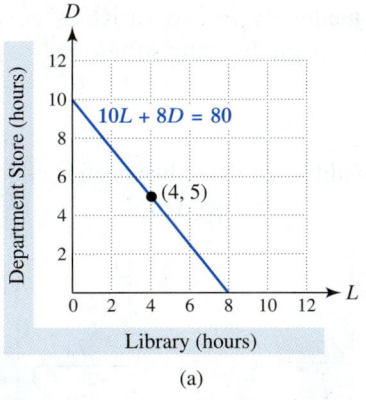

Figure 4.13 A Student's Earnings

There are many situations in which the student can make more than $80. For example, the **test point** $(6, 6)$ lies above the line in Figure 4.13(b), indicating that, if the student works 6 hours at both the library and the department store, then the pay is more than $80. In fact, any point *above* the line results in pay *greater than* $80. The region **above** the line is described by the inequality

$$10L + 8D > 80.$$

The test point (**6**, **6**) represents earnings of $108 and *satisfies* this inequality because

$$10(6) + 8(6) > 80$$

is a true statement. Similarly, any point *below* the line gives an ordered pair (L, D) that results in earnings *less than* $80. The point (**2**, **3**) in Figure 4.13(b) lies below the line and represents earnings of $44. That is,

$$10(2) + 8(3) < 80.$$

The region **below** the line is described by the inequality

$$10L + 8D < 80.$$

Solutions to One Inequality

Any linear equation in two variables can be written in standard form as

$$Ax + By = C,$$

where A, B, and C are constants. When the equals sign is replaced with $<$, $>$, $\leq$, or $\geq$, a **linear inequality in two variables** results. Examples of linear *equations* in two variables include

$$2x + 3y = 10 \quad \text{and} \quad y = \frac{1}{2}x - 5,$$

and so examples of linear *inequalities* in two variables include

$$2x + 3y < 10 \quad \text{and} \quad y \geq \frac{1}{2}x - 5.$$

A *solution* to a linear inequality in two variables is an ordered pair (x, y) that makes the inequality a true statement. The *solution set* is the set of all solutions to the inequality. The solution set to an inequality in two variables is typically a region in the xy-plane, which means that there are infinitely many solutions.

EXAMPLE 1   Writing a linear inequality

Write a linear inequality that describes each shaded region.

(a)

(b)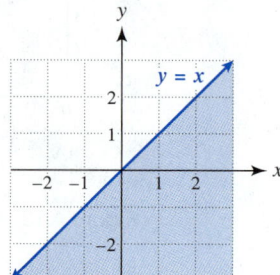

Solution

(a) The shaded region is bounded by the line $x = 2$. The *dashed* line indicates that the line is not included in the solution set. Only points with x-coordinates **greater than** 2 are shaded. Thus every point in the shaded region satisfies $x > 2$.

(b) The solution set includes all points that are on or **below** the line $y = x$. An inequality that describes this region is $y \leq x$, which can also be written as $-x + y \leq 0$.

Now Try Exercises 25, 27

TECHNOLOGY NOTE: Shading an Inequality

Graphing calculators can be used to shade a solution set to an inequality. The left-hand screen shows how to enter the equation from Example 1(b), and the right-hand screen shows the resulting graph.

CALCULATOR HELP

To shade an inequality, see Appendix A (page AP-7).

EXAMPLE 2   Graphing a linear inequality

Shade the solution set for each inequality.
(a) $y \leq 1$ **(b)** $x + y < 3$ **(c)** $-x + 2y \geq 2$

Solution

(a) First graph the horizontal line $y = 1$, where a solid line indicates that the line is included in the solution set. Next decide whether to shade above or below this line. The inequality $y \leq 1$ indicates that a solution (x, y) must have a y-coordinate less than or equal to 1. There are no restrictions on the x-coordinate. The shaded region **below** the line $y = 1$ in Figure 4.14(a) depicts all ordered pairs (x, y) satisfying $y \leq 1$. The horizontal line $y = 1$ is solid because it is included in the solution set.

(b) Graph the line $x + y = 3$, as shown in Figure 4.14(b). Because the inequality is $<$ and not $\leq$, the line is not included and is dashed rather than solid. To decide whether to shade above or below this dashed line, select a *test point* in either region. For example, the point $(0, 0)$ satisfies the inequality $x + y < 3$, because $0 + 0 < 3$ is a true statement. Therefore shade the region that does contain $(0, 0)$, which is *below* the line.

(c) Graph $-x + 2y = 2$ as a solid line, as shown in Figure 4.14(c). To decide whether to shade above or below the solid line, use the test point $(0, 0)$ again. (Note that other test points can be used.) The point $(0, 0)$ does not satisfy the inequality $-x + 2y \geq 2$ because $-0 + 2(0) \geq 2$ is a false statement. Therefore shade the region that does not contain $(0, 0)$, which is *above* the line.

(a)

(b)

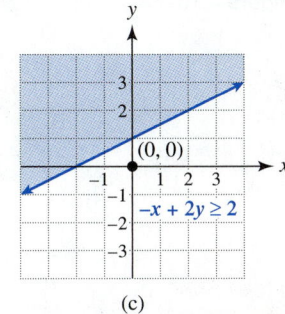
(c)

Figure 4.14

Now Try Exercises 33, 39, 41

NOTE: In both parts (b) and (c) of Example 2, the test point $(0, 0)$ was used to determine which region should be shaded. However, *any* point that is not on the solid or dashed line can be used as a test point. When the line does not pass through the origin, it is often convenient to use $(0, 0)$ as a test point because substituting 0 for both x and y in an inequality results in a very simple computation.

GRAPHING A LINEAR INEQUALITY

1. Replace the inequality symbol with an equals sign and graph the resulting line. If the inequality is $<$ or $>$, use a dashed line, and if it is $\leq$ or $\geq$, use a solid line.
2. Pick a test point that does *not* lie on the line. Substitute this point in the given inequality. Determine whether the resulting statement is true or false.
3. If the statement is true, shade the region containing the test point. If the statement is false, shade the region not containing the test point.

Solutions to Systems of Inequalities

Sometimes a solution set must satisfy two inequalities. The point (2, 1) satisfies both of the inequalities in the system of inequalities given by

$$x > 1$$
$$y < 2.$$

Figure 4.15(a) shows the solution set to $x > 1$ in blue, and Figure 4.15(b) shows the solution set to $y < 2$ in red. The two regions are shaded together in Figure 4.15(c) where the blue and red regions intersect to form a purple region. The solution set to the system of inequalities includes all points in the purple region, as shown in Figure 4.15(d). Points in this region satisfy *both* inequalities.

(a)

(b)

(c)

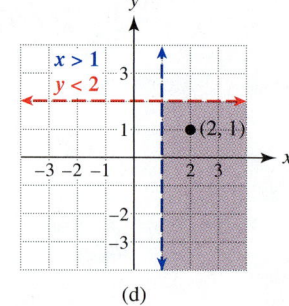

(d)

Figure 4.15

EXAMPLE 3 **Graphing a system of linear inequalities**

Shade the solution set to the system of inequalities.

$$x > -1$$
$$x + y \leq 1$$

Solution

In this example, we start by graphing the solution set to each inequality. The solution set to $x > -1$ is the blue region to the right of the dashed vertical line $x = -1$, as shown in Figure 4.16(a). The solution set to $x + y \leq 1$ includes the solid line $x + y = 1$ and the red region that lies below it, as shown in Figure 4.16(b). For a point to satisfy the *system* of inequalities it must satisfy *both* inequalities. Therefore the solution set is the *intersection* of the blue and red regions, shown as the purple region in Figure 4.16(c). Note that the test point (0, 0), located in the shaded region, satisfies both inequalities.

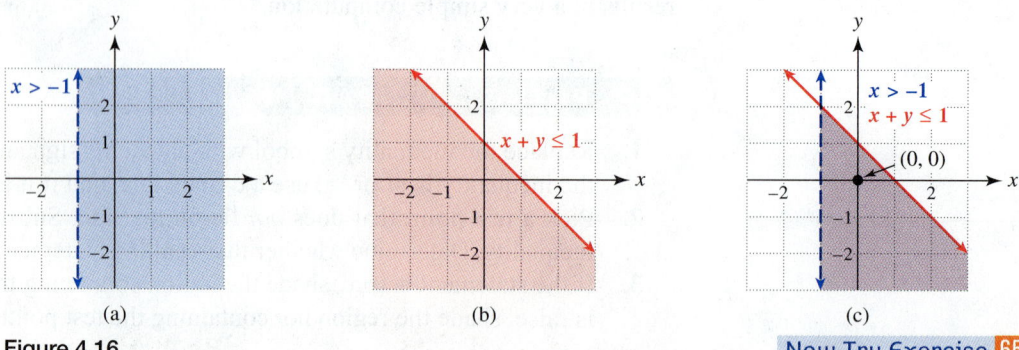

(a)   (b)   (c)

Figure 4.16

Now Try Exercise 65

EXAMPLE 4 **Graphing a system of linear inequalities**

Shade the solution set to the system of inequalities.

$$x + 2y < -2$$
$$2x + y \geq 2$$

Solution

In this example we use test points to determine the solution set. We start by graphing the dashed line $x + 2y = -2$ and the solid line $2x + y = 2$, as shown in Figure 4.17(a). Note that these two lines divide the xy-plane into 4 regions, numbered 1, 2, 3, and 4. If we let $(0, 0)$ be a test point, it does not satisfy either inequality. Therefore we do not shade region 2, which contains $(0, 0)$. However, there are still 3 possible regions. If we try the test point $(4, -4)$ in region 4, it satisfies both the given inequalities.

$$4 + 2(-4) < -2 \qquad \text{A true statement}$$
$$2(4) + (-4) \geq 2 \qquad \text{A true statement}$$

Thus we shade region 4, as shown in Figure 4.17(b). Once a test point is found that makes both inequalities true, there is no need to check test points in other regions.

CRITICAL THINKING

Does the solution set in Figure 4.17(b) include the point of intersection, $(2, -2)$? Explain your reasoning.

 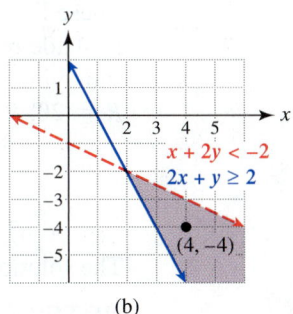

(a)   (b)

Figure 4.17

Now Try Exercise 61

MAKING CONNECTIONS

Solving for *y* and shading inequalities

Another way to solve the system of inequalities in Example 4 that does not involve test points is to solve each inequality for y to obtain

$$y < -\frac{1}{2}x - 1$$

$$y \geq -2x + 2.$$

The solution set is the region in Figure 4.17(b) that lies *below* the line $y = -\frac{1}{2}x - 1$ and *above and including* the line $y = -2x + 2$.

Applications

▶ **REAL-WORLD CONNECTION**   The next two examples illustrate applications of inequalities.

EXAMPLE **Manufacturing radios and CD players**

A business manufactures radios and CD players. Because every CD player contains a radio, it must produce at least as many radios as CD players. In addition, the total number of radios and CD players produced each day cannot exceed 50 because of limited resources. Shade the region that shows the numbers of radios R and CD players P that can be produced within these restrictions. Label the horizontal axis R and the vertical axis P.

Solution
Because the company must produce *at least* as many radios R as CD players P, we have $R \geq P$, which can also be written as $P \leq R$. The total number of radios and CD players *cannot exceed* 50 so $R + P \leq 50$. To shade the solution set for

$$P \leq R$$
$$R + P \leq 50,$$

we first graph the lines $P = R$ and $R + P = 50$, as shown in Figure 4.18(a). Because the number of radios and CD players cannot be negative, the graph includes only quadrant I. These lines divide this quadrant into four regions, and we can determine the correct region to shade by selecting one test point from each region. The region containing the test point satisfying both inequalities is the one to be shaded. For example, the test point (**20**, **10**) with $R = 20$, $P = 10$ satisfies both inequalities.

$$10 \leq 20 \qquad \text{A true statement; } P \leq R$$
$$20 + 10 \leq 50 \qquad \text{A true statement; } R + P \leq 50$$

The solution set is shaded in Figure 4.18(b).

NOTE: An alternative solution is to write the inequalities as $P \leq R$ and $P \leq -R + 50$. Then the solution set lies *below both lines*. This region is shaded in Figure 4.18(b).

(a)

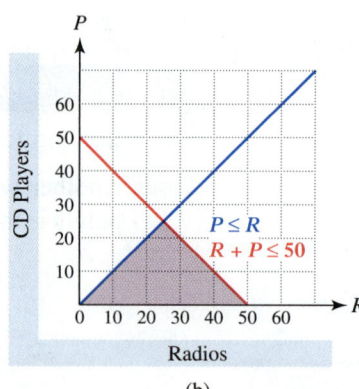

(b)

Figure 4.18

Now Try Exercise **71**

The next example discusses the application from the beginning of this section.

EXAMPLE 6   **Finding weight–height combinations**

Figure 4.19 shows a shaded region containing recommended weights *w* for heights *h*.

Figure 4.19

(a) What does this graph indicate about someone who is 68 inches tall and weighs 150 pounds?

(b) The shaded region in Figure 4.19 is determined by the following system of inequalities.

$$25h - 7w \leq 800$$
$$5h - w \geq 170$$

Verify that $h = 68$ and $w = 150$ satisfies the system of inequalities.

(c) What do ordered pairs (w, h) to the left of the shaded region indicate?

CRITICAL THINKING

In Example 6 what do ordered pairs (w, h) to the right of the shaded region represent? Explain your answer.

Solution

(a) The point (150, 68) lies in the shaded region. Therefore someone who is 68 inches tall and weighs 150 pounds falls within the recommended guidelines.

(b) Both inequalities are satisfied by $h = 68$, $w = 150$.

$$25(\mathbf{68}) - 7(\mathbf{150}) = 650 \leq 800$$
$$5(\mathbf{68}) - \mathbf{150} = 190 \geq 170$$

(c) To the left of the shaded region are ordered pairs (w, h) that represent smaller weights and larger heights. This region corresponds to people who weigh less than recommended.

Now Try Exercise 75

4.4 PUTTING IT ALL TOGETHER

The following table summarizes important concepts related to linear inequalities in two variables.

Concept	Explanation	Examples
Linear Inequality in Two Variables	An inequality that can be written as $$Ax + By < C,$$ where < can also be ≤, >, or ≥	$3x + y \geq 10$, $-x + 3y < 5$, $y \leq 5 - x$, and $x > 5$

continued on next page

continued from previous page

Concept	Explanation	Examples
Solution	A solution (x, y) makes the inequality a true statement.	The point $(0, 0)$ satisfies $$2x - y < 2,$$ so it is a solution to the inequality.
Solution Set	The set of all solutions Usually a region in the xy-plane	The solution set to $x + y > 2$ is all points above the line $x + y = 2$. *(graph showing $x + y > 2$)*
System of Linear Inequalities in Two Variables	Solutions to systems must satisfy both inequalities. The solution set usually includes infinitely many solutions.	The point $(0, 0)$ is a solution to $$x + y \leq \ \ 2$$ $$2x - y > -4,$$ because both inequalities are true when $x = 0$ and $y = 0$. *(graph showing $2x - y > -4$ and $x + y \leq 2$ with point $(0, 0)$)*

4.4 Exercises

CONCEPTS

1. Describe the graph of the solution set to $y \leq k$ for some number k.

2. Describe the graph of the solution set to $x > k$ for some number k.

3. Describe the graph of the solution set to $y \geq x$.

4. When graphing the solution set to a linear inequality, one way to determine which region to shade is to use a _____ point.

5. When graphing a linear inequality containing either $<$ or $>$, use a _____ line.

6. When graphing a linear inequality containing either $\leq$ or $\geq$, use a _____ line.

7. When graphing the linear inequality $Ax + By < C$, a first step is to graph the line _____.

8. A solution to a system of two inequalities must make (both inequalities/one inequality) true.

9. If two shaded regions represent the solution sets for two inequalities in a system, then the solution set for the system is where these two shaded regions _____.

10. If a test point is found that satisfies both inequalities in a system, do other test points still need to be checked?

SOLUTIONS TO LINEAR INEQUALITIES

Exercises 11–22: Determine whether the test point is a solution to the linear inequality.

11. $(3, 1), x > 2$

12. $(-3, 4), x \le -3$

13. $(0, 0), y \ge 2$

14. $(0, 0), y < -3$

15. $(5, 4), y \ge x$

16. $(-1, 2), y < x$

17. $(3, 0), y < x - 1$

18. $(0, 5), y > 2x + 4$

19. $(-2, 6), x + y \le 4$

20. $(2, -4), x - y \ge 7$

21. $(-1, -1), 2x + y \ge -1$

22. $(0, 1), -x - 5y \ge -1$

Exercises 23–30: Write a linear inequality that describes the shaded region.

23.

24.

25.

26.

27.

28.

29.

30.

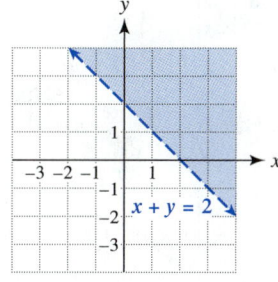

Exercises 31–42: Shade the solution set to the inequality.

31. $x \le -1$

32. $x > 3$

33. $y < -2$

34. $y \ge 0$

35. $y > x$

36. $y \le x$

37. $y \ge 3x$

38. $y < -2x$

39. $x + y \le 1$

40. $x + y \ge -2$

41. $2x - y > 2$

42. $-x - y < 1$

Exercises 43–48: Determine if the test point is a solution to the system of linear inequalities.

43. $(3, 1)$
$$x - y < 3$$
$$x + y > 3$$

44. $(0, 0)$
$$x - 2y < 1$$
$$2x - y > -1$$

45. $(-2, 3)$
$$3x - 2y \ge 1$$
$$-x + 3y > 3$$

46. $(1, 2)$
$$2x - 2y < 5$$
$$x - y > -1$$

47. $(4, -2)$
$$x - 2y \ge 8$$
$$-2x - 5y > 0$$

48. $(-1, -2)$
$$x + y < 0$$
$$-2x - 3y \le -1$$

Exercises 49–52: The graphs of two equations are shown with four test points labeled. Use these points to decide which region should be shaded to solve the given system of inequalities.

49. $x \le 2$
     $x + y \ge 2$

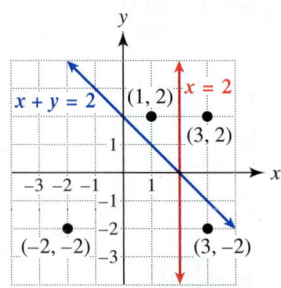

50. $y \ge 1$
     $2x - y \ge 3$

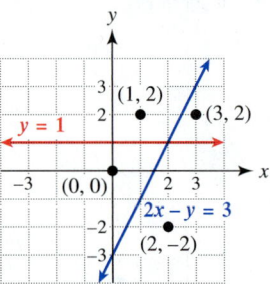

51. $x + y \le 3$
     $y \le 2x$

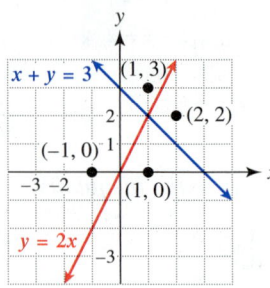

52. $y \le \ x$
     $y \ge -x$

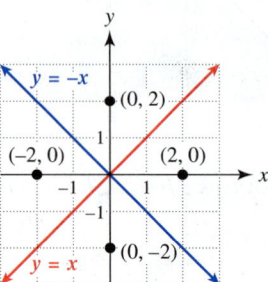

Exercises 53–70: Shade the solution set to the system of inequalities.

53. $x > 2$
     $y < 3$

54. $x \le -1$
     $y \ge \ 3$

55. $x \le -2$
     $y < 2x$

56. $y > 2$
     $y \ge -x$

57. $y \le x$
     $y > -x$

58. $y \le \frac{1}{2}x$
     $y \ge -2x$

59. $x + y \le 3$
     $-x + y \le 1$

60. $x + y > 2$
     $x - y < 2$

61. $2x + y > -3$
      $x + y \le -1$

62. $-x + y \ge \ 3$
     $2x - y \ge -2$

63. $2x + y \ge -3$
      $x + y > -1$

64. $-x + y \ge \ 2$
     $3x - y \ge -2$

65. $y > -2$
     $x + 2y \le -4$

66. $x \ge 2$
     $3y < x - 3$

67. $x + 2y > -4$
     $2x + \ y \le \ 3$

68. $x + 3y \ge 3$
     $3x - 2y \ge 6$

69. $3x + 4y \le 12$
     $5x + 3y \ge 15$

70. $-2x + \ y \ge \ 6$
     $x - 2y \ge -4$

APPLICATIONS

71. *Radios and CDs*   (Refer to Example 5.) A business manufactures at least two radios for each CD player. The total number of radios and CD players must be less than 90. Shade the region that represents the number of radios R and CD players P that can be produced within these restrictions. Put P on the horizontal axis.

72. *Working on Two Projects*   An employee is required to spend more time on project X than on project Y. The employee can work at most 40 hours on these two projects. Shade the region in the *xy*-plane that represents the number of hours that the employee can spend on each project.

73. *Maximum Heart Rate*   When exercising, people often try to maintain target heart rates that are a percentage of their maximum heart rate. Maximum heart rate R is $R = 220 - A$, where A is the person's age and R is the heart rate in beats per minute.
   (a) Find R for a person 20 years old; 70 years old.
   (b) Sketch a graph of $R \le 220 - A$. Assume that A is between 20 and 70 and put A on the horizontal axis of your graph.
   (c) Interpret your graph.

74. *Target Heart Rate*   (Refer to the preceding exercise.) A target heart rate T that is half a person's maximum heart rate is given by $T = 110 - \frac{1}{2}A$, where A is a person's age.
   (a) What is T for a person 30 years old? 50 years old?
   (b) Sketch a graph of the system of inequalities.

$$T \ge 110 - \frac{1}{2}A$$
$$T \le 220 - A$$

   Assume that A is between 20 and 60.
   (c) Interpret your graph.

75. *Height and Weight*   Use Figure 4.19 in Example 6 to determine the range of recommended weights for a person who is 74 inches tall.

76. *Height and Weight*   Use Figure 4.19 in Example 6 to determine the range of recommended heights for a person who weighs 150 pounds.

WRITING ABOUT MATHEMATICS

77. What is the solution set to the following system of inequalities? Explain your reasoning.

$$y > x$$
$$y < x - 1$$

78. Write down a system of linear inequalities whose solution set is the entire xy-plane. Explain your reasoning.

CHECKING BASIC CONCEPTS

SECTIONS 4.3 AND 4.4

1. Use elimination to solve the system of equations.

$$2x + 3y = 5$$
$$x - 7y = -6$$

2. Use elimination to solve each system of equations. How many solutions are there in each case?

(a) $\quad x + y = -1$
$\quad\quad x - 2y = 2$

(b) $\quad 5x - 6y = 4$
$\quad\quad -5x + 6y = 1$

(c) $\quad x - 3y = 0$
$\quad\quad 2x - 6y = 0$

3. Solve the system of equations symbolically, graphically, and numerically.

$$-2x + y = 0$$
$$y = 2x$$

4. Shade the solution set to each inequality.

(a) $y < -1$　　**(b)** $x + y < 1$

5. Shade the solution set to the given system of inequalities.

$$x \le -1$$
$$-2x + y > -3$$

6. *Large Cities in the United States* The combined population of New York and Chicago was 11 million people in 2005. The population of New York exceeded the population of Chicago by 5 million people.

(a) Let x be the population of New York and y be the population of Chicago. Write a system of equations whose solution gives the population of each city in 2005.

(b) Solve the system of equations.

CHAPTER 4 　 SUMMARY

SECTION 4.1 ■ SOLVING SYSTEMS OF LINEAR EQUATIONS GRAPHICALLY AND NUMERICALLY

System of Linear Equations

Solution	An ordered pair (x, y) that satisfies *both* equations
Solution Set	The set of all solutions
Graphical Solution	Graph each equation. A point of intersection is a solution. (Sometimes determining the exact answer when estimating from a graph may be difficult.)
Numerical Solution	Solve each equation for y and make a table for each equation. A solution occurs when two y-values are equal for a given x-value.

Example: The ordered pair $(3, 1)$ is the solution to the following system.

$$x + y = 4 \qquad 3 + 1 = 4 \text{ is a true statement.}$$
$$x - y = 2 \qquad 3 - 1 = 2 \text{ is a true statement.}$$

A Graphical Solution

The point of intersection, (3, 1), is the solution to the system of equations.

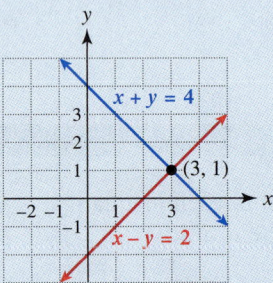

A Numerical Solution

The ordered pair (**3**, **1**) is the solution. When $x = 3$, both y-values equal 1.

x	1	2	3	4
$y = 4 - x$	3	2	1	0
$y = x - 2$	-1	0	1	2

SECTION 4.2 ■ SOLVING SYSTEMS OF LINEAR EQUATIONS BY SUBSTITUTION

Method of Substitution   This method can be used to solve a system of equations symbolically and always gives the exact solution, provided one exists.

Example: $-2x + y = -3$

$\qquad\qquad x + y = 3$

STEP 1: Solve one of the equations for a convenient variable.

$$x + y = 3 \quad \text{becomes} \quad y = 3 - x.$$

STEP 2: Substitute this result in the other equation and then solve.

$$-2x + (3 - x) = -3 \qquad \text{Substitute } (3 - x) \text{ for } y.$$
$$-3x = -6 \qquad \text{Combine like terms; subtract 3.}$$
$$x = 2 \qquad \text{Divide each side by } -3.$$

STEP 3: Find the value of the other variable. Because $y = 3 - x$ and $x = 2$, it follows that $y = 3 - 2 = 1$.

STEP 4: Check to determine that (2, 1) is the solution.

$$-2(2) + (1) \overset{?}{=} -3 \qquad \text{A true statement}$$
$$2 + 1 \overset{?}{=} 3 \qquad \text{A true statement}$$

The solution (2, 1) checks.

Types of Systems of Linear Equations   Can have no solutions, one solution, or infinitely many solutions

No solutions   Inconsistent (parallel lines)
One solution   Consistent (independent equations)
Infinitely many solutions Consistent (dependent equations)

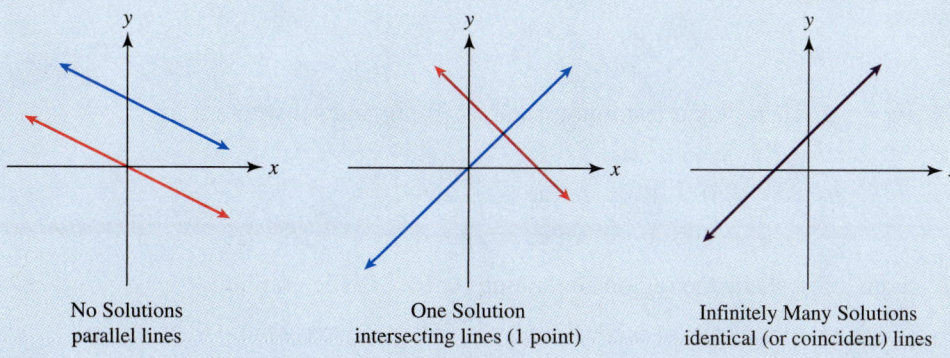

No Solutions
parallel lines

One Solution
intersecting lines (1 point)

Infinitely Many Solutions
identical (or coincident) lines

SECTION 4.3 ■ SOLVING SYSTEMS OF LINEAR EQUATIONS BY ELIMINATION

Method of Elimination   This method can be used to solve a system of linear equations symbolically and always gives the exact solution, provided one exists.

Example: $x + 3y = 1$

$\underline{-x + \ y = 3}$

$4y = 4, \quad \text{or} \quad y = 1$ Add and solve for y.

Substitute $y = 1$ in either of the given equations: $x + 3(1) = 1$ implies that $x = -2$, so $(-2, 1)$ is the solution.

NOTE: To eliminate a variable, it may be necessary to multiply one or both equations by a constant before adding.

Recognizing Types of Systems

No solutions   Final equation is always false, such as $0 = 1$.
One solution   Final equation has one solution, such as $x = 1$.
Infinitely many solutions   Final equation is always true, such as $0 = 0$.

SECTION 4.4 ■ SYSTEMS OF LINEAR INEQUALITIES

Graphing a Linear Inequality in Two Variables

1. Replace the inequality symbol with an equals sign and graph the resulting line. If the inequality is $<$ or $>$ use a dashed line, and if it is $\leq$ or $\geq$ use a solid line.
2. Pick a *test point* that does *not* lie on the line. Substitute this point in the given inequality. Determine whether the resulting statement is true or false.
3. If the statement is true, shade the region containing the test point. If the statement is false, shade the region on the other side of the line.

Solving a System of Linear Inequalities

1. Perform Step 1 above for each inequality in the system.
2. Pick a test point from one region and substitute it in the given inequalities.
3. If the resulting statements are true, shade the region containing the test point. If not, pick a test point from a different region and substitute it in the given inequalities. Repeat this step until the region to be shaded is found.

Example: $x \leq 1$

$x + y \leq 2$

Graph the lines $x = 1$ and $x + y = 2$. Then pick a test point, such as $(0, 0)$, and substitute it in each inequality.

$$0 \leq 1 \qquad \text{A true statement}$$
$$0 + 0 \leq 2 \qquad \text{A true statement}$$

Because $(0, 0)$ satisfies *both* inequalities, shade the region containing $(0, 0)$. See the graph.

NOTE: When shading the solution set to a *system* of inequalities, you may need to try more than one test point.

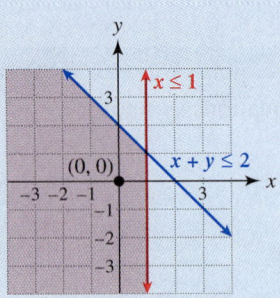

NOTE: An alternative way to determine the solution set is to shade the region to the *left* of the line $x = 1$ (because $x \leq 1$) and *below* the line $y = -x + 2$ (because $y \leq -x + 2$.)

CHAPTER 4   REVIEW EXERCISES

SECTION 4.1

Exercises 1 and 2: Determine graphically the x-value for the equation when $y = 3$.

1. $y = 2x - 3$ **2.** $y = \frac{3}{2}x$

Exercises 3–6: Determine which ordered pair is a solution to the system of equations.

3. $(0, 1), (1, 2)$
$x + 2y = 5$
$x - y = -1$

4. $(5, 2), (4, 0)$
$2x - y = 8$
$x + 3y = 11$

5. $(2, 2), (4, 3)$
$\frac{1}{2}x = y - 1$
$2x = 3y - 1$

6. $(2, -4), (-1, 2)$
$5x - 2y = 18$
$y = -2x$

Exercises 7 and 8: The graphs for two equations are shown. Use the intersection-of-graphs method to identify the solution to both equations. Then check your result.

7.

8.

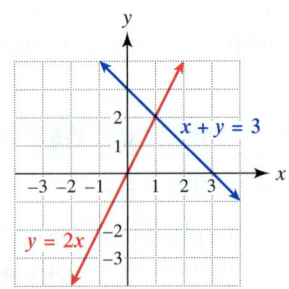

Exercises 9 and 10: A table for two equations is given. Identify the solution to both equations.

9.

x	1	2	3	4
$y = 3x$	3	6	9	12
$y = 6$	6	6	6	6

10.

x	-1	0	1	2
$y = 2x - 1$	-3	-1	1	3
$y = 2 - x$	3	2	1	0

Exercises 11–16: Solve the system of linear equations graphically.

11. $\begin{aligned} y &= -3 \\ x + y &= 1 \end{aligned}$ 12. $\begin{aligned} x &= 1 \\ x - y &= -1 \end{aligned}$

13. $\begin{aligned} 2x + y &= 3 \\ -x + y &= 0 \end{aligned}$ 14. $\begin{aligned} y &= 2x \\ 2x + y &= 4 \end{aligned}$

15. $\begin{aligned} x + 2y &= 3 \\ 2x + y &= 3 \end{aligned}$ 16. $\begin{aligned} -3x - y &= 7 \\ 2x + 3y &= -7 \end{aligned}$

SECTION 4.2

Exercises 17–22: Use the method of substitution to solve the system of linear equations.

17. $\begin{aligned} x + y &= 8 \\ y &= 3x \end{aligned}$ 18. $\begin{aligned} x - 2y &= 22 \\ y &= -5x \end{aligned}$

19. $\begin{aligned} 2x + y &= 5 \\ -3x + y &= 0 \end{aligned}$ 20. $\begin{aligned} 3x - y &= 5 \\ x - y &= -5 \end{aligned}$

21. $\begin{aligned} x + 3y &= 1 \\ -2x + 2y &= 6 \end{aligned}$ 22. $\begin{aligned} 3x - 2y &= -4 \\ 2x - y &= -4 \end{aligned}$

Exercises 23–26: The graphs of two equations are shown.
  (a) State the number of solutions to the system of equations.
  (b) Is the system consistent or inconsistent? If the system is consistent, state whether the equations are dependent or independent.

23.

24.

25.

26.
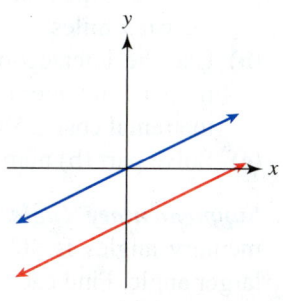

Exercises 27–30: Use the method of substitution to solve the system of linear equations. Then solve the system graphically. Note that these systems may have no solutions, one solution, or infinitely many solutions.

27. $\begin{aligned} x + y &= 2 \\ y &= -x \end{aligned}$ 28. $\begin{aligned} x + y &= -2 \\ x + y &= 3 \end{aligned}$

29. $\begin{aligned} -x + 2y &= 2 \\ x - 2y &= -2 \end{aligned}$ 30. $\begin{aligned} -x - y &= -2 \\ 2x - y &= 1 \end{aligned}$

SECTION 4.3

Exercises 31 and 32: Use the graph to solve the system of equations. Then use the elimination method to verify your answer.

31. $\begin{aligned} x + y &= 3 \\ x - y &= 1 \end{aligned}$ 32. $\begin{aligned} 2x + 3y &= 4 \\ x - 2y &= -5 \end{aligned}$

   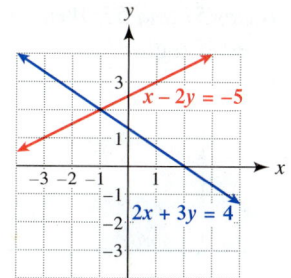

Exercises 33–40: Use the elimination method to solve the system of equations.

33. $\begin{aligned} x + y &= 10 \\ x - y &= 12 \end{aligned}$ 34. $\begin{aligned} 2x - y &= 2 \\ 3x + y &= 3 \end{aligned}$

35. $\begin{aligned} -2x + 2y &= -1 \\ x - 3y &= -3 \end{aligned}$ 36. $\begin{aligned} 2x - 5y &= 0 \\ 2x + 4y &= 9 \end{aligned}$

37. $\begin{aligned} 2a + b &= 3 \\ -3a - 2b &= -1 \end{aligned}$ 38. $\begin{aligned} a - 3b &= 2 \\ 3a + b &= 26 \end{aligned}$

39. $\begin{aligned} 5r + 3t &= -1 \\ -2r - 5t &= -11 \end{aligned}$ 40. $\begin{aligned} 5r + 2t &= 5 \\ 3r - 7t &= 3 \end{aligned}$

Exercises 41 and 42: Solve the system of equations (a) symbolically, (b) graphically, and (c) numerically.

41. $\begin{aligned} 3x + y &= 6 \\ x - y &= -2 \end{aligned}$ 42. $\begin{aligned} 2x + y &= 3 \\ -x + 2y &= -4 \end{aligned}$

Exercises 43–46: Use elimination to determine whether the system of equations has no solutions, one solution, or infinitely many solutions.

43. $x - y = 5$
$-x + y = -5$

44. $3x - 3y = 0$
$-x + y = 0$

45. $-2x + y = 3$
$2x - y = 3$

46. $-2x + y = 2$
$3x - y = 3$

SECTION 4.4

Exercises 47–50: Determine whether the test point is a solution to the linear inequality.

47. $(5, -3)$ $y \leq 2$

48. $(-1, 3)$ $x > -1$

49. $(1, 2)$ $x + y < -2$

50. $(1, -4)$ $2x - 3y \geq 2$

Exercises 51 and 52: Write a linear inequality that describes the shaded region.

51.

52.

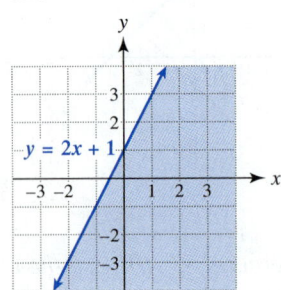

Exercises 53–58: Shade the solution set for the inequality.

53. $x \leq 1$

54. $y > 2$

55. $y > 3x$

56. $x \geq 2y$

57. $y < x + 1$

58. $2x + y \geq -2$

Exercises 59 and 60: Determine whether the test point is a solution to the system of linear inequalities.

59. $(1, -2)$
$x - 2y > 3$
$2x + y < 3$

60. $(4, -3)$
$x - y \geq 1$
$4x + 3y \leq 4$

Exercises 61 and 62: The graphs of two equations are shown at the top of the next column with four test points labeled. Use these points to decide which region should be shaded to solve the system of inequalities.

61. $y \leq 1$
$2x + y \geq -1$

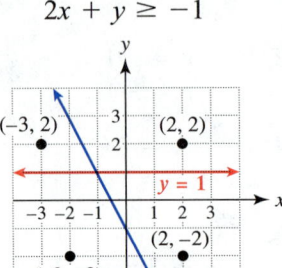

62. $y \geq x$
$x + y \geq 2$

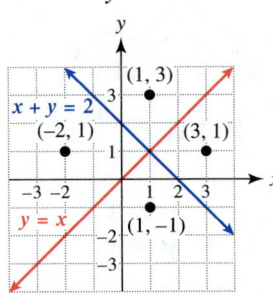

Exercises 63–68: Shade the solution set for the system of inequalities.

63. $x > -1$
$y < -2$

64. $y \leq x$
$y \geq -2x$

65. $x + y \leq 3$
$y \geq -x$

66. $2x + y < 3$
$y > x$

67. $\frac{1}{2}x + y \geq 2$
$x - 2y \leq 0$

68. $2x - y < 3$
$4x + 2y > -6$

APPLICATIONS

69. *Motor Vehicle Fatalities* The number of motor vehicle deaths increased by 13.75 times from 1912 to 2003. There were 39,525 more deaths in 2003 than in 1912. Find the number of motor vehicle deaths in each of the two years. Note that the number of motor vehicles on the road increased from 1 million to 230 million between 1912 and 2003. (**Source:** Department of Health and Human Services.)

70. *Lung Cancer* In 2005, 185,000 new cases of lung cancer were reported. There were 20,000 more new cases for men than for women. How many new cases of lung cancer were there for men and for women? (**Source:** American Cancer Society.)

71. *Renting a Car* A rental car costs $40 plus $0.20 per mile that it is driven.
(**a**) Write an equation that gives the cost C of driving the car x miles.
(**b**) Use the intersection-of-graphs method to determine the number of miles that the car is driven if the rental cost is $90.
(**c**) Solve part (b) numerically with a table of values.

72. *Supplementary Angles* The smaller of two supplementary angles is 30° less than the measure of the larger angle. Find each angle.

73. *Triangle* In an isosceles triangle, the measures of the two smaller angles are equal and their sum is 40° more than the larger angle.

 (a) Let x be the measure of each of the two smaller angles and y be the measure of the larger angle. Write a system of linear equations whose solution gives the measures of these angles.

 (b) Use the method of substitution to solve the system.

 (c) Use the method of elimination to solve the system.

74. *Dimensions of a Garden* A rectangular garden has 88 feet of fencing around it. The garden is 4 feet longer than it is wide. Find the dimensions of the garden.

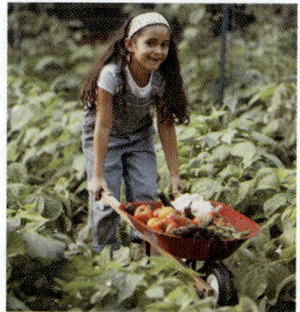

75. *Room Prices* Ten rooms are rented at rates of $80 and $120 per night. The total collected for the 10 rooms is $920.

 (a) Write a system of linear equations whose solution gives the number of each type of room rented. Be sure to state what each variable represents.

 (b) Solve the system of equations.

76. *Mixture Problem* One type of candy sells for $2 per pound, and another type sells for $3 per pound. An order for 18 pounds of candy costs $47. How much of each type of candy was bought?

77. *Burning Calories* An athlete burns 9 calories per minute on a stationary bicycle and 11 calories per minute on a stair climber. In a 60-minute workout the athlete burns 590 calories. How many minutes does the athlete spend on each type of exercise equipment?
(**Source:** Runner's World.)

78. *River Current* A riverboat travels 140 miles downstream in 10 hours, and the return trip takes 14 hours. What is the speed of the current?

79. *Garage Dimensions* The blue graph shown in the figure gives possible dimensions for a rectangular garage with perimeter 80 feet. The red graph shows possible dimensions for a garage that has width W two-thirds of its length L.

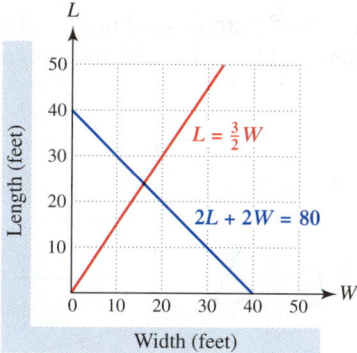

 (a) Use the graph to estimate the dimensions of a garage with perimeter 80 feet and width two-thirds its length.

 (b) Solve this problem symbolically.

80. *Wheels and Trailers* A business manufactures at least two wheels for each trailer it makes. The total number of trailers and wheels manufactured cannot exceed 30 per week. Shade the region that represents numbers of wheels W and trailers T that can be produced each week within these restrictions. Label the horizontal axis W and the vertical axis T.

81. *Target Heart Rate* A target heart rate T that is 70% of a person's maximum heart rate is approximated by $T = 150 - 0.7A$, where A is a person's age.

 (a) What is T for a person 20 years old? 60 years old?

 (b) Sketch a graph of $T \geq 150 - 0.7A$. Assume that A is between 20 and 60.

 (c) Interpret this graph.

CHAPTER 4   TEST 🎓Pass the Test Video solutions to all test exercises

1. Determine which ordered pair is a solution to the system of equations.

$$(3, -1), (1, 2)$$
$$3x + 2y = 7$$
$$2x - y = 0$$

2. The graphs for two equations are shown. Use the intersection-of-graphs method to identify the solution. Then check your solution.

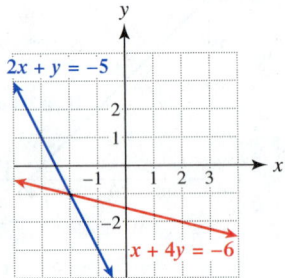

3. A table for two equations is given. Identify the solution to both equations.

x	-2	-1	0	1
$y = 2x$	-4	-2	0	2
$y = 3x + 1$	-5	-2	1	4

4. Solve the system of equations graphically.

$$x + 2y = 4$$
$$x + y = 1$$

5. Use the method of substitution to solve the system of linear equations.

$$3x + 2y = 9$$
$$y = 3x$$

6. Use the method of substitution to solve the system of linear equations. How many solutions are there? Is the system consistent or inconsistent?

(a) $x + 3y = 5$
     $3x - 2y = 4$

(b) $-x + \frac{1}{2}y = 12$
     $2x - y = -4$

Exercises 7 and 8: The graphs of two equations are shown.
(a) State the number of solutions to the system of equations.
(b) Is the system consistent or inconsistent? If the system is consistent, state whether the equations are dependent or independent.

7.

8.
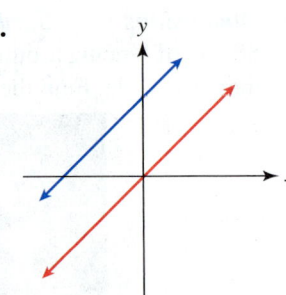

Exercises 9–12: Use the elimination method to solve the system of equations. Note that these systems may have no solutions, one solution, or infinitely many solutions.

9. $x + 2y = 5$
    $3x - 2y = -17$

10. $2x - 2y = 3$
    $-x + y = 5$

11. $x - 2y = 3$
    $-3x + 6y = -9$

12. $4x + 3y = 5$
    $3x - 2y = -9$

13. Write a linear inequality that describes the shaded region.

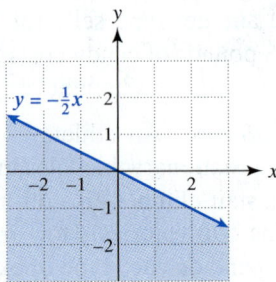

14. Determine whether the test point $(4, -3)$ is a solution to the system of linear inequalities.

$$2x + y > 3$$
$$x - y \geq 7$$

Exercises 15 and 16: Shade the solution set for the given inequality.

15. $x \leq 4$

16. $x + y > 2$

Exercises 17 and 18: Shade the solution set for the given system of inequalities.

17. $x > 2$
$y < 2x$

18. $2x + y \leq 3$
$x - y \geq 0$

19. *IRS Collections*   In 2003 and 2004, the IRS collected a total of $3.9 trillion in taxes. The IRS collected $0.1 trillion more in 2004 than in 2003. How much did the IRS collect in each year?   (*Source:* Internal Revenue Service.)

20. *Jogging Speed*   An athlete jogs at 6 miles per hour and at 9 miles per hour for a total time of 1 hour, covering a distance of 7 miles. How long does the athlete jog at each speed?

CHAPTER 4   EXTENDED AND DISCOVERY EXERCISES

*Exercises 1–4: **Plant Growth**   Before doing these exercises, read the introduction to this chapter. Then refer to the figure shown here.*

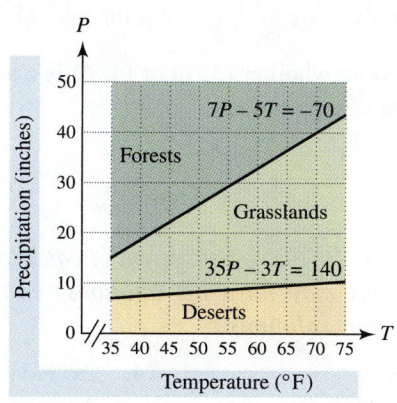

1. The equation of the line that separates grasslands and forests is

$$7P - 5T = -70.$$

Write an inequality that describes temperatures and amounts of precipitation that correspond to forested regions. (Include the line.)

2. The equation of the line that separates grasslands and deserts is

$$35P - 3T = 140.$$

Write an inequality that describes temperatures and amounts of precipitation that correspond to desert regions. (Include the line.)

3. Using the information from Exercises 1 and 2, write a system of inequalities that describes temperatures and amounts of precipitation that correspond to grassland regions.

4. Cheyenne, Wyoming, has an average annual temperature of about 50°F and an average annual precipitation of about 14 inches. Use the graph to predict the type of plant growth you might expect near Cheyenne. Then check to determine whether $T = 50$ and $P = 14$ satisfy the proper inequalities.

*Exercises 5–8: **Numerical Solutions**   If a solution to a linear equation does not appear in a table of values, can you still find it? Use only the table to solve the equation. Then explain how you got your answer.*

5. $2x + 1 = 0$

x	-2	-1	0	1	2
$y = 2x + 1$	-3	-1	1	3	5

6. $4x + 3 = 5$

x	-2	-1	0	1	2
$y = 4x + 3$	-5	-1	3	7	11

7. $\frac{1}{2}x + 3 = 3.75$

x	-2	-1	0	1	2
$y = \frac{1}{2}x + 3$	2	2.5	3	3.5	4

8. $3x - 1 = 0$

x	-2	-1	0	1	2
$y = 3x - 1$	-7	-4	-1	2	5

CHAPTERS 1–4   CUMULATIVE REVIEW EXERCISES

1. Write 120 as a product of prime numbers.

2. Evaluate the expression by hand.
   (a) $2^3 \div \frac{5+7}{9-3}$ (b) $-\frac{2}{5} \cdot (5 - 25)$

3. Classify the number as rational or irrational.
   (a) -6.9 (b) $\sqrt{14}$

4. Insert $>$ or $<$ to make each statement true.
   (a) -5 _____ $|-5|$ (b) $|7|$ _____ $|-1|$

5. State the property of real numbers illustrated by the equation $3 \cdot (2 \cdot 7) = (3 \cdot 2) \cdot 7$.

6. Use properties of real numbers to evaluate $30 \cdot 102$ mentally.

7. Simplify the expression.
   (a) $5x^2 - x^2$ (b) $3 - 2x + 7x - 5$

Exercises 8 and 9: Solve the equation.

8. $5(2x + 1) = 7 + x$ 9. $1 - (x + 1) = x - 1$

10. Determine whether the equation $2(5x + 1) = 10x - 3$ has no solutions, one solution, or infinitely many solutions.

11. Find four consecutive integers whose sum is 50.

12. Find the area of a rectangle having a 36-inch length and a 1-foot width.

13. Solve the formula $W = 3x - 7y$ for x.

14. Solve the inequality $3 - (2x - 7) \le 8x$.

15. Use the table of xy-values to make a line graph.

x	-2	-1	0	1	2
y	-2	2	0	3	-1

16. Graph the equation.
    (a) $y = -2x + 2$ (b) $3y + 2x = 6$

17. Find the x- and y-intercepts for the graph of $4x - y = 8$.

18. Sketch a line with the given slope that passes through the given point.
    (a) $m = -2$, $(1, 1)$ (b) $m = \frac{4}{3}$, $(-3, 2)$

19. Write the slope-intercept form for the line.
    (a)   (b)

    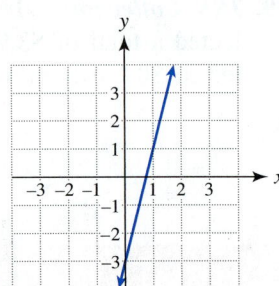

20. Find the slope–intercept form of the line passing through $(4, 2)$ and $(-4, 4)$.

21. Write the slope–intercept form of the line perpendicular to $2x - 6y = 7$, passing through $(1, 1)$.

22. Determine whether $(3, 1)$ or $(4, 4)$ is a solution to the system of equations.

$$3x - y = \ \ 8$$
$$2x + y = 12$$

Exercises 23–26: Solve the system of equations. Note that these systems may have no solutions, one solution, or infinitely many solutions.

23. $\begin{aligned} x - y &= 4 \\ -2x - y &= 1 \end{aligned}$ 24. $\begin{aligned} 2x + 3y &= 4 \\ -4x - 6y &= 7 \end{aligned}$

25. $\begin{aligned} 3x - \ \ 4y &= \ \ \ 8 \\ -15x + 20y &= -40 \end{aligned}$ 26. $\begin{aligned} 7x + 2y &= -3 \\ -5x - 3y &= -1 \end{aligned}$

Exercises 27 and 28: Shade the solution set for the system of inequalities.

27. $\begin{aligned} y &< x + 1 \\ x + y &\le 3 \end{aligned}$ 28. $\begin{aligned} 3x + \ \ y &\ge 6 \\ x - 3y &\le 3 \end{aligned}$

29. **Rods to Feet**   There are 16.5 feet in 1 rod. Write a formula that converts R rods to F feet.

30. **Temperature Change**   Find the temperature change for a package of frozen carrots that is removed from a freezer at $-11°C$ and placed in water at $83°C$.

31. **Cost of a Digital Camera**   A 7% sales tax on a digital camera amounts to $17.15. Find the cost of the digital camera.

32. *Tuition Increase* If tuition is currently $145 per credit and it is going to be increased by 9%, what will the new tuition per credit be?

33. *Gasoline Consumption* Write an equation in slope–intercept form that models the number of gallons G of gas in a truck's tank after x hours if the tank initially contains 30 gallons of gas and the truck uses 3 gallons every hour.

34. *Bank Loans* An individual has two low-interest loans totaling $2400. One loan charges 5% interest and the other charges 6% interest. If the interest for one year is $132, how much money is borrowed at each rate?

CHAPTER 5

Polynomials and Exponents

5.1 Rules for Exponents

5.2 Addition and
Subtraction
of Polynomials

5.3 Multiplication of
Polynomials

5.4 Special Products

5.5 Integer Exponents
and the Quotient Rule

5.6 Division of
Polynomials

Digital images were first sent between New York and London by cable in the early 1920s. Unfortunately the transmission time was 3 hours and the quality was poor. Digital photography was developed further by NASA in the 1960s because ordinary pictures were subject to interference when transmitted through space. Digital pictures remain crystal clear even if they travel millions of miles. The following digital picture shows the planet Mars.

Digital images comprise tiny units called pixels, which are represented in a computer or camera by numbers. As a result, mathematics plays an important role in digital images. In this chapter we illustrate some of the ways mathematics is used to describe digital pictures. We also use mathematics to model heart rate, computer sales, motion of the planets, and interest on money.

> *If you want to do something, do it!*
>
> —PLAUTUS

Source: NASA. (Photograph reprinted with permission.)

5.1 RULES FOR EXPONENTS

Review of Bases and Exponents ▪ Zero Exponents ▪ The Product Rule ▪ Power Rules

A LOOK INTO MATH ▷

Exponents occur throughout mathematics. Regardless of how many mathematics courses you plan to take, exponents are sure to play an important role in each. Because exponents are so important, this section is essential for your becoming successful in mathematics. It takes practice, so set aside some extra time.

Review of Bases and Exponents

The expression 5^3 is an exponential expression with *base* **5** and *exponent* **3**. Its value is

$$5 \cdot 5 \cdot 5 = 125.$$

In general, b^n is an exponential expression with base b and exponent n. If n is a natural number, it indicates the number of times the base b is to be multiplied with itself.

$$\text{Exponent} \searrow$$
$$b^n = \underbrace{b \cdot b \cdot b \cdots \cdot b}_{n \text{ times}}$$
$$\text{Base} \nearrow$$

When evaluating expressions, evaluate exponents *before* performing addition, subtraction, multiplication, division, or negation. The following may be helpful when evaluating expressions.

EVALUATING EXPRESSIONS

When evaluating expressions, use the following order of operations.

1. Evaluate exponents.
2. Perform negation.
3. Do multiplication and division from left to right.
4. Do addition and subtraction from left to right.

EXAMPLE 1 **Evaluating exponential expressions**

Evaluate each expression.

(a) $1 + \dfrac{2^4}{4}$ **(b)** $3\left(\frac{1}{3}\right)^2$ **(c)** -2^4 **(d)** $(-2)^4$

Solution

(a) Evaluate the exponent first.

$$1 + \frac{2^4}{4} = 1 + \frac{\overbrace{2 \cdot 2 \cdot 2 \cdot 2}^{4 \text{ factors}}}{4} = 1 + \frac{16}{4} = 1 + 4 = 5$$

(b) $3\left(\frac{1}{3}\right)^2 = 3\left(\underbrace{\frac{1}{3} \cdot \frac{1}{3}}_{2 \text{ factors}}\right) = 3 \cdot \frac{1}{9} = \frac{3}{9} = \frac{1}{3}$

(c) Because exponents are evaluated before negation is performed,

4 factors

$$-2^4 = -(2 \cdot 2 \cdot 2 \cdot 2) = -16.$$

4 factors

(d) $(-2)^4 = (-2)(-2)(-2)(-2) = 16.$ Now Try Exercises 11, 13, 15, 25

NOTE: Parts (c) and (d) of Example 1 appear to be very similar. However, the placement of the negation sign inside the parentheses in part (d) means that the base for the exponential expression is -2. In part (c) no parentheses are used, indicating that the base of the exponential expression is 2. In general, operations within parentheses should be evaluated *before* using the order of operations.

TECHNOLOGY NOTE: Evaluating Exponents

Exponents can often be evaluated on calculators by using the ^ key. The four expressions from Example 1 are evaluated with a calculator and the results are shown in the two figures. When evaluating the last two expressions on your calculator, remember to use the negation key rather than the subtraction key.

Zero Exponents

So far we have discussed the meaning of natural number exponents. What if an exponent is equal to 0? What does 2^0 equal? To answer these questions consider Table 5.1, which shows values for decreasing powers of 2. Note that each time the power of 2 decreases by 1, the resulting value is divided by 2. For this pattern to continue, we need to define 2^0 to be 1 because if we divide 2 by 2, the result is 1.

This discussion suggests that $2^0 = 1$, and is generalized as follows.

TABLE 5.1   Powers of 2

Power of 2	Value
2^3	8
2^2	4
2^1	2
2^0	?

ZERO EXPONENT

For any nonzero real number b,

$$b^0 = 1.$$

The expression 0^0 is undefined.

EXAMPLE 2   Evaluating zero exponents

Evaluate each expression. Assume that all variables represent nonzero numbers.

(a) 7^0 (b) $3\left(\frac{4}{9}\right)^0$ (c) $\left(\frac{x^2 y^5}{3z}\right)^0$

Solution

(a) $7^0 = 1$

(b) $3\left(\frac{4}{9}\right)^0 = 3(1) = 3$. (Note that the exponent 0 does not apply to 3.)

(c) All variables are nonzero, so the expression inside the parentheses is also nonzero. Thus $\left(\frac{x^2 y^5}{3z}\right)^0 = 1$.

Now Try Exercises 17, 35, 79

The Product Rule

We can calculate products of exponential expressions *provided their bases are the same*. For example,

$$4^3 \cdot 4^2 = \underbrace{(4 \cdot 4 \cdot 4) \cdot (4 \cdot 4)}_{5 \text{ factors}} = 4^5.$$

The expression $4^3 \cdot 4^2$ has a total of $3 + 2 = 5$ factors of 4, so the result is $4^{3+2} = 4^5$. To multiply exponential expressions with the *same* base, we add exponents and the base does not change. We generalize this discussion as the following rule.

THE PRODUCT RULE

For any real number a and natural numbers m and n,

$$a^m \cdot a^n = a^{m+n}.$$

NOTE: The product $2^4 \cdot 3^5$ cannot be simplified by using the product rule because the exponential expressions have different bases: **2** and **3**.

EXAMPLE 3 Using the product rule

Multiply and simplify.

(a) $2^3 \cdot 2^2$ (b) $x^4 x^5$ (c) $2x^2 \cdot 5x^6$ (d) $x^3(2x + 3x^2)$

Solution

(a) $2^3 \cdot 2^2 = 2^{3+2} = 2^5 = 32$

(b) $x^4 x^5 = x^{4+5} = x^9$

(c) Begin by applying the commutative property of multiplication to write the product in a more convenient order.

$$2x^2 \cdot 5x^6 = 2 \cdot 5 \cdot x^2 \cdot x^6 = 10x^{2+6} = 10x^8$$

(d) To simplify this expression, first apply a distributive property.

$$x^3(2x + 3x^2) = x^3 \cdot 2x + x^3 \cdot 3x^2 = 2x^4 + 3x^5$$

Exponent is 1.

Now Try Exercises 29, 33, 39, 83

If an exponent does not appear on an expression, it is assumed to be 1. For example, x can be written as x^1 and $(x + y)$ can be written as $(x + y)^1$.

EXAMPLE 4 **Applying the product rule**

Multiply and simplify.
(a) $x \cdot x^3$ **(b)** $(a + b)(a + b)^4$

Solution
(a) Begin by writing x as x^1. Then $x^1 \cdot x^3 = x^{1+3} = x^4$.
(b) First write $(a + b)$ as $(a + b)^1$. Then

$$(a + b)^1 \cdot (a + b)^4 = (a + b)^{1+4} = (a + b)^5.$$

Now Try Exercises 27, 75

VISUALIZING EXPONENTS (OPTIONAL) We can visualize exponents by stacking wooden blocks having different lengths. For example, the product $2 \cdot 2 \cdot 2$ can be thought of as three blocks with length 2, as illustrated in Figure 5.1. Because they all have the same length (same base), we can stack them and the result represents 2^3.

Figure 5.1 Visualizing $2 \cdot 2 \cdot 2 = 2^3$

In Figure 5.2 the expression $3^2 \cdot 3^3$ can be simplified by stacking blocks of length 3. Note that the stack is 5 blocks high, so the result represents 3^5.

Figure 5.2 Visualizing $3^2 \cdot 3^3 = 3^5$

However, Figure 5.3 reveals that $2^2 \cdot 3^3$ cannot be stacked properly because the blocks have different lengths (bases).

Blocks do not stack properly.

Figure 5.3 Visualizing $2^2 \cdot 3^3 = ?$

Power Rules

How should $(4^3)^2$ be evaluated? To answer this question consider

$$(4^3)^2 = \underbrace{4^3 \cdot 4^3}_{\text{2 factors}} = 4^{\overset{3+3=3\cdot 2}{3+3}} = 4^6.$$

Similarly,

$$(a^5)^3 = \underbrace{a^5 \cdot a^5 \cdot a^5}_{\text{3 factors}} = a^{\overbrace{5+5+5}^{5+5+5=5\cdot3}} = a^{15}.$$

This discussion suggests that to raise a power to a power, we multiply the exponents.

RAISING A POWER TO A POWER

For any real number a and natural numbers m and n,

$$(a^m)^n = a^{mn}.$$

EXAMPLE 5 Raising a power to a power

Simplify the expression.
(a) $(3^2)^4$ **(b)** $(a^3)^2$

Solution
(a) $(3^2)^4 = 3^{2\cdot4} = 3^8$ **(b)** $(a^3)^2 = a^{3\cdot2} = a^6$

Now Try Exercises **43, 45**

To decide how to simplify the expression $(2x)^3$, consider

$$(2x)^3 = \underbrace{2x \cdot 2x \cdot 2x}_{\text{3 factors}} = \underbrace{(2 \cdot 2 \cdot 2)}_{\text{3 factors}} \cdot \underbrace{(x \cdot x \cdot x)}_{\text{3 factors}} = 2^3x^3.$$

To raise a product to a power, we raise each factor to the power, which we generalize as follows.

RAISING A PRODUCT TO A POWER

For any real numbers a and b and natural number n,

$$(ab)^n = a^n b^n.$$

EXAMPLE 6 Raising a product to a power

Simplify the expression.
(a) $(3z)^2$ **(b)** $(-2x^2)^3$ **(c)** $4(x^2y^3)^5$ **(d)** $(-2^2a^5)^3$

Solution
(a) $(3z)^2 = 3^2z^2 = 9z^2$
(b) $(-2x^2)^3 = (-2)^3(x^2)^3 = -8x^6$
(c) $4(x^2y^3)^5 = 4(x^2)^5(y^3)^5 = 4x^{10}y^{15}$
(d) $(-2^2a^5)^3 = (-4a^5)^3 = (-4)^3(a^5)^3 = -64a^{15}$

Now Try Exercises **49, 51, 55**

The following equation illustrates a third power rule.

$$\left(\frac{2}{3}\right)^4 = \frac{2}{3} \cdot \frac{2}{3} \cdot \frac{2}{3} \cdot \frac{2}{3} = \frac{2 \cdot 2 \cdot 2 \cdot 2}{3 \cdot 3 \cdot 3 \cdot 3} = \frac{2^4}{3^4}$$

That is, to raise a quotient to a power, raise both the numerator and the denominator to the power, which we generalize as follows.

RAISING A QUOTIENT TO A POWER

For any real numbers a and b and natural number n,

$$\left(\frac{a}{b}\right)^n = \frac{a^n}{b^n}. \qquad b \neq 0$$

EXAMPLE 7   Raising a quotient to a power

Simplify the expression.

(a) $\left(\dfrac{2}{3}\right)^3$ (b) $\left(\dfrac{a}{b}\right)^9$ (c) $\left(\dfrac{a+b}{5}\right)^2$

Solution

(a) $\left(\dfrac{2}{3}\right)^3 = \dfrac{2^3}{3^3} = \dfrac{8}{27}$ (b) $\left(\dfrac{a}{b}\right)^9 = \dfrac{a^9}{b^9}$

(c) Because the numerator is an expression with more than one term, we must place parentheses around it before raising it to the power 2.

$$\left(\frac{a+b}{5}\right)^2 = \frac{(a+b)^2}{5^2} = \frac{(a+b)^2}{25}$$

Now Try Exercises 63, 65, 67

MAKING CONNECTIONS

Raising a Sum or Difference to a Power

Although there are power rules for products and quotients, there are not similar rules for sums and differences. In general, $(a+b)^n \neq a^n + b^n$ and $(a-b)^n \neq a^n - b^n$. For example, $(3+4)^2 = 7^2 = 49$ but $3^2 + 4^2 = 9 + 16 = 25$. Similarly, $(4-1)^3 = 3^3 = 27$ but $4^3 - 1^3 = 64 - 1 = 63$.

Simplification of some expressions may require the application of more than one rule of exponents. This is demonstrated in the next example.

EXAMPLE 8   Combining rules for exponents

Simplify the expression.

(a) $(2a)^2(3a)^3$ (b) $\left(\dfrac{a^2 b^3}{c}\right)^4$ (c) $(2x^3 y)^2(-4x^2 y^3)^3$

Solution

(a) $(2a)^2(3a)^3 = 2^2 a^2 \cdot 3^3 a^3$ 　　　　Raising a product to a power

　　　　$= 4 \cdot 27 \cdot a^2 \cdot a^3$ 　　　　Evaluate powers; commutative property

　　　　$= 108 a^5$ 　　　　Product rule

(b) $\left(\dfrac{a^2 b^3}{c}\right)^4 = \dfrac{(a^2)^4 (b^3)^4}{c^4}$ Raising a quotient to a power; raising a product to a power

$= \dfrac{a^8 b^{12}}{c^4}$ Raising a power to a power

(c) $(2x^3 y)^2 (-4x^2 y^3)^3 = 2^2 (x^3)^2 y^2 (-4)^3 (x^2)^3 (y^3)^3$ Raising a product to a power

$= 4x^6 y^2 (-64) x^6 y^9$ Raising a power to a power

$= 4(-64) x^6 x^6 y^2 y^9$ Commutative property

$= -256 x^{12} y^{11}$ Product rule

Now Try Exercises 61, 73

EXAMPLE 9   **Calculating growth of an investment**

If a parcel of property increases in value by about 11% each year for 20 years, then its value will double three times.

(a) Write an exponential expression that represents "doubling three times."

(b) If the property is initially worth \$25,000, how much will it be worth if it doubles 3 times?

Solution

(a) Doubling three times is represented by 2^3.

(b) $2^3 (25,000) = 8(25,000) = \$200,000$ Now Try Exercise 95

5.1 PUTTING IT ALL TOGETHER

The following table summarizes properties of exponents.

Concept	Explanation	Examples
Bases and Exponents	In the expression b^n, b is the base and n is the exponent. If n is a natural number, then $$b^n = \underbrace{b \cdot b \cdots \cdot b}_{n \text{ times}}.$$	2^3 has base 2 and exponent 3. $9^1 = 9$, $3^2 = 3 \cdot 3 = 9$, $4^3 = 4 \cdot 4 \cdot 4 = 64$, and $-6^2 = -(6 \cdot 6) = -36$
Zero Exponents	$b^0 = 1$ for any nonzero number b.	$5^0 = 1$, $x^0 = 1$, and $(xy^3)^0 = 1$
The Product Rule	$a^m \cdot a^n = a^{m+n}$, where m and n are natural numbers.	$2^4 \cdot 2^3 = 2^{4+3} = 2^7$, $x \cdot x^2 \cdot x^6 = x^{1+2+6} = x^9$, and $(x+1) \cdot (x+1)^2 = (x+1)^3$
Raising a Power to a Power	$(a^m)^n = a^{mn}$, where m and n are natural numbers.	$(2^4)^2 = 2^{4 \cdot 2} = 2^8$, $(x^2)^5 = x^{2 \cdot 5} = x^{10}$, and $(a^4)^3 = a^{4 \cdot 3} = a^{12}$

Concept	Explanation	Examples
Raising a Product to a Power	$(ab)^n = a^n b^n$, where n is a natural number.	$(3x)^3 = 3^3 x^3 = 27x^3$, $(x^2 y)^4 = (x^2)^4 y^4 = x^8 y^4$, and $(-xy)^6 = (-x)^6 y^6 = x^6 y^6$
Raising a Quotient to a Power	$\left(\dfrac{a}{b}\right)^n = \dfrac{a^n}{b^n}$, where $b \neq 0$ and n is a natural number.	$\left(\dfrac{x}{y}\right)^5 = \dfrac{x^5}{y^5}$ and $\left(\dfrac{a^2 b}{d^3}\right)^4 = \dfrac{(a^2)^4 b^4}{(d^3)^4} = \dfrac{a^8 b^4}{d^{12}}$

5.1 Exercises

CONCEPTS

1. In the expression b^n, b is the _____ and n is the _____.

2. The expression $b^0 =$ _____ for any nonzero number b.

3. Write $\frac{1}{2} \cdot \frac{1}{2} \cdot \frac{1}{2}$ by using exponents.

4. Write $x \cdot x \cdot x \cdot x$ by using exponents.

5. $-1^2 =$ _____

6. $(-1)^2 =$ _____

7. $a^m \cdot a^n =$ _____

8. $(a^m)^n =$ _____

9. $(ab)^n =$ _____

10. $\left(\dfrac{a}{b}\right)^n =$ _____

PROPERTIES OF EXPONENTS

Exercises 11–26: Evaluate the expression.

11. 8^2

12. 4^3

13. $(-2)^3$

14. $(-3)^4$

15. -2^3

16. -3^4

17. 6^0

18. $(-0.5)^0$

19. $2 \cdot 4^2$

20. $-3 \cdot 2^4$

21. $1 + 5^2$

22. $5^2 - 4^2$

23. $\dfrac{4^2}{2}$

24. $\left(\dfrac{-4}{2}\right)^2$

25. $4 \cdot \dfrac{1}{2^3}$

26. $\dfrac{2}{2^3} - \dfrac{1}{2^3}$

Exercises 27–86: Simplify the expression. Assume that all variables represent nonzero numbers.

27. $3 \cdot 3^2$

28. $x^3 \cdot x$

29. $4^2 \cdot 4^6$

30. $5^3 \cdot 5^3$

31. $2^3 \cdot 2^2$

32. $10^4 \cdot 10^3$

33. $x^3 \cdot x^6$

34. $a^5 \cdot a^2$

35. $z^0 z^4$

36. $a^2 a^3 a^1$

37. $x^2 x^2 x^2$

38. $y^7 y^3 y^0$

39. $4x^2 \cdot 5x^5$

40. $-2y^6 \cdot 5y^2$

41. $3(-xy^3)(x^2 y)$

42. $(a^2 b^3)(-ab^2)$

43. $(2^3)^2$

44. $(10^3)^4$

45. $(n^3)^4$

46. $(z^7)^3$

47. $x(x^3)^2$

48. $(z^3)^2(5z^5)$

49. $(-7b)^2$

50. $(-4z)^3$

51. $(ab)^3$

52. $(xy)^8$

53. $(2x^2)^2$

54. $(3a^2)^4$

55. $(-4b^2)^3$

56. $(-3r^4 t^3)^2$

57. $(x^2 y^3)^7$

58. $(rt^2)^5$

59. $(y^3)^2(x^4 y)^3$

60. $(ab^3)^2(ab)^3$

61. $(a^2b)^2(a^2b^2)^3$

62. $(x^3y)(x^2y^4)^2$

63. $\left(\frac{1}{3}\right)^3$

64. $\left(\frac{5}{2}\right)^2$

65. $\left(\frac{a}{b}\right)^5$

66. $\left(\frac{x}{2}\right)^4$

67. $\left(\frac{x-y}{3}\right)^3$

68. $\left(\frac{4}{x+y}\right)^2$

69. $\left(\frac{5}{a+b}\right)^2$

70. $\left(\frac{a-b}{2}\right)^3$

71. $\left(\frac{2x}{5}\right)^3$

72. $\left(\frac{3y}{2}\right)^4$

73. $\left(\frac{3x^2}{5y^4}\right)^3$

74. $\left(\frac{a^2b^3}{3}\right)^5$

75. $(x+y)(x+y)^3$

76. $(a-b)^2(a-b)$

77. $(a+b)^2(a+b)^3$

78. $(x-y)^5(x-y)^4$

79. $6(x^4y^6)^0$

80. $\left(\frac{xy}{z^2}\right)^0$

81. $a(a^2+2b^2)$

82. $x^3(3x-5y^4)$

83. $3a^3(4a^2+2b)$

84. $2x^2(5-4y^3)$

85. $(r+t)(rt)$

86. $(x-y)(x^2y^3)$

87. Thinking Generally Students sometimes mistakenly apply the "rule" $a^m \cdot b^n \overset{?}{=} (ab)^{m+n}$. In general, this equation is *not true*. Find values for a, b, m, and n with $a \neq b$ and $m \neq n$ that will make this equation true.

88. Thinking Generally Students sometimes mistakenly apply the "rule" $(a+b)^n \overset{?}{=} a^n + b^n$. In general, this equation is *not true*. Find values for a, b, and n with $a \neq b$ that will make this equation true.

APPLICATIONS

89. *Area of a Rectangle* Find the area of the rectangle.

90. *Area of a Square* Find the area of the square whose sides have length $2ab$.

91. *Volume of a Box* A box has the dimensions shown. Write a simplified expression for its volume.

92. *Volume of a Cube* Each side of a cube is $3a^2$. Write a simplified expression for its volume.

93. *Area of a Circle* Write a simplified expression for the area of the circle shown.

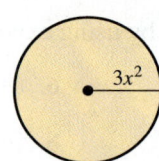

94. *Volume of a Sphere* The volume V of a sphere with radius r is given by $V = \frac{4}{3}\pi r^3$. Find a simplified expression for the volume of a sphere with radius $2b$.

95. *Compound Interest* If P dollars are deposited in an account that pays 5% annual interest, then the amount of money in the account after 3 years is $P(1 + 0.05)^3$. Find the amount when $P = \$1000$.

96. *Compound Interest* If P dollars are deposited in an account that pays 9% annual interest, then the amount of money in the account after 4 years is $P(1 + 0.09)^4$. Find the amount when $P = \$500$.

Exercises 97–100: (Refer to the discussion about visualizing exponents found in this section.) Make a sketch of blocks that represents the given exponential equation.

97. $4^2 \cdot 4^1 = 4^3$

98. $2^3 \cdot 2^3 = 2^6$

99. $2^3 \cdot 3^2 \cdot 3^1 \cdot 2^2 = 2^5 \cdot 3^3$

100. $4^2 \cdot 2^2 \cdot 4^2 = 4^5$ (*Hint:* Write 2^2 with a different base.)

WRITING ABOUT MATHEMATICS

101. Are the expressions $(4x)^2$ and $4x^2$ equal? Explain your answer.

102. Are the expressions $3^3 \cdot 2^3$ and 6^6 equal? Explain your answer.

5.2 ADDITION AND SUBTRACTION OF POLYNOMIALS

Monomials and Polynomials ■ Addition of Polynomials ■ Subtraction of Polynomials ■ Evaluating Polynomial Expressions

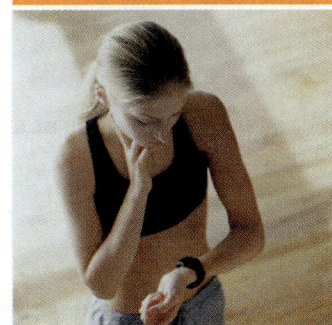

A LOOK INTO MATH ▷ If you have ever exercised strenuously and then taken your pulse immediately afterward, you may have discovered that your pulse slowed quickly at first and then gradually leveled off. A typical scatterplot of this phenomenon is shown in Figure 5.4(a). These data points cannot be modeled accurately with a line, so a new expression, called a *polynomial,* is needed to model them. A graph of this new expression is shown in Figure 5.4(b) and discussed in Exercise 79.   (***Source:*** V. Thomas, *Science and Sport.*)

(a)

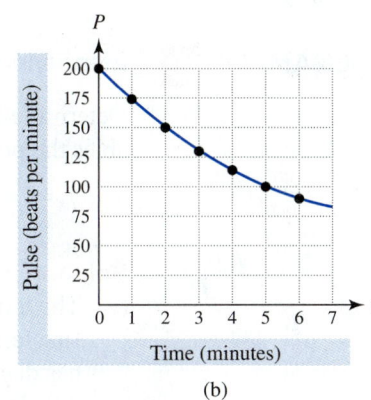

(b)

Figure 5.4 Heart Rate After Exercising

Monomials and Polynomials

A **monomial** is a number, a variable, or a product of numbers and variables raised to natural number powers. Examples of monomials include

$$-3, \quad xy^2, \quad 5a^2, \quad -z^3, \quad \text{and} \quad -\frac{1}{2}xy^3.$$

A monomial may contain more than one variable, but monomials do not contain division by variables. For example, the expression $\frac{3}{z}$ is not a monomial. If an expression contains addition or subtraction signs, it is *not* a monomial.

   The **degree of a monomial** is the sum of the exponents of the variables. If the monomial has only one variable, its degree is the exponent of that variable. Remember, when a variable does not have a written exponent, the exponent is implied to be 1. A nonzero number has degree 0, and the number 0 has *undefined* degree. The number in a monomial is called the **coefficient of the monomial**. Table 5.2 contains the degree and coefficient of several monomials.

TABLE 5.2   Properties of Monomials

Monomial	-5	$6a^3b$	$-xy$	$7y^3$
Degree	0	4	2	3
Coefficient	-5	6	-1	7

A **polynomial** is the sum of one or more monomials. Each monomial is called a *term* of the polynomial. Addition or subtraction signs separate terms. The expression $2x^2 - 3x + 5$ is a **polynomial in one variable** with three terms. Examples of polynomials in one variable include

$$-2x, \quad 3x + 1, \quad 4y^2 - y + 7, \quad \text{and} \quad x^5 - 3x^3 + x - 7.$$

These polynomials have 1, 2, 3, and 4 terms, respectively. A polynomial with *two terms* is called a **binomial**, and a polynomial with *three terms* is called a **trinomial**.

A polynomial can have more than one variable, as in

$$x^2y^2, \quad 2xy^2 + 5x^2y - 1, \quad \text{and} \quad a^2 + 2ab + b^2.$$

Note that all variables in a polynomial are raised to natural number powers. The **degree of a polynomial** is the degree of the term (or monomial) with highest degree.

EXAMPLE 1   **Identifying properties of polynomials**

Determine whether the expression is a polynomial. If it is, state how many terms and variables the polynomial contains and its degree.

(a) $7x^2 - 3x + 1$ **(b)** $5x^3 - 3x^2y^3 + xy^2 - 2y^3$ **(c)** $4x^2 + \dfrac{5}{x + 1}$

Solution
(a) The expression $7x^2 - 3x + 1$ is a polynomial with three terms and one variable. The first term $7x^2$ has degree 2 because the exponent on the variable is 2. The second term $-3x$ has degree 1 because the exponent on the variable is implied to be 1. The third term 1 has degree 0 because it is a nonzero number. The term with highest degree is $7x^2$, so the polynomial has degree **2**.
(b) The expression $5x^3 - 3x^2y^3 + xy^2 - 2y^3$ is a polynomial with four terms and two variables. The first term has degree 3 because the exponent on the variable is 3. The second term has degree 5 because the *sum* of the exponents on the variables is 5. Likewise, the third term has degree 3 and the fourth term has degree 3. The term with highest degree is $-3x^2y^3$, so the polynomial has degree **2** + **3** = 5.
(c) The expression $4x^2 + \dfrac{5}{x+1}$ is not a polynomial because it contains division by the polynomial $x + 1$.

Now Try Exercises **21, 23, 25**

Addition of Polynomials

Suppose that we have 2 identical rectangles with length L and width W, as illustrated in Figure 5.5. Then the area of one rectangle is LW and the total area is

$$LW + LW.$$

This area is equivalent to 2 times LW, which can be expressed as $2LW$, or

$$LW + LW = 2LW.$$

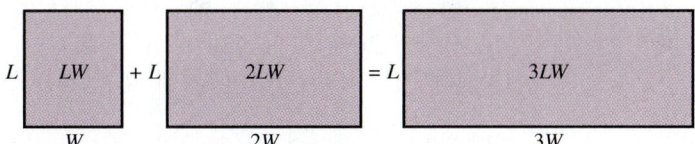

Figure 5.5 Adding $LW + LW$

If two monomials contain the same variables raised to the same powers, we call them **like terms**. We can add or subtract *like* terms but not *unlike* terms. The terms LW and $2LW$ are like terms and can be combined geometrically, as shown in Figure 5.6. If we joined one of the small rectangles with area LW and a larger rectangle with area $2LW$, then the total area is $3LW$.

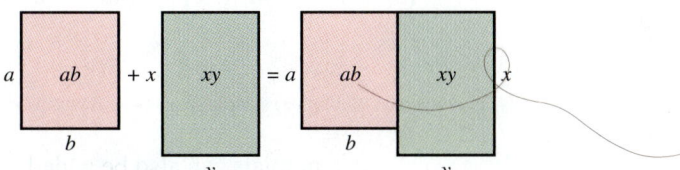

Figure 5.6 Adding $LW + 2LW$

The *distributive property* justifies combining like terms.

$$1LW + 2LW = (1 + 2)LW = 3LW$$

The rectangles shown in Figure 5.7 have areas of ab and xy. Their combined area is their sum, $ab + xy$. However, because these monomials are unlike terms, they cannot be combined into one term.

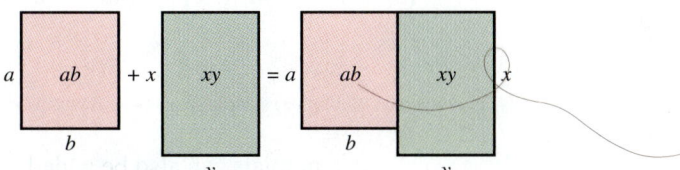

Figure 5.7 Unlike terms: $ab + xy$

EXAMPLE 2 **Adding like terms**

State whether each pair of expressions contains like terms or unlike terms. If they are like terms, add them.
(a) $5x^2, -x^2$ **(b)** $7a^2b, 10ab^2$ **(c)** $4rt^2, \frac{1}{2}rt^2$

Solution
(a) The terms $5x^2$ and $-x^2$ have the same variable raised to the same power, so they are like terms. To add like terms add their coefficients. Note that the coefficient of $-x^2$ is -1.

$$5x^2 + (-x^2) = \big(5 + (-1)\big)x^2 \qquad \text{Distributive property}$$
$$= 4x^2 \qquad\qquad\qquad \text{Add.}$$

(b) The terms $7a^2b$ and $10ab^2$ have the same variables, but these variables are not raised to the same power. They are unlike terms and therefore cannot be added.

(c) The terms $4rt^2$ and $\frac{1}{2}rt^2$ have the same variables raised to the same powers, so they are like terms.

$$4rt^2 + \frac{1}{2}rt^2 = \left(4 + \frac{1}{2}\right)rt^2 \qquad \text{Distributive property}$$

$$= \frac{9}{2}rt^2 \qquad \text{Add.} \qquad \boxed{\text{Now Try Exercises } 33, 35, 39}$$

To add two polynomials, add like terms, as illustrated in the next example.

EXAMPLE 3   **Adding polynomials**

Add each pair of polynomials by combining like terms.
(a) $(3x + 4) + (-4x + 2)$
(b) $(y^2 - 2y + 1) + (3y^2 + y + 11)$
(c) $(3a^3 - 4a + 1) + (4a^3 + a^2 - 5)$

Solution
(a) $(3x + 4) + (-4x + 2) = 3x + (-4x) + 4 + 2$
$$= (3 - 4)x + (4 + 2)$$
$$= -x + 6$$
(b) $(y^2 - 2y + 1) + (3y^2 + y + 11) = y^2 + 3y^2 - 2y + y + 1 + 11$
$$= (1 + 3)y^2 + (-2 + 1)y + (1 + 11)$$
$$= 4y^2 - y + 12$$

NOTE: With practice the first two steps can be done mentally.

(c) $(3a^3 - 4a + 1) + (4a^3 + a^2 - 5) = 3a^3 + 4a^3 + a^2 - 4a + 1 - 5$
$$= (3 + 4)a^3 + a^2 - 4a - 4$$
$$= 7a^3 + a^2 - 4a - 4$$

$\boxed{\text{Now Try Exercises } 41, 43, 47}$

Polynomials can also be added vertically, as demonstrated in the next example.

EXAMPLE 4   **Adding polynomials vertically**

Simplify $(3x^2 - 3x + 5) + (-x^2 + x - 6)$.

Solution
Write the polynomials in a vertical format and then add each column of like terms.

$$
\begin{array}{r}
3x^2 - 3x + 5 \\
\underline{-x^2 + x - 6} \\
2x^2 - 2x - 1
\end{array}
\qquad \text{Add.}
$$

Regardless of the method used, the same answer should be obtained. However, adding vertically requires that *like terms be placed in the same column.* $\boxed{\text{Now Try Exercise } 53}$

Subtraction of Polynomials

To subtract one integer from another, add the first integer with the *additive inverse* or *opposite* of the second integer. For example, $3 - 5$ is evaluated as follows.

$$3 - 5 = 3 + (-5) \qquad \text{Add the opposite.}$$

$$= -2 \qquad \text{Simplify.}$$

Similarly, to subtract one polynomial from another, add the first polynomial and the *opposite* of the second polynomial. To find the opposite of a polynomial, simply negate each term. Table 5.3 lists some polynomials and their opposites.

CRITICAL THINKING

What is the result when a polynomial and its opposite are added?

TABLE 5.3   Opposites of Polynomials

Polynomial	Opposite
$2x - 4$	$-2x + 4$
$-x^2 - 2x + 9$	$x^2 + 2x - 9$
$6x^3 - 12$	$-6x^3 + 12$
$-3x^4 - 2x^2 - 8x + 3$	$3x^4 + 2x^2 + 8x - 3$

EXAMPLE 5   Subtracting polynomials

Simplify each expression.
(a) $(3x - 4) - (5x + 1)$
(b) $(5x^2 + 2x - 3) - (6x^2 - 7x + 9)$
(c) $(6x^3 + x^2) - (-3x^3 - 9)$

Solution
(a) To subtract $(5x + 1)$ from $(3x - 4)$, we add the opposite of $(5x + 1)$, or $(-5x - 1)$.

$$(3x - 4) - (5x + 1) = (3x - 4) + (-5x - 1)$$
$$= (3 - 5)x + (-4 - 1)$$
$$= -2x - 5$$

(b) The opposite of $(6x^2 - 7x + 9)$ is $(-6x^2 + 7x - 9)$.

$$(5x^2 + 2x - 3) - (6x^2 - 7x + 9) = (5x^2 + 2x - 3) + (-6x^2 + 7x - 9)$$
$$= (5 - 6)x^2 + (2 + 7)x + (-3 - 9)$$
$$= -x^2 + 9x - 12$$

(c) The opposite of $(-3x^3 - 9)$ is $(3x^3 + 9)$.

$$(6x^3 + x^2) - (-3x^3 - 9) = (6x^3 + x^2) + (3x^3 + 9)$$
$$= (6 + 3)x^3 + x^2 + 9$$
$$= 9x^3 + x^2 + 9$$

Now Try Exercises 63, 65, 69

NOTE: Some students prefer to subtract one polynomial from another by noting that a subtraction sign in front of parentheses changes the signs of all of the terms within the parentheses. For example, part (a) of the previous example could be worked as follows.

$$(3x - 4) - (5x + 1) = 3x - 4 - 5x - 1$$
$$= (3 - 5)x + (-4 - 1)$$
$$= -2x - 5$$

EXAMPLE 6   Subtracting polynomials vertically

Simplify $(5x^2 - 2x + 7) - (-3x^2 + 3)$.

Solution

To subtract one polynomial from another vertically, simply add the first polynomial and the opposite of the second polynomial. No x-term occurs in the second polynomial, so insert $0x$.

$$5x^2 - 2x + 7$$
$$\underline{3x^2 + 0x - 3} \qquad \text{Opposite of } -3x^2 + 3 \text{ is } 3x^2 - 3 \text{ or } 3x^2 + 0x - 3.$$
$$8x^2 - 2x + 4 \qquad \text{Add like terms in each column.} \qquad \text{Now Try Exercise } \boxed{77}$$

Evaluating Polynomial Expressions

Frequently, monomials and polynomials represent formulas that may be evaluated. We illustrate such applications in the next two examples.

EXAMPLE 7   Writing and evaluating a monomial

Write the monomial that represents the volume of the box having a square bottom, as shown in Figure 5.8. Find the volume of the box if $x = 3$ feet and $y = 2$ feet.

CRITICAL THINKING

Write an expression that gives the volume of six identical cubes having sides of length L.

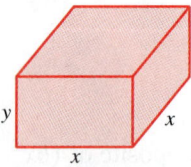

Figure 5.8

Solution

The volume of a box is found by multiplying the length, width, and height together. Because the length and width are both x and the height is y, the monomial xxy represents the volume of the box. This can be written x^2y. To calculate the volume let $x = 3$ and $y = 2$ in the monomial x^2y.

$$x^2y = 3^2 \cdot 2 = 9 \cdot 2 = 18 \text{ cubic feet} \qquad \text{Now Try Exercise } \boxed{81}$$

EXAMPLE 8   **Modeling sales of personal computers**

Worldwide sales of personal computers have increased dramatically in recent years, as illustrated in Figure 5.9. The polynomial

$$0.7868x^2 + 12x + 79.5$$

approximates the number of computers sold in millions, where $x = 0$ corresponds to 1997, $x = 1$ to 1998, and so on. Estimate the number of personal computers sold in 2002 by using both the graph and the polynomial.   (**Source:** International Data Corporation.)

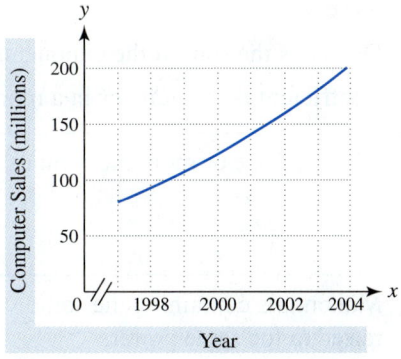

Figure 5.9 Worldwide Computer Sales

Solution

From the graph shown in Figure 5.10, it appears that personal computer sales were slightly more than 150 million, or about 160 million, in 2002. The year 2002 corresponds to $x = 5$ in the given polynomial, so substitute 5 for x and evaluate the resulting expression.

$$0.7868x^2 + 12x + 79.5 = 0.7868(5)^2 + 12(5) + 79.5$$
$$\approx 159 \text{ million}$$

The graph and the polynomial give similiar results.

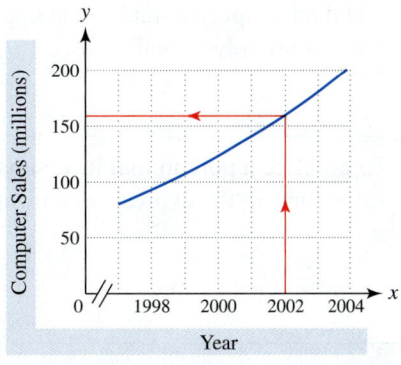

Figure 5.10 Worldwide Computer Sales

Now Try Exercise 80

5.2 PUTTING IT ALL TOGETHER

In this section we discussed monomials and polynomials, including how to add, subtract, and evaluate them. The following table summarizes several important concepts related to these topics.

Concept	Explanation	Examples
Monomial	A number, variable, or product of numbers and variables raised to natural number powers Degree is the sum of the exponents. Coefficient is the number in a monomial.	$4x^2y$ Degree: 3, coefficient: 4 $-6x^2$ Degree: 2, coefficient: -6 $-a^4$ Degree: 4, coefficient: -1 x Degree: 1, coefficient: 1 -8 Degree: 0, coefficient: -8
Polynomial	A sum of one or more monomials	$4x^2 + 8xy^2 + 3y^2$ Trinomial $-9x^4 + 100$ Binomial $-3x^2y^3$ Monomial
Like Terms	Monomials containing the same variables raised to the same powers	$10x$ and $-2x$, $4x^2$ and $3x^2$ $5ab^2$ and $-ab^2$, $5z$ and $\frac{1}{2}z$
Addition of Polynomials	To add polynomials combine like terms by applying the distributive property.	$(x^2 + 3x + 1) + (2x^2 - 2x + 7)$ $= (1 + 2)x^2 + (3 - 2)x + (1 + 7)$ $= 3x^2 + x + 8$ $3xy + 5xy = (3 + 5)xy = 8xy$
Opposite of a Polynomial	To obtain the opposite of a polynomial negate each term.	*Polynomial* *Opposite* $-2x^2 + x - 6$ $2x^2 - x + 6$ $a^2 - b^2$ $-a^2 + b^2$ $-3x - 18$ $3x + 18$
Subtraction of Polynomials	To subtract one polynomial from another, add the first polynomial to the opposite of the second polynomial.	$(x^2 + 3x) - (2x^2 - 5x)$ $= (x^2 + 3x) + (-2x^2 + 5x)$ $= (1 - 2)x^2 + (3 + 5)x$ $= -x^2 + 8x$

Evaluating a Polynomial	To evaluate a polynomial in x, substitute a value for x in the expression and simplify.	To evaluate the polynomial
		    $3x^2 - 2x + 1$ for $x = \mathbf{2}$,
		substitute 2 for x and simplify.
		    $3(\mathbf{2})^2 - 2(\mathbf{2}) + 1 = 9$

5.2 Exercises

CONCEPTS

1. A ____ is a number, a variable, or a product of numbers and variables raised to a natural number power.

2. The coefficient of $-3xy$ is ____.

3. A ____ is a monomial or a sum of monomials.

4. The ____ of a monomial is the sum of the exponents of the variables.

5. A polynomial with two terms is called a ____.

6. A polynomial with three terms is called a ____.

7. $4x^3 + x^2$ has ____ terms and its degree is ____.

8. To add two polynomials, combine ____ terms.

9. To subtract two polynomials, add the first polynomial to the ____ of the second polynomial.

10. Polynomials can be added horizontally or ____.

PROPERTIES OF POLYNOMIALS

Exercises 11–18: Identify the degree and coefficient of the monomial.

11. $3x^2$

12. y

13. $-ab$

14. $2xy$

15. $-5rt$

16. $8x^2y^5$

17. -6

18. $\frac{1}{2}$

Exercises 19–30: Determine whether the expression is a polynomial. If it is, state how many terms and variables the polynomial contains. Then state its degree.

19. $-x$

20. $7z$

21. $4x^2 - 5x + 9$

22. $x^3 - 9$

23. $x + \dfrac{1}{x}$

24. $\dfrac{5}{xy + 1}$

25. $3x^2y - xy^3$

26. $a^3 + 3a^2b + 3ab^2$

27. $3x^{-2}y^{-3}$

28. $5^2a^3b^4$

29. -2^3a^4bc

30. $-7y^{-1}z^{-3}$

Exercises 31–40: State whether the given pair of expressions are like terms. If they are like terms, add them.

31. $5x, -4x$

32. $x^2, 8x^2$

33. $x^3, -6x^3$

34. $4xy, -9xy$

35. $9x, -xy$

36. $5x^2y, -3xy^2$

37. ab, ba

38. $rt^2, -2t^2r$

39. $7xy^2, -3xy^2$

40. a, b

ADDITION OF POLYNOMIALS

Exercises 41–52: Add the polynomials.

41. $(3x + 5) + (-4x + 4)$

42. $(-x + 5) + (2x - 5)$

43. $(3x^2 + 4x + 1) + (x^2 + 4x - 6)$

44. $(-x^2 - x + 7) + (2x^2 + 3x - 1)$

45. $(a^3 - 6) + (4a^3 + 7)$

46. $(2b^4 - 3b^2) + (-3b^4 + b^2)$

47. $(y^3 + 3y^2 - 5) + (3y^3 + 4y - 4)$

48. $(4z^4 + z^2 - 10) + (-z^4 + 4z - 5)$

49. $(-xy + 5) + (5xy - 4)$

50. $(2a^2 + b^2) + (3a^2 - 5b^2)$

51. $(a^3b^2 + a^2b^3) + (a^2b^3 - a^3b^2)$

52. $(a^2 + ab + b^2) + (a^2 - ab + b^2)$

Exercises 53–56: Add the polynomials vertically.

53. $\begin{aligned} 4x^2 - 2x + 1 \\ 5x^2 + 3x - 7 \end{aligned}$

54. $\begin{aligned} 8x^2 + 3x + 5 \\ -x^2 - 3x - 9 \end{aligned}$

55. $\begin{aligned} -x^2 + x \\ 2x^2 - 8x - 1 \end{aligned}$

56. $\begin{aligned} a^3 - 3a^2b + 3ab^2 - b^3 \\ a^3 + 3a^2b + 3ab^2 + b^3 \end{aligned}$

SUBTRACTION OF POLYNOMIALS

Exercises 57–62: Write the opposite of the polynomial.

57. $5x^2$

58. $17x + 12$

59. $3a^2 - a + 4$

60. $-b^3 + 3b$

61. $-2t^2 - 3t + 4$

62. $7t^2 + t - 10$

Exercises 63–74: Subtract the polynomials.

63. $(3x + 1) - (-x + 3)$

64. $(-2x + 5) - (x + 7)$

65. $(-x^2 + 6x + 8) - (2x^2 + x - 2)$

66. $(2y^2 + 3y - 2) - (y^2 - y - 4)$

67. $(a^2 - 2a) - (4a^2 + 3a)$

68. $(7b^3 + 3b) - (-3b^3 - b)$

69. $(z^3 - 2z^2 - z) - (4z^2 + 5z + 1)$

70. $(3z^4 - z) - (-z^4 + 4z^2 - 5)$

71. $(4xy + x^2y^2) - (xy - x^2y^2)$

72. $(a^2 + b^2) - (-a^2 + b^2)$

73. $(ab^2) - (ab^2 + a^3b)$

74. $(x^2 + 3xy + 4y^2) - (x^2 - xy + 4y^2)$

Exercises 75–78: (Refer to Example 6.) Subtract the polynomials vertically.

75. $(x^2 + 2x - 3) - (2x^2 + 7x + 1)$

76. $(5x^2 - 9x - 1) - (x^2 - x + 3)$

77. $(3x^3 - 2x) - (5x^3 + 4x + 2)$

78. $(a^2 + 3ab + 2b^2) - (a^2 - 3ab + 2b^2)$

APPLICATIONS

79. *Exercise and Heart Rate*   The polynomial given by $1.6t^2 - 28t + 200$ calculates the heart rate shown in Figure 5.4(b) where t represents the elapsed time in minutes since exercise stopped.
   (a) What is the heart rate when the athlete first stops exercising?
   (b) What is the heart rate after 5 minutes?
   (c) Describe what happens to the heart rate after exercise stops.

80. *Cellular Phone Subscribers*   In the early years of cellular phone technology—from 1986 through 1991—the number of subscribers in millions could be modeled by the polynomial $0.163x^2 - 0.146x + 0.205$, where $x = 1$ corresponds to 1986, $x = 2$ to 1987, and so on. The graph illustrates this growth.

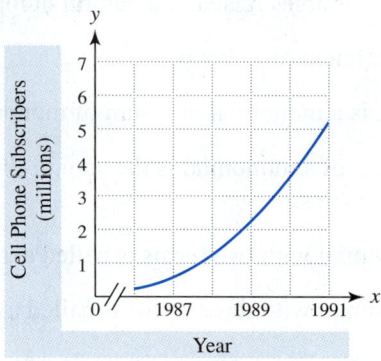

   (a) Use the graph to estimate the number of cellular phone subscribers in 1990.
   (b) Use the polynomial to estimate the number of cellular phone subscribers in 1990.
   (c) Do your answers from parts (a) and (b) agree?

81. *Areas of Squares*   Write a monomial that equals the sum of the areas of the squares. Then calculate this sum for $z = 10$ inches.

82. *Areas of Rectangles*   Find a monomial that equals the sum of the areas of the three rectangles. Find this sum for $a = 5$ yards and $b = 3$ yards.

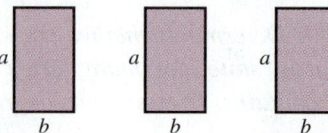

83. *Area of a Figure*   Find a polynomial that equals the area of the figure. Calculate its area for $x = 6$ feet.

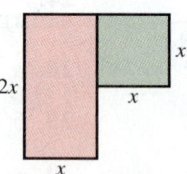

84. *Area of a Rectangle*   Write a polynomial that gives the area of the rectangle. Calculate its area for $x = 3$ feet.

85. *Areas of Circles*   Write a polynomial that gives the sum of the areas of two circles, one with radius x and the other with radius y. Find this sum for $x = 2$ feet and $y = 3$ feet. Leave your answer in terms of π.

86. *Squares and Circles*   Write a polynomial that gives the sum of the areas of a square having sides of length x and a circle having diameter x. Approximate this sum to the nearest hundredth of a square foot for $x = 6$ feet.

87. *World Population*   The table lists actual and projected world population P in billions for selected years t.

t	1974	1987	1999	2012
P	4	5	6	7

Source: U.S. Census Bureau.

(a) Find the slope of each line segment connecting consecutive data points in the table. Can these data be modeled with a line? Explain.

(b) Does the polynomial $0.077t - 148$ give good estimates for the world population in year t? Explain how you decided.

88. *Price of a Stamp*   The table lists the price P of a first-class postage stamp for selected years t.

t	1963	1975	1987	2002	2007
P	5¢	13¢	25¢	37¢	41¢

Source: U.S. Postal Service.

(a) Does the polynomial $0.835t - 1635$ model the data in the table exactly?

(b) Does it give approximations that are within 1.5¢ of the actual values?

WRITING ABOUT MATHEMATICS

89. Explain what the terms monomial, binomial, trinomial, and polynomial mean. Give an example of each.

90. Explain how to determine the degree of a polynomial having one variable. Give an example.

91. Explain how to obtain the opposite of a polynomial. Give an example.

92. Explain how to subtract two polynomials. Give an example.

CHECKING BASIC CONCEPTS
SECTIONS 5.1 AND 5.2

1. Evaluate each expression.
   (a) -5^2 **(b)** $3^2 - 2^3$

2. Simplify each expression.
   (a) $10^3 \cdot 10^5$ **(b)** $(3x^2)(-4x^5)$
   (c) $(a^3 b)^2$ **(d)** $\left(\dfrac{x}{z^3}\right)^4$

3. State the number of terms and variables in the polynomial $5x^3 y - 2x^2 y + 5$. What is its degree?

4. A box has a rectangular bottom twice as long as it is wide.

(a) If the bottom has width W and the box has height H, write a monomial that gives the volume of the box.

(b) Find the volume of the box for $W = 12$ inches and $H = 10$ inches.

5. Simplify each expression.
   (a) $(2a^2 + 3a - 1) + (a^2 - 3a + 7)$
   (b) $(4z^3 + 5z) - (2z^3 - 2z + 8)$
   (c) $(x^2 + 2xy + y^2) - (x^2 - 2xy + y^2)$

6. Evaluate $5x^2 - 7x$ for $x = -2$.

5.3   MULTIPLICATION OF POLYNOMIALS

Multiplying Monomials ■ Review of the Distributive Properties ■
Multiplying Monomials and Polynomials ■ Multiplying Polynomials

A LOOK INTO MATH ▷   The study of polynomials dates back to Babylonian civilization in about 1800–1600 B.C. Many eighteenth-century mathematicians devoted their entire careers to the study of polynomials. Polynomials still play an important role in mathematics, often being used to approximate unknown quantities. In this section we discuss the basics of multiplying polynomials.   (**Source:** *Historical Topics for the Mathematics Classroom, Thirty-first Yearbook,* NCTM.)

Multiplying Monomials

A monomial is a number, a variable, or a product of numbers and variables raised to natural number powers. To multiply monomials, we often use the product rule for exponents.

EXAMPLE   **Multiplying monomials**

Multiply.
(a) $-5x^2 \cdot 4x^3$ **(b)** $(7xy^4)(x^3y^2)$

Solution
(a) $\begin{aligned} -5x^2 \cdot 4x^3 &= (-5)(4)x^2x^3 \qquad &\text{Commutative property} \\ &= -20x^{2+3} &\text{The product rule} \\ &= -20x^5 &\text{Simplify.} \end{aligned}$

(b) $\begin{aligned} (7xy^4)(x^3y^2) &= 7xx^3y^4y^2 \qquad &\text{Commutative property} \\ &= 7x^{1+3}y^{4+2} &\text{The product rule} \\ &= 7x^4y^6 &\text{Simplify.} \end{aligned}$ **Now Try Exercises** 11, 15

Review of the Distributive Properties

Distributive properties are used frequently for multiplying monomials and polynomials. For all real numbers a, b, and c,

$$a(b + c) = ab + ac \quad \text{and}$$
$$a(b - c) = ab - ac.$$

The first distributive property above can be visualized geometrically. For example,

$$3(x + 2) = 3x + 6$$

is illustrated in Figure 5.11. The dimensions of the large rectangle are 3 by $x + 2$, and its area is $3(x + 2)$. The areas of the two small rectangles, $3x$ and 6, equal the area of the large rectangle. Therefore $3(x + 2) = 3x + 6$.

In the next example we use the distributive properties to multiply expressions.

Figure 5.11 Area: $3x + 6$

EXAMPLE   **Using distributive properties**

Multiply.
(a) $2(3x + 4)$ **(b)** $(3x^2 + 4)5$ **(c)** $-x(3x - 6)$

Solution

(a) $2(3x + 4) = \mathbf{2} \cdot 3x + \mathbf{2} \cdot 4 = 6x + 8$

(b) $(3x^2 + 4)\mathbf{5} = 3x^2 \cdot \mathbf{5} + 4 \cdot \mathbf{5} = 15x^2 + 20$

(c) $-x(3x - 6) = -x \cdot 3x + x \cdot 6 = -3x^2 + 6x$

Now Try Exercises 17, 21, 23

Multiplying Monomials and Polynomials

A monomial consists of one term, whereas a polynomial consists of one or more terms separated by $+$ or $-$ signs. To multiply a monomial by a polynomial, we apply the distributive properties and the product rule.

EXAMPLE 3   **Multiplying monomials and polynomials**

Multiply.
(a) $9x(2x^2 - 3)$ (b) $(5x - 8)x^2$
(c) $-7(2x^2 - 4x + 6)$ (d) $4x^3(x^4 + 9x^2 - 8)$

Solution

(a) $\quad 9x(2x^2 - 3) = \mathbf{9x} \cdot 2x^2 - \mathbf{9x} \cdot 3$ Distributive property
$\qquad\qquad\qquad = 18x^3 - 27x$ The product rule

(b) $\quad (5x - 8)x^2 = 5x \cdot x^2 - 8 \cdot x^2$ Distributive property
$\qquad\qquad\qquad = 5x^3 - 8x^2$ The product rule

(c) $-7(2x^2 - 4x + 6) = -7 \cdot 2x^2 + 7 \cdot 4x - 7 \cdot 6$ Distributive property
$\qquad\qquad\qquad\qquad = -14x^2 + 28x - 42$ Simplify.

(d) $4x^3(x^4 + 9x^2 - 8) = 4x^3 \cdot x^4 + 4x^3 \cdot 9x^2 - 4x^3 \cdot 8$ Distributive property
$\qquad\qquad\qquad\qquad = 4x^7 + 36x^5 - 32x^3$ The product rule

Now Try Exercises 25, 27, 29, 31

We can also multiply monomials and polynomials that contain more than one variable.

EXAMPLE 4   **Multiplying monomials and polynomials**

Multiply.
(a) $2xy(7x^2y^3 - 1)$ (b) $-ab(a^2 - b^2)$

Solution

(a) $\quad 2xy(7x^2y^3 - 1) = \mathbf{2xy} \cdot 7x^2y^3 - \mathbf{2xy} \cdot 1$ Distributive property
$\qquad\qquad\qquad\quad = 14xx^2yy^3 - 2xy$ Commutative property
$\qquad\qquad\qquad\quad = 14x^3y^4 - 2xy$ The product rule

(b) $\quad -ab(a^2 - b^2) = -ab \cdot a^2 + ab \cdot b^2$ Distributive property
$\qquad\qquad\qquad\quad = -aa^2b + abb^2$ Commutative property
$\qquad\qquad\qquad\quad = -a^3b + ab^3$ The product rule

Now Try Exercises 33, 37

Multiplying Polynomials

Monomials, binomials, and trinomials are examples of polynomials. Recall that a monomial has one term, a binomial has two terms, and a trinomial has three terms. In the next example we multiply two binomials, using both geometric and symbolic techniques.

EXAMPLE 5   **Multiplying binomials**

Multiply $(x + 4)(x + 2)$
(a) geometrically and   **(b)** symbolically.

Solution

(a) To multiply $(x + 4)(x + 2)$ geometrically, draw a rectangle $x + 4$ long and $x + 2$ wide, as shown in Figure 5.12(a). The area of this rectangle equals length times width, or $(x + 4)(x + 2)$. The large rectangle can be divided into four smaller rectangles, which have areas of x^2, $4x$, $2x$, and 8, as shown in Figure 5.12(b). Thus

$$(x + 4)(x + 2) = x^2 + 4x + 2x + 8$$
$$= x^2 + 6x + 8.$$

(b) To multiply $(x + 4)(x + 2)$ symbolically, apply the distributive property.

$$(x + 4)(x + 2) = (x + 4)(x) + (x + 4)(2)$$
$$= x \cdot x + 4 \cdot x + x \cdot 2 + 4 \cdot 2$$
$$= x^2 + 4x + 2x + 8$$
$$= x^2 + 6x + 8$$

Now Try Exercise **41**

The distributive properties used in part (b) of the previous example show that if we want to multiply $(x + 4)$ by $(x + 2)$, we should multiply every term in $x + 4$ by every term in $x + 2$.

$$(x + 4)(x + 2) = x^2 + 2x + 4x + 8$$
$$= x^2 + 6x + 8$$

NOTE: This process of multiplying binomials is sometimes called *FOIL*. This acronym may be used to remind us to multiply the first terms (*F*), outside terms (*O*), inside terms (*I*), and last terms (*L*).

Multiply the *First terms* to obtain x^2. $(x + 4)(x + 2)$

Multiply the *Outside terms* to obtain $2x$. $(x + 4)(x + 2)$

Multiply the *Inside terms* to obtain $4x$. $(x + 4)(x + 2)$

Multiply the *Last terms* to obtain 8. $(x + 4)(x + 2)$

The following statement summarizes how to multiply two polynomials in general.

MULTIPLYING POLYNOMIALS

The product of two polynomials may be found by multiplying every term in the first polynomial by every term in the second polynomial and then combining like terms.

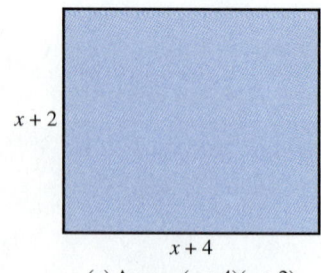

$x + 2$

$x + 4$

(a) Area $= (x + 4)(x + 2)$

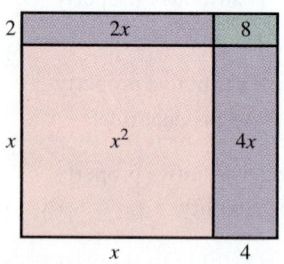

2	$2x$	8
x	x^2	$4x$
	x	4

(b) Area $= x^2 + 4x + 2x + 8$

Figure 5.12

EXAMPLE 6   **Multiplying binomials**

Multiply. Draw arrows to show how each term is found.

(a) $(3x + 2)(x + 1)$ **(b)** $(1 - x)(1 + 2x)$ **(c)** $(4x - 3)(x^2 - 2x)$

Solution

(a) $(3x + 2)(x + 1) = 3x \cdot x + 3x \cdot 1 + 2 \cdot x + 2 \cdot 1$

$ = 3x^2 + 3x + 2x + 2$

$ = 3x^2 + 5x + 2$

(b) $(1 - x)(1 + 2x) = 1 \cdot 1 + 1 \cdot 2x - x \cdot 1 - x \cdot 2x$

$ = 1 + 2x - x - 2x^2$

$ = 1 + x - 2x^2$

(c) $(4x - 3)(x^2 - 2x) = 4x \cdot x^2 - 4x \cdot 2x - 3 \cdot x^2 + 3 \cdot 2x$

$ = 4x^3 - 8x^2 - 3x^2 + 6x$

$ = 4x^3 - 11x^2 + 6x$ **Now Try Exercises** 53, 55, 61

EXAMPLE 7   **Multiplying polynomials**

Multiply.

(a) $(2x + 3)(x^2 + x - 1)$ **(b)** $(a - b)(a^2 + ab + b^2)$
(c) $(x^4 + 2x^2 - 5)(x^2 + 1)$

Solution

(a) Multiply every term in $(2x + 3)$ by every term in $(x^2 + x - 1)$.

$(2x + 3)(x^2 + x - 1) = 2x \cdot x^2 + 2x \cdot x - 2x \cdot 1 + 3 \cdot x^2 + 3 \cdot x - 3 \cdot 1$

$ = 2x^3 + 2x^2 - 2x + 3x^2 + 3x - 3$

$ = 2x^3 + 5x^2 + x - 3$

(b) $(a - b)(a^2 + ab + b^2) = a \cdot a^2 + a \cdot ab + a \cdot b^2 - b \cdot a^2 - b \cdot ab - b \cdot b^2$

$ = a^3 + a^2b + ab^2 - a^2b - ab^2 - b^3$

$ = a^3 - b^3$

(c) $(x^4 + 2x^2 - 5)(x^2 + 1) = x^4 \cdot x^2 + x^4 \cdot 1 + 2x^2 \cdot x^2 + 2x^2 \cdot 1 - 5 \cdot x^2 - 5 \cdot 1$

$ = x^6 + x^4 + 2x^4 + 2x^2 - 5x^2 - 5$

$ = x^6 + 3x^4 - 3x^2 - 5$ **Now Try Exercises** 65, 69, 71

Polynomials can be multiplied vertically in a manner similar to multiplication of real numbers. For example, multiplication of 123 times 12 is performed as follows.

$$
\begin{array}{r}
1\ 2\ 3 \\
\times\ 1\ 2 \\
\hline
2\ 4\ 6 \\
1\ 2\ 3 \\
\hline
1\ 4\ 7\ 6 \\
\end{array}
$$

A similar method can be used to multiply polynomials vertically.

EXAMPLE 8   **Multiplying polynomials vertically**

Multiply $2x^2 - 4x + 1$ and $x + 3$.

Solution

Write the polynomials vertically. Then multiply every term in the first polynomial by each term in the second polynomial. Arrange the results so that *like terms are in the same column*.

$$
\begin{array}{r}
2x^2 - 4x + 1 \\
x + 3 \\
\hline
6x^2 - 12x + 3 \\
2x^3 - 4x^2 + x \\
\hline
2x^3 + 2x^2 - 11x + 3
\end{array}
$$

Multiply top row by 3.
Multiply top row by x.
Add each column.   **Now Try Exercise** 73

MAKING CONNECTIONS

Vertical and Horizontal Formats

Whether you decide to add, subtract, or multiply polynomials vertically or horizontally, remember that the same answer is obtained either way.

EXAMPLE 9   **Finding the volume of a box**

A box has a width 3 inches less than its height and a length 4 inches more than its height.
(a) If h represents the height of the box, write a polynomial that represents the volume of the box.
(b) Use this polynomial to calculate the volume of the box if $h = 10$ inches.

Solution

(a) If h is the height, then $h - 3$ is the width and $h + 4$ is the length, as illustrated in Figure 5.13. Its volume equals the product of these three expressions.

$$
h(h - 3)(h + 4) = (h^2 - 3h)(h + 4)
$$
$$
= h^2 \cdot h + h^2 \cdot 4 - 3h \cdot h - 3h \cdot 4
$$
$$
= h^3 + 4h^2 - 3h^2 - 12h
$$
$$
= h^3 + h^2 - 12h
$$

(b) If $h = 10$, then the volume is

$$
10^3 + 10^2 - 12(10) = 1000 + 100 - 120 = 980 \text{ cubic inches.}
$$

Now Try Exercise 81

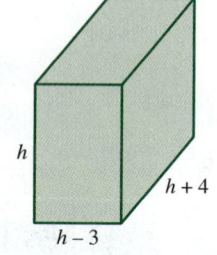

h

$h + 4$

$h - 3$

Figure 5.13

5.3 PUTTING IT ALL TOGETHER

The following table summarizes multiplication of polynomials.

Concept	Explanation	Examples
Distributive Properties	For all real numbers a, b, and c, $a(b + c) = ab + ac$ and $a(b - c) = ab - ac$.	$5(x + 3) = 5x + 15$, $3(x - 6) = 3x - 18$, and $-2x(3 - 5x^3) = -6x + 10x^4$

Concept	Explanation	Examples
Multiplying Polynomials	The product of two polynomials may be found by multiplying every term in the first polynomial by every term in the second polynomial and then combining like terms.	$3x(5x^2 + 2x - 7)$ $= 3x \cdot 5x^2 + 3x \cdot 2x - 3x \cdot 7$ $= 15x^3 + 6x^2 - 21x$ $(x + 2)(7x - 3)$ $= x \cdot 7x - x \cdot 3 + 2 \cdot 7x - 2 \cdot 3$ $= 7x^2 - 3x + 14x - 6$ $= 7x^2 + 11x - 6$

5.3 Exercises

CONCEPTS

1. The equation $x^2 \cdot x^3 = x^5$ illustrates what rule of exponents?

2. The equation $3(x - 2) = 3x - 6$ illustrates what property?

3. A monomial has ____ term(s), a binomial has ____ term(s), and a trinomial has ____ term(s).

4. The product of two polynomials may be found by multiplying every ____ in the first polynomial by every ____ in the second polynomial and then combining like terms.

5. Are $(x + y)(x + y)$ and $x^2 + y^2$ equivalent?

6. Polynomials can be multiplied horizontally or ____.

MULTIPLICATION OF MONOMIALS

Exercises 7–16: Multiply.

7. $x^2 \cdot x^5$

8. $-a \cdot a^5$

9. $-3a \cdot 4a$

10. $7x \cdot 5x$

11. $4x^3 \cdot 5x^2$

12. $6b^6 \cdot 3b^5$

13. $xy^2 \cdot 4xy$

14. $3ab \cdot ab^2$

15. $(-3xy^2)(4x^2y)$

16. $(-r^2t^2)(-r^3t)$

MULTIPLICATION OF MONOMIALS AND POLYNOMIALS

Exercises 17–38: Multiply and simplify the expression.

17. $3(x + 4)$

18. $-7(4x - 1)$

19. $-5(9x + 1)$

20. $10(1 - 6x)$

21. $(4 - z)z$

22. $3z(1 - 5z)$

23. $-y(5 + 3y)$

24. $(2y - 8)2y$

25. $3x(5x^2 - 4)$

26. $-6x(2x^3 + 1)$

27. $(6x - 6)x^2$

28. $(1 - 2x^2)3x^2$

29. $-8(4t^2 + t + 1)$

30. $7(3t^2 - 2t - 5)$

31. $n^2(-5n^2 + n - 2)$

32. $6n^3(2 - 4n + n^2)$

33. $xy(x + y)$

34. $ab(2a - 3b)$

35. $x^2(x^2y - xy^2)$

36. $2y^2(xy - 5)$

37. $-ab(a^3 - 2b^3)$

38. $5rt(r^2 + 2rt + t^2)$

MULTIPLICATION OF POLYNOMIALS

Exercises 39–44: (Refer to Example 5.) Multiply the given expression (a) geometrically and (b) symbolically.

39. $x(x + 3)$

40. $2x(x + 5)$

41. $(x + 2)(x + 2)$

42. $(x + 1)(x + 3)$

43. $(x + 3)(x + 6)$

44. $(x + 5)(x + 2)$

Exercises 45–72: Multiply and simplify the expression.

45. $(x + 3)(x + 5)$ **46.** $(x - 4)(x - 7)$

47. $(x - 8)(x - 9)$ **48.** $(x + 10)(x + 10)$

49. $(3z - 2)(2z - 5)$ **50.** $(z + 6)(2z - 1)$

51. $(8b - 1)(8b + 1)$ **52.** $(3t + 2)(3t - 2)$

53. $(10y + 7)(y - 1)$ **54.** $(y + 6)(2y + 7)$

55. $(5 - 3a)(1 - 2a)$ **56.** $(4 - a)(5 + 3a)$

57. $(1 - 3x)(1 + 3x)$ **58.** $(10 - x)(5 - 2x)$

59. $(x - 1)(x^2 + 1)$ **60.** $(x + 2)(x^2 - x)$

61. $(x^2 + 4)(4x - 3)$ **62.** $(3x^2 - 1)(3x^2 + 1)$

63. $(2n + 1)(n^2 + 3)$ **64.** $(2 - n^2)(1 + n^2)$

65. $(m + 1)(m^2 + 3m + 1)$

66. $(m - 2)(m^2 - m + 5)$

67. $(3x - 2)(2x^2 - x + 4)$

68. $(5x + 4)(x^2 - 3x + 2)$

69. $(x + 1)(x^2 - x + 1)$

70. $(x - 2)(x^2 + 4x + 4)$

71. $(4b^2 + 3b + 7)(b^2 + 3)$

72. $(-3a^2 - 2a + 1)(3a^2 - 3)$

Exercises 73–78: Multiply the polynomials vertically.

73. $(x + 2)(x^2 - 3x + 1)$

74. $(2y - 3)(3y^2 - 2y - 2)$

75. $(a - 2)(a^2 + 2a + 4)$

76. $(b - 3)(b^2 + 3b + 9)$

77. $(3x^2 - x + 1)(2x^2 + 1)$

78. $(2x^2 - 3x - 5)(2x^2 + 3)$

79. **Thinking Generally**   If a polynomial with m terms and a polynomial with n terms are multiplied, how many terms are there in the product before like terms are combined?

80. **Thinking Generally**   When a polynomial with m terms is multiplied by a second polynomial, the product contains k terms before like terms are combined. How many terms does the second polynomial contain?

APPLICATIONS

81. *Volume of a Box*   (Refer to Example 9.) A box has a width 4 inches less than its height and a length 2 inches more than its height.
   (a) If h is the height of the box, write a polynomial that represents the volume of the box.
   (b) Use this polynomial to calculate the volume for $h = 25$ inches.

82. *Surface Area of a Box*   Use the drawing of the box to write a polynomial that represents each of the following.

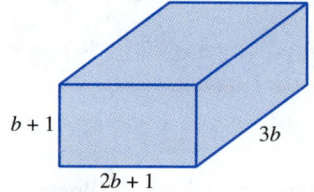

   (a) The area of its bottom
   (b) The area of its front
   (c) The area of its right side
   (d) The total area of its six sides

83. *Perimeter of a Pen*   A rectangular pen for a pet has a perimeter of 100 feet. If one side of the pen has length x, then its area is given by $x(50 - x)$.
   (a) Multiply this expression.
   (b) Evaluate the expression obtained in part (a) for $x = 25$.

84. *Rectangular Garden*   A rectangular garden has a perimeter of 500 feet.
   (a) If one side of the garden has length x, then write a polynomial expression that gives its area. Multiply this expression completely.
   (b) Evaluate the expression for $x = 50$ and interpret your answer.

85. *Surface Area of a Cube*   Write a polynomial that represents the total area of the six sides of the cube having edges with length $x + 1$.

86. *Surface Area of a Sphere* The surface area of a sphere with radius r is $4\pi r^2$. Write a polynomial that gives the surface area of a sphere with radius $x + 2$. Leave your answer in terms of π.

87. *Toy Rocket* A toy rocket is shot straight up into the air. Its height h in feet above the ground after t seconds is represented by the expression $t(64 - 16t)$.
  (a) Multiply this expression.
  (b) Evaluate the expression obtained in part (a) and the given expression for $t = 2$.
  (c) Are your answers in part (b) the same? Should they be the same?

88. *Toy Rocket on the Moon* (Refer to the preceding exercise.) If the same toy rocket were flown on the moon, then its height h in feet after t seconds would be $t(64 - \frac{5}{2}t)$.
  (a) Multiply this expression.
  (b) Evaluate the expression obtained in part (a) and the given expression for $t = 2$. Did the rocket go higher on the moon?

WRITING ABOUT MATHEMATICS

89. Explain how the acronym FOIL relates to multiplying two binomials, such as $x + 3$ and $2x + 1$.

90. Does the FOIL method work for multiplying a binomial and a trinomial? Explain.

91. Explain in words how to multiply any two polynomials. Give an example.

92. Give two properties of real numbers that are used for multiplying $3x(5x^2 - 3x + 2)$. Explain your answer.

GROUP ACTIVITY
WORKING WITH REAL DATA

Directions: Form a group of 2 to 4 people. Select someone to record the group's responses for this activity. All members of the group should work cooperatively to answer the questions. If your instructor asks for your results, each member of the group should be prepared to respond.

Biology Some types of worms have a remarkable ability to live without moisture. The following table from one study shows the number of worms W surviving after x days without moisture.

x (days)	0	20	40	80	120	160
W (worms)	50	48	45	36	20	3

Source: D. Brown and P. Rothery, *Models in Biology*.

  (a) Use the equation $W = -0.0014x^2 - 0.076x + 50$ to find W for each x-value in the table.
  (b) Discuss how well this equation approximates the data.
  (c) Use this equation to estimate the number of worms on day 60 and on day 180. Which answer is most accurate? Explain.

5.4 SPECIAL PRODUCTS

Product of a Sum and Difference ■ Squaring Binomials ■ Cubing Binomials

A LOOK INTO MATH ▷ Polynomials are often used to approximate real-world phenomena in applications. Polynomials have played an important role in the development of everyday products such as computers, cell phones, and automobiles. Even digital images in computers and interest calculations at a bank make use of polynomials. In this section we discuss how to multiply some special types of binomials.

Product of a Sum and Difference

Products of the form $(a + b)(a - b)$ occur frequently in mathematics. Other examples include

$$(x + y)(x - y) \quad \text{and} \quad (2r + 3t)(2r - 3t).$$

These products can always be multiplied by using the techniques discussed in Section 5.3. However, there is a faster way to multiply these special products.

$$(a + b)(a - b) = a \cdot a - a \cdot b + b \cdot a - b \cdot b$$
$$= a^2 - ab + ba - b^2$$
$$= a^2 - b^2$$

In words, the product of a sum of two numbers and their difference equals the difference of their squares. We generalize this method as follows.

PRODUCT OF A SUM AND DIFFERENCE

For any real numbers a and b,

$$(a + b)(a - b) = a^2 - b^2.$$

EXAMPLE 1   Finding products of sums and differences

Multiply.
(a) $(x + y)(x - y)$ **(b)** $(z - 2)(z + 2)$
(c) $(2r + 3t)(2r - 3t)$ **(d)** $(5m^2 - 4n^2)(5m^2 + 4n^2)$

Solution
(a) If we let $a = x$ and $b = y$, then we can apply the rule

$$(a + b)(a - b) = a^2 - b^2.$$

Thus

$$(x + y)(x - y) = (x)^2 - (y)^2$$
$$= x^2 - y^2.$$

(b) Because the expressions $(z + 2)(z - 2)$ and $(z - 2)(z + 2)$ are equal by the commutative property, we can apply the formula for the product of a sum and difference.

$$(z - 2)(z + 2) = (z)^2 - (2)^2$$
$$= z^2 - 4$$

(c) Let $a = 2r$ and $b = 3t$. Then the product can be evaluated as follows.
$$(2r + 3t)(2r - 3t) = (2r)^2 - (3t)^2$$
$$= 4r^2 - 9t^2$$

(d) $(5m^2 - 4n^2)(5m^2 + 4n^2) = (5m^2)^2 - (4n^2)^2 = 25m^4 - 16n^4$

Now Try Exercises 11, 17, 23

The next example demonstrates how to multiply mentally some products of numbers.

EXAMPLE 2   **Finding a product**

Use the product of a sum and difference to find $22 \cdot 18$.

Solution
Because $22 = 20 + 2$ and $18 = 20 - 2$, rewrite and evaluate $22 \cdot 18$ as follows.

$$22 \cdot 18 = (20 + 2)(20 - 2) \qquad \text{Product of a sum and difference}$$
$$= 20^2 - 2^2 \qquad (a + b)(a - b) = a^2 - b^2$$
$$= 400 - 4 \qquad \text{Evaluate exponents.}$$
$$= 396 \qquad \text{Subtract.} \qquad \text{Now Try Exercise } 27$$

Squaring Binomials

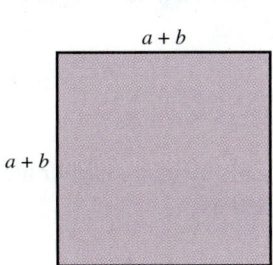

$a + b$

$a + b$

Area = $(a + b)^2$

Figure 5.14

Because each side of the square shown in Figure 5.14 has length $(a + b)$, its area equals
$$(a + b)(a + b),$$
which can be written as $(a + b)^2$. We can multiply this expression as follows.
$$(a + b)^2 = (a + b)(a + b)$$
$$= a^2 + ab + ba + b^2$$
$$= a^2 + 2ab + b^2$$

This result is illustrated geometrically in Figure 5.15, where the area of the large square is $(a + b)^2$. This area can also be found by adding the areas of the four small rectangles.
$$a^2 + ab + ba + b^2 = a^2 + 2ab + b^2$$

The geometric and symbolic results are the same. Note that to obtain the middle term, $2ab$, we can multiply the two terms in the binomial and *double* the result.

A similar product that is also the square of a binomial can be calculated as
$$(a - b)^2 = (a - b)(a - b)$$
$$= a^2 - ab - ba + b^2$$
$$= a^2 - 2ab + b^2.$$

The results are summarized as follows.

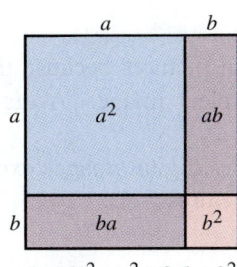

a b

a a^2 ab

b ba b^2

$(a + b)^2 = a^2 + 2ab + b^2$

Figure 5.15

SQUARING A BINOMIAL

For any real numbers a and b,
$$(a + b)^2 = a^2 + 2ab + b^2 \qquad \text{and}$$
$$(a - b)^2 = a^2 - 2ab + b^2.$$

That is, the square of a binomial equals the square of the first term, plus (or minus) twice the product of the two terms, plus the square of the last term.

NOTE: $(a + b)^2 \neq a^2 + b^2$. Do not forget the middle term when squaring a binomial.

EXAMPLE 3   Squaring a binomial

Multiply.
(a) $(x + 3)^2$ **(b)** $(2x - 5)^2$ **(c)** $(1 - 5y)^2$ **(d)** $(7a^2 + 3b)^2$

Solution
(a) If we let $a = x$ and $b = 3$, then we can apply the formula
$$(a + b)^2 = a^2 + 2ab + b^2.$$
Thus
$$(x + 3)^2 = (x)^2 + 2(x)(3) + (3)^2$$
$$= x^2 + 6x + 9.$$

(b) Applying the formula $(a - b)^2 = a^2 - 2ab + b^2$ with $a = 2x$ and $b = 5$ gives
$$(2x - 5)^2 = (2x)^2 - 2(2x)(5) + (5)^2$$
$$= 4x^2 - 20x + 25.$$

(c) $(1 - 5y)^2 = (1)^2 - 2(1)(5y) + (5y)^2 = 1 - 10y + 25y^2$

(d) $(7a^2 + 3b)^2 = (7a^2)^2 + 2(7a^2)(3b) + (3b)^2 = 49a^4 + 42a^2b + 9b^2$

Now Try Exercises 33, 37, 41, 47

MAKING CONNECTIONS

Multiplying Binomial and Special Products

If you forget these special products, you can still multiply polynomials by using earlier techniques. For example, the binomial in Example 3(b) can be multiplied as
$$(2x - 5)^2 = (2x - 5)(2x - 5)$$
$$= 2x \cdot 2x - 2x \cdot 5 - 5 \cdot 2x + 5 \cdot 5$$
$$= 4x^2 - 10x - 10x + 25$$
$$= 4x^2 - 20x + 25.$$

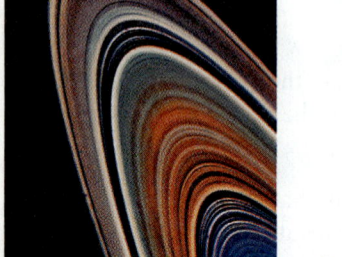

Figure 5.16 Digital Picture

▶ **REAL-WORLD CONNECTION**   NASA first developed digital pictures because they were easy to transmit through space and because they provided clear images. A digital image supplied by NASA is shown in Figure 5.16.

Today, digital cameras are readily available, and the Internet uses digital images exclusively. The next example shows how polynomials relate to digital pictures.

EXAMPLE 4   Calculating the size of a digital picture

Figure 5.17

A digital picture comprises tiny square units called *pixels*. Shading individual pixels creates a picture. A simplified version of a digital picture of the letter T is shown in Figure 5.17. The dimensions of this picture are 3 pixels by 3 pixels with a 1-pixel border.
(a) Suppose that a square digital picture, including its border, is $x + 2$ pixels by $x + 2$ pixels. Find a polynomial that gives the total number of pixels in the picture, including the border.
(b) Let $x = 3$ and evaluate the polynomial. Does it agree with Figure 5.17?

Solution

(a) The total number of pixels equals $(x + 2)$ times $(x + 2)$, or $(x + 2)^2$.

$$(x + 2)^2 = x^2 + 4x + 4$$

(b) For $x = 3$, the polynomial evaluates to $3^2 + 4 \cdot 3 + 4 = 25$, the total number of pixels. This result agrees with Figure 5.17, which has a total of $5 \cdot 5 = 25$ pixels with a 3×3–pixel picture of the letter T inside.   **Now Try Exercise 90**

Cubing Binomials

Figure 5.18
Volume $= (x + 2)^3$

To calculate the volume of the cube shown in Figure 5.18, we find the product of its length, width, and height. Because all sides have the same measure, its volume is $(x + 2)^3$. That is, the volume equals the **cube** of $x + 2$.

To multiply the expression $(x + 2)^3$, we proceed as follows.

$(x + 2)^3 = (x + 2)(x + 2)^2$ $a^3 = a \cdot a^2$

$= (x + 2)(x^2 + 4x + 4)$ Square the binomial.

$= x \cdot x^2 + x \cdot 4x + x \cdot 4 + 2 \cdot x^2 + 2 \cdot 4x + 2 \cdot 4$ Multiply the polynomials.

$= x^3 + 4x^2 + 4x + 2x^2 + 8x + 8$ Simplify terms.

$= x^3 + 6x^2 + 12x + 8$ Add like terms.

EXAMPLE 5

CRITICAL THINKING

Suppose that a student is convinced that the expressions

$(x + y)^3$ and $x^3 + y^3$

are equal. How could you convince the student otherwise?

Cubing a binomial

Multiply $(2z - 3)^3$.

Solution

$(2z - 3)^3 = (2z - 3)(2z - 3)^2$ $a^3 = a \cdot a^2$

$= (2z - 3)(4z^2 - 12z + 9)$ Square the binomial.

$= 8z^3 - 24z^2 + 18z - 12z^2 + 36z - 27$ Multiply the polynomials.

$= 8z^3 - 36z^2 + 54z - 27$ Add like terms.

NOTE: $(2z - 3)^3 \neq (2z)^3 - (3)^3 = 8z^3 - 27$.   **Now Try Exercise 55**

EXAMPLE 6

Calculating interest

If a savings account pays x percent annual interest, where x is expressed as a decimal, then after 3 years a sum of money will grow by a factor of $(1 + x)^3$.
(a) Multiply this expression.
(b) Evaluate the expression for $x = 0.10$ (or 10%), and interpret the result.

Solution

(a) $(1 + x)^3 = (1 + x)(1 + x)^2$ $a^3 = a \cdot a^2$

$= (1 + x)(1 + 2x + x^2)$ Square the binomial.

$= 1 + 2x + x^2 + x + 2x^2 + x^3$ Multiply the polynomials.

$= 1 + 3x + 3x^2 + x^3$ Add like terms.

(b) Let $x = \mathbf{0.1}$ in the expression $1 + 3x + 3x^2 + x^3$. Then

$$1 + 3(\mathbf{0.1}) + 3(\mathbf{0.1})^2 + (\mathbf{0.1})^3 = 1.331.$$

The sum of money will increase by a factor of 1.331. For example, if $1000 are deposited in this account, it will grow to $1331 after 3 years.   Now Try Exercise 85

5.4 PUTTING IT ALL TOGETHER

The following table summarizes some special products of polynomials.

Concept	Explanation	Examples
Product of a Sum and Difference	For any real numbers x and y, $$(x + y)(x - y) = x^2 - y^2.$$	$(x + 6)(x - 6) = x^2 - 36,$ $(2x - 3)(2x + 3) = 4x^2 - 9,$ and $(x^2 + y^2)(x^2 - y^2) = x^4 - y^4$
Squaring a Binomial	For all real numbers x and y, $$(x + y)^2 = x^2 + 2xy + y^2$$ and $$(x - y)^2 = x^2 - 2xy + y^2.$$	$(x + 4)^2 = x^2 + 8x + 16,$ $(5x - 2)^2 = 25x^2 - 20x + 4,$ $(1 - 7x)^2 = 1 - 14x + 49x^2,$ and $(x^2 + y^2)^2 = x^4 + 2x^2y^2 + y^4$
Cubing a Binomial	Multiply the binomial by its square.	$(x + 3)^3$ $= (x + 3)(x + 3)^2$ $= (x + 3)(x^2 + 6x + 9)$ $= x^3 + 6x^2 + 9x + 3x^2 + 18x + 27$ $= x^3 + 9x^2 + 27x + 27$

5.4 Exercises

CONCEPTS

1. $(a + b)(a - b) =$ _____

2. Are the expressions $(3z - 1)(3z + 1)$ and $9z^2 - 1$ equal for all real numbers z? Explain.

3. $(a + b)^2 =$ _____

4. $(a - b)^2 =$ _____

5. Are the expressions $(x + y)^2$ and $x^2 + y^2$ equal for all real numbers x and y? Explain.

6. Are the expressions $(r - t)^2$ and $r^2 - t^2$ equal for all real numbers r and t? Explain.

7. Are the expressions $(z + 5)^3$ and $z^3 + 5^3$ equal for all real numbers z? Explain.

8. $(a + b)^3 = (a + b) \cdot$ _____

PRODUCT OF A SUM AND DIFFERENCE

Exercises 9–26: Multiply.

9. $(x - 3)(x + 3)$ **10.** $(x + 6)(x - 6)$

11. $(x + 1)(x - 1)$ **12.** $(x - 7)(x + 7)$

13. $(4x - 1)(4x + 1)$ **14.** $(10x + 3)(10x - 3)$

15. $(1 + 2a)(1 - 2a)$ **16.** $(4 - 9b)(4 + 9b)$

17. $(2x + 3y)(2x - 3y)$ **18.** $(5r - 6t)(5r + 6t)$

19. $(ab - 5)(ab + 5)$ **20.** $(2xy + 7)(2xy - 7)$

21. $(ab + 4)(ab - 4)$ **22.** $(2 - 3yz)(2 + 3yz)$

23. $(a^2 - b^2)(a^2 + b^2)$ **24.** $(3x^2 + y^2)(3x^2 - y^2)$

25. $(x^3 - y^3)(x^3 + y^3)$

26. $(2a^4 + b^4)(2a^4 - b^4)$

Exercises 27–32: (Refer to Example 2.) Use the product of a sum and a difference to evaluate the expression.

27. $101 \cdot 99$ **28.** $52 \cdot 48$

29. $23 \cdot 17$ **30.** $29 \cdot 31$

31. $90 \cdot 110$ **32.** $38 \cdot 42$

SQUARING BINOMIALS

Exercises 33–48: Multiply.

33. $(x + 1)^2$ **34.** $(a + 3)^2$

35. $(a - 2)^2$ **36.** $(x - 7)^2$

37. $(2x + 3)^2$ **38.** $(7x - 2)^2$

39. $(3b + 5)^2$ **40.** $(7t + 10)^2$

41. $\left(\frac{3}{4}a - 4\right)^2$ **42.** $\left(\frac{1}{5}a + 1\right)^2$

43. $(1 - b)^2$ **44.** $(1 - 4a)^2$

45. $(5 + y^3)^2$ **46.** $(9 - 5x^2)^2$

47. $(a^2 + b)^2$ **48.** $(x^3 - y^3)^2$

CUBING BINOMIALS

Exercises 49–58: Multiply.

49. $(a + 1)^3$ **50.** $(b + 4)^3$

51. $(x - 2)^3$ **52.** $(y - 7)^3$

53. $(2x + 1)^3$ **54.** $(4z + 3)^3$

55. $(6u - 1)^3$ **56.** $(5v + 3)^3$

57. $t(t + 2)^3$ **58.** $a(2a - b)^3$

MULTIPLICATION OF POLYNOMIALS

Exercises 59–76: Multiply, using any appropriate method.

59. $4(5x + 9)$ **60.** $(2x + 1)(3x - 5)$

61. $(x - 5)(x + 7)$ **62.** $(x + 10)(x + 10)$

63. $(3x - 5)^2$ **64.** $(x - 3)(x + 9)$

65. $(5x + 3)(5x + 4)$ **66.** $-x^3(x^2 - x + 1)$

67. $(4b - 5)(4b + 5)$ **68.** $(x + 5)^3$

69. $-5x(4x^2 - 7x + 2)$ **70.** $(4x^2 - 5)(4x^2 + 5)$

71. $(4 - a)^3$ **72.** $2x(x - 3)^3$

73. $x(x + 3)^2$ **74.** $(x - 1)^2(x + 1)$

75. $(x + 2)(x - 2)(x + 1)(x - 1)$

76. $(x - y)(x + y)(x^2 + y^2)$

77. Thinking Generally Multiply $(a^n + b^n)(a^n - b^n)$.

78. Thinking Generally Multiply $(a^n + b^n)^2$.

APPLICATIONS

Exercises 79–82: Do each part and verify that your answers are the same.

  (a) *Find the area of the large square by multiplying its length and width.*

  (b) *Find the sum of the areas of the smaller rectangles inside the large square.*

79.

80.

81.

82.

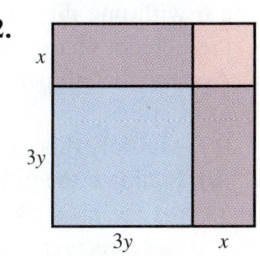

Exercises 83 and 84: Find a polynomial that represents the following.

    (a) *The outside surface area given by the six sides of the cube*

    (b) *The volume of the cube*

83.

$x + 5$

$x + 5$

$x + 5$

84.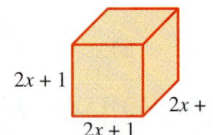

$2x + 1$

$2x + 1$

$2x + 1$

85. *Compound Interest* (Refer to Example 6.) If a sum of money is deposited in a savings account that is paying x percent annual interest (expressed as a decimal), then this sum of money increases by a factor of $(1 + x)^2$ after 2 years.

    (a) Multiply this expression.

    (b) Evaluate the polynomial expression found in part (a) for an annual interest rate of 10%, or $x = 0.10$, and interpret the answer.

86. *Compound Interest* If a sum of money is deposited in a savings account that is paying x percent annual interest, then this sum of money increases by a factor of $\left(1 + \frac{1}{100}x\right)^3$ after 3 years.

    (a) Multiply this expression.

    (b) Evaluate the polynomial expression found in part (a) for an annual interest rate of 8%, or $x = 8$, and interpret the answer.

87. *Probability* If there is an x percent chance of rain on each of two consecutive days, then the expression $(1 - x)^2$ gives the percent chance of no rain either day. Assume that all percentages are expressed as decimals.

    (a) Multiply this expression.

    (b) Evaluate the polynomial expression found in part (a) for a 50% chance of rain, or $x = 0.50$, and interpret the answer.

88. *Probability* If there is an x percent chance of rolling a 6 with one die, then the expression $(1 - x)^3$ gives the percent chance of not rolling a 6 with three dice. Assume that all percentages are expressed as decimals or fractions.

    (a) Multiply this expression.

    (b) Evaluate the polynomial expression found in part (a) for a $16.\overline{6}\%$ chance of rolling a 6, or $x = \frac{1}{6}$, and interpret the answer.

89. *Swimming Pool* A square swimming pool has an 8-foot–wide sidewalk around it.

    (a) If the sides of the pool have length z, find a polynomial that gives the area of the sidewalk.

    (b) Evaluate the polynomial in part (a) for $z = 60$ and interpret the answer.

90. *Digital Picture* (Refer to Example 4.) Suppose that a digital picture, including its border, is $x + 2$ pixels by $x + 2$ pixels and that the actual picture inside the border is $x - 2$ pixels by $x - 2$ pixels.

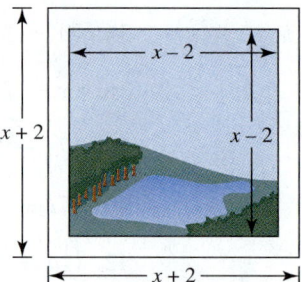

    (a) Find a polynomial that gives the number of pixels in the border.

    (b) Evaluate the polynomial in part (a) for $x = 5$.

    (c) Sketch a digital picture of the letter H with $x = 5$. Does the picture agree with the answer in part (b)?

    (d) Digital pictures typically have large values for x. If a picture has $x = 500$, find the total number of pixels in its border.

WRITING ABOUT MATHEMATICS

91. Explain why $(a + b)^2$ does not equal $a^2 + b^2$ in general for real numbers a and b.

92. Explain how to find the cube of a binomial.

CHECKING BASIC CONCEPTS
SECTIONS 5.3 AND 5.4

1. Multiply each expression.
   (a) $(-3xy^4)(5x^2y)$ (b) $-x(6 - 4x)$
   (c) $3ab(a^2 - 2ab + b^2)$

2. Multiply each expression.
   (a) $(x + 3)(4x - 3)$
   (b) $(x^2 - 1)(2x^2 + 2)$
   (c) $(x + y)(x^2 - xy + y^2)$

3. Multiply each expression.
   (a) $(5x + 2)(5x - 2)$ (b) $(x + 3)^2$
   (c) $(2 - 7x)^2$ (d) $(t + 2)^3$

4. *Surface Area*   A sphere with radius r has surface area $A = 4\pi r^2$.
   (a) Find an expression for the surface area for $r = a + 7$.
   (b) Find the surface area for $a = 8$. Leave your answer in terms of π.

5. Complete each part and verify that your answers are the same.
   (a) Find the area of the large square by squaring the length of one of its sides.
   (b) Find the sum of the areas of the smaller rectangles inside the large square.

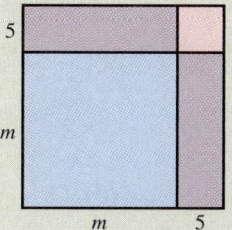

5.5 INTEGER EXPONENTS AND THE QUOTIENT RULE

Negative Integers as Exponents ▪ The Quotient Rule ▪ Other Rules for Exponents ▪ Scientific Notation

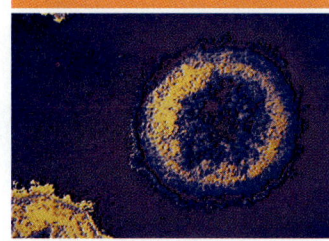

A LOOK INTO MATH ▷   Technology has brought with it the need for both very small and very large numbers. The size of an average virus (shown to the left) is 5 millionths of a centimeter, whereas the distance to the nearest star, Alpha Centauri, is 25 trillion miles. Exponents are often used to represent such numbers. In this section we discuss properties of integer exponents and some of their applications.   (**Source:** C. Ronan, *The Natural History of the Universe.*)

Negative Integers as Exponents

So far we have defined exponents that are whole numbers. For example,

$$5^0 = 1 \quad \text{and} \quad 2^3 = 2 \cdot 2 \cdot 2 = 8.$$

We can also define exponents that are negative integers as follows.

NEGATIVE INTEGER EXPONENTS

Let a be a nonzero real number and n be a positive integer. Then

$$a^{-n} = \frac{1}{a^n}.$$

That is, a^{-n} is the reciprocal of a^n.

In the next example we evaluate negative exponents.

EXAMPLE 1   Evaluating negative exponents

Simplify each expression.

(a) 2^{-3} **(b)** 7^{-1} **(c)** x^{-2} **(d)** $(x+y)^{-8}$

Solution

(a) Because $a^{-n} = \dfrac{1}{a^n}$, $2^{-3} = \dfrac{1}{2^3} = \dfrac{1}{2 \cdot 2 \cdot 2} = \dfrac{1}{8}$.

(b) $7^{-1} = \dfrac{1}{7^1} = \dfrac{1}{7}$

(c) $x^{-2} = \dfrac{1}{x^2}$

(d) $(x+y)^{-8} = \dfrac{1}{(x+y)^8}$

Now Try Exercises 13, 25, 32(b)

CALCULATOR HELP

To use the fraction feature (Frac), see Appendix A (page AP-2).

TECHNOLOGY NOTE: Negative Exponents

Calculators can be used to evaluate negative exponents. The figure shows how a graphing calculator evaluates the expressions in parts (a) and (b) of Example 1.

```
2^(-3)▶Frac
              1/8
7^(-1)▶Frac
              1/7
```

The rules for exponents discussed in this chapter so far also apply to expressions having negative exponents. For example, we can apply the product rule, $a^m \cdot a^n = a^{m+n}$, as follows.

$$\overset{\text{Add}}{2^{-3} \cdot 2^2 = 2^{-3+2} = 2^{-1} = \dfrac{1}{2}}$$

We can check this result by evaluating the expression without using the product rule.

$$2^{-3} \cdot 2^2 = \dfrac{1}{2^3} \cdot 2^2 = \dfrac{1}{8} \cdot 4 = \dfrac{4}{8} = \dfrac{1}{2}$$

EXAMPLE 2   Using the product rule with negative exponents

Evaluate each expression.
(a) $5^2 \cdot 5^{-4}$ **(b)** $3^{-2} \cdot 3^{-1}$

Solution

(a) $5^2 \cdot 5^{-4} \overset{\text{Add}}{=} 5^{2+(-4)} = 5^{-2} = \dfrac{1}{5^2} = \dfrac{1}{25}$

(b) $3^{-2} \cdot 3^{-1} = 3^{-2+(-1)} = 3^{-3} = \dfrac{1}{3^3} = \dfrac{1}{27}$

Now Try Exercise 15

EXAMPLE 3 **Using the rules of exponents**

Simplify the expression. Write the answer using positive exponents.

(a) $x^2 \cdot x^{-5}$ **(b)** $(y^3)^{-4}$ **(c)** $(rt)^{-5}$ **(d)** $(ab)^{-3}(a^{-2}b)^3$

Solution

(a) Using the product rule, $a^m \cdot a^n = a^{m+n}$, gives

$$x^2 \cdot x^{-5} = x^{2+(-5)} = x^{-3} = \frac{1}{x^3}.$$

(b) Using the power rule, $(a^m)^n = a^{mn}$, gives

$$(y^3)^{-4} = y^{3(-4)} = y^{-12} = \frac{1}{y^{12}}.$$

(c) Using the power rule, $(ab)^n = a^n b^n$, gives

$$(rt)^{-5} = r^{-5}t^{-5} = \frac{1}{r^5} \cdot \frac{1}{t^5} = \frac{1}{r^5 t^5}.$$

This expression could also be simplified as follows.

$$(rt)^{-5} = \frac{1}{(rt)^5} = \frac{1}{r^5 t^5}$$

(d) $(ab)^{-3}(a^{-2}b)^3 = a^{-3}b^{-3}a^{-6}b^3$

$$= a^{-3+(-6)}b^{-3+3}$$

$$= a^{-9}b^0$$

$$= \frac{1}{a^9} \cdot 1$$

$$= \frac{1}{a^9}$$

Now Try Exercises 27, 33, 35(a)

The Quotient Rule

Consider the division problem

$$\frac{3^4}{3^2} = \frac{3 \cdot 3 \cdot 3 \cdot 3}{3 \cdot 3} = \frac{3}{3} \cdot \frac{3}{3} \cdot 3 \cdot 3 = 1 \cdot 1 \cdot 3^2 = 3^2.$$

Because there are two more 3s in the numerator than in the denominator, the result is $3^{4-2} = 3^2$. That is, to divide exponential expressions having the *same base*, subtract the exponent of the denominator from the exponent of the numerator and keep the same base. This rule is called the *quotient rule,* which we express in symbols as follows.

THE QUOTIENT RULE

For any nonzero number a and integers m and n,

$$\frac{a^m}{a^n} = a^{m-n}.$$

EXAMPLE 4   **Using the quotient rule**

Simplify each expression. Write the answer using positive exponents.

(a) $\dfrac{4^3}{4^5}$ (b) $\dfrac{6a^7}{3a^4}$ (c) $\dfrac{xy^7}{x^2y^5}$

Solution

(a) $\dfrac{4^3}{4^5} = 4^{3-5} = 4^{-2} = \dfrac{1}{4^2} = \dfrac{1}{16}$ ⌐ Subtract

(b) $\dfrac{6a^7}{3a^4} = \dfrac{6}{3} \cdot \dfrac{a^7}{a^4} = 2a^{7-4} = 2a^3$

(c) $\dfrac{xy^7}{x^2y^5} = \dfrac{x^1}{x^2} \cdot \dfrac{y^7}{y^5} = x^{1-2}y^{7-5} = x^{-1}y^2 = \dfrac{y^2}{x}$

Now Try Exercises 19(b), 37(b), 39(a)

MAKING CONNECTIONS

The Quotient Rule and Simplifying Quotients

Some quotients can be simplified mentally. Because

$$\frac{x^5}{x^3} = \frac{x \cdot x \cdot x \cdot x \cdot x}{x \cdot x \cdot x},$$

the quotient $\frac{x^5}{x^3}$ has five factors of x in the numerator and three factors of x in the denominator. There are two more factors of x in the numerator than in the denominator, $5 - 3 = 2$, so this expression simplifies to x^2. Similarly,

$$\frac{x^3}{x^5} = \frac{x \cdot x \cdot x}{x \cdot x \cdot x \cdot x \cdot x}$$

has two more factors of x in the denominator than in the numerator. This expression simplifies to $\frac{1}{x^2}$. Use this technique to simplify the expressions

$$\frac{z^7}{z^4}, \quad \frac{a^5}{a^8}, \quad \text{and} \quad \frac{x^6y^2}{x^3y^7}.$$

Other Rules for Exponents

Other rules can be used to simplify expressions with negative exponents.

QUOTIENTS AND NEGATIVE EXPONENTS

The following three rules hold for any nonzero real numbers a and b and positive integers m and n.

1. $\dfrac{1}{a^{-n}} = a^n$ **2.** $\dfrac{a^{-n}}{b^{-m}} = \dfrac{b^m}{a^n}$ **3.** $\left(\dfrac{a}{b}\right)^{-n} = \left(\dfrac{b}{a}\right)^n$

We demonstrate the validity of these rules as follows.

1. $\dfrac{1}{a^{-n}} = \dfrac{1}{\frac{1}{a^n}} = 1 \cdot \dfrac{a^n}{1} = a^n$

2. $\dfrac{a^{-n}}{b^{-m}} = \dfrac{\frac{1}{a^n}}{\frac{1}{b^m}} = \dfrac{1}{a^n} \cdot \dfrac{b^m}{1} = \dfrac{b^m}{a^n}$

3. $\left(\dfrac{a}{b}\right)^{-n} = \dfrac{a^{-n}}{b^{-n}} = \dfrac{\frac{1}{a^n}}{\frac{1}{b^n}} = \dfrac{1}{a^n} \cdot \dfrac{b^n}{1} = \dfrac{b^n}{a^n} = \left(\dfrac{b}{a}\right)^n$

In the next example we apply these rules.

EXAMPLE 5   **Working with quotients and negative exponents**

Simplify each expression. Write the answer using positive exponents.

(a) $\dfrac{1}{2^{-5}}$ **(b)** $\dfrac{3^{-3}}{4^{-2}}$ **(c)** $\dfrac{5x^{-4}y^2}{10x^2y^{-4}}$ **(d)** $\left(\dfrac{2}{z^2}\right)^{-4}$

Solution

(a) $\dfrac{1}{2^{-5}} = 2^5 = 2 \cdot 2 \cdot 2 \cdot 2 \cdot 2 = 32$

(b) $\dfrac{3^{-3}}{4^{-2}} = \dfrac{4^2}{3^3} = \dfrac{16}{27}$

(c) $\dfrac{5x^{-4}y^2}{10x^2y^{-4}} = \dfrac{y^2y^4}{2x^2x^4} = \dfrac{y^6}{2x^6}$

(d) $\left(\dfrac{2}{z^2}\right)^{-4} = \left(\dfrac{z^2}{2}\right)^4 = \dfrac{z^8}{2^4} = \dfrac{z^8}{16}$

Now Try Exercises 21(b), 23, 41(a), 47

Scientific Notation

Powers of 10 are important because they are used in science to express numbers that are either very small or very large in absolute value. Table 5.4 on the next page lists some powers of 10. Note that if the power of 10 decreases by 1, the result decreases by a factor of $\frac{1}{10}$, or equivalently, the decimal point is moved one place to the left. Table 5.5 on the next page shows the values of some important powers of 10.

TECHNOLOGY NOTE:
Powers of 10

Calculators make use of scientific notation, as illustrated in the accompanying figure. The letter E denotes a power of 10. That is,

$2.5\text{E}13 = 2.5 \times 10^{13}$ and
$5\text{E}{-6} = 5 \times 10^{-6}$.

```
25000000000000
              2.5E13
.000005
          5E-6
```

NOTE: The calculator has been set in *scientific mode*.

TABLE 5.4 Powers of 10

Power of 10	Value
10^3	1000
10^2	100
10^1	10
10^0	1
10^{-1}	$\frac{1}{10} = 0.1$
10^{-2}	$\frac{1}{100} = 0.01$
10^{-3}	$\frac{1}{1000} = 0.001$

TABLE 5.5 Important Powers of 10

Number	Value
10^3	Thousand
10^6	Million
10^9	Billion
10^{12}	Trillion
10^{-1}	Tenth
10^{-2}	Hundredth
10^{-3}	Thousandth
10^{-6}	Millionth

Recall that numbers written in decimal notation are sometimes said to be in *standard form*. Decimal numbers that are either very large or very small in absolute value can be expressed in *scientific notation*.

▶ **REAL-WORLD CONNECTION** The distance to the nearest star, Alpha Centauri, is 25 trillion miles. This distance can be written in scientific notation as 2.5×10^{13} because

$$25{,}000{,}000{,}000{,}000 = 2.5 \times 10^{13}.$$
$$\underbrace{\qquad\qquad}_{\text{13 decimal places}}$$

The 10^{13} indicates that the decimal point in 2.5 should be moved **13** places to the **right**.

A typical virus is about 5 millionths of a centimeter in diameter, which can be written in scientific notation as 5×10^{-6} because

$$0.000005 = 5 \times 10^{-6} \text{ cm}.$$
$$\underbrace{\qquad\quad}_{\text{6 decimal places}}$$

The 10^{-6} indicates that the decimal point in 5 should be moved **6** places to the **left**.

The following definition provides a more complete explanation of scientific notation.

SCIENTIFIC NOTATION

A real number a is in **scientific notation** when a is written in the form $b \times 10^n$, where $1 \le |b| < 10$ and n is an integer.

EXAMPLE 6 Converting scientific notation to standard form

Write each number in standard form.
(a) 5.23×10^4 **(b)** 8.1×10^{-3} **(c)** 6×10^{-2}

Solution
(a) The positive exponent 4 indicates that the decimal point in 5.23 is to be moved 4 places to the *right*.

$$5.23 \times 10^4 = 5.2\,3\,0\,0. = 52{,}300$$
$$\underbrace{\quad\ \ \ }_{1\ 2\ 3\ 4}$$

(b) A negative exponent -3 indicates that the decimal point in 8.1 is to be moved 3 places to the *left*.

$$8.1 \times 10^{-3} = 0.\underset{1\ 2\ 3}{\underbrace{0\ 0\ 8}}.1 = 0.0081$$

(c) $6 \times 10^{-2} = 0.\underset{1\ 2}{\underbrace{0\ 6}}. = 0.06$

Now Try Exercises 59, 61

The following steps can be used for writing a positive number a in scientific notation.

WRITING A POSITIVE NUMBER IN SCIENTIFIC NOTATION

For a positive, rational number a expressed as a decimal, if $1 \le a < 10$, then $a = a \times 10^0$. Otherwise, use the following process to write a in scientific notation.

1. Move the decimal point in a until it becomes a number b such that $1 \le b < 10$.
2. Count the number of places that the decimal point was moved. Let this positive integer be n.
3. If the decimal point was moved to the *left*, then $a = b \times 10^n$. If the decimal point was moved to the *right*, then $a = b \times 10^{-n}$.

NOTE: The scientific notation for a negative number a is the opposite of the scientific notation of $|a|$. For example, $450 = 4.5 \times 10^2$ and $-450 = -4.5 \times 10^2$.

EXAMPLE 7 Writing a number in scientific notation

Write each number in scientific notation.
(a) 281,000,000   (U.S. population in 2000)
(b) 0.001   (Approximate time in seconds for sound to travel one foot)

Solution
(a) Move the assumed decimal point in 281,000,000 eight places to the *left* to obtain 2.81.

$$2.8\,1\,\underset{1\ 2\ 3\ 4\ 5\ 6\ 7\ 8}{\underbrace{0\ 0\ 0\ 0\ 0\ 0}}\,0.$$

The scientific notation for 281,000,000 is 2.81×10^8.
(b) Move the decimal point in 0.001 three places to the *right* to obtain 1.

$$0.\underset{1\ 2\ 3}{\underbrace{0\ 0\ 1}}.$$

The scientific notation for 0.001 is 1×10^{-3}.

Now Try Exercises 73, 77

MAKING CONNECTIONS

Scientific Notation

When a positive number a is expressed in scientific notation, a negative exponent on 10 indicates that $a < 1$, thus a is relatively small, and a positive exponent on 10 indicates that $a \geq 10$, thus a is relatively large. This can be helpful when converting from scientific notation to standard notation or vice versa. For example, to write the number 3.4×10^7 in standard form, we move the decimal point in 3.4 seven places. Because the exponent on 10 is positive, the resulting number should be relatively large. Moving the decimal point to the right results in 34,000,000. To express the number 0.00087 in scientific notation, we move the decimal point four places to obtain 8.7. Because the number 0.00087 is relatively small, the exponent on 10 will be negative. The resulting scientific notation is 8.7×10^{-4}.

CALCULATOR HELP

To display numbers in scientific notation, see Appendix A (page AP-2).

```
(6*10^4)(3*10^3)
                1.8E8
(6*10^4)/(3*10^3
)
                2E1
```

Figure 5.19

Numbers in scientific notation can be multiplied by applying properties of real numbers and properties of exponents.

$$(6 \times 10^4) \cdot (3 \times 10^3) = (6 \cdot 3) \times (10^4 \cdot 10^3) \quad \text{Properties of real numbers}$$
$$= 18 \times 10^7 \quad \text{Product rule}$$
$$= 1.8 \times 10^8 \quad \text{Scientific notation}$$

Division can also be performed with scientific notation.

$$\frac{6 \times 10^4}{3 \times 10^3} = \frac{6}{3} \times \frac{10^4}{10^3} \quad \text{Property of fractions}$$
$$= 2 \times 10^1 \quad \text{Quotient rule}$$

These results are supported in Figure 5.19, where the calculator is in scientific mode.

In the next example we show how to use scientific notation in an application.

EXAMPLE 8   Analyzing the cost of advertising

In 2005, a total of 2.46×10^{11} was spent on advertising in the United States. At that time the population of the United States was 2.98×10^8. Determine how much was spent per person on advertising.

Solution

To determine the amount spent per person, divide $\$2.46 \times 10^{11}$ by 2.98×10^8.

$$\frac{2.46 \times 10^{11}}{2.98 \times 10^8} = \frac{2.46}{2.98} \times 10^{11-8} \approx 0.826 \times 10^3 = 826$$

In 2005, about $826 was spent on advertising for every person in the United States.

Now Try Exercise 93

CRITICAL THINKING

Estimate the number of seconds that you have been alive. Write your answer in scientific notation.

5.5 PUTTING IT ALL TOGETHER

In this section we discussed negative integer exponents and scientific notation. The following table summarizes several important concepts related to these topics. Assume that a and b are nonzero real numbers and that m and n are integers.

Concept	Explanation	Examples
Negative Integer Exponents	$a^{-n} = \dfrac{1}{a^n}$	$2^{-4} = \dfrac{1}{2^4} = \dfrac{1}{16}$, $\quad a^{-8} = \dfrac{1}{a^8}$, and $(xy)^{-2} = \dfrac{1}{(xy)^2} = \dfrac{1}{x^2 y^2}$
Quotient Rule	$\dfrac{a^m}{a^n} = a^{m-n}$	$\dfrac{7^2}{7^4} = 7^{2-4} = 7^{-2} = \dfrac{1}{7^2} = \dfrac{1}{49}$ and $\dfrac{x^6}{x^3} = x^{6-3} = x^3$
Quotients and Negative Integer Exponents	1. $\dfrac{1}{a^{-n}} = a^n$ 2. $\dfrac{a^{-n}}{b^{-m}} = \dfrac{b^m}{a^n}$ 3. $\left(\dfrac{a}{b}\right)^{-n} = \left(\dfrac{b}{a}\right)^n$	1. $\dfrac{1}{5^{-2}} = 5^2 = 25$ 2. $\dfrac{x^{-4}}{y^{-2}} = \dfrac{y^2}{x^4}$ 3. $\left(\dfrac{2}{3}\right)^{-3} = \left(\dfrac{3}{2}\right)^3 = \dfrac{3^3}{2^3} = \dfrac{27}{8}$
Scientific Notation	Write a as $b \times 10^n$, where $1 \le \lvert b \rvert < 10$ and n is an integer.	$23{,}500 = 2.35 \times 10^4$, $0.0056 = 5.6 \times 10^{-3}$, and $1000 = 1 \times 10^3$

5.5 Exercises

CONCEPTS

1. $a^{-n} = $ _____

2. $2^{-3} = $ _____

3. $\dfrac{1}{a^{-n}} = $ _____

4. $\dfrac{1}{2^{-3}} = $ _____

5. $\dfrac{a^m}{a^n} = $ _____

6. $\dfrac{2^5}{2^2} = $ _____

7. $\dfrac{a^{-n}}{b^{-m}} = $ _____

8. $\dfrac{2^{-3}}{5^{-2}} = $ _____

9. $\left(\dfrac{a}{b}\right)^{-n} = $ _____

10. $\left(\dfrac{3}{5}\right)^{-2} = $ _____

11. To write a number a in scientific notation as $b \times 10^n$, the number b must satisfy _____.

12. Are 0.005 and 5×10^3 equivalent numbers?

NEGATIVE EXPONENTS

Exercises 13–24: Simplify the expression.

13. (a) 4^{-1} (b) $\left(\dfrac{1}{3}\right)^{-2}$

14. (a) 6^{-2} (b) 2.5^{-1}

15. (a) $2^3 \cdot 2^{-2}$ (b) $10^4 \cdot 10^{-2}$

16. (a) $3^{-4} \cdot 3^2$ (b) $10^{-1} \cdot 10^{-2}$

17. (a) $3^{-2} \cdot 3^{-1} \cdot 3^{-1}$ (b) $(2^3)^{-1}$

18. (a) $2^{-3} \cdot 2^5 \cdot 2^{-4}$ (b) $(3^{-2})^{-2}$

19. (a) $(3^2 4^3)^{-1}$ (b) $\dfrac{4^5}{4^2}$

20. (a) $(2^{-2} 3^2)^{-2}$ (b) $\dfrac{5^5}{5^3}$

21. (a) $\dfrac{1^9}{1^7}$ (b) $\dfrac{1}{4^{-3}}$

22. (a) $\dfrac{-6^4}{6}$ (b) $\dfrac{1}{6^{-2}}$

23. (a) $\dfrac{5^{-2}}{5^{-4}}$ (b) $\left(\dfrac{2}{7}\right)^{-2}$

24. (a) $\dfrac{7^{-3}}{7^{-1}}$ (b) $\left(\dfrac{3}{4}\right)^{-3}$

Exercises 25–48: Simplify the expression. Write the answer using positive exponents.

25. (a) x^{-1} (b) a^{-4}

26. (a) y^{-2} (b) z^{-7}

27. (a) $x^{-2} \cdot x^{-1} \cdot x$ (b) $a^{-5} \cdot a^{-2} \cdot a^{-1}$

28. (a) $y^{-3} \cdot y^4 \cdot y^{-5}$ (b) $b^5 \cdot b^{-3} \cdot b^{-6}$

29. (a) $x^2 y^{-3} x^{-5} y^6$ (b) $(xy)^{-3}$

30. (a) $a^{-2} b^{-6} b^3 a^{-1}$ (b) $(ab)^{-1}$

31. (a) $(2t)^{-4}$ (b) $(x+1)^{-7}$

32. (a) $(8c)^{-2}$ (b) $(a+b)^{-9}$

33. (a) $(a^{-2})^{-4}$ (b) $(rt^3)^{-2}$

34. (a) $(4x^3)^{-3}$ (b) $(xy^{-3})^{-2}$

35. (a) $(ab)^2 (a^2)^{-3}$ (b) $\dfrac{x^4}{x^2}$

36. (a) $(x^3)^{-2} (xy)^4 y^{-5}$ (b) $\dfrac{y^9}{y^5}$

37. (a) $\dfrac{a^{10}}{a^{-3}}$ (b) $\dfrac{4z}{2z^4}$

38. (a) $\dfrac{b^5}{b^{-2}}$ (b) $\dfrac{12x^2}{24x^7}$

39. (a) $\dfrac{-4xy^5}{6x^3y^2}$ (b) $\dfrac{x^{-4}}{x^{-1}}$

40. (a) $\dfrac{12a^6b^2}{8ab^3}$ (b) $\dfrac{y^{-2}}{y^{-7}}$

41. (a) $\dfrac{10b^{-4}}{5b^{-5}}$ (b) $\left(\dfrac{a}{b}\right)^3$

42. (a) $\dfrac{8a^{-2}}{2a^{-3}}$ (b) $\left(\dfrac{2x}{y}\right)^5$

43. (a) $\dfrac{1}{y^{-5}}$ (b) $\dfrac{4}{2t^{-3}}$

44. (a) $\dfrac{1}{z^{-6}}$ (b) $\dfrac{5}{10b^{-5}}$

45. (a) $\dfrac{1}{(xy)^{-2}}$ (b) $\dfrac{1}{(a^2b)^{-3}}$

46. (a) $\dfrac{1}{(ab)^{-1}}$ (b) $\dfrac{1}{(rt^4)^{-2}}$

47. (a) $\left(\dfrac{a}{b}\right)^{-2}$ (b) $\left(\dfrac{u}{4v}\right)^{-1}$

48. (a) $\left(\dfrac{2x}{y}\right)^{-3}$ (b) $\left(\dfrac{5u}{3v}\right)^{-2}$

49. **Thinking Generally** For positive integers m and n show that $\dfrac{a^n}{a^m} = \dfrac{1}{a^{m-n}}$.

50. **Thinking Generally** For positive integers m and n show that $\dfrac{a^{-n}}{a^{-m}} = a^{m-n}$.

SCIENTIFIC NOTATION

Exercises 51–56: (Refer to Table 5.5.) Write the value of the power of 10 in words.

51. 10^3 52. 10^6

53. 10^9 54. 10^{-1}

55. 10^{-2} 56. 10^{-6}

Exercises 57–68: Write the expression in standard form.

57. 2×10^3

58. 5×10^2

59. 4.5×10^4

60. 7.1×10^6

61. 8×10^{-3}

62. 9×10^{-1}

63. 4.56×10^{-4}

64. 9.4×10^{-2}

65. 3.9×10^7

66. 5.27×10^6

67. -5×10^5

68. -9.5×10^3

Exercises 69–80: Write the number in scientific notation.

69. 2000

70. 11,000

71. 567

72. 9300

73. 12,000,000

74. 600,000

75. 0.004

76. 0.0008

77. 0.000895

78. 0.0123

79. -0.05

80. -0.934

Exercises 81–88: Evaluate the expression. Write the answer in standard form.

81. $(5 \times 10^3)(3 \times 10^2)$

82. $(2.1 \times 10^2)(2 \times 10^4)$

83. $(-3 \times 10^{-3})(5 \times 10^2)$

84. $(4 \times 10^2)(1 \times 10^3)(5 \times 10^{-4})$

85. $\dfrac{4 \times 10^5}{2 \times 10^2}$

86. $\dfrac{9 \times 10^2}{3 \times 10^6}$

87. $\dfrac{8 \times 10^{-6}}{4 \times 10^{-3}}$

88. $\dfrac{6.3 \times 10^2}{2 \times 10^{-3}}$

APPLICATIONS

89. *Light-year* The distance that light travels in 1 year is called a *light-year*. Light travels at 1.86×10^5 miles per second, and there are about 3.15×10^7 seconds in 1 year.

   (a) Estimate the number of miles in 1 light-year.

   (b) Except for the sun, Alpha Centauri is the nearest star, and its distance is 4.27 light-years from Earth. Estimate its distance in miles. Write your answer in scientific notation.

90. *Milky Way* It takes 2×10^8 years for the sun to make one orbit around the Milky Way galaxy. Write this number in standard form.

91. *Speed of the Sun* (Refer to the two previous exercises.) Assume that the sun's orbit in the Milky Way galaxy is circular with a diameter of 10^5 light-years. Estimate how many miles the sun travels in 1 year.

92. *Distance to the Moon* The moon is about 240,000 miles from Earth.

   (a) Write this number in scientific notation.

   (b) If a rocket traveled at 4×10^4 miles per hour, how long would it take for it to reach the moon?

93. *Gross Domestic Product* The gross domestic product (GDP) is the total national output of goods and services valued at market prices *within* the United States. The GDP of the United States in 2005 was $12,460,000,000,000. (*Source: Bureau of Economic Analysis.*)

   (a) Write this number in scientific notation.

   (b) In 2005, the U.S. population was 2.98×10^8. On average, how many dollars of goods and services were produced by each individual?

94. *Average Family Net Worth*   A family refers to a group of two or more people related by birth, marriage, or adoption who reside together. In 2000, the average family net worth was $280,000, and there were about 7.2×10^7 families. Calculate the total family net worth in the United States in 2000. (***Source:*** U.S. Census Bureau.)

WRITING ABOUT MATHEMATICS

95. Explain what a negative exponent is and how it is different from a positive exponent. Give an example.

96. Explain why scientific notation is helpful for writing some numbers.

GROUP ACTIVITY
WORKING WITH REAL DATA

Directions: Form a group of 2 to 4 people. Select a person to record the group's responses for this activity. All members of the group should work cooperatively to answer the questions. If your instructor asks for your results, each member of the group should be prepared to respond.

Water in a Lake   East Battle Lake in Minnesota covers an area of about 1950 acres or 8.5×10^7 square feet, and its average depth is about 3.2×10^1 feet.
(a) Estimate the cubic feet of water in the lake. (*Hint:* volume = area × average depth.)
(b) One cubic foot of water equals about 7.5 gallons. How many gallons of water are in this lake?

(c) The population of the United States is about 3.02×10^8, and the average American uses about 1.5×10^2 gallons of water per day. Could this lake supply the American population with water for 1 day?

5.6 DIVISION OF POLYNOMIALS

Division by a Monomial ▪ Division by a Polynomial

A LOOK INTO MATH ▷   The study of polynomials has occupied the minds of mathematicians for centuries. During the sixteenth century, Girolamo Cardano and other Italian mathematicians discovered how to solve higher degree polynomial equations. In this section we demonstrate how to divide polynomials. Division is often needed to factor polynomials and to solve polynomial equations. (***Source:*** H. Eves, *An Introduction to the History of Mathematics.*)

Division by a Monomial

To add two fractions with like denominators, we use the property

$$\frac{a}{d} + \frac{b}{d} = \frac{a+b}{d}.$$

Girolamo Cardano (1501–1576)

For example, $\frac{1}{7} + \frac{3}{7} = \frac{1+3}{7} = \frac{4}{7}$.
     To divide a polynomial by a monomial we use the same property, only in reverse. That is,

$$\frac{a+b}{d} = \frac{a}{d} + \frac{b}{d}.$$

Note that each term in the numerator is divided by the monomial in the denominator. The next example shows how to divide a polynomial by a monomial.

EXAMPLE 1   Dividing a polynomial by a monomial

Divide.

(a) $\dfrac{a^5 + a^3}{a^2}$ (b) $\dfrac{5x^4 + 10x}{10x}$ (c) $\dfrac{3y^2 + 2y - 12}{6y}$

Solution

(a) $\dfrac{a^5 + a^3}{a^2} = \dfrac{a^5}{a^2} + \dfrac{a^3}{a^2} = a^{5-2} + a^{3-2} = a^3 + a$

(b) $\dfrac{5x^4 + 10x}{10x} = \dfrac{5x^4}{10x} + \dfrac{10x}{10x} = \dfrac{5}{10} \cdot \dfrac{x^4}{x} + \dfrac{10x}{10x} = \dfrac{1}{2}x^3 + 1$

(c) $\dfrac{3y^2 + 2y - 12}{6y} = \dfrac{3y^2}{6y} + \dfrac{2y}{6y} - \dfrac{12}{6y} = \dfrac{1}{2}y + \dfrac{1}{3} - \dfrac{2}{y}$ Now Try Exercises 15, 17

MAKING CONNECTIONS

Division and Simplification

A common mistake made when dividing expressions is to "cancel" incorrectly. Note in Example 1(b) that

$$\frac{5x^4 + 10x}{10x} \neq 5x^4 + \frac{10x}{10x}.$$

The monomial must be divided into *every* term in the numerator.

When dividing two natural numbers, we can check our work by multiplying. For example, $\frac{10}{5} = 2$, and we can check this result by finding the product $5 \cdot 2 = 10$. Similarly, to check

$$\frac{a^5 + a^3}{a^2} = a^3 + a$$

we can multiply a^2 and $a^3 + a$.

$$a^2(a^3 + a) = a^2 \cdot a^3 + a^2 \cdot a \qquad \text{Distributive property}$$
$$= a^5 + a^3 \qquad \text{It checks.}$$

EXAMPLE 2   Dividing and checking

Divide the expression $\frac{8x^3 - 4x^2 + 6x}{2x^2}$ and then check the result.

Solution

Be sure to divide $2x^2$ into *every* term in the numerator.

$$\frac{8x^3 - 4x^2 + 6x}{2x^2} = \frac{8x^3}{2x^2} - \frac{4x^2}{2x^2} + \frac{6x}{2x^2} = 4x - 2 + \frac{3}{x}$$

Check:

$$2x^2\left(4x - 2 + \frac{3}{x}\right) = 2x^2 \cdot 4x - 2x^2 \cdot 2 + 2x^2 \cdot \frac{3}{x}$$
$$= 8x^3 - 4x^2 + 6x \qquad \text{Now Try Exercise 13}$$

EXAMPLE 3   Finding the length of a rectangle

The rectangle in Figure 5.20 has an area $A = x^2 + 2x$ and width x. Write an expression for its length L in terms of x.

Figure 5.20

Solution
The area A of a rectangle equals length L times width W, or $A = LW$. Solving for L gives

$$L = \frac{A}{W}.$$

Thus to find the length of the given rectangle, divide the area by the width.

$$L = \frac{x^2 + 2x}{x} = \frac{x^2}{x} + \frac{2x}{x} = x + 2$$

The length of the rectangle is $x + 2$. The answer checks because $x(x + 2) = x^2 + 2x$.

Now Try Exercise **43**

Division by a Polynomial

To enable you to understand division by a polynomial better, we first need to review some terminology related to long division of natural numbers. To compute $271 \div 4$, we complete long division as follows.

$$
\begin{array}{r}
\text{Quotient} \longrightarrow 67 \text{ R } 3 \longleftarrow \text{Remainder} \\
\text{Divisor} \longrightarrow 4\overline{)271} \longleftarrow \text{Dividend} \\
\underline{24} \\
31 \\
\underline{28} \\
3
\end{array}
$$

To check this result, we find the product of the quotient and divisor and then add the remainder. Because $67 \cdot 4 + 3 = 271$, the answer checks. The quotient and remainder can also be expressed as $67\frac{3}{4}$. Division of polynomials is similar to long division of natural numbers.

EXAMPLE 4   Dividing polynomials

Divide $\frac{6x^2 + 13x + 3}{3x + 2}$ and check.

Solution
Begin by dividing the first term of $3x + 2$ into the first term of $6x^2 + 13x + 3$. That is, divide $3x$ into $6x^2$ to obtain $2x$. Then find the product of $2x$ and $3x + 2$, or $6x^2 + 4x$, place it below $6x^2 + 13x$, and subtract. Bring down the 3.

$$
\begin{array}{r}
2x \\
3x + 2\overline{)6x^2 + 13x + 3} \\
\underline{6x^2 + 4x} \\
9x + 3
\end{array}
$$

$\dfrac{6x^2}{3x} = 2x$

$2x(3x + 2) = 6x^2 + 4x$

Subtract: $13x - 4x = 9x$. Bring down the 3.

In the next step, divide $3x$ into the first term of $9x + 3$ to obtain 3. Then find the product of 3 and $3x + 2$, or $9x + 6$, place it below $9x + 3$, and subtract.

$$
\begin{array}{r}
2x + 3 \\
3x + 2 \overline{)6x^2 + 13x + 3} \\
\underline{6x^2 + 4x} \\
9x + 3 \\
\underline{9x + 6} \\
-3
\end{array}
\qquad
\begin{array}{l}
\dfrac{9x}{3x} = 3 \\[3mm]
\\
3(3x + 2) = 9x + 6 \\
\text{Subtract: } 3 - 6 = -3.
\end{array}
$$

The quotient is $2x + 3$ with remainder -3. This result can also be written as

$$
2x + 3 + \frac{-3}{3x + 2}, \qquad \text{Quotient} + \frac{\text{Remainder}}{\text{Divisor}}
$$

in the same manner that 67 R 3 can be written as $67\frac{3}{4}$.

Check polynomial division by adding the remainder to the product of the divisor and the quotient. That is,

$$
(\text{Divisor})(\text{Quotient}) + \text{Remainder} = \text{Dividend}.
$$

For this example, the equation becomes

$$
\begin{aligned}
(3x + 2)(2x + 3) + (-3) &= 3x \cdot 2x + 3x \cdot 3 + 2 \cdot 2x + 2 \cdot 3 - 3 \\
&= 6x^2 + 9x + 4x + 6 - 3 \\
&= 6x^2 + 13x + 3. \qquad \text{It checks.}
\end{aligned}
$$

Now Try Exercise **23**

EXAMPLE **5**   **Dividing polynomials having a missing term**

Simplify $(3x^3 + 2x - 4) \div (x - 2)$.

Solution
Because the dividend does not have an x^2-term, insert $0x^2$ as a "place holder." Then begin by dividing x into $3x^3$ to obtain $3x^2$.

$$
\begin{array}{r}
3x^2 \\
x - 2 \overline{)3x^3 + 0x^2 + 2x - 4} \\
\underline{3x^3 - 6x^2} \\
6x^2 + 2x
\end{array}
\qquad
\begin{array}{l}
\dfrac{3x^3}{x} = 3x^2 \\[3mm]
\\
3x^2(x - 2) = 3x^3 - 6x^2 \\
\text{Subtract: } 0x^2 - (-6x^2) = 6x^2. \text{ Bring down } 2x.
\end{array}
$$

In the next step, divide x into $6x^2$.

$$
\begin{array}{r}
3x^2 + 6x \\
x - 2 \overline{)3x^3 + 0x^2 + 2x - 4} \\
\underline{3x^3 - 6x^2} \\
6x^2 + 2x \\
\underline{6x^2 - 12x} \\
14x - 4
\end{array}
\qquad
\begin{array}{l}
\dfrac{6x^2}{x} = 6x \\[3mm]
\\
6x(x - 2) = 6x^2 - 12x \\
\text{Subtract: } 2x - (-12x) = 14x. \text{ Bring down } -4.
\end{array}
$$

Now divide x into $14x$.

$$\begin{array}{r} 3x^2 + 6x + \mathbf{14} \\ x - 2\overline{)3x^3 + 0x^2 + 2x - 4} \\ \underline{3x^3 - 6x^2} \\ 6x^2 + 2x \\ \underline{6x^2 - 12x} \\ \mathbf{14x} - 4 \\ \underline{14x - 28} \\ 24 \end{array}$$

$\dfrac{14x}{x} = 14$

$14(x - 2) = 14x - 28$

Subtract: $-4 - (-28) = 24$.

The quotient is $3x^2 + 6x + 14$ with remainder 24. This result can also be written as

$$3x^2 + 6x + 14 + \frac{24}{x - 2}.$$

Now Try Exercise **31**

EXAMPLE 6   Dividing with a quadratic divisor

Divide $x^3 - 3x^2 + 3x + 2$ by $x^2 + 1$.

Solution
Begin by writing $x^2 + 1$ as $x^2 + 0x + 1$.

$$\begin{array}{r} x - 3 \\ x^2 + 0x + 1\overline{)x^3 - 3x^2 + 3x + 2} \\ \underline{x^3 + 0x^2 + x} \\ -3x^2 + 2x + 2 \\ \underline{-3x^2 + 0x - 3} \\ 2x + 5 \end{array}$$

The quotient is $x - 3$ with remainder $2x + 5$. This result can also be written as

$$x - 3 + \frac{2x + 5}{x^2 + 1}.$$

Now Try Exercise **35**

5.6 PUTTING IT ALL TOGETHER

In this section we discussed division of polynomials by monomials and other polynomials. The following table summarizes several important concepts related to division of polynomials.

Concept	Explanation	Examples
Division by a Monomial	Use the property $$\frac{a + b}{d} = \frac{a}{d} + \frac{b}{d}.$$ Be sure to divide the denominator into every term in the numerator.	$\dfrac{2x^3 + 4x}{2x^2} = \dfrac{2x^3}{2x^2} + \dfrac{4x}{2x^2} = x + \dfrac{2}{x}$ and $\dfrac{a^2 - 2a}{4a} = \dfrac{a^2}{4a} - \dfrac{2a}{4a} = \dfrac{a}{4} - \dfrac{1}{2}$

Concept	Explanation	Examples
Division by a Polynomial	Is done similarly to the way long division of natural numbers is performed	Divide $x^2 + 3x + 3$ by $x + 1$. $$\begin{array}{r} x + 2 \\ x + 1 \overline{)\ x^2 + 3x + 3} \\ \underline{x^2 + x} \\ 2x + 3 \\ \underline{2x + 2} \\ 1 \end{array}$$ The quotient is $x + 2$ with remainder 1, which can be expressed as $$x + 2 + \frac{1}{x + 1}.$$
Checking a Result	Dividend = (Divisor)(Quotient) + Remainder	When $x^2 + 3x + 3$ is divided by $x + 1$, the quotient is $x + 2$ with remainder 1. Thus $$(x + 1)(x + 2) + 1 = x^2 + 3x + 3,$$ and the answer checks.

5.6 Exercises

MyMathLab

PRACTICE   WATCH   DOWNLOAD   READ   REVIEW

CONCEPTS

1. $\frac{a + b}{d} = $ _____

2. $\frac{a + b - c}{d} = $ _____

3. When dividing a polynomial by a monomial, the monomial must be divided into every _____ of the polynomial.

4. Are the expressions $\frac{5x^2 + 2x}{2x}$ and $5x^2 + 1$ equal?

5. Are the expressions $\frac{5x^2 + 2x}{2x}$ and $\frac{5x^2}{2x}$ equal?

6. Because $\frac{37}{9} = 4$ with remainder 1, it follows that $37 = $ _____ $\cdot$ _____ $+$ _____.

7. Because $2x^3 - x + 5$ divided by $x + 1$ equals $2x^2 - 2x + 1$ with remainder 4, it follows that $2x^3 - x + 5 = $ _____ $\cdot$ _____ $+$ _____.

8. When dividing $2x^3 + 3x - 1$ by $x - 1$, insert _____ into the dividend as a "place holder" for the missing x^2-term.

DIVISION BY A MONOMIAL

Exercises 9–14: Divide and check.

9. $\frac{6x^2}{3x}$

10. $\frac{-5x^2}{10x^4}$

11. $\frac{z^4 + z^3}{z}$

12. $\frac{t^3 - t}{t}$

13. $\frac{a^5 - 6a^3}{2a^3}$

14. $\frac{b^4 - 4b}{4b^2}$

Exercises 15–22: Divide.

15. $\frac{4x - 7x^4}{x^2}$

16. $\frac{1 + 6x^4}{3x^3}$

17. $\frac{9x^4 - 3x + 6}{3x}$

18. $\frac{y^3 - 4y + 6}{y}$

19. $\dfrac{12y^4 - 3y^2 + 6y}{3y^2}$ **20.** $\dfrac{2x^2 - 6x + 9}{12x}$

21. $\dfrac{15m^4 - 10m^3 + 20m^2}{5m^2}$ **22.** $\dfrac{n^8 - 8n^6 + 4n^4}{2n^5}$

Exercises 23–28: Divide and check.

23. $\dfrac{2x^2 - 3x + 1}{x - 2}$ **24.** $\dfrac{4x^2 - x + 3}{x + 2}$

25. $\dfrac{x^2 + 2x + 1}{x + 1}$ **26.** $\dfrac{4x^2 - 4x + 1}{2x - 1}$

27. $\dfrac{x^3 - x^2 + x - 2}{x - 1}$ **28.** $\dfrac{2x^3 + 3x^2 + 3x - 1}{2x + 1}$

Exercises 29–40: Divide.

29. $\dfrac{4x^3 - 3x^2 + 7x + 3}{4x + 1}$ **30.** $\dfrac{10x^3 - x^2 - 17x - 7}{5x + 2}$

31. $\dfrac{x^3 - x + 2}{x - 2}$ **32.** $\dfrac{6x^3 + 8x^2 + 4}{3x + 4}$

33. $(3x^3 + 2) \div (x - 1)$

34. $(-3x^3 + 8x^2 + x) \div (3x + 4)$

35. $(x^3 + 3x^2 + 1) \div (x^2 + 1)$

36. $(x^4 - x^3 + x^2 - x + 1) \div (x^2 - 1)$

37. $\dfrac{x^3 + 1}{x^2 - x + 1}$ **38.** $\dfrac{4x^3 + 3x + 2}{2x^2 - x + 1}$

39. $\dfrac{x^3 + 8}{x + 2}$ **40.** $\dfrac{x^4 - 16}{x - 2}$

41. **Thinking Generally**   If the quotient in a polynomial division problem is an integer, what must be true about the degrees of the dividend and divisor?

42. **Thinking Generally**   If the quotient in a polynomial division problem is a polynomial of degree 1, what must be true about the degrees of the dividend and divisor?

APPLICATIONS

43. *Area of a Rectangle*   The area A of a rectangle is $8x^2$, and one of its sides has length $2x$. Find an expression for the length L of the other side.

44. *Area of a Rectangle*   The area A of a rectangle is $x^2 - 1$, and one of its sides has length $x + 1$. Find the width W of the other side.

45. *Volume of a Box*   The volume V of a box is $2x^3 + 4x^2$, and the area of its bottom is $2x^2$. Find the height of the box in terms of x. Make a possible sketch of the box, and label the length of each side.

46. *Area of a Triangle*   A triangle has height h and area $A = 2h^2 - 4h$. Find its base b in terms of h. Make a possible sketch of the triangle, and label the height and base. (*Hint:* $A = \frac{1}{2}bh$.)

WRITING ABOUT MATHEMATICS

47. Suppose that one polynomial is divided into another polynomial and the remainder is 0. What does the product of the divisor and quotient equal? Explain.

48. A student simplifies the expression $\dfrac{4x^3 - 1}{4x^2}$ to $x - 1$. Explain the student's error.

CHECKING BASIC CONCEPTS
SECTIONS 5.5 AND 5.6

1. Simplify each expression. Write the result with positive exponents.

   (a) 9^{-2} (b) $\dfrac{3x^{-3}}{6x^4}$ (c) $(4ab^{-4})^{-2}$

2. Simplify each expression. Write the result with positive exponents.

   (a) $\dfrac{1}{z^{-5}}$ (b) $\dfrac{x^{-3}}{y^{-6}}$ (c) $\left(\dfrac{3}{x^2}\right)^{-3}$

3. Write each expression in scientific notation.
   (a) 45,000   (b) 0.000234   (c) 0.01

4. Write each expression in standard form.
   (a) 4.71×10^4 (b) 6×10^{-3}

5. Simplify $\dfrac{25a^4 - 15a^3}{5a^3}$.

6. Divide $3x^2 - x - 4$ by $x - 1$. State the quotient and remainder.

7. *Distance to the Sun*   The distance to the sun is approximately 93 million miles.
   (a) Write this distance in scientific notation.
   (b) Light travels at 1.86×10^5 miles per second. How long does it take for the sun's light to reach Earth?

CHAPTER 5   SUMMARY

SECTION 5.1 ■ RULES FOR EXPONENTS

Bases and Exponents   The expression b^n has base b, exponent n, and equals the expression $\underbrace{b \cdot b \cdot b \cdot \,\cdots\, \cdot b}_{n \text{ times}}$, when n is a natural number.

Example: 2^3 has base 2, exponent 3, and equals $2 \cdot 2 \cdot 2 = 8$.

Evaluating Expressions   When evaluating expressions, evaluate exponents *before* performing addition, subtraction, multiplication, division, or negation. In general, operations within parentheses should be evaluated *before* using the order of operations.

1. Evaluate exponents.
2. Perform negation.
3. Do multiplication and division from left to right.
4. Do addition and subtraction from left to right.

Example: $-3^2 + 3 \cdot 4 = -9 + 3 \cdot 4 = -9 + 12 = 3$

Zero Exponents   For any nonzero number b, $b^0 = 1$. Note that 0^0 is undefined.

Examples: $5^0 = 1$ and $\left(\dfrac{x}{y}\right)^0 = 1$, where x and y are nonzero.

Product Rule   For any real number a and natural numbers m and n,

$$a^m \cdot a^n = a^{m+n}.$$

Examples: $3^4 \cdot 3^2 = 3^6$ and $x^3 x^2 x^4 = x^9$

Power Rules   For any real numbers a and b and natural numbers m and n,

$$(a^m)^n = a^{mn}, \quad (ab)^n = a^n b^n, \quad \text{and} \quad \left(\frac{a}{b}\right)^n = \frac{a^n}{b^n}, b \neq 0.$$

Examples: $(x^2)^3 = x^6, \quad (3x)^4 = 3^4 x^4 = 81x^4, \quad \text{and} \quad \left(\frac{2}{y}\right)^3 = \frac{2^3}{y^3} = \frac{8}{y^3}$

SECTION 5.2 ■ ADDITION AND SUBTRACTION OF POLYNOMIALS

Terms Related to Polynomials

Monomial	A number, variable, or product of numbers and variables raised to natural number powers
Degree of a Monomial	Sum of the exponents of the variables
Coefficient of a Monomial	The number in a monomial

Example: The monomial $-3x^2 y^3$ has degree 5 and coefficient -3.

Polynomial	A monomial or a sum of monomials
Term of a Polynomial	Each monomial is a term of the polynomial.
Binomial	A polynomial with two terms
Trinomial	A polynomial with three terms
Degree of a Polynomial	The degree of the term with highest degree
Opposite of a Polynomial	The opposite is found by negating each term.

Example: $2x^3 - 4x + 5$ is a trinomial with degree 3. Its opposite is $-2x^3 + 4x - 5$.

Like Terms	Two monomials with the same variables raised to the same powers

Examples: $3xy^2$ and $-xy^2$ are like terms.

$5x^3$ and $3x^3$ are like. terms.

$5x^2$ and $5x$ are unlike terms.

Addition of Polynomials   Combine like terms, using the distributive property.

Example: $(2x^2 - 4x) + (-x^2 - x) = (2 - 1)x^2 + (-4 - 1)x$

$$= x^2 - 5x$$

Subtraction of Polynomials   Add the first polynomial to the opposite of the second polynomial.

Example: $(4x^4 - 5x) - (7x^4 + 6x) = (4x^4 - 5x) + (-7x^4 - 6x)$

$$= (4 - 7)x^4 + (-5 - 6)x$$

$$= -3x^4 - 11x$$

SECTION 5.3 ■ MULTIPLICATION OF POLYNOMIALS

Multiplication of Monomials   Use the commutative property and the product rule.

Examples: $-2x^3 \cdot 3x^2 = -2 \cdot 3 \cdot x^3 \cdot x^2 = -6x^5$

$(2xy^2)(3x^2y^3) = 2 \cdot 3 \cdot x \cdot x^2 \cdot y^2 \cdot y^3 = 6x^3y^5$

$\uparrow$———— Assumed exponent of 1

Distributive Properties

$$a(b + c) = ab + ac \quad \text{and} \quad a(b - c) = ab - ac$$

Examples: $4x(3x + 6) = 4x \cdot 3x + 4x \cdot 6 = 12x^2 + 24x$

$ab(a^2 - b^2) = ab \cdot a^2 - ab \cdot b^2 = a^3b - ab^3$

Multiplication of Monomials and Polynomials   Apply the distributive properties. Be sure to multiply every term in the polynomial by the monomial.

Example: $-2x^2(4x^2 - 5x - 3) = -8x^4 + 10x^3 + 6x^2$

Multiplication of Polynomials   The product of two polynomials may be found by multiplying every term in the first polynomial by every term in the second polynomial. Be sure to combine like terms.

Examples: $(x + 3)(2x - 5) = 2x^2 - 5x + 6x - 15$

$= 2x^2 + x - 15$

$(2x + 1)(x^2 - 5x + 2) = 2x^3 - 10x^2 + 4x + x^2 - 5x + 2$

$= 2x^3 - 9x^2 - x + 2$

SECTION 5.4 ■ SPECIAL PRODUCTS

Product of a Sum and Difference

$$(a + b)(a - b) = a^2 - b^2$$

Examples: $(x + 4)(x - 4) = x^2 - 16$

$(2r - 3t)(2r + 3t) = (2r)^2 - (3t)^2 = 4r^2 - 9t^2$

Squaring Binomials

$$(a + b)^2 = a^2 + 2ab + b^2 \quad \text{and} \quad (a - b)^2 = a^2 - 2ab + b^2$$

Examples: $(2x + 1)^2 = (2x)^2 + 2(2x)1 + 1^2 = 4x^2 + 4x + 1$

$(z^2 - 2)^2 = (z^2)^2 - 2z^2(2) + 2^2 = z^4 - 4z^2 + 4$

Cubing Binomials   To multiply $(a + b)^3$ write it as $(a + b)(a + b)^2$.

Example: $(x + 4)^3 = (x + 4)(x + 4)^2$

$= (x + 4)(x^2 + 8x + 16)$ Square the binomial.

$= x^3 + 8x^2 + 16x + 4x^2 + 32x + 64$ Distributive property

$= x^3 + 12x^2 + 48x + 64$ Combine like terms.

SECTION 5.5 ■ INTEGER EXPONENTS AND THE QUOTIENT RULE

Negative Integers as Exponents   For any nonzero real number a and positive integer n,

$$a^{-n} = \frac{1}{a^n}.$$

Examples: $5^{-2} = \dfrac{1}{5^2}$ and $x^{-4} = \dfrac{1}{x^4}$

The Quotient Rule   For any nonzero real number a and integers m and n,

$$\frac{a^m}{a^n} = a^{m-n}.$$

Examples: $\dfrac{6^4}{6^2} = 6^{4-2} = 6^2 = 36$ and $\dfrac{xy^3}{x^4y^2} = x^{1-4}y^{3-2} = x^{-3}y^1 = \dfrac{y}{x^3}$

Other Rules   For any nonzero real numbers a and b and positive integers m and n,

$$\frac{1}{a^{-n}} = a^n, \qquad \frac{a^{-n}}{b^{-m}} = \frac{b^m}{a^n}, \qquad \text{and} \qquad \left(\frac{a}{b}\right)^{-n} = \left(\frac{b}{a}\right)^n.$$

Examples: $\dfrac{1}{4^{-3}} = 4^3$, $\dfrac{x^{-3}}{y^{-2}} = \dfrac{y^2}{x^3}$,   and $\left(\dfrac{4}{5}\right)^{-2} = \left(\dfrac{5}{4}\right)^2$

Scientific Notation   A real number a written as $b \times 10^n$, where $1 \le |b| < 10$ and n is an integer

Examples: $2.34 \times 10^3 = 2340$ Move the decimal point 3 places to the right.

$2.34 \times 10^{-3} = 0.00234$ Move the decimal point 3 places to the left.

SECTION 5.6 ■ DIVISION OF POLYNOMIALS

Division of a Polynomial by a Monomial   Divide the monomial into *every* term of the polynomial.

Example: $\dfrac{5x^3 - 10x^2 + 15x}{5x} = \dfrac{5x^3}{5x} - \dfrac{10x^2}{5x} + \dfrac{15x}{5x} = x^2 - 2x + 3$

Division of a Polynomial by a Polynomial   Division of polynomials is performed similarly to long division of natural numbers.

Example: Divide $2x^3 + 4x^2 - 3x + 1$ by $x + 1$.

$$
\begin{array}{r}
2x^2 + 2x - 5 \\
x + 1 \overline{)\,2x^3 + 4x^2 - 3x + 1} \\
\underline{2x^3 + 2x^2 } \\
2x^2 - 3x \\
\underline{2x^2 + 2x} \\
-5x + 1 \\
\underline{-5x - 5} \\
6
\end{array}
$$

The quotient is $2x^2 + 2x - 5$ with remainder 6, which can be written as

$$2x^2 + 2x - 5 + \frac{6}{x + 1}.$$

CHAPTER 5   REVIEW EXERCISES

SECTION 5.1

Exercises 1–6: Evaluate the expression.

1. 5^3

2. -3^4

3. $4(-2)^0$

4. $3 + 3^2 - 3^0$

5. $\dfrac{-5^2}{5}$

6. $\left(\dfrac{-5}{5}\right)^2$

Exercises 7–22: Simplify the expression.

7. $6^2 \cdot 6^3$

8. $10^5 \cdot 10^7$

9. $z^4 \cdot z^5$

10. $y^2 \cdot y \cdot y^3$

11. $5x^2 \cdot 6x^7$

12. $(ab^3)(a^3b)$

13. $(2^5)^2$

14. $(m^4)^5$

15. $(ab)^3$

16. $(x^2y^3)^4$

17. $(xy)^3(x^2y^4)^2$

18. $(a^2b^9)^0$

19. $(r - t)^4(r - t)^5$

20. $(a + b)^2(a + b)^4$

21. $\left(\dfrac{3}{x - y}\right)^2$

22. $\left(\dfrac{x + y}{2}\right)^3$

SECTION 5.2

Exercises 23 and 24: Identify the degree and coefficient of the monomial.

23. $6x^7$

24. $-x^2y^3$

Exercises 25–28: Determine whether the expression is a polynomial. If it is, state how many terms and variables the polynomial contains. Then state its degree.

25. $8y$

26. $8x^3 - 3x^2 + x - 5$

27. $a^2 + 2ab + b^2$

28. $\dfrac{1}{xy}$

29. Add the polynomials vertically.

$$3x^2 + 4x + 8$$
$$\underline{2x^2 - 5x - 5}$$

30. Write the opposite of $6x^2 - 3x - 7$.

Exercises 31–36: Simplify.

31. $(4x - 3) + (-x + 7)$

32. $(3x^2 - 1) - (5x^2 + 12)$

33. $(x^2 + 5x + 6) - (3x^2 - 4x + 1)$

34. $(a^3 + 4a^2) + (a^3 - 5a^2 + 7a)$

35. $(xy + y^2) + (4y^2 - 4xy)$

36. $(7x^2 + 2xy + y^2) - (7x^2 - 2xy + y^2)$

SECTION 5.3

Exercises 37–50: Multiply and simplify.

37. $-x^2 \cdot x^3$

38. $-(r^2t^3)(rt)$

39. $-3(2t - 5)$

40. $2y(1 - 6y)$

41. $6x^3(3x^2 + 5x)$

42. $-x(x^2 - 2x + 9)$

43. $-ab(a^2 - 2ab + b^2)$

44. $(a - 2)(a + 5)$

45. $(8x - 3)(x + 2)$

46. $(2x - 1)(1 - x)$

47. $(y^2 + 1)(2y + 1)$

48. $(y^2 - 1)(2y^2 + 1)$

49. $(z + 1)(z^2 - z + 1)$

50. $(4z - 3)(z^2 - 3z + 1)$

Exercises 51 and 52: Multiply the expression
    (a) *geometrically and*
    (b) *symbolically.*

51. $z(z + 1)$

52. $2x(x + 2)$

SECTION 5.4

Exercises 53–68: Multiply.

53. $(z + 2)(z - 2)$

54. $(5z - 9)(5z + 9)$

55. $(1 - 3y)(1 + 3y)$

56. $(5x + 4y)(5x - 4y)$

57. $(rt + 1)(rt - 1)$

58. $(2m^2 - n^2)(2m^2 + n^2)$

59. $(x + 1)^2$

60. $(4x + 3)^2$

61. $(y - 3)^2$

62. $(2y - 5)^2$

63. $(4 + a)^2$

64. $(4 - a)^2$

65. $(x^2 + y^2)^2$

66. $(xy - 2)^2$

67. $(z + 5)^3$

68. $(2z - 1)^3$

Exercises 69 and 70: Use the product of sum and a difference to evaluate the expression.

69. $59 \cdot 61$

70. $22 \cdot 18$

SECTION 5.5

Exercises 71–76: Simplify the expression.

71. 9^{-1}

72. 3^{-2}

73. $4^3 \cdot 4^{-2}$

74. $10^{-6} \cdot 10^3$

75. $\dfrac{1}{6^{-2}}$

76. $\dfrac{5^7}{5^9}$

Exercises 77–92: Simplify the expression. Write the answer using positive exponents.

77. z^{-2}

78. y^{-4}

79. $a^{-4} \cdot a^2$

80. $x^2 \cdot x^{-5} \cdot x$

81. $(2t)^{-2}$

82. $(ab^2)^{-3}$

83. $(xy)^{-2}(x^{-2}y)^{-1}$

84. $\dfrac{x^6}{x^2}$

85. $\dfrac{4x}{2x^4}$

86. $\dfrac{20x^5y^3}{30xy^6}$

87. $\left(\dfrac{a}{b}\right)^5$

88. $\dfrac{4}{t^{-4}}$

89. $\left(\dfrac{x}{3}\right)^{-3}$

90. $\dfrac{2}{(ab)^{-1}}$

91. $\left(\dfrac{x}{y}\right)^{-2}$

92. $\left(\dfrac{3u}{2v}\right)^{-1}$

Exercises 93–96: Write the expression in standard form.

93. 6×10^2

94. 5.24×10^4

95. 3.7×10^{-3}

96. 6.234×10^{-2}

Exercises 97–100: Write the number in scientific notation.

97. $10,000$

98. $56,100,000$

99. 0.000054

100. 0.001

Exercises 101 and 102: Evaluate the expression. Write the result in standard form.

101. $(4 \times 10^2)(6 \times 10^4)$ **102.** $\dfrac{8 \times 10^3}{4 \times 10^4}$

SECTION 5.6

Exercises 103–110: Divide and check.

103. $\dfrac{5x^2 + 3x}{3x}$

104. $\dfrac{6b^4 - 4b^2 + 2}{2b^2}$

105. $\dfrac{3x^2 - x + 2}{x - 1}$

106. $\dfrac{9x^2 - 6x - 2}{3x + 2}$

107. $\dfrac{4x^3 - 11x^2 - 7x - 1}{4x + 1}$

108. $\dfrac{2x^3 - x^2 - 1}{2x - 1}$

109. $\dfrac{x^3 - x^2 - x + 1}{x^2 + 1}$

110. $\dfrac{x^4 + 3x^3 + 8x^2 + 7x + 5}{x^2 + x + 1}$

APPLICATIONS

111. *Heart Rate* An athlete starts running and continues for 10 seconds. The polynomial $t^2 + 60$ calculates the heart rate of the athlete in beats per minute t seconds after beginning the run, where $t \leq 10$.
   (a) What is the athlete's heart rate when the athlete first starts to run?
   (b) What is the athlete's heart rate after 10 seconds?
   (c) What happens to the athlete's heart rate while the athlete is running?

112. *Areas of Rectangles* Find a monomial equal to the sum of the areas of the rectangles. Calculate this sum for $x = 3$ feet and $y = 4$ feet.

113. *Area of a Rectangle* Write a polynomial that gives the area of the rectangle. Calculate its area for $z = 6$ inches.

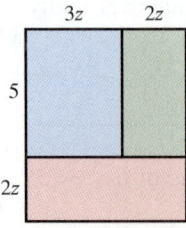

114. *Area of a Square* Find the area of the square whose sides have length x^2y.

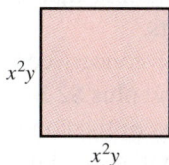

115. *Compound Interest* If P dollars are deposited in an account that pays 6% annual interest, then the amount of money after 3 years is given by $P(1 + 0.06)^3$. Find this amount when $P = \$700$.

116. *Volume of a Sphere* The expression for the volume of a sphere with radius r is $\frac{4}{3}\pi r^3$. Find a polynomial that gives the volume of a sphere with radius $x + 2$. Leave your answer in terms of π.

117. *Height Reached by a Baseball* A baseball is hit straight up. Its height h in feet above the ground after t seconds is given by $t(96 - 16t)$.
  (a) Multiply this expression.
  (b) Evaluate both the expression in part (a) and the given expression for $t = 2$. Interpret the result.

118. *Rectangular Building* A rectangular building has a perimeter of 1200 feet.
  (a) If one side of the building has length L, write a polynomial expression that gives its area. (Be sure to multiply your expression.)
  (b) Evaluate the expression in part (a) for $L = 50$ and interpret the answer.

119. *Geometry* Complete each part and verify that your answers are equal.

  (a) Find the area of the large square by multiplying its length and width.
  (b) Find the sum of the areas of the smaller rectangles inside the large square.

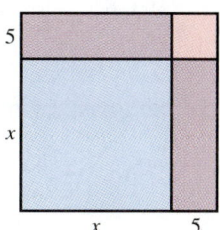

120. *Digital Picture* A digital picture, including its border, is $x + 4$ pixels by $x + 4$ pixels, and the actual picture inside the border is $x - 4$ pixels by $x - 4$ pixels.

  (a) Find a polynomial that gives the number of pixels in the border.
  (b) Let $x = 100$ and evaluate the polynomial.

121. *Federal Debt* In 1990, the federal debt held by the public was $2.19 trillion, and the population of the United States was 249 million. Use scientific notation to approximate the national debt per person. (*Source:* U.S. Department of the Treasury.)

122. *Alcohol Consumption* In 2004, about 233 million people in the United States were aged 14 or older. They consumed, on average, 2.23 gallons of alcohol per person. Use scientific notation to estimate the total number of gallons of alcohol consumed by this age group. (*Source:* Department of Health and Human Services.)

CHAPTER 5 TEST 📹 Pass the Test Video solutions to all test exercises

1. Is $5x^2 - 3xy - 7y^3$ a polynomial? If it is, state how many terms and variables the polynomial contains. Then state its degree.

2. Write the opposite of $-x^3 + 4x - 8$.

Exercises 3–6: Simplify.

3. $(-3x + 4) + (7x + 2)$

4. $(y^3 - 2y + 6) - (4y^3 + 5)$

5. $(5x^2 - x + 3) - (4x^2 - 2x + 10)$

6. $(a^3 + 5ab) + (3a^3 - 3ab)$

7. Evaluate each expression by hand.
   (a) $-4^2 + 10$ **(b)** 8^{-2} **(c)** $\dfrac{1}{2^{-3}}$
   (d) $-3x^0$

Exercises 8–15: Write the given expression with positive exponents.

8. $6y^4 \cdot 4y^7$

9. $(a^2b^3)^2(ab^2)$

10. $x^7 \cdot x^{-3}$

11. $(a^{-1}b^2)^{-3}$

12. $ab(a^2 - b^2)$

13. $\left(\dfrac{3a^2}{2b^{-3}}\right)^{-2}$

14. $\dfrac{12xy^4}{6x^2y}$

15. $\left(\dfrac{2}{a+b}\right)^4$

Exercises 16–21: Multiply and simplify.

16. $3x^2(4x^3 - 6x + 1)$ **17.** $(z - 3)(2z + 4)$

18. $(7y^2 - 3)(7y^2 + 3)$ **19.** $(3x - 2)^2$

20. $(m + 3)^3$

21. $(y + 2)(y^2 - 2y + 3)$

22. Evaluate $78 \cdot 82$ using the product of a sum and a difference.

23. Write 6.1×10^{-3} in standard form.

24. Write 5410 in scientific notation.

Exercises 25 and 26: Divide.

25. $\dfrac{9x^3 - 6x^2 + 3x}{3x^2}$ **26.** $\dfrac{x^3 + x^2 - x + 1}{x + 2}$

27. *Concert Tickets*   Tickets for a concert are sold for $20 each.
   (a) Write a polynomial that gives the revenue from selling t tickets.
   (b) Putting on the concert costs management $2000 to hire the band plus $2 for each ticket sold. What is the total cost of the concert if t tickets are sold?
   (c) Subtract the polynomial that you found in part (b) from the polynomial that you found in part (a). What does this polynomial represent?

28. *Areas of Rectangles*   Find a polynomial representing the sum of the areas of two identical rectangles that have width $2x$ and length $3x$. Calculate this sum for $x = 10$ feet.

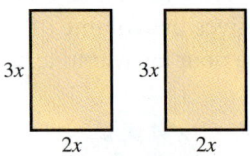

29. *Volume of a Box*   Write a polynomial that represents the volume of the box. Be sure to multiply your answer completely.

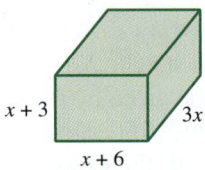

30. *Height Reached by a Golf Ball*   When a golf ball is hit into the air its height in feet above the ground after t seconds is given by $t(88 - 16t)$.
   (a) Multiply this expression.
   (b) Evaluate the expression in part (a) for $t = 3$. Interpret the result.

CHAPTER 5   EXTENDED AND DISCOVERY EXERCISES

Exercises 1–6: Arithmetic and Scientific Notation   The product $(4 \times 10^3) \times (2 \times 10^2)$ can be evaluated as

$$(4 \times 2) \times (10^3 \times 10^2) = 8 \times 10^5,$$

and the quotient $(4 \times 10^3) \div (2 \times 10^2)$ can be evaluated as

$$\frac{4 \times 10^3}{2 \times 10^2} = \frac{4}{2} \times \frac{10^3}{10^2} = 2 \times 10^1.$$

How would you evaluate $(4 \times 10^3) + (2 \times 10^2)$? How would you evaluate $(4 \times 10^3) - (2 \times 10^2)$? Make a conjecture as to how numbers in scientific notation should be added and subtracted. Try your method on these problems and then check your answers with a calculator set in scientific mode. Does your method work?

1. $(4 \times 10^3) + (3 \times 10^3)$

2. $(5 \times 10^{-2}) - (2 \times 10^{-2})$

3. $(1.2 \times 10^4) - (3 \times 10^3)$

4. $(2 \times 10^2) + (6 \times 10^1)$

5. $(2 \times 10^{-1}) + (4 \times 10^{-2})$

6. $(2 \times 10^{-3}) - (5 \times 10^{-2})$

*Exercises 7 and 8: **Constructing a Box**   A box is constructed from a rectangular piece of metal by cutting squares from the corners and folding up the sides. The square, cutout corners are x inches by x inches.*

7. Suppose that the dimensions of the metal piece are 20 inches by 30 inches.
   (a) Write a polynomial that gives the volume of the box.
   (b) Find the volume of the box for $x = 4$ inches.

8. Suppose that the metal piece is square with sides of length 25 inches.
   (a) Write a polynomial expression that gives the outside surface area of the box. (Assume that the box does not have a top.)
   (b) Find this area for $x = 3$ inches.

*Exercises 9–12: **Calculators and Polynomials**   A graphing calculator can be used to help determine whether two polynomial expressions in one variable are equal. For example, suppose that a student believes that $(x + 2)^2$ and $x^2 + 4$ are equal. Then the first two calculator tables shown demonstrate that the two expressions are not equal except for $x = 0$.*

X	Y1
-3	1
-2	0
-1	1
0	4
1	9
2	16
3	25

Y1=(X+2)²

X	Y1
-3	13
-2	8
-1	5
0	4
1	5
2	8
3	13

Y1=X²+4

The next two calculator tables support the fact that $(x + 1)^2$ and $x^2 + 2x + 1$ are equal for all x.

X	Y1
-3	4
-2	1
-1	0
0	1
1	4
2	9
3	16

Y1=(X+1)²

X	Y1
-3	4
-2	1
-1	0
0	1
1	4
2	9
3	16

Y1=X²+2X+1

Use a graphing calculator to determine whether the first expression is equal to the second expression. If the expressions are not equal, multiply the first expression and simplify it.

9. $3x(4 - 5x)$, $12x - 5x$

10. $(x - 1)^2$, $x^2 - 1$

11. $(x - 1)(x^2 + x + 1)$, $x^3 - 1$

12. $(x - 2)^3$, $x^3 - 8$

CHAPTERS 1–5   CUMULATIVE REVIEW EXERCISES

Exercises 1 and 2: Evaluate each expression by hand.

1. **(a)** $18 - 2 \cdot 5$ **(b)** $42 \div 7 + 2$

2. **(a)** $21 - (-8)$ **(b)** $-\frac{7}{3} \div \left(-\frac{14}{9}\right)$

Exercises 3 and 4: Solve the equation. Note that these equations may have no solutions, one solution, or infinitely many solutions.

3. **(a)** $(x - 3) + x = 4 + x$
   (b) $2(5x - 4) = 1 + 10x$

4. **(a)** $2 + 6x = 2(3x + 1)$
   (b) $11x - 9 = -31$

5. Find the average speed of a car that travels 306 miles in 4 hours 30 minutes.

6. Find three consecutive numbers whose sum is -114.

7. Write each value as a fraction in lowest terms.
   (a) 42%   **(b)** 0.076

8. Solve the formula $Wx = 4x + Y$ for x.

9. Graph the equation $4x - 5y = 20$.

10. Sketch a line with slope $-\frac{2}{3}$ that passes through the point $(1, 1)$.

11. Write the slope–intercept form for the line shown.

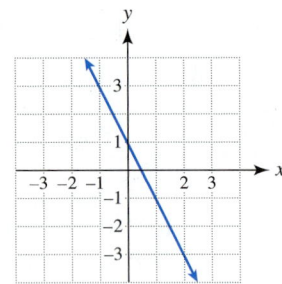

12. Find the x- and y-intercepts for the graph of the equation $2y = 3x - 6$.

Exercises 13 and 14: Write the slope–intercept form of a line that satisfies the given information.

13. Parallel to $3x - 6y = 7$, passing through $(2, -3)$

14. Passing through $(-2, -5)$ and $(1, 4)$

Exercises 15–18: Solve the system of equations. Note that these systems may have no solutions, one solution, or infinitely many solutions.

15. $4x + 3y = -6$
$8x + 6y = 12$

16. $x - 3y = 5$
$3x + y = 5$

17. $x + 4y = -8$
$-3x - 12y = 24$

18. $x - 5y = 30$
$2x + y = -6$

Exercises 19 and 20: Shade the solution set for the system of inequalities.

19. $x + y < 3$
$y \geq x + 2$

20. $x - 2y > 4$
$3x + y < 6$

21. Simplify the expression.
  (a) $3x^2 \cdot 5x^3$ **(b)** $(x^3y)^2(x^4y^5)$

22. Simplify.
  (a) $(5x^2 - 3x + 4) - (3x^2 - 2x + 1)$
  (b) $(7a^3 - 4a^2 - 5) + (5a^3 + 4a^2 + a)$

23. Multiply and simplify.
  (a) $(2x + 3)(x - 7)$ **(b)** $(y + 3)(y^2 - 3y - 1)$
  (c) $(4x + 7)(4x - 7)$ **(d)** $(5a + 3)^2$

24. Simplify the expression. Write the answer using positive exponents.
  (a) $x^{-5} \cdot x^3 \cdot x$ **(b)** $\left(\dfrac{2}{x^3}\right)^{-3}$
  (c) $\dfrac{3x^2y^{-1}}{6x^{-2}y}$ **(d)** $(xy^{-2})^3(x^{-2}y)^{-2}$

25. Write $24{,}000{,}000{,}000$ in scientific notation.

26. Write 4.71×10^{-7} in standard form.

27. Divide.
  (a) $\dfrac{8x^3 - 2x}{2x}$ **(b)** $\dfrac{2x^2 + x - 14}{x + 3}$

28. *Price Decrease*   If the price of a computer is reduced from \$1200 to \$900, find the percent change.

29. *Mixing an Acid Solution*   How many milliliters of a 3% acid solution should be added to 400 milliliters of a 6% acid solution to dilute it to a 5% acid solution?

30. *Surface Area of a Box*   Use the drawing of the box to write a polynomial that represents the area of each of the following.

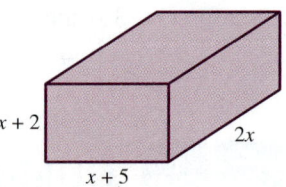

  (a) The bottom
  (b) The front
  (c) The right side
  (d) All six sides

CHAPTER **6**

Factoring Polynomials and Solving Equations

Have you ever noticed that cars are designed so that their exteriors are curved and smooth? This characteristic is especially true for solar cars. In fact, a side view of a solar car often resembles the cross section of an airplane wing, as illustrated in the accompanying figure. The reason for this design is to reduce drag from air resistance. Reducing drag increases fuel efficiency and gives the car a smoother ride in the wind.

Views of a Solar Car and a Standard Car

Mathematics plays an important role in the design of cars. Cubic polynomials called *cubic splines* are used extensively by engineers to obtain a smooth shape for new cars. Although the topic of cubic splines is covered in advanced mathematics and engineering courses, in this chapter we introduce many concepts necessary for understanding polynomials and polynomial equations, which are used in the design of cars.

What we see depends mainly on what we look for.

—JOHN LUBBOCK

Source: R. Burden and J. Faires, *Numerical Analysis.*

6.1 INTRODUCTION TO FACTORING

Common Factors ■ Factoring by Grouping

Polynomials are frequently used in applications to approximate data and to model such things as changes in the weather, sales of a new video game, and the growth of young children. As a result, scientists and mathematicians commonly solve equations involving polynomials. One way to solve these equations is to use **factoring**. When factoring a polynomial, we usually write it as a *product* of lower degree polynomials. For example, because $x(x-1) = x^2 - x$, we say that the polynomial $x^2 - x$ can be *factored* as $x(x-1)$. Note that $x^2 - x$ is a polynomial of degree 2, whereas x and $x - 1$ are both degree 1 polynomials. In this section we introduce two basic methods of factoring polynomials.

Common Factors

When factoring a polynomial, we first look for factors that are common to each term. By applying a distributive property we can often write a polynomial as a product. For example, each term in the polynomial $8x^2 + 6x$ has a factor of $2x$ because

$$8x^2 = \mathbf{2x} \cdot \mathbf{4x} \quad \text{and} \quad 6x = \mathbf{2x} \cdot \mathbf{3}.$$

Therefore by the distributive property,

$$8x^2 + 6x = \mathbf{2x}(\mathbf{4x} + \mathbf{3}).$$

Thus the product $2x(4x + 3)$ equals $8x^2 + 6x$. We check this result by multiplying.

$$2x(4x + 3) = 2x \cdot 4x + 2x \cdot 3$$
$$= 8x^2 + 6x \qquad \text{It checks.}$$

This factorization is shown visually in Figure 6.1, where possible dimensions for a rectangle with an area of $8x^2 + 6x$ are $2x$ by $4x + 3$.

Figure 6.1 $8x^2 + 6x = 2x(4x + 3)$

The expressions $8x^2 + 6x$ and $2x(4x + 3)$ are equal for all values of x. Figure 6.2 also illustrates this fact with a partial table of values for each expression.

X	Y1
-3	54
-2	20
-1	2
0	0
1	14
2	44
3	90

Y1≣8X^2+6X

(a)

X	Y1
-3	54
-2	20
-1	2
0	0
1	14
2	44
3	90

Y1≣2X(4X+3)

(b)

Figure 6.2 $8x^2 + 6x = 2x(4x + 3)$

EXAMPLE 1 **Factoring an expression**

Factor the expression and sketch a rectangle that illustrates the factorization.
(a) $10x + 6$ **(b)** $6x^2 + 15x$

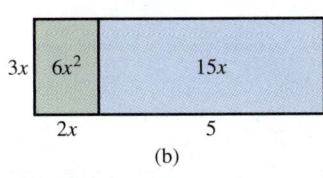

Figure 6.3

Solution
(a) Each term in the polynomial $10x + 6$ has a factor of 2 because

$$10x = 2 \cdot 5x \quad \text{and} \quad 6 = 2 \cdot 3.$$

By the distributive property, $10x + 6 = 2(5x + 3)$. This factorization is illustrated visually in Figure 6.3(a).

(b) Each term in the polynomial $6x^2 + 15x$ has a factor of $3x$ because

$$6x^2 = 3x \cdot 2x \quad \text{and} \quad 15x = 3x \cdot 5.$$

By the distributive property, $6x^2 + 15x = 3x(2x + 5)$. This factorization is illustrated visually in Figure 6.3(b).   **Now Try Exercises 9, 11**

EXAMPLE 2 **Finding common factors**

Factor.
(a) $15x^2 + 10x$ **(b)** $6y^3 - 2y^2$ **(c)** $3z^3 + 9z^2 - 6z$ **(d)** $2x^2y^2 + 4xy^3$

Solution
(a) In the expression $15x^2 + 10x$, the terms $15x^2$ and $10x$ both contain a common factor of $5x$ because

$$15x^2 = 5x \cdot 3x \quad \text{and} \quad 10x = 5x \cdot 2.$$

Therefore this polynomial can be factored as

$$15x^2 + 10x = 5x(3x + 2).$$

(b) In the expression $6y^3 - 2y^2$, the terms $6y^3$ and $2y^2$ both contain a common factor of $2y^2$ because

$$6y^3 = 2y^2 \cdot 3y \quad \text{and} \quad 2y^2 = 2y^2 \cdot 1.$$

Therefore this polynomial can be factored as

$$6y^3 - 2y^2 = 2y^2(3y - 1).$$

(c) In the expression $3z^3 + 9z^2 - 6z$, the terms $3z^3$, $9z^2$, and $6z$ all contain a common factor of $3z$ because

$$3z^3 = 3z \cdot z^2, \quad 9z^2 = 3z \cdot 3z, \quad \text{and} \quad 6z = 3z \cdot 2.$$

Therefore this polynomial can be factored as

$$3z^3 + 9z^2 - 6z = 3z(z^2 + 3z - 2).$$

(d) In the expression $2x^2y^2 + 4xy^3$, the terms $2x^2y^2$ and $4xy^3$ both contain a common factor of $2xy^2$ because

$$2x^2y^2 = 2xy^2 \cdot x \quad \text{and} \quad 4xy^3 = 2xy^2 \cdot 2y.$$

Thus $2x^2y^2 + 4xy^3 = 2xy^2(x + 2y)$.   **Now Try Exercises 13, 15, 17, 19**

MAKING CONNECTIONS

Checking Common Factors with Multiplication

When factoring we can check our work by multiplying. For example, if we are uncertain whether the equation

$$6y^3 - 2y^2 = 2y^2(3y - 1)$$

is correct, we can apply the distributive property to the right side of the above equation to obtain

$$2y^2(3y - 1) = 2y^2 \cdot 3y - 2y^2 \cdot 1$$
$$= 6y^3 - 2y^2. \qquad \text{It checks.}$$

In most situations we factor out the *greatest common factor* (GCF). For example, the polynomial $12b^3 + 8b^2$ has a common factor of $2b$. We could factor this polynomial as

$$12b^3 + 8b^2 = 2b(6b^2 + 4b).$$

However, we can factor out $4b^2$ instead.

$$12b^3 + 8b^2 = 4b^2(3b + 2)$$

Because $4b^2$ is the common factor with the greatest (integer) coefficient and highest degree, we say that $4b^2$ is the **greatest common factor** of $12b^3 + 8b^2$. In Examples 1 and 2 we factored out the greatest common factor for each expression.

When finding the greatest common factor for a polynomial, it is often helpful first to completely factor each term of the polynomial. To completely factor a term, write its coefficient as the product of prime numbers and write any powers of variables as repeated multiplication. For example, the complete factorization of $24x^4$ is $2 \cdot 2 \cdot 2 \cdot 3 \cdot x \cdot x \cdot x \cdot x$ and the complete factorization of $18xy^3$ is $2 \cdot 3 \cdot 3 \cdot x \cdot y \cdot y \cdot y$. The next example shows how to find the greatest common factor for a polynomial by using the complete factorization of each term.

EXAMPLE 3   Finding the greatest common factor

Find the greatest common factor for each expression. Then factor the expression.
(a) $9x^2 + 6x$ **(b)** $4z^4 + 8z^2$ **(c)** $8a^2b^3 - 16a^3b^2$

Solution
(a) Because

$$9x^2 = 3 \cdot 3 \cdot x \cdot x \quad \text{and}$$
$$6x = 3 \cdot 2 \cdot x,$$

both terms have common factors of **3** and **x**. The GCF is the product of these two factors, or $3 \cdot x = \mathbf{3x}$. Thus the expression $9x^2 + 6x$ can be factored as $\mathbf{3x}(3x + 2)$. Note that the term $3x$ inside the parentheses is the product of the (black) factors of $9x^2$ that were not part of the greatest common factor and the term 2 inside the parentheses is the product of the (black) factors of $6x$ that were not part of the greatest common factor.

(b) Because

$$4z^4 = \mathbf{2} \cdot \mathbf{2} \cdot \mathbf{z} \cdot z \cdot z \cdot z \quad \text{and}$$
$$8z^2 = \mathbf{2} \cdot \mathbf{2} \cdot 2 \cdot \mathbf{z} \cdot \mathbf{z},$$

both terms have common factors of $\mathbf{2}, \mathbf{2}, \mathbf{z}$, and $\mathbf{z}$. The GCF is the product of these four factors, or $2 \cdot 2 \cdot z \cdot z = \mathbf{4z^2}$. Thus the expression $4z^4 + 8z^2$ can be factored as $\mathbf{4z^2}(z^2 + 2)$.

(c) Because

$$8a^2b^3 = \mathbf{2} \cdot \mathbf{2} \cdot \mathbf{2} \cdot \mathbf{a} \cdot \mathbf{a} \cdot \mathbf{b} \cdot \mathbf{b} \cdot b \quad \text{and}$$
$$16a^3b^2 = \mathbf{2} \cdot \mathbf{2} \cdot \mathbf{2} \cdot 2 \cdot \mathbf{a} \cdot \mathbf{a} \cdot a \cdot \mathbf{b} \cdot \mathbf{b},$$

both terms have common factors of $\mathbf{2}, \mathbf{2}, \mathbf{2}, \mathbf{a}, \mathbf{a}, \mathbf{b}$, and $\mathbf{b}$. The GCF is the product of these seven factors, or $2 \cdot 2 \cdot 2 \cdot a \cdot a \cdot b \cdot b = \mathbf{8a^2b^2}$. Here $8a^2b^3 - 16a^3b^2$ can be factored as $\mathbf{8a^2b^2}(b - 2a)$.   Now Try Exercises 21, 23, 35

NOTE: With practice, you may find that you can determine the GCF mentally without factoring each term as was done in Example 3.

In the next example, we factor an expression that occurs in a scientific application.

EXAMPLE **4**   **Modeling the flight of a golf ball**

If a golf ball is hit upward at 66 feet per second (45 miles per hour), then its height in feet after t seconds is approximated by $66t - 16t^2$. Factor this expression.

Solution
The GCF for $66t$ and $16t^2$ is $2t$ because

$$66t = \mathbf{2t} \cdot \mathbf{33} \quad \text{and} \quad 16t^2 = \mathbf{2t} \cdot \mathbf{8t}.$$

Therefore this polynomial can be factored as

$$66t - 16t^2 = \mathbf{2t(33 - 8t)}. \quad\quad \text{Now Try Exercise } 77$$

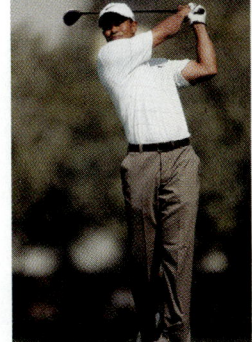

Factoring by Grouping

Factoring by grouping is a technique that makes use of the associative and distributive properties. The next example illustrates one step in this factoring technique.

EXAMPLE **5**   **Factoring out binomials**

Factor.
(a) $5x(x + 3) + 6(x + 3)$ **(b)** $x^2(2x - 5) - 4x(2x - 5)$

Solution
(a) Each term in the expression $5x(x + 3) + 6(x + 3)$ contains the binomial $(x + 3)$. Therefore the distributive property can be used to factor this expression.

$$\mathbf{5x(x + 3) + 6(x + 3)} = \mathbf{(5x + 6)(x + 3)}$$

(b) Each term in $x^2(2x - 5) - 4x(2x - 5)$ contains the binomial $(2x - 5)$. Therefore the distributive property can be used to factor this expression.

$$x^2(2x - 5) - 4x(2x - 5) = (x^2 - 4x)(2x - 5)$$
$$= x(x - 4)(2x - 5)$$

Now Try Exercises 39, 43

Now consider the polynomial

$$x^3 + x^2 + 2x + 2.$$

We can factor this polynomial by first grouping it into two binomials.

$(x^3 + x^2) + (2x + 2)$	Associative property
$x^2(x + 1) + 2(x + 1)$	Factor out common factors.
$(x^2 + 2)(x + 1)$	Factor out $(x + 1)$.

When factoring by grouping, we factor out a common factor more than once.

The first step in factoring a four-term polynomial by grouping requires the use of the associative property to write the polynomial as the *sum* of two binomials. However, this property must be applied carefully to avoid sign errors. The terms of a polynomial are separated by addition $(+)$ or subtraction $(-)$ symbols. There are three of these symbols in a four-term polynomial. The next two examples illustrate that the middle arithmetic symbol in a four-term polynomial determines how the associative property is applied.

EXAMPLE 6 **Factoring by grouping when the middle symbol is (+)**

Factor each polynomial.
(a) $2x^3 - 4x^2 + 3x - 6$ **(b)** $3x + 3y + ax + ay$

Solution
(a) Use the associative property to write the polynomial as the *sum* of two binomials.

$2x^3 - 4x^2 + 3x - 6 = (2x^3 - 4x^2) + (3x - 6)$	Associative property
$= 2x^2(x - 2) + 3(x - 2)$	Factor out common factors.
$= (2x^2 + 3)(x - 2)$	Factor out $(x - 2)$.

(b) Group the polynomial into the *sum* of two binomials.

$3x + 3y + ax + ay = (3x + 3y) + (ax + ay)$	Associative property
$= 3(x + y) + a(x + y)$	Factor out common factors.
$= (3 + a)(x + y)$	Factor out $(x + y)$.

Now Try Exercises 45, 63

EXAMPLE 7 **Factoring by grouping when the middle symbol is (−)**

Factor each polynomial.
(a) $3y^3 - y^2 - 9y + 3$ **(b)** $z^3 + 4z^2 - 5z - 20$

Solution

(a) Begin by changing the middle subtraction to addition by adding the opposite of $9y$. Then apply the associative property to write the result as the *sum* of two binomials.

$$\begin{aligned}
3y^3 - y^2 - 9y + 3 &= 3y^3 - y^2 + (-9y) + 3 & &\text{Add the opposite of } 9y. \\
&= (3y^3 - y^2) + (-9y + 3) & &\text{Associative property} \\
&= y^2(3y - 1) - 3(3y - 1) & &\text{Factor out common factors.} \\
&= (y^2 - 3)(3y - 1) & &\text{Factor out } (3y - 1).
\end{aligned}$$

Note that in the third step, -3 was factored from the second binomial.

(b) Begin by changing the middle subtraction to addition by adding the opposite of $5z$. Then apply the associative property to write the result as the *sum* of two binomials.

$$\begin{aligned}
z^3 + 4z^2 - 5z - 20 &= z^3 + 4z^2 + (-5z) - 20 & &\text{Add the opposite of } 5z. \\
&= (z^3 + 4z^2) + (-5z - 20) & &\text{Associative property} \\
&= z^2(z + 4) - 5(z + 4) & &\text{Factor out common factors.} \\
&= (z^2 - 5)(z + 4) & &\text{Factor out } (z + 4).
\end{aligned}$$

Now Try Exercises 53, 59

When factoring some polynomials, it may be necessary to factor out the greatest common factor before completing other factoring techniques such as factoring by grouping. In the next example the GCF is factored out before grouping is applied.

EXAMPLE 8 **Factoring out the GCF before grouping**

Completely factor each polynomial.
(a) $6x^3 - 12x^2 - 3x + 6$ **(b)** $2x^5 - 8x^4 + 6x^3 - 24x^2$

Solution

(a) The GCF of $6x^3 - 12x^2 - 3x + 6$ is 3, so factor out 3 before factoring the remaining polynomial by grouping.

$$\begin{aligned}
6x^3 - 12x^2 - 3x + 6 &= 3(2x^3 - 4x^2 - x + 2) & &\text{Factor out the GCF.} \\
&= 3\big(2x^3 - 4x^2 + (-x) + 2\big) & &\text{Add the opposite of } x. \\
&= 3\big((2x^3 - 4x^2) + (-x + 2)\big) & &\text{Associative property} \\
&= 3\big(2x^2(x - 2) - 1(x - 2)\big) & &\text{Factor out common factors.} \\
&= 3(2x^2 - 1)(x - 2) & &\text{Factor out } (x - 2).
\end{aligned}$$

(b) The GCF of $2x^5 - 8x^4 + 6x^3 - 24x^2$ is $2x^2$, so factor out $2x^2$ before factoring the remaining polynomial by grouping.

$$\begin{aligned}
2x^5 - 8x^4 + 6x^3 - 24x^2 &= 2x^2(x^3 - 4x^2 + 3x - 12) & &\text{Factor out the GCF.} \\
&= 2x^2\big((x^3 - 4x^2) + (3x - 12)\big) & &\text{Associative property} \\
&= 2x^2\big(x^2(x - 4) + 3(x - 4)\big) & &\text{Factor out common factors.} \\
&= 2x^2(x^2 + 3)(x - 4) & &\text{Factor out } (x - 4).
\end{aligned}$$

Now Try Exercises 65, 69

6.1 PUTTING IT ALL TOGETHER

In this section we introduced basic concepts used to factor polynomials. They are summarized in the following table.

Concept	Explanation	Examples
Common Factor	Factor out a monomial common to each term in a polynomial.	$6z^2 - 6z = 6z(z - 1)$ $4y^3 - 6y^2 = 2y^2(2y - 3)$ $5x^3 - 10x^2 + 15x = 5x(x^2 - 2x + 3)$ $2a^3b^3 - 4a^2b^3 = 2a^2b^3(a - 2)$
Greatest Common Factor (GCF)	The common factor with the greatest (integer) coefficient and highest degree	The GCF of $10x^4 + 15x^2$ is $5x^2$. Common factors include $1, 5, x, 5x, x^2$, and $5x^2$. However, $5x^2$ is *the greatest common factor*.
Factoring by Grouping	Factoring by grouping is a method that can be used to factor *four terms* into a product of two binomials. It makes use of the associative and distributive properties.	$2x^3 + 3x^2 + 2x + 3$ $\quad = (2x^3 + 3x^2) + (2x + 3)$ $\quad = x^2(2x + 3) + 1(2x + 3)$ $\quad = (x^2 + 1)(2x + 3)$ $4x^3 - 24x^2 - 3x + 18$ $\quad = 4x^3 - 24x^2 + (-3x) + 18$ $\quad = (4x^3 - 24x^2) + (-3x + 18)$ $\quad = 4x^2(x - 6) - 3(x - 6)$ $\quad = (4x^2 - 3)(x - 6)$

6.1 Exercises

CONCEPTS

1. When you write a polynomial as a product of two or more polynomials, it is called _____.

2. A common factor in the expression $ab + ac$ is _____.

3. When factoring, we can check our work by _____.

4. To completely factor a term, write the coefficient as the product of _____ numbers and write any powers of variables as repeated _____.

5. The _____ of a polynomial is the common factor with the greatest coefficient and highest degree.

6. Factoring by _____ is a method that can be used to factor four terms into a product of two binomials by using the associative and distributive properties.

7. Identify four common factors of $2x^2 + 4x$.

8. Identify the greatest common factor (GCF) of the expression $2x^2 + 4x$.

COMMON FACTORS

Exercises 9–12: Factor the expression. Then make a sketch of a rectangle that illustrates this factorization.

9. $2x + 4$

10. $6 + 3x$

11. $z^2 + 4z$

12. $2z^2 + 10z$

Exercises 13–20: Factor the expression.

13. $3x^2 + 9x$

14. $10y^2 + 2y$

15. $4y^3 - 2y^2$

16. $6x^4 + 9x^2$

17. $2z^3 + 8z^2 - 4z$

18. $5x^4 - 15x^3 - 10x^2$

19. $6x^2y - 3xy^2$

20. $7x^3y^3 + 14x^2y^2$

Exercises 21–38: Identify the greatest common factor. Then factor the expression.

21. $6x - 18x^2$

22. $16x^2 - 24x^3$

23. $8y^3 - 12y^2$

24. $12y^3 - 8y^2 + 4y$

25. $6z^3 + 3z^2 + 9z$

26. $16z^3 - 24z^2 - 36z$

27. $x^4 - 5x^3 - 4x^2$

28. $2x^4 + 8x^2$

29. $5y^5 + 10y^4 - 15y^3 + 10y^2$

30. $7y^4 - 14y^3 - 21y^2 + 7y$

31. $xy + xz$

32. $ab - bc$

33. $ab^2 - a^2b$

34. $4x^2y + 6xy^2$

35. $5x^2y^4 + 10x^3y^3$

36. $3r^3t^3 - 6r^4t^2$

37. $a^2b + ab^2 + ab$

38. $6ab^2 - 9ab + 12b^2$

FACTORING BY GROUPING

Exercises 39–44: Factor.

39. $x(x + 1) - 2(x + 1)$

40. $5x(3x - 2) + 2(3x - 2)$

41. $(z + 5)z + (z + 5)4$

42. $3y^2(y - 2) + 5(y - 2)$

43. $4x^3(x - 5) - 2x(x - 5)$

44. $8x^2(x + 3) + (x + 3)$

Exercises 45–64: Factor by grouping.

45. $x^3 + 2x^2 + 3x + 6$

46. $x^3 + 6x^2 + x + 6$

47. $2y^3 + y^2 + 2y + 1$

48. $4y^3 + 10y^2 + 2y + 5$

49. $2z^3 - 6z^2 + 5z - 15$

50. $15z^3 - 5z^2 + 6z - 2$

51. $4t^3 - 20t^2 + 3t - 15$

52. $4t^3 - 12t^2 + 3t - 9$

53. $9r^3 + 6r^2 - 6r - 4$

54. $3r^3 + 12r^2 - 2r - 8$

55. $7x^3 + 21x^2 - 2x - 6$

56. $6x^3 + 3x^2 - 10x - 5$

57. $2y^3 - 7y^2 - 4y + 14$

58. $y^3 - 5y^2 - 3y + 15$

59. $z^3 - 4z^2 - 7z + 28$

60. $12z^3 - 18z^2 - 10z + 15$

61. $2x^4 - 3x^3 + 4x - 6$

62. $x^4 + x^3 + 5x + 5$

63. $ax + bx + ay + by$

64. $ax - bx + ay - by$

Exercises 65–74: Completely factor the polynomial.

65. $3x^3 + 6x^2 + 3x + 6$

66. $5x^3 - 5x^2 + 5x - 5$

67. $6y^4 - 24y^3 - 2y^2 + 8y$

68. $6x^4 - 12x^3 + 3x^2 - 6x$

69. $x^5 + 2x^4 - 3x^3 - 6x^2$

70. $y^6 + 3y^5 - 2y^4 - 6y^3$

71. $4x^5 + 2x^4 - 12x^3 - 6x^2$

72. $18y^5 + 27y^4 + 12y^3 + 18y^2$

73. $x^3y + x^2y^2 - 2x^2y - 2xy^2$

74. $6x^3y - 3x^2y^2 + 18x^2y - 9xy^2$

75. Thinking Generally   Factor a from $ax^2 + bx + c$.

76. Thinking Generally   Factor c from $ax^2 + bx + c$.

APPLICATIONS

77. *Flight of a Golf Ball*   The height of a golf ball in feet after t seconds is given by $80t - 16t^2$.
   (a) Identify the greatest common factor.
   (b) Factor this expression.

78. *Flight of a Golf Ball*   Repeat the previous exercise if the height of a golf ball in feet after t seconds is given by $128t - 16t^2$.

79. *Volume of a Box*   A box is constructed by cutting out square corners of a rectangular piece of cardboard and folding up the sides. If the cutout corners have sides with length x, then the volume of the box is given by the polynomial $4x^3 - 60x^2 + 200x$.

(a) Find the volume of the box when $x = 3$ inches.
(b) Factor out the greatest common factor for this expression.

80. *Volume of a Box*   (Refer to the preceding exercise.) A box is constructed from a square piece of metal that is 20 inches on a side.
(a) If the square corners of length x are cut out, write a polynomial that gives the volume of the box.
(b) Evaluate the polynomial when $x = 4$ inches.
(c) Factor out the greatest common factor for this polynomial expression.

WRITING ABOUT MATHEMATICS

81. Use an example to explain the difference between a common factor and the greatest common factor.

82. Use an example to explain how to factor a polynomial by grouping. What two properties of real numbers did you use?

6.2 FACTORING TRINOMIALS I ($x^2 + bx + c$)

Review of the FOIL Method ■ Factoring Trinomials Having a Leading Coefficient of 1

A LOOK INTO MATH ▷   In Section 6.1 we discussed two types of factoring: common factors and grouping. In this section we introduce methods for factoring certain types of trinomials. These techniques are frequently used in mathematics classes that you might take in the future. We use factoring later in this chapter to solve equations.

Review of the FOIL Method

A **trinomial** is a polynomial that has three terms. We begin by reviewing products of binomials that result in trinomials.

$$(x + 2)(x + 3) = x \cdot x + x \cdot 3 + 2 \cdot x + 2 \cdot 3$$
$$= x^2 + 5x + 6$$

Note that the first term, x^2, in the trinomial results from multiplying the *first* terms of each binomial. The middle term, $5x$, results from adding the product of the *outside* terms and the product of the *inside* terms. Finally the last term, 6, results from multiplying the *last* terms of each binomial. We discussed this method of multiplying binomials, called FOIL, in Section 5.3 and illustrate it as follows.

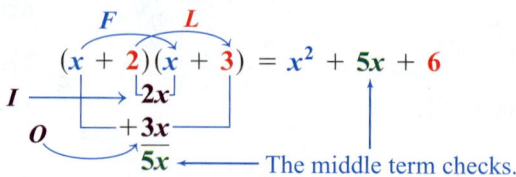

Factoring Trinomials Having a Leading Coefficient of 1

Any trinomial of degree 2 in the variable x can be written in *standard form* as $ax^2 + bx + c$, where a, b, and c are constants. The constant a is called the **leading coefficient** of the trinomial. In this section we focus on trinomials where $a = 1$ and b and c are integers.

Recall that the binomials $(x + m)$ and $(x + n)$ are multiplied as follows.

$$(x + m)(x + n) = x^2 + nx + mx + mn$$
$$= x^2 + (m + n)x + mn$$

Note that the coefficient of the x-term is the sum of m and n and that the constant (or third) term is the product of m and n. Thus to factor a trinomial in the form $x^2 + bx + c$, we start by finding two numbers, m and n, such that when they are multiplied $m \cdot n = c$ and when they are added $m + n = b$. We illustrate this statement with $x^2 + 6x + 8$.

Standard Form	*Example*
$ax^2 + bx + c$	$x^2 + 6x + 8$
$m \cdot n = c$	$m \cdot n = 8$
$m + n = b$	$m + n = 6$

To determine possible values for m and n, we list factor pairs for 8 and search for a pair whose sum is 6, as in Table 6.1.

Because $2 \cdot 4 = 8$ and $2 + 4 = 6$, we let $m = \mathbf{2}$ and $n = \mathbf{4}$. We then factor the given trinomial as

$$x^2 + 6x + 8 = (x + \mathbf{2})(x + \mathbf{4}).$$

TABLE 6.1

Factor Pairs for 8

Factors	1, 8	2, 4
Sum	9	6

Note that, if you can find this factor pair mentally, making a table is not necessary.

We check the result by multiplying the two binomials.

$$(x + 2)(x + 4) = x^2 + 6x + 8$$

$2x$
$+4x$
$\overline{6x}$ ——— The middle term checks.

FACTORING $x^2 + bx + c$

To factor the trinomial $x^2 + bx + c$, find two numbers m and n that satisfy

$$m \cdot n = c \quad \text{and} \quad m + n = b.$$

Then $x^2 + bx + c = (x + m)(x + n)$.

EXAMPLE 1 **Factoring a trinomial having only positive coefficients**

Factor each trinomial.
(a) $x^2 + 7x + 12$ **(b)** $x^2 + 13x + 30$ **(c)** $z^2 + 9z + 20$

Solution
(a) To factor $x^2 + 7x + 12$ we need to find a factor pair for 12 whose sum is 7. To do so we make Table 6.2 as shown on the next page.

TABLE 6.2 **Factor Pairs for 12**

Factors	1, 12	2, 6	3, 4
Sum	13	8	7

The required factor pair is **3** and **4** because $3 \cdot 4 = 12$ and $3 + 4 = 7$. Therefore the given trinomial can be factored as

$$x^2 + 7x + 12 = (x + \mathbf{3})(x + \mathbf{4}).$$

(b) To factor $x^2 + 13x + 30$ we need to find a factor pair for 30 whose sum is 13. The required pair is **3** and **10**. Thus

$$x^2 + 13x + 30 = (x + \mathbf{3})(x + \mathbf{10}).$$

(c) To factor $z^2 + 9z + 20$ we need to find a factor pair for 20 whose sum is 9. The required pair is 4 and 5. Thus

$$z^2 + 9z + 20 = (z + 4)(z + 5).$$

Now Try Exercises 21, 23, 25

In the next example, the coefficients of the middle terms are negative.

ЄXAMPLE 2   **Factoring trinomials having a negative middle coefficient**

Factor each trinomial.
(a) $x^2 - 7x + 10$ **(b)** $x^2 - 8x + 15$ **(c)** $y^2 - 9y + 18$

Solution
(a) To factor $x^2 - 7x + 10$ we need to find a factor pair for 10 whose sum equals -7. To have a positive product *and* a negative sum, *both* numbers must be negative, as shown in Table 6.3.

TABLE 6.3 **Factor Pairs for 10**

Factors	$-1, -10$	$-2, -5$
Sum	-11	-7

The required pair is **-2** and **-5** because $-2 \cdot (-5) = 10$ and $-2 + (-5) = -7$. Therefore the given trinomial can be factored as

$$x^2 - 7x + 10 = (x - \mathbf{2})(x - \mathbf{5}).$$

(b) To factor $x^2 - 8x + 15$ we need to find a factor pair for 15 whose sum is -8. The required pair is **-3** and **-5**. Thus

$$x^2 - 8x + 15 = (x - \mathbf{3})(x - \mathbf{5}).$$

(c) To factor $y^2 - 9y + 18$ we need to find a factor pair for 18 whose sum is -9. The required pair is -3 and -6. Thus

$$y^2 - 9y + 18 = (y - 3)(y - 6).$$

Now Try Exercises 29, 31, 33

In Examples 1 and 2 the coefficient of the last term was always positive. In the next example, this coefficient is negative and the coefficient of the middle term is either positive or negative.

EXAMPLE 3 Factoring trinomials having a negative constant term

Factor each trinomial.
(a) $x^2 - 3x - 4$ **(b)** $x^2 + 7x - 8$ **(c)** $t^2 - 2t - 24$

Solution

(a) To factor $x^2 - 3x - 4$ we need to find a factor pair for -4 whose sum is -3. To have a negative product one factor must be positive and the other factor must be negative, as shown in Table 6.4.

TABLE 6.4 Factor Pairs for -4

Factors	$-1, 4$	$1, -4$	$-2, 2$
Sum	3	-3	0

The required pair is **1** and **-4** because $1 \cdot (-4) = -4$ and $1 + (-4) = -3$. Therefore the given trinomial can be factored as

$$x^2 - 3x - 4 = (x + \mathbf{1})(x - \mathbf{4}),$$

which can be checked by multiplying $(x + 1)(x - 4)$.

(b) To factor $x^2 + 7x - 8$ we need to find a factor pair for -8 whose sum is 7. The required pair is **-1** and **8**. Thus

$$x^2 + 7x - 8 = (x - \mathbf{1})(x + \mathbf{8}).$$

(c) To factor $t^2 - 2t - 24$ we need to find a factor pair for -24 whose sum is -2. The required pair is -6 and 4. Thus

$$t^2 - 2t - 24 = (t - 6)(t + 4).$$ Now Try Exercises 39, 55

A polynomial with integer coefficients that cannot be factored by using integer coefficients is called a **prime** polynomial. The next example illustrates that some trinomials of the form $x^2 + bx + c$ cannot be factored into the product of two binomials.

EXAMPLE 4 Discovering that a trinomial is prime

Factor each trinomial.
(a) $x^2 + 9x + 12$ **(b)** $x^2 + 5x - 4$

Solution

(a) To factor $x^2 + 9x + 12$, we need to find a factor pair for 12 whose sum is 9. Table 6.5 reveals that no such factor pair exists.

TABLE 6.5 Factor Pairs for 12

Factors	$1, 12$	$2, 6$	$3, 4$
Sum	13	8	7

The trinomial $x^2 + 9x + 12$ is prime.

(b) At first glance it may appear that the required factor pair is 4 and 1 because $4 \cdot 1 = 4$ and $4 + 1 = 5$. However, it is important to pay close attention to the signs of the coefficients. To factor $x^2 + 5x - 4$, we need to find a factor pair for -4 whose sum is 5. No such factor pair exists. The trinomial $x^2 + 5x - 4$ is prime.   Now Try Exercises 19, 43

When factoring some trinomials, it may be necessary to factor out the greatest common factor before attempting to factor the trinomial into the product of two binomials. The next example illustrates this process.

EXAMPLE 5 Factoring out the GCF before factoring further

Factor each trinomial completely.
(a) $7x^2 + 35x + 42$ **(b)** $2x^4 - 4x^3 - 6x^2$

Solution
(a) Because the GCF of $7x^2 + 35x + 42$ is 7, factor out **7** before factoring the remaining trinomial.

$$7x^2 + 35x + 42 = \mathbf{7}(x^2 + 5x + 6)$$

To factor $x^2 + 5x + 6$, we need to find a factor pair for 6 whose sum is 5. The required pair is 2 and 3. Thus

$$7x^2 + 35x + 42 = 7(x + 2)(x + 3).$$

(b) Because the GCF of $2x^4 - 4x^3 - 6x^2$ is $2x^2$, factor out $\mathbf{2x^2}$ before factoring the remaining trinomial.

$$2x^4 - 4x^3 - 6x^2 = \mathbf{2x^2}(x^2 - 2x - 3)$$

To factor $x^2 - 2x - 3$, we need to find a factor pair for -3 whose sum is -2. The required pair is -3 and 1. Thus

$$2x^4 - 4x^3 - 6x^2 = 2x^2(x - 3)(x + 1)$$

Now Try Exercises 65, 69

CRITICAL THINKING

A cube has a surface area of $6x^2 + 24x + 24$. What is the length of each side?

MAKING CONNECTIONS

The Signs in the Binomial Factors

If a trinomial of the form $x^2 + bx + c$ can be factored, the signs of the coefficients in the trinomial can be used to determine the signs in the binomial factors. If c is positive, the binomial factors must have the same signs. If c is negative, the binomial factors must have opposite signs. If b and c represent positive numbers, this can be summarized as follows.

Form of the Trinomial	*Signs in the Binomial Factors*
$x^2 + bx + c$	$(+)(+)$
$x^2 - bx + c$	$(-)(-)$
$x^2 + bx - c$	$(-)(+)$
$x^2 - bx - c$	$(-)(+)$

EXAMPLE 6 Finding the dimensions of a rectangle

Find one possibility for the dimensions of a rectangle that has an area of $x^2 + 6x + 5$.

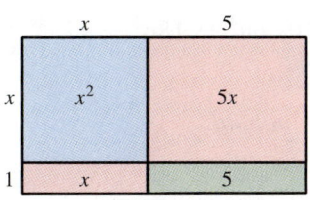

Figure 6.4 Area = $x^2 + 6x + 5$

Solution

The area of a rectangle equals length times width. If we can factor $x^2 + 6x + 5$, then the factors can represent its length and width. Because

$$x^2 + 6x + 5 = (x + 1)(x + 5),$$

one possibility for the rectangle's dimensions is width $x + 1$ and length $x + 5$, as illustrated in Figure 6.4.

Now Try Exercise **83**

6.2 PUTTING IT ALL TOGETHER

In this section we discussed factoring trinomials of the form $x^2 + bx + c$. The following table summarizes these techniques.

Concept	Explanation	Examples
Factoring Trinomials of the Form $x^2 + bx + c$	Find two numbers m and n that satisfy $mn = c$ and $m + n = b$. Then $$x^2 + bx + c = (x + m)(x + n).$$	$x^2 + 9x + 20 = (x + 4)(x + 5)$ because $4 \cdot 5 = 20$ and $4 + 5 = 9$. $x^2 + x - 6 = (x - 2)(x + 3)$ because $-2 \cdot 3 = -6$ and $-2 + 3 = 1$. $x^2 - 8x + 12 = (x - 6)(x - 2)$ because $-6 \cdot (-2) = 12$ and $-6 + (-2) = -8$.

6.2 Exercises

CONCEPTS

1. In the trinomial $x^2 + bx + c$, the leading coefficient is _____.

2. Multiply $(x + m)(x + n)$. What is the coefficient of the x-term? What is the constant term?

3. To factor $x^2 + bx + c$, start by finding two numbers m and n that satisfy _____ = c and _____ = b.

4. A trinomial with integer coefficients that cannot be factored using integer coefficients is _____.

5. List all positive integer pairs that have a product of 12.

6. List all negative integer pairs that have a product of 30.

Exercises 7–14: Find the integer pair that has the given product and sum.

7. Product: 28   Sum: 11

8. Product: 35   Sum: 12

9. Product: -30 Sum: -7

10. Product: -100 Sum: 21

11. Product: -50 Sum: 5

12. Product: -15 Sum: -2

13. Product: 28   Sum: -11

14. Product: 80   Sum: -42

FACTORING TRINOMIALS

Exercises 15–60: Factor the trinomial. If the trinomial cannot be factored, write "prime."

15. $x^2 + 3x + 2$ **16.** $x^2 + 5x + 4$

17. $y^2 + 4y + 4$ **18.** $y^2 + 8y + 7$

19. $z^2 + 3z + 7$ **20.** $z^2 + 4z + 5$

21. $x^2 + 8x + 15$ **22.** $x^2 + 9x + 14$

23. $m^2 + 13m + 36$ **24.** $m^2 + 15m + 36$

25. $n^2 + 20n + 100$ **26.** $n^2 + 52n + 100$

27. $x^2 - 6x + 5$ **28.** $x^2 - 6x + 8$

29. $y^2 - 7y + 12$ **30.** $y^2 - 12y + 27$

31. $z^2 - 13z + 40$ **32.** $z^2 - 15z + 54$

33. $a^2 - 16a + 63$ **34.** $a^2 - 82a + 81$

35. $y^2 - 6y + 10$ **36.** $y^2 - 2y + 3$

37. $b^2 - 30b + 125$ **38.** $b^2 - 19b + 90$

39. $x^2 + 13x - 90$ **40.** $x^2 + 15x - 100$

41. $m^2 + 4m - 45$ **42.** $m^2 + 4m - 60$

43. $a^2 + 16a - 63$ **44.** $a^2 + 13a - 42$

45. $n^2 + 10n - 200$ **46.** $n^2 + 2n - 120$

47. $x^2 + 22x - 23$ **48.** $x^2 + 18x - 19$

49. $a^2 + 4a - 32$ **50.** $a^2 + 9a - 36$

51. $b^2 - b - 20$ **52.** $b^2 - b - 12$

53. $m^2 - 14m - 22$ **54.** $m^2 - 11m - 24$

55. $x^2 - x - 72$ **56.** $x^2 - 2x - 80$

57. $y^2 - 15y - 34$ **58.** $y^2 - 10y - 39$

59. $z^2 - 5z - 66$ **60.** $z^2 - 6z - 55$

Exercises 61–70: Factor the trinomial completely.

61. $5x^2 - 10x - 40$ **62.** $2x^2 + 8x - 10$

63. $y^3 - 7y^2 + 10y$ **64.** $z^3 + 9z^2 + 20z$

65. $3a^3 + 21a^2 + 18a$ **66.** $5b^3 - 5b^2 - 60b$

67. $2x^3 - 6x^2 + 8x$ **68.** $4y^3 - 20y^2 + 32y$

69. $2m^4 - 10m^3 - 28m^2$ **70.** $6n^4 - 18n^3 + 12n^2$

Exercises 71–78: Factor each trinomial.

71. $5 + 6x + x^2$
   (*Hint:* Write the expression in standard form.)

72. $8 + 6x + x^2$ **73.** $3 - 4x + x^2$

74. $10 - 7x + x^2$

75. $12 + 4x - x^2$
   (*Hint:* Write $(m - x)(n + x)$ and find m and n.)

76. $28 + 3x - x^2$ **77.** $32 - 4x - x^2$

78. $40 - 3x - x^2$

79. **Thinking Generally**   Factor the trinomial expression $x^2 + (k + 1)x + k$.

80. **Thinking Generally**   Factor the trinomial expression $x^2 + (k - 2)x - 2k$.

GEOMETRY

81. A square has an area of $x^2 + 2x + 1$. Find the length of a side. Make a sketch of the square.

82. A square has an area of $x^2 + 6x + 9$. Find the length of a side. Make a sketch of the square.

83. A rectangle has an area of $x^2 + 3x + 2$. Find one possibility for its width and length. Make a sketch of the rectangle.

84. A rectangle has an area of $x^2 + 9x + 8$. Find one possibility for its width and length. Make a sketch of the rectangle.

85. A cube has a surface area of $6x^2 + 12x + 6$. Find the length of a side. (*Hint:* First factor out the GCF.)

86. A cube has a surface area of $6x^2 + 36x + 54$. Find the length of a side.

87. Write a polynomial in factored form that represents the total area of the figure.

88. Write a polynomial in factored form that represents the total area of the figure.

89. Explain how to determine whether a trinomial has been factored correctly. Give an example.

90. Factoring $x^2 + bx + c$ involves finding two integers m and n such that $mn = c$ and $m + n = b$. When you are finding m and n, is it better first to determine values of m and n so that the product is c or the sum is b? Explain your reasoning.

CHECKING BASIC CONCEPTS

SECTIONS 6.1 AND 6.2

1. What is the greatest common factor for the expression $8x^3 - 12x^2 + 24x$?

2. Factor $12z^3 - 18z^2$.

3. Factor each expression completely.
(a) $6y(y - 2) + 5(y - 2)$
(b) $2x^3 + x^2 + 10x + 5$
(c) $4z^3 - 12z^2 + 4z - 12$

4. Factor each trinomial completely.
(a) $x^2 + 6x + 8$ (b) $x^2 - x - 42$

(c) $a^2 + 3a - 5$ (d) $4a^3 + 20a^2 + 24a$

5. Write a polynomial in factored form that represents the total area of the figure.

6.3 FACTORING TRINOMIALS II ($ax^2 + bx + c$)

Factoring Trinomials by Grouping ▪ Factoring with FOIL in Reverse

A LOOK INTO MATH ▷

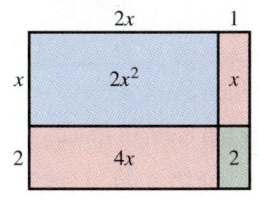

$2x^2 + 5x + 2 = (x + 2)(2x + 1)$

Figure 6.5

The sum of the areas of the four small rectangles shown in Figure 6.5 is $2x^2 + 5x + 2$. Note that the length of the large rectangle is $2x + 1$ and that its width is $x + 2$. One way to determine these dimensions is to factor $2x^2 + 5x + 2$. However, this trinomial has a leading coefficient of 2. Thus far we have factored only trinomials that have a leading coefficient of 1. In this section we discuss two methods used to factor trinomials in the form $ax^2 + bx + c$, where $a \neq 1$. In Example 1(a), we demonstrate how to factor the trinomial $2x^2 + 5x + 2$ as $(x + 2)(2x + 1)$.

Factoring Trinomials by Grouping

To factor the polynomial given by $x^2 + 6x + 5$ we find two numbers, m and n, such that $mn = 5$ and $m + n = 6$. For this trinomial we let $m = 1$ and $n = 5$, which gives

$$x^2 + 6x + 5 = (x + 1)(x + 5).$$

To factor the polynomial $2x^2 + 7x + 3$, which has a leading coefficient of 2, we must find two numbers m and n such that $mn = 2 \cdot 3 = 6$ and $m + n = 7$. One solution is

$m = 1$ and $n = 6$. Using grouping, we can now factor this trinomial by writing $7x$ as the sum $1x + 6x$.

$$
\begin{aligned}
2x^2 + 7x + 3 &= 2x^2 + \overset{\lceil 7x \rceil}{x + 6x} + 3 & &\text{Write } 7x \text{ as } x + 6x. \\
&= (2x^2 + x) + (6x + 3) & &\text{Associative property} \\
&= x(2x + 1) + 3(2x + 1) & &\text{Factor out common factors.} \\
&= (x + 3)(2x + 1) & &\text{Factor out } (2x + 1).
\end{aligned}
$$

This technique of factoring trinomials by grouping is summarized as follows.

FACTORING $ax^2 + bx + c$ BY GROUPING

To factor $ax^2 + bx + c$ perform the following steps. (Assume that a, b, and c are integers and have no common factors.)

1. Find two numbers, m and n, such that $mn = ac$ and $m + n = b$.
2. Write the trinomial as $ax^2 + mx + nx + c$.
3. Use grouping to factor this expression into two binomials.

EXAMPLE 1   Factoring $ax^2 + bx + c$ by grouping

Factor each trinomial.
(a) $2x^2 + 5x + 2$ **(b)** $3z^2 + z - 2$ **(c)** $10t^2 - 11t + 3$

Solution

(a) To factor $2x^2 + 5x + 2$ we need to find m and n that satisfy $mn = 2 \cdot 2 = 4$ and $m + n = 5$. Table 6.6 shows that two such numbers are $m = 1$ and $n = 4$.

TABLE 6.6

Factor Pairs for 4

Factors	1, 4	2, 2
Sum	5	4

$$
\begin{aligned}
2x^2 + 5x + 2 &= 2x^2 + \overset{\lceil 5x \rceil}{x + 4x} + 2 & &\text{Write } 5x \text{ as } x + 4x. \\
&= (2x^2 + x) + (4x + 2) & &\text{Associative property} \\
&= x(2x + 1) + 2(2x + 1) & &\text{Factor out common factors.} \\
&= (x + 2)(2x + 1) & &\text{Factor out } (2x + 1).
\end{aligned}
$$

(b) To factor $3z^2 + 1z - 2$ we need to find m and n that satisfy $mn = 3 \cdot (-2) = -6$ and $m + n = 1$. Two such numbers are $m = 3$ and $n = -2$.

$$
\begin{aligned}
3z^2 + z - 2 &= 3z^2 + 3z - 2z - 2 & &\text{Write } z \text{ as } 3z - 2z. \\
&= (3z^2 + 3z) + (-2z - 2) & &\text{Associative property} \\
&= 3z(z + 1) - 2(z + 1) & &\text{Factor out common factors.} \\
&= (3z - 2)(z + 1) & &\text{Factor out } (z + 1).
\end{aligned}
$$

(c) To factor $10t^2 - 11t + 3$ we need to find m and n that satisfy $mn = 10 \cdot 3 = 30$ and $m + n = -11$. Two such numbers are $m = -5$ and $n = -6$.

$$
\begin{aligned}
10t^2 - 11t + 3 &= 10t^2 - 5t - 6t + 3 & &\text{Write } -11t \text{ as } -5t - 6t. \\
&= (10t^2 - 5t) + (-6t + 3) & &\text{Associative property} \\
&= 5t(2t - 1) - 3(2t - 1) & &\text{Factor out common factors.} \\
&= (5t - 3)(2t - 1) & &\text{Factor out } (2t - 1).
\end{aligned}
$$

Now Try Exercises **11, 21, 33**

MAKING CONNECTIONS

Different Ways to Factor by Grouping

In Example 1(c) we could have written $-11t$ as $-6t - 5t$, rather than $-5t - 6t$. Then the factoring could have been written as

$$\begin{aligned}
10t^2 - 11t + 3 &= 10t^2 - 6t - 5t + 3 \\
&= (10t^2 - 6t) + (-5t + 3) \\
&= 2t(5t - 3) - 1(5t - 3) \\
&= (2t - 1)(5t - 3),
\end{aligned}$$

which gives the same result.

In the previous section we showed that some trinomials of the form $x^2 + bx + c$ are prime and cannot be factored into the product of two binomials with integer coefficients. The next example illustrates that some trinomials of the form $ax^2 + bx + c$, with $a \neq 1$, may also be prime.

EXAMPLE 2   Discovering that a trinomial is prime

Factor each trinomial.
(a) $3x^2 + 5x + 4$ **(b)** $2x^2 - 8x - 3$

Solution
(a) To factor $3x^2 + 5x + 4$, we need to find integers m and n such that $mn = 3 \cdot 4 = 12$ and $m + n = 5$. Because the middle term is positive, we consider only positive factors of 12. Table 6.7 reveals that no such integers exist.

TABLE 6.7   **Factor Pairs for 12**

Factors	1, 12	2, 6	3, 4
Sum	13	8	7

The trinomial $3x^2 + 5x + 4$ is prime.
(b) To factor $2x^2 - 8x - 3$, we must find integers m and n such that $mn = 2 \cdot (-3) = -6$ and $m + n = -8$. Table 6.8 reveals that no such integers exist.

TABLE 6.8   **Factor Pairs for −6**

Factors	−1, 6	−2, 3	2, −3	1, −6
Sum	5	1	−1	−5

The trinomial $2x^2 - 8x - 3$ is prime.   Now Try Exercises 13, 23

Although some trinomials may look as though they can be factored using the process discussed in Example 1, it is important to remember to first factor out the greatest common factor whenever possible. In the next example we factor out the GCF before factoring the trinomial further.

EXAMPLE 3   **Factoring out the GCF before factoring further**

Factor each trinomial completely.
(a) $15x^2 - 50x - 40$ **(b)** $4x^3 - 22x^2 + 30x$

Solution

(a) Because the GCF of $15x^2 - 50x - 40$ is 5, factor out 5 before factoring the remaining trinomial.

$$15x^2 - 50x - 40 = 5(3x^2 - 10x - 8)$$

To factor $3x^2 - 10x - 8$, we need numbers m and n such that $mn = 3 \cdot (-8) = -24$ and $m + n = -10$. Two such numbers are -12 and 2.

$$
\begin{aligned}
15x^2 - 50x - 40 &= 5(3x^2 - 12x + 2x - 8) && \text{Write } -10x \text{ as } -12x + 2x. \\
&= 5\big((3x^2 - 12x) + (2x - 8)\big) && \text{Associative property} \\
&= 5\big(3x(x - 4) + 2(x - 4)\big) && \text{Factor out common factors.} \\
&= 5(3x + 2)(x - 4) && \text{Factor out } (x - 4).
\end{aligned}
$$

(b) Because the GCF of $4x^3 - 22x^2 + 30x$ is $2x$, factor out $2x$ before factoring the remaining trinomial.

$$4x^3 - 22x^2 + 30x = 2x(2x^2 - 11x + 15)$$

To factor the trinomial $2x^2 - 11x + 15$, we need to find numbers m and n such that $mn = 2 \cdot 15 = 30$ and $m + n = -11$. Two such numbers are -6 and -5.

$$
\begin{aligned}
4x^3 - 22x^2 + 30x &= 2x(2x^2 - 6x - 5x + 15) && \text{Write } -11x \text{ as } -6x - 5x. \\
&= 2x\big((2x^2 - 6x) + (-5x + 15)\big) && \text{Associative property} \\
&= 2x\big(2x(x - 3) - 5(x - 3)\big) && \text{Factor out common factors.} \\
&= 2x(2x - 5)(x - 3) && \text{Factor out } (x - 3).
\end{aligned}
$$

Now Try Exercises 51, 55

Factoring with FOIL in Reverse

Rather than factoring a trinomial by grouping, we can use FOIL in reverse to determine its binomial factors. For example, the factors of $2x^2 + 5x + 2$ are two binomials:

$$2x^2 + 5x + 2 = (\underline{\quad} + \underline{\quad})(\underline{\quad} + \underline{\quad}),$$

where the expressions to be placed in the four blanks are yet to be found. By the FOIL method, we know that the product of the first terms of these two binomials is $2x^2$. Because $2x^2 = 2x \cdot x$, we can write

$$2x^2 + 5x + 2 = (\underline{2x} + \underline{\quad})(\underline{x} + \underline{\quad}).$$

By FOIL, the product of the last terms in these binomials must be 2. Because $2 = 1 \cdot 2$, we could put 1 and 2 in the blanks. However, we must be sure to place them correctly so that the product of the outside terms plus the product of the inside terms is $5x$.

$$(2x + 1)(x + 2) = 2x^2 + 5x + 2$$

$1x$

$+4x$

$\overline{5x}$ ⟵——Middle term checks.

If we had interchanged the 1 and 2, we would have obtained an incorrect result.

$$(2x + 2)(x + 1) = 2x^2 + 4x + 2$$

Middle term is *not* 5x.

In the next example we factor expressions of the form $ax^2 + bx + c$, where $a \neq 1$. We use FOIL in reverse and *trial and error* to find the correct factors.

EXAMPLE 4 **Factoring the form $ax^2 + bx + c$**

Factor each trinomial.
(a) $3x^2 + 5x + 2$ **(b)** $6x^2 + 7x - 3$ **(c)** $6 + 4y^2 - 11y$

Solution

(a) To factor $3x^2 + 5x + 2$ we start by finding factors of $3x^2$, which include **3x** and **x**, so we write

$$3x^2 + 5x + 2 = (\mathbf{3x} + \underline{\quad})(x + \underline{\quad}).$$

The factors of the last term 2 are **1** and **2**. (Because the middle term is positive, we do not consider -1 and -2.) If we place values of 1 and 2 in the binomial as follows, the middle term becomes $7x$ rather than $5x$.

$$(\mathbf{3x} + \mathbf{1})(x + \mathbf{2}) = 3x^2 + 7x + 2$$

Middle term is *not* 5x.

By reversing the positions of 1 and 2, we obtain a correct factorization.

$$(\mathbf{3x} + \mathbf{2})(x + \mathbf{1}) = 3x^2 + 5x + 2$$

Middle term checks.

(b) To factor $6x^2 + 7x - 3$, we start by finding factors of $6x^2$, which include $2x$ and $3x$ or $6x$ and x. Factors of the last term, -3, include -1 and 3 or 1 and -3. The following two factorizations give incorrect results.

$$(2x + 1)(3x - 3) = 6x^2 - 3x - 3$$

Middle term is *not* 7x.

$$(6x + 1)(x - 3) = 6x^2 - 17x - 3$$

Middle term is *not* 7x.

To obtain a middle term of $7x$ we use the following arrangement.

$$(3x - 1)(2x + 3) = 6x^2 + 7x - 3$$

Middle term checks.

To find the correct factorization we often need to try more than once.

(c) To factor $6 + 4y^2 - 11y$, we start by writing the trinomial in standard form as $4y^2 - 11y + 6$. Then we find possible factors of the first term, $4y^2$, which include $2y$ and $2y$ or $4y$ and y. Factors of the last term, 6, include either -1 and -6 or -2 and -3. (Because the middle term is negative, we do not use the positive factors of 1 and 6 or 2 and 3.) To obtain a middle term of $-11y$ we use the following arrangement.

$$(4y - 3) \ (y - 2) = 4y^2 - 11y + 6$$

$$-3y$$
$$-8y$$
$$\overline{-11y} \longleftarrow \text{Middle term checks.}$$

Now Try Exercises 15, 29, 65

MAKING CONNECTIONS

The Signs in the Binomial Factors

Let a, b, and c represent positive integers. If a trinomial of the form $ax^2 + bx + c$ can be factored, the signs in the binomial factors can be summarized as follows.

Form of the Trinomial	*Signs in the Binomial Factors*
$ax^2 + bx + c$	$(\ + \ )(\ + \ )$
$ax^2 - bx + c$	$(\ - \ )(\ - \ )$
$ax^2 + bx - c$	$(\ - \ )(\ + \ )$
$ax^2 - bx - c$	$(\ - \ )(\ + \ )$

In the next example we demonstrate how to factor trinomials having a negative leading coefficient. This task is sometimes accomplished by first factoring out -1.

EXAMPLE 5   **Factoring trinomials having a negative leading coefficient**

Factor each trinomial.
(a) $-6x^2 + 17x - 5$ **(b)** $1 - x - 2x^2$

Solution
(a) One way to factor $-6x^2 + 17x - 5$ is to start by factoring out -1. Then we can apply FOIL in reverse.

Opposite of $-6x^2 + 17x - 5$

$$\begin{aligned} -6x^2 + 17x - 5 &= -1(6x^2 - 17x + 5) && \text{Factor out } -1. \\ &= -1(3x - 1)(2x - 5) && \text{Factor the trinomial.} \\ &= -(3x - 1)(2x - 5) && \text{Rewrite.} \end{aligned}$$

(b) To factor the trinomial $1 - x - 2x^2$, write it in standard form, factor out -1, and then apply FOIL in reverse.

$$\begin{aligned} 1 - x - 2x^2 &= -2x^2 - x + 1 && \text{Standard form} \\ &= -1(2x^2 + x - 1) && \text{Factor out } -1. \\ &= -(2x - 1)(x + 1) && \text{Factor the trinomial.} \end{aligned}$$

Now Try Exercises 67, 71

6.3 PUTTING IT ALL TOGETHER

In this section we discussed factoring trinomials in the form $ax^2 + bx + c$ by using grouping and FOIL in reverse. The following table summarizes these two methods.

Concept	Explanation	Examples
Factoring Trinomials by Grouping	To factor $ax^2 + bx + c$, find two numbers, m and n, such that $mn = ac$ and $m + n = b$. Then write $$ax^2 + mx + nx + c.$$ Use grouping to factor this expression into two binomials.	For $3x^2 + 10x + 8$ let $m = 6$ and $n = 4$ because $mn = 24$ and $m + n = 10$. $$\begin{aligned} 3x^2 + 10x + 8 &= 3x^2 + 6x + 4x + 8 \\ &= (3x^2 + 6x) + (4x + 8) \\ &= 3x(x + 2) + 4(x + 2) \\ &= (3x + 4)(x + 2) \end{aligned}$$
Factoring Trinomials by Using FOIL in Reverse	To factor $ax^2 + bx + c$, find factors of ax^2 and of c. Choose and arrange these factors in two binomials so that the middle term is bx.	For $3x^2 + 10x + 8$ the factors of $3x^2$ are $3x$ and x. The positive factors of 8 are either 2 and 4, or 1 and 8. A middle term of $10x$ can be obtained as follows. $$(3x + 4)(x + 2) = 3x^2 + 10x + 8$$ $4x$ $+6x$ $10x \leftarrow$ Middle term checks.
Choosing Signs When Factoring $ax^2 + bx + c$	For $ax^2 + bx + c$, with $a > 0$, **1.** If $c > 0$ and $b > 0$, use $(\_ + \_)(\_ + \_)$. **2.** If $c > 0$ and $b < 0$, use $(\_ - \_)(\_ - \_)$. **3.** If $c < 0$, use $(\_ - \_)(\_ + \_)$ or $(\_ + \_)(\_ - \_)$.	**1.** $2x^2 + 7x + 3 = (2x + 1)(x + 3)$ $(c = 3, b = 7)$ **2.** $2x^2 - 7x + 3 = (2x - 1)(x - 3)$ $(c = 3, b = -7)$ **3.** $2x^2 + 5x - 3 = (2x - 1)(x + 3)$ $(c = -3, b = 5)$ $2x^2 - 5x - 3 = (2x + 1)(x - 3)$ $(c = -3, b = -5)$

6.3 Exercises

MyMathLab   PRACTICE WATCH DOWNLOAD READ REVIEW

CONCEPTS

1. To factor the polynomial $ax^2 + bx + c$ by grouping, you first find two numbers, m and n, such that $mn = $ _____ and $m + n = $ _____.

2. To factor the polynomial $ax^2 + bx + c$ with FOIL in reverse, you first find possible factors for _____ and for _____.

3. If $3x^2 + 5x + 2$ is factored, the result is the product $(3x \_\_ 2)(x \_\_ 1)$.

4. If $3x^2 - x - 2$ is factored, the result is the product $(3x \_\_ 2)(x \_\_ 1)$.

5. If $3x^2 - 5x + 2$ is factored, the result is the product $(3x \_\_ 2)(x \_\_ 1)$.

6. If $3x^2 + x - 2$ is factored, the result is the product $(3x \_\_ 2)(x \_\_ 1)$.

7. If $4x^2 + 11x + 6$ is factored, the result is the product $(4x + \_\_)(\_\_ + 2)$.

8. If $4x^2 - 5x - 6$ is factored, the result is the product $(x - \_\_)(\_\_ + 3)$.

9. If $4x^2 + 4x - 3$ is factored, the result is the product $(2x - \_\_)(\_\_ + 3)$.

10. If $4x^2 - 8x + 3$ is factored, the result is the product $(2x - \_\_)(\_\_ - 3)$.

FACTORING TRINOMIALS

Exercises 11–50: Factor the trinomial. If the trinomial cannot be factored, write "prime."

11. $2x^2 + 7x + 3$

12. $2x^2 + 3x + 1$

13. $3y^2 + 2y + 4$

14. $2y^2 + 5y + 1$

15. $3x^2 + 4x + 1$

16. $3x^2 + 10x + 3$

17. $6x^2 + 11x + 3$

18. $6x^2 + 17x + 5$

19. $5x^2 - 11x + 2$

20. $7x^2 - 8x + 1$

21. $2y^2 - 7y + 5$

22. $2y^2 - 11y + 12$

23. $3m^2 - 11m - 6$

24. $5m^2 - 7m - 2$

25. $7z^2 - 37z + 10$

26. $3z^2 - 11z + 6$

27. $3t^2 - 7t - 6$

28. $8t^2 - 6t - 9$

29. $15r^2 + r - 6$

30. $12r^2 + r - 6$

31. $24m^2 - 23m - 12$

32. $24m^2 + 29m - 4$

33. $25x^2 + 5x - 2$

34. $30x^2 + 7x - 2$

35. $6x^2 + 11x - 2$

36. $12x^2 + 28x - 5$

37. $15y^2 - 7y + 2$

38. $14y^2 - 5y + 1$

39. $21n^2 + 4n - 1$

40. $21n^2 + 10n + 1$

41. $14y^2 + 23y + 3$

42. $28y^2 + 25y + 3$

43. $28z^2 - 25z + 3$

44. $15z^2 - 19z + 6$

45. $30x^2 - 29x + 6$

46. $50x^2 - 55x + 12$

47. $20a^2 + 18a - 5$

48. $40a^2 + 21a - 2$

49. $18t^2 + 23t - 6$

50. $33t^2 + 7t - 10$

Exercises 51–60: Factor the trinomial completely.

51. $12a^2 + 12a - 9$

52. $21b^2 - 14b - 56$

53. $12y^3 - 11y^2 + 2y$

54. $10z^3 + 19z^2 + 6z$

55. $24x^3 - 30x^2 + 9x$

56. $8y^3 - 16y^2 + 6y$

57. $8x^4 - 6x^3 + 2x^2$

58. $10y^3 + 15y^2 - 5y$

59. $28x^4 + 56x^3 + 21x^2$

60. $20y^4 + 42y^3 - 20y^2$

61. Thinking Generally Factor the trinomial expression $3x^2 + (3k + 1)x + k$.

62. Thinking Generally Factor the trinomial expression $3x^2 + (3k - 2)x - 2k$.

Exercises 63–72: Factor.

63. $2 + 15x + 7x^2$

64. $3 + 16x + 5x^2$

65. $2 - 5x + 2x^2$

66. $5 - 6x + x^2$

67. $3 - 2x - 8x^2$

68. $5 - 3x - 2x^2$

69. $-2x^2 - 7x + 15$

70. $-5x^2 - 19x + 4$

71. $-5x^2 + 14x + 3$

72. $-6x^2 + 17x + 14$

73. A rectangle has an area of $6x^2 + 7x + 2$. Find possible dimensions for this rectangle. Make a sketch of the rectangle.

74. A rectangle has an area of $2x^2 + 5x + 3$. Find possible dimensions for the rectangle. Make a sketch of the rectangle.

75. Write a polynomial in factored form that represents the total area of the figure.

76. Write a polynomial in factored form that represents the total area of the figure.

WRITING ABOUT MATHEMATICS

77. Explain how the sign of the third term in the trinomial $ax^2 + bx + c$ affects how it is factored.

78. Explain the steps to be used to factor $ax^2 + bx + c$ by grouping.

GROUP ACTIVITY
WORKING WITH REAL DATA

Directions: Form a group of 2 to 4 people. Select someone to record the group's responses for this activity. All members of the group should work cooperatively to answer the questions. If your instructor asks for your results, each member of the group should be prepared to respond.

AIDS Cases From 1993 to 2003 the cumulative number N of AIDS cases in thousands can be approximated by $N = -2x^2 + 76x + 430$, where $x = 0$ corresponds to the year 1993.

Year	1993	1995	1997	1999	2001	2003
Cases	422	565	677	762	844	930

Source: U.S. Department of Health and Human Services.

(a) Use the equation to find N for each year in the table.

(b) Discuss how well this equation approximates the data.

(c) Rewrite the equation with the right side completely factored.

(d) Use your equation from part (c) to find N for each year in the table. Do your answers agree with those found in part (a)?

6.4 SPECIAL TYPES OF FACTORING

Difference of Two Squares ▪ Perfect Square Trinomials ▪ Sum and Difference of Two Cubes

A LOOK INTO MATH ▷ Some polynomials can be factored by using special methods. These methods are used to solve equations and simplify expressions. In this section we discuss some of these methods.

Difference of Two Squares

In Section 5.4 we showed that

$$(a - b)(a + b) = a^2 - b^2.$$

We can use this equation to factor a difference of two squares. For example, to factor the expression $x^2 - 25$ we can write it in the form $a^2 - b^2$, where $a = x$ and $b = 5$. Then the equation

$$a^2 - b^2 = (a - b)(a + b)$$

becomes

$$x^2 - 25 = (x - 5)(x + 5).$$

DIFFERENCE OF TWO SQUARES

For any real numbers a and b,

$$a^2 - b^2 = (a - b)(a + b).$$

In the next example we apply this method to other expressions.

EXAMPLE 1   **Factoring the difference of two squares**

Factor each difference of two squares.
(a) $x^2 - 36$ **(b)** $4x^2 - 9$ **(c)** $100 - 16t^2$ **(d)** $49y^2 - 64z^2$

Solution
(a) Substitute x for a and 6 for b. Then the equation

$$a^2 - b^2 = (a - b)(a + b)$$

becomes

$$x^2 - 6^2 = (x - 6)(x + 6).$$

(b) The expression $4x^2 - 9$ can be written as $(2x)^2 - 3^2$. Thus

$$4x^2 - 9 = (2x - 3)(2x + 3).$$

(c) The expression $100 - 16t^2$ can be written as $(10)^2 - (4t)^2$. Thus

$$100 - 16t^2 = (10 - 4t)(10 + 4t).$$

(d) The expression $49y^2 - 64z^2$ can be written as $(7y)^2 - (8z)^2$. Thus

$$49y^2 - 64z^2 = (7y - 8z)(7y + 8z).$$

Now Try Exercises 15, 17, 23, 27

MAKING CONNECTIONS

Sum of Squares versus Difference of Squares

The sum of two squares, $a^2 + b^2$, cannot be factored by using real numbers. However, the difference of two squares, $a^2 - b^2$, can be factored. For example, $x^2 + 4$ cannot be factored, but $x^2 - 4$ can be factored as $(x - 2)(x + 2)$.

Perfect Square Trinomials

In Section 5.4 we also showed how to compute $(a + b)^2$ and $(a - b)^2$:

$$(a + b)^2 = a^2 + 2ab + b^2 \quad \text{and}$$
$$(a - b)^2 = a^2 - 2ab + b^2.$$

The expressions $a^2 + 2ab + b^2$ and $a^2 - 2ab + b^2$ are called **perfect square trinomials.** If we can recognize a perfect square trinomial, we can use the following formulas to factor it.

PERFECT SQUARE TRINOMIALS

For any real numbers a and b,

$$a^2 + 2ab + b^2 = (a + b)^2 \quad \text{and}$$

$$a^2 - 2ab + b^2 = (a - b)^2.$$

When factoring a trinomial as a perfect square trinomial, we must first verify that the middle term is correct. This technique is demonstrated in the next example.

EXAMPLE 2 **Factoring perfect square trinomials**

If possible, factor each trinomial as a perfect square trinomial.
(a) $x^2 + 10x + 25$ (b) $4x^2 - 4x + 1$
(c) $9z^2 + 18z + 4$ (d) $x^2 - 4xy + 4y^2$

Solution
(a) To factor $x^2 + 10x + 25$ we let $a^2 = x^2$ and $b^2 = 5^2$ so that $a = x$ and $b = 5$. To be a perfect square trinomial, the middle term must be $2ab$. That is, the middle term must be *twice* the product of x and 5. So

$$2ab = 2 \cdot x \cdot 5 = 10x,$$

which is the middle term in the given expression. Thus

$$a^2 + 2ab + b^2 = (a + b)^2 \quad \text{becomes} \quad x^2 + 10x + 25 = (x + 5)^2.$$

(b) To factor $4x^2 - 4x + 1$ we let $a^2 = (2x)^2$ and $b^2 = 1^2$ so that $a = 2x$ and $b = 1$. The middle term must be $2ab$. So

$$2ab = 2 \cdot 2x \cdot 1 = 4x,$$

which is the middle term in the given expression, except for the subtraction sign. Thus

$$a^2 - 2ab + b^2 = (a - b)^2 \quad \text{becomes} \quad 4x^2 - 4x + 1 = (2x - 1)^2.$$

(c) To factor $9z^2 + 18z + 4$ we let $a^2 = (3z)^2$ and $b^2 = 2^2$ so that $a = 3z$ and $b = 2$. The middle term must be twice the product of $3z$ and 2. So

$$2ab = 2 \cdot 3z \cdot 2 = 12z,$$

which is *not* the middle term of $18z$. We cannot factor this expression as a perfect square trinomial.

(d) To factor $x^2 - 4xy + 4y^2$ we let $a^2 = x^2$ and $b^2 = (2y)^2$ so that $a = x$ and $b = 2y$. The middle term must be twice the product of x and $2y$. So

$$2ab = 2 \cdot x \cdot 2y = 4xy,$$

which is the middle term, except for the subtraction sign. Thus

$$a^2 - 2ab + b^2 = (a - b)^2 \quad \text{becomes} \quad x^2 - 4xy + 4y^2 = (x - 2y)^2.$$

Now Try Exercises 35, 39, 41, 49

MAKING CONNECTIONS

Special Factoring and General Techniques

If you do not recognize a polynomial as the difference of two squares or a perfect square trinomial, you can still factor the polynomial by using the methods discussed in earlier sections.

Sum and Difference of Two Cubes

The sum or difference of two cubes may be factored—a result of the two equations

$$(a + b)(a^2 - ab + b^2) = a^3 + b^3 \quad \text{and}$$
$$(a - b)(a^2 + ab + b^2) = a^3 - b^3.$$

These equations can be verified by multiplying the left side to obtain the right side. For example, multiplying the polynomials on the left side of the first equation results in

$$(a + b)(a^2 - ab + b^2) = a \cdot a^2 - a \cdot ab + a \cdot b^2 + b \cdot a^2 - b \cdot ab + b \cdot b^2$$
$$= a^3 - a^2b + ab^2 + a^2b - ab^2 + b^3$$
$$= a^3 + b^3.$$

SUM AND DIFFERENCE OF TWO CUBES

For any real numbers a and b,

Opposite Signs

$$a^3 + b^3 = (a + b)(a^2 - ab + b^2) \quad \text{and}$$
$$a^3 - b^3 = (a - b)(a^2 + ab + b^2).$$

Opposite Signs

Any binomial whose terms can be expressed as cubes can be factored as a sum or difference of cubes. We demonstrate this method in the next example.

EXAMPLE 3   Factoring the sum and difference of two cubes

Factor each polynomial.
(a) $z^3 + 8$ **(b)** $x^3 - 27$ **(c)** $8x^3 - 1$

Solution
(a) To factor $z^3 + 8$ we let $a^3 = z^3$ and $b^3 = 2^3$ so that $a = z$ and $b = 2$. Then

$$a^3 + b^3 = (a + b)(a^2 - ab + b^2)$$

becomes

$$z^3 + 2^3 = (z + 2)(z^2 - z \cdot 2 + 2^2)$$
$$= (z + 2)(z^2 - 2z + 4).$$

(b) To factor $x^3 - 27$ we let $a^3 = x^3$ and $b^3 = 3^3$ so that $a = x$ and $b = 3$. Then

$$a^3 - b^3 = (a - b)(a^2 + ab + b^2)$$

becomes

$$x^3 - 3^3 = (x - 3)(x^2 + x \cdot 3 + 3^2)$$
$$= (x - 3)(x^2 + 3x + 9).$$

(c) To factor $8x^3 - 1$ we let $a^3 = (2x)^3$ and $b^3 = 1^3$ so that $a = 2x$ and $b = 1$. Then

$$(2x)^3 - 1^3 = (2x - 1)((2x)^2 + 2x \cdot 1 + 1^2)$$
$$= (2x - 1)(4x^2 + 2x + 1). \quad \text{Now Try Exercises } 55, 63$$

In this section we have discussed special methods for factoring polynomials that can be identified as the difference of two squares, perfect square trinomials, the sum of two cubes, or the difference of two cubes. The next example demonstrates how to recognize and factor such polynomials.

EXAMPLE 4   **Recognizing polynomials that can be factored with special methods**

Factor each polynomial.
(a) $8x^3 + 27$ **(b)** $4y^2 - 20y + 25$ **(c)** $9z^2 - 64$

Solution
(a) Because this polynomial has only two terms, it cannot be factored as a perfect square trinomial. Since it is not a difference, it cannot be factored as the difference of two squares. We will try to factor the polynomial as the sum of two cubes. To factor $8x^3 + 27$, we note that $8x^3 = (2x)^3$ and $27 = 3^3$. Then

$$8x^3 + 27 = (2x)^3 + 3^3$$
$$= (2x + 3)((2x)^2 - 2x \cdot 3 + 3^2)$$
$$= (2x + 3)(4x^2 - 6x + 9).$$

(b) Because this polynomial has three terms, we will try to factor it as a perfect square trinomial. To factor $4y^2 - 20y + 25$, we let $a^2 = (2y)^2$ and $b^2 = 5^2$ so that $a = 2y$ and $b = 5$. The middle term must be twice the product of $2y$ and 5 so that $2ab = 2 \cdot 2y \cdot 5 = 20y$. This is the correct middle term except for the subtraction sign. Thus

$$a^2 - 2ab + b^2 = (a - b)^2 \quad \text{becomes} \quad 4y^2 - 20y + 25 = (2y - 5)^2.$$

(c) Because this polynomial is a difference of two terms that appear to be square terms, we will try to factor the polynomial as the difference of two squares. The expression $9z^2 - 64$ can be written as $(3z)^2 - 8^2$. Thus

$$9z^2 - 64 = (3z - 8)(3z + 8).$$

Now Try Exercises 29, 45, 65

When using the special factoring methods discussed in this section, it is important to remember to first factor out the greatest common factor whenever possible. In the next example we factor out the GCF before factoring further.

EXAMPLE 5   **Factoring out the GCF before factoring further**

Factor each polynomial completely.
(a) $27x^3 + 72x^2 + 48x$ **(b)** $18a^3 - 8ab^2$

Solution
(a) Because the GCF of $27x^3 + 72x^2 + 48x$ is $3x$, factor out $3x$ before factoring the remaining trinomial.

$$27x^3 + 72x^2 + 48x = 3x(9x^2 + 24x + 16)$$

The expression $9x^2 + 24x + 16$ is a perfect square trinomial with $a^2 = (\mathbf{3x})^2$ and $b^2 = \mathbf{4}^2$ so that $a = \mathbf{3x}$ and $b = \mathbf{4}$. Thus

$$a^2 + 2ab + b^2 = (a + b)^2 \quad \text{becomes} \quad 9x^2 + 24x + 16 = (\mathbf{3x + 4})^2.$$

As a result, $27x^3 + 72x^2 + 48x = 3x(3x + 4)^2$.
(b) Because the GCF of $18a^3 - 8ab^2$ is $2a$, factor out $2a$ before factoring the remaining polynomial.

$$18a^3 - 8ab^2 = 2a(9a^2 - 4b^2)$$

The expression $9a^2 - 4b^2$ is the difference of two squares and can be written as $(\mathbf{3a})^2 - (\mathbf{2b})^2$. Thus

$$18a^3 - 8ab^2 = 2a(\mathbf{3a - 2b})(\mathbf{3a + 2b}).$$

Now Try Exercises 69, 77

6.4 PUTTING IT ALL TOGETHER

In this section we discussed some special types of factoring, which are summarized in the following table.

Factoring	Explanation	Examples
Difference of Two Squares	$a^2 - b^2 = (a - b)(a + b)$ **NOTE:** The *sum* of two squares, $a^2 + b^2$, cannot be factored by using real numbers.	$x^2 - 49 = (x - 7)(x + 7)$ $81 - z^2 = (9 - z)(9 + z)$ $4r^2 - 25t^2 = (2r - 5t)(2r + 5t)$ $16a^2 + b^2$ cannot be factored.
Perfect Square Trinomial	$a^2 + 2ab + b^2 = (a + b)^2$ $a^2 - 2ab + b^2 = (a - b)^2$ Be sure to verify that the given middle term equals $2ab$ before factoring.	$m^2 + 2m + 1 = (m + 1)^2$ $25y^2 - 30y + 9 = (5y - 3)^2$ $36r^2 + 12rt + t^2 = (6r + t)^2$ $x^2 + 5x + 4$ is *not* a perfect square trinomial because $2ab = 2 \cdot x \cdot 2 = 4x \neq 5x.$

Factoring	Explanation	Examples
Sum and Difference of Two Cubes	$a^3 + b^3 = (a + b)(a^2 - ab + b^2)$ $a^3 - b^3 = (a - b)(a^2 + ab + b^2)$	$y^3 + 27 = (y + 3)(y^2 - y \cdot 3 + 3^2)$ $ = (y + 3)(y^2 - 3y + 9)$ $27r^3 - 64t^3$ $ = (3r - 4t)((3r)^2 + 3r \cdot 4t + (4t)^2)$ $ = (3r - 4t)(9r^2 + 12rt + 16t^2)$

6.4 Exercises

CONCEPTS

1. $a^2 - b^2 = $ _____

2. The expression $a^2 + b^2$ (can/cannot) be factored by using real numbers.

3. If the expression $36x^2 - 49y^2$ is written in the form $a^2 - b^2$, then $a = $ _____ and $b = $ _____.

4. $a^2 + 2ab + b^2 = $ _____

5. $a^2 - 2ab + b^2 = $ _____

6. $x^2 + $ _____ $+ 9$ is a perfect square trinomial.

7. $4r^2 - $ _____ $+ 25t^2$ is a perfect square trinomial.

8. $a^3 + b^3 = $ _____

9. $a^3 - b^3 = $ _____

10. If the expression $8x^3 + 27y^3$ is written in the form $a^3 + b^3$, then $a = $ _____ and $b = $ _____.

11. $y^3 - 8 = (y \underline{} 2)(y^2 \underline{} 2y + 4)$

12. $64z^3 + 27 = (4z \underline{} 3)(16z^2 \underline{} 12z + 9)$

FACTORING THE DIFFERENCE OF TWO SQUARES

Exercises 13–30: Factor.

13. $x^2 - 1$

14. $x^2 - 16$

15. $z^2 - 100$

16. $z^2 - 81$

17. $4y^2 - 1$

18. $9y^2 - 16$

19. $36z^2 - 25$

20. $49z^2 - 64$

21. $9 - x^2$

22. $25 - x^2$

23. $1 - 9y^2$

24. $49 - 16y^2$

25. $4a^2 - 9b^2$

26. $16a^2 - b^2$

27. $36m^2 - 25n^2$

28. $49m^2 - 100n^2$

29. $81r^2 - 49t^2$

30. $625r^2 - 121t^2$

FACTORING PERFECT SQUARE TRINOMIALS

Exercises 31–52: Factor as a perfect square trinomial whenever possible.

31. $x^2 + 8x + 16$

32. $x^2 + 4x + 4$

33. $z^2 + 12z + 25$

34. $z^2 - 18z + 36$

35. $x^2 - 6x + 9$

36. $x^2 - 10x + 25$

37. $9y^2 + 6y + 1$

38. $16y^2 + 8y + 1$

39. $4z^2 - 4z + 1$

40. $25z^2 - 12z + 1$

41. $9t^2 + 16t + 4$

42. $4t^2 + 12t + 9$

43. $9x^2 + 30x + 25$

44. $25x^2 + 60x + 36$

45. $4a^2 - 36a + 81$

46. $9a^2 - 60a + 100$

47. $x^2 + 2xy + y^2$

48. $x^2 - 6xy + 9y^2$

49. $r^2 - 10rt + 25t^2$

50. $15r^2 + 10rt + t^2$

51. $4y^2 - 10yz + 9z^2$

52. $25y^2 - 20yz + 4z^2$

FACTORING SUMS AND DIFFERENCES OF TWO CUBES

Exercises 53–66: Factor.

53. $z^3 + 1$

54. $z^3 + 8$

55. $x^3 + 64$

56. $x^3 + 125$

57. $y^3 - 8$

58. $y^3 - 27$

59. $n^3 - 1$

60. $n^3 - 64$

61. $8x^3 + 1$

62. $27x^3 - 1$

63. $m^3 - 64n^3$

64. $m^3 + 8n^3$

65. $8x^3 + 125y^3$

66. $27x^3 + 64y^3$

GENERAL FACTORING USING SPECIAL METHODS

Exercises 67–84: Factor the expression completely.

67. $4x^2 - 16$

68. $12x^2 - 60x + 75$

69. $2y^2 - 28y + 98$

70. $y^3 - 9y$

71. $5z^3 + 40$

72. $4z^3 + 36z^2 + 100z$

73. $x^3y - xy^3$

74. $8m^3 - 8$

75. $2m^3 - 10m^2 + 18m$

76. $2a^3b - 18ab^3$

77. $700x^4 - 63x^2y^2$

78. $135r^3 - 5t^3$

79. $16a^3 + 2b^3$

80. $192x^2y^2 - 3y^4$

81. $4b^4 + 24b^3 + 36b^2$

82. $2y^4 + 24y^3 + 72y^2$

83. $500r^3 - 32t^3$

84. $8r^3 - 64t^3$

GEOMETRY

85. A square has an area of $4x^2 + 12x + 9$. Find the length of a side. Make a sketch of the square.

86. A square has an area of $9x^2 + 30x + 25$. Find the length of a side. Make a sketch of the square.

WRITING ABOUT MATHEMATICS

87. Explain how factoring $x^3 + y^3$ is different from factoring $x^3 - y^3$.

88. Using the techniques discussed in this section, can you factor the expression $4x^2 + 9y^2$ into two binomials? Explain your reasoning.

CHECKING BASIC CONCEPTS

SECTIONS 6.3 AND 6.4

1. Factor each trinomial.
  (a) $2x^2 - 5x - 12$ **(b)** $6x^2 + 17x - 14$

2. Factor completely when possible.
  (a) $3y^2 + 4y - 2$ **(b)** $6y^3 - 10y^2 - 4y$

3. Write a polynomial in factored form that represents the total area of the figure.

4. Factor each polynomial.
  (a) $z^2 - 64$ **(b)** $9r^2 - 4t^2$

5. Factor each trinomial.
  (a) $x^2 + 12x + 36$ **(b)** $9a^2 - 12ab + 4b^2$

6. Factor.
  (a) $m^3 - 27$ **(b)** $125n^3 + 27$

7. Factor completely.
  (a) $16x^2 - 4$ **(b)** $3y^4 + 24y$

6.5 SUMMARY OF FACTORING

Guidelines for Factoring Polynomials ■ Factoring Polynomials

A LOOK INTO MATH ▷ A well-known line from the movie *Forrest Gump* states that "Life is like a box of chocolates. You never know what you're gonna get." When we factor a polynomial, the techniques needed to complete the factoring are usually not spelled out as part of the problem— "You never know what you're gonna get." In this section we discuss general guidelines that can be used to factor polynomials.

Guidelines for Factoring Polynomials

The following guidelines can be used to factor polynomials in general.

FACTORING POLYNOMIALS

STEP 1: Factor out the greatest common factor, if possible.

STEP 2: A. If the polynomial has *four terms*, try factoring by grouping.
   B. If the polynomial is a *binomial*, try one of the following.

   1. $a^2 - b^2 = (a - b)(a + b)$ Difference of two squares
   2. $a^3 - b^3 = (a - b)(a^2 + ab + b^2)$ Difference of two cubes
   3. $a^3 + b^3 = (a + b)(a^2 - ab + b^2)$ Sum of two cubes

   C. If the polynomial is a *trinomial*, check for a perfect square.

   1. $a^2 + 2ab + b^2 = (a + b)^2$ Perfect square trinomial
   2. $a^2 - 2ab + b^2 = (a - b)^2$ Perfect square trinomial

   Otherwise, apply FOIL in reverse or apply grouping, as described in Sections 6.2 and 6.3.

STEP 3: Check to make sure that the polynomial is *completely* factored.

NOTE: Always perform Step 1 first. Factoring out the greatest common factor usually makes it easier to factor the resulting polynomial. After a polynomial has been factored, remember to perform Step 3 so that you are sure the given polynomial is completely factored.

Factoring Polynomials

In the first example, we apply Step 1 to a polynomial with a common factor.

EXAMPLE **1**   **Factoring out a common factor**

Factor $5x^3 - 10x^2 + 15x$.

Solution

STEP 1: The greatest common factor is **$5x$**.

$$5x^3 - 10x^2 + 15x = 5x(x^2 - 2x + 3)$$

STEP 2C: The trinomial $x^2 - 2x + 3$ is prime and cannot be factored further.

STEP 3: The completely factored polynomial is $5x(x^2 - 2x + 3)$.   Now Try Exercise

When factoring polynomials completely, it is often necessary to apply more than one factoring technique. In several of the next examples we factor polynomials that require more than one method of factoring.

EXAMPLE 2 Factoring a difference of squares

Factor $3x^4 - 48x^2$.

Solution
STEP 1: The greatest common factor is $3x^2$.
$$3x^4 - 48x^2 = 3x^2(x^2 - 16)$$

STEP 2B: The binomial $x^2 - 16$ can be factored as a difference of squares.
$$3x^2(x^2 - 16) = 3x^2(x - 4)(x + 4)$$

STEP 3: The completely factored polynomial is $3x^2(x - 4)(x + 4)$. Now Try Exercise **37**

EXAMPLE 3 Factoring a perfect square trinomial

Factor $36y^3 - 24y^2 + 4y$.

Solution
STEP 1: The greatest common factor is $4y$.
$$36y^3 - 24y^2 + 4y = 4y(9y^2 - 6y + 1)$$

STEP 2C: We can factor $9y^2 - 6y + 1$ as a perfect square trinomial.
$$4y(9y^2 - 6y + 1) = 4y(3y - 1)(3y - 1)$$

STEP 3: The completely factored polynomial is $4y(3y - 1)^2$. Now Try Exercise **39**

MAKING CONNECTIONS

Factoring Polynomials

We can often determine how a polynomial should be factored by considering the number of terms in the polynomial. This is summarized as follows.

Type of Polynomial	*Factoring Technique*
4-term Polynomial	Grouping
Trinomial	Perfect square trinomial FOIL in reverse or grouping
Binomial	Difference of squares Sum or difference of cubes

EXAMPLE 4   **Factoring a sum of cubes**

Factor $27z^3 + 64$.

Solution

STEP 1: There are no common factors.

STEP 2B: The binomial $27z^3 + 64$ can be written as $(3z)^3 + 4^3$ and can be factored as a sum of cubes.

$$27z^3 + 64 = (3z + 4)\big((3z)^2 - 3z \cdot 4 + 4^2\big)$$
$$= (3z + 4)(9z^2 - 12z + 16)$$

NOTE: The trinomial $9z^2 - 12z + 16$ cannot be factored further.

STEP 3: The completely factored polynomial is $(3z + 4)(9z^2 - 12z + 16)$.

Now Try Exercise 29

EXAMPLE 5   **Factoring a trinomial**

Factor $14x^4 + 7x^3 - 42x^2$.

Solution

STEP 1: The greatest common factor is $7x^2$.

$$14x^4 + 7x^3 - 42x^2 = 7x^2(2x^2 + x - 6)$$

STEP 2C: We can factor $2x^2 + x - 6$ using FOIL in reverse.

$$7x^2(2x^2 + x - 6) = 7x^2(2x - 3)(x + 2)$$

STEP 3: The completely factored polynomial is $7x^2(2x - 3)(x + 2)$.

Now Try Exercise 33

EXAMPLE 6   **Factoring by grouping**

Factor $15x^3 + 10x^2 - 60x - 40$.

Solution

STEP 1: The greatest common factor is 5.

$$15x^3 + 10x^2 - 60x - 40 = 5(3x^3 + 2x^2 - 12x - 8)$$

STEP 2A: Because the resulting polynomial has four terms, we apply grouping.

$$5(3x^3 + 2x^2 - 12x - 8) = 5\big((3x^3 + 2x^2) + (-12x - 8)\big) \quad \text{Associative property}$$
$$= 5\big(x^2(3x + 2) - 4(3x + 2)\big) \quad \text{Factor out common factors.}$$
$$= 5(x^2 - 4)(3x + 2) \quad \text{Factor out } (3x + 2).$$

STEP 2B: The binomial $x^2 - 4$ can now be factored as a difference of squares.

$$5(x^2 - 4)(3x + 2) = 5(x - 2)(x + 2)(3x + 2)$$

STEP 3: The completely factored polynomial is $5(x - 2)(x + 2)(3x + 2)$.

Now Try Exercise 31

EXAMPLE 7 Factoring a polynomial having two variables

Factor $18x^3y - 8xy^3$.

Solution

STEP 1: The greatest common factor is $2xy$.

$$18x^3y - 8xy^3 = 2xy(9x^2 - 4y^2)$$

STEP 2B: The binomial $9x^2 - 4y^2$ can be written as $(\mathbf{3x})^2 - (\mathbf{2y})^2$ and can be factored as a difference of squares.

$$9x^2 - 4y^2 = (\mathbf{3x - 2y})(\mathbf{3x + 2y})$$

STEP 3: The completely factored polynomial is $2xy(3x - 2y)(3x + 2y)$.

Now Try Exercise 53

EXAMPLE 8 Applying several techniques

Factor $3x^5 - 3x^3 - 24x^2 + 24$.

Solution

STEP 1: The greatest common factor is 3.

$$3x^5 - 3x^3 - 24x^2 + 24 = 3(x^5 - x^3 - 8x^2 + 8)$$

STEP 2A: The resulting four-term polynomial can be factored by grouping.

$$
\begin{aligned}
3(x^5 - x^3 - 8x^2 + 8) &= 3\big((x^5 - x^3) + (-8x^2 + 8)\big) && \text{Associative property} \\
&= 3\big(x^3(x^2 - 1) - 8(x^2 - 1)\big) && \text{Factor out common factors.} \\
&= 3(\mathbf{x^3 - 8})(\mathbf{x^2 - 1}) && \text{Factor out } x^2 - 1.
\end{aligned}
$$

STEP 2B: Both binomials in this expression can be factored further. The binomial $\mathbf{x^3 - 8}$ can be factored as a difference of cubes and the binomial $\mathbf{x^2 - 1}$ can be factored as a difference of squares.

$$3(\mathbf{x^3 - 8})(\mathbf{x^2 - 1}) = 3(\mathbf{x - 2})(\mathbf{x^2 + 2x + 4})(\mathbf{x - 1})(\mathbf{x + 1})$$

NOTE: The trinomial $x^2 + 2x + 4$ cannot be factored further.

STEP 3: The completely factored polynomial is $3(x - 2)(x^2 + 2x + 4)(x - 1)(x + 1)$.

Now Try Exercise 43

6.5 PUTTING IT ALL TOGETHER

The following table provides a summary of factoring rules that may be helpful when applying the guidelines for factoring polynomials presented in this section.

Concept	Explanation	Examples
Greatest Common Factor	Factor out the greatest common factor, or monomial, that occurs in each term.	$2x^2 - 4x + 10 = 2(x^2 - 2x + 5)$ $3x^3 + 6x = 3x(x^2 + 2)$ $7xy - x^2y = xy(7 - x)$

Concept	Explanation	Examples
Factoring by Grouping	Use the associative and distributive properties to factor a polynomial with four terms.	$\begin{aligned} x^3 - 3x^2 + 2x - 6 &= (x^3 - 3x^2) + (2x - 6) \\ &= x^2(x - 3) + 2(x - 3) \\ &= (x^2 + 2)(x - 3) \end{aligned}$
Factoring Binomials	Use the difference of squares, the difference of cubes, or the sum of cubes.	$9x^2 - 4 = (3x - 2)(3x + 2)$ $x^3 - 27 = (x - 3)(x^2 + 3x + 9)$ $x^3 + 27 = (x + 3)(x^2 - 3x + 9)$
Factoring Trinomials	Use FOIL in reverse or grouping.	$x^2 + 5x - 6 = (x + 6)(x - 1)$ Check middle term: $-x + 6x = 5x$. $\begin{aligned} 4x^2 + 4x - 3 &= (4x^2 - 2x) + (6x - 3) \\ &= 2x(2x - 1) + 3(2x - 1) \\ &= (2x + 3)(2x - 1) \end{aligned}$

6.5 Exercises

CONCEPTS

1. What do the letters GCF mean?

2. A good first step for factoring polynomials is to factor out the _____.

3. If a polynomial has four terms, what factoring method might be appropriate?

4. Can $x^2 + 1$ be factored? Explain.

5. Can $x^3 + 1$ be factored? Explain.

6. The last step for factoring is to be sure the polynomial is _____ factored.

WARM UP

Exercises 7–22: Factor completely, if possible.

7. $4x - 2$ **8.** $x^2 + 3x$

9. $2y^2 - 4y + 4$ **10.** $5y^2 - 25y + 10$

11. $z^2 - 4$ **12.** $9z^2 - 25$

13. $a^3 + 8$ **14.** $8a^3 - 1$

15. $4b^2 - 12b + 9$ **16.** $b^2 + 4b + 4$

17. $m^2 + 9$ **18.** $4m^2 + 49$

19. $x^3 - x^2 + 5x - 5$ **20.** $3x^3 + 6x^2 + x + 2$

21. $y^2 - 5y + 4$ **22.** $y^2 - 3y - 10$

GENERAL FACTORING

Exercises 23–60: Factor completely.

23. $x^3 + 4x^2 - 9x - 36$ **24.** $6x^2 - 19x + 15$

25. $8a^3 - 64$ **26.** $ab^2 - 4a$

27. $12x^4 - 18x^3 + 4x^2 - 6x$

28. $3x^2y + 24xy + 48y$

29. $54t^4 + 16t$ **30.** $3t^3 + 18t^2 - 48t$

31. $2r^3 + 6r^2 - 2r - 6$

32. $3r^4 + 3r^3 - 24r - 24$ **33.** $6z^4 - 21z^3 - 45z^2$

34. $3x^4y + 24xy^4$ **35.** $12b^4 - 10b^3 + 2b^2$

36. $6a^4b + 4a^3b + 18a^2b + 12ab$

37. $6y^2z - 24z^3$ **38.** $6y^3z - 48z^4$

39. $3x^2y - 30xy + 75y$ **40.** $8x^3 + y^3$

41. $27m^3 - 8n^3$ **42.** $45m^3 - 69m^2 + 12m$

43. $3x^5 - 12x^3 - 3x^2 + 12$

44. $8x^3 - 8$ **45.** $5a^2 - 27a - 18$

46. $2a^2 - 6ab + 3a - 9b$

47. $3rt^2 + 33rt + 90r$ **48.** $9t^2 + 24t + 16$

49. $9b^3 + 6b^2 + 12b + 8$ **50.** $5b^3 - 55b^2 - 60b$

51. $6n^3 + 2n^2 - 10n$ **52.** $7n^4 + 28n^3 - 63n^2$

53. $4x^2 - 36y^2$ **54.** $64x^2 - 25y^2$

55. $2a^3 - 16a^2 + 32a$ **56.** $24a^3 + 72a^2 + 54a$

57. $32xy^3 + 4x$ **58.** $24x^3 - 4x^2 - 160x$

59. $8b^4 + 24b^3 - 2b^2 - 6b$

60. $3z^3 - 6z^2 - 27z + 54$

GEOMETRY

61. *Dimensions of a Square*   If three identical squares have a total area of $27x^2 + 18x + 3$, find a possible length of one side of one of the squares.

62. *Dimensions of a Cube*   If three identical cubes have a total volume of $3x^3 + 18x^2 + 36x + 24$, find a possible length of one side of one of the cubes.

WRITING ABOUT MATHEMATICS

63. Explain how the number of terms in a polynomial can help determine what method should be used to factor it.

64. Describe a method for determining whether a polynomial has been factored correctly.

6.6 SOLVING EQUATIONS BY FACTORING I (QUADRATICS)

The Zero-Product Property ■ Solving Quadratic Equations ■ Applications

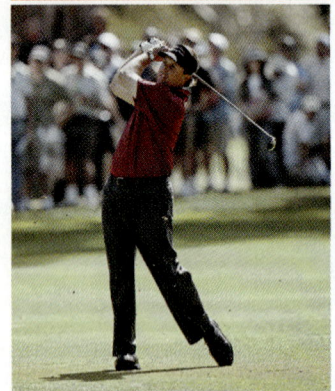

A LOOK INTO MATH ▷

If a golf ball is hit upward at 132 feet per second, or 90 miles per hour, then its height h in feet above the ground after t seconds is given by $h = 132t - 16t^2$. The expression $132t - 16t^2$ is an example of a *quadratic polynomial*. To determine the elapsed time between when the ball is hit and when it strikes the ground (or when $h = 0$) we solve the *quadratic equation*

$$132t - 16t^2 = 0.$$

One method for solving this equation is by factoring. (See Example 4.) In this section we discuss how to use factoring to solve a variety of equations.

The Zero-Product Property

To solve equations we often use the **zero-product property**, which states that if the product of two numbers is 0, then at least one of the numbers must be 0.

ZERO-PRODUCT PROPERTY

For all real numbers a and b, if $ab = 0$, then $a = 0$ or $b = 0$ (or both).

NOTE: The zero-product property works only for 0. If $ab = 1$, then it does *not* follow that $a = 1$ or $b = 1$. For example, $a = \frac{1}{3}$ and $b = 3$ satisfy the equation $ab = 1$.

After factoring an expression, we can use the zero-product property to solve an equation. The left side of the equation

$$3t^2 - 9t = 0$$

may be factored to obtain

$$3t(t - 3) = 0.$$

Note that the product of $3t$ and $t - 3$ is 0. By the zero-product property, either

$$3t = 0 \quad \text{or} \quad t - 3 = 0.$$

Solving each equation for t results in

$$t = 0 \quad \text{or} \quad t = 3.$$

These values can be checked by substituting them into the given equation $3t^2 - 9t = 0$.

$$3(0)^2 - 9(0) = 0 \qquad \text{Let } t = 0. \text{ It checks.}$$
$$3(3)^2 - 9(3) = 0 \qquad \text{Let } t = 3. \text{ It checks.}$$

The t–values of 0 and 3 are called **zeros** of the polynomial $3t^2 - 9t$, because when either is substituted in this polynomial, the result is 0.

EXAMPLE 1 | Applying the zero-product property

Solve each equation.
(a) $x(x - 1) = 0$ **(b)** $2z^2 = 0$
(c) $(t + 3)(t + 2) = 0$ **(d)** $x(x - 2)(2x + 1) = 0$

Solution
(a) By the zero-product property, $x(x - 1) = 0$ when $x = 0$ or $x - 1 = 0$. The solutions are 0 and 1.
(b) $2z^2 = 2 \cdot z \cdot z$ and $2 \neq 0$, so $2z^2 = 0$ when $z = 0$.
(c) $(t + 3)(t + 2) = 0$ implies that $t + 3 = 0$ or $t + 2 = 0$. The solutions to the equation are -3 and -2.
(d) We apply the zero-product property to $x(x - 2)(2x + 1) = 0$. Thus $x = 0$ or $x - 2 = 0$ or $2x + 1 = 0$. The solutions are $-\frac{1}{2}$, 0, and 2.

Now Try Exercises 12, 13, 17, 21

Solving Quadratic Equations

Any **quadratic polynomial** in the variable x can be written as $ax^2 + bx + c$, with $a \neq 0$. Any **quadratic equation** in the variable x can be written as $ax^2 + bx + c = 0$, with $a \neq 0$. This form of quadratic equation is called the **standard form** of a quadratic equation. For example, $x^2 + 2x - 3$ is a quadratic polynomial and $x^2 + 2x - 3 = 0$ is a quadratic equation written in standard form. When a quadratic polynomial in the variable x is in standard form and we read it from left to right, the terms contain descending powers of x. In other words, the first term contains x^2, the second term contains x, and the third term is a constant (the exponent on x is 0).

To solve a quadratic equation we often use factoring and the zero-product property. This method is summarized by the following steps. Although it is not necessary to label each step in the solution to a quadratic equation, it is important to keep these steps in mind.

> ### SOLVING QUADRATIC EQUATIONS
>
> To solve a quadratic equation by factoring, follow these steps.
>
> **STEP 1:** If necessary, write the equation in standard form as $ax^2 + bx + c = 0$.
>
> **STEP 2:** Factor the left side of the equation using any method.
>
> **STEP 3:** Apply the zero-product property.
>
> **STEP 4:** Solve each of the resulting equations. Check any solutions.

EXAMPLE **2**   ### Solving equations by factoring

Solve each quadratic equation. Check your answers.

(a) $x^2 + 2x = 0$ **(b)** $y^2 = 16$ **(c)** $z^2 - 3z + 2 = 0$ **(d)** $2x^2 = 5 - 9x$

Solution

(a) Because $x^2 + 2x = 0$ is in standard form, we begin by factoring out the GCF of x.

$x^2 + 2x = 0$	Given equation
$x(x + 2) = 0$	Factor out x. (Step 2)
$x = 0$ or $x + 2 = 0$	Zero-product property (Step 3)
$x = 0$ or $x = -2$	Solve for x. (Step 4)

To check these values, substitute -2 and 0 for x in the given equation.

$$(-2)^2 + 2(-2) \stackrel{?}{=} 0 \qquad (0)^2 + 2(0) \stackrel{?}{=} 0 \quad \text{Substitute } -2 \text{ and } 0.$$
$$0 = 0 \qquad\qquad\qquad 0 = 0 \quad \text{Both answers check.}$$

Therefore the solutions are -2 and 0.

(b) To write $y^2 = 16$ in standard form we begin by subtracting 16 from each side to obtain 0 on the right side.

$y^2 = 16$	Given equation
$y^2 - 16 = 0$	Subtract 16. (Step 1)
$(y - 4)(y + 4) = 0$	Difference of squares (Step 2)
$y - 4 = 0$ or $y + 4 = 0$	Zero-product property (Step 3)
$y = 4$ or $y = -4$	Solve for y. (Step 4)

To check these values, substitute -4 and 4 for y in the given equation.

$$(-4)^2 \stackrel{?}{=} 16 \qquad (4)^2 \stackrel{?}{=} 16 \quad \text{Substitute } -4 \text{ and } 4.$$
$$16 = 16 \qquad\quad 16 = 16 \quad \text{Both answers check.}$$

The solutions are -4 and 4.

(c) We begin by factoring the left side of the equation, $z^2 - 3z + 2$.

$z^2 - 3z + 2 = 0$	Given equation
$(z - 1)(z - 2) = 0$	Factor. (Step 2)
$z - 1 = 0$ or $z - 2 = 0$	Zero-product property (Step 3)
$z = 1$ or $z = 2$	Solve for z. (Step 4)

To check these values, substitute 1 and 2 for z in the given equation.

$$1^2 - 3(1) + 2 \overset{?}{=} 0 \qquad 2^2 - 3(2) + 2 \overset{?}{=} 0 \qquad \text{Substitute 1 and 2.}$$
$$0 = 0 \qquad\qquad\qquad 0 = 0 \qquad\qquad \text{Both answers check.}$$

The solutions are 1 and 2.

(d) We write $2x^2 = 5 - 9x$ in standard form by adding -5 and $9x$ to each side.

$$\begin{array}{ll}
2x^2 = 5 - 9x & \text{Given equation} \\
2x^2 + 9x - 5 = 0 & \text{Add } -5 \text{ and } 9x. \text{ (Step 1)} \\
(2x - 1)(x + 5) = 0 & \text{Factor. (Step 2)} \\
2x - 1 = 0 \quad \text{or} \quad x + 5 = 0 & \text{Zero-product property (Step 3)} \\
x = \dfrac{1}{2} \quad \text{or} \quad x = -5 & \text{Solve for } x. \text{ (Step 4)}
\end{array}$$

To check these values, substitute -5 and $\frac{1}{2}$ for x in the given equation.

$$2(-5)^2 \overset{?}{=} 5 - 9(-5) \qquad 2\left(\frac{1}{2}\right)^2 \overset{?}{=} 5 - 9\left(\frac{1}{2}\right) \qquad \text{Substitute } -5 \text{ and } \tfrac{1}{2}.$$
$$50 = 50 \qquad\qquad\qquad \frac{1}{2} = \frac{1}{2} \qquad\qquad \text{Both answers check.}$$

The solutions are -5 and $\frac{1}{2}$.

Now Try Exercises **25, 33, 41, 47**

EXAMPLE 3 Solving an equation by factoring

Solve $6x^2 - x = 12$.

Solution

We cannot solve the equation $6x^2 - x = 12$ by factoring out the common factor of x in $6x^2 - x$ and setting each factor equal to 12. Instead we apply the zero-product property by first writing the given equation in the standard form: $ax^2 + bx + c = 0$.

$$\begin{array}{ll}
6x^2 - x = 12 & \text{Given equation} \\
6x^2 - x - 12 = 0 & \text{Subtract 12. (Step 1)} \\
(2x - 3)(3x + 4) = 0 & \text{Factor. (Step 2)} \\
2x - 3 = 0 \quad \text{or} \quad 3x + 4 = 0 & \text{Zero-product property (Step 3)} \\
x = \dfrac{3}{2} \quad \text{or} \quad x = -\dfrac{4}{3} & \text{Solve for } x. \text{ (Step 4)}
\end{array}$$

The solutions are $-\frac{4}{3}$ and $\frac{3}{2}$.

Now Try Exercise **49**

MAKING CONNECTIONS

Equations and Expressions

The words "equation" and "expression" occur frequently in mathematics. However, they are *not* interchangeable. We often want to *solve* equations to find the values of the variable that make the equation true. We *factor* and *simplify* expressions. An equation is a statement that two expressions are equal. For example, $2x^2 - 3 = 5x + 1$ is an equation where $2x^2 - 3$ and $5x + 1$ are each expressions

Applications

▶ **REAL-WORLD CONNECTION**   To solve application problems we often need to solve equations. The next example illustrates how to solve the application presented in the introduction to this section.

EXAMPLE **4**   **Modeling the flight of a golf ball**

If a golf ball is hit upward at 132 feet per second, or 90 miles per hour, then its height h in feet after t seconds is $h = 132t - 16t^2$. After how long does the golf ball strike the ground?

Solution

The golf ball strikes the ground when its height is 0.

$$132t - 16t^2 = 0 \qquad \text{Let } h = 0.$$
$$4t(33 - 4t) = 0 \qquad \text{Factor out } 4t.$$
$$4t = 0 \quad \text{or} \quad 33 - 4t = 0 \qquad \text{Zero-product property}$$
$$t = 0 \quad \text{or} \quad -4t = -33 \qquad \text{Divide by 4; subtract 33.}$$
$$t = 0 \quad \text{or} \quad t = \frac{33}{4} \qquad \text{Solve for } t.$$

The ball strikes the ground after $\frac{33}{4} = 8.25$ seconds. The solution of 0 is not used in this problem because it corresponds to the time when the ball is hit.   Now Try Exercise 65(a)

▶ **REAL-WORLD CONNECTION**   When you try to stop a car, the greater the speed the greater is the stopping distance. In fact, if you drive twice as fast, the braking distance will be about four times as much. And if you drive three times faster, the braking distance will be about nine times as much.

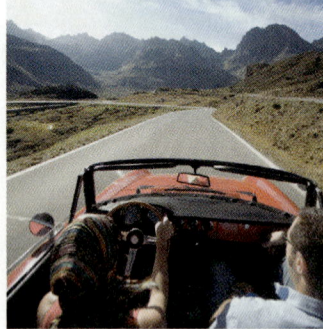

EXAMPLE **5**   **Modeling braking distance**

The braking distance D in feet required to stop a car traveling at x miles per hour on dry, level pavement can be approximated by $D = \frac{1}{11}x^2$.   (**Source:** L. Haefner.)
(a) Calculate the braking distance for a car traveling 70 miles per hour.
(b) If the braking distance is 44 feet, calculate the speed of the car.
(c) If you have a calculator available, use it to solve part (b) numerically with a table of values.

Solution
(a) If $x = \mathbf{70}$, then $D = \frac{1}{11}(\mathbf{70})^2 = \frac{4900}{11} \approx 445$ feet.
(b) *Symbolic Solution*   Let $D = 44$ in the given equation and solve.

$$\frac{1}{11}x^2 = 44 \qquad \text{Let } D = 44.$$
$$\frac{1}{11}x^2 - 44 = 0 \qquad \text{Subtract 44.}$$
$$x^2 - 484 = 0 \qquad \text{Multiply by 11.}$$
$$(x - 22)(x + 22) = 0 \qquad \text{Difference of two squares}$$
$$\qquad\qquad\qquad\qquad \textbf{\textit{Note:}} \; 22^2 = 484$$
$$x - 22 = 0 \quad \text{or} \quad x + 22 = 0 \qquad \text{Zero-product property}$$
$$x = 22 \quad \text{or} \quad x = -22 \qquad \text{Solve for } x.$$

CALCULATOR HELP

To make a table of values, see Appendix A (page AP-2).

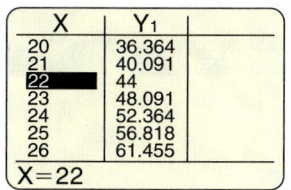

Figure 6.6

The car is traveling at (approximately) 22 miles per hour. (Note that $x = -22$ has no physical meaning in this problem.)

(c) *Numerical Solution*   Let $Y_1 = X^2/11$ and make a table of values. Scroll through the table, as shown in Figure 6.6, to where $x = 22$ when $y_1 = 44$. Thus the solution is 22 miles per hour.   Now Try Exercise **67**

▶ **REAL-WORLD CONNECTION**   Quadratic equations sometimes are used in applications involving rectangular shapes. The next application involves finding the dimensions of a digital photograph.

EXAMPLE **6**   **Finding the dimensions of a digital photograph**

A small digital photograph is 20 pixels longer than it is wide, as illustrated in Figure 6.7. It has a total of 2400 pixels. Find the dimensions of this photograph.

x

2400 pixels

$x + 20$

Figure 6.7 Dimensions of a Photograph

Solution

From Figure 6.7 the rectangular photograph has an area of 2400 pixels.

$$x(x + 20) = 2400 \qquad \text{Area} = \text{width} \times \text{length}$$
$$x^2 + 20x = 2400 \qquad \text{Distributive property}$$
$$x^2 + 20x - 2400 = 0 \qquad \text{Subtract 2400.}$$
$$(x - 40)(x + 60) = 0 \qquad \text{Factor.}$$
$$x - 40 = 0 \quad \text{or} \quad x + 60 = 0 \qquad \text{Zero-product property}$$
$$x = 40 \quad \text{or} \quad x = -60 \qquad \text{Solve for } x.$$

The valid solution is 40. Thus the dimensions of the photograph are 40 pixels by $40 + 20 = 60$ pixels.   Now Try Exercise **71**

CRITICAL THINKING

Are the two solutions to $x(2x + 1) = 1$ found by letting $x = 1$ or $2x + 1 = 1$? Explain your reasoning. What are the solutions to this equation?

6.6 PUTTING IT ALL TOGETHER

In this section we discussed solving quadratic equations by using factoring and the zero-product property. These topics are summarized in the following table.

Concept	Explanation	Examples
Zero-Product Property	If the product of two or more expressions is 0, then at least one of the expressions must equal 0.	$ab = 0$ implies that $a = 0$ or $b = 0$. $x(x + 1)$ implies that $x = 0$ or $x + 1 = 0$. $z(z - 1)(z + 2) = 0$ implies that $z = 0 \quad \text{or} \quad z - 1 = 0 \quad \text{or} \quad z + 2 = 0.$

continued on next page

continued from previous page

Concept	Explanation	Examples
Solving Quadratic Equations by Factoring	1. Write the equation as $$ax^2 + bx + c = 0.$$ 2. Factor the left side of this equation. 3. Apply the zero-product property. 4. Solve each resulting equation. Check any solutions.	$2x^2 + 11x = 6$ $2x^2 + 11x - 6 = 0$ Step 1 $(2x - 1)(x + 6) = 0$ Step 2 $2x - 1 = 0$ or $x + 6 = 0$ Step 3 $x = \dfrac{1}{2}$ or $x = -6$ Step 4

6.6 Exercises

MyMathLab

PRACTICE   WATCH   DOWNLOAD   READ   REVIEW

CONCEPTS

1. If $ab = 0$, then either $a =$ _____ or $b =$ _____.

2. Can the zero-product property be used to state that if $(x - 1)(x - 2) = 3$, then either $x - 1 = 3$ or $x - 2 = 3$? Explain your answer.

3. If $2x(x + 6) = 0$, then either _____ or _____.

4. What is a good first step when you are solving the equation $4x^2 + 1 = 4x$ by factoring?

5. What is the next step when you are solving the equation $(x + 5)(x - 4) = 0$?

6. Factoring is an important method for _____ equations.

7. Any quadratic equation in the variable x can be written in standard form as _____.

8. Standard form for $x^2 + 1 = 6x$ is _____.

9. Because $2(4) - 8 = 0$, the value 4 is called a _____ of the polynomial $2x - 8$.

10. What is the zero of the polynomial $3x - 6$?

ZERO-PRODUCT PROPERTY

Exercises 11–22: Solve the equation.

11. $xy = 0$

12. $m^2 = 0$

13. $2x(x + 8) = 0$

14. $x(x + 10) = 0$

15. $(y - 1)(y - 2) = 0$

16. $(y + 4)(y - 3) = 0$

17. $(2z - 1)(4z - 3) = 0$

18. $(6z + 5)(z - 7) = 0$

19. $(1 - 3n)(3 - 7n) = 0$

20. $(5 - n)(5 + n) = 0$

21. $x(x - 5)(x - 8) = 0$

22. $x(x + 1)(x - 6) = 0$

SOLVING QUADRATIC EQUATIONS

Exercises 23–58: Solve and check.

23. $x^2 - x = 0$

24. $2x^2 + 4x = 0$

25. $z^2 - 5z = 0$

26. $6z^2 - 3z = 0$

27. $10y^2 + 15y = 0$

28. $2y^2 + 3y = 0$

29. $x^2 - 1 = 0$

30. $x^2 - 9 = 0$

31. $4n^2 - 1 = 0$

32. $9n^2 - 4 = 0$

33. $z^2 + 3z + 2 = 0$

34. $z^2 - 2z - 3 = 0$

35. $x^2 - 12x + 35 = 0$

36. $x^2 - x - 20 = 0$

37. $2b^2 + 3b - 2 = 0$

38. $3b^2 + b - 2 = 0$

39. $6y^2 + 19y + 10 = 0$

40. $4y^2 - 25y - 21 = 0$

41. $x^2 = 25$

42. $x^2 = 81$

43. $t^2 = 5t$

44. $10t^2 = -5t$

45. $3m^2 = -9m$

46. $4m^2 = 9$

47. $x^2 = 5x + 6$

48. $2x^2 + 3x = 14$

49. $12z^2 + 11z = 15$

50. $12z^2 = 5 - 4z$

51. $t(t + 1) = 2$

52. $t(t - 7) = -12$

53. $x(2x + 5) = 3$

54. $x(3x + 2) = 5$

55. $12x^2 + 12x = -3$

56. $20x^2 - 8 = 6x$

57. $30y^2 + 50y + 20 = 0$

58. $30y^2 - 25y + 5 = 0$

GEOMETRY

59. *Dimensions of a Square*   A square has an area of 144 square feet. Find the length of a side.

60. *Dimensions of a Cube*   A cube has a surface area of 96 square feet. Find the length of a side.

61. *Radius of a Circle*   The numerical difference between the area and the circumference of a circle is 8π. Find the radius of the circle. (*Hint:* First factor out π in your equation.)

62. *Dimensions of a Rectangle*   A rectangle is 5 feet longer than it is wide and has an area of 126 square feet. What are its dimensions?

Exercises 63 and 64: ***Pythagorean Theorem***   *Suppose that a right triangle has legs a and b and hypotenuse c, as illustrated in the figure. Then these values satisfy* $a^2 + b^2 = c^2$.

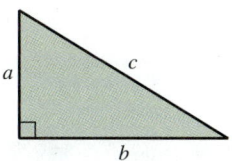

Use the Pythagorean Theorem as described above to find the value of x in the figure.

63.

64.

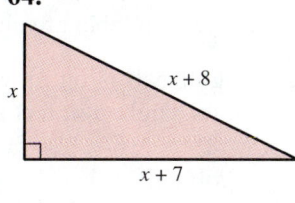

APPLICATIONS

65. *Flight of a Golf Ball*   (Refer to Example 4.) The height h in feet of a golf ball after t seconds is given by $h = 96t - 16t^2$.

(a) How long did it take for the ball to hit the ground?

(b) Make a table of h for $t = 0, 1, 2, \ldots, 6$. After how many seconds did the golf ball reach its maximum height?

66. *Flight of a Baseball*   The height h in feet of a baseball after t seconds is given by $h = -16t^2 + 88t + 3$. At what values of t is the height of the baseball 75 feet?

67. *Braking Distance*   (Refer to Example 5.) The braking distance D in feet required to stop a car traveling x miles per hour on dry, level pavement can be approximated by $D = \frac{1}{11}x^2$.

(a) Calculate the braking distance for 30 miles per hour and 60 miles per hour. How do your answers compare?

(b) If the braking distance is 33 feet, estimate the speed of the car.

(c) If you have a calculator, use it to solve part (b) numerically. Do your answers agree?

68. *Braking Distance*   The braking distance D in feet required to stop a car traveling x miles per hour on wet, level pavement is approximated by $D = \frac{1}{9}x^2$.

(a) Calculate the braking distance for 36 miles per hour and 72 miles per hour. How do your answers compare?

(b) If the braking distance is 49 feet, estimate the speed of the car.

(c) If you have a calculator, use it to solve part (b) numerically. Do your answers agree?

69. *Women in the Workforce*   The number of women W in the workforce in millions can be estimated by the equation $W = \frac{19}{3125}x^2 + \frac{11}{2}$, where $x = 0$ corresponds to 1900, $x = 10$ to 1910, and so on until $x = 100$ corresponds to 2000. (*Source:* U.S. Census Bureau.)

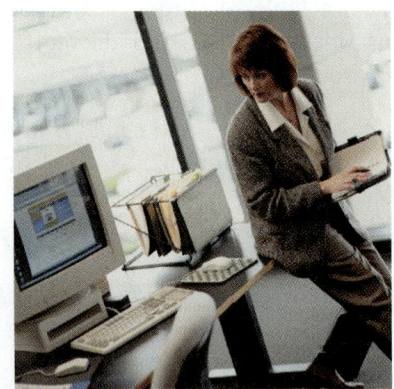

(a) How many women were in the workforce in 1930 and in 2000?

(b) Use a table of values to estimate the year in which 45 million women were in the workforce.

70. *AIDS Deaths*   The cumulative number of AIDS deaths D from 1993 to 2003 can be modeled by

$$D = -1808x^2 + 71{,}154x + 359{,}262,$$

where $x = 1$ corresponds to 1993, $x = 2$ to 1994, and so on until $x = 11$ corresponds to 2003. (*Source:* U.S. Department of Health and Human Services.)

(a) Estimate the cumulative number of AIDS deaths in 2002.

(b) Use a table of values to estimate the year in which the cumulative number of AIDS deaths reached 721,000.

71. *Digital Photographs*   (Refer to Example 6.) A digital photograph is 10 pixels longer than it is wide and has a total area of 2000 pixels. Find the dimensions of this rectangular photograph.

72. *Dimensions of a Building*   The rectangular floor of a shed has a length 4 feet longer than its width, and its area is 140 square feet. Let x be the width of the floor.

(a) Write a quadratic equation whose solution gives the width of the floor.

(b) Solve this equation.

WRITING ABOUT MATHEMATICS

73. List four steps for solving a quadratic equation by factoring.

74. Explain why factoring is important.

CHECKING BASIC CONCEPTS

SECTIONS 6.5 AND 6.6

1. Factor out the greatest common factor.
(a) $9a^2 - 18a + 27$ (b) $7xy^2 + 28x$

2. Factor completely.
(a) $6z^4 - 28z^3 + 16z^2$ (b) $2r^2t^2 - 18r^2$

3. Factor completely.
(a) $36x^3 - 48x^2 + 16x$
(b) $24b^3 - 81$

4. Solve each quadratic equation.
(a) $4y^2 - 6y = 0$ (b) $5z^2 + 2z = 3$

5. Solve $x^2 + 2x - 3 = 0$ symbolically and numerically with a table of values.

6. If a golf ball is hit upward at 60 miles per hour, then its height h in feet after t seconds is given by $h = 88t - 16t^2$. Use factoring to determine when the golf ball strikes the ground.

6.7 SOLVING EQUATIONS BY FACTORING II (HIGHER DEGREE)

Polynomials Having Common Factors ∎ Special Types of Polynomials

A LOOK INTO MATH ▷ In this section we discuss factoring polynomials having higher degree. Polynomials of degree 2 or higher are often used in applications. For example, the polynomial

$$0.0013x^3 - 0.085x^2 + 1.6x + 12$$

models natural gas consumption in the United States (in trillions of cubic feet) where $x = 0$ corresponds to 1960, $x = 10$ to 1970, and so on until $x = 40$ corresponds to 2000, as shown in Figure 6.8. In Exercises 75–78 we discuss further the consumption of natural gas. (**Source:** Department of Energy.)

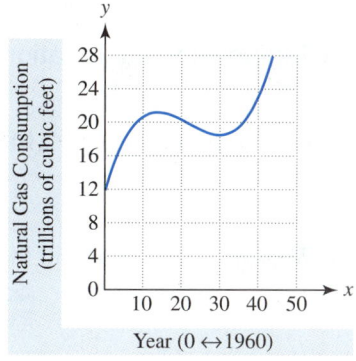

Figure 6.8 Natural Gas Consumption

Polynomials Having Common Factors

The first step in factoring a polynomial is to factor out the greatest common factor (GCF). For example, to factor $x^3 - x$ we start by factoring out the GCF, which is x. The resulting expression can be factored as the difference of two squares.

$$x^3 - x = x(x^2 - 1) \qquad \text{Factor out } x.$$
$$= x(x - 1)(x + 1) \qquad \text{Difference of two squares}$$

EXAMPLE 1 **Factoring trinomials with common factors**

Factor each trinomial completely.
(a) $-4x^2 + 28x - 40$ **(b)** $10x^3 + 28x^2 - 6x$

Solution
(a) Each term in $-4x^2 + 28x - 40$ has a factor of -4, so we start by factoring out -4. (In this example, we factor out the *negative* of the GCF to obtain a positive leading coefficient.)

$$-4x^2 + 28x - 40 = -4(x^2 - 7x + 10) \qquad \text{Factor out } -4.$$
$$= -4(x - 5)(x - 2) \qquad \text{Factor the trinomial.}$$

(b) We start by factoring out the GCF for $10x^3 + 28x^2 - 6x$, which is **$2x$**.

$$10x^3 + 28x^2 - 6x = \mathbf{2x}(5x^2 + 14x - 3) \qquad \text{Factor out } 2x.$$
$$= 2x(5x - 1)(x + 3) \qquad \text{Factor the trinomial.}$$

Now Try Exercises 11, 19

Many equations involving higher degree polynomials can be solved using the four-step process discussed in Section 6.6. It is important to remember that the zero-product property applies to *all* factors that contain the variable. In the next example we apply the zero-product property to three factors, resulting in three solutions.

EXAMPLE 2   **Solving polynomial equations**

Solve each equation.
(a) $x^3 - x^2 - 6x = 0$ **(b)** $4x^4 + 10x^3 = 6x^2$

Solution
(a) We start by factoring out the GCF, which is x.

$$\begin{aligned} x^3 - x^2 - 6x &= 0 && \text{Given equation} \\ \mathbf{x}(x^2 - x - 6) &= 0 && \text{Factor out } x. \\ \mathbf{x}(\mathbf{x - 3})(\mathbf{x + 2}) &= 0 && \text{Factor the trinomial.} \end{aligned}$$

$$x = 0 \quad \text{or} \quad x - 3 = 0 \quad \text{or} \quad x + 2 = 0 \qquad \text{Zero-product property}$$
$$x = 0 \quad \text{or} \quad x = 3 \quad \text{or} \quad x = -2 \qquad \text{Solve for } x.$$

The solutions are -2, 0, and 3.

(b) We start by subtracting $6x^2$ from each side to obtain 0 on one side of the equation.

$$\begin{aligned} 4x^4 + 10x^3 &= 6x^2 && \text{Given equation} \\ 4x^4 + 10x^3 - 6x^2 &= 0 && \text{Subtract } 6x^2. \\ 2x^2(2x^2 + 5x - 3) &= 0 && \text{Factor out the GCF, } 2x^2. \\ 2x^2(2x - 1)(x + 3) &= 0 && \text{Factor the trinomial.} \end{aligned}$$

$$2x \cdot x = 0 \quad \text{or} \quad 2x - 1 = 0 \quad \text{or} \quad x + 3 = 0 \qquad \text{Zero-product property}$$

$$x = 0 \quad \text{or} \quad x = \frac{1}{2} \quad \text{or} \quad x = -3 \qquad \text{Solve for } x.$$

The solutions are -3, 0, and $\frac{1}{2}$.

Now Try Exercises 55, 61

▶ **REAL-WORLD CONNECTION** The corners of a square piece of metal are cut out to form a box, as shown in Figure 6.9. This square piece of metal has sides with length 10 inches, and the cutout corners are squares with length x. The outside surface area A of this box, including the bottom and the sides but *not* the top, is $A = 100 - 4x^2$. (See Critical Thinking in the margin on the next page.)

CRITICAL THINKING

Calculate the outside surface area A of the box shown in Figure 6.9 two different ways.

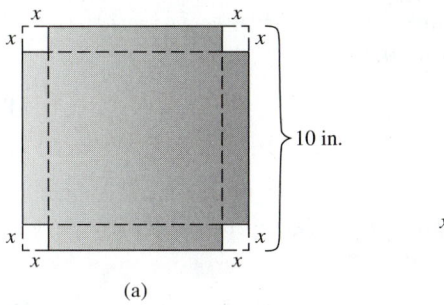

(a)   (b)

Figure 6.9

EXAMPLE **3** **Finding a dimension of a box**

Find the value of x in Figure 6.9 if its outside surface area A is 84 square inches.

Solution

Let $A = 84$ in the equation $A = 100 - 4x^2$ and solve for x.

$$100 - 4x^2 = 84 \qquad A = 84$$

$$16 - 4x^2 = 0 \qquad \text{Subtract 84.}$$

$$4(4 - x^2) = 0 \qquad \text{Factor out 4.}$$

$$4(2 - x)(2 + x) = 0 \qquad \text{Difference of squares}$$

$$2 - x = 0 \quad \text{or} \quad 2 + x = 0 \qquad \text{Zero-product property}$$

$$x = 2 \quad \text{or} \quad x = -2 \qquad \text{Solve for } x.$$

If squares measuring 2 inches on a side are cut out of each corner, the surface area of the box is 84 square inches. (Note that $x = -2$ has no physical meaning in this problem.)

Now Try Exercise **73**

Special Types of Polynomials

Some types of polynomials of higher degree can be factored by using methods that we have already presented.

EXAMPLE **4** **Factoring higher degree polynomials**

Factor each polynomial completely.
(a) $x^4 - 16$ **(b)** $y^4 + 5y^2 + 4$ **(c)** $x^4 + 2x^2y^2 + y^4$ **(d)** $r^4 - t^4$

Solution
(a) We view this polynomial as the difference of two squares, where $x^4 = (x^2)^2$. Then we factor twice.

$$x^4 - 16 = (x^2)^2 - 4^2 \qquad \text{Rewrite.}$$

$$= (x^2 - 4)(x^2 + 4) \qquad \text{Difference of squares}$$

$$= (x - 2)(x + 2)(x^2 + 4) \qquad \text{Difference of squares}$$

Note that $x^2 + 4$ does not factor.

(b) Because $a^2 + 5a + 4 = (a + 1)(a + 4)$, we let $a = y^2$ and then factor the given trinomial.

$$y^4 + 5y^2 + 4 = (y^2)^2 + 5(y^2) + 4$$
$$= (y^2 + 1)(y^2 + 4)$$

Note that neither $y^2 + 1$ nor $y^2 + 4$ can be factored further.

(c) Because $a^2 + 2ab + b^2 = (a + b)^2$, we let $a = x^2$ and $b = y^2$ and then factor the given trinomial.

$$x^4 + 2x^2y^2 + y^4 = (x^2)^2 + 2x^2y^2 + (y^2)^2$$
$$= (x^2 + y^2)^2$$

(d) Because $a^2 - b^2 = (a - b)(a + b)$, we let $a = r^2$ and $b = t^2$ and then factor the given binomial.

$$r^4 - t^4 = (r^2)^2 - (t^2)^2$$
$$= (r^2 - t^2)(r^2 + t^2)$$
$$= (r - t)(r + t)(r^2 + t^2)$$

Note that $r^2 + t^2$ cannot be factored.   Now Try Exercises 35, 41, 43

EXAMPLE 5   **Solving an equation**

Solve $x^5 - 81x = 0$.

Solution

We start by factoring out the common factor of x.

$x^5 - 81x = 0$	Given equation
$x(x^4 - 81) = 0$	Factor out x.
$x(x^2 - 9)(x^2 + 9) = 0$	Difference of two squares
$x(x - 3)(x + 3)(x^2 + 9) = 0$	Difference of two squares
$x = 0$ or $x - 3 = 0$ or $x + 3 = 0$ or $x^2 + 9 = 0$	Zero-product property
$x = 0$ or $x = 3$ or $x = -3$	Solve for x.

Note that $x^2 + 9 = 0$ has no real number solutions because the square of a number plus 9 is never 0. The solutions are -3, 0, and 3.   Now Try Exercise 59

6.7   PUTTING IT ALL TOGETHER

The following table summarizes factoring higher degree polynomials and solving higher degree polynomial equations.

Concept	Explanation	Examples
Common Factors	A first step when factoring polynomials is to factor out the GCF.	$x^3 - 4x^2 - 5x = x(x^2 - 4x - 5)$ $= x(x - 5)(x + 1)$ $4x^4 - 16x^2 = 4x^2(x^2 - 4)$ $= 4x^2(x - 2)(x + 2)$

Concept	Explanation	Examples
Factoring Trinomials	Some higher degree trinomials can be factored by using the same methods that we used to factor quadratic trinomials. **NOTE:** $a^2 + b^2$ cannot be factored.	$x^4 + 6x^2 + 5 = (x^2 + 5)(x^2 + 1)$ $4y^4 - 25 = (2y^2 - 5)(2y^2 + 5)$ $2z^4 - 32 = 2(z^4 - 16)$ $= 2(z^2 - 4)(z^2 + 4)$ $= 2(z - 2)(z + 2)(z^2 + 4)$
Solving Equations by Factoring	Use factoring and the zero-product property to solve polynomial equations.	$3x^3 - 12x^2 + 9x = 0$ $3x(x^2 - 4x + 3) = 0$ $3x(x - 3)(x - 1) = 0$ $3x = 0$ or $x - 3 = 0$ or $x - 1 = 0$ $x = 0$ or $x = 3$ or $x = 1$ The solutions are 0, 1, and 3.

6.7 Exercises

CONCEPTS

1. When you are factoring polynomials, a good first step is to factor out the _____.

2. When you are solving an equation by factoring, the _____ property is used.

3. The zero-product property applies to all _____ containing the variable.

4. Because $a^2 - 2ab + b^2 = (a - b)^2$, it follows that $x^4 - 2x^2y^2 + y^4 =$_____.

5. Because $x^2 + 3x + 2 = (x + 1)(x + 2)$, it follows that $z^4 + 3z^2 + 2 =$ _____.

6. Is $x^4 - 1 = (x^2 - 1)(x^2 + 1)$ factored completely? Explain.

7. When you are solving the equation $x^3 = x$, what is a good first step?

8. When you are solving $x^4 - x^2 = 0$, what is a good first step?

FACTORING POLYNOMIALS

Exercises 9–20: Factor the polynomial completely.

9. $5x^2 - 5x - 30$

10. $3x^2 - 15x + 12$

11. $-4y^2 - 32y - 48$

12. $-7y^2 + 14y + 21$

13. $-20z^2 - 110z - 50$

14. $-12z^2 - 54z + 30$

15. $60 - 64t - 28t^2$

16. $18 - 45t - 27t^2$

17. $r^3 - r$

18. $r^3 + 2r^2 - 3r$

19. $3x^3 + 3x^2 - 18x$

20. $6x^3 - 26x^2 - 20x$

Exercises 21–46: Factor completely.

21. $72z^3 + 12z^2 - 24z$

22. $6z^3 - 4z^2 - 42z$

23. $x^4 - 4x^2$

24. $4x^4 - 36x^2$

25. $t^4 + t^3 - 2t^2$

26. $t^4 + 5t^3 - 24t^2$

27. $x^4 - 5x^2 + 6$

28. $x^4 - 3x^2 - 10$

29. $2x^4 + 7x^2 + 3$

30. $3x^4 - 8x^2 + 5$

31. $y^4 + 6y^2 + 9$

32. $y^4 - 10y^2 + 25$

33. $x^4 - 9$

34. $x^4 - 25$

35. $x^4 - 81$

36. $4x^4 - 64$

37. $z^5 + 2z^4 + z^3$

38. $6z^5 - 47z^4 + 35z^3$

39. $2x^2 + xy - y^2$

40. $2x^2 + 5xy + 2y^2$

41. $a^4 - 2a^2b^2 + b^4$ **42.** $a^3 + 2a^2b + ab^2$

43. $x^3 - xy^2$ **44.** $2x^2y - 2y^3$

45. $4x^3 + 4x^2y + xy^2$ **46.** $x^2y - 6xy^2 + 9y^3$

SOLVING EQUATIONS

Exercises 47–50: Do the following.

47. **(a)** Factor $x^3 - 4x$.
(b) Solve $x^3 - 4x = 0$.

48. **(a)** Factor $4x^3 - 16x$.
(b) Solve $4x^3 - 16x = 0$.

49. **(a)** Factor $2y^3 - 6y^2 - 36y$.
(b) Solve $2y^3 - 6y^2 - 36y = 0$.

50. **(a)** Factor $z^4 - 13z^2 + 36$.
(b) Solve $z^4 - 13z^2 + 36 = 0$.

Exercises 51–72: Solve.

51. $3x^2 + 33x + 72 = 0$ **52.** $4x^2 - 16x - 20 = 0$

53. $25x^2 = 50x + 75$ **54.** $10x^2 = 20x + 80$

55. $y^3 - 3y^2 - 4y = 0$ **56.** $y^3 - 3y^2 + 2y = 0$

57. $3z^3 + 6z^2 = 72z$ **58.** $4z^3 = 4z^2 + 24z$

59. $x^4 - 36x^2 = 0$ **60.** $4x^4 = 100x^2$

61. $r^4 + 6r^3 = 7r^2$ **62.** $r^4 + 30r^2 = 11r^3$

63. $x^4 - 13x^2 = -36$ **64.** $x^4 - 17x^2 + 16 = 0$

65. $x^4 + 1 = 2x^2$ **66.** $x^4 - 8x^2 + 16 = 0$

67. $a^4 = 81$ **68.** $b^3 = -8$

69. $x^3 - 2x^2 - x + 2 = 0$

70. $x^3 - x^2 + 4x - 4 = 0$

71. $x^3 - 5x^2 + x - 5 = 0$

72. $3x^3 + 2x^2 - 27x - 18 = 0$

APPLICATIONS

73. *Dimensions of a Box*   (Refer to Example 3.) A box is made from a rectangular piece of metal with length 20 inches and width 15 inches. The box has no top.

(a) What are the limitations on the size of x? Explain. your answer.

(b) Write an expression that gives the outside surface area of the box. (*Hint:* Consider the size of the metal sheet and how much was cut out.)

(c) If the outside surface area of the box is 275 square inches, find x.

74. *Dimensions of a Box*   Refer to the previous exercise.
(a) Find a polynomial that gives the volume of the box for a given x.
(b) Factor your polynomial completely.
(c) What are the zeros of your polynomial? What do they represent in this problem?

Exercises 75–78: **U.S. Natural Gas Consumption**   *(Refer to the introduction to this section.) The polynomial*

$$0.0013x^3 - 0.085x^2 + 1.6x + 12$$

models natural gas consumption in trillions of cubic feet, where $x = 0$ corresponds to 1960, $x = 1$ to 1961, and so on.

75. How much natural gas was consumed in 1990?

76. In which year (between 1970 and 1990) was natural gas consumption about 20.4 trillion cubic feet?

77. Explain any difficulties encountered when you try to solve the equation

$$0.0013x^3 - 0.085x^2 + 1.6x + 12 = 23.2.$$

78. How might you solve this equation without factoring? If you were to find the solution to this equation, what would it represent?

WRITING ABOUT MATHEMATICS

79. Compare factoring the polynomial $x^2 + 6x + 5$ with factoring the polynomial $z^4 + 6z^2 + 5$.

80. Suppose that a polynomial can be factored. Explain how its factors can be used to find the zeros of the polynomial. Give an example.

CHECKING BASIC CONCEPTS
SECTION 6.7

1. Factor the trinomial completely.
   (a) $3x^2 - 6x - 24$ **(b)** $-10y^2 + 5y + 5$

2. Factor the binomial completely.
   (a) $z^4 - 25$ **(b)** $7t^4 - 7$

3. Factor.
   (a) $x^4 - 8x^2 + 16$ **(b)** $2y^3 + 17y^2 - 30y$

4. Solve $t^4 + t^3 = 12t^2$.

5. Solve $x^3 - 3x^2 + 2x - 6 = 0$.

CHAPTER 6   SUMMARY

SECTION 6.1 ■ INTRODUCTION TO FACTORING

Terms Related to Factoring Polynomials

Factoring	Writing a polynomial as a product, usually of lower degree polynomials
Common Factor	An expression that is a factor of each term in a polynomial

Example: Some common factors of $4x^4 + 8x^2$ are $2x$, x^2, and $4x^2$.

Greatest Common Factor (GCF)	The common factor with the greatest (integer) coefficient and the highest degree

Example: The GCF of $4x^4 + 8x^2$ is $4x^2$.

$$4x^4 + 8x^2 = 4x^2(x^2 + 2)$$

Factoring by Grouping	Used to factor a four-term polynomial into two binomials

Example: $x^3 + 5x^2 + 3x + 15 = x^2(x + 5) + 3(x + 5)$
$$= (x^2 + 3)(x + 5)$$

SECTION 6.2 ■ FACTORING TRINOMIALS I ($x^2 + bx + c$)

Review of the FOIL Method   A method used for multiplying two binomials

First Terms	$(2x + 3)(5x + 4)$:	$2x \cdot 5x = 10x^2$
Outside Terms	$(2x + 3)(5x + 4)$:	$2x \cdot 4 = 8x$
Inside Terms	$(2x + 3)(5x + 4)$:	$3 \cdot 5x = 15x$
Last Terms	$(2x + 3)(5x + 4)$:	$3 \cdot 4 = 12$

The product is the sum of these four terms:

$$(2x + 3)(5x + 4) = 10x^2 + 8x + 15x + 12 = 10x^2 + 23x + 12.$$

Factoring Trinomials Having a Leading Coefficient of 1   To factor the trinomial $x^2 + bx + c$, find two numbers, m and n, that satisfy

$$m \cdot n = c \quad \text{and} \quad m + n = b.$$

Then $x^2 + bx + c = (x + m)(x + n)$.

Example: Because $-3 \cdot 5 = -15$ and $-3 + 5 = 2$,

$$x^2 + 2x - 15 = (x - 3)(x + 5).$$

SECTION 6.3 ■ FACTORING TRINOMIALS II ($ax^2 + bx + c$)

Factoring Trinomials by Grouping   To factor $ax^2 + bx + c$, perform the following steps. (Assume that a, b, and c are integers and have no factor in common.)

1. Find two numbers, m and n, such that $mn = ac$ and $m + n = b$.
2. Write the trinomial as $ax^2 + mx + nx + c$.
3. Use grouping to factor this expression into two binomials.

Example: To factor $3x^2 + 10x - 8$, find two numbers whose product is -24 and whose sum is 10. These two numbers are $m = 12$ and $n = -2$, so write $10x$ as $12x - 2x$.

$$\begin{aligned} 3x^2 + 10x - 8 &= 3x^2 + 12x - 2x - 8 \\ &= (3x^2 + 12x) + (-2x - 8) \\ &= 3x(x + 4) - 2(x + 4) \\ &= (3x - 2)(x + 4) \end{aligned}$$

Factoring with FOIL in Reverse   Use trial and error and FOIL in reverse to find the factors of a trinomial.

Example: To factor $3x^2 + 10x - 8$, first find factors of $3x^2$.

$$(3x + \underline{})(x + \underline{})$$

Then place factors of -8 so that the resulting middle term is $10x$.

$$(3x + \underline{-2}) \cdot (x + \underline{4})$$

$-2x$

$12x$

$10x$ ⟵ Middle term checks.

SECTION 6.4 ■ SPECIAL TYPES OF FACTORING

Difference of Two Squares

$$a^2 - b^2 = (a - b)(a + b)$$

Examples: $x^2 - 16 = (x - 4)(x + 4)$ $(a = x, b = 4)$
$4r^2 - 9t^2 = (2r - 3t)(2r + 3t)$ $(a = 2r, b = 3t)$

Perfect Square Trinomials

$$a^2 + 2ab + b^2 = (a + b)^2 \quad \text{and} \quad a^2 - 2ab + b^2 = (a - b)^2$$

Examples: $4x^2 + 4x + 1 = (2x)^2 + 2(2x)1 + 1^2 = (2x + 1)^2$ $(a = 2x, b = 1)$

$x^2 - 10x + 25 = x^2 - 2 \cdot x \cdot 5 + 5^2 = (x - 5)^2$ $(a = x, b = 5)$

Sums and Differences of Two Cubes

$a^3 + b^3 = (a + b)(a^2 - ab + b^2)$ and $a^3 - b^3 = (a - b)(a^2 + ab + b^2)$

Examples: $x^3 + 8 = (x + 2)(x^2 - 2x + 4)$ $(a = x, b = 2)$

$27x^3 - 1 = (3x - 1)(9x^2 + 3x + 1)$ $(a = 3x, b = 1)$

SECTION 6.5 ■ SUMMARY OF FACTORING

Guidelines for Factoring Polynomials The following guidelines can be used to factor polynomials in general.

STEP 1: Factor out the greatest common factor, if possible.

STEP 2: **A.** If the polynomial has *four terms*, try factoring by grouping.

B. If the polynomial is a *binomial,* try one of the following.

1. $a^2 - b^2 = (a - b)(a + b)$ Difference of two squares
2. $a^3 - b^3 = (a - b)(a^2 + ab + b^2)$ Difference of two cubes
3. $a^3 + b^3 = (a + b)(a^2 - ab + b^2)$ Sum of two cubes

C. If the polynomial is a *trinomial*, check for a perfect square.

1. $a^2 + 2ab + b^2 = (a + b)^2$ Perfect square trinomial
2. $a^2 - 2ab + b^2 = (a - b)^2$ Perfect square trinomial

Otherwise, apply FOIL in reverse or apply grouping, as described in Sections 6.2 and 6.3.

STEP 3: Check to make sure that the polynomial is *completely* factored.

Examples: $12x^3 - 12x^2 + 3x = 3x(4x^2 - 4x + 1) = 3x(2x - 1)^2$ Steps 1, 2C, and 3

$9x^3 - 6x^2 + 18x - 12 = 3(3x^3 - 2x^2 + 6x - 4) = 3(x^2 + 2)(3x - 2)$ Steps 1, 2A, and 3

$16x^3 - 100x = 4x(4x^2 - 25) = 4x(2x - 5)(2x + 5)$ Steps 1, 2B, and 3

SECTION 6.6 ■ SOLVING EQUATIONS BY FACTORING I (QUADRATICS)

Zero-Product Property

For any real numbers a and b, if $ab = 0$, then $a = 0$ or $b = 0$ (or both).
The zero-product property is used to solve equations.

Examples: $xy = 0$ implies that $x = 0$ or $y = 0$.

$(x + 5)(x - 3) = 0$ implies $x + 5 = 0$ or $x - 3 = 0$.

Zero of a Polynomial A number a is a zero of a polynomial if the result is 0 when a is substituted in that polynomial.

Example: The number -2 is a zero of $x^2 - 4$ because $(-2)^2 - 4 = 0$.

Solving Quadratic Equations by Factoring   To solve a quadratic equation by factoring, follow these steps.

STEP 1: If necessary, use algebra to write the equation as $ax^2 + bx + c = 0$.

STEP 2: Factor the left side of the equation using any method.

STEP 3: Apply the zero-product property.

STEP 4: Solve each of the resulting equations. Check any solutions.

Example:

$$x^2 + 7x = 8 \quad \text{Given equation}$$
$$x^2 + 7x - 8 = 0 \quad \text{Step 1}$$
$$(x + 8)(x - 1) = 0 \quad \text{Step 2}$$
$$x + 8 = 0 \quad \text{or} \quad x - 1 = 0 \quad \text{Step 3}$$
$$x = -8 \quad \text{or} \quad x = 1 \quad \text{Step 4}$$

SECTION 6.7 ■ SOLVING EQUATIONS BY FACTORING II (HIGHER DEGREE)

Factoring Polynomials of Higher Degree   The distributive property and the techniques for factoring quadratic polynomials can also be applied to polynomials of higher degree.

Examples: $10r^3 + 15r = 5r(2r^2 + 3)$ (To check, multiply the right side.)

Because $2x^2 + x - 1 = (x + 1)(2x - 1)$,
it follows that $2z^4 + z^2 - 1 = (z^2 + 1)(2z^2 - 1)$.

Solving Equations by Factoring   Use algebra to obtain 0 on one side of the equation. Factor the other side and apply the zero-product property.

Example:

$$x^3 = 4x$$
$$x^3 - 4x = 0 \quad \text{Subtract } 4x.$$
$$x(x^2 - 4) = 0 \quad \text{Factor out the GCF, } x.$$
$$x(x - 2)(x + 2) = 0 \quad \text{Difference of squares}$$
$$x = 0 \quad \text{or} \quad x - 2 = 0 \quad \text{or} \quad x + 2 = 0 \quad \text{Zero-product property}$$
$$x = 0 \quad \text{or} \quad x = 2 \quad \text{or} \quad x = -2 \quad \text{Solve for } x.$$

CHAPTER 6   REVIEW EXERCISES

SECTION 6.1

Exercises 1–4: Identify the greatest common factor for the expression and then factor the expression.

1. $8z^3 - 4z^2$

2. $6x^4 + 3x^3 - 12x^2$

3. $9xy + 15yz^2$

4. $a^2b^3 + a^3b^2$

Exercises 5–12: Use grouping to factor the given polynomial completely.

5. $x(x + 2) - 3(x + 2)$

6. $y^2(x - 5) + 3y(x - 5)$ **7.** $z^3 - 2z^2 + 5z - 10$

8. $t^3 + t^2 + 8t + 8$ **9.** $x^3 - 3x^2 + 6x - 18$

10. $ax + bx - ay - by$ **11.** $x^5 + 3x^4 - 2x^3 - 6x^2$

12. $2y^4 + 6y^3 + 2y^2 + 6y$

SECTION 6.2

Exercises 13–16: Find an integer pair that has the given product and sum.

13. Product: 20   Sum: 9

14. Product: -21 Sum: 4

15. Product: 36   Sum: -13

16. Product: -100 Sum: -21

Exercises 17–26: Factor the trinomial completely.

17. $x^2 - x - 12$ **18.** $x^2 + 10x + 24$

19. $x^2 + 6x - 16$ **20.** $x^2 - x - 42$

21. $x^2 + 2x - 3$ **22.** $x^2 + 22x + 120$

23. $2x^3 + 6x^2 - 20x$ **24.** $x^4 - 3x^3 - 28x^2$

25. $10 - 7x + x^2$ **26.** $24 + 2x - x^2$

SECTION 6.3

Exercises 27–36: Factor the trinomial completely.

27. $9x^2 + 3x - 2$ **28.** $2x^2 + 3x - 5$

29. $3x^2 + 14x + 15$ **30.** $35x^2 - 2x - 1$

31. $24x^2 - 7x - 5$ **32.** $4x^2 + 33x - 27$

33. $12x^3 + 48x^2 + 21x$ **34.** $8x^4 + 14x^3 - 30x^2$

35. $12 - 5x - 2x^2$ **36.** $1 + 3x - 10x^2$

SECTION 6.4

Exercises 37–50: Factor completely.

37. $z^2 - 4$ **38.** $9z^2 - 64$

39. $36 - y^2$ **40.** $100a^2 - 81b^2$

41. $x^2 + 14x + 49$ **42.** $x^2 - 10x + 25$

43. $4x^2 - 12x + 9$ **44.** $9x^2 + 48x + 64$

45. $8t^3 - 1$ **46.** $27r^3 + 8t^3$

47. $2x^3 - 50x$ **48.** $24x^3 + 81$

49. $2x^3 + 28x^2 + 98x$ **50.** $2x^4 - 128x$

SECTION 6.5

Exercises 51–58: Factor completely.

51. $9y^2 - 6y + 6$ **52.** $yz^2 - 9y$

53. $x^4 + 7x^3 - 4x^2 - 28x$ **54.** $12x^3 + 36x^2 + 27x$

55. $3ab^3 - 24a$ **56.** $5x^3 + 20x$

57. $24x^3 - 6xy^2$ **58.** $x^3y + 27y$

SECTION 6.6

Exercises 59–70: Solve the equation.

59. $mn = 0$ **60.** $y^2 = 0$

61. $(4x - 3)(x + 9) = 0$

62. $(1 - 4x)(6 + 5x) = 0$

63. $z(z - 1)(z - 2) = 0$ **64.** $z^2 - 7z = 0$

65. $y^2 - 64 = 0$ **66.** $y^2 + 9y + 14 = 0$

67. $x^2 = x + 6$ **68.** $10x^2 + 11x = 6$

69. $t(t - 14) = 72$ **70.** $t(2t - 1) = 10$

SECTION 6.7

Exercises 71–80: Factor completely.

71. $5x^2 - 15x - 50$ **72.** $-3x^2 - 6x + 45$

73. $y^3 - 4y$ **74.** $3y^3 + 6y^2 - 9y$

75. $2z^4 + 14z^3 + 20z^2$ **76.** $8z^4 - 32z^2$

77. $x^4 - 6x^2 + 9$ **78.** $2x^4 - 15x^2 - 27$

79. $a^2 + 10ab + 25b^2$ **80.** $x^3 - xy^2$

Exercises 81–88: Solve.

81. $16x^2 - 72x - 40 = 0$

82. $2x^3 - 11x^2 + 15x = 0$ **83.** $t^3 = 25t$

84. $t^4 - 7t^3 + 12t^2 = 0$ **85.** $z^4 + 16 = 8z^2$

86. $z^4 - 256 = 0$ **87.** $y^3 = -64$

88. $y^3 - y^2 - y + 1 = 0$

APPLICATIONS

89. A square has area $9x^2 + 42x + 49$. Find the length of a side. Make a sketch of the square.

90. A rectangle has area $x^2 + 6x + 5$. Find possible dimensions for the rectangle. Make a sketch of the rectangle.

91. A cube has surface area $6x^2 + 12x + 6$. Find the length of a side.

92. Write a polynomial in factored form that represents the total area of the rectangle.

93. Write a polynomial in factored form that represents the total area of the rectangle.

94. *Radius of a Circle*   The area and the circumference of a circle are numerically equal. Find the radius of the circle.

95. *Dimensions of a Shed*   The floor of a rectangular shed is 7 feet longer than it is wide and has an area of 120 square feet. What are its dimensions?

96. *Flight of a Ball*   A ball is hit upward. Its height h in feet after t seconds is given by $h = -16t^2 + 80t + 4$. At what times is the ball 100 feet in the air?

97. *Stopping Distance*   The distance D in feet that it takes to stop a car traveling x miles per hour on wet, level pavement can be approximated by $D = \frac{1}{9}x^2 + \frac{11}{3}x$.
   (a) Estimate the distance required for the car to stop when it is traveling 45 miles per hour.
   (b) If the stopping distance is 80 feet, what is the speed of the car?
   (c) If you have a calculator, use it to solve part (b) numerically with a table of values. Do your answers agree?

98. *Revenue*   A company makes tops for the boxes of pickup trucks. The total revenue R in dollars from selling the tops for p dollars each is given by $R = p(200 - p)$, where $p \le 200$.
   (a) Find R when $p = \$100$.
   (b) Find p when $R = \$7500$.
   (c) If you have a calculator, use it to solve part (b) numerically with a table of values. Do your answers agree?

99. *Airline Passengers*   The number N of worldwide airline passengers in millions from 1950 to 2000 is approximated by

$$N = 0.68y^2 + 3.8y + 24,$$

where $y = 0$ corresponds to 1950, $y = 1$ to 1951, and so on until $y = 50$ corresponds to 2000.
   (a) Estimate the number of airline passengers in 1970.
   (b) Use a table of values to estimate the year in which the number of airline passengers reached 544 million.

100. *Digital Photographs*   A digital photograph is 30 pixels longer than it is wide and has a total area of 4000 pixels. Find the dimensions of this photograph.

101. *Dimensions of a Box*   A box is made from a rectangular piece of metal with length 50 inches and width 40 inches by cutting out square corners of length x and folding up the sides.
   (a) Write an expression that gives the surface area of the inside of the box.
   (b) If the surface area of the box is 1900 square inches, find x.

CHAPTER 6 TEST ⊙Pass the Test Video solutions to all test exercises

Exercises 1 and 2: Identify the greatest common factor for the expression. Then factor the expression.

1. $4x^2y - 20xy^2 + 12xy$ **2.** $9a^3b^2 + 3a^2b^2$

Exercises 3 and 4: Factor by grouping.

3. $ay + by + az + bz$ **4.** $3x^3 + x^2 - 15x - 5$

Exercises 5–8: Factor the trinomial.

5. $y^2 + 4y - 12$ **6.** $4x^2 + 20x + 25$

7. $4z^2 - 19z + 12$ **8.** $21 - 17t + 2t^2$

Exercises 9–14: Factor completely.

9. $6x^3 + 3x^2 - 3x$ **10.** $2z^4 - 12z^2 - 54$

11. $36y^3 - 100y$ **12.** $7x^4 + 56x$

13. $16a^4 + 24a^3 + 9a^2$ **14.** $2b^4 - 32$

Exercises 15–20: Solve the equation.

15. $x^2 - 16 = 0$ **16.** $y^2 = y + 20$

17. $9z^2 + 16 = 24z$ **18.** $x(x - 5) = 66$

19. $y^3 = 9y$ **20.** $x^4 - 5x^2 + 4 = 0$

21. A square has area $9x^2 + 30x + 25$. Find the length of a side in terms of x.

22. Write a polynomial in factored form that represents the total area of the rectangle.

23. *Braking Distance* The braking distance D in feet required for a car traveling at x miles per hour to stop on dry, level pavement can be modeled by $D = \frac{1}{11}x^2$.
(a) Calculate the distance required for the car to stop when it is traveling 55 miles per hour.
(b) If the braking distance is 99 feet, estimate the speed of the car.

24. *Flight of a Ball* A ball is thrown upward. Its height h in feet after t seconds is given by $h = -16t^2 + 48t + 4$. At what times is the ball 36 feet in the air?

CHAPTER 6 EXTENDED AND DISCOVERY EXERCISES

Exercises 1–6: Difference of Two Squares The difference of two squares can be factored by using

$$a^2 - b^2 = (a - b)(a + b).$$

This equation can also be used in some situations where an expression may not appear to be the difference of two squares. For example, because $(\sqrt{3})^2 = 3$, $x^2 - 3$ can be written and then factored as

$$x^2 - 3 = x^2 - (\sqrt{3})^2$$
$$= (x - \sqrt{3})(x + \sqrt{3}).$$

Use this concept to factor the following expressions as the difference of two squares.

1. $x^2 - 5$ **2.** $y^2 - 7$

3. $3z^2 - 25$ **4.** $7t^2 - 2$

5. $x - 4$ for $x \geq 0$ (*Hint:* $(\sqrt{x})^2 = x$.)

6. $x - 7$ for $x \geq 0$

Exercises 7–12: Solving Equations (Refer to Exercises 1–6.) Solve the equation by factoring it as the difference of two squares.

7. $x^2 - 3 = 0$ **8.** $y^2 - 7 = 0$

9. $3x^2 - 25 = 0$ **10.** $7x^2 - 11 = 0$

11. $x^4 - 9 = 0$ **12.** $x^4 - 25 = 0$

CHAPTERS 1–6   CUMULATIVE REVIEW EXERCISES

1. Write 144 as a product of prime numbers.

2. Evaluate $-2x + 3y$ for $x = -2$ and $y = 4$.

Exercises 3 and 4: Evaluate by hand and then simplify to lowest terms.

3. $\frac{3}{5} \cdot \frac{15}{21}$ 4. $\frac{4}{5} - \frac{1}{10}$

Exercises 5 and 6: Evaluate by hand.

5. $26 - 3 \cdot 6 \div 2$ 6. $-2^2 + \frac{3+2}{8+2}$

7. Solve $4t - 7 = 25$ and check.

8. Complete the table. Then use the table to solve the equation $2x + 3 = 5$.

x	-2	-1	0	1	2
$2x + 3$					

9. Translate the sentence "Triple a number decreased by 5 equals the number decreased by 7" into an equation using the variable n. Then solve the equation.

10. Convert 5.7% to fraction and decimal notation.

11. Convert 0.123 to a percentage.

12. Solve $P = 2W + 2L$ for W.

13. Solve $5 - 3z < -1$.

14. Make a scatterplot having the following five points: $(-2, 3)$, $(-1, 2)$, $(0, -1)$, $(1, 1)$, and $(2, 2)$.

Exercises 15 and 16: Graph the given equation. Determine any intercepts.

15. $y = 3x - 2$ 16. $y = -2$

17. Identify the x-intercept and the y-intercept. Then write the slope–intercept form of the line.

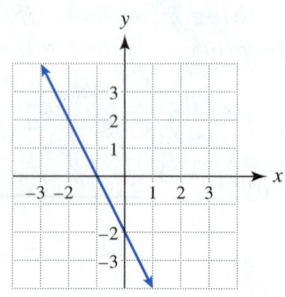

Exercises 18 and 19: Find the slope–intercept form for the line satisfying the given conditions.

18. Perpendicular to $2x - 3y = -6$ and passing through the point $(1, 2)$

19. Passing through the points $(-2, 1)$ and $(1, 5)$

20. The graphs of two equations are shown. Use the graphs to identify the solution to the system of equations. Then check your answer.

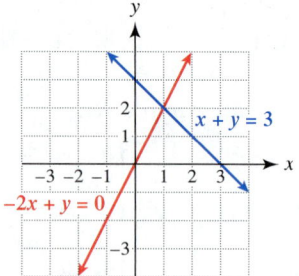

Exercises 21 and 22: Solve the system of equations.

21. $\begin{aligned} y &= -1 \\ 2x + y &= 1 \end{aligned}$ 22. $\begin{aligned} 5x + y &= -5 \\ -x + 2y &= 12 \end{aligned}$

Exercises 23 and 24: The graphs of two linear equations are shown.

(a) State the number of solutions to the system of equations.

(b) Is the system consistent or inconsistent? If the system is consistent, state whether the equations are dependent or independent.

23.   24.

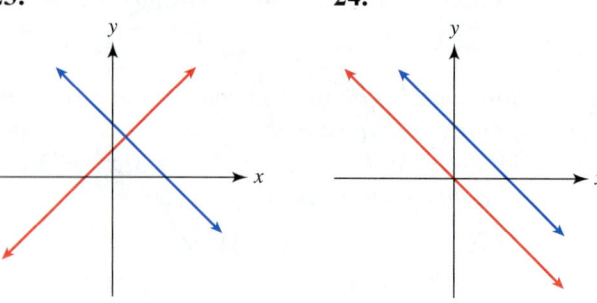

Exercises 25 and 26: Shade the solution set to the given inequality.

25. $x \le 2$ 26. $2x + 3y \ge 6$

Exercises 27 and 28: Shade the solution set to the system of inequalities.

27. $x > -1$
$\quad y < x$

28. $2x - y \le 4$
$\quad x + 2y \ge 2$

Exercises 29–34: Simplify the expression.

29. -2^4

30. $(xy)^0$

31. $(xy)^4(x^3y^{-4})^2$

32. $7x^3(-2x^2 + 3x)$

33. $(7x - 2)(3x + 5)$

34. $(x - 7)^2$

Exercises 35–38: Simplify and write the expression using positive exponents.

35. $a^{-4} \cdot a^2$

36. $(2t^3)^{-2}$

37. $(xy)^{-3}(x^{-1}y^2)^{-1}$

38. $\left(\dfrac{2x}{y^{-2}}\right)^5$

39. Write 6.23×10^{-3} in standard form.

40. Write 543,000 in scientific notation.

Exercises 41 and 42: Divide and check.

41. $\dfrac{6x^3 + 12x^2}{3x}$

42. $\dfrac{3x^3 - x + 1}{x^2 + 1}$

Exercises 43 and 44: Factor by grouping.

43. $2y^2(x + 2) - 5(x + 2)$ **44.** $t^3 + 6t^2 + t + 6$

Exercises 45–52: Factor completely.

45. $x^2 + 3x - 28$

46. $6y^2 + y - 12$

47. $25x^2 - 4y^2$

48. $64x^2 - 16x + 1$

49. $27t^3 - 8$

50. $-4x^2 + 4x + 24$

51. $x^4 - 12x^2 + 27$

52. $x^3y - x^2y^2$

Exercises 53–56: Solve the equation.

53. $y^4 = 25y^2$

54. $8z^2 + 8z - 16 = 0$

55. $4z^3 = 49z$

56. $x^4 - 18x^2 + 81 = 0$

APPLICATIONS

57. *Shoveling the Driveway* Two people are shoveling snow from a driveway. The first person shovels 10 square feet per minute, while the second person shovels 8 square feet per minute.

(a) Write and simplify an expression that gives the total square feet that the two people shovel in x minutes.

(b) How many minutes would it take for them to clear a driveway with an area of 900 square feet?

58. *Running Distance* A person runs the following distances over three days: $1\frac{3}{4}$ miles, $2\frac{1}{2}$ miles, and $2\frac{2}{3}$ miles. How far does the person run altogether?

59. *Cost of a Car* A 7% sales tax on a car amounted to $1470. Find the cost of the car.

60. *Burning Calories* An athlete can burn 12 calories per minute while cross-country skiing and 9 calories per minute while running at 5 miles per hour. If the athlete burns 615 calories in 60 minutes, how long is spent on each activity?

61. *Renting a Car* A rental car costs $20 plus $0.25 per mile driven.

(a) Write an equation that gives the cost C of driving the car x miles.

(b) Determine the number of miles that the car is driven if the rental cost is $100.

62. *Triangle* In an isosceles triangle, the measures of the two smaller angles are equal and their sum is 20° more than the larger angle.

(a) Let x be the measure of one of the two smaller angles and y be the measure of the larger angle. Write a system of linear equations whose solution gives the measures of these angles.

(b) Solve your system.

63. *Radios and CD Players* A business manufactures at least three radios for each CD player. The total number of radios and CD players cannot exceed 90 per week. Shade the region that represents numbers of radios R and CD players P that can be produced each week within these restrictions.

64. *Areas of Rectangles* Find a monomial equal to the sum of the areas of the rectangles. Calculate this sum for $x = 2$ yards and $y = 3$ yards.

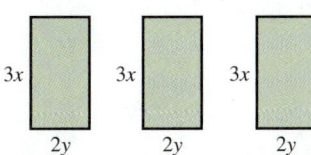

65. *Volume of a Cube* Find an expression for the volume of a cube whose sides have length $2xy^2$.

66. *Area of a Square*   Complete each part and verify that your answers are the same.
  (a) Find the area of the large square by multiplying its length and width
  (b) Find the sum of the areas of the four smaller rectangles inside the large square.

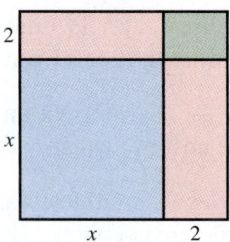

67. *Area of a Square*   A square has area $x^2 + 12x + 36$. Find the length of a side.

68. *Flight of a Golf Ball*   A golf ball is hit upward. Its height h in feet after t seconds is given by the formula $h = 64t - 16t^2$.
  (a) How long does it take for the ball to hit the ground?
  (b) At what times is the ball 48 feet in the air?

CHAPTER 7

Rational Expressions

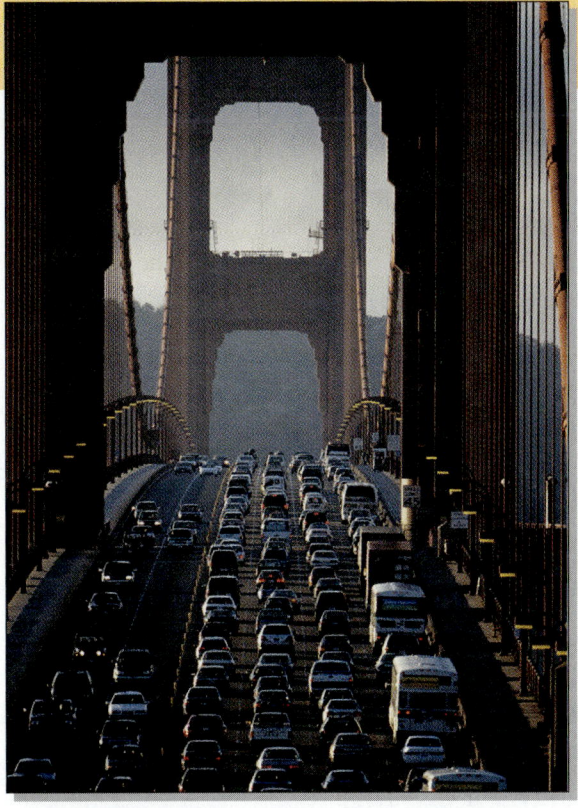

One of the most significant problems facing the U.S. transportation system is chronic highway congestion. According to a newly released "Highway Statistics," Americans drove about 3 trillion miles in 2005, nearly a 25% increase over 1995. There were 241.2 million vehicles registered in the United States in 2005. Our ability to keep traffic moving smoothly and safely is key to keeping our economy strong, and traffic congestion costs motorists more than $80 billion annually in wasted time and fuel. People spend billions of hours stuck in traffic.

Traffic congestion is subject to a nonlinear effect. If the amount of traffic doubles on a highway, the wait in traffic may more than double. At times, only a slight increase in the traffic rate can result in a dramatic increase in the time spent waiting. (See Example 6 in Section 7.1.) Mathematics can describe this effect by using rational expressions, which are used frequently to diagnose and solve traffic and highway engineering problems. In this chapter we introduce rational expressions and some of their applications.

There is nothing wrong with making mistakes. Just don't respond with encores.

—ANONYMOUS

Sources: Philip J. Longman, "American Gridlock." *U.S. News & World Report*, May 28, 2001; U.S. Department of Transportation, 2006.

7.1 INTRODUCTION TO RATIONAL EXPRESSIONS

Basic Concepts ■ Simplifying Rational Expressions ■ Applications

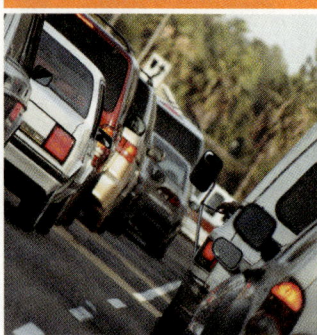

A LOOK INTO MATH ▷ Have you ever been moving smoothly in traffic and noticed how it can come to a halt all of a sudden? Mathematics shows that if the number of cars on a road increases even slightly, then the movement of traffic can slow dramatically. To understand this process better, we need to discuss rational expressions.

Basic Concepts

In Chapters 5 and 6 we discussed polynomials. Examples of polynomials include

$$3, \quad 2x, \quad x^2 + 4, \quad \text{and} \quad x^3 - 1.$$

In this chapter we discuss *rational expressions*. Rational expressions can be written as quotients (fractions) of two polynomials. Examples of rational expressions include

$$\frac{3}{2x}, \quad \frac{2x}{x^2 + 4}, \quad \frac{x^2 + 4}{3}, \quad \text{and} \quad \frac{x^3 - 1}{x^2 + 4}.$$

> ### RATIONAL EXPRESSION
>
> A rational expression can be written as $\frac{P}{Q}$, where P and Q are polynomials. A rational expression is defined whenever $Q \neq 0$.

We can evaluate polynomials for different values of a variable. For example, for $x = 2$ the polynomial $x^2 - 3x + 1$ evaluates to

$$(2)^2 - 3(2) + 1 = -1.$$

Rational expressions can be evaluated similarly.

EXAMPLE 1   Evaluating rational expressions

If possible, evaluate each expression for the given value of the variable.

(a) $\dfrac{1}{x + 1}$ $x = 2$ **(b)** $\dfrac{y^2}{2y - 1}$ $y = -4$

(c) $\dfrac{5z + 8}{z^2 - 2z + 1}$ $z = 1$ **(d)** $\dfrac{2 - x}{x - 2}$ $x = -3$

Solution

(a) If $x = 2$, then $\frac{1}{x + 1}$ becomes $\frac{1}{2 + 1} = \frac{1}{3}$.

(b) If $y = -4$, then $\frac{y^2}{2y - 1}$ becomes $\frac{(-4)^2}{2(-4) - 1} = -\frac{16}{9}$.

(c) If $z = 1$, then $\frac{5z + 8}{z^2 - 2z + 1}$ becomes $\frac{5(1) + 8}{1^2 - 2(1) + 1}$, or $\frac{13}{0}$, which is undefined because division by 0 is not possible.

(d) If $x = -3$, then $\frac{2 - x}{x - 2}$ becomes $\frac{2 - (-3)}{-3 - 2} = \frac{5}{-5} = -1$.

Now Try Exercises **11, 15, 17, 21**

Division by 0 is undefined. As a result, rational expressions are different from polynomials because they are undefined whenever their denominators are 0. For example, the expression in Example 1(c) is undefined when $z = 1$.

EXAMPLE 2 **Determining when a rational expression is undefined**

Find all values of the variable for which each expression is undefined.

(a) $\dfrac{1}{x}$ (b) $\dfrac{4t}{t - 3}$ (c) $\dfrac{1 - 6r}{r^2 - 4}$ (d) $\dfrac{4}{x^2 + 1}$

Solution

(a) A rational expression is undefined when its denominator is 0. Thus $\frac{1}{x}$ is undefined when $x = 0$.

(b) The expression $\frac{4t}{t - 3}$ is undefined when its denominator, $t - 3$, is 0, or when $t = 3$.

(c) The expression $\frac{1 - 6r}{r^2 - 4}$ is undefined when its denominator, $r^2 - 4$, is 0. Here

$$r^2 - 4 = (r - 2)(r + 2) = 0$$

implies that the denominator is 0 when $r = -2$ or $r = 2$.

(d) In the expression $\frac{4}{x^2 + 1}$ the denominator, $x^2 + 1$, is never 0 because any real number squared plus 1 is always greater than or equal to 1. Thus this rational expression is defined for all real numbers x.   Now Try Exercises 29, 31, 35, 37

Simplifying Rational Expressions

In Chapter 1, we used the *basic principle of fractions*,

$$\frac{a \cdot c}{b \cdot c} = \frac{a}{b}.$$

When we use this basic principle, the fraction $\frac{8}{12}$, for example, simplifies to

$$\frac{8}{12} = \frac{2 \cdot 4}{3 \cdot 4} = \frac{2}{3}.$$

EXAMPLE 3 **Simplifying fractions**

Simplify each fraction by applying the basic principle of fractions.

(a) $\dfrac{5}{10}$ (b) $-\dfrac{36}{48}$

Solution

(a) $\dfrac{5}{10} = \dfrac{1 \cdot 5}{2 \cdot 5} = \dfrac{1}{2}$ (b) $-\dfrac{36}{48} = -\dfrac{3 \cdot 12}{4 \cdot 12} = -\dfrac{3}{4}$ Now Try Exercises 43, 47

We can also apply this basic principle to rational expressions. For example,

$$\frac{x(x - 1)}{4(x - 1)} = \frac{x}{4},$$

provided that $x \neq 1$.

NOTE: This simplification is not valid when $x = 1$ because the expression is undefined for this x-value. When simplifying rational expressions, *we assume that values of the variable making a rational expression undefined are excluded.*

BASIC PRINCIPLE OF RATIONAL EXPRESSIONS

The following property can be used to simplify rational expressions, where P, Q, and R are polynomials.

$$\frac{P \cdot R}{Q \cdot R} = \frac{P}{Q} \qquad Q \text{ and } R \text{ are nonzero.}$$

NOTE: $\frac{P \cdot R}{Q \cdot R} = \frac{P}{Q} \cdot \frac{R}{R} = \frac{P}{Q} \cdot 1 = \frac{P}{Q}$, provided that $Q \neq 0$ and $R \neq 0$.

Like fractions, rational expressions can be written in *lowest terms*. For example, the rational expression $\frac{x^2 - 1}{x^2 + 2x + 1}$ can be written in lowest terms by factoring the numerator and the denominator and then applying the basic principle of rational expressions.

$$\frac{x^2 - 1}{x^2 + 2x + 1} = \frac{(x - 1)(x + 1)}{(x + 1)(x + 1)} \qquad \text{Factor the numerator and the denominator.}$$

$$= \frac{x - 1}{x + 1} \qquad \text{Apply } \frac{PR}{QR} = \frac{P}{Q} \text{ with } R = x + 1.$$

Because the basic principle of rational expressions cannot be applied further to $\frac{x-1}{x+1}$, we say that this expression is written in **lowest terms**.

EXAMPLE 4 Simplifying rational expressions

Simplify each expression.

(a) $\dfrac{8y}{4y^2}$ **(b)** $\dfrac{2x + 6}{3x + 9}$ **(c)** $\dfrac{(z + 1)(z - 5)}{(z - 5)(z + 3)}$ **(d)** $\dfrac{x^2 - 9}{2x^2 + 7x + 3}$

Solution

(a) Factor out the greatest common factor, $4y$, in the numerator and the denominator. Then apply the basic principle of rational expressions.

$$\frac{8y}{4y^2} = \frac{2 \cdot 4y}{y \cdot 4y} = \frac{2}{y}$$

(b) Use the distributive property and then the basic principle of rational expressions.

$$\frac{2x + 6}{3x + 9} = \frac{2(x + 3)}{3(x + 3)} = \frac{2}{3}$$

(c) Use the commutative property to rewrite the denominator of the expression.

$$\frac{(z + 1)(z - 5)}{(z - 5)(z + 3)} = \frac{(z + 1)(z - 5)}{(z + 3)(z - 5)} = \frac{z + 1}{z + 3}$$

(d) Start by factoring the numerator and the denominator.

$$\frac{x^2 - 9}{2x^2 + 7x + 3} = \frac{(x - 3)(x + 3)}{(2x + 1)(x + 3)} = \frac{x - 3}{2x + 1}$$

Now Try Exercises 55, 59, 65, 83

MAKING CONNECTIONS

Expressions and Equations

Expressions and equations are different concepts. An expression does not contain an equals sign, whereas an equation is a statement that two expressions are equal and always contains an equals sign. For example,

$$\frac{x}{x+4} \quad \text{and} \quad \frac{2}{x}$$

are two rational *expressions*, and

$$\frac{x}{x+4} = \frac{2}{x}$$

is a rational *equation*. In this section we evaluate and simplify rational expressions. Later, we will *solve* rational equations by finding x-values that make an equation true.

A negative sign can be placed in a fraction in a number of ways. For example,

$$-\frac{5}{7} = \frac{-5}{7} = \frac{5}{-7}$$

illustrates three fractions that are equal. This property can also be applied to rational expressions, as demonstrated in the next example.

EXAMPLE 5   Distributing a negative sign

Simplify each expression.

(a) $\dfrac{-x-6}{2x+12}$ **(b)** $\dfrac{10-z}{z-10}$ **(c)** $-\dfrac{5-x}{x-5}$

Solution

(a) Factor -1 out of the numerator and 2 out of the denominator.

$$\frac{-x-6}{2x+12} = \frac{-1(x+6)}{2(x+6)} = -\frac{1}{2}$$

(b) Factor -1 out of the numerator.

$$\frac{10-z}{z-10} = \frac{-1(-10+z)}{z-10} = \frac{-1(z-10)}{z-10} = -1$$

(c) Rewrite the expression with the negative sign in the numerator and then apply the distributive property. Be sure to include parentheses around the numerator.

$$-\frac{5-x}{x-5} = \frac{-(5-x)}{x-5} = \frac{-5+x}{x-5} = \frac{x-5}{x-5} = 1$$

The same answer can be obtained by distributing the negative sign over the denominator.

$$-\frac{5-x}{x-5} = \frac{5-x}{-(x-5)} = \frac{5-x}{-x+5} = \frac{5-x}{5-x} = 1$$

Now Try Exercises 67, 71, 75

The result for Example 5(b) becomes more obvious if we substitute a number for z. For example, if we let $z = 6$, then

$$\frac{10 - z}{z - 10} = \frac{10 - 6}{6 - 10} = \frac{4}{-4} = -1.$$

MAKING CONNECTIONS

Negative Signs and Rational Expressions

In general, $(b - a)$ equals $-1(a - b)$. Thus if $a \neq b$, then $\frac{b - a}{a - b} = -1$.

Applications

▶ **REAL-WORLD CONNECTION**   The next example is based on the introduction to this chapter and illustrates modeling traffic flow with a rational expression.

EXAMPLE 6   **Modeling traffic flow**

Suppose that 10 cars per minute can pass through a construction zone. If traffic arrives *randomly* at an average rate of x cars per minute, the average time T in minutes spent waiting in line and passing through the construction zone is given by

$$T = \frac{1}{10 - x},$$

where $x < 10$.   (**Source:** N. Garber and L. Hoel, *Traffic and Highway Engineering.*)

(a) Complete Table 7.1 by finding T for each value of x.

TABLE 7.1   Waiting in Traffic

x (cars/minute)	5	7	9	9.5	9.9	9.99
T (minutes)						

(b) Interpret the results.

Solution

(a) When $x = 5$ cars per minute, then $T = \frac{1}{10 - 5} = \frac{1}{5}$ minute. Other values are found similarly, as shown in Table 7.2.

TABLE 7.2   Waiting in Traffic

x (cars/minute)	5	7	9	9.5	9.9	9.99
T (minutes)	$\frac{1}{5}$	$\frac{1}{3}$	1	2	10	100

(b) As the average traffic rate increases from 9 cars per minute to 9.9 cars per minute, the time needed to pass through the construction zone increases from 1 minute to 10 minutes. As x nears 10 cars per minute, a small increase in x increases the waiting time dramatically.

Now Try Exercise **101**

This nonlinear effect for traffic congestion in Example 6 is shown in Figure 7.1, where points from Table 7.2 have been plotted and a curve passing through them has been sketched. A vertical dashed line was also sketched at $x = 10$. This dashed line is called a **vertical asymptote** and indicates that the rational expression is undefined at this value of x. Near the vertical asymptote, the waiting time T increases dramatically for small increases in x. The graph of T does not intersect or cross this vertical asymptote.

Figure 7.1

TECHNOLOGY NOTE: Making Tables

Table 7.2 can also be created with a graphing calculator by using the Ask feature, as illustrated in the following displays.

CALCULATOR HELP

To make a table of values, see Appendix A (page AP-3).

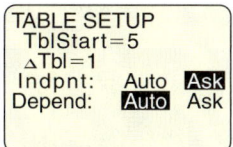

AN APPLICATION INVOLVING PROBABILITY (OPTIONAL)   If 10 marbles, one blue and nine red, are placed in a jar, then the *probability*, or *likelihood*, of picking the blue marble at random is 1 *chance* in 10, or $\frac{1}{10}$. The probability of drawing a red marble at random is 9 chances in 10, or $\frac{9}{10}$. **Probability** is a real number between 0 and 1. A probability of 0, or 0%, indicates that an event is impossible, whereas a probability of 1, or 100%, indicates that an event is certain. Rational expressions are often used to describe probability.

EXAMPLE 7   Calculating probability

Suppose that n balls, numbered 1 to n, are placed in a container and only one ball has the winning number.

(a) What is the probability of drawing the winning ball at random?
(b) Calculate this probability for $n = 100, 1000,$ and $10,000$.
(c) What happens to the probability of drawing the winning ball as the number of balls increases?

Solution

(a) There is 1 chance in n of drawing the winning ball, so the probability is $\frac{1}{n}$.
(b) For $n = 100, 1000,$ and $10,000$, the probabilities are $\frac{1}{100}, \frac{1}{1000},$ and $\frac{1}{10,000}$.
(c) As the number of balls increases, the probability of picking the winning ball decreases.

Now Try Exercise 105

7.1 PUTTING IT ALL TOGETHER

In this section we discussed some basic concepts of rational expressions. They are summarized in the following table.

Concept	Explanation	Examples
Rational Expression	An expression of the form $\frac{P}{Q}$, where P and Q are polynomials with $Q \neq 0$	$\frac{1}{x}, \frac{x-3}{2x^2-1}, \frac{2x+9}{5x}$, and $\frac{x^2+3x-5}{1}$
Undefined Values for Rational Expressions	If the *denominator* of a rational expression is 0, then the rational expression is undefined for that value of the variable.	$\frac{1}{x-3}$ is undefined when $x = 3$. $\frac{5y}{y^2-1}$ is undefined when $y = 1$ or when $y = -1$.
Basic Principle of Rational Expressions	Factor the numerator and the denominator completely. Then apply $$\frac{P \cdot R}{Q \cdot R} = \frac{P}{Q}.$$	$\frac{4xy^2}{6xy^3} = \frac{2(2xy^2)}{3y(2xy^2)} = \frac{2}{3y}$ $\frac{4x(x-4)}{(x+1)(x-4)} = \frac{4x}{x+1}$

7.1 Exercises

MathXL PRACTICE   WATCH   DOWNLOAD   READ   REVIEW

CONCEPTS

1. A rational expression can be written as _____, where P and Q are _____ with $Q \neq 0$.

2. Is $\frac{x}{2x^2+1}$ a rational expression? Why or why not?

3. A rational expression is undefined whenever _____.

4. The rational expression $\frac{1}{x-a}$ is undefined whenever $x =$ _____.

5. The basic principle of fractions states that a fraction can be simplified by using $\frac{a \cdot c}{b \cdot c} =$ _____.

6. The basic principle of rational expressions can be used to simplify _____ expressions.

7. The fraction $\frac{18}{60}$ simplifies to _____.

8. The expression $\frac{3(x+1)}{2x(x+1)}$ simplifies to _____.

9. The expression $\frac{x-a}{a-x}$ simplifies to _____.

10. The expression $-\frac{x-1}{x+1}$ simplifies to $\frac{?}{x+1}$.

EVALUATING RATIONAL EXPRESSIONS

Exercises 11–24: If possible, evaluate the expression for the given value of the variable.

11. $\frac{3}{x}$ $x = -7$

12. $\frac{3}{x+3}$ $x = 0$

13. $-\dfrac{x}{x-5}$ $x = -4$

14. $-\dfrac{4x}{5x+1}$ $x = 1$

15. $\dfrac{y+1}{y^2}$ $y = -2$

16. $\dfrac{3y-1}{y^2+1}$ $y = -1$

17. $\dfrac{7z}{z^2-4}$ $z = -2$

18. $\dfrac{5}{z^2-3z+2}$ $z = -1$

19. $\dfrac{5}{3t+6}$ $t = -2$

20. $\dfrac{4t}{2t+5}$ $t = -\dfrac{5}{2}$

21. $\dfrac{4-x}{x-4}$ $x = -2$

22. $\dfrac{x-7}{7-x}$ $x = 4$

23. $-\dfrac{6-x}{x-6}$ $x = 0$

24. $\dfrac{8-2x}{2x-8}$ $x = -5$

Exercises 25–28: Complete the table for the given expression. If a value is undefined, place a dash in the table.

25.

x	-2	-1	0	1	2
$\dfrac{x}{x+1}$					

26.

x	-2	-1	0	1	2
$\dfrac{2x}{3x-1}$					

27.

x	-2	-1	0	1	2
$\dfrac{3x}{2x^2+1}$					

28.

x	-2	-1	0	1	2
$\dfrac{2x-1}{x^2-1}$					

Exercises 29–42: Find any values of the variable that make the expression undefined.

29. $-\dfrac{8}{x}$ **30.** $\dfrac{7}{x+1}$

31. $\dfrac{4}{z-3}$ **32.** $\dfrac{7-z}{z-7}$

33. $\dfrac{4y}{5y+4}$ **34.** $\dfrac{3+y}{3y-7}$

35. $\dfrac{5t+2}{t^2+1}$ **36.** $\dfrac{8t}{t^2+25}$

37. $\dfrac{8x}{x^2-25}$ **38.** $\dfrac{x+4}{x^2-36}$

39. $\dfrac{x^2+3x+2}{x^2+5x+6}$ **40.** $\dfrac{2x-1}{x^2-7x+10}$

41. $\dfrac{8z^2+z+1}{2z^2-7z+5}$ **42.** $\dfrac{4n^2+17n-15}{3n^2-8n+4}$

SIMPLIFYING RATIONAL EXPRESSIONS

Exercises 43–50: Simplify the fraction to lowest terms.

43. $\dfrac{12}{18}$ **44.** $\dfrac{24}{32}$

45. $\dfrac{24}{48}$ **46.** $-\dfrac{22}{33}$

47. $-\dfrac{6}{15}$ **48.** $\dfrac{8}{22}$

49. $-\dfrac{25}{75}$ **50.** $-\dfrac{36}{42}$

Exercises 51–54: First simplify the fraction, then simplify the rational expression.

51. (a) $\dfrac{8}{16}$ (b) $\dfrac{x+2}{2x+4}$

52. (a) $\dfrac{6}{9}$ (b) $\dfrac{2x+6}{3x+9}$

53. (a) $\dfrac{7-3}{3-7}$ (b) $\dfrac{7-x}{x-7}$

54. (a) $-\dfrac{8-5}{5-8}$ (b) $-\dfrac{x-5}{5-x}$

Exercises 55–92: Simplify the expression.

55. $\dfrac{5x^4}{10x^6}$

56. $\dfrac{6y^2}{9y}$

57. $\dfrac{8xy^3}{6x^2y^2}$

58. $\dfrac{36x^2y^5}{6x^5y}$

59. $\dfrac{x + 4}{2x + 8}$

60. $\dfrac{5x - 10}{x - 2}$

61. $\dfrac{3z - 9}{5z - 15}$

62. $\dfrac{4z + 8}{10 + 5z}$

63. $\dfrac{(x + 1)(x - 1)}{(x + 6)(x - 1)}$

64. $\dfrac{(2x + 1)(x + 9)}{(4x - 3)(x + 9)}$

65. $\dfrac{(5y + 3)(2y - 1)}{(2y - 1)(y + 2)}$

66. $\dfrac{(4y - 1)(5y + 7)}{(5y + 7)(1 - 4y)}$

67. $\dfrac{x - 7}{7 - x}$

68. $\dfrac{5 - x}{x - 5}$

69. $\dfrac{a - b}{b - a}$

70. $\dfrac{2t - 3r}{3r - 2t}$

71. $\dfrac{-6 - x}{18 + 3x}$

72. $\dfrac{-2x - 6}{x + 3}$

73. $\dfrac{x + 1}{-2x - 2}$

74. $\dfrac{3x + 21}{-7 - x}$

75. $-\dfrac{9 - x}{x - 9}$

76. $-\dfrac{4 - 2x}{x - 2}$

77. $\dfrac{(3x + 5)(x - 1)}{(3x - 5)(1 - x)}$

78. $\dfrac{(2 - x)(x - 2)}{(x - 2)(2 - x)}$

79. $\dfrac{n^2 - n}{n^2 - 5n}$

80. $\dfrac{3n^2 - 4n}{n^2 + 4n}$

81. $\dfrac{x^2 - 3x}{6x - 18}$

82. $\dfrac{4x^2 + 16x}{5x^2 + 20x}$

83. $\dfrac{z^2 - 3z + 2}{z^2 - 4z + 3}$

84. $\dfrac{z^2 - 3z - 10}{z^2 - 2z - 8}$

85. $\dfrac{2x^2 + 7x - 4}{6x^2 + x - 2}$

86. $\dfrac{5x^2 + 3x - 2}{5x^2 + 13x - 6}$

87. $\dfrac{x - 3}{3x^2 - 11x + 6}$

88. $\dfrac{2x - 1}{4x^2 + 6x - 4}$

89. $-\dfrac{a - 9}{9 - a}$

90. $-\dfrac{-b - 6}{b + 6}$

91. $\dfrac{-2x - 1}{4x + 2}$

92. $\dfrac{4x + 3}{-8x - 6}$

Exercises 93–100: Do the following.
  (a) *Decide whether you are given an expression or an equation.*
  (b) *If you are given an expression, simplify it. If you are given an equation, solve it.*

93. $x + 1 = 7$

94. $x^2 - 4 = 0$

95. $\dfrac{x}{x(x + 1)}$

96. $\dfrac{x - 2}{(x - 2)(x - 8)}$

97. $\dfrac{x^2 - 4}{x + 2}$

98. $\dfrac{x - 4}{8 - 2x}$

99. $\dfrac{x}{2(1 + 3)} = 1$

100. $\dfrac{x}{2} + 2 = \dfrac{8}{2}$

APPLICATIONS

101. *Modeling Traffic Flow*   (Refer to Example 6.) Five vehicles per minute can pass through a construction zone. If the traffic arrives randomly at an average rate of x vehicles per minute, the average time T in minutes spent waiting in line and passing through the construction zone is given by $T = \dfrac{1}{5 - x}$ for $x < 5$. (*Source:* N. Garber.)
  (a) Evaluate T for $x = 3$ and interpret the result.
  (b) Complete the table and interpret the results.

x	2	4	4.5	4.9	4.99
T					

102. *Standing in Line*   A worker at a poolside store can serve 20 customers per hour. If children arrive randomly at an average rate of x per hour, then the average number of children N waiting in line is given by $N = \dfrac{x^2}{400 - 20x}$ for $x < 20$. (*Source:* N. Garber.)
  (a) Complete the table.

x	5	10	18	19	20
N					

  (b) Compare the number of children waiting in line if the average rate increases from 18 to 19 children per hour.

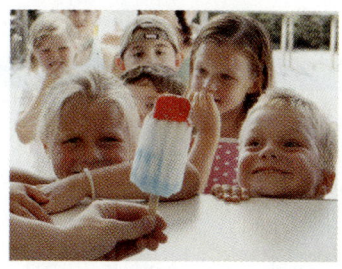

103. *Probability*   Suppose that a coin is flipped. What is the probability that a head appears?

104. *Probability*   A die shows the numbers 1, 2, 3, 4, 5, and 6. If each number has an equal chance of appearing on any given roll, what is the probability that a 2 or 4 appears?

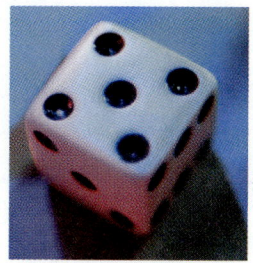

105. *Probability*   (Refer to Example 7.) Suppose that there are n balls in a container and that three balls have a winning number. If a ball is drawn randomly, do each of the following.
  (a) Write a rational expression that gives the probability of drawing a winning ball.
  (b) Write a rational expression that gives the probability of not drawing a winning ball. Evaluate your expression for $n = 100$ and interpret the result.

106. *Surface Area of a Cylinder*   If a cylindrical container has a volume of π cubic feet, then its surface area S in square feet (excluding the top and bottom) is given by $S = \frac{2\pi}{r}$, where r is the radius of the cylinder.

  (a) Calculate S when $r = \frac{1}{2}$ foot.
  (b) What happens to this surface area when r becomes large? Sketch this situation.

(c) What happens to the surface area when r becomes small (nearly 0)? Sketch this situation.

107. *Distance and Time*   A car is traveling at 60 miles per hour.
  (a) How long does it take the car to travel 360 miles?
  (b) Write a rational expression that gives the time that it takes the car to travel M miles.

108. *Distance and Time*   A bicyclist rides uphill at 10 miles per hour for 5 miles and then rides downhill at 20 miles per hour for 5 miles. What is the bicyclist's average speed? (*Hint:* Average speed equals distance divided by time.)

Exercises 109 and 110: *Traffic Flow*   *(Refer to Example 6.) The figure shows a graph of the waiting time T in minutes at a construction zone when cars are arriving randomly at an average rate of x cars per minute.*
  (a) *Give the equation of the vertical asymptote.*
  (b) *Explain how the graph relates to traffic flow.*

109.

110.

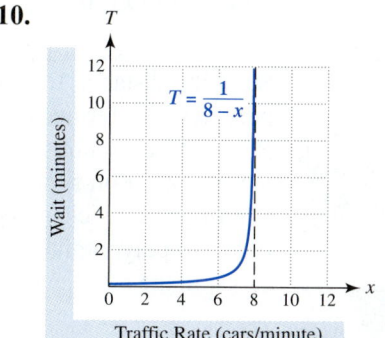

WRITING ABOUT MATHEMATICS

111. Explain what a rational expression is. When is a rational expression undefined?

112. Does the rational expression $\frac{5x + 2}{10x + 4}$ equal $\frac{5x}{10x} + \frac{2}{4}$? Explain your answer.

GROUP ACTIVITY
WORKING WITH REAL DATA

Directions: Form a group of 2 to 4 people. Select someone to record the group's responses for this activity. All members of the group should work cooperatively to answer the questions. If your instructor asks for your results, each member of the group should be prepared to respond.

Students Per Computer   In the early years of microcomputers, school districts could not afford to buy a computer for every student. As the price of computers decreased, more and more school districts have been able to attain this goal. The following table lists numbers of students per computer during these early years.

Year	1983	1985	1987	1989
Students/Computer	125	50	32	22

Year	1991	1993	1995	1997
Students/Computer	18	14	10	6

Source: Quality Education Data, Inc.

(a) Make a scatterplot of the data. Would a straight line model the data accurately? Explain.

(b) Discuss how well the formula

$$S = \frac{125}{1 + 0.7(y - 1983)}, \quad y \geq 1983$$

models these data, where S represents the students per computer and y represents the year.

(c) In what year does the formula reveal that there were about 17 students per computer?

7.2   MULTIPLICATION AND DIVISION OF RATIONAL EXPRESSIONS

Review of Multiplication and Division of Fractions ■
Multiplication of Rational Expressions ■ Division of Rational Expressions

A LOOK INTO MATH ▷

Stopping distance for a car can vary depending on the road conditions. If the road is slippery, it takes farther to stop. Also, it takes farther to stop if the car is traveling downhill. Rational expressions are frequently used by highway engineers to estimate the stopping distance of a car on slippery surfaces or on hills.

In previous chapters we reviewed how to add, subtract, multiply, and divide real numbers and polynomials. In this section we show how to multiply and divide rational expressions; in the next section we discuss addition and subtraction of rational expressions.

Review of Multiplication and Division of Fractions

To multiply two fractions we use the property

$$\frac{a}{b} \cdot \frac{c}{d} = \frac{ac}{bd}.$$

In the next example we review multiplication of fractions. (See Section 1.2.)

EXAMPLE 1 **Multiplying fractions**

Multiply and simplify your answers to lowest terms.

(a) $\frac{3}{7} \cdot \frac{4}{5}$ (b) $2 \cdot \frac{3}{4}$ (c) $\frac{4}{21} \cdot \frac{7}{8}$

Solution

(a) $\frac{3}{7} \cdot \frac{4}{5} = \frac{12}{35}$ (b) $2 \cdot \frac{3}{4} = \frac{2}{1} \cdot \frac{3}{4} = \frac{6}{4} = \frac{3}{2}$

(c) $\frac{4}{21} \cdot \frac{7}{8} = \frac{7 \cdot 4}{21 \cdot 8} = \frac{1}{3} \cdot \frac{1}{2} = \frac{1}{6}$

Now Try Exercises 9, 11, 13

To divide two fractions we "invert and multiply." That is, we change a division problem to a multiplication problem by using

$$\frac{a}{b} \div \frac{c}{d} = \frac{a}{b} \cdot \frac{d}{c}.$$

EXAMPLE 2 **Dividing fractions**

Divide and simplify your answers to lowest terms.

(a) $\frac{1}{3} \div \frac{5}{7}$ (b) $\frac{4}{5} \div 8$ (c) $\frac{8}{9} \div \frac{10}{3}$

Solution

(a) $\frac{1}{3} \div \frac{5}{7} = \frac{1}{3} \cdot \frac{7}{5} = \frac{7}{15}$

(b) $\frac{4}{5} \div 8 = \frac{4}{5} \cdot \frac{1}{8} = \frac{4}{40} = \frac{1}{10}$

(c) $\frac{8}{9} \div \frac{10}{3} = \frac{8}{9} \cdot \frac{3}{10} = \frac{24}{90} = \frac{4 \cdot 6}{15 \cdot 6} = \frac{4}{15}$

Now Try Exercises 17, 19, 21

Multiplication of Rational Expressions

Multiplying rational expressions is similar to multiplying fractions.

> ### PRODUCTS OF RATIONAL EXPRESSIONS
>
> To multiply two rational expressions multiply the numerators and multiply the denominators. That is,
>
> $$\frac{A}{B} \cdot \frac{C}{D} = \frac{AC}{BD},$$
>
> where B and D are nonzero.

EXAMPLE 3 **Multiplying rational expressions**

Multiply and simplify to lowest terms. Leave your answers in factored form.

(a) $\frac{3}{x} \cdot \frac{2x - 5}{x - 1}$

(b) $\frac{x - 1}{4x} \cdot \frac{x + 3}{x - 1}$

(c) $\frac{x^2 - 4}{x + 3} \cdot \frac{x + 3}{x + 2}$

(d) $\frac{4}{x^2 + 3x + 2} \cdot \frac{x^2 + 2x + 1}{8}$

Solution

(a) $\dfrac{3}{x} \cdot \dfrac{2x-5}{x-1} = \dfrac{3(2x-5)}{x(x-1)}$ Multiply the numerators and the denominators.

(b) $\dfrac{x-1}{4x} \cdot \dfrac{x+3}{x-1} = \dfrac{(x-1)(x+3)}{4x(x-1)}$ Multiply the numerators and the denominators.

$\qquad\qquad = \dfrac{(x+3)(x-1)}{4x(x-1)}$ Commutative property

$\qquad\qquad = \dfrac{x+3}{4x}$ Simplify.

(c) $\dfrac{x^2-4}{x+3} \cdot \dfrac{x+3}{x+2} = \dfrac{(x-2)(x+2)}{x+3} \cdot \dfrac{x+3}{x+2}$ Factor.

$\qquad\qquad = \dfrac{(x-2)(x+2)(x+3)}{(x+3)(x+2)}$ Multiply the numerators and the denominators.

$\qquad\qquad = \dfrac{(x-2)(x+2)(x+3)}{(x+2)(x+3)}$ Commutative property

$\qquad\qquad = x - 2$ Simplify.

(d) $\dfrac{4}{x^2+3x+2} \cdot \dfrac{x^2+2x+1}{8} = \dfrac{4(x^2+2x+1)}{8(x^2+3x+2)}$ Multiply the numerators and the denominators.

$\qquad\qquad = \dfrac{4(x+1)(x+1)}{8(x+2)(x+1)}$ Factor.

$\qquad\qquad = \dfrac{x+1}{2(x+2)}$ Simplify.

Now Try Exercises 33, 35, 45, 49

▶ **REAL-WORLD CONNECTION**   The next example illustrates a rational expression that is used in highway design.

EXAMPLE 4   **Estimating stopping distance**

If a car is traveling at 60 miles per hour on a slippery road, then its stopping distance D in feet can be calculated by

$$D = \dfrac{3600}{30} \cdot \dfrac{1}{x},$$

where x is the coefficient of friction between the tires and the road and $0 < x \le 1$. The more slippery the road is, the smaller the value of x.   (**Source:** L. Haefner, *Introduction to Transportation Systems.*)

(a) Multiply and simplify the formula for D.
(b) Compare the stopping distance on an icy road with $x = 0.1$ and on dry pavement with $x = 0.4$.

Solution

(a) Because

$$\dfrac{3600}{30} \cdot \dfrac{1}{x} = \dfrac{3600}{30x} = \dfrac{120 \cdot 30}{x \cdot 30} = \dfrac{120}{x},$$

it follows that $D = \frac{120}{x}$.

(b) When $x = \mathbf{0.1}, D = \frac{120}{\mathbf{0.1}} = 1200$ feet, and when $x = \mathbf{0.4}, D = \frac{120}{\mathbf{0.4}} = 300$ feet. When a car is moving at 60 miles per hour, an icy road can increase the stopping distance by 900 feet, or by a factor of 4.   Now Try Exercise 77

Division of Rational Expressions

Dividing rational expressions is similar to dividing fractions.

QUOTIENTS OF RATIONAL EXPRESSIONS

To divide two rational expressions multiply by the reciprocal of the divisor. That is,

$$\frac{A}{B} \div \frac{C}{D} = \frac{A}{B} \cdot \frac{D}{C},$$

where B, C, and D are nonzero.

EXAMPLE 5 Dividing rational expressions

Divide and simplify to lowest terms.

(a) $\dfrac{5}{2x} \div \dfrac{10}{x-4}$ **(b)** $\dfrac{x^2-9}{x^2+4} \div (x-3)$ **(c)** $\dfrac{x^2-x}{x^2-x-2} \div \dfrac{x}{x-2}$

Solution

(a)

$$\frac{5}{2x} \div \frac{10}{x-4} = \frac{5}{2x} \cdot \frac{x-4}{10} \qquad \text{Invert and multiply.}$$

$$= \frac{5(x-4)}{20x} \qquad \text{Multiply the numerators and the denominators.}$$

$$= \frac{x-4}{4x} \qquad \text{Simplify. Note that } \tfrac{5}{20} = \tfrac{1}{4}.$$

(b)

$$\frac{x^2-9}{x^2+4} \div (x-3) = \frac{x^2-9}{x^2+4} \cdot \frac{1}{x-3} \qquad \text{Invert and multiply.}$$

$$= \frac{x^2-9}{(x^2+4)(x-3)} \qquad \text{Multiply the numerators and the denominators.}$$

$$= \frac{(x+3)(x-3)}{(x^2+4)(x-3)} \qquad \text{Factor the numerator.}$$

$$= \frac{x+3}{x^2+4} \qquad \text{Simplify.}$$

(c)

$$\frac{x^2-x}{x^2-x-2} \div \frac{x}{x-2} = \frac{x^2-x}{x^2-x-2} \cdot \frac{x-2}{x} \qquad \text{Invert and multiply.}$$

$$= \frac{(x^2-x)(x-2)}{(x^2-x-2)x} \qquad \text{Multiply the numerators and the denominators.}$$

$$= \frac{x(x-1)(x-2)}{x(x+1)(x-2)} \qquad \text{Factor numerator and denominator.}$$

$$= \frac{x-1}{x+1} \qquad \text{Simplify.}$$

Now Try Exercises 53, 61, 69

7.2   PUTTING IT ALL TOGETHER

In this section we showed how to multiply and divide rational expressions. The following table summarizes these methods.

Concept	Explanation	Examples
Basic Principle of Rational Expressions	$\frac{PR}{QR} = \frac{P}{Q}$, where Q and R are nonzero polynomials. Apply this principle to simplify rational expressions to lowest terms.	$\frac{(x+2)(2x-3)}{(x-1)(2x-3)} = \frac{x+2}{x-1}$
Multiplication of Rational Expressions	Multiply the numerators and multiply the denominators: $$\frac{A}{B} \cdot \frac{C}{D} = \frac{AC}{BD}.$$ Then simplify the result to lowest terms.	$\frac{4x}{x-1} \cdot \frac{x-1}{x+1} = \frac{4x(x-1)}{(x-1)(x+1)}$ $= \frac{4x}{x+1}$
Division of Rational Expressions	Multiply the first expression by the reciprocal of the second expression: $$\frac{A}{B} \div \frac{C}{D} = \frac{A}{B} \cdot \frac{D}{C}.$$ Then simplify the result to lowest terms.	$\frac{x+1}{x-3} \div \frac{x+1}{x-5} = \frac{x+1}{x-3} \cdot \frac{x-5}{x+1}$ $= \frac{(x-5)(x+1)}{(x-3)(x+1)}$ $= \frac{x-5}{x-3}$

7.2   Exercises

MyMathLab

 PRACTICE   WATCH   DOWNLOAD   READ   REVIEW

CONCEPTS

1. $\frac{2}{3} \cdot \frac{5}{7} = $ _____

2. $\frac{2}{3} \div \frac{5}{7} = $ _____

3. $\frac{A}{B} \cdot \frac{C}{D} = $ _____

4. $\frac{A}{B} \div \frac{C}{D} = $ _____

5. Simplify $\frac{(x+7)(x+2)}{(x+1)(x+2)}$.

6. $\frac{AC}{BC} = $ _____

7. Does $\frac{x+2}{x}$ equal $\frac{x}{x} + 2$?

8. To divide $\frac{1}{x}$ by $\frac{y}{z}$ multiply _____ by _____.

REVIEW OF FRACTIONS

Exercises 9–24: Simplify to lowest terms.

9. $\frac{1}{2} \cdot \frac{4}{5}$

10. $\frac{6}{7} \cdot \frac{7}{18}$

11. $\frac{3}{7} \cdot 4$

12. $5 \cdot \frac{4}{5}$

13. $\frac{5}{4} \cdot \frac{8}{15}$

14. $\frac{3}{10} \cdot \frac{5}{9}$

15. $\frac{1}{3} \cdot \frac{2}{3} \cdot \frac{9}{11}$

16. $\frac{2}{5} \cdot \frac{10}{11} \cdot \frac{1}{4}$

17. $\frac{2}{3} \div \frac{1}{6}$

18. $\frac{5}{7} \div \frac{5}{8}$

19. $\frac{8}{9} \div \frac{5}{3}$

20. $\frac{7}{3} \div 6$

21. $8 \div \frac{4}{5}$

22. $\frac{1}{2} \div \frac{5}{4} \div \frac{2}{5}$

23. $\frac{4}{5} \div \frac{2}{3} \div \frac{1}{2}$

24. $\frac{7}{20} \div \frac{14}{5}$

MULTIPLYING RATIONAL EXPRESSIONS

Exercises 25–32: Simplify the expression.

25. $\dfrac{x+5}{x+5}$

26. $\dfrac{2x-3}{2x-3}$

27. $\dfrac{(z+1)(z+2)}{(z+4)(z+2)}$

28. $\dfrac{(2z-7)(z+5)}{(3z+5)(z+5)}$

29. $\dfrac{8y(y+7)}{12y(y+7)}$

30. $\dfrac{6(y+1)}{12(y+1)}$

31. $\dfrac{x(x+2)(x+3)}{x(x-2)(x+3)}$

32. $\dfrac{2(x+1)(x-1)}{4(x+1)(x-1)}$

Exercises 33–52: Multiply and simplify to lowest terms. Leave your answers in factored form.

33. $\dfrac{8}{x}\cdot\dfrac{x+1}{x}$

34. $\dfrac{7}{2x}\cdot\dfrac{x}{x-1}$

35. $\dfrac{8+x}{x}\cdot\dfrac{x-3}{x+8}$

36. $\dfrac{5x^2+x}{2x-1}\cdot\dfrac{1}{x}$

37. $\dfrac{z+3}{z+4}\cdot\dfrac{z+4}{z-7}$

38. $\dfrac{2z+1}{3z}\cdot\dfrac{3z}{z+2}$

39. $\dfrac{5x+1}{3x+2}\cdot\dfrac{3x+2}{5x+1}$

40. $\dfrac{x+1}{x+3}\cdot\dfrac{x+3}{x+1}$

41. $\dfrac{(t+1)^2}{t+2}\cdot\dfrac{(t+2)^2}{t+1}$

42. $\dfrac{(t-1)^2}{(t+5)^2}\cdot\dfrac{t+5}{t-1}$

43. $\dfrac{x^2}{x^2+4}\cdot\dfrac{x+4}{x}$

44. $\dfrac{x-1}{x^2}\cdot\dfrac{x^2}{x^2+1}$

45. $\dfrac{z^2-1}{z^2-4}\cdot\dfrac{z-2}{z+1}$

46. $\dfrac{z^2-9}{z-5}\cdot\dfrac{z-5}{z+3}$

47. $\dfrac{y^2-2y}{y^2-1}\cdot\dfrac{y+1}{y-2}$

48. $\dfrac{y^2-4y}{y+1}\cdot\dfrac{y+1}{y-4}$

49. $\dfrac{2x^2-x-3}{3x^2-8x-3}\cdot\dfrac{3x+1}{2x-3}$

50. $\dfrac{6x^2+11x-2}{3x^2+11x-4}\cdot\dfrac{3x-1}{6x-1}$

51. $\dfrac{(x-3)^3}{x^2-2x+1}\cdot\dfrac{x-1}{(x-3)^2}$

52. $\dfrac{x^2+4x+4}{x^2-2x+1}\cdot\dfrac{(x-1)^2}{(x+2)^2}$

Exercises 53–74: Divide and simplify to lowest terms. Leave your answers in factored form.

53. $\dfrac{2}{x}\div\dfrac{2x+3}{x}$

54. $\dfrac{6}{2x}\div\dfrac{x+2}{2x}$

55. $\dfrac{x-2}{3x}\div\dfrac{2-x}{6x}$

56. $\dfrac{x+1}{2x-1}\div\dfrac{x+1}{x}$

57. $\dfrac{z+2}{z+1}\div\dfrac{z+2}{z-1}$

58. $\dfrac{z+7}{z-4}\div\dfrac{z+7}{z-4}$

59. $\dfrac{3y+4}{2y+1}\div\dfrac{3y+4}{y+2}$

60. $\dfrac{y+5}{y-2}\div\dfrac{y}{y+3}$

61. $\dfrac{t^2-1}{t^2+1}\div\dfrac{t+1}{4}$

62. $\dfrac{4}{2t^3}\div\dfrac{8}{t^2}$

63. $\dfrac{y^2-9}{y^2-25}\div\dfrac{y+3}{y+5}$

64. $\dfrac{y+1}{y-4}\div\dfrac{y^2-1}{y^2-16}$

65. $\dfrac{2x^2-4x}{2x-1}\div\dfrac{x-2}{2x-1}$

66. $\dfrac{x-4}{x^2+x}\div\dfrac{5}{x+1}$

67. $\dfrac{2z^2-5z-3}{z^2+z-20}\div\dfrac{z-3}{z-4}$

68. $\dfrac{z^2+12z+27}{z^2-5z-14}\div\dfrac{z+3}{z+2}$

69. $\dfrac{t^2-1}{t^2+5t-6}\div(t+1)$

70. $\dfrac{t^2-2t-3}{t^2-5t-6}\div(t-3)$

71. $\dfrac{a-b}{a+b}\div\dfrac{a-b}{2a+3b}$

72. $\dfrac{x^3-y^3}{x^2-y^2}\div\dfrac{x^2+xy+y^2}{x-y}$

73. $\dfrac{x-y}{x^2+2xy+y^2}\div\dfrac{1}{(x+y)^2}$

74. $\dfrac{a^2-b^2}{4a^2-9b^2}\div\dfrac{a-b}{2a+3b}$

75. Thinking Generally Simplify $\dfrac{a-b}{b-c}\cdot\dfrac{c-b}{b-a}$.

76. Thinking Generally Simplify $\dfrac{a-b}{b-c}\div\dfrac{b-a}{a-b}$.

APPLICATIONS

77. *Stopping on Slippery Roads* (Refer to Example 4.) If a car is traveling at 30 miles per hour on a slippery road, then its stopping distance D in feet can be calculated by

$$D = \frac{900}{30} \cdot \frac{1}{x},$$

where x is the coefficient of friction between the tires and the road and $0 < x \le 1$. (**Source:** L. Haefner.)
(a) Multiply and simplify the formula for D.
(b) Compare the stopping distance on an icy road with $x = 0.1$ and on dry pavement with $x = 0.4$.

78. *Stopping on Hills* If a car is traveling at 50 miles per hour on a hill with wet pavement, then its stopping distance D is given by

$$D = \frac{2500}{30} \cdot \frac{1}{x + 0.3},$$

where x equals the slope of the hill. (**Source:** L. Haefner.)
(a) Multiply and simplify the formula for D.
(b) Compare the stopping distance for an uphill slope of $x = 0.1$ to a downhill slope of $x = -0.1$.

79. *Probability* Suppose that one jar holds n balls and that a second jar holds $n + 1$ balls. Each jar contains one winning ball.

(a) The probability, or chance, of drawing the winning ball from the first jar and *not* drawing it from the second jar is

$$\frac{1}{n} \cdot \frac{n}{n + 1}.$$

Simplify this expression.
(b) Find this probability for $n = 99$.

80. *U.S. AIDS Cases* The cumulative number of AIDS cases C in the United States from 1982 to 1994 can be modeled by $C = 3200x^2 + 1586$, and the cumulative number of AIDS deaths D from 1982 to 1994 can be modeled by $D = 1900x^2 + 619$. In these equations $x = 0$ corresponds to 1982, $x = 1$ to 1983, and so on until $x = 12$ corresponds to 1994. (**Source:** U.S. Department of Health.)
(a) Write the rational expression $\frac{D}{C}$ in terms of x.
(b) Evaluate your expression for $x = 4, 7$, and 10. Round your answers to the nearest thousandth. Interpret the results.
(c) Explain what the rational expression $\frac{D}{C}$ represents.

WRITING ABOUT MATHEMATICS

81. Explain how to multiply two rational expressions.

82. Explain how to divide two rational expressions.

CHECKING BASIC CONCEPTS
SECTIONS 7.1 AND 7.2

1. If possible, evaluate the expression $\frac{3}{x^2 - 1}$ for $x = -1$ and $x = 3$.

2. Simplify to lowest terms.
(a) $\frac{6x^3y^2}{15x^2y^3}$ **(b)** $\frac{5x - 15}{x - 3}$ **(c)** $\frac{x^2 - x - 6}{x^2 + x - 12}$

3. Multiply and simplify to lowest terms.
(a) $\frac{4}{3x} \cdot \frac{2x}{6}$ **(b)** $\frac{2x + 4}{x^2 - 1} \cdot \frac{x + 1}{x + 2}$

4. Divide and simplify to lowest terms.
(a) $\frac{7}{3z^2} \div \frac{14}{5z^3}$ **(b)** $\frac{x^2 + x}{x - 3} \div \frac{x}{x - 3}$

5. *Waiting in Line* Customers are waiting in line at a department store. They arrive randomly at an average rate of x per minute. If the clerk can wait on 2 customers per minute, then the average time in minutes spent waiting in line is given by $T = \frac{1}{2 - x}$ for $x < 2$. (**Source:** N. Garber, *Traffic and Highway Engineering*.)
(a) Complete the table.

x	0.5	1.0	1.5	1.9
T				

(b) What happens to the waiting time as x increases but remains less than 2?

7.3   ADDITION AND SUBTRACTION WITH LIKE DENOMINATORS

Review of Addition and Subtraction of Fractions ■ Rational Expressions Having Like Denominators

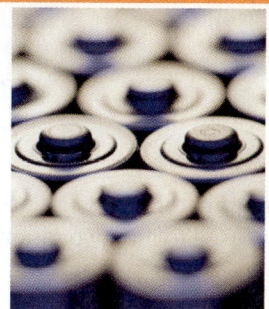

When companies manufacture a large number of items, quality control is important. For example, suppose that a company makes batteries. Because it is not practical to check every battery to make sure that it works properly, inspectors often check a random sample of batteries. By using mathematics and rational expressions, this technique helps determine the likelihood that all the batteries are good. (See Example 6.)

In this section we discuss methods for adding and subtracting rational expressions having like denominators. These methods are similar to the ones used to add and subtract fractions having like denominators.

Review of Addition and Subtraction of Fractions

In Section 1.2 we demonstrated how the property

$$\frac{a}{c} + \frac{b}{c} = \frac{a+b}{c}$$

can be used to add fractions having like denominators. For example,

$$\frac{3}{7} + \frac{2}{7} = \frac{3+2}{7} = \frac{5}{7}.$$

To subtract two fractions having like denominators the property

$$\frac{a}{c} - \frac{b}{c} = \frac{a-b}{c}$$

is used. For example,

$$\frac{2}{5} - \frac{4}{5} = \frac{2-4}{5} = -\frac{2}{5}.$$

EXAMPLE 1   Adding and subtracting fractions having like denominators

Simplify each expression to lowest terms.

(a) $\frac{3}{8} + \frac{4}{8}$ **(b)** $\frac{5}{9} + \frac{1}{9}$ **(c)** $\frac{12}{5} - \frac{7}{5}$ **(d)** $\frac{23}{20} - \frac{13}{20}$

Solution

(a) $\frac{3}{8} + \frac{4}{8} = \frac{3+4}{8} = \frac{7}{8}$

(b) $\frac{5}{9} + \frac{1}{9} = \frac{5+1}{9} = \frac{6}{9} = \frac{2}{3}$

(c) $\frac{12}{5} - \frac{7}{5} = \frac{12-7}{5} = \frac{5}{5} = 1$

(d) $\frac{23}{20} - \frac{13}{20} = \frac{23-13}{20} = \frac{10}{20} = \frac{1}{2}$

Now Try Exercises 15, 19

TECHNOLOGY NOTE: Arithmetic of Fractions

Many calculators have the capability to perform addition and subtraction of fractions, as illustrated in the following figures. Compare these results with those from Example 1.

```
(3/8)+(4/8)▶Frac
               7/8
(5/9)+(1/9)▶Frac
               2/3
```

```
(12/5)−(7/5)▶Fra
c
                1
(23/20)−(13/20)▶
Frac
              1/2
```

Rational Expressions Having Like Denominators

Addition and subtraction of rational expressions having like denominators are similar to addition and subtraction of fractions. The following property can be used to add two rational expressions having like denominators.

SUMS OF RATIONAL EXPRESSIONS

To add two rational expressions having like denominators, add their numerators. Keep the same denominator.

$$\frac{A}{C} + \frac{B}{C} = \frac{A + B}{C} \qquad C \text{ is nonzero.}$$

When we add rational expressions with like denominators, we add the numerators. Then we combine like terms and simplify the resulting expression by applying the basic principle of rational expressions. For example, we can add $\frac{2x}{x + 1}$ and $\frac{1 - x}{x + 1}$ as follows.

$$\frac{2x}{x + 1} + \frac{1 - x}{x + 1} = \frac{2x + 1 - x}{x + 1} \qquad \text{Add the numerators.}$$

$$= \frac{2x - x + 1}{x + 1} \qquad \text{Commutative property}$$

$$= \frac{x + 1}{x + 1} \qquad \text{Combine like terms.}$$

$$= 1 \qquad \text{Simplify.}$$

It is important to understand that the expressions $\frac{2x}{x + 1} + \frac{1 - x}{x + 1}$ and 1 are *equivalent expressions*. That is, they are equal for *every* value of x except -1, for which the first expression is undefined.

In the next example, we add rational expressions with like denominators and simplify the result to lowest terms.

EXAMPLE 2 Adding rational expressions having like denominators

Add and simplify to lowest terms.

(a) $\dfrac{3}{b} + \dfrac{2}{b}$

(b) $\dfrac{z}{z + 2} + \dfrac{2}{z + 2}$

(c) $\dfrac{x - 1}{x^2 + x} + \dfrac{1}{x^2 + x}$

(d) $\dfrac{t^2 + t}{t - 1} + \dfrac{1 - 3t}{t - 1}$

Solution

(a) $\dfrac{3}{b} + \dfrac{2}{b} = \dfrac{3+2}{b} = \dfrac{5}{b}$ Add the numerators.

(b) $\dfrac{z}{z+2} + \dfrac{2}{z+2} = \dfrac{z+2}{z+2}$ Add the numerators.

$\quad\quad = 1$ Simplify.

(c) $\dfrac{x-1}{x^2+x} + \dfrac{1}{x^2+x} = \dfrac{x-1+1}{x^2+x}$ Add the numerators.

$\quad\quad = \dfrac{x}{x(x+1)}$ Factor the denominator.

$\quad\quad = \dfrac{1}{x+1}$ Simplify.

(d) $\dfrac{t^2+t}{t-1} + \dfrac{1-3t}{t-1} = \dfrac{t^2+t+1-3t}{t-1}$ Add the numerators.

$\quad\quad = \dfrac{t^2-2t+1}{t-1}$ Combine like terms.

$\quad\quad = \dfrac{(t-1)(t-1)}{t-1}$ Factor the numerator.

$\quad\quad = t-1$ Simplify.

Now Try Exercises 25, 32, 39, 41

EXAMPLE 3   Adding rational expressions having two variables

Add and simplify to lowest terms.

(a) $\dfrac{4}{xy} + \dfrac{5}{xy}$ (b) $\dfrac{a}{a^2-b^2} + \dfrac{b}{a^2-b^2}$ (c) $\dfrac{1}{x-y} + \dfrac{-1}{y-x}$

Solution

(a) $\dfrac{4}{xy} + \dfrac{5}{xy} = \dfrac{4+5}{xy} = \dfrac{9}{xy}$ Add the numerators.

(b) $\dfrac{a}{a^2-b^2} + \dfrac{b}{a^2-b^2} = \dfrac{a+b}{a^2-b^2}$ Add the numerators.

$\quad\quad = \dfrac{a+b}{(a-b)(a+b)}$ Factor the denominator.

$\quad\quad = \dfrac{1}{a-b}$ Simplify.

(c) First write $\dfrac{1}{x-y} + \dfrac{-1}{y-x}$ with a common denominator. Note that, if we multiply the second term by 1, written in the form $\dfrac{-1}{-1}$, it becomes

$$\dfrac{-1}{y-x} \cdot \dfrac{-1}{-1} = \dfrac{(-1)(-1)}{(y-x)(-1)} = \dfrac{1}{-y+x} = \dfrac{1}{x-y}.$$

Thus the given sum can be simplified as follows.

$$\frac{1}{x - y} + \frac{-1}{y - x} = \frac{1}{x - y} + \frac{1}{x - y}$$ Rewrite the second term.

$$= \frac{2}{x - y}$$ Add the numerators.

Now Try Exercises **57, 59, 61**

Next we consider subtraction of rational expressions having like denominators.

DIFFERENCES OF RATIONAL EXPRESSIONS

To subtract two rational expressions having like denominators, subtract their numerators. Keep the same denominator.

$$\frac{A}{C} - \frac{B}{C} = \frac{A - B}{C}$$ C is nonzero.

Subtraction of rational expressions with like denominators is similar to addition except that instead of adding numerators, we subtract them. For example, the expressions $\frac{3x}{x - 4}$ and $\frac{2x}{x - 4}$ have like denominators and can be subtracted as follows.

$$\frac{3x}{x - 4} - \frac{2x}{x - 4} = \frac{3x - 2x}{x - 4}$$ Subtract the numerators.

$$= \frac{x}{x - 4}$$ Combine like terms.

In the next example, we subtract rational expressions having like denominators and simplify the result to lowest terms.

EXAMPLE **4**   **Subtracting rational expressions having like denominators**

Subtract and simplify to lowest terms.

(a) $\dfrac{a + 1}{a} - \dfrac{1}{a}$ **(b)** $\dfrac{2y}{3y - 1} - \dfrac{3y}{3y - 1}$ **(c)** $\dfrac{1 + x}{2x^2 + 5x - 3} - \dfrac{-2}{2x^2 + 5x - 3}$

Solution

(a) $\dfrac{a + 1}{a} - \dfrac{1}{a} = \dfrac{a + 1 - 1}{a} = \dfrac{a}{a} = 1$

(b) $\dfrac{2y}{3y - 1} - \dfrac{3y}{3y - 1} = \dfrac{2y - 3y}{3y - 1} = \dfrac{-y}{3y - 1}$ or $-\dfrac{y}{3y - 1}$

(c) $\dfrac{1 + x}{2x^2 + 5x - 3} - \dfrac{-2}{2x^2 + 5x - 3} = \dfrac{1 + x - (-2)}{2x^2 + 5x - 3}$

$$= \dfrac{x + 3}{(2x - 1)(x + 3)}$$

$$= \dfrac{1}{2x - 1}$$

Now Try Exercises **27, 33, 51**

If the numerator of the second fraction in a difference has more than one term, it is important to put parentheses around the second numerator.

$$\frac{x+1}{2x+1} - \frac{3-x}{2x+1} = \frac{x+1-(3-x)}{2x+1}$$ Subtract the numerators; insert parentheses.

$$= \frac{x+1-3-(-x)}{2x+1}$$ Distributive property

$$= \frac{x+1-3+x}{2x+1}$$ Double negative property

$$= \frac{2x-2}{2x+1}$$ Combine like terms.

NOTE: If parentheses were not inserted in the previous calculation, the numerator would be

$$x+1-3-x = -2,$$

which would give an incorrect result.

EXAMPLE 5 **Subtracting rational expressions having like denominators**

Subtract and simplify to lowest terms.

(a) $\dfrac{2x}{x+1} - \dfrac{x-1}{x+1}$ **(b)** $\dfrac{x+y}{3y} - \dfrac{x-y}{3y}$

Solution

(a) $\dfrac{2x}{x+1} - \dfrac{x-1}{x+1} = \dfrac{2x-(x-1)}{x+1}$ Subtract the numerators.

$$= \frac{2x-x+1}{x+1}$$ Distributive property

$$= \frac{x+1}{x+1}$$ Simplify the numerator.

$$= 1$$ Simplify.

(b) $\dfrac{x+y}{3y} - \dfrac{x-y}{3y} = \dfrac{x+y-(x-y)}{3y}$ Subtract the numerators.

$$= \frac{x+y-x+y}{3y}$$ Distributive property

$$= \frac{2y}{3y}$$ Simplify the numerator.

$$= \frac{2}{3}$$ Simplify.

Now Try Exercises **37, 63**

▶ **REAL-WORLD CONNECTION** The introduction to this section discusses how quality control uses rational expressions. The next example illustrates one way this can occur.

EXAMPLE 6   Analyzing quality control

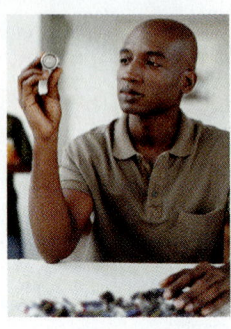

A container holds a mixture of size A and size AA batteries. In this container, there is a total of n batteries, including 2 defective A batteries and 4 defective AA batteries. If a battery is picked at random by a quality control inspector, then the probability, or chance, of one of the defective batteries being chosen is given by the expression $\frac{2}{n} + \frac{4}{n}$.
(a) Simplify this expression.   **(b)** Interpret the result.

Solution
(a) Because the denominators are the same, we simply add the numerators.

$$\frac{2}{n} + \frac{4}{n} = \frac{2+4}{n} = \frac{6}{n}$$

(b) There are 6 in n chances that a defective battery is chosen.   Now Try Exercise **73**

7.3 PUTTING IT ALL TOGETHER

In this section we discussed how to add and subtract rational expressions having like denominators. After adding or subtracting rational expressions, be sure to factor the numerator and the denominator completely. Then simplify the result to lowest terms by applying the basic principle of rational expressions. The following table summarizes this discussion, where A, B, and C represent polynomials and C is nonzero.

Concept	Explanation	Examples
Addition of Rational Expressions	$\dfrac{A}{C} + \dfrac{B}{C} = \dfrac{A+B}{C}$	$\dfrac{x}{x+1} + \dfrac{1-x}{x+1} = \dfrac{x+1-x}{x+1} = \dfrac{1}{x+1}$ $\dfrac{2x}{x^2-1} + \dfrac{x}{x^2-1} = \dfrac{2x+x}{x^2-1} = \dfrac{3x}{x^2-1}$
Subtraction of Rational Expressions	$\dfrac{A}{C} - \dfrac{B}{C} = \dfrac{A-B}{C}$ If B consists of more than one term, put parentheses around B and apply the distributive property.	$\dfrac{2x}{x^2-4} - \dfrac{x+2}{x^2-4} = \dfrac{2x-(x+2)}{x^2-4}$ $= \dfrac{2x-x-2}{x^2-4}$ $= \dfrac{x-2}{(x+2)(x-2)}$ $= \dfrac{1}{x+2}$

7.3 Exercises

CONCEPTS

1. To add two rational expressions having like denominators, _____ their _____. The _____ do not change.

2. To subtract two rational expressions having like denominators, _____ their _____. The _____ do not change.

3. $\frac{2}{5} + \frac{1}{5} =$ _____

4. $\frac{6}{7} - \frac{2}{7} =$ _____

5. $\frac{A}{C} + \frac{B}{C} =$ _____

6. $\frac{A}{C} - \frac{B}{C} =$ _____

7. $5 - (x + 1) =$ _____

8. $\frac{5}{x - 1} - \frac{x + 1}{x - 1} =$ _____

9. $3x - (2x - 5) =$ _____

10. $\frac{3x}{7} - \frac{2x - 5}{7} =$ _____

ADDITION AND SUBTRACTION OF FRACTIONS

Exercises 11–24: Simplify to lowest terms.

11. $\frac{1}{2} + \frac{1}{2}$

12. $\frac{3}{7} + \frac{2}{7}$

13. $\frac{4}{5} + \frac{2}{5}$

14. $\frac{3}{11} + \frac{5}{11}$

15. $\frac{1}{6} + \frac{5}{6}$

16. $\frac{3}{10} + \frac{5}{10}$

17. $\frac{4}{7} - \frac{1}{7}$

18. $\frac{5}{13} - \frac{7}{13}$

19. $\frac{7}{8} - \frac{3}{8}$

20. $\frac{9}{16} - \frac{5}{16}$

21. $\frac{11}{12} - \frac{5}{12}$

22. $\frac{7}{24} - \frac{3}{24}$

23. $\frac{7}{15} + \frac{4}{15} - \frac{1}{15}$

24. $\frac{11}{36} - \frac{5}{36} + \frac{1}{36}$

ADDITION AND SUBTRACTION OF RATIONAL EXPRESSIONS

Exercises 25–70: Simplify to lowest terms.

25. $\frac{2}{x} + \frac{1}{x}$

26. $\frac{9}{x} - \frac{7}{x}$

27. $\frac{7 + 2x}{4x} - \frac{7}{4x}$

28. $\frac{x - 1}{5x} + \frac{2x + 1}{5x}$

29. $\frac{y + 3}{y - 3} + \frac{2y - 12}{y - 3}$

30. $\frac{5 - y}{y + 2} + \frac{y}{y + 2}$

31. $\frac{x}{x - 3} - \frac{3}{x - 3}$

32. $\frac{2x}{2x + 1} + \frac{1}{2x + 1}$

33. $\frac{5z}{4z + 3} - \frac{z}{4z + 3}$

34. $\frac{z}{2z + 1} - \frac{1 - z}{2z + 1}$

35. $\frac{t + 5}{t + 6} + \frac{t + 7}{t + 6}$

36. $\frac{t + 1}{t - 4} + \frac{1 - t}{t - 4}$

37. $\frac{5x}{2x + 3} - \frac{3x - 3}{2x + 3}$

38. $\frac{x}{5 - x} - \frac{2x - 5}{5 - x}$

39. $\frac{x - 4}{x^2 - x} + \frac{4}{x^2 - x}$

40. $\frac{2x - 2}{4x^2 - 1} + \frac{1}{4x^2 - 1}$

41. $\frac{z^2 - 1}{z - 2} + \frac{3 - 3z}{z - 2}$

42. $\frac{x^2 + 2}{x + 1} + \frac{3x}{x + 1}$

43. $\frac{x^2 + 4x - 1}{4x + 2} - \frac{x^2 - 4x - 5}{4x + 2}$

44. $\frac{2x^2 - x + 5}{x^2 - 9} - \frac{x^2 - x + 14}{x^2 - 9}$

45. $\frac{3y}{5} + \frac{2y - 5}{5}$

46. $\frac{3y - 22}{11} + \frac{8y}{11}$

47. $\frac{x + y}{4} + \frac{x - y}{4}$

48. $\frac{x + y}{4} - \frac{x - y}{4}$

49. $\frac{z^2 + 4}{z - 2} - \frac{4z}{z - 2}$

50. $\frac{z^2 + 2z}{z + 1} + \frac{1}{z + 1}$

51. $\frac{2x^2 - 5x}{2x + 1} - \frac{3}{2x + 1}$

52. $\frac{2x^2}{x + 2} + \frac{9x + 10}{x + 2}$

53. $\frac{3n}{2n^2 - n + 5} + \frac{4n}{2n^2 - n + 5}$

54. $\frac{n}{n^2 + n + 1} - \frac{1}{n^2 + n + 1}$

55. $\frac{1}{x + 3} + \frac{2}{x + 3} + \frac{3}{x + 3}$

56. $\frac{x}{2x - 5} - \frac{1}{2x - 5} + \frac{2x + 1}{2x - 5}$

57. $\dfrac{8}{ab} + \dfrac{1}{ab}$

58. $\dfrac{6}{xy} + \dfrac{9}{xy}$

59. $\dfrac{x}{(x+y)^2} + \dfrac{y}{(x+y)^2}$

60. $\dfrac{x-2y}{x^2-y^2} + \dfrac{y}{x^2-y^2}$

61. $\dfrac{8}{a-b} + \dfrac{-8}{b-a}$

62. $\dfrac{6}{x-y} + \dfrac{6}{y-x}$

63. $\dfrac{a+b}{4a} - \dfrac{a-b}{4a}$

64. $\dfrac{x-y}{5x} - \dfrac{x+y}{5x}$

65. $\dfrac{x}{x+y} + \dfrac{y}{x+y}$

66. $\dfrac{x}{x^2-y^2} - \dfrac{y}{x^2-y^2}$

67. $\dfrac{a^2}{a+b} - \dfrac{b^2}{a+b}$

68. $\dfrac{a^2}{a+b} + \dfrac{2ab+b^2}{a+b}$

69. $\dfrac{4x^2}{2x+3y} - \dfrac{9y^2}{2x+3y}$

70. $\dfrac{x^3}{x^2+xy+y^2} - \dfrac{y^3}{x^2+xy+y^2}$

71. **Thinking Generally** If $\dfrac{2}{3+x} + \dfrac{3}{3+x}$ equals $\dfrac{5}{10}$, what must be true about x?

72. **Thinking Generally** If $\dfrac{8}{6+x} - \dfrac{4}{3+y}$ equals 0, what must be true about x and y?

APPLICATIONS

73. *Quality Control*   (Refer to Example 6.) A container holds a total of $n+1$ batteries. In this container, there are 6 defective B batteries, 5 defective C batteries, and 3 defective D batteries. If a battery is chosen at random by a quality control inspector, the probability, or chance, of one of the defective batteries being chosen is

$$\dfrac{6}{n+1} + \dfrac{5}{n+1} + \dfrac{3}{n+1}.$$

(a) Simplify this expression.
(b) Evaluate the simplified expression for $n = 99$ and interpret the result.

74. *Intensity of a Light Bulb*   The farther a person is from a light bulb, the less intense its light is. The equation $I = \dfrac{19}{4d^2}$ approximates the intensity of light from a 60-watt light bulb at a distance of d meters, where I is measured in watts per square meter.
(*Source:* R. Weidner.)

(a) Find I for $d = 2$ meters and interpret the result.
(b) The intensity of light from a 100-watt light bulb is about $I = \dfrac{32}{4d^2}$. Find an expression for the sum of the intensities of light from a 100-watt bulb and a 60-watt bulb.

WRITING ABOUT MATHEMATICS

75. Explain how to add two rational expressions having like denominators. Give an example.

76. Explain how to subtract two rational expressions having like denominators. Give an example.

7.4   ADDITION AND SUBTRACTION WITH UNLIKE DENOMINATORS

Finding Least Common Multiples ▪ Review of Fractions Having Unlike Denominators ▪ Rational Expressions Having Unlike Denominators

A LOOK INTO MATH ▷

The sum and difference of rational expressions frequently occur in the design of electrical devices, such as cell phones and televisions. In Section 7.3, we added and subtracted rational expressions having like denominators. Although the denominators of rational expressions are often unlike, rational expressions can still be added and subtracted after a common denominator is found. One way to find the least common denominator for a sum or difference of rational expressions is to find the least common multiple of the denominators. In the following subsection we show how to find the least common multiple.

Finding Least Common Multiples

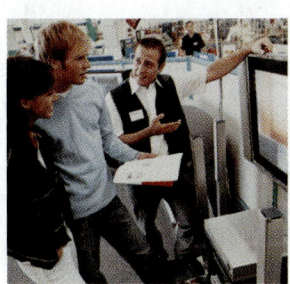

▶ **REAL-WORLD CONNECTION**   Two friends work part-time at a store. The first person works every sixth day, and the second person works every eighth day. If they both work today, how many days will pass before they work on the same day again?

We can answer this question by listing the days that each person works.

First person: 6,   12,   18,   **24**,   30,   36,   42,   **48**,   54

Second person: 8,   16,   **24**,   32,   40,   **48**,   56,   64

After 24 days, the two friends work on the same day. The next time is after 48 days. The numbers 24 and 48 are *common multiples* of 6 and 8. (Find another.) However, 24 is the *least* common multiple (LCM) of 6 and 8.

Another way to find the least common multiple of 6 and 8 is to factor each number into prime numbers.

$$6 = 2 \cdot \mathbf{3} \quad \text{and} \quad 8 = \mathbf{2 \cdot 2 \cdot 2}$$

To find the least common multiple, first list each factor the greatest number of times that it occurs in either factorization. Then find the product of these numbers. For this example, the factor 2 occurs three times in the factorization of 8 and only once in the factorization of 6, so list 2 three times. The factor 3 appears only once in the factorization of 6 so list it once:

2, 2, 2, 3.

The least common multiple is their product: $\mathbf{2 \cdot 2 \cdot 2 \cdot 3} = 24$.

This same procedure can also be used to find the least common multiple for two or more polynomials.

FINDING THE LEAST COMMON MULTIPLE

The least common multiple (LCM) of two or more polynomials can be found as follows.

STEP 1: Factor each polynomial completely.

STEP 2: List each factor the greatest number of times that it occurs in any factorization.

STEP 3: Find the product of this list of factors. The result is the LCM.

EXAMPLE 1   Finding least common multiples

Find the least common multiple of each pair of expressions.
(a) $2x, 5x^2$ **(b)** $x^2 - x, x - 1$
(c) $x + 2, x - 3$ **(d)** $x^2 + 2x + 1, x^2 + 3x + 2$

Solution
(a) STEP 1: Factor $2x$ and $5x^2$ completely.

$$2x = \mathbf{2} \cdot \boldsymbol{x} \quad \text{and} \quad 5x^2 = \mathbf{5} \cdot \boldsymbol{x} \cdot \boldsymbol{x}$$

STEP 2: In either factorization, the factor 2 occurs at most once, the factor 5 occurs at most once, and the factor x occurs at most twice. The list of factors is: $\mathbf{2}, \mathbf{5}, \boldsymbol{x}, \boldsymbol{x}$.

STEP 3: The LCM equals the product

$$\mathbf{2} \cdot \mathbf{5} \cdot \boldsymbol{x} \cdot \boldsymbol{x} = 10x^2.$$

(b) STEP 1: Factor $x^2 - x$ and $x - 1$ completely. Note that $x - 1$ cannot be factored.

$$x^2 - x = x(x - 1) \quad \text{and} \quad x - 1 = x - 1$$

STEP 2: Both factors, x and $x - 1$, occur at most once in either factorization. The list of factors is $x, (x - 1)$.

STEP 3: The LCM is the product $x(x - 1)$, or $x^2 - x$.

(c) STEP 1: Neither $x + 2$ nor $x - 3$ can be factored.

STEP 2: The list of factors is $(x + 2), (x - 3)$.

STEP 3: The LCM is the product $(x + 2)(x - 3)$, or $x^2 - x - 6$.

(d) STEP 1: Factor $x^2 + 2x + 1$ and $x^2 + 3x + 2$ completely.

$$x^2 + 2x + 1 = (x + 1)(x + 1) \quad \text{and} \quad x^2 + 3x + 2 = (x + 1)(x + 2)$$

STEP 2: In either factorization, the factor $(x + 1)$ occurs at most twice and the factor $(x + 2)$ occurs at most once. The list is $(x + 1), (x + 1), (x + 2)$.

STEP 3: The LCM is the product $(x + 1)^2(x + 2)$.   Now Try Exercises 17, 21, 33, 37

Review of Fractions Having Unlike Denominators

Before we can find the sum $\frac{1}{2} + \frac{1}{3}$ by hand, we need to rewrite these fractions by using their least common denominator. The least common denominator (LCD) of $\frac{1}{2}$ and $\frac{1}{3}$ corresponds to the least common multiple of 2 and 3, which is 6. As a result, we rewrite these fractions as

$$\frac{1}{2} \cdot \frac{3}{3} = \frac{3}{6} \quad \text{and} \quad \frac{1}{3} \cdot \frac{2}{2} = \frac{2}{6}.$$

Their sum is $\frac{1}{2} + \frac{1}{3} = \frac{3}{6} + \frac{2}{6} = \frac{5}{6}$.

EXAMPLE 2   **Adding and subtracting fractions having unlike denominators**

Simplify each expression.

(a) $\frac{3}{10} + \frac{4}{15}$ **(b)** $\frac{7}{8} - \frac{1}{6}$

Solution

(a) The LCD for $\frac{3}{10}$ and $\frac{4}{15}$ equals the LCM of 10 and 15, which is 30. We rewrite these fractions as

$$\frac{3}{10} \cdot \frac{3}{3} = \frac{9}{30} \quad \text{and} \quad \frac{4}{15} \cdot \frac{2}{2} = \frac{8}{30}.$$

Their sum is

$$\frac{3}{10} + \frac{4}{15} = \frac{9}{30} + \frac{8}{30} = \frac{17}{30}.$$

(b) The LCD for $\frac{7}{8}$ and $\frac{1}{6}$ is the LCM of 8 and 6, which is 24. We rewrite these fractions as

$$\frac{7}{8} \cdot \frac{3}{3} = \frac{21}{24} \quad \text{and} \quad \frac{1}{6} \cdot \frac{4}{4} = \frac{4}{24}.$$

Their difference is

$$\frac{7}{8} - \frac{1}{6} = \frac{21}{24} - \frac{4}{24} = \frac{17}{24}.$$ Now Try Exercises 51, 53

Rational Expressions Having Unlike Denominators

The first step in adding or subtracting rational expressions having unlike denominators is to rewrite each expression by using the least common denominator. Then the sum or difference can be found by using the techniques discussed in Section 7.3.

NOTE: The LCD of two or more rational expressions equals the LCM of their denominators.

To add or subtract rational expressions, we often rewrite a rational expression with a different denominator. This technique is demonstrated in the next example and is used in future examples.

EXAMPLE 3 **Rewriting rational expressions**

Rewrite each rational expression so it has the given denominator D.

(a) $\dfrac{3}{2x}$, $D = 8x^2$ **(b)** $\dfrac{1}{x+1}$, $D = x^2 - 1$

Solution

(a) We need to write $\dfrac{3}{2x}$ so that it is equivalent to $\dfrac{?}{8x^2}$. Because $8x^2 = 2x \cdot \mathbf{4x}$, we can multiply $\dfrac{3}{2x}$ by 1 in the form $\dfrac{4x}{4x}$ as follows.

$$\frac{3}{2x} \cdot \frac{\mathbf{4x}}{\mathbf{4x}} = \frac{12x}{8x^2} \qquad \text{Multiply rational expressions.}$$

(b) We must write $\dfrac{1}{x+1}$ so that it is equivalent to $\dfrac{?}{x^2-1}$. Because $x^2 - 1 = (x+1)(\mathbf{x-1})$, we can multiply $\dfrac{1}{x+1}$ by 1 in the form $\dfrac{x-1}{x-1}$ as follows.

$$\frac{1}{x+1} \cdot \frac{\mathbf{x-1}}{\mathbf{x-1}} = \frac{x-1}{x^2-1} \qquad \text{Multiply rational expressions.}$$

Now Try Exercises 43, 45

EXAMPLE 4 **Adding rational expressions having unlike denominators**

Find each sum and leave your answer in factored form.

(a) $\dfrac{5}{8y} + \dfrac{7}{4y^2}$ **(b)** $\dfrac{1}{x-1} + \dfrac{1}{x+1}$ **(c)** $\dfrac{x}{x^2+2x+1} + \dfrac{1}{x+1}$

Solution

(a) First find the LCM for $8y$ and $4y^2$.

$$8y = \mathbf{2 \cdot 2 \cdot 2} \cdot y \quad \text{and} \quad 4y^2 = 2 \cdot 2 \cdot \mathbf{y \cdot y}$$

Thus the LCM is $\mathbf{2 \cdot 2 \cdot 2} \cdot \mathbf{y \cdot y} = 8y^2$. Because

$$8y^2 = 8y \cdot \mathbf{y} \quad \text{and} \quad 8y^2 = 4y^2 \cdot \mathbf{2},$$

we multiply the first expression by $\dfrac{y}{y}$ and the second expression by $\dfrac{2}{2}$.

$$\frac{5}{8y} + \frac{7}{4y^2} = \frac{5}{8y} \cdot \frac{y}{y} + \frac{7}{4y^2} \cdot \frac{2}{2} \qquad \text{Rewrite by using the LCD.}$$

$$= \frac{5y}{8y^2} + \frac{14}{8y^2} \qquad \text{Multiply the fractions.}$$

$$= \frac{5y+14}{8y^2} \qquad \text{Add the numerators.}$$

(b) The LCM for $x - 1$ and $x + 1$ is their product, $(x - 1)(x + 1)$.

$$\frac{1}{x - 1} + \frac{1}{x + 1} = \frac{1}{x - 1} \cdot \frac{x + 1}{x + 1} + \frac{1}{x + 1} \cdot \frac{x - 1}{x - 1}$$ Rewrite by using the LCD.

$$= \frac{x + 1}{(x - 1)(x + 1)} + \frac{x - 1}{(x + 1)(x - 1)}$$ Multiply the fractions.

$$= \frac{x + 1 + x - 1}{(x - 1)(x + 1)}$$ Add the numerators.

$$= \frac{2x}{(x - 1)(x + 1)}$$ Simplify the numerator.

(c) First find the LCM for $x^2 + 2x + 1$ and $x + 1$. Because

$$x^2 + 2x + 1 = (x + 1)(x + 1),$$

their LCM is $(x + 1)(x + 1) = (x + 1)^2$.

$$\frac{x}{x^2 + 2x + 1} + \frac{1}{x + 1} = \frac{x}{(x + 1)^2} + \frac{1}{x + 1} \cdot \frac{x + 1}{x + 1}$$ Rewrite by using the LCD.

$$= \frac{x}{(x + 1)^2} + \frac{x + 1}{(x + 1)^2}$$ Multiply the fractions.

$$= \frac{2x + 1}{(x + 1)^2}$$ Add the numerators.

Now Try Exercises 59, 71, 77

Subtraction of rational expressions is performed in a manner similar to addition and is illustrated in the next example.

EXAMPLE 5 **Subtracting rational expressions having unlike denominators**

Simplify each expression. Write your answer in lowest terms and leave it in factored form.

(a) $\dfrac{5}{z} - \dfrac{z}{z - 1}$ **(b)** $\dfrac{5}{x + 1} - \dfrac{1}{x^2 - 1}$

(c) $\dfrac{x}{x^2 - 2x} - \dfrac{1}{x^2 + 2x}$ **(d)** $\dfrac{3}{x - 1} - \dfrac{3}{x^2 - x} + \dfrac{5}{x}$

Solution
(a) The LCD is $z(z - 1)$.

$$\frac{5}{z} - \frac{z}{z - 1} = \frac{5}{z} \cdot \frac{z - 1}{z - 1} - \frac{z}{z - 1} \cdot \frac{z}{z}$$ Rewrite by using the LCD.

$$= \frac{5(z - 1)}{z(z - 1)} - \frac{z^2}{z(z - 1)}$$ Multiply the fractions.

$$= \frac{5(z - 1) - z^2}{z(z - 1)}$$ Add the numerators.

$$= \frac{-z^2 + 5z - 5}{z(z - 1)}$$ Simplify the numerator.

(b) The LCD is $(x - 1)(x + 1)$. Note that $x^2 - 1 = (x - 1)(x + 1)$.

$$\frac{5}{x + 1} - \frac{1}{x^2 - 1} = \frac{5}{x + 1} \cdot \frac{x - 1}{x - 1} - \frac{1}{(x - 1)(x + 1)}$$ Rewrite by using the LCD.

$$= \frac{5(x - 1) - 1}{(x - 1)(x + 1)}$$ Subtract the numerators.

$$= \frac{5x - 5 - 1}{(x - 1)(x + 1)}$$ Distributive property

$$= \frac{5x - 6}{(x - 1)(x + 1)}$$ Simplify the numerator.

(c) Start by factoring each denominator. Because

$$x^2 - 2x = x(x - 2) \quad \text{and} \quad x^2 + 2x = x(x + 2),$$

the LCD is $x(x - 2)(x + 2)$.

$$\frac{x}{x^2 - 2x} - \frac{1}{x^2 + 2x} = \frac{x}{x(x - 2)} \cdot \frac{x + 2}{x + 2} - \frac{1}{x(x + 2)} \cdot \frac{x - 2}{x - 2}$$ Rewrite by using the LCD.

$$= \frac{x(x + 2)}{x(x - 2)(x + 2)} - \frac{x - 2}{x(x - 2)(x + 2)}$$ Multiply the rational expressions.

$$= \frac{x(x + 2) - (x - 2)}{x(x - 2)(x + 2)}$$ Subtract the numerators.

$$= \frac{x^2 + 2x - x + 2}{x(x - 2)(x + 2)}$$ Distributive property

$$= \frac{x^2 + x + 2}{x(x - 2)(x + 2)}$$ Simplify the numerator.

(d) The given expression contains three rational expressions. Begin by finding the LCM of the three denominators: $x - 1$, $x^2 - x$, and x. Because $x^2 - x = x(x - 1)$, the LCM is $x(x - 1)$.

$$\frac{3}{x - 1} - \frac{3}{x^2 - x} + \frac{5}{x} = \frac{3}{x - 1} \cdot \frac{x}{x} - \frac{3}{x(x - 1)} + \frac{5}{x} \cdot \frac{x - 1}{x - 1}$$ Rewrite by using the LCD.

$$= \frac{3x}{x(x - 1)} - \frac{3}{x(x - 1)} + \frac{5(x - 1)}{x(x - 1)}$$ Multiply the expressions.

$$= \frac{3x - 3 + 5x - 5}{x(x - 1)}$$ Combine the expressions.

$$= \frac{8x - 8}{x(x - 1)}$$ Simplify the numerator.

$$= \frac{8(x - 1)}{x(x - 1)}$$ Factor the numerator.

$$= \frac{8}{x}$$ Simplify to lowest terms.

Now Try Exercises 69, 83, 85, 87

CRITICAL THINKING

Find the reciprocal of the sum $x + \frac{1}{x}$.

► **REAL-WORLD CONNECTION** Sums of rational expressions occur in applications involving electricity. The flow of electricity through a wire can be compared to the flow of water through a hose. Voltage is the force "pushing" the electricity and corresponds to water pressure in a hose. Resistance is the opposition to the flow of electricity and corresponds to the diameter of a hose. More resistance results in less flow of electricity. An ordinary light bulb is an example of a resistor in an electrical circuit. Resistance is often measured in units called *ohms*. For example, a standard 60-watt light bulb has a resistance of about 200 ohms.

Suppose that two light bulbs are wired in parallel so that electricity can flow through either light bulb, as illustrated in Figure 7.2. If the individual resistances of the light bulbs are R and S, then the total resistance of the circuit is given by the *reciprocal* of the sum

Figure 7.2

$$\frac{1}{R} + \frac{1}{S}.$$

(**Source:** R. Weidner and R. Sells, *Elementary Classical Physics, Vol. 2.*)

EXAMPLE **6**   **Modeling electrical resistance**

Add $\frac{1}{R} + \frac{1}{S}$, and then find the reciprocal of the result.

Solution

The LCD for $\frac{1}{R}$ and $\frac{1}{S}$ is RS.

$$\frac{1}{R} + \frac{1}{S} = \frac{1}{R} \cdot \frac{S}{S} + \frac{1}{S} \cdot \frac{R}{R} \qquad \text{Rewrite by using the LCD.}$$

$$= \frac{S}{RS} + \frac{R}{RS} \qquad \text{Multiply the fractions.}$$

$$= \frac{S + R}{RS} \qquad \text{Add the numerators.}$$

In general, the reciprocal of $\frac{a}{b}$ is $\frac{b}{a}$, so the reciprocal of $\frac{S + R}{RS}$ is $\frac{RS}{S + R}$. This final expression can be used to find the total resistance of the circuit.   Now Try Exercise **103**

7.4   PUTTING IT ALL TOGETHER

In this section we discussed how to add and subtract rational expressions having unlike denominators. The following table summarizes this discussion.

Concept	Explanation	Examples
Least Common Multiple (LCM)	1. Factor each polynomial completely. 2. List each factor the greatest number of times that it occurs in any factorization. 3. The LCM is the product of this list.	1. $x^2 - 2x = x(x - 2)$ $x^2 - 6x + 8 = (x - 2)(x - 4)$ 2. $x, (x - 2), (x - 4)$ 3. LCM $= x(x - 2)(x - 4)$

Concept	Explanation	Examples
Least Common Denominator (LCD)	The LCD of two or more rational expressions equals the LCM of their denominators.	The LCD of $\frac{2}{x^2 - 2x}$ and $\frac{3}{x^2 - 6x + 8}$ is $x(x - 2)(x - 4)$ because the LCM of $x^2 - 2x$ and $x^2 - 6x + 8$ is $x(x - 2)(x - 4)$, as shown in the preceding numbered list.
Addition and Subtraction of Rational Expressions Having Unlike Denominators	First rewrite each expression by using the LCD. Then add or subtract the expressions. Finally, write your answer in lowest terms.	The LCM of x and $x + 2$ is $x(x + 2)$. $$\frac{2}{x} + \frac{5}{x + 2} = \frac{2}{x} \cdot \frac{x + 2}{x + 2} + \frac{5}{x + 2} \cdot \frac{x}{x}$$ $$= \frac{2x + 4}{x(x + 2)} + \frac{5x}{x(x + 2)}$$ $$= \frac{7x + 4}{x(x + 2)}$$

7.4 Exercises

CONCEPTS

1. Give a common multiple of 6 and 9 that is not the *least* common multiple.

2. The LCM of x and y is _____.

3. To rewrite $\frac{3}{4}$ having denominator 12, multiply $\frac{3}{4}$ by the fraction _____.

4. To rewrite $\frac{4}{x - 1}$ having denominator $x^2 - 1$, multiply $\frac{4}{x - 1}$ by the rational expression _____.

5. What is the LCD for $\frac{1}{4}$ and $\frac{5}{6}$?

6. What is the LCD for $\frac{1}{2x}$ and $\frac{5}{6x^2}$?

LEAST COMMON MULTIPLES

Exercises 7–14: Find the least common multiple.

7. 4, 6

8. 6, 9

9. 2, 3

10. 5, 4

11. 10, 15

12. 8, 12

13. 24, 36

14. 32, 40

Exercises 15–38: Find the least common multiple. Leave your answer in factored form.

15. $4x, 6x$

16. $6x, 9x$

17. $5x, 10x^2$

18. $4x^2, 12x$

19. $x, x + 1$

20. $4x, x - 1$

21. $2x + 1, x + 3$

22. $5x + 3, x + 9$

23. $4x^2, 9x^3$

24. $12x^3, 15x^5$

25. $x^2 - x, x^2 + x$

26. $x^2 + 2x, x^2$

27. $(x + 1)^2, x + 1$

28. $(x - 8)^2, (x - 8)(x + 1)$

29. $(2x - 1)^3, (2x - 1)(x + 3)$

30. $(x - 4)(x + 4), (x - 4)(x + 3)$

31. $4x^2 - 1, 2x + 1$

32. $x^2 + 4x + 3, x + 3$

33. $x^2 - 1, x + 1$

34. $x^2 - 4, x - 2$

35. $x^2 + 4, 4x$ **36.** $4x^2 + x, x$

37. $2x^2 + 7x + 6, x^2 + 5x + 6$

38. $x^2 - 3x + 2, x^2 + 2x - 3$

ADDITION AND SUBTRACTION OF RATIONAL EXPRESSIONS

Exercises 39–50: Rewrite the rational expression so it has the given denominator D. For example, if the fraction $\frac{1}{4}$ is written with denominator $D = 8$, it becomes $\frac{2}{8}$.

39. $\frac{1}{3}$, $D = 9$ **40.** $\frac{3}{4}$, $D = 24$

41. $\frac{5}{7}$, $D = 21$ **42.** $\frac{4}{5}$, $D = 30$

43. $\frac{1}{4x}$, $D = 8x^3$ **44.** $\frac{5}{3x}$, $D = 9x^2$

45. $\frac{1}{x + 2}$, $D = x^2 - 4$ **46.** $\frac{3}{x - 3}$, $D = x^2 - 9$

47. $\frac{1}{x + 1}$, $D = x^2 + x$ **48.** $\frac{3}{x - 3}$, $D = x^2 - 3x$

49. $\frac{2x}{x + 1}$, $D = x^2 + 2x + 1$

50. $\frac{x}{2x - 1}$, $D = 2x^2 + 11x - 6$

Exercises 51–102: Simplify the expression. Write your answer in lowest terms and leave it in factored form.

51. $\frac{4}{5} + \frac{1}{2}$ **52.** $\frac{3}{8} + \frac{1}{4}$

53. $\frac{5}{9} - \frac{1}{3}$ **54.** $\frac{7}{10} - \frac{3}{15}$

55. $\frac{4}{25} + \frac{2}{5}$ **56.** $\frac{7}{9} - \frac{1}{3}$

57. $\frac{1}{5} + \frac{3}{4} - \frac{1}{2}$ **58.** $\frac{6}{7} - \frac{8}{9} + \frac{2}{3}$

59. $\frac{1}{3x} + \frac{3}{4x}$ **60.** $\frac{4}{2x^2} + \frac{7}{3x}$

61. $\frac{5}{z^2} - \frac{7}{z^3}$ **62.** $\frac{8}{z} - \frac{3}{2z}$

63. $\frac{1}{x} - \frac{1}{y}$ **64.** $\frac{1}{xy} - \frac{4}{y}$

65. $\frac{a}{b} + \frac{b}{a}$ **66.** $\frac{3}{x} - \frac{4}{y}$

67. $\frac{1}{2x + 4} + \frac{3}{x + 2}$ **68.** $\frac{1}{5x - 10} - \frac{x}{x - 2}$

69. $\frac{2}{t - 2} - \frac{1}{t}$ **70.** $\frac{7}{2t} + \frac{1}{t + 5}$

71. $\frac{5}{n - 1} + \frac{n}{n + 1}$ **72.** $\frac{4n}{3n - 2} + \frac{n}{n + 1}$

73. $\frac{3}{x - 3} + \frac{6}{3 - x}$ **74.** $\frac{x}{x - 8} + \frac{x}{8 - x}$

75. $\frac{1}{5k - 1} + \frac{1}{1 - 5k}$ **76.** $\frac{4}{4 - 3k} + \frac{3k}{3k - 4}$

77. $\frac{2x}{(x - 1)^2} + \frac{4}{x - 1}$ **78.** $\frac{5}{(x + 5)} - \frac{x}{(x + 5)^2}$

79. $\frac{2y}{y(2y - 1)} + \frac{1}{2y - 1}$ **80.** $\frac{5y}{y(y + 1)} - \frac{5}{y + 1}$

81. $\frac{1}{x + 2} - \frac{1}{x^2 + 2x}$ **82.** $\frac{1}{x - 3} - \frac{2}{x^2 - 3x}$

83. $\frac{3}{x - 2} - \frac{1}{x^2 - 4}$ **84.** $\frac{x}{9 - x^2} - \frac{1}{3 - x}$

85. $\frac{2}{x^2 - 3x} - \frac{1}{x^2 + 3x}$ **86.** $\frac{3}{x^2 + 4x} - \frac{2}{x^2 - 4x}$

87. $\frac{1}{x - 2} - \frac{1}{x + 2} + \frac{1}{x}$ **88.** $\frac{1}{x^2} - \frac{2}{x} + \frac{2}{x - 1}$

89. $\frac{x}{x^2 + 4x + 4} + \frac{1}{x + 2}$

90. $\frac{1}{x^2 - 3x - 4} - \frac{1}{x + 1}$

91. $\frac{x}{(x + 1)(x + 2)} - \frac{1}{(x + 2)(x + 3)}$

92. $\frac{2x}{(x - 1)(x - 2)} - \frac{5}{x - 2}$

93. $\frac{1}{a + b} - \frac{1}{a - b}$

94. $\frac{x}{x^2 - y^2} - \frac{1}{x + y}$

95. $\frac{r}{r - t} + \frac{t}{t - r} - 1$

96. $\frac{1}{x} + \frac{2}{x^2 - 2x} + \frac{5}{x - 2}$

97. $\dfrac{1}{2a} + \dfrac{1}{3a} + \dfrac{1}{4a}$

98. $\dfrac{1}{b} + \dfrac{1}{b+1} + \dfrac{1}{b+2}$

99. $\dfrac{2}{x-y} + \dfrac{3}{y-x} + \dfrac{1}{x-y}$

100. $\dfrac{a}{a-b} + \dfrac{b}{b-a} + \dfrac{3}{b}$

101. $\dfrac{3}{x-3} - \dfrac{3}{x^2-3x} - \dfrac{6}{x(x-3)}$

102. $\dfrac{3}{2a-4} + \dfrac{5}{2a} - \dfrac{3}{a^2-2a}$

APPLICATIONS

103. *Electricity*   In Example 6, we showed that the expressions

$$\frac{1}{R} + \frac{1}{S} \quad \text{and} \quad \frac{S+R}{RS}$$

are equivalent. Evaluate both expressions by using $R = 120$ and $S = 200$. Are your answers the same?

104. *Intensity of a Light Bulb*   The formula $I = \dfrac{32}{4d^2}$ approximates the intensity of light from a 100-watt light bulb at a distance of d meters, where I is in watts per square meter. For light from a 40-watt bulb the equation for its intensity becomes $I = \dfrac{16}{5d^2}$.
(*Source:* R. Weidner.)
(a) Find an expression for the sum of the intensities of light from the two light bulbs.
(b) Find the combined intensity of their light at $d = 5$ meters.

105. *Photography*   A lens in a camera has a focal length, which is important for focusing the camera. If an object is at a distance D from the lens that has a focal length F, then to be in focus the distance S between the lens and the film should satisfy the equation

$$\frac{1}{S} = \frac{1}{F} - \frac{1}{D},$$

as illustrated in the figure. Write the difference $\frac{1}{F} - \frac{1}{D}$ as one term.

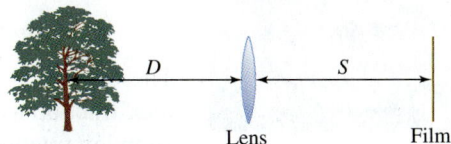

Lens   Film

106. *Geometry*   Find the sum of the areas of the two rectangles shown in the figure. Write your answer in factored form.

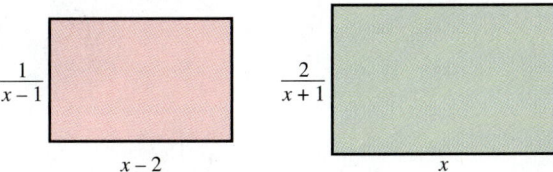

WRITING ABOUT MATHEMATICS

107. Explain how to find the least common multiple of two polynomials.

108. Explain how to subtract two rational expressions having unlike denominators.

CHECKING BASIC CONCEPTS
SECTIONS 7.3 AND 7.4

1. Simplify each expression.
(a) $\dfrac{x}{x+2} + \dfrac{2}{x+2}$ **(b)** $\dfrac{2}{3x} - \dfrac{x}{3x}$
(c) $\dfrac{z^2+z}{z+2} + \dfrac{z}{z+2}$

2. Find the least common multiple of each pair of expressions.
(a) $3x, 5x$ **(b)** $4x, x^2+x$
(c) $x+1, x-1$

3. Simplify each expression.
(a) $\dfrac{1}{x+1} + \dfrac{5}{x}$ **(b)** $\dfrac{5}{x-3} + \dfrac{1}{3-x}$
(c) $\dfrac{-4}{4x+2} - \dfrac{x+2}{2x+1}$

4. Simplify the expression $\dfrac{a}{a-b} - \dfrac{b}{a+b}$.

7.5 COMPLEX FRACTIONS

Basic Concepts ■ Simplifying Complex Fractions

If two and a half pizzas are cut so that each piece equals one-eighth of a pizza, then there are 20 pieces of pizza. This problem can be written as

$$\frac{2\frac{1}{2}}{\frac{1}{8}} \quad \text{or} \quad \frac{2 + \frac{1}{2}}{\frac{1}{8}}.$$

These expressions are examples of *complex* fractions. Typically, we want to rewrite a complex fraction as a standard fraction in the form $\frac{a}{b}$. In this section we show how to simplify complex fractions.

Basic Concepts

A **complex fraction** is a rational expression that contains fractions in its numerator, denominator, or both. Examples of complex fractions include

$$\frac{1 + \frac{1}{2}}{1 - \frac{1}{3}}, \quad \frac{5z}{\frac{7}{z} - \frac{4}{2z}}, \quad \text{and} \quad \frac{\frac{2x}{3} + \frac{1}{x}}{x - \frac{1}{x + 1}}.$$

The expression $\frac{a}{b} \div \frac{c}{d}$ can be written as the complex fraction

$$\frac{\frac{a}{b}}{\frac{c}{d}}.$$

Because $\frac{a}{b} \div \frac{c}{d} = \frac{a}{b} \cdot \frac{d}{c}$, we can simplify some complex fractions by multiplying the numerator and the reciprocal of the denominator. This "invert and multiply" strategy is summarized by the following.

SIMPLIFYING BASIC COMPLEX FRACTIONS

For any real numbers a, b, c, and d,

$$\frac{\frac{a}{b}}{\frac{c}{d}} = \frac{a}{b} \cdot \frac{d}{c},$$

where b, c, and d are nonzero.

Simplifying Complex Fractions

There are two basic methods of simplifying a complex fraction. The first is to simplify both the numerator and the denominator and then divide the resulting two fractions. The second is to multiply the numerator and denominator by the least common denominator of all fractions within the complex fraction.

SIMPLIFYING THE NUMERATOR AND DENOMINATOR   The following steps outline Method I for simplifying complex fractions.

SIMPLIFYING COMPLEX FRACTIONS (Method I)

To simplify a complex fraction, perform the following steps.

STEP 1: Write the numerator as a single fraction; write the denominator as a single fraction.

STEP 2: Divide the denominator into the numerator by multiplying the numerator and the reciprocal of the denominator.

STEP 3: Simplify the result to lowest terms.

Sometimes a complex fraction already has both its numerator and denominator written as single fractions. As a result, we can start with Step 2. This situation is shown in Example 1.

EXAMPLE 1   Simplifying basic complex fractions

Simplify each complex fraction.

(a) $\dfrac{\frac{3}{4}}{\frac{9}{8}}$ (b) $\dfrac{3\frac{3}{4}}{1\frac{1}{2}}$ (c) $\dfrac{\frac{x}{4}}{\frac{3}{2y}}$ (d) $\dfrac{\frac{(x-1)^2}{4}}{\frac{x-1}{8}}$

Solution

(a) We can skip Step 1. To simplify $\frac{3}{4} \div \frac{9}{8}$, multiply $\frac{3}{4}$ by the reciprocal of $\frac{9}{8}$, or $\frac{8}{9}$.

$$\frac{\frac{3}{4}}{\frac{9}{8}} = \frac{3}{4} \cdot \frac{8}{9} \qquad \text{Invert and multiply (Step 2).}$$

$$= \frac{24}{36} \qquad \text{Multiply the fractions.}$$

$$= \frac{2}{3} \qquad \text{Simplify (Step 3).}$$

(b) Start by writing $3\frac{3}{4}$ and $1\frac{1}{2}$ as improper fractions.

$$\frac{3\frac{3}{4}}{1\frac{1}{2}} = \frac{\frac{15}{4}}{\frac{3}{2}} \qquad \text{Write as improper fractions (Step 1).}$$

$$= \frac{15}{4} \cdot \frac{2}{3} \qquad \text{Invert and multiply (Step 2).}$$

$$= \frac{5}{2} \qquad \text{Multiply and simplify (Step 3).}$$

(c) We can skip Step 1. To simplify $\frac{x}{4} \div \frac{3}{2y}$, multiply $\frac{x}{4}$ by the reciprocal of $\frac{3}{2y}$, or $\frac{2y}{3}$.

$$\frac{\dfrac{x}{4}}{\dfrac{3}{2y}} = \frac{x}{4} \cdot \frac{2y}{3} \qquad \text{Invert and multiply (Step 2).}$$

$$= \frac{2xy}{12} \qquad \text{Multiply the fractions.}$$

$$= \frac{xy}{6} \qquad \text{Simplify (Step 3).}$$

(d) We can skip Step 1. To simplify $\frac{(x-1)^2}{4} \div \frac{x-1}{8}$, multiply $\frac{(x-1)^2}{4}$ by the reciprocal of $\frac{x-1}{8}$, or $\frac{8}{x-1}$.

$$\frac{\dfrac{(x-1)^2}{4}}{\dfrac{x-1}{8}} = \frac{(x-1)^2}{4} \cdot \frac{8}{x-1} \qquad \text{Invert and multiply (Step 2).}$$

$$= \frac{8(x-1)(x-1)}{4(x-1)} \qquad \text{Multiply the fractions.}$$

$$= 2(x-1) \qquad \text{Simplify (Step 3).}$$

Now Try Exercises **15, 17, 21, 31**

EXAMPLE 2 **Simplifying complex fractions**

Simplify. Write your answer in lowest terms.

(a) $\dfrac{\dfrac{1}{a}+\dfrac{1}{b}}{\dfrac{1}{a}-\dfrac{1}{b}}$ **(b)** $\dfrac{x-\dfrac{4}{x}}{x+\dfrac{4}{x}}$ **(c)** $\dfrac{\dfrac{1}{x}+\dfrac{2}{x-1}}{\dfrac{2}{x}-\dfrac{1}{x-1}}$ **(d)** $\dfrac{\dfrac{1}{x}+\dfrac{1}{y}}{\dfrac{1}{x^2}-\dfrac{1}{y^2}}$

Solution

(a) In Step 1, write the numerator as one fraction by using the LCD, ab.

$$\frac{1}{a}+\frac{1}{b} = \frac{b}{ab}+\frac{a}{ab} = \frac{b+a}{ab}$$

Write the denominator as one fraction by using the LCD, ab.

$$\frac{1}{a}-\frac{1}{b} = \frac{b}{ab}-\frac{a}{ab} = \frac{b-a}{ab}$$

Finally, use these results to simplify the given complex fraction.

$$\frac{\dfrac{1}{a} + \dfrac{1}{b}}{\dfrac{1}{a} - \dfrac{1}{b}} = \frac{\dfrac{b+a}{ab}}{\dfrac{b-a}{ab}}$$

Simplify the numerator and the denominator (Step 1).

$$= \frac{b+a}{ab} \cdot \frac{ab}{b-a}$$

Invert and multiply (Step 2).

$$= \frac{ab(b+a)}{ab(b-a)}$$

Multiply.

$$= \frac{b+a}{b-a}$$

Simplify (Step 3).

(b) The LCD for both the numerator and the denominator is x.

$$\frac{x - \dfrac{4}{x}}{x + \dfrac{4}{x}} = \frac{\dfrac{x^2}{x} - \dfrac{4}{x}}{\dfrac{x^2}{x} + \dfrac{4}{x}}$$

Write by using the LCD (Step 1).

$$= \frac{\dfrac{x^2 - 4}{x}}{\dfrac{x^2 + 4}{x}}$$

Combine terms.

$$= \frac{x^2 - 4}{x} \cdot \frac{x}{x^2 + 4}$$

Invert and multiply (Step 2).

$$= \frac{x(x^2 - 4)}{x(x^2 + 4)}$$

Multiply.

$$= \frac{x^2 - 4}{x^2 + 4}$$

Simplify (Step 3).

(c) The LCD for the numerator and the denominator is $x(x-1)$.

$$\frac{\dfrac{1}{x} + \dfrac{2}{x-1}}{\dfrac{2}{x} - \dfrac{1}{x-1}} = \frac{\dfrac{x-1}{x(x-1)} + \dfrac{2x}{x(x-1)}}{\dfrac{2(x-1)}{x(x-1)} - \dfrac{x}{x(x-1)}}$$

Write by using the LCD (Step 1).

$$= \frac{\dfrac{x-1+2x}{x(x-1)}}{\dfrac{2(x-1)-x}{x(x-1)}}$$

Combine terms.

$$= \frac{\dfrac{3x-1}{x(x-1)}}{\dfrac{x-2}{x(x-1)}}$$

Simplify.

$$= \frac{3x-1}{x(x-1)} \cdot \frac{x(x-1)}{x-2}$$

Invert and multiply (Step 2).

$$= \frac{3x-1}{x-2}$$

Multiply and simplify (Step 3).

(d) The LCD for the numerator is xy, and the LCD for the denominator is x^2y^2.

$$\frac{\dfrac{1}{x} + \dfrac{1}{y}}{\dfrac{1}{x^2} - \dfrac{1}{y^2}} = \frac{\dfrac{y}{xy} + \dfrac{x}{xy}}{\dfrac{y^2}{x^2y^2} - \dfrac{x^2}{x^2y^2}}$$ Write by using the LCD (Step 1).

$$= \frac{\dfrac{y + x}{xy}}{\dfrac{y^2 - x^2}{x^2y^2}}$$ Combine terms.

$$= \frac{y + x}{xy} \cdot \frac{x^2y^2}{y^2 - x^2}$$ Invert and multiply (Step 2).

$$= \frac{x^2y^2(y + x)}{xy(y^2 - x^2)}$$ Multiply.

$$= \frac{x^2y^2(y + x)}{xy(y - x)(y + x)}$$ Factor the denominator (Step 3).

$$= \frac{xy}{y - x}$$ Simplify.

Now Try Exercises 33, 39, 43, 47

MULTIPLYING BY THE LCD   Method II for simplifying a complex fraction is to multiply the numerator and denominator by the least common denominator of all the fractions in *both* the numerator and the denominator.

The following steps outline Method II for simplifying complex fractions.

SIMPLIFYING COMPLEX FRACTIONS (Method II)

To simplify a complex fraction, perform the following steps.

STEP 1: Find the LCD of all fractions within the complex fraction.

STEP 2: Multiply the numerator and the denominator of the complex fraction by the LCD.

STEP 3: Simplify the result to lowest terms.

We can use this method to simplify the complex fraction in Example 2(a). Because the LCD for the numerator and the denominator is ab, we multiply the complex fraction by 1, expressed in the form $\frac{ab}{ab}$. This method is equivalent to multiplying the numerator and the denominator by ab.

$$\frac{\dfrac{1}{a} + \dfrac{1}{b}}{\dfrac{1}{a} - \dfrac{1}{b}} = \frac{\left(\dfrac{1}{a} + \dfrac{1}{b}\right)ab}{\left(\dfrac{1}{a} - \dfrac{1}{b}\right)ab}$$ Multiply by $\dfrac{ab}{ab} = 1$.

$$= \frac{\dfrac{ab}{a} + \dfrac{ab}{b}}{\dfrac{ab}{a} - \dfrac{ab}{b}}$$ Distributive property

$$= \frac{b + a}{b - a}$$ Simplify.

EXAMPLE 3 **Simplifying complex fractions**

Simplify.

(a) $\dfrac{2z}{\dfrac{4}{z} + \dfrac{3}{z}}$ **(b)** $\dfrac{\dfrac{1}{x - 3}}{\dfrac{1}{x} + \dfrac{3}{x - 3}}$ **(c)** $\dfrac{\dfrac{1}{a} - \dfrac{1}{b}}{\dfrac{1}{2b^2} - \dfrac{1}{2a^2}}$

Solution

(a) The LCD for the numerator *and* the denominator is z, so multiply each by z (Step 1).

$$\frac{2z}{\dfrac{4}{z} + \dfrac{3}{z}} = \frac{(2z)z}{\left(\dfrac{4}{z} + \dfrac{3}{z}\right)z}$$ Multiply by $\frac{z}{z}$ (Step 2).

$$= \frac{2z^2}{\dfrac{4z}{z} + \dfrac{3z}{z}}$$ Distributive property

$$= \frac{2z^2}{4 + 3}$$ Simplify (Step 3).

$$= \frac{2z^2}{7}$$ Add.

(b) The LCD is the product $x(x - 3)$ (Step 1).

$$\frac{\dfrac{1}{x - 3}}{\dfrac{1}{x} + \dfrac{3}{x - 3}} = \frac{\dfrac{1}{x - 3}}{\left(\dfrac{1}{x} + \dfrac{3}{x - 3}\right)} \cdot \frac{x(x - 3)}{x(x - 3)}$$ Multiply by 1 (Step 2).

$$= \frac{\dfrac{x(x - 3)}{x - 3}}{\dfrac{x(x - 3)}{x} + \dfrac{3x(x - 3)}{x - 3}}$$ Distributive property

$$= \frac{x}{(x - 3) + 3x}$$ Simplify the fractions (Step 3).

$$= \frac{x}{4x - 3}$$ Simplify the denominator.

(c) The LCD for the numerator *and* the denominator is $2a^2b^2$ (Step 1).

$$\frac{\dfrac{1}{a} - \dfrac{1}{b}}{\dfrac{1}{2b^2} - \dfrac{1}{2a^2}} = \frac{\dfrac{1}{a} - \dfrac{1}{b}}{\dfrac{1}{2b^2} - \dfrac{1}{2a^2}} \cdot \frac{\mathbf{2a^2b^2}}{\mathbf{2a^2b^2}}$$ Multiply by 1 (Step 2).

$$= \frac{\left(\dfrac{1}{a} - \dfrac{1}{b}\right)2a^2b^2}{\left(\dfrac{1}{2b^2} - \dfrac{1}{2a^2}\right)2a^2b^2}$$ Multiply the fractions.

$$= \frac{\dfrac{2a^2b^2}{a} - \dfrac{2a^2b^2}{b}}{\dfrac{2a^2b^2}{2b^2} - \dfrac{2a^2b^2}{2a^2}}$$ Distributive property

$$= \frac{2ab^2 - 2a^2b}{a^2 - b^2}$$ Simplify the fractions (Step 3).

$$= \frac{2ab(b - a)}{(a - b)(a + b)}$$ Factor.

$$= -\frac{2ab}{a + b}$$ Simplify.

Now Try Exercises 35, 37, 51

CRITICAL THINKING

Are the expressions $\dfrac{\dfrac{a}{b}}{\dfrac{a}{b} + 1}$

and $\dfrac{1}{1 + 1}$ equal? Explain.

Are the expressions $\dfrac{\dfrac{a}{b} + 1}{\dfrac{a}{b}}$

and $1 + \dfrac{b}{a}$ equal? Explain.

7.5 PUTTING IT ALL TOGETHER

In this section we discussed how to simplify complex fractions to the form $\frac{a}{b}$. The following table summarizes this discussion.

Concept	Explanation	Examples
Complex Fraction	A rational expression that contains fractions in its numerator, denominator, or both	$\dfrac{3 + \dfrac{1}{x + 1}}{3 - \dfrac{1}{x + 1}}$ and $\dfrac{\dfrac{x}{y} - \dfrac{y}{x}}{\dfrac{x}{y} + \dfrac{y}{x}}$
Simplifying Basic Complex Fractions	$\dfrac{\dfrac{a}{b}}{\dfrac{c}{d}} = \dfrac{a}{b} \cdot \dfrac{d}{c}$	$\dfrac{\dfrac{2}{x}}{\dfrac{4}{x - 1}} = \dfrac{2}{x} \cdot \dfrac{x - 1}{4} = \dfrac{x - 1}{2x}$

Concept	Explanation	Examples
Method I: Simplifying the Numerator and Denominator First	Combine the terms in the numerator, combine the terms in the denominator, and then invert and multiply.	$\dfrac{\dfrac{1}{x}+\dfrac{3}{x}}{\dfrac{5}{y}-\dfrac{4}{y}}=\dfrac{\dfrac{4}{x}}{\dfrac{1}{y}}=\dfrac{4}{x}\cdot\dfrac{y}{1}=\dfrac{4y}{x}$
Method II: Multiplying the Numerator and Denominator by the LCD	Multiply the numerator *and* the denominator by the LCD of *all* fractions within the expression.	$\dfrac{\dfrac{2}{x}+\dfrac{1}{y}}{\dfrac{4}{y}-\dfrac{1}{x}}=\dfrac{\left(\dfrac{2}{x}+\dfrac{1}{y}\right)xy}{\left(\dfrac{4}{y}-\dfrac{1}{x}\right)xy}$ $=\dfrac{2y+x}{4x-y}$ Note that the LCD is xy.

7.5 Exercises

CONCEPTS

1. $\dfrac{\frac{1}{2}}{\frac{3}{4}} = $ _____

2. $\dfrac{\frac{a}{b}}{\frac{c}{d}} = $ _____

3. A complex fraction is a rational expression that contains _____ in its numerator, denominator, or both.

4. Write "the quantity x plus one half divided by the quantity x minus one half" as a complex fraction.

5. What operation does the fraction bar represent?

6. Write the expression $\frac{x}{2} \div \frac{1}{x-1}$ as a complex fraction.

7. Write the expression $\frac{a}{b} \div \frac{c}{d}$ as a complex fraction.

8. What is the LCD for $\frac{1}{x+2}$ and $\frac{1}{x}$?

SIMPLIFYING COMPLEX FRACTIONS

Exercises 9–14: For the complex fraction, determine the LCD of all the fractions appearing in both the numerator and the denominator.

9. $\dfrac{\dfrac{x}{5}-\dfrac{1}{6}}{\dfrac{2}{15}-3x}$

10. $\dfrac{\dfrac{1}{2}-\dfrac{1}{x}}{\dfrac{1}{2}+\dfrac{1}{x}}$

11. $\dfrac{\dfrac{2}{x+1}-x}{\dfrac{2}{x-1}+x}$

12. $\dfrac{\dfrac{1}{4x}-\dfrac{4}{x}}{\dfrac{1}{2x}+\dfrac{1}{3x}}$

13. $\dfrac{\dfrac{1}{2x-1}-\dfrac{1}{2x+1}}{\dfrac{x+1}{x}}$

14. $\dfrac{\dfrac{1}{4x^2}-\dfrac{1}{2x^3}}{\dfrac{1}{x-1}+\dfrac{1}{x-1}}$

Exercises 15–54: Simplify the complex fraction.

15. $\dfrac{\frac{2}{3}}{\frac{5}{6}}$

16. $\dfrac{\frac{8}{9}}{\frac{5}{4}}$

17. $\dfrac{2\frac{1}{2}}{1\frac{3}{4}}$

18. $\dfrac{1\frac{2}{3}}{3\frac{1}{2}}$

19. $\dfrac{1\frac{1}{2}}{2\frac{1}{3}}$

20. $\dfrac{2\frac{1}{5}}{2\frac{1}{10}}$

21. $\dfrac{\frac{r}{t}}{\frac{2r}{t}}$

22. $\dfrac{\frac{8}{p}}{\frac{4}{p}}$

23. $\dfrac{\dfrac{6}{x}}{\dfrac{2}{y}}$

24. $\dfrac{\dfrac{3}{14x}}{\dfrac{6}{7x}}$

25. $\dfrac{\dfrac{6}{m-2}}{\dfrac{2}{m-2}}$

26. $\dfrac{\dfrac{3}{n+1}}{\dfrac{6}{n+1}}$

27. $\dfrac{\dfrac{p+1}{p}}{\dfrac{p+2}{p}}$

28. $\dfrac{\dfrac{2p}{2p+5}}{\dfrac{1}{4p+10}}$

29. $\dfrac{\dfrac{5}{z^2-1}}{\dfrac{z}{z^2-1}}$

30. $\dfrac{\dfrac{z}{z-2}}{\dfrac{z}{z-2}}$

31. $\dfrac{\dfrac{y}{y^2-9}}{\dfrac{1}{y+3}}$

32. $\dfrac{\dfrac{2y}{2y-1}}{\dfrac{1}{4y^2-1}}$

33. $\dfrac{x-\dfrac{1}{x}}{x+\dfrac{1}{x}}$

34. $\dfrac{4-\dfrac{1}{x}}{4+\dfrac{1}{x}}$

35. $\dfrac{x}{\dfrac{2}{x}+\dfrac{1}{x}}$

36. $\dfrac{5x}{1+\dfrac{1}{x}}$

37. $\dfrac{\dfrac{3}{x+1}}{\dfrac{4}{x+1}-\dfrac{1}{x+1}}$

38. $\dfrac{\dfrac{5}{2x-3}-\dfrac{4}{2x-3}}{\dfrac{7}{2x-3}+\dfrac{8}{2x-3}}$

39. $\dfrac{\dfrac{1}{m^2n}+\dfrac{1}{mn^2}}{\dfrac{1}{m^2n}-\dfrac{1}{mn^2}}$

40. $\dfrac{\dfrac{3}{x-1}-\dfrac{2}{x}}{\dfrac{3}{x-1}+\dfrac{2}{x}}$

41. $\dfrac{\dfrac{1}{2x}+\dfrac{1}{y}}{\dfrac{1}{y}-\dfrac{1}{2x}}$

42. $\dfrac{\dfrac{3}{x}-\dfrac{2}{y}}{\dfrac{3}{x}+\dfrac{2}{y}}$

43. $\dfrac{\dfrac{1}{ab}+\dfrac{1}{a}}{\dfrac{1}{ab}-\dfrac{1}{b}}$

44. $\dfrac{\dfrac{1}{a}+\dfrac{2}{3b}}{\dfrac{1}{a}-\dfrac{5}{2b}}$

45. $\dfrac{\dfrac{2}{q}-\dfrac{1}{q+1}}{\dfrac{1}{q+1}}$

46. $\dfrac{\dfrac{5}{p}+\dfrac{4}{p-5}}{\dfrac{5}{p}-\dfrac{5}{p-5}}$

47. $\dfrac{\dfrac{1}{x+1}+\dfrac{1}{x+2}}{\dfrac{1}{x+1}-\dfrac{1}{x+2}}$

48. $\dfrac{\dfrac{1}{x-3}-\dfrac{1}{x+3}}{1-\dfrac{1}{x^2-9}}$

49. $\dfrac{\dfrac{1}{2x-1}-\dfrac{1}{2x+1}}{\dfrac{x+1}{x}}$

50. $\dfrac{\dfrac{1}{4x^2}-\dfrac{1}{x^3}}{\dfrac{1}{x-1}+\dfrac{1}{x-1}}$

51. $\dfrac{\dfrac{1}{ab^2}-\dfrac{1}{a^2b}}{\dfrac{1}{b}-\dfrac{1}{a}}$

52. $\dfrac{\dfrac{1}{x^2}-\dfrac{1}{y^2}}{\dfrac{1}{x}-\dfrac{1}{y}}$

53. $\dfrac{1}{a^{-1}+b^{-1}}$

54. $\dfrac{a^2-b^2}{a^{-2}-b^{-2}}$

APPLICATIONS

55. *Annuity* If P dollars are deposited every 2 weeks in an account paying an annual interest rate r expressed as a decimal, then the amount A in the account after 2 years can be approximated by

$$\left(P\left(1+\dfrac{r}{26}\right)^{52}-P\right)\div\dfrac{r}{26}.$$

Write this expression as a complex fraction.

56. *Annuity* (Continuation of the preceding exercise) Use a calculator to evaluate the expression when $r = 0.026$ (2.6%) and $P = \$100$. Interpret the result.

57. *Resistance in Electricity*   Light bulbs are often wired so that electricity can flow through either bulb, as illustrated in the accompanying figure.

In this way, if one bulb burns out, the other bulb still works. If two light bulbs have resistances T and S, their combined resistance R is

$$R = \dfrac{1}{\dfrac{1}{T} + \dfrac{1}{S}}.$$

Simplify this formula.

58. *Resistance in Electricity*   (Refer to the preceding exercise.) Evaluate the formula

$$R = \dfrac{1}{\dfrac{1}{T} + \dfrac{1}{S}}$$

when $T = 100$ and $S = 200$.

WRITING ABOUT MATHEMATICS

59. A student simplifies a complex fraction as shown below. Explain the student's mistake and how you would simplify the complex fraction correctly.

$$\dfrac{\dfrac{1}{x} + 1}{\dfrac{1}{x}} \overset{?}{=} \dfrac{\dfrac{1}{x}}{\dfrac{1}{x}} + 1 = 2$$

60. Explain one method for simplifying a complex fraction.

7.6   RATIONAL EQUATIONS AND FORMULAS

Solving Rational Equations ▪ Rational Expressions and Equations ▪ Graphical and Numerical Solutions ▪ Solving a Formula for a Variable ▪ Applications

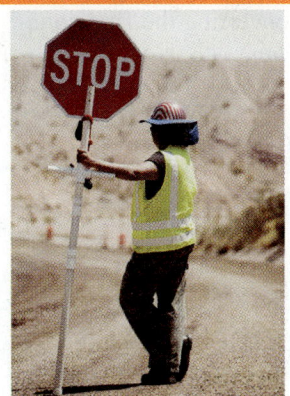

A LOOK INTO MATH ▷   In Section 7.1 we demonstrated that, if cars arrive randomly at a construction site at an average rate of x cars per minute and if 10 cars per minute can pass through the site, then the average time T that a driver spends waiting in line and passing through the construction site is

$$T = \frac{1}{10 - x},$$

where T is in minutes and $x < 10$. If the highway department wants to limit the average wait for a car to $\frac{1}{2}$ minute or less, then mathematics can be used to determine the corresponding value of x. However, before solving this problem, we discuss rational equations. (**Source:** N. Garber and L. Hoel, *Traffic and Highway Engineering.*)

Solving Rational Equations

▶ **REAL-WORLD CONNECTION**   If an equation contains one or more rational expressions, it is called a **rational equation**. Rational equations occur in mathematics whenever a rational expression is set equal to a constant. For example, the rational expression

$$\frac{1}{10 - x}$$

from the introduction can be used to estimate the average time that drivers wait to get through a construction site. If the wait is $\frac{1}{2}$ minute, we determine x by solving the *rational equation*

$$\frac{1}{10 - x} = \frac{1}{2}.$$

To solve this equation, we multiply each side by the LCD: $2(10 - x)$.

$$\frac{2(10 - x)}{10 - x} = \frac{2(10 - x)}{2} \qquad \text{Multiply by the LCD.}$$

$$2 = 10 - x \qquad \text{Simplify.}$$

$$x = 8 \qquad \text{Add } x; \text{ subtract 2.}$$

The average wait is $\frac{1}{2}$ minute when cars arrive randomly at an average rate of 8 cars per minute. In general, if a rational equation is in the form

$$\frac{a}{b} = \frac{c}{d},$$

we can multiply each side of this equation by the common denominator bd to obtain

$$\frac{a(bd)}{b} = \frac{c(bd)}{d},$$

which simplifies to $ad = cb$. This technique can be used to solve some types of basic rational equations.

SOLVING BASIC RATIONAL EQUATIONS

The equations

$$\frac{a}{b} \diagdown \frac{c}{d} \quad \text{and} \quad ad = bc$$

are equivalent, provided that b and d are nonzero. Note that converting the first equation to the second equation is sometimes called *cross multiplying*.

EXAMPLE 1   **Solving rational equations**

Solve each equation.

(a) $\dfrac{5}{3} = \dfrac{4}{x}$ **(b)** $\dfrac{x + 1}{5} = \dfrac{3x}{2}$ **(c)** $\dfrac{4}{3x - 4} = x$ **(d)** $\dfrac{1}{x} + \dfrac{2}{x} = \dfrac{3}{7}$

Solution

(a)

$$\frac{5}{3} = \frac{4}{x} \qquad \text{Given equation}$$

$$5x = 12 \qquad \text{Cross multiply.}$$

$$x = \frac{12}{5} \qquad \text{Divide by 5.}$$

The solution is $\frac{12}{5}$.

(b)

$$\frac{x+1}{5} = \frac{3x}{2} \qquad \text{Given equation}$$

$$2(x+1) = 15x \qquad \text{Cross multiply.}$$

$$2x + 2 = 15x \qquad \text{Distributive property}$$

$$-13x = -2 \qquad \text{Subtract 2 and } 15x.$$

$$x = \frac{2}{13} \qquad \text{Divide by } -13.$$

The solution is $\frac{2}{13}$.

(c)

$$\frac{4}{3x-4} = \frac{x}{1} \qquad \text{Write } x \text{ as } \frac{x}{1}.$$

$$x(3x-4) = 4 \cdot 1 \qquad \text{Cross multiply.}$$

$$3x^2 - 4x = 4 \qquad \text{Distributive property}$$

$$3x^2 - 4x - 4 = 0 \qquad \text{Subtract 4.}$$

$$(3x+2)(x-2) = 0 \qquad \text{Factor.}$$

$$3x + 2 = 0 \quad \text{or} \quad x - 2 = 0 \qquad \text{Zero-product property}$$

$$x = -\frac{2}{3} \quad \text{or} \quad x = 2 \qquad \text{Solve each equation.}$$

The solutions are $-\frac{2}{3}$ and 2.

(d)

$$\frac{1}{x} + \frac{2}{x} = \frac{3}{7} \qquad \text{Given equation}$$

$$\frac{3}{x} = \frac{3}{7} \qquad \text{Add the rational expressions.}$$

$$3x = 21 \qquad \text{Cross multiply.}$$

$$x = 7 \qquad \text{Divide by 3.}$$

The solution is 7.   Now Try Exercises 9, 13, 33, 43

Another technique for solving rational equations is to multiply each side by the least common denominator. Unlike cross multiplying, this technique can always be used.

EXAMPLE 2   **Multiplying by the LCD**

Solve each equation. Check your answer.

(a) $\dfrac{1}{x-1} - \dfrac{1}{x} = \dfrac{1}{9x}$ **(b)** $\dfrac{1}{x-1} + \dfrac{1}{x+1} = \dfrac{12}{x^2-1}$

Solution

(a) Start by multiplying each term by the LCD, $9x(x-1)$.

$$\frac{1}{x-1} - \frac{1}{x} = \frac{1}{9x} \qquad \text{Given equation}$$

$$\frac{9x(x-1)}{x-1} - \frac{9x(x-1)}{x} = \frac{9x(x-1)}{9x} \qquad \text{Multiply each term by the LCD.}$$

$$9x - 9(x-1) = x - 1 \qquad \text{Simplify each rational expression.}$$

$$9 = x - 1 \qquad \text{Distributive property}$$

$$x = 10 \qquad \text{Add 1 and rewrite.}$$

Check:

$$\frac{1}{10 - 1} - \frac{1}{10} \stackrel{?}{=} \frac{1}{9(10)}$$ — Substitute 10 for x.

$$\frac{10}{90} - \frac{9}{90} \stackrel{?}{=} \frac{1}{90}$$ — LCD $= 90$

$$\frac{1}{90} = \frac{1}{90}$$ — The answer checks.

(b) Start by multiplying each term by the LCD, $x^2 - 1 = (x - 1)(x + 1)$.

$$\frac{1}{x - 1} + \frac{1}{x + 1} = \frac{12}{x^2 - 1}$$ — Given equation

$$\frac{(x - 1)(x + 1)}{x - 1} + \frac{(x - 1)(x + 1)}{x + 1} = \frac{12(x - 1)(x + 1)}{(x - 1)(x + 1)}$$ — Multiply each term by the LCD.

$$(x + 1) + (x - 1) = 12$$ — Simplify each rational expression.

$$2x = 12$$ — Combine like terms.

$$x = 6$$ — Divide by 2.

Check:

$$\frac{1}{6 - 1} + \frac{1}{6 + 1} \stackrel{?}{=} \frac{12}{6^2 - 1}$$ — Substitute 6 for x.

$$\frac{1}{5} + \frac{1}{7} \stackrel{?}{=} \frac{12}{35}$$ — Simplify.

$$\frac{7}{35} + \frac{5}{35} \stackrel{?}{=} \frac{12}{35}$$ — The LCD is 35.

$$\frac{12}{35} = \frac{12}{35}$$ — The answer checks.

Now Try Exercises 51, 57

In Example 2 we checked our answers. Although the answer checked in both cases, it may not always check. When multiplying a rational equation by the LCD, it is possible to obtain *extraneous solutions* that do not satisfy the *given* equation. This situation is demonstrated in Example 3. Before solving Example 3, we present a step-by-step strategy for solving rational equations.

STEPS FOR SOLVING A RATIONAL EQUATION

STEP 1: Find the LCD of the terms in the equation.

STEP 2: Multiply each side of the equation by the LCD.

STEP 3: Simplify each term.

STEP 4: Solve the resulting equation.

STEP 5: Check each answer in the *given* equation. Reject any value that makes a denominator equal 0.

NOTE: If the rational equation is in the form $\frac{a}{b} = \frac{c}{d}$, or "fraction equals fraction," cross multiplying to obtain $ad = bc$ may be helpful. However, be sure to check your answers.

EXAMPLE **3** Solving an equation having an extraneous solution

If possible, solve $\dfrac{1}{x - 2} + \dfrac{1}{x + 2} = \dfrac{4}{x^2 - 4}$.

Solution

The LCD is $(x - 2)(x + 2) = x^2 - 4$ (Step 1).

$$\frac{1}{x - 2} + \frac{1}{x + 2} = \frac{4}{x^2 - 4}$$ Given equation

$$\frac{(x - 2)(x + 2)}{x - 2} + \frac{(x - 2)(x + 2)}{x + 2} = \frac{4(x^2 - 4)}{x^2 - 4}$$ Multiply by the LCD (Step 2).

$$(x + 2) + (x - 2) = 4$$ Simplify each term (Step 3).

$$2x = 4$$ Combine like terms (Step 4).

$$x = 2$$ Divide by 2.

Check: $$\frac{1}{2 - 2} + \frac{1}{2 + 2} \stackrel{?}{=} \frac{4}{2^2 - 4}$$ Substitute 2 for x (Step 5).

Note that both sides of the equation are undefined because it is not possible to divide by 0. Therefore 2 is not a solution; rather, it is an *extraneous solution*. That is, there are no solutions to the given equation.

Now Try Exercise **55**

Rational Expressions and Equations

Rational expressions and rational equations are not the same concepts. For example,

$$\frac{3}{x} \quad \text{and} \quad \frac{x}{x^2 - x}$$

are both rational expressions. Neither one contains an equals sign. We often simplify or evaluate rational expressions. By factoring the denominator and applying the basic principle of rational expressions, we can simplify the second expression.

$$\frac{x}{x^2 - x} = \frac{x}{x(x - 1)}$$ Factor the denominator.

$$= \frac{1}{x - 1}$$ Simplify.

Thus the expressions $\frac{x}{x^2 - x}$ and $\frac{1}{x - 1}$ are equal for every value of x (except 0 and 1). Expressions can also be evaluated. For example, if we replace x with 3, the expression $\frac{3}{x}$ equals $\frac{3}{3}$, or 1.

When two expressions are set equal, an equation is formed that is typically true for only a limited number of x-values. For example, $\frac{3}{x} = \frac{1}{x - 1}$ is a rational equation that is true for only one x-value.

$$\frac{3}{x} = \frac{1}{x - 1}$$ Given equation

$$3(x - 1) = x \cdot 1$$ Cross multiply.

$$3x - 3 = x$$ Distributive property

$$2x = 3$$ Add 3; subtract x.

$$x = \frac{3}{2}$$ Divide each side by 2.

The only solution is $\frac{3}{2}$. (Check this solution.) If we replace x with any value other than $\frac{3}{2}$, the equation $\frac{3}{x} = \frac{1}{x - 1}$ is a false statement.

EXAMPLE **4**   Identifying expressions and equations

Determine whether you are given an expression or an equation. If it is an expression, simplify it and then evaluate it for $x = 5$. If it is an equation, solve it.

(a) $\dfrac{x - 1}{x + 1} = \dfrac{x}{x + 3}$ **(b)** $\dfrac{x^2 - 2x}{x - 1} + \dfrac{1}{x - 1}$

Solution
(a) There is an equals sign, so it is an equation.

$$\dfrac{x - 1}{x + 1} = \dfrac{x}{x + 3} \qquad \text{Given equation}$$

$$(x - 1)(x + 3) = x(x + 1) \qquad \text{Cross multiply.}$$

$$x^2 + 2x - 3 = x^2 + x \qquad \text{Multiply.}$$

$$2x - 3 = x \qquad \text{Subtract } x^2.$$

$$x = 3 \qquad \text{Subtract } x; \text{ add 3.}$$

Check: $\dfrac{3 - 1}{3 + 1} \overset{?}{=} \dfrac{3}{3 + 3}$ Replace x with 3 in given equation.

$\dfrac{1}{2} = \dfrac{1}{2}$ The answer checks.

The solution to the equation is 3.

(b) There is no equals sign, so it is an expression. The common denominator is $x - 1$, so add the numerators.

$$\dfrac{x^2 - 2x}{x - 1} + \dfrac{1}{x - 1} = \dfrac{x^2 - 2x + 1}{x - 1} \qquad \text{Add the numerators.}$$

$$= \dfrac{(x - 1)(x - 1)}{x - 1} \qquad \text{Factor the numerator.}$$

$$= x - 1 \qquad \text{Simplify.}$$

The given expression simplifies to $x - 1$. When x equals 5, the expression evaluates to $5 - 1$, or 4.   Now Try Exercises 65, 69

MAKING CONNECTIONS

Expressions versus Equations

1. If a problem does not contain an equals sign, you are probably adding, subtracting, multiplying, dividing, or otherwise simplifying an expression. Your answer will be an expression, not a value for x.
2. If the problem has an equals sign, it is an equation to be solved. Your answer will be a value (or values) for x that makes the equation a true statement. One strategy for solving a rational equation is to multiply each side of the equation by the LCD. Be sure to check all your answers.

Graphical and Numerical Solutions

Like other types of equations, rational equations can also be solved graphically and numerically. Graphs of rational expressions are not lines. They are typically curves that can be graphed by plotting several points and then sketching a graph or by using a graphing calculator.

EXAMPLE 5 Solving a rational equation graphically and numerically

Solve $\frac{2}{x} = x + 1$ graphically and numerically.

Solution

Graphical Solution   To solve $\frac{2}{x} = x + 1$, graph $y_1 = \frac{2}{x}$ and $y_2 = x + 1$. The graph of y_2 is a line with slope 1 and y-intercept 1. To graph y_1 make a table of values, as shown in Table 7.3. Then plot the points and connect them with a smooth curve. Note that y_1 is undefined for $x = 0$. Generally, it is a good idea to plot at least three points on each side of an asymptote. These points were plotted and the curves sketched in Figure 7.3. Note that the graphs of y_1 and y_2 intersect at $(-2, -1)$ and $(1, 2)$. The solutions to the equation are -2 and 1. Check these solutions.

TABLE 7.3

x	$\frac{2}{x}$
-3	$-\frac{2}{3}$
-2	-1
-1	-2
0	—
1	2
2	1
3	$\frac{2}{3}$

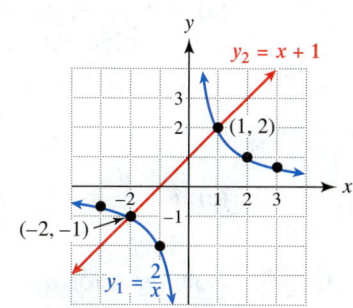

Figure 7.3

Numerical Solution   Make a table of values for $y_1 = \frac{2}{x}$ and $y_2 = x + 1$, as shown in Table 7.4. Note that $\frac{2}{x} = x + 1$ when $x = -2$ or $x = 1$.

TABLE 7.4

x	-3	-2	-1	0	1	2	3
$y_1 = \frac{2}{x}$	$-\frac{2}{3}$	-1	-2	—	2	1	$\frac{2}{3}$
$y_2 = x + 1$	-2	-1	0	1	2	3	4

Now Try Exercise **77**

A graphing calculator can be used efficiently to create Figure 7.3 and Table 7.4, as illustrated in Figure 7.4 on the next page.

CRITICAL THINKING

If you solve Example 5 symbolically, what will be the result? Verify your answer.

CALCULATOR HELP

To make a table or graph, or to enter a formula, see Appendix A (page AP-3 or AP-5).

(a)

(b)

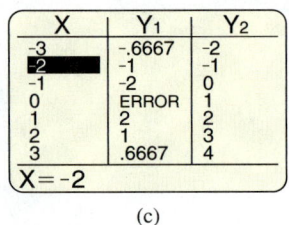
(c)

Figure 7.4

Solving a Formula for a Variable

▶ **REAL-WORLD CONNECTION**   Formulas in applications often involve both the use of rational expressions and the need to solve an equation for a variable. This idea is illustrated in the next two examples.

EXAMPLE 6   Finding time in a distance problem

If a person travels at a speed, or rate, r for time t, then the distance d traveled is $d = rt$.
(a) How far does a person travel in 2 hours when traveling at 60 miles per hour?
(b) Solve the formula $d = rt$ for t.
(c) How long does it take a person to go 250 miles when traveling at 40 miles per hour?

Solution
(a) $d = rt = 60 \cdot 2 = 120$ miles
(b) To solve $d = rt$ for t divide each side by r to obtain

$$\frac{d}{r} = \frac{rt}{r} \quad \text{or} \quad \frac{d}{r} = t.$$

(c) $t = \dfrac{d}{r} = \dfrac{250}{40} = \dfrac{25}{4} = 6.25$ hours, or 6 hours and 15 minutes.   Now Try Exercise 89

EXAMPLE 7   Solving a formula for a variable

Solve each equation for the specified variable.
(a) $A = \frac{bh}{2}$ for b **(b)** $P = \frac{nRT}{V}$ for V **(c)** $S = 2\pi rh + \pi r^2$ for h

Solution
(a) First, multiply each side by 2.

$$A = \frac{bh}{2} \qquad \text{Given equation}$$

$$2A = bh \qquad \text{Multiply each side by 2.}$$

$$\frac{2A}{h} = b \qquad \text{Divide each side by } h.$$

(b) Begin by multiplying each side by V.

$$P = \frac{nRT}{V} \qquad \text{Given equation}$$

$$PV = nRT \qquad \text{Multiply each side by } V.$$

$$V = \frac{nRT}{P} \qquad \text{Divide each side by } P.$$

(c) Begin by subtracting πr^2 from each side.

$$S = 2\pi r\boldsymbol{h} + \pi r^2 \qquad \text{Given equation}$$

$$S - \pi r^2 = 2\pi r\boldsymbol{h} \qquad \text{Subtract } \pi r^2 \text{ from each side.}$$

$$\frac{S - \pi r^2}{2\pi r} = \boldsymbol{h} \qquad \text{Divide each side by } 2\pi r.$$

Now Try Exercises 91, 93, 94

Applications

▶ **REAL-WORLD CONNECTION**   Rational equations sometimes occur in time and rate problems, as demonstrated in the next two examples.

EXAMPLE 8   ### Mowing a lawn

Two people are mowing a large lawn. One person has a riding mower, and the other person has a push mower. The person with the riding mower can cut the lawn alone in 4 hours, and the person with the push mower can cut the lawn alone in 9 hours. How long does it take for them, working together, to cut the lawn?

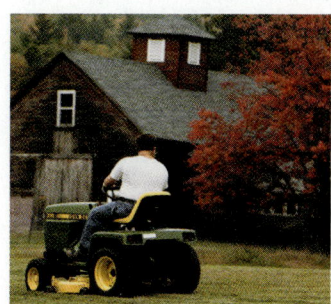

Solution

The first person can cut the entire lawn in 4 hours, so this person can cut $\frac{1}{4}$ of the lawn in 1 hour, $\frac{2}{4}$ of the lawn in 2 hours, and in general, $\frac{t}{4}$ of the lawn in t hours. The second person can cut the lawn in 9 hours so (using similar reasoning) this person can cut $\frac{t}{9}$ of the lawn in t hours. Together they can cut

$$\frac{t}{4} + \frac{t}{9}$$

of the lawn in t hours. The job is complete when the fraction of the lawn cut reaches 1. To find out how long this task takes, solve the equation

$$\frac{t}{4} + \frac{t}{9} = 1.$$

Begin by multiplying each side by the LCD, or 36.

$$\frac{t}{4} + \frac{t}{9} = 1 \qquad \text{Equation to be solved}$$

$$\frac{36t}{4} + \frac{36t}{9} = 36(1) \qquad \text{Multiply each term by 36.}$$

$$9t + 4t = 36 \qquad \text{Simplify.}$$

$$13t = 36 \qquad \text{Combine like terms.}$$

$$t = \frac{36}{13} \qquad \text{Divide each side by 13.}$$

Working together they can cut the lawn in $\frac{36}{13} \approx 2.8$ hours.   Now Try Exercise 105

CRITICAL THINKING

If one person can mow a lawn in x hours and another person can mow it in y hours, how long does it take them, working together, to mow the lawn?

EXAMPLE 9   Solving a distance problem

Suppose that the winner of a 600-mile car race finishes 12 minutes ahead of the second-place finisher. If the winner averages 5 miles per hour faster than the second racer, find the average speed of each racer.

Solution

Here we apply the four-step method for solving an application problem.

STEP 1: *Identify any variables.*

$$x: \qquad \text{speed of slower car in miles per hour}$$
$$x + 5: \qquad \text{speed of faster car in miles per hour}$$

STEP 2: *Write an equation.* To determine the time required for each car to finish the race we use the equation $t = \frac{d}{r}$. Because the race is 600 miles, the time for the slower car is $\frac{600}{x}$ and the time for the faster car is $\frac{600}{x+5}$. The difference between these times is 12 minutes, or $\frac{12}{60} = \frac{1}{5}$ hour. Thus to determine x, we solve

$$\frac{600}{x} - \frac{600}{x+5} = \frac{1}{5}.$$

NOTE: Speeds are in miles per *hour*, so time must be in hours not minutes.

STEP 3: *Solve the equation.* Start by multiplying by the LCD, $5x(x+5)$.

$$\frac{600}{x} - \frac{600}{x+5} = \frac{1}{5} \qquad \text{Equation to be solved}$$

$$\frac{600 \cdot 5x(x+5)}{x} - \frac{600 \cdot 5x(x+5)}{x+5} = \frac{1 \cdot 5x(x+5)}{5} \qquad \begin{array}{l}\text{Multiply each term} \\ \text{by the LCD.}\end{array}$$

$$3000(x+5) - 3000x = x(x+5) \qquad \text{Simplify.}$$

$$3000x + 15{,}000 - 3000x = x^2 + 5x \qquad \text{Distributive property}$$

$$15{,}000 = x^2 + 5x \qquad \text{Combine like terms.}$$

$$x^2 + 5x - 15{,}000 = 0 \qquad \text{Rewrite the equation.}$$

$$(x - 120)(x + 125) = 0 \qquad \text{Factor.}$$

$$x - 120 = 0 \quad \text{or} \quad x + 125 = 0 \qquad \text{Zero-product property}$$

$$x = 120 \quad \text{or} \quad x = -125 \qquad \text{Solve.}$$

The slower car travels at 120 miles per hour, and the faster car travels 5 miles per hour faster, or 125 miles per hour. (The solution -125 has no meaning in this problem.)

STEP 4: *Check your answer.* The slower car travels 600 miles at 120 miles per hour, which takes $\frac{600}{120} = 5$ hours. The faster car travels 600 miles at 125 miles per hour, which requires $\frac{600}{125} = 4.8$ hours. The difference between their times is $5 - 4.8 = 0.2$ hour, or $0.2 \times 60 = 12$ minutes, so the answer checks.   **Now Try Exercise 107**

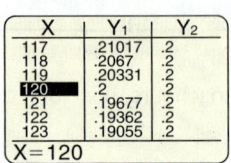

TECHNOLOGY NOTE: Solving Rational Equations Numerically

Tables can be used to find solutions to rational equations. The displays in the margin show the positive solution of 120 from Example 9. Note the use of parentheses for entering the formula for Y_1.

7.6 PUTTING IT ALL TOGETHER

In this section we discussed two methods of solving rational equations. The following table summarizes these methods.

Concept	Explanation	Examples
Solving the Equation $\dfrac{a}{b} = \dfrac{c}{d}$	Cross multiply to obtain $ad = bc$.	$\dfrac{5}{2x} = \dfrac{3}{6}$ is equivalent to $6x = 30$. $\dfrac{x}{2} = \dfrac{8}{x}$ is equivalent to $x^2 = 16$. (Cross multiplication works only for equations having *one* rational expression on each side: "fraction equals fraction.")
Multiplying by the LCD	**STEP 1:** Find the LCD. **STEP 2:** Multiply each term by the LCD. **STEP 3:** Simplify each term. **STEP 4:** Solve the resulting equation. **STEP 5:** Check each possible solution.	Solve $\dfrac{1}{x} - \dfrac{2}{3x} = \dfrac{5}{6}$. **STEP 1:** The LCD is $6x$. **STEP 2:** $\dfrac{1(6x)}{x} - \dfrac{2(6x)}{3x} = \dfrac{5(6x)}{6}$ **STEP 3:** $6 - 4 = 5x$ **STEP 4:** $\dfrac{2}{5} = x$ **STEP 5:** $\dfrac{1}{\frac{2}{5}} - \dfrac{2}{3\left(\frac{2}{5}\right)} = \dfrac{5}{2} - \dfrac{5}{3} = \dfrac{5}{6}$ It checks.

7.6 Exercises

PRACTICE   WATCH   DOWNLOAD   READ   REVIEW

CONCEPTS

1. If an equation contains one or more rational expressions, it is called a _____ equation.

2. Give an example of a rational expression and an example of a rational equation.

3. The equation $\frac{a}{b} = \frac{c}{d}$ is equivalent to _____, provided that _____ and _____ are nonzero.

4. Are the equations $\frac{2}{x-1} = 5$ and $5(x-1) = 2$ equivalent provided that $x \neq 1$?

5. One way to solve the equation $\frac{5}{3x} + \frac{3}{4x} = 1$ is to multiply each side by the LCD, or _____.

6. To solve the equation $T = \frac{R}{SV}$ for V, multiply each side by the variable _____ and then divide each side by the variable _____.

SOLVING RATIONAL EQUATIONS

Exercises 7–64: Solve and check your answer.

7. $\dfrac{x}{2} = \dfrac{3}{4}$

8. $\dfrac{2x}{3} = \dfrac{2}{5}$

9. $\dfrac{3}{z} = \dfrac{6}{5}$

10. $\dfrac{2}{7} = \dfrac{1}{z}$

11. $\dfrac{12}{7} = \dfrac{2}{t}$

12. $\dfrac{10}{t} = \dfrac{5}{7}$

13. $\dfrac{3y}{4} = \dfrac{7y}{2}$

14. $\dfrac{y}{6} = \dfrac{5y}{3}$

15. $\dfrac{2}{3} = \dfrac{1}{2x + 1}$

16. $\dfrac{1}{x + 4} = \dfrac{3}{5}$

17. $\dfrac{5}{2x} = \dfrac{8}{x + 2}$

18. $\dfrac{1}{x - 1} = \dfrac{5}{3x}$

19. $\dfrac{1}{z - 1} = \dfrac{2}{z + 1}$

20. $\dfrac{4}{z + 3} = \dfrac{2}{z - 2}$

21. $\dfrac{3}{n + 5} = \dfrac{2}{n - 5}$

22. $\dfrac{4}{3n + 2} = \dfrac{1}{n - 1}$

23. $\dfrac{m}{m - 1} = \dfrac{5}{4}$

24. $\dfrac{5m}{2m - 1} = \dfrac{3}{2}$

25. $\dfrac{5x}{5 - x} = \dfrac{1}{3}$

26. $\dfrac{x + 2}{3x} = \dfrac{4}{3}$

27. $\dfrac{6}{5 - 2x} = 2$

28. $\dfrac{x + 1}{x} = 6$

29. $\dfrac{2x}{2x + 1} = \dfrac{-1}{2x + 1}$

30. $\dfrac{x}{x - 4} = \dfrac{4}{x - 4}$

31. $\dfrac{1}{1 - x} = \dfrac{3}{1 + x}$

32. $\dfrac{2x}{1 - 2x} = \dfrac{1}{2}$

33. $\dfrac{1}{z + 2} = -z$

34. $\dfrac{1}{z - 2} = \dfrac{z}{3}$

35. $\dfrac{-1}{2x + 5} = \dfrac{x}{3}$

36. $\dfrac{x}{2} = \dfrac{1}{3x + 5}$

37. $\dfrac{x}{2} + \dfrac{x}{4} = 3$

38. $\dfrac{x}{4} - \dfrac{x}{3} = 1$

39. $\dfrac{3x}{4} - \dfrac{x}{2} = 1$

40. $\dfrac{2x}{3} + \dfrac{x}{3} = 6$

41. $\dfrac{4}{t + 1} + \dfrac{1}{t + 1} = -1$

42. $\dfrac{2}{t - 5} - \dfrac{5}{t - 5} = 3$

43. $\dfrac{1}{x} + \dfrac{2}{x} = \dfrac{1}{2}$

44. $\dfrac{1}{2x} - \dfrac{2}{x} = -3$

45. $\dfrac{2}{x - 1} + 1 = \dfrac{4}{x^2 - 1}$

46. $\dfrac{1}{x} + 2 = \dfrac{1}{x^2 + x}$

47. $\dfrac{1}{x + 2} = \dfrac{4}{4 - x^2} - 1$

48. $\dfrac{1}{x - 3} + 1 = \dfrac{6}{x^2 - 9}$

49. $\dfrac{5}{4z} - \dfrac{2}{3z} = 1$

50. $\dfrac{3}{z + 1} - \dfrac{1}{z + 1} = 2$

51. $\dfrac{4}{y - 1} + \dfrac{1}{y} = \dfrac{6}{5}$

52. $\dfrac{6}{y + 1} + \dfrac{6}{y} = 5$

53. $\dfrac{1}{2x} - \dfrac{1}{x + 3} = 0$

54. $\dfrac{2}{x} - \dfrac{6}{2x - 1} = -1$

55. $\dfrac{1}{x - 1} + \dfrac{1}{x + 1} = \dfrac{2}{x^2 - 1}$

56. $\dfrac{1}{2x + 1} + \dfrac{1}{2x - 1} = \dfrac{2}{4x^2 - 1}$

57. $\dfrac{1}{x - 2} + \dfrac{1}{x + 2} = \dfrac{6}{x^2 - 4}$

58. $\dfrac{2}{x + 3} - \dfrac{1}{x - 3} = \dfrac{1}{x^2 - 9}$

59. $\dfrac{1}{p + 1} + \dfrac{1}{p + 2} = \dfrac{1}{p^2 + 3p + 2}$

60. $\dfrac{1}{p - 1} - \dfrac{1}{p + 3} = \dfrac{1}{p^2 + 2p - 3}$

61. $\dfrac{1}{x - 2} + \dfrac{3}{2x - 4} = \dfrac{6}{3x - 6}$

62. $\dfrac{4}{x + 1} - \dfrac{4}{2x + 2} = \dfrac{1}{(x + 1)^2}$

63. $\dfrac{1}{r^2 - r - 2} + \dfrac{2}{r^2 - 2r} = \dfrac{1}{r^2 + r}$

64. $\dfrac{3}{r^2 - 1} + \dfrac{1}{r^2 + r} = \dfrac{3}{r^2 - r}$

*Exercises 65–72: **Expressions and Equations**   (Refer to Example 4.) Determine whether you are given an expression or an equation. If it is an expression, simplify (if possible) and evaluate it for x = 2. If it is an equation, solve it.*

65. $\dfrac{1}{x} - \dfrac{1 - x}{x}$

66. $\dfrac{1}{x} - x = 0$

67. $\dfrac{1}{2x} - \dfrac{1}{4x} = \dfrac{1}{8}$

68. $\dfrac{1}{2x} - \dfrac{1}{4x}$

69. $\dfrac{x + 1}{x - 1} = \dfrac{2x - 3}{2x - 5}$

70. $\dfrac{2x - 1}{4x + 1} = \dfrac{x + 1}{2x - 1}$

71. $\dfrac{4x + 4}{x + 2} + \dfrac{x^2}{x + 2}$

72. $\dfrac{x^2 - 2}{x - 2} - \dfrac{x}{x - 2}$

GRAPHICAL AND NUMERICAL SOLUTIONS

Exercises 73–76: Use the graph to solve the given equation. Check your answers.

73. $\dfrac{1}{x} = 4x$

74. $\dfrac{x}{x-1} = x$

75. $\dfrac{x-1}{x} = -2x$

76. $\dfrac{3}{x^2-1} = 1$

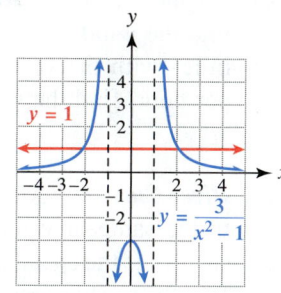

Exercises 77–84: (Refer to Example 5.) Solve the equation
(a) *graphically and*
(b) *numerically.*

77. $\dfrac{3}{x} = x + 2$

78. $-\dfrac{2}{x} = 1 - x$

79. $\dfrac{3x}{2} = \dfrac{1}{2}x - 1$

80. $\dfrac{x}{3} = 2 - \dfrac{2}{3}x$

81. $\dfrac{3}{x-1} = 3$

82. $\dfrac{2}{x+5} = 1$

 83. $\dfrac{4}{x^2} = 1$

84. $\dfrac{-18}{x^2} = -2$

 Exercises 85–88: Solve the rational equation graphically to the nearest thousandth.

85. $\dfrac{1}{\pi x - 2} + \dfrac{\sqrt{2}}{2.1x} = 1.3$

86. $\dfrac{6}{x^2 - 4} + \dfrac{\pi}{2x} = \dfrac{3}{2}$

87. $\dfrac{1}{x^3} - \dfrac{\pi}{4} = 0$

88. $\dfrac{1}{2\pi x^2} - \dfrac{x}{2} = 1$

SOLVING AN EQUATION FOR A VARIABLE

Exercises 89–100: (Refer to Examples 6 and 7.) Solve the equation for the specified variable.

89. $m = \dfrac{F}{a}$ for a

90. $m = \dfrac{2K}{v^2}$ for K

91. $I = \dfrac{V}{R + r}$ for r

92. $\dfrac{1}{T} = \dfrac{r}{R - r}$ for R

93. $h = \dfrac{2A}{b}$ for b

94. $h = \dfrac{2A}{b_1 + b_2}$ for b_1

95. $\dfrac{3}{k} = \dfrac{z}{z + 5}$ for z

96. $\dfrac{5}{r} = \dfrac{t + r}{t}$ for t

97. $T = \dfrac{ab}{a + b}$ for b

98. $A = \dfrac{2b}{a - b}$ for b

99. $\dfrac{3}{k} = \dfrac{1}{x} - \dfrac{2}{y}$ for x

100. $\dfrac{1}{R} = \dfrac{1}{R_1} + \dfrac{1}{R_2}$ for R_1

APPLICATIONS

101. *Waiting in Line*   (Refer to the introduction for this section.) Solve the equation

$$\frac{1}{10 - x} = 1$$

to determine the traffic rate x in cars per minute corresponding to an average waiting time of 1 minute.

102. *Waiting in Line*   At a post office customers arrive randomly at an average rate of x people per minute. The clerk can wait on 4 customers per minute. The average time T in minutes spent waiting in line is given by

$$T = \frac{1}{4 - x},$$

where $x < 4$.
(a) Evaluate T for $x = 3, 3.9$, and 3.99. What happens to the waiting time as the arrival rate nears 4 people per minute?
(b) Find x when the waiting time is 5 minutes.

103. *Shoveling a Sidewalk*   It takes an older employee 4 hours to shovel the snow from a sidewalk, but a younger employee can shovel the same sidewalk in 3 hours. How long will it take them to clear the walk if they work together?

104. *Pumping Water*   One pump can empty a pool in 5 days, whereas a second pump can empty the pool in 7 days. How long will it take the two pumps, working together, to empty the pool?

105. *Painting a House*   One painter can paint a house in 8 days, yet a more experienced painter can paint the house in 4 days. How long will it take the two painters, working together, to paint the house?

106. *Working Together*   Suppose that one person can mow a lawn in x hours and that a second person can mow the same lawn in y hours.
   (a) Show that it takes $\frac{xy}{x + y}$ hours for the two people, working together, to mow the lawn.
   (b) Use this expression to solve Example 8.

107. *Bicycle Race*   The winner of a 6-mile bicycle race finishes 2 minutes ahead of a teammate and travels, on average, 2 miles per hour faster than the teammate. Find the average speed of each racer.

108. *Freeway Travel*   Two drivers travel 150 miles on a freeway and then stop at a wayside rest area. The first driver travels 5 miles per hour faster and arrives $\frac{1}{7}$ hour ahead of the second. Find the average speed of each car.

109. *Braking Distance*   If a car is traveling *downhill* at 30 miles per hour on wet pavement, then the braking distance B in feet for this car is given by

$$B = \frac{30}{0.3 + m},$$

where $m < 0$ is the slope of the hill.   (**Source:** L. Haefner, *Introduction to Transportation Systems.*)
   (a) Find the braking distance for $m = -0.05$ and interpret the result.
   (b) Find m if the braking distance is 150 feet.

110. *Slippery Roads*   If a car is traveling at 30 miles per hour on a level road, then its braking distance in feet is $\frac{30}{x}$, where x is the coefficient of friction between

the road and the tires. The variable x is positive and satisfies $x \leq 1$. The closer the value of x is to 0, the more slippery is the road.   (**Source:** L. Haefner.)
   (a) Evaluate the expression for $x = 1$, 0.5, and 0.1. Interpret the results.
   (b) Find x for a braking distance of 150 feet.

111. *River Current*   A boat can travel 36 miles upstream in the same time that it can travel 54 miles downstream. If the speed of the current is 3 miles per hour, find the speed of the boat without a current.

112. *Airplane Speed*   An airplane can travel 380 miles into the wind in the same time that it can travel 420 miles with the wind. If the wind speed is 10 miles per hour, find the speed of the airplane without any wind.

113. *Airplane Speed*   An airplane can travel 450 miles into the wind in the same time that it can travel 750 miles with the wind. If the wind speed is 50 miles per hour, find the speed of the airplane without any wind.

114. *River Current*   A boat can travel 114 miles upstream in the same time that it can travel 186 miles downstream. If the speed of the current is 6 miles per hour, find the speed of the boat without a current.

115. *Running and Walking*   An athlete runs 10 miles and then walks home. The trip home takes 1 hour longer than it took to run that distance. If the athlete runs 5 miles per hour faster than she walks, what are her average running and walking speeds?

116. *Speed Limit*   A person drives 390 miles on a stretch of road. Half the distance is driven traveling 5 miles per hour below the speed limit, and half the distance is driven traveling 5 miles per hour above the speed limit. If the time spent traveling at the slower speed exceeds the time spent traveling at the faster speed by 24 minutes, find the speed limit.

117. *Unknown Number*   Find a number n that makes the expression $\frac{2 + n}{8 - n}$ equal 1.

118. *Unknown Number*   Find a number n that makes the expression $\frac{2n + 4}{4 - 3n}$ equal $-\frac{9}{11}$.

119. *Unknown Number*   Find a number n that makes the expression $\frac{4 + 6n}{4 - 4n}$ equal $-\frac{49}{31}$.

120. *Highway Curves* To make a highway curve safe, highway engineers often bank it, as shown in the figure. If a curve is designed for a speed of 50 miles per hour and is banked with positive slope m, then a minimum radius R in feet for the curve is given by

$$R = \frac{2500}{15m + 2}.$$

(**Source:** N. Garber and L. Hoel, *Traffic and Highway Engineering.*)

(a) Find R for $m = 0.1$. Interpret the result.
(b) If $R = 500$, find m. Interpret the result.

WRITING ABOUT MATHEMATICS

121. Do all rational equations have solutions? Explain.

122. Why is it important to check your answer when solving rational equations? Explain.

CHECKING BASIC CONCEPTS
SECTIONS 7.5 AND 7.6

1. Simplify each complex fraction.

(a) $\dfrac{\dfrac{x}{3}}{\dfrac{2x}{5}}$

(b) $\dfrac{\dfrac{2}{2x} - \dfrac{1}{3x}}{6x}$

(c) $\dfrac{\dfrac{1}{a} - \dfrac{1}{b}}{\dfrac{1}{a} + \dfrac{1}{b}}$

(d) $\dfrac{\dfrac{1}{r^2} - \dfrac{1}{t^2}}{\dfrac{2}{r} - \dfrac{2}{t}}$

2. Solve each equation. Check your answer.

(a) $\dfrac{1}{2x} = \dfrac{3}{x + 1}$

(b) $\dfrac{x}{2x + 3} = \dfrac{4}{5}$

3. Solve each equation. Check your answer.

(a) $\dfrac{1}{2x} + \dfrac{3}{2x} = 1$

(b) $\dfrac{3}{x + 1} - \dfrac{2}{x} = -2$

4. Solve the equation and check your answer.

$$\frac{1}{x - 1} = \frac{2}{x^2 - 1} - \frac{1}{2}$$

5. Solve each equation for the specified variable.

(a) $\dfrac{ax}{2} - 3y = b$ for x

(b) $\dfrac{1}{2m - 1} = \dfrac{k}{m}$ for m

6. *Braking Distance* If a car is traveling *uphill* at 60 miles per hour on wet pavement, then the braking distance D in feet for this car is given by the rational expression

$$D = \frac{120}{0.3 + m},$$

where $m > 0$ is the slope of the hill. (**Source:** L. Haefner, *Introduction to Transportation Systems.*)

(a) Find D for $m = 0.1$ and interpret the result.
(b) Find the slope of the road if D is 200 feet.

7.7 PROPORTIONS AND VARIATION

Proportions ■ Direct Variation ■ Inverse Variation ■ Joint Variation

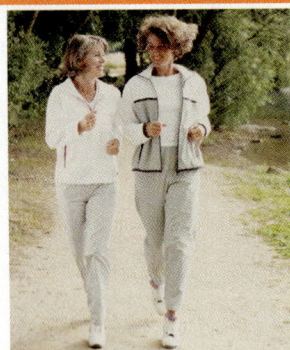

A LOOK INTO MATH ▷   Proportions frequently are used to solve applications. The following are a few examples.

- If someone earns $120 per day, then that person can earn $600 in 5 days.
- If a car goes 280 miles on 10 gallons of gas, then it can go 560 miles on 20 gallons of gas.
- If a person walks 1 mile in 20 minutes, then that person can walk $\frac{1}{2}$ mile in 10 minutes.

In this section we discuss several applications of proportions.

Proportions

▶ **REAL-WORLD CONNECTION**   A **ratio** is a comparison of two quantities, expressed as a quotient. For example, a math class might have 7 boys for every 8 girls. Thus the boy–girl ratio in this class is *7 to 8*, or $\frac{7}{8}$. In mathematics, ratios are typically expressed as fractions.

Ratios and proportions are sometimes used to find how much space remains for music on a compact disc (CD). A 700-megabyte CD can store about 80 minutes of music. Suppose that some music has been recorded on the CD and that 256 megabytes are still available. Using ratios and proportions, we can find how many more minutes of music could be recorded. A **proportion** is a statement that two *ratios* are equal.   (*Source:* Maxell Corporation.)

Let x represent the number of minutes available on a CD. Then **80** minutes are to **700** megabytes as x minutes are to **256** megabytes. By setting the *ratios* $\frac{80}{700}$ and $\frac{x}{256}$ equal to each other, we obtain the *proportion*

$$\frac{80}{700} = \frac{x}{256}. \qquad \frac{\text{Minutes}}{\text{Megabytes}} = \frac{\text{Minutes}}{\text{Megabytes}}$$

Solving this equation for x gives

$$700x = 80(256) \qquad \text{Cross multiply.}$$

$$x = \frac{80 \cdot 256}{700} \approx 29.3 \text{ minutes.} \qquad \text{Divide by 700.}$$

About 29 minutes are available to record on the CD.

MAKING CONNECTIONS

Proportions and Fractional Parts

We could have solved the preceding problem by noting that the fraction of the CD still available for recording music is $\frac{256}{700}$. So $\frac{256}{700}$ of 80 minutes is

$$\frac{256}{700} \cdot 80 \approx 29.3 \text{ minutes.}$$

EXAMPLE 1   Calculating the water content in snow

Six inches of light, fluffy snow are equivalent to about half an inch of rain in terms of water content. If 15 inches of this type of snow fall, estimate the water content.

Solution

Let x be the equivalent amount of rain. Then **6** inches of snow are to $\frac{1}{2}$ inch of rain as **15** inches of snow are to x inches of rain, which can be written as the proportion

$$\frac{6}{\frac{1}{2}} = \frac{15}{x}. \qquad\qquad \frac{\text{Snow}}{\text{Rain}} = \frac{\text{Snow}}{\text{Rain}}$$

Solving this equation gives

$$6x = \frac{15}{2} \quad\text{or}\quad x = \frac{15}{12} = 1.25.$$

Thus 15 inches of light, fluffy snow are equivalent to about 1.25 inches of rain.

Now Try Exercise **57**

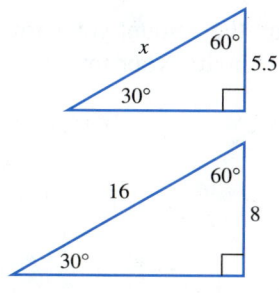

Figure 7.5

Proportions frequently occur in geometry when we work with similar figures. Two triangles are similar if the measures of their corresponding angles are equal. Corresponding sides of similar triangles are proportional. Figure 7.5 shows two right triangles that are similar because each has angles of 30°, 60°, and 90°.

We can find the length of side x by using proportions. Side x is to 16 as 5.5 is to 8, which can be written as the proportion

$$\frac{x}{16} = \frac{5.5}{8}. \qquad\qquad \frac{\text{Hypotenuse}}{\text{Hypotenuse}} = \frac{\text{Shorter leg}}{\text{Shorter leg}}$$

Solving yields the equation

$$8x = 5.5(16) \qquad\qquad \text{Cross multiply.}$$
$$x = 11. \qquad\qquad \text{Divide by 8.}$$

NOTE: Proportions can be set up in different ways and still produce the correct result. For example, we could say that x is to 5.5 in the smaller triangle as 16 is to 8 in the larger triangle.

$$\frac{x}{5.5} = \frac{16}{8} \qquad\qquad \frac{\text{Hypotenuse}}{\text{Shorter leg}} = \frac{\text{Hypotenuse}}{\text{Shorter leg}}$$

Solving, we obtain $8x = 5.5(16)$, or $x = 11$, which is the same answer.

EXAMPLE 2 Calculating the height of a tree

A 6-foot-tall person casts a 4-foot-long shadow. If a nearby tree casts a 36-foot-long shadow, estimate the height of the tree. See Figure 7.6.

6 feet

4 feet   36 feet

NOT TO SCALE

Figure 7.6

Figure 7.7

Solution

The triangles shown in Figure 7.7 are similar because the measures of the corresponding angles are equal. Therefore their sides are proportional. Let h be the height of the tree.

$$\frac{h}{6} = \frac{36}{4} \qquad \frac{\text{Height}}{\text{Height}} = \frac{\text{Shadow length}}{\text{Shadow length}}$$

$$4h = 6(36) \qquad \text{Cross multiply.}$$

$$h = \frac{6(36)}{4} \qquad \text{Divide each side by 4.}$$

$$h = 54 \qquad \text{Simplify.}$$

The tree is 54 feet tall.

Now Try Exercise 56

Direct Variation

▶ **REAL-WORLD CONNECTION** If your wage is $12 per hour, the amount you earn is proportional to the number of hours that you work. If you work H hours, your total pay P satisfies the equation

$$\frac{P}{H} = \frac{12}{1}, \qquad \frac{\text{Pay}}{\text{Hours}}$$

or, equivalently,

$$P = 12H.$$

We say that your pay P is *directly proportional* to the number of hours H worked. The constant of proportionality is 12.

DIRECT VARIATION

Let x and y denote two quantities. Then y is **directly proportional** to x, or y **varies directly** with x, if there is a nonzero number k such that

$$y = kx.$$

The number k is called the **constant of proportionality**, or the **constant of variation**.

The following 4-step process can be helpful when solving a variation application. This process is used to solve Examples 3 and 4 and can also be used to solve other types of variation problems.

SOLVING A VARIATION APPLICATION

When solving a variation problem, the following steps can be used.

STEP 1: Write the general equation for the type of variation problem that you are solving.

STEP 2: Substitute given values in this equation so the constant of variation k is the only unknown value in the equation. Solve for k.

STEP 3: Substitute the value of k in the general equation in Step 1.

STEP 4: Use this equation to find the requested quantity.

EXAMPLE 3   Solving a direct variation problem

Let y be directly proportional to x, or vary directly with x. Suppose $y = 7$ when $x = 5$. Find y when $x = 11$.

Solution

STEP 1: The general equation for direct variation is $y = kx$.

STEP 2: Substitute **7** for y and **5** for x in $y = k\boldsymbol{x}$. Solve for k.

$$\boldsymbol{7} = k(\boldsymbol{5}) \qquad \text{Let } y = 7 \text{ and } x = 5.$$

$$\frac{\boldsymbol{7}}{\boldsymbol{5}} = k \qquad \text{Divide each side by 5.}$$

STEP 3: Replace k with $\frac{7}{5}$ in the equation $y = k\boldsymbol{x}$ to obtain $y = \frac{7}{5}x$.

STEP 4: To find y, let $x = \boldsymbol{11}$. Then $y = \frac{7}{5}(11) = \frac{77}{5} = 15.4$.   Now Try Exercise 31

EXAMPLE 4   Solving a direct variation application

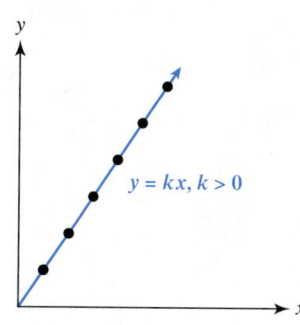

The strength of a beam of wood varies directly with its width. A beam that is 2.5 inches wide can support 800 pounds. How much weight can a similar beam support if its width is 3.2 inches?

Solution

STEP 1: The general equation for direct variation is $y = kx$, where y is the weight and x is the width.

STEP 2: Substitute **800** for y and **2.5** for x in $y = k\boldsymbol{x}$. Solve for k.

$$\boldsymbol{800} = k(\boldsymbol{2.5}) \qquad \text{Let } y = 800 \text{ and } x = 2.5.$$

$$\frac{\boldsymbol{800}}{\boldsymbol{2.5}} = k \qquad \text{Divide each side by 2.5.}$$

$$k = 320 \qquad \text{Simplify; rewrite the equation.}$$

STEP 3: Replace k with **320** in $y = k\boldsymbol{x}$ to obtain $y = \boldsymbol{320}x$.

STEP 4: Find the weight y by substituting **3.2** for x in $y = 320x$.

$$y = 320(\boldsymbol{3.2}) = 1024 \text{ pounds} \qquad \text{Now Try Exercise 59}$$

The graph of $y = kx$ is a line passing through the origin, as illustrated in Figure 7.8. Sometimes data in a scatterplot indicate that two quantities are directly proportional. The constant of proportionality k corresponds to the slope of the graph, which can be negative.

$y = kx, k > 0$

Figure 7.8 Direct Variation

EXAMPLE 5   Modeling college tuition

Table 7.5 lists the tuition for taking various numbers of credits.
(a) A scatterplot of the data is shown in Figure 7.9. Could the data be modeled using a line?

TABLE 7.5

Credits	Tuition
3	$189
5	$315
8	$504
11	$693
17	$1071

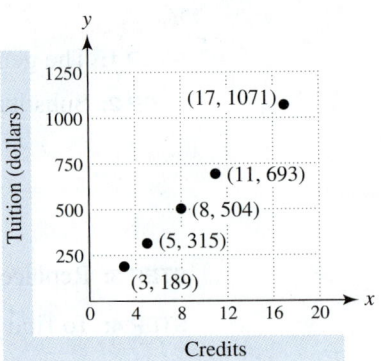

Figure 7.9

(b) Explain why tuition is directly proportional to the number of credits taken.
(c) Find the constant of proportionality. Interpret your result.
(d) Predict the cost of taking 16 credits.

Solution
(a) The data are linear and suggest a line passing through the origin.
(b) Because the data can be modeled by a line passing through the origin, tuition is directly proportional to the number of credits taken. Hence doubling the credits will double the tuition and tripling the credits will triple the tuition.
(c) The slope of the line equals the constant of proportionality k. If we use the first and last data points $(3, 189)$ and $(17, 1071)$, the slope is

$$k = \frac{1071 - 189}{17 - 3} = 63.$$

That is, tuition is $63 per credit. If we graph the line $y = 63x$, it models the data, as shown in Figure 7.10. This graph can also be created with a graphing calculator.
(d) If y represents tuition and x represents the credits taken, **16** credits would cost

$$y = 63(16) = \$1008.$$

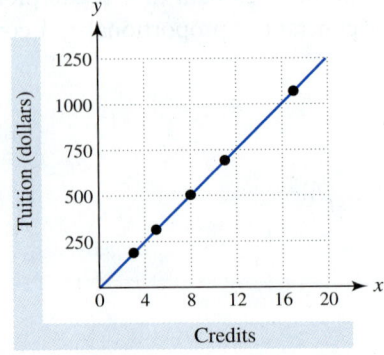

Figure 7.10

Now Try Exercise 63

Ratios and the Constant of Proportionality

The constant of proportionality in Example 5 can also be found by calculating the ratios $\frac{y}{x}$, where y is the tuition and x is the credits taken. Note that each ratio in the table is 63 because the equation $y = 63x$ is equivalent to the equation $\frac{y}{x} = \mathbf{63}$.

x	3	5	8	11	17
y	189	315	504	693	1071
$\frac{y}{x}$	**63**	**63**	**63**	**63**	**63**

Inverse Variation

▶ **REAL-WORLD CONNECTION**   When two quantities vary inversely, an increase in one quantity results in a decrease in the second quantity. For example, at 30 miles per hour a car travels 120 miles in 4 hours, whereas at 60 miles per hour the car travels 120 miles in 2 hours. Doubling the speed (or rate) decreases the travel time by half. Distance equals rate times time, so $d = rt$. Thus

$$120 = rt, \quad \text{or equivalently,} \quad t = \frac{120}{r}.$$

We say that the time t to travel 120 miles is *inversely proportional* to the speed or rate r. The constant of proportionality or constant of variation is 120.

Let x and y denote two quantities. Then y is **inversely proportional** to x, or y **varies inversely** with x, if there is a nonzero number k such that

$$y = \frac{k}{x}.$$

The data shown in Figure 7.11 represent inverse variation and are modeled by $y = \frac{k}{x}$. Note that, as x increases, y decreases. We assume k is positive.

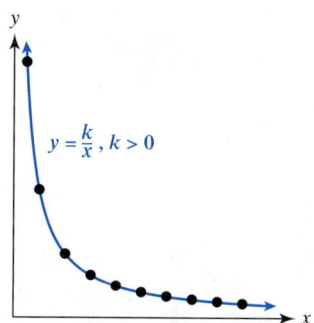

$y = \frac{k}{x}, k > 0$

Figure 7.11 Inverse Variation

EXAMPLE **6** Solving an inverse variation problem

Let y be inversely proportional to x, or vary inversely with x. Suppose $y = 5$ when $x = 6$. Find y when $x = 21$.

Solution

STEP 1: The general equation for inverse variation is $y = \frac{k}{x}$.

STEP 2: Because $y = 5$ when $x = 6$, substitute **5** for y and **6** for x in $y = \frac{k}{x}$. Solve for k.

$$5 = \frac{k}{6} \qquad \text{Let } y = 5 \text{ and } x = 6.$$

$$30 = k \qquad \text{Multiply each side by 6.}$$

STEP 3: Replace k with **30** in the equation $y = \frac{k}{x}$ to obtain $y = \frac{30}{x}$.

STEP 4: To find y, let $x = 21$. Then $y = \frac{30}{21} = \frac{10}{7}$. Now Try Exercise **37**

▶ **REAL-WORLD CONNECTION**   A wrench is commonly used to loosen a nut on a bolt. See Figure 7.12. If the nut is difficult to loosen, a wrench with a longer handle is often helpful.

Figure 7.12

EXAMPLE **7** Illustrating inverse variation with a wrench

Table 7.6 lists the force F necessary to loosen a particular nut with wrenches of different lengths L.

(a) Make a scatterplot of the data and discuss the graph. Are the data linear?

(b) Explain why the force F is inversely proportional to the handle length L. Find k so that $F = \frac{k}{L}$ models the data.

(c) Predict the force needed to loosen the nut with a 15-inch wrench.

Solution

(a) The scatterplot shown in Figure 7.13 reveals that the data are nonlinear. As the length L of the wrench increases, the force F necessary to loosen the nut decreases.

TABLE 7.6

L (inches)	F (pounds)
6	12
8	9
12	6
18	4
24	3

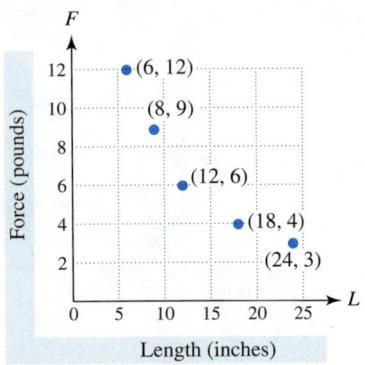

Figure 7.13

(b) If F is inversely proportional to L, then $F = \frac{k}{L}$, or $FL = k$. That is, the product of F and L equals the constant of proportionality k. In Table 7.6, the product of F and L always equals 72 for each data point. Thus F is inversely proportional to L with constant of proportionality $k = 72$.

(c) If $L = 15$, then $F = \frac{72}{15} = 4.8$. A wrench with a 15-inch handle requires a force of 4.8 pounds to loosen the nut.   **Now Try Exercise** 67

TECHNOLOGY NOTE: Scatterplots and Graphs

A graphing calculator can be used to create scatterplots and graphs. A scatterplot of the data in Table 7.6 is shown in the first figure. In the second figure the data and the equation $y = \frac{72}{x}$ are graphed. Note that each tick mark represents 5 units.

CALCULATOR HELP

To make a scatterplot see Appendix A (page AP-4).

Example 7 demonstrates that the product of the variables in an inverse variation equation equals a constant. This result is true in general. If we multiply each side of $y = \frac{k}{x}$ by x, we obtain $xy = k$. That is, the product of the variables in inverse variation equals the constant of variation k. In the next example, we use this fact to determine if a table of data represents inverse variation.

EXAMPLE 8   Analyzing data

Determine whether the data in each table represent direct variation, inverse variation, or neither.

(a)

x	4	5	10	20
y	40	32	16	8

(b)

x	2	5	9	11
y	18	45	81	99

(c)

x	2	4	6	8
y	8	20	30	56

Solution

(a) As x increases, y decreases. Because $xy = 160$ for each data point in the table, the equation $y = \frac{160}{x}$ models the data. The data represent inverse variation.

(b) As $\frac{y}{x} = 9$ for each data point in the table, the equation $y = 9x$ models the data in the table. These data represent direct variation.

(c) Neither the product xy nor the ratio $\frac{y}{x}$ are constant for the data in the table. Therefore these data represent neither direct variation nor inverse variation.

Now Try Exercises 43, 45, 47

Joint Variation

▶ **REAL-WORLD CONNECTION**   In many applications a quantity depends on more than one variable. In *joint variation* a quantity varies with the product of more than one variable. For example, the formula for the area A of a rectangle is given by

$$A = WL,$$

where W and L are the width and length, respectively. Thus the area of a rectangle varies jointly with the width and length.

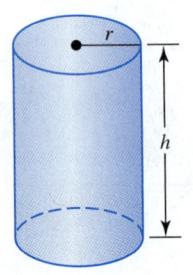

Figure 7.14

JOINT VARIATION

Let x, y, and z denote three quantities. Then z **varies jointly** with x and y if there is a nonzero number k such that

$$z = kxy.$$

Sometimes joint variation can involve a power of a variable. For example, the volume V of a cylinder is given by $V = \pi r^2 h$, where r is its radius and h is its height, as illustrated in Figure 7.14. In this case we say that the volume varies jointly with the height and the *square* of the radius. The constant of variation is $k = \pi$.

EXAMPLE 9   Finding the strength of a rectangular beam

Figure 7.15

The strength S of a rectangular beam varies jointly as its width w and the square of its thickness t. See Figure 7.15. If a beam 3 inches wide and 5 inches thick supports 750 pounds, how much can a similar beam 2 inches wide and 6 inches thick support?

Solution
The strength of a beam is modeled by $S = kwt^2$, where k is a constant of variation. We can find k by substituting $S = 750$, $w = 3$, and $t = 5$ in the formula.

$$750 = k \cdot 3 \cdot 5^2 \qquad \text{Substitute in } S = kwt^2.$$

$$k = \frac{750}{3 \cdot 5^2} \qquad \text{Solve for } k; \text{ rewrite.}$$

$$= 10 \qquad \text{Simplify.}$$

Thus $S = 10wt^2$ models the strength of this type of beam. When $w = 2$ and $t = 6$, the beam can support

$$S = 10 \cdot 2 \cdot 6^2 = 720 \text{ pounds.} \qquad \text{Now Try Exercise } 81$$

CRITICAL THINKING

Compare the increased strength of a beam if the width doubles and if the thickness doubles. What happens to the strength of a beam if both the width and thickness triple?

7.7 PUTTING IT ALL TOGETHER

In this section we introduced some basic concepts of proportion and variation. They are summarized in the following table.

Concept	Explanation	Examples
Proportion	A statement that two ratios are equal	$\dfrac{8}{17} = \dfrac{49}{x}$ and $\dfrac{x}{6} = \dfrac{3}{14}$
Direct Variation	Two quantities x and y vary according to the equation $y = kx$, where k is a nonzero constant. The constant of proportionality (or variation) is k.	$y = 4x$ or $\dfrac{y}{x} = 4$ $\begin{array}{\|c\|c\|c\|c\|} \hline x & 1 & 2 & 4 \\ \hline y & 4 & 8 & 16 \\ \hline \end{array}$ Note that if x doubles, then y also doubles.
Inverse Variation	Two quantities x and y vary according to the equation $y = \frac{k}{x}$, where k is a nonzero constant. The constant of proportionality (or variation) is k.	$y = \dfrac{3}{x}$ or $xy = 3$ $\begin{array}{\|c\|c\|c\|c\|} \hline x & 1 & 3 & 6 \\ \hline y & 3 & 1 & \frac{1}{2} \\ \hline \end{array}$ Note that if x doubles, then y decreases by half.
Joint Variation	Three quantities x, y, and z vary according to the equation $z = kxy$, where k is a nonzero constant.	The area A of a triangle varies jointly as b and h according to the equation $A = \frac{1}{2}bh$, where b is its base and h is its height. The constant of variation is $k = \frac{1}{2}$

7.7 Exercises

CONCEPTS

1. What is a proportion?

2. If 5 is to 6 as x is to 7, write a proportion that allows you to find x.

3. Suppose that y is directly proportional to x. If x doubles, what happens to y?

4. Suppose that y is inversely proportional to x. If x doubles, what happens to y?

5. If y varies directly with x, then $\frac{y}{x}$ equals a _____.

6. If y varies inversely with x, then xy equals a _____.

7. Would the food bill B generally vary directly or inversely with the number of people N being fed? Explain your reasoning.

8. Would the time T needed to paint a building vary directly or inversely with the number of painters N working on the job? Explain your reasoning.

PROPORTIONS

Exercises 9–20: Solve the proportion.

9. $\dfrac{x}{24} = \dfrac{5}{8}$ **10.** $\dfrac{x}{5} = \dfrac{3}{7}$

11. $\dfrac{14}{x} = \dfrac{2}{3}$ **12.** $\dfrac{4}{9} = \dfrac{9}{x}$

13. $\dfrac{3}{16} = \dfrac{h}{256}$ **14.** $\dfrac{20}{a} = \dfrac{15}{4}$

15. $\dfrac{3}{4} = \dfrac{2x}{7}$ **16.** $\dfrac{7}{3z} = \dfrac{5}{4}$

17. $\dfrac{x}{6} = \dfrac{8}{3x}$ **18.** $\dfrac{4}{x} = \dfrac{4x}{9}$

19. $\dfrac{x}{7} = \dfrac{7}{4x}$ **20.** $\dfrac{2}{3x} = \dfrac{27x}{8}$

21. Thinking Generally Solve $\dfrac{a}{b} = \dfrac{c}{d}$ for b.

22. Thinking Generally Solve $\dfrac{a+b}{c^2} = \dfrac{1}{2}$ for b.

Exercises 23–30: Do the following.
(a) *Write a proportion that models the situation.*
(b) *Solve the proportion for x.*

23. 5 is to 8, as 9 is to x

24. x is to 11, as 7 is to 4

25. A triangle has sides of 4, 7, and 10. In a similar triangle the shortest side is 8 and the longest side is x.

26. A rectangle has sides of 5 and 12. In a similar rectangle the longer side is 10 and the shorter side is x.

27. If you earn \$98 in 7 hours, then you can earn x dollars in 11 hours.

28. If 14 gallons of gasoline contain 1.4 gallons of ethanol, then 22 gallons of gasoline contain x gallons of ethanol.

29. If 3 cassette tapes can record 180 minutes of music, then 7 cassette tapes can record x minutes.

30. If a gas pump fills a 25-gallon tank in 6 minutes, it can fill a 14-gallon tank in x minutes.

VARIATION

Exercises 31–36: **Direct Variation** *Suppose that y is directly proportional to x.*
(a) *Use the given information to find the constant of proportionality k.*
(b) *Then use $y = kx$ to find y for $x = 6$.*

31. $y = 4$ when $x = 2$ **32.** $y = 5$ when $x = 10$

33. $y = 3$ when $x = 2$ **34.** $y = 11$ when $x = 55$

35. $y = -60$ when $x = 8$

36. $y = -17$ when $x = 68$

Exercises 37–42: **Inverse Variation** *Suppose that y is inversely proportional to x.*
(a) *Use the given information to find the constant of proportionality k.*
(b) *Then use $y = \dfrac{k}{x}$ to find y for $x = 8$.*

37. $y = 6$ when $x = 4$ **38.** $y = 2$ when $x = 24$

39. $y = 80$ when $x = \frac{1}{2}$ **40.** $y = \frac{1}{4}$ when $x = 32$

41. $y = 20$ when $x = 20$ **42.** $y = \frac{8}{3}$ when $x = 12$

Exercises 43–48: **Joint Variation** *Let z vary jointly with x and y.*
(a) *Find the constant of variation k.*
(b) *Use $z = kxy$ to find z when $x = 5$ and $y = 7$.*

43. $z = 6$ when $x = 3$ and $y = 8$

44. $z = 135$ when $x = 2.5$ and $y = 9$

45. $z = 5775$ when $x = 25$ and $y = 21$

46. $z = 1530$ when $x = 22.5$ and $y = 4$

47. $z = 25$ when $x = \frac{1}{2}$ and $y = 5$

48. $z = 12$ when $x = \frac{1}{4}$ and $y = 12$

Exercises 49–54: (Refer to Example 8.)
(a) *Determine whether the data represent direct variation, inverse variation, or neither.*
(b) *If the data represent either direct or inverse variation, find an equation that models the data.*
(c) *Graph the equation and the data when possible.*

49.

x	2	3	4	5
y	3	4.5	6	7.5

50.

x	10	20	30	40
y	12	6	5	4

51.

x	3	6	9	12
y	12	6	4	3

52.

x	2	6	10	14
y	105	35	21	15

53.

x	4	6	12	20
y	10	20	30	40

54.

x	1	5	9	15
y	6	30	54	90

Exercises 55–60: Use the graph to determine whether the data represent direct variation, inverse variation, or neither. Find the constant of variation whenever possible.

55.

56.

57.

58.

59.

60.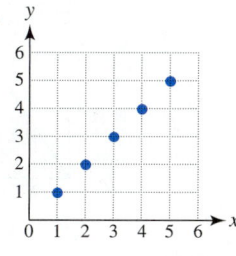

APPLICATIONS

61. *Recording Music*   A 750-megabyte CD can record 85 minutes of music. How many minutes can be recorded on 420 megabytes?

62. *Height of a Tree*   (Refer to Example 2.) A 5-foot person casts an 8-foot shadow, and a nearby tree casts a 30-foot shadow. Estimate the height of the tree.

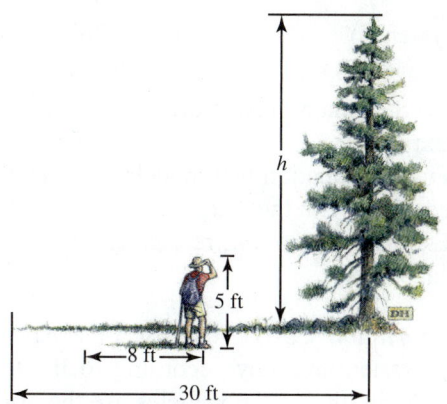

63. *Water Content in Snow*   (Refer to Example 1.) Eight inches of heavy, wet snow are equivalent to an inch of rain. Estimate the water content in 13 inches of heavy, wet snow.

64. *Wages*   If a person working for an hourly wage earns $143 in 13 hours, how much will that person earn in 15 hours?

65. *Strength of a Beam*   (Refer to Example 4.) The strength of a metal beam varies directly with its width. A beam that is 6.2 inches wide can support 2800 pounds. How much weight can a similar beam support if it is 4.7 inches wide?

66. *Strength of a Beam*   The strength of a wood beam varies inversely with its length. A beam that is 33 feet long can support 1200 pounds. How much weight can a similar beam support if it is 23 feet long?

67. *Making Fudge*   If $2\frac{2}{3}$ cups of sugar can make 14 pieces of fudge, how much sugar is needed to make 49 pieces of fudge?

68. *Making Coffee*   If 6 tablespoons of coffee grounds make 10 cups of coffee, how many tablespoons of coffee grounds are needed to make 35 cups of coffee?

69. *Rolling Resistance of Cars*   If you were to try to push a car, you would experience *rolling resistance*. This resistance equals the force necessary to keep the car moving slowly in neutral gear. The following table shows the rolling resistance R for passenger cars of different gross weights W.   (***Source:*** N. Garber and L. Hoel, *Traffic and Highway Engineering*.)

W (pounds)	2000	2500	3000	3500
R (pounds)	24	30	36	42

(a) Do the data represent direct or inverse variation? Explain.
(b) Find an equation that models the data. Graph the equation with the data.
(c) Estimate the rolling resistance of a 3200-pound car.

70. *Transportation Costs*   The use of a particular toll bridge varies inversely according to the toll. If the toll is $0.50, then 8000 vehicles use the bridge. Estimate the number of users if the toll is $0.80.   (***Source:*** N. Garber.)

71. *Flow of Water*   The gallons of water G flowing in 1 minute through a hose with a cross-sectional area A are shown in the table.

A (square inch)	0.2	0.3	0.4	0.5
G (gallons)	5.4	8.1	10.8	13.5

(a) Do the data represent direct or inverse variation? Explain.
(b) Find an equation that models the data. Graph the equation with the data.
(c) Interpret the constant of variation k.

72. *Hooke's Law*   The table shows the distance D that a spring stretches when a weight W is hung on it.

W (pounds)	2	6	9	15
D (inches)	1.6	4.8	7.2	12

(a) Do the data represent direct or inverse variation? Explain.
(b) Find an equation that models the data.
(c) How far will the spring stretch if an 11-pound weight is hung on it, as depicted in the figure?

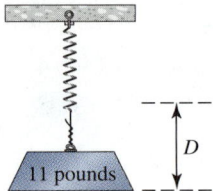

73. *Tightening Lug Nuts*   (Refer to Example 7.) When a tire is mounted on a car, the lug nuts should not be over-tightened. The following table shows the maximum force used with wrenches of different lengths.

L (inches)	8	10	16
F (pounds)	150	120	75

Source: Tires Plus.

(a) Model the data, using the equation $F = \frac{k}{L}$.
(b) How much force should be used with a wrench 15 inches long?

74. *Ozone and UV Radiation*   Ozone in the atmosphere filters out approximately 90% of the harmful ultraviolet (UV) rays from the sun. Depletion of the ozone layer increases the amount of UV radiation reaching Earth's surface. An increase in UV radiation is associated with skin cancer. The following graph shows the percentage increase y in UV radiation for a decrease in the ozone layer of x percent.   (***Source:*** R. Turner, D. Pearce, and I. Bateman, *Environmental Economics*.)

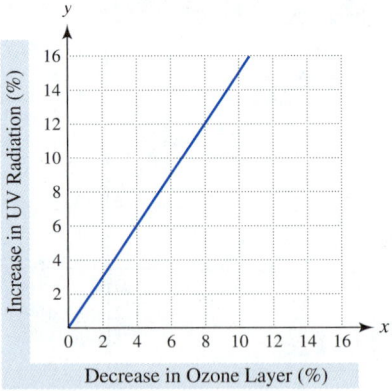

(a) Does this graph represent direct or inverse variation?

(b) Find an equation for the line in the graph.

(c) Estimate the percentage increase in UV radiation if the ozone layer decreases by 5%.

75. *Air Temperature and Altitude*   In the first 6 miles of Earth's atmosphere, air cools as the altitude increases. The following graph shows the temperature change y in degrees Fahrenheit at an altitude of x miles. (**Source:** A. Miller and R. Anthes, *Meteorology.*)

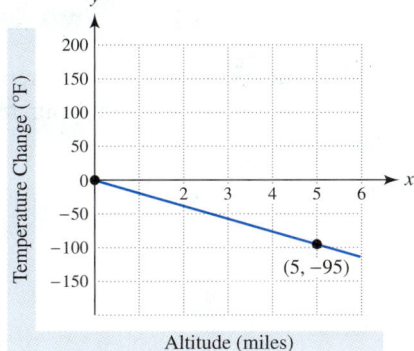

(a) Does this graph represent direct variation or inverse variation?

(b) Find an equation that models the data in the graph.

(c) Is the constant of proportionality k positive or negative? Interpret k.

(d) Find the change in air temperature 2.5 miles high.

76. *Cost of Tuition*   (Refer to Example 5.) The cost of tuition is directly proportional to the number of credits taken. If 6 credits cost $483, find the cost of 13 credits. What does the constant of proportionality represent?

77. *Electrical Resistance*   The electrical resistance of a wire is directly proportional to its length. If a 30-foot-long wire has a resistance of 3 ohms, find the resistance of an 18-foot-long wire.

78. *Resistance and Current*   The current that flows through an electrical circuit is inversely proportional to the resistance. When the resistance R is 180 ohms, the current I is 0.6 amp. Find the current when the resistance is 54 ohms.

79. *Joint Variation*   The variable z varies jointly as the second power of x and the third power of y. Write a formula for z if $z = 31.9$ when $x = 2$ and $y = 2.5$.

80. *Wind Power*   The electrical power generated by a windmill varies jointly with the square of the diameter of the area swept out by the blades and the cube of the wind velocity. If a windmill with a 10-foot diameter and a 16-mile-per-hour wind generates 15,392 watts, how much power would be generated if the blades swept out an area 12 feet in diameter and the wind speed was 15 miles per hour?

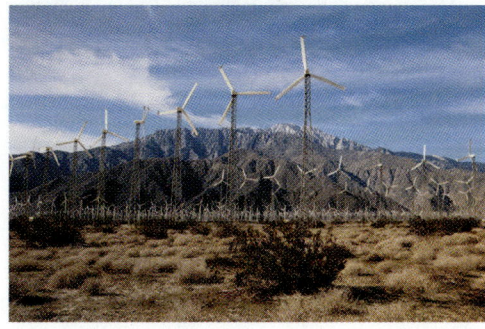

81. *Strength of a Beam*   (Refer to Example 9.) If a wood beam 5 inches wide and 3 inches thick supports 300 pounds, how much can a similar beam 5 inches wide and 2 inches thick support?

82. *Carpeting*   The cost of carpet for a rectangular room varies jointly as its width and length. If a room 11 feet wide and 14 feet long costs $539 to carpet, find the cost to carpet a room 17 feet by 19 feet. Interpret the constant of variation k.

83. *Weight on the Moon*   The weight of a person on the moon is directly proportional to the weight of the person on Earth. If a 175-pound person weighs 28 pounds on the moon, how much will a 220-pound person weigh on the moon?

84. *Weight Near Earth*   The weight W of a person near Earth is inversely proportional to the square of the person's distance d from the *center* of Earth. If a person weighs 200 pounds when $d = 4000$ miles, how much does the same person weigh when $d = 7000$ miles? (*Note:* The radius of Earth is about 4000 miles.)

85. *Ohm's Law*   The voltage V in an electrical circuit varies jointly with the amperage I and resistance R. If $V = 220$ when $I = 10$ and $R = 22$, find V when $I = 15$ and $R = 50$.

86. *Revenue*   The revenue R from selling x items at price p varies jointly with x and p. If $R = \$24,000$ when $x = 3000$ and $p = \$8$, find the number of items sold x when $R = \$30,000$ and $p = \$6$.

WRITING ABOUT MATHEMATICS

87. Explain what it means for a quantity y to be directly proportional to a quantity x.

88. Explain what it means for a quantity y to be inversely proportional to a quantity x.

CHECKING BASIC CONCEPTS
SECTION 7.7

1. Solve each proportion.

(a) $\dfrac{x}{9} = \dfrac{2}{5}$ **(b)** $\dfrac{4}{3} = \dfrac{5}{b}$

2. Write a proportion that models each situation. Then solve it.
   (a) 4 is to 6 as 8 is to x
   (b) If 2 compact discs can record 148 minutes of music, then 5 compact discs can record x minutes of music.

3. Suppose that y is inversely proportional to x. If $y = 4$ when $x = 15$, find the constant of proportionality k. Find y when $x = 10$.

4. Decide whether the data in the table represent direct or inverse variation. Explain your reasoning. Find the constant of variation.

(a)

x	2	4	6	8
y	3	6	9	12

(b)

x	2	4	6	8
y	12	6	4	3

5. *Wages*   If a person working for an hourly wage earns $272 in 17 hours, how much will the person earn in 10 hours?

CHAPTER 7   SUMMARY
SECTION 7.1 ■ INTRODUCTION TO RATIONAL EXPRESSIONS

Rational Expression
A rational expression can be written in the form $\frac{P}{Q}$, where P and Q are polynomials, and is defined whenever $Q \neq 0$.

Example: $\frac{x^2}{x-5}$ is a rational expression that is defined for all real numbers except $x = 5$.

Simplifying a Rational Expression   To simplify a rational expression, factor the numerator and the denominator. Then apply the basic principle of rational expressions,

$$\frac{PR}{QR} = \frac{P}{Q}.$$

Examples: $\dfrac{x^2 - 4}{x^2 - 3x + 2} = \dfrac{(x+2)(x-2)}{(x-1)(x-2)} = \dfrac{x+2}{x-1}$, $\dfrac{x-3}{3-x} = -\dfrac{x-3}{x-3} = -1$

SECTION 7.2 ■ MULTIPLICATION AND DIVISION OF RATIONAL EXPRESSIONS

Multiplying Rational Expressions To multiply two rational expressions, multiply the numerators and multiply the denominators.

$$\frac{A}{B} \cdot \frac{C}{D} = \frac{AC}{BD}$$

Example: $\dfrac{3}{x-1} \cdot \dfrac{4}{x+1} = \dfrac{12}{(x-1)(x+1)}$

Dividing Rational Expressions To divide two rational expressions, multiply by the reciprocal of the divisor.

$$\frac{A}{B} \div \frac{C}{D} = \frac{A}{B} \cdot \frac{D}{C} = \frac{AD}{BC}$$

Example: $\dfrac{x+2}{x^2+3x} \div \dfrac{x+2}{x} = \dfrac{x+2}{x(x+3)} \cdot \dfrac{x}{x+2} = \dfrac{x(x+2)}{x(x+3)(x+2)} = \dfrac{1}{x+3}$

SECTION 7.3 ■ ADDITION AND SUBTRACTION WITH LIKE DENOMINATORS

Addition of Rational Expressions Having Like Denominators To add two rational expressions having like denominators, add their numerators. Keep the same denominator.

$$\frac{A}{C} + \frac{B}{C} = \frac{A+B}{C}$$

Example: $\dfrac{5x}{x+4} + \dfrac{1}{x+4} = \dfrac{5x+1}{x+4}$

Subtraction of Rational Expressions Having Like Denominators To subtract two rational expressions having like denominators, subtract their numerators. Keep the same denominator.

$$\frac{A}{C} - \frac{B}{C} = \frac{A-B}{C}$$

Example: $\dfrac{6}{2x+1} - \dfrac{2}{2x+1} = \dfrac{4}{2x+1}$

SECTION 7.4 ■ ADDITION AND SUBTRACTION WITH UNLIKE DENOMINATORS

Finding Least Common Multiples The least common multiple (LCM) of two or more polynomials can be found as follows.

STEP 1: Factor each polynomial completely.

STEP 2: List each factor the greatest number of times that it occurs in any factorization.

STEP 3: Find the product of this list of factors. It is the LCM.

Example: $4x^2(x + 1) = 2 \cdot 2 \cdot x \cdot x \cdot (x + 1)$
$2x(x^2 - 1) = 2 \cdot x \cdot (x + 1) \cdot (x - 1)$

Listing each factor the greatest number of times and multiplying gives

$$2 \cdot 2 \cdot x \cdot x \cdot (x + 1)(x - 1) = 4x^2(x^2 - 1),$$

which is the LCM of $4x^2(x + 1)$ and $2x(x^2 - 1)$.

Finding the Least Common Denominator   The least common denominator (LCD) is the least common multiple (LCM) of the denominators.

Example: From the preceding example, the LCD for $\frac{1}{4x^2(x + 1)}$ and $\frac{1}{2x(x^2 - 1)}$ is $4x^2(x^2 - 1)$.

Addition and Subtraction of Rational Expressions Having Unlike Denominators   First write each rational expression by using the LCD. Then add or subtract the resulting rational expressions.

Example: $\dfrac{1}{x - 1} - \dfrac{1}{x} = \dfrac{x}{x(x - 1)} - \dfrac{x - 1}{x(x - 1)} = \dfrac{x - (x - 1)}{x(x - 1)} = \dfrac{1}{x(x - 1)}$

Note that the LCD is $x(x - 1)$.

SECTION 7.5 ■ COMPLEX FRACTIONS

Complex Fractions   A complex fraction is a rational expression that contains fractions in its numerator, denominator, or both. The following equation can be used to simplify some complex fractions.

$$\frac{\dfrac{a}{b}}{\dfrac{c}{d}} = \frac{a}{b} \cdot \frac{d}{c}$$

Example: $\dfrac{\dfrac{x}{3}}{\dfrac{x}{x - 1}} = \dfrac{x}{3} \cdot \dfrac{x - 1}{x} = \dfrac{x(x - 1)}{3x} = \dfrac{x - 1}{3}$

Simplifying Complex Fractions

Method I   Combine terms in the numerator, combine terms in the denominator, and simplify the resulting expression.

Method II   Multiply the numerator and denominator by the LCD for both and simplify the resulting expression.

Example: *Method I* $\dfrac{\dfrac{1}{a} - \dfrac{1}{b}}{\dfrac{1}{a} + \dfrac{1}{b}} = \dfrac{\dfrac{b - a}{ab}}{\dfrac{b + a}{ab}} = \dfrac{b - a}{ab} \cdot \dfrac{ab}{b + a} = \dfrac{b - a}{b + a}$

Method II   The least common denominator is *ab*.

$$\frac{\dfrac{1}{a} - \dfrac{1}{b}}{\dfrac{1}{a} + \dfrac{1}{b}} = \frac{\left(\dfrac{1}{a} - \dfrac{1}{b}\right)ab}{\left(\dfrac{1}{a} + \dfrac{1}{b}\right)ab} = \frac{\dfrac{ab}{a} - \dfrac{ab}{b}}{\dfrac{ab}{a} + \dfrac{ab}{b}} = \frac{b - a}{b + a}$$

SECTION 7.6 ■ RATIONAL EQUATIONS AND FORMULAS

Solving Rational Equations   One way to solve the equation $\frac{a}{b} = \frac{c}{d}$ is to cross multiply to obtain $ad = bc$. (Check each answer.) A general way to solve rational equations is to follow these steps.

STEP 1: Find the LCD of the terms in the equation.

STEP 2: Multiply each side of the equation by the LCD.

STEP 3: Simplify each term.

STEP 4: Solve the resulting equation.

STEP 5: Check each answer in the *given* equation. Reject any value that makes a denominator equal 0.

Examples: $\frac{1}{2x} = \frac{2}{x + 3}$ implies that $x + 3 = 4x$, or $x = 1$. This answer checks.

To solve $\frac{5}{x} - \frac{1}{3x} = \frac{7}{3}$ multiply each term by the LCD, $3x$:

$$\frac{5(3x)}{x} - \frac{3x}{3x} = \frac{7(3x)}{3},$$

which simplifies to $15 - 1 = 7x$, or $x = 2$. This answer checks.

Solving for a Variable   Many formulas contain more than one variable. To solve for a particular variable, use the rules of algebra to isolate the variable.

Example: To solve $S = \frac{2\pi}{r}$ for r, multiply each side by r to obtain $Sr = 2\pi$ and then divide each side by S to obtain $r = \frac{2\pi}{S}$.

SECTION 7.7 ■ PROPORTIONS AND VARIATION

Proportions   A proportion is a statement that two ratios are equal.

Example: $\frac{5}{x} = \frac{4}{7}$

Similar Triangles   Two triangles are similar if the measures of their corresponding angles are equal. Corresponding sides of similar triangles are proportional.

Example: A right triangle has legs of lengths 3 and 4. A similar right triangle has a shorter leg with length 6. Its longer leg can be found by solving the proportion $\frac{3}{6} = \frac{4}{x}$ to obtain $x = 8$.

Direct Variation   A quantity y is *directly proportional* to a quantity x, or y *varies directly* with x, if there is a nonzero constant k such that $y = kx$. The number k is called the *constant of proportionality* or the *constant of variation*.

Example: If y varies directly with x, then the ratios $\frac{y}{x}$ always equal k. The following data satisfy $\frac{y}{x} = 4$, so the constant of variation is 4. Thus $y = 4x$.

x	1	2	3	4
y	4	8	12	16

Inverse Variation   A quantity y is *inversely proportional* to a quantity x, or y *varies inversely* with x, if there is a nonzero constant k such that $y = \frac{k}{x}$.

Example: If y varies inversely with x, then the products xy always equal k. The following data satisfy $xy = 12$, so the constant of variation is 12. Thus $y = \frac{12}{x}$.

x	1	2	4	6
y	12	6	3	2

Joint Variation   The quantity z *varies jointly* with x and y if $z = kxy$, $k \neq 0$.

Example: The area A of a rectangle varies jointly with the width W and length L because $A = LW$. Note that $k = 1$ in this example.

CHAPTER 7   REVIEW EXERCISES

SECTION 7.1

Exercises 1–4: If possible, evaluate the expression for the given value of x.

1. $\dfrac{3}{x-3}$ $x = -2$

2. $\dfrac{4x}{5-x^2}$ $x = 3$

3. $\dfrac{-x}{7-x}$ $x = 7$

4. $\dfrac{4x}{x^2-3x+2}$ $x = 2$

5. Complete the table for the rational expression. If a value is undefined, place a dash in the table.

x	-2	-1	0	1	2
$\frac{3x}{x-1}$					

6. Find the x-values that make $\dfrac{8}{x^2-4}$ undefined.

Exercises 7–12: Simplify to lowest terms.

7. $\dfrac{25x^3y^4}{15x^5y}$

8. $\dfrac{x^2-36}{x+6}$

9. $\dfrac{x-9}{9-x}$

10. $\dfrac{x^2-5x}{5x}$

11. $\dfrac{2x^2+5x-3}{2x^2+x-1}$

12. $\dfrac{3x^2+10x-8}{3x^2+x-2}$

SECTION 7.2

Exercises 13–16: Multiply and write in lowest terms.

13. $\dfrac{x-3}{x+1} \cdot \dfrac{2x+2}{x-3}$

14. $\dfrac{2x+5}{(x+5)(x-1)} \cdot \dfrac{x-1}{2x+5}$

15. $\dfrac{z+3}{z-4} \cdot \dfrac{z-4}{(z+3)^2}$

16. $\dfrac{x^2}{x^2-4} \cdot \dfrac{x+2}{x}$

Exercises 17–22: Divide and write in lowest terms.

17. $\dfrac{x+1}{2x} \div \dfrac{3x+3}{5x}$

18. $\dfrac{4}{x^3} \div \dfrac{x+1}{2x^2}$

19. $\dfrac{x-5}{x+2} \div \dfrac{2x-10}{x+2}$

20. $\dfrac{x^2-6x+5}{x^2-25} \div \dfrac{x-1}{x+5}$

21. $\dfrac{x^2-y^2}{x+y} \div \dfrac{x-y}{x+y}$

22. $\dfrac{a^3-b^3}{a+b} \div \dfrac{a-b}{2a+2b}$

SECTION 7.3

Exercises 23–30: Add or subtract and write in lowest terms.

23. $\dfrac{2}{x + 10} + \dfrac{8}{x + 10}$

24. $\dfrac{9}{x - 1} - \dfrac{8}{x - 1}$

25. $\dfrac{x + 2y}{2x} + \dfrac{x - 2y}{2x}$

26. $\dfrac{x}{x + 3} + \dfrac{3}{x + 3}$

27. $\dfrac{x}{x^2 - 1} - \dfrac{1}{x^2 - 1}$

28. $\dfrac{2x}{x^2 - 25} + \dfrac{10}{x^2 - 25}$

29. $\dfrac{3}{xy} - \dfrac{1}{xy}$

30. $\dfrac{x + y}{2y} + \dfrac{x - y}{2y}$

SECTION 7.4

Exercises 31–36: Find the least common multiple for the expressions. Leave your answer in factored form.

31. $3x, 5x$

32. $5x^2, 10x$

33. $x, x - 5$

34. $10x^2, x^2 - x$

35. $x^2 - 1, (x + 1)^2$

36. $x^2 - 4x, x^2 - 16$

Exercises 37–42: Rewrite the rational expression by using the given denominator D.

37. $\dfrac{3}{8}, D = 24$

38. $\dfrac{4}{3x}, D = 12x$

39. $\dfrac{3x}{x - 2}, D = x^2 - 4$

40. $\dfrac{2}{x + 1}, D = x^2 + x$

41. $\dfrac{3}{5x}, D = 5x^2 - 5x$

42. $\dfrac{2x}{2x - 3}, D = 2x^2 + x - 6$

Exercises 43–54: Simplify the expression.

43. $\dfrac{5}{8} + \dfrac{1}{6}$

44. $\dfrac{3}{4x} + \dfrac{1}{x}$

45. $\dfrac{5}{9x} - \dfrac{2}{3x}$

46. $\dfrac{7}{x - 1} - \dfrac{3}{x}$

47. $\dfrac{1}{x + 1} + \dfrac{1}{x - 1}$

48. $\dfrac{4}{3x^2} - \dfrac{3}{2x}$

49. $\dfrac{1 + x}{3x} - \dfrac{3}{2x}$

50. $\dfrac{x}{x^2 - 1} - \dfrac{1}{x - 1}$

51. $\dfrac{2}{x - y} - \dfrac{3}{x + y}$

52. $\dfrac{2}{x} - \dfrac{1}{2x} + \dfrac{2}{3x}$

53. $\dfrac{3}{2y} + \dfrac{1}{2x}$

54. $\dfrac{x}{y - x} + \dfrac{y}{x - y}$

SECTION 7.5

Exercises 55–64: Simplify the complex fraction.

55. $\dfrac{\frac{3}{4}}{\frac{7}{11}}$

56. $\dfrac{\frac{x}{5}}{\frac{2x}{7}}$

57. $\dfrac{\frac{m}{n}}{\frac{2m}{n^2}}$

58. $\dfrac{\frac{3}{p - 1}}{\frac{1}{p + 1}}$

59. $\dfrac{\frac{3}{m - 1}}{\frac{2m - 2}{m + 1}}$

60. $\dfrac{\frac{2}{2n + 1}}{\frac{8}{2n - 1}}$

61. $\dfrac{\frac{1}{2x} - \frac{1}{3x}}{\frac{2}{3x} - \frac{1}{6x}}$

62. $\dfrac{\frac{2}{xy} - \frac{1}{y}}{\frac{2}{xy} + \frac{1}{y}}$

63. $\dfrac{\frac{1}{x} - \frac{1}{x + 1}}{\frac{x}{x + 1}}$

64. $\dfrac{\frac{2}{x - 1} - \frac{1}{x + 1}}{\frac{1}{x^2 - 1}}$

SECTION 7.6

Exercises 65–70: Solve and check your answer.

65. $\dfrac{x}{5} = \dfrac{4}{7}$

66. $\dfrac{4}{x} = \dfrac{3}{2}$

67. $\dfrac{3}{z + 1} = \dfrac{1}{2z}$

68. $\dfrac{x + 2}{x} = \dfrac{3}{5}$

69. $\dfrac{1}{x + 1} = \dfrac{2}{x - 2}$

70. $\dfrac{x}{3} = \dfrac{-1}{x + 4}$

Exercises 71–84: If possible, solve. Check your answer.

71. $\dfrac{1}{5x} + \dfrac{3}{5x} = \dfrac{1}{5}$

72. $\dfrac{1}{x-1} + \dfrac{2x}{x-1} = 1$

73. $\dfrac{1}{x} + \dfrac{2}{3x} = \dfrac{1}{3}$

74. $\dfrac{1}{x+3} + \dfrac{2x}{x+3} = \dfrac{3}{2}$

75. $\dfrac{5}{x} - \dfrac{3}{x+1} = \dfrac{1}{2}$

76. $\dfrac{1}{x-1} - \dfrac{1}{x+1} = \dfrac{1}{4}$

77. $\dfrac{4}{p} - \dfrac{5}{p+2} = 0$

78. $\dfrac{1}{x-3} - \dfrac{1}{x+3} = \dfrac{1}{x^2-9}$

79. $\dfrac{1}{x+1} = \dfrac{-x}{x+1}$

80. $\dfrac{2}{x} = \dfrac{2}{x^2+x} - 4$

81. $\dfrac{2}{x^2-2x} + \dfrac{1}{x^2-4} = \dfrac{1}{x^2+2x}$

82. $\dfrac{3}{x^2-3x} - \dfrac{1}{x^2-9} = \dfrac{1}{x^2+3x}$

83. $\dfrac{1}{x^2} - \dfrac{5}{x^2+4x} = \dfrac{1}{x^2+4x}$

84. $\dfrac{5}{x^2-1} - \dfrac{1}{x^2+2x+1} = \dfrac{3}{x^2-1}$

Exercises 85 and 86: Solve for the specified variable.

85. $\dfrac{1}{a} + \dfrac{2}{b} = \dfrac{3}{c}$ for b

86. $y = \dfrac{x}{x-1}$ for x

SECTION 7.7

Exercises 87 and 88: Solve the proportion.

87. $\dfrac{x}{6} = \dfrac{1}{3}$

88. $\dfrac{5}{x} = \dfrac{7}{3}$

Exercises 89 and 90: **Proportions**   *Do the following.*
  (a) *Write a proportion that models the situation.*
  (b) *Solve the proportion for x.*

89. A rectangle has sides of 6 and 13. In a similar rectangle the longer side is 20 and the shorter side is x.

90. If you earn \$341 in 11 hours, then you can earn x dollars in 8 hours.

Exercises 91 and 92: **Direct Variation**   *Suppose that y is directly proportional to x.*
  (a) *Use the given information to find the constant of proportionality k.*
  (b) *Then use $y = kx$ to find y for $x = 5$.*

91. $y = 8$ when $x = 2$

92. $y = 21$ when $x = 7$

Exercises 93 and 94: **Inverse Variation**   *Suppose that y is inversely proportional to x.*
  (a) *Use the given information to find the constant of proportionality k.*
  (b) *Then use $y = \frac{k}{x}$ to find y for $x = 5$.*

93. $y = 2.5$ when $x = 4$

94. $y = 7$ when $x = 3$

95. **Joint Variation**   Suppose that z varies jointly with x and y. If $z = 483$ when $x = 23$ and $y = 7$, find the constant of variation k.

96. **Joint Variation**   Suppose that z varies jointly with x and the square of y. If $z = 891$ when $x = 22$ and $y = 3$, find z when $x = 10$ and $y = 4$.

Exercises 97 and 98: Do the following.
  (a) *Determine whether the data represent direct or inverse variation.*
  (b) *Find an equation that models the data.*
  (c) *Graph the data and your equation.*

97.

x	2	3	4	5
y	30	20	15	12

98.

x	2	4	6	8
y	6	12	18	24

Exercises 99 and 100: Use the graph to determine whether the data represent direct or inverse variation. Find the constant of variation.

99.

100.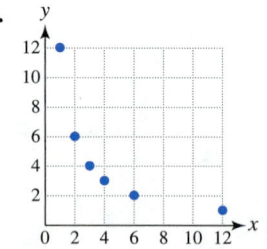

APPLICATIONS

101. *Modeling Traffic Flow*   Fifteen vehicles per minute can pass through an intersection. If vehicles arrive randomly at an average rate of x per minute, the average waiting time T in minutes is given by $T = \frac{1}{15 - x}$ for $x < 15$.   (*Source:* N. Garber, *Traffic and Highway Engineering.*)

  (a) Find T when $x = 10$ and interpret the result.

  (b) Complete the table.

x	5	10	13	14	14.9
T					

  (c) What happens to the waiting time as the traffic rate x approaches 15 vehicles per minute?

102. *Distance and Time*   A car traveled at 50 miles per hour for 150 miles and then traveled at 75 miles per hour for 150 miles. What was the car's average speed?

103. *Emptying a Swimming Pool*   A large pump can empty a swimming pool in 100 hours, whereas a small pump can empty the pool in 160 hours. How long will it take to empty the pool if both pumps are used?

104. *Jogging*   Two athletes jog 10 miles. One of the athletes jogs 2 miles per hour faster and finishes 10 minutes ahead of the other athlete. Find the average speed of each athlete.

105. *River Current*   A boat can travel 16 miles upstream in the same time that it can travel 48 miles downstream. If the speed of the current is 4 miles per hour, find the boat's speed.

106. *Height of a Tree*   A 5-foot-tall person has a 6-foot-long shadow, and a nearby tree has a 40-foot-long shadow. Estimate the tree's height.

107. *Transportation Costs*   Use of a toll road varies inversely with the toll. If the toll is $0.25, then 400 vehicles use the road. Estimate the number of users for a toll of $0.50.

108. *Cost of Carpet*   The cost of carpet is directly proportional to the amount of carpet purchased. If 17 square yards cost $612, find the cost of 13 square yards.

109. *Tightening a Bolt*   The torque exerted on a nut by a wrench is inversely proportional to the length of the wrench's handle. Suppose that a 12-inch wrench can be used to tighten a nut by using 30 pounds of force. How much force is necessary to tighten the same nut by using a 10-inch wrench?

110. *Water Content in Snow*   Twenty inches of extremely dry, powdery snow are equivalent to an inch of rain. Estimate the water content in 32 inches of this type of snow.

111. *Cold Water Survival*   When a person falls through the ice on a lake, the cold water removes body heat 25 times faster than air at the same temperature. To survive, a person has between 30 and 90 seconds to get out of the water. How long could a person remain in air at the same temperature?

112. *Strength of a Beam*   The strength of a wood beam varies inversely with its length. A beam that is 18 feet long can support 900 pounds. How much weight can a similar beam support if its length is 21 feet?

113. *Wind Power*   The electric power generated by a windmill varies jointly with the square of the diameter of the area swept out by the blades and the cube of the wind velocity. If a windmill with 6-foot-diameter blades and a 20-mile-per-hour wind generates 10,823 watts, how much power would be generated if the blades were 10 feet in diameter and the wind speed were 12 miles per hour?

114. *Strength of a Beam*   The strength of a beam varies jointly as its width w and the square of its thickness t. If a beam 8 inches wide and 5 inches thick supports 650 pounds, how much can a similar beam 6 inches wide and 6 inches thick support?

CHAPTER 7   TEST *Pass the Test* **Video solutions to all test exercises**

1. Evaluate the expression $\dfrac{3x}{2x-1}$ for $x = 3$.

2. Find any x-value that makes $\dfrac{x-1}{x+2}$ undefined.

Exercises 3 and 4: Simplify the expression.

3. $\dfrac{x^2 - 25}{x - 5}$

4. $\dfrac{3x^2 - 15x}{3x}$

Exercises 5–12: Simplify the expression. Write your answer in lowest terms.

5. $\dfrac{x-2}{x+4} \cdot \dfrac{3x+12}{x-2}$

6. $\dfrac{z+1}{z+3} \cdot \dfrac{2z+6}{z+1}$

7. $\dfrac{x+1}{5x} \div \dfrac{2x+2}{x-1}$

8. $\dfrac{2}{x^2} \div \dfrac{x+3}{3x}$

9. $\dfrac{x}{x+4} + \dfrac{3x+1}{x+4}$

10. $\dfrac{4t+1}{2t-3} - \dfrac{3t-6}{2t-3}$

11. $\dfrac{1}{y^2+y} - \dfrac{y-1}{y^2-y}$

12. $\dfrac{1}{xy} + \dfrac{x}{y} - \dfrac{1}{y^2}$

Exercises 13 and 14: Simplify the complex fraction.

13. $\dfrac{\dfrac{a}{3b}}{\dfrac{5a}{b^2}}$

14. $\dfrac{1 + \dfrac{1}{p-1}}{1 - \dfrac{1}{p-1}}$

Exercises 15–22: Solve the equation and check your answer.

15. $\dfrac{2}{7} = \dfrac{5}{x}$

16. $\dfrac{x+3}{2x} = 1$

17. $\dfrac{1}{2x} + \dfrac{2}{5x} = \dfrac{9}{10}$

18. $\dfrac{1}{x-1} + \dfrac{2}{x+2} = \dfrac{3}{2}$

19. $\dfrac{1}{x^2-1} - \dfrac{4}{x+1} = \dfrac{3}{x-1}$

20. $\dfrac{1}{x^2 - 4x} + \dfrac{2}{x^2 - 16} = \dfrac{2}{x^2 + 4x}$

21. $\dfrac{x}{2x-1} = \dfrac{1-x}{2x-1}$

22. $\dfrac{x}{x-5} + \dfrac{x}{x+5} = \dfrac{10x}{x^2-25}$

Exercises 23 and 24: Solve the equation for the specified variable.

23. $y = \dfrac{2}{3x-5}$ for x

24. $\dfrac{a+b}{ab} = 1$ for b

25. Suppose that y is directly proportional to x.
 (a) If $y = 14$ when $x = 4$, find k so that $y = kx$.
 (b) Then use $y = kx$ to find y for $x = 6$.

26. Use the table to determine whether y varies directly or inversely with x. Find the constant of variation.

x	2	4	8	16
y	16	8	4	2

27. *Emptying a Swimming Pool*   It takes a large pump 40 hours to empty a swimming pool, whereas a small pump can empty the pool in 60 hours. How long will it take to empty the pool if both pumps are used?

28. *Height of a Building*   A 5-foot-tall post has a 4-foot-long shadow, and a nearby building has a 54-foot-long shadow. Estimate the height of the building.

29. *Standing in Line*   A department store clerk can wait on 30 customers per hour. If people arrive randomly at an average rate of x per hour, then the average number of customers N waiting in line is given by $N = \dfrac{x^2}{900 - 30x}$, for $x < 30$. Evaluate the expression for $x = 24$ and interpret your result.

CHAPTER 7   EXTENDED AND DISCOVERY EXERCISES

1. *Graph of a Rational Function*   A car wash can clean 15 cars per hour. If cars arrive randomly at an average rate of x per hour, the average number N of cars waiting in line is given by

$$N = \frac{x^2}{225 - 15x},$$

where $x < 15$.   (**Source:** N. Garber and L. Hoel, *Traffic and Highway Engineering.*)

(a) Complete the table.

x	3	9	12	13	14
N					

(b) For what value of x is N undefined?

(c) Plot the points from the table. Then graph $x = 15$ as a vertical, dashed line.

(d) Sketch a graph of N that passes through these points. Do not allow your graph to cross the vertical, dashed line, called an asymptote.

(e) Use the graph to explain why a small increase in x can sometimes lead to a long wait.

(f) Explain what happens over a long period of time if the arrival rate x exceeds 15 cars per hour. (*Hint:* The formula is not valid for $x \geq 15$.)

Exercises 2–6: Graphing Rational Functions   *Complete the following.*

(a) *Use the given equation to complete the table of values for y.*

x	−4	−3	−2	−1	0	1	2	3	4
y									

(b) *Determine any x-value that will make the expression undefined.*

(c) *Sketch a dashed, vertical line (asymptote) in the xy-plane at any undefined values of x.*

(d) *Plot the points from the table.*

(e) *Sketch a graph of the equation. Do not let your graph cross the vertical, dashed line.*

2. $y = \dfrac{1}{x - 1}$ **3.** $y = \dfrac{1}{x + 1}$

4. $y = \dfrac{4}{x^2 + 1}$ **5.** $y = \dfrac{x}{x + 1}$

6. $y = \dfrac{x}{x - 1}$

CHAPTERS 1–7   CUMULATIVE REVIEW EXERCISES

1. Evaluate $\pi r^2 h$ when $r = 2$ and $h = 6$.

2. Translate the phrase "two less than twice a number" into an algebraic expression using the variable x.

3. Find the reciprocal of $-\frac{8}{5}$.

Exercises 4 and 5: Evaluate and simplify to lowest terms.

4. $\frac{1}{2} \div \frac{5}{4}$ **5.** $\frac{5}{8} + \frac{1}{8}$

6. Simplify $\dfrac{4x}{9y} \div \dfrac{6x}{3y}$.

Exercises 7 and 8: Simplify the expression.

7. $-2 + 7x + 4 - 5x$

8. $-4(4 - y) + (5 - 3y)$

9. Solve $-2x + 11 = 13$ and check the solution.

10. Convert 0.045 to a percentage.

11. Solve $V = 6LW$ for W.

12. Solve $-3x + 1 \geq x$.

Exercises 13 and 14: Graph the equation. Determine any intercepts.

13. $2x - 3y = 6$ **14.** $x = 1$

15. Sketch a line with slope $m = 3$ passing through $(-2, -1)$. Write its slope–intercept form.

16. The table lists points located on a line. Write the slope–intercept form of the line.

x	−2	−1	0	1
y	−5	−3	−1	1

Exercises 17 and 18: Find the slope–intercept form for the line satisfying the given conditions.

17. Parallel to $y = -\frac{2}{3}x + 1$ passing through $(2, -1)$

18. Passing through $(-1, 2)$ and $(2, 4)$

19. A fish population is initially 2000 and is increasing at 200 per month. Write an equation that gives the number of fish N after x months.

20. Determine which ordered pair is a solution to the system of equations: $(2, -6)$ or $(1, -2)$.

$$4x + y = 2$$
$$x - 4y = 9$$

21. A table listing solutions for the equations $y = 2x$ and $y = 3 - x$ is given. Identify the solution to both equations.

x	−1	0	1	2
y = 2x	−2	0	2	4
y = 3 − x	4	3	2	1

22. Solve the system of equations.

$$-3r - t = 2$$
$$2r + t = -4$$

Exercises 23 and 24: Determine if the system of linear equations has no solutions, one solution, or infinitely many solutions.

23. $2x - y = 5$
    $-2x + y = -5$

24. $4x - 6y = 12$
    $-6x + 9y = 18$

25. Solve the system of equations symbolically, graphically, and numerically.

$$-2x + y = 0$$
$$x + y = 3$$

26. Use the graph to identify one solution to $2x + y < 1$. *Answers may vary.* Then shade the solution set.

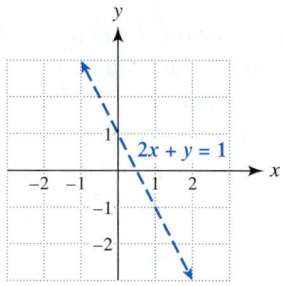

Exercises 27–32: Simplify the expression.

27. $(5x^2 - 3) + (-x^2 + 4)$

28. $3z^2 \cdot 5z^6$ 29. $(ab)^3$

30. $(2y - 3)(5y + 2)$ 31. $(x^2 - y^2)^2$

32. $(2t + 5)^2$

Exercises 33–36: Simplify and write the expression using positive exponents.

33. $2^{-4} \cdot 2^5$ 34. $\dfrac{1}{3^{-2}}$

35. $(3x^2)^{-3}$ 36. $\dfrac{4x^2}{2x^4}$

37. Write 0.00123 in scientific notation.

38. Divide and check: $\dfrac{2x^2 - x + 3}{x - 1}$.

Exercises 39 and 40: Identify the greatest common factor for each expression. Then factor the expression.

39. $20z^3 - 15z^2$ 40. $12x^2y + 15xy^2$

Exercises 41–46: Factor completely.

41. $6 + 13x - 5x^2$ 42. $9z^2 - 4$

43. $t^2 + 16t + 64$ 44. $x^3 - 16x$

45. $x^3 + 2x^2 - 99x$ 46. $a^2 + 6ab + 9b^2$

Exercises 47–50: Solve the equation.

47. $(2x - 7)(x + 5) = 0$ 48. $2x^2 - 4x = 0$

49. $y^2 + 5y - 14 = 0$ 50. $x^3 = 4x$

51. Evaluate $\frac{1}{x-2}$ at $x = 3$. For what value of x is the expression undefined?

52. Simplify $\frac{x^2 + 2x + 1}{x^2 - 1}$ to lowest terms.

Exercises 53–56: Simplify completely.

53. $\frac{x}{x+1} + \frac{1}{x+1}$

54. $\frac{1}{x-1} - \frac{1}{x^2 - 1}$

55. $\frac{3x}{4y} \cdot \frac{y}{9x^2}$

56. $\frac{x}{x^2 - 4} \div \frac{2x}{x-2}$

57. Simplify $\dfrac{1 + \frac{2}{x}}{1 - \frac{2}{x}}$.

Exercises 58–60: Solve.

58. $\frac{5}{x} = \frac{7}{8}$

59. $\frac{4}{3x} - \frac{3}{4x} = 1$

60. $\frac{1}{x-1} + \frac{2}{x+2} = \frac{3}{2}$

61. Solve $z = 3x - 2y$ for x.

62. Suppose that y is directly proportional to x and that $y = 7$ when $x = 14$. Find y when $x = 11$.

63. Suppose that y is inversely proportional to x and that $y = 8$ when $x = 20$. Find y when $x = 2$.

64. Determine whether the data in the table represent direct or inverse variation. Find an equation that models the data.

x	1	2	4	10
y	20	10	5	2

APPLICATIONS

65. *Shoveling the Driveway*   Two people are shoveling snow from a driveway. The first person shovels 12 square feet per minute, while the second person shovels 9 square feet per minute.
(a) Write a simplified expression that gives the total square feet that the two people shovel in x minutes.

(b) How long would it take them to clear a driveway with 1890 square feet?

66. *Burning Calories*   An athlete can burn 10 calories per minute while running and 4 calories per minute while walking. If the athlete burns 450 calories in 60 minutes, how long is spent on each activity?

67. *Head Start*   In 1980 there were 376 thousand children participating in the federal Head Start program. In 2000 this number increased to 858 thousand students.   (*Source:* Department of Health and Human Services.)
(a) Find the slope of the line passing through the points (1980, 376) and (2000, 858).
(b) Interpret the slope as a rate of change.
(c) If trends continue, estimate the number of participants in 2010.

68. *Triangle*   In an isosceles triangle, the measures of the two smaller angles are equal and their sum is 32° more than the largest angle.
(a) Let x be the measure of one of the two smaller angles and y be the measure of the largest angle. Write a system of linear equations whose solution gives the measures of these angles.
(b) Solve your system.

69. *Energy*   The U.S. consumption of natural gas G in trillions of cubic feet from 1990 to 2000 can be approximated by

$$G = -0.036x^2 + 0.76x + 18.7,$$

where $x = 0$ corresponds to 1990, $x = 1$ to 1991, and so on.   (*Source:* Department of Energy.)
(a) Evaluate this polynomial when $x = 5$ and interpret the answer.
(b) In what year was the natural gas consumption 22.7 trillion cubic feet?

70. *Strength of a Beam*   The strength of a wood beam varies inversely with its length. A beam that is 10 feet long can support 1100 pounds. How much weight can a similar beam support if it is 22 feet long?

CHAPTER 8

Introduction to Functions

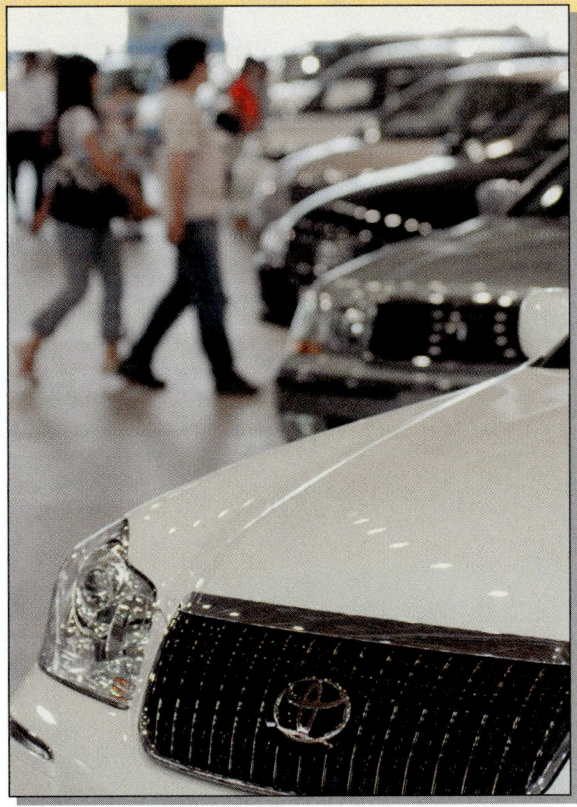

Every day our society creates enormous amounts of data, and mathematics is an important tool for summarizing those data and discovering trends. For example, the table shows the number of Toyota vehicles sold in the United States for selected years.

Year	1998	1999	2000	2001	2002
Vehicles (millions)	1.4	1.5	1.6	1.7	1.8

Source: Autodata.

These data contain an obvious pattern: Sales increased by 0.1 million each year. A scatterplot of these data and a line that models this situation are shown in the figure. In this chapter you will learn how to model these data with a function. (See Section 8.2, Example 6.)

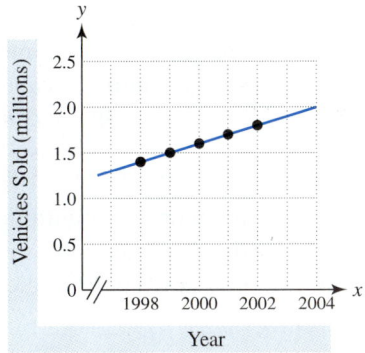

No great thing is created suddenly.

—EPICTETUS

8.1 FUNCTIONS AND THEIR REPRESENTATIONS

Basic Concepts ■ Representations of a Function ■ Definition of a Function ■ Identifying a Function ■ Graphing Calculators (Optional)

A LOOK INTO MATH ▷

In earlier chapters we showed how to use numbers to describe data. For example, instead of simply saying that it is *hot* outside, we might use the number 102°F to describe the temperature. We also showed that data can be modeled with formulas and graphs. Formulas and graphs are sometimes used to represent *functions*, which are important in mathematics. In this section we introduce functions and their representations.

Basic Concepts

▶ **REAL-WORLD CONNECTION** Functions are used to calculate many important quantities. For example, suppose that a person works for $7 per hour. Then we could use a function f to calculate the amount of money the person earned after working x hours simply by multiplying the *input* x by 7. The result y is called the *output*. This concept is shown visually in the following diagram.

For each valid input x, a function computes *exactly one* output y, which may be represented by the ordered pair (x, y). If the input is 5 hours, f outputs $7 \cdot 5 = \$35$; if the input is 8 hours, f outputs $7 \cdot 8 = \$56$. These results can be represented by the ordered pairs (5, 35) and (8, 56). Sometimes an input may not be valid. For example, if $x = -3$, there is no reasonable output because a person cannot work -3 hours.

We say that *y is a function of x* because the output y is determined by and *depends* on the input x. As a result, y is called the *dependent variable* and x is the *independent variable*. To emphasize that y is a function of x, we use the notation $y = f(x)$. The symbol $f(x)$ does not represent multiplication of a variable f and a variable x. The notation $y = f(x)$ is called *function notation*, is read "y equals f of x," and means that function f with input x produces output y. For example, if $x = 3$ hours, $y = f(3) = \$21$.

FUNCTION NOTATION

The notation $y = f(x)$ is called **function notation**. The **input** is x, the **output** is y, and the *name* of the function is f.

$$\text{Name}$$
$$\downarrow$$
$$y = f(x)$$
$$\nearrow \qquad \nwarrow$$
$$\text{Output} \qquad \text{Input}$$

The variable y is called the **dependent variable** and the variable x is called the **independent variable**. The expression $f(4) = 28$ is read "f of 4 equals 28" and indicates that f outputs 28 when the input is 4. A function computes *exactly one* output for each valid input. The letters f, g, and h are often used to denote names of functions.

Input x

$9 = g(4)$
Output y

$y = x + 5$

Figure 8.1 Function Machine

▶ **REAL-WORLD CONNECTION**   Functions can be used to compute a variety of quantities. For example, suppose that a boy has a sister who is exactly 5 years older than he is. If the age of the boy is x, then a function g can calculate the age of his sister by adding 5 to x. Thus $g(4) = 4 + 5 = 9$, $g(10) = 10 + 5 = 15$, and in general $g(x) = x + 5$. That is, function g adds 5 to every input x to obtain the output $y = g(x)$.

Functions can be represented by an input–output machine, as illustrated in Figure 8.1. This machine represents function g and receives input $x = 4$, adds 5 to this value, and then outputs $g(4) = 4 + 5 = 9$.

Representations of a Function

▶ **REAL-WORLD CONNECTION**   A function f forms a relation between inputs x and outputs y that can be represented verbally, numerically, symbolically, and graphically. Functions can also be represented with diagrams. We begin by considering a function f that converts yards to feet.

TABLE 8.1

x (yards)	y (feet)
1	3
2	6
3	9
4	12
5	15
6	18
7	21

VERBAL REPRESENTATION (WORDS)   To convert x yards to y feet we must multiply x by 3. Therefore, if function f computes the number of feet in x yards, a **verbal representation** of f is "Multiply the input x in yards by 3 to obtain the output y in feet."

NUMERICAL REPRESENTATION (TABLE OF VALUES)   A function f that converts yards to feet is shown in Table 8.1, where $y = f(x)$.

A *table of values* is called a **numerical representation** of a function. Many times it is impossible to list all valid inputs x in a table. On the one hand, if a table does not contain every x-input, it is a *partial* numerical representation. On the other hand, a *complete* numerical representation includes *all* valid inputs. Table 8.1 is a partial numerical representation of f because many valid inputs, such as $x = 10$ or $x = 5.3$, are not shown in it. Note that for each valid input x there is exactly one output y. *For a function, inputs are not listed more than once in a table.*

SYMBOLIC REPRESENTATION (FORMULA)   A *formula* provides a **symbolic representation** of a function. The computation performed by f to convert x yards to y feet is expressed by $y = 3x$. A formula for f is $f(x) = 3x$, where $y = f(x)$. We say that function f is *defined by* or *given by* $f(x) = 3x$. Thus $f(2) = 3 \cdot 2 = 6$.

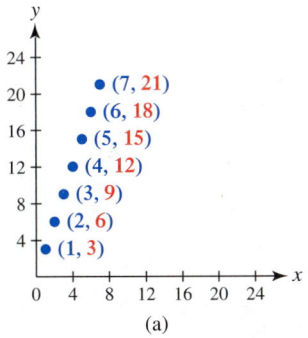

(a)

Figure 8.2

GRAPHICAL REPRESENTATION (GRAPH)   A **graphical representation**, or **graph**, visually associates an x-input with a y-output. The ordered pairs

$$(1, 3), (2, 6), (3, 9), (4, 12), (5, 15), (6, 18), \text{ and } (7, 21)$$

from Table 8.1 are plotted in Figure 8.2(a). This scatterplot suggests a line for the graph f. For each real number x there is exactly one real number y determined by $y = 3x$. If we restrict inputs to $x \geq 0$ and plot all ordered pairs $(x, 3x)$, then a line with no breaks will appear, as shown in Figure 8.2(b) on the next page.

(Figure 8.2)

(a) Function

(b) Function

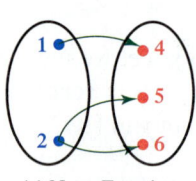

(c) Not a Function

Figure 8.3

MAKING CONNECTIONS

Functions, Points, and Graphs

If $f(a) = b$, then the point (a, b) lies on the graph of f. Conversely, if the point (a, b) lies on the graph of f, then $f(a) = b$. Thus each point on the graph of f can be written in the form $(a, f(a))$.

DIAGRAMMATIC REPRESENTATION (DIAGRAM)   Functions may be represented by **diagrams**. Figure 8.3(a) is a diagram of a function, where an arrow is used to identify the output y associated with input x. For example, input **2** results in output **6**, which is written in function notation as $f(2) = 6$. That is, **2** yards are equivalent to **6** feet. Figure 8.3(b) shows a function f even though $f(1) = 4$ and $f(2) = 4$. Although two inputs for f have the same output, each valid input has exactly one output. In contrast, Figure 8.3(c) shows a relation that is not a function because input 2 results in two different outputs, 5 and 6.

MAKING CONNECTIONS

Four Representations of a Function

Symbolic Representation $f(x) = x + 1$

Numerical Representation

x	y
-2	-1
-1	0
0	1
1	2
2	3

Graphical Representation

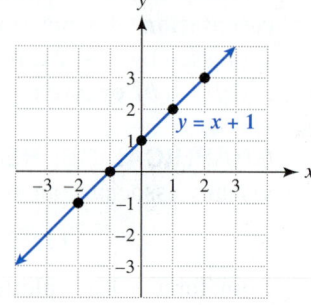

Verbal Representation f adds 1 to an input x to produce an output y.

▶ **REAL-WORLD CONNECTION**   In the next example we calculate sales tax by evaluating different representations of a function.

EXAMPLE 1   Calculating sales tax

Let a function f compute a sales tax of 7% on a purchase of x dollars. Use the given representation to evaluate $f(2)$.

(a) *Verbal Representation*   Multiply a purchase of x dollars by 0.07 to obtain a sales tax of y dollars.

(b) *Numerical Representation*   Shown in Table 8.2

(c) *Symbolic Representation* $f(x) = 0.07x$

(d) *Graphical Representation*   Shown in Figure 8.4

(e) *Diagrammatic Representation*   Shown in Figure 8.5

TABLE 8.2

x	$f(x)$
$1.00	$0.07
$2.00	$0.14
$3.00	$0.21
$4.00	$0.28

Figure 8.4 Sales Tax of 7%

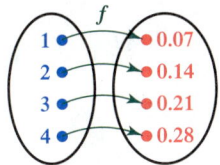

Figure 8.5

Solution

(a) Multiply the input 2 by 0.07 to obtain 0.14. The sales tax on $2.00 is $0.14.

(b) From Table 8.2, $f(2) = \$0.14$.

(c) Because $f(x) = 0.07x, f(2) = 0.07(2) = \textbf{0.14}$, or $0.14.

(d) To evaluate $f(2)$ with a graph, first find 2 on the x-axis. Then move vertically upward until you reach the graph of f. The point on the graph may be estimated as $(2, \textbf{0.14})$, meaning that $f(2) = \textbf{0.14}$ (see Figure 8.6). Note that it may not be possible to find the exact answer from a graph. For example, one might estimate $f(2)$ to be 0.13 or 0.15 instead of 0.14.

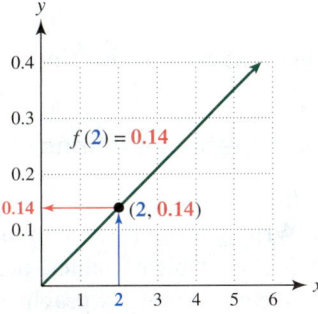

Figure 8.6

(e) In Figure 8.5, follow the arrow from **2** to **0.14**. Thus $f(2) = \textbf{0.14}$.

Now Try Exercises   19, 31, 51, 57, 59

EXAMPLE **2**   **Evaluating symbolic representations (formulas)**

Evaluate each function f at the given value of x.
(a) $f(x) = 3x - 7$ $\qquad x = -2$
(b) $f(x) = \dfrac{x}{x + 2}$ $\qquad x = 0.5$
(c) $f(x) = \sqrt{x - 1}$ $\qquad x = 10$

Solution
(a) $f(-2) = 3(-2) - 7 = -6 - 7 = -13$
(b) $f(0.5) = \dfrac{0.5}{0.5 + 2} = \dfrac{0.5}{2.5} = 0.2$
(c) $f(10) = \sqrt{10 - 1} = \sqrt{9} = 3$ Now Try Exercises 21, 29

▶ **REAL-WORLD CONNECTION**   There are many examples of functions. To give more meaning to a function, sometimes we change both its name and its input variable. For instance, if we know the radius r of a circle, we can calculate its circumference by using $C(r) = 2\pi r$. The next example illustrates how functions are used in physical therapy.

EXAMPLE **3**   **Computing crutch length**

People who sustain leg injuries often require crutches. A proper crutch length can be estimated without using trial and error. The function L, given by $L(t) = 0.72t + 2$, outputs an appropriate crutch length in inches for a person t inches tall.   (**Source:** *Journal of the American Physical Therapy Association.*)
(a) Find $L(60)$ and interpret the result.
(b) If one person is 70 inches tall and another person is 71 inches tall, what should be the difference in their crutch lengths?

Solution
(a) $L(60) = 0.72(60) + 2 = 45.2$. Thus a person 60 inches tall needs crutches that are about 45.2 inches long.
(b) From the formula, $L(t) = 0.72t + 2$, we can see that each 1-inch increase in t results in a 0.72-inch increase in L. For example,

$$L(71) - L(70) = 53.12 - 52.4 = 0.72.$$ Now Try Exercise 77

In the next example we find a formula and then sketch a graph of a function.

EXAMPLE **4**   **Finding representations of a function**

Let function f square the input x and then subtract 1 to obtain the output y.
(a) Write a formula, or symbolic representation, for f.
(b) Make a table of values, or numerical representation, for f. Use $x = -2, -1, 0, 1, 2$.
(c) Sketch a graph, or graphical representation, of f.

Solution
(a) *Symbolic Representation*   If we square x and then subtract 1, we obtain $x^2 - 1$. Thus a formula for f is $f(x) = x^2 - 1$.

TABLE 8.3

x	$f(x)$
-2	3
-1	0
0	-1
1	0
2	3

(b) *Numerical Representation*   Make a table of values for $f(x)$, as shown in Table 8.3. For example,

$$f(-2) = (-2)^2 - 1 = 4 - 1 = 3.$$

(c) *Graphical Representation*   To obtain a graph of $f(x) = x^2 - 1$, plot the points from Table 8.3 and then connect them with a smooth curve, as shown in Figure 8.7. Note that we need to plot enough points so that we can determine the overall shape of the graph.

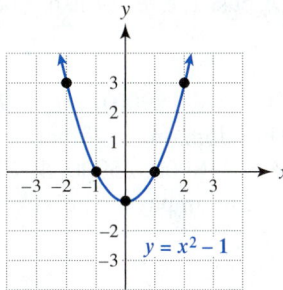

Figure 8.7

Now Try Exercise **61**

Definition of a Function

A function is a fundamental concept in mathematics. Its definition should allow for all representations of a function. *A function receives an input x and produces exactly one output y,* which can be expressed as an ordered pair:

$$(x, y).$$

Input   Output

A **relation** is a set of ordered pairs, and a function is a special type of relation.

> **FUNCTION**
>
> A **function** f is a set of ordered pairs (x, y), where each x-value corresponds to exactly one y-value.

The **domain** of f is the set of all x-values, and the **range** of f is the set of all y-values. For example, a function f that converts 1, 2, 3, and 4 yards to feet could be expressed as

$$f = \{(1, 3), (2, 6), (3, 9), (4, 12)\}.$$

The domain of f is $D = \{1, 2, 3, 4\}$, and the range of f is $R = \{3, 6, 9, 12\}$.

MAKING CONNECTIONS

Relations and Functions

A relation can be thought of as a set of input–output pairs. A function is a special type of relation whereby each input results in exactly one output.

▶ **REAL-WORLD CONNECTION**   In the next example, we see how education can improve a person's chances for earning a higher income.

EXAMPLE **5**   Computing average income

The function f computes the average 2004 individual income in dollars by educational attainment. This function is defined by $f(N) = 18{,}900$, $f(H) = 25{,}900$, $f(B) = 45{,}400$, and $f(M) = 62{,}300$, where N denotes no diploma, H a high school diploma, B a bachelor's degree, and M a master's degree.   (*Source:* U.S. Census Bureau.)

(a) Write f as a set of ordered pairs.
(b) Give the domain and range of f.
(c) Discuss the relationship between education and income.

Solution

(a) $f = \{(N, 18900), (H, 25900), (B, 45400), (M, 62300)\}$.
(b) The domain of function f is $D = \{N, H, B, M\}$, and the range of function f is $R = \{18900, 25900, 45400, 62300\}$.
(c) Education pays—the greater the educational attainment, the greater are annual earnings.

Now Try Exercise **37**

EXAMPLE **6**   Finding the domain and range graphically

Use the graphs of f shown in Figures 8.8 and 8.9 to find each function's domain and range.

(a)

Figure 8.8

(b)

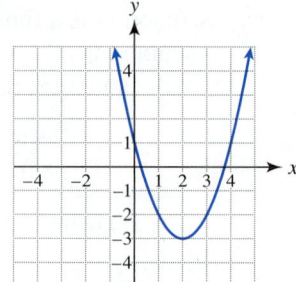

Figure 8.9

Solution

(a) The domain is the set of all x-values that correspond to points on the graph of f. Figure 8.10 shows that the domain D includes all x-values satisfying $-3 \le x \le 3$. (Recall that the symbol $\le$ is read "less than or equal to.") Because the graph is a semicircle with no breaks, the domain includes all real numbers between and including -3 and 3. The range R is the set of y-values that correspond to points on the graph of f. Thus R includes all y-values satisfying $0 \le y \le 3$.

Figure 8.10

(b) The arrows on the ends of the graph in Figure 8.9 indicate that the graph extends indefinitely left and right, as well as upward. Thus D includes all real numbers. The smallest y-value on the graph is $y = -3$, which occurs when $x = 2$. Thus the range is $y \geq -3$. (Recall that the symbol $\geq$ is read "greater than or equal to.")

Now Try Exercises 79, 83

Symbolic, numerical, and graphical representations of three common functions are shown in Figure 8.11. Note that their graphs are not lines. Use the graphs to find the domain and range of each function.

Absolute value: $f(x) = |x|$

x	-2	-1	0	1	2
$|x|$	2	1	0	1	2

Square: $f(x) = x^2$

x	-2	-1	0	1	2
x^2	4	1	0	1	4

Square root: $f(x) = \sqrt{x}$

x	0	1	4	9
$\sqrt{x}$	0	1	2	3

(a)

(b)

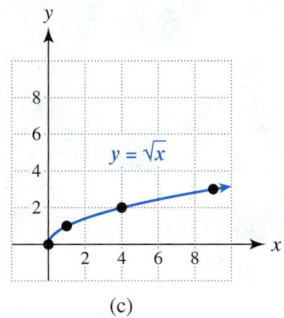

(c)

Figure 8.11

CRITICAL THINKING

Suppose that a car travels at 50 miles per hour to a city that is 250 miles away. Sketch a graph of a function f that gives the distance y traveled after x hours. Identify the domain and range of f.

The domain of a function is the set of all valid inputs. To determine the domain of a function from a formula, we must determine x-values for which the formula is defined. This concept is demonstrated in the next example.

EXAMPLE 7 Finding the domain of a function

Use $f(x)$ to find the domain of f.

(a) $f(x) = 5x$ **(b)** $f(x) = \dfrac{1}{x - 2}$ **(c)** $f(x) = \sqrt{x}$

Solution

(a) Because we can always multiply a real number x by 5, $f(x) = 5x$ is defined for all real numbers. Thus the domain of f includes all real numbers.

(b) Because we cannot divide by 0, input $x = 2$ is not valid for $f(x) = \frac{1}{x - 2}$. The expression for $f(x)$ is defined for all other values of x. Thus the domain of f includes all real numbers except 2, or $x \neq 2$.

(c) Because square roots of negative numbers are not real numbers, the inputs for $f(x) = \sqrt{x}$ cannot be negative. Thus the domain of f includes all nonnegative numbers, or $x \geq 0$.

Now Try Exercises 89, 93, 97

Identifying a Function

Recall that for a function each valid input x produces exactly one output y. In the next three examples we demonstrate techniques for identifying a function.

EXAMPLE 8

Determining whether a set of ordered pairs is a function

The set S of ordered pairs (x, y) represents the monthly average temperature y in degrees Fahrenheit for the month x in Washington, D.C. Determine whether S is a function.

$S = \{$(January, 33), (February, 37), (March, 45), (April, 53), (May, 66), (June, 73),
      (July, 77), (August, 77), (September, 70), (October, 51), (November, 48),
      (December, 37)$\}$

(**Source:** A. Miller and J. Thompson, *Elements of Meteorology*.)

Solution

The input x is the month and the output y is the monthly average temperature. The set S *is* a function because each month x is paired with exactly one monthly average temperature y. Note that even though an average temperature of 37°F occurs in both February and December, S is nonetheless a function.

Now Try Exercise 121

EXAMPLE 9

Determining whether a table of values represents a function

TABLE 8.4

Determine whether Table 8.4 represents a function.

x	y
1	−4
2	8
3	2
1	5
4	−6

Solution

The table does not represent a function because input $x = 1$ produces two outputs: −4 and 5. That is, the following two ordered pairs both belong to this relation.

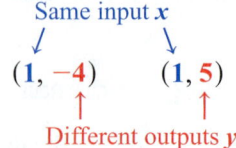

Same input x

$(1, -4)$ $(1, 5)$

Different outputs y

Now Try Exercise 123

VERTICAL LINE TEST   To determine whether a graph represents a function, we must be convinced that it is impossible for an input x to have two or more outputs y. If two distinct points have the same x-coordinate on a graph, then the graph cannot represent a function. For example, the ordered pairs $(-1, 1)$ and $(-1, -1)$ could not lie on the graph of a function because input -1 results in *two* outputs: 1 and -1. When the points $(-1, 1)$ and $(-1, -1)$ are plotted, they lie on the same vertical line, as shown in Figure 8.12(a). A graph passing through these points intersects the vertical line twice, as illustrated in Figure 8.12(b).

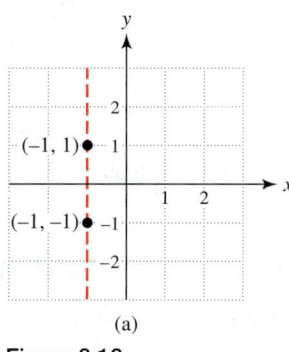

(a)   (b)

Figure 8.12

To determine whether a graph represents a function, visualize vertical lines moving across the *xy*-plane. If each vertical line intersects the graph *at most once*, then it is a graph of a function. This test is called the **vertical line test**. Note that the graph in Figure 8.12(b) fails the vertical line test and therefore does not represent a function.

VERTICAL LINE TEST

If every vertical line intersects a graph at no more than one point, then the graph represents a function.

EXAMPLE 10 **Determining whether a graph represents a function**

Determine whether the graphs shown in Figures 8.13 and 8.14 represent functions.

(a)

Figure 8.13

(b)

Figure 8.14

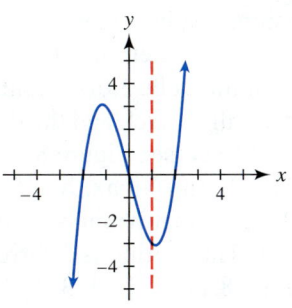

Figure 8.15 Passes Vertical Line Test

Solution

(a) Any vertical line will cross the graph at most once, as depicted in Figure 8.15. Therefore the graph *does* represent a function.

(b) The graph *does not* represent a function because there exist vertical lines that can intersect the graph twice. One such line is shown in Figure 8.16 on the next page.

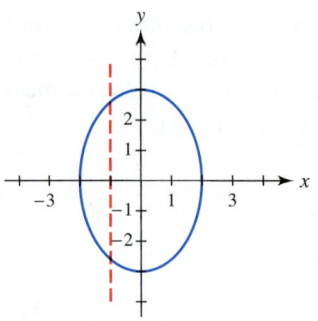

Figure 8.16 Fails Vertical Line Test

Now Try Exercises 113, 115

Graphing Calculators (Optional)

Graphing calculators provide several features beyond those found on scientific calculators. Graphing calculators have keys that can be used to create tables, scatterplots, and graphs.

▶ **REAL-WORLD CONNECTION** The **viewing rectangle**, or **window**, on a graphing calculator is similar to the viewfinder in a camera. A camera cannot take a picture of an entire scene. The camera must be centered on some object and can photograph only a portion of a scene. A camera can capture different views of the same scene by zooming in and out, as can graphing calculators. The xy-plane is infinite, but the calculator screen can show only a finite, rectangular region of the xy-plane. The viewing rectangle must be specified by setting minimum and maximum values for both the x- and y-axes before a graph can be drawn.

Figure 8.17

We use the following terminology regarding the size of a viewing rectangle. **Xmin** is the minimum x-value along the x-axis, and **Xmax** is the maximum x-value. Similarly, **Ymin** is the minimum y-value along the y-axis, and **Ymax** is the maximum y-value. Most graphs show an x-scale and a y-scale with tick marks on the respective axes. Sometimes the distance between consecutive tick marks is 1 unit, but at other times it might be 5 or 10 units. The distance represented by consecutive tick marks on the x-axis is called **Xscl**, and the distance represented by consecutive tick marks on the y-axis is called **Yscl** (see Figure 8.17).

This information about the viewing rectangle can be written as [Xmin, Xmax, Xscl] by [Ymin, Ymax, Yscl]. For example, $[-10, 10, 1]$ by $[-10, 10, 1]$ means that Xmin $= -10$, Xmax $= 10$, Xscl $= 1$, Ymin $= -10$, Ymax $= 10$, and Yscl $= 1$. This setting is referred to as the **standard viewing rectangle**. The window in Figure 8.17 is $[-3, 3, 1]$ by $[-3, 3, 1]$.

EXAMPLE 11 **Setting the viewing rectangle**

Show the viewing rectangle $[-2, 3, 0.5]$ by $[-100, 200, 50]$ on your calculator. How many tick marks are there on the positive x-axis?

Solution

The window setting and viewing rectangle are displayed in Figure 8.18. Note that in Figure 8.18(b) there are 6 tick marks on the positive x-axis because its length is 3 units and the distance between consecutive tick marks is 0.5 unit.

CALCULATOR HELP

To set a viewing rectangle, see Appendix A (page AP-4).

(a)

$[-2, 3, 0.5]$ by $[-100, 200, 50]$
(b)

Figure 8.18

Now Try Exercise 129

SCATTERPLOTS Many graphing calculators have the capability to create scatterplots. The next example illustrates how to make a scatterplot with a graphing calculator.

NOTE: Refer to Appendix A for help with keystrokes.

EXAMPLE 12 **Making a scatterplot with a graphing calculator**

Plot the points $(-2, -2)$, $(-1, 3)$, $(1, 2)$, and $(2, -3)$ in $[-4, 4, 1]$ by $[-4, 4, 1]$.

Solution

We entered the points $(-2, -2)$, $(-1, 3)$, $(1, 2)$, and $(2, -3)$, as shown in Figure 8.19(a), using the STAT EDIT feature. The variable L1 represents the list of x-values, and the variable L2 represents the list of y-values. In Figure 8.19(b) we set the graphing calculator to make a scatterplot with the STATPLOT feature, and in Figure 8.19(c) the points have been plotted. If you have a different model of calculator, you may need to consult your owner's manual.

CALCULATOR HELP

To make a scatterplot, see Appendix A (page AP-4).

(a)

(b)

$[-4, 4, 1]$ by $[-4, 4, 1]$
(c)

Figure 8.19

Now Try Exercise 133

GRAPHS AND TABLES We can also use graphing calculators to create graphs and tables, usually more efficiently and reliably than with pencil-and-paper techniques. However, a graphing calculator uses the same techniques that we might use to sketch a graph. For example, one way to sketch a graph of $y = 2x - 1$ is first to make a table of values, as shown in Table 8.5.

TABLE 8.5

x	−1	0	1	2
y	−3	−1	1	3

We can plot these points in the xy-plane, as shown in Figure 8.20(a). Next we might connect the points, as shown in Figure 8.20(b).

(a) Plotting Points

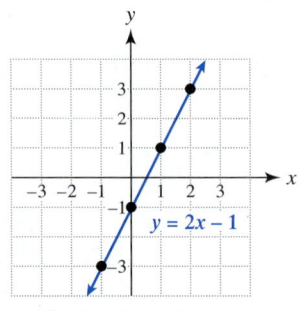
(b) Graphing a Line

Figure 8.20

In a similar manner a graphing calculator plots numerous points and connects them to make a graph. To create a similar graph with a graphing calculator, we enter the formula $Y_1 = 2X - 1$, set an appropriate viewing rectangle, and graph, as shown in Figures 8.21(a) and (b). A table of values can also be generated, as illustrated in Figure 8.21(c).

CALCULATOR HELP

To make a graph, see Appendix A (page AP-5). To make a table, see Appendix A (page AP-3).

$[-10, 10, 1]$ by $[-10, 10, 1]$

(a)

(b)

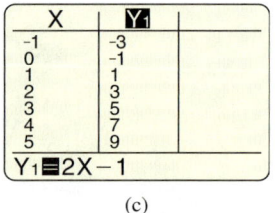
(c)

Figure 8.21

8.1   PUTTING IT ALL TOGETHER

One important concept in mathematics is that of a function. A function calculates exactly one output for each valid input and produces input–output ordered pairs in the form (x, y). A function typically computes something, such as area, speed, or sales tax. The following table summarizes some concepts related to functions.

Concept	Explanation	Examples
Function	A set of ordered pairs (x, y), where each x-value corresponds to exactly one y-value	$f = \{(1, 3), (2, 3), (3, 1)\}$ $f(x) = 2x$ A graph of $y = x + 2$ A table of values for $y = 4x$
Independent Variable	The *input* variable for a function	*Function* *Independent Variable* $f(x) = 2x$ x $A(r) = \pi r^2$ r $V(s) = s^3$ s
Dependent Variable	The *output* variable of a function. There is exactly one output for each valid input.	*Function* *Dependent Variable* $y = f(x)$ y $T = F(r)$ T $V = g(r)$ V
Domain and Range of a Function	The domain D is the set of all valid inputs. The range R is the set of all outputs.	For $S = \{(-1, 0), (3, 4), (5, 0)\}$, $D = \{-1, 3, 5\}$ and $R = \{0, 4\}$. For $f(x) = \frac{1}{x}$ the domain includes all real numbers except 0, or $x \neq 0$.
Vertical Line Test	If every vertical line intersects a graph at no more than one point, the graph represents a function.	This graph does *not* pass this test and thus does not represent a function.
Scatterplot	A scatterplot results when individual points are plotted in the xy-plane.	

A function can be represented verbally, symbolically, numerically, and graphically. The following table summarizes these four representations.

continued on next page

continued from previous page

Type of Representation	Explanation	Comments
Verbal	Precise word description of what is computed	May be oral or written Must be stated *precisely*
Symbolic	Mathematical formula	Efficient and concise way of representing a function (e.g., $f(x) = 2x - 3$)
Numerical	List of specific inputs and their outputs	May be in the form of a table or an explicit set of ordered pairs
Graphical, diagrammatic	Shows inputs and outputs visually	No words, formulas, or tables Many types of graphs and diagrams are possible.

8.1 Exercises

MyMathLab

 PRACTICE   WATCH   DOWNLOAD   READ   REVIEW

CONCEPTS

1. The notation $y = f(x)$ is called _____ notation.

2. The notation $y = f(x)$ is read _____.

3. The notation $f(x) = x^2 + 1$ is a _____ representation of a function.

4. A table of values is a _____ representation of a function.

5. The set of valid inputs for a function is the _____.

6. The set of outputs for a function is the _____.

7. A function computes _____ output for each valid input.

8. What is the vertical line test used for?

9. Name four types of representations for a function.

10. If $f(3) = 4$, the point _____ is on the graph of f. If $(3, 6)$ is on the graph of f, $f(\underline{\quad}) = \underline{\quad}$.

11. **Thinking Generally**   If $f(a) = b$, the point _____ is on the graph of f.

12. **Thinking Generally**   If (c, d) is on the graph of g, then $g(c) = \underline{\quad}$.

13. **Thinking Generally**   If a is in the domain of f, then $f(a)$ represents how many outputs?

14. **Thinking Generally**   If $f(x) = x$ for every x in the domain of f, then the domain and range of f are _____.

Exercises 15–18: Determine whether the phrase describes a function.

15. Calculating the square of a number

16. Determining your age to the nearest whole number

17. Listing the students who passed a given math exam

18. Listing the children of parent x

REPRESENTING AND EVALUATING FUNCTIONS

Exercises 19–30: Evaluate $f(x)$ at the given values of x.

19. $f(x) = 4x - 2$ $x = -1, 0$

20. $f(x) = 5 - 3x$ $x = -4, 2$

21. $f(x) = \sqrt{x}$ $x = 0, \frac{9}{4}$

22. $f(x) = \sqrt[3]{x}$ $x = -1, 27$

23. $f(x) = x^2$ $x = -5, \frac{3}{2}$

24. $f(x) = x^3$ $\qquad$ $x = -2, 0.1$

25. $f(x) = 3$ $\qquad$ $x = -8, \frac{7}{3}$

26. $f(x) = 100$ $\qquad$ $x = -\pi, \frac{1}{3}$

27. $f(x) = 5 - x^3$ $\qquad$ $x = -2, 3$

28. $f(x) = x^2 + 5$ $\qquad$ $x = -\frac{1}{2}, 6$

29. $f(x) = \dfrac{2}{x + 1}$ $\qquad$ $x = -5, 4$

30. $f(x) = \dfrac{x}{x - 4}$ $\qquad$ $x = -3, 1$

Exercises 31–36: Do the following.
  (a) *Write a formula for the function described.*
  (b) *Evaluate the function for input 10 and interpret the results.*

31. Function I computes the number of inches in x yards.

32. Function M computes the number of miles in x feet.

33. Function A computes the area of a circle with radius r.

34. Function C computes the circumference of a circle with radius r.

35. Function A computes the square feet in x acres. (*Hint:* There are 43,560 square feet in one acre.)

36. Function K computes the number of kilograms in x pounds. (*Hint:* There are about 2.2 pounds in one kilogram.)

Exercises 37–40: Write each function f as a set of ordered pairs. Give the domain and range of f.

37. $f(1) = 3, f(2) = -4, f(3) = 0$

38. $f(-1) = 4, f(0) = 6, f(1) = 4$

39. $f(a) = b, f(c) = d, f(e) = a, f(d) = b$

40. $f(a) = 7, f(b) = 7, f(c) = 7, f(d) = 7$

Exercises 41–50: Sketch a graph of f.

41. $f(x) = -x + 3$ $\qquad$ **42.** $f(x) = -2x + 1$

43. $f(x) = 2x$ $\qquad$ **44.** $f(x) = \frac{1}{2}x - 2$

45. $f(x) = 4 - x$ $\qquad$ **46.** $f(x) = 6 - 3x$

47. $f(x) = x^2$ $\qquad$ **48.** $f(x) = \sqrt{x}$

49. $f(x) = \sqrt{x + 1}$ $\qquad$ **50.** $f(x) = \frac{1}{2}x^2 - 1$

Exercises 51–56: Use the graph of f to evaluate the given expressions.

51. $f(0)$ and $f(2)$ $\qquad$ **52.** $f(-2)$ and $f(2)$

 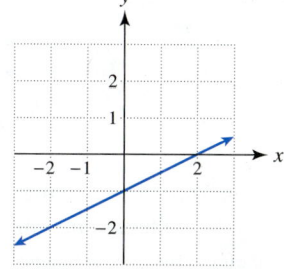

53. $f(-2)$ and $f(1)$ $\qquad$ **54.** $f(-1)$ and $f(0)$

 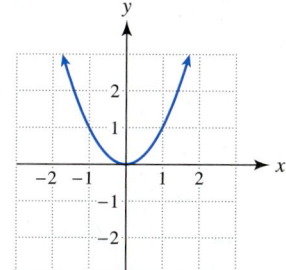

55. $f(1)$ and $f(2)$ $\qquad$ **56.** $f(-1)$ and $f(4)$

 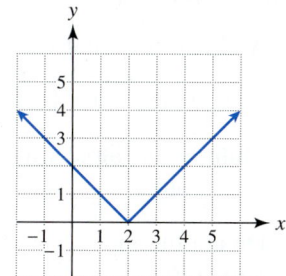

Exercises 57 and 58: Use the table to evaluate the given expressions.

57. $f(0)$ and $f(2)$

x	0	1	2	3	4
$f(x)$	5.5	4.3	3.7	2.5	1.9

58. $f(-10)$ and $f(5)$

x	-10	-5	0	5	10
$f(x)$	23	96	-45	-33	23

Exercises 59 and 60: Use the diagram to evaluate f(1990).
Interpret your answer.

59. The function f computes average fuel efficiency of new U.S. passenger cars in miles per gallon during year x.   (**Source:** Department of Transportation.)

60. The function f computes average cost of tuition at public colleges and universities during academic year x.
(**Source:** The College Board.)

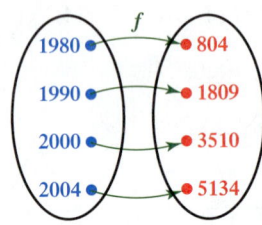

Exercises 61–64: Express the verbal representation for the function f numerically, symbolically, and graphically. Let $x = -3, -2, -1, \ldots, 3$ for the numerical representation (table), and let $-3 \le x \le 3$ for the graph.

61. Add 5 to the input x to obtain the output y.

62. Square the input x to obtain the output y.

63. Multiply the input x by 5 and then subtract 2 to obtain the output y.

64. Divide the input x by 2 and then add 3 to obtain the output y.

Exercises 65–68: Express the function f numerically and graphically. Let $x = -3, -2, -1, \ldots, 3$ for your numerical representation and use the standard window for your graph.

65. $f(x) = \sqrt{x + 3}$ **66.** $f(x) = x^3 - \frac{1}{2}x^2$

67. $f(x) = \dfrac{5 - x}{5 + x}$ **68.** $f(x) = |2 - x| + \sqrt[3]{x}$

Exercises 69–74: Give a verbal representation for $f(x)$.

69. $f(x) = x - \frac{1}{2}$ **70.** $f(x) = \frac{3}{4}x$

71. $f(x) = \dfrac{x}{3}$ **72.** $f(x) = x^2 + 1$

73. $f(x) = \sqrt{x - 1}$ **74.** $f(x) = 1 - 3x$

75. *Cost of Driving*   In 2005, the average cost of driving a new car in the United States was about 50 cents per mile. Symbolically, graphically, and numerically represent a function f that computes the cost in dollars of driving x miles. For the numerical representation (table) let $x = 10, 20, 30, \ldots, 70$.   (**Source:** Associated Press.)

76. *Federal Income Taxes*   In 2005, the lowest U.S. income tax rate was 15 percent. Symbolically, graphically, and numerically represent a function f that computes the tax on a taxable income of x dollars. For the numerical representation (table) let $x = 1000, 2000, 3000, \ldots, 7000$, and for the graphical representation let $0 \le x \le 30{,}000$.   (**Source:** Internal Revenue Service.)

77. *Home Prices*   The average price P of a single-family home in thousands of dollars from 1990 to 2000 can be approximated by $P(x) = 5.7(x - 1990) + 150$, where x is the year. Evaluate $P(1995)$ and interpret the result.   (**Source:** U.S. Census Bureau.)

78. *Median Family Income*   The median income I of a family in thousands of dollars from 2000 to 2005 can be approximated by $I(x) = 1.4(x - 2000) + 51$, where x is the year. Evaluate $P(2003)$ and interpret the result.   (**Source:** U.S. Census Bureau.)

IDENTIFYING DOMAINS AND RANGES

Exercises 79–86: Use the graph of f to estimate its domain and range.

79.

80.

81.

82.

83.

84.

85.

86.

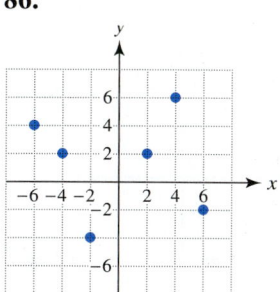

Exercises 87 and 88: Use the diagram to find the domain and range of f.

87.

88.

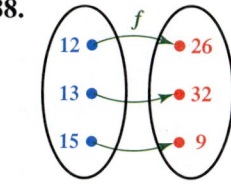

Exercises 89–102: Find the domain.

89. $f(x) = 10x$

90. $f(x) = 5 - x$

91. $f(x) = x^2 - 3$

92. $f(x) = \frac{1}{2}x^2$

93. $f(x) = \dfrac{3}{x - 5}$

94. $f(x) = \dfrac{x}{x + 1}$

95. $f(x) = \dfrac{2x}{x^2 + 1}$

96. $f(x) = \dfrac{6}{1 - x}$

97. $f(x) = \sqrt{x - 1}$

98. $f(x) = |x|$

99. $f(x) = |x - 5|$

100. $f(x) = \sqrt{2 - x}$

101. $f(x) = \dfrac{1}{x}$

102. $f(x) = 1 - 3x^2$

103. *Accidental Deaths* Function f computes the number of accidental deaths y per 100,000 people during year x. (**Source:** Department of Health and Human Services.)

$$f = \{(1910, 84.4), (1930, 80.5), (1950, 60.3),$$
$$(1970, 56.2), (1990, 36.9), (2000, 35.5)\}$$

(a) Evaluate $f(1950)$ and interpret the result.
(b) Identify the domain and range of f.
(c) Describe the trend in accidental deaths from 1910 through 2000.

104. *Motor Vehicle Registrations* The following table lists motor vehicle registrations y in millions during year x. Let $y = f(x)$.

x	1920	1940	1960	1980	2000
$f(x)$	9	32	74	156	216

Source: American Automobile Manufacturers Association.

(a) Evaluate $f(1940)$ and interpret the result.
(b) Identify the domain and range of f.
(c) Represent f with a diagram.

IDENTIFYING A FUNCTION

Exercises 105–108: Determine whether the diagram could represent a function.

105.

106.

107.

108.

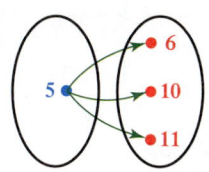

109. *Average Precipitation* The table lists the monthly average precipitation P in Las Vegas, Nevada, where $x = 1$ corresponds to January and $x = 12$ corresponds to December.

x (month)	1	2	3	4	5	6
P (inches)	0.5	0.4	0.4	0.2	0.2	0.1

x (month)	7	8	9	10	11	12
P (inches)	0.4	0.5	0.3	0.2	0.4	0.3

Source: J. Williams.

(a) Determine the value of P during May.
(b) Is P a function of x? Explain.
(c) If $P = 0.4$, find x.

110. *Wind Speeds* The table lists the monthly average wind speed W in Louisville, Kentucky, where $x = 1$ corresponds to January and $x = 12$ corresponds to December.

x (month)	1	2	3	4	5	6
W (mph)	10.4	12.7	10.4	10.4	8.1	8.1

x (month)	7	8	9	10	11	12
W (mph)	6.9	6.9	6.9	8.1	9.2	9.2

Source: J. Williams.

(a) Determine the month with the highest average wind speed.
(b) Is W a function of x? Explain.
(c) If $W = 6.9$, find x.

Exercises 111–120: Determine whether the graph represents a function. If it does, identify the domain and range.

111.

112.

113.

114.

115.

116.

117.

118.

119.

120.

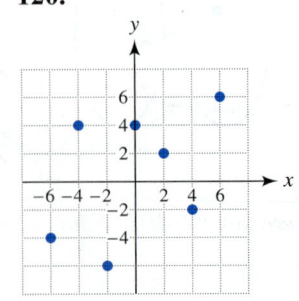

Exercises 121–124: Determine whether S is a function.

121. $S = \{(1, 2), (4, 5), (7, 8), (5, 4), (2, 2)\}$

122. $S = \{(4, 7), (-2, 1), (3, 8), (4, 9)\}$

123. *S* is given by the table.

x	5	10	5
y	2	1	0

124. *S* is given by the table.

x	−3	−2	−1
y	10	10	10

GRAPHICAL INTERPRETATION

Exercises 125 and 126: The graph represents the distance that a person is from home while walking on a straight path. The x-axis represents time and the y-axis represents distance. Interpret the graph.

125.   **126.**

GRAPHING CALCULATORS

 Exercises 127–132: Show the given viewing rectangle on your graphing calculator. Predict the number of tick marks on the positive x-axis and the positive y-axis.

127. Standard viewing rectangle

128. [−12, 12, 2] by [−8, 8, 2]

129. [0, 100, 10] by [−50, 50, 10]

130. [−30, 30, 5] by [−20, 20, 5]

131. [1980, 1995, 1] by [12000, 16000, 1000]

132. [1900, 1990, 10] by [1700, 2800, 100]

Exercises 133–138: Use your calculator to make a scatterplot of the relation after determining an appropriate viewing rectangle.

133. $\{(4, 3), (-2, 1), (-3, -3), (5, -2)\}$

134. $\{(5, 5), (2, 0), (-2, 7), (2, -8), (-1, -5)\}$

135. $\{(20, 40), (-25, -15), (-20, 25), (15, -25)\}$

136. $\{(-13, 12), (3, 10), (-15, -4), (12, -9)\}$

137. $\{(100, -100), (50, 200), (-150, -140),$ $(-30, 80)\}$

138. $\{(-125, 75), (45, 65), (-53, -67), (150, -80)\}$

WRITING ABOUT MATHEMATICS

139. Give an example of a function. Identify the domain and range of your function.

140. Explain in your own words what a function is. How is a function different from other relations?

141. Explain how to evaluate a function by using a graph. Give an example.

142. Give one difficulty that may occur when you use a table of values to evaluate a function.

GROUP ACTIVITY
WORKING WITH REAL DATA

Directions: Form a group of 2 to 4 people. Select someone to record the group's responses for this activity. All members of the group should work cooperatively to answer the questions. If your instructor asks for your results, each member of the group should be prepared to respond.

Large Tires on Vehicles Manufacturers are installing fewer small rims in favor of larger rims. One advantage of bigger rims is that they are better for braking. The following table lists the increase in 16-inch rim sales as a percentage of new vehicle sales for selected years from 1996 to 2002.

Sales of 16-Inch Rims

Year	1996	1998	2000	2002
Percentage	28	36	43	51

Source: USA Today.

continued on next page

continued from previous page

1. Plot the data in a scatterplot.

2. Calculate the slope between consecutive points in the table. Interpret each slope.

3. Find the equation of a line that models the data. (*Hint:* You may want to use the point–slope form, $y - y_1 = m(x - x_1)$.)

4. Does your line model the data exactly? Could a line model these data exactly? Explain.

5. Use your equation to write a function P that models these data.

6. Evaluate $P(1999)$ and interpret your result.

8.2 LINEAR FUNCTIONS

Basic Concepts ▪ **Representations of Linear Functions** ▪ **Modeling Data with Linear Functions** ▪ **The Midpoint Formula (Optional)**

A LOOK INTO MATH ▷ In mathematics, functions are used to model real-world phenomena, such as electricity, weather, and the economy. Because there are so many different applications of mathematics, a wide assortment of functions has been created. In fact, new functions are invented every day for use in business, education, and government. In this section we discuss an important type of function called a *linear function*.

Basic Concepts

▶ **REAL-WORLD CONNECTION** Suppose that the air conditioner is turned on when the temperature inside a house is 80°F. The resulting temperatures are listed in Table 8.6 for various elapsed times. Note that for each 1-hour increase in elapsed time, the temperature decreases by 2°F.

TABLE 8.6 House Temperature

1-hour increase

Elapsed Time (hours)	0	1	2	3	4	5
Temperature (°F)	80	78	76	74	72	70

2°F decrease

A scatterplot is shown in Figure 8.22, which suggests that a line models these data.

CALCULATOR HELP

To make a scatterplot, see Appendix A (page AP-4).

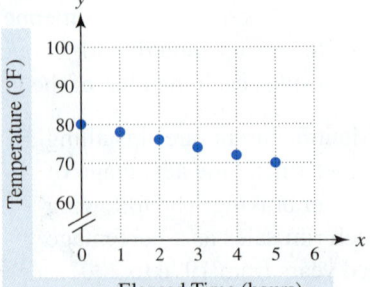

Figure 8.22 Temperature in a Home

Over this 5-hour period, the air conditioner lowers the temperature by 2°F for each hour that it runs. The temperature is found by multiplying the elapsed time x by -2 and adding the initial temperature of 80°F. This situation is modeled by $f(x) = -2x + 80$. For example,

$$f(\mathbf{2.5}) = -2(\mathbf{2.5}) + 80 = \mathbf{75}$$

means that the temperature is **75**°F after the air conditioner has run for **2.5** hours. A graph of $f(x) = -2x + 80$, where $0 \leq x \leq 5$, is shown in Figure 8.23. We call f a *linear function* because its graph is a *line*. If a function is *not* a linear function, then it is a **nonlinear function.**

Figure 8.23 A Linear Function

LINEAR FUNCTION

A function f defined by $f(x) = ax + b$, where a and b are constants, is a **linear function**.

For $f(x) = \mathbf{-2}x + \mathbf{80}$, we have $a = \mathbf{-2}$ and $b = \mathbf{80}$. The constant a represents the rate at which the air conditioner cools the building, and the constant b represents the initial temperature.

▶ **REAL-WORLD CONNECTION**   In general, a linear function defined by $f(x) = ax + b$ changes by a units for each unit increase in x. This **rate of change** is an increase if $a > 0$ and a decrease if $a < 0$. For example, if new carpet costs \$20 per square yard, then the linear function defined by $C(x) = 20x$ gives the cost of buying x square yards of carpet. The value of $a = 20$ gives the cost (rate of change) for each additional square yard of carpet. For function C, the value of b is 0 because it costs \$0 to buy 0 square yards of carpet.

NOTE: If f is a linear function, then $f(\mathbf{0}) = a(\mathbf{0}) + b = b$. Thus b can be found by evaluating $f(x)$ at $x = 0$.

EXAMPLE 1   **Identifying linear functions**

Determine whether f is a linear function. If f is a linear function, find values for a and b so that $f(x) = ax + b$.
(a) $f(x) = 4 - 3x$ **(b)** $f(x) = 8$ **(c)** $f(x) = 2x^2 + 8$

Solution
(a) Let $a = -3$ and $b = 4$. Then $f(x) = -3x + 4$, and f is a linear function.
(b) Let $a = 0$ and $b = 8$. Then $f(x) = 0x + 8$, and f is a linear function.
(c) Function f is not linear because its formula contains x^2. The formula for a linear function cannot contain an x with an exponent other than 1.   Now Try Exercises 9, 11, 13

EXAMPLE 2   **Determining linear functions**

Use each table of values to determine whether $f(x)$ could represent a linear function. If f could be linear, write a formula for f in the form $f(x) = ax + b$.

(a)

x	0	1	2	3
$f(x)$	10	15	20	25

(b)

x	-2	0	2	4
$f(x)$	4	2	0	-2

(c)

x	0	1	2	3
$f(x)$	1	2	4	7

Solution

(a) For each unit increase in x, $f(x)$ increases by 5 units so $f(x)$ could be linear with $a = 5$. Because $f(0) = 10$, $b = 10$. Thus $f(x) = 5x + 10$.

(b) For each 2-unit increase in x, $f(x)$ decreases by 2 units. Equivalently, each unit increase in x results in a 1-unit decrease in $f(x)$, so $f(x)$ could be linear with $a = -1$. Because $f(0) = 2$, $b = 2$. Thus $f(x) = -x + 2$.

(c) Each unit increase in x does not result in a constant change in $f(x)$. Thus $f(x)$ does not represent a linear function.   Now Try Exercises 21, 23, 25

Representations of Linear Functions

The graph of a linear function is a line. To graph a linear function f, we usually start by making a table of values and then plot three or more points. We can then sketch the graph of f by drawing a line through these points, as demonstrated in the next example.

EXAMPLE 3   **Graphing a linear function by hand**

Sketch a graph of $f(x) = x - 1$. Use the graph to evaluate $f(-2)$.

Solution

Begin by making a table of values containing at least three points. Pick convenient values of x, such as $x = -1, 0, 1$.

$$f(-1) = -1 - 1 = -2$$
$$f(0) = 0 - 1 = -1$$
$$f(1) = 1 - 1 = 0$$

Display the results, as shown in Table 8.7.

Plot the points $(-1, -2)$, $(0, -1)$, and $(1, 0)$. Then sketch a line through the points to obtain the graph of f. A graph of a line results when *infinitely* many points are plotted, as shown in Figure 8.24(a).

To evaluate $f(-2)$, first find $x = -2$ on the x-axis. See Figure 8.24(b). Then move downward to the graph of f. By moving across to the y-axis, we see that the corresponding y-value is -3. Thus $f(-2) = -3$.

TABLE 8.7

x	y
-1	-2
0	-1
1	0

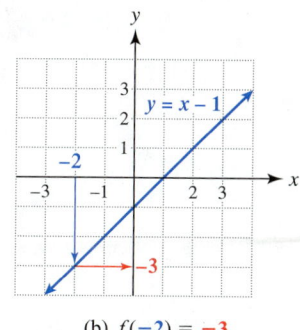

(a)

(b) $f(-2) = -3$

Figure 8.24

Now Try Exercises 35, 53

CRITICAL THINKING

Two points determine a line. Why is it a good idea to plot at least three points when graphing a linear function by hand?

EXAMPLE **4**   Representing a linear function

A linear function is given by $f(x) = -3x + 2$.
(a) Give a verbal representation of f.
(b) Make a numerical representation (table) of f by letting $x = -1, 0, 1$.
(c) Plot the points listed in the table from part (b). Then sketch a graph of $y = f(x)$.

Solution

TABLE 8.8

x	$f(x)$
-1	5
0	2
1	-1

(a) *Verbal Representation*   Multiply the input x by -3 and then add 2 to obtain the output.

(b) *Numerical Representation*   Evaluate the formula $f(x) = -3x + 2$ at $x = -1, 0, 1$, which results in Table 8.8. Note that $f(-1) = 5, f(0) = 2,$ and $f(1) = -1$.

(c) *Graphical Representation*   To make a graph of f by hand without a graphing calculator, plot the points $(-1, 5), (0, 2),$ and $(1, -1)$ from Table 8.8. Then draw a line passing through these points, as shown in Figure 8.25.

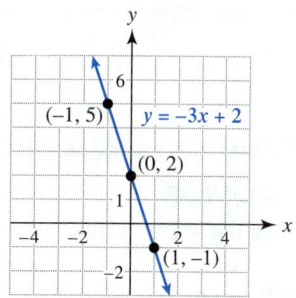

Figure 8.25

Now Try Exercise 67

In the next example a graphing calculator is used to create a graph and table.

EXAMPLE 5   **Using a graphing calculator**

Give numerical and graphical representations of $f(x) = \frac{1}{2}x - 2$.

Solution

Numerical Representation   To make a numerical representation, construct the table for $Y_1 = .5X - 2$, starting at $x = -3$ and incrementing by 1, as shown in Figure 8.26(a). (Other tables are possible.)

Graphical Representation   Graph Y_1 in the standard viewing rectangle, as shown in Figure 8.26(b). (Other viewing rectangles may be used.)

CALCULATOR HELP

To make a table, see Appendix A (page AP-3). To make a graph, see Appendix A (page AP-5).

$[-10, 10, 1]$ by $[-10, 10, 1]$

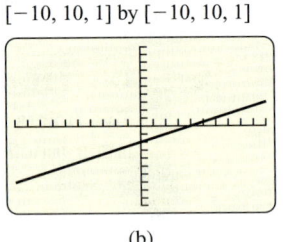

X	Y1
-3	-3.5
-2	-3
-1	-2.5
0	-2
1	-1.5
2	-1
3	-.5

$Y_1 \blacksquare .5X - 2$

(a)   (b)

Figure 8.26

Now Try Exercise 71

MAKING CONNECTIONS

Mathematics in Newspapers

Think of the mathematics that you see in newspapers. Often percentages are described *verbally*, numbers are displayed in *tables*, and data are shown in *graphs*. Seldom are *formulas* given, which is an important reason not to study only symbolic representations.

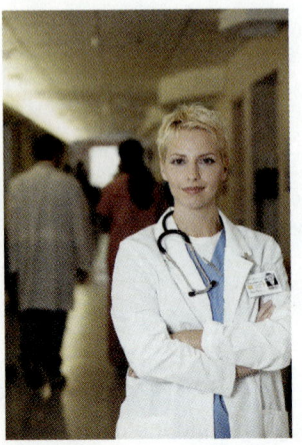

Modeling Data with Linear Functions

▶ **REAL-WORLD CONNECTION**   A distinguishing feature of a linear function is that when the input x increases by 1 unit, the output $f(x) = ax + b$ always changes by an amount equal to a. For example, the number of doctors in the private sector from 1970 to 2000 can be modeled by

$$f(x) = 15{,}260x - 29{,}729{,}000,$$

where x is the year. The value $a = 15{,}260$ indicates that the number of doctors has increased, on average, by 15,260 each year.   (***Source:*** American Medical Association.)

The following are other examples of quantities that are modeled by linear functions. Try to determine the value of the constant a.

- The wages earned by an individual working x hours at \$8 per hour
- The distance traveled by light in x seconds if the speed of light is 186,000 miles per second
- The cost of tuition and fees when registering for x credits if each credit costs \$80 and the fees are fixed at \$50

When we are modeling data with a linear function defined by $f(x) = ax + b$, the following concepts are helpful to determine a and b.

MODELING DATA WITH A LINEAR FUNCTION

The formula $f(x) = ax + b$ may be interpreted as follows.

$$f(x) \quad = \quad ax \quad + \quad b$$

(New amount) = (Change) + (Fixed amount)

When x represents time, *change* equals (rate of change) × (time).

$$f(x) \quad = \quad a \quad \times \quad x \quad + \quad b$$

(Future amount) = (Rate of change) × (Time) + (Initial amount)

▶ **REAL-WORLD CONNECTION**   These concepts are applied in the next three examples.

EXAMPLE 6   Modeling car sales

Table 8.9 shows numbers of Toyota vehicles sold in the United States. (Refer to the introduction to this chapter.)

TABLE 8.9   Toyota Vehicles Sold (millions)

Year	1998	1999	2000	2001	2002
Vehicles	1.4	1.5	1.6	1.7	1.8

Source: Autodata.

(a) What were the sales in 1998?
(b) What was the annual increase in sales?
(c) Find a linear function f that models these data. Let $x = 0$ correspond to 1998, $x = 1$ to 1999, and so on.
(d) Use f to predict sales in 2004.

Solution

(a) In 1998, 1.4 million vehicles were sold.
(b) Sales have increased by 0.1 million (100,000) vehicles per year. Because this rate of change is the same each year, we can model the data *exactly* with a linear function.
(c) Initial sales in 1998 ($x = 0$) were 1.4 million vehicles, and each year sales increased by 0.1 million vehicles. Thus

$$f(x) \quad = \quad 0.1 \quad \times \quad x \quad + \quad 1.4$$

(Future sales) = (Rate of change in sales) × (Time) + (Initial sales),

or $f(x) = 0.1x + 1.4$.

(d) Because $x = 6$ corresponds to 2004, evaluate $f(6)$.

$$f(6) = 0.1(6) + 1.4 = 2 \text{ million vehicles}$$

Note that the actual sales in 2004 were 2 million vehicles, so the estimate is accurate.

Now Try Exercise **111**

In the next example we model tuition and fees.

EXAMPLE **7**   Modeling the cost of tuition and fees

Suppose that tuition costs $80 per credit and that student fees are fixed at $50. Find a formula for a linear function that models tuition and fees.

Solution

Total cost is found by multiplying $80 (rate, or cost per credit) by the number of credits x and then adding the fixed fees (fixed amount) of $50. Thus $f(x) = 80x + 50$.

Now Try Exercise **62**

In the next example we consider a simple linear function that models the speed of a car.

EXAMPLE **8**   Modeling with a constant function

A car travels on a freeway with its speed recorded at regular intervals, as listed in Table 8.10.

TABLE 8.10   Speed of a Car

Elapsed Time (hours)	0	1	2	3	4
Speed (miles per hour)	70	70	70	70	70

(a) Discuss the speed of the car during this time interval.
(b) Find a formula for a function f that models these data.
(c) Sketch a graph of f together with the data.

Solution
(a) The speed of the car appears to be constant at 70 miles per hour.
(b) Because the speed is constant, the rate of change is 0. Thus

$$f(x) \quad = \quad 0x \quad + \quad 70$$
(Future speed) = (Change in speed) + (Initial speed)

and $f(x) = 70$.

(c) Because $y = f(x)$, graph $y = 70$ with the data points

$$(0, 70), (1, 70), (2, 70), (3, 70), \text{ and } (4, 70)$$

to obtain Figure 8.27.

Figure 8.27 Speed of a Car

Now Try Exercise **99**

The function defined by $f(x) = 70$ is an example of a *constant function*. A **constant function** *is a linear function* with $a = 0$ and can be written as $f(x) = b$. Regardless of the input, a constant function always outputs the same value, b. Its graph is a horizontal line.

▶ **REAL-WORLD CONNECTION**   The following are two applications of constant functions.

- A thermostat calculates a constant function regardless of the weather outside by maintaining a set temperature.
- A cruise control in a car calculates a constant function by maintaining a fixed speed, regardless of the type of road or terrain.

The Midpoint Formula (Optional)

▶ **REAL-WORLD CONNECTION**   A common way to make estimations is to average data items. For example, in 1990 about 30% of all music was recorded on compact discs. By 2000, this percentage had increased to about 90%. (**Source:** Recording Industry Association of America.) To estimate the percentage of music recorded on compact discs in 1995, we could average the 1990 and 2000 percentages.

$$\frac{30\% + 90\%}{2} = 60\%$$

This technique is sometimes referred to as finding the *midpoint*. The **midpoint** of a line segment is the unique point on the line segment that is equidistant from the endpoints.

On the real number line, the midpoint of the line segment connecting the numbers x_1 and x_2 equals their average. For example, if a line segment on the real number line has endpoints -3 and 5, as shown in Figure 8.28, then the coordinate of the midpoint is

Figure 8.28

$$M = \frac{x_1 + x_2}{2} = \frac{-3 + 5}{2} = 1.$$

The midpoint of a line segment in the xy-plane can be found in a similar way. Figure 8.29(a) shows the midpoint on the line segment connecting the points (x_1, y_1) and (x_2, y_2). The x-coordinate of M is equal to the average of x_1 and x_2, and the y-coordinate of M is equal to the average of y_1 and y_2. For example, the line segment with endpoints $(-2, 1)$ and $(4, -3)$ is shown in Figure 8.29(b). The coordinates of the midpoint are

$$M = \left(\frac{-2 + 4}{2}, \frac{1 + (-3)}{2} \right) = (1, -1).$$

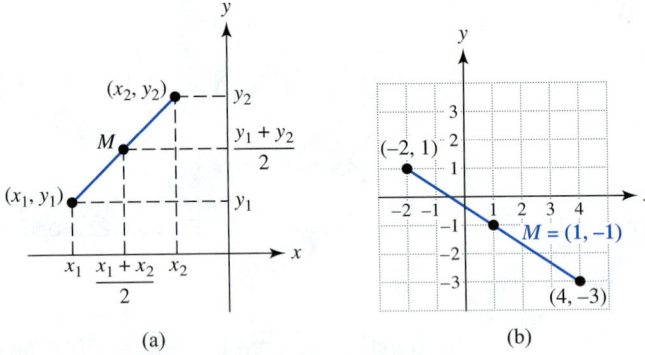

(a)   (b)

Figure 8.29

This discussion is summarized as follows.

MIDPOINT FORMULA IN THE xy-PLANE

The midpoint of the line segment with endpoints (x_1, y_1) and (x_2, y_2) in the xy-plane is

$$\left(\frac{x_1 + x_2}{2}, \frac{y_1 + y_2}{2} \right).$$

EXAMPLE 9   Finding the midpoint

Find the midpoint of the line segment connecting the points $(-3, -2)$ and $(4, 1)$.

Solution
In the midpoint formula let $(-3, -2)$ be (x_1, y_1) and $(4, 1)$ be (x_2, y_2).

$$M = \left(\frac{x_1 + x_2}{2}, \frac{y_1 + y_2}{2} \right)$$

$$= \left(\frac{-3 + 4}{2}, \frac{-2 + 1}{2} \right)$$

$$= \left(\frac{1}{2}, -\frac{1}{2} \right)$$

The midpoint of the line segment is $\left(\frac{1}{2}, -\frac{1}{2} \right)$.

Now Try Exercise **83**

▶ **REAL-WORLD CONNECTION**   In the next example we use the midpoint formula to estimate the number of radio stations in 2000.

EXAMPLE 10   **Estimating the number of radio stations**

The number of radio stations in 2000 is missing from a data table. However, the table states that there were about 12,600 radio stations in 1999 and about 13,000 radio stations in 2001. Use the midpoint formula to estimate the number of radio stations in 2000.   (*Source: The New York Times Almanac, 2006.*)

Solution

In the midpoint formula let (**1999**, **12600**) be (x_1, y_1) and let (**2001**, **13000**) be (x_2, y_2).

$$M = \left(\frac{x_1 + x_2}{2}, \frac{y_1 + y_2}{2} \right)$$

$$= \left(\frac{1999 + 2001}{2}, \frac{12{,}600 + 13{,}000}{2} \right)$$

$$= (2000, 12800)$$

The midpoint formula estimates that in **2000** there were about **12,800** radio stations.

Now Try Exercise **95**

NOTE: An estimate obtained from the midpoint formula is equal to an estimate obtained from a linear function whose graph passes through the endpoints of the line segment. This fact is illustrated in the next example.

EXAMPLE 11   **Comparing estimations**

The graph of a linear function f passes through the points $(0, 4)$ and $(2, -6)$.
(a) Determine $f(1)$ by finding a formula for $f(x)$.
(b) Determine $f(1)$ by finding the midpoint of the line segment connecting $(0, 4)$ and $(2, -6)$.
(c) Compare your answers for parts (a) and (b).

Solution

(a) The graph of f is a line that passes through $(0, 4)$ and $(2, -6)$. The slope of this line is

$$m = \frac{-6 - 4}{2 - 0} = -5$$

and the y-intercept is **4**. Thus $f(x) = -5x + 4$ and $f(1) = -5(1) + 4 = -1$.

(b) $M = \left(\frac{0 + 2}{2}, \frac{4 + (-6)}{2} \right) = (1, -1)$, so the midpoint formula gives $f(1) = -1$.

(c) The answers are equal. Because the graph of f is a line, the midpoint lies on the graph of f and the midpoint formula gives an exact value for $f(1)$.   Now Try Exercise **93**

8.2 PUTTING IT ALL TOGETHER

A linear function is a relatively simple function. A clear understanding of linear functions is essential to the understanding of functions in general.

Concept	Explanation	Examples
Linear Function	Can be represented by $f(x) = ax + b$	$f(x) = 2x - 6$, $a = 2$ and $b = -6$ $f(x) = 10$, $a = 0$ and $b = 10$
Constant Function	Can be represented by $f(x) = b$	$f(x) = -7$, $b = -7$ $f(x) = 22$, $b = 22$
Rate of Change for a Linear Function	The output of a linear function changes by a constant amount for each unit increase in the input.	$f(x) = -3x + 8$ decreases 3 units for each unit increase in x. $f(x) = 5$ neither increases nor decreases. The rate of change is 0.
Midpoint Formula	The midpoint of the line segment connecting (x_1, y_1) and (x_2, y_2) is $$\left(\frac{x_1 + x_2}{2}, \frac{y_1 + y_2}{2} \right).$$	The midpoint of the line segment connecting $(-2, 3)$ and $(4, 5)$ is $$\left(\frac{-2 + 4}{2}, \frac{3 + 5}{2} \right) = (1, 4).$$

The following table summarizes symbolic, verbal, numerical, and graphical representations of a linear function.

Type of Representation	Comments	Example
Symbolic	Mathematical formula in the form $f(x) = ax + b$	$f(x) = 2x + 1$, where $a = 2$ and $b = 1$
Verbal	Multiply the input x by a and add b.	Multiply the input x by 2 and then add 1 to obtain the output.
Numerical (table of values)	For each unit increase in x in the table, the output of $f(x) = ax + b$ changes by an amount equal to a.	1-unit increase 	x	0	1	2	 	$f(x)$	1	3	5	 2-unit increase
Graphical	The graph of a linear function is a line. Plot at least 3 points and then sketch the line.	$y = 2x + 1$

8.2 Exercises

CONCEPTS

1. The formula for a linear function is $f(x) = $ ____.

2. The formula for a constant function is $f(x) = $ ____.

3. The graph of a linear function is a ____.

4. The graph of a constant function is a ____ line.

5. If $f(x) = 7x + 5$, each time x increases by 1 unit, $f(x)$ increases by ____ units.

6. If $f(x) = 5$, each time x increases by 1 unit, $f(x)$ increases by ____ units.

7. If $C(x) = 2x$ calculates the cost in dollars of buying x square feet of carpet, what does 2 represent in the formula? Interpret the fact that the point (10, 20) lies on the graph of C.

8. If $G(x) = 100 - 4x$ calculates the number of gallons of water in a tank after x minutes, what does -4 represent in the formula? Interpret the fact that the point (5, 80) lies on the graph of G.

IDENTIFYING LINEAR FUNCTIONS

Exercises 9–16: Determine whether f is a linear function. If f is linear, give values for a and b so that f may be expressed as $f(x) = ax + b$.

9. $f(x) = \frac{1}{2}x - 6$

10. $f(x) = x$

11. $f(x) = \frac{5}{2} - x^2$

12. $f(x) = \sqrt{x} + 3$

13. $f(x) = -9$

14. $f(x) = 1.5 - 7.3x$

15. $f(x) = -9x$

16. $f(x) = \dfrac{1}{x}$

Exercises 17–20: Determine whether the graph represents a linear function.

17.

18.

19.

20.

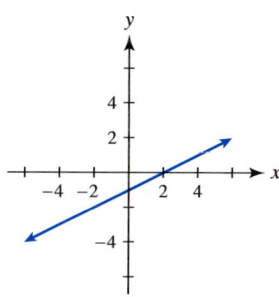

Exercises 21–28: (Refer to Example 2.) Use the table to determine whether $f(x)$ could represent a linear function. If it could, write $f(x)$ in the form $f(x) = ax + b$.

21.

x	0	1	2	3
$f(x)$	-6	-3	0	3

22.

x	0	2	4	6
$f(x)$	-2	2	6	10

23.

x	-2	0	2	4
$f(x)$	6	3	0	-3

24.

x	0	3	6	9
$f(x)$	8	4	2	1

25.

x	-2	-1	0	1
$f(x)$	-5	0	20	40

26.

x	-2	-1	0	1
$f(x)$	6	3	0	-3

27.

x	1	2	3	4
$f(x)$	0	2	4	6

28.

x	1	2	3	4
$f(x)$	0	1	3	7

EVALUATING LINEAR FUNCTIONS

Exercises 29–34: Evaluate $f(x)$ at the given values of x.

29. $f(x) = 4x$ $x = -4, 5$

30. $f(x) = -2x + 1$ $x = -2, 3$

31. $f(x) = 5 - x$ $x = -\frac{2}{3}, 3$

32. $f(x) = \frac{1}{2}x - \frac{1}{4}$ $x = 0, \frac{1}{2}$

33. $f(x) = -22$ $x = -\frac{3}{4}, 13$

34. $f(x) = 9x - 7$ $x = -1.2, 2.8$

Exercises 35–40: Use the graph of f to evaluate the given expressions.

35. $f(-1)$ and $f(0)$

36. $f(-2)$ and $f(2)$

37. $f(-2)$ and $f(4)$

38. $f(0)$ and $f(3)$

39. $f(-3)$ and $f(1)$

40. $f(1.5)$ and $f(0.5\pi)$

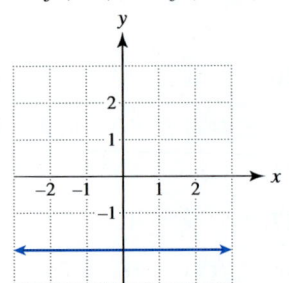

Exercises 41–44: Use the verbal description to write a formula for $f(x)$. Then evaluate $f(3)$.

41. Multiply the input by 6.

42. Multiply the input by -3 and add 7.

43. Divide the input by 6 and subtract $\frac{1}{2}$.

44. Output 8.7 for every input.

REPRESENTING LINEAR FUNCTIONS

Exercises 45–48: Match $f(x)$ with its graph (a.–d.).

45. $f(x) = 3x$ **46.** $f(x) = -2x$

47. $f(x) = x - 2$ **48.** $f(x) = 2x + 1$

a.   **b.**

c.   **d.**

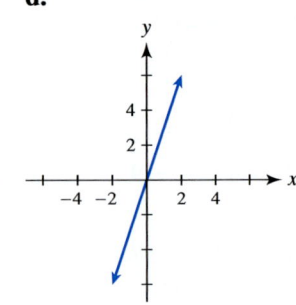

Exercises 49–58: Sketch a graph of $y = f(x)$.

49. $f(x) = 2$ **50.** $f(x) = -1$

51. $f(x) = -\frac{1}{2}x$ **52.** $f(x) = 2x$

53. $f(x) = x + 1$ **54.** $f(x) = x - 2$

55. $f(x) = 3x - 3$ **56.** $f(x) = -2x + 1$

57. $f(x) = 3 - x$ **58.** $f(x) = \frac{1}{4}x + 2$

Exercises 59–64: Write a symbolic representation (formula) for a linear function f that calculates the following.

59. The number of pounds in x ounces

60. The number of dimes in x dollars

61. The distance traveled by a car moving at 65 miles per hour for t hours

62. The long-distance phone bill *in dollars* for calling t minutes at 10 cents per minute and a fixed fee of $4.95

63. The total number of hours in a day during day x

64. The total cost of downhill skiing x times with a $500 season pass

65. **Thinking Generally** For each 1-unit increase in x, $y = ax + b$ with $a > 0$, y increases by _____ units.

66. **Thinking Generally** For each 1-unit decrease in $y = cx + d$ with $c < 0$, x increases by _____ units.

Exercises 67–70: Do the following.
(a) Give a verbal representation of f.
(b) Make a numerical representation (table) of f for $x = -2, 0, 2$.
(c) Plot the points listed in the table from part (b). Then sketch a graph of f.

67. $f(x) = -2x + 1$ **68.** $f(x) = 1 - x$

69. $f(x) = \frac{1}{2}x - 1$ **70.** $f(x) = \frac{3}{4}x$

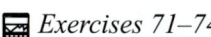 *Exercises 71–74: Do the following.*
(a) Make a numerical representation (table) of f for $x = -3, -2, -1, \ldots, 3$.
(b) Graph f in the window $[-6, 6, 1]$ by $[-4, 4, 1]$.

71. $f(x) = \frac{1}{3}x + \sqrt{2}$ **72.** $f(x) = -\frac{2}{3}x - \sqrt{3}$

73. $f(x) = \frac{x + 2}{5}$ **74.** $f(x) = \frac{2 - 3x}{7}$

MODELING

Exercises 75–78: Match the situation with the graph (a.–d.) that models it best, where x-values represent time from 1995 to 2005.

75. The cost of college tuition

76. The cost of 1 gigabyte of computer memory

77. The distance between Chicago and Denver

78. The total distance traveled by a satellite orbiting Earth if the satellite was launched in 1995.

a.   **b.**

c.   **d.**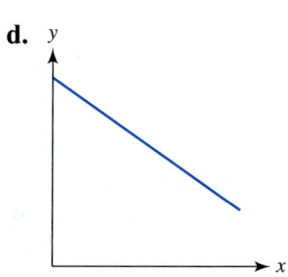

MIDPOINT FORMULA

Exercises 79–82: Find the midpoint of the line segment shown.

79.

80.

81.   **82.**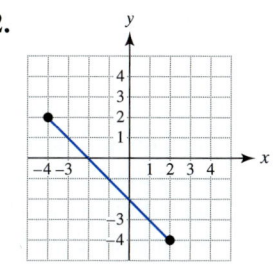

Exercises 83–92: Find the midpoint of the line segment connecting the given points.

83. $(-9, -3), (-7, 1)$ **84.** $(7, -2), (-5, 8)$

85. $\left(\frac{1}{2}, \frac{1}{3}\right), \left(-\frac{5}{2}, -\frac{2}{3}\right)$ **86.** $\left(-\frac{3}{5}, -\frac{1}{4}\right), \left(\frac{1}{10}, \frac{1}{2}\right)$

87. $(-0.3, 0.1), (0.7, 0.4)$

88. $(0.8, -0.4), (0.9, -0.1)$

89. $(2000, 5), (2010, 13)$ **90.** $(2005, 9), (2011, 3)$

91. Thinking Generally $(a, -b)$, $(3a, 5b)$

92. Thinking Generally $(-a, b)$, $(a, -b)$

Exercises 93 and 94: (Refer to Example 11.)   The graph of a linear function f passes through the two given points.
(a) *Determine f(2) by finding a formula for f(x).*
(b) *Determine f(2) by finding the midpoint of the line segment connecting the given points.*
(c) *Compare your answers for parts (a) and (b).*

93. $(0, 5)$, $(4, -3)$ **94.** $(0, 2)$, $(4, 10)$

Exercises 95–98: Use the midpoint formula to make the requested estimation.

95. *U.S. Life Expectancy*   The life expectancy of a female born in 1990 was 78.8 years and the life expectancy of a female born in 2004 rose to 80.4 years. Estimate the life expectancy of a female born in 1997. (The actual value was 79.4 years). (***Source:*** Centers for Disease Control.)

96. *U.S. Life Expectancy*   The life expectancy of a male born in 1990 was 71.8 years and the life expectancy of a male born in 2004 rose to 75.2 years. Estimate the life expectancy of a male born in 1997. (The actual value was 73.6 years.) (***Source:*** Centers for Disease Control.)

97. *U.S. Population*   The population of the United States in 1970 was 205 million and in 1990 it was 249 million. Estimate the population in 1980. (The actual value was 228 million.) (***Source:*** U.S. Census Bureau.)

98. *Distance Traveled*   A car is moving at a constant speed on an Interstate highway. After 1 hour the car passes the 103-mile marker and after 5 hours the car passes the 391-mile marker. What mile marker does the car pass after 3 hours?

APPLICATIONS

99. *Thermostat*   Let $y = f(x)$ describe the temperature y of a room that is kept at 70°F for x hours.
(a) Represent f symbolically and graphically over a 24-hour period for $0 \le x \le 24$.
(b) Construct a table of f for $x = 0, 4, 8, 12, \ldots, 24$.
(c) What type of function is f?

100. *Cruise Control*   Let $y = f(x)$ describe the speed y of an automobile after x minutes if the cruise control is set at 60 miles per hour.

(a) Represent f symbolically and graphically over a 15-minute period for $0 \le x \le 15$.
(b) Construct a table of f for $x = 0, 1, 2, \ldots, 6$.
(c) What type of function is f?

101. *Distance*   A car is initially 50 miles south of the Minnesota–Iowa border traveling south on Interstate 35. Distances D between the car and the border are recorded in the table for various elapsed times t. Find a linear function given by $D(t) = at + b$ that models these data.

t (hours)	0	2	3	5
D (miles)	50	170	230	350

102. *Estimating the Weight of a Bass*   Sometimes the weight of a fish can be estimated by measuring its length. The table lists typical weights of bass having various lengths.

Length (inches)	12	14	16	18	20	22
Weight (pounds)	1.0	1.7	2.5	3.6	5.0	6.6

Source: Minnesota Department of Natural Resources.

(a) Let x be the length and y be the weight. Make a line graph of the data.
(b) Could the data be modeled accurately with a linear function? Explain your answer.

103. *Solid Waste*   In 1960, the average American disposed of 2.7 pounds of garbage per day, whereas in 2003 this amount was 4.3 pounds per day. (***Source:*** Environmental Protection Agency.)

(a) Find a linear function f that calculates the amount of garbage disposed of by a person in 1960 after x days. Find $f(60)$ and interpret the result.

(b) Repeat part (a) for 2003 with a linear function g.

104. *Rain Forests*   Rain forests are forests that grow in regions receiving more than 70 inches of rain per year. The world is losing about 49 million acres of rain forest each year. (*Source: New York Times Almanac.*)

(a) Find a linear function f that calculates the change in the acres of rain forest in millions in x years.

(b) Evaluate $f(7)$ and interpret the result.

105. *Mobile Malware*   As the number of smart mobile telephones has increased, so has the number of malicious software viruses that affect these devices. In 2006 the total number N of *malware* viruses detected from January to July could be modeled by the formula $N(x) = 16x + 136$, where $x = 1$ corresponds to January, $x = 2$ to February, and so on. (*Source:* Hypponen, Mikko, "Malware Goes Mobile," *Scientific American*, November 2006.)

(a) Use the window [1, 7, 1] by [0, 300, 100] to graph f.

(b) Discuss any trend in the graph.

(c) Evaluate $f(3)$ and interpret your result.

(d) What does the number 16 in the formula indicate?

106. *Temperature and Volume*   If a sample of a gas, such as helium, is heated, it will expand. The formula $V(T) = 0.147T + 40$ calculates the volume V in cubic inches of a sample of gas at temperature T in degrees Celsius.

(a) Evaluate $V(0)$ and interpret the result.

(b) If the temperature increases by $10°C$, by how much does the volume increase?

(c) What is the volume of the gas when the temperature is $100°C$?

107. *Temperature and Volume*   (Refer to the preceding exercise.) A sample of gas at $0°C$ has a volume V of 137 cubic centimeters, which increases in volume by 0.5 cubic centimeter for every $1°C$ increase in temperature T.

(a) Write a formula $V(T) = aT + b$ that gives the volume of the gas at temperature T.

(b) Find the volume of the gas when $T = 50°C$.

108. *Cost*   To make a music video it costs $750 to rent a studio plus $5 for each copy produced.

(a) Write a formula $C(x) = ax + b$ that calculates the cost of producing x videos.

(b) Find the cost of producing 2500 videos.

109. *Weight Lifting*   Lifting weights can increase a person's muscle mass. Each additional pound of muscle burns an extra 40 calories per day. Write a linear function that models the number of calories burned each day by x pounds of muscle. By burning an extra 3500 calories a person can lose 1 pound of fat. How many pounds of muscle are needed to burn 1 pound of fat in 30 days? (*Source: Runner's World.*)

110. *Number of Doctors*   The number N of doctors in the private sector from 1970 to 2000 can be approximated by $N(t) = 15{,}260t + 333{,}200$, where t is in years and $t = 0$ corresponds to 1970.

(a) Evaluate $N(0)$ and $N(30)$ and interpret each result.

(b) Explain what the numbers 15,260 and 333,200 represent in the formula.

111. *Baseball*   The table shows the average length of a major league baseball game in minutes for various years.

Year	2000	2001	2002
Length (minutes)	180	176	172

Source: Elias Sports Bureau.

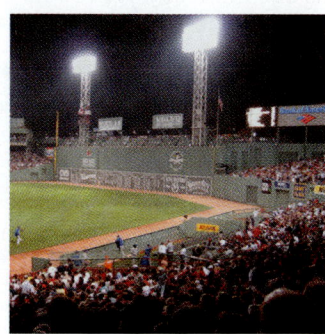

(a) What was the average length in 2000?

(b) By how many minutes did the average length change each year?

(c) Find a linear function f that models these data. Let $x = 0$ correspond to 2000.

(d) Use f to predict the average length in 2004.

112. ***Wal-Mart Sales***   The table shows Wal-Mart's share as a percentage of overall U.S. retail sales for various years. (This percentage excludes restaurants and motor vehicles.)

Year	1998	1999	2000	2001	2002
Share (%)	6	6.5	7	7.5	8

Source: Commerce Department, Wal-Mart.

(a) What was Wal-Mart's share in 1998?
(b) By how much (percent) did Wal-Mart's share increase each year?

(c) Find a linear function f that models these data. Let $x = 0$ correspond to 1998.
(d) Use f to estimate Wal-Mart's share in 2005.

WRITING ABOUT MATHEMATICS

113. Explain how you can determine whether a function is linear by using its
(a) symbolic representation,
(b) graphical representation, and
(c) numerical representation.

114. Describe one way to determine whether a set of data points can be modeled by a linear function.

CHECKING BASIC CONCEPTS
SECTIONS 8.1 AND 8.2

1. Find a formula and sketch a graph for a function that squares the input x and then subtracts 1.

2. Use the graph below to do the following.
(a) Find the domain and range of f.
(b) Evaluate $f(0)$ and $f(2)$.
(c) Is f a linear function? Explain.

3. Determine whether f is a linear function.
(a) $f(x) = 4x - 2$
(b) $f(x) = 2\sqrt{x} - 5$
(c) $f(x) = -7$
(d) $f(x) = 9 - 2x + 5x$

4. Graph $f(x) = 4 - 3x$. Evaluate $f(-2)$.

5. Find a formula for a linear function that models the data.

x	0	1	2	3	4
$f(x)$	-1	$-\frac{1}{2}$	0	$\frac{1}{2}$	1

6. The median age in the United States from 1970 to 2000 can be approximated by

$$f(x) = 0.264x + 27.7,$$

where $x = 0$ corresponds to 1970, $x = 1$ to 1971, and so on.
(a) Evaluate $f(20)$ and interpret the result.
(b) Interpret the numbers 0.264 and 27.7.

7. Find the midpoint of the line segment connecting the points $(-3, 4)$ and $(5, -6)$.

8.3 COMPOUND INEQUALITIES

Basic Concepts ▪ Symbolic Solutions and Number Lines ▪
Numerical and Graphical Solutions ▪ Interval Notation

A LOOK INTO MATH ▷ A person weighing 143 pounds and needing to purchase a life vest for white-water rafting is not likely to find one designed exactly for this weight. Life vests are manufactured to support a range of body weights. A vest approved for weights between 100 and 160 pounds might be appropriate for this person. In other words, if a person's weight is w, this life vest is safe if $w \geq 100$ *and* $w \leq 160$. This example illustrates the concept of a *compound inequality*.

Basic Concepts

A **compound inequality** consists of two inequalities joined by the words *and* or *or*. The following are two examples of compound inequalities.

$$2x \geq -3 \quad \textbf{and} \quad 2x < 5 \qquad (1)$$
$$x + 2 \geq 3 \quad \textbf{or} \quad x - 1 < -5 \qquad (2)$$

If a compound inequality contains the word *and*, a solution must satisfy *both* inequalities. For example, 1 is a solution to the first compound inequality because

$$2(\mathbf{1}) \geq -3 \quad \text{and} \quad 2(\mathbf{1}) < 5$$
$$\text{True} \qquad\qquad \text{True}$$

are both true statements.

If a compound inequality contains the word *or*, a solution must satisfy *at least one* of the two inequalities. Thus 5 is a solution to the second compound inequality because the first statement is true.

$$\mathbf{5} + 2 \geq 3 \quad \text{or} \quad \mathbf{5} - 1 < -5$$
$$\text{True} \qquad\qquad \text{False}$$

Note that 5 does not need to satisfy both statements for this compound inequality to be true.

EXAMPLE 1 Determining solutions to compound inequalities

Determine whether the given x-values are solutions to the compound inequalities.
(a) $x + 1 < 9$ and $2x - 1 > 8$ $\qquad x = 5, -5$
(b) $5 - 2x \leq -4$ or $5 - 2x \geq 4$ $\qquad x = 2, -3$

Solution
(a) Substitute $x = 5$ in the given compound inequality.

$$\mathbf{5} + 1 < 9 \quad \text{and} \quad 2(\mathbf{5}) - 1 > 8$$
$$\text{True} \qquad\qquad\quad \text{True}$$

Both inequalities are true, so 5 is a solution. Now substitute $x = -5$.

$$\mathbf{-5} + 1 < 9 \quad \text{and} \quad 2(\mathbf{-5}) - 1 > 8$$
$$\text{True} \qquad\qquad\quad\;\; \text{False}$$

To be a solution both inequalities must be true, so -5 is not a solution.

(b) Substitute $x = 2$ into the given compound inequality.

$$5 - 2(2) \leq -4 \quad \text{or} \quad 5 - 2(2) \geq 4$$

False   **False**

Neither inequality is true, so 2 is not a solution. Now substitute $x = -3$.

$$5 - 2(-3) \leq -4 \quad \text{or} \quad 5 - 2(-3) \geq 4$$

False   **True**

At least one of the two inequalities is true, so -3 is a solution.

Now Try Exercises **7, 9**

CRITICAL THINKING

Graph the following inequalities and discuss your results.

1. $x < 2$ and $x > 5$

2. $x > 2$ or $x < 5$

Symbolic Solutions and Number Lines

We can use a number line to graph solutions to compound inequalities, such as

$$x \leq 6 \quad \text{and} \quad x > -4.$$

The solution set for $x \leq 6$ is shaded to the left of 6, with a bracket placed at $x = 6$, as shown in Figure 8.30. The solution set for $x > -4$ can be shown by shading a different number line to the right of -4 and placing a left parenthesis at -4. Because the inequalities are connected by *and*, the solution set consists of all numbers that are shaded on *both* number lines. The final number line represents the *intersection* of the two solution sets. That is, the solution set includes real numbers where the graphs "overlap." For any two sets A and B, the **intersection** of A and B, denoted $A \cap B$, is defined by

$$A \cap B = \{x \mid x \text{ is an element of } A \text{ and an element of } B\}.$$

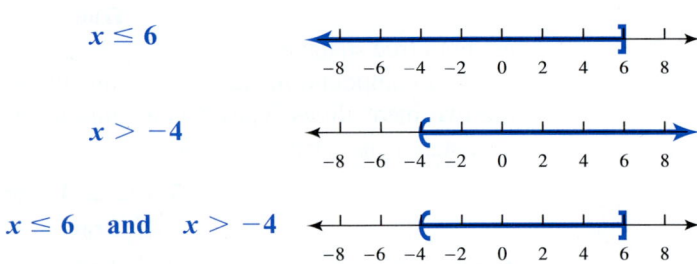

Figure 8.30

NOTE: A bracket, either [or], is used when an inequality contains $\leq$ or $\geq$. A parenthesis, either (or), is used when an inequality contains $<$ or $>$. This notation makes clear whether an endpoint is included in the inequality.

EXAMPLE 2   Solving a compound inequality containing "and"

Solve $2x + 4 > 8$ and $5 - x < 9$. Graph the solution.

Solution

First solve each linear inequality separately.

$$2x + 4 > 8 \quad \text{and} \quad 5 - x < 9$$
$$2x > 4 \quad \text{and} \quad -x < 4$$
$$x > 2 \quad \text{and} \quad x > -4$$

Graph the two inequalities on two different number lines. On a third number line, shade solutions that appear on both of the first two number lines. As shown in Figure 8.31, the solution set is $\{x \mid x > 2\}$.

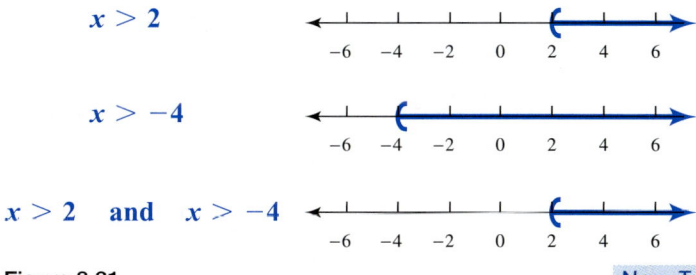

$x > 2$

$x > -4$

$x > 2$ **and** $x > -4$

Figure 8.31

Now Try Exercise **41**

Sometimes a compound inequality containing the word *and* can be combined into a three-part inequality. For example, rather than writing

$$x > 5 \quad \text{and} \quad x \leq 10,$$

we could write the **three-part inequality**

$$5 < x \leq 10.$$

This three-part inequality is represented by the number line shown in Figure 8.32.

$5 < x \leq 10$

Figure 8.32

EXAMPLE 3 Solving three-part inequalities

Solve each inequality. Write the solution set in set-builder notation. (To review set-builder notation, see page 142.)

(a) $4 < t + 2 \leq 8$ **(b)** $-3 \leq 3z \leq 6$ **(c)** $-\dfrac{5}{2} < \dfrac{1 - m}{2} < 4$

Solution

(a) To solve a three-part inequality, isolate the variable by applying properties of inequalities to each part of the inequality.

$$4 < t + 2 \leq 8 \qquad \text{Given three-part inequality}$$
$$4 - \mathbf{2} < t + 2 - \mathbf{2} \leq 8 - \mathbf{2} \qquad \text{Subtract 2 from each part.}$$
$$2 < t \leq 6 \qquad \text{Simplify each part.}$$

The solution set in set-builder notation is $\{t \mid 2 < t \leq 6\}$.

(b) To simplify, divide each part by 3.

$$-3 \leq 3z \leq 6 \qquad \text{Given three-part inequality}$$
$$\frac{-3}{\mathbf{3}} \leq \frac{3z}{\mathbf{3}} \leq \frac{6}{\mathbf{3}} \qquad \text{Divide each part by 3.}$$
$$-1 \leq z \leq 2 \qquad \text{Simplify each part.}$$

The solution set in set-builder notation is $\{z \mid -1 \leq z \leq 2\}$.

(c) Multiply each part by 2 to clear fractions.

$$-\frac{5}{2} < \frac{1-m}{2} < 4 \qquad \text{Given three-part inequality}$$

$$2 \cdot \left(-\frac{5}{2}\right) < 2 \cdot \left(\frac{1-m}{2}\right) < 2 \cdot 4 \qquad \text{Multiply each part by 2.}$$

$$-5 < 1 - m < 8 \qquad \text{Simplify each part.}$$

$$-5 - 1 < 1 - m - 1 < 8 - 1 \qquad \text{Subtract 1 from each part.}$$

$$-6 < -m < 7 \qquad \text{Simplify each part.}$$

$$-1 \cdot (-6) > -1 \cdot (-m) > -1 \cdot 7 \qquad \begin{array}{l}\text{Multiply each part by } -1;\\ \textit{reverse} \text{ inequality symbols.}\end{array}$$

$$6 > m > -7 \qquad \text{Simplify each part.}$$

$$-7 < m < 6 \qquad \text{Rewrite inequality.}$$

The solution set in set-builder notation is $\{m \mid -7 < m < 6\}$.

NOTE: Either $6 > m > -7$ or $-7 < m < 6$ is a correct way to write a three-part inequality. *However*, we usually write the smaller number on the left side and the larger number on the right side. Now Try Exercises **57, 61, 77**

▶ **REAL-WORLD CONNECTION** Three-part inequalities occur frequently in applications. In the next example we find altitudes at which the air temperature is in a certain range.

EXAMPLE 4 Solving a three-part inequality

If the ground-level temperature is $80°F$, the air temperature x miles above Earth's surface is cooler and can be modeled by $T(x) = 80 - 19x$. Find the altitudes at which the air temperature ranges from $42°F$ down to $23°F$. (*Source:* A. Miller and R. Anthes, *Meteorology.*)

Solution
We write and solve the three-part inequality $23 \le T(x) \le 42$.

$$23 \le 80 - 19x \le 42 \qquad \text{Substitute for } T(x).$$

$$-57 \le -19x \le -38 \qquad \text{Subtract 80 from each part.}$$

$$\frac{-57}{-19} \ge x \ge \frac{-38}{-19} \qquad \text{Divide by } -19; \textit{reverse} \text{ inequality symbols.}$$

$$3 \ge x \ge 2 \qquad \text{Simplify.}$$

$$2 \le x \le 3 \qquad \text{Rewrite inequality.}$$

The air temperature ranges from $42°F$ to $23°F$ for altitudes between 2 and 3 miles. Now Try Exercise **109**

MAKING CONNECTIONS

Writing Three-Part Inequalities

The inequality $-2 < x < 1$ means that $x > -2$ *and* $x < 1$. A three-part inequality should *not* be used when *or* connects a compound inequality. Writing $x < -2$ or $x > 1$ as $1 < x < -2$ is **incorrect** because it states that x must be both greater than 1 *and* less than -2. It is impossible for any value of x to satisfy this statement.

We can also solve compound inequalities containing the word *or*. To write the solution to such an inequality we sometimes use *union* notation. For any two sets A and B, the **union** of A and B, denoted $A \cup B$, is defined by

$$A \cup B = \{x \mid x \text{ is an element of } A \text{ or an element of } B\}.$$

If the solution to an inequality is $\{x \mid x < 1\}$ or $\{x \mid x \geq 3\}$, then it can also be written as

$$\{x \mid x < 1\} \cup \{x \mid x \geq 3\}.$$

That is, we can replace the word *or* with the $\cup$ symbol.

EXAMPLE 5   Solving a compound inequality containing "or"

Solve $x + 2 < -1$ or $x + 2 > 1$.

Solution
We first solve each linear inequality.

$$
\begin{array}{lll}
x + 2 < -1 & \text{or} \quad x + 2 > 1 & \text{Given compound inequality} \\
x < -3 & \text{or} \quad x > -1 & \text{Subtract 2.}
\end{array}
$$

We can graph the simplified inequalities on different number lines, as shown in Figure 8.33. A solution must satisfy at least one of the two inequalities. Thus the solution set for the compound inequality results from taking the *union* of the first two number lines. We can write the solution, using set-builder notation, as $\{x \mid x < -3\} \cup \{x \mid x > -1\}$ or as $\{x \mid x < -3 \text{ or } x > -1\}$.

$x < -3$

$x > -1$

$x < -3 \quad$ or $\quad x > -1$

Figure 8.33   Now Try Exercise **45**

CRITICAL THINKING
Carbon dioxide is emitted when human beings breathe. In one study of college students, the amount of carbon dioxide exhaled in grams per hour was measured during lectures and exams. The average amount exhaled during lectures L satisfied $25.33 \leq L \leq 28.17$, whereas the average amount exhaled during exams E satisfied $36.58 \leq E \leq 40.92$. What do these results indicate? Explain.   (***Source:*** T. Wang, *ASHRAE Trans.*)

Numerical and Graphical Solutions

Compound inequalities can also be solved graphically and numerically, as illustrated in the next example.

EXAMPLE 6   Solving a compound inequality numerically and graphically

Tuition and fees at private colleges and universities from 1980 to 2000 can be modeled by $f(x) = 575(x - 1980) + 3600$. Estimate when the average tuition and fees ranged from $8200 to $10,500.

Solution

Numerical Solution Let $Y_1 = 575(X - 1980) + 3600$. Make a table of values, as shown in Figure 8.34(a). In 1988, the average tuition and fees were \$8200 and in 1992 they were \$10,500. Thus from 1988 to 1992 the average tuition and fees ranged from \$8200 to \$10,500.

Graphical Solution Graph $Y_1 = 575(X - 1980) + 3600$, $Y_2 = 8200$, and $Y_3 = 10,500$. We must find x-values so that $y_2 \leq y_1 \leq y_3$. Figures 8.34(b) and (c) show that y_1 is between y_2 and y_3 when $1988 \leq x \leq 1992$.

CALCULATOR HELP

To find a point of intersection, see Appendix A (page AP-7).

[1980, 2000, 1] by [3000, 12000, 3000]   [1980, 2000, 1] by [3000, 12000, 3000]

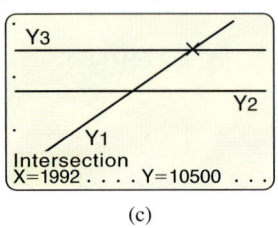

X	Y₁
1987	7625
1988	8200
1989	8775
1990	9350
1991	9925
1992	10500
1993	11075

X=1988

(a)   (b)   (c)

Figure 8.34

Now Try Exercise **103(b)**

Interval Notation

Figure 8.35

The solution set in Example 4 was $\{x \mid 2 \leq x \leq 3\}$. This solution set can be graphed on a number line, as shown in Figure 8.35.

A convenient notation for number line graphs is called **interval notation**. Instead of drawing the entire number line as in Figure 8.35, the solution set can be expressed as [2, 3] in interval notation. Because the solution set includes the endpoints 2 and 3, brackets are used. A solution set that includes all real numbers satisfying $-2 < x < 3$ can be expressed as $(-2, 3)$. Parentheses indicate that the endpoints are *not* included. The interval $0 \leq x < 4$ is represented by $[0, 4)$.

Table 8.11 provides some examples of interval notation. The symbol ∞ refers to **infinity**, and it does not represent a real number. The interval $(5, \infty)$ represents $x > 5$, which has no maximum x-value, so ∞ is used for the right endpoint. The symbol $-\infty$ may be used similarly and denotes **negative infinity**.

TABLE 8.11 Interval Notation

Inequality	Interval Notation	Number Line Graph
$-1 < x < 3$	$(-1, 3)$	
$-3 < x \leq 2$	$(-3, 2]$	
$-2 \leq x \leq 2$	$[-2, 2]$	
$x < -1$ or $x > 2$	$(-\infty, -1) \cup (2, \infty)$ ($\cup$ is the union symbol.)	
$x > -1$	$(-1, \infty)$	
$x \leq 2$	$(-\infty, 2]$	

> ## MAKING CONNECTIONS
>
> ### Points and Intervals
>
> The expression $(1, 2)$ may represent a point in the *xy*-plane or the interval $1 < x < 2$. To alleviate confusion, phrases such as "the point $(1, 2)$" or "the interval $(1, 2)$" are used.

EXAMPLE 7 Writing inequalities in interval notation

Write each expression in interval notation.
(a) $-2 \le x < 5$ **(b)** $x \ge 3$ **(c)** $x < -5$ or $x \ge 2$
(d) $\{x \mid x > 0 \text{ and } x \le 3\}$ **(e)** $\{x \mid x \le 1 \text{ or } x \ge 3\}$

Solution
(a) $[-2, 5)$ **(b)** $[3, \infty)$ **(c)** $(-\infty, -5) \cup [2, \infty)$
(d) $(0, 3]$ **(e)** $(-\infty, 1] \cup [3, \infty)$

Now Try Exercises **13, 17, 23, 27, 37**

EXAMPLE 8 Solving an inequality

Solve $2x + 1 \le -1$ or $2x + 1 \ge 3$. Write the solution in interval notation.

Solution
First solve each inequality.

$$
\begin{array}{llll}
2x + 1 \le -1 & \text{or} & 2x + 1 \ge 3 & \text{Given compound inequality} \\
2x \le -2 & \text{or} & 2x \ge 2 & \text{Subtract 1.} \\
x \le -1 & \text{or} & x \ge 1 & \text{Divide by 2.}
\end{array}
$$

The solution set may be written as $(-\infty, -1] \cup [1, \infty)$.

Now Try Exercise **53**

8.3 PUTTING IT ALL TOGETHER

The following table summarizes some important topics related to compound inequalities.

Concept	Explanation	Examples
Compound Inequality	Two inequalities joined by *and* or *or*	$x < -1$ or $x > 2$; $2x \ge 10$ and $x + 2 < 6$
Three-Part Inequality	Can be used to write some types of compound inequalities involving *and*	$x > -2$ and $x \le 3$ is equivalent to $-2 < x \le 3$.
Interval Notation	Notation used to write sets of real numbers rather than using number lines or inequalities See Table 8.11.	$-2 \le z \le 4$ is equivalent to $[-2, 4]$. $x < 4$ is equivalent to $(-\infty, 4)$. $x \le -2$ or $x > 0$ is equivalent to $(-\infty, -2] \cup (0, \infty)$.

continued on next page

continued from previous page

The following table lists basic strategies for solving compound inequalities.

Type of Inequality	Method to Solve Inequality
Solving a Compound Inequality Containing *and*	**STEP 1:** First solve each inequality individually. **STEP 2:** The solution set includes values that satisfy *both* inequalities from Step 1.
Solving a Compound Inequality Containing *or*	**STEP 1:** First solve each inequality individually. **STEP 2:** The solution set includes values that satisfy *at least one* of the inequalities from Step 1.
Solving a Three-Part Inequality	Work on all three parts at the same time. Be sure to perform the same step on each part. Continue until the variable is isolated in the middle part.

8.3 Exercises

CONCEPTS

1. Give an example of a compound inequality containing the word *and*.

2. Give an example of a compound inequality containing the word *or*.

3. Is 1 a solution to $x > 3$ and $x \le 5$?

4. Is 1 a solution to $x < 3$ or $x \ge 5$?

5. Is the compound inequality $x \ge -5$ and $x \le 5$ equivalent to $-5 \le x \le 5$?

6. Name three ways to solve a compound inequality.

Exercises 7–12: Determine whether the given values of x are solutions to the compound inequality.

7. $x - 1 < 5$ and $2x > 3$ $x = 2, x = 6$

8. $2x + 1 \ge 4$ and $1 - x \le 3$ $x = -2, x = 3$

9. $3x < -5$ or $2x \ge 3$ $x = 0, x = 3$

10. $x + 1 \le -4$ or $x + 1 \ge 4$ $x = -5, x = 2$

11. $2 - x > -5$ and $2 - x \le 4$ $x = -3, x = 0$

12. $x + 5 \ge 6$ or $3x \le 3$ $x = -1, x = 1$

INTERVAL NOTATION

Exercises 13–38: Write the inequality in interval notation.

13. $2 \le x \le 10$

14. $-1 < x < 5$

15. $5 < x \le 8$

16. $-\frac{1}{2} \le x \le \frac{5}{6}$

17. $x < 4$

18. $x \le -3$

19. $x > -2$

20. $x \ge 6$

21. $x \ge -2$ and $x < 5$

22. $x \le 6$ and $x \ge 2$

23. $x \le 8$ and $x > -8$

24. $x \ge -4$ and $x < 3$

25. $x \ge 6$ or $x > 3$

26. $x \le -4$ or $x < -3$

27. $x \le -2$ or $x \ge 4$

28. $x \le -1$ or $x > 6$

29. $x < 1$ or $x \ge 5$

30. $x < -3$ or $x > 3$

31.

![number line from -6 to 6 with shaded region from -2 to 4]

32.

(number line from −6 to 6, arrow starting at 2 going right)

33.

(number line from −6 to 6, arrow from −2 going left)

34.

(number line from −6 to 6, segment from −4 to 4)

35. $\{x \mid x < 4\}$ **36.** $\{x \mid -1 \le x < 4\}$

37. $\{x \mid x < 1 \text{ or } x > 2\}$ **38.** $\{x \mid -\infty < x < \infty\}$

SYMBOLIC SOLUTIONS

Exercises 39–46: Solve the compound inequality. Graph the solution set on a number line.

39. $x \le 3$ and $x \ge -1$ **40.** $x \ge 5$ and $x > 6$

41. $2x < 5$ and $2x > -4$

42. $2x + 1 < 3$ and $x - 1 \ge -5$

43. $x \le -1$ or $x \ge 2$

44. $2x \le -6$ or $x \ge 6$

45. $5 - x > 1$ or $x + 3 \ge -1$

46. $1 - 2x > 3$ or $2x - 4 \ge 4$

Exercises 47–56: Solve the compound inequality. Write your answer in interval notation.

47. $x - 3 \le 4$ and $x + 5 \ge -1$

48. $2z \ge -10$ and $z < 8$

49. $3t - 1 > -1$ and $2t - \frac{1}{2} > 6$

50. $2(x + 1) < 8$ and $-2(x - 4) > -2$

51. $x - 4 \ge -3$ or $x - 4 \le 3$

52. $1 - 3n \ge 6$ or $1 - 3n \le -4$

53. $-x < 1$ or $5x + 1 < -10$

54. $7x - 6 > 0$ or $-\frac{1}{2}x \le 6$

55. $1 - 7x < -48$ and $3x + 1 \le -9$

56. $3x - 4 \le 8$ or $4x - 1 \le 13$

Exercises 57–78: Solve the three-part inequality. Write your answer in interval notation.

57. $-2 \le t + 4 < 5$ **58.** $5 < t - 7 < 10$

59. $-\frac{5}{8} \le y - \frac{3}{8} < 1$ **60.** $-\frac{1}{2} < y - \frac{3}{2} < \frac{1}{2}$

61. $-27 \le 3x \le 9$ **62.** $-4 < 2y < 22$

63. $\frac{1}{2} < -2y \le 8$ **64.** $-16 \le -4x \le 8$

65. $-4 < 5z + 1 \le 6$ **66.** $-3 \le 3z + 6 < 9$

67. $3 \le 4 - n \le 6$ **68.** $-1 < 3 - n \le 1$

69. $-1 < 2z - 1 < 3$ **70.** $2 \le 4z + 5 \le 6$

71. $-2 \le 5 - \frac{1}{3}m < 2$ **72.** $-\frac{3}{2} < 4 - 2m < \frac{7}{2}$

73. $100 \le 10(5x - 2) \le 200$

74. $-15 < 5(x - 1990) < 30$

75. $-3 < \dfrac{3z + 1}{4} < 1$ **76.** $-3 < \dfrac{z - 1}{2} < 5$

77. $-\dfrac{5}{2} \le \dfrac{2 - m}{4} \le \dfrac{1}{2}$ **78.** $\dfrac{4}{5} \le \dfrac{4 - 2m}{10} \le 2$

NUMERICAL AND GRAPHICAL SOLUTIONS

Exercises 79–82: Use the table to solve the three-part inequality. Write your answer in interval notation.

79. $-3 \le 3x \le 6$ **80.** $-5 \le 2x - 1 \le 1$

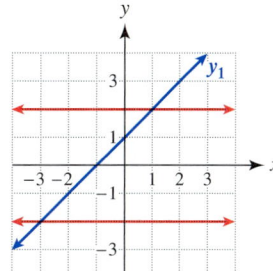

X	Y1
-2	-6
-1	-3
0	0
1	3
2	6
3	9
4	12

Y1■3X

X	Y1
-4	-9
-3	-7
-2	-5
-1	-3
0	-1
1	1
2	3

Y1■2X−1

81. $-1 < 1 - x < 2$ **82.** $-2 \le -2x < 4$

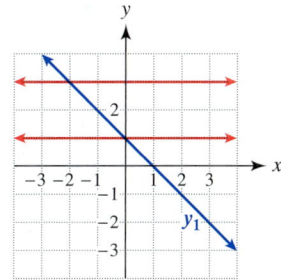

X	Y1
-2	3
-1	2
0	1
1	0
2	-1
3	-2
4	-3

Y1■1−X

X	Y1
-3	6
-2	4
-1	2
0	0
1	-2
2	-4
3	-6

Y1■-2X

Exercises 83–86: Use the graph to solve the compound inequality. Write your answer in interval notation.

83. $-2 \le y_1 \le 2$ **84.** $1 \le y_1 < 3$

(graph with line y_1)   (graph with line y_1)

85. $y_1 < -2$ or $y_1 > 2$ **86.** $y_1 \le -2$ or $y_1 \ge 4$

     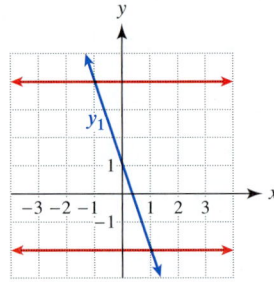

87. *Distance*   The function f, shown in the figure, gives the distance y in miles between a car and the city of Omaha, Nebraska, after x hours, where $0 \le x \le 6$.
  (a) Is the car moving toward or away from Omaha? Explain.
  (b) Determine the times when the car is 100 miles or 200 miles from Omaha.
  (c) When is the car from 100 to 200 miles from Omaha?
  (d) When is the car's distance from Omaha greater than or equal to 200 miles?

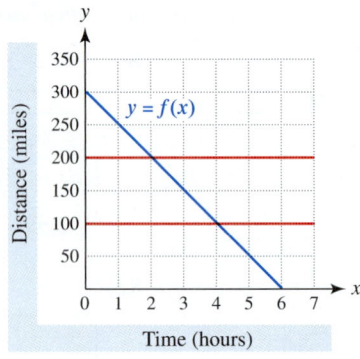

88. Use the figure to solve each equation or inequality. Let the domains of y_1, y_2, and y_3 be $0 \le x \le 5$.

  (a) $y_1 = y_2$ **(b)** $y_2 = y_3$
  (c) $y_1 \le y_2 \le y_3$ **(d)** $y_2 < y_3$

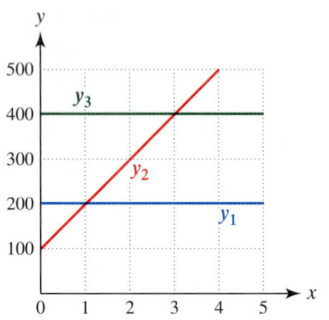

Exercises 89–94: Solve numerically or graphically. Write your answer in interval notation.

89. $-2 \le 2x - 4 \le 4$

90. $-1 \le 1 - x \le 3$

91. $x + 1 < -1$ or $x + 1 > 1$

92. $2x - 1 < -3$ or $2x - 1 > 5$

93. $95 \le 25(x - 2000) + 45 \le 295$

94. $42 \le -13(x - 2005) + 120 \le 94$

USING MORE THAN ONE METHOD

Exercises 95–100: Solve symbolically, graphically, and numerically. Write the solution set in interval notation.

95. $4 \le 5x - 1 \le 14$

96. $-4 < 2x < 4$

97. $4 - x \ge 1$ or $4 - x < 3$

98. $x + 3 \ge -2$ or $x + 3 \le 1$

99. $2x + 1 < 3$ or $2x + 1 \ge 7$

100. $3 - x \le 4$ or $3 - x > 8$

101. Thinking Generally   Solve $c < x + b \le d$ for x.

102. Thinking Generally   Solve $c \le ax + b \le d$ for x, if $a < 0$.

APPLICATIONS

103. *Home Prices*   The average price P of a single-family home in thousands of dollars from 1990 to 2000 can be modeled by $P(x) = 5.7(x - 1990) + 150$, where x is the year.   (*Source:* U.S. Census Bureau.) Determine the years when the average price ranged from \$178,500 to \$207,000
  (a) symbolically,
  (b) graphically or numerically.

104. *Median Family Income*   The median income I of a family in thousands of dollars from 2000 to 2005 can be modeled by $I(x) = 1.4(x - 2000) + 51$, where x is the year. (*Source:* U.S. Census Bureau.) Determine the years when the median income ranged from \$52,400 to \$56,600
  (a) symbolically,
  (b) graphically or numerically.

105. *Altitude and Dew Point*   If the dew point D on the ground is 60°F, then the dew point x miles high is given by $D(x) = 60 - 5.8x$. Find the altitudes where the dew point ranges from 57.1°F to 51.3°F. (*Source:* A. Miller.)

106. *Medicare Costs*   Medicare costs in billions of dollars may be modeled by $f(x) = 18x - 35,750$, where $1995 \le x \le 2007$. Estimate the years when Medicare costs were from 250 to 340 billion dollars. (*Source:* Office of Management and Budget.)

107. *Geometry*   For what values of x is the perimeter of the rectangle from 40 to 60 feet?

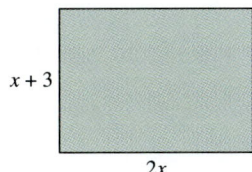

$x + 3$

$2x$

108. *Geometry*   A rectangle is three times as long as it is wide. If the perimeter ranges from 100 to 160 inches, what values for the width are possible?

109. *Altitude and Temperature*   If the air temperature at ground level is 70°F, the air temperature x miles high is given by $T(x) = 70 - 19x$. Determine the altitudes at which the air temperature is from 41.5°F to 22.5°F.   (*Source:* A. Miller and R. Anthes, *Meteorology*.)

110. *Distance*   A car's distance in miles from a rest stop after x hours is given by $f(x) = 70x + 50$.
   (a) Construct a table for f for $x = 4, 5, 6, \dots, 10$ and use the table to solve the inequality $470 \le f(x) \le 680$. Explain your result.
   (b) Solve the inequality in part (a) symbolically.

111. *Temperature*   The formula $F = \frac{9}{5}C + 32$ may be used to convert Celsius temperatures to Fahrenheit temperatures. The greatest temperature ranges on Earth are recorded in Siberia where the temperature has varied from −90°F to 98°F. Find this temperature range in degrees Celsius.

112. *Temperature*   The formula $C = \frac{5}{9}(F - 32)$ may be used to convert Fahrenheit temperatures to Celsius temperatures. If the Celsius temperature ranges from 5°C to 20°C, use this formula to find the corresponding temperature range in degrees Fahrenheit.

WRITING ABOUT MATHEMATICS

113. Suppose that the solution set for a compound inequality can be written as $x < -3$ or $x > 2$. A student writes it as $2 < x < -3$. Is the student's three-part inequality correct? Explain your answer.

114. How can you determine whether an x-value is a solution to a compound inequality containing the word *and*? Give an example. Repeat the question for a compound inequality connected by the word *or*.

8.4   OTHER FUNCTIONS AND THEIR PROPERTIES

Expressing Domain and Range in Interval Notation ■ Absolute Value Function ■ Polynomial Functions ■ Rational Functions (Optional) ■ Operations on Functions

A LOOK INTO MATH ▷

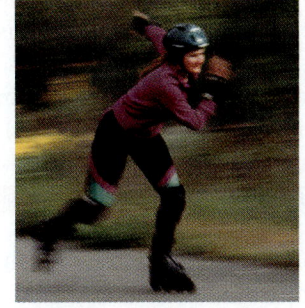

An important application of mathematics is to model and summarize data. Many times when data are plotted they do not lie on a line and, as a result, cannot be modeled with a linear function, as described in Section 8.2. For example, Table 8.12 on the next page lists the heart rate in beats per minute (bpm) of an athlete after exercise has stopped. Note that the heart rate does not decrease by the same amount during each 2-minute interval. Instead, it decreases faster at first and then begins to level off, indicating that the data are *nonlinear*.

A graph of these data is shown in Figure 8.36 on the next page. To model such data we can use **nonlinear functions** whose graphs are *not* lines. Because graphs of nonlinear data can have a wide variety of shapes, it has been necessary for mathematicians to create many different types of functions. In this section we discuss some of these functions and use a polynomial function to model the data in Table 8.12. (See Example 7.)

TABLE 8.12 Heart Rate of an Athlete

Time (minutes)	Heart Rate (bpm)
0	175
2	139
4	111
6	91
8	79
10	75

Source: V. Thomas, *Science and Sport.*

Figure 8.36 Heart Rate of an Athlete

Expressing Domain and Range in Interval Notation

The set of all valid inputs for a function is called the *domain*, and the set of all outputs from a function is called the *range*. For example, all real numbers are valid inputs for $f(x) = x^2$. Rather than writing "the set of all real numbers" for the domain of f, we can use *interval notation* to express the domain as $(-\infty, \infty)$. The symbol ∞ represents infinity and is not a real number. Because $x^2 \geq 0$ for every real number x, the output from $f(x) = x^2$ is never negative. Therefore the range of f is $[0, \infty)$, which denotes all nonnegative real numbers, or $x \geq 0$. Note that 0 is in the range of f because $f(0) = 0$, and a bracket "[" is used to indicate that 0 is included in the range of f.

EXAMPLE 1   **Writing domains in interval notation**

Write the domain for each function in interval notation.

(a) $f(x) = 4x$ **(b)** $g(t) = \sqrt{t - 1}$ **(c)** $h(v) = \dfrac{1}{v + 3}$

Solution

(a) The expression $4x$ is defined for all real numbers x. Thus the domain of f is $(-\infty, \infty)$.

(b) The square root $\sqrt{t - 1}$ is defined only when $t - 1$ is *not* negative. Thus the domain of g includes all real numbers satisfying $t - 1 \geq 0$ or $t \geq 1$. In interval notation this inequality is written as $[1, \infty)$.

(c) The expression $\frac{1}{v + 3}$ is defined except when $v + 3 = 0$ or $v = -3$. Thus the domain of h includes all real numbers except -3 and can be written as $(-\infty, -3) \cup (-3, \infty)$. Parentheses are used because -3 is not included in the domain of h.

Now Try Exercises 13, 21, 25

In the next example, we determine the domain and range of a function from its graph. Note that dots placed at each end of a graph indicate that the endpoints are included.

EXAMPLE 2 **Writing the domain and range in interval notation**

Use the graph of f in Figure 8.37 to write its domain and range in interval notation.

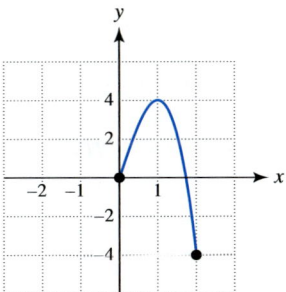

Figure 8.37

Solution

Because dots are placed at $(0, 0)$ and $(2, -4)$, the endpoints are included in the graph of f. Thus the graph in Figure 8.37 includes x-values from $x = 0$ to $x = 2$. In interval notation, the domain of f is $[0, 2]$. The range of f includes y-values from -4 to 4 and can be expressed in interval notation as $[-4, 4]$.   Now Try Exercise **39**

Absolute Value Function

In Chapter 1 we discussed the absolute value of a number. We can define a function called the **absolute value function** as $f(x) = |x|$. We evaluate f as follows.

$$f(11) = |11| = 11, \quad f(-4) = |-4| = 4, \quad \text{and} \quad f(-\pi) = |-\pi| = \pi$$

To graph $y = |x|$ we begin by making a table of values, as shown in Table 8.13. Next we plot these points and sketch the graph, as shown in Figure 8.38. Note that the graph is V-shaped and never lies below the x-axis because the absolute value of a number cannot be negative.

TABLE 8.13

x	y
-2	2
-1	1
0	0
1	1
2	2

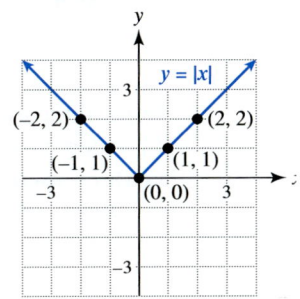

Figure 8.38 Absolute Value

Because valid inputs for $f(x) = |x|$ include all real numbers, the domain of the absolute value function is $(-\infty, \infty)$. The graph of the absolute value function never lies below the x-axis, so the range includes all real numbers greater than or equal to 0 and can be written as $[0, \infty)$.

EXAMPLE 3 Graphing a function

Sketch a graph of $f(x) = |x - 1|$. Write its domain and range in interval notation.

Solution
To graph $f(x) = |x - 1|$ we begin by making a table of values, as shown in Table 8.14. We then plot the points and sketch the V-shaped graph shown in Figure 8.39. Note that the graph of $f(x) = |x - 1|$ has the same shape as the graph of $y = |x|$, except that the graph of $y = f(x)$ is translated 1 unit to the right. The domain of f is $(-\infty, \infty)$, and the range of f is $[0, \infty)$.

TABLE 8.14

x	y
-1	2
0	1
1	0
2	1
3	2

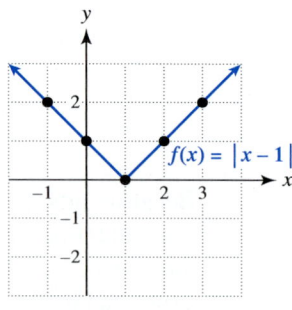

Figure 8.39

Now Try Exercise **81**

Polynomial Functions

In Chapter 5 we introduced polynomials and defined their degrees. The following expressions are examples of polynomials in one variable.

$$1 - 5x, \quad 3t^2 - 5t + 1, \quad \text{and} \quad z^3 + 5$$

(The exponents on variables in polynomials must be nonnegative integers.) Recall that the *degree* of a polynomial in one variable equals the largest exponent on the variable. Thus the degree of $1 - 5x$ is 1, the degree of $3t^2 - 5t + 1$ is 2, and the degree of $z^3 + 5$ is 3.
The equations

$$f(x) = 1 - 5x, \quad g(t) = 3t^2 - 5t + 1, \quad \text{and} \quad h(z) = z^3 + 5$$

define three **polynomial functions in one variable**. Function f is a **linear function** because it has degree 1, function g is a **quadratic function** because it has degree 2, and function h is a **cubic function** because it has degree 3. The domain of any polynomial function is the set of *all* real numbers.

EXAMPLE 4 Identifying polynomial functions

Determine whether $f(x)$ represents a polynomial function. If possible, identify the type of polynomial function and its degree.

(a) $f(x) = 5x^3 - x + 10$ **(b)** $f(x) = x^{-2.5} + 1$

(c) $f(x) = 1 - 2x$ **(d)** $f(x) = \dfrac{3}{x - 1}$

Solution
(a) The expression $5x^3 - x + 10$ is a cubic polynomial, so $f(x)$ represents a cubic polynomial function. It has degree 3.
(b) $f(x)$ does not represent a polynomial function because the variables in a polynomial must have *nonnegative integer* exponents.
(c) $f(x) = 1 - 2x$ represents a polynomial function that is linear. It has degree 1.
(d) $f(x)$ does not represent a polynomial function because $\frac{3}{x-1}$ is not a polynomial.

Now Try Exercises 43, 47, 51, 53

EXAMPLE 5   **Evaluating a polynomial function graphically and symbolically**

A graph of $f(x) = 4x - x^3$ is shown in Figure 8.40. Evaluate $f(-1)$ graphically and check your result symbolically.

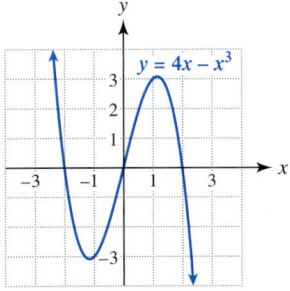

Figure 8.40

Solution
Graphical Evaluation   To calculate $f(-1)$ graphically, find -1 on the x-axis and move down until the graph of f is reached. Then move horizontally to the y-axis, as shown in Figure 8.41. Thus, when $x = -1$, $y = -3$ and $f(-1) = -3$.

Symbolic Evaluation   Evaluation of $f(x) = 4x - x^3$ is performed as follows.

$$f(\mathbf{-1}) = 4(\mathbf{-1}) - (\mathbf{-1})^3 = -4 - (-1) = -3$$

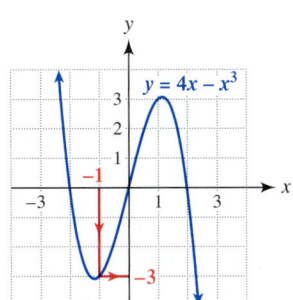

Figure 8.41 $f(-1) = -3$ Now Try Exercise 73

EXAMPLE 6   **Evaluating a polynomial function symbolically**

Evaluate $f(x)$ at the given value of x.
(a) $f(x) = -3x^4 - 2$, $x = 2$ (b) $f(x) = -2x^3 - 4x^2 + 5$, $x = -3$

Solution
(a) Be sure to evaluate exponents before multiplying.

$$f(\mathbf{2}) = -3(\mathbf{2})^4 - 2 = -3 \cdot 16 - 2 = -50$$

(b) $f(\mathbf{-3}) = -2(\mathbf{-3})^3 - 4(\mathbf{-3})^2 + 5 = -2(-27) - 4(9) + 5 = 23$

Now Try Exercises 59, 64

▶ **REAL-WORLD CONNECTION**   In the next example we discuss further the application presented in the introduction to this section.

EXAMPLE 7   **Modeling the heart rate of an athlete**

The quadratic function $P(t) = t^2 - 20t + 175$ models an athlete's pulse, or heart rate, in beats per minute t minutes after exercise has stopped, where $0 \leq t \leq 10$. See Table 8.12 and Figure 8.36 in the introduction to this section.

(a) Use P to determine the heart rate when $t = 0, 4$, and 10. Do your values agree with those shown in Table 8.12?

(b) Evaluate $P(20)$. Is this value valid?

Solution

(a) Start by evaluating $P(0), P(4)$, and $P(10)$.

$$P(0) = 0^2 - 20(0) + 175 = \mathbf{175}$$
$$P(4) = 4^2 - 20(4) + 175 = \mathbf{111}$$
$$P(10) = 10^2 - 20(10) + 175 = \mathbf{75}$$

These values agree with those shown in Table 8.12.

(b) $P(20) = 20^2 - 20(20) + 175 = 175$. Unless the athlete started to exercise again, 175 beats per minute is probably not valid. Note that valid inputs for the formula are limited to $0 \leq t \leq 10$. When t is not in this interval, the results may not be meaningful.

Now Try Exercise 107

Rational Functions (Optional)

A rational expression is formed when a polynomial is divided by a polynomial. For example, the expressions

$$\frac{2x - 1}{x}, \quad \frac{5}{x^2 + 1}, \quad \text{and} \quad \frac{2x - 5}{x^2 - 9}$$

are rational expressions. Rational expressions can be used to define *rational functions*.

> **RATIONAL FUNCTION**
>
> Let $p(x)$ and $q(x)$ be polynomials. Then a **rational function** is given by
>
> $$f(x) = \frac{p(x)}{q(x)}.$$
>
> The domain of f includes all x-values such that $q(x) \neq 0$.

From this definition, it follows that

$$f(x) = \frac{2x - 1}{x}, \quad g(x) = \frac{5}{x^2 + 1}, \quad \text{and} \quad h(x) = \frac{2x - 5}{x^2 - 9}$$

define rational functions. The domain of f includes all real numbers except 0, the domain of g includes all real numbers because $x^2 + 1 \neq 0$ for any x-value, and the domain of h includes all real numbers except ± 3.

EXAMPLE **8**   **Identifying the domains of rational functions**

Write the domain of each function in interval notation.

(a) $f(x) = \dfrac{1}{x + 2}$ **(b)** $g(x) = \dfrac{2x}{x^2 - 3x + 2}$ **(c)** $h(t) = \dfrac{4}{t^3 - t}$

Solution

(a) The domain of f includes all x-values except when the denominator equals 0.

$$x + 2 = 0 \qquad \text{Set the denominator equal to 0.}$$

$$x = -2 \qquad \text{Subtract 2.}$$

Thus $f(-2)$ is undefined and -2 must be excluded from the domain of f. In interval notation the domain of f is $(-\infty, -2) \cup (-2, \infty)$.

(b) The domain of g includes all real numbers except when $x^2 - 3x + 2 = 0$.

$$x^2 - 3x + 2 = 0 \qquad \text{Set the denominator equal to 0.}$$

$$(x - 1)(x - 2) = 0 \qquad \text{Factor.}$$

$$x = 1 \quad \text{or} \quad x = 2 \qquad \text{Zero-product property}$$

Because $g(1)$ and $g(2)$ are both undefined, 1 and 2 must be excluded from the domain of g. In interval notation the domain of g is $(-\infty, 1) \cup (1, 2) \cup (2, \infty)$.

(c) The domain of h includes all real numbers except when $t^3 - t = 0$.

$$t^3 - t = 0 \qquad \text{Set the denominator equal to 0.}$$

$$t(t^2 - 1) = 0 \qquad \text{Factor out } t.$$

$$t(t - 1)(t + 1) = 0 \qquad \text{Difference of squares}$$

$$t = 0 \quad \text{or} \quad t = 1 \quad \text{or} \quad t = -1 \qquad \text{Zero-product property}$$

In interval notation the domain of h is $(-\infty, -1) \cup (-1, 0) \cup (0, 1) \cup (1, \infty)$.

Now Try Exercises **27, 33, 35**

Like other functions, rational functions have graphs. To graph a rational function by hand, we usually start by making a table of values, as demonstrated in the next example. Because the graphs of rational functions are typically nonlinear, it is a good idea to plot at least three points on each side of an x-value where the formula is undefined.

EXAMPLE **9**   **Graphing a rational function**

Graph $f(x) = \frac{1}{x}$. State the domain of f.

Solution

Make a table of values for $f(x) = \frac{1}{x}$, as shown in Table 8.15. Notice that $x = 0$ is not in the domain of f, and a dash can be used to denote this undefined value. The domain of f is all real numbers such that $x \neq 0$. Start by picking three x-values on each side of 0.

TABLE 8.15

x	-2	-1	$-\frac{1}{2}$	0	$\frac{1}{2}$	1	2
$\frac{1}{x}$	$-\frac{1}{2}$	-1	-2	—	2	1	$\frac{1}{2}$

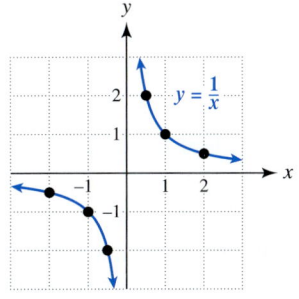

Figure 8.42

Plot the points shown in Table 8.15 and then connect the points with a smooth curve, as shown in Figure 8.42. Because $f(0)$ is undefined, the graph of $f(x) = \frac{1}{x}$ does not cross the line $x = 0$, the y-axis.

Now Try Exercise **89**

In the next example we evaluate a rational function in two ways.

EXAMPLE 10   **Evaluating a rational function symbolically and graphically**

Use the formula for $f(x)$ and Figure 8.43 to evaluate $f(-1), f(1),$ and $f(2)$.

(a) $f(x) = \dfrac{2x}{x - 1}$

(b)

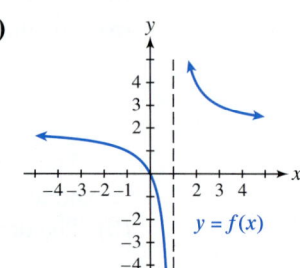

Figure 8.43

Solution

(a) *Symbolic Evaluation*

$$f(-1) = \frac{2(-1)}{-1 - 1} = 1$$

$$f(1) = \frac{2(1)}{1 - 1} = \frac{2}{0}, \quad \text{which is undefined. Input 1 is not in the domain of } f.$$

$$f(2) = \frac{2(2)}{2 - 1} = 4$$

(b) *Graphical Evaluation*   To evaluate $f(-1)$ graphically, find $x = -1$ on the x-axis and move upward to the graph of f. The y-value is 1 at the point of intersection, so $f(-1) = 1$, as shown in Figure 8.44(a). In Figure 8.44(b) the vertical line $x = 1$ is called a *vertical asymptote*. Because the graph of f does not intersect this line, $f(1)$ is undefined. Figure 8.44(c) reveals that $f(2) = 4$.

(a)

(b)

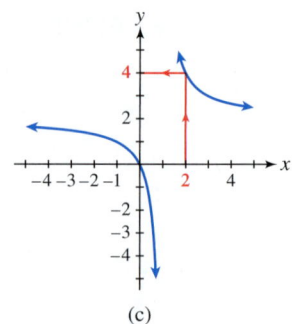

(c)

Figure 8.44

Now Try Exercise 77

A **vertical asymptote** is a vertical line that typically occurs in (but is not part of) the graph of a rational function when the denominator of the rational expression is 0 but the numerator is *not* 0. The graph of a rational function *never* crosses a vertical asymptote. In Figure 8.42, the vertical asymptote is the *y*-axis or $x = 0$.

NOTE: In Figures 8.42 and 8.43 the *y*-values on the graph become very large (approach ∞) or become very small (approach −∞) for *x*-values near the vertical asymptote. This characteristic is true in general for vertical asymptotes.

TECHNOLOGY NOTE: Asymptotes, Dot Mode, and Decimal Windows

When rational functions are graphed on graphing calculators, pseudo-asymptotes often occur because the calculator is simply connecting dots to draw a graph. The accompanying figures show the graph of $y = \frac{2}{x-2}$ in connected mode, dot mode, and with a *decimal*, or *friendly*, *window*. In dot mode, pixels in the calculator screen are not connected. With dot mode (and sometimes with a decimal window) pseudo-asymptotes do not appear. To learn more about these features, consult your owner's manual.

CALCULATOR HELP

To set a calculator in dot mode or to set a decimal window, see Appendix A (page AP-10).

$[-6, 6, 1]$ by $[-4, 4, 1]$

Connected Mode

$[-6, 6, 1]$ by $[-4, 4, 1]$

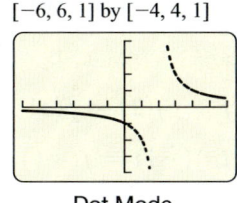

Dot Mode

$[-4.7, 4.7, 1]$ by $[-3.1, 3.1, 1]$

Decimal Window

Operations on Functions

▶ **REAL-WORLD CONNECTION** Frequently, it is necessary to add numbers or variables. Similarly, there may be a need to add two functions. For example, suppose that a student works road construction during summer break. On the taxable portion of this income, there might be a 6% state income tax and a 15% federal income tax. Suppose that *x* is the student's taxable income for *both* state and federal taxes. Then $S(x) = 0.06x$ calculates the state income tax and $F(x) = 0.15x$ calculates the federal income tax. Thus if $x = \$5000$, then the *total income tax T* is

$$T(\mathbf{5000}) = S(\mathbf{5000}) + F(\mathbf{5000})$$
$$= 0.06(\mathbf{5000}) + 0.15(\mathbf{5000})$$
$$= \$300 + \$750$$
$$= \$1050.$$

That is, to find the total tax, we add state and federal taxes. This statement can be written in *function notation* for a taxable income *x* as

$$T(x) = S(x) + F(x).$$

Given two functions, *f* and *g*, we define the sum, $f + g$, difference, $f - g$, product, *fg*, and quotient, *f/g*, as follows.

OPERATIONS ON FUNCTIONS

If $f(x)$ and $g(x)$ are both defined, then the sum, difference, product, and quotient of two functions f and g are defined by

$$(f + g)(x) = f(x) + g(x) \qquad \text{Sum}$$
$$(f - g)(x) = f(x) - g(x) \qquad \text{Difference}$$
$$(fg)(x) = f(x) \cdot g(x) \qquad \text{Product}$$
$$\left(\frac{f}{g}\right)(x) = \frac{f(x)}{g(x)}, \text{ where } g(x) \neq 0. \qquad \text{Quotient}$$

EXAMPLE 11 Performing arithmetic on functions

Use $f(x) = x^2$ and $g(x) = 2x - 4$ to evaluate each of the following.

(a) $(f + g)(3)$ (b) $(fg)(-1)$ (c) $\left(\dfrac{f}{g}\right)(0)$ (d) $(f/g)(2)$

Solution

(a) $(f + g)(3) = f(3) + g(3) = 3^2 + (2 \cdot 3 - 4) = 9 + 2 = 11$

(b) $(fg)(-1) = f(-1) \cdot g(-1) = (-1)^2 \cdot (2 \cdot (-1) - 4) = 1 \cdot (-6) = -6$

(c) $\left(\dfrac{f}{g}\right)(0) = \dfrac{f(0)}{g(0)} = \dfrac{0^2}{2 \cdot 0 - 4} = \dfrac{0}{-4} = 0$

(d) Note that $(f/g)(2)$ is equivalent to $\left(\dfrac{f}{g}\right)(2)$.

$$(f/g)(2) = \frac{f(2)}{g(2)} = \frac{2^2}{2 \cdot 2 - 4} = \frac{4}{0},$$

which is not possible. Thus $(f/g)(2)$ is undefined.   Now Try Exercise **101**

In the next example, we find the sum, difference, product, and quotient of two functions for a general x.

EXAMPLE 12 Performing arithmetic on functions

Use $f(x) = 4x - 5$ and $g(x) = 3x + 1$ to evaluate each of the following.

(a) $(f + g)(x)$ (b) $(f - g)(x)$ (c) $(fg)(x)$ (d) $\left(\dfrac{f}{g}\right)(x)$

Solution

(a) $(f + g)(x) = f(x) + g(x) = (4x - 5) + (3x + 1) = 7x - 4$

(b) $(f - g)(x) = f(x) - g(x) = (4x - 5) - (3x + 1) = x - 6$

(c) $(fg)(x) = f(x) \cdot g(x) = (4x - 5)(3x + 1) = 12x^2 - 11x - 5$

(d) $\left(\dfrac{f}{g}\right)(x) = \dfrac{f(x)}{g(x)} = \dfrac{4x - 5}{3x + 1}$

Now Try Exercise **105**

8.4 PUTTING IT ALL TOGETHER

The following table summarizes some important topics from this section.

Concept	Comments	Examples
Writing Domain and Range in Interval Notation	Interval notation can be used to specify the domain and range of a function.	If $f(x) = x^2 + 1$, the domain of f is $(-\infty, \infty)$, and the range of f is $[1, \infty)$.
Absolute Value Function	Defined by $$f(x) = \lvert x \rvert$$ and has a V-shaped graph	$f(-5) = \lvert -5 \rvert = 5$ $f(0) = \lvert 0 \rvert = 0$ $f(4) = \lvert 4 \rvert = 4$
Polynomial Function in One Variable	Can be defined by a polynomial; its degree equals the largest exponent of the variable.	Because $x^3 - 4x^2 + 6$ is a polynomial with degree 3, $$f(x) = x^3 - 4x^2 + 6$$ defines a polynomial function of degree 3 and is called a cubic function.
Rational Function	A rational function can be written as $$f(x) = \frac{p(x)}{q(x)},$$ where $p(x)$ and $q(x)$ are polynomials. Note that $q(x) \neq 0$.	Because $2x - 3$ and $x + 1$ are polynomials, $$f(x) = \frac{2x - 3}{x + 1}$$ defines a rational function. Because $f(x)$ is undefined at $x = -1$, the domain of f is $(-\infty, -1) \cup (-1, \infty)$.
Operations on Functions	$(f + g)(x) = f(x) + g(x)$ Sum $(f - g)(x) = f(x) - g(x)$ Difference $(fg)(x) = f(x)g(x)$ Product $\left(\dfrac{f}{g}\right)(x) = \dfrac{f(x)}{g(x)}, g(x) \neq 0$ Quotient	Let $f(x) = x^2$ and $g(x) = 1 - x^2$. $(f + g)(x) = f(x) + g(x)$ $\quad = x^2 + (1 - x^2)$ $\quad = 1$ $(f - g)(x) = f(x) - g(x)$ $\quad = x^2 - (1 - x^2)$ $\quad = 2x^2 - 1$ $(fg)(x) = f(x)g(x)$ $\quad = x^2(1 - x^2)$ $\quad = x^2 - x^4$ $\left(\dfrac{f}{g}\right)(x) = \dfrac{f(x)}{g(x)}$ $\quad = \dfrac{x^2}{1 - x^2}, x \neq -1, x \neq 1$

8.4 Exercises

CONCEPTS

1. The set of all valid inputs for a function is called its
   _____.

2. The set of all outputs for a function is called its _____.

3. The inequality $x \geq 1$ can be written in interval notation as _____.

4. The inequality $x < -1$ can be written in interval
   notation as _____.

5. The set of all real numbers can be written in interval
   notation as _____.

6. If the domain of a function includes all real numbers
   except 5, then its domain can be written in interval
   notation as _____.

7. The graph of the _____ function is V-shaped.

8. The degree of a polynomial in one variable equals the
   largest _____ of the variable.

9. A quadratic function has degree _____.

10. If a function is linear, then its degree is _____.

11. If $f(x) = \frac{x}{2x + 1}$, then f is a _____ function.

12. If $f(x) = \frac{x}{2x + 1}$, then the domain of f includes all
    real numbers except _____.

DOMAIN AND RANGE

*Exercises 13–24: Write the domain and the range of the
function in interval notation. (Hint: You may want to consider the graph of the function.)*

13. $f(x) = -2x$

14. $f(x) = -\frac{1}{4}x + 1$

15. $g(t) = \frac{2}{3}t - 3$

16. $g(t) = 9t$

17. $h(z) = z^2 + 2$

18. $h(z) = z^2 - 1$

19. $f(z) = -z^2$

20. $f(z) = -\frac{1}{4}z^2$

21. $g(x) = \sqrt{x + 1}$

22. $g(x) = \sqrt{x - 2}$

23. $h(x) = |x - 1|$

24. $h(x) = |2x|$

*Exercises 25–36: Write the domain of the rational func-
tion in interval notation.*

25. $f(x) = \dfrac{1}{x - 1}$

26. $f(x) = \dfrac{6}{x}$

27. $f(x) = \dfrac{x}{6 - 3x}$

28. $f(x) = \dfrac{3x}{2x - 4}$

29. $g(t) = \dfrac{2}{t^2 - 4}$

30. $g(t) = \dfrac{5}{1 - t^2}$

31. $g(t) = \dfrac{5t}{t^2 - 2t}$

32. $g(t) = \dfrac{-t}{2t^2 - 3t}$

33. $h(z) = \dfrac{2 - z}{z^3 - 1}$

34. $h(z) = \dfrac{z + 1}{z^3 - z^2}$

35. $f(x) = \dfrac{4}{x^2 - 2x - 3}$

36. $f(x) = \dfrac{1}{x^2 + 4x - 5}$

*Exercises 37–42: A graph of a function is shown. Write
the domain and range of the function in interval notation.*

37.

38.

39.

40.

41.

42.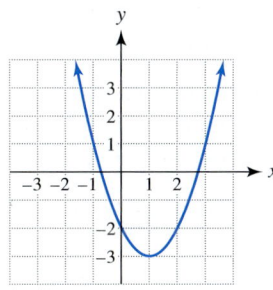

66. $g(t) = \dfrac{2}{3 - t}$ $t = 10, t = 3$

67. $g(t) = -\dfrac{t}{t + 1}$ $t = 5, t = -1$

68. $g(t) = -\dfrac{2 - t}{4t}$ $t = 4, t = -1$

69. $g(t) = \dfrac{t^2}{t^2 - t}$ $t = -5, t = 1$

70. $g(t) = \dfrac{t - 3}{t^2 - 3t + 2}$ $t = -2, t = 1$

IDENTIFYING POLYNOMIAL FUNCTIONS

Exercises 43–54: Determine whether $f(x)$ represents a polynomial function. If possible, identify the degree and type of polynomial function.

43. $f(x) = 5x - 11$ **44.** $f(x) = 9 - x$

45. $f(x) = x^3$ **46.** $f(x) = x^2 + 3$

47. $f(x) = \dfrac{6}{x + 5}$ **48.** $f(x) = |x|$

49. $f(x) = 1 + 2x - x^2$ **50.** $\frac{1}{4}x^3 - x$

51. $f(x) = 5x^{-2}$ **52.** $f(x) = x^2 + x^{-1}$

53. $f(x) = x^4 + 2x^2$ **54.** $f(x) = x^5 - 3x^3$

Exercises 71–78: If possible, use the graph to evaluate each expression. Then use the formula for $f(x)$ to check your results.

71. $f(0)$ and $f(1)$ **72.** $f(-1)$ and $f(2)$

     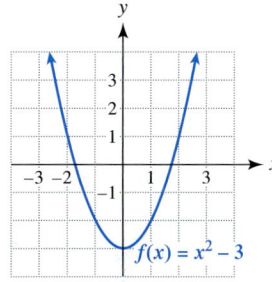

EVALUATING FUNCTIONS

Exercises 55–70: If possible, evaluate $g(t)$ for the given values of t.

55. $g(t) = |4t|$ $t = 3, t = 0$

56. $g(t) = |t + 12|$ $t = 18, t = -15$

57. $g(t) = |t - 2|$ $t = 1, t = -\frac{3}{4}$

58. $g(t) = |2t + 1|$ $t = 2, t = -\frac{1}{2}$

59. $g(t) = t^2 - t - 6$ $t = 3, t = -3$

60. $g(t) = 3t^2 - 2t$ $t = -2, t = 4$

61. $g(t) = 2t^3 - t$ $t = 2, t = -2$

62. $g(t) = \frac{1}{3}t^3$ $t = 1, t = -3$

63. $g(t) = t^2 - 2t - 6$ $t = 0, t = -3$

64. $g(t) = 2t^3 - t^2 + 4$ $t = 2, t = -1$

65. $g(t) = \dfrac{1}{t}$ $t = 11, t = -7$

73. $f(-1)$ and $f(2)$ **74.** $f(0)$ and $f(-2)$

     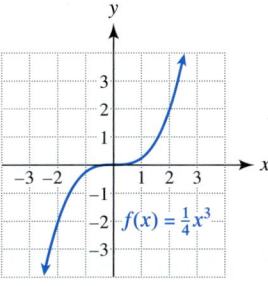

75. $f(-2)$ and $f(2)$ **76.** $f(-1)$ and $f(0)$

77. $f(-3)$ and $f(-1)$

78. $f(0)$ and $f(1)$

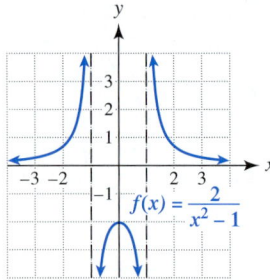

Exercises 79–98: Graph $y = f(x)$.

79. $f(x) = |2x|$

80. $f(x) = \left|\frac{1}{2}x\right|$

81. $f(x) = |x + 2|$

82. $f(x) = |x - 2|$

83. $f(x) = 1 - 2x$

84. $f(x) = \frac{1}{2}x + 1$

85. $f(x) = \frac{1}{2}x^2$

86. $f(x) = x^2 - 2$

87. $f(x) = \dfrac{1}{x - 1}$

88. $f(x) = \dfrac{1}{x + 1}$

89. $f(x) = \dfrac{1}{2x}$

90. $f(x) = \dfrac{2}{x}$

91. $f(x) = \dfrac{1}{x + 2}$

92. $f(x) = \dfrac{1}{x - 2}$

93. $f(x) = \dfrac{4}{x^2 + 1}$

94. $f(x) = \dfrac{6}{x^2 + 2}$

95. $f(x) = \dfrac{3}{2x - 3}$

96. $f(x) = \dfrac{1}{3x + 2}$

97. $f(x) = \dfrac{1}{x^2 - 1}$

98. $f(x) = \dfrac{4}{4 - x^2}$

OPERATIONS ON FUNCTIONS

Exercises 99–102: Use $f(x)$ and $g(x)$ to evaluate each of the following.

   (a) $(f + g)(3)$ *(b)* $(f - g)(-2)$

   (c) $(fg)(5)$ *(d)* $(f/g)(0)$

99. $f(x) = 5x, g(x) = x + 1$

100. $f(x) = x^2 + 2, g(x) = -2x$

101. $f(x) = 2x - 1, g(x) = 4x^2$

102. $f(x) = x^2 - 1, g(x) = x + 2$

Exercises 103–106: Use $f(x)$ and $g(x)$ to find each of the following.

   (a) $(f + g)(x)$ *(b)* $(f - g)(x)$

   (c) $(fg)(x)$ *(d)* $(f/g)(x)$

103. $f(x) = x + 1, g(x) = x + 2$

104. $f(x) = -3x, g(x) = x - 1$

105. $f(x) = 1 - x, g(x) = x^2$

106. $f(x) = x^2 + 4, g(x) = 6x$

APPLICATIONS

107. *Heart Rate of an Athlete*   The following table lists the heart rate of an athlete running a 100-meter race. The race lasts 10 seconds.

Time (seconds)	0	2	4	6	8	10
Heart Rate (bpm)	90	100	113	127	143	160

   (a) Does $P(t) = 0.2t^2 + 5t + 90$ model the data in the table exactly? Explain.
   (b) Does P provide a reasonable model for the athlete's heart rate?
   (c) Does $P(12)$ have significance in this situation? What should be the domain of P?

108. *Heart Rate of an Athlete*   The following table lists an athlete's heart rate after the athlete finishes exercising strenuously.

Time (minutes)	0	2	4	6
Heart Rate (bpm)	180	137	107	90

   (a) Does $P(t) = \frac{5}{3}t^2 - 25t + 180$ model the data in the table exactly? Explain.
   (b) Does P provide a reasonable model for the athlete's heart rate?
   (c) Does $P(12)$ have significance in this situation? What should be the domain of P?

109. *Burning Calories*   The following table lists the calories burned by a 140-pound person walking at 4 miles per hour for various lengths of time.

Time (minutes)	6	12	18	24
Calories	32	64	96	128

Source: Healthstats.com.

(a) Would it be reasonable to assume that the point (0, 0) could be included in the table? Why or why not?

(b) What type of polynomial function would model these data? Explain.

(c) Find a function C that models the calories burned in t minutes.

110. **Burning Calories** The following table lists the calories burned by a 180-pound person shopping at a mall.

Time (minutes)	10	30	50	80
Calories	30	90	150	240

Source: Healthstats.com.

(a) Would it be reasonable to assume that the point (0, 0) could be included in the table? Why or why not?

(b) What type of polynomial function would model these data? Explain.

(c) Find a function C that models the calories burned in t minutes.

Exercises 111–114: **Rational Models** *Match the physical situation with the graph (a.–d.) of the rational function that models it best.*

111. A population of fish that increases and then levels off

112. An insect population that dies out

113. The length of a ticket line as the rate at which people arrive in line increases

114. The wind speed during a day that is initially calm, becomes windy, and then is calm again

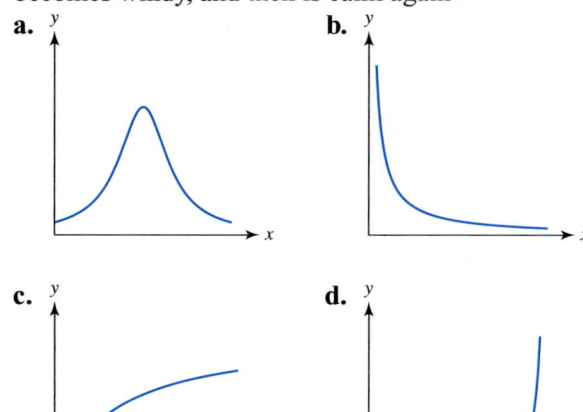

115. **Modeling a Train Track Curve** When curves are designed for train tracks, sometimes the outer rail is elevated, or banked, so that a locomotive and its cars can safely negotiate the curve at a higher speed than if the tracks were level. Suppose that a circular curve with a radius of r feet is being designed for a train traveling 60 miles per hour. Then $f(r) = \frac{2540}{r}$ calculates the proper elevation y in inches for the outer rail, where $y = f(r)$. See the accompanying figure.
(*Source:* L. Haefner, *Introduction to Transportation Systems.*)

(a) Evaluate $f(300)$ and interpret the result.

(b) Graph f in the window [0, 600, 100] by [0, 50, 10].

(c) Discuss how the elevation of the outer rail changes as the radius r increases.

116. **Grade of a Highway** The *grade x* of a hill is a measure of the steepness and corresponds to the slope of the highway. The accompanying figure shows a grade of $x = \frac{10}{100} = 10\%$ because the road rises 10 feet for every 100 feet of horizontal run. The braking distance D in feet for a car traveling 60 miles per hour on a wet uphill grade is given by

$$D(x) = \frac{3600}{30x + 9}.$$

(*Source:* N. Garber and L. Hoel, *Traffic and Highway Engineering.*)

(a) Evaluate $D(0.05)$ and interpret the result.

(b) Find the grade x for a braking distance of 300 feet.

WRITING ABOUT MATHEMATICS

117. Name two functions. Give their formulas, sketch their graphs, and state their domains and ranges.

118. Explain the difference between the domain and the range of a function.

CHECKING BASIC CONCEPTS
SECTIONS 8.3 AND 8.4

1. **(a)** Is 3 a solution to the compound inequality
$x + 2 < 4$ or $2x - 1 \geq 3$?
(b) Is 3 a solution to the compound inequality
$x + 2 < 4$ and $2x - 1 \geq 3$?

2. Solve the following compound inequalities. Write your answer in interval notation.
(a) $-5 \leq 2x + 1 \leq 3$
(b) $1 - x \leq -2$ or $1 - x \geq 2$
(c) $-2 < \dfrac{4 - 3x}{2} \leq 6$

3. Write the domain of each function in interval notation.
(a) $f(x) = x^2$
(b) $g(t) = \dfrac{1}{t - 1}$
(c) $h(z) = \sqrt{z}$

4. Use the graph of f to do the following.
(a) Write the domain and range of f in interval notation.
(b) Evaluate $f(0)$ and $f(-2)$.

5. Graph $f(x) = |x - 3|$.

8.5 ABSOLUTE VALUE EQUATIONS AND INEQUALITIES

Absolute Value Equations ■ Absolute Value Inequalities

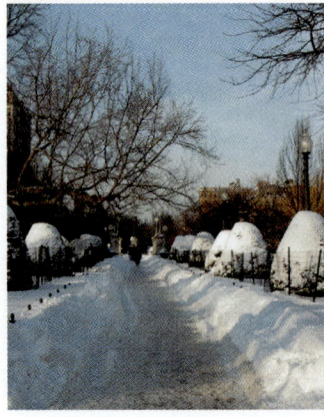

A LOOK INTO MATH ▷

Monthly average temperatures can vary greatly from one month to another, whereas yearly average temperatures remain fairly constant from one year to the next. In Boston, Massachusetts, the yearly average temperature is 50°F, but monthly average temperatures can vary from 28°F to 72°F. Because 50°F − 28°F = 22°F and 72°F − 50°F = 22°F, the monthly average temperatures are always within 22°F of the yearly average temperature. If T represents a monthly average temperature, we can model this situation by using the absolute value inequality

$$|T - 50| \leq 22.$$

The absolute value is necessary because a monthly average temperature T can be either greater than or less than 50°F by as much as 22°F. In this section we discuss absolute value equations and inequalities. (**Source:** A. Miller and J. Thompson, *Elements of Meteorology.*)

Absolute Value Equations

An equation that contains an absolute value is an **absolute value equation**. Examples are

$$|x| = 2, \quad |2x - 1| = 5, \quad \text{and} \quad |5 - 3x| - 3 = 1.$$

Consider the absolute value equation $|x| = 2$. This equation has *two* solutions: 2 and -2 because $|2| = 2$ and $|-2| = 2$. We can also demonstrate this result with a table of values or a graph. Refer to Table 8.16: $|x| = 2$ when $x = -2$ or $x = 2$. In Figure 8.45 the graph of $y_1 = |x|$ intersects the graph of $y_2 = 2$ at the points $(-2, 2)$ and $(2, 2)$. The x-values at these points of intersection correspond to the solutions -2 and 2.

TABLE 8.16

x	$|x|$
-2	2
-1	1
0	0
1	1
2	2

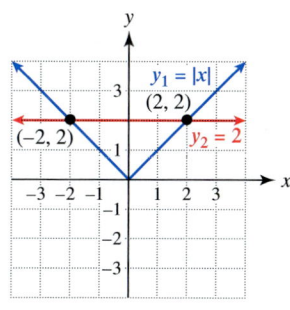

Figure 8.45

We generalize this discussion in the following manner.

SOLVING $|x| = k$

1. If $k > 0$, then $|x| = k$ is equivalent to $x = k$ or $x = -k$.
2. If $k = 0$, then $|x| = k$ is equivalent to $x = 0$.
3. If $k < 0$, then $|x| = k$ has no solutions.

EXAMPLE 1   Solving absolute value equations

Solve each equation.
(a) $|x| = 20$ **(b)** $|x| = -5$

Solution
(a) The solutions are -20 and 20.
(b) There are no solutions because $|x|$ is never negative.

Now Try Exercises 17, 19

EXAMPLE 2   Solving an absolute value equation

Solve $|2x - 5| = 3$ symbolically.

Solution
If $|2x - 5| = 3$, then either $2x - 5 = 3$ or $2x - 5 = -3$. Solve each equation separately.

$2x - 5 = 3$ or	$2x - 5 = -3$	Equations to be solved
$2x = 8$ or	$2x = 2$	Add 5.
$x = 4$ or	$x = 1$	Divide by 2.

The solutions are 1 and 4.

Now Try Exercise 25

A table of values can be used to solve the equation $|2x - 5| = 3$ from Example 2. Table 8.17 shows that $|2x - 5| = 3$ when $x = 1$ or $x = 4$.

Numerical Solution

TABLE 8.17

x	0	1	2	3	4	5	6
$|2x - 5|$	5	3	1	1	3	5	7

Graphical Solution

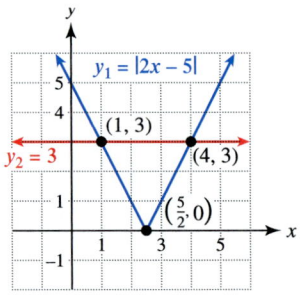

Figure 8.46

This equation can also be solved by graphing $y_1 = |2x - 5|$ and $y_2 = 3$. To graph y_1, first plot some of the points from Table 8.17. Its graph is V-shaped, as shown in Figure 8.46. Note that the x-coordinate of the "point" or vertex of the V can be found by solving the equation $2x - 5 = 0$ to obtain $\frac{5}{2}$. The graph of y_1 intersects the graph of y_2 at the points $(1, 3)$ and $(4, 3)$, giving the solutions **1** and **4**, so the graphical solutions agree with the numerical and symbolic solutions.

This discussion leads to the following result.

ABSOLUTE VALUE EQUATIONS

If $k > 0$, then

$$|ax + b| = k$$

is equivalent to

$$ax + b = k \quad \text{or} \quad ax + b = -k.$$

EXAMPLE **3**   Solving absolute value equations

Solve.
(a) $|5 - x| - 2 = 8$ **(b)** $\left|\frac{1}{2}(x - 6)\right| = \frac{3}{4}$

Solution
(a) Start by adding 2 to each side to obtain $|5 - x| = 10$. This new equation is satisfied by the solution from either of the following equations.

$5 - x = 10$ or $5 - x = -10$	Equations to be solved
$-x = 5$ or $-x = -15$	Subtract 5.
$x = -5$ or $x = 15$	Multiply by -1.

The solutions are -5 and 15.

(b) This equation is satisfied by the solution from either of the following equations.

$$\frac{1}{2}(x - 6) = \frac{3}{4} \quad \text{or} \quad \frac{1}{2}(x - 6) = -\frac{3}{4} \qquad \text{Equations to be solved}$$

$$2(x - 6) = 3 \quad \text{or} \quad 2(x - 6) = -3 \qquad \text{Multiply by 4 to clear fractions.}$$

$$2x - 12 = 3 \quad \text{or} \quad 2x - 12 = -3 \qquad \text{Distributive property}$$

$$2x = 15 \quad \text{or} \qquad 2x = 9 \qquad \text{Add 12.}$$

$$x = \frac{15}{2} \quad \text{or} \qquad x = \frac{9}{2} \qquad \text{Divide by 2.}$$

The solutions are $\frac{9}{2}$ and $\frac{15}{2}$. **Now Try Exercises 27, 29**

The next example illustrates absolute value equations that have either no solutions or one solution.

EXAMPLE 4 Solving absolute value equations

Solve.
(a) $|2x - 1| = -2$ **(b)** $|4 - 2x| = 0$

Solution
(a) Because an absolute value is never negative, there are no solutions. Figure 8.47 shows that the graph of $y_1 = |2x - 1|$ never intersects the graph of $y_2 = -2$.
(b) If $|y| = 0$, then $y = 0$. Thus $|4 - 2x| = 0$ when $4 - 2x = 0$ or when $x = 2$. The solution is 2. **Now Try Exercises 31, 32**

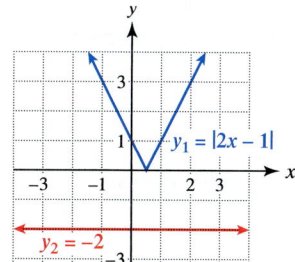

Figure 8.47

Sometimes an equation can have an absolute value on each side. An example would be $|2x| = |x - 3|$. In this situation either $2x = x - 3$ (the two expressions are equal) or $2x = -(x - 3)$ (the two expressions are opposites).

These concepts are summarized as follows.

> **SOLVING $|ax + b| = |cx + d|$**
>
> Let a, b, c, and d be constants. Then $|ax + b| = |cx + d|$ is equivalent to
> $$ax + b = cx + d \quad \text{or} \quad ax + b = -(cx + d).$$

EXAMPLE 5 Solving an absolute value equation

Solve $|2x| = |x - 3|$.

Solution
Solve the following compound equality.

$$2x = x - 3 \quad \text{or} \quad 2x = -(x - 3)$$
$$x = -3 \quad \text{or} \quad 2x = -x + 3$$
$$3x = 3$$
$$x = 1$$

The solutions are -3 and 1. **Now Try Exercise 35**

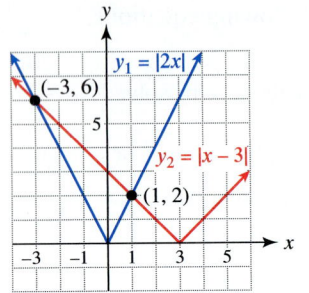

Figure 8.48

Example 5 is solved graphically in Figure 8.48. The graphs of $y_1 = |2x|$ and $y_2 = |x - 3|$ are V-shaped and intersect at $(-3, 6)$ and $(1, 2)$. The solutions are -3 and 1.

Absolute Value Inequalities

We can solve absolute value inequalities graphically. For example, to solve $|x| < 3$, let $y_1 = |x|$ and $y_2 = 3$ (see Figure 8.49). Their graphs intersect at $(-3, 3)$ and $(3, 3)$. The graph of y_1 is *below* the graph of y_2 for x-values *between*, but not including, $x = -3$ and $x = 3$. The solution set is $\{x \mid -3 < x < 3\}$ and is shaded on the x-axis.

Other absolute value inequalities can be solved graphically in a similar way. In Figure 8.50 the solutions to $|2x - 1| = 3$ are -1 and 2. The V-shaped graph of $y_1 = |2x - 1|$ is below the horizontal line $y_2 = 3$ when $-1 < x < 2$. Thus $|2x - 1| < 3$ whenever $-1 < x < 2$. The solution set is shaded on the x-axis.

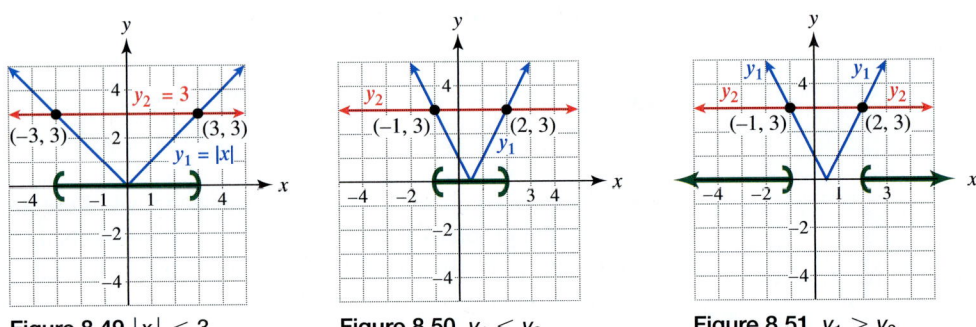

Figure 8.49 $|x| < 3$ **Figure 8.50** $y_1 < y_2$ **Figure 8.51** $y_1 > y_2$

In Figure 8.51 the V-shaped graph of $y_1 = |2x - 1|$ is above the horizontal line $y_2 = 3$ both to the left of -1 and to the right of 2. That is, $|2x - 1| > 3$ whenever $x < -1$ or $x > 2$. The solution set is shaded on the x-axis.

This discussion is summarized as follows.

ABSOLUTE VALUE INEQUALITIES

Let the solutions to $|ax + b| = k$ be c and d, where $c < d$ and $k > 0$.

1. $|ax + b| < k$ is equivalent to $c < x < d$.
2. $|ax + b| > k$ is equivalent to $x < c$ or $x > d$.

Similar statements can be made for inequalities involving $\leq$ or $\geq$.

EXAMPLE 6   **Solving absolute value equations and inequalities**

Solve each absolute value equation and inequality.
(a) $|2 - 3x| = 4$ **(b)** $|2 - 3x| < 4$ **(c)** $|2 - 3x| > 4$

Solution

(a) The given equation is equivalent to the following equations.

$$2 - 3x = 4 \quad \text{or} \quad 2 - 3x = -4 \qquad \text{Equations to be solved}$$

$$-3x = 2 \quad \text{or} \quad -3x = -6 \qquad \text{Subtract 2.}$$

$$x = -\frac{2}{3} \quad \text{or} \quad x = 2 \qquad \text{Divide by } -3.$$

The solutions are $-\frac{2}{3}$ and 2.

(b) Solutions to $|2 - 3x| < 4$ include x-values **between**, but not including, $-\frac{2}{3}$ and 2. Thus the solution set is $\left\{ x \mid -\frac{2}{3} < x < 2 \right\}$, or in interval notation, $\left(-\frac{2}{3}, 2 \right)$.

(c) Solutions to $|2 - 3x| > 4$ include x-values to the **left** of $x = -\frac{2}{3}$ **or** to the **right** of $x = 2$. Thus the solution set is $\left\{ x \mid x < -\frac{2}{3} \text{ or } x > 2 \right\}$, or in interval notation, $\left(-\infty, -\frac{2}{3} \right) \cup (2, \infty)$.

Now Try Exercise **43**

EXAMPLE 7 Solving an absolute value inequality

Solve $\left| \frac{2x - 5}{3} \right| > 3$. Write the solution set in interval notation.

Solution

Start by solving $\left| \frac{2x - 5}{3} \right| = 3$ as follows.

$$\frac{2x - 5}{3} = 3 \quad \text{or} \quad \frac{2x - 5}{3} = -3 \qquad \text{Equations to be solved}$$

$$2x - 5 = 9 \quad \text{or} \quad 2x - 5 = -9 \qquad \text{Multiply by 3.}$$

$$2x = 14 \quad \text{or} \quad 2x = -4 \qquad \text{Add 5.}$$

$$x = 7 \quad \text{or} \quad x = -2 \qquad \text{Divide by 2.}$$

Because the inequality symbol is $>$, the solution set is $x < -2$ or $x > 7$, or in interval notation, $(-\infty, -2) \cup (7, \infty)$.

Now Try Exercise **71**

The results from Examples 6 and 7 can be generalized as follows.

INEQUALITIES AND ABSOLUTE VALUES

If $k > 0$ and $y = f(x)$, then

$$|y| < k \text{ is equivalent to } -k < y < k \text{ and}$$

$$|y| > k \text{ is equivalent to } y < -k \text{ or } y > k.$$

Similar statements can be made for inequalities involving $\leq$ and $\geq$.

In the next example, we use the fact that $-k \le y \le k$ is equivalent to $|y| \le k$.

EXAMPLE 8 Analyzing error

An engineer is designing a circular cover for a container. The diameter d of the cover is to be 4.25 inches and must be accurate to within 0.01 inch. Write an absolute value inequality that gives acceptable values for d.

Solution

The diameter d must satisfy $4.24 \le d \le 4.26$. Subtracting 4.25 from each part gives

$$-0.01 \le d - 4.25 \le 0.01,$$

which is equivalent to $|d - 4.25| \le 0.01$.   Now Try Exercise 113

EXAMPLE 9 Modeling temperature in Boston

CALCULATOR HELP

To graph an absolute value, see Appendix A (page AP-7).

[0, 100, 10] by [0, 70, 10]

(a)

[0, 100, 10] by [0, 70, 10]

(b)

Figure 8.52

In the introduction to this section we discussed how the inequality $|T - 50| \le 22$ models the range for the monthly average temperatures T in Boston, Massachusetts.

(a) Solve this inequality and interpret the result.

(b) Give graphical support for part (a).

Solution

(a) *Symbolic Solution*   Start by solving $|T - 50| = 22$.

$$T - 50 = 22 \quad \text{or} \quad T - 50 = -22 \qquad \text{Equations to be solved}$$
$$T = 72 \quad \text{or} \qquad\quad T = 28 \qquad \text{Add 50 to each side.}$$

Thus the solution set to $|T - 50| \le 22$ is $\{T \mid 28 \le T \le 72\}$. Monthly average temperatures in Boston vary from $28°F$ to $72°F$.

(b) *Graphical Solution*   The graphs of $y_1 = |x - 50|$ and $y_2 = 22$ intersect at $(28, 22)$ and $(72, 22)$, as shown in Figures 8.52(a) and (b). The V-shaped graph of y_1 intersects the horizontal graph of y_2, or is below it, when $28 \le x \le 72$. Thus the solution set is $\{T \mid 28 \le T \le 72\}$. This result agrees with the symbolic result.   Now Try Exercise 105

Sometimes the solution set to an absolute value inequality can be either the empty set or the set of all real numbers. These two situations are illustrated in the next example.

EXAMPLE 10 Solving absolute value inequalities

Solve if possible.

(a) $|2x - 5| > -1$ **(b)** $|5x - 1| + 3 \le 2$

Solution

(a) Because the absolute value of an expression cannot be negative, $|2x - 5|$ is greater than -1 for every x-value. The solution set is all real numbers, or $(-\infty, \infty)$.

(b) Subtracting 3 from each side results in $|5x - 1| \le -1$. Because the absolute value is always greater than or equal to 0, no x-values satisfy this inequality. There are no solutions.   Now Try Exercises 63, 65

8.5 PUTTING IT ALL TOGETHER

The absolute value function is given by $f(x) = |x|$, and its graph is V-shaped. Its domain (set of valid inputs) includes all real numbers, and its range (outputs) includes all nonnegative real numbers. The following table summarizes methods for solving absolute value equations and inequalities involving $<$ and $>$ symbols. Inequalities containing $\leq$ and $\geq$ symbols are solved similarly.

Problem	Symbolic Solution	Example
$\|ax + b\| = k, k > 0$	Solve the equations $$ax + b = k$$ and $$ax + b = -k.$$	If $\|x - 2\| = 3$, then $$x - 2 = 3 \quad \text{or} \quad x - 2 = -3;$$ $$x = 5 \quad \text{or} \quad x = -1.$$
$\|ax + b\| < k, k > 0$	If the solutions to $$\|ax + b\| = k$$ are c and d, $c < d$, then the solutions to $$\|ax + b\| < k$$ satisfy $$c < x < d.$$	The solutions to $$\|x - 2\| < 3$$ satisfy $$-1 < x < 5.$$
$\|ax + b\| > k, k > 0$	If the solutions to $$\|ax + b\| = k$$ are c and d, $c < d$, then the solutions to $$\|ax + b\| > k$$ satisfy $$x < c \quad \text{or} \quad x > d.$$	The solutions to $$\|x - 2\| > 3$$ satisfy $$x < -1 \quad \text{or} \quad x > 5.$$

8.5 Exercises

CONCEPTS

1. Give an example of an absolute value equation.

2. Give an example of an absolute value inequality.

3. Is -3 a solution to $|x| = 3$?

4. Is -4 a solution to $|x| > 3$?

5. Is $|x| = 5$ equivalent to $x = -5$ or $x = 5$?

6. Is $|x| < 3$ equivalent to $x < -3$ or $x > 3$? Explain.

7. How many times does the graph of $y = |2x - 1|$ intersect the graph of $y = 5$?

8. How many times does the graph of $y = |2x - 1|$ intersect the graph of $y = -5$?

CHECKING SOLUTIONS

Exercises 9–14: Determine whether the given values of x are solutions to the absolute value equation or inequality.

9. $|2x - 5| = 1$ $x = -3, x = 3$

10. $|5 - 6x| = 1$ $x = 1, x = 0$

11. $|7 - 4x| \leq 5$ $x = -2, x = 2$

12. $|2 + x| < 2$ $x = -4, x = -1$

13. $|7x + 4| > -1$ $x = -\frac{4}{7}, x = 2$

14. $|12x + 3| \geq 3$ $x = -\frac{1}{4}, x = 2$

Exercises 15 and 16: Use the graph to solve the equation.

15. $y_1 = 2$ **16.** $y_1 = 3$

 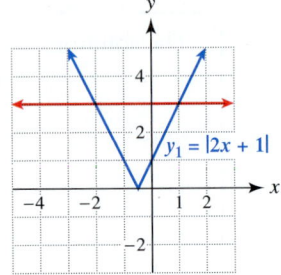

SYMBOLIC SOLUTIONS

Exercises 17–40: Solve the absolute value equation.

17. $|x| = 7$ **18.** $|x| = 4$

19. $|x| = -6$ **20.** $|x| = 0$

21. $|4x| = 9$ **22.** $|-3x| = 7$

23. $|-2x| - 6 = 2$ **24.** $|5x| + 1 = 5$

25. $|2x + 1| = 11$ **26.** $|1 - 3x| = 4$

27. $|-2x + 3| + 3 = 4$ **28.** $|6x + 2| - 2 = 6$

29. $\left|\frac{1}{2}x - 1\right| = 5$ **30.** $\left|6 - \frac{3}{4}x\right| = 3$

31. $|2x - 6| = -7$ **32.** $\left|1 - \frac{2}{3}x\right| = 0$

33. $\left|\frac{2}{3}z - 1\right| - 3 = 8$ **34.** $|1 - 2z| + 5 = 10$

35. $|z - 1| = |2z|$ **36.** $|2z + 3| = |2 - z|$

37. $|3t + 1| = |2t - 4|$ **38.** $\left|\frac{1}{2}t - 1\right| = \left|3 - \frac{3}{2}t\right|$

39. $\left|\frac{1}{4}x\right| = \left|3 + \frac{1}{4}x\right|$ **40.** $|2x - 1| = |2x + 2|$

Exercises 41–44: Solve each equation or inequality.

41. (a) $|2x| = 8$ (b) $|2x| < 8$
      (c) $|2x| > 8$

42. (a) $|3x - 9| = 6$
      (b) $|3x - 9| \leq 6$
      (c) $|3x - 9| \geq 6$

43. (a) $|5 - 4x| = 3$ (b) $|5 - 4x| \leq 3$
      (c) $|5 - 4x| \geq 3$

44. (a) $\left|\frac{x - 5}{2}\right| = 2$ (b) $\left|\frac{x - 5}{2}\right| < 2$

      (c) $\left|\frac{x - 5}{2}\right| > 2$

Exercises 45–76: Solve the absolute value inequality. Write your answer in interval notation.

45. $|x| \leq 3$ **46.** $|x| < 2$

47. $|k| > 4$ **48.** $|k| \geq 5$

49. $|t| \leq -3$ **50.** $|t| < -1$

51. $|z| > 0$ **52.** $|2z| \geq 0$

53. $|2x| > 7$ **54.** $|-12x| < 30$

55. $|-4x + 4| < 16$ **56.** $|-5x - 8| > 2$

57. $2|x + 5| \geq 8$ **58.** $-3|x - 1| \geq -9$

59. $|8 - 6x| - 1 \leq 2$ **60.** $4 - \left|\frac{2x}{3}\right| < -7$

61. $5 + \left|\frac{2 - x}{3}\right| \leq 9$ **62.** $\left|\frac{x + 3}{5}\right| \leq 12$

63. $|2x - 1| \leq -3$ **64.** $|x + 6| \geq -5$

65. $|x + 1| - 1 > -3$ **66.** $-2|1 - 7x| \geq 2$

67. $|2z - 4| \leq -1$ **68.** $|4 - z| \leq 0$

69. $|3z - 1| > -3$ **70.** $|2z| \geq -2$

71. $\left|\frac{2 - t}{3}\right| \geq 5$ **72.** $\left|\frac{2t + 3}{5}\right| \geq 7$

73. $|t - 1| \leq 0.1$ **74.** $|t - 2| \leq 0.01$

75. $|b - 10| > 0.5$ **76.** $|b - 25| \geq 1$

NUMERICAL AND GRAPHICAL SOLUTIONS

Exercises 77 and 78: Use the table of $y = |ax + b|$ to solve each equation or inequality. Write your answers in interval notation for parts (b) and (c).

77. (a) $y = 2$ **(b)** $y < 2$ **(c)** $y > 2$

x	-2	-1	0	1	2	3	4
y	3	2	1	0	1	2	3

78. (a) $y = 6$ **(b)** $y \leq 6$ **(c)** $y \geq 6$

x	-12	-6	0	6	12	18	24
y	9	6	3	0	3	6	9

Exercises 79 and 80: Use the graph of y_1 to solve each equation or inequality. Write your answers in interval notation for parts (b) and (c).

79. (a) $y_1 = 1$ **(b)** $y_1 \leq 1$ **(c)** $y_1 \geq 1$

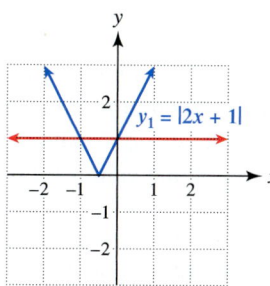

80. (a) $y_1 = 3$ **(b)** $y_1 < 3$ **(c)** $y_1 > 3$

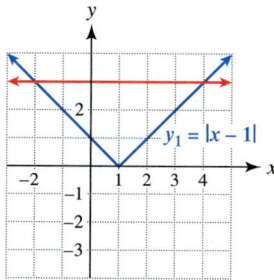

Exercises 81–90: Solve the inequality graphically. Write your answer in interval notation.

81. $|x| \geq 1$ **82.** $|x| < 2$

83. $|x - 1| \leq 3$ **84.** $|x + 5| \geq 2$

85. $|4 - 2x| > 2$ **86.** $|1.5x - 3| \geq 6$

87. $|10 - 3x| < 4$ **88.** $|7 - 4x| \leq 2.5$

89. $|8.1 - x| > -2$ **90.** $\left|\dfrac{5x - 9}{2}\right| \leq -1$

USING MORE THAN ONE METHOD

Exercises 91–94: Solve the absolute value inequality
     (a) *symbolically,*
     (b) *graphically, and*
     (c) *numerically.*
Write your answer in set-builder notation.

91. $|3x| \leq 9$ **92.** $|5 - x| \geq 3$

93. $|2x - 5| > 1$ **94.** $|-8 - 4x| < 6$

WRITING ABSOLUTE VALUE EQUATIONS

Exercises 95–102: Write each compound inequality as an absolute value inequality. Do not simplify Exercises 99–102.

95. $-4 \leq x \leq 4$ **96.** $-0.1 < y < 0.1$

97. $y < -2$ or $y > 2$ **98.** $-0.1 \leq x \leq 0.1$

99. $-0.3 \leq 2x + 1 \leq 0.3$

100. $4x < -5$ or $4x > 5$

101. $\pi x \leq -7$ or $\pi x \geq 7$

102. $-0.9 \leq x - \sqrt{2} \leq 0.9$

103. Thinking Generally If $a \neq 0$ and $k > 0$, then the graph of $y = |ax + b|$ intersects the graph of $y = k$ at _____ points.

104. Thinking Generally If a and k are positive, then the solution set to $|ax + b| < k$ is _____.

APPLICATIONS

Exercises 105–108: **Average Temperatures** *(Refer to Example 9.) The given inequality models the range for the monthly average temperatures T in degrees Fahrenheit at the location specified.*
     (a) *Solve the inequality.*
     (b) *Give a possible interpretation of the inequality.*

105. $|T - 43| \leq 24$, Marquette, Michigan

106. $|T - 62| \leq 19$, Memphis, Tennessee

107. $|T - 10| \leq 36$, Chesterfield, Canada

108. $|T - 61.5| \leq 12.5$, Buenos Aires, Argentina

109. *Highest Elevations*   The table lists the highest elevation in each continent.

Continent	Elevation (feet)
Asia	29,028
S. America	22,834
N. America	20,320
Africa	19,340
Europe	18,510
Antarctica	16,066
Australia	7,310

Source: National Geographic.

(a) Calculate the average A of these elevations.
(b) Which continents have their highest elevations within 1000 feet of A?
(c) Which continents have their highest elevations within 5000 feet of A?

110. *Distance*   Suppose that two cars, both traveling at a constant speed of 60 miles per hour, approach each other on a straight highway.
(a) If they are initially 4 miles apart, sketch a graph of the distance between the cars after x minutes, where $0 \le x \le 4$. (*Hint:* 60 miles per hour = 1 mile per minute.)
(b) Write an absolute value equation whose solution gives the times when the cars are 2 miles apart.
(c) Solve your equation from part (b).

111. *Error in Measurements*   Products are often manufactured to be a given size or shape to within a certain tolerance. For instance, if an aluminum can is supposed to have a diameter of 2.5 inches, either 2.501 inches or 2.499 inches might be acceptable. If the maximum error in the diameter of the can is restricted to 0.002 inch, an acceptable diameter d must satisfy the absolute value inequality

$$|d - 2.5| \le 0.002.$$

Solve this inequality for d and interpret the result.

112. *Error in Measurements*   (Refer to the preceding exercise.) Suppose that a person can operate a stopwatch accurately to within 0.02 second. If a runner's

time in the 400-meter dash is recorded as 51.57 seconds, write an absolute value inequality that gives the possible values for the actual time t.

113. *Error in Measurements*   A circular lid is being designed for a container. The diameter d of the lid is to be 3.8 inches and must be accurate to within 0.03 inch. Write an absolute value inequality that gives acceptable values for d.

114. *Manufacturing a Tire*   An engineer is designing a tire for a truck. The diameter d of the tire is to be 36 inches and the circumference must be accurate to within 0.1 inch. Write an absolute value inequality that gives acceptable values for d.

115. *Relative Error*   If a quantity is measured to be x and the true value is t, then the relative error in the measurement is $\left|\frac{x-t}{t}\right|$. If the true measurement is $t = 20$ and you want the relative error to be less than 0.05 (5%), what values for x are possible?

116. *Relative Error*   (Refer to the preceding exercise.) The volume V of a box is 50 cubic inches. How accurately must you measure the volume of the box for the relative error to be less than 3%?

WRITING ABOUT MATHEMATICS

117. If $a \ne 0$, how many solutions are there to the equation $|ax + b| = k$ when
(a) $k > 0$,   **(b)** $k = 0$, and   **(c)** $k < 0$?
Explain each answer.

118. Suppose that you know two solutions to the equation $|ax + b| = k$. How can you use these solutions to solve the inequalities $|ax + b| < k$ and $|ax + b| > k$? Give an example.

CHECKING BASIC CONCEPTS
SECTION 8.5

Write answers in interval notation whenever possible.

1. Solve $\left|\frac{3}{4}x - 1\right| - 3 = 5$.

2. Solve the absolute value equation and inequalities.
(a) $|3x - 6| = 8$
(b) $|3x - 6| < 8$
(c) $|3x - 6| > 8$

3. Solve the inequality $|-2(3 - x)| < 6$. Then solve $|-2(3 - x)| \geq 6$.

4. Use the graph to solve the equation and inequalities.

(a) $|2x - 4| = 2$ (b) $|2x - 4| \leq 2$
(c) $|2x - 4| \geq 2$

CHAPTER 8   SUMMARY

SECTION 8.1 ■ FUNCTIONS AND THEIR REPRESENTATIONS

Function   A function is a set of ordered pairs (x, y), where each x-value corresponds to exactly one y-value. A function takes a valid input x and computes exactly one output y, forming the ordered pair (x, y).

Domain and Range of a Function   The domain D is the set of all valid inputs, or x-values, and the range R is the set of all outputs, or y-values.

Examples: $f = \{(1, 2), (2, 3), (3, 3)\}$ has $D = \{1, 2, 3\}$ and $R = \{2, 3\}$.

$f(x) = x^2$ has domain all real numbers and range $y \geq 0$. (See the graph below.)

Function Notation $y = f(x)$ and is read "y equals f of x."

Example: $f(x) = \frac{2x}{x - 1}$ implies that $f(3) = \frac{2 \cdot 3}{3 - 1} = \frac{6}{2} = 3$. Thus the point $(3, 3)$ is on the graph of f.

Function Representations   A function can be represented symbolically, numerically, graphically, or verbally.

Symbolic Representation (Formula) $f(x) = x^2$

Numerical Representation (Table)

Graphical Representation (Graph)

x	y
-2	4
-1	1
0	0
1	1
2	4

Verbal Representation (Words) f computes the square of the input x.

Vertical Line Test   If every vertical line intersects a graph at most once, then the graph represents a function.

SECTION 8.2 ■ LINEAR FUNCTIONS

Linear Function   A linear function can be represented by $f(x) = ax + b$. Its graph is a (straight) line. For each unit increase in x, $f(x)$ changes by an amount equal to a.

Example: $f(x) = 2x - 1$ represents a linear function with $a = 2$ and $b = -1$.

Numerical Representation

Graphical Representation

Each 1-unit increase in x results in a 2-unit increase in $f(x)$.

NOTE: A numerical representation is a table of values of $f(x)$.

Modeling Data with Linear Functions   When data have a constant rate of change, they can be modeled by $f(x) = ax + b$. The constant a represents the *rate of change*, and the constant b represents the *initial amount* or the value when $x = 0$. That is,

$$f(x) = (\text{Rate of change})x + (\text{Initial amount}).$$

Example: In the following table, the y-values decrease by 3 units for each unit increase in x. When $x = 0$, $y = 4$. Thus the data are modeled by $f(x) = -3x + 4$.

x	-2	-1	0	1	2
y	10	7	4	1	-2

Midpoint Formula   The midpoint of the line segment connecting (x_1, y_1) and (x_2, y_2) is

$$\left(\frac{x_1 + x_2}{2}, \frac{y_1 + y_2}{2} \right).$$

Example: The midpoint of the line segment connecting $(-5, 8)$ and $(9, 4)$ is

$$\left(\frac{-5 + 9}{2}, \frac{8 + 4}{2} \right) = (2, 6).$$

SECTION 8.3 ■ COMPOUND INEQUALITIES

Compound Inequality   Two inequalities connected by *and* or *or*.

Examples: For $x + 1 < 3$ *or* $x + 1 > 6$, a solution satisfies *at least* one of the inequalities.
For $2x + 1 < 3$ *and* $1 - x > 6$, a solution satisfies *both* inequalities.

Three-Part Inequality   A compound inequality in the form $x > a$ and $x < b$ can be written as the three-part inequality $a < x < b$.

Example: $1 \le x < 7$ means $x \ge 1$ *and* $x < 7$.

Interval Notation   Can be used to identify intervals on the real number line.

Examples: $-2 < x \le 3$ is equivalent to $(-2, 3]$.
$x < 5$ is equivalent to $(-\infty, 5)$.
Real numbers are denoted $(-\infty, \infty)$.

SECTION 8.4 ■ OTHER FUNCTIONS AND THEIR PROPERTIES

Domain and Range in Interval Notation   The domain and range of a function can often be expressed in interval notation.

Example: The domain of $f(x) = x^2 - 2$ is all real numbers, or $(-\infty, \infty)$, and its range is real numbers greater than or equal to -2, or $[-2, \infty)$.

Absolute Value Function   The domain of $f(x) = |x|$ is $(-\infty, \infty)$, and the range is $[0, \infty)$.

Polynomial Functions   If $f(x)$ is a polynomial, then f is a polynomial function. The degree of a polynomial function (in one variable) equals the largest exponent of a variable. The graphs of polynomial functions with degree greater than 1 are not lines. The domain of a polynomial function is $(-\infty, \infty)$.

Examples: $f(x) = 4x - 1$ defines a linear function with degree 1.
$g(x) = 4x^2 + x - 4$ defines a quadratic function with degree 2.
$h(x) = x^3 + 0.7x - 1$ defines a cubic function with degree 3.

Rational Functions   If $f(x) = \dfrac{p(x)}{q(x)}$, where $p(x)$ and $q(x)$ are polynomials, f is a rational function. The domain of a rational function includes all real numbers, except x-values that make the denominator 0.

Examples: $f(x) = \frac{1}{x}$ has domain $(-\infty, 0) \cup (0, \infty)$, or $x \ne 0$.
$g(x) = \frac{x}{x^2 - 9}$ has domain $(-\infty, -3) \cup (-3, 3) \cup (3, \infty)$, or $x \ne -3, x \ne 3$.

Operations on Functions   If $f(x)$ and $g(x)$ are both defined, then the sum, difference, product, and quotient of two functions f and g are defined by

$$(f + g)(x) = f(x) + g(x) \qquad \text{Sum}$$
$$(f - g)(x) = f(x) - g(x) \qquad \text{Difference}$$
$$(fg)(x) = f(x) \cdot g(x) \qquad \text{Product}$$
$$\left(\frac{f}{g}\right)(x) = \frac{f(x)}{g(x)}, \text{ where } g(x) \neq 0. \qquad \text{Quotient}$$

Example: Let $f(x) = x^2 - 1$ and $g(x) = x^2 + 1$.

$$(f + g)(x) = f(x) + g(x) = (x^2 - 1) + (x^2 + 1) = 2x^2$$
$$(f - g)(x) = f(x) - g(x) = (x^2 - 1) - (x^2 + 1) = -2$$
$$(fg)(x) = f(x) \cdot g(x) = (x^2 - 1)(x^2 + 1) = x^4 - 1$$
$$\left(\frac{f}{g}\right)(x) = \frac{f(x)}{g(x)} = \frac{x^2 - 1}{x^2 + 1}$$

SECTION 8.5 ■ ABSOLUTE VALUE EQUATIONS AND INEQUALITIES

Absolute Value Equations   The graph of $y = |ax + b|$, $a \neq 0$, is V-shaped and intersects the horizontal line $y = k$ twice if $k > 0$. In this case there are two solutions to the equation $|ax + b| = k$ determined by $ax + b = k$ or $ax + b = -k$.

Example: The equation $|2x - 1| = 5$ has two solutions.

Symbolic Solution

$2x - 1 = 5$	or	$2x - 1 = -5$
$2x = 6$	or	$2x = -4$	Add 1.
$x = 3$	or	$x = -2$	Divide by 2.

Graphical Solution

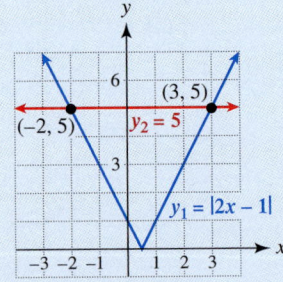

The solutions are -2 and 3.

Numerical Solution

x	-3	-2	-1	0	1	2	3
$|2x - 1|$	7	5	3	1	1	3	5

The solutions are -2 and 3.

Absolute Value Inequalities   If the solutions to $|ax + b| = k$ are c and d with $c < d$, then the solution set for $|ax + b| < k$ is $\{x \mid c < x < d\}$, and the solution set for $|ax + b| > k$ is $\{x \mid x < c \text{ or } x > d\}$.

Example: The solutions to the equation $|2x - 1| = 5$ are -2 and 3, so the solution set for $|2x - 1| < 5$ is $\{x \mid -2 < x < 3\}$, and the solution set for $|2x - 1| > 5$ is $\{x \mid x < -2 \text{ or } x > 3\}$.

If $k > 0$ and $y = f(x)$, then

$$|y| < k \text{ is equivalent to } -k < y < k \text{ and}$$
$$|y| > k \text{ is equivalent to } y < -k \text{ or } y > k.$$

Examples: $|3 - x| < 5$ is equivalent to $-5 < 3 - x < 5$ and

$|3 - x| > 5$ is equivalent to $3 - x < -5$ or $3 - x > 5$.

CHAPTER 8   REVIEW EXERCISES

SECTION 8.1

Exercises 1–4: Evaluate $f(x)$ for the given values of x.

1. $f(x) = 3x - 1$ $x = -2, \frac{1}{3}$

2. $f(x) = 5 - 3x^2$ $x = -3, 1$

3. $f(x) = \sqrt{x} - 2$ $x = 0, 9$

4. $f(x) = 5$ $x = -5, \frac{7}{5}$

Exercises 5 and 6: Do the following.
  (a) *Write a symbolic representation (formula) for the function described.*
  (b) *Evaluate the function for input 5 and interpret the result.*

5. Function P computes the number of pints in q quarts.

6. Function f computes 3 less than 4 times a number x.

7. If $f(3) = -2$, then the point _____ lies on the graph of f.

8. If $(4, -6)$ lies on the graph of f, then $f(\_\_) = \_\_\_$.

Exercises 9–12: Sketch a graph of $y = f(x)$.

9. $f(x) = -2x$ **10.** $f(x) = \frac{1}{2}x - \frac{3}{2}$

11. $f(x) = x^2 - 1$ **12.** $f(x) = \sqrt{x + 1}$

Exercises 13 and 14: Use the graph of f to evaluate the given expressions.

13. $f(0)$ and $f(-3)$ **14.** $f(-2)$ and $f(1)$

 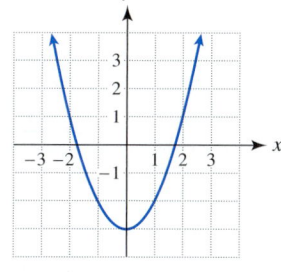

15. Use the table to evaluate $f(-1)$ and $f(3)$.

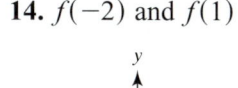

x	-1	1	3	5
$f(x)$	7	3	-1	-5

16. A function f is represented verbally by "Multiply the input x by 3 and then subtract 2." Give numerical, symbolic, and graphical representations for f. Let $x = -3, -2, -1, \ldots, 3$ in the table of values, and let $-3 \le x \le 3$ for the graph.

Exercises 17 and 18: Use the graph of f to estimate its domain and range.

17.   **18.**

 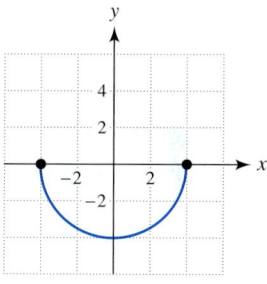

Exercises 19 and 20: Does the graph represent a function?

19.   **20.**

 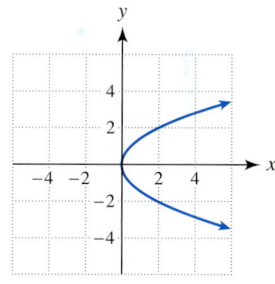

Exercises 21 and 22: Find the domain and range of S. Then state whether S defines a function.

21. $S = \{(-3, 4), (-1, 4), (2, 3), (4, -1)\}$

22. $S = \{(-1, 5), (0, 3), (1, -2), (-1, 2), (2, 4)\}$

Exercises 23–30: Find the domain.

23. $f(x) = -3x + 7$ **24.** $f(x) = \sqrt{x}$

25. $f(x) = \frac{3}{x}$ **26.** $f(x) = x^2 + 2$

27. $f(x) = \sqrt{5 - x}$ **28.** $f(x) = \frac{x}{x + 2}$

29. $f(x) = |2x + 1|$ **30.** $f(x) = x^3$

SECTION 8.2

Exercises 31 and 32: Does the graph represent a linear function?

31.

32.

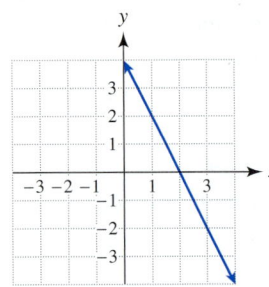

Exercises 33 and 34: Determine whether f is a linear function. If f is linear, give values for a and b so that f may be expressed as $f(x) = ax + b$.

33. $f(x) = -4x + 5$ **34.** $f(x) = \sqrt{x}$

Exercises 35 and 36: Use the table to determine whether $f(x)$ could represent a linear function. If it could, write the formula for f in the form $f(x) = ax + b$.

35.

x	0	2	4	6
$f(x)$	-3	0	3	6

36.

x	-1	0	1	2
$f(x)$	-5	0	10	15

37. Evaluate $f(x) = \frac{1}{2}x + 3$ at $x = -4$.

38. Use the graph to evaluate $f(-2)$ and $f(1)$.

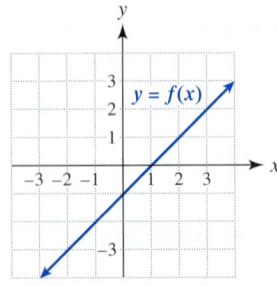

Exercises 39–42: Sketch a graph of $y = f(x)$.

39. $f(x) = x + 1$ **40.** $f(x) = 1 - 2x$

41. $f(x) = -\frac{1}{3}x$ **42.** $f(x) = -1$

43. Write a symbolic representation (formula) for a linear function H that calculates the number of hours in x days. Evaluate $H(2)$ and interpret the result.

44. Let $f(x) = \sqrt{x + 2} - x^2$.
   (a) Make a numerical representation (table) for f with $x = 1, 2, 3, \ldots, 7$.
   (b) Graph f in the standard window. What is the domain of f?

Exercises 45 and 46: Find the midpoint of the line segment connecting the given points.

45. $(-5, 3), (6, -9)$ **46.** $\left(\frac{2}{3}, -\frac{3}{4}\right), \left(\frac{1}{6}, \frac{3}{2}\right)$

SECTION 8.3

Use interval notation whenever possible for the remaining exercises.

Exercises 47–50: Solve the compound inequality. Graph the solution set on a number line.

47. $x + 1 \leq 3$ and $x + 1 \geq -1$

48. $2x + 7 < 5$ and $-2x \geq 6$

49. $5x - 1 \leq 3$ or $1 - x < -1$

50. $3x + 1 > -1$ or $3x + 1 < 10$

51. Use the table to solve $-2 \leq 2x + 2 \leq 4$.

x	-3	-2	-1	0	1	2	3
$2x + 2$	-4	-2	0	2	4	6	8

52. Use the following figure to solve each equation and inequality.
   (a) $y_1 = y_2$ (b) $y_2 = y_3$
   (c) $y_1 \leq y_2 \leq y_3$ (d) $y_2 < y_3$

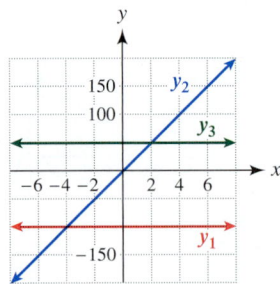

53. The graphs of y_1 and y_2 are shown at the top of the next page. Solve each equation and inequality.
   (a) $y_1 = y_2$ (b) $y_1 < y_2$
   (c) $y_1 > y_2$

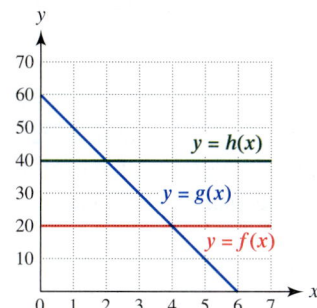

54. The graphs of three linear functions f, g, and h are shown in the following figure. Solve each equation and inequality.
(a) $f(x) = g(x)$ (b) $g(x) = h(x)$
(c) $f(x) < g(x) < h(x)$

Exercises 55–60: Write the inequality in interval notation.

55. $-3 \le x \le \frac{2}{3}$ **56.** $-6 < x \le 45$

57. $x < \frac{7}{2}$ **58.** $x \ge 1.8$

59. $x > -3$ and $x < 4$ **60.** $x < 4$ or $x > 10$

Exercises 61–66: Solve the three-part inequality. Write the solution set in interval notation.

61. $-4 < x + 1 < 6$ **62.** $20 \le 2x + 4 \le 60$

63. $-3 < 4 - \frac{1}{3}x < 7$ **64.** $2 \le \frac{1}{2}x - 2 \le 12$

65. $-3 \le \dfrac{4 - 5x}{3} - 2 < 3$

66. $30 \le \dfrac{2x - 6}{5} - 4 < 50$

SECTION 8.4

Exercises 67 and 68: Write the domain and the range of the function.

67. $f(t) = \frac{1}{2}t^2$ **68.** $f(x) = |x + 2|$

69. Write the domain of $f(x) = \frac{x + 1}{2x - 8}$.

70. Write the domain and range of the function shown in the graph.

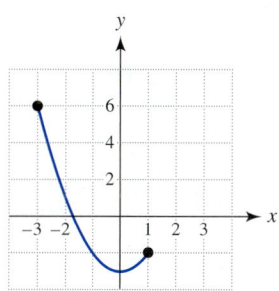

Exercises 71–74: Determine whether $f(x)$ represents a polynomial function. If possible, identify the degree and type of polynomial function.

71. $f(x) = 1 + 2x - 3x^2$

72. $f(x) = 5 + 7x$

73. $f(x) = x^3 + 2x$ **74.** $f(x) = |2x - 1|$

Exercises 75 and 76: If possible, evaluate $g(t)$ for the given values of t.

75. $g(t) = |1 - 4t|$ $t = 3, t = -\frac{1}{4}$

76. $g(t) = \dfrac{4}{4 - t^2}$ $t = 3, t = -2$

Exercises 77–80: Graph $y = f(x)$.

77. $f(x) = |x + 3|$ **78.** $f(x) = x^2 + 1$

79. $f(x) = \dfrac{1}{x}$ **80.** $f(x) = -3x$

81. Use $f(x) = 2x^2 - 3x$ and $g(x) = 2x - 3$ to find each of the following.
(a) $(f + g)(3)$ (b) $(fg)(3)$

82. Use $f(x) = x^2 - 1$ and $g(x) = x - 1$ to find each of the following.
(a) $(f - g)(x)$ (b) $(f/g)(x)$

SECTION 8.5

Exercises 83–86: Determine whether the given values of x are solutions to the absolute value equation or inequality.

83. $|12x - 24| = 24$ $x = -3; x = 2$

84. $|5 - 3x| > 3$ $x = \frac{4}{3}; x = 0$

85. $|3x - 6| \le 6$ $x = -3; x = 4$

86. $|2 + 3x| + 4 < 11$ $x = -3; x = \frac{2}{3}$

87. Use the accompanying table to solve each equation or inequality.
(a) $y_1 = 2$ (b) $y_1 < 2$ (c) $y_1 > 2$

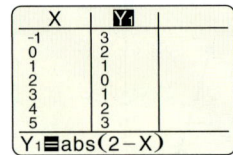

88. Use the graph of $y = |2x + 2|$ to solve each equation or inequality.
(a) $|2x + 2| = 4$ (b) $|2x + 2| \le 4$
(c) $|2x + 2| \ge 4$

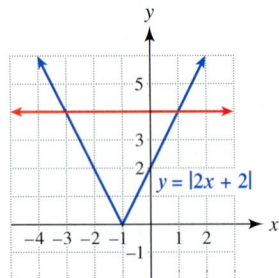

Exercises 89–94: Solve the absolute value equation.

89. $|x| = 22$ **90.** $|2x - 9| = 7$

91. $\left|4 - \frac{1}{2}x\right| = 17$ **92.** $\frac{1}{3}|3x - 1| + 1 = 9$

93. $|2x - 5| = |5 - 3x|$

94. $|-3 + 3x| = |-2x + 6|$

Exercises 95 and 96: Solve each absolute value equation or inequality.

95. (a) $|x + 1| = 7$
(b) $|x + 1| \le 7$
(c) $|x + 1| \ge 7$

96. (a) $|1 - 2x| = 6$
(b) $|1 - 2x| \le 6$
(c) $|1 - 2x| \ge 6$

Exercises 97–104: Solve the absolute value inequality.

97. $|x| > 3$ **98.** $|-5x| < 20$

99. $|4x - 2| \le 14$ **100.** $\left|1 - \frac{4}{5}x\right| \ge 3$

101. $|t - 4.5| \le 0.1$ **102.** $-2|13t - 5| \ge -4$

103. $|5 - 4x| > -5$ **104.** $|2t - 3| \le 0$

Exercises 105 and 106: Solve the inequality graphically.

105. $|2x| \ge 3$ **106.** $\left|\frac{1}{2}x - 1\right| \le 2$

Exercises 107 and 108: Write each compound inequality as an absolute value inequality.

107. $-0.05 \le x \le 0.05$

108. $5x - 1 < -4$ or $5x - 1 > 4$

APPLICATIONS

109. *Age at First Marriage*   The median age at the first marriage for men from 1890 to 1960 can be modeled by $f(x) = -0.0492x + 119.1$, where x is the year.   (*Source:* National Center of Health Statistics.)
(a) Find the median age in 1910.
(b) Graph f in [1885, 1965, 10] by [22, 26, 1]. What happened to the median age?
(c) What is the slope of the graph of f? Interpret the slope as a rate of change.

110. *Marriages*   From 1980 to 2000 the number of U.S. marriages in millions could be modeled by the formula $f(x) = 2.4$, where x is the year.
(a) Estimate the number of marriages in 1991.
(b) What information does f give about the number of marriages during this time period?

111. *Fat Grams*   A cup of milk contains 8 grams of fat.
(a) Give a formula for $f(x)$ that calculates the number of fat grams in x cups of milk.
(b) What is the slope of the graph of f?
(c) Interpret the slope as a rate of change.

112. *Birth Rate*   The U.S. birth rate per 1000 people from 1990 through 1997 is shown in the table.

Year	1990	1991	1992	1993
Birth Rate	16.7	16.3	16.0	15.7

Year	1994	1995	1996	1997
Birth Rate	15.3	14.8	14.7	14.5

Source: U.S. Census Bureau.

(a) Make a scatterplot of the data.
(b) Model the data with $f(x) = mx + b$, where x is the year. Answers may vary.
(c) Use f to estimate the birth rate in 2000.

113. *Unhealthy Air Quality*   The Environmental Protection Agency (EPA) monitors air quality in U.S. cities. The function f gives the annual number of days with unhealthy air quality in Los Angeles, California, for selected years.

x	1995	1997	1999	2000	2003
$f(x)$	113	60	56	87	88

Source: Environmental Protection Agency.

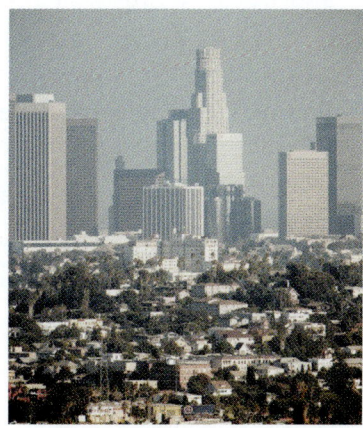

(a) Find $f(1995)$ and interpret your result.
(b) Identify the domain and range of f.
(c) Discuss the trend of air pollution in Los Angeles.

114. *Temperature Scales*   The following table shows equivalent temperatures in degrees Celsius and degrees Fahrenheit.

°C	−40	0	15	35	100
°F	−40	32	59	95	212

(a) Plot the data. Let the x-axis correspond to the Celsius temperature and the y-axis correspond to the Fahrenheit temperature. What type of relation exists between the data?
(b) Find $f(x) = ax + b$ so that f receives the Celsius temperature x as input and outputs the corresponding Fahrenheit temperature. Interpret the slope of the graph of f.
(c) If the temperature is 20°C, what is this temperature in degrees Fahrenheit?

115. *High Temperatures*   The formula
$$f(x) = -1.466x^2 + 20.25x + 9$$
models the monthly average high temperatures in degrees Fahrenheit at Columbus, Ohio. In this formula, let $x = 1$ be January, $x = 2$ be February, and so on.   (*Source:* J. Williams, *The Weather Almanac.*)
(a) What is the average high temperature in May?
(b) Make a table of $f(x)$, starting at $x = 1$ and incrementing by 1. During what month is the average temperature the greatest?
(c) Graph f in [1, 12, 1] by [30, 90, 10] and interpret the graph.

116. *Geometry*   A rectangle is 5 feet longer than twice its width. If the rectangle has a perimeter of 88 feet, what are the dimensions of the rectangle?

117. *Temperature*   The formula $F = \frac{9}{5}C + 32$ may be used to convert Fahrenheit temperature to Celsius temperature. The temperature range at Houghton Lake, Michigan, has varied between −48°F and 107°F. Find this temperature range in Celsius.

118. *Error in Measurements*   A square garden is being fenced along its 160-foot perimeter. If the length L of the fence must be within 1 foot of the garden's perimeter, write an absolute value inequality that gives acceptable values for L. Solve your inequality.

119. *Average Precipitation*   The average rainfall in Houston, Texas, is 3.9 inches per month. Each month's average A is within 1.7 inches of 3.9 inches. (*Source:* J. Williams, *The Weather Almanac 1995.*)
(a) Write an absolute value inequality that models this situation.
(b) Solve the inequality.

120. *Relative Error*   If a quantity is measured to be T and the actual value is A, then the relative error in this measurement is $\left|\frac{T - A}{A}\right|$. If $A = 35$ and the relative error is to be less than 0.08 (8%), what values for T are possible?

CHAPTER 8   TEST *Pass the Test* Video solutions to all test exercises

1. Evaluate $f(4)$ if $f(x) = 3x^2 - \sqrt{x}$. Give a point on the graph of f.

2. Write a symbolic representation (formula) for a function C that calculates the cost of buying x pounds of candy at \$4 per pound. Evaluate $C(5)$ and interpret your result.

3. Sketch a graph of $y = f(x)$.
   (a) $f(x) = -2x + 1$ (b) $f(x) = x^2 + 1$
   (c) $f(x) = \sqrt{x + 3}$ (d) $f(x) = |x + 1|$

4. Use the graph of f to evaluate $f(-3)$ and $f(0)$. Determine the domain and range of f.

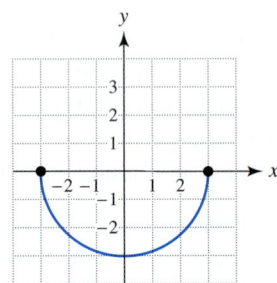

5. A function f is represented verbally by "Square the input x and then subtract 5." Give symbolic, numerical, and graphical representations of f. Let $x = -3, -2, -1, \ldots, 3$ in the numerical representation (table) and let $-3 \le x \le 3$ for the graph.

6. Determine whether the graph represents a function. Explain your reasoning.

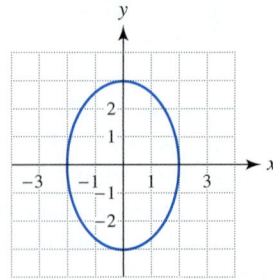

7. Find the domain of function f.
   (a) $f = \{(-2, 3), (-1, 5), (0, 3), (5, 7)\}$

   (b) $f(x) = \frac{3}{4}x - 5$ (c) $f(x) = \sqrt{x + 4}$

   (d) $f(x) = 2x^2 - 1$ (e) $f(x) = \frac{3x}{5 - x}$

8. Determine if $f(x) = 6 - 8x$ is a linear function. If it is, write it in the form $f(x) = ax + b$.

9. Graph the solution set to $2x + 6 < 2$ and $-3x \ge 3$ on a number line.

10. Use the table to solve the compound inequality $-3x < -3$ or $-3x > 6$. Write your answer in interval notation.

x	-3	-2	-1	0	1	2	3
$-3x$	9	6	3	0	-3	-6	-9

11. Use the following figure to solve the equations and inequalities. Write your answers for parts (c) and (d) in interval notation.
    (a) $y_1 = y_2$ (b) $y_2 = y_3$
    (c) $y_1 \le y_2 \le y_3$ (d) $y_2 < y_3$

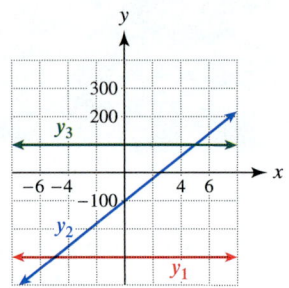

12. Solve $-2 < 2 + \frac{1}{2}x < 2$ and write the solution set in interval notation.

13. Solve the equation $|2 - \frac{1}{3}x| = 6$.

14. Solve each inequality. Write your answer in interval notation.
    (a) $|x| \le 5$ (b) $|x| > 0$

    (c) $|2x + 7| \le -2$ (d) $|5x| > 10$

    (e) $-2|2 - 5x| + 1 \le 5$ (f) $|3 - x| < 5$

15. Solve each equation or inequality. Write your answers for parts (b) and (c) in interval notation.
    (a) $|1 - 5x| = 3$ (b) $|1 - 5x| \le 3$
    (c) $|1 - 5x| \ge 3$

16. Write the domain and range of $f(x) = |2x - 5|$ in interval notation.

17. Determine whether $f(x) = 1 - 2x + x^3$ represents a polynomial function. If possible, identify the degree and type of polynomial function.

18. Evaluate $h(t) = -\frac{4t}{5 - t}$ at $t = -2$. Write the domain of h in interval notation.

19. Let $f(x) = x^2 + 1$ and $g(x) = 2x$. Find each of the following.
  (a) $(f - g)(-2)$ **(b)** $(fg)(x)$

20. *Drinking Fluids and Exercise*   To determine the number of ounces of fluid that a person should drink in a day, divide his or her weight in pounds by 2 and then add 0.4 ounce for every minute of exercise.
  (a) Write a function that gives the fluid requirements for a person weighing 150 pounds and exercising x minutes a day.

(b) If a 150-pound runner needs 89 ounces of fluid each day, determine the runner's daily minutes of exercise.

21. *Heart Rate of an Athlete*   The following table lists the heart rate or pulse of an athlete running a 400-meter race. The race lasts 50 seconds.

Time (seconds)	0	20	30	50
Heart Rate (bpm)	100	134	150	180

  (a) Does $P(t) = -\frac{1}{300}t^2 + \frac{53}{30}t + 100$ model the data in the table exactly? Explain.
  (b) Does $P(60)$ have significance in this situation? What should be the domain of P?

CHAPTER 8   EXTENDED AND DISCOVERY EXERCISES

1. *Developing a Model*   Two identical cylindrical tanks, A and B, each contain 100 gallons of water. Tank A has a pump that begins removing water at a constant rate of 8 gallons per minute. Tank B has a plug removed from its bottom and water begins to flow out—faster at first and then more slowly.
  (a) Assuming that the tanks become empty at the same time, sketch a graph that models the amount of water in each tank. Explain your graphs.
  (b) Which tank is half empty first? Explain.

 2. *Modeling Real Data*   Per capita personal incomes in the United States from 1990 through 2000 are listed in the following table.

Year	1990	1991	1992	1993
Income	$18,666	$19,091	$20,105	$20,800

Year	1994	1995	1996	2000
Income	$21,809	$23,359	$24,436	$29,676

Source: Department of Commerce.

  (a) Make a scatterplot of the data.
  (b) Find a function f that models the data. Explain your reasoning.
  (c) Use f to estimate per capita personal income in 1998.

3. *Weight of a Small Fish*   The figure shows a graph of a function f that models the weight in milligrams of a small fish, *Lebistes reticulatus*, during the first 14 weeks of its life.   (*Source:* D. Brown and P. Rothery, *Models in Biology*.)

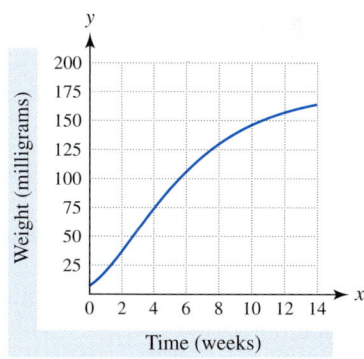

  (a) Estimate the weight of the fish when it hatches, at 6 weeks, and at 12 weeks.
  (b) If (x_1, y_1) and (x_2, y_2) are points on the graph of a function, the *average rate of change of f from x_1 to x_2* is given by $\frac{y_2 - y_1}{x_2 - x_1}$. Approximate the average rates of change of f from hatching to 6 weeks and from 6 weeks to 12 weeks.
  (c) Interpret these rates of change.
  (d) During which time period does the fish gain weight the fastest?

4. *Recording Music* A compact disc (CD) can hold approximately 600 million bytes. One *byte* is capable of storing one letter of the alphabet. For example, the word "function" requires 8 bytes to store in computer memory. One million bytes is commonly referred to as a *megabyte* (MB). Recording music requires an enormous amount of memory. The accompanying table lists the megabytes x needed to record y seconds (sec) of music.

x (MB)	0.129	0.231	0.415	0.491
y (sec)	6.01	10.74	19.27	22.83

x (MB)	0.667	1.030	1.160	1.260
y (sec)	31.00	49.00	55.25	60.18

Source: Gateway 2000 System CD.

(a) Make a scatterplot of the data.
(b) What type of relationship seems to exist between x and y? Why does this relationship seem reasonable?
(c) Find the slope–intercept form of a line that models the data. Interpret the slope of this line as a rate of change. Answers may vary.
(d) Check your answer in part (c) by graphing the line and data in the same graph.
(e) Write a linear equation whose solution gives the megabytes needed to record 120 seconds of music.
(f) Solve the equation in part (e) graphically or symbolically.

CHAPTERS 1–8   CUMULATIVE REVIEW EXERCISES

1. Write 120 as a product of prime numbers.

2. Translate the sentence "Triple a number decreased by 4 equals the number" into an equation using the variable n. Solve the equation.

Exercises 3 and 4: Simplify completely.

3. $\frac{1}{4} \div \frac{3}{4} - \frac{1}{2}$ **4.** $12 - 3^2 \div 3 \cdot 2$

5. Solve $-3(3 - x) - 6 = 2x$.

6. Convert 0.075 to a percentage.

7. Solve $A = \frac{1}{3}(2a - b)$ for a.

8. Solve $5 - 3t < 1 - t$. Write the solution set in set-builder notation.

Exercises 9 and 10: Graph the equation.

9. $3x - 2y = -6$ **10.** $y = |2x - 4|$

11. Write the slope–intercept form of a line that passes through the points $(-4, 4)$ and $(2, 1)$.

12. Solve the system of equations. Write your answer as an ordered pair.
$$2x + 3y = 5$$
$$3x - 2y = 1$$

Exercises 13 and 14: Shade the solution set for the given inequality.

13. $x \geq -1$ **14.** $x - y \geq 2$

15. Shade the solution set for the system of inequalities.
$$x - y \geq 1$$
$$2x + y \leq 0$$

16. Simplify $(3x^2 + 2x - 4) - (4x^2 + 5)$.

Exercises 17–20: Simplify and use positive exponents.

17. $\dfrac{x^{-4}}{x^{-3}}$ **18.** $(3b^{-3})(2b^4)$

19. $3(2t)^3$ **20.** $\left(\dfrac{2x^3}{x^2 y^{-1}}\right)^{-2}$

Exercises 21–24: Multiply the expression.

21. $2x^2(x^3 - 4x^2 - 5)$

22. $(5x + 1)(2x - 7)$

23. $(y - 3)(y + 3)$

24. $(x - 4y)^2$

25. Write 2.5×10^4 in standard form.

26. Write 0.028 in scientific notation.

Exercises 27 and 28: Divide.

27. $\dfrac{6x^3 - 4x^2 + 8x}{2x}$

28. $(3x^3 + 2x^2 + 1) \div (x - 1)$

Exercises 29–34: Factor completely.

29. $10x^2y^3 - 15x^3y^2$ **30.** $x^3 + 3x^2 - x - 3$

31. $2z^2 + z - 3$ **32.** $16x^2 - 25$

33. $a^3 - 8$ **34.** $z^4 + 7z^2 + 6$

Exercises 35–38: Solve the equation.

35. $x(x + 5) = 0$ **36.** $4x^2 = 0$

37. $2x^2 + 5x = 3$ **38.** $x^3 = x$

39. Simplify $\dfrac{x^2 + 4x + 4}{x + 2}$ to lowest terms.

40. If possible, evaluate $\dfrac{x^2 + 1}{x - 1}$ for $x = -2$ and $x = 1$.

Exercises 41 and 42: Simplify to lowest terms.

41. $\dfrac{x^2 - 3x + 2}{x + 2} \div \dfrac{x - 1}{2x + 4}$

42. $\dfrac{1}{x + 3} + \dfrac{2}{x + 1}$

43. Solve $\dfrac{x + 2}{x - 1} = \dfrac{2}{3}$.

44. Suppose that y is directly proportional to x and that $y = 5$ when $x = 10$. Find y when $x = 20$.

45. Evaluate $f(x) = x^2 - 4x$ for $x = -2$.

46. Graph $f(x) = x^2 - 2$ and identify the domain and range of f. Write your answer in interval notation.

47. Solve $x - 2 < 3$ or $x - 2 > 6$. Write your answer in interval notation.

48. Solve $|2x - 4| = 6$.

49. Solve $|2x - 4| \le 6$. Write your answer in interval notation.

50. Solve $|x - 4| > 2$. Write your answer in interval notation.

APPLICATIONS

51. *Modeling Motion*   The table lists distance d in miles traveled by a car for various elapsed times t in hours. Find an equation that models these data.

t (hours)	2	3	4	6
d (miles)	144	216	288	432

52. *Rainfall*   At noon 2 inches of rain had fallen. For the next 6 hours rain fell at $\frac{1}{4}$ inch per hour.
   (a) Find an equation in the form $y = mx + b$ that calculates the number of inches of rain that fell x hours past noon.
   (b) What is the slope of the graph of f?
   (c) Interpret the slope as a rate of change.
   (d) How much rain had fallen by 4 P.M.?

53. *Interest*   A total of $5000 are deposited in two accounts paying 5% and 7% annual interest. If total interest received at the end of the year is $308, determine how much is invested at each interest rate.

54. *Height of a Building*   An 8-foot-tall stop sign casts a 5-foot-long shadow, while a nearby building casts a 65-foot-long shadow. Find the height of the building.

55. *Shoveling the Driveway*   Two people are shoveling snow from a driveway. The first person shovels 9 square feet per minute, while the second person shovels 11 square feet per minute.
   (a) Write a simplified expression that gives the total number of square feet the two people can shovel in x minutes.
   (b) How many minutes would it take for them to clear a driveway with 1000 square feet?

56. *Cost of a Television*   A 5% sales tax on a television amounts to $82.50. Find the cost of the television.

Systems of Linear Equations

I n 1940, a physicist named John Atanasoff at Iowa State University needed to solve 29 equations with 29 variables simultaneously. This task was too difficult to do by hand, so he and a graduate student invented the first fully electronic digital computer. Thus the desire to solve a mathematical problem led to one of the most important inventions of the twentieth century. Today people can solve thousands of equations with thousands of variables. Solutions to such equations have resulted in better airplanes, cars, electronic devices, weather forecasts, and medical equipment.

Equations are also widely used in biology. The following table contains the weight W, neck size N, and chest size C for three black bears. Suppose that park rangers find a bear with a neck size of 22 inches and a chest size of 38 inches. Can they use the data in the table to estimate the bear's weight? Using systems of linear equations, they can answer this question. See Example 7 and Exercise 58 in Section 9.2.

W (pounds)	N (inches)	C (inches)
80	16	26
344	28	45
416	31	54
?	22	38

Education is what survives when what has been learned has been forgotten.

—B. F. SKINNER

Sources: A. Tucker, *Fundamentals of Computing*; M. Triola, *Elementary Statistics*; Minitab, Inc.

9.1 SYSTEMS OF LINEAR EQUATIONS IN THREE VARIABLES

Basic Concepts ▪ Solving Linear Systems with Substitution and Elimination ▪ Modeling Data ▪ Systems of Equations with No Solutions ▪ Systems of Equations with Infinitely Many Solutions

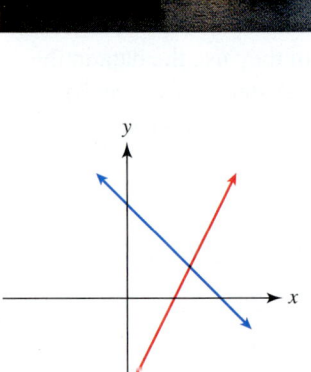

In Chapter 4, we described how to solve systems of linear equations in two variables. In applications, systems commonly have many variables. Large linear systems are used in the design of electrical circuits, bridges, and ships. They also are used in business, economics, and psychology. Because of the enormous amount of work needed to solve large systems, technology is usually used to find approximate solutions. In this section we discuss symbolic methods for finding solutions of linear systems in three variables. These methods provide the basis for understanding how technology can solve large linear systems.

Basic Concepts

When we solve a linear system in two variables, we can express a solution as an ordered pair (x, y). A linear equation in two variables can be represented graphically by a line. A system of two linear equations with a unique solution can be represented graphically by two lines intersecting at a point, as shown in Figure 9.1.

When solving linear systems in three variables, we often use the variables x, y, and z. A solution is expressed as an **ordered triple** (x, y, z) rather than an ordered pair (x, y). For example, if the ordered triple $(1, 2, 3)$ is a solution, $x = 1$, $y = 2$, and $z = 3$ satisfy each equation. A linear equation in three variables can be represented by a flat plane in space. If the solution is unique, we can represent a linear system of three equations in three variables graphically by three planes intersecting at a single point, as illustrated in Figure 9.2.

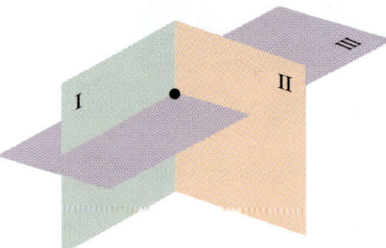

Figure 9.1

Figure 9.2

NOTE: The three planes in Figure 9.2 all intersect at right angles. In general, three planes can intersect at a point even if they are not at right angles to each other.

CRITICAL THINKING

The following figures of three planes in space represent a system of three linear equations in three variables. How many solutions are there in each case? Explain.

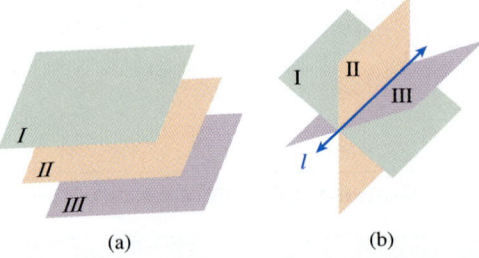

(a)   (b)

The next example shows how to check whether an ordered triple is a solution to a system of linear equations in three variables.

EXAMPLE 1   Checking for solutions to a system of three equations

Determine whether $(4, 2, -1)$ or $(-1, 0, 3)$ is a solution to the system.

$$2x - 3y + \ z = 1$$
$$x - 2y + 2z = 5$$
$$2y + \ z = 3$$

Solution
To check $(4, 2, -1)$, substitute $x = 4$, $y = 2$, and $z = -1$ in each equation.

$$2(4) - 3(2) + (-1) \overset{?}{=} 1 \qquad \text{True}$$
$$4 - 2(2) + 2(-1) \overset{?}{=} 5 \qquad \text{False}$$
$$2(2) + (-1) \overset{?}{=} 3 \qquad \text{True}$$

The ordered triple $(4, 2, -1)$ *does not satisfy all three equations, so it is not a solution.* Next, substitute $x = -1$, $y = 0$, and $z = 3$ in each equation.

$$2(-1) - 3(0) + 3 \overset{?}{=} 1 \qquad \text{True}$$
$$-1 - 2(0) + 2(3) \overset{?}{=} 5 \qquad \text{True}$$
$$2(0) + 3 \overset{?}{=} 3 \qquad \text{True}$$

The ordered triple $(-1, 0, 3)$ *satisfies all three equations, so it is a solution.*

Now Try Exercise 9

In the next example we set up a system of three equations in three variables that involves the angles of a triangle. You are asked to solve this system in Exercise 51.

EXAMPLE 2   Setting up a system of equations

The measure of the largest angle in a triangle is 40° greater than the sum of the two smaller angles and 90° more than the smallest angle. Set up a system of three linear equations in three variables whose solution gives the measure of each angle.

Solution
Let x, y, and z be the measures of the three angles from largest to smallest. Because the sum of the measures of the angles in a triangle equals 180°, we have

$$x + y + z = 180.$$

The measure of the largest angle x is 40° greater than the sum of the measures of the two smaller angles $y + z$, so

$$x - (y + z) = 40 \quad \text{or} \quad x - y - z = 40.$$

The measure of the largest angle x is 90° more than the measure of the smallest angle z, so

$$x - z = 90.$$

Thus the required system of equations can be written as follows.

$$x + y + z = 180$$
$$x - y - z = 40$$
$$x \qquad - z = 90$$

Now Try Exercise 49(a)

▶ **REAL-WORLD CONNECTION**   In the next example we show how a linear system involving three equations and three variables can be used to model a real-world situation. We solve this system of equations in Example 7.

EXAMPLE **3**   **Modeling real data with a linear system**

The Bureau of Land Management studies antelope populations in Wyoming. It monitors the number of adult antelope, the number of fawns each spring, and the severity of the winter. The first two columns of Table 9.1 contain counts of fawns and adults for three representative winters. The third column shows the severity of each winter. The severity of the winter is measured from 1 to 5, with 1 being mild and 5 being severe.

TABLE 9.1

Fawns (F)	Adults (A)	Winter (W)
405	870	3
414	848	2
272	684	5
?	750	4

We want to use the data in the first three rows of the table to estimate the number of fawns F in the fourth row when the number of adults is 750 and the severity of the winter is 4. To do so, we use the formula

$$F = a + bA + cW,$$

where a, b, and c are constants. Write a system of linear equations whose solution gives appropriate values for a, b, and c.

Solution
From the first row in the table, when $F = 405$, $A = 870$, and $W = 3$, the formula

$$F = a + bA + cW$$

becomes

$$405 = a + b(870) + c(3).$$

Similarly, $F = 414$, $A = 848$, and $W = 2$ gives

$$414 = a + b(848) + c(2),$$

and $F = 272$, $A = 684$, and $W = 5$ yields

$$272 = a + b(684) + c(5).$$

To find values for a, b, and c we can solve the following system of linear equations.

$$405 = a + 870b + 3c$$
$$414 = a + 848b + 2c$$
$$272 = a + 684b + 5c$$

We can also write these equations as a linear system in the following form.

$$a + 870b + 3c = 405$$
$$a + 848b + 2c = 414$$
$$a + 684b + 5c = 272$$

Finding values for a, b, and c will allow us to use the formula $F = a + bA + cW$ to predict the number of fawns F when the number of adults A is 750 and the severity of the winter W is 4 (see Example 7).   **Now Try Exercise** 53(a)

NOTE: Linear systems of two equations can have no solutions, one solution, or infinitely many solutions. The same is true for larger linear systems. In the following subsection we focus on linear systems having one solution.

Solving Linear Systems with Substitution and Elimination

When solving systems of linear equations with more than two variables, we usually use both substitution and elimination. However, in the next example we use only substitution to solve a particular type of linear system in three variables.

EXAMPLE 4   **Using substitution to solve a linear system of equations**

Solve the following system.

$$2x - y + z = 7$$
$$3y - z = 1$$
$$z = 2$$

Solution

The last equation gives us the value of z immediately. We can substitute $z = 2$ in the second equation and determine y.

$3y - z = 1$	Second equation
$3y - 2 = 1$	Substitute $z = 2$.
$3y = 3$	Add 2 to each side.
$y = 1$	Divide each side by 3.

Knowing that $y = 1$ and $z = 2$ allows us to find x by using the first equation.

$2x - y + z = 7$	First equation
$2x - 1 + 2 = 7$	Let $y = 1$ and $z = 2$.
$2x = 6$	Simplify and subtract 1.
$x = 3$	Divide each side by 2.

Thus $x = 3$, $y = 1$, and $z = 2$ and the solution is $(3, 1, 2)$.   **Now Try Exercise** 13

In the next example we use elimination and substitution to solve a system of linear equations. This four-step method is summarized by the following.

SOLVING A LINEAR SYSTEM IN THREE VARIABLES

STEP 1: Eliminate one variable, such as x, from two of the equations.

STEP 2: Use the two resulting equations in two variables to eliminate one of the variables, such as y. Solve for the remaining variable, z.

STEP 3: Substitute z in one of the two equations from Step 2. Solve for the unknown variable y.

STEP 4: Substitute values for y and z in one of the given equations and find x. The solution is (x, y, z).

EXAMPLE 5   Solving a linear system in three variables

Solve the following system.

$$
\begin{aligned}
x - y + 2z &= 6 \\
2x + y - 2z &= -3 \\
-x - 2y + 3z &= 7
\end{aligned}
$$

Solution

STEP 1: We begin by eliminating the variable x from the second and third equations. To eliminate x from the second equation, we multiply the first equation by -2 and then add it to the second equation. To eliminate x from the third equation, we add the first and third equations.

$-2x + 2y - 4z = -12$	First equation times -2	$x - y + 2z = 6$	First equation
$\underline{2x + y - 2z = -3}$	Second equation	$\underline{-x - 2y + 3z = 7}$	Third equation
$3y - 6z = -15$	Add.	$-3y + 5z = 13$	Add.

STEP 2: Take the two resulting equations from Step 1 and eliminate either variable. Here we add the two equations to eliminate the variable y.

$$
\begin{aligned}
3y - 6z &= -15 \\
\underline{-3y + 5z} &= \underline{13} \\
-z &= -2 \qquad \text{Add the equations.} \\
z &= 2 \qquad \text{Multiply by } -1.
\end{aligned}
$$

STEP 3: Now we can use substitution to find the values of x and y. We let $z = 2$ in either equation used in Step 2 to find y.

$$
\begin{aligned}
3y - 6z &= -15 \\
3y - 6(2) &= -15 \qquad \text{Substitute } z = 2. \\
3y - 12 &= -15 \qquad \text{Multiply.} \\
3y &= -3 \qquad \text{Add 12.} \\
y &= -1 \qquad \text{Divide by 3.}
\end{aligned}
$$

STEP 4: Finally, we substitute $y = -1$ and $z = 2$ in one of the given equations to find x.

$$x - y + 2z = 6 \qquad \text{First given equation}$$
$$x - (-1) + 2(2) = 6 \qquad \text{Let } y = -1 \text{ and } z = 2.$$
$$x + 1 + 4 = 6 \qquad \text{Simplify.}$$
$$x = 1 \qquad \text{Subtract 5.}$$

The solution is $(1, -1, 2)$. Check this solution.   `Now Try Exercise` **25**

▶ **REAL-WORLD CONNECTION**   In the next example we solve a system of linear equations to determine the number of tickets sold at a play.

EXAMPLE **6**   ## Finding the number of tickets sold

One thousand tickets were sold for a play, which generated $3800 in revenue. The prices of the tickets were $3 for children, $4 for students, and $5 for adults. There were 100 fewer student tickets sold than adult tickets. Find the number of each type of ticket sold.

Solution

Let x be the number of tickets sold to children, y be the number of tickets sold to students, and z be the number of tickets sold to adults. The total number of tickets sold was 1000, so

$$x + y + z = 1000.$$

Each child's ticket cost $3, so the revenue generated from selling x tickets was $3x$. Similarly, the revenue generated from students was $4y$, and the revenue from adults was $5z$. Total ticket sales were $3800, so

$$3x + 4y + 5z = 3800.$$

The equation $z - y = 100$, or $y - z = -100$, must also be satisfied, as 100 fewer tickets were sold to students than adults.

To find the price of a ticket, we need to solve the following system of linear equations.

$$x + y + z = 1000$$
$$3x + 4y + 5z = 3800$$
$$y - z = -100$$

STEP 1: We begin by eliminating the variable x from the second equation. To do so, we multiply the first equation by -3 and add the second equation.

$$\begin{array}{ll} -3x - 3y - 3z = -3000 & \text{First given equation times } -3 \\ \underline{3x + 4y + 5z = 3800} & \text{Second equation} \\ y + 2z = 800 & \text{Add.} \end{array}$$

STEP 2: We then multiply the resulting equation from Step 1 by -1 and add the third equation to eliminate y.

$$\begin{array}{ll} -y - 2z = -800 & \text{Equation from Step 1 times } -1 \\ \underline{y - z = -100} & \text{Third equation} \\ -3z = -900 & \text{Add the equations.} \\ z = 300 & \text{Divide by } -3. \end{array}$$

STEP 3: To find y, we can substitute $z = \mathbf{300}$ in the third equation.

$$
\begin{array}{ll}
y - z = -100 & \text{Third equation} \\
y - \mathbf{300} = -100 & \text{Let } z = 300. \\
y = \mathbf{200} & \text{Add 300.}
\end{array}
$$

STEP 4: Finally, we substitute $y = \mathbf{200}$ and $z = \mathbf{300}$ in the first equation.

$$
\begin{array}{ll}
x + y + z = 1000 & \text{First equation} \\
x + \mathbf{200} + \mathbf{300} = 1000 & \text{Let } y = 200 \text{ and } z = 300. \\
x = \mathbf{500} & \text{Subtract 500.}
\end{array}
$$

Thus **500** tickets were sold to children, **200** to students, and **300** to adults.

Now Try Exercise **47**

Modeling Data

In the next example we solve the system of equations that we discussed in Example 3.

EXAMPLE 7 Predicting fawns in the spring

Solve the following linear system for a, b, and c.

$$
\begin{aligned}
a + 870b + 3c &= 405 \\
a + 848b + 2c &= 414 \\
a + 684b + 5c &= 272
\end{aligned}
$$

Then use $F = a + bA + cW$ to predict the number of fawns when there are 750 adults and the severity of the winter is 4.

Solution

STEP 1: We begin by eliminating the variable a from the second and third equations. To do so, we add the second and third equations to the first equation times -1.

$$
\begin{array}{l}
-a - 870b - 3c = -405 \\
\underline{a + 848b + 2c = 414} \\
{-22b} - c = 9
\end{array}
\qquad
\begin{array}{ll}
-a - 870b - 3c = -405 & \text{First times } -1 \\
\underline{a + 684b + 5c = 272} & \text{Second, third equations} \\
-186b + 2c = -133 & \text{Add.}
\end{array}
$$

STEP 2: We use the two resulting equations from Step 1 to eliminate c. To do so, we multiply $-22b - c = 9$ by 2 and add it to the other equation.

$$
\begin{array}{lll}
-44b - 2c = 18 & & (-22b - c = 9) \text{ times } 2 \\
\underline{-186b + 2c = -133} & & \\
-230b = -115 & & \text{Add the equations.} \\
b = 0.5 & & \text{Divide by } -230.
\end{array}
$$

STEP 3: To find c, we substitute $b = \mathbf{0.5}$ in either equation used in Step 2.

$$
\begin{array}{ll}
-44b - 2c = 18 & \\
-44(\mathbf{0.5}) - 2c = 18 & \text{Let } b = 0.5. \\
-22 - 2c = 18 & \text{Multiply.} \\
-2c = 40 & \text{Add 22.} \\
c = \mathbf{-20} & \text{Divide by } -2.
\end{array}
$$

STEP 4: Finally, we substitute $b = \mathbf{0.5}$ and $c = \mathbf{-20}$ in one of the given equations to find a.

$$a + 870b + 3c = 405 \qquad \text{First given equation}$$
$$a + 870(\mathbf{0.5}) + 3(\mathbf{-20}) = 405 \qquad \text{Let } b = 0.5 \text{ and } c = -20.$$
$$a + 435 - 60 = 405 \qquad \text{Multiply.}$$
$$a = \mathbf{30} \qquad \text{Solve for } a.$$

The solution is $a = \mathbf{30}$, $b = \mathbf{0.5}$, and $c = \mathbf{-20}$. Thus we may write

$$F = a + bA + cW$$
$$= \mathbf{30} + \mathbf{0.5}A - \mathbf{20}W.$$

If there are 750 adults and the winter has a severity of 4, this model predicts

$$F = 30 + 0.5(750) - 20(4)$$
$$= 325 \text{ fawns.} \qquad \text{Now Try Exercise } \boxed{53}$$

CRITICAL THINKING

Give reasons why the coefficient for A is positive and the coefficient for W is negative in the formula

$$F = 30 + 0.5A - 20W.$$

Systems of Equations with No Solutions

It is possible for a system of three linear equations in three variables to be inconsistent and have no solutions. If we apply substitution and elimination to this type of system, we arrive at a contradiction. This case is demonstrated in the next example.

EXAMPLE 8 Recognizing an inconsistent system

Solve the system, if possible.

$$x + y + z = 4$$
$$-x + y + z = 2$$
$$y + z = 1$$

Solution

STEP 1: If we add the first two equations, we can eliminate x. The variable x is already eliminated from the third equation.

$$x + y + z = 4 \qquad \text{First equation}$$
$$\underline{-x + y + z = 2} \qquad \text{Second equation}$$
$$2y + 2z = 6 \qquad \text{Add.}$$

STEP 2: If we multiply the third *given* equation by -2 and add it to the resulting equation in Step 1, we arrive at a contradiction.

$$-2y - 2z = -2 \qquad \text{Third equation times } -2$$
$$\underline{2y + 2z = 6} \qquad \text{Equation from Step 1}$$
$$\mathbf{0} = \mathbf{4} \qquad \text{Add equations.}$$

Because $\mathbf{0} = \mathbf{4}$ is a contradiction, there are no solutions to the given system of equations.

Now Try Exercise $\boxed{31}$

Systems of Equations with Infinitely Many Solutions

It is possible for a system of linear equations in three variables to have infinitely many solutions. If we apply substitution and elimination to this type of system, we arrive at an identity. This case is demonstrated in the next example.

EXAMPLE 9   Solving a system with infinitely many solutions

Solve the system.

$$x + y + z = 2$$
$$x - y + z = 4$$
$$3x - y + 3z = 10$$

Solution

STEP 1: To eliminate y from the second equation, add the first equation to the second. To eliminate y from the third equation, add the first equation to the third equation.

$x + y + z = 2$	First equation		$x + y + z = 2$	First equation
$x - y + z = 4$	Second equation		$3x - y + 3z = 10$	Third equation
$\overline{2x + 2z = 6}$	Add.		$\overline{4x + 4z = 12}$	Add.

STEP 2: If we multiply the first resulting equation in Step 1 by -2 and add it to the second resulting equation in Step 1, we arrive at an identity.

$$-4x - 4z = -12 \qquad (2x + 2z = 6) \text{ times } -2$$
$$\underline{4x + 4z = 12} \qquad \text{Second equation from Step 1}$$
$$0 = 0 \qquad \text{Add. (Identity)}$$

The variable x can be written in terms of z by solving $2x + 2z = 6$ for x.

$$2x + 2z = 6 \qquad \text{Equation from Step 1}$$
$$2x = 6 - 2z \qquad \text{Subtract } 2z.$$
$$x = 3 - z \qquad \text{Divide by 2.}$$

STEP 3: To find y in terms of z, substitute $3 - z$ for x in the first *given* equation.

$$x + y + z = 2 \qquad \text{First equation}$$
$$(3 - z) + y + z = 2 \qquad \text{Let } x = 3 - z.$$
$$y = -1 \qquad \text{Solve for } y.$$

All solutions have the form $(3 - z, -1, z)$, where z can be any real number. For example, if $z = 1$, then $(2, -1, 1)$ is one of infinitely many solutions to the system of equations.   **Now Try Exercise 33**

9.1 PUTTING IT ALL TOGETHER

In this section we discussed how to solve a system of three linear equations in three variables. Systems of linear equations can have no solutions, one solution, or infinitely many solutions. The following table summarizes some of the important concepts presented in this section.

Concept	Explanation
System of Linear Equations in Three Variables	The following is a system of three linear equations in three variables. $$\begin{aligned} x - 2y + z &= 0 \\ -x + y + z &= 4 \\ -y + 4z &= 10 \end{aligned}$$
Solution to a Linear System in Three Variables	The solution to a linear system in three variables is an ordered triple, expressed as (x, y, z). The solution to the preceding system is $(1, 2, 3)$ because substituting $x = 1$, $y = 2$, and $z = 3$ in each equation results in a true statement. We can check solutions this way. $$\begin{aligned} (1) - 2(2) + (3) &= 0 \qquad \text{True} \\ -(1) + (2) + (3) &= 4 \qquad \text{True} \\ - (2) + 4(3) &= 10 \qquad \text{True} \end{aligned}$$
Solving a Linear System with Substitution and Elimination	**STEP 1:** Eliminate one variable, such as x, from two of the equations. **STEP 2:** Use the two resulting equations in two variables to eliminate one of the variables, such as y. Solve for the remaining variable z. **STEP 3:** Substitute z in one of the two equations from Step 2. Solve for the unknown variable y. **STEP 4:** Substitute values for y and z in one of the given equations and find x. The solution is (x, y, z).

9.1 Exercises

MyMathLab

CONCEPTS

1. Can a system of three linear equations and three variables have exactly two solutions? Explain.

2. Give an example of a system of three linear equations in three variables.

3. Does the ordered triple $(1, 2, 3)$ satisfy the equation $x + y + z = 6$?

4. Does $(3, 4)$ represent a solution to the equation $x + y + z = 7$? Explain.

5. To solve uniquely for two variables, how many equations do you usually need?

6. To solve uniquely for three variables, how many equations do you usually need?

7. If a contradiction occurs while solving a system of linear equations, then there are _____ solution(s).

8. If an identity occurs while solving a linear system, how many solutions are there?

SOLVING LINEAR SYSTEMS

Exercises 9–12: Determine which ordered triple is a solution to the linear system.

9. $(1, 2, 3)$, $(0, 2, 4)$
$$\begin{aligned} x + y + z &= 6 \\ x - y - z &= -4 \\ -x - y + z &= 0 \end{aligned}$$

10. $(-1, 0, 2)$, $(0, 4, 4)$
$$\begin{aligned} 2x + y - 3z &= -8 \\ x - 3y + 2z &= -4 \\ 3x - 2y + z &= -4 \end{aligned}$$

11. $(1, 0, 3), (-1, 1, 2)$

$$3x - 2y + z = -3$$
$$-x + 3y - 2z = 0$$
$$x + 4y + 2z = 7$$

12. $\left(\frac{1}{2}, \frac{3}{2}, -\frac{1}{2}\right), (-1, 0, -2)$

$$x + 3y - 4z = 7$$
$$-x + 5y + 3z = \frac{11}{2}$$
$$3x - 2y - 7z = 2$$

Exercises 13–18: (Refer to Example 4.) Use substitution to solve the system of linear equations. Check your solution.

13. $x + y - z = 1$
$2y + z = -1$
$z = 1$

14. $2x + y - 3z = 1$
$y + 4z = 0$
$z = -1$

15. $-x - 3y + z = -2$
$2y + 3z = 3$
$z = 2$

16. $3x + 2y - 3z = -4$
$-y + 2z = 4$
$z = 0$

17. $a - b + 2c = 3$
$-3b + c = 4$
$c = -2$

18. $5a + 2b - 3c = 10$
$5b - 2c = -4$
$c = 3$

Exercises 19–44: Solve the system, if possible.

19. $x + y - z = 11$
$-x + 2y + 3z = -1$
$2z = 4$

20. $x + 2y - 3z = -7$
$-2x + y + z = -1$
$3z = 9$

21. $x + y - z = -2$
$-x + z = 1$
$y + 2z = 3$

22. $x + y - 3z = 11$
$-2x + y + 2z = 1$
$-3y + 3z = -21$

23. $x + y - 2z = -7$
$y + z = -1$
$-y + 3z = 9$

24. $2x + 3y + z = 5$
$y + 2z = 4$
$-2y + z = 2$

25. $x + 2y + 2z = 1$
$x + y + z = 0$
$-x - 2y + 3z = -11$

26. $x + y - z = 0$
$x - 3y + z = -2$
$x - y + 3z = 8$

27. $x + y + z = 5$
$y + z = 6$
$x + z = 3$

28. $x + y + z = 0$
$x - y - z = 6$
$-x + y - z = 4$

29. $x + 2y + 3z = 24$
$-x + y + 2z = 1$
$x + y - 2z = 9$

30. $5x - 15y + z = 22$
$-10x + 12y - 2z = -8$
$4x - 2y - 3z = 9$

31. $x + y + z = 2$
$x - y + z = 1$
$x + z = 3$

32. $4x - y + 3z = 3$
$2x + y + z = 2$
$x - y + z = 1$

33. $x + y + z = 6$
$x - y + z = 2$
$-x + 5y - z = 6$

34. $x - y + z = 3$
$2x - y + z = 2$
$-x - y + z = 5$

35. $2x + y + z = 3$
$2x - y - z = 9$
$x + y - z = 0$

36. $x + 3y + z = -8$
$x - 2y = 11$
$2y - z = -16$

37. $2x + 6y - 2z = 47$
$2x + y + 3z = -28$
$-x + y + z = -\frac{7}{2}$

38. $x + y + 2z = 23$
$3x - y + 3z = 8$
$2x + 2y + z = 13$

39. $x + 3y - 4z = \frac{13}{2}$
$-2x + 3y - z = \frac{1}{2}$
$3x + z = 4$

40. $x - 2y + z = \frac{9}{2}$
$4x - y + 3z = 9$
$x + 2y = -\frac{3}{2}$

41. $2x - 2y + z = 4$
$x - y + z = 1$
$x - y + z = 3$

42. $x - 2y + 3z = 1$
$x + 3y + z = 2$
$3x + 4y + 5z = 7$

43. $x + y + z = 5$
$x - y + z = 3$
$2x + y + 2z = 9$

44. $x + y - 2z = 0$
$x + 2y - 4z = -1$
$y - 2z = -1$

Exercises 45 and 46: **Thinking Generally** *Solve the system of linear equations, if $a \neq 0$.*

45. $x + y + z = a$
$x + y + z = 2a$
$-x + y + z = 0$

46. $x + y + z = a$
$-x - y - z = -a$
$y - z = 0$

APPLICATIONS

47. *Finding Costs*   The accompanying table shows the costs of purchasing different combinations of hamburgers, fries, and soft drinks.

Hamburgers	Fries	Soft Drinks	Total Cost
1	2	4	$10
1	4	6	$15
0	3	2	$6

(a) Let x be the cost of a hamburger, y the cost of fries, and z be the cost of a soft drink. Write a system of three linear equations that represents the data in the table.

(b) Solve the system and interpret your answer.

48. *Cost of CDs*   The accompanying table shows the total cost of purchasing combinations of differently priced CDs. The types of CDs are labeled A, B, and C.

A	B	C	Total Cost
1	1	1	$37
3	2	1	$69
1	1	4	$82

(a) Let x be the cost of a CD of type A, y be the cost of a CD of type B, and z be the cost of a CD of type C. Write a system of three linear equations that represents the data in the table.

(b) Solve the system and interpret your answer.

49. *Geometry*   The largest angle in a triangle is 55° more than the smallest angle. The sum of the measures of the two smaller angles is 10° more than the measure of the largest angle.

(a) Let x, y, and z be the measures of the three angles from largest to smallest. Write a system of three linear equations whose solution gives the measure of each angle.

(b) Solve the system.

(c) Check your solution.

50. *Geometry*   The perimeter of a triangle is 90 inches. The longest side is 20 inches longer than the shortest side and 10 inches longer than the remaining side.

(a) Let x, y, and z be the lengths of the three sides from largest to smallest. Write a system of three linear equations whose solution gives the lengths of each side.

(b) Solve the system.

(c) Check your solution.

51. *Geometry*   (Refer to Example 2.) Solve the system to find the measure of each angle.

$$x + y + z = 180$$
$$x - y - z = 40$$
$$x \qquad - z = 90$$

52. *Loan Mixture*   A student takes out a total of $5000 in three loans: one subsidized, one unsubsidized, and one from the parents of the student. The subsidized loan is $200 more than the combined total of the unsubsidized and parent loans. The unsubsidized loan is twice the amount of the parent loan. Find the amount of each loan.

53. *Predicting Fawns*   (Refer to Examples 3 and 7.) The accompanying table shows counts for fawns and adult deer and the severity of the winter. These data may be modeled by the equation $F = a + bA + cW$.

Fawns (F)	Adults (A)	Winter (W)
525	600	4
365	400	2
805	900	5
?	500	3

(a) Use the first three rows to write a system of three linear equations in three variables whose solution gives values for a, b, and c.

(b) Solve the system.

(c) Predict the number of fawns when there are 500 adults and the winter has severity 3.

54. *Business Production*   A business has three machines that manufacture containers. Together they make 100 containers per day, whereas the two fastest machines can make 80 containers per day. The fastest machine makes 34 more containers per day than the slowest machine.

(a) Let x, y, and z be the number of containers that the machines make from fastest to slowest. Write a system of three equations whose solution gives the number of containers that each machine can make.

(b) Solve the system.

55. *Mixture Problem* One type of lawn fertilizer consists of a mixture of nitrogen, N, phosphorus, P, and potassium, K. An 80-pound sample contains 8 more pounds of nitrogen and phosphorus than potassium. There is 9 times as much potassium as phosphorus.
  (a) Write a system of three equations whose solution gives the amount of nitrogen, phosphorus, and potassium in this sample.
  (b) Solve the system.

56. *Predicting Home Prices* Selling prices of homes can depend on several factors such as size and age. The accompanying table shows the selling price for three homes. In this table, price P is given in thousands of dollars, age A in years, and home size S in thousands of square feet. These data may be modeled by the equation $P = a + bA + cS$.

Price (P)	Age (A)	Size (S)
190	20	2
320	5	3
50	40	1

  (a) Write a system of linear equations whose solution gives a, b, and c.
  (b) Solve the system.
  (c) Predict the price of a home that is 10 years old and has 2500 square feet.

57. *Investment Mixture* A sum of $30,000 was invested in three mutual funds. In one year the first fund grew by 8%, the second by 10%, and the third by 15%. Total earnings were $3550. The amount invested in the third fund was $2000 less than the combined amount invested in the other two funds. Use a linear system of equations to determine the amount invested in each fund.

58. *Football Tickets* A total of 2500 tickets were sold at a football game. Prices were $2 for children, $3 for students, and $5 for adults. Twice as many tickets were sold to students as children and ticket revenues were $7250. Use a system of linear equations to determine how many of each type of ticket were sold.

WRITING ABOUT MATHEMATICS

59. In the previous section we solved problems with two variables; to obtain a unique solution, we needed two linear equations. In this section we solved problems with three variables; to obtain a unique solution, we needed three equations. Try to generalize these results. In the design of aircraft, problems commonly involve 100,000 variables. How many equations are required to solve such problems? Can such problems be solved by hand? If not, how are such problems solved? Explain your answers.

60. In Exercise 56 the price of a home was estimated by its age and size. What other factors might affect the price of a home? Explain how these factors might affect the number of variables and equations in the linear system.

GROUP ACTIVITY
WORKING WITH REAL DATA

Directions: Form a group of 2 to 4 people. Select someone to record the group's responses for this activity. All of the members of the group should work cooperatively to answer the questions. If your instructor asks for your results, each member of the group should be prepared to respond.

CEO Salaries In 2002, hourly wages for the top three CEOs (chief executive officers) of U.S. corporations were calculated based on their working 14 hours per day for 365 days. Together the three CEOs made $44,000 per hour. The top CEO earned $2,000 more per hour than the combined hourly wages of the next two top CEOs, and the top CEO earned $7,000 more per hour than the next top CEO. (*Source: USA Today*, March 31, 2003.)

  (a) Let x, y, and z be the hourly wages in *thousands* of dollars for the top CEOs from greatest to least. Write a system of equations whose solution gives these hourly wages.
  (b) Solve the system. Interpret the answer.
  (c) The average American worker earned $16.23 per hour in 2002. How many hours did the average American work to earn an amount equal to one hour of work by the top CEO in 2002?

9.2 MATRIX SOLUTIONS OF LINEAR SYSTEMS

Representing Systems of Linear Equations with Matrices ■ Gauss–Jordan Elimination ■ Using Technology to Solve Systems of Linear Equations (Optional)

A LOOK INTO MATH ▷

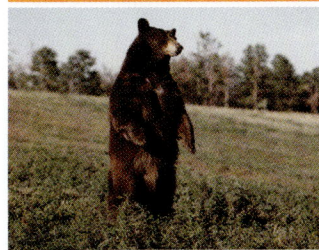

Suppose that the size of a bear's head and its overall length are known. Can its weight be estimated from these variables? Can a bear's weight be estimated if its neck size and chest size are known? In this section we show that systems of linear equations can be used to make such estimates.

In the previous section we solved systems of three linear equations in three variables by using elimination and substitution. In real life, systems of equations often contain thousands of variables. To solve a large system of equations, we need an efficient method. Long before the invention of the computer, Carl Fredrich Gauss (1777–1855) developed a method called *Gaussian elimination* to solve systems of linear equations. Even though it was developed more than 150 years ago, it is still used today in modern computers and calculators. In this section we introduce the Gauss–Jordan method, which is based on Gaussian elimination.

Representing Systems of Linear Equations with Matrices

Arrays of numbers are used frequently in many different real-world situations. Spreadsheets often make use of arrays. A **matrix** is a rectangular array of numbers. Each number in a matrix is called an **element**. The following are examples of *matrices* (plural of *matrix*), with their dimensions written below them.

$$\begin{bmatrix} 2 & 0 \\ 3 & 1 \end{bmatrix} \quad \begin{bmatrix} -1.2 & 5 & 0 \\ 1 & 0 & 1 \\ 4 & -5 & 7 \end{bmatrix} \quad \begin{bmatrix} 3 & -6 & 0 & \sqrt{3} \\ 1 & 4 & 0 & 9 \\ -3 & 1 & 1 & 18 \\ -10 & -4 & 5 & -1 \end{bmatrix} \quad \begin{bmatrix} 4 & 2 \\ 0 & 1 \\ 1 & 0 \end{bmatrix} \quad \begin{bmatrix} 1 & 5 & -1 \\ 3 & 4 & 2 \end{bmatrix}$$

$$\boxed{2 \times 2} \qquad \boxed{3 \times 3} \qquad\qquad\qquad \boxed{4 \times 4} \qquad\qquad\qquad \boxed{3 \times 2} \qquad \boxed{2 \times 3}$$

$$\boxed{\text{rows} \times \text{columns}}$$

▶ **REAL-WORLD CONNECTION** The dimension of a matrix is stated much like the dimensions of a rectangular room. We might say that a room is m feet long and n feet wide. Similarly, the **dimension of a matrix** is $m \times n$ (m by n), if it has m rows and n columns. For example, the last matrix in the preceding group has a dimension of 2×3 because it has 2 rows and 3 columns. If the number of rows equals the number of columns, the matrix is a **square matrix**. The first three matrices in that group are square matrices.

Matrices can be used to represent a system of linear equations. For example, if we have the system of equations

$$\begin{aligned} 3x - y + 2z &= 7 \\ x - 2y + z &= 0 \\ 2x + 5y - 7z &= -9, \end{aligned}$$

we can represent the system with the following **augmented matrix**. Note how the coefficients of the variables were placed in the matrix. A vertical line is positioned in the matrix where the equals signs occur in the system. The rows and columns are labeled, and the elements of the **main diagonal** of the augmented matrix are circled. The matrix has dimension 3×4.

Main diagonal

$$\begin{bmatrix} 3 & -1 & 2 & 7 \\ 1 & -2 & 1 & 0 \\ 2 & 5 & -7 & -9 \end{bmatrix} \begin{matrix} \leftarrow \text{Row 1} \\ \leftarrow \text{Row 2} \\ \leftarrow \text{Row 3} \end{matrix}$$

Column 1   Column 2   Column 3   Column 4

EXAMPLE 1   Representing a linear system

Represent each linear system with an augmented matrix. State the dimension of the matrix.

(a) $\begin{aligned} x - 2y &= 9 \\ 6x + 7y &= 16 \end{aligned}$

(b) $\begin{aligned} x - 3y + 7z &= 4 \\ 2x + 5y - z &= 15 \\ 2x + y \phantom{{}+ 7z} &= 8 \end{aligned}$

Solution

(a) This system can be represented by the following 2×3 augmented matrix.

$$\begin{bmatrix} 1 & -2 & 9 \\ 6 & 7 & 16 \end{bmatrix}$$

(b) This system can be represented by the following 3×4 augmented matrix.

$$\begin{bmatrix} 1 & -3 & 7 & 4 \\ 2 & 5 & -1 & 15 \\ 2 & 1 & 0 & 8 \end{bmatrix}$$

Now Try Exercises 11, 13

Gauss–Jordan Elimination

A convenient matrix form for representing a system of linear equations is **reduced row–echelon form**. The following matrices are examples of reduced row–echelon form. Note that there are 1s on the main diagonal with 0s above and below the 1s.

$$\begin{bmatrix} 1 & 0 & 3 \\ 0 & 1 & -2 \end{bmatrix} \qquad \begin{bmatrix} 1 & 0 & 0 & 3 \\ 0 & 1 & 0 & 1 \\ 0 & 0 & 1 & -1 \end{bmatrix} \qquad \begin{bmatrix} 1 & 0 & 0 & 8 \\ 0 & 1 & 0 & 2 \\ 0 & 0 & 1 & 3 \end{bmatrix}$$

If an augmented matrix representing a linear system is in reduced row–echelon form, we can usually determine the solution easily.

EXAMPLE 2   Determining a solution from reduced row–echelon form

Each matrix represents a system of linear equations. Find the solution.

(a) $\begin{bmatrix} 1 & 0 & 0 & 2 \\ 0 & 1 & 0 & -3 \\ 0 & 0 & 1 & 5 \end{bmatrix}$

(b) $\begin{bmatrix} 1 & 0 & 10 \\ 0 & 1 & -4 \end{bmatrix}$

Solution

(a) The top row represents $1x + 0y + 0z = 2$ or $x = 2$. The second and third rows tell us that $y = -3$ and $z = 5$. The solution is $(2, -3, 5)$.

(b) The system involves two equations in two variables. The solution is $(10, -4)$.

Now Try Exercise **19**

We can use a numerical method called **Gauss–Jordan elimination** to solve a linear system. It makes use of the following matrix row transformations.

MATRIX ROW TRANSFORMATIONS

For any augmented matrix representing a system of linear equations, the following row transformations result in an equivalent system of linear equations.

1. Any two rows may be interchanged.
2. The elements of any row may be multiplied by a nonzero constant.
3. Any row may be changed by adding to (or subtracting from) its elements a nonzero multiple of the corresponding elements of another row.

Gauss–Jordan elimination can be used to transform an augmented matrix into reduced row–echelon form. Its objective is to use these matrix row transformations to obtain a matrix that has the following reduced row–echelon form, where (a, b) represents the solution.

$$\begin{bmatrix} 1 & 0 & | & a \\ 0 & 1 & | & b \end{bmatrix}$$

This method is illustrated in the next example.

EXAMPLE 3 Transforming a matrix into reduced row–echelon form

Use Gauss–Jordan elimination to transform the augmented matrix of the linear system into reduced row–echelon form. Find the solution.

$$x + y = 5$$
$$-x + y = 1$$

Solution

Both the linear system and the augmented matrix are shown.

$$\begin{array}{cc} \textit{Linear System} & \textit{Augmented Matrix} \\[6pt] \begin{array}{r} x + y = 5 \\ -x + y = 1 \end{array} & \begin{bmatrix} 1 & 1 & | & 5 \\ -1 & 1 & | & 1 \end{bmatrix} \end{array}$$

First, we want to obtain a 0 in the second row, where the -1 is highlighted. To do so, we add row 1 to row 2 and place the result in row 2. This step is denoted $R_2 + R_1$ and eliminates the x-variable from the second equation.

$$\begin{array}{r} x + \ y = 5 \\ 2y = 6 \end{array} \qquad R_2 + R_1 \rightarrow \begin{bmatrix} 1 & 1 & | & 5 \\ 0 & 2 & | & 6 \end{bmatrix}$$

To obtain a 1 where the 2 in the second row is located, we multiply the second row by $\frac{1}{2}$, denoted $\frac{1}{2}R_2$.

$$
\begin{aligned}
x + y &= 5 \\
y &= 3
\end{aligned}
\qquad \frac{1}{2}R_2 \rightarrow \quad
\begin{bmatrix}
1 & 1 & | & 5 \\
0 & 1 & | & 3
\end{bmatrix}
$$

Next, we need to obtain a 0 where the 1 is highlighted. We do so by subtracting row 2 from row 1 and placing the result in row 1, denoted $R_1 - R_2$.

$$
\begin{aligned}
x &= 2 \\
y &= 3
\end{aligned}
\qquad R_1 - R_2 \rightarrow \quad
\begin{bmatrix}
1 & 0 & | & 2 \\
0 & 1 & | & 3
\end{bmatrix}
$$

This matrix is in reduced row–echelon form. The solution is $(2, 3)$.   Now Try Exercise **21**

In the next example, we use Gauss–Jordan elimination to solve a system with three linear equations and three variables. To do so we transform the matrix into the following reduced row–echelon form, where (a, b, c) represents the solution.

$$
\begin{bmatrix}
1 & 0 & 0 & | & a \\
0 & 1 & 0 & | & b \\
0 & 0 & 1 & | & c
\end{bmatrix}
$$

EXAMPLE **4**   **Transforming a matrix to reduced row–echelon form**

Use Gauss–Jordan elimination to transform the augmented matrix of the linear system into reduced row–echelon form. Find the solution.

$$
\begin{aligned}
x + y + 2z &= 1 \\
-x \phantom{{}+y} + z &= -2 \\
2x + y + 5z &= -1
\end{aligned}
$$

Solution
The linear system and the augmented matrix are both shown.

Linear System   **Augmented Matrix**

$$
\begin{aligned}
x + y + 2z &= 1 \\
-x \phantom{{}+y} + z &= -2 \\
2x + y + 5z &= -1
\end{aligned}
\qquad
\begin{bmatrix}
1 & 1 & 2 & | & 1 \\
-1 & 0 & 1 & | & -2 \\
2 & 1 & 5 & | & -1
\end{bmatrix}
$$

First, we want to put 0s in the second and third rows, where the -1 and 2 are highlighted. To obtain a 0 in the first position of the second row, we add row 1 to row 2 and place the result in row 2, denoted $R_2 + R_1$. To obtain a 0 in the first position of the third row, we subtract 2 times row 1 from row 3 and place the result in row 3, denoted $R_3 - 2R_1$. Row 1 does not change. These steps eliminate the x-variable from the second and third equations.

$$
\begin{aligned}
x + y + 2z &= 1 \\
y + 3z &= -1 \\
-y + z &= -3
\end{aligned}
\qquad
\begin{matrix}
\\
R_2 + R_1 \rightarrow \\
R_3 - 2R_1 \rightarrow
\end{matrix}
\begin{bmatrix}
1 & 1 & 2 & | & 1 \\
0 & 1 & 3 & | & -1 \\
0 & -1 & 1 & | & -3
\end{bmatrix}
$$

To eliminate the y-variable in row 1, we subtract row 2 from row 1. To eliminate the y-variable from row 3, we add row 2 to row 3.

$$
\begin{aligned}
x \quad\; - \; z &= \; 2 \\
y + 3z &= -1 \\
4z &= -4
\end{aligned}
\qquad
\begin{array}{l}
R_1 - R_2 \rightarrow \\
\\
R_3 + R_2 \rightarrow
\end{array}
\left[\begin{array}{ccc|c}
1 & 0 & -1 & 2 \\
0 & 1 & 3 & -1 \\
0 & 0 & 4 & -4
\end{array}\right]
$$

To obtain a 1 in row 3, where the highlighted 4 is located, we multiply row 3 by $\frac{1}{4}$.

$$
\begin{aligned}
x \quad\; - \; z &= \; 2 \\
y + 3z &= -1 \\
z &= -1
\end{aligned}
\qquad
\begin{array}{l}
\\
\\
\frac{1}{4}R_3 \rightarrow
\end{array}
\left[\begin{array}{ccc|c}
1 & 0 & -1 & 2 \\
0 & 1 & 3 & -1 \\
0 & 0 & 1 & -1
\end{array}\right]
$$

For the matrix to be in reduced row–echelon form, we need 0s in the highlighted locations. We first add row 3 to row 1 and then subtract 3 times row 3 from row 2.

$$
\begin{aligned}
x &= \; 1 \\
y &= \; 2 \\
z &= -1
\end{aligned}
\qquad
\begin{array}{l}
R_1 + R_3 \rightarrow \\
R_2 - 3R_3 \rightarrow \\
\\
\end{array}
\left[\begin{array}{ccc|c}
1 & 0 & 0 & 1 \\
0 & 1 & 0 & 2 \\
0 & 0 & 1 & -1
\end{array}\right]
$$

This matrix is now in reduced row–echelon form. The solution is $(1, 2, -1)$.

Now Try Exercise **31**

CRITICAL THINKING

An *inconsistent* system of linear equations has no solutions, and a system of *dependent* linear equations has infinitely many solutions. Suppose that an augmented matrix row reduces to either of the following matrices. Explain what each matrix indicates about the given system of linear equations.

$$
\left[\begin{array}{ccc|c}
1 & 0 & 0 & 2 \\
0 & 1 & 0 & 3 \\
0 & 0 & 0 & 1
\end{array}\right]
\qquad
\left[\begin{array}{ccc|c}
1 & 0 & 1 & 2 \\
0 & 1 & 2 & 3 \\
0 & 0 & 0 & 0
\end{array}\right]
$$

In the next example we find the amounts invested in three mutual funds.

EXAMPLE **5** ### Determining investment amounts in mutual funds

A total of $8000 was invested in three funds that grew at a rate of 5%, 10%, and 20% over 1 year. After 1 year, the combined value of the three funds had grown by $1200. Five times as much money was invested at 20% as at 10%. Find the amount invested in each fund.

Solution
Let x be the amount invested at 5%, y be the amount invested at 10%, and z be the amount invested at 20%. The total amount invested was $8000, so

$$x + y + z = 8000.$$

The growth in the first mutual fund, paying 5% of x, is given by $0.05x$. Similarly, the growths in the other mutual funds are given by $0.10y$ and $0.20z$. As the total growth was $1200, we can write

$$0.05x + 0.10y + 0.20z = 1200.$$

Multiplying each side of this equation by 20 to eliminate decimals results in

$$x + 2y + 4z = 24{,}000.$$

Five times as much was invested at 20% as at 10%, so $z = 5y$, or $5y - z = 0$.

These three equations can be written as a system of linear equations and as an augmented matrix.

Linear System   **Augmented Matrix**

$$\begin{aligned} x + y + z &= 8{,}000 \\ x + 2y + 4z &= 24{,}000 \\ 5y - z &= \phantom{24{,}00}0 \end{aligned} \qquad \left[\begin{array}{ccc|c} 1 & 1 & 1 & 8{,}000 \\ 1 & 2 & 4 & 24{,}000 \\ 0 & 5 & -1 & 0 \end{array}\right]$$

A 0 can be obtained in the highlighted position by subtracting row 1 from row 2.

$$\begin{aligned} x + y + z &= 8{,}000 \\ y + 3z &= 16{,}000 \\ 5y - z &= \phantom{16{,}00}0 \end{aligned} \qquad R_2 - R_1 \rightarrow \left[\begin{array}{ccc|c} 1 & 1 & 1 & 8{,}000 \\ 0 & 1 & 3 & 16{,}000 \\ 0 & 5 & -1 & 0 \end{array}\right]$$

Zeros can be obtained in the highlighted positions by subtracting row 2 from row 1 and by subtracting 5 times row 2 from row 3.

$$\begin{aligned} x - 2z &= -8{,}000 \\ y + 3z &= 16{,}000 \\ -16z &= -80{,}000 \end{aligned} \qquad \begin{aligned} R_1 - R_2 &\rightarrow \\ R_3 - 5R_2 &\rightarrow \end{aligned} \left[\begin{array}{ccc|c} 1 & 0 & -2 & -8{,}000 \\ 0 & 1 & 3 & 16{,}000 \\ 0 & 0 & -16 & -80{,}000 \end{array}\right]$$

To obtain a 1 in the highlighted position, multiply row 3 by $-\frac{1}{16}$.

$$\begin{aligned} x - 2z &= -8{,}000 \\ y + 3z &= 16{,}000 \\ z &= 5{,}000 \end{aligned} \qquad -\frac{1}{16}R_3 \rightarrow \left[\begin{array}{ccc|c} 1 & 0 & -2 & -8{,}000 \\ 0 & 1 & 3 & 16{,}000 \\ 0 & 0 & 1 & 5{,}000 \end{array}\right]$$

To obtain a 0 in each of the highlighted positions, add twice row 3 to row 1 and subtract three times row 3 from row 2.

$$\begin{aligned} x &= 2{,}000 \\ y &= 1{,}000 \\ z &= 5{,}000 \end{aligned} \qquad \begin{aligned} R_1 + 2R_3 &\rightarrow \\ R_2 - 3R_3 &\rightarrow \end{aligned} \left[\begin{array}{ccc|c} 1 & 0 & 0 & 2{,}000 \\ 0 & 1 & 0 & 1{,}000 \\ 0 & 0 & 1 & 5{,}000 \end{array}\right]$$

Thus $2000 was invested at 5%, $1000 at 10%, and $5000 at 20%.   **Now Try Exercise 63**

Using Technology to Solve Systems of Linear Equations (Optional)

▶ **REAL-WORLD CONNECTION**   Examples 4 and 5 involve a lot of arithmetic. Trying to solve a large system of equations by hand is an enormous—if not impossible—task. In the real world, people use technology to solve large systems. In the next example we solve the linear systems from Examples 3 and 4 with a graphing calculator.

EXAMPLE 6   Using technology

Use a graphing calculator to solve the following systems of equations.

(a) $\begin{aligned} x + y &= 5 \\ -x + y &= 1 \end{aligned}$ (b) $\begin{aligned} x + y + 2z &= 1 \\ -x + z &= -2 \\ 2x + y + 5z &= -1 \end{aligned}$

Solution

(a) Enter the 2×3 augmented matrix from Example 3 in a graphing calculator, as shown in Figure 9.3(a). Then transform the matrix into reduced row–echelon form (rref), as shown in Figure 9.3(b). The solution is (2, 3).

(a)   (b)

Figure 9.3

(b) Enter the 3×4 augmented matrix from Example 4, as shown in Figure 9.4(a). (The fourth column of A can be seen by scrolling right.) Then transform the matrix into reduced row–echelon form (rref), as shown in Figure 9.4(b). The solution is $(1, 2, -1)$.

CALCULATOR HELP

To enter a matrix and put it in reduced row–echelon form, see Appendix A (pages AP-8 and AP-9).

(a)   (b)

Figure 9.4

Now Try Exercises 39, 41

▶ **REAL-WORLD CONNECTION**   In the next example we use technology to solve an application.

EXAMPLE 7   Modeling the weight of male bears

The data shown in Table 9.2 give the weight W, head length H, and overall length L of three bears. These data can be modeled with the equation $W = a + bH + cL$, where a, b, and c are constants that we need to determine. (**Sources:** M. Triola, *Elementary Statistics*; Minitab, Inc.)

TABLE 9.2

W (pound)	H (inches)	L (inches)
362	16	72
300	14	68
147	11	52

(a) Set up a system of equations whose solution gives values for constants a, b, and c.
(b) Solve the system.
(c) Predict the weight of a bear with $H = 13$ inches and $L = 65$ inches.

Solution
(a) Substitute each row of Table 9.2 in the equation $W = a + bH + cL$.

$$362 = a + b(16) + c(72)$$
$$300 = a + b(14) + c(68)$$
$$147 = a + b(11) + c(52)$$

Rewrite this system as

$$a + 16b + 72c = 362$$
$$a + 14b + 68c = 300$$
$$a + 11b + 52c = 147$$

and represent it as the augmented matrix

$$A = \begin{bmatrix} 1 & 16 & 72 & 362 \\ 1 & 14 & 68 & 300 \\ 1 & 11 & 52 & 147 \end{bmatrix}.$$

(a)

(b)

Figure 9.5

(b) Enter A and put it in reduced row–echelon form, as shown in Figures 9.5(a) and (b), respectively. The solution is $a = -374$, $b = 19$, and $c = 6$.
(c) For $W = a + bH + cL$, use

$$W = -374 + 19H + 6L$$

to predict the weight of a bear with head length $H = 13$ and overall length $L = 65$.

$$W = -374 + 19(13) + 6(65) = 263 \text{ pounds}$$

Now Try Exercise **58**

9.2   PUTTING IT ALL TOGETHER

A matrix is a rectangular array of numbers. An augmented matrix can be used to represent any system of linear equations. One common method for solving a system of linear equations is Gauss–Jordan elimination. Matrix row operations can be used to transform an augmented matrix to reduced row–echelon form. Technology can be used to solve systems of linear equations efficiently. The following table summarizes augmented matrices and reduced row–echelon form.

Concept	Explanation
Augmented Matrix	A linear system can be represented by an augmented matrix. The following matrix has dimension 3×4. **Linear System** $$x + 2y - z = 6$$ $$-2x + y - z = 7$$ $$2x + 3z = -11$$ **Augmented Matrix** $$\begin{bmatrix} 1 & 2 & -1 & 6 \\ -2 & 1 & -1 & 7 \\ 2 & 0 & 3 & -11 \end{bmatrix}$$
Reduced Row–Echelon Form	The following augmented matrix is in reduced row–echelon form, which results from transforming the preceding system to reduced row–echelon form. There are 1s along the main diagonal and 0s elsewhere in the first three columns. The solution to the linear system is $(-1, 2, -3)$. $$\begin{bmatrix} 1 & 0 & 0 & -1 \\ 0 & 1 & 0 & 2 \\ 0 & 0 & 1 & -3 \end{bmatrix}$$

9.2 Exercises

CONCEPTS

1. What is a matrix?

2. Give an example of a matrix and state its dimension.

3. Give an example of an augmented matrix and state its dimension.

4. If an augmented matrix is used to solve a system of three linear equations in three variables, what will be its dimension?

5. Give an example of a matrix that is in reduced row–echelon form.

6. Identify the elements on the main diagonal in the augmented matrix.
$$\begin{bmatrix} 4 & -6 & -1 & 3 \\ 6 & 2 & -2 & 9 \\ 7 & 5 & -3 & 1 \end{bmatrix}$$

DIMENSIONS OF MATRICES AND AUGMENTED MATRICES

Exercises 7–10: State the dimension of the matrix.

7. $\begin{bmatrix} 3 & -3 & 7 \\ 2 & 6 & -2 \\ 4 & 2 & 5 \end{bmatrix}$

8. $\begin{bmatrix} -2 & 3 & 0 \\ 1 & -8 & 4 \end{bmatrix}$

9. $\begin{bmatrix} 1 & 7 \\ 0 & 2 \\ 2 & -5 \end{bmatrix}$

10. $\begin{bmatrix} 4 & 2 & -3 & -1 \\ 4 & -3 & 2 & -7 \\ 14 & 6 & 4 & 0 \end{bmatrix}$

Exercises 11–14: Represent the linear system as an augmented matrix.

11. $x - 3y = 1$
$-x + 3y = -1$

12. $4x + 2y = -5$
$5x + 8y = 2$

13. $2x - y + 2z = -4$
$x - 2y = 2$
$-x + y - 2z = -6$

14. $3x - 2y + z = 5$
$-x + 2z = -4$
$x - 2y + z = -1$

Exercises 15–20: Write the system of linear equations that the augmented matrix represents. Use the variables x, y, and z.

15. $\begin{bmatrix} 1 & 2 & | & -6 \\ 5 & -1 & | & 4 \end{bmatrix}$ **16.** $\begin{bmatrix} 1 & -5 & | & 7 \\ 0 & -3 & | & 6 \end{bmatrix}$

17. $\begin{bmatrix} 1 & -1 & 2 & | & 6 \\ 2 & 1 & -2 & | & 1 \\ -1 & 2 & -1 & | & 3 \end{bmatrix}$ **18.** $\begin{bmatrix} 3 & -1 & 2 & | & -1 \\ 2 & -2 & 2 & | & 4 \\ 1 & 7 & -2 & | & 2 \end{bmatrix}$

19. $\begin{bmatrix} 1 & 0 & 0 & | & 4 \\ 0 & 1 & 0 & | & -2 \\ 0 & 0 & 1 & | & 7 \end{bmatrix}$ **20.** $\begin{bmatrix} 1 & 0 & 0 & | & 6 \\ 0 & 1 & 0 & | & -2 \\ 0 & 0 & 1 & | & 4 \end{bmatrix}$

GAUSSIAN ELIMINATION

Exercises 21–38: Use Gauss–Jordan elimination to find the solution. Write the solution as an ordered pair or ordered triple and check the solution.

21. $x + y = 4$
$x + 3y = 10$

22. $x - 3y = -7$
$2x + y = 0$

23. $2x + 3y = 3$
$-2x + 2y = 7$

24. $x + 3y = -14$
$2x + 5y = -24$

25. $x - y = 5$
$x + 3y = -1$

26. $x + 4y = 1$
$3x - 2y = 10$

27. $4x - 8y = -10$
$x + y = 2$

28. $x - 7y = -16$
$4x + 10y = 50$

29. $x + y + z = 6$
$2y - z = 1$
$y + z = 5$

30. $x + y + z = 3$
$x + y - z = 2$
$y + z = 2$

31. $x + 2y + 3z = 6$
$-x + 3y + 4z = 0$
$x + y - 2z = -6$

32. $2x - 4y + 2z = 10$
$-x + 3y - 4z = -19$
$2x - y - 6z = -28$

33. $x + y + z = 0$
$2x + y + 2z = -1$
$x + y = 0$

34. $x + y - 2z = 5$
$x + 2y - 2z = 4$
$-x - y + z = -4$

35. $x + y + z = 3$
$-x - z = -2$
$x + y + 2z = 4$

36. $x + 2y - z = 3$
$-x - y + z = 0$
$x + 2y = 5$

37. $x + 2y + z = 3$
$2x + y - z = -6$
$-x - y + 2z = 5$

38. $x + y + z = -3$
$x - y - z = -1$
$-2x + y + 4z = 4$

*Exercises 39–48: **Technology**   Use a graphing calculator to solve the system of linear equations.*

39. $x + 4y = 13$
$5x - 3y = -50$

40. $9x - 11y = 7$
$5x + 6y = 16$

41. $2x - y + 3z = 9$
$-4x + 5y + 2z = 12$
$2x + 7z = 23$

42. $3x - 2y + 4z = 29$
$2x + 3y - 7z = -14$
$5x - y + 11z = 59$

43. $6x + 2y + z = 4$
$-2x + 4y + z = -3$
$2x - 8y = -2$

44. $-x - 9y + 2z = -28.5$
$2x - y + 4z = -17$
$x - y + 8z = -9$

45. $4x + 3y + 12z = -9.25$
$15y + 8z = -4.75 + x$
$7z = -5.5 - 6y$

46. $5x + 4y = 13.3 + z$
$7y + 9z = 16.9 - x$
$x - 3y + 4z = -4.1$

47. $1.2x - 0.9y + 2.7z = 5.37$
$3.1x - 5.1y + 7.2z = 14.81$
$1.8y + 6.38 = 3.6z - 0.2x$

48. $11x + 13y - 17z = 380$
$5x - 14y - 19z = 24$
$-21y + 46z = -676 + 7x$

Exercises 49–54: (Refer to the Critical Thinking box in this section.) Row-reduce the matrix associated with the given system to determine whether the system of linear equations is inconsistent or dependent.

49. $x + 2y = 4$
$-2x - 4y = -8$

50. $x - 5y = 4$
$-2x + 10y = 8$

51. $x + y + z = 3$
$x + y - z = 1$
$x + y = 3$

52. $x + y + z = 5$
$x - y - z = 8$
$2x + 2y + 2z = 6$

53. $\begin{aligned} x + 2y + 3z &= 14 \\ 2x - 3y - 2z &= -10 \\ 3x - y + z &= 4 \end{aligned}$ **54.** $\begin{aligned} x + 2y + 3z &= 6 \\ -x + 3y + 4z &= 6 \\ 5y + 7z &= 12 \end{aligned}$

Exercises 55 and 56: **Thinking Generally**   *The matrix represents a linear system of equations. Solve the system, if a and b are nonzero constants.*

55. $\begin{bmatrix} a & 0 & 0 & | & 1 \\ 0 & b & 0 & | & 1 \\ 0 & 0 & ab & | & 2 \end{bmatrix}$ **56.** $\begin{bmatrix} a & 0 & 0 & | & 1 \\ 0 & b & 0 & | & 2 \\ 0 & 0 & 0 & | & a \end{bmatrix}$

APPLICATIONS

57. *Weight of a Bear*   Use the results of Example 7 to estimate the weight of a bear with a head length of 12 inches and an overall length of 60 inches.

58. *Weight of a Bear*   (Refer to Example 7.) Head length and overall length are not the only variables that can be used to estimate the weight of a bear. The data in the accompanying table list the weight W, neck size N, and chest size C of three bears. These data can be modeled by $W = a + bN + cC$.   (**Sources:** M. Triola, *Elementary Statistics*; Minitab, Inc.)

W (pounds)	N (inches)	C (inches)
80	16	26
344	28	45
416	31	54

(a) Set up a system of equations whose solution gives values for the constants a, b, and c.
(b) Solve this system. Round each value to the nearest tenth.
(c) Predict the weight of a bear with neck size $N = 22$ inches and chest size $C = 38$ inches.

59. *Garbage and Household Size*   A larger household produces more garbage, on average, than a smaller household. If we know the amount of metal M and plastic P waste produced each week, we can estimate the household size H from $H = a + bM + cP$. The table contains representative data for three households.   (**Source:** M. Triola, *Elementary Statistics*.)

H (people)	M (pounds)	P (pounds)
3	2.00	1.40
2	1.50	0.65
6	4.00	3.40

(a) Set up a system of equations whose solution gives values for the constants a, b, and c.
(b) Solve this system with a graphing calculator.
(c) Predict the size of a household that produces 3 pounds of metal waste and 2 pounds of plastic waste each week.

60. *Old Faithful Geyser*   In Yellowstone National Park, Old Faithful Geyser has been a favorite attraction for decades. Although this geyser erupts about every 80 minutes, this time interval varies, as do the duration and height of the eruptions. The accompanying table shows the height H, duration D, and time interval T for three eruptions.   (**Source:** National Park Service.)

H (feet)	D (seconds)	T (minutes)
160	276	94
125	203	84
140	245	79

(a) Assume that these data can be modeled by $H = a + bD + cT$. Set up a system of equations whose solution gives values for the constants a, b, and c.
(b) Solve this system by using a graphing calculator. Round each value to the nearest thousandth.
(c) Use this equation to estimate H when $D = 220$ and $T = 81$.

61. *Jogging Speeds*   A runner in preparation for a marathon jogs at 5, 6, and 8 miles per hour. The runner travels a total distance of 12.5 miles in 2 hours and jogs the same length of time at 5 miles per hour and at 8 miles per hour. How long does the runner jog at each speed?

62. *Mixture Problem*   Three types of candy that cost $2, $3, and $4 per pound are to be mixed to produce a 5-pound bag of candy that costs $14.50. If there are to be equal amounts of the $3-per-pound candy and the $4-per-pound candy, how much of each type of candy should be included in the mixture?

63. *Interest and Investments*   (Refer to Example 5.) A total of $3000 is invested at 5%, 8%, and 12% annual interest. The interest earned after 1 year equals $285. The amount invested at 12% is triple the amount invested at 5%. Find the amount invested at each rate.

64. *Geometry*   The measure of the largest angle in a triangle is twice the measure of the smallest angle. The remaining angle is 10° less than the largest angle. Find the measure of each angle.

WRITING ABOUT MATHEMATICS

65. Explain what the dimension of a matrix means. What is the difference between a matrix that has dimension 3×4 and one that has dimension 4×3?

66. Discuss the advantages of using technology to transform an augmented matrix to reduced row–echelon form. Are there any disadvantages? Explain.

CHECKING BASIC CONCEPTS
SECTIONS 9.1 AND 9.2

1. Determine which ordered triple is the solution to the system of equations: $(5, -4, 0)$, $(1, 3, -1)$.

$$
\begin{aligned}
x - y + 7z &= -9 \\
2x - 2y + 5z &= -9 \\
-x + 3y - 2z &= 10
\end{aligned}
$$

2. Solve the system of equations by using elimination and substitution.

$$
\begin{aligned}
x - y + z &= 2 \\
2x - 3y + z &= -1 \\
-x + y + z &= 4
\end{aligned}
$$

3. Use an augmented matrix to represent the system of equations. Solve the system by using

   (a) Gauss–Jordan elimination and

 (b) technology.

$$
\begin{aligned}
x + 2y + z &= 1 \\
x + y + z &= -1 \\
y + z &= 1
\end{aligned}
$$

9.3 DETERMINANTS

Calculation of Determinants ■ Area of Regions ■ Cramer's Rule

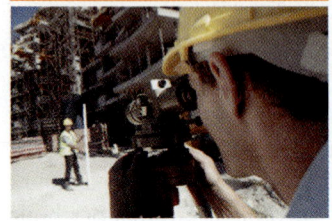

A LOOK INTO MATH ▷   Surveyors commonly calculate the areas of parcels of land. To do so, they frequently divide the land into triangular regions. When the coordinates of the vertices of a triangle are known, determinants may be used to find the area of the triangle. *A determinant is a real number that can be calculated for any square matrix.* In this section we use determinants to find areas and to solve systems of linear equations.

Calculation of Determinants

▶ **REAL-WORLD CONNECTION**   The concept of determinants originated with the Japanese mathematician Seki Kowa (1642–1708), who used them to solve systems of linear equations. Later, Gottfried Leibniz (1646–1716) formally described determinants and also used them to solve systems of linear equations.   (**Source:** *Historical Topics for the Mathematical Classroom*, NCTM.)

We begin by defining a determinant of a 2×2 matrix.

DETERMINANT OF A 2 × 2 MATRIX

The **determinant** of

$$A = \begin{bmatrix} a & b \\ c & d \end{bmatrix}$$

is a *real number* defined by

$$\det A = ad - cb.$$

EXAMPLE 1 Calculating determinants

Find det A for each 2×2 matrix.

(a) $A = \begin{bmatrix} 1 & 2 \\ 3 & 4 \end{bmatrix}$ **(b)** $A = \begin{bmatrix} -1 & -3 \\ 2 & -8 \end{bmatrix}$

Solution
(a) The determinant is calculated as follows.

$$\det A = \det \begin{bmatrix} 1 & 2 \\ 3 & 4 \end{bmatrix} = (1)(4) - (3)(2) = -2$$

(b) Similarly,

$$\det A = \det \begin{bmatrix} -1 & -3 \\ 2 & -8 \end{bmatrix} = (-1)(-8) - (2)(-3) = 14.$$

Now Try Exercise **5**

We can use determinants of 2×2 matrices to find determinants of 3×3 matrices. This method is called **expansion of a determinant by minors**.

DETERMINANT OF A 3 × 3 MATRIX

$$\det A = \det \begin{bmatrix} a_1 & b_1 & c_1 \\ a_2 & b_2 & c_2 \\ a_3 & b_3 & c_3 \end{bmatrix}$$

$$= a_1 \cdot \det \begin{bmatrix} b_2 & c_2 \\ b_3 & c_3 \end{bmatrix} - a_2 \cdot \det \begin{bmatrix} b_1 & c_1 \\ b_3 & c_3 \end{bmatrix} + a_3 \cdot \det \begin{bmatrix} b_1 & c_1 \\ b_2 & c_2 \end{bmatrix}$$

The 2×2 matrices in this equation are called **minors**.

EXAMPLE 2   **Calculating 3 × 3 determinants**

Evaluate det A.

(a) $A = \begin{bmatrix} 2 & 1 & -1 \\ -1 & 3 & 2 \\ 4 & -3 & -5 \end{bmatrix}$ (b) $A = \begin{bmatrix} 5 & -2 & 4 \\ 0 & 2 & 1 \\ -1 & 4 & -4 \end{bmatrix}$

Solution

(a) We evaluate the determinant as follows.

$$\det \begin{bmatrix} 2 & 1 & -1 \\ -1 & 3 & 2 \\ 4 & -3 & -5 \end{bmatrix} = 2 \cdot \det \begin{bmatrix} 3 & 2 \\ -3 & -5 \end{bmatrix} - (-1) \cdot \det \begin{bmatrix} 1 & -1 \\ -3 & -5 \end{bmatrix}$$

$$+ 4 \cdot \det \begin{bmatrix} 1 & -1 \\ 3 & 2 \end{bmatrix}$$

$$= 2(-9) + 1(-8) + 4(5)$$

$$= -6$$

(b) We evaluate the determinant as follows.

$$\det \begin{bmatrix} 5 & -2 & 4 \\ 0 & 2 & 1 \\ -1 & 4 & -4 \end{bmatrix} = 5 \cdot \det \begin{bmatrix} 2 & 1 \\ 4 & -4 \end{bmatrix} - (0) \cdot \det \begin{bmatrix} -2 & 4 \\ 4 & -4 \end{bmatrix}$$

$$+ (-1) \cdot \det \begin{bmatrix} -2 & 4 \\ 2 & 1 \end{bmatrix}$$

$$= 5(-12) - 0(-8) + (-1)(-10)$$

$$= -50$$

Now Try Exercise 13

Many graphing calculators can evaluate the determinant of a matrix, as illustrated in the next example, where we evaluate the determinants from Example 2.

EXAMPLE 3   **Using technology to find determinants**

Find each determinant of A, using a graphing calculator.

(a) $A = \begin{bmatrix} 2 & 1 & -1 \\ -1 & 3 & 2 \\ 4 & -3 & -5 \end{bmatrix}$ (b) $A = \begin{bmatrix} 5 & -2 & 4 \\ 0 & 2 & 1 \\ -1 & 4 & -4 \end{bmatrix}$

Solution

(a) Begin by entering the matrix and then evaluate the determinant, as shown in Figure 9.6. The result is det $A = -6$, which agrees with our earlier calculation.

CALCULATOR HELP

To find a determinant, see Appendix A (page AP-9).

(a)

(b)

Figure 9.6

(b) The determinant of A evaluates to -50 (see Figure 9.7).

(a)   (b)

Figure 9.7

Now Try Exercise **21**

Area of Regions

▶ **REAL-WORLD CONNECTION**   A determinant may be used to find the area of a triangle. For example, if a triangle has vertices (a_1, a_2), (b_1, b_2), and (c_1, c_2), its area equals the absolute value of D, where

$$D = \frac{1}{2}\det\begin{bmatrix} a_1 & b_1 & c_1 \\ a_2 & b_2 & c_2 \\ 1 & 1 & 1 \end{bmatrix}.$$

If the vertices are entered in the columns of D counterclockwise as they appear in the xy-plane, D will be positive.   (**Source:** W. Taylor, *The Geometry of Computer Graphics.*)

EXAMPLE 4   Computing the area of a triangular parcel of land

A triangular parcel of land is shown in Figure 9.8. If all units are miles, find the area of the parcel of land by using a determinant.

Solution

The vertices of the triangular parcel of land are $(2, 2)$, $(5, 4)$, and $(3, 8)$. The area of the triangle is

$$D = \frac{1}{2}\det\begin{bmatrix} 2 & 5 & 3 \\ 2 & 4 & 8 \\ 1 & 1 & 1 \end{bmatrix} = \frac{1}{2} \cdot 16 = 8.$$

Figure 9.8

The area of the triangle is 8 square miles.

Now Try Exercise **27**

CRITICAL THINKING

Suppose that you are given three distinct points in the xy-plane and $D = 0$. What must be true about the three points?

Cramer's Rule

▶ **REAL-WORLD CONNECTION**   Determinants were developed independently by Gabriel Cramer (1704–1752). His work, published in 1750, provided a method called **Cramer's rule** for solving systems of linear equations.

CRAMER'S RULE FOR LINEAR SYSTEMS IN TWO VARIABLES

The solution to the system of linear equations

$$a_1x + b_1y = c_1$$

$$a_2x + b_2y = c_2$$

is given by $x = \frac{E}{D}$ and $y = \frac{F}{D}$, where

$$E = \det\begin{bmatrix} c_1 & b_1 \\ c_2 & b_2 \end{bmatrix}, \quad F = \det\begin{bmatrix} a_1 & c_1 \\ a_2 & c_2 \end{bmatrix}, \quad \text{and} \quad D = \det\begin{bmatrix} a_1 & b_1 \\ a_2 & b_2 \end{bmatrix} \neq 0.$$

NOTE: If $D = 0$, the system has either no solutions or infinitely many solutions.

EXAMPLE 5   Using Cramer's rule

Use Cramer's rule to solve the following linear systems.
(a) $3x - 4y = 18$ **(b)** $-4x + 9y = -24$
     $7x + 5y = -1$ $6x + 17y = -25$

Solution

(a) $E = \det\begin{bmatrix} c_1 & b_1 \\ c_2 & b_2 \end{bmatrix} = \det\begin{bmatrix} 18 & -4 \\ -1 & 5 \end{bmatrix} = (18)(5) - (-1)(-4) = \mathbf{86}$

$F = \det\begin{bmatrix} a_1 & c_1 \\ a_2 & c_2 \end{bmatrix} = \det\begin{bmatrix} 3 & 18 \\ 7 & -1 \end{bmatrix} = (3)(-1) - (7)(18) = \mathbf{-129}$

$D = \det\begin{bmatrix} a_1 & b_1 \\ a_2 & b_2 \end{bmatrix} = \det\begin{bmatrix} 3 & -4 \\ 7 & 5 \end{bmatrix} = (3)(5) - (7)(-4) = \mathbf{43}$

Because $x = \frac{E}{D} = \frac{86}{43} = 2$ and $y = \frac{F}{D} = \frac{-129}{43} = -3$, the solution is $(2, -3)$.

(b) $E = \det\begin{bmatrix} c_1 & b_1 \\ c_2 & b_2 \end{bmatrix} = \det\begin{bmatrix} -24 & 9 \\ -25 & 17 \end{bmatrix} = (-24)(17) - (-25)(9) = \mathbf{-183}$

$F = \det\begin{bmatrix} a_1 & c_1 \\ a_2 & c_2 \end{bmatrix} = \det\begin{bmatrix} -4 & -24 \\ 6 & -25 \end{bmatrix} = (-4)(-25) - (6)(-24) = \mathbf{244}$

$D = \det\begin{bmatrix} a_1 & b_1 \\ a_2 & b_2 \end{bmatrix} = \det\begin{bmatrix} -4 & 9 \\ 6 & 17 \end{bmatrix} = (-4)(17) - (6)(9) = \mathbf{-122}$

Because $x = \frac{E}{D} = \frac{-183}{-122} = 1.5$ and $y = \frac{F}{D} = \frac{244}{-122} = -2$, the solution is $(1.5, -2)$.

Now Try Exercise **33**

Cramer's rule can be applied to systems that have any number of linear equations. Cramer's rule for three linear equations is discussed in the Extended and Discovery Exercises at the end of this chapter.

▶ **REAL-WORLD CONNECTION**   In applications, equations with hundreds of variables are routinely solved. Such systems of equations could be solved with Cramer's rule.

However, using Cramer's rule and the expansion of a determinant with minors to solve a linear system of equations with n variables requires at least

$$1 \cdot 2 \cdot 3 \cdot 4 \cdot \cdots \cdot n \cdot (n + 1)$$

multiplication operations. To solve a linear system involving only 25 variables would require about

$$1 \cdot 2 \cdot 3 \cdot 4 \cdot \cdots \cdot 25 \cdot 26 \approx 4 \times 10^{26}$$

multiplication operations. Supercomputers can perform about 1 trillion (1×10^{12}) multiplication operations per second. This would take about

$$\frac{4 \times 10^{26}}{1 \times 10^{12}} = 4 \times 10^{14} \text{ seconds.}$$

With $60 \times 60 \times 24 \times 365 = 31{,}536{,}000$ seconds in a year, 4×10^{14} seconds equals

$$\frac{4 \times 10^{14}}{31{,}536{,}000} \approx 12{,}700{,}000 \text{ years!}$$

Modern software packages do *not* use Cramer's rule for three or more variables.

9.3 PUTTING IT ALL TOGETHER

The determinant of a square matrix A is a real number, denoted det A. Cramer's rule is a method that uses determinants to solve systems of linear equations. The following table summarizes important topics from this section.

Concept	Explanation
Determinant of a 2 × 2 Matrix	The determinant of a 2×2 matrix A is given by $$\det A = \det \begin{bmatrix} a & b \\ c & d \end{bmatrix} = ad - cb.$$
Determinant of a 3 × 3 Matrix	The determinant of a 3×3 matrix A is given by $$\det A = \det \begin{bmatrix} a_1 & b_1 & c_1 \\ a_2 & b_2 & c_2 \\ a_3 & b_3 & c_3 \end{bmatrix}$$ $$= a_1 \cdot \det \begin{bmatrix} b_2 & c_2 \\ b_3 & c_3 \end{bmatrix} - a_2 \cdot \det \begin{bmatrix} b_1 & c_1 \\ b_3 & c_3 \end{bmatrix}$$ $$+ a_3 \cdot \det \begin{bmatrix} b_1 & c_1 \\ b_2 & c_2 \end{bmatrix}$$

continued on next page

continued from previous page

Concept	Explanation
Area of a Triangle	If a triangle has vertices (a_1, a_2), (b_1, b_2), and (c_1, c_2), its area equals the absolute value of D, where $$D = \frac{1}{2}\det\begin{bmatrix} a_1 & b_1 & c_1 \\ a_2 & b_2 & c_2 \\ 1 & 1 & 1 \end{bmatrix}.$$
Cramer's Rule for Linear Systems in Two Variables	The solution to the linear system $$a_1x + b_1y = c_1$$ $$a_2x + b_2y = c_2$$ is given by $x = \frac{E}{D}$ and $y = \frac{F}{D}$, where $$E = \det\begin{bmatrix} c_1 & b_1 \\ c_2 & b_2 \end{bmatrix}, \quad F = \det\begin{bmatrix} a_1 & c_1 \\ a_2 & c_2 \end{bmatrix}, \quad \text{and}$$ $$D = \det\begin{bmatrix} a_1 & b_1 \\ a_2 & b_2 \end{bmatrix} \neq 0.$$ **NOTE:** If $D = 0$, then the system has either no solutions or infinitely many solutions.

9.3 Exercises

MyMathLab   Math XL PRACTICE   WATCH   DOWNLOAD   READ   REVIEW

CONCEPTS

1. We can find the determinant of a _____ matrix.

2. If we find a determinant, the answer is a _____.

3. Cramer's rule can be used to solve a _____ .

4. If the first column of a matrix is all 0s, then its determinant equals _____.

CALCULATING DETERMINANTS

Exercises 5–20: Evaluate det A by hand where A is the given matrix.

5. $\begin{bmatrix} 1 & -2 \\ 3 & -8 \end{bmatrix}$

6. $\begin{bmatrix} 5 & -1 \\ 3 & 7 \end{bmatrix}$

7. $\begin{bmatrix} -3 & 7 \\ 8 & -1 \end{bmatrix}$

8. $\begin{bmatrix} 0 & -7 \\ -3 & 1 \end{bmatrix}$

9. $\begin{bmatrix} 23 & 4 \\ 6 & -13 \end{bmatrix}$

10. $\begin{bmatrix} 44 & -51 \\ -9 & 32 \end{bmatrix}$

11. $\begin{bmatrix} 1 & -1 & 2 \\ 0 & 1 & -3 \\ 0 & -4 & 7 \end{bmatrix}$

12. $\begin{bmatrix} 2 & -1 & -5 \\ -1 & 4 & -2 \\ 0 & 1 & 4 \end{bmatrix}$

13. $\begin{bmatrix} 2 & -1 & 0 \\ 1 & -2 & 6 \\ 0 & 1 & 8 \end{bmatrix}$

14. $\begin{bmatrix} 0 & 1 & -4 \\ 3 & -6 & 10 \\ 4 & -2 & 7 \end{bmatrix}$

15. $\begin{bmatrix} -1 & 3 & 5 \\ 3 & -3 & 5 \\ 2 & -3 & 7 \end{bmatrix}$ 16. $\begin{bmatrix} 6 & -1 & 9 \\ 7 & 0 & -3 \\ 2 & 5 & -1 \end{bmatrix}$

17. $\begin{bmatrix} 5 & 0 & 0 \\ 0 & -2 & 0 \\ 0 & 0 & 5 \end{bmatrix}$ 18. $\begin{bmatrix} 1 & 2 & 3 \\ 2 & 4 & 6 \\ 3 & 6 & 9 \end{bmatrix}$

19. $\begin{bmatrix} 0 & 2 & -3 \\ 0 & 3 & -9 \\ 0 & 5 & 9 \end{bmatrix}$ 20. $\begin{bmatrix} 3 & -1 & 2 \\ 0 & 5 & 7 \\ 0 & 0 & -1 \end{bmatrix}$

 Exercises 21–24: Use technology to calculate det A, where A is the given matrix.

21. $\begin{bmatrix} 2 & -5 & 13 \\ 10 & 15 & -10 \\ 17 & -19 & 22 \end{bmatrix}$ 22. $\begin{bmatrix} 1.6 & 3.1 & 5.7 \\ 2.1 & 6.7 & 8.1 \\ -0.4 & -0.8 & -3.1 \end{bmatrix}$

23. $\begin{bmatrix} 17 & 0 & 4 \\ -9 & 14 & 1.5 \\ 13 & 67 & -11 \end{bmatrix}$ 24. $\begin{bmatrix} 121 & 45 & -56 \\ -45 & 87 & 32 \\ -14 & -34 & 67 \end{bmatrix}$

Exercises 25 and 26: **Thinking Generally**   *Find det A, if a, b, and c are nonzero constants.*

25. $A = \begin{bmatrix} a & 0 & 0 \\ 0 & b & 0 \\ 0 & 0 & c \end{bmatrix}$ 26. $A = \begin{bmatrix} 0 & 0 & 0 \\ a & b & 0 \\ 0 & 0 & c \end{bmatrix}$

CALCULATING AREA

Exercises 27–32: Find the area of the figure by using a determinant. Assume that units are feet.

27.

28.

29.

30.

31.

32.
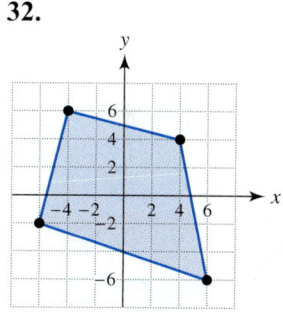

CRAMER'S RULE

Exercises 33–38: Solve the system of equations by using Cramer's rule.

33. $5x + 3y = 4$
$\quad 6x - 4y = 20$

34. $-5x + 4y = -5$
$\quad 4x + 4y = -32$

35. $\quad 7x - 5y = -3$
$\quad -4x + 6y = -8$

36. $-4x - 9y = -17$
$\quad 8x + 4y = 9$

37. $\quad 8x = 3y - 61$
$\quad -x = 4y - 23$

38. $15y = -188 - 22x$
$\quad 23y = -173 - 16x$

WRITING ABOUT MATHEMATICS

39. Suppose that one row of a 3×3 matrix A is all 0s. What is the value of det A? Give an example and explain your answer.

40. Estimate how long it might take a supercomputer to solve a linear system of equations with 15 variables, using Cramer's rule and expansion of a determinant by minors. Explain your calculations. (*Hint:* See the discussion at the end of this section.)

CHECKING BASIC CONCEPTS
SECTION 9.3

1. Evaluate det A.

(a) $A = \begin{bmatrix} -3 & 4 \\ -2 & 3 \end{bmatrix}$

(b) $A = \begin{bmatrix} 1 & -2 & 3 \\ 5 & 1 & 1 \\ 0 & 2 & -1 \end{bmatrix}$

2. Use Cramer's rule to solve the system of linear equations.

$$2x - \;\;y = -14$$
$$3x - 4y = -36$$

3. Find the area of a triangle that has the vertices $(-1, 2)$, $(5, 6)$, and $(2, -3)$.

CHAPTER 9   SUMMARY

SECTION 9.1 ■ SYSTEMS OF LINEAR EQUATIONS IN THREE VARIABLES

Solution to a System of Linear Equations in Three Variables   An ordered triple (x, y, z) that satisfies *every* equation

Example: $x - y + 2z = 3$
$$2x - y + \;\;z = 5$$
$$x + y + \;\;z = 6$$

The solution is $(\mathbf{3}, \mathbf{2}, \mathbf{1})$ because these values for (x, y, z) satisfy all three equations.

$$\mathbf{3} \;- \mathbf{2} + 2(\mathbf{1}) = 3 \qquad \text{True}$$
$$2(\mathbf{3}) - \mathbf{2} + \;\;\mathbf{1} \;= 5 \qquad \text{True}$$
$$\mathbf{3} \;+ \mathbf{2} + \;\;\mathbf{1} \;= 6 \qquad \text{True}$$

Elimination and Substitution   Systems of linear equations in *three* variables can be solved by elimination and substitution, using the following steps.

STEP 1: Eliminate one variable, such as x, from two of the given equations.

STEP 2: Use the two resulting equations in two variables to eliminate one of the variables, such as y. Solve for the remaining variable z.

STEP 3: Substitute z in one of the two equations from Step 2. Solve for the unknown variable y.

STEP 4: Substitute values for y and z in one of the given equations. Then find x.
The solution is (x, y, z).

SECTION 9.2 ■ MATRIX SOLUTIONS OF LINEAR SYSTEMS

Matrix A rectangular array of numbers is a matrix. If a matrix has m rows and n columns, it has dimension $m \times n$.

Example: Matrix $A = \begin{bmatrix} 3 & -1 & 7 \\ 0 & 6 & -2 \end{bmatrix}$ has dimension 2×3.

Augmented Matrix Any linear system can be represented with an augmented matrix.

Linear System

$4x - 3y = 5$
$x + 2y = 4$

Augmented Matrix

$\begin{bmatrix} 4 & -3 & | & 5 \\ 1 & 2 & | & 4 \end{bmatrix}$

Gauss–Jordan Elimination A numerical method that uses matrix row transformations to tranform a matrix into reduced row–echelon form

Example: The matrix $\begin{bmatrix} 4 & -3 & | & 5 \\ 1 & 2 & | & 4 \end{bmatrix}$ reduces to $\begin{bmatrix} 1 & 0 & | & 2 \\ 0 & 1 & | & 1 \end{bmatrix}$.

The solution to the system is $(2, 1)$.

SECTION 9.3 ■ DETERMINANTS

Determinant for a 2 × 2 Matrix A determinant is a *real number*. The determinant of a 2×2 matrix is

$$\det A = \det \begin{bmatrix} a & b \\ c & d \end{bmatrix} = ad - cb.$$

Example: $\det \begin{bmatrix} 2 & 3 \\ 4 & 5 \end{bmatrix} = (2)(5) - (4)(3) = -2$

Determinant for a 3 × 3 Matrix

$$\det A = \det \begin{bmatrix} a_1 & b_1 & c_1 \\ a_2 & b_2 & c_2 \\ a_3 & b_3 & c_3 \end{bmatrix}$$

$$= a_1 \cdot \det \begin{bmatrix} b_2 & c_2 \\ b_3 & c_3 \end{bmatrix} - a_2 \cdot \det \begin{bmatrix} b_1 & c_1 \\ b_3 & c_3 \end{bmatrix} + a_3 \cdot \det \begin{bmatrix} b_1 & c_1 \\ b_2 & c_2 \end{bmatrix}$$

Example: $\det \begin{bmatrix} 2 & 3 & 2 \\ 3 & 7 & -3 \\ 0 & 0 & -1 \end{bmatrix} = 2 \det \begin{bmatrix} 7 & -3 \\ 0 & -1 \end{bmatrix} - 3 \det \begin{bmatrix} 3 & 2 \\ 0 & -1 \end{bmatrix} + 0$

$$= 2(-7) - 3(-3) = -5$$

Cramer's rule uses determinants to solve linear systems of equations. Determinants can also be used to find areas of triangles. See Putting It All Together for Section 9.3.

CHAPTER 9   REVIEW EXERCISES

SECTION 9.1

1. Is $(3, -4, 5)$ a solution for $x + y + z = 4$?

2. Decide whether either ordered triple is a solution: $(1, -1, 2)$ or $(1, 0, 5)$.
$$2x - 3y + z = 7$$
$$-x - y + 3z = 6$$
$$3x - 2y + z = 7$$

Exercises 3–8: Use elimination and substitution to solve the system of linear equations, if possible.

3.
$$x - y - 2z = -11$$
$$-x + 2y + 3z = 16$$
$$3z = 6$$

4.
$$x + y = 4$$
$$-2x + y + 3z = -2$$
$$x - 2y + 5z = -26$$

5.
$$2x - y = -5$$
$$x + 2y + z = 7$$
$$-2x + y + z = 7$$

6.
$$2x + 3y + z = 6$$
$$-x + 2y + 2z = 3$$
$$x + y + 2z = 4$$

7.
$$x - y + 3z = 2$$
$$2x + y + 4z = 3$$
$$x + 2y + z = 5$$

8.
$$x - y + 3z = 3$$
$$x + y - z = 1$$
$$x + z = 2$$

SECTION 9.2

Exercises 9–12: Write the system of linear equations as an augmented matrix. Then use Gauss–Jordan elimination to solve the system, writing the solution as an ordered triple. Check your solution.

9.
$$x + y + z = -6$$
$$x + 2y + z = -8$$
$$y + z = -5$$

10.
$$x + y + z = -3$$
$$-x + y = 5$$
$$y + z = -1$$

11.
$$x + 2y - z = 1$$
$$-x + y - 2z = 5$$
$$2y + z = 10$$

12.
$$2x + 2y - 2z = -14$$
$$-2x - 3y + 2z = 12$$
$$x + y - 4z = -22$$

 Exercises 13 and 14: **Technology**   *Use a graphing calculator to solve the system of linear equations.*

13.
$$3x - 2y + 6z = -17$$
$$-2x - y + 5z = 20$$
$$4y + 7z = 30$$

14.
$$19x - 13y - 7z = 7.4$$
$$22x + 33y - 8z = 110.5$$
$$10x - 56y + 9z = 23.7$$

SECTION 9.3

Exercises 15–18: Evaluate det A, if A is the given matrix.

15. $\begin{bmatrix} 6 & -5 \\ -4 & 2 \end{bmatrix}$

16. $\begin{bmatrix} 0 & -6 \\ 5 & 9 \end{bmatrix}$

17. $\begin{bmatrix} 3 & -5 & -3 \\ 1 & 4 & 7 \\ 0 & -3 & 1 \end{bmatrix}$

18. $\begin{bmatrix} -2 & -1 & -7 \\ 2 & 1 & -3 \\ 3 & -5 & 8 \end{bmatrix}$

 Exercises 19 and 20: Use technology to calculate det A, where A is the given matrix.

19. $\begin{bmatrix} 22 & -45 & 3 \\ 15 & -12 & -93 \\ 5 & 81 & -21 \end{bmatrix}$

20. $\begin{bmatrix} 0.5 & -7.3 & 9.6 \\ 0.1 & 3.1 & 9.2 \\ -0.5 & -1.9 & 5.4 \end{bmatrix}$

Exercises 21 and 22: Use a determinant to find the area of the triangle. Assume that the units are feet.

21.

22.
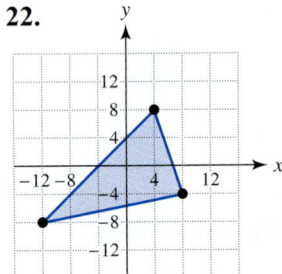

Exercises 23–26: Use Cramer's rule to solve the system.

23.
$$7x + 6y = 8$$
$$5x - 8y = 18$$

24.
$$-2x + 5y = 25$$
$$3x + 4y = -3$$

25.
$$3x - 6y = 1.5$$
$$7x - 5y = 8$$

26.
$$-5x + 4y = -47$$
$$6x - 7y = 63$$

APPLICATIONS

27. *Pedestrian Fatalities*   Forty-seven percent of pedestrian fatalities occur on Friday and Saturday nights. The combined total of pedestrian fatalities in 1994 and 2004 was 10,130. There were 848 more fatalities in 1994 than in 2004. Find the number of pedestrian fatalities during each year.   (*Source:* National Highway Traffic Safety Administration.)

28. *Tickets*   Tickets for a football game cost $8 and $12. If 480 tickets were sold for total receipts of $4620, how many of each type of ticket were sold?

29. *Determining Costs*   The accompanying table shows the costs for purchasing different combinations of malts, cones, and ice cream bars.

Malts	Cones	Bars	Total Cost
1	3	5	$14
1	2	4	$11
0	1	3	$5

(a) Let m be the cost of a malt, c the cost of a cone, and b the cost of an ice cream bar. Write a system of equations that represents the data in the table.
(b) Solve this system.

30. *Geometry*   The largest angle in a triangle is $20°$ more than the sum of the two smaller angles. The measure of the largest angle is $85°$ more than the smallest angle. Find the measure of each angle in the triangle.

31. *Mixture Problem*   Three types of candy that cost $1.50, $2.00, and $2.50 per pound are to be mixed to produce 12 pounds of candy worth $26.00. If there is to be 2 pounds more of the $2.50 candy than the $2.00 candy, how much of each type of candy should be used in the mixture?

32. *Estimating the Chest Size of a Bear*   The accompanying table shows the chest size C, weight W, and overall length L of three bears. These data can be modeled with the formula $C = a + bW + cL$.   (*Sources:* M. Triola, *Elementary Statistics*; Minitab, Inc.)

C (inches)	W (pounds)	L (inches)
40	202	63
50	365	70
55	446	77

(a) Set up a system of linear equations whose solution gives values for the constants a, b, and c.
(b) Solve this system. Round each value to the nearest thousandth.
(c) Predict the chest size of a bear weighing 300 pounds and having a length of 68 inches.

CHAPTER 9   TEST   Pass the Test   Video solutions to all test exercises

1. Can a system of three equations and three variables have exactly three solutions? Explain.

2. Determine which ordered triple is a solution to the linear system.

$$(-4, 3, 3), (1, -2, -2)$$
$$x - 3y + 4z = -1$$
$$2x + y - 3z = 6$$
$$x - y + z = 1$$

Exercises 3–5: Use substitution and elimination to solve the system, if possible.

3.
$$x + 3y = 2$$
$$-2x + y + z = 5$$
$$y + z = -3$$

4.
$$x + y - z = 1$$
$$2x - 3y + z = 0$$
$$x - 4y + 2z = 2$$

5.
$$x - y + 2z = 5$$
$$-x + y - 3z = -8$$
$$x - 2y + 2z = 3$$

6. Consider the system of linear equations.

$$2x - 4y = -10$$
$$-3x - 2y = 7$$

(a) Write the system of linear equations as an augmented matrix.
(b) Use Gauss–Jordan elimination to solve the system, writing the solution as an ordered pair.

7. Consider the system of linear equations.

$$x + y + z = 2$$
$$x - y - z = 3$$
$$2x + 2y + z = 6$$

(a) Write the system as an augmented matrix.
(b) Use Gauss–Jordan elimination to solve the system, writing the solution as an ordered triple.

8. Evaluate $\det A$ if $A = \begin{bmatrix} 3 & 2 & -1 \\ 6 & 2 & -6 \\ 0 & 8 & -3 \end{bmatrix}$.

9. Solve the system of equations with Cramer's rule.

$$5x - 3y = 7$$
$$-4x + 2y = 11$$

10. *Jogging Speeds*   A runner in preparation for a race jogs at 6, 7, and 9 miles per hour and travels a total of 7.1 miles in 1 hour. The runner jogs 12 minutes longer at 6 miles per hour than at 9 miles per hour. How many minutes did the runner jog at each speed?

11. *Geometry*   The largest angle in a triangle is 50° more than the smallest angle. The sum of the measures of the smaller two angles is 10° more than the largest angle. Find the measure of each angle.

12. *Weight of a Bear*   The data in the accompanying table list the weight W, neck size N, and chest size C of three bears. These data can be modeled by $W = a + bN + cC$.

W (pounds)	N (inches)	C (inches)
168	20	25
270	24	40
405	30	50

(a) Set up a system of equations whose solution gives values for the constants a, b, and c.
(b) Solve this system.
(c) Predict the weight W of a bear with neck size $N = 26$ and chest size $C = 44$.

CHAPTER 9   EXTENDED AND DISCOVERY EXERCISES

CRAMER'S RULE

Exercises 1–6: Cramer's rule can be applied to systems of three equations in three variables. For the system of equations

$$a_1 x + b_1 y + c_1 z = d_1$$
$$a_2 x + b_2 y + c_2 z = d_2$$
$$a_3 x + b_3 y + c_3 z = d_3,$$

the solution can be written as follows.

$$D = \det \begin{bmatrix} a_1 & b_1 & c_1 \\ a_2 & b_2 & c_2 \\ a_3 & b_3 & c_3 \end{bmatrix}, \quad E = \det \begin{bmatrix} d_1 & b_1 & c_1 \\ d_2 & b_2 & c_2 \\ d_3 & b_3 & c_3 \end{bmatrix}$$

$$F = \det \begin{bmatrix} a_1 & d_1 & c_1 \\ a_2 & d_2 & c_2 \\ a_3 & d_3 & c_3 \end{bmatrix}, \quad G = \det \begin{bmatrix} a_1 & b_1 & d_1 \\ a_2 & b_2 & d_2 \\ a_3 & b_3 & d_3 \end{bmatrix}$$

If $D \neq 0$, a unique solution exists and is given by

$$x = \frac{E}{D}, \quad y = \frac{F}{D}, \quad z = \frac{G}{D}.$$

Use Cramer's rule to solve the equations.

1.
$$x + y + z = 6$$
$$2x + y + 2z = 9$$
$$y + 3z = 9$$

2.
$$y + z = 1$$
$$2x - y - z = -1$$
$$x + y - z = 3$$

3.
$$x + z = 2$$
$$x + y = 0$$
$$y + 2z = 1$$

4.
$$x + y + 2z = 1$$
$$-x - 2y - 3z = -2$$
$$y - 3z = 5$$

5.
$$x + 2z = 7$$
$$-x + y + z = 5$$
$$2x - y + 2z = 6$$

6.
$$x + 2y + 3z = -1$$
$$2x - 3y - z = 12$$
$$x + 4y - 2z = -12$$

MATRICES AND ROAD MAPS

Exercises 7–12: Adjacency Matrix   A matrix A can be used to represent a map showing distances between cities. Let a_{ij} denote the number in row i and column j of a matrix A. Now consider the following map illustrating freeway distances in miles between four cities. Each city has been assigned a number. For example, there is a direct route from Denver, Colorado (city 1), to Colorado Springs, Colorado (city 2), of approximately 60 miles.

Therefore $a_{12} = 60$ in the accompanying matrix A at the top of the next column. (Note that a_{12} is the number in row 1 and column 2.) The distance from Colorado Springs to Denver is also 60 miles, so $a_{21} = 60$. As there is no direct freeway connection between Las Vegas, Nevada (city 4), and Colorado Springs (city 2), we let $a_{24} = a_{42} = *$. The matrix A is called an **adjacency matrix**. (**Source:** S. Baase, *Computer Algorithms: Introduction to Design and Analysis.*)

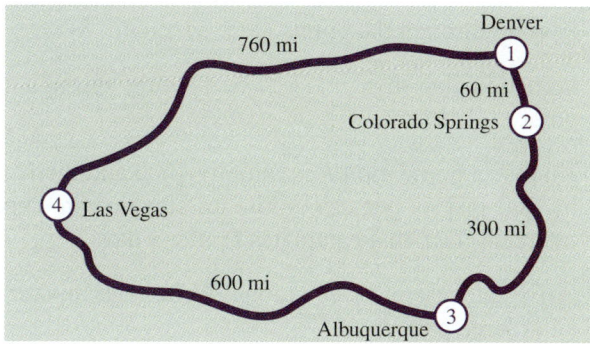

$$A = \begin{bmatrix} 0 & 60 & * & 760 \\ 60 & 0 & 300 & * \\ * & 300 & 0 & 600 \\ 760 & * & 600 & 0 \end{bmatrix}$$

7. Explain how to use A to find the freeway distance from Denver to Las Vegas.

8. Explain how to use A to find the freeway distance from Denver to Albuquerque.

9. If a map shows 20 cities, what would be the dimension of the adjacency matrix? How many elements would there be in this matrix?

10. Why are there only zeros on the main diagonal of A?

11. What does $a_{14} + a_{41}$ equal?

12. What does $a_{11} + a_{44}$ equal?

Exercises 13 and 14: (Refer to Exercises 7–12.) Find an adjacency matrix A for the road map.

13.

14.

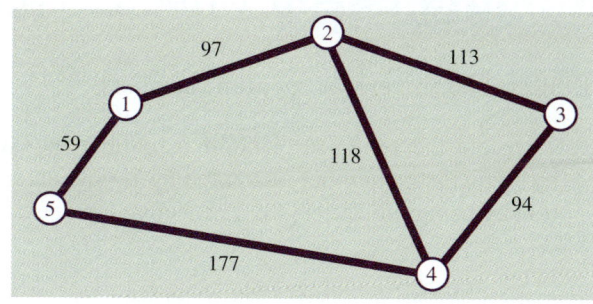

Exercises 15 and 16: (Refer to Exercises 7–12.) Sketch a road map represented by the adjacency matrix A. Is your answer unique?

15. $A = \begin{bmatrix} 0 & 30 & 20 & 5 \\ 30 & 0 & 15 & * \\ 20 & 15 & 0 & 25 \\ 5 & * & 25 & 0 \end{bmatrix}$

16. $A = \begin{bmatrix} 0 & 5 & * & 13 & 20 \\ 5 & 0 & 5 & * & * \\ * & 5 & 0 & 13 & * \\ 13 & * & 13 & 0 & 10 \\ 20 & * & * & 10 & 0 \end{bmatrix}$

SOLVING AN EQUATION IN FOUR VARIABLES

17. *Weight of a Bear* In Section 9.2 we estimated the weight of a bear by using two variables. We may be able to make more accurate estimates by using three variables. The accompanying table shows the weight W, neck size N, overall length L, and chest size C for four bears. (**Sources:** M. Triola, *Elementary Statistics*; Minitab, Inc.)

W (pounds)	N (inches)	L (inches)	C (inches)
125	19	57.5	32
316	26	65	42
436	30	72	48
514	30.5	75	54

(a) We can model the data in the table with the equation $W = a + bN + cL + dC$, where a, b, c, and d are constants. To do so, represent a system of linear equations by a 4×5 augmented matrix whose solution gives values for a, b, c, and d.

(b) Solve the system with a graphing calculator. Round each value to the nearest thousandth.

(c) Predict the weight of a bear with $N = 24$, $L = 63$, and $C = 39$. Interpret the result.

CHAPTERS 1–9   CUMULATIVE REVIEW EXERCISES

1. Write 360 as a product of prime numbers.

2. Translate the sentence "Double a number increased by 7 equals the number decreased by 2" into an equation by using the variable n. Then solve the equation.

Exercises 3 and 4: Simplify to lowest terms.

3. $\frac{2}{3} + \frac{4}{7} \cdot \frac{21}{28}$

4. $\frac{3}{5} \div \frac{6}{5} - \frac{2}{3}$

Exercises 5 and 6: Evaluate by hand.

5. $30 - 4 \div 2 \cdot 6$

6. $\dfrac{3^2 - 2^3}{20 - 5 \cdot 2}$

7. Solve $2(x + 1) - 6x = x - 4$.

8. Solve $4 - 3x = -2$ graphically. Check your answer.

9. Convert 124% to fraction and decimal notation.

10. If $A = 30$ square miles and $h = 10$ miles, use the formula $A = \frac{1}{2}bh$ to find b.

11. Solve $A = \frac{h}{2}(a + b)$ for b.

12. Solve $6t - 1 < 3 - t$.

13. Make a scatterplot with the points $(-1, 2)$, $(1, -2)$, $(0, 3)$, $(-2, 0)$, and $(2, 3)$.

Exercises 14–16: Graph the equation and determine any intercepts.

14. $y = -\frac{1}{2}x + 2$

15. $-3x + 4y = 12$

16. $x = -2$

17. Use the graph to identify the x-intercept and the y-intercept. Then write the slope–intercept form of the line.

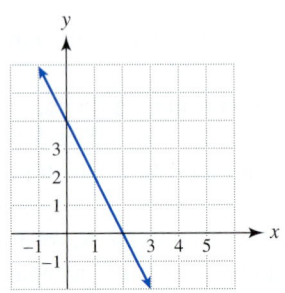

18. Write the slope–intercept form of a line that passes through the point $(-1, 3)$ with slope $m = -2$.

Exercises 19 and 20: Find the slope–intercept form of the line satisfying the given conditions.

19. Passing through the points $(-3, 5)$ and $(2, 8)$

20. Perpendicular to $x + 2y = 5$, passing through the point $(-1, 1)$

21. An insect population P is initially 4000 and increases by 500 insects per day. Write an equation in slope–intercept form that calculates P after x days.

22. Solve the system of equations. Write your answer as an ordered pair.
$$\begin{aligned} -2a + b &= -5 \\ 4a - 3b &= 0 \end{aligned}$$

Exercises 23 and 24: Shade the solution set.

23. $y \geq -1$

24. $3x - 2y \leq 6$

Exercises 25 and 26: Shade the solution set for the system.

25. $y \geq x - 1$
    $y \leq 2x$

26. $x - y < 3$
    $2x + y \geq 6$

Exercises 27–34: Simplify and use positive exponents, when appropriate, to write the expression.

27. $5 - 3^4$

28. $4(4^{-2})(3^{-1})(3^4)$

29. $\dfrac{2^{-4}}{4^{-2}}$

30. $(8t^{-3})(3t^2)(t^5)$

31. $2(rt)^4$

32. $(2t^3)^{-2}$

33. $(4a^2b^3)^2(2ab)^{-3}$

34. $\left(\dfrac{2a^{-1}}{ab^{-2}}\right)^{-3}$

Exercises 35–39: Multiply the expression.

35. $2a^2(a^2 - 2a + 3)$

36. $(5x + 1)(x - 7)$

37. $(y - 3)(2y + 3)$

38. $(a + b)(a - b)$

39. $(2x + 3y)^2$

40. Write 2,130,000 in scientific notation.

Exercises 41 and 42: Divide.

41. $\dfrac{4x^3 - 8x^2 + 6x}{2x}$

42. $(x^4 - 9x^3 + 23x^2 - 17x + 11) \div (x - 5)$

Exercises 43–50: Factor completely.

43. $10ab^2 - 25a^3b^5$ **44.** $y^3 - 3y^2 + 2y - 6$

45. $6z^2 + 7z - 3$ **46.** $4z^2 - 9$

47. $4y^2 - 20y + 25$ **48.** $a^3 - 27$

49. $4z^4 - 17z^2 + 15$ **50.** $2a^3b + a^2b^2 - ab^3$

Exercises 51–54: Solve the equation.

51. $(x - 1)(x + 2) = 0$ **52.** $x^2 - 9x = 0$

53. $6y^2 - 7y = 3$ **54.** $x^3 = 4x$

Exercises 55 and 56: Simplify to lowest terms.

55. $\dfrac{x^2 - 16}{x + 4}$ **56.** $\dfrac{2x^2 - 11x - 6}{6x^2 - 5x - 4}$

Exercises 57–60: Simplify and write in lowest terms.

57. $\dfrac{x^2 - 3x + 2}{x + 7} \div \dfrac{x - 2}{2x + 14}$

58. $\dfrac{x}{2x + 3} + \dfrac{x + 3}{2x + 3}$ **59.** $\dfrac{5x}{x^2 - 1} - \dfrac{3}{x + 1}$

60. $\dfrac{\dfrac{2}{x} - \dfrac{2}{y}}{\dfrac{2}{x} + \dfrac{2}{y}}$

Exercises 61–63: Solve the equation.

61. $\dfrac{x + 2}{5} = \dfrac{x}{4}$ **62.** $\dfrac{1}{3x} + \dfrac{5}{2x} = 2$

63. $\dfrac{1}{x - 2} + \dfrac{2}{x + 2} = \dfrac{1}{x^2 - 4}$

64. Suppose that y is inversely proportional to x and that $y = 25$ when $x = 4$. Find y for $x = 10$.

65. Evaluate $f(x) = 1 - 4x$ for $x = -3$.

66. Graph $f(x) = x^2 - 2$ and identify the domain and range of f. Write your answer in interval notation.

Exercises 67 and 68: Use the graph or table to evaluate $f(0)$ and $f(-2)$.

67.

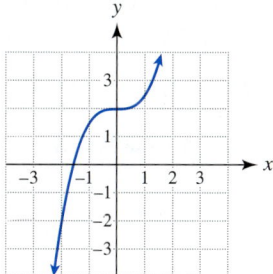

68.

x	-2	-1	0	1	2
$f(x)$	-4	-1	2	5	8

Exercises 69 and 70: Solve the inequality symbolically. Write the solution set in interval notation.

69. $\dfrac{4x - 9}{6} > \dfrac{1}{2}$

70. $\dfrac{2}{3}z - 2 \le \dfrac{1}{4}z - (2z + 2)$

Exercises 71 and 72: Solve the compound inequality. Graph the solution set on a number line.

71. $x + 2 > 1$ and $2x - 1 \le 9$

72. $4x + 7 < 1$ or $3x + 2 \ge 11$

Exercises 73 and 74: Solve the three-part inequality. Write the solution set in interval notation.

73. $-7 \le 2x - 3 \le 5$ **74.** $-8 \le -\frac{1}{2}x - 3 \le 5$

75. Use the graph to solve the equation and inequalities.
(a) $y_1 = 2$ (b) $y_1 \le 2$
(c) $y_1 \ge 2$

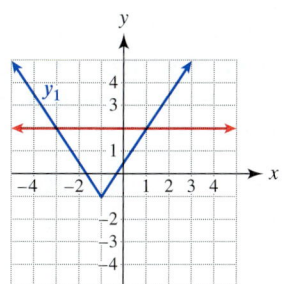

76. Solve the absolute value equation $\left|\frac{2}{3}x - 4\right| = 8$.

Exercises 77 and 78: Solve the absolute value inequality. Write the solution set in interval notation.

77. $|3x + 5| > 13$

78. $-3|2t - 11| \geq -9$

79. Use elimination and substitution to solve the system of linear equations.
$$2x + 3y - z = 3$$
$$3x - y + 4z = 10$$
$$2x + y - 2z = -1$$

80. Write the system of linear equations as an augmented matrix. Then use Gauss–Jordan elimination to solve the system. Write the solution as an ordered triple.
$$x + y - z = 4$$
$$-x - y - z = 0$$
$$x - 2y + z = -9$$

81. Evaluate $\det A$.
$$A = \begin{bmatrix} 4 & -2 \\ 1 & 3 \end{bmatrix}$$

APPLICATIONS

82. *Graphical Model*   An athlete rides a bicycle on a straight road away from home at 20 miles per hour for 1.5 hours. The athlete then turns around and rides home at 15 miles per hour. Sketch a graph that depicts the athlete's distance d from home after t hours.

83. *Ticket Prices*   The price of admission to a baseball game is $15 for children and $25 for adults. If a group of 8 people pays $170 to enter the game, find the number of children and the number of adults in the group.

84. *Working Together*   One person can shovel the snow from a sidewalk in 2 hours and another person can shovel the same sidewalk in 1.5 hours. How long will it take for them to shovel the snow from the sidewalk if they work together?

85. *Flight of a Golf Ball*   If a golf ball is hit upward with a velocity of 88 feet per second (60 miles per hour), then its height h in feet after t seconds can be approximated by
$$h(t) = 88t - 16t^2.$$

(a) What is the height of the golf ball after 2 seconds?

(b) After how long does the golf ball strike the ground?

86. *Height of a Building*   A 7-foot-tall stop sign casts a 4-foot-long shadow, while a nearby building casts a 35-foot-long shadow. Find the height of the building.

87. *Traffic Flow*   Twenty vehicles per minute can pass through an intersection. If vehicles arrive randomly at an average rate of x per minute, the average waiting time T in minutes is $T = \frac{1}{20 - x}$ for $x < 20$.   (*Source:* N. Garber, *Traffic and Highway Engineering.*)

(a) Evaluate the expression for $x = 15$ and interpret the result.

(b) Complete the table.

x	5	10	15	19	19.9
T					

(c) What happens to the waiting time as the traffic rate approaches 20 vehicles per minute?

88. *Error in Measurements*   An apprentice carpenter measures the length of a board to be L feet. If the actual measurement of the board is A feet, then the relative error in this measurement is $\left|\frac{L - A}{A}\right|$. If $A = 74$ inches and the relative error is to be less than or equal to 0.005 (0.5%), what values for L are possible?

89. *Determining Costs*   The accompanying table shows the costs for purchasing different combinations of burgers, fries, and malts.

Burgers	Fries	Malts	Total Cost
4	3	4	$23
1	2	1	$7
3	1	2	$13

(a) Let b be the cost of a burger, f be the cost of an order of fries, and m be the cost of a malt. Write a system of three linear equations that represents the data in the table.

(b) Solve this system and find the price of each item.

CHAPTER 10

Radical Expressions and Functions

Throughout history, people have created new numbers. Often these new numbers were met with resistance and regarded as being imaginary or unreal. The number 0 was not invented at the same time as the natural numbers. There was no Roman numeral for 0, which is one reason why our calendar started with A.D. 1 and, as a result, the twenty-first century began in 2001. No doubt there were skeptics during the time of the Roman Empire who questioned why anyone needed a number to represent nothing. Negative numbers also met strong resistance. After all, how could anyone possibly have −6 apples?

In this chapter we describe a new number system called *complex numbers*, which involve square roots of negative numbers. The Italian mathematician Cardano (1501–1576) was one of the first mathematicians to work with complex numbers and called them useless. René Descartes (1596–1650) originated the term *imaginary number*, which is associated with complex numbers. However, today complex numbers are used in many applications, such as electricity, fiber optics, and the design of airplanes. We are privileged to study in a period of days what took people centuries to discover.

Bear in mind that the wonderful things you learn in schools are the work of many generations, produced by enthusiastic effort and infinite labor in every country.

—ALBERT EINSTEIN

Source: Historical Topics for the Mathematics Classroom, Thirty-first Yearbook, NCTM.

10.1 RADICAL EXPRESSIONS AND RATIONAL EXPONENTS

Radical Notation ■ Rational Exponents ■ Properties of Rational Exponents

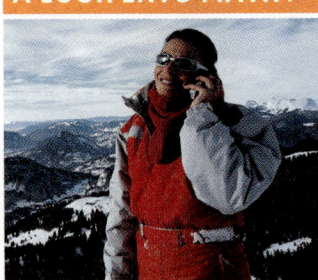

A LOOK INTO MATH ▷

Cellular phone technology has become a part of everyday life. In order to have cellular phone coverage, transmission towers, or cellular sites, are spread throughout a region. To estimate the minimum broadcasting distance for each cellular site, *radical expressions* are needed. (See Example 4.) In this section we discuss radical expressions and rational exponents and show how to manipulate them symbolically.   (**Source:** C. Smith, *Practical Cellular & PCS Design.*)

Radical Notation

Recall the definition of the square root of a number a.

> **SQUARE ROOT**
>
> The number b is a *square root* of a if $b^2 = a$.

EXAMPLE 1   Finding square roots

Find the square roots of 100.

Solution

The square roots of 100 are **10** *and* **−10** because $10^2 = 100$ and $(-10)^2 = 100$.

Now Try Exercise **1**

Every positive number a has two square roots, one positive and one negative. Recall that the *positive* square root is called the *principal square root* and is denoted $\sqrt{a}$. The *negative square root* is denoted $-\sqrt{a}$. To identify both square roots we write $\pm\sqrt{a}$. The symbol $\pm$ is read "plus or minus." The symbol $\sqrt{}$ is called the **radical sign**. The expression under the radical sign is called the **radicand**, and an expression containing a radical sign is called a **radical expression**. Examples of radical expressions include

$$\sqrt{6}, \quad 5 + \sqrt{x+1}, \quad \text{and} \quad \sqrt{\frac{3x}{2x-1}}.$$

> **MAKING CONNECTIONS**
>
> **Expressions and Equations**
>
> Expressions and equations are *different* mathematical concepts. An expression does not contain an equals sign, whereas an equation *always* contains an equals sign. *An equation is a statement that two expressions are equal.* For example,
>
> $$\sqrt{x+1} \quad \text{and} \quad \sqrt{5-x}$$
>
> are two different expressions, and
>
> $$\sqrt{x+1} = \sqrt{5-x}$$
>
> is an equation. We often *solve an equation*, but we *do not solve an expression*. Instead, we simplify and evaluate expressions.

In the next example we show how to find the principal square root of an expression.

EXAMPLE **2**   Finding principal square roots

Evaluate each square root.

(a) $\sqrt{25}$ **(b)** $\sqrt{0.49}$ **(c)** $\sqrt{\frac{4}{9}}$ **(d)** $\sqrt{c^2}, c > 0$

Solution
(a) Because $5 \cdot 5 = 25$, the principal, or *positive*, square root of 25 is $\sqrt{25} = 5$.
(b) Because $(0.7)(0.7) = 0.49$, the principal square root of 0.49 is $\sqrt{0.49} = 0.7$.
(c) Because $\frac{2}{3} \cdot \frac{2}{3} = \frac{4}{9}$, the principal square root of $\frac{4}{9}$ is $\sqrt{\frac{4}{9}} = \frac{2}{3}$.
(d) The principal square root of c^2 is $\sqrt{c^2} = c$, as c is positive.

Now Try Exercises 11, 13, 15, 17

EXAMPLE **3**   Approximating a square root

Approximate $\sqrt{17}$ to the nearest thousandth.

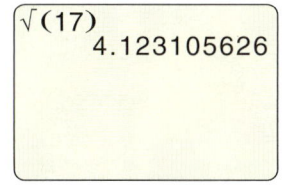

Figure 10.1

Solution
Figure 10.1 shows that $\sqrt{17} \approx 4.123$, rounded to the nearest thousandth. This result means that $4.123 \times 4.123 \approx 17$.

NOTE: Calculators often give decimal approximations rather than exact answers when evaluating radical expressions.

Now Try Exercise 35

▶ **REAL-WORLD CONNECTION**   In the next example we use the principal square root to estimate the minimum transmission distance for a cellular site.

EXAMPLE **4**   Estimating cellular phone transmission distance

Figure 10.2

If the ground is level, a cellular transmission tower will broadcast its signal in roughly a circular pattern, whose radius can be altered by changing the strength of its signal. See Figure 10.2. Suppose that a city has an area of 50 square miles and that there are 10 identical transmission towers spread evenly throughout the city. Estimate a *minimum* transmission radius R for each tower. (Note that a larger radius would probably be necessary to adequately cover the city.)   (**Source:** C. Smith.)

Solution
The circular area A covered by one transmission tower is $A = \pi R^2$. The total area covered by 10 towers is $10\pi R^2$, which must equal *at least* 50 square miles.

$$10\pi R^2 = 50$$
$$R^2 = \frac{50}{10\pi} \qquad \text{Divide by } 10\pi.$$
$$R^2 = \frac{5}{\pi} \qquad \text{Simplify.}$$
$$R = \sqrt{\frac{5}{\pi}} \approx 1.26 \qquad \text{Take principal square root.}$$

Each transmission tower must broadcast with a minimum radius of about 1.26 miles.

Now Try Exercise 117

Another common radical expression is the cube root of a number a, denoted $\sqrt[3]{a}$.

> ### CUBE ROOT
>
> The number b is a *cube root* of a if $b^3 = a$.

Although the square root of a negative number is *not* a real number, the cube root of a negative number is a negative real number. *Every real number has one real cube root.*

EXAMPLE 5   Finding cube roots

Evaluate the cube root. Approximate your answer to the nearest hundredth when appropriate.

(a) $\sqrt[3]{8}$ (b) $\sqrt[3]{-27}$ (c) $\sqrt[3]{\frac{1}{64}}$ (d) $\sqrt[3]{d^6}$ (e) $\sqrt[3]{16}$

$\sqrt[3]{(16)}$
            2.5198421

Figure 10.3

CALCULATOR HELP

To calculate a cube root, see Appendix A (page AP-1).

Solution

(a) $\sqrt[3]{8} = 2$ because $2^3 = 2 \cdot 2 \cdot 2 = 8$.

(b) $\sqrt[3]{-27} = -3$ because $(-3)^3 = (-3)(-3)(-3) = -27$.

(c) $\sqrt[3]{\frac{1}{64}} = \frac{1}{4}$ because $\left(\frac{1}{4}\right)^3 = \frac{1}{4} \cdot \frac{1}{4} \cdot \frac{1}{4} = \frac{1}{64}$.

(d) $\sqrt[3]{d^6} = d^2$ because $(d^2)^3 = d^2 \cdot d^2 \cdot d^2 = d^{2+2+2} = d^6$.

(e) $\sqrt[3]{16}$ is not an integer. Figure 10.3 shows that $\sqrt[3]{16} \approx 2.52$.

Now Try Exercises **19, 21, 23, 27, 37**

NOTE: $\sqrt[3]{-b} = -\sqrt[3]{b}$ for any real number b. That is, the cube root of a negative is the negative of the cube root. For example, $\sqrt[3]{-8} = -\sqrt[3]{8} = -2$.

We can generalize square roots and cube roots to include *n*th roots of a number a. The number b is an **nth root** of a if $b^n = a$, where n is a positive integer. For example, $2^5 = 32$ and so the 5th root of 32 is 2 and can be written as $\sqrt[5]{32} = 2$.

> ### THE NOTATION $\sqrt[n]{a}$
>
> The equation $\sqrt[n]{a} = b$ means that $b^n = a$, where n is a natural number called the **index**. If n is odd, we are finding an **odd root** and if n is even, we are finding an **even root**.
>
> 1. If $a > 0$, then $\sqrt[n]{a}$ is a positive number.
>
> 2. If $a < 0$ and
>   (a) n is odd, then $\sqrt[n]{a}$ is a negative number.
>   (b) n is even, then $\sqrt[n]{a}$ is *not* a real number.

If $a > 0$ and n is even, then a has two real *n*th roots: one positive and one negative. In this case, the positive root is denoted $\sqrt[n]{a}$ and called the **principal nth root of a**. For example, $(-3)^4 = 81$ *and* $3^4 = 81$, but $\sqrt[4]{81} = 3$ in the same way *principal square roots* are calculated.

EXAMPLE 6   **Finding nth roots**

Find each root, if possible.
(a) $\sqrt[4]{16}$ **(b)** $\sqrt[5]{-32}$ **(c)** $\sqrt[4]{-81}$

Solution
(a) $\sqrt[4]{16} = 2$ because $2^4 = 2 \cdot 2 \cdot 2 \cdot 2 = 16$.
(b) $\sqrt[5]{-32} = -2$ because $(-2)^5 = (-2)(-2)(-2)(-2)(-2) = -32$.
(c) An *even* root of a *negative* number is *not* a real number.

Now Try Exercises 29, 31, 33

Consider the calculations

$$\sqrt{3^2} = \sqrt{9} = \mathbf{3}, \quad \sqrt{(-4)^2} = \sqrt{16} = \mathbf{4}, \quad \text{and} \quad \sqrt{(-6)^2} = \sqrt{36} = \mathbf{6}.$$

In general, the expression $\sqrt{x^2}$ equals $|x|$. Graphical support is shown in Figure 10.4, where the graphs of $Y_1 = \sqrt{(X^\wedge 2)}$ and $Y_2 = \text{abs}(X)$ appear to be identical.

CRITICAL THINKING

Evaluate $\sqrt[6]{(-2)^6}$ and $\sqrt[3]{(-2)^3}$. Now simplify $\sqrt[n]{x^n}$ when n is even and when n is odd.

$[-6, 6, 1]$ by $[-4, 4, 1]$ $[-6, 6, 1]$ by $[-4, 4, 1]$

          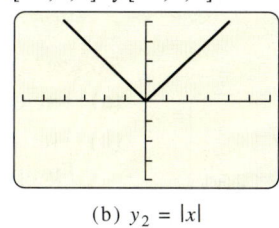

(a) $y_1 = \sqrt{x^2}$ (b) $y_2 = |x|$

Figure 10.4

THE EXPRESSION $\sqrt{x^2}$

For every real number x, $\sqrt{x^2} = |x|$.

EXAMPLE 7   **Simplifying expressions**

Write each expression in terms of an absolute value.
(a) $\sqrt{(-3)^2}$ **(b)** $\sqrt{(x + 1)^2}$ **(c)** $\sqrt{z^2 - 4z + 4}$

Solution
(a) $\sqrt{x^2} = |x|$, so $\sqrt{(-3)^2} = |-3| = 3$
(b) $\sqrt{(x + 1)^2} = |x + 1|$
(c) $\sqrt{z^2 - 4z + 4} = \sqrt{(z - 2)^2} = |z - 2|$

Now Try Exercises 103, 107, 109

Rational Exponents

When m and n are integers, the product rule states that $a^m a^n = a^{m+n}$. This rule can be extended to include exponents that are fractions. For example,

$$4^{1/2} \cdot 4^{1/2} = 4^{1/2 + 1/2} = 4^1 = 4.$$

That is, if we multiply the expression $4^{1/2}$ by itself, the result is 4. Because we also know that $\sqrt{4} \cdot \sqrt{4} = 4$, this discussion suggests that $4^{1/2} = \sqrt{4}$ and motivates the following definition.

THE EXPRESSION $a^{1/n}$

If n is an integer greater than 1 and a is a real number, then
$$a^{1/n} = \sqrt[n]{a}.$$

NOTE: If $a < 0$ and n is an even positive integer, then $a^{1/n}$ is not a real number.

In the next two examples we show how to interpret rational exponents.

EXAMPLE **8** **Interpreting rational exponents**

Write each expression in radical notation. Then evaluate the expression and round to the nearest hundredth when appropriate.
(a) $36^{1/2}$ **(b)** $23^{1/5}$ **(c)** $x^{1/3}$ **(d)** $(5x)^{1/2}$

Solution
(a) The exponent $\frac{1}{2}$ indicates a square root. Thus $36^{1/2} = \sqrt{36}$, which evaluates to 6.
(b) The exponent $\frac{1}{5}$ indicates a fifth root. Thus $23^{1/5} = \sqrt[5]{23}$, which is not an integer. Figure 10.5 shows this expression approximated in both exponential and radical notation. In either case $23^{1/5} \approx 1.87$.
(c) The exponent $\frac{1}{3}$ indicates a cube root, so $x^{1/3} = \sqrt[3]{x}$.
(d) The exponent $\frac{1}{2}$ indicates a square root, so $(5x)^{1/2} = \sqrt{5x}$.

Now Try Exercises **41, 49, 63, 67**

```
23^(1/5)
          1.872171231
5ˣ√(23)
          1.872171231
```

Figure 10.5

CALCULATOR HELP

To calculate other roots, see Appendix A (page AP-1).

Suppose that we want to define the expression $8^{2/3}$. On the one hand, using properties of exponents we have
$$8^{1/3} \cdot 8^{1/3} = 8^{1/3+1/3} = 8^{2/3}.$$

On the other hand, we have
$$8^{1/3} \cdot 8^{1/3} = \sqrt[3]{8} \cdot \sqrt[3]{8} = 2 \cdot 2 = 4.$$

Thus $8^{2/3} = 4$, and that value is obtained whether we interpret $8^{2/3}$ as either
$$8^{2/3} = (8^{1/3})^2 = (\sqrt[3]{8})^2 = 2^2 = 4$$
or
$$8^{2/3} = (8^2)^{1/3} = \sqrt[3]{8^2} = \sqrt[3]{64} = 4.$$

This result suggests the following definition.

THE EXPRESSION $a^{m/n}$

If m and n are positive integers with $\frac{m}{n}$ in lowest terms, then
$$a^{m/n} = \sqrt[n]{a^m} = (\sqrt[n]{a})^m.$$

NOTE: If $a < 0$ and n is an even integer, then $a^{m/n}$ is *not* a real number.

The exponent $\frac{m}{n}$ indicates that we either take the nth root and then calculate the mth power of the result or calculate the mth power and then take the nth root. For example, $4^{3/2}$ means that we can either take the square root of 4 and then cube the result or we cube 4 and then take the square root of the result. In either case, the result is the same: $4^{3/2} = 8$. This concept is illustrated in the next example.

EXAMPLE 9   **Interpreting rational exponents**

Write each expression in radical notation. Evaluate the expression by hand when possible.

(a) $(-27)^{2/3}$ **(b)** $12^{3/5}$

Solution

(a) The exponent $\frac{2}{3}$ indicates that we either take the cube root of -27 and then square it or that we square -27 and then take the cube root. Thus

$$(-27)^{2/3} = (\sqrt[3]{-27})^2 = (-3)^2 = 9$$

or

$$(-27)^{2/3} = \sqrt[3]{(-27)^2} = \sqrt[3]{729} = 9.$$

Same result

(b) The exponent $\frac{3}{5}$ indicates that we either take the fifth root of 12 and then cube it or that we cube 12 and then take the fifth root. Thus

$$12^{3/5} = (\sqrt[5]{12})^3 \quad \text{or} \quad 12^{3/5} = \sqrt[5]{12^3}.$$

This result cannot be evaluated by hand.   Now Try Exercises 45, 55

TECHNOLOGY NOTE: Rational Exponents

When evaluating expressions with rational (fractional) exponents, be sure to put parentheses around the fraction. For example, most calculators will evaluate 8^(2/3) and 8^2/3 differently. The accompanying figure shows evaluation of $8^{2/3}$ input correctly, 8^(2/3), as 4 but shows evaluation of $8^{2/3}$ input incorrectly, 8^2/3, as $\frac{8^2}{3} = 21.\overline{3}$.

Correct → | 8^(2/3)
| 4
Incorrect → | 8^2/3
| 21.33333333

From properties of exponents we know that $a^{-n} = \frac{1}{a^n}$, where n is a positive integer. We now define this property for negative rational exponents.

THE EXPRESSION $a^{-m/n}$

If m and n are positive integers with $\frac{m}{n}$ in lowest terms, then

$$a^{-m/n} = \frac{1}{a^{m/n}}, \qquad a \neq 0.$$

EXAMPLE 10   Interpreting negative rational exponents

Write each expression in radical notation and then evaluate.

(a) $64^{-1/3}$ **(b)** $81^{-3/4}$

Solution

(a) $64^{-1/3} = \dfrac{1}{64^{1/3}} = \dfrac{1}{\sqrt[3]{64}} = \dfrac{1}{4}.$

(b) $81^{-3/4} = \dfrac{1}{81^{3/4}} = \dfrac{1}{(\sqrt[4]{81})^3} = \dfrac{1}{3^3} = \dfrac{1}{27}.$

Now Try Exercises 57, 59

▶ **REAL-WORLD CONNECTION**   In the next example we use a formula from biology that involves a rational exponent.

EXAMPLE 11   Analyzing stepping frequency

When smaller (four-legged) animals walk, they tend to take faster, shorter steps, whereas larger animals tend to take slower, longer steps. If an animal is h feet high at the shoulder, then the number N of steps per second that the animal takes *while walking* can be estimated by $N(h) = 1.6h^{-1/2}$. Use a calculator to estimate N for an elephant that is 10 feet high at the shoulder.   (**Source:** C. Pennycuick, *Newton Rules Biology.*)

Solution

$$N(10) = 1.6(10)^{-1/2} = \frac{1.6}{\sqrt{10}} \approx 0.51$$

The elephant takes about $\frac{1}{2}$ step per second while walking or about 1 step every 2 seconds.

Now Try Exercise 119

Properties of Rational Exponents

Any rational number can be written as a ratio of two integers. That is, if p is a rational number, then $p = \frac{m}{n}$, where m and n are integers. Properties for integer exponents also apply to rational exponents, with one exception. If n is even in the expression $a^{m/n}$ and $\frac{m}{n}$ is written in lowest terms, then a must be nonnegative (not negative) for the result to be a real number.

PROPERTIES OF EXPONENTS

Let p and q be rational numbers written in lowest terms. For all real numbers a and b for which the expressions are real numbers the following properties hold.

1. $a^p \cdot a^q = a^{p+q}$ Product rule for exponents

2. $a^{-p} = \dfrac{1}{a^p}, \quad \dfrac{1}{a^{-p}} = a^p$ Negative exponents

3. $\left(\dfrac{a}{b}\right)^{-p} = \left(\dfrac{b}{a}\right)^{p}$ Negative exponents for quotients

4. $\dfrac{a^p}{a^q} = a^{p-q}$ Quotient rule for exponents

5. $(a^p)^q = a^{pq}$ Power rule for exponents

6. $(ab)^p = a^p b^p$ Power rule for products

7. $\left(\dfrac{a}{b}\right)^p = \dfrac{a^p}{b^p}$ Power rule for quotients

In the next two examples we apply these properties.

EXAMPLE 12   **Applying properties of exponents**

Write each expression using rational exponents and simplify. Write the answer with a positive exponent. Assume that all variables are positive numbers.

(a) $\sqrt{x} \cdot \sqrt[3]{x}$ **(b)** $\sqrt[3]{27x^2}$ **(c)** $\dfrac{\sqrt[4]{16x}}{\sqrt[3]{x}}$ **(d)** $\left(\dfrac{x^2}{81}\right)^{-1/2}$

Solution

(a) $\sqrt{x} \cdot \sqrt[3]{x} = x^{1/2} \cdot x^{1/3}$ Use rational exponents.

$\qquad\qquad = x^{1/2+1/3}$ Product rule for exponents

$\qquad\qquad = x^{5/6}$ Simplify: $\frac{1}{2} + \frac{1}{3} = \frac{3}{6} + \frac{2}{6} = \frac{5}{6}$.

(b) $\sqrt[3]{27x^2} = (27x^2)^{1/3}$ Use rational exponents.

$\qquad\qquad = 27^{1/3}(x^2)^{1/3}$ Power rule for products

$\qquad\qquad = 3x^{2/3}$ Simplify; power rule for exponents

(c) $\dfrac{\sqrt[4]{16x}}{\sqrt[3]{x}} = \dfrac{(16x)^{1/4}}{x^{1/3}}$ Use rational exponents.

$\qquad\qquad = \dfrac{16^{1/4}x^{1/4}}{x^{1/3}}$ Power rule for products

$\qquad\qquad = 16^{1/4}x^{1/4-1/3}$ Quotient rule for exponents

$\qquad\qquad = 2x^{-1/12}$ Simplify: $\frac{1}{4} - \frac{1}{3} = \frac{3}{12} - \frac{4}{12} = -\frac{1}{12}$.

$\qquad\qquad = \dfrac{2}{x^{1/12}}$ Negative exponents

(d) $\left(\dfrac{x^2}{81}\right)^{-1/2} = \left(\dfrac{81}{x^2}\right)^{1/2}$ Negative exponents for quotients

$\qquad\qquad = \dfrac{(81)^{1/2}}{(x^2)^{1/2}}$ Power rule for quotients

$\qquad\qquad = \dfrac{9}{x}$ Simplify; power rule for exponents

Now Try Exercises 81, 89, 90, 97

EXAMPLE 13   Applying properties of exponents

Write each expression with positive rational exponents and simplify, if possible.

(a) $\sqrt[3]{\sqrt{x+1}}$ **(b)** $\sqrt[5]{c^{15}}$ **(c)** $\dfrac{y^{-1/2}}{x^{-1/3}}$ **(d)** $\sqrt{x}(\sqrt{x}-1)$

Solution

(a) $\sqrt[3]{\sqrt{x+1}} = \left((x+1)^{1/2}\right)^{1/3} = (x+1)^{1/6}$

(b) $\sqrt[5]{c^{15}} = c^{15/5} = c^3$

(c) $\dfrac{y^{-1/2}}{x^{-1/3}} = \dfrac{x^{1/3}}{y^{1/2}}$

(d) $\sqrt{x}(\sqrt{x}-1) = x^{1/2}(x^{1/2}-1) = x^{1/2}x^{1/2} - x^{1/2} = x - x^{1/2}$

Now Try Exercises   **91, 95, 101, 115**

10.1   PUTTING IT ALL TOGETHER

Properties of radicals and rational exponents are summarized in the following table.

Concept	Explanation	Examples
*n*th Root of a Real Number	An *n*th root of a real number a is b if $b^n = a$ and the (principal) *n*th root is denoted $\sqrt[n]{a}$. If $a < 0$ and n is even, $\sqrt[n]{a}$ is not a real number.	The square roots of 25 are 5 and -5. The principal square root is $\sqrt{25} = 5$. $\sqrt[3]{-125} = -5$ because $(-5)^3 = -125$.
Rational Exponents	If m and n are positive integers with $\frac{m}{n}$ in lowest terms, $$a^{m/n} = \sqrt[n]{a^m} = \left(\sqrt[n]{a}\right)^m.$$ If $a < 0$ and n is even, $a^{m/n}$ is not a real number.	$8^{4/3} = \left(\sqrt[3]{8}\right)^4 = 2^4 = 16$ $(-27)^{3/4} = (\sqrt[4]{-27})^3$ is *not* a real number.
Properties of Exponents	Let p and q be rational numbers. **1.** $a^p \cdot a^q = a^{p+q}$ **2.** $a^{-p} = \dfrac{1}{a^p}, \dfrac{1}{a^{-p}} = a^p$ **3.** $\left(\dfrac{a}{b}\right)^{-p} = \left(\dfrac{b}{a}\right)^p$ **4.** $\dfrac{a^p}{a^q} = a^{p-q}$ **5.** $(a^p)^q = a^{pq}$ **6.** $(ab)^p = a^p b^p$ **7.** $\left(\dfrac{a}{b}\right)^p = \dfrac{a^p}{b^p}$	**1.** $2^{1/3} \cdot 2^{2/3} = 2^{1/3+2/3} = 2^1 = 2$ **2.** $2^{-1/2} = \dfrac{1}{2^{1/2}}, \dfrac{1}{3^{-1/4}} = 3^{1/4}$ **3.** $\left(\dfrac{3}{4}\right)^{-4/5} = \left(\dfrac{4}{3}\right)^{4/5}$ **4.** $\dfrac{7^{2/3}}{7^{1/3}} = 7^{2/3-1/3} = 7^{1/3}$ **5.** $(8^{2/3})^{1/2} = 8^{(2/3)\cdot(1/2)} = 8^{1/3} = 2$ **6.** $(2x)^{1/3} = 2^{1/3}x^{1/3}$ **7.** $\left(\dfrac{x}{y}\right)^{1/6} = \dfrac{x^{1/6}}{y^{1/6}}$

10.1 Exercises

CONCEPTS

1. What are the square roots of 9?

2. What is the principal square root of 9?

3. What is the cube root of 8?

4. Does every real number have a cube root?

5. If $b^n = a$ and $b > 0$, then $\sqrt[n]{a} =$ _____.

6. Use a rational exponent to write $\sqrt{x}$.

7. Use a rational exponent to write $\sqrt[3]{a^4}$.

8. Write $a^{1/n}$ in radical notation.

9. Write $a^{m/n}$ in radical notation.

10. Does $(a^{1/3})^2 = a^{2/3}$?

RADICAL EXPRESSIONS AND RATIONAL EXPONENTS

Exercises 11–34: Evaluate the expression by hand, if possible. Variables represent any real number.

11. $\sqrt{9}$

12. $\sqrt{121}$

13. $\sqrt{0.36}$

14. $\sqrt{0.64}$

15. $\sqrt{\frac{16}{25}}$

16. $\sqrt{\frac{9}{49}}$

17. $\sqrt{x^2}, x > 0$

18. $\sqrt{(x-1)^2}, x > 1$

19. $\sqrt[3]{27}$

20. $\sqrt[3]{64}$

21. $\sqrt[3]{-64}$

22. $-\sqrt[3]{-1}$

23. $\sqrt[3]{\frac{8}{27}}$

24. $\sqrt[3]{-\frac{1}{125}}$

25. $-\sqrt[3]{x^9}$

26. $\sqrt[3]{(x+1)^6}$

27. $\sqrt[3]{(2x)^6}$

28. $\sqrt[3]{27x^3}$

29. $\sqrt[4]{81}$

30. $\sqrt[5]{-1}$

31. $\sqrt[5]{-243}$

32. $\sqrt[4]{625}$

33. $\sqrt[4]{-16}$

34. $\sqrt[6]{-64}$

Exercises 35–48: Approximate to the nearest hundredth.

35. $-\sqrt{5}$

36. $\sqrt{11}$

37. $\sqrt[3]{5}$

38. $\sqrt[3]{-13}$

39. $\sqrt[5]{-7}$

40. $\sqrt[4]{6}$

41. $16^{1/5}$

42. $7^{1/4}$

43. $5^{1/3}$

44. $11^{1/2}$

45. $9^{3/5}$

46. $13^{5/4}$

47. $4^{-3/7}$

48. $2^{-3/4}$

Exercises 49–68: Write each expression in radical notation. Evaluate the expression by hand when possible.

49. $9^{1/2}$

50. $100^{1/2}$

51. $8^{1/3}$

52. $(-8)^{1/3}$

53. $\left(\frac{4}{9}\right)^{1/2}$

54. $\left(\frac{64}{27}\right)^{1/3}$

55. $(-8)^{2/3}$

56. $(16)^{3/2}$

57. $\left(\frac{1}{8}\right)^{-1/3}$

58. $\left(\frac{1}{81}\right)^{-1/4}$

59. $16^{-3/4}$

60. $(32)^{-2/5}$

61. $(4^{1/2})^{-3}$

62. $(8^{1/3})^{-2}$

63. $z^{1/4}$

64. $b^{1/3}$

65. $y^{-2/5}$

66. $z^{-3/4}$

67. $(3x)^{1/3}$

68. $(5x)^{1/2}$

Exercises 69–76: Use a positive rational exponent to write the expression.

69. $\sqrt{y}$

70. $\sqrt{z+1}$

71. $\sqrt{x} \cdot \sqrt{x}$

72. $\sqrt[3]{x} \cdot \sqrt[3]{x}$

73. $\sqrt[3]{8x^2}$

74. $\sqrt[3]{27z}$

75. $\dfrac{\sqrt{49x}}{\sqrt[3]{x^2}}$

76. $\dfrac{\sqrt[3]{8x}}{\sqrt[4]{x^3}}$

*Exercises 77 and 78: **Thinking Generally** Simplify the expression. Assume that a and b are positive integers.*

77. $(b^{1/a}b^{2/a})^a$

78. $b^{(b-1)/b} \cdot \sqrt[b]{b}$

Exercises 79–102: Simplify the expression. Assume that all variables are positive.

79. $(x^2)^{3/2}$

80. $(y^4)^{1/2}$

81. $\sqrt[3]{x^3 y^6}$

82. $\sqrt{16x^4}$

83. $\sqrt{\dfrac{y^4}{x^2}}$

84. $\sqrt[3]{\dfrac{x^{12}}{z^6}}$

85. $\sqrt[3]{y^3} \cdot \sqrt[3]{y^2}$

86. $\left(\dfrac{x^6}{81}\right)^{1/4}$

87. $\left(\dfrac{x^6}{27}\right)^{2/3}$

88. $\left(\dfrac{1}{x^8}\right)^{-1/4}$

89. $\left(\dfrac{x^2}{y^6}\right)^{-1/2}$

90. $\dfrac{\sqrt{x}}{\sqrt[3]{27x^6}}$

91. $\sqrt{\sqrt{y}}$

92. $\sqrt{\sqrt[3]{(3x)^2}}$

93. $(a^{-1/2})^{4/3}$

94. $(x^{-3/2})^{2/3}$

95. $\dfrac{(k^{1/2})^{-3}}{(k^2)^{1/4}}$

96. $\dfrac{(b^{3/4})^4}{(b^{4/5})^{-5}}$

97. $\sqrt{b} \cdot \sqrt[4]{b}$

98. $\sqrt[3]{t} \cdot \sqrt[5]{t}$

99. $p^{1/2}\left(p^{3/2} + p^{1/2}\right)$

100. $d^{3/4}(d^{1/4} - d^{-1/4})$

101. $\sqrt[3]{x}\left(\sqrt{x} - \sqrt[3]{x^2}\right)$

102. $\tfrac{1}{2}\sqrt{x}\left(\sqrt{x} + \sqrt[4]{x^2}\right)$

Exercises 103–116: Simplify the expression. Assume that all variables are real numbers.

103. $\sqrt{(-4)^2}$

104. $\sqrt{9^2}$

105. $\sqrt{y^2}$

106. $\sqrt{z^4}$

107. $\sqrt{(x-5)^2}$

108. $\sqrt{(2x-1)^2}$

109. $\sqrt{x^2 - 2x + 1}$

110. $\sqrt{4x^2 + 4x + 1}$

111. $\sqrt[4]{y^4}$

112. $\sqrt[4]{x^8 z^4}$

113. $\sqrt[4]{x^{12}}$

114. $\sqrt[6]{x^6}$

115. $\sqrt[5]{x^5}$

116. $\sqrt[5]{32(x+4)^5}$

APPLICATIONS

117. *Cellular Phone Technology*   (Refer to Example 4.) Suppose that a city has an area of 65 square miles and 15 cellular transmission towers spread evenly throughout it. Estimate a minimum radius R for each tower.

118. *Musical Tones*   One octave on a piano contains 12 keys (including both the black and white keys). The frequency of each successive key increases by a factor of $2^{1/12}$. For example, middle C is two keys below the first D above it. Therefore the frequency of this D is

$$2^{1/12} \cdot 2^{1/12} = 2^{1/6} \approx 1.12$$

times the frequency of middle C.

(a) If two tones are one octave apart, how do their frequencies compare?

(b) The A tone below middle C has a frequency of 220 cycles per second. Middle C is 3 keys above this A note. Estimate the frequency of middle C.

119. *Animal Stepping Frequency*   (Refer to Example 11.) Use the formula $N(h) = 1.6h^{-1/2}$ to estimate the stepping frequency of a dog that is 2.5 feet high at the shoulders.   (*Source:* C. Pennycuick.)

120. *Animal Pulse Rate*   According to one model, an animal's heart rate varies according to its weight. The formula $N(w) = 885w^{-1/2}$ gives an estimate for the average number N of beats per minute for an animal that weighs w pounds. Use the formula to estimate the heart rate for a horse that weighs 800 pounds. (*Source:* C. Pennycuick.)

Exercises 121 and 122: Heron's Formula   Suppose the lengths of the sides of a triangle are a, b, and c as illustrated in the figure.

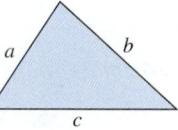

If the **semiperimeter** of the triangle is $s = \frac{1}{2}(a + b + c)$, then the area of the triangle is

$$A = \sqrt{s(s - a)(s - b)(s - c)}.$$

Find the area of the triangle with the given sides.

121. $a = 3, b = 4, c = 5$

122. $a = 5, b = 9, c = 10$

WRITING ABOUT MATHEMATICS

123. Try to calculate $\sqrt{-7}$, $\sqrt[4]{-56}$, and $\sqrt[6]{-10}$ with a calculator. Describe what happens when you evaluate an even root of a negative number. Does the same difficulty occur when you evaluate an odd root of a negative number? Try to evaluate $\sqrt[3]{-7}$, $\sqrt[5]{-56}$, and $\sqrt[7]{-10}$. Explain.

124. Explain the difference between a root and a positive integer power of a number. Give examples.

10.2 SIMPLIFYING RADICAL EXPRESSIONS

Product Rule for Radical Expressions ■ Quotient Rule for Radical Expressions

A LOOK INTO MATH ▷

Radical expressions often occur in biology. For example, heavier birds tend to have larger wings. The relationship between the weight of a bird and the surface area of its wings can be modeled by a radical expression. (See Exercise 111.) In this section we discuss performing arithmetic operations with radical expressions.

Product Rule for Radical Expressions

Consider the following examples of multiplying radical expressions. The equations

$$\sqrt{4} \cdot \sqrt{25} = 2 \cdot 5 = 10 \quad \text{and} \quad \sqrt{4 \cdot 25} = \sqrt{100} = 10$$

imply that

$$\sqrt{4} \cdot \sqrt{25} = \sqrt{4 \cdot 25} \quad \text{(see Figure 10.6(a))}.$$

Similarly, the equations

$$\sqrt[3]{8} \cdot \sqrt[3]{27} = 2 \cdot 3 = 6 \quad \text{and} \quad \sqrt[3]{8 \cdot 27} = \sqrt[3]{216} = 6$$

imply that

$$\sqrt[3]{8} \cdot \sqrt[3]{27} = \sqrt[3]{8 \cdot 27} \quad \text{(see Figure 10.6(b))}.$$

These examples suggest that *the product of the roots is equal to the root of the product.*

(a)

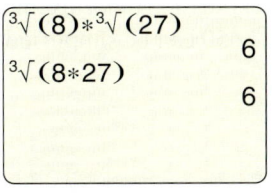

(b)

Figure 10.6

PRODUCT RULE FOR RADICAL EXPRESSIONS

Let a and b be real numbers, where $\sqrt[n]{a}$ and $\sqrt[n]{b}$ are both defined. Then

$$\sqrt[n]{a} \cdot \sqrt[n]{b} = \sqrt[n]{a \cdot b}.$$

NOTE: The product rule only works when the radicals have the *same index*. For example, the product $\sqrt{2} \cdot \sqrt[3]{4}$ cannot be simplified because the indexes are 2 and 3. (However, by using rational exponents, we can simplify this product. See Example 5(b).)

We apply the product rule in the next two examples.

EXAMPLE 1   Multiplying radical expressions

Multiply each radical expression.

(a) $\sqrt{5} \cdot \sqrt{20}$ **(b)** $\sqrt[3]{-3} \cdot \sqrt[3]{9}$ **(c)** $\sqrt[4]{\frac{1}{3}} \cdot \sqrt[4]{\frac{1}{9}} \cdot \sqrt[4]{\frac{1}{3}}$

Solution

(a) $\sqrt{5} \cdot \sqrt{20} = \sqrt{5 \cdot 20} = \sqrt{100} = 10$

(b) $\sqrt[3]{-3} \cdot \sqrt[3]{9} = \sqrt[3]{-3 \cdot 9} = \sqrt[3]{-27} = -3$

(c) The product rule can also be applied to three or more factors. Thus

$$\sqrt[4]{\frac{1}{3}} \cdot \sqrt[4]{\frac{1}{9}} \cdot \sqrt[4]{\frac{1}{3}} = \sqrt[4]{\frac{1}{3} \cdot \frac{1}{9} \cdot \frac{1}{3}} = \sqrt[4]{\frac{1}{81}} = \frac{1}{3}$$

because $\frac{1}{3} \cdot \frac{1}{3} \cdot \frac{1}{3} \cdot \frac{1}{3} = \frac{1}{81}$.   **Now Try Exercises** 13, 15, 21

EXAMPLE 2   Multiplying radical expressions containing variables

Multiply each radical expression. Assume that all variables are positive.

(a) $\sqrt{x} \cdot \sqrt{x^3}$ **(b)** $\sqrt[3]{2a} \cdot \sqrt[3]{5a}$ **(c)** $\sqrt{11} \cdot \sqrt{xy}$ **(d)** $\sqrt[5]{\frac{2x}{y}} \cdot \sqrt[5]{\frac{16y}{x}}$

Solution

(a) $\sqrt{x} \cdot \sqrt{x^3} = \sqrt{x \cdot x^3} = \sqrt{x^4} = x^2$

(b) $\sqrt[3]{2a} \cdot \sqrt[3]{5a} = \sqrt[3]{2a \cdot 5a} = \sqrt[3]{10a^2}$

(c) $\sqrt{11} \cdot \sqrt{xy} = \sqrt{11xy}$

(d) $\sqrt[5]{\frac{2x}{y}} \cdot \sqrt[5]{\frac{16y}{x}} = \sqrt[5]{\frac{2x}{y} \cdot \frac{16y}{x}}$ Product rule

$\qquad\qquad = \sqrt[5]{\frac{32xy}{xy}}$ Multiply fractions.

$\qquad\qquad = \sqrt[5]{32}$ Simplify.

$\qquad\qquad = 2$ $2^5 = 32$ **Now Try Exercises** 23, 51, 57, 61

An integer a is a **perfect nth power** if there exists an integer b such that $b^n = a$. Thus 36 is a **perfect square** because $6^2 = 36$. Similarly, 8 is a **perfect cube** because $2^3 = 8$, and 81 is a *perfect fourth power* because $3^4 = 81$.

The product rule for radicals can be used to simplify radical expressions. For example, because the largest perfect square factor of 50 is 25, the expression $\sqrt{50}$ can be simplified as

$$\sqrt{50} = \sqrt{25 \cdot 2} = \sqrt{25} \cdot \sqrt{2} = 5\sqrt{2}.$$

This procedure is generalized as follows.

SIMPLIFYING RADICALS (nth ROOTS)

STEP 1: Determine the largest perfect nth power factor of the radicand.

STEP 2: Use the product rule to factor out and simplify this perfect nth power.

EXAMPLE 3 **Simplifying radical expressions**

Simplify each expression.

(a) $\sqrt{300}$ **(b)** $\sqrt[3]{16}$ **(c)** $\sqrt{54}$ **(d)** $\sqrt[4]{512}$

Solution

(a) First note that $300 = 100 \cdot 3$ and that 100 is the largest perfect square factor of 300.

$$\sqrt{300} = \sqrt{100} \cdot \sqrt{3} = 10\sqrt{3}$$

(b) The largest perfect cube factor of 16 is 8. Thus $\sqrt[3]{16} = \sqrt[3]{8} \cdot \sqrt[3]{2} = 2\sqrt[3]{2}$.

(c) $\sqrt{54} = \sqrt{9} \cdot \sqrt{6} = 3\sqrt{6}$

(d) $\sqrt[4]{512} = \sqrt[4]{256} \cdot \sqrt[4]{2} = 4\sqrt[4]{2}$ because $4^4 = 256$.   Now Try Exercises 73, 75, 77, 79

NOTE: To simplify a cube root of a negative number we factor out the negative of the largest perfect cube factor. For example, $-16 = -8 \cdot 2$, so $\sqrt[3]{-16} = \sqrt[3]{-8} \cdot \sqrt[3]{2} = -2\sqrt[3]{2}$. This procedure can be used with any odd root of a negative number.

EXAMPLE 4 **Simplifying radical expressions**

Simplify each expression. Assume that all variables are positive.

(a) $\sqrt{25x^4}$ **(b)** $\sqrt{32n^3}$ **(c)** $\sqrt[3]{-16x^3y^5}$ **(d)** $\sqrt[3]{2a} \cdot \sqrt[3]{4a^2b}$

Solution

(a) $\sqrt{25x^4} = 5x^2$ Perfect square: $(5x^2)^2 = 25x^4$

(b) $\sqrt{32n^3} = \sqrt{(16n^2)2n}$ $16n^2$ is the largest perfect square factor.

$\qquad\qquad = \sqrt{16n^2} \cdot \sqrt{2n}$ Product rule

$\qquad\qquad = 4n\sqrt{2n}$ $(4n)^2 = 16n^2$

(c) $\sqrt[3]{-16x^3y^5} = \sqrt[3]{(-8x^3y^3)2y^2}$ $8x^3y^3$ is the largest perfect cube factor.

$\qquad\qquad = \sqrt[3]{-8x^3y^3} \cdot \sqrt[3]{2y^2}$ Product rule

$\qquad\qquad = -2xy\sqrt[3]{2y^2}$ $(-2xy)^3 = -8x^3y^3$

(d) $\sqrt[3]{2a} \cdot \sqrt[3]{4a^2b} = \sqrt[3]{(2a)(4a^2b)}$ Product rule

$\qquad\qquad = \sqrt[3]{(8a^3)b}$ $8a^3$ is the largest perfect cube factor.

$\qquad\qquad = \sqrt[3]{8a^3} \cdot \sqrt[3]{b}$ Product rule

$\qquad\qquad = 2a\sqrt[3]{b}$ $(2a)^3 = 8a^3$

Now Try Exercises 45, 85, 89, 91

The product rule for radical expressions cannot be used if the radicals do not have the same indexes. In this case we use rational exponents, as illustrated in the next example.

EXAMPLE 5 **Multiplying radicals with different indexes**

Simplify each expression. Write your answer in radical notation.

(a) $\sqrt{5} \cdot \sqrt[4]{5}$ **(b)** $\sqrt{2} \cdot \sqrt[3]{4}$ **(c)** $\sqrt[3]{x} \cdot \sqrt[4]{x}$

Solution

(a) The given expression is $\sqrt{5} \cdot \sqrt[4]{5}$. Because $\sqrt{5} = 5^{1/2}$ and $\sqrt[4]{5} = 5^{1/4}$,

$$\sqrt{5} \cdot \sqrt[4]{5} = 5^{1/2} \cdot 5^{1/4} = 5^{1/2+1/4} = 5^{3/4}.$$

In radical notation, $5^{3/4} = \sqrt[4]{5^3} = \sqrt[4]{125}$.

(b) The given expression is $\sqrt{2} \cdot \sqrt[3]{4}$. Because $\sqrt[3]{4} = \sqrt[3]{2^2} = 2^{2/3}$,

$$\sqrt{2} \cdot \sqrt[3]{4} = 2^{1/2} \cdot 2^{2/3} = 2^{1/2+2/3} = 2^{7/6}.$$

In radical notation, $2^{7/6} = \sqrt[6]{2^7} = \sqrt[6]{2^6 \cdot 2^1} = \sqrt[6]{2^6} \cdot \sqrt[6]{2} = 2\sqrt[6]{2}$.

(c) $\sqrt[3]{x} \cdot \sqrt[4]{x} = x^{1/3} \cdot x^{1/4} = x^{7/12} = \sqrt[12]{x^7}$　　　Now Try Exercises　101, 103, 107

Quotient Rule for Radical Expressions

Consider the following examples of dividing radical expressions. The equations

$$\sqrt{\frac{4}{9}} = \sqrt{\frac{2}{3} \cdot \frac{2}{3}} = \frac{2}{3} \quad \text{and} \quad \frac{\sqrt{4}}{\sqrt{9}} = \frac{2}{3}$$

imply that

$$\sqrt{\frac{4}{9}} = \frac{\sqrt{4}}{\sqrt{9}} \quad \text{(see Figure 10.7)}.$$

These examples suggest that *the root of a quotient is equal to the quotient of the roots.*

Figure 10.7

CALCULATOR HELP

To use the Frac feature, see Appendix A (page AP-2).

> ### QUOTIENT RULE FOR RADICAL EXPRESSIONS
>
> Let a and b be real numbers, where $\sqrt[n]{a}$ and $\sqrt[n]{b}$ are both defined and $b \neq 0$. Then
>
> $$\sqrt[n]{\frac{a}{b}} = \frac{\sqrt[n]{a}}{\sqrt[n]{b}}.$$

EXAMPLE 6　**Simplifying quotients**

Simplify each radical expression. Assume that all variables are positive.

(a) $\sqrt[3]{\frac{5}{8}}$　　**(b)** $\sqrt[4]{\frac{x}{16}}$　　**(c)** $\sqrt{\frac{16}{y^2}}$

Solution

(a) $\sqrt[3]{\frac{5}{8}} = \frac{\sqrt[3]{5}}{\sqrt[3]{8}} = \frac{\sqrt[3]{5}}{2}$　　**(b)** $\sqrt[4]{\frac{x}{16}} = \frac{\sqrt[4]{x}}{\sqrt[4]{16}} = \frac{\sqrt[4]{x}}{2}$

(c) $\sqrt{\frac{16}{y^2}} = \frac{\sqrt{16}}{\sqrt{y^2}} = \frac{4}{y}$　　because $y > 0$.

　　　Now Try Exercises　25, 27, 29

EXAMPLE 7   **Simplifying quotients**

Simplify each radical expression. Assume that all variables are positive.

(a) $\dfrac{\sqrt{40}}{\sqrt{10}}$ (b) $\dfrac{\sqrt[3]{2}}{\sqrt[3]{16}}$ (c) $\dfrac{\sqrt{x^2 y}}{\sqrt{y}}$

Solution

(a) $\dfrac{\sqrt{40}}{\sqrt{10}} = \sqrt{\dfrac{40}{10}} = \sqrt{4} = 2$

(b) $\dfrac{\sqrt[3]{2}}{\sqrt[3]{16}} = \sqrt[3]{\dfrac{2}{16}} = \sqrt[3]{\dfrac{1}{8}} = \dfrac{1}{2}$ because $\dfrac{1}{2} \cdot \dfrac{1}{2} \cdot \dfrac{1}{2} = \dfrac{1}{8}$.

(c) $\dfrac{\sqrt{x^2 y}}{\sqrt{y}} = \sqrt{\dfrac{x^2 y}{y}} = \sqrt{x^2} = x$ because $x > 0$.

Now Try Exercises **33, 39, 41**

NOTE: When we simplify a radical *expression* we do *not* set the expression equal to 0 and solve the equation. Instead we write a sequence of equivalent expressions.

MAKING CONNECTIONS

Rules for Radical Expressions and Rational Exponents

The rules for radical expressions are a result of the properties of rational exponents.

$$\sqrt[n]{a \cdot b} = \sqrt[n]{a} \cdot \sqrt[n]{b} \quad \text{is equivalent to} \quad (a \cdot b)^{1/n} = a^{1/n} \cdot b^{1/n}.$$

$$\sqrt[n]{\dfrac{a}{b}} = \dfrac{\sqrt[n]{a}}{\sqrt[n]{b}} \quad \text{is equivalent to} \quad \left(\dfrac{a}{b}\right)^{1/n} = \dfrac{a^{1/n}}{b^{1/n}}.$$

EXAMPLE 8   **Simplifying radical expressions**

Simplify each radical expression. Assume that all variables are positive.

(a) $\sqrt[4]{\dfrac{16x^3}{y^4}}$ (b) $\sqrt{\dfrac{5a^2}{8}} \cdot \sqrt{\dfrac{5a^3}{2}}$

Solution

(a) To simplify this expression we first use the quotient rule for radical expressions and then apply the product rule for radical expressions.

$$\sqrt[4]{\dfrac{16x^3}{y^4}} = \dfrac{\sqrt[4]{16x^3}}{\sqrt[4]{y^4}} \qquad \text{Quotient rule}$$

$$= \dfrac{\sqrt[4]{16}\,\sqrt[4]{x^3}}{\sqrt[4]{y^4}} \qquad \text{Product rule}$$

$$= \dfrac{2\,\sqrt[4]{x^3}}{y} \qquad \text{Evaluate 4th roots.}$$

(b) To simplify this expression, we use both the product and quotient rules.

$$\sqrt{\frac{5a^2}{8}} \cdot \sqrt{\frac{5a^3}{2}} = \sqrt{\frac{25a^5}{16}} \qquad \text{Product rule}$$

$$= \frac{\sqrt{25a^5}}{\sqrt{16}} \qquad \text{Quotient rule}$$

$$= \frac{\sqrt{25a^4 \cdot a}}{\sqrt{16}} \qquad \text{Factor out largest perfect square.}$$

$$= \frac{\sqrt{25a^4} \cdot \sqrt{a}}{\sqrt{16}} \qquad \text{Product rule}$$

$$= \frac{5a^2\sqrt{a}}{4} \qquad (5a^2)^2 = 25a^4$$

Now Try Exercises 95, 97

EXAMPLE 9 **Simplifying products and quotients of roots**

Simplify each expression. Assume all radicands are positive.

(a) $\sqrt{x-3} \cdot \sqrt{x+3}$ **(b)** $\dfrac{\sqrt[3]{x^2+3x+2}}{\sqrt[3]{x+1}}$

Solution

(a) Start by applying the product rule for radical expressions.

$$\sqrt{x-3} \cdot \sqrt{x+3} = \sqrt{(x-3)(x+3)} \qquad \text{Product rule}$$

$$= \sqrt{x^2 - 9} \qquad \text{Multiply binomials.}$$

NOTE: The expression $\sqrt{x^2 - 9}$ is *not* simplified further. It is important to realize that

$$\sqrt{x^2 - 9} \neq \sqrt{x^2} - \sqrt{9} = x - 3.$$

For example, $\sqrt{5^2 - 3^2} = \sqrt{16} = 4$, but $\sqrt{5^2} - \sqrt{3^2} = 5 - 3 = 2$.

(b) Start by applying the quotient rule for radical expressions.

$$\frac{\sqrt[3]{x^2+3x+2}}{\sqrt[3]{x+1}} = \sqrt[3]{\frac{x^2+3x+2}{x+1}} \qquad \text{Quotient rule}$$

$$= \sqrt[3]{\frac{(x+1)(x+2)}{x+1}} \qquad \text{Factor trinomial.}$$

$$= \sqrt[3]{x+2} \qquad \text{Simplify quotient.}$$

Now Try Exercises 63, 67

10.2 PUTTING IT ALL TOGETHER

In this section we discussed how to simplify radical expressions. Results are summarized in the following table.

Procedure	Explanation	Examples
Product Rule for Radical Expressions	Let a and b be real numbers, where $\sqrt[n]{a}$ and $\sqrt[n]{b}$ are both defined. Then $$\sqrt[n]{a} \cdot \sqrt[n]{b} = \sqrt[n]{a \cdot b}.$$	$$\sqrt{2} \cdot \sqrt{32} = \sqrt{64} = 8$$ $$\sqrt{500} = \sqrt{100} \cdot \sqrt{5} = 10\sqrt{5}$$
Quotient Rule for Radical Expressions	Let a and b be real numbers, where $\sqrt[n]{a}$ and $\sqrt[n]{b}$ are both defined and $b \neq 0$. Then $$\sqrt[n]{\frac{a}{b}} = \frac{\sqrt[n]{a}}{\sqrt[n]{b}}.$$	$$\frac{\sqrt{60}}{\sqrt{15}} = \sqrt{\frac{60}{15}} = \sqrt{4} = 2 \quad \text{and}$$ $$\sqrt[3]{\frac{x^2}{-27}} = \frac{\sqrt[3]{x^2}}{\sqrt[3]{-27}} = \frac{\sqrt[3]{x^2}}{-3} = -\frac{\sqrt[3]{x^2}}{3}$$
Simplifying Radicals (nth roots)	**STEP 1:** Find the largest nth power factor of the radicand. **STEP 2:** Factor out and simplify this perfect nth power.	$$\sqrt{12} = \sqrt{4 \cdot 3} = \sqrt{4} \cdot \sqrt{3} = 2\sqrt{3}$$ $$\sqrt[3]{81} = \sqrt[3]{27 \cdot 3} = \sqrt[3]{27} \cdot \sqrt[3]{3} = 3\sqrt[3]{3}$$ $$\sqrt[4]{x^5} = \sqrt[4]{x^4 \cdot x} = \sqrt[4]{x^4} \cdot \sqrt[4]{x} = x\sqrt[4]{x},$$ provided $x \geq 0$.

10.2 Exercises

MyMathLab   Math XL PRACTICE   WATCH   DOWNLOAD   READ   REVIEW

CONCEPTS

1. Does $\sqrt{2} \cdot \sqrt{3}$ equal $\sqrt{6}$?

2. Does $\sqrt{5} \cdot \sqrt[3]{5}$ equal 5?

3. $\sqrt[3]{a} \cdot \sqrt[3]{b} = $ _____

4. $\dfrac{\sqrt{a}}{\sqrt{b}} = \sqrt{?}$ **5.** $\dfrac{\sqrt[n]{a}}{\sqrt[n]{b}} = \sqrt[n]{?}$

6. $\dfrac{\sqrt{3}}{\sqrt{27}} = $ _____

7. Does $\sqrt{50} = \sqrt{25} + \sqrt{25}$?

8. Does $\sqrt{50} = 5\sqrt{2}$?

9. Is $\sqrt[3]{3}$ equal to 1? Explain.

10. Is 64 a perfect cube? Explain.

MULTIPLYING AND DIVIDING

Exercises 11–62: Simplify the expression. Assume that all variables are positive.

11. $\sqrt{3} \cdot \sqrt{3}$ **12.** $\sqrt{2} \cdot \sqrt{18}$

13. $\sqrt{2} \cdot \sqrt{50}$ **14.** $\sqrt[3]{-2} \cdot \sqrt[3]{-4}$

15. $\sqrt[3]{4} \cdot \sqrt[3]{16}$ **16.** $\sqrt[3]{x} \cdot \sqrt[3]{x^2}$

17. $\sqrt{\dfrac{9}{25}}$ **18.** $\sqrt[3]{\dfrac{x}{8}}$

19. $\sqrt{\dfrac{1}{2}} \cdot \sqrt{\dfrac{1}{8}}$ **20.** $\sqrt{\dfrac{5}{3}} \cdot \sqrt{\dfrac{1}{3}}$

21. $\sqrt[3]{\dfrac{2}{3}} \cdot \sqrt[3]{\dfrac{4}{3}} \cdot \sqrt[3]{\dfrac{1}{3}}$ **22.** $\sqrt[4]{\dfrac{8}{3}} \cdot \sqrt[4]{\dfrac{4}{9}} \cdot \sqrt[4]{\dfrac{8}{3}}$

23. $\sqrt{x^3} \cdot \sqrt{x^3}$ **24.** $\sqrt{z} \cdot \sqrt{z^7}$

25. $\sqrt[3]{\dfrac{7}{27}}$

26. $\sqrt[3]{\dfrac{9}{64}}$

27. $\sqrt[4]{\dfrac{x}{81}}$

28. $\sqrt[4]{\dfrac{16x}{81}}$

29. $\sqrt{\dfrac{9}{z^2}}$

30. $\sqrt{\dfrac{x^2}{81}}$

31. $\sqrt{\dfrac{x}{2}} \cdot \sqrt{\dfrac{x}{8}}$

32. $\sqrt{\dfrac{4}{y}} \cdot \sqrt{\dfrac{y}{5}}$

33. $\dfrac{\sqrt{45}}{\sqrt{5}}$

34. $\dfrac{\sqrt{7}}{\sqrt{28}}$

35. $\sqrt[3]{-4} \cdot \sqrt[3]{-16}$

36. $\sqrt[3]{9} \cdot \sqrt[3]{3}$

37. $\sqrt[4]{9} \cdot \sqrt[4]{9}$

38. $\sqrt[5]{16} \cdot \sqrt[5]{-2}$

39. $\dfrac{\sqrt[5]{64}}{\sqrt[5]{-2}}$

40. $\dfrac{\sqrt[4]{324}}{\sqrt[4]{4}}$

41. $\dfrac{\sqrt{a^2b}}{\sqrt{b}}$

42. $\dfrac{\sqrt{4xy^2}}{\sqrt{x}}$

43. $\dfrac{\sqrt[3]{54}}{\sqrt[3]{2}}$

44. $\dfrac{\sqrt[3]{x^3y^7}}{\sqrt[3]{y^4}}$

45. $\sqrt{4x^4}$

46. $\sqrt[3]{-8y^3}$

47. $\sqrt[3]{-5a^6}$

48. $\sqrt{9x^2y}$

49. $\sqrt[4]{16x^4y}$

50. $\sqrt[3]{8xy^3}$

51. $\sqrt{3x} \cdot \sqrt{12x}$

52. $\sqrt{6x^5} \cdot \sqrt{6x}$

53. $\sqrt[3]{8x^6y^3z^9}$

54. $\sqrt{16x^4y^6}$

55. $\sqrt[4]{\dfrac{3}{4}} \cdot \sqrt[4]{\dfrac{27}{4}}$

56. $\sqrt[5]{\dfrac{4}{-9}} \cdot \sqrt[5]{\dfrac{8}{-27}}$

57. $\sqrt[3]{12} \cdot \sqrt[3]{ab}$

58. $\sqrt{5x} \cdot \sqrt{5z}$

59. $\sqrt[4]{25z} \cdot \sqrt[4]{25z}$

60. $\sqrt[5]{3z^2} \cdot \sqrt[5]{7z}$

61. $\sqrt[5]{\dfrac{7a}{b^2}} \cdot \sqrt[5]{\dfrac{b^2}{7a^6}}$

62. $\sqrt[3]{\dfrac{8m}{n}} \cdot \sqrt[3]{\dfrac{n^4}{m^2}}$

Exercises 63–68: Use properties of polynomials to simplify the expression. Assume all radicands are positive.

63. $\sqrt{x + 4} \cdot \sqrt{x - 4}$

64. $\sqrt[3]{x - 1} \cdot \sqrt[3]{x^2 + x + 1}$

65. $\sqrt[3]{a + 1} \cdot \sqrt[3]{a^2 - a + 1}$

66. $\sqrt{b - 1} \cdot \sqrt{b + 1}$

67. $\dfrac{\sqrt{x^2 + 2x + 1}}{\sqrt{x + 1}}$

68. $\dfrac{\sqrt{x^2 - 4x + 4}}{\sqrt{x - 2}}$

Exercises 69–74: Complete the equation.

69. $\sqrt{500} = \underline{\quad} \sqrt{5}$

70. $\sqrt{28} = \underline{\quad} \sqrt{7}$

71. $\sqrt{8} = \underline{\quad} \sqrt{2}$

72. $\sqrt{99} = \underline{\quad} \sqrt{11}$

73. $\sqrt{45} = \underline{\quad} \sqrt{5}$

74. $\sqrt{243} = \underline{\quad} \sqrt{3}$

Exercises 75–98: Simplify the radical expression by factoring out the largest perfect nth power. Assume that all variables are positive.

75. $\sqrt{200}$

76. $\sqrt{72}$

77. $\sqrt[3]{81}$

78. $\sqrt[3]{256}$

79. $\sqrt[4]{64}$

80. $\sqrt[5]{27 \cdot 81}$

81. $\sqrt[5]{-64}$

82. $\sqrt[3]{-81}$

83. $\sqrt{b^5}$

84. $\sqrt{t^3}$

85. $\sqrt{8n^3}$

86. $\sqrt{32a^2}$

87. $\sqrt{12a^2b^5}$

88. $\sqrt{20a^3b^2}$

89. $\sqrt[3]{-125x^4y^5}$

90. $\sqrt[3]{-81a^5b^2}$

91. $\sqrt[3]{5t} \cdot \sqrt[3]{125t}$

92. $\sqrt[4]{4bc^3} \cdot \sqrt[4]{64ab^3c^2}$

93. $\sqrt[4]{\dfrac{9t^5}{r^8}} \cdot \sqrt[4]{\dfrac{9r}{5t}}$

94. $\sqrt[5]{\dfrac{4t^6}{r}} \cdot \sqrt[5]{\dfrac{8t}{r^6}}$

95. $\sqrt[3]{\dfrac{27x^2}{y^3}}$

96. $\sqrt[4]{\dfrac{32x^8}{z^4}}$

97. $\sqrt{\dfrac{7a^2}{27}} \cdot \sqrt{\dfrac{7a}{3}}$

98. $\sqrt{\dfrac{8a}{125}} \cdot \sqrt{\dfrac{2a}{5}}$

Exercises 99 and 100: **Thinking Generally**   *Assume that m and n are positive integers and that each expression exists.*

99. Simplify $\left(\sqrt[mn]{a^m b^m}\right)^n$ so that it equals ab.

100. Simplify $\left(\sqrt[mn]{a^n b^m}\right) \cdot \left(\sqrt[mn]{a^m b^n}\right)$ so that it equals $\sqrt[m]{ab} \cdot \sqrt[n]{ab}$.

Exercises 101–110: Simplify the expression. Assume that all variables are positive and write your answer in radical notation.

101. $\sqrt{3} \cdot \sqrt[3]{3}$

102. $\sqrt{5} \cdot \sqrt[3]{5}$

103. $\sqrt[4]{8} \cdot \sqrt[3]{4}$

104. $\sqrt[5]{16} \cdot \sqrt{2}$

105. $\sqrt[4]{27} \cdot \sqrt[3]{9} \cdot \sqrt{3}$

106. $\sqrt[5]{16} \cdot \sqrt[3]{16}$

107. $\sqrt[4]{x^3} \cdot \sqrt[3]{x}$

108. $\sqrt[4]{x^3} \cdot \sqrt{x}$

109. $\sqrt[4]{rt} \cdot \sqrt[3]{r^2t}$

110. $\sqrt[3]{a^3b^2} \cdot \sqrt{a^2b}$

APPLICATIONS

111. *Bird Wings* Heavier birds tend to have larger wings than lighter birds do. For some birds the relationship between the surface area A of the bird's wings in square inches and its weight W in pounds can be modeled by $A = 100\sqrt[3]{W^2}$. (**Source:** C. Pennycuick, *Newton Rules Biology.*)

    (a) Find the area of the wings when the weight is 8 pounds.

    (b) Write this formula with rational exponents.

112. *Orbits and Distance* Johannes Kepler (1571–1630) discovered a relationship between a planet's distance D from the sun and the time T it takes to orbit the sun. This formula is $T = \sqrt{D^3}$, where T is in Earth years and $D = 1$ corresponds to the distance between Earth and the sun, or 93,000,000 miles.

    (a) Neptune is 30 times farther from the sun than Earth ($D = 30$). Estimate the number of years required for Neptune to orbit the sun.

    (b) Write this formula with rational exponents.

WRITING ABOUT MATHEMATICS

113. Explain what it means for a positive integer to be a *perfect square* or a *perfect cube*. List the positive integers that are perfect squares and less than 101. List the positive integers that are perfect cubes and less than 220.

114. Explain how the product and quotient rules for radical expressions are the result of properties of rational exponents.

CHECKING BASIC CONCEPTS

SECTIONS 10.1 AND 10.2

1. Find the following.
    (a) The square roots of 49
    (b) The principal square root of 49

2. Evaluate.
    (a) $\sqrt[3]{-8}$ **(b)** $-\sqrt[4]{81}$

3. Write the expression in radical notation.
    (a) $x^{3/2}$ **(b)** $x^{2/3}$ **(c)** $x^{-2/5}$

4. Simplify $\sqrt{(x-1)^2}$ for any real number x.

5. Simplify each expression. Assume that all variables are positive.
    (a) $(64^{-3/2})^{1/3}$ **(b)** $\sqrt{5} \cdot \sqrt{20}$
    (c) $\sqrt[3]{-8x^4y}$ **(d)** $\sqrt{\dfrac{4b}{5}} \cdot \sqrt{\dfrac{4b^3}{5}}$

6. Simplify $\sqrt[3]{7} \cdot \sqrt{7}$.

10.3 OPERATIONS ON RADICAL EXPRESSIONS

Addition and Subtraction ▪ Multiplication ▪ Rationalizing the Denominator

A LOOK INTO MATH ▷ So far we have discussed how to add, subtract, multiply, and divide numbers and variables. In this section we extend these operations to radical expressions. In doing so, we use many of the techniques discussed in Sections 10.1 and 10.2.

Addition and Subtraction

We can add $2x^2$ and $5x^2$ to obtain $7x^2$ because they are *like* terms. That is,

$$2x^2 + 5x^2 = (2 + 5)x^2 = 7x^2.$$

We can also add and subtract *like radicals*. **Like radicals** *have the same index and the same radicand*. For example, we can add $3\sqrt{2}$ and $5\sqrt{2}$ because they are like radicals.

$$3\sqrt{2} + 5\sqrt{2} = (3 + 5)\sqrt{2} = 8\sqrt{2}$$

EXAMPLE 1   Adding like radicals

If possible, add the expressions and simplify.
(a) $10\sqrt{11} + 4\sqrt{11}$ **(b)** $5\sqrt[3]{6} + \sqrt[3]{6}$
(c) $4 + 5\sqrt{3}$ **(d)** $\sqrt{7} + \sqrt{11}$

Solution
(a) These terms are like radicals because they have the same index, 2, and the same radicand, 11.

$$10\sqrt{11} + 4\sqrt{11} = (10 + 4)\sqrt{11} = 14\sqrt{11}$$

(b) These terms are like radicals because they have the same index, 3, and the same radicand, 6. Note that the coefficient on the second term is understood to be 1.

$$5\sqrt[3]{6} + 1\sqrt[3]{6} = (5 + 1)\sqrt[3]{6} = 6\sqrt[3]{6}$$

(c) The expression $4 + 5\sqrt{3}$ can be written as $4\sqrt{1} + 5\sqrt{3}$. These terms cannot be added because they are not like radicals.

   NOTE: $4 + 5\sqrt{3} \neq 9\sqrt{3}$

(d) The expression $\sqrt{7} + \sqrt{11}$ contains unlike radicals that *cannot* be added.

Now Try Exercises 19, 21, 23, 25

Sometimes two radical expressions that are not alike can be added by changing them to like radicals. For example, $\sqrt{20}$ and $\sqrt{5}$ are unlike radicals. However,

$$\sqrt{20} = \sqrt{4 \cdot 5} = \sqrt{4} \cdot \sqrt{5} = 2\sqrt{5},$$

so

$$\sqrt{20} + \sqrt{5} = 2\sqrt{5} + 1\sqrt{5} = 3\sqrt{5}.$$

We cannot combine $x + x^2$ because they are unlike terms. Similarly, we cannot combine $\sqrt{2} + \sqrt{5}$ because they are unlike radicals. When combining radicals, the first step is to see if we can write pairs of terms as like radicals, as demonstrated in the next example.

EXAMPLE 2   Finding like radicals

Write each pair of terms as like radicals, if possible.
(a) $\sqrt{45}, \sqrt{20}$ **(b)** $\sqrt{27}, \sqrt{5}$ **(c)** $5\sqrt[3]{16}, 4\sqrt[3]{54}$

Solution

(a) The expressions $\sqrt{45}$ and $\sqrt{20}$ are unlike radicals. However, they can be changed to like radicals as follows.

$$\sqrt{45} = \sqrt{9 \cdot 5} = \sqrt{9} \cdot \sqrt{5} = 3\sqrt{5} \quad \text{and}$$

$$\sqrt{20} = \sqrt{4 \cdot 5} = \sqrt{4} \cdot \sqrt{5} = 2\sqrt{5}$$

The expressions $3\sqrt{5}$ and $2\sqrt{5}$ are like radicals.

(b) Because $\sqrt{27} = \sqrt{9 \cdot 3} = \sqrt{9} \cdot \sqrt{3} = 3\sqrt{3}$, the given expressions $\sqrt{27}$ and $\sqrt{5}$ are unlike radicals and cannot be written as like radicals.

(c) $5\sqrt[3]{16} = 5\sqrt[3]{8 \cdot 2} = 5\sqrt[3]{8} \cdot \sqrt[3]{2} = 5 \cdot 2 \cdot \sqrt[3]{2} = 10\sqrt[3]{2} \quad$ and

$4\sqrt[3]{54} = 4\sqrt[3]{27 \cdot 2} = 4\sqrt[3]{27} \cdot \sqrt[3]{2} = 4 \cdot 3 \cdot \sqrt[3]{2} = 12\sqrt[3]{2}$

The expressions $10\sqrt[3]{2}$ and $12\sqrt[3]{2}$ are like radicals.

Now Try Exercises **9, 11, 13**

We use these techniques to add radical expressions in the next two examples.

EXAMPLE 3 Adding radical expressions

Add the expressions and simplify.

(a) $\sqrt{12} + 7\sqrt{3}$ **(b)** $\sqrt[3]{16} + \sqrt[3]{2}$ **(c)** $3\sqrt{2} + \sqrt{8} + \sqrt{18}$

Solution

(a)
$$\sqrt{12} + 7\sqrt{3} = \sqrt{4 \cdot 3} + 7\sqrt{3}$$
$$= \sqrt{4} \cdot \sqrt{3} + 7\sqrt{3}$$
$$= 2\sqrt{3} + 7\sqrt{3}$$
$$= 9\sqrt{3}$$

(b)
$$\sqrt[3]{16} + \sqrt[3]{2} = \sqrt[3]{8 \cdot 2} + \sqrt[3]{2}$$
$$= \sqrt[3]{8} \cdot \sqrt[3]{2} + \sqrt[3]{2}$$
$$= 2\sqrt[3]{2} + 1\sqrt[3]{2}$$
$$= (2 + 1)\sqrt[3]{2}$$
$$= 3\sqrt[3]{2}$$

(c)
$$3\sqrt{2} + \sqrt{8} + \sqrt{18} = 3\sqrt{2} + \sqrt{4 \cdot 2} + \sqrt{9 \cdot 2}$$
$$= 3\sqrt{2} + \sqrt{4} \cdot \sqrt{2} + \sqrt{9} \cdot \sqrt{2}$$
$$= 3\sqrt{2} + 2\sqrt{2} + 3\sqrt{2}$$
$$= 8\sqrt{2}$$

Now Try Exercises **29, 31, 43**

EXAMPLE 4 Adding radical expressions

Add the expressions and simplify. Assume that all variables are positive.

(a) $\sqrt[4]{32} + 3\sqrt[4]{2}$ **(b)** $-2\sqrt{4x} + \sqrt{x}$ **(c)** $3\sqrt{3k} + 5\sqrt{12k} + 9\sqrt{48k}$

Solution

(a) Because $\sqrt[4]{32} = \sqrt[4]{16 \cdot 2} = \sqrt[4]{16} \cdot \sqrt[4]{2} = 2\sqrt[4]{2}$, we can add and simplify as follows.

$$\sqrt[4]{\mathbf{32}} + 3\sqrt[4]{2} = \mathbf{2\sqrt[4]{2}} + 3\sqrt[4]{2} = 5\sqrt[4]{2}$$

(b) Note that $\sqrt{4x} = \sqrt{4} \cdot \sqrt{x} = 2\sqrt{x}$.

$$-2\sqrt{4x} + \sqrt{x} = -2(2\sqrt{x}) + \sqrt{x} = \mathbf{-4}\sqrt{x} + \mathbf{1}\sqrt{x} = \mathbf{-3}\sqrt{x}$$

(c) Note that $\sqrt{12k} = \sqrt{4} \cdot \sqrt{3k} = 2\sqrt{3k}$ and that $\sqrt{48k} = \sqrt{16} \cdot \sqrt{3k} = 4\sqrt{3k}$.

$$3\sqrt{3k} + 5\sqrt{12k} + 9\sqrt{48k} = 3\sqrt{3k} + 5(2\sqrt{3k}) + 9(4\sqrt{3k})$$

$$= (3 + 10 + 36)\sqrt{3k}$$

$$= 49\sqrt{3k} \qquad \text{Now Try Exercises } \boxed{45, 47, 49}$$

Subtraction of radical expressions is similar to addition, as illustrated in the next three examples.

EXAMPLE 5   Subtracting like radicals

Simplify the expressions.

(a) $5\sqrt{7} - 3\sqrt{7}$ **(b)** $8\sqrt[3]{5} - 3\sqrt[3]{5} + \sqrt[3]{11}$ **(c)** $5\sqrt{z} + \sqrt[3]{z} - 2\sqrt{z}$

Solution

(a) $\qquad \mathbf{5}\sqrt{\mathbf{7}} - \mathbf{3}\sqrt{\mathbf{7}} = (\mathbf{5} - \mathbf{3})\sqrt{\mathbf{7}} = \mathbf{2}\sqrt{\mathbf{7}}$

(b) $\mathbf{8}\sqrt[3]{\mathbf{5}} - \mathbf{3}\sqrt[3]{\mathbf{5}} + \sqrt[3]{\mathbf{11}} = (\mathbf{8} - \mathbf{3})\sqrt[3]{\mathbf{5}} + \sqrt[3]{\mathbf{11}} = \mathbf{5}\sqrt[3]{\mathbf{5}} + \sqrt[3]{\mathbf{11}}$

(c) $\quad 5\sqrt{z} + \sqrt[3]{z} - 2\sqrt{z} = 5\sqrt{z} - 2\sqrt{z} + \sqrt[3]{z} \qquad$ Commutative property

$$= (5 - 2)\sqrt{z} + \sqrt[3]{z} \qquad \text{Distributive property}$$

$$= 3\sqrt{z} + \sqrt[3]{z} \qquad \text{Subtract.}$$

NOTE: We cannot combine $3\sqrt{z} + \sqrt[3]{z}$ because their indexes are different. That is, one term is a square root and the other is a cube root.   Now Try Exercises $\boxed{33, 35, 39}$

EXAMPLE 6   Subtracting radical expressions

Subtract and simplify. Assume that all variables are positive.

(a) $3\sqrt[3]{xy^2} - 2\sqrt[3]{xy^2}$ **(b)** $\sqrt{16x^3} - \sqrt{x^3}$ **(c)** $\sqrt[3]{\dfrac{5x}{27}} - \dfrac{\sqrt[3]{5x}}{6}$

Solution

(a) $\qquad 3\sqrt[3]{xy^2} - 2\sqrt[3]{xy^2} = (3 - 2)\sqrt[3]{xy^2} = \sqrt[3]{xy^2}$

(b) $\qquad \sqrt{\mathbf{16x^3}} - \sqrt{\mathbf{x^3}} = \sqrt{\mathbf{16x^2}} \cdot \sqrt{\mathbf{x}} - \sqrt{\mathbf{x^2}} \cdot \sqrt{\mathbf{x}} \qquad$ Factor out perfect squares.

$$= 4x\sqrt{x} - x\sqrt{x} \qquad \text{Simplify.}$$

$$= (4x - x)\sqrt{x} \qquad \text{Distributive property}$$

$$= 3x\sqrt{x} \qquad \text{Subtract.}$$

(c)

$$\sqrt[3]{\dfrac{5x}{27}} - \dfrac{\sqrt[3]{5x}}{6} = \dfrac{\sqrt[3]{5x}}{\sqrt[3]{27}} - \dfrac{\sqrt[3]{5x}}{6} \qquad \text{Quotient rule for radical expressions}$$

$$= \dfrac{\sqrt[3]{5x}}{3} - \dfrac{\sqrt[3]{5x}}{6} \qquad \text{Evaluate } \sqrt[3]{27} = 3.$$

$$= \dfrac{2\sqrt[3]{5x}}{6} - \dfrac{\sqrt[3]{5x}}{6} \qquad \text{Find a common denominator.}$$

$$= \dfrac{2\sqrt[3]{5x} - \sqrt[3]{5x}}{6} \qquad \text{Subtract numerators.}$$

$$= \dfrac{\sqrt[3]{5x}}{6} \qquad \text{Simplify.}$$

Now Try Exercises 55, 59, 61

EXAMPLE 7   Subtracting radical expressions

Subtract and simplify. Assume that all variables are positive.

(a) $\dfrac{5\sqrt{2}}{3} - \dfrac{2\sqrt{2}}{4}$ **(b)** $\sqrt[4]{81a^5b^6} - \sqrt[4]{16ab^2}$ **(c)** $3\sqrt[3]{\dfrac{n^5}{27}} - 2\sqrt[3]{n^2}$

Solution

(a)

$$\dfrac{5\sqrt{2}}{3} - \dfrac{2\sqrt{2}}{4} = \dfrac{5\sqrt{2}}{3} \cdot \dfrac{4}{4} - \dfrac{2\sqrt{2}}{4} \cdot \dfrac{3}{3} \qquad \text{LCD is 12.}$$

$$= \dfrac{20\sqrt{2}}{12} - \dfrac{6\sqrt{2}}{12} \qquad \text{Multiply fractions.}$$

$$= \dfrac{14\sqrt{2}}{12} \qquad \text{Subtract numerators.}$$

$$= \dfrac{7\sqrt{2}}{6} \qquad \text{Simplify.}$$

(b) $\sqrt[4]{81a^5b^6} - \sqrt[4]{16ab^2} = \sqrt[4]{81a^4b^4} \cdot \sqrt[4]{ab^2} - \sqrt[4]{16} \cdot \sqrt[4]{ab^2}$ Factor out perfect powers.

$$= 3ab\sqrt[4]{ab^2} - 2\sqrt[4]{ab^2} \qquad \text{Simplify.}$$

$$= (3ab - 2)\sqrt[4]{ab^2} \qquad \text{Distributive property}$$

(c) $3\sqrt[3]{\dfrac{n^5}{27}} - 2\sqrt[3]{n^2} = 3\sqrt[3]{\dfrac{n^3}{27}} \cdot \sqrt[3]{n^2} - 2\sqrt[3]{n^2}$ Factor out perfect cube.

$$= \dfrac{3\sqrt[3]{n^3}}{\sqrt[3]{27}} \cdot \sqrt[3]{n^2} - 2\sqrt[3]{n^2} \qquad \text{Quotient rule}$$

$$= n\sqrt[3]{n^2} - 2\sqrt[3]{n^2} \qquad \text{Simplify.}$$

$$= (n - 2)\sqrt[3]{n^2} \qquad \text{Distributive property}$$

Now Try Exercises 65, 73, 75

Radicals often occur in geometry. In the next example we find the perimeter of a triangle by adding radical expressions.

EXAMPLE **Finding the perimeter of a triangle**

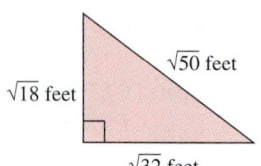

Figure 10.8

Find the *exact* perimeter of the right triangle shown in Figure 10.8. Then approximate your answer to the nearest hundredth of a foot.

Solution

The sum of the lengths of the sides of the triangle is

$$\sqrt{18} + \sqrt{32} + \sqrt{50} = 3\sqrt{2} + 4\sqrt{2} + 5\sqrt{2} = 12\sqrt{2}.$$

The perimeter is $12\sqrt{2} \approx 16.97$ feet.   Now Try Exercise 117

Multiplication

Some types of radical expressions can be multiplied like binomials. For example, because

$$(x + 1)(x + 2) = x^2 + 3x + 2,$$

we have $(\sqrt{x} + 1)(\sqrt{x} + 2) = (\sqrt{x})^2 + 2\sqrt{x} + 1\sqrt{x} + 2 = x + 3\sqrt{x} + 2,$

provided that $x \geq 0$. The next example demonstrates this technique.

EXAMPLE 9   **Multiplying radical expressions**

Multiply and simplify.

(a) $(\sqrt{b} - 4)(\sqrt{b} + 5)$ **(b)** $(4 + \sqrt{3})(4 - \sqrt{3})$

Solution

(a) This expression can be multiplied and then simplified.

$$(\sqrt{b} - 4)(\sqrt{b} + 5) = \sqrt{b} \cdot \sqrt{b} + 5\sqrt{b} - 4\sqrt{b} - 4 \cdot 5$$
$$= b + \sqrt{b} - 20$$

Compare this product with $(b - 4)(b + 5) = b^2 + b - 20$.

(b) This expression is in the form $(a + b)(a - b)$, which equals $a^2 - b^2$.

$$(4 + \sqrt{3})(4 - \sqrt{3}) = (4)^2 - (\sqrt{3})^2$$
$$= 16 - 3$$
$$= 13$$ Now Try Exercises 77, 79

NOTE: Example 9(b) illustrates a special case for multiplying radicals. In general,

$$(\sqrt{a} + \sqrt{b})(\sqrt{a} - \sqrt{b}) = (\sqrt{a})^2 - (\sqrt{b})^2 = a - b,$$

provided a and b are nonnegative.

Rationalizing the Denominator

In mathematics it is common to write expressions without radicals in the denominator. Quotients containing radical expressions in the numerator or denominator can appear to be different but actually be equal. For example, $\frac{1}{\sqrt{3}}$ and $\frac{\sqrt{3}}{3}$ represent the same real number even though they look like they are unequal. To show that they are equal, we multiply the first quotient by 1 in the form $\frac{\sqrt{3}}{\sqrt{3}}$:

$$\frac{1}{\sqrt{3}} \cdot \frac{\sqrt{3}}{\sqrt{3}} = \frac{1 \cdot \sqrt{3}}{\sqrt{3} \cdot \sqrt{3}} = \frac{\sqrt{3}}{3}.$$

If the denominator of a quotient contains only one term with one square root, then we can rationalize the denominator by multiplying the numerator and denominator by this square root. For example, the denominator of $\frac{1}{\sqrt{3}}$ contains one term, which is $\sqrt{3}$. Therefore, we multiplied the $\frac{1}{\sqrt{3}}$ by $\frac{\sqrt{3}}{\sqrt{3}}$ to rationalize the denominator.

NOTE: $\sqrt{b} \cdot \sqrt{b} = \sqrt{b^2} = b$ for any *positive* number b.

One way to *standardize* radical expressions is to remove any radical expressions from the denominator. This process is called **rationalizing the denominator**. Exercise 119 suggests one reason why people rationalized denominators before calculators were invented. The next example demonstrates how to rationalize the denominator of several quotients.

EXAMPLE 10   Rationalizing the denominator

Rationalize each denominator. Assume that all variables are positive.

(a) $\dfrac{1}{\sqrt{2}}$ **(b)** $\dfrac{3}{5\sqrt{3}}$ **(c)** $\sqrt{\dfrac{x}{24}}$ **(d)** $\dfrac{xy}{\sqrt{y^3}}$

Solution
(a) We start by multiplying this expression by 1 in the form $\frac{\sqrt{2}}{\sqrt{2}}$:

$$\frac{1}{\sqrt{2}} \cdot \frac{\sqrt{2}}{\sqrt{2}} = \frac{\sqrt{2}}{\sqrt{4}} = \frac{\sqrt{2}}{2}.$$

Note that the expression $\frac{\sqrt{2}}{2}$ does not have a radical in the denominator.

(b) We multiply this expression by 1 in the form $\frac{\sqrt{3}}{\sqrt{3}}$:

$$\frac{3}{5\sqrt{3}} \cdot \frac{\sqrt{3}}{\sqrt{3}} = \frac{3\sqrt{3}}{5\sqrt{9}} = \frac{3\sqrt{3}}{5 \cdot 3} = \frac{\sqrt{3}}{5}.$$

(c) Because $\sqrt{24} = \sqrt{4} \cdot \sqrt{6} = 2\sqrt{6}$, we start by simplifying the expression.

$$\sqrt{\frac{x}{24}} = \frac{\sqrt{x}}{\sqrt{24}} = \frac{\sqrt{x}}{2\sqrt{6}}$$

To rationalize the denominator, we multiply this expression by 1 in the form $\frac{\sqrt{6}}{\sqrt{6}}$:

$$\frac{\sqrt{x}}{2\sqrt{6}} = \frac{\sqrt{x}}{2\sqrt{6}} \cdot \frac{\sqrt{6}}{\sqrt{6}} = \frac{\sqrt{6x}}{12}.$$

(d) Because $\sqrt{y^3} = \sqrt{y^2} \cdot \sqrt{y} = y\sqrt{y}$, we start by simplifying the expression.

$$\frac{xy}{\sqrt{y^3}} = \frac{xy}{y\sqrt{y}} = \frac{x}{\sqrt{y}}$$

To rationalize the denominator, we multiply by 1 in the form $\frac{\sqrt{y}}{\sqrt{y}}$:

$$\frac{x}{\sqrt{y}} \cdot \frac{\sqrt{y}}{\sqrt{y}} = \frac{x\sqrt{y}}{y}.$$

Now Try Exercises 89, 93, 95, 97

When the denominator is either a sum or difference containing a square root, we multiply the numerator and denominator by the *conjugate* of the denominator. In this case, the **conjugate** of the denominator is found by changing a + sign to a − sign or vice versa. Table 10.1 lists examples of conjugates.

TABLE 10.1

Expression	$1 + \sqrt{2}$	$\sqrt{3} - 2$	$\sqrt{x} + 7$	$\sqrt{a} - \sqrt{b}$
Conjugate	$1 - \sqrt{2}$	$\sqrt{3} + 2$	$\sqrt{x} - 7$	$\sqrt{a} + \sqrt{b}$

In the next two examples we illustrate this method to rationalize a denominator.

EXAMPLE 11   Using a conjugate to rationalize the denominator

Rationalize the denominator of $\frac{1}{1 + \sqrt{2}}$.

Solution
From Table 10.1, the conjugate of $1 + \sqrt{2}$ is $1 - \sqrt{2}$.

$$\frac{1}{1 + \sqrt{2}} = \frac{1}{1 + \sqrt{2}} \cdot \frac{(1 - \sqrt{2})}{(1 - \sqrt{2})} \quad \text{Multiply numerator and denominator by the conjugate.}$$

$$= \frac{1 - \sqrt{2}}{(1)^2 - (\sqrt{2})^2} \quad (a+b)(a-b) = a^2 - b^2$$

$$= \frac{1 - \sqrt{2}}{1 - 2} \quad \text{Simplify.}$$

$$= \frac{1 - \sqrt{2}}{-1} \quad \text{Subtract.}$$

$$= \frac{1}{-1} - \frac{\sqrt{2}}{-1} \quad \frac{a-b}{c} = \frac{a}{c} - \frac{b}{c}$$

$$= -1 + \sqrt{2} \quad \text{Simplify.} \quad \text{Now Try Exercise } 99$$

EXAMPLE **12**   Rationalizing the denominator

Rationalize the denominator.

(a) $\dfrac{3 + \sqrt{5}}{2 - \sqrt{5}}$ **(b)** $\dfrac{\sqrt{x}}{\sqrt{x} - 2}$

Solution

(a) The conjugate of the denominator is **$2 + \sqrt{5}$**.

$$\frac{3 + \sqrt{5}}{2 - \sqrt{5}} = \frac{(3 + \sqrt{5})}{(2 - \sqrt{5})} \cdot \frac{(2 + \sqrt{5})}{(2 + \sqrt{5})} \qquad \text{Multiply by 1.}$$

$$= \frac{6 + 3\sqrt{5} + 2\sqrt{5} + (\sqrt{5})^2}{(2)^2 - (\sqrt{5})^2} \qquad \text{Multiply.}$$

$$= \frac{11 + 5\sqrt{5}}{4 - 5} \qquad \text{Combine terms.}$$

$$= -11 - 5\sqrt{5} \qquad \text{Simplify.}$$

(b) The conjugate of the denominator is **$\sqrt{x} + 2$**.

$$\frac{\sqrt{x}}{\sqrt{x} - 2} = \frac{\sqrt{x}}{(\sqrt{x} - 2)} \cdot \frac{(\sqrt{x} + 2)}{(\sqrt{x} + 2)} \qquad \text{Multiply by 1.}$$

$$= \frac{x + 2\sqrt{x}}{x - 4} \qquad \text{Multiply.}$$

Now Try Exercises **103, 107**

EXAMPLE **13**   Rationalizing a denominator having a cube root

Rationalize the denominator of $\dfrac{5}{\sqrt[3]{x}}$.

Solution

The expression $\dfrac{5}{\sqrt[3]{x}}$ is equal to $\dfrac{5}{x^{1/3}}$. To rationalize the denominator, $x^{1/3}$, we can multiply it by $x^{2/3}$ because $x^{1/3} \cdot x^{2/3} = x^{1/3 + 2/3} = x^1$.

$$\frac{5}{x^{1/3}} = \frac{5}{x^{1/3}} \cdot \frac{x^{2/3}}{x^{2/3}} \qquad \text{Multiply by 1.}$$

$$= \frac{5x^{2/3}}{x^{1/3 + 2/3}} \qquad \text{Product rule}$$

$$= \frac{5\sqrt[3]{x^2}}{x} \qquad \text{Add; write in radical notation.}$$

Now Try Exercise **113**

10.3 PUTTING IT ALL TOGETHER

In this section we discussed how to add, subtract, and multiply radical expressions. Rationalization of the denominator is a technique that can sometimes be helpful when dividing rational expressions. The following table summarizes important topics in this section.

Concept	Explanation	Examples
Like Radicals	Like radicals have the same index and the same radicand.	$7\sqrt{5}$ and $3\sqrt{5}$ are like radicals. $5\sqrt[3]{ab}$ and $\sqrt[3]{ab}$ are like radicals. $\sqrt[3]{5}$ and $\sqrt[3]{4}$ are unlike radicals. $\sqrt[3]{7}$ and $\sqrt{7}$ are unlike radicals.
Adding and Subtracting Radical Expressions	Combine like radicals when adding or subtracting. We cannot combine unlike radicals such as $\sqrt{2}$ and $\sqrt{5}$. But sometimes we can rewrite radicals and then combine.	$6\sqrt{13} + \sqrt{13} = (6 + 1)\sqrt{13} = 7\sqrt{13}$ $\begin{aligned}\sqrt{40} - \sqrt{10} &= \sqrt{4} \cdot \sqrt{10} - \sqrt{10}\\ &= 2\sqrt{10} - \sqrt{10}\\ &= \sqrt{10}\end{aligned}$
Multiplying Radical Expressions	Radical expressions can sometimes be multiplied like binomials.	$(\sqrt{a} - 5)(\sqrt{a} + 5) = a - 25$ and $(\sqrt{x} - 3)(\sqrt{x} + 1) = x - 2\sqrt{x} - 3$
Conjugate	The conjugate is found by changing a $+$ sign to a $-$ sign or vice versa.	$\begin{array}{cc}\textit{Expression} & \textit{Conjugate}\\ \sqrt{x} + 7 & \sqrt{x} - 7\\ \sqrt{a} - 2\sqrt{b} & \sqrt{a} + 2\sqrt{b}\end{array}$
Rationalizing a Denominator Having One Term	Write the quotient without a radical expression in the denominator.	To rationalize $\frac{5}{\sqrt{7}}$, multiply the expression by 1 in the form $\frac{\sqrt{7}}{\sqrt{7}}$: $\frac{5}{\sqrt{7}} \cdot \frac{\sqrt{7}}{\sqrt{7}} = \frac{5\sqrt{7}}{\sqrt{49}} = \frac{5\sqrt{7}}{7}.$
Rationalizing a Denominator Having Two Terms Containing Square Roots	Multiply the numerator and denominator by the conjugate of the denominator.	$\begin{aligned}\frac{1}{2 - \sqrt{3}} &= \frac{1}{2 - \sqrt{3}} \cdot \frac{(2 + \sqrt{3})}{(2 + \sqrt{3})}\\ &= \frac{2 + \sqrt{3}}{4 - 3} = 2 + \sqrt{3}\end{aligned}$

10.3 Exercises

MyMathLab PRACTICE WATCH DOWNLOAD READ REVIEW

CONCEPTS

1. $\sqrt{a} + \sqrt{a} =$ _____

2. $\sqrt[3]{b} + \sqrt[3]{b} + \sqrt[3]{b} =$ _____

3. You cannot simplify $\sqrt[3]{4} + \sqrt[3]{7}$ because they are not _____ radicals.

4. Can you simplify $4\sqrt{15} - 3\sqrt{15}$? Explain.

5. Does $6 + 3\sqrt{5}$ equal $9\sqrt{5}$?

6. To rationalize the denominator of $\frac{2}{\sqrt{7}}$, multiply this expression by _____.

7. What is the conjugate of $\sqrt{t} - 5$?

8. To rationalize the denominator of $\frac{1}{5 - \sqrt{2}}$, multiply this expression by _____.

LIKE RADICALS

Exercises 9–18: (Refer to Example 2.) Write the terms as like radicals, if possible. Assume that all variables are positive.

9. $\sqrt{12}, \sqrt{24}$

10. $\sqrt{18}, \sqrt{27}$

11. $\sqrt{7}, \sqrt{28}, \sqrt{63}$

12. $\sqrt{200}, \sqrt{300}, \sqrt{500}$

13. $\sqrt[3]{16}, \sqrt[3]{-54}$

14. $\sqrt[3]{80}, \sqrt[3]{10}$

15. $\sqrt{x^2 y}, \sqrt{4y^2}$

16. $\sqrt{x^5 y^3}, \sqrt{9xy}$

17. $\sqrt[3]{8xy}, \sqrt[3]{x^4 y^4}$

18. $\sqrt[3]{64x^4}, \sqrt[3]{-8x}$

ADDITION AND SUBTRACTION OF RADICALS

Exercises 19–76: If possible, simplify the expression. Assume that all variables are positive.

19. $2\sqrt{3} + 7\sqrt{3}$

20. $8\sqrt{7} + 2\sqrt{7}$

21. $4\sqrt[3]{5} + 2\sqrt[3]{5}$

22. $\sqrt[3]{13} + 3\sqrt[3]{13}$

23. $7 + 4\sqrt{7}$

24. $8 - 4\sqrt{3}$

25. $2\sqrt{3} + 3\sqrt{2}$

26. $\sqrt{6} + \sqrt{17}$

27. $\sqrt{3} + \sqrt[3]{3}$

28. $\sqrt{6} + \sqrt[4]{6}$

29. $\sqrt[3]{16} + 3\sqrt[3]{2}$

30. $\sqrt[3]{24} + \sqrt[3]{81}$

31. $\sqrt{2} + \sqrt{18} + \sqrt{32}$

32. $2\sqrt{3} + \sqrt{12} + \sqrt{27}$

33. $11\sqrt{11} - 5\sqrt{11}$

34. $9\sqrt{5} + \sqrt{2} - \sqrt{5}$

35. $\sqrt{x} + \sqrt{x} - \sqrt{y}$

36. $\sqrt{xy^2} - \sqrt{x}$

37. $\sqrt[3]{z} + \sqrt[3]{z}$

38. $\sqrt[3]{y} - \sqrt[3]{y}$

39. $2\sqrt[3]{6} - 7\sqrt[3]{6}$

40. $18\sqrt[3]{3} + 3\sqrt[3]{3}$

41. $\sqrt[3]{y^6} - \sqrt[3]{y^3}$

42. $2\sqrt{20} + 7\sqrt{5} + 3\sqrt{2}$

43. $3\sqrt{28} + 3\sqrt{7}$

44. $9\sqrt{18} - 2\sqrt{8}$

45. $\sqrt[4]{48} + 4\sqrt[4]{3}$

46. $\sqrt[4]{32} + \sqrt[4]{16}$

47. $\sqrt{9x} + \sqrt{16x}$

48. $-3\sqrt{x} + 5\sqrt{x}$

49. $3\sqrt{2k} + \sqrt{8k} + \sqrt{18k}$

50. $3\sqrt{k} + 2\sqrt{4k} + \sqrt{9k}$

51. $\sqrt{44} - 4\sqrt{11}$

52. $\sqrt[4]{5} + 2\sqrt[4]{5}$

53. $2\sqrt[3]{16} + \sqrt[3]{2} - \sqrt{2}$

54. $5\sqrt[3]{x} - 3\sqrt[3]{x}$

55. $\sqrt[3]{xy} - 2\sqrt[3]{xy}$

56. $3\sqrt{x^3} - \sqrt{x}$

57. $\sqrt{4x + 8} + \sqrt{x + 2}$

58. $\sqrt{2a + 1} + \sqrt{8a + 4}$

59. $\sqrt{25x^3} - \sqrt{x^3}$

60. $\sqrt{36x^5} - \sqrt{25x^5}$

61. $\sqrt[3]{\frac{7x}{8}} - \frac{\sqrt[3]{7x}}{3}$

62. $\sqrt[3]{\frac{8x^2}{27}} - \sqrt[3]{\frac{x^2}{8}}$

63. $\frac{4\sqrt{3}}{3} + \frac{\sqrt{3}}{6}$

64. $\frac{8\sqrt{5}}{7} + \frac{4\sqrt{5}}{2}$

65. $\frac{15\sqrt{8}}{4} - \frac{2\sqrt{2}}{5}$

66. $\frac{23\sqrt{11}}{2} - \frac{\sqrt{44}}{8}$

67. $2\sqrt[4]{64} - \sqrt[4]{324} + \sqrt[4]{4}$

68. $2\sqrt[3]{16} - 5\sqrt[3]{54} + 10\sqrt[3]{2}$

69. $5\sqrt[4]{x^5} - \sqrt[4]{x}$

70. $20\sqrt[3]{b^4} - 4\sqrt[3]{b}$

71. $\sqrt{64x^3} - \sqrt{x} + 3\sqrt{x}$

72. $2\sqrt{3z} + 3\sqrt{12z} + 3\sqrt{48z}$

73. $\sqrt[4]{81a^5b^5} - \sqrt[4]{ab}$ **74.** $\sqrt[4]{xy^5} - \sqrt[4]{x^5y}$

75. $5\sqrt[3]{\dfrac{n^4}{125}} - 2\sqrt[3]{n}$ **76.** $\sqrt[3]{\dfrac{8x}{27}} - \dfrac{2\sqrt[3]{x}}{3}$

MULTIPLYING BINOMIALS CONTAINING RADICALS

Exercises 77–88: Multiply and simplify.

77. $(\sqrt{x} - 3)(\sqrt{x} + 2)$

78. $(2\sqrt{x} + 1)(\sqrt{x} + 4)$

79. $(3 + \sqrt{7})(3 - \sqrt{7})$

80. $(5 - \sqrt{5})(5 + \sqrt{5})$

81. $(11 - \sqrt{2})(11 + \sqrt{2})$

82. $(6 + \sqrt{3})(6 - \sqrt{3})$

83. $(\sqrt{x} + 8)(\sqrt{x} - 8)$

84. $(\sqrt{ab} - 3)(\sqrt{ab} + 3)$

85. $(\sqrt{ab} - \sqrt{c})(\sqrt{ab} + \sqrt{c})$

86. $(\sqrt{2x} + \sqrt{3y})(\sqrt{2x} - \sqrt{3y})$

87. $(\sqrt{x} - 7)(\sqrt{x} + 8)$

88. $(\sqrt{ab} - 1)(\sqrt{ab} - 2)$

RATIONALIZING THE DENOMINATOR

Exercises 89–116: Rationalize the denominator.

89. $\dfrac{1}{\sqrt{7}}$ **90.** $\dfrac{1}{\sqrt{23}}$

91. $\dfrac{4}{\sqrt{3}}$ **92.** $\dfrac{8}{\sqrt{2}}$

93. $\dfrac{5}{3\sqrt{5}}$ **94.** $\dfrac{6}{11\sqrt{3}}$

95. $\sqrt{\dfrac{b}{12}}$ **96.** $\sqrt{\dfrac{5b}{72}}$

97. $\dfrac{rt}{2\sqrt{r^3}}$ **98.** $\dfrac{m^2n}{2\sqrt{m^5}}$

99. $\dfrac{1}{3 - \sqrt{2}}$ **100.** $\dfrac{1}{\sqrt{3} - 2}$

101. $\dfrac{\sqrt{2}}{\sqrt{5} + 2}$ **102.** $\dfrac{\sqrt{3}}{\sqrt{3} + 2}$

103. $\dfrac{\sqrt{7} - 2}{\sqrt{7} + 2}$ **104.** $\dfrac{\sqrt{3} - 1}{\sqrt{3} + 1}$

105. $\dfrac{1}{\sqrt{7} - \sqrt{6}}$ **106.** $\dfrac{1}{\sqrt{8} - \sqrt{7}}$

107. $\dfrac{\sqrt{z}}{\sqrt{z} - 3}$ **108.** $\dfrac{2\sqrt{z}}{2 - \sqrt{z}}$

109. $\dfrac{\sqrt{a} + \sqrt{b}}{\sqrt{a} - \sqrt{b}}$ **110.** $\dfrac{\sqrt{x} - 2\sqrt{y}}{\sqrt{x} + 2\sqrt{y}}$

111. $\dfrac{1}{\sqrt{x + 1} - \sqrt{x}}$ **112.** $\dfrac{1}{\sqrt{a + 1} + \sqrt{a}}$

113. $\dfrac{3}{\sqrt[3]{x}}$ **114.** $\dfrac{6}{5\sqrt[3]{x}}$

115. $\dfrac{1}{\sqrt[3]{x^2}}$ **116.** $\dfrac{2}{\sqrt[3]{(x - 2)^2}}$

GEOMETRY

117. *Perimeter*   (Refer to Example 8.) Find the exact perimeter of the right triangle. Then approximate your answer.

118. *Perimeter*   Find the exact perimeter of the rectangle. Then approximate your answer.

WRITING ABOUT MATHEMATICS

119. Suppose that a student knows that $\sqrt{3} \approx 1.73205$ and does not have a calculator. Which of the following expressions would be easier to evaluate by hand? Why?

$$\dfrac{1}{\sqrt{3}} \quad \text{or} \quad \dfrac{\sqrt{3}}{3}$$

120. A student simplifies an expression *incorrectly*:

$$\sqrt{8} + \sqrt[3]{16} \overset{?}{=} \sqrt{4 \cdot 2} + \sqrt[3]{8 \cdot 2}$$
$$\overset{?}{=} \sqrt{4} \cdot \sqrt{2} + \sqrt[3]{8} \cdot \sqrt[3]{2}$$
$$\overset{?}{=} 2\sqrt{2} + 2\sqrt[3]{2}$$
$$\overset{?}{=} 4\sqrt{4}$$
$$\overset{?}{=} 8.$$

Explain any errors that the student made. What would you do differently?

Directions: Form a group of 2 to 4 people. Select someone to record the group's responses for this activity. All members of the group should work cooperatively to answer the questions. If your instructor asks for your results, each member of the group should be prepared to respond.

Designing a Paper Cup   A paper drinking cup is being designed in the shape shown in the accompanying figure. The amount of paper needed to manufacture the cup is determined by the surface area S of the cup, which is given by

$$S = \pi r \sqrt{r^2 + h^2},$$

where r is the radius and h is the height.

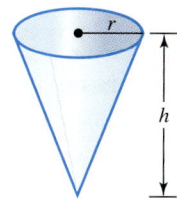

(a) Approximate S to the nearest hundredth when $r = 1.5$ inches and $h = 4$ inches.
(b) Could the formula for S be simplified as follows?

$$\pi r \sqrt{r^2 + h^2} \stackrel{?}{=} \pi r (\sqrt{r^2} + \sqrt{h^2})$$
$$\stackrel{?}{=} \pi r (r + h)$$

Try evaluating this second formula with $r = 1.5$ inches and $h = 4$ inches. Is your answer the same as in part (a)? Explain.

(c) Discuss why evaluating real-world formulas correctly is important.
(d) In general, does $\sqrt{a + b}$ equal $\sqrt{a} + \sqrt{b}$? Justify your answer by completing the following table. Approximate answers to the nearest hundredth when appropriate.

a	b	$\sqrt{a + b}$	$\sqrt{a} + \sqrt{b}$
0	4		
4	0		
5	4		
9	7		
4	16		
25	100		

10.4 RADICAL FUNCTIONS

The Square Root Function ■ The Cube Root Function ■ Power Functions ■ Modeling with Power Functions (Optional)

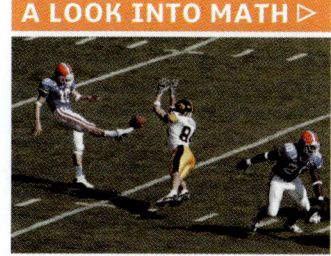

A good punter can kick a football so that it has a long *hang time*. Hang time is the length of time that the football is in the air. Long hang time gives the kicking team time to run down the field and stop the punt return. Using square roots, we can derive a function that calculates hang time. (See the discussion before Example 3.) First, we discuss the square root function.

The Square Root Function

The square root function is given by $f(x) = \sqrt{x}$. The domain of the square root function is all nonnegative real numbers because we have not defined the square root of a negative number. Table 10.2 lists three points that lie on the graph of $f(x) = \sqrt{x}$. In Figure 10.9 these points are plotted and the graph of $y = \sqrt{x}$ has been sketched. The graph does not appear to the left of the origin because $f(x) = \sqrt{x}$ is undefined for negative inputs.

TABLE 10.2

x	$\sqrt{x}$
0	0
1	1
4	2

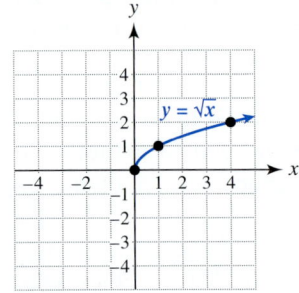

Figure 10.9 Square Root Function

TECHNOLOGY NOTE: Square Roots of Negative Numbers

If a table of values for $y_1 = \sqrt{x}$ includes both negative and positive values for x, then many calculators give error messages when x is negative, as shown in the accompanying figure.

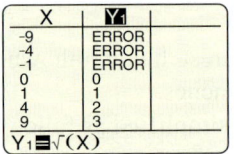

EXAMPLE 1   Evaluating functions involving square roots

If possible, evaluate $f(1)$ and $f(-2)$ for each $f(x)$.
(a) $f(x) = \sqrt{2x - 1}$ **(b)** $f(x) = \sqrt{4 - x^2}$

Solution
(a) $f(\mathbf{1}) = \sqrt{2(\mathbf{1}) - 1} = \sqrt{1} = 1$
$f(\mathbf{-2}) = \sqrt{2(\mathbf{-2}) - 1} = \sqrt{-5}$, which does not equal a real number.
(b) $f(\mathbf{1}) = \sqrt{4 - (\mathbf{1})^2} = \sqrt{3}$; $f(\mathbf{-2}) = \sqrt{4 - (\mathbf{-2})^2} = \sqrt{0} = 0$

Now Try Exercises 11, 13

In the next example we find a formula, a table of values, and a graph for a function.

EXAMPLE 2   Finding symbolic, numerical, and graphical representations

A function f takes the square root of x and then multiplies the result by 2. Give symbolic, numerical, and graphical representations of f.

Solution
First find a formula for f. The square root of x multiplied by 2 equals $2\sqrt{x}$, so $f(x) = 2\sqrt{x}$. A table of values is shown in Table 10.3. A curve is sketched through these points to obtain the graph of f in Figure 10.10. The domain of f includes all nonnegative numbers.

Symbolic Representation	Numerical Representation	Graphical Representation

$$f(x) = 2\sqrt{x}$$

TABLE 10.3

x	$2\sqrt{x}$
0	0
1	2
4	4
9	6

$f(x) = 2\sqrt{x}$

Figure 10.10

Now Try Exercise **25**

▶ **REAL-WORLD CONNECTION**   To derive a function that calculates the hang time of a football we need two facts from physics. First, when a ball is kicked into the air, half the hang time of the ball is spent going up and the other half is spent coming down. Second, the time t in seconds required for a ball to fall from a height of h feet is modeled by the equation $16t^2 = h$. Solving this equation for t gives half the hang time.

$$16t^2 = h \qquad \text{Given equation}$$

$$t^2 = \frac{h}{16} \qquad \text{Divide by 16.}$$

$$t = \sqrt{\frac{h}{16}} \qquad \text{Take the principal square root.}$$

$$t = \frac{\sqrt{h}}{4} \qquad \text{Simplify.}$$

Half the hang time is $\frac{\sqrt{h}}{4}$, so the total hang time T in seconds is given by

$$T(h) = 2 \cdot \frac{\sqrt{h}}{4} = \frac{\sqrt{h}}{2},$$

where h is the maximum height of the ball.

In the next example we use this formula to calculate hang time.

EXAMPLE 3   Calculating hang time

If a football is kicked 50 feet into the air, estimate the hang time. Does the hang time double for a football kicked 100 feet into the air?

Solution

A football kicked 50 feet into the air has a hang time of $T(50) = \frac{\sqrt{50}}{2} \approx 3.5$ seconds. For 100 feet, the hang time is $T(100) = \frac{\sqrt{100}}{2} = 5$ seconds. The hang time does not double.

Now Try Exercise **79**

CRITICAL THINKING

How high would a football have to be kicked to have twice the hang time of a football kicked 50 feet into the air?

EXAMPLE 4   **Finding the domain of a function**

Let $f(x) = \sqrt{x - 1}$.

(a) Find the domain of f. Write your answer in interval notation.

(b) Graph $y = f(x)$ and compare it to the graph of $y = \sqrt{x}$.

Solution

(a) For $f(x)$ to be defined, $x - 1$ cannot be negative. Thus valid inputs for x must satisfy

$$x - 1 \geq 0 \quad \text{or} \quad x \geq 1.$$

The domain is $[1, \infty)$.

(b) Table 10.4 lists points that lie on the graph of $y = \sqrt{x - 1}$. Note in Figure 10.11 that the graph appears only when $x \geq 1$. This graph is similar to $y = \sqrt{x}$ (see Figure 10.9) except that it is shifted one unit to the right.

TABLE 10.4

x	$\sqrt{x - 1}$
1	0
2	1
5	2

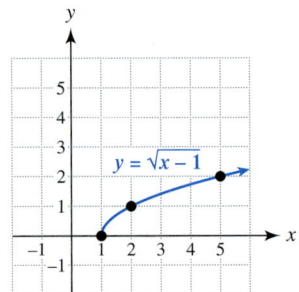

Figure 10.11

Now Try Exercises 33, 47

EXAMPLE 5   **Finding the domains of functions**

Find the domain of each function. Write your answer in interval notation.

(a) $f(x) = \sqrt{4 - 2x}$ **(b)** $g(x) = \sqrt{x^2 + 1}$

Solution

(a) To determine when $f(x)$ is defined, we must solve the inequality $4 - 2x \geq 0$.

$$4 - 2x \geq 0 \qquad \text{Inequality to be solved}$$
$$4 \geq 2x \qquad \text{Add } 2x \text{ to each side.}$$
$$2 \geq x \qquad \text{Divide each side by 2.}$$

The domain is $(-\infty, 2]$.

(b) Regardless of the value of x, the expression $x^2 + 1$ is always positive because $x^2 \geq 0$. Thus $g(x)$ is defined for all real numbers, and its domain is $(-\infty, \infty)$.

Now Try Exercises 41, 43

MAKING CONNECTIONS

Domains of Functions and Their Graphs

In Example 5, the domains of f and g were found *symbolically*. Notice that the graph of f does not appear to the right of $x = 2$ because the domain of f is $(-\infty, 2]$, whereas the graph of g appears for all values of x because the domain of g is $(-\infty, \infty)$.

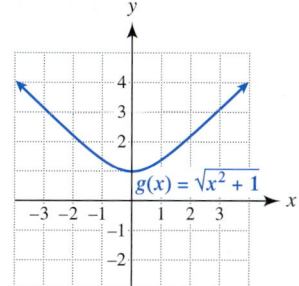

The Cube Root Function

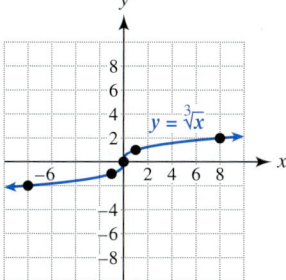

Figure 10.12 Cube Root Function

In Section 10.1 we discussed the cube root of a number. We can define the **cube root function** by $f(x) = \sqrt[3]{x}$. Cube roots are defined for both positive and negative numbers, so *the domain of the cube root function includes all real numbers*. Table 10.5 lists points that lie on the graph of the cube root function. Figure 10.12 shows a graph of $y = \sqrt[3]{x}$.

TABLE 10.5 Cube Root Function

x	-27	-8	-1	0	1	8	27
$\sqrt[3]{x}$	-3	-2	-1	0	1	2	3

MAKING CONNECTIONS

Domains of the Square Root and Cube Root Functions

$f(x) = \sqrt{x}$ equals a real number for any nonnegative x. Thus $D = [0, \infty)$.

$f(x) = \sqrt[3]{x}$ equals a real number for any x. Thus $D = (-\infty, \infty)$.

Examples: $\sqrt{4} = 2$ but $\sqrt{-4}$ is *not* a real number. $\sqrt[3]{8} = 2$ and $\sqrt[3]{-8} = -2$.

EXAMPLE 6   **Evaluating functions involving cube roots**

Evaluate $f(1)$ and $f(-3)$ for each $f(x)$.
(a) $f(x) = \sqrt[3]{x^2 - 1}$ **(b)** $f(x) = \sqrt[3]{2 - x^2}$

Solution
(a) $f(\mathbf{1}) = \sqrt[3]{\mathbf{1}^2 - 1} = \sqrt[3]{0} = 0$; $f(\mathbf{-3}) = \sqrt[3]{(\mathbf{-3})^2 - 1} = \sqrt[3]{8} = 2$
(b) $f(\mathbf{1}) = \sqrt[3]{2 - \mathbf{1}^2} = \sqrt[3]{1} = 1$; $f(\mathbf{-3}) = \sqrt[3]{2 - (\mathbf{-3})^2} = \sqrt[3]{-7}$ or $-\sqrt[3]{7}$

Now Try Exercises **15, 19**

EXAMPLE 7   Finding symbolic, numerical, and graphical representations

A function f takes the cube root of twice x. Give symbolic, numerical, and graphical representations of f.

Solution
First find a formula for f. Twice x is $2x$, so $f(x) = \sqrt[3]{2x}$. A table of values is shown in Table 10.6. A curve is sketched through these points to obtain the graph of f in Figure 10.13.

Symbolic Representation

$$f(x) = \sqrt[3]{2x}$$

Numerical Representation

TABLE 10.6

x	$\sqrt[3]{2x}$
-4	-2
$-\frac{1}{2}$	-1
0	0
$\frac{1}{2}$	1
4	2

Graphical Representation

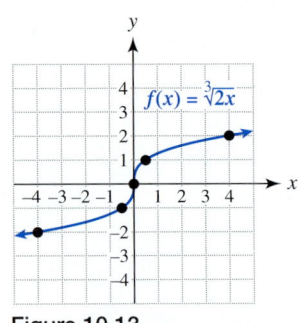

Figure 10.13

Now Try Exercise 29

Power Functions

Power functions are a generalization of root functions. Examples of power functions include

$$f(x) = x^{1/2}, \quad g(x) = x^{2/3}, \quad \text{and} \quad h(x) = x^{-3/5}.$$

The exponents for power functions can be rational numbers. Any rational number can be written in lowest terms as $\frac{m}{n}$, where m and n are integers. When n is an odd integer, then $f(x) = x^{m/n}$ is defined for all real numbers. When n is even, $f(x) = x^{m/n}$ is defined only for nonnegative real numbers.

POWER FUNCTION

If a function f can be represented by

$$f(x) = x^p,$$

where p is a rational number, then f is a **power function**. If $p = \frac{1}{n}$, where $n \geq 2$ is an integer, then f is also a **root function**, which is given by

$$f(x) = \sqrt[n]{x}.$$

EXAMPLE 8   Evaluating power functions

If possible, evaluate $f(x)$ at the given value of x.
(a) $f(x) = x^{0.75}$ at $x = 16$ **(b)** $f(x) = x^{1/4}$ at $x = -81$

Solution

(a) $0.75 = \frac{3}{4}$, so $f(x) = x^{3/4}$. Thus $f(\mathbf{16}) = \mathbf{16}^{3/4} = (16^{1/4})^3 = 2^3 = 8$.

   NOTE: $16^{1/4} = \sqrt[4]{16} = 2$.

(b) $f(-81) = (-81)^{1/4} = \sqrt[4]{-81}$, which is undefined. There is no real number a such that $a^4 = -81$ because a^4 is never negative.   Now Try Exercises **59, 63**

▶ **REAL-WORLD CONNECTION**   The surface area of the skin covering the human body is influenced by both the height and weight of a person. A taller person tends to have a larger surface area, as does a heavier person. In the next example, we use a power function from biology to model this situation.

EXAMPLE **9**   Modeling surface area of the human body

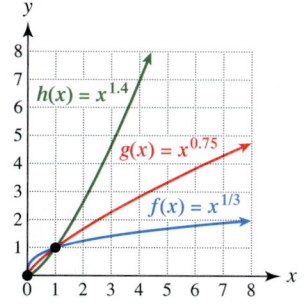

The surface area of a person who is 66 inches tall and weighs w pounds can be estimated by $S(w) = 327w^{0.425}$, where S is in square inches.   (***Source:*** H. Lancaster, Quantitative *Methods in Biological and Medical Sciences*.)

(a) Find S if this person weighs 130 pounds.

(b) If the person gains 20 pounds, by how much does the person's surface area increase?

Solution

(a) $S(130) = 327(130)^{0.425} \approx 2588$ square inches

(b) Because $S(150) = 327(150)^{0.425} \approx 2750$ square inches, the surface area of the person increases by about $2750 - 2588 = 162$ square inches.   Now Try Exercise **87**

In the next example we investigate the graph of $y = x^p$ for different values of p.

EXAMPLE **10**   Graphing power functions

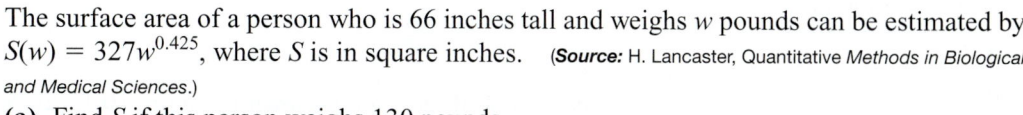

The graphs of three power functions,

$$f(x) = x^{1/3}, \quad g(x) = x^{0.75}, \quad \text{and} \quad h(x) = x^{1.4},$$

are shown in Figure 10.14. Discuss how the value of p affects the graph of $y = x^p$ when $x > 1$ and when $0 < x < 1$.

Figure 10.14 Power Functions

Solution

First note that $g(x)$ and $h(x)$ can be written as $g(x) = x^{3/4}$ and $h(x) = x^{7/5}$. When $x > 1$, $h(x) > g(x) > f(x)$ and the graphs increase (rise) faster for larger values of p. When $0 < x < 1$, $h(x) < g(x) < f(x)$. Thus smaller values of p result in larger y-values when $0 < x < 1$. All three graphs appear to intersect at the points $(0, 0)$ and $(1, 1)$.

Now Try Exercise **67**

Modeling with Power Functions (Optional)

▶ **REAL-WORLD CONNECTION**   Allometry is the study of the relative sizes of different characteristics of an organism. For example, the weight of a bird is related to the surface area of its wings: Heavier birds tend to have larger wings. Allometric relations are often modeled with $f(x) = kx^p$, where k and p are constants.   (***Source:*** C. Pennycuick, *Newton Rules Biology*.)

EXAMPLE   Modeling surface area of wings

The surface area A of a bird's wings with weight w is shown in Table 10.7.

TABLE 10.7   Weight and Wing Size

w (kilograms)	0.5	2.0	3.5	5.0
A (square meters)	0.069	0.175	0.254	0.325

CALCULATOR HELP

To make a scatterplot, see Appendix A (page AP-4).

[0, 6, 1] by [0, 0.4, 0.1]

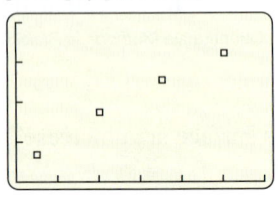

(a)

[0, 6, 1] by [0, 0.4, 0.1]

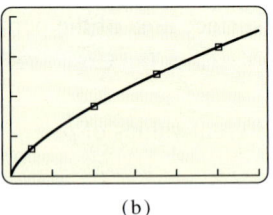

(b)

Figure 10.15

(a) Make a scatterplot of the data. Discuss any trends in the data.
(b) Biologists modeled the data with $A(w) = kw^{2/3}$, where k is a constant. Find k.
(c) Graph A and the data in the same viewing rectangle.
(d) Estimate the area of the wings of a 3-kilogram bird.

Solution
(a) A scatterplot of the data is shown in Figure 10.15(a). As the weight of a bird increases so does the surface area of its wings.
(b) To determine k, substitute one of the data points into $A(w)$. We use (**2.0, 0.175**).

$$A(w) = kw^{2/3} \qquad \text{Given formula}$$
$$\mathbf{0.175} = k(\mathbf{2})^{2/3} \qquad \begin{array}{l}\text{Let } w = 2 \text{ and } A(w) = 0.175 \\ \text{(any data point could be used).}\end{array}$$
$$k = \frac{0.175}{2^{2/3}} \qquad \text{Solve for } k; \text{ rewrite equation.}$$
$$k \approx 0.11 \qquad \text{Approximate } k.$$

Thus $A(w) = 0.11w^{2/3}$.
(c) The data and graph of $Y_1 = 0.11X^{\wedge}(2/3)$ are shown in Figure 10.15(b). Note that the graph appears to pass through each data point.
(d) $A(\mathbf{3}) = 0.11(\mathbf{3})^{2/3} \approx 0.23$ square meter

Now Try Exercise **83**

10.4   PUTTING IT ALL TOGETHER

In this section we discussed some new functions involving radicals and rational exponents. Properties of these functions are summarized in the following table.

Function	Explanation
Square Root and Cube Root	$f(x) = \sqrt{x}$ and $g(x) = \sqrt[3]{x}$ The cube root function g is defined for all inputs, whereas the square root function f is defined only for nonnegative inputs.

Function	Explanation
Square Root and Cube Root (*continued*)	Square Root Function $\quad$ Cube Root Function
Power and Root	$f(x) = x^p$, where p is a rational number, is a power function. If $p = \frac{1}{n}$, where $n \geq 2$ is an integer, f is also a root function, given by $f(x) = \sqrt[n]{x}$. ***Examples:*** $f(x) = x^{5/3}$ $\qquad$ Power function $\qquad\qquad\quad g(x) = x^{1/4} = \sqrt[4]{x}$ $\quad$ Root and power function

10.4 Exercises

  PRACTICE $\quad$ WATCH $\quad$ DOWNLOAD $\quad$ READ $\quad$ REVIEW

CONCEPTS

1. Sketch a graph of the square root function.

2. Sketch a graph of the cube root function.

3. What is the domain of the square root function?

4. What is the domain of the cube root function?

5. Give a symbolic representation for a power function.

6. Give a symbolic representation for a root function.

7. What is the domain of $f(x) = \sqrt[4]{x}$?

8. What is the domain of $f(x) = \sqrt[5]{x}$?

ROOT FUNCTIONS

Exercises 9–24: If possible, evaluate the function at the given value(s) of the variable.

9. $f(x) = \sqrt{x-1}$ $\qquad\qquad x = 10, 0$

10. $f(x) = \sqrt{4-3x}$ $\qquad\quad x = -4, 1$

11. $f(x) = \sqrt{3-3x}$ $\qquad\quad x = -1, 5$

12. $f(x) = \sqrt{x-5}$ $\qquad\qquad x = -1, 5$

13. $f(x) = \sqrt{x^2 - x}$ $\qquad\quad x = -4, 3$

14. $f(x) = \sqrt{2x^2 - 3}$ $\qquad\quad x = -1, 2$

15. $f(x) = \sqrt[3]{x^2 - 8}$ $\qquad\quad x = -3, 4$

16. $f(x) = \sqrt[3]{2x^2}$ $\qquad\qquad x = -2, 2$

17. $f(x) = \sqrt[3]{x - 9}$ $\qquad\quad x = 1, 10$

18. $f(x) = \sqrt[3]{5x - 2}$ $\qquad\quad x = -5, 2$

19. $f(x) = \sqrt[3]{3 - x^2}$ $\qquad\quad x = -2, 3$

20. $f(x) = \sqrt[3]{-1 - x^2}$ $\qquad\quad x = 0, 3$

21. $T(h) = \frac{1}{2}\sqrt{h}$ $\qquad\qquad h = 64$

22. $L(k) = 2\sqrt{k + 2}$ $\qquad\quad k = 23$

23. $f(x) = \sqrt{x + 5} + \sqrt{x}$ $\quad x = 4$

24. $f(x) = \dfrac{\sqrt{x - 5} - \sqrt{x}}{2}$ $\qquad x = 9$

Exercises 25–32: Give symbolic, numerical, and graphical representations for the function f.

25. Function f takes the square root of x and then adds 1 to the result.

26. Function f takes the square root of x and then subtracts 2 from the result.

27. Function f takes the square root of three times x.

28. Function f takes the square root of the quantity x plus 1.

29. Function f takes the cube root of x and then multiplies the result by 2.

30. Function f takes the cube root of 4 times x.

31. Function f takes the cube root of the quantity x minus 1.

32. Function f takes the cube root of x and then adds 1.

Exercises 33–46: Find the domain of f. Write your answer in interval notation.

33. $f(x) = \sqrt{x + 2}$ **34.** $f(x) = \sqrt{x - 1}$

35. $f(x) = \sqrt{x - 2}$ **36.** $f(x) = \sqrt{x + 1}$

37. $f(x) = \sqrt{2x - 4}$ **38.** $f(x) = \sqrt{4x + 2}$

39. $f(x) = \sqrt{1 - x}$ **40.** $f(x) = \sqrt{6 - 3x}$

41. $f(x) = \sqrt{8 - 5x}$ **42.** $f(x) = \sqrt{3 - 2x}$

43. $f(x) = \sqrt{3x^2 + 4}$ **44.** $f(x) = \sqrt{1 + 2x^2}$

45. $f(x) = \dfrac{1}{\sqrt{2x + 1}}$ **46.** $f(x) = \dfrac{1}{\sqrt{x - 1}}$

Exercises 47–52: Graph the equation. Compare the graph to either $y = \sqrt{x}$ or $y = \sqrt[3]{x}$.

47. $y = \sqrt{x} + 2$ **48.** $y = \sqrt{x} - 1$

49. $y = \sqrt{x + 2}$ **50.** $y = \sqrt[3]{x} + 2$

51. $y = \sqrt[3]{x + 2}$ **52.** $y = \sqrt[3]{x} - 1$

POWER FUNCTIONS

Exercises 53–58: Use a radical to write f(x).

53. $f(x) = x^{1/2}$ **54.** $f(x) = x^{1/3}$

55. $f(x) = x^{2/3}$ **56.** $f(x) = x^{3/4}$

57. $f(x) = x^{-1/5}$ **58.** $f(x) = x^{-2/5}$

Exercises 59–66: If possible, evaluate f(x) at the given values of x. When appropriate, approximate the answer to the nearest hundredth.

59. $f(x) = x^{5/2}$ $x = 4, x = 5$

60. $f(x) = x^{-3/4}$ $x = 1, x = 3$

61. $f(x) = x^{-7/5}$ $x = -32, x = 10$

62. $f(x) = x^{4/3}$ $x = -8, x = 27$

63. $f(x) = x^{1/4}$ $x = 256, x = -10$

64. $f(x) = x^{3/4}$ $x = 16, x = -1$

65. $f(x) = x^{2/5}$ $x = 32, x = -32$

66. $f(x) = x^{5/6}$ $x = -5, x = 64$

Exercises 67–70: Graph f and g in the window [0, 6, 1] by [0, 6, 1]. Which function is greater when x > 1?

67. $f(x) = x^{1/5}, g(x) = x^{1/3}$

68. $f(x) = x^{4/5}, g(x) = x^{5/4}$

69. $f(x) = x^{1.2}, g(x) = x^{0.45}$

70. $f(x) = x^{-1.4}, g(x) = x^{1.4}$

Exercises 71 and 72: **Thinking Generally**   *Let $0 < q < p$, where p and q are rational numbers.*

71. For $x > 1$, is $x^p > x^q$ or $x^p < x^q$?

72. For $0 < x < 1$, is $x^p > x^q$ or $x^p < x^q$?

Exercises 73 and 74: **Operations on Functions**   *For the given f(x) and g(x), evaluate each expression.*

    (a) $(f + g)(2)$ **(b)** $(f - g)(x)$
    (c) $(fg)(x)$ **(d)** $(f/g)(x)$

73. $f(x) = \sqrt{8x}, g(x) = \sqrt{2x}$

74. $f(x) = \sqrt{9x + 18}, g(x) = \sqrt{4x + 8}$

Exercises 75–78: **Graphical Interpretation**   *Match the situation with the graph of the power function (a.–d.) on the next page that models it best.*

75. Amount of water in a barrel that is initially full and has a hole near the bottom

76. The average weight of a type of bird as its wing span increases (*Hint*: The exponent is greater than 1.)

77. Money made after x hours by a person who works for a fixed hourly wage

78. Rapid growth of an insect population that eventually slows down

a.

b.

c.

d.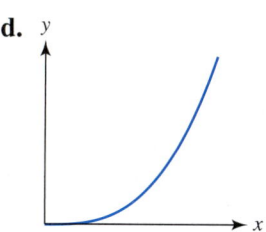

APPLICATIONS

79. *Jumping*   (Refer to Example 3.) If a person jumps 4 feet off the ground, estimate how long the person is in the air.

80. *Hang Time*   (Refer to Example 3.) Find the hang time for a golf ball hit 80 feet into the air.

81. *Aging More Slowly*   Albert Einstein in his theory of relativity showed that if a person travels at nearly the speed of light, then time slows down significantly. Suppose that there are twins; one remains on Earth and the other leaves in a very fast spaceship having velocity v. If the twin on Earth ages T_0 years, then according to Einstein the twin in the spaceship ages T years, where

$$T(v) = T_0\sqrt{1 - (v/c)^2}.$$

In this formula c represents the speed of light, which is 186,000 miles per second.

(a) Evaluate T when $v = 0.8c$ (eight-tenths the speed of light) and $T_0 = 10$ years. (*Hint:* Simplify $\frac{v}{c}$ without using 186,000 miles per second.)

(b) Interpret your result.

82. *Increasing Your Weight*   (Refer to Exercise 81.) Albert Einstein also showed that the weight (mass) of an object increases when traveling near the speed of light. If a person's weight on Earth is W_0, then the same person's weight W in a spaceship traveling at velocity v is

$$W(v) = \frac{W_0}{\sqrt{1 - (v/c)^2}}.$$

(a) Evaluate W when $v = 0.6c$ (six-tenths the speed of light) and $W_0 = 220$ pounds (100 kilograms).

(b) Interpret your result.

83. *Modeling Wing Size*   (Refer to Example 11.) The surface area of wings for a species of bird with weight w is shown in the table.

w (kilograms)	2	3	4
A (square meters)	0.254	0.333	0.403

(a) Make a scatterplot of the data.

(b) The data can be modeled with $A(w) = kw^{2/3}$. Find k.

(c) Graph A and the data in the same viewing rectangle. Does the graph of A pass through the data points?

(d) Estimate the area of the wings of a 2.5-kilogram bird.

84. *Pulse Rate in Animals*   The following table lists typical pulse rates R in beats per minute (bpm) for animals with various weights W in pounds. (*Source:* C. Pennycuick.)

W (pounds)	20	150	500	1500
R (beats per minute)	198	72	40	23

(a) Describe what happens to the pulse rate as the size of the animal increases.

(b) Plot the data in [0, 1600, 400] by [0, 220, 20].

(c) If $R = kW^{-1/2}$, find k.

(d) Find R when $W = 700$, and interpret the result.

85. *Modeling Wing Span*   (Refer to Example 11.) Biologists have found that the weight W of a bird and the length L of its wing span are related by $L = kW^{1/3}$, where k is a constant. The following table lists L and W for one species of bird.   (***Source:*** C. Pennycuick.)

W (kilograms)	0.1	0.4	0.8	1.1
L (meters)	0.422	0.670	0.844	0.938

(a) Use the data to approximate the value of k.
(b) Graph L and the data in the same viewing rectangle. What happens to L as W increases?
(c) Find the wing span of a 0.7-kilogram bird.
(d) Find L when $W = 0.65$, and interpret the result.

86. *Design of Open Channels*   To protect cities from flooding during heavy rains, open channels are sometimes constructed to handle runoff. The rate R at which water flows through the channel is modeled by $R = k\sqrt{m}$, where m is the slope of the channel and k is a constant determined by the shape of the channel.
(***Source:*** N. Garber and L. Hoel, *Traffic and Highway Design.*)
(a) Suppose that a channel has a slope of $m = 0.01$ (or 1%) and a runoff rate of $R = 340$ cubic feet per second (cfs). Find k.
(b) If the slope of the channel increases to $m = 0.04$ (or 4%), what happens to R? Be specific.

Exercises 87 and 88: (Refer to Example 9.) The surface area of the skin of a person who is 70 inches tall and weighs w pounds can be estimated by $S(w) = 342w^{0.425}$, where S is in square inches. Evaluate each of the following.

87. $S(150)$ **88.** $S(200)$

WRITING ABOUT MATHEMATICS

89. Explain why a root function is an example of a power function.

90. Discuss the shape of the graph of $y = x^p$ as p increases. Assume that p is a positive rational number and that x is a positive real number.

CHECKING BASIC CONCEPTS
SECTIONS 10.3 AND 10.4

1. Simplify each expression.
  (a) $\sqrt{3} \cdot \sqrt{12}$ (b) $\dfrac{\sqrt[3]{81}}{\sqrt[3]{3}}$
  (c) $\sqrt{36x^6}, x > 0$

2. Simplify each expression.
  (a) $5\sqrt{6} + 2\sqrt{6} + \sqrt{7}$
  (b) $8\sqrt[3]{x} - 3\sqrt[3]{x}$ (c) $\sqrt{9x} - \sqrt{4x}$

3. Simplify each expression.
  (a) $\sqrt[3]{xy^4} - \sqrt[3]{x^4y}$
  (b) $(4 - \sqrt{2})(4 + \sqrt{2})$

4. Rationalize the denominator of $\dfrac{6}{2\sqrt{6}}$.

5. Rationalize the denominator of $\dfrac{2}{\sqrt{5} - 1}$.

6. Sketch a graph of each function and then evaluate $f(-1)$, if possible.
  (a) $f(x) = \sqrt{x}$
  (b) $f(x) = \sqrt[3]{x}$
  (c) $f(x) = \sqrt{x^2}$

7. Evaluate $f(x) = 0.2x^{2/3}$ when $x = 64$.

8. Find the domain of $f(x) = \sqrt{x - 4}$. Write your answer in interval notation.

10.5 EQUATIONS INVOLVING RADICAL EXPRESSIONS

Solving Radical Equations ▪ The Distance Formula ▪ Solving the Equation $x^n = k$

A LOOK INTO MATH ▷

In Section 10.4 we showed that for some types of birds there is a relationship between weight and the size of the wings—heavier birds tend to have larger wings. This relationship can sometimes be modeled by $A = 100\sqrt[3]{W^2}$, where W is weight in pounds and A is area in square inches. Suppose that we want to estimate the weight of a bird whose wings have an area of 600 square inches. To do so, we would need to solve the equation

$$600 = 100\sqrt[3]{W^2}$$

for W. This equation contains a radical expression. In this section we explain how to solve this type of equation.   (***Source:*** C. Pennycuick, *Newton Rules Biology.*)

Solving Radical Equations

Many times, equations contain either radical expressions or rational exponents. Examples include

$$\sqrt{x} = 6, \quad 5x^{1/2} = 1, \quad \text{and} \quad \sqrt[3]{x-1} = 3.$$

One strategy for solving an equation containing a square root is to isolate the square root (if necessary) and then square each side of the equation. This technique is an example of the *power rule for solving equations.*

> **POWER RULE FOR SOLVING EQUATIONS**
>
> If each side of an equation is raised to the same positive integer power, then any solutions to the given equation are among the solutions to the new equation. That is, the solutions to the equation $a = b$ are among the solutions to $a^n = b^n$.

We must check our solutions when applying the power rule. For example, consider the equation $2x = 6$. If we square each side, we obtain $4x^2 = 36$. Solving this second equation gives $x^2 = 9$, or $x = \pm 3$. Here, 3 is a solution to both equations, but -3 is an **extraneous solution** that satisfies the second equation but not the given equation. That is, $4(-3)^2 = 36$ is a *true* statement, but $2(-3) = 6$ is a *false* statement. Thus -3 is *not* a solution to the given equation.

We illustrate the power rule in the next example.

EXAMPLE 1 **Solving a radical equation symbolically**

Solve $\sqrt{2x-1} = 3$. Check your solution.

Solution
Begin by squaring each side of the equation.

$$
\begin{array}{ll}
\sqrt{2x-1} = 3 & \text{Given equation} \\
(\sqrt{2x-1})^2 = 3^2 & \text{Square each side.} \\
2x - 1 = 9 & \text{Simplify.} \\
2x = 10 & \text{Add 1.} \\
x = 5 & \text{Divide by 2.}
\end{array}
$$

To check this answer, we substitute $x = \mathbf{5}$ in the given equation.

$$\sqrt{2(\mathbf{5}) - 1} \overset{?}{=} 3$$

$$3 = 3 \qquad \text{It checks.} \qquad \text{Now Try Exercise } \boxed{21}$$

NOTE: To simplify $(\sqrt{2x - 1})^2$ in Example 1, we used the fact that

$$(\sqrt{a})^2 = \sqrt{a} \cdot \sqrt{a} = a.$$

The following steps can be used to solve a radical equation.

SOLVING A RADICAL EQUATION

STEP 1: Isolate a radical term on one side of the equation.

STEP 2: Apply the power rule by raising each side of the equation to the power equal to the index of the isolated radical term.

STEP 3: Solve the equation. If it still contains a radical, repeat Steps 1 and 2.

STEP 4: Check your answers by substituting each result in the *given* equation.

NOTE: In Sections 10.1–10.3, we *simplified expressions*. In this section we *solve equations*. Equations contain equals signs and when we solve an equation, we try to find values of the variable that make the equation a true statement.

In the next example we apply these steps to a radical equation.

EXAMPLE $\boxed{2}$ **Isolating the radical term**

Solve $\sqrt{4 - x} + 5 = 8$.

Solution

STEP 1: To isolate the radical term, we subtract 5 from each side of the equation.

$$\sqrt{4 - x} + 5 = 8 \qquad \text{Given equation}$$

$$\sqrt{4 - x} = 3 \qquad \text{Subtract 5.}$$

STEP 2: The isolated term involves a square root, so we must square each side.

$$(\sqrt{4 - x})^2 = (3)^2 \qquad \text{Square each side.}$$

STEP 3: Next we solve the resulting equation. (It is not necessary to repeat Steps 1 and 2 because the resulting equation does not contain any radical expressions.)

$$4 - x = 9 \qquad \text{Simplify.}$$

$$-x = 5 \qquad \text{Subtract 4.}$$

$$x = -5 \qquad \text{Multiply by } -1.$$

STEP 4: To check this answer, we substitute $x = -5$ in the given equation.

$$\sqrt{4 - (-5)} + 5 \overset{?}{=} 8$$

$$\sqrt{9} + 5 \overset{?}{=} 8$$

$$8 = 8 \qquad \text{It checks.} \qquad \text{Now Try Exercise } \boxed{23}$$

Example 3 shows that we must check our answers when squaring each side of an equation to identify extraneous solutions.

EXAMPLE 3

Solving a radical equation

Solve $\sqrt{3x + 3} = 2x - 1$. Check your results and then solve the equation graphically.

Solution

Symbolic Solution   Begin by squaring each side of the equation.

$$\sqrt{3x + 3} = 2x - 1 \qquad \text{Given equation}$$
$$(\sqrt{3x + 3})^2 = (2x - 1)^2 \qquad \text{Square each side.}$$
$$3x + 3 = 4x^2 - 4x + 1 \qquad \text{Multiply.}$$
$$0 = 4x^2 - 7x - 2 \qquad \text{Subtract } 3x + 3.$$
$$0 = (4x + 1)(x - 2) \qquad \text{Factor.}$$
$$x = -\frac{1}{4} \quad \text{or} \quad x = 2 \qquad \text{Solve for } x.$$

To check these values, substitute $x = -\frac{1}{4}$ and $x = 2$ in the given equation.

$$\sqrt{3\left(-\frac{1}{4}\right) + 3} \stackrel{?}{=} 2\left(-\frac{1}{4}\right) - 1$$
$$\sqrt{2.25} \stackrel{?}{=} -1.5$$
$$1.5 \neq -1.5 \qquad \text{It does not check.}$$

Thus $-\frac{1}{4}$ is an *extraneous solution*. Next substitute $x = 2$ in the given equation.

$$\sqrt{3 \cdot 2 + 3} \stackrel{?}{=} 2 \cdot 2 - 1$$
$$\sqrt{9} \stackrel{?}{=} 3$$
$$3 = 3 \qquad \text{It checks.}$$

The only solution is 2.

Graphical Solution   The solution 2 is supported graphically in Figure 10.16, where the graphs of $Y_1 = \sqrt{(3X + 3)}$ and $Y_2 = 2X - 1$ intersect at the point $(2, 3)$. *Note that the graphical solution does not give an extraneous solution.*   Now Try Exercises 27, 79(a), (b)

$[-5, 5, 1]$ by $[-5, 5, 1]$

Intersection
X=2   Y=3

Figure 10.16

CALCULATOR HELP

To find a point of intersection, see Appendix A (page AP-7).

Example 3 demonstrates that *checking a solution is essential when you are squaring each side of an equation.* Squaring may introduce extraneous solutions, which are solutions to the resulting equation but are not solutions to the given equation.

CRITICAL THINKING

Will a numerical solution give extraneous solutions?

When an equation contains two or more terms with square roots, it may be necessary to square each side of the equation more than once. In these situations, isolate one of the square roots and then square each side of the equation. If a radical term remains after simplifying, repeat these steps. We apply this technique in the next example.

EXAMPLE 4   Squaring twice

Solve $\sqrt{2x} - 1 = \sqrt{x + 1}$.

Solution
Begin by squaring each side of the equation.

$$\sqrt{2x} - 1 = \sqrt{x + 1} \qquad \text{Given equation}$$
$$(\sqrt{2x} - 1)^2 = (\sqrt{x + 1})^2 \qquad \text{Square each side.}$$
$$(\sqrt{2x})^2 - 2(\sqrt{2x})(1) + 1^2 = x + 1 \qquad (a - b)^2 = a^2 - 2ab + b^2$$
$$2x - 2\sqrt{2x} + 1 = x + 1 \qquad \text{Simplify.}$$
$$2x - 2\sqrt{2x} = x \qquad \text{Subtract 1.}$$
$$x = 2\sqrt{2x} \qquad \text{Subtract } x \text{ and add } 2\sqrt{2x}.$$
$$x^2 = 4(2x) \qquad \text{Square each side again.}$$
$$x^2 - 8x = 0 \qquad \text{Subtract } 8x.$$
$$x(x - 8) = 0 \qquad \text{Factor.}$$
$$x = 0 \quad \text{or} \quad x = 8 \qquad \text{Solve.}$$

To check these answers, substitute $x = 0$ and $x = 8$ in the given equation.

$$\sqrt{2 \cdot 0} - 1 \stackrel{?}{=} \sqrt{0 + 1} \qquad\qquad \sqrt{2 \cdot 8} - 1 \stackrel{?}{=} \sqrt{8 + 1}$$
$$-1 \neq 1 \quad \text{It does not check.} \qquad\qquad 3 = 3 \quad \text{It checks.}$$

The only solution is 8.   Now Try Exercise 43

In the next example we apply the power rule to an equation that contains a cube root.

EXAMPLE 5   Solving an equation containing a cube root

Solve $\sqrt[3]{4x - 7} = 4$.

Solution
STEP 1: The cube root term is already isolated, so we proceed to Step 2.

STEP 2: Because the index is 3, we cube each side of the equation.

$$\sqrt[3]{4x - 7} = 4 \qquad \text{Given equation}$$
$$(\sqrt[3]{4x - 7})^3 = (4)^3 \qquad \text{Cube each side.}$$

STEP 3: We solve the resulting equation.

$$4x - 7 = 64 \qquad \text{Simplify.}$$
$$4x = 71 \qquad \text{Add 7 to each side.}$$
$$x = \frac{71}{4} \qquad \text{Divide each side by 4.}$$

STEP 4: To check this answer, we substitute $x = \frac{71}{4}$ in the given equation.

$$\sqrt[3]{4\left(\frac{71}{4}\right) - 7} \stackrel{?}{=} 4$$
$$\sqrt[3]{64} \stackrel{?}{=} 4$$
$$4 = 4 \qquad \text{It checks.} \qquad \text{Now Try Exercise 31}$$

▶ **REAL-WORLD CONNECTION**   In the next example we solve the equation presented in the introduction to this section that relates a bird's weight and wing surface area.

EXAMPLE 6   **Finding the weight of a bird**

Solve the equation $600 = 100\sqrt[3]{W^2}$ to determine the weight in pounds of a bird having wings with an area of 600 square inches.

Solution
Begin by dividing each side of the equation by 100 to isolate the radical term.

$$\frac{600}{100} = \sqrt[3]{W^2} \qquad \text{Divide each side by 100.}$$

$$(6)^3 = (\sqrt[3]{W^2})^3 \qquad \text{Cube each side.}$$

$$216 = W^2 \qquad \text{Simplify.}$$

$$\sqrt{216} = W \qquad \text{Take principal square root, } W > 0.$$

$$W \approx 14.7 \qquad \text{Approximate.}$$

The weight of the bird is approximately 14.7 pounds.   Now Try Exercise 117

TECHNOLOGY NOTE: Graphing Radical Expressions

The equation in Example 6 can be solved graphically. Sometimes it is more convenient to use rational exponents than radical notation. Thus $y = 100\sqrt[3]{W^2}$ can be entered as $Y_1 = 100X^\wedge(2/3)$. (Be sure to include parentheses around the 2/3.) The accompanying figure shows y_1 intersecting the line $y_2 = 600$ near the point (14.7, 600), which supports our symbolic result.

[0, 20, 5] by [0, 800, 100]

In the next example we solve an equation that would be difficult to solve symbolically, but an *approximate* solution can be found graphically.

EXAMPLE 7   **Solving an equation with rational exponents**

Solve $x^{2/3} = 3 - x^2$ graphically.

Solution
Graph $Y_1 = X^\wedge(2/3)$ and $Y_2 = 3 - X^\wedge2$. Their graphs intersect near $(-1.34, 1.21)$ and $(1.34, 1.21)$, as shown in Figure 10.17. Thus the solutions are given by $x \approx \pm 1.34$.

$[-6, 6, 1]$ by $[-4, 4, 1]$ $[-6, 6, 1]$ by $[-4, 4, 1]$

          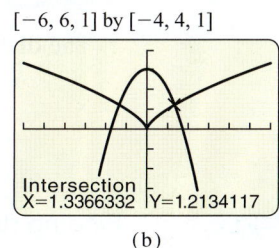

(a)   (b)

Figure 10.17   Now Try Exercise 73

The Distance Formula

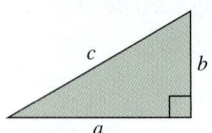

Figure 10.18
$a^2 + b^2 = c^2$

▶ **REAL-WORLD CONNECTION**   One of the most famous theorems in mathematics is the **Pythagorean theorem**. It states that if a right triangle has legs a and b with hypotenuse c (see Figure 10.18), then

$$a^2 + b^2 = c^2.$$

For example, if the legs of a right triangle are $a = 3$ and $b = 4$, the hypotenuse is $c = 5$ because $3^2 + 4^2 = 5^2$. Also, if the sides of a triangle satisfy $a^2 + b^2 = c^2$, it is a right triangle.

EXAMPLE 8   Applying the Pythagorean theorem

A rectangular television screen has a width of 20 inches and a height of 15 inches. Find the diagonal of the television. Why is it called a 25-inch television?

Figure 10.19

Solution
In Figure 10.19, let $a = \mathbf{20}$ and $b = \mathbf{15}$. Then the diagonal of the television corresponds to the hypotenuse of a right triangle with legs of 20 inches and 15 inches.

$$c^2 = a^2 + b^2 \qquad \text{Pythagorean theorem}$$
$$c = \sqrt{a^2 + b^2} \qquad \text{Take the principal square root, } c > 0.$$
$$c = \sqrt{20^2 + 15^2} \qquad \text{Substitute } a = 20 \text{ and } b = 15.$$
$$c = 25 \qquad \text{Simplify.}$$

A 25-inch television has a diagonal of 25 inches.   **Now Try Exercise 123**

The Pythagorean theorem can be used to determine the distance between two points. Suppose that a line segment has endpoints (x_1, y_1) and (x_2, y_2), as illustrated in Figure 10.20. The lengths of the legs of the right triangle are $x_2 - x_1$ and $y_2 - y_1$. The distance d is the hypotenuse of a right triangle. Applying the Pythagorean theorem, we have

$$d^2 = (x_2 - x_1)^2 + (y_2 - y_1)^2.$$

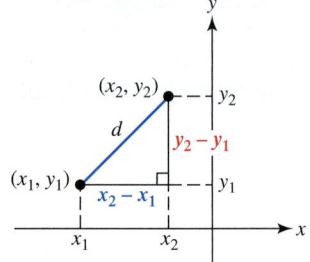

Figure 10.20

Distance is nonnegative, so we let d be the principal square root and obtain

$$d = \sqrt{(x_2 - x_1)^2 + (y_2 - y_1)^2}.$$

DISTANCE FORMULA

The **distance** d between the points (x_1, y_1) and (x_2, y_2) in the xy-plane is

$$d = \sqrt{(x_2 - x_1)^2 + (y_2 - y_1)^2}.$$

EXAMPLE 9   Finding distance between points

Find the distance between the points $(-2, 3)$ and $(1, -4)$.

Solution

Start by letting $(x_1, y_1) = (-2, 3)$ and $(x_2, y_2) = (1, -4)$. Then substitute these values into the distance formula.

$$\begin{aligned} d &= \sqrt{(x_2 - x_1)^2 + (y_2 - y_1)^2} && \text{Distance formula} \\ &= \sqrt{(1 - (-2))^2 + (-4 - 3)^2} && \text{Substitute.} \\ &= \sqrt{9 + 49} && \text{Simplify.} \\ &= \sqrt{58} && \text{Add.} \\ &\approx 7.62 && \text{Approximate.} \end{aligned}$$

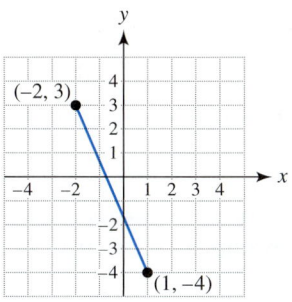

Figure 10.21

The distance between the points, as shown in Figure 10.21, is exactly $\sqrt{58}$ units, or about 7.62 units. Note that we would obtain the same result if we let $(x_1, y_1) = (1, -4)$ and $(x_2, y_2) = (-2, 3)$.

Now Try Exercise **109**

NOTE: In Example 9, $\sqrt{9 + 49} \neq \sqrt{9} + \sqrt{49} = 3 + 7 = 10$. In general, for any a and b, $\sqrt{a^2 + b^2} \neq a + b$.

▶ **REAL-WORLD CONNECTION**   In the next example the distance formula is applied to road construction.

EXAMPLE **10**   Designing a highway curve

Figure 10.22 shows a circular highway curve joining a straight section of road. A surveyor is trying to locate the x-coordinate of the *point of curvature PC* where the two sections of the highway meet. The distance between the surveyor and PC should be 400 feet. Estimate the x-coordinate of PC if x is positive.

Figure 10.22

Solution

In Figure 10.22, the distance between the points $(0, 75)$ and $(x, 300)$ is 400 feet. We can apply the distance formula and solve for x.

$$\begin{aligned} d &= \sqrt{(x_2 - x_1)^2 + (y_2 - y_1)^2} && \text{Distance formula} \\ 400 &= \sqrt{(x - 0)^2 + (300 - 75)^2} && \text{Substitute.} \\ 400^2 &= (x - 0)^2 + (300 - 75)^2 && \text{Square each side.} \\ 160{,}000 &= x^2 + 50{,}625 && \text{Simplify.} \\ x^2 &= 109{,}375 && \text{Solve for } x^2. \\ x &= \sqrt{109{,}375} && \text{Take the principal square root, } x > 0. \\ x &\approx 330.7 && \text{Approximate.} \end{aligned}$$

The x-coordinate is about 330.7.

Now Try Exercise **113**

Solving the Equation $x^n = k$

The equation $x^n = k$, where n is a *positive integer*, can be solved by taking the nth root of each side of the equation. The following technique allows us to find all *real* solutions to this equation.

SOLVING THE EQUATION $x^n = k$

Take the nth root of each side of $x^n = k$ to obtain $\sqrt[n]{x^n} = \sqrt[n]{k}$.

1. If n is *odd*, then $\sqrt[n]{x^n} = x$ and the equation becomes $x = \sqrt[n]{k}$.
2. If n is *even* and $k > 0$, then $\sqrt[n]{x^n} = |x|$ and the equation becomes $|x| = \sqrt[n]{k}$.

   (If $k < 0$, there are no real solutions.)

To understand this technique better, consider the following examples. First let $x^3 = 8$, so that n is odd. Taking the cube root of each side gives

$$\sqrt[3]{x^3} = \sqrt[3]{8}, \quad \text{which is equivalent to} \quad x = \sqrt[3]{8}, \quad \text{or} \quad x = 2.$$

Next let $x^2 = 4$, so that n is even. Taking the square root of each side gives

$$\sqrt{x^2} = \sqrt{4}, \quad \text{which is equivalent to} \quad |x| = \sqrt{4}, \quad \text{or} \quad |x| = 2.$$

The solutions to $|x| = 2$ are -2 or 2, which can be written ± 2.

EXAMPLE 11 **Solving the equation $x^n = k$**

Solve each equation.
(a) $x^3 = -64$ (b) $x^2 = 12$ (c) $2(x - 1)^4 = 32$

Solution
(a) Taking the cube root of each side of $x^3 = -64$ gives

$$\sqrt[3]{x^3} = \sqrt[3]{-64} \quad \text{or} \quad x = -4.$$

(b) Taking the square root of each side of $x^2 = 12$ gives

$$\sqrt{x^2} = \sqrt{12} \quad \text{or} \quad |x| = \sqrt{12}.$$

The equation $|x| = \sqrt{12}$ is equivalent to $x = \pm\sqrt{12}$.

(c) First divide each side of $2(x - 1)^4 = 32$ by 2 to isolate the power of $x - 1$.

$(x - 1)^4 = 16$	Divide each side by 2.
$\sqrt[4]{(x - 1)^4} = \sqrt[4]{16}$	Take the 4th root of each side.
$\lvert x - 1 \rvert = 2$	Simplify: $\sqrt[4]{y^4} = \lvert y \rvert$.
$x - 1 = -2 \text{ or } x - 1 = 2$	Solve the absolute value equation.
$x = -1 \text{ or } x = 3$	Add 1 to each side.

Now Try Exercises 47, 55, 67

▶ **REAL-WORLD CONNECTION** In some parts of the United States, wind power is used to generate electricity. Suppose that the diameter of the circular path created by the blades of a wind-powered generator is 8 feet. Then the wattage W generated by a wind velocity of v miles per hour is modeled by

$$W(v) = 2.4v^3.$$

If the wind blows at 10 miles per hour, the generator can produce about

$$W(\mathbf{10}) = 2.4 \cdot \mathbf{10}^3 = 2400 \text{ watts.}$$

(**Source:** *Conquering the Sciences*, Sharp Electronics.)

EXAMPLE **12**   Modeling a wind generator

The formula $W(v) = 2.4v^3$ is used to calculate the watts generated when there is a wind velocity of v miles per hour.
(a) Find a function f that calculates the wind velocity when W watts are being produced.
(b) If the wattage doubles, has the wind velocity also doubled? Explain.

Solution
(a) Given W we need a formula to find v, so solve $W = 2.4v^3$ for v.

$$W = 2.4v^3 \qquad \text{Given formula}$$

$$\frac{W}{2.4} = v^3 \qquad \text{Divide by 2.4.}$$

$$\sqrt[3]{\frac{W}{2.4}} = \sqrt[3]{v^3} \qquad \text{Take the cube root of each side.}$$

$$v = \sqrt[3]{\frac{W}{2.4}} \qquad \text{Simplify and rewrite equation.}$$

Thus $f(W) = \sqrt[3]{\frac{W}{2.4}}$.

(b) Suppose that the power generated is 1000 watts. Then the wind speed is

$$f(1000) = \sqrt[3]{\frac{1000}{2.4}} \approx 7.5 \text{ miles per hour.}$$

If the power doubles to 2000 watts, then the wind speed is

$$f(2000) = \sqrt[3]{\frac{2000}{2.4}} \approx 9.4 \text{ miles per hour.}$$

Thus for the wattage to double, the wind speed does not need to double.

Now Try Exercise **131**

10.5 PUTTING IT ALL TOGETHER

In this section we focused on equations that contain either radical expressions or rational exponents. A good strategy for solving an equation containing radical expressions *symbolically* is to raise each side of the equation to the same positive integer power. However, checking answers is important to eliminate extraneous solutions. Note that these extraneous solutions do not occur when an equation is being solved graphically or numerically.

Concept	Description	Example
Power Rule for Solving Equations	If each side of an equation is raised to the same positive integer power, any solutions to the given equation are among the solutions to the new equation.	$\sqrt{2x} = x$ $2x = x^2$ Square each side. $x^2 - 2x = 0$ Rewrite equation. $x = 0 \quad \text{or} \quad x = 2$ Factor and solve. Be sure to check any solutions.

continued on next page

continued from previous page

Concept	Description	Example
Pythagorean Theorem	If c is the hypotenuse of a right triangle and a and b are its legs, then $a^2 + b^2 = c^2$.	If the sides of the right triangle are $a = 5$, $b = 12$, and $c = 13$, then they satisfy $a^2 + b^2 = c^2$ or $5^2 + 12^2 = 13^2$.
Distance Formula	The distance d between the points (x_1, y_1) and (x_2, y_2) is $$d = \sqrt{(x_2 - x_1)^2 + (y_2 - y_1)^2}.$$	The distance between the points $(2, 3)$ and $(-3, 4)$ is $$d = \sqrt{(-3 - 2)^2 + (4 - 3)^2}$$ $$= \sqrt{(-5)^2 + (1)^2} = \sqrt{26}.$$
Solving the Equation $x^n = k$, Where n Is a Positive Integer	Take the nth root of each side to obtain $\sqrt[n]{x^n} = \sqrt[n]{k}$. Then, 1. $x = \sqrt[n]{k}$, if n is odd. 2. $x = \pm \sqrt[n]{k}$, if n is even and $k \geq 0$.	1. n odd: If $x^5 = 32$, then $x = \sqrt[5]{32} = 2$. 2. n even: If $x^4 = 81$, then $x = \pm \sqrt[4]{81} = \pm 3$.

Equations involving radical expressions can be solved symbolically, numerically, and graphically. All symbolic solutions must be checked; *numerical and graphical methods do not give extraneous solutions.* These concepts are illustrated for the equation $\sqrt{x + 2} = x$, where $Y_1 = \sqrt{(X + 2)}$ and $Y_2 = X$.

Symbolic Solution

$$\sqrt{x + 2} = x$$
$$x + 2 = x^2$$
$$x^2 - x - 2 = 0$$
$$(x - 2)(x + 1) = 0$$
$$x = 2 \text{ or } x = -1$$

Check: $\sqrt{2 + 2} = 2$
$\sqrt{-1 + 2} \neq -1$

The only solution is 2.

Numerical Solution

X	Y₁	Y₂
-2	0	-2
-1	1	-1
0	1.4142	0
1	1.7321	1
■	2	2
3	2.2361	3
4	2.4495	4

X=2

$y_1 = y_2$ when $x = 2$, so 2 is a solution. Note that -1 is *not* a solution.

Graphical Solution

$[-4.7, 4.7, 1]$ by $[-3.1, 3.1, 1]$

Intersection
X=2   Y=2

The graphs intersect at $(2, 2)$. Note that there is no point of intersection when $x = -1$.

10.5 Exercises

CONCEPTS

1. What is a good first step for solving $\sqrt{4x - 1} = 5$?

2. What is a good first step for solving $\sqrt[3]{x + 1} = 6$?

3. Can an equation involving rational exponents have more than one solution?

4. When you square each side of an equation to solve for an unknown, what must you do with any answers?

5. What is the Pythagorean theorem used for?

6. If the legs of a right triangle are 3 and 4, what is the length of the hypotenuse?

7. What formula can you use to find the distance d between two points?

8. Write the equation $\sqrt{x} + \sqrt[4]{x^3} = 2$ with rational exponents.

SIMPLIFYING RADICALS

Exercises 9–16: Simplify. Assume radicands of square roots are positive.

9. $\sqrt{2} \cdot \sqrt{2}$

10. $\sqrt{5} \cdot \sqrt{5}$

11. $\sqrt{x} \cdot \sqrt{x}$

12. $\sqrt{2x} \cdot \sqrt{2x}$

13. $(\sqrt{2x + 1})^2$

14. $(\sqrt{7x})^2$

15. $(\sqrt[3]{5x^2})^3$

16. $(\sqrt[3]{2x - 5})^3$

SYMBOLIC SOLUTIONS

Exercises 17–46: Solve the equation symbolically. Check your results.

17. $\sqrt{x} = 8$

18. $\sqrt{3z} = 6$

19. $\sqrt[4]{x} = 3$

20. $\sqrt[3]{x - 4} = 2$

21. $\sqrt{2t + 4} = 4$

22. $\sqrt{y + 4} = 3$

23. $\sqrt{x + 1} - 3 = 4$

24. $\sqrt{2x + 5} + 2 = 5$

25. $2\sqrt{x - 2} + 1 = 5$

26. $-\sqrt{x + 7} - 1 = -7$

27. $\sqrt{x + 6} = x$

28. $\sqrt{z + 6} = z$

29. $\sqrt[3]{x} = 3$

30. $\sqrt[3]{x + 10} = 4$

31. $\sqrt[3]{2z - 4} = -2$

32. $\sqrt[3]{z - 1} = -3$

33. $\sqrt[4]{t + 1} = 2$

34. $\sqrt[4]{5t} = 5$

35. $\sqrt{5z - 1} = \sqrt{z + 1}$

36. $y = \sqrt{y + 1} + 1$

37. $\sqrt{1 - x} = 1 - x$

38. $\sqrt[3]{4x} = x$

39. $\sqrt{b^2 - 4} = b - 2$

40. $\sqrt{b^2 - 2b + 1} = b$

41. $\sqrt{1 - 2x} = x + 7$

42. $\sqrt{4 - y} = y - 2$

43. $\sqrt{x} = \sqrt{x - 5} + 1$

44. $\sqrt{x - 1} = \sqrt{x + 4} - 1$

45. $\sqrt{2t - 2} + \sqrt{t} = 7$

46. $\sqrt{x + 1} - \sqrt{x - 6} = 1$

SOLVING THE EQUATION $x^n = k$

Exercises 47–68: (Refer to Example 11.) Solve.

47. $x^2 = 49$

48. $x^2 = 9$

49. $2z^2 = 200$

50. $3z^2 = 48$

51. $(t + 1)^2 = 16$

52. $(t - 5)^2 = 81$

53. $(4 - 2x)^2 = 100$

54. $(3x - 6)^2 = 25$

55. $b^3 = 64$

56. $a^3 = 1000$

57. $2t^3 = -128$

58. $3t^3 = -81$

59. $(x + 1)^3 = 8$

60. $(4 - x)^3 = -1$

61. $(2 - 5z)^3 = -125$

62. $(2x + 4)^3 = 125$

63. $x^4 = 16$

64. $x^4 = 7$

65. $x^5 = 12$

66. $x^5 = -32$

67. $2(x + 2)^4 = 162$

68. $\frac{1}{2}(x - 1)^5 = 16$

GRAPHICAL SOLUTIONS

Exercises 69–78: Solve graphically. Approximate solutions to the nearest hundredth when appropriate.

69. $\sqrt[3]{x + 5} = 2$

70. $\sqrt[3]{x} + \sqrt{x} = 3.43$

71. $\sqrt{2x - 3} = \sqrt{x} - \frac{1}{2}$

72. $x^{4/3} - 1 = 2$

73. $x^{5/3} = 2 - 3x^2$ **74.** $x^{3/2} = \sqrt{x + 2} - 2$

75. $z^{1/3} - 1 = 2 - z$ **76.** $z^{3/2} - 2z^{1/2} - 1 = 0$

77. $\sqrt{y + 2} + \sqrt{3y + 2} = 2$

78. $\sqrt{x + 1} - \sqrt{x - 1} = 4$

USING MORE THAN ONE METHOD

Exercises 79–82: Solve the equation
   (a) symbolically,
   (b) graphically, and
   (c) numerically.

79. $2\sqrt{x} = 8$ **80.** $\sqrt[3]{5 - x} = 2$

81. $\sqrt{6z - 2} = 8$ **82.** $\sqrt{y + 4} = \dfrac{y}{3}$

SOLVING AN EQUATION FOR A VARIABLE

Exercises 83–86: Solve for the indicated variable.

83. $T = 2\pi\sqrt{\dfrac{L}{32}}$ for L

84. $Z = \sqrt{L^2 + R^2}$ for R

85. $r = \sqrt{\dfrac{A}{\pi}}$ for A

86. $F = \dfrac{1}{2\pi\sqrt{LC}}$ for C

PYTHAGOREAN THEOREM

Exercises 87–94: If the sides of a triangle are a, b, and c and they satisfy $a^2 + b^2 = c^2$, the triangle is a right triangle. Determine whether the triangle with the given sides is a right triangle.

87. $a = 6$ $b = 8$ $c = 10$

88. $a = 5$ $b = 12$ $c = 13$

89. $a = \sqrt{5}$ $b = \sqrt{9}$ $c = \sqrt{14}$

90. $a = 4$ $b = 5$ $c = 7$

91. $a = 7$ $b = 24$ $c = 25$

92. $a = 1$ $b = \sqrt{3}$ $c = 2$

93. $a = 8$ $b = 8$ $c = 16$

94. $a = 11$ $b = 60$ $c = 61$

Exercises 95–98: Find the length of the missing side in the right triangle.

95.

96.

97.

98.

Exercises 99–104: A right triangle has legs a and b with hypotenuse c. Find the length of the missing side.

99. $a = 3, b = 4$ **100.** $a = 4, b = 7$

101. $a = \sqrt{3}, c = 8$ **102.** $a = \sqrt{6}, c = \sqrt{10}$

103. $b = 48, c = 50$ **104.** $b = 10, c = 26$

DISTANCE FORMULA

Exercises 105–108: Find the length of the line segment.

105.

106.

107.

108.

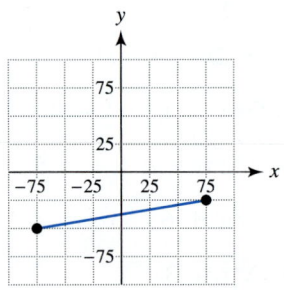

Exercises 109–112: Find the distance between the points.

109. $(-1, 2)$, $(4, 10)$ **110.** $(5, -40)$, $(-6, 20)$

111. $(0, -3)$, $(4, 0)$ **112.** $(3, 9)$, $(-4, 2)$

Exercises 113–116: (Refer to Example 10.) Find x if the distance between the points is d. Assume that $x \geq 0$.

113. $(x, 3)$, $(0, 6)$ $d = 5$

114. $(x, -1)$, $(6, 11)$ $d = 13$

115. $(x, -5)$, $(62, 6)$ $d = 61$

116. $(x, 3)$, $(12, -4)$ $d = 25$

APPLICATIONS

Exercises 117 and 118: **Weight of a Bird** *(Refer to Example 6.) Estimate the weight W of a bird having wings of area A. Let $A = 100\sqrt[3]{W^2}$.*

117. $A = 400$ square inches

118. $A = 1000$ square inches

Exercises 119–122: **Distance to the Horizon** *Because of Earth's curvature, a person can see a limited distance to the horizon. The higher the location of the person, the farther that person can see. The distance D in miles to the horizon can be estimated by $D(h) = 1.22\sqrt{h}$, where h is the height of the person above the ground in feet.*

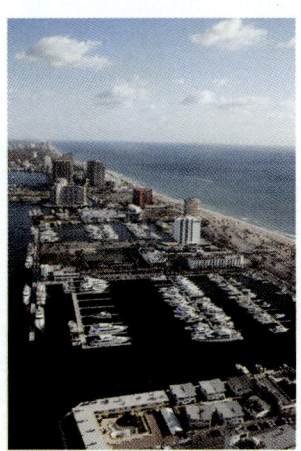

119. Find D for a 6-foot-tall person standing on level ground.

120. Find D for a person on top of Mount Everest with a height of 29,028 feet.

121. How high does a person need to be to see 20 miles?

122. How high does a plane need to fly for the pilot to be able to see 100 miles?

123. *Diagonal of a Television*   (Refer to Example 8.) A rectangular television screen is 11.4 inches by 15.2 inches. Find the diagonal of the television set.

124. *Dimensions of a Television*   The height of a television with a 13-inch diagonal is $\frac{3}{4}$ of its width. Find the width and height of the television set.

125. *DVD and Picture Dimensions*   If the picture shown on a television set is h units high and w units wide, the *aspect ratio* of the picture is $\frac{w}{h}$ (see the accompanying figure). Digital video discs support the newer aspect ratio of $\frac{16}{9}$ rather than the older ratio of $\frac{4}{3}$. If the width of a picture with an aspect ratio of $\frac{16}{9}$ is 29 inches, approximate the height and diagonal of the rectangular picture. (*Source:* J. Taylor, *DVD Demystified.*)

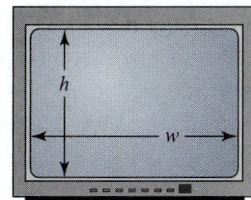

126. *Flood Control*   The spillway capacity of a dam is important in flood control. Spillway capacity Q in cubic feet of water per second flowing over the spillway depends on the width W and the depth D of the spillway, as illustrated in the accompanying figure. If W and D are measured in feet, capacity can be modeled by $Q = 3.32WD^{3/2}$.   (*Source:* D. Callas, Project Director, *Snapshots of Applications in Mathematics.*)

  (a) Find the capacity of a spillway with $W = 20$ feet and $D = 5$ feet.

  (b) A spillway with a width of 30 feet is to have a capacity of $Q = 2690$ cubic feet per second. Estimate to the nearest foot the appropriate depth of the spillway.

127. *Sky Diving* When sky divers initially fall from an airplane, their velocity v in miles per hour after free falling d feet can be approximated by $v = \frac{60}{11}\sqrt{d}$. (Because of air resistance, they will eventually reach a terminal velocity.) How far do sky divers need to fall to attain the following velocities? (These values for d represent minimum distances.)

(a) 60 miles per hour   **(b)** 100 miles per hour

128. *Guy Wire* A guy wire attached to the top of a 30-foot-long pole is anchored 10 feet from the base of the pole, as illustrated in the figure. Find the length of the guy wire to the nearest tenth of a foot.

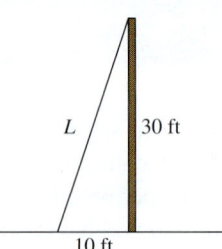

129. *Skid Marks* Vehicles involved in accidents often leave skid marks, which can be used to determine how fast a vehicle was traveling. To determine this speed, officials often use a test vehicle to compare skid marks on the same section of road. Suppose that a vehicle in a crash left skid marks D feet long and that a test vehicle traveling at v miles per hour leaves skid marks d feet long. Then the speed V of the vehicle involved in the crash is given by

$$V = v\sqrt{\frac{D}{d}}.$$

(**Source:** N. Garber and L. Hoel, *Traffic and Highway Engineering.*)

(a) Find V if $v = 30$ mph, $D = 285$ feet, and $d = 178$ feet. Interpret your result.

(b) A test vehicle traveling at 45 mph leaves skid marks 255 feet long. How long would the skid marks be for a vehicle traveling 60 miles per hour?

130. *Highway Curves* If a circular curve without any banking has a radius of R feet, the speed limit L in miles per hour for the curve is $L = 1.5\sqrt{R}$. (**Source:** N. Garber.)

(a) Find the speed limit for a curve having a radius of 400 feet.

(b) If the radius of a curve doubles, what happens to the speed limit?

(c) A curve with a 40-mile-per-hour speed limit is being designed. What should be its radius?

131. *Wind Power* (Refer to Example 12.) If a wind-powered generator has blades that create a circular path with a diameter of 10 feet, then the wattage W generated by a wind velocity of v miles per hour is modeled by $W(v) = 3.8v^3$.

(a) If the wind velocity doubles, what happens to the wattage generated?

(b) Solve $W = 3.8v^3$ for v.

(c) If the wind generator is producing 30,400 watts, find the wind speed.

132. *Height and Weight* Suppose that the weight of a person is directly proportional to the cube of the person's height. If one person weighs twice as much as a (similarly proportioned) second person, by what factor is the heavier person's height greater than the shorter person's height?

133. *45°–45° Right Triangle* Suppose that the legs of a right triangle with angles of 45° and 45° both have length a, as depicted in the accompanying figure. Find the length of the hypotenuse.

134. *30°–60° Right Triangle* In a right triangle with angles of 30° and 60°, the shortest side is half the length of the hypotenuse (see the accompanying figure). If the hypotenuse has length c, find the length of the other two sides, a and b, in terms of c.

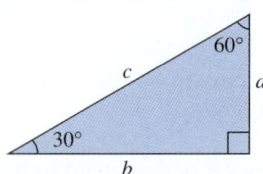

WRITING ABOUT MATHEMATICS

135. A student solves an equation *incorrectly* as follows.

$$\sqrt{3 - x} = \sqrt{x} - 1$$
$$(\sqrt{3 - x})^2 \stackrel{?}{=} (\sqrt{x})^2 - (1)^2$$
$$3 - x \stackrel{?}{=} x - 1$$
$$-2x \stackrel{?}{=} -4$$
$$x \stackrel{?}{=} 2$$

(a) How could you convince the student that the answer is wrong?

(b) Discuss where any errors were made.

136. When each side of an equation is squared, you must check your results. Explain why.

GROUP ACTIVITY
WORKING WITH REAL DATA

Directions: Form a group of 2 to 4 people. Select someone to record the group's responses for this activity. All members of the group should work cooperatively to answer the questions. If your instructor asks for your results, each member of the group should be prepared to respond.

Simple Pendulum Gravity is responsible for an object falling toward Earth. The farther the object falls, the faster it is moving when it hits the ground. For each second that an object falls, its speed increases by a constant amount, called the *acceleration due to gravity*, denoted g. One way to calculate the value of g is to use a simple pendulum. See the accompanying figure.

The time T for a pendulum to swing back and forth once is called its *period* and is given by

$$T = 2\pi\sqrt{\frac{L}{g}},$$

where L equals the length of the pendulum. The table lists the periods of pendulums with different lengths.

L (feet)	0.5	1.0	1.5
T (seconds)	0.78	1.11	1.36

(a) Solve the formula for g.
(b) Use the table to determine the value of g. (*Note:* The units for g are feet per second per second.)
(c) Interpret your result.

10.6 COMPLEX NUMBERS

Basic Concepts ■ Addition, Subtraction, and Multiplication ■ Powers of i ■ Complex Conjugates and Division

A LOOK INTO MATH ▷ Mathematics is both applied and theoretical. A common misconception is that abstract or theoretical mathematics is unimportant in today's world. Many new ideas with great practical importance were first developed as abstract concepts with no particular application in mind. For example, complex numbers, which are related to square roots of negative numbers, started as an abstract concept to solve equations. Today complex numbers are used in many sophisticated applications, such as the design of electrical circuits, ships, and airplanes. Even the *fractal image* shown to the left would not have been discovered without complex numbers.

Basic Concepts

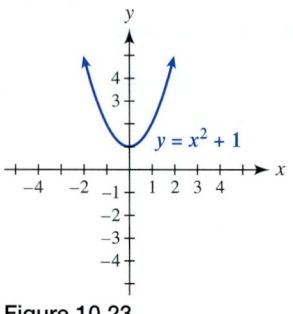

Figure 10.23

A graph of $y = x^2 + 1$ is shown in Figure 10.23. There are no x-intercepts, so the equation $x^2 + 1 = 0$ has no real number solutions.

If we try to solve $x^2 + 1 = 0$ by subtracting 1 from each side, the result is $x^2 = -1$. Because $x^2 \geq 0$ for any real number x, there are no real solutions. However, mathematicians have invented solutions.

$$x^2 = -1$$
$$x = \pm \sqrt{-1} \qquad \text{Solve for } x.$$

We now define a number called the **imaginary unit**, denoted i.

PROPERTIES OF THE IMAGINARY UNIT i

$$i = \sqrt{-1} \quad \text{and} \quad i^2 = -1$$

By creating the number i, the solutions to the equation $x^2 + 1 = 0$ are i and $-i$. Using the real numbers and the imaginary unit i, we can define a new set of numbers called the *complex numbers*. A **complex number** can be written in **standard form**, as $a + bi$, where a and b are real numbers. The **real part** is a and the **imaginary part** is b. Every real number a is also a complex number because it can be written $a + 0i$. A complex number $a + bi$ with $b \neq 0$ is an **imaginary number**. A complex number $a + bi$ with $a = 0$ and $b \neq 0$ is sometimes called a **pure imaginary number**. Examples include $4i$ and $-2i$. Table 10.8 lists several complex numbers with their real and imaginary parts.

TABLE 10.8

Complex Number: $a + bi$	$-3 + 2i$	5	$-3i$	$-1 + 7i$	$-5 - 2i$	$4 + 6i$
Real Part: a	-3	5	0	-1	-5	4
Imaginary Part: b	2	0	-3	7	-2	6

Figure 10.24 shows how different sets of numbers are related. Note that *the set of complex numbers contains the set of real numbers*.

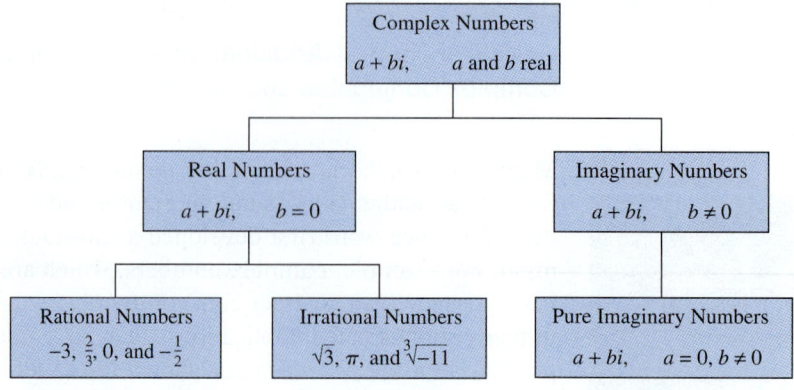

Figure 10.24

Using the imaginary unit i, we may write the square root of a negative number as a complex number. For example, $\sqrt{-2} = i\sqrt{2}$, and $\sqrt{-4} = i\sqrt{4} = 2i$. This method is summarized as follows.

CALCULATOR HELP

To set your calculator in $a + bi$ mode, see Appendix A (page AP-11).

THE EXPRESSION $\sqrt{-a}$

If $a > 0$, then $\sqrt{-a} = i\sqrt{a}$.

NOTE: Although it is standard for a complex number to be expressed as $a + bi$, we often write $\sqrt{2}i$ as $i\sqrt{2}$ so that it is clear that i is *not under the square root*. Similarly $\frac{1}{2}\sqrt{2}i$ is sometimes written $\frac{1}{2}i\sqrt{2}$ or $\frac{i\sqrt{2}}{2}$.

EXAMPLE 1 Writing the square root of a negative number

Write each square root using the imaginary unit i.

(a) $\sqrt{-25}$ **(b)** $\sqrt{-7}$ **(c)** $\sqrt{-20}$

Solution

(a) $\sqrt{-25} = i\sqrt{25} = 5i$ **(b)** $\sqrt{-7} = i\sqrt{7}$

(c) $\sqrt{-20} = i\sqrt{20} = i\sqrt{4}\sqrt{5} = 2i\sqrt{5}$

Now Try Exercises **13, 15, 19**

Addition, Subtraction, and Multiplication

Addition can be defined for complex numbers in a similar manner to how we add binomials. For example,

$$(-3 + 2x) + (2 - x) = (-3 + 2) + (2x - x) = -1 + x.$$

ADDITION AND SUBTRACTION To add the complex numbers $(-3 + 2i)$ and $(2 - i)$ add the real parts and then add the imaginary parts.

$$(-3 + 2i) + (2 - i) = (-3 + 2) + (2i - i)$$
$$= (-3 + 2) + (2 - 1)i$$
$$= -1 + i$$

This same process works for subtraction.

$$(6 - 3i) - (2 + 5i) = (6 - 2) + (-3i - 5i)$$
$$= (6 - 2) + (-3 - 5)i$$
$$= 4 - 8i$$

This method is summarized as follows.

SUM OR DIFFERENCE OF COMPLEX NUMBERS

Let $a + bi$ and $c + di$ be two complex numbers. Then

$$(a + bi) + (c + di) = (a + c) + (b + d)i \qquad \text{Sum}$$

and

$$(a + bi) - (c + di) = (a - c) + (b - d)i. \qquad \text{Difference}$$

EXAMPLE 2   Adding and subtracting complex numbers

Write each sum or difference in standard form.
(a) $(-7 + 2i) + (3 - 4i)$ **(b)** $3i - (5 - i)$

Solution
(a) $(-7 + 2i) + (3 - 4i) = (-7 + 3) + (2 - 4)i = -4 - 2i$
(b) $3i - (5 - i) = 3i - 5 + i = -5 + (3 + 1)i = -5 + 4i$

Now Try Exercises 23, 29

TECHNOLOGY NOTE: Complex Numbers

Many calculators can perform arithmetic with complex numbers. The figure shows a calculator display for the results in Example 2.

CALCULATOR HELP

To access the imaginary unit i, see Appendix A (page AP-11).

```
(-7+2i)+(3-4i)
               -4-2i
3i-(5-i)
               -5+4i
```

MULTIPLICATION   We multiply two complex numbers in the same way that we multiply binomials and then we apply the property $i^2 = -1$. For example,

$$(2 - 3x)(1 + 4x) = 2 + 8x - 3x - 12x^2 = 2 + 5x - 12x^2.$$

In the next example we find the product of $2 - 3i$ and $1 + 4i$ in a similar manner.

EXAMPLE 3   Multiplying complex numbers

Write each product in standard form.
(a) $(2 - 3i)(1 + 4i)$ **(b)** $(5 - 2i)(5 + 2i)$

Solution
(a) Multiply the complex numbers like binomials.

$$(2 - 3i)(1 + 4i) = (2)(1) + (2)(4i) - (3i)(1) - (3i)(4i)$$
$$= 2 + 8i - 3i - 12i^2$$
$$= 2 + 5i - 12(-1)$$
$$= 14 + 5i$$

```
(2-3i)(1+4i)
               14+5i
(5-2i)(5+2i)
               29
```

(b)
$$(5 - 2i)(5 + 2i) = (5)(5) + (5)(2i) - (2i)(5) - (2i)(2i)$$
$$= 25 + 10i - 10i - 4i^2$$
$$= 25 - 4(-1)$$
$$= 29$$

Figure 10.25

These results are supported in Figure 10.25.

Now Try Exercises 31, 35

Powers of i

An interesting pattern appears when powers of i are calculated.

$$i^1 = i$$
$$i^2 = -1$$
$$i^3 = i^2 \cdot i = -1 \cdot i = -i$$
$$i^4 = i^2 \cdot i^2 = (-1)(-1) = 1$$
$$i^5 = i^4 \cdot i = (1)i = i$$
$$i^6 = i^4 \cdot i^2 = (1)(-1) = -1$$
$$i^7 = i^4 \cdot i^3 = (1)(-i) = -i$$
$$i^8 = i^4 \cdot i^4 = (1)(1) = 1$$

The powers of i cycle with the pattern i, -1, $-i$, and 1. These examples suggest the following method for calculating powers of i.

POWERS OF i

The value of i^n can be found by dividing n (a positive integer) by 4. If the remainder is r, then

$$i^n = i^r.$$

Note that $i^0 = 1$, $i^1 = i$, $i^2 = -1$, and $i^3 = -i$.

EXAMPLE 4 Calculating powers of i

Evaluate each expression.
(a) i^9 **(b)** i^{19} **(c)** i^{40}

Solution
(a) When 9 is divided by 4, the result is 2 with remainder **1**. Thus $i^9 = i^1 = i$.
(b) When 19 is divided by 4, the result is 4 with remainder **3**. Thus $i^{19} = i^3 = -i$.
(c) When 40 is divided by 4, the result is 10 with remainder **0**. Thus $i^{40} = i^0 = 1$.

Now Try Exercises 49, 51, 55

Complex Conjugates and Division

The **complex conjugate** of $a + bi$ is $a - bi$. To find the conjugate, we change the sign of the imaginary part b. Table 10.9 contains some complex numbers and their conjugates.

TABLE 10.9 Complex Conjugates

Number	$2 + 5i$	$6 - 3i$	$-2 + 7i$	$-1 - i$	5	$-4i$
Conjugate	$2 - 5i$	$6 + 3i$	$-2 - 7i$	$-1 + i$	5	$4i$

NOTE: The product of a complex number and its conjugate is a real number. That is,

$$(a + bi)(a - bi) = a^2 + b^2.$$

For example, $(3 + 4i)(3 - 4i) = 3^2 + 4^2 = 25$, which is a real number.

This property of complex conjugates is used to divide two complex numbers. To convert the quotient $\frac{2 + 3i}{3 - i}$ into standard form $a + bi$, we multiply the numerator and the denominator by the complex conjugate of the *denominator*, which is $3 + i$. The next example illustrates this method.

EXAMPLE 5   **Dividing complex numbers**

Write each quotient in standard form.

(a) $\dfrac{2 + 3i}{3 - i}$ **(b)** $\dfrac{4}{2i}$

Solution

(a) Multiply the numerator and denominator by $3 + i$.

$$\frac{2 + 3i}{3 - i} = \frac{(2 + 3i)(3 + i)}{(3 - i)(3 + i)} \qquad \text{Multiply by 1.}$$

$$= \frac{2(3) + (2)(i) + (3i)(3) + (3i)(i)}{(3)(3) + (3)(i) - (i)(3) - (i)(i)} \qquad \text{Multiply.}$$

$$= \frac{6 + 2i + 9i + 3i^2}{9 + 3i - 3i - i^2} \qquad \text{Simplify.}$$

$$= \frac{6 + 11i + 3(-1)}{9 - (-1)} \qquad i^2 = -1$$

$$= \frac{3 + 11i}{10} \qquad \text{Simplify.}$$

$$= \frac{3}{10} + \frac{11}{10}i \qquad \frac{a + bi}{c} = \frac{a}{c} + \frac{b}{c}i$$

(b) Multiply the numerator and denominator by $-2i$.

$$\frac{4}{2i} = \frac{(4)(-2i)}{(2i)(-2i)} \qquad \text{Multiply by 1.}$$

$$= \frac{-8i}{-4i^2} \qquad \text{Simplify.}$$

$$= \frac{-8i}{-4(-1)} \qquad i^2 = -1$$

$$= \frac{-8i}{4} \qquad \text{Simplify.}$$

$$= -2i \qquad \text{Divide.}$$

```
(2+3i)/(3-i)▶Fra
c
        3/10+11/10i
4/(2i)
              -2i
```

Figure 10.26

These results are supported in Figure 10.26.

Now Try Exercises 71, 73

10.6 PUTTING IT ALL TOGETHER

In this section we discussed complex numbers and how to perform arithmetic operations with them. Complex numbers allow us to solve equations that could not be solved only with real numbers. The following table summarizes the important concepts in the section.

Concept	Explanation	Examples
Complex Numbers	A complex number can be expressed as $a + bi$, where a and b are real numbers. The imaginary unit i satisfies $i = \sqrt{-1}$ and $i^2 = -1$. As a result, we can write $\sqrt{-a} = i\sqrt{a}$ if $a > 0$.	The real part of $5 - 3i$ is 5 and the imaginary part is -3. $\sqrt{-13} = i\sqrt{13}$ and $\sqrt{-9} = 3i$
Addition, Subtraction, and Multiplication	To add (subtract) complex numbers, add (subtract) the real parts and then add (subtract) the imaginary parts. Multiply complex numbers in a similar manner to how *FOIL* is used to multiply binomials. Then apply the property $i^2 = -1$.	$(3 + 6i) + (-1 + 2i)$ Sum $\quad = (3 + -1) + (6 + 2)i$ $\quad = 2 + 8i$ $(2 - 5i) - (1 + 4i)$ Difference $\quad = (2 - 1) + (-5 - 4)i$ $\quad = 1 - 9i$ $(-1 + 2i)(3 + i)$ Product $\quad = (-1)(3) + (-1)(i) + (2i)(3) + (2i)(i)$ $\quad = -3 - i + 6i + 2i^2$ $\quad = -3 + 5i + 2(-1)$ $\quad = -5 + 5i$
Complex Conjugates	The conjugate of $a + bi$ is $a - bi$.	The conjugate of $3 - 5i$ is $3 + 5i$. The conjugate of $2i$ is $-2i$.
Division	To simplify a quotient, multiply the numerator and denominator by the complex conjugate of the *denominator*. Then simplify the expression and write it in standard form as $a + bi$.	$\dfrac{10}{1 + 2i} = \dfrac{10(1 - 2i)}{(1 + 2i)(1 - 2i)}$ Quotient $\quad = \dfrac{10 - 20i}{5}$ $\quad = 2 - 4i$

10.6 Exercises

MyMathLab   PRACTICE   WATCH   DOWNLOAD   READ   REVIEW

CONCEPTS

1. Give an example of a complex number that is not a real number.

2. Can you give an example of a real number that is not a complex number? Explain.

3. $\sqrt{-1} =$ _____

4. $i^2 =$ _____

5. $\sqrt{-a} =$ _____, if $a > 0$.

6. The complex conjugate of $10 + 7i$ is _____.

7. The standard form for a complex number is _____.

8. Write $\frac{2 + 4i}{2}$ in standard form.

9. The real part of $4 - 5i$ is _____.

10. The imaginary part of $4 - 5i$ is _____.

11. The imaginary part of -7 is _____.

12. The number $7i$ is called a(n) _____ imaginary number.

COMPLEX NUMBERS

Exercises 13–22: Use i to write the expression.

13. $\sqrt{-5}$ **14.** $\sqrt{-21}$

15. $\sqrt{-100}$ **16.** $\sqrt{-49}$

17. $\sqrt{-144}$ **18.** $\sqrt{-64}$

19. $\sqrt{-12}$ **20.** $\sqrt{-8}$

21. $\sqrt{-18}$ **22.** $\sqrt{-48}$

Exercises 23–46: Write the expression in standard form.

23. $(5 + 3i) + (-2 - 3i)$

24. $(1 - i) + (5 - 7i)$

25. $2i + (-8 + 5i)$ **26.** $-3i + 5i$

27. $(2 - 7i) - (1 + 2i)$ **28.** $(1 + 8i) - (3 + 9i)$

29. $5i - (10 - 2i)$ **30.** $-3(4 - 3i)$

31. $(3 + 2i)(-1 + 5i)$ **32.** $(1 + i) - (1 - i)$

33. $4(5 - 3i)$ **34.** $(1 + 2i)(-6 - i)$

35. $(5 + 4i)(5 - 4i)$ **36.** $(3 + 5i)(3 - 5i)$

37. $(-4i)(5i)$ **38.** $(-6i)(-4i)$

39. $3i + (2 - 3i) - (1 - 5i)$

40. $4 - (5 - 7i) + (3 + 7i)$

41. $(2 + i)^2$ **42.** $(-1 + 2i)^2$

43. $2i(-3 + i)$ **44.** $5i(1 - 9i)$

45. $i(1 + i)^2$ **46.** $2i(1 - i)^2$

Exercises 47 and 48: **Thinking Generally**   *Write in standard form.*

47. $(a + 3bi)(a - 3bi)$ **48.** $(a + bi) - (a - bi)$

Exercises 49–56: (Refer to Example 4.) Simplify.

49. i^{11} **50.** i^{50}

51. i^{21} **52.** i^{103}

53. i^{58} **54.** i^{61}

55. i^{64} **56.** i^{28}

Exercises 57–64: Write the complex conjugate.

57. $3 + 4i$ **58.** $1 - 4i$

59. $-6i$ **60.** -10

61. $5 - 4i$ **62.** $7 + 2i$

63. -1 **64.** $19i$

Exercises 65–78: Write the expression in standard form.

65. $\dfrac{2}{1 + i}$ **66.** $\dfrac{-6}{2 - i}$

67. $\dfrac{3i}{5 - 2i}$ **68.** $\dfrac{-8}{2i}$

69. $\dfrac{8 + 9i}{5 + 2i}$ **70.** $\dfrac{3 - 2i}{1 + 4i}$

71. $\dfrac{5 + 7i}{1 - i}$ **72.** $\dfrac{-7 + 4i}{3 - 2i}$

73. $\dfrac{2-i}{i}$ **74.** $\dfrac{3+2i}{-i}$

75. $\dfrac{1}{i}+\dfrac{1}{2i}$ **76.** $\dfrac{3}{4i}+\dfrac{2}{i}$

77. $\dfrac{1}{-1+i}-\dfrac{2}{i}$ **78.** $-\dfrac{3}{2i}-\dfrac{2}{1+i}$

APPLICATIONS

Exercises 79 and 80: **Corrosion in Airplanes** *Corrosion in the metal surface of an airplane can be difficult to detect visually. One test used to locate it involves passing an alternating current through a small area on the plane's surface. If the current varies from one region to another, it may indicate that corrosion is occurring. The impedance Z (or opposition to the flow of electricity) of the metal is related to the voltage V and current I by the equation $Z = \frac{V}{I}$, where Z, V, and I are complex numbers. Calculate Z for the given values of V and I.* (**Source:** Society for Industrial and Applied Mathematics.)

79. $V = 40 + 70i, I = 2 + 3i$

80. $V = 10 + 20i, I = 3 + 7i$

WRITING ABOUT MATHEMATICS

81. A student multiplies $(2 + 3i)(4 - 5i)$ *incorrectly* to obtain $8 - 15i$. What is the student's mistake?

82. A student divides $\frac{6 - 10i}{3 + 2i}$ *incorrectly* to obtain $2 - 5i$. What is the student's mistake?

CHECKING BASIC CONCEPTS
SECTIONS 10.5 AND 10.6

1. Solve each equation. Check your answers.
  (a) $\sqrt{2x - 4} = 2$ **(b)** $\sqrt[3]{x - 1} = 3$
  (c) $\sqrt{3x} = 1 + \sqrt{x + 1}$

2. Find the distance between $(-3, 5)$ and $(2, -7)$.

3. A 16-inch diagonal television set has a rectangular picture with a width of 12.8 inches. Find the height of the picture.

4. Solve $(x + 1)^4 = 16$.

5. Use i to write each expression.
  (a) $\sqrt{-64}$ **(b)** $\sqrt{-17}$

6. Simplify each expression.
  (a) $(2 - 3i) + (1 - i)$
  (b) $4i - (2 + i)$
  (c) $(3 - 2i)(1 + i)$
  (d) $\dfrac{3}{2 - 2i}$

CHAPTER 10   SUMMARY
SECTION 10.1 ■ RADICAL EXPRESSIONS AND RATIONAL EXPONENTS

Radicals and Radical Notation

Square Root b is a square root of a if $b^2 = a$.

Principal Square Root $\sqrt{a} = b$ if $b^2 = a$ and $b \geq 0$.

  Examples: $\sqrt{16} = 4$, $-\sqrt{9} = -3$,
        and $\pm\sqrt{36} = \pm 6$

Cube Root

b is a cube root of a if $b^3 = a$.

Examples: $\sqrt[3]{27} = 3$, $\sqrt[3]{-8} = -2$

nth Root

b is an nth root of a if $b^n = a$.

Example: $\sqrt[4]{16} = 2$ because $2^4 = 16$.

NOTE: An *even* root of a *negative* number is not a real number. Also, $\sqrt[n]{a}$ denotes the *principal* nth root.

Absolute Value

The expressions $|x|$ and $\sqrt{x^2}$ are equivalent.

Example: $\sqrt{(x + y)^2} = |x + y|$

The Expression $a^{1/n}$

$a^{1/n} = \sqrt[n]{a}$ if n is an integer greater than 1.

Examples: $5^{1/2} = \sqrt{5}$ and $64^{1/3} = \sqrt[3]{64} = 4$

The Expression $a^{m/n}$

$a^{m/n} = \sqrt[n]{a^m}$ or $a^{m/n} = (\sqrt[n]{a})^m$

Examples: $8^{2/3} = \sqrt[3]{8^2} = \sqrt[3]{64} = 4$ and

$$8^{2/3} = (\sqrt[3]{8})^2 = (2)^2 = 4$$

Properties of Exponents

Product Rule

$a^p a^q = a^{p+q}$

Negative Exponents

$a^{-p} = \dfrac{1}{a^p},\ \dfrac{1}{a^{-p}} = a^p$

Negative Exponents for Quotients

$\left(\dfrac{a}{b}\right)^{-p} = \left(\dfrac{b}{a}\right)^{p}$

Quotient Rule for Exponents

$\dfrac{a^p}{a^q} = a^{p-q}$

Power Rule for Exponents

$(a^p)^q = a^{pq}$

Power Rule for Products

$(ab)^p = a^p b^p$

Power Rule for Quotients

$\left(\dfrac{a}{b}\right)^p = \dfrac{a^p}{b^p}$

SECTION 10.2 ■ SIMPLIFYING RADICAL EXPRESSIONS

Product Rule for Radical Expressions   Provided each expression is defined,

$$\sqrt[n]{a} \cdot \sqrt[n]{b} = \sqrt[n]{a \cdot b}.$$

Example: $\sqrt[3]{3} \cdot \sqrt[3]{9} = \sqrt[3]{27} = 3$

Perfect nth Power   An integer a is a perfect nth power if $b^n = a$ for some integer b.

Examples: 25 is a perfect square, 8 is a perfect cube, and 16 is a perfect fourth power.

Quotient Rule for Radical Expressions   Provided each expression is defined,

$$\sqrt[n]{\frac{a}{b}} = \frac{\sqrt[n]{a}}{\sqrt[n]{b}}.$$

Example: $\dfrac{\sqrt[3]{24}}{\sqrt[3]{3}} = \sqrt[3]{\dfrac{24}{3}} = \sqrt[3]{8} = 2$

SECTION 10.3 ■ OPERATIONS ON RADICAL EXPRESSIONS

Addition and Subtraction   Combine like radicals.

Examples: $2\sqrt[3]{4} + 3\sqrt[3]{4} = 5\sqrt[3]{4}$ and $\sqrt{5} - 2\sqrt{5} = -\sqrt{5}$

Multiplication   Sometimes radical expressions can be multiplied like binomials.

Examples: $(4 - \sqrt{2})(2 + \sqrt{2}) = 8 + 4\sqrt{2} - 2\sqrt{2} - 2 = 6 + 2\sqrt{2}$

$(5 - \sqrt{3})(5 + \sqrt{3}) = (5)^2 - (\sqrt{3})^2 = 25 - 3 = 22$ because

$(a - b)(a + b) = a^2 - b^2.$

Rationalizing the Denominator   One technique is to multiply the numerator and denominator by the conjugate of the denominator if the denominator is a binomial containing square roots.

Examples: $\dfrac{1}{4 + \sqrt{2}} = \dfrac{1}{(4 + \sqrt{2})} \cdot \dfrac{(4 - \sqrt{2})}{(4 - \sqrt{2})} = \dfrac{4 - \sqrt{2}}{(4)^2 - (\sqrt{2})^2} = \dfrac{4 - \sqrt{2}}{14}$

$\dfrac{4}{\sqrt{7}} = \dfrac{4}{\sqrt{7}} \cdot \dfrac{\sqrt{7}}{\sqrt{7}} = \dfrac{4\sqrt{7}}{7}$

SECTION 10.4 ■ RADICAL FUNCTIONS

The Square Root Function   The square root function is denoted $f(x) = \sqrt{x}$. Its domain is $\{x \mid x \geq 0\}$ and its graph is shown in the figure.

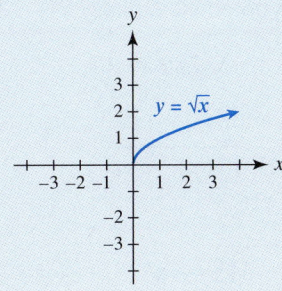

The Cube Root Function

The cube root function is denoted $f(x) = \sqrt[3]{x}$. Its domain is all real numbers and its graph is shown in the figure.

Power Functions

If a function can be defined by $f(x) = x^p$, where p is a rational number, then it is a power function.

Examples: $f(x) = x^{4/5}$ and $g(x) = x^{2.3}$

SECTION 10.5 ■ EQUATIONS INVOLVING RADICAL EXPRESSIONS

Power Rule for Solving Radical Equations

The solutions to $a = b$ are among the solutions to $a^n = b^n$, where n is a positive integer.

Example: The solutions to the equation $\sqrt{3x + 3} = 2x - 1$ are among the solutions to $3x + 3 = (2x - 1)^2$.

Solving Radical Equations

STEP 1: Isolate a radical term on one side of the equation.

STEP 2: Apply the power rule by raising each side of the equation to the power equal to the index of the isolated radical term.

STEP 3: Solve the equation. If it still contains a radical, repeat Steps 1 and 2.

STEP 4: Check your answers by substituting each result in the *given* equation.

Example: To isolate the radical in $\sqrt{x + 1} + 4 = 6$ subtract 4 from each side to obtain $\sqrt{x + 1} = 2$. Next square each side, which gives $x + 1 = 4$ or $x = 3$. Checking verifies that 3 is a solution.

Pythagorean Theorem

If a right triangle has legs a and b with hypotenuse c, then
$$a^2 + b^2 = c^2.$$

Example: If a right triangle has legs 8 and 15, then the hypotenuse equals
$$c = \sqrt{8^2 + 15^2} = \sqrt{289} = 17.$$

The Distance Formula

The distance d between (x_1, y_1) and (x_2, y_2) is
$$d = \sqrt{(x_2 - x_1)^2 + (y_2 - y_1)^2}.$$

Example: The distance between $(-1, 3)$ and $(4, 5)$ is
$$d = \sqrt{\left(4 - (-1)\right)^2 + (5 - 3)^2} = \sqrt{25 + 4} = \sqrt{29}.$$

Solving the Equation $x^n = k$ Let n be a positive integer.

Take the nth root of each side of $x^n = k$ to obtain $\sqrt[n]{x^n} = \sqrt[n]{k}$.

1. If n is *odd*, then $\sqrt[n]{x^n} = x$ and the equation becomes $x = \sqrt[n]{k}$.
2. If n is *even* and $k \geq 0$, then $\sqrt[n]{x^n} = |x|$ and the equation becomes $|x| = \sqrt[n]{k}$.

Examples: $x^3 = -27$ implies that $x = \sqrt[3]{-27} = -3$.
$x^4 = 81$ implies that $|x| = \sqrt[4]{81} = 3$ or $x = \pm 3$.

SECTION 10.6 ■ COMPLEX NUMBERS

Complex Numbers

Imaginary Unit	$i = \sqrt{-1}$ and $i^2 = -1$
Standard Form	$a + bi$, where a and b are real numbers
	Examples: $4 + 3i$, $5 - 6i$, 8, and $-2i$
Real Part	The real part of $a + bi$ is a.
	Example: The real part of $3 - 2i$ is 3.
Imaginary Part	The imaginary part of $a + bi$ is b.
	Example: The imaginary part of $2 - i$ is -1.
Arithmetic Operations	Arithmetic operations are similar to arithmetic operations on binomials.

Examples: $(2 + 2i) + (3 - i) = 5 + i,$
$(1 - i) - (1 - 2i) = i,$
$(1 - i)(1 + i) = 1^2 - i^2 = 1 - (-1) = 2,$ and
$$\frac{2}{1 - i} = \frac{2}{1 - i} \cdot \frac{1 + i}{1 + i} = \frac{2 + 2i}{2} = 1 + i$$

Powers of i   The value of i^n equals i^r, where r is the remainder when n is divided by 4. Note that $i^0 = 1$, $i^1 = i$, $i^2 = -1$, and $i^3 = -i$.

Example: $i^{21} = i^1 = i$ because when 21 is divided by 4 the remainder is 1.

CHAPTER 10   REVIEW EXERCISES

SECTION 10.1

Exercises 1–12: Simplify the expression.

1. $\sqrt{4}$
2. $\sqrt{36}$
3. $\sqrt{9x^2}$
4. $\sqrt{(x-1)^2}$
5. $\sqrt[3]{-64}$
6. $\sqrt[3]{-125}$
7. $\sqrt[3]{x^6}$
8. $\sqrt[3]{27x^3}$
9. $\sqrt[4]{16}$
10. $\sqrt[5]{-1}$
11. $\sqrt[4]{x^8}$
12. $\sqrt[5]{(x+1)^5}$

Exercises 13–16: Write the expression in radical notation.

13. $14^{1/2}$
14. $(-5)^{1/3}$
15. $\left(\dfrac{x}{y}\right)^{3/2}$
16. $(xy)^{-2/3}$

Exercises 17–20: Evaluate the expression.

17. $(-27)^{2/3}$

18. $16^{1/4}$

19. $16^{3/2}$

20. $81^{3/4}$

Exercises 21–24: Simplify the expression. Assume that all variables are positive.

21. $(z^3)^{2/3}$

22. $(x^2 y^4)^{1/2}$

23. $\left(\dfrac{x^2}{y^6}\right)^{3/2}$

24. $\left(\dfrac{x^3}{y^6}\right)^{-1/3}$

SECTION 10.2

Exercises 25–40: Simplify the expression. Assume that all variables are positive.

25. $\sqrt{2} \cdot \sqrt{32}$

26. $\sqrt[3]{-4} \cdot \sqrt[3]{2}$

27. $\sqrt[3]{x^4} \cdot \sqrt[3]{x^2}$

28. $\dfrac{\sqrt{80}}{\sqrt{20}}$

29. $\sqrt[3]{-\dfrac{x}{8}}$

30. $\sqrt{\dfrac{1}{3}} \cdot \sqrt{\dfrac{1}{3}}$

31. $\sqrt{48}$

32. $\sqrt{54}$

33. $\sqrt[3]{\dfrac{3}{x}} \cdot \sqrt[3]{\dfrac{9}{x^2}}$

34. $\sqrt{32 a^3 b^2}$

35. $\sqrt{3xy} \cdot \sqrt{27xy}$

36. $\sqrt[3]{-25z^2} \cdot \sqrt[3]{-5z^2}$

37. $\sqrt{x^2 + 2x + 1}$

38. $\sqrt[4]{\dfrac{2a^2}{b}} \cdot \sqrt[4]{\dfrac{8a^3}{b^3}}$

39. $2\sqrt{x} \cdot \sqrt[3]{x}$

40. $\sqrt[3]{rt} \cdot \sqrt[3]{r^2 t^4}$

SECTION 10.3

Exercises 41–50: Simplify the expression. Assume that all variables are positive.

41. $3\sqrt{3} + \sqrt{3}$

42. $\sqrt[3]{x} + 2\sqrt[3]{x}$

43. $3\sqrt[3]{5} - 6\sqrt[3]{5}$

44. $\sqrt[4]{y} - 2\sqrt[4]{y}$

45. $2\sqrt{12} + 7\sqrt{3}$

46. $3\sqrt{18} - 2\sqrt{2}$

47. $7\sqrt[3]{16} - \sqrt[3]{2}$

48. $\sqrt{4x + 4} + \sqrt{x + 1}$

49. $\sqrt{4x^3} - \sqrt{x}$

50. $\sqrt[3]{ab^4} + 2\sqrt[3]{a^4 b}$

Exercises 51–56: Multiply and simplify.

51. $(1 + \sqrt{2})(3 + \sqrt{2})$

52. $(7 - \sqrt{5})(1 + \sqrt{3})$

53. $(3 + \sqrt{6})(3 - \sqrt{6})$

54. $(10 - \sqrt{5})(10 + \sqrt{5})$

55. $(\sqrt{a} + \sqrt{2b})(\sqrt{a} - \sqrt{2b})$

56. $(\sqrt{xy} - 1)(\sqrt{xy} + 2)$

Exercises 57–62: Rationalize the denominator.

57. $\dfrac{4}{\sqrt{5}}$

58. $\dfrac{r}{2\sqrt{t}}$

59. $\dfrac{1}{\sqrt{2} + 3}$

60. $\dfrac{2}{5 - \sqrt{7}}$

61. $\dfrac{1}{\sqrt{8} - \sqrt{7}}$

62. $\dfrac{\sqrt{a} - \sqrt{b}}{\sqrt{a} + \sqrt{b}}$

SECTION 10.4

Exercises 63 and 64: Graph the equation.

63. $y = \sqrt{x}$

64. $y = \sqrt[3]{x}$

Exercises 65 and 66: Write $f(x)$ in radical notation and evaluate $f(4)$.

65. $f(x) = x^{1/2}$

66. $f(x) = x^{2/7}$

Exercises 67 and 68: Graph the equation. Compare the graph to either $y = \sqrt{x}$ or $y = \sqrt[3]{x}$.

67. $y = \sqrt{x} - 2$

68. $y = \sqrt[3]{x} - 1$

Exercises 69–72: Find the domain of f. Write your answer in interval notation.

69. $f(x) = \sqrt{x - 1}$

70. $f(x) = \sqrt{6 - 2x}$

71. $f(x) = \sqrt{x^2 + 1}$

72. $f(x) = \dfrac{1}{\sqrt{x + 2}}$

SECTION 10.5

Exercises 73–78: Solve. Check your answer.

73. $\sqrt{x + 2} = x$

74. $\sqrt{2x - 1} = \sqrt{x + 3}$

75. $\sqrt[3]{x - 1} = 2$

76. $\sqrt[3]{3x} = 3$

77. $\sqrt{2x} = x - 4$

78. $\sqrt{x + 1} = \sqrt{x + 2}$

Exercises 79 and 80: Solve graphically. Approximate solutions to the nearest hundredth when appropriate.

79. $\sqrt[3]{2x - 1} = 2$

80. $x^{2/3} = 3 - x$

Exercises 81 and 82: A right triangle has legs a and b with hypotenuse c. Find the length of the missing side.

81. $a = 4, b = 7$

82. $a = 5, c = 8$

Exercises 83 and 84: Find the exact distance between the given points.

83. $(-2, 3), (2, -2)$

84. $(2, -3), (-4, 1)$

Exercises 85–94: Solve.

85. $x^2 = 121$

86. $2z^2 = 32$

87. $(x - 1)^2 = 16$

88. $x^3 = 64$

89. $(x - 1)^3 = 8$

90. $(2x - 1)^3 = 27$

91. $x^4 = 256$

92. $x^5 = -1$

93. $(x - 3)^5 = -32$

94. $3(x + 1)^4 = 3$

SECTION 10.6

Exercises 95–100: Write the complex expression in standard form.

95. $(1 - 2i) + (-3 + 2i)$

96. $(1 + 3i) - (3 - i)$

97. $(1 - i)(2 + 3i)$

98. $\dfrac{3 + i}{1 - i}$

99. $\dfrac{i(4 + i)}{2 - 3i}$

100. $(1 - i)^2(1 + i)$

APPLICATIONS

101. *Hang Time* (Refer to Example 3, Section 10.4.) A football is punted and has a hang time T of 4.6 seconds. Use the formula $T(h) = \frac{1}{2}\sqrt{h}$ to estimate the height h that it was kicked to the nearest foot.

102. *Baseball Diamond* The four bases of a baseball diamond form a square that is 90 feet on a side. Find the distance from home plate to second base.

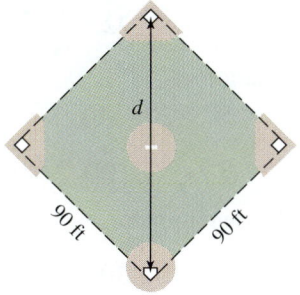

103. *Falling Time* The time T in seconds for an object to fall from a height of h feet is given by $T = \frac{1}{4}\sqrt{h}$. If a person steps off a 10-foot-high board into a swimming pool, how long is the person in the air?

104. *Geometry* A cube has sides of length $\sqrt{5}$.
 (a) Find the area of one side of the cube.
 (b) Find the volume of the cube.
 (c) Find the length of the diagonal of one of the sides.
 (d) Find the distance from A to B in the figure.

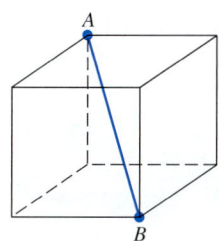

105. *Pendulum* The time required for a pendulum to swing back and forth is called its *period* (see the figure). The period T of a pendulum does not depend on its weight, only on its length L and gravity. It is given by $T = 2\pi\sqrt{\frac{L}{32.2}}$, where T is in seconds and L is in feet. Estimate the length of a pendulum with a period of 1 second.

106. *Pendulum* (Refer to Exercise 105.) If a pendulum were on the moon, its period could be calculated by $T = 2\pi\sqrt{\frac{L}{5.1}}$. Estimate the length of a pendulum with a period of 1 second on the moon. Compare your answer to that for Exercise 105.

107. *Population Growth* In 1790 the population of the United States was 4 million, and by 2000 it had grown to 281 million. The *average* annual percentage growth in the population r (expressed as a decimal) can be determined by the polynomial equation $281 = 4(1 + r)^{210}$. Solve this equation for r and interpret the result.

108. Highway Curves   If a circular highway curve is banked with a slope of $m = \frac{1}{10}$ (see the accompanying figure) and has a radius of R feet, then the speed limit L in miles per hour for the curve is given by

$$L = \sqrt{3.75R}.$$

(**Source:** N. Garber and L. Hoel, *Traffic and Highway Engineering.*)

(a) Find the speed limit if R is 500 feet.

(b) With no banking, the speed limit is given by $L = 1.5\sqrt{R}$. Find the speed limit for a curve with no banking and a radius of 500 feet. How does banking affect the speed limit? Does this result agree with your intuition?

109. Geometry   Find the length of a side of a square if the square has an area of 7 square feet.

110. Surface Area of a Cone   The surface area of a cone having radius r and height h is given by the formula $S = \pi r\sqrt{r^2 + h^2}$. See the accompanying figure. Estimate the surface area if $r = 11$ inches and $h = 60$ inches.

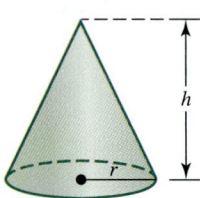

111. Radioactive Carbon Dating   Living plants and animals have a constant amount of radioactive carbon in their cells, which comes from the carbon dioxide they breathe. When a plant or animal dies, the exchange of oxygen and carbon dioxide halts and the amount of radioactive carbon starts to decrease. The fraction of radioactive carbon remaining t years after death is given by $2^{-t/5700}$. Find the fraction left after each time period.

(a) 5700 years   (b) 20,000 years

CHAPTER 10   TEST   🎓Pass the Test   Video solutions to all test exercises

Exercises 1–6: Simplify the expression.

1. $\sqrt[3]{-27}$

2. $\sqrt{(z + 1)^2}$

3. $\sqrt{25x^4}$

4. $\sqrt[3]{8z^6}$

5. $\sqrt[4]{16x^4y^5}; x > 0, y > 0$

6. $(\sqrt{3} - \sqrt{2})(\sqrt{3} + \sqrt{2})$

Exercises 7 and 8: Write the expression in radical notation.

7. $7^{2/5}$

8. $\left(\dfrac{x}{y}\right)^{-2/3}$

Exercises 9 and 10: Evaluate the expression by hand.

9. $(-8)^{4/3}$

10. $36^{-3/2}$

Exercises 11 and 12: Use rational exponents to write the expression.

11. $\sqrt[3]{x^4}$

12. $\sqrt{x} \cdot \sqrt[5]{x}$

13. Find the domain of $f(x) = \sqrt{4 - x}$. Write your answer in interval notation.

14. Sketch a graph of $y = \sqrt{x} + 3$.

Exercises 15–22: Simplify the expression. Assume that all variables are positive.

15. $(2z^{1/2})^3$

16. $\left(\dfrac{y^2}{z^3}\right)^{-1/3}$

17. $\sqrt{3} \cdot \sqrt{27}$

18. $\dfrac{\sqrt{y^3}}{\sqrt{4y}}$

19. $7\sqrt{7} - 3\sqrt{7} + \sqrt{5}$

20. $7\sqrt[3]{x} - \sqrt[3]{x}$

21. $4\sqrt{18} + \sqrt{8}$

22. $\dfrac{\sqrt[3]{32}}{\sqrt[3]{4}}$

23. Solve each equation.

(a) $\sqrt{x} - 2 = 5$

(b) $\sqrt[3]{x} + 1 = 2$

(c) $(x - 1)^3 = 8$

(d) $\sqrt{2x + 2} = x - 11$

24. Rationalize the denominator.

(a) $\dfrac{2}{3\sqrt{7}}$

(b) $\dfrac{1}{1 + \sqrt{5}}$

25. Solve $\sqrt{3x} - x + 1 = \sqrt[3]{x - 1}$ graphically. Round solutions to the nearest hundredth.

26. One leg of a right triangle has length 7 and the hypotenuse has length 13. Find the length of the third side.

27. Find the distance between $(-3, 5)$ and $(-1, 7)$.

Exercises 28–31: Write the complex expression in standard form.

28. $(-5 + i) + (7 - 20i)$

29. $(3i) - (6 - 5i)$

30. $\left(\dfrac{1}{2} - i\right)\left(\dfrac{1}{2} + i\right)$

31. $\dfrac{2i}{5 + 2i}$

32. *Volume of a Sphere* The volume V of a sphere is given by $V = \frac{4}{3}\pi r^3$, where r is its radius.
(a) Solve the equation for r.
(b) Find r if $V = 50$ cubic inches.

33. *Wing Span of a Bird* The wing span L of a bird with weight W can sometimes be modeled by $L = 27.4W^{1/3}$, where L is in inches and W is in pounds. Use this formula to estimate the weight of a bird that has a wing span of 30 inches. (*Source:* C. Pennycuick, *Newton Rules Biology.*)

CHAPTER 10   EXTENDED AND DISCOVERY EXERCISES

 1. *Moons of Jupiter* The accompanying table lists the orbital distances and periods of several moons of Jupiter. Let x be the distance and y be the period. These data can be modeled by a power function of the form $f(x) = 0.0002x^{m/n}$. Use trial and error to find the value of the fraction $\frac{m}{n}$. Graph f and the data in the same viewing rectangle.

Moon	Distance (10^3 kilometers)	Period (days)
Metis	128	0.29
Almathea	181	0.50
Thebe	222	0.67
Io	422	1.77
Europa	671	3.55
Ganymede	1070	7.16
Callisto	1883	16.69

Source: M. Zeilik, *Introductory Astronomy and Astrophysics.*

2. *Modeling Wood in a Tree* In forestry the volume of timber in a given area of forest is often estimated. To make such estimates, scientists have developed formulas to find the amount of wood contained in a tree with height h in feet and diameter d in inches. One study concluded that the volume V of wood in cubic feet in a tree is given by $V = kh^{1.12}d^{1.98}$,

where k is a constant. Note that the diameter is measured 4.5 feet above the ground. (*Source:* B. Ryan, B. Joiner, and T. Ryan, *Minitab Handbook.*)
(a) A tree with an 11-inch diameter and a 47-foot height has a volume of 11.4 cubic feet. Approximate the constant k.
(b) Estimate the volume of wood in the same type of tree with $d = 20$ inches and $h = 105$ feet.

3. *Area of Skin* The surface area of the skin covering the human body is a function of more than one variable. Both height and weight influence the surface area of a person's body. Hence a taller person tends to have a larger surface area, as does a heavier person. A formula to determine the area of a person's skin in square inches is $S = 15.7w^{0.425}h^{0.725}$, where w is weight in pounds and h is height in inches. (*Source:* H. Lancaster, *Quantitative Methods in Biological and Medical Sciences.*)
(a) Use S to estimate the area of a person's skin who is 65 inches tall and weighs 154 pounds.
(b) If a person's weight doubles, what happens to the area of the person's skin? Explain.
(c) If a person's height doubles, what happens to the area of the person's skin? Explain.

4. *Minimizing Cost* A natural gas line running along a river is to be connected from point A to a cabin on the other bank located at point D, as illustrated in the figure on the next page. The width of the river is 500 feet, and the distance from point A to point C is 1000 feet.

The cost of running the pipe along the shoreline is $30 per foot, and the cost of running it underwater is $50 per foot. The cost of connecting the gas line from A to D is to be minimized.

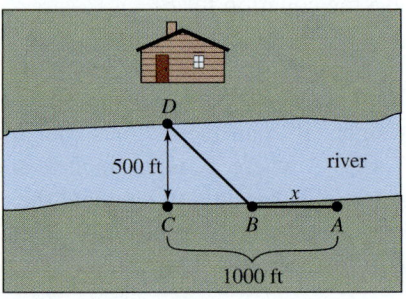

(a) Write an expression that gives the cost of running the line from A to B if the distance between these points is x feet.

(b) Find the distance from B to D in terms of x.

(c) Write an expression that gives the cost of running the line from B to D.

(d) Use your answer from parts (a) and (c) to write an expression that gives the cost of running the line from A to B to D.

 (e) Graph your expression from part (d) in the window [0, 1000, 100] by [40000, 60000, 5000] to determine the value of x that minimizes the cost of the line going from A to D. What is the minimum cost?

CHAPTERS 1–10   CUMULATIVE REVIEW EXERCISES

1. Evaluate $S = 4\pi r^2$ when $r = 3$.

2. Identify the domain and range of the function given by $f = \{(-1, 2), (0, 4), (1, 2)\}$.

3. Simplify each expression. Write the result using positive exponents.
   (a) $\left(\dfrac{ab^2}{b^{-1}}\right)^{-3}$ (b) $\dfrac{(x^2y)^3}{x^2(y^2)^{-3}}$ (c) $(rt)^2(r^2t)^3$

4. Write 0.00043 in scientific notation.

5. Find $f(3)$ if $f(x) = \dfrac{x}{x-2}$. What is the domain of f?

6. Find the domain of $f(x) = \sqrt[3]{2x}$.

Exercises 7 and 8: Graph the equation by hand.

7. $y = 1 - 2x$ 8. $y = \sqrt{x+1}$

9. Sketch a graph of $f(x) = x^2 - 4$.
   (a) Identify the domain and range.
   (b) Evaluate $f(-2)$.
   (c) Identify the x-intercepts.
   (d) Solve the equation $f(x) = 0$.

10. Find slope–intercept form of the line that passes through $(-1, 2)$ and is perpendicular to $y = -2x$.

11. Let f be a linear function. Find a formula for $f(x)$.

x	-2	-1	1	2
$f(x)$	7	4	-2	-5

12. Sketch a graph of a line passing through $(-1, 1)$ with slope $-\frac{1}{2}$.

13. Solve $5x - (3 - x) = \frac{1}{2}x$.

14. Solve $2x - 5 \le 4 - x$.

15. Solve $|x - 2| \le 3$.

16. Solve $-1 \le 1 - 2x \le 6$.

17. Solve each system symbolically if possible.
   (a) $2x - y = 4$ (b) $3x - 4y = 2$
   $\ x + y = 8$ $\ x - \frac{4}{3}y = 1$

18. Shade the solution set in the xy-plane.
$$x + 2y \le 2$$
$$-x + 3y \ge 3$$

19. Solve the system.
$$\begin{aligned} x + 2y - z &= 6 \\ x - 3y + z &= -2 \\ x + y + z &= 6 \end{aligned}$$

20. Calculate $\det \begin{bmatrix} 4 & 2 & -1 \\ 2 & 1 & 0 \\ 0 & -2 & 1 \end{bmatrix}$.

Exercises 21–24: Multiply the expression.

21. $4x(4 - x^3)$ 22. $(x - 4)(x + 4)$

23. $(5x + 3)(x - 2)$ 24. $(4x + 9)^2$

Exercises 25–30: Factor the expression.

25. $9x^2 - 16$

26. $x^2 - 4x + 4$

27. $15x^3 - 9x^2$

28. $12x^2 - 5x - 3$

29. $r^3 - 1$

30. $x^3 - 3x^2 + 5x - 15$

Exercises 31 and 32: Solve each equation.

31. $x^2 - 3x + 2 = 0$

32. $x^3 = 4x$

Exercises 33 and 34: Simplify the expression.

33. $\dfrac{x^2 + 3x + 2}{x - 3} \div \dfrac{x + 1}{2x - 6}$

34. $\dfrac{2}{x - 1} + \dfrac{5}{x}$

Exercises 35–42: Simplify the expression. Assume all variables are positive.

35. $\sqrt{36x^2}$

36. $\sqrt[3]{64}$

37. $16^{-3/2}$

38. $\sqrt[4]{625}$

39. $\sqrt{2x} \cdot \sqrt{8x}$

40. $\sqrt{x} \cdot \sqrt[4]{x}$

41. $\dfrac{\sqrt[3]{16x^4}}{\sqrt[3]{2x}}$

42. $4\sqrt{12x} - 2\sqrt{3x}$

43. Multiply $(2x + \sqrt{3})(x - \sqrt{3})$.

44. Find the domain of $f(x) = \sqrt{1 - x}$.

45. Graph $f(x) = 3\sqrt[3]{x}$.

46. Find the distance between $(-2, 3)$ and $(1, 2)$.

47. Write $(1 - i)(2 + 3i)$ in standard form.

48. The lengths of the legs of a right triangle are 5 and 12. What is the length of the hypotenuse?

Exercises 49–52: Solve symbolically.

49. $2\sqrt{x + 3} = x$

50. $\sqrt[3]{x - 1} = 3$

51. $\sqrt{x + 4} = 2\sqrt{x + 5}$

52. $\frac{1}{3}x^4 = 27$

Exercises 53 and 54: Solve graphically and approximate your answer to the nearest hundredth.

53. $\sqrt[3]{x^2 - 2} + x = \sqrt{x}$

54. $x - \sqrt[3]{x} = \sqrt{x + 2}$

APPLICATIONS

55. *Calculating Water Flow*   The gallons G of water in a tank after t minutes are given by $G(t) = 300 - 15t$. Interpret the slope of the graph of G as a rate of change.

56. *Distance*   A person jogs away from home at 8 miles per hour for 1 hour, rests in a park for 2 hours, and then walks back home at 4 miles per hour. Sketch a graph that models the distance y that the person is from home after x hours.

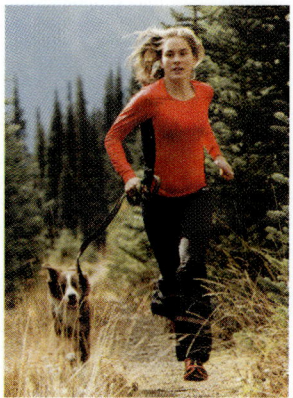

57. *Investment*   Suppose $2000 are deposited in two accounts paying 5% and 4% annual interest. If the total interest after 1 year is $93, how much is invested at each rate?

58. *Geometry*   A rectangular box with a square base has a volume of 256 cubic inches. If its height is 4 inches less than the length of an edge of the base, find the dimensions of the box.

59. *Height of a Building*   A 5-foot 3-inch person casts a shadow that is 7.5 feet while a nearby building casts a 32-foot shadow. Find the height of the building.

60. *Angles in a Triangle*   The measure of the largest angle in a triangle is 20° less than the sum of the measures of the two smaller angles. The sum of the measures of the two larger angles is 90° greater than the measure of the smaller angle. Find the measure of each angle.

CHAPTER 11

Quadratic Functions and Equations

W hat size television should you buy? Should you buy a 32-inch screen or a 50-inch screen? According to a home entertainment article in *Money* magazine, the answer depends on how far you sit from your television. The farther you sit from your television, the larger the television that is recommended. For example, if you sit only 6 feet from the screen, then a 32-inch television would be adequate; if you sit 10 feet from the screen, then a 50-inch television is more appropriate.

We can use a *quadratic function* to calculate the size S of the television screen needed for a person who sits x feet from the screen, where $6 \le x \le 15$. This function is given by

$$S(x) = -0.227x^2 + 8.155x - 8.8.$$

You might want to use S to determine the size of screen recommended when you sit 8 feet from a television. (See Exercises 89 and 90 in Section 11.1.) Quadratic functions are a special type of polynomial function that occur frequently in applications involving economics, road construction, falling objects, geometry, and modeling real-world data. In this chapter we discuss quadratic functions and equations.

There is no branch of mathematics, however abstract, which may not some day be applied to the real world.

—Nikolai Lobachevsky

Source: *Money*, January 2007, p. 107; hdguru.com.

11.1 QUADRATIC FUNCTIONS AND THEIR GRAPHS

Graphs of Quadratic Functions ■ Basic Transformations of Graphs ■
More About Graphing Quadratic Functions (Optional) ■ Min–Max Applications

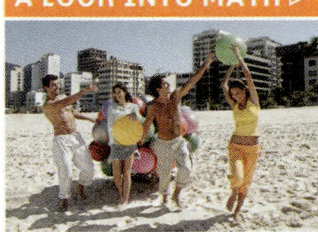

Suppose that a hotel is considering giving a group discount on room rates. The regular price is $80, but for each room rented the price decreases by $2. On the one hand, if the hotel rents one room, it makes only $78. On the other hand, if the hotel rents 40 rooms, the rooms are all free and the hotel makes nothing. Is there an optimal number of rooms between 1 and 40 that should be rented to maximize the revenue? In this section we use quadratic functions and their graphs to answer this question.

Graphs of Quadratic Functions

In Section 8.4 we discussed how a quadratic function could be represented by a polynomial of degree 2. We now give an alternative definition of a quadratic function.

QUADRATIC FUNCTION

A **quadratic function** can be written in the form

$$f(x) = ax^2 + bx + c,$$

where a, b, and c are constants with $a \neq 0$.

The graph of *any* quadratic function is a *parabola*. Recall that a parabola is a ∪-shaped graph that opens either upward or downward. The graph of the simple quadratic function $y = x^2$ is a parabola that opens upward, with its *vertex* located at the origin, as shown in Figure 11.1(a). The **vertex** is the *lowest* point on the graph of a parabola that opens upward and the *highest* point on the graph of a parabola that opens downward. A parabola opening downward is shown in Figure 11.1(b). Its vertex is the point (0, 2) and is the highest point on the graph. If we were to fold the *xy*-plane along the *y*-axis, the left and right sides of the graph would match. That is, the graph is symmetric with respect to the *y*-axis. In this case the *y*-axis is the **axis of symmetry** for the graph. Figure 11.1(c) shows a parabola that opens upward with vertex (2, −1) and axis of symmetry $x = 2$.

(a)

(b)

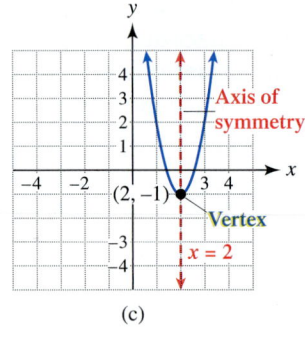

(c)

Figure 11.1

INCREASING AND DECREASING   Suppose that the graph of the equation $y = x^2$ shown in Figure 11.2 represents a valley. If we walk from *left to right*, the valley "goes down" and then "goes up." Mathematically, we say that the graph of $y = x^2$ is *decreasing* when $x \leq 0$ and *increasing* when $x \geq 0$. The vertex represents the point at which the graph switches from decreasing to increasing. In Figure 11.1(b), the graph increases when $x \leq 0$ and decreases when $x \geq 0$, and in Figure 11.1(c) the graph decreases when $x \leq 2$ and increases when $x \geq 2$.

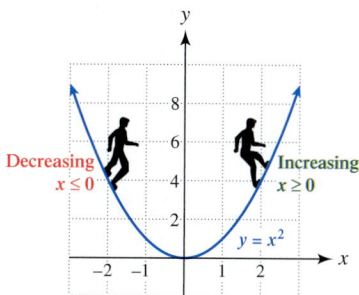

Figure 11.2

NOTE: When determining where a graph is increasing and where it is decreasing, we must "walk" along the graph *from left to right*. (We read English from left to right, which might help you remember.)

EXAMPLE 1   **Graphing quadratic functions**

Graph each quadratic function. Identify the vertex and the axis of symmetry. Then state where the graph is increasing and where it is decreasing.
(a) $f(x) = x^2 - 1$ **(b)** $f(x) = -(x + 1)^2$ **(c)** $f(x) = x^2 + 4x + 3$

Solution
(a) Begin by making a convenient table of values (see Table 11.1). Then plot the points and sketch a smooth ∪-shaped curve that opens upward, as shown in Figure 11.3. The lowest point on this graph is $(0, -1)$, which is the vertex. The axis of symmetry is the vertical line $x = 0$, which passes through the vertex and coincides with the y-axis. Note also the symmetry of the y-values in Table 11.1 about the point $(0, -1)$. This graph is decreasing when $x \leq 0$ and increasing when $x \geq 0$.

TABLE 11.1

x	$y = x^2 - 1$
-2	3
-1	0
0 (Vertex →)	-1
1	0
2	3

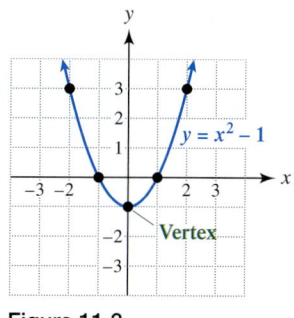

Figure 11.3

(b) Make a table of values (see Table 11.2). Plot the points and sketch a smooth ∩-shaped curve opening downward, as shown in Figure 11.4. The highest point on this graph is $(-1, 0)$, which is the vertex. The axis of symmetry is the vertical line $x = -1$, which passes through the vertex. This graph is increasing when $x \leq -1$ and decreasing when $x \geq -1$.

TABLE 11.2

x	$y = -(x + 1)^2$	
-3	-4	
-2	-1	
Vertex → -1	0	Equal
0	-1	
1	-4	

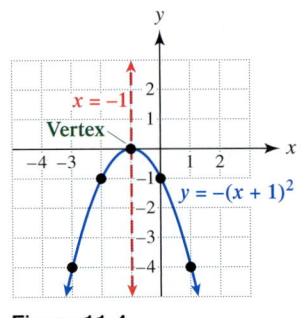

Figure 11.4

(c) Table 11.3 lists points on the graph of f, which is shown in Figure 11.5. The vertex is $(-2, -1)$ and the axis of symmetry is $x = -2$. The graph is decreasing when $x \leq -2$ and increasing when $x \geq -2$.

TABLE 11.3

x	$y = x^2 + 4x + 3$	
-5	8	
-4	3	
-3	0	
Vertex → -2	-1	Equal
-1	0	
0	3	
1	8	

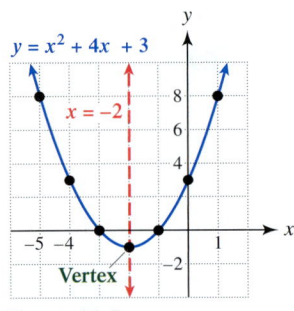

Figure 11.5

Now Try Exercises 15, 17, 21, 23

THE VERTEX FORMULA   Rather than finding a vertex by graphing, we can use the following formula. This formula can be derived by completing the square, a technique discussed in the next section.

VERTEX FORMULA

The x-coordinate of the vertex of the graph of $y = ax^2 + bx + c$, $a \neq 0$, is given by

$$x = -\frac{b}{2a}.$$

To find the y-coordinate of the vertex, substitute this x-value in the equation.

NOTE: The equation of the axis of symmetry for $f(x) = ax^2 + bx + c$ is $x = -\frac{b}{2a}$, and the vertex is the point $\left(-\frac{b}{2a}, f\left(-\frac{b}{2a}\right)\right)$.

We apply the vertex formula in the next example.

EXAMPLE 2 Finding the vertex of a parabola

Find the vertex for the graph of $f(x) = 2x^2 - 4x + 1$. Support your answer graphically.

Solution

For $f(x) = 2x^2 - 4x + 1$, $a = \mathbf{2}$ and $b = \mathbf{-4}$. The x-coordinate of the vertex is

$$x = -\frac{b}{2a} = -\frac{(-4)}{2(2)} = \mathbf{1}.$$

To find the y-coordinate of the vertex, substitute $x = \mathbf{1}$ in the given formula.

$$f(\mathbf{1}) = 2(\mathbf{1})^2 - 4(\mathbf{1}) + 1 = \mathbf{-1}.$$

Thus the vertex is located at $(\mathbf{1}, \mathbf{-1})$, which is supported by Figure 11.6.

Now Try Exercise **39**

[−4.7, 4.7, 1] by [−3.1, 3.1, 1]

Figure 11.6

The vertex formula can help us find maximum y-values on a graph.

EXAMPLE 3 Finding a maximum y-value

Find the maximum y-value on the graph of $f(x) = -x^2 + 4x - 3$. Then state where the graph of f is increasing and where it is decreasing.

Solution

For $f(x) = -x^2 + 4x - 3$, $a = \mathbf{-1}$ and $b = \mathbf{4}$. The graph of $f(x) = -x^2 + 4x - 3$ is a parabola that *opens downward* because $a < 0$. The highest point on the graph of f is the vertex and the y-coordinate of the vertex corresponds to the maximum y-coordinate. (A graph of f is shown in Figure 11.7, but it is not necessary to graph f.) The x-coordinate of the vertex is

$$x = -\frac{b}{2a} = -\frac{4}{2(-1)} = \mathbf{2}.$$

The corresponding y-coordinate of the vertex is $f(\mathbf{2}) = -(\mathbf{2})^2 + 4(\mathbf{2}) - 3 = \mathbf{1}$. The maximum y-value is $\mathbf{1}$. Because the graph opens downward, f is increasing when $x \leq 2$ and decreasing when $x \geq 2$.

Now Try Exercise **61**

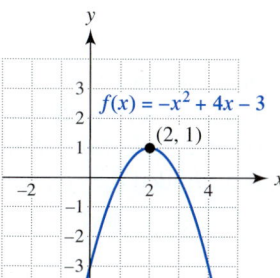

Figure 11.7 Maximum y-value: 1

Basic Transformations of Graphs

THE GRAPH OF $y = ax^2$, $a > 0$ First, graph $y_1 = \frac{1}{2}x^2$, $y_2 = x^2$, and $y_3 = 2x^2$, as shown in Figure 11.8(a) on the next page. Note that $a = \frac{1}{2}$, $a = 1$, and $a = 2$, respectively, and that as a increases, the resulting parabola becomes narrower. The graph of $y_1 = \frac{1}{2}x^2$ is wider than the graph of $y_2 = x^2$, and the graph of $y_3 = 2x^2$ is narrower than the graph of $y_2 = x^2$. In general, the graph of $y = ax^2$ is wider than the graph of $y = x^2$ when $0 < a < 1$ and narrower than the graph of $y = x^2$ when $a > 1$. When $a > 0$, the graph of $y = ax^2$ never lies *below* the x-axis.

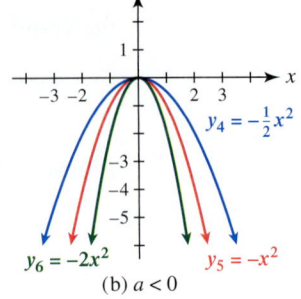

(a) $a > 0$ (b) $a < 0$

Figure 11.8 $y = ax^2$

THE GRAPH OF $y = ax^2$, $a < 0$ When $a < 0$, the graph of $y = ax^2$ never lies *above* the x-axis because, for any input x, the product $ax^2 \leq 0$. The graphs of $y_4 = -\frac{1}{2}x^2$, $y_5 = -x^2$, and $y_6 = -2x^2$ are shown in Figure 11.8(b). The graph of $y_4 = -\frac{1}{2}x^2$ is wider than the graph of $y_5 = -x^2$ and the graph of $y_6 = -2x^2$ is narrower than the graph of $y_5 = -x^2$.

> ### THE GRAPH OF $y = ax^2$
>
> The graph of $y = ax^2$ is a parabola with the following characteristics.
>
> 1. The vertex is $(0, 0)$, and the axis of symmetry is given by $x = 0$.
> 2. It opens upward if $a > 0$ and opens downward if $a < 0$.
> 3. It is wider than the graph of $y = x^2$, if $0 < |a| < 1$. It is narrower than the graph of $y = x^2$, if $|a| > 1$.

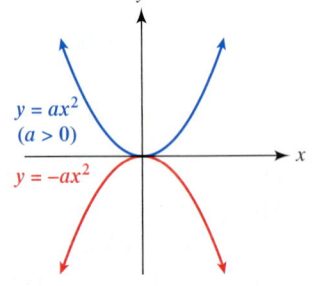

Figure 11.9 Reflection Across the x-Axis

REFLECTIONS OF $y = ax^2$ In Figure 11.8 the graph of $y_1 = \frac{1}{2}x^2$ can be *transformed* into the graph of $y_4 = -\frac{1}{2}x^2$ by *reflecting* it across the x-axis. The graph of y_4 is a **reflection** of the graph of y_1 across the x-axis. In general, *the graph of $y = -ax^2$ is a reflection of the graph of $y = ax^2$ across the x-axis*, as shown in Figure 11.9. That is, if we folded the xy-plane along the x-axis the two graphs would match.

EXAMPLE 4   Graphing $y = ax^2$

Compare the graph of $g(x) = -3x^2$ to the graph of $f(x) = x^2$. Then graph both functions on the same coordinate axes.

Solution

Both graphs are parabolas. However, the graph of g opens downward and is narrower than the graph of f. Their graphs are shown in Figure 11.10.   Now Try Exercise **47**

Figure 11.10

More About Graphing Quadratic Functions (Optional)

The graph of a quadratic function is a parabola, and any quadratic function f can be written as $f(x) = ax^2 + bx + c$, where a, b, and c are constants. As a result, the values of a, b, and c

determine both the shape and position of the parabola in the xy-plane. In this subsection, we summarize some of the effects that these constants have on the graph of f.

THE EFFECTS OF a   The effects of a on the graph of $y = ax^2$ were discussed extensively in the previous subsection, and these effects can be generalized to include the graph of $y = ax^2 + bx + c$.

1. **Width:** The graph of $f(x) = ax^2 + bx + c$ is wider than the graph of $y = x^2$ if $0 < |a| < 1$ and narrower if $|a| > 1$. See Figures 11.11(a) and (b).
2. **Opening:** The graph of $f(x) = ax^2 + bx + c$ opens upward if $a > 0$ and downward if $a < 0$. See Figure 11.11(c).

      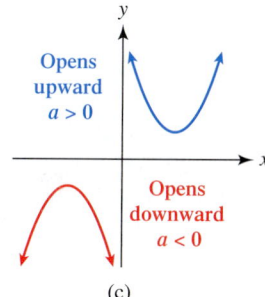

(a)   (b)   (c)

Figure 11.11 The Effects of a

THE EFFECTS OF c   The value of c affects the vertical placement of the parabola in the xy-plane.

1. **y-intercept:** Because $f(0) = a(0)^2 + b(0) + c = c$, it follows that the y-intercept for the graph of f is c. See Figure 11.12(a).
2. **Vertical Shifts:** The graph of $f(x) = ax^2 + bx + c$ is shifted vertically c units compared to the graph of $y = ax^2 + bx$. If $c < 0$, the shift is downward; if $c > 0$, the shift is upward. See Figures 11.12(b) and (c). The parabolas in both figures have identical shapes.

      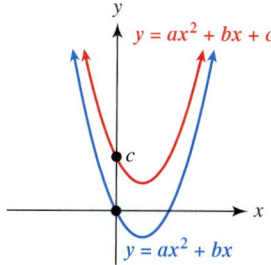

(a) y-intercept: c (b) Shifted downward: $c < 0$ (c) Shifted upward: $c > 0$

Figure 11.12 The Effects of c

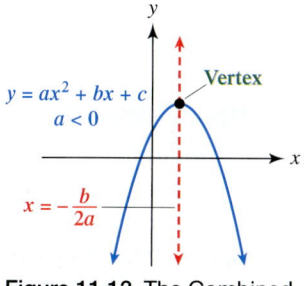

Figure 11.13 The Combined Effects of a and b

THE COMBINED EFFECTS OF a AND b   The combined values of a and b determine the x-coordinate of the vertex and the equation of the axis of symmetry.

1. **Vertex:** The x-coordinate of the vertex is $-\frac{b}{2a}$.
2. **Axis of Symmetry:** The axis of symmetry is given by $x = -\frac{b}{2a}$.

Figure 11.13 illustrates these concepts.

EXAMPLE **5** **Analyzing the graph of $y = ax^2 + bx + c$**

Let $f(x) = -\frac{1}{2}x^2 + x + \frac{3}{2}$.

(a) Does the graph of f open upward or downward? Is this graph wider or narrower than the graph of $y = x^2$?

(b) Find the axis of symmetry and the vertex.

(c) Find the y-intercept and any x-intercepts.

(d) Sketch a graph of f.

Solution

(a) If $f(x) = -\frac{1}{2}x^2 + x + \frac{3}{2}$, then $a = -\frac{1}{2}$, $b = 1$, and $c = \frac{3}{2}$. Because $a = -\frac{1}{2} < 0$, the parabola opens downward. Also, because $0 < |a| < 1$, the graph is wider than the graph of $y = x^2$.

(b) The axis of symmetry is $x = -\frac{b}{2a} = -\dfrac{1}{2\left(-\frac{1}{2}\right)} = 1$, or $x = 1$. Because

$$f(\mathbf{1}) = -\frac{1}{2}(\mathbf{1})^2 + (\mathbf{1}) + \frac{3}{2} = -\frac{1}{2} + 1 + \frac{3}{2} = \mathbf{2},$$

the vertex is $(\mathbf{1}, \mathbf{2})$.

(c) The y-intercept equals c, or $\frac{3}{2}$. To find x-intercepts we let $y = 0$ and solve for x.

$$\begin{array}{ll} -\dfrac{1}{2}x^2 + x + \dfrac{3}{2} = 0 & \text{Equation to be solved} \\[2mm] x^2 - 2x - 3 = 0 & \text{Multiply by } -2; \text{ clear fractions.} \\[2mm] (x + 1)(x - 3) = 0 & \text{Factor.} \\[2mm] x + 1 = 0 \text{ or } x - 3 = 0 & \text{Zero-product property} \\[2mm] x = -1 \quad \text{or} \quad x = 3 & \text{Solve.} \end{array}$$

The x-intercepts are -1 and 3.

(d) Start by plotting the vertex and intercepts, as shown in Figure 11.14. Then sketch a smooth, ∩-shaped graph that connects these points.

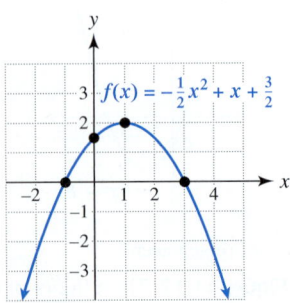

Figure 11.14

Now Try Exercise **67**

Min–Max Applications

Sometimes when a quadratic function f is used to model real-world data, the vertex provides important information. The reason is that the y-coordinate of the vertex gives either the maximum value of $f(x)$ or the minimum value of $f(x)$. For example, Figure 11.15(a) shows a parabola that opens upward. The minimum y-value on this graph is 1 and occurs at the vertex $(2, 1)$. Similarly, Figure 11.15(b) shows a parabola that opens downward. The maximum y-value on this graph is 3 and occurs at the vertex $(-1, 3)$.

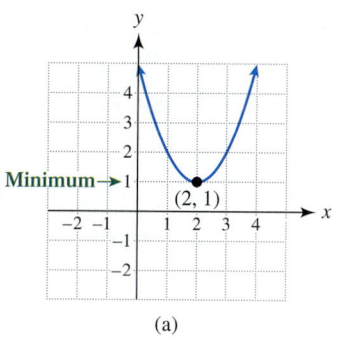

(a)   (b)

Figure 11.15

▶ **REAL-WORLD CONNECTION**   In the next example we demonstrate finding a maximum height reached by a baseball.

EXAMPLE 6   Finding maximum height

A baseball is hit into the air and its height h in feet after t seconds can be calculated by $h(t) = -16t^2 + 96t + 3$.
(a) What is the height of the baseball when it is hit?
(b) Determine the maximum height of the baseball.

Solution
(a) The baseball is hit when $t = \mathbf{0}$, so $h(\mathbf{0}) = -16(\mathbf{0})^2 + 96(\mathbf{0}) + 3 = 3$ feet.
(b) The graph of h opens downward because $a = -16 < 0$. Thus the maximum height of the baseball occurs at the vertex. To find the vertex, we apply the vertex formula with $a = \mathbf{-16}$ and $b = \mathbf{96}$ because $h(t) = \mathbf{-16}t^2 + \mathbf{96}t + 3$.

$$t = -\frac{b}{2a} = -\frac{\mathbf{96}}{2(\mathbf{-16})} = \mathbf{3} \text{ seconds}$$

The maximum height of the baseball occurs at $t = \mathbf{3}$ seconds and is

$$h(\mathbf{3}) = -16(\mathbf{3})^2 + 96(\mathbf{3}) + 3 = 147 \text{ feet.} \quad \boxed{\text{Now Try Exercise } 79}$$

▶ **REAL-WORLD CONNECTION**   In the next example we answer the question presented in the introduction to this section.

EXAMPLE 7   Maximizing revenue

A hotel is considering giving the following group discount on room rates. The regular price for a room is \$80, but for each room rented the price decreases by \$2. A graph of the revenue received from renting x rooms is shown in Figure 11.16 on the next page.

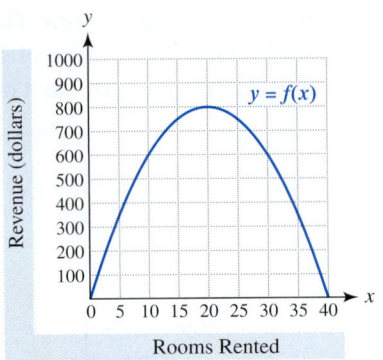

Figure 11.16

(a) Interpret the graph.
(b) What is the maximum revenue? How many rooms should be rented to receive the maximum revenue?
(c) Write a formula for $y = f(x)$ whose graph is shown in Figure 11.16.
(d) Use $f(x)$ to determine symbolically the maximum revenue and the number of rooms that should be rented.

Solution

(a) The revenue increases at first, reaches a maximum, which corresponds to the vertex, and then decreases.
(b) In Figure 11.16 the vertex is (20, 800). Thus the maximum revenue of $800 occurs when 20 rooms are rented.
(c) If x rooms are rented, the price for each room is $80 - 2x$. The revenue equals the number of rooms rented times the price of each room. Thus $f(x) = x(80 - 2x)$.
(d) First, multiply $x(80 - 2x)$ to obtain $80x - 2x^2$ and then let $f(x) = -2x^2 + 80x$. The x-coordinate of the vertex is

$$x = -\frac{b}{2a} = -\frac{80}{2(-2)} = 20.$$

The y-coordinate is $f(20) = -2(20)^2 + 80(20) = 800$. These calculations verify our results in part (b).   **Now Try Exercise** 83

TECHNOLOGY NOTE: Locating a Vertex

Some graphing calculators can locate a vertex on a parabola with either the MAXIMUM or MINIMUM utility. The maximum for the graph in Example 7 is found in the accompanying figure. This utility is typically more accurate than the TRACE utility.

CALCULATOR HELP

To find a minimum or maximum, see Appendix A (page AP-12).

[0, 50, 10] by [0, 1000, 100]

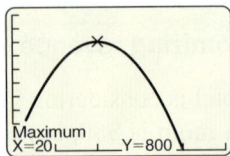

▶ **REAL-WORLD CONNECTION**   In the next example we minimize a quadratic function that models percentages of births by cesarean section (C-section).

EXAMPLE **8**   Analyzing percentage of births by cesarean section

The percentage P of births performed by cesarean section between 1991 and 2001 is modeled by the formula $P(t) = 0.105t^2 - 1.08t + 23.6$, where $t = 1$ corresponds to 1991, $t = 2$ to 1992, and so on.   (*Source:* The National Center for Health Statistics.)

(a) Estimate the year when the percentage of births by cesarean section was minimum.

(b) What is this minimum percentage?

Solution

(a) The graph of P is a parabola that opens upward because $a = 0.105 > 0$. Therefore the t-coordinate of the vertex represents the year when the percentage of births done by cesarean section was minimum, or

$$t = -\frac{b}{2a} = -\frac{-1.08}{2(0.105)} \approx 5.1.$$

Because $t = 5$ corresponds to 1995, the minimum percentage occurred during 1995.

(b) In 1995, this percentage was about $P(5) = 0.105(5)^2 - 1.08(5) + 23.6 \approx 20.8\%$.

Now Try Exercise **88**

11.1 PUTTING IT ALL TOGETHER

The following table summarizes some of the important topics in this section.

Concept	Explanation	Examples
Quadratic Function	Can be written as $$f(x) = ax^2 + bx + c, a \neq 0$$	$f(x) = x^2 + x - 2$ and $g(x) = -2x^2 + 4$ $(b = 0)$
Graph of a Quadratic Function	Its graph is a parabola that opens upward if $a > 0$ and downward if $a < 0$. The value of $\lvert a \rvert$ affects the width of the parabola. The vertex can be used to determine the maximum or minimum output of a quadratic function.	The graph of $y = -\frac{1}{4}x^2$ opens downward, and is wider than the graph of $y = x^2$ as shown in the figure. Each graph has its vertex at $(0, 0)$.

continued on next page

continued from previous page

Concept	Explanation	Examples
Vertex of a Parabola	The x-coordinate of the vertex for the function $f(x) = ax^2 + bx + c$ with $a \neq 0$ is given by $$x = -\frac{b}{2a}.$$ The y-coordinate of the vertex is found by substituting this x-value in the equation. Hence the vertex is $\left(-\frac{b}{2a}, f\left(-\frac{b}{2a}\right)\right)$.	If $f(x) = -2x^2 + 8x - 7$, then $$x = -\frac{8}{2(-2)} = 2$$ and $$f(2) = -2(2)^2 + 8(2) - 7 = 1.$$ The vertex is $(2, 1)$. The graph of f opens downward because $a < 0$.

Like other functions that we have studied, quadratic functions also have symbolic, numerical, and graphical representations. The following gives these representations for the quadratic function f that *squares the input x and then subtracts* 1.

Symbolic Representation

$f(x) = x^2 - 1$

Numerical Representation

x	$f(x) = x^2 - 1$
-2	3
-1	0
0	-1
1	0
2	3

Equal

Graphical Representation

11.1 Exercises

CONCEPTS

1. The graph of a quadratic function is called a _____.

2. If a parabola opens upward, what is the lowest point on the parabola called?

3. If a parabola is symmetric with respect to the y-axis, the y-axis is called the _____.

4. The vertex of $y = x^2$ is _____.

5. Sketch a parabola that opens downward with a vertex of $(1, 2)$.

6. If $y = ax^2 + bx + c$, the x-coordinate of the vertex is given by $x =$ _____.

7. Compared to the graph of $y = x^2$, the graph of $y = 2x^2$ is (wider/narrower).

8. The graph of $y = -x^2$ is similar to the graph of $y = x^2$ except that it is _____ across the x-axis.

9. Any quadratic function can be written in the form $f(x) =$ _____.

10. If a parabola opens downward, the point with the largest y-value is called the _____.

GRAPHS OF QUADRATIC FUNCTIONS

Exercises 11–14: Use the given graph of f to evaluate the expressions.

11. $f(-2)$ and $f(0)$

12. $f(-2)$ and $f(2)$

13. $f(-3)$ and $f(1)$

14. $f(-1)$ and $f(2)$

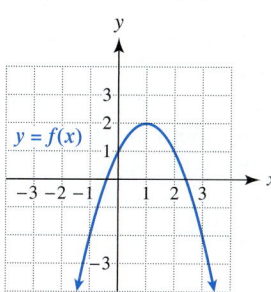

Exercises 15–18: Identify the vertex, axis of symmetry, and whether the parabola opens upward or downward. State where the graph is increasing and where it is decreasing.

15.

16.

17.

18.

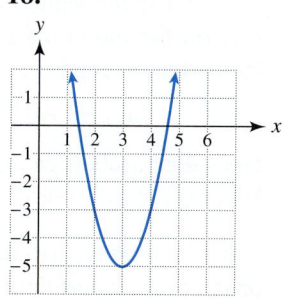

Exercises 19–38: Do the following for the given $f(x)$.
  (a) Graph $y = f(x)$.
  (b) Identify the vertex and axis of symmetry.
  (c) Evaluate $f(-2)$ and $f(3)$.

19. $f(x) = \frac{1}{2}x^2$ **20.** $f(x) = -3x^2$

21. $f(x) = x^2 - 2$ **22.** $f(x) = x^2 - 1$

23. $f(x) = -3x^2 + 1$ **24.** $f(x) = \frac{1}{2}x^2 + 2$

25. $f(x) = (x - 1)^2$ **26.** $f(x) = (x + 2)^2$

27. $f(x) = -(x + 2)^2$ **28.** $f(x) = -(x - 1)^2$

29. $f(x) = x^2 + x - 2$ **30.** $f(x) = x^2 - 2x + 2$

31. $f(x) = 2x^2 - 3$ **32.** $f(x) = 1 - 2x^2$

33. $f(x) = 2x - x^2$ **34.** $f(x) = x^2 + 2x - 8$

35. $f(x) = -2x^2 + 4x - 1$

36. $f(x) = -\frac{1}{2}x^2 + 2x - 3$

37. $f(x) = \frac{1}{4}x^2 - x + 5$ **38.** $f(x) = 3 - 6x - 4x^2$

Exercises 39–46: Find the vertex of the parabola.

39. $f(x) = x^2 - 4x - 2$

40. $f(x) = 2x^2 + 6x - 3$

41. $f(x) = -\frac{1}{3}x^2 - 2x + 1$

42. $f(x) = 5 - 4x + x^2$ **43.** $f(x) = 3 - 2x^2$

44. $f(x) = \frac{1}{4}x^2 - 3x - 2$

45. $f(x) = -0.3x^2 + 0.6x + 1.1$

46. $f(x) = 25 - 10x + 20x^2$

Exercises 47–54: Graph $f(x)$. Compare the graph to $y = x^2$.

47. $f(x) = -x^2$ **48.** $f(x) = -2x^2$

49. $f(x) = 2x^2$ **50.** $f(x) = 3x^2$

51. $f(x) = \frac{1}{4}x^2$ **52.** $f(x) = \frac{1}{2}x^2$

53. $f(x) = -\frac{1}{2}x^2$ **54.** $f(x) = -\frac{3}{2}x^2$

Exercises 55–60: Find the minimum y-value on the graph of $y = f(x)$. Then state where the graph of f is increasing and where it is decreasing.

55. $f(x) = x^2 + 2x - 1$ **56.** $f(x) = x^2 + 6x + 2$

57. $f(x) = x^2 - 5x$ **58.** $f(x) = x^2 - 3x$

59. $f(x) = 2x^2 + 2x - 3$ **60.** $f(x) = 3x^2 - 3x + 7$

Exercises 61–66: Find the maximum y-value on the graph of $y = f(x)$. Then state where the graph of f is increasing and where it is decreasing.

61. $f(x) = -x^2 + 2x + 5$

62. $f(x) = -x^2 + 4x - 3$

63. $f(x) = 4x - x^2$ **64.** $f(x) = 6x - x^2$

65. $f(x) = -2x^2 + x - 5$

66. $f(x) = -5x^2 + 15x - 2$

Exercises 67–72: (Refer to Example 5.) Use the given $f(x)$ to complete the following.
   (a) Does the graph of f open upward or downward? Is this graph wider, narrower, or the same as the graph of $y = x^2$?
   (b) Find the axis of symmetry and the vertex.
   (c) Find the y-intercept and any x-intercepts.
   (d) Sketch a graph of f.

67. $f(x) = \frac{1}{2}x^2 + x - \frac{3}{2}$ **68.** $f(x) = -x^2 + 4x + 5$

69. $f(x) = 2x - x^2$ **70.** $f(x) = x - 2x^2$

71. $f(x) = 2x^2 + 2x - 4$ **72.** $f(x) = \frac{1}{2}x^2 - \frac{1}{2}x - 1$

73. *Numbers* Find two positive numbers whose sum is 20 and whose product is maximum.

74. **Thinking Generally** Find two positive numbers whose sum is k and whose product is maximum.

APPLICATIONS

Exercises 75–78: Quadratic Models Match the physical situation with the graph (a.–d.) that models it best.

75. The height y of a stone thrown from ground level after x seconds.

76. The number of people attending a popular movie x weeks after its opening.

77. The temperature after x hours in a house when the furnace quits and a repair person fixes it.

78. U.S. population from 1800 to the present.

a.

b.

c.

d.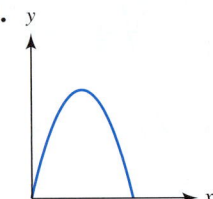

79. *Height Reached by a Baseball*   (Refer to Example 6.) A baseball is hit into the air, and its height h in feet after t seconds is given by $h(t) = -16t^2 + 64t + 2$.
   (a) What is the height of the baseball when it is hit?
   (b) After how many seconds does the baseball reach its maximum height?
   (c) Determine the maximum height of the baseball.

80. *Height Reached by a Golf Ball*   A golf ball is hit into the air, and its height h in feet after t seconds is given by $h(t) = -16t^2 + 128t$.
   (a) What is the height of the golf ball when it is hit?
   (b) After how many seconds does the golf ball reach its maximum height?
   (c) Determine the maximum height of the golf ball.

81. *Height Reached by a Baseball*   Suppose that a baseball is thrown upward with an initial velocity of 66 feet per second (45 miles per hour) and it is released 6 feet above the ground. Its height h after t seconds is given by $h(t) = -16t^2 + 66t + 6$. After how many seconds does the baseball reach a maximum height? Estimate this height.

82. *Throwing a Baseball on the Moon*   (Refer to Exercise 81.) If the same baseball were thrown the same way on the moon, its height h above the moon's surface after t seconds would be

$$h(t) = -2.55t^2 + 66t + 6.$$

Does the baseball go higher on the moon or on Earth? What is the difference in these two heights?

83. *Concert Tickets*   (Refer to Example 7.) An agency is promoting concert tickets by offering a group-discount rate. The regular price is $100 and for each ticket bought the price decreases by $1. (One ticket costs $99, two tickets cost $98 *each*, and so on.)

(a) A graph of the revenue received from selling x tickets is shown in the figure. When is revenue increasing and when is it decreasing?

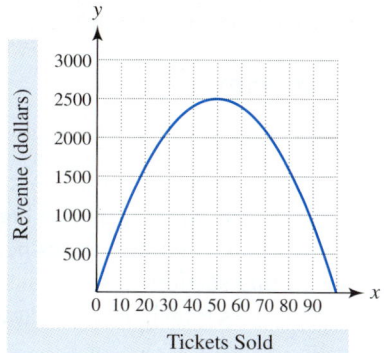

Tickets Sold

(b) What is the maximum revenue? How many tickets should be sold to maximize revenue?

(c) Write a formula for $y = f(x)$ whose graph is shown in the figure.

(d) Use $f(x)$ to determine symbolically the maximum revenue and the number of tickets that should be sold to maximize revenue.

84. *Maximizing Revenue*   The regular price for a round-trip ticket to Las Vegas, Nevada, charged by an airline charter company is $300. For a group rate the company will reduce the price of each ticket by $1.50 for every passenger on the flight.

(a) Write a formula $f(x)$ that gives the revenue from selling x tickets.

(b) Determine how many tickets should be sold to maximize the revenue. What is the maximum revenue?

85. *Maximizing Area*   A farmer is fencing a rectangular area for cattle using a straight portion of a river as one side of the rectangle as illustrated in the figure. Note that there is no fence along the river. If the farmer has 1200 feet of fence, find the dimensions for the rectangular area that gives a maximum area for the cattle.

River

86. *Maximizing Area*   A rectangular pen being constructed for a pet requires 60 feet of fence.

(a) Write a formula $f(x)$ that gives the area of the pen if one side of the pen has length x.

(b) Find the dimensions of the pen that give the largest area. What is the largest area?

87. *Seedling Growth*   The effect of temperature on the growth of melon seedlings was studied, where the seedlings were grown at different temperatures, and their heights were measured after a fixed period of time. The findings of this study can be modeled by

$$f(x) = -0.095x^2 + 5.4x - 52.2,$$

where x is the temperature in degrees Celsius and the output $f(x)$ gives the resulting average height in centimeters.   (*Source:* R. Pearl, "The growth of *Cucumis melo* seedlings at different temperatures.")

(a) Graph f in [20, 40, 5] by [0, 30, 5].

(b) Estimate graphically the temperature that resulted in the greatest height for the melon seedlings.

(c) Solve part (b) symbolically.

88. *Game Length*   The quadratic function defined by

$$L(x) = -2x^2 + 8000x - 7,999,820$$

approximates the length of the average major league baseball game in minutes during year x, where $1998 \leq x \leq 2002$.   (*Source:* Elias Sports Bureau.)

(a) Determine the year when games were the longest.

(b) How long did the average game last?

Exercises 89 and 90: Large-Screen Televisions (Refer to the introduction to this chapter.) Use the formula

$$S(x) = -0.227x^2 + 8.155x - 8.8$$

to estimate the recommended screen size in inches when viewers sit x feet from the screen.

89. $x = 8$ feet   90. $x = 12$ feet

WRITING ABOUT MATHEMATICS

91. If $f(x) = ax^2 + bx + c$, explain how the values of a and c affect the graph of f.

92. Suppose that a quantity Q is modeled by the formula $Q(x) = ax^2 + bx + c$ with $a < 0$. Explain how to find the x-value that maximizes $Q(x)$. How do you find the maximum value of $Q(x)$?

11.2 PARABOLAS AND MODELING

Vertical and Horizontal Translations ■ Vertex Form ■ Modeling with Quadratic Functions (Optional)

A LOOK INTO MATH ▷ A taxiway used by an airplane to exit a runway often contains curves. A curve that is too sharp for the speed of the plane is a safety hazard. The scatterplot shown in Figure 11.17 gives an appropriate radius R of a curve designed for an airplane taxiing x miles per hour. The data are nonlinear because they do not lie on a line. In this section we explain how a quadratic function may be used to model such data. First we discuss translations of parabolas.

Figure 11.17

Vertical and Horizontal Translations

The graph of $y = x^2$ is a parabola opening upward with vertex $(0, 0)$. Suppose that we graph $y_1 = x^2$, $y_2 = x^2 + 1$, and $y_3 = x^2 - 2$ in the same xy-plane, as calculated for Table 11.4 and shown in Figure 11.18. All three graphs have the same shape. However, compared to the graph of $y_1 = x^2$, the graph of $y_2 = x^2 + 1$ is shifted *upward* **1** unit and the graph of $y_3 = x^2 - 2$ is shifted *downward* **2** units. Such shifts are called **translations** because they do not change the shape of a graph—only its position.

TABLE 11.4

x	$y_1 = x^2$	$y_2 = x^2 + 1$	$y_3 = x^2 - 2$
-2	4	5	2
-1	1	2	-1
0	0	1	-2
1	1	2	-1
2	4	5	2

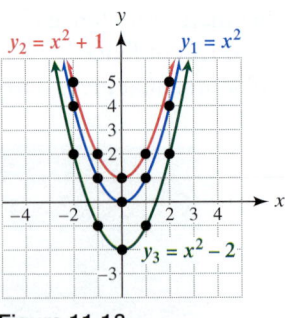

Figure 11.18

Next, suppose that we graph $y_1 = x^2$ and $y_2 = (x - 1)^2$ in the same xy-plane. Compare Tables 11.5 and 11.6. Note that the y-values are equal when the x-value for y_2 is 1 unit *larger* than the x-value for y_1. For example, $y_1 = 4$ when $x = -2$ and $y_2 = 4$ when $x = -1$. Thus the graph of $y_2 = (x - 1)^2$ has the same shape as the graph of $y_1 = x^2$, except that it is translated *horizontally to the right* 1 unit, as illustrated in Figure 11.19.

TABLE 11.5

x	$y_1 = x^2$
-2	4
-1	1
0	0
1	1
2	4

TABLE 11.6

x	$y_2 = (x - 1)^2$
-1	4
0	1
1	0
2	1
3	4

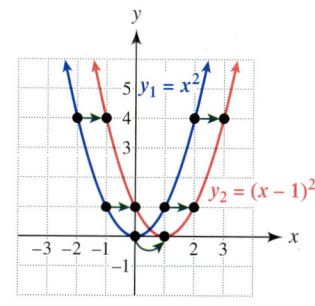

Figure 11.19

The graphs $y_1 = x^2$ and $y_2 = (x + 2)^2$ are shown in Figure 11.20. Note that Tables 11.7 and 11.8 show their y-values to be equal when the x-value for y_2 is 2 units *smaller* than the x-value for y_1. As a result, the graph of $y_2 = (x + 2)^2$ has the same shape as the graph of $y_1 = x^2$ except that it is translated *horizontally to the left* 2 units.

TABLE 11.7

x	$y_1 = x^2$
-2	4
-1	1
0	0
1	1
2	4

TABLE 11.8

x	$y_2 = (x + 2)^2$
-4	4
-3	1
-2	0
-1	1
0	4

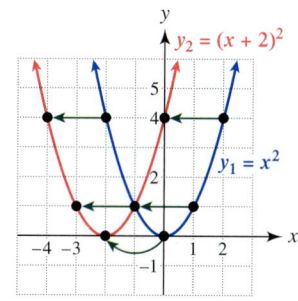

Figure 11.20

These results are summarized as follows.

VERTICAL AND HORIZONTAL TRANSLATIONS OF PARABOLAS

Let h and k be positive numbers.

To graph	*shift the graph of $y = x^2$ by k units*
$y = x^2 + k$	upward.
$y = x^2 - k$	downward.

To graph	*shift the graph of $y = x^2$ by h units*
$y = (x - h)^2$	right.
$y = (x + h)^2$	left.

The next example demonstrates this method.

EXAMPLE **1** **Translating the graph $y = x^2$**

Sketch the graph of the equation and identify the vertex.
(a) $y = x^2 + 2$ **(b)** $y = (x + 3)^2$ **(c)** $y = (x - 2)^2 - 3$

Solution
(a) The graph of $y = x^2 + 2$ is similar to the graph of $y = x^2$ except that it has been translated upward 2 units, as shown in Figure 11.21(a). The vertex is $(0, 2)$.
(b) The graph of $y = (x + 3)^2$ is similar to the graph of $y = x^2$ except that it has been translated *left* 3 units, as shown in Figure 11.21(b). The vertex is $(-3, 0)$.

> **NOTE:** If you are thinking that the graph should be shifted right (instead of left) 3 units, try graphing $y = (x + 3)^2$ on a graphing calculator.

(c) The graph of $y = (x - 2)^2 - 3$ is similar to the graph of $y = x^2$ except that it has been translated downward 3 units *and* right 2 units, as shown in Figure 11.21(c). The vertex is $(2, -3)$.

(a)

(b)

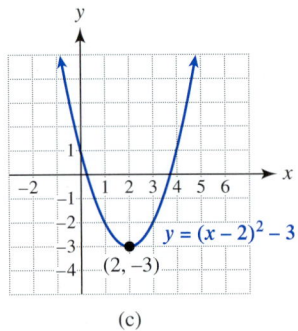
(c)

Figure 11.21

Now Try Exercises **9, 13, 21**

Vertex Form

Suppose that a parabola has the equation $y = ax^2 + bx + c$ with vertex (h, k). We can write this equation in a different form called the *vertex form* by transforming the graph of $y = x^2$. The vertex for $y = x^2$ is $(0, 0)$ so we need to translate it h units horizontally and k units vertically. Thus $y = (x - \boldsymbol{h})^2 + \boldsymbol{k}$ has vertex $(\boldsymbol{h}, \boldsymbol{k})$. For the graph of our new equation to open correctly and have the same shape as $y = ax^2 + bx + c$, we must be sure that their leading coefficients are identical. That is, the graph of $y = a(x - h)^2 + k$ is identical to $y = ax^2 + bx + c$, provided that the vertex for the second equation is (h, k). This discussion is summarized as follows.

> **VERTEX FORM**
>
> The **vertex form** of the equation of a parabola with vertex (h, k) is
> $$y = a(x - h)^2 + k,$$
> where $a \neq 0$ is a constant. If $a > 0$, the parabola opens upward; if $a < 0$, the parabola opens downward.

NOTE: Vertex form is sometimes called **standard form for a parabola with a vertical axis.**

In the next three examples we demonstrate the graphing of parabolas in vertex form, finding their equations, and writing vertex forms of equations.

EXAMPLE 2   Graphing parabolas in vertex form

Compare the graph of $y = f(x)$ to the graph of $y = x^2$. Then sketch a graph of $y = f(x)$ and $y = x^2$ in the same xy-plane.

(a) $f(x) = \frac{1}{2}(x - 5)^2 + 2$ **(b)** $f(x) = -3(x + 5)^2 - 3$

Solution

(a) Compared to the graph of $y = x^2$, the graph of $y = f(x)$ is translated 5 units right and 2 units upward. The vertex for $f(x)$ is $(5, 2)$, whereas the vertex of $y = x^2$ is $(0, 0)$. Because $a = \frac{1}{2}$, the graph of $y = f(x)$ opens upward and is wider than the graph of $y = x^2$. These graphs are shown in Figure 11.22(a).

(b) Compared to the graph of $y = x^2$, the graph of $y = f(x)$ is translated 5 units left and 3 units downward. The vertex for $f(x)$ is $(-5, -3)$. Because $a = -3$, the graph of $y = f(x)$ opens downward and is narrower than the graph of $y = x^2$. These graphs are shown in Figure 11.22(b).

(a)

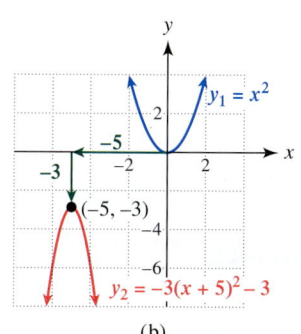

(b)

Figure 11.22

Now Try Exercises 29, 31

EXAMPLE 3   Finding equations of parabolas

Write the vertex form of a parabola with $a = 2$ and vertex $(-2, 1)$. Then express this equation in the form $y = ax^2 + bx + c$.

Solution

The vertex form of a parabola is $y = a(x - h)^2 + k$, where the vertex is (h, k). For $a = \mathbf{2}, h = \mathbf{-2}$, and $k = \mathbf{1}$, the equation becomes

$$y = \mathbf{2}(x - (\mathbf{-2}))^2 + \mathbf{1} \quad \text{or} \quad y = 2(x + 2)^2 + 1.$$

To write this equation in the form $y = ax^2 + bx + c$, do the following.

$$
\begin{aligned}
y &= 2(x^2 + 4x + 4) + 1 && \text{Multiply } (x + 2)^2. \\
&= (2x^2 + 8x + 8) + 1 && \text{Distributive property} \\
&= 2x^2 + 8x + 9 && \text{Add.}
\end{aligned}
$$

The equivalent equation is $y = 2x^2 + 8x + 9$.

Now Try Exercise 35

COMPLETING THE SQUARE TO FIND THE VERTEX   If we are given the equation $y = x^2 + 4x + 2$, can we write it in an equivalent vertex form? The answer is *yes*, and in this process we use the **completing the square method**. In Section 5.4 we learned that $(a + b)^2 = a^2 + 2ab + b^2$. If we apply this result to $\left(x + \frac{b}{2}\right)^2$, we obtain the following.

$$\left(x + \frac{b}{2}\right)^2 = x^2 + 2x\left(\frac{b}{2}\right) + \left(\frac{b}{2}\right)^2$$

$$= x^2 + bx + \left(\frac{b}{2}\right)^2$$

CRITICAL THINKING

Use the figure to *complete the square* for $x^2 + 8x$.

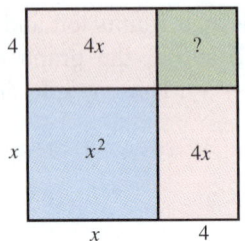

Thus we can always complete the square for $x^2 + bx$ by adding $\left(\frac{b}{2}\right)^2$. For example, if $y = x^2 + 8x + 2$, then $b = 8$ and so $\left(\frac{b}{2}\right)^2 = \left(\frac{8}{2}\right)^2 = 16$. If we simply add 16 to the right side of this equation to complete the square, the resulting equation will *not be equivalent* to the given equation. To avoid this situation, we *add and subtract* 16 on the right side of the equation. Adding 16 and then subtracting 16 is equivalent to adding 0, which does not change the equation.

$$
\begin{aligned}
y &= x^2 + \mathbf{8}x + 2 && \text{Given equation} \\
&= (x^2 + 8x + \mathbf{16}) - \mathbf{16} + 2 && \text{Add and subtract 16.} \\
&= (x + 4)^2 - 14 && \text{Perfect square trinomial}
\end{aligned}
$$

Thus $y = x^2 + 8x + 2$ and $y = (x + 4)^2 - 14$ are equivalent equations. The vertex for this parabola is $(-4, -14)$. Note that we added *and* subtracted 16, so the right side of the equation did not change in value.

EXAMPLE 4   Writing vertex form

Write each equation in vertex form. Identify the vertex.
(a) $y = x^2 - 6x - 1$ **(b)** $y = x^2 + 3x + 4$ **(c)** $y = 2x^2 + 4x - 1$

Solution
(a) Because $\left(\frac{b}{2}\right)^2 = \left(\frac{-6}{2}\right)^2 = \mathbf{9}$, add and subtract 9 on the right side.

$$
\begin{aligned}
y &= x^2 - \mathbf{6}x - 1 && \text{Given equation} \\
&= (x^2 - 6x + \mathbf{9}) - \mathbf{9} - 1 && \text{Add and subtract 9.} \\
&= (x - 3)^2 - 10 && \text{Perfect square trinomial}
\end{aligned}
$$

The vertex is $(3, -10)$.

(b) Because $\left(\frac{b}{2}\right)^2 = \left(\frac{3}{2}\right)^2 = \frac{9}{4}$, add and subtract $\frac{9}{4}$ on the right side.

$$
\begin{aligned}
y &= x^2 + \mathbf{3}x + 4 && \text{Given equation} \\
&= \left(x^2 + 3x + \frac{\mathbf{9}}{\mathbf{4}}\right) - \frac{\mathbf{9}}{\mathbf{4}} + 4 && \text{Add and subtract } \frac{9}{4}. \\
&= \left(x + \frac{3}{2}\right)^2 + \frac{7}{4} && \text{Perfect square trinomial}
\end{aligned}
$$

The vertex is $\left(-\frac{3}{2}, \frac{7}{4}\right)$.

(c) This equation is slightly different because the leading coefficient is 2 rather than 1. Start by factoring 2 from the first two terms on the right side.

$$y = 2x^2 + 4x - 1 \qquad \text{Given equation}$$
$$= 2(x^2 + \mathbf{2}x) - 1 \qquad \text{Factor out 2.}$$
$$= 2(x^2 + 2x + \mathbf{1} - \mathbf{1}) - 1 \qquad \left(\tfrac{b}{2}\right)^2 = \left(\tfrac{2}{2}\right)^2 = 1$$
$$= 2(x^2 + 2x + 1) - 2 - 1 \qquad \text{Distributive property } (2 \cdot -1)$$
$$= 2(x + 1)^2 - 3 \qquad \text{Perfect square trinomial; add.}$$

The vertex is $(-1, -3)$.

Now Try Exercises 47, 51, 55

Modeling with Quadratic Functions (Optional)

▶ **REAL-WORLD CONNECTION** In the introduction to this section we discussed airport taxiway curves designed for airplanes. The data previously shown in Figure 11.17 are listed in Table 11.9.

TABLE 11.9

x (mph)	10	20	30	40	50	60
R (ft)	50	200	450	800	1250	1800

Source: Federal Aviation Administration.

A second scatterplot of the data is shown in Figure 11.23. The data may be modeled by $R(x) = ax^2$ for some value a. To illustrate this relation, graph R for different values of a. In Figures 11.24–11.26, R has been graphed for $a = 2$, -1, and $\frac{1}{2}$, respectively. When $a > 0$ the parabola opens upward and when $a < 0$ the parabola opens downward. Larger values of $|a|$ make a parabola narrower, whereas smaller values of $|a|$ make the parabola wider. Through trial and error, $a = \frac{1}{2}$ gives a good fit to the data, so $R(x) = \frac{1}{2}x^2$ models the data.

CALCULATOR HELP

To make a scatterplot, see Appendix A (page AP-4).

$[-70, 70, 10]$ by $[-2000, 2000, 500]$

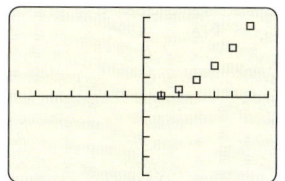

Figure 11.23

$[-70, 70, 10]$ by $[-2000, 2000, 500]$

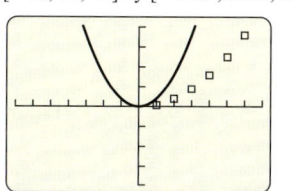

Figure 11.24 $a = 2$

$[-70, 70, 10]$ by $[-2000, 2000, 500]$

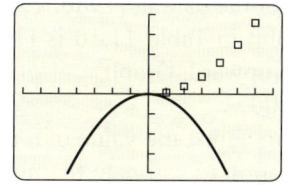

Figure 11.25 $a = -1$

$[-70, 70, 10]$ by $[-2000, 2000, 500]$

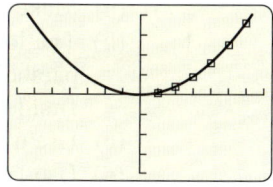

Figure 11.26 $a = \frac{1}{2}$

This value of a can also be found *symbolically*, as demonstrated in the next example.

EXAMPLE 5 **Modeling safe taxiway speed**

Find a value for the constant a symbolically so that $R(x) = ax^2$ models the data in Table 11.9 on the previous page. Check your result by making a table of values for $R(x)$.

Solution

When $x = 10$ miles per hour, the curve radius is $R(x) = 50$ feet. Therefore

$$R(10) = 50 \quad \text{or} \quad a(10)^2 = 50.$$

Solving for a gives

$$a = \frac{50}{10^2} = \frac{1}{2}.$$

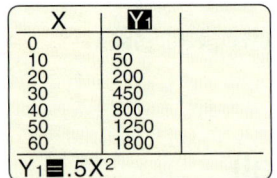

Figure 11.27

To be sure that $R(x) = \frac{1}{2}x^2$ is correct, construct a table, as shown in Figure 11.27. Its values agree with those in Table 11.9.

Now Try Exercise 69

▶ **REAL-WORLD CONNECTION**   In 1981 the first cases of AIDS were reported in the United States. Since then, AIDS has become one of the most devastating diseases of recent times. Table 11.10 lists the *cumulative* number of AIDS cases in the United States for various years. For example, between 1981 and 1990, a total of 199,608 AIDS cases were reported.

A scatterplot of these data is shown in Figure 11.28. To model these nonlinear data, we want to find (the right half of) a parabola with the shape illustrated in Figure 11.29. We do so in the next example.

TABLE 11.10

Year	AIDS Cases
1981	425
1984	11,106
1987	71,414
1990	199,608
1993	417,835
1996	609,933

Source: Department of Health and Human Services.

Figure 11.28 AIDS Cases

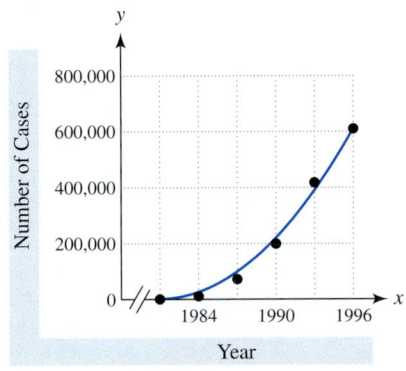

Figure 11.29 Modeling AIDS Cases

EXAMPLE 6 **Modeling AIDS cases**

Use the data in Table 11.10 to complete the following.
(a) Make a scatterplot of the data in $[1980, 1997, 2]$ by $[-10000, 800000, 100000]$.
(b) The lowest data point in Table 11.10 is $(1981, 425)$. Let this point be the vertex of a parabola that opens upward. Graph $y = a(x - 1981)^2 + 425$ together with the data by first letting $a = 1000$.
(c) Use trial and error to adjust the value of a until the graph models the data.
(d) Use your final equation to estimate the number of AIDS cases in 1992. Compare it to the known value of 338,786.

Solution

(a) A scatterplot of the data is shown in Figure 11.30.

[1980, 1997, 2] by
[−10000, 800000, 100000]

[1980, 1997, 2] by
[−10000, 800000, 100000]

[1980, 1997, 2] by
[−10000, 800000, 100000]

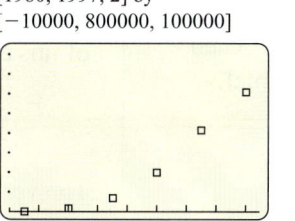

Figure 11.30

Figure 11.31 $a = 1000$

Figure 11.32 $a = 2700$

(b) A graph of $y = 1000(x - 1981)^2 + 425$ is shown in Figure 11.31. To have a better fit of the data, a larger value for a is needed.

(c) Figure 11.32 shows the effect of adjusting the value of a to 2700. This value provides a reasonably good fit. (Note that you may decide on a slightly different value for a.)

(d) If $a = 2700$, the modeling equation becomes

$$y = 2700(x - 1981)^2 + 425.$$

To estimate the number of AIDS cases in 1992, substitute $x = \mathbf{1992}$ to obtain

$$y = 2700(\mathbf{1992} - 1981)^2 + 425 = 327{,}125.$$

This number is about 12,000 less than the known value of 338,786.

Now Try Exercise **71**

11.2 PUTTING IT ALL TOGETHER

The following table summarizes some of the important topics in this section.

Concept	Explanation	Examples
Translations of Parabolas	Compared to the graph $y = x^2$, the graph of $y = x^2 + k$ is shifted vertically k units and the graph of $y = (x - h)^2$ is shifted horizontally h units.	Compared to the graph of $y = x^2$, the graph of $y = x^2 - 4$ is shifted downward 4 units. Compared to the graph of $y = x^2$, the graph of $y = (x - 4)^2$ is shifted right 4 units and the graph of $y = (x + 4)^2$ is shifted left 4 units.
Vertex Form	The vertex form of the equation of a parabola with vertex (h, k) is $$y = a(x - h)^2 + k,$$ where $a \neq 0$ is a constant. If $a > 0$, the parabola opens upward; if $a < 0$, the parabola opens downward.	The graph of $y = 3(x + 2)^2 - 7$ has a vertex of $(-2, -7)$ and opens upward because $3 > 0$.

continued on next page

continued from previous page

Concept	Explanation	Examples
Completing the Square Method	To complete the square to obtain the vertex form, add and subtract $\left(\frac{b}{2}\right)^2$ on the right side of $y = x^2 + bx + c$. Then factor the perfect square trinomial.	If $y = x^2 + 10x - 3$, then add and subtract $\left(\frac{b}{2}\right)^2 = \left(\frac{10}{2}\right)^2 = 25$ on the right side of this equation. $$y = (x^2 + 10x + 25) - 25 - 3$$ $$= (x + 5)^2 - 28$$ The vertex is $(-5, -28)$.

11.2 Exercises

MyMathLab

PRACTICE   WATCH   DOWNLOAD   READ   REVIEW

CONCEPTS

1. Compared to the graph of $y = x^2$, the graph of $y =$ _____ is shifted upward 2 units.

2. Compared to the graph of $y = x^2$, the graph of $y =$ _____ is shifted to the right 2 units.

3. The vertex of $y = (x - 1)^2 + 2$ is _____.

4. The vertex of $y = (x + 1)^2 - 2$ is _____.

5. A quadratic function f may be written either in the form _____ or _____.

6. The vertex form of a parabola is given by _____ and its vertex is _____.

7. The graph of $y = -x^2$ is a parabola that opens _____.

8. The x-coordinate of the vertex of $y = ax^2 + bx + c$ is $x =$ _____.

GRAPHS OF PARABOLAS

Exercises 9–28: Do the following.
  (a) Sketch a graph of the equation.
  (b) Identify the vertex.
  (c) Compare the graph to the graph of $y = x^2$. (State any transformations used.)

9. $f(x) = x^2 - 4$ 10. $f(x) = x^2 - 1$

11. $f(x) = 2x^2 + 1$ 12. $f(x) = \frac{1}{2}x^2 + 1$

13. $f(x) = (x - 3)^2$ 14. $f(x) = (x + 1)^2$

15. $f(x) = -x^2$ 16. $f(x) = -(x + 2)^2$

17. $f(x) = 2 - x^2$ 18. $f(x) = (x - 1)^2$

19. $f(x) = (x + 2)^2$ 20. $f(x) = (x - 2)^2 - 3$

21. $f(x) = (x + 1)^2 - 2$

22. $f(x) = (x - 3)^2 + 1$

23. $f(x) = (x - 1)^2 + 2$

24. $f(x) = \frac{1}{2}(x + 3)^2 - 3$

25. $f(x) = 2(x - 5)^2 - 4$

26. $f(x) = -3(x + 4)^2 + 5$

27. $f(x) = -\frac{1}{2}(x + 3)^2 + 1$

28. $f(x) = 2(x - 5)^2 + 10$

Exercises 29–32: Compare the graph of $y = f(x)$ to the graph of $y = x^2$. Then sketch a graph of $y = f(x)$ and $y = x^2$ in the same xy-plane.

29. $f(x) = \frac{1}{2}(x - 1)^2 - 2$

30. $f(x) = 2(x + 2)^2 - 1$

31. $f(x) = -2(x + 1)^2 + 3$

32. $f(x) = -\frac{1}{2}(x - 2)^2 + 2$

Exercises 33 and 34: Graph the equation in a window that shows the vertex and all intercepts.

33. $f(x) = -0.4x^2 + 6x - 10$

34. $f(x) = 3x^2 - 40x + 50$

VERTEX FORM

Exercises 35–38: (Refer to Example 3.) Write the vertex form of a parabola that satisfies the conditions given. Then write the equation in the form $y = ax^2 + bx + c$.

35. Vertex $(3, 4)$ and $a = 3$

36. Vertex $(-1, 3)$ and $a = -5$

37. Vertex $(5, -2)$ and $a = -\frac{1}{2}$

38. Vertex $(-2, -6)$ and $a = \frac{3}{4}$

Exercises 39–42: Write the vertex form of a parabola that satisfies the conditions given. Assume that $a = \pm 1$.

39. Opens upward, vertex $(1, 2)$

40. Opens downward, vertex $(-1, -2)$

41. Opens downward, vertex $(0, -3)$

42. Opens upward, vertex $(5, -4)$

Exercises 43–46: Write the vertex form of the parabola shown in the graph. Assume that $a = \pm 1$.

43.

44.

45.

46.
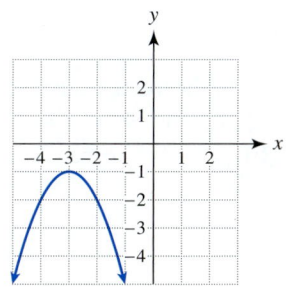

Exercises 47–60: (Refer to Example 4.) Write the equation in vertex form. Identify the vertex.

47. $y = x^2 + 2x - 3$ **48.** $y = x^2 + 4x + 1$

49. $y = x^2 - 4x + 5$ **50.** $y = x^2 - 8x + 10$

51. $y = x^2 + 3x - 2$ **52.** $y = x^2 + 5x - 4$

53. $y = x^2 - 7x + 1$ **54.** $y = x^2 - 3x + 5$

55. $y = 3x^2 + 6x - 1$ **56.** $y = 2x^2 + 4x - 9$

57. $y = 2x^2 - 3x$ **58.** $y = 3x^2 - 7x$

59. $y = -2x^2 - 8x + 5$ **60.** $y = -3x^2 + 6x + 1$

MODELING DATA

Exercises 61–64: Find a value for the constant a so that $f(x) = ax^2$ models the data. If you are uncertain about your value for a, check it by making a table of values.

61.

x	1	2	3
y	2	8	18

62.

x	-2	0	2
y	6	0	6

63.

x	2	4	6	8
y	1.2	4.8	10.8	19.2

64.

x	5	10	15	20
y	17.5	70	157.5	280

 Exercises 65–68: **Modeling Quadratic Data** *(Refer to Example 6.) Find a quadratic function expressed in vertex form that models the data in the given table.*

65.

x	1	2	3	4
y	-3	-1	5	15

66.

x	-2	-1	0	1	2
y	5	2	-7	-22	-43

67.

x	1980	1990	2000	2010
y	6	55	210	450

68.

x	1990	1995	2000	2005
y	10	60	205	470

Year	1982	1986	1990	1994
Deaths (thousands)	1	25	123	300

(**Source:** Department of Health and Human Services.)

(a) Determine $f(x) = a(x - h)^2 + k$, so that f models these data.

(b) Estimate the number of AIDS deaths in 1992 and compare it to the actual value of 202 thousand.

69. *Braking Distance*   The table lists approximate braking distances D in feet for cars traveling at x miles per hour on dry, level pavement.

x	12	24	36	48
D	12	48	108	192

(a) Make a scatterplot of the data.
(b) Find a function given by $D(x) = ax^2$ that models these data.

72. *Head Start Enrollment*   The table lists numbers of students *in thousands* enrolled in Head Start for selected years.

Year	1966	1980	1995
Students (thousands)	733	376	750

70. *Health Care Costs*   The table lists approximate *annual* percentage increases in the cost of health insurance premiums between 1992 and 2000.

Year	1992	1994	1996	1998	2000
Increase	11%	4%	1%	4%	11%

Source: Kaiser Family Foundation.

(a) Determine $f(x) = a(x - h)^2 + k$, so that f models these data.
(b) Estimate Head Start enrollment in 1990 and compare it to the actual value of 541 thousand.

WRITING ABOUT MATHEMATICS

73. Explain how to find the vertex of $y = x^2 + bx + c$ by completing the square.

(a) Make a scatterplot of the data.
(b) Find a function given by $C(x) = a(x - h)^2 + k$ that models these data. (Answers may vary.)

74. If $f(x) = a(x - h)^2 + k$, explain how the values of a, h, and k affect the graph of $y = f(x)$.

71. *U.S. AIDS Deaths*   (Refer to Example 6.) The table at the top of the next column lists cumulative numbers of AIDS deaths in *thousands* for selected years.

CHECKING BASIC CONCEPTS

SECTIONS 11.1 AND 11.2

1. Graph each quadratic function. Identify the vertex and axis of symmetry.
(a) $f(x) = x^2 - 2$
(b) $f(x) = x^2 - 2x - 2$

2. Compare the graph of $y_1 = 2x^2$ to the graph of $y_2 = -\frac{1}{2}x^2$.

3. Find the maximum y-value on the graph of $y = -3x^2 + 12x - 5$. State where the graph is increasing and where it is decreasing.

4. Sketch a graph of $y = f(x)$. Compare the graph of f to the graph of $y = x^2$.
(a) $f(x) = (x - 1)^2 + 2$
(b) $f(x) = -(x + 3)^2$

5. Write the vertex form for each equation.
(a) $y = x^2 + 14x - 7$
(b) $y = 4x^2 + 8x - 2$

11.3 QUADRATIC EQUATIONS

Basics of Quadratic Equations ▪ The Square Root Property ▪ Completing the Square ▪ Solving an Equation for a Variable ▪ Applications of Quadratic Equations

A LOOK INTO MATH ▷

In Section 11.2 we modeled curves on airport taxiways by using $R(x) = \frac{1}{2}x^2$. In this formula x represented the airplane's speed in miles per hour, and R represented the radius of the curve in feet. This formula may be used to determine the speed limit for a curve with a radius of **650** feet by solving the *quadratic equation*

$$\frac{1}{2}x^2 = 650.$$

In this section we demonstrate techniques for solving this and other quadratic equations.

Basics of Quadratic Equations

Any quadratic function f can be represented by $f(x) = ax^2 + bx + c$ with $a \neq 0$. Examples of quadratic functions include

$$f(x) = 2x^2 - 1, \quad g(x) = -\frac{1}{3}x^2 + 2x, \quad \text{and} \quad h(x) = x^2 + 2x - 1.$$

Quadratic functions can be used to write quadratic equations. Examples of quadratic equations include

$$2x^2 - 1 = 0, \quad -\frac{1}{3}x^2 + 2x = 0, \quad \text{and} \quad x^2 + 2x - 1 = 0.$$

QUADRATIC EQUATION

A **quadratic equation** is an equation that can be written as

$$ax^2 + bx + c = 0,$$

where a, b, and c are constants with $a \neq 0$.

Solutions to the quadratic equation $ax^2 + bx + c = 0$ correspond to x-intercepts of the graph of $y = ax^2 + bx + c$. Because the graph of a quadratic function is either $\cup$-shaped or $\cap$-shaped, it can intersect the x-axis zero, one, or two times, as illustrated in Figure 11.33. Hence a quadratic equation can have zero, one, or two real solutions.

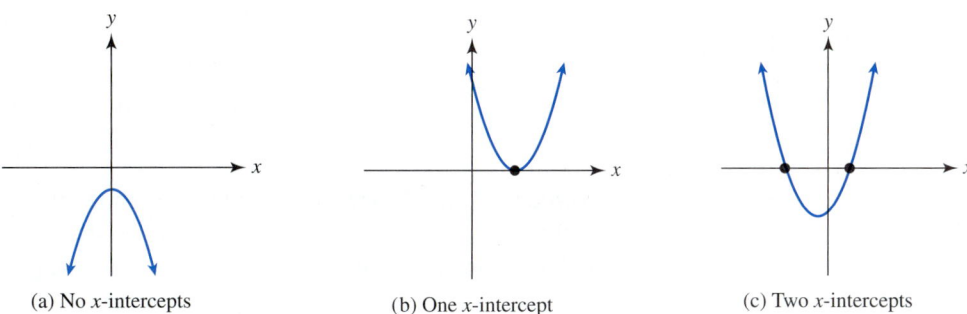

(a) No x-intercepts   (b) One x-intercept   (c) Two x-intercepts

Figure 11.33

We have already solved quadratic equations by factoring, graphing, and constructing tables. In the next example we apply these three techniques to quadratic equations that have no real solutions, one real solution, and two real solutions.

EXAMPLE **1**   ### Solving quadratic equations

Solve each quadratic equation. Support your results numerically and graphically.
(a) $2x^2 + 1 = 0$ (No real solutions)
(b) $x^2 + 4 = 4x$ (One real solution)
(c) $x^2 - 6x + 8 = 0$ (Two real solutions)

Solution
(a) *Symbolic Solution*

$$2x^2 + 1 = 0 \qquad \text{Given equation}$$
$$2x^2 = -1 \qquad \text{Subtract 1.}$$
$$x^2 = -\frac{1}{2} \qquad \text{Divide by 2.}$$

This equation has no real-number solutions because $x^2 \geq 0$ for all real numbers x.

Numerical and Graphical Solution   The points in Table 11.11 for $y = 2x^2 + 1$ are plotted in Figure 11.34 and connected with a parabolic graph. The graph of $y = 2x^2 + 1$ has no x-intercepts, indicating that there are no real solutions.

TABLE 11.11

x	y
-2	9
-1	3
0	1
1	3
2	9

For all x, $y \neq 0$.

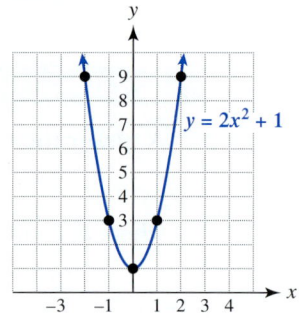

Figure 11.34 No Solutions

(b) *Symbolic Solution*

$$x^2 + 4 = 4x \qquad \text{Given equation}$$
$$x^2 - 4x + 4 = 0 \qquad \text{Subtract } 4x \text{ from each side.}$$
$$(x - 2)(x - 2) = 0 \qquad \text{Factor.}$$
$$x - 2 = 0 \quad \text{or} \quad x - 2 = 0 \qquad \text{Zero-product property}$$
$$x = 2 \qquad \text{There is one solution.}$$

Numerical and Graphical Solution   Because the given quadratic equation is equivalent to $x^2 - 4x + 4 = 0$, we let $y = x^2 - 4x + 4$. The points in Table 11.12 are plotted in Figure 11.35 and connected with a parabolic graph. The graph of $y = x^2 - 4x + 4$ has one x-intercept, 2. Note that in Table 11.12, $y = 0$ when $x = 2$, indicating that the equation has one solution.

TABLE 11.12

x	y
0	4
1	1
2	**0**
3	1
4	4

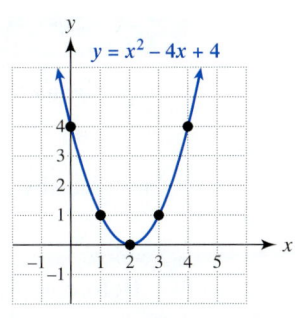

Figure 11.35 One Solution

(c) *Symbolic Solution*

$$
\begin{aligned}
x^2 - 6x + 8 &= 0 && \text{Given equation} \\
(x - 2)(x - 4) &= 0 && \text{Factor.} \\
x - 2 = 0 \quad \text{or} \quad x - 4 &= 0 && \text{Zero-product property} \\
x = 2 \quad \text{or} \quad x &= 4 && \text{There are two solutions.}
\end{aligned}
$$

Numerical and Graphical Solution   The points in Table 11.13 for $y = x^2 - 6x + 8$ are plotted in Figure 11.36 and connected with a parabolic graph. The graph of $y = x^2 - 6x + 8$ has two x-intercepts, 2 and 4, indicating two solutions. Note that in Table 11.13 $y = 0$ when $x = 2$ or $x = 4$.

TABLE 11.13

x	y
0	8
1	3
2	**0**
3	−1
4	**0**
5	3
6	8

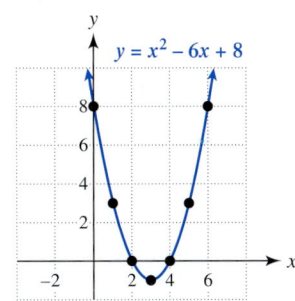

Figure 11.36 Two Solutions

Now Try Exercises 29, 37, 39

The Square Root Property

The **square root property** is used to solve quadratic equations that have no x-terms. The following is an example of the square root property.

$$x^2 = 25 \quad \text{is equivalent to} \quad x = \pm 5.$$

The equation $x = \pm 5$ (read "x equals plus or minus 5") indicates that either $x = 5$ or $x = -5$. Each value is a solution because $(5)^2 = 25$ and $(-5)^2 = 25$.

We can derive this result in general for $k \geq 0$.

$$
\begin{aligned}
x^2 &= k && \text{Given quadratic equation} \\
\sqrt{x^2} &= \sqrt{k} && \text{Take the square root of each side.} \\
|x| &= \sqrt{k} && \sqrt{x^2} = |x| \text{ for all } x. \\
x &= \pm\sqrt{k} && |x| = b \text{ implies } x = \pm b, b \geq 0.
\end{aligned}
$$

This result is summarized by the *square root property*.

SQUARE ROOT PROPERTY

Let k be a nonnegative number. Then the solutions to the equation

$$x^2 = k$$

are given by $x = \pm \sqrt{k}$. If $k < 0$, then this equation has no real solutions.

Before applying the square root property in the next two examples, we review a quotient property of square roots. If a and b are positive numbers, then

$$\sqrt{\frac{a}{b}} = \frac{\sqrt{a}}{\sqrt{b}}.$$

For example, $\sqrt{\frac{25}{36}} = \frac{\sqrt{25}}{\sqrt{36}} = \frac{5}{6}$.

EXAMPLE 2   Using the square root property

Solve each equation.
(a) $x^2 = 7$ **(b)** $16x^2 - 9 = 0$ **(c)** $(x - 4)^2 = 25$

Solution
(a) $x^2 = 7$ is equivalent to $x = \pm \sqrt{7}$ by the square root property. The solutions are $\sqrt{7}$ and $-\sqrt{7}$.

(b)

$16x^2 - 9 = 0$	Given equation
$16x^2 = 9$	Add 9.
$x^2 = \dfrac{9}{16}$	Divide by 16.
$x = \pm\sqrt{\dfrac{9}{16}}$	Square root property
$x = \pm\dfrac{3}{4}$	Simplify.

The solutions are $\frac{3}{4}$ and $-\frac{3}{4}$.

(c)

$(x - 4)^2 = 25$	Given equation
$(x - 4) = \pm\sqrt{25}$	Square root property
$x - 4 = \pm 5$	Simplify.
$x = 4 \pm 5$	Add 4.
$x = 9$ or $x = -1$	Evaluate $4 + 5$ and $4 - 5$.

The solutions are 9 and -1.

Now Try Exercises 51, 53, 57

▶ **REAL-WORLD CONNECTION**   If an object is dropped from a height of h feet, its distance d above the ground after t seconds is given by

$$d(t) = h - 16t^2.$$

This formula can be used to estimate the time it takes for a falling object to hit the ground.

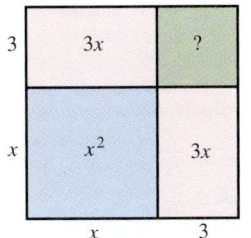

EXAMPLE 3   Modeling a falling object

A toy falls 30 feet from a window. How long does the toy take to hit the ground?

Solution
The height of the window above the ground is 30 feet so let $d(t) = 30 - 16t^2$. The toy strikes the ground when the distance d above the ground equals 0.

$$30 - 16t^2 = 0 \qquad \text{Equation to solve for } t$$
$$-16t^2 = -30 \qquad \text{Subtract 30.}$$
$$t^2 = \frac{30}{16} \qquad \text{Divide by } -16.$$
$$t = \pm\sqrt{\frac{30}{16}} \qquad \text{Square root property}$$
$$t = \pm\frac{\sqrt{30}}{4} \qquad \text{Simplify.}$$

Time cannot be negative in this problem, so the appropriate solution is $t = \frac{\sqrt{30}}{4} \approx 1.4$. The toy hits the ground after about 1.4 seconds.   **Now Try Exercise 115**

Completing the Square

In Section 11.2 we used the *method of completing the square* to find the vertex of a parabola. This method can also be used to solve quadratic equations. Because

$$x^2 + bx + \left(\frac{b}{2}\right)^2 = \left(x + \frac{b}{2}\right)^2,$$

we can solve a quadratic equation in the form $x^2 + bx = d$, where b and d are constants, by adding $\left(\frac{b}{2}\right)^2$ to each side and then factoring the resulting perfect square trinomial.

In the equation $x^2 + 6x = 7$ we have $b = 6$, so we add $\left(\frac{6}{2}\right)^2 = 9$ to each side.

$$x^2 + 6x = 7 \qquad \text{Given equation}$$
$$x^2 + 6x + 9 = 7 + 9 \qquad \text{Add 9 to each side.}$$
$$(x + 3)^2 = 16 \qquad \text{Perfect square trinomial}$$
$$x + 3 = \pm 4 \qquad \text{Square root property}$$
$$x = -3 \pm 4 \qquad \text{Add } -3 \text{ to each side.}$$
$$x = 1 \quad \text{or} \quad x = -7 \qquad \text{Simplify } -3 + 4 \text{ and } -3 - 4.$$

The solutions are 1 and -7. Note that the left side of the equation becomes a perfect square trinomial. We show how to create one in the next example.

CRITICAL THINKING

Use the figure to *complete the square* for $x^2 + 6x$.

3	$3x$	?
x	x^2	$3x$
	x	3

EXAMPLE 4   Creating a perfect square trinomial

Find the term that should be added to $x^2 - 10x$ to form a perfect square trinomial.

Solution
The coefficient of the x-term is -10, so we let $b = -10$. To complete the square we divide b by 2 and then square the result.

$$\left(\frac{b}{2}\right)^2 = \left(\frac{-10}{2}\right)^2 = 25$$

If we add **25** to $x^2 - 10x$, a perfect square trinomial is formed.

$$x^2 - \mathbf{10}x + \mathbf{25} = (x - 5)^2$$ Now Try Exercise 67

Completing the square can be used to solve quadratic equations when a trinomial does not factor easily, as illustrated in the next two examples.

EXAMPLE 5   Completing the square when the leading coefficient is 1

Solve the equation $x^2 - 4x + 2 = 0$.

Solution
Start by writing the equation in the form $x^2 + bx = d$.

$x^2 - 4x + 2 = 0$	Given equation
$x^2 - \mathbf{4}x = -2$	Subtract 2.
$x^2 - \mathbf{4}x + \mathbf{4} = -2 + \mathbf{4}$	Add $\left(\frac{b}{2}\right)^2 = \left(\frac{-4}{2}\right)^2 = 4$.
$(x - 2)^2 = 2$	Perfect square trinomial; add.
$x - 2 = \pm\sqrt{2}$	Square root property
$x = 2 \pm \sqrt{2}$	Add 2.

The solutions are $2 + \sqrt{2} \approx 3.41$ and $2 - \sqrt{2} \approx 0.59$.   Now Try Exercise 73

EXAMPLE 6   Completing the square when the leading coefficient is not 1

Solve the equation $2x^2 + 7x - 5 = 0$.

Solution
Start by writing the equation in the form $x^2 + bx = d$. That is, add 5 to each side and then divide the equation by 2 so that the leading coefficient of the x^2-term becomes 1.

$2x^2 + 7x - 5 = 0$	Given equation
$2x^2 + 7x = 5$	Add 5.
$x^2 + \dfrac{\mathbf{7}}{\mathbf{2}}x = \dfrac{5}{2}$	Divide by 2.
$x^2 + \dfrac{\mathbf{7}}{\mathbf{2}}x + \dfrac{\mathbf{49}}{\mathbf{16}} = \dfrac{5}{2} + \dfrac{\mathbf{49}}{\mathbf{16}}$	Add $\left(\frac{b}{2}\right)^2 = \left(\frac{7}{4}\right)^2 = \frac{49}{16}$.
$\left(x + \dfrac{7}{4}\right)^2 = \dfrac{89}{16}$	Perfect square trinomial; add.
$x + \dfrac{7}{4} = \pm\dfrac{\sqrt{89}}{4}$	Square root property
$x = -\dfrac{7}{4} \pm \dfrac{\sqrt{89}}{4}$	Add $-\frac{7}{4}$.
$x = \dfrac{-7 \pm \sqrt{89}}{4}$	Combine fractions.

CRITICAL THINKING

What happens if you try to solve

$$2x^2 - 13 = 1$$

by completing the square? What method could you use to solve this problem?

The solutions are $\dfrac{-7 + \sqrt{89}}{4} \approx 0.61$ and $\dfrac{-7 - \sqrt{89}}{4} \approx -4.1$.   Now Try Exercise 81

Figure 11.37

Solving an Equation for a Variable

We often need to solve an equation or formula for a variable. For example, the formula $V = \frac{1}{3}\pi r^2 h$ calculates the volume of the cone shown in Figure 11.37. Let's say that we know the volume V is 120 cubic inches and the height h is 15 inches. We can then find the radius of the cone by solving the equation for r.

$$V = \frac{1}{3}\pi r^2 h \qquad \text{Solve the equation for } r.$$

$$3V = \pi r^2 h \qquad \text{Multiply by 3.}$$

$$\frac{3V}{\pi} = r^2 h \qquad \text{Divide by } \pi.$$

$$\frac{3V}{\pi h} = r^2 \qquad \text{Divide by } h.$$

$$r = \pm\sqrt{\frac{3V}{\pi h}} \qquad \text{Square root property; rewrite.}$$

Because $r \geq 0$, we use the positive or *principal square root*. Thus for $V = \mathbf{120}$ cubic inches and $h = \mathbf{15}$ inches,

$$r = \sqrt{\frac{3(\mathbf{120})}{\pi(\mathbf{15})}} = \sqrt{\frac{24}{\pi}} \approx 2.8 \text{ inches.}$$

EXAMPLE 7 Solving equations for variables

Solve each equation for the specified variable.
(a) $s = -\frac{1}{2}gt^2 + h$, for t **(b)** $d^2 = x^2 + y^2$, for y

Solution
(a) Begin by subtracting h from each side of the equation.

$$s = -\tfrac{1}{2}gt^2 + h \qquad \text{Solve the equation for } t.$$

$$s - h = -\tfrac{1}{2}gt^2 \qquad \text{Subtract } h.$$

$$-2(s - h) = gt^2 \qquad \text{Multiply by } -2.$$

$$\frac{2h - 2s}{g} = t^2 \qquad \text{Divide by } g; \text{ simplify.}$$

$$t = \pm\sqrt{\frac{2h - 2s}{g}} \qquad \text{Square root property; rewrite.}$$

(b) Begin by subtracting x^2 from each side of the equation.

$$d^2 = x^2 + y^2 \qquad \text{Solve the equation for } y.$$

$$d^2 - x^2 = y^2 \qquad \text{Subtract } x^2.$$

$$y = \pm\sqrt{d^2 - x^2} \qquad \text{Square root property; rewrite.}$$

Now Try Exercises 107, 108

Applications of Quadratic Equations

▶ **REAL-WORLD CONNECTION**   In the introduction to this section we discussed how the solution to $\frac{1}{2}x^2 = 650$ would give a safe speed limit for a curve with a radius of 650 feet on an airport taxiway. We solve this problem in the next example.   (**Source:** FAA.)

EXAMPLE 8   Finding a safe speed limit

Solve $\frac{1}{2}x^2 = 650$ and interpret any solutions.

Solution
Use the square root property to solve this problem.

$$\frac{1}{2}x^2 = 650 \qquad \text{Given equation}$$

$$x^2 = 1300 \qquad \text{Multiply by 2.}$$

$$x = \pm\sqrt{1300} \qquad \text{Square root property}$$

The solutions are $\sqrt{1300} \approx 36$ and $-\sqrt{1300} \approx -36$. The solution of $x \approx 36$ indicates that a safe speed limit for a curve with a radius of 650 feet should be 36 miles per hour. (The negative solution has no physical meaning in this problem.)   **Now Try Exercise 113**

▶ **REAL-WORLD CONNECTION**   In applications, solving a quadratic equation either graphically or numerically is often easier than solving it symbolically. We do so in the next example.

EXAMPLE 9   Modeling numbers of Internet users

Use of the Internet in Western Europe has increased dramatically. Figure 11.38 shows a scatterplot of online users in Western Europe, together with a graph of a function f that models the data. The function f is given by

$$f(x) = 0.976x^2 - 4.643x + 0.238,$$

where the output is in millions of users. In this formula $x = 6$ corresponds to 1996, $x = 7$ to 1997, and so on, until $x = 12$ represents 2002.   (**Source:** Nortel Networks.)

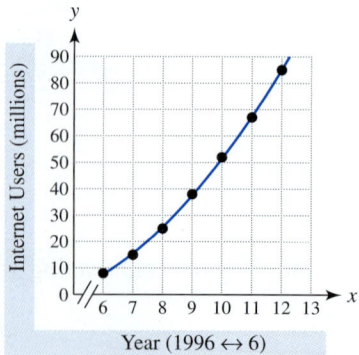

Figure 11.38 Internet Usage in Western Europe

(a) Evaluate $f(10)$ and interpret the result.

(b) Graph f and estimate the year when the number of Internet users reached 85 million. Compare your result with Figure 11.38.

(c) Solve part (b) numerically.

Solution

(a) Substituting $x = \mathbf{10}$ in the formula gives

$$f(\mathbf{10}) = 0.976(\mathbf{10})^2 - 4.643(\mathbf{10}) + 0.238 \approx 51.4.$$

Because $x = 10$ corresponds to 2000, there were about 51.4 million Internet users in Western Europe in 2000.

(b) Graph $Y_1 = .976X^2 - 4.643X + .238$ and $Y_2 = 85$, as shown in Figure 11.39. Their graphs intersect near $x = 12$, which corresponds to 2002 and agrees with Figure 11.38.

(c) The table for y_1, shown in Figure 11.40, reveals $y_1 \approx 85$ when $x = 12$.

CALCULATOR HELP

To find a point of intersection, see Appendix A (page AP-7).

[5, 13, 1] by [0, 100, 10]

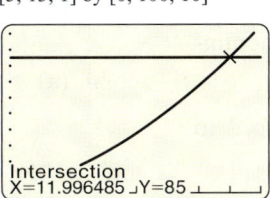

Intersection
X=11.996485 Y=85

Figure 11.39

X	Y₁	Y₂
9	37.507	85
10	51.408	85
11	67.261	85
12	85.066	85
13	104.82	85
14	126.53	85
15	150.19	85

X=12

Figure 11.40

Now Try Exercise 124

11.3 PUTTING IT ALL TOGETHER

Quadratic equations can be expressed in the form $ax^2 + bx + c = 0$, $a \neq 0$. They can have no real solutions, one real solution, or two real solutions and may be solved symbolically, graphically, and numerically. Symbolic techniques for solving quadratic equations include factoring, the square root property, and completing the square. We discussed factoring extensively in Chapter 6, so the following table summarizes only the square root property and the method of completing the square.

Technique	Description	Examples
Square Root Property	If $k \geq 0$, the solutions to the equation $x^2 = k$ are $\pm\sqrt{k}$.	$x^2 = 100$ is equivalent to $x = \pm 10$ and $x^2 = 13$ is equivalent to $x = \pm\sqrt{13}$. $x^2 = -2$ has no real solutions.

continued on next page

continued from previous page

Technique	Description	Examples
Method of Completing the Square	To solve an equation in the form $x^2 + bx = d$, add $\left(\frac{b}{2}\right)^2$ to each side of the equation. Factor the resulting perfect square trinomial and solve for x by applying the square root property.	To solve $x^2 + 8x = 3$, add $\left(\frac{8}{2}\right)^2 = 16$ to each side. $x^2 + 8x + 16 = 3 + 16$ Add 16 to each side. $(x + 4)^2 = 19$ Perfect square trinomial $x + 4 = \pm\sqrt{19}$ Square root property $x = -4 \pm \sqrt{19}$ Add -4.

11.3 Exercises

CONCEPTS

1. Give an example of a quadratic equation. How many real solutions can a quadratic equation have?

2. Is a quadratic equation a linear equation or a nonlinear equation?

3. Name three symbolic methods that can be used to solve a quadratic equation.

4. Sketch a graph of a quadratic function that has two x-intercepts and opens downward.

5. Sketch a graph of a quadratic function that has no x-intercepts and opens upward.

6. If the graph of $y = ax^2 + bx + c$ intersects the x-axis twice, how many solutions does the equation $ax^2 + bx + c = 0$ have? Explain.

7. Solve $x^2 = 64$. What property did you use?

8. To solve $x^2 + bx = 6$ by completing the square, what value should be added to each side of the equation?

Exercises 9–16: Is the given equation quadratic?

9. $x^2 - 3x + 1 = 0$ 10. $2x^2 - 3 = 0$

11. $3x + 1 = 0$ 12. $x^3 - 3x^2 + x = 0$

13. $-3x^2 + x = 16$ 14. $x^2 - 1 = 4x$

15. $x^2 = \sqrt{x} + 1$ 16. $\dfrac{1}{x - 1} = 5$

Exercises 17–20: Approximate to the nearest hundredth.

17. (a) $1 \pm \sqrt{7}$ (b) $-2 \pm \sqrt{11}$

18. (a) $\pm\dfrac{\sqrt{3}}{2}$ (b) $\pm\dfrac{2\sqrt{5}}{7}$

19. (a) $\dfrac{3 \pm \sqrt{13}}{5}$ (b) $\dfrac{-5 \pm \sqrt{6}}{9}$

20. (a) $\dfrac{2}{5} \pm \dfrac{\sqrt{5}}{5}$ (b) $-\dfrac{3}{7} \pm \dfrac{\sqrt{3}}{7}$

SOLVING QUADRATIC EQUATIONS

Exercises 21–24: A graph of $y = ax^2 + bx + c$ is given. Use this graph to solve $ax^2 + bx + c = 0$, if possible.

21.

22.

23.

24.
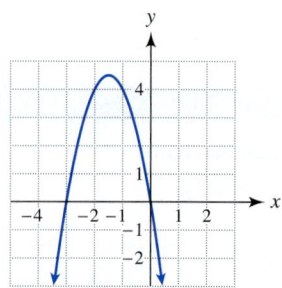

Exercises 25–28: A table of $y = ax^2 + bx + c$ is given. Use this table to solve $ax^2 + bx + c = 0$.

25.

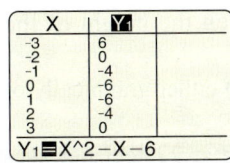

X	Y₁
-3	6
-2	0
-1	-4
0	-6
1	-6
2	-4
3	0

$Y_1 \blacksquare X^2 - X - 6$

26.

X	Y₁
-6	0
-4	-16
-2	-24
0	-24
2	-16
4	0
6	24

$Y_1 \blacksquare X^2 + 2X - 24$

27.

X	Y₁
-2	9
-1.5	4
-1	1
-.5	0
0	1
.5	4
1	9

$Y_1 \blacksquare 4X^2 + 4X + 1$

28.

X	Y₁
-4	-2.5
-3	0
-2	1.5
-1	2
0	1.5
1	0
2	-2.5

$Y_1 \blacksquare -X^2/2 - X + 3/2$

Exercises 29–40: Solve the quadratic equation. Support your results numerically and graphically.

29. $x^2 - 4x - 5 = 0$

30. $x^2 - x - 6 = 0$

31. $x^2 + 2x = 3$

32. $x^2 + 4x = 5$

33. $x^2 = 9$

34. $x^2 = 4$

35. $4x^2 - 4x - 3 = 0$

36. $2x^2 + x = 1$

37. $x^2 + 2x = -1$

38. $4x^2 - 4x + 1 = 0$

39. $x^2 + 2 = 0$

40. $-4x^2 - 1 = 2$

Exercises 41–50: Solve by factoring.

41. $x^2 + 2x - 35 = 0$

42. $2x^2 - 7x + 3 = 0$

43. $6x^2 - x - 1 = 0$

44. $x^2 + 4x + 6 = -3x$

45. $2x^2 + x + 3 = 6x$

46. $6x^2 - x = 15$

47. $25x^2 - 350 = 125x$

48. $20x^2 + 150 = 130x$

49. $2(5x^2 + 9) = 27x$

50. $15(3x^2 + x) = 10$

Exercises 51–62: Use the square root property to solve.

51. $x^2 = 144$

52. $4x^2 - 5 = 0$

53. $5x^2 - 64 = 0$

54. $3x^2 = 7$

55. $(x + 1)^2 = 25$

56. $(x + 4)^2 = 9$

57. $(x - 1)^2 = 64$

58. $(x - 3)^2 = 0$

59. $(2x - 1)^2 = 5$

60. $(5x + 3)^2 = 7$

61. $10(x - 5)^2 = 50$

62. $7(3x + 1)^2 = 14$

COMPLETING THE SQUARE

Exercises 63–66: To solve by completing the square, what value should you add to each side of the equation?

63. $x^2 + 4x = -3$

64. $x^2 - 6x = 4$

65. $x^2 - 5x = 4$

66. $x^2 + 3x = 1$

Exercises 67–70: (Refer to Example 4.) Find the term that should be added to the expression to form a perfect square trinomial. Write the resulting perfect square trinomial in factored form.

67. $x^2 - 8x$

68. $x^2 - 5x$

69. $x^2 + 9x$

70. $x^2 + x$

Exercises 71–86: Solve by completing the square.

71. $x^2 - 2x = 24$

72. $x^2 - 2x + \frac{1}{2} = 0$

73. $x^2 + 6x - 2 = 0$

74. $x^2 - 16x = 5$

75. $x^2 - 3x = 5$

76. $x^2 + 5x = 2$

77. $x^2 - 5x + 1 = 0$

78. $x^2 - 9x + 7 = 0$

79. $x^2 - 4 = 2x$

80. $x^2 + 1 = 7x$

81. $2x^2 - 3x = 4$

82. $3x^2 + 6x - 5 = 0$

83. $4x^2 - 8x - 7 = 0$

84. $25x^2 - 20x - 1 = 0$

85. $36x^2 + 18x + 1 = 0$

86. $12x^2 + 8x - 2 = 0$

Exercises 87–96: Solve by any method.

87. $3x^2 + 12x = 36$

88. $6x^2 + 9x = 27$

89. $x^2 + 4x = -2$

90. $x^2 + 6x + 3 = 0$

91. $3x^2 - 4 = 2$

92. $-2x^2 + 3 = 1$

93. $-6x^2 + 70 = 16x$

94. $-15x^2 + 25x + 10 = 0$

95. $-3x(x - 8) = 6$

96. $-2x(4 - x) = 8$

Exercises 97 and 98: **Thinking Generally** *Solve for x. Assume a and c are positive.*

97. $ax^2 - c = 0$

98. $ax^2 + bx = 0$

SOLVING EQUATIONS BY MORE THAN ONE METHOD

Exercises 99–104: Solve the quadratic equation
  (a) *symbolically,*
  (b) *graphically, and*
  (c) *numerically.*

99. $x^2 - 3x - 18 = 0$ **100.** $\frac{1}{2}x^2 + 2x - 6 = 0$

101. $x^2 - 8x + 15 = 0$ **102.** $2x^2 + 3 = 7x$

103. $4(x^2 + 35) = 48x$ **104.** $4x(2 - x) = -5$

SOLVING AN EQUATION FOR A VARIABLE

Exercises 105–112: Solve for the specified variable.

105. $x = y^2 - 1$ for y **106.** $x = 9y^2$ for y

107. $K = \frac{1}{2}mv^2$ for v **108.** $c^2 = a^2 + b^2$ for b

109. $E = \frac{k}{r^2}$ for r **110.** $W = I^2 R$ for I

111. $LC = \dfrac{1}{(2\pi f)^2}$ for f **112.** $F = \dfrac{KmM}{r^2}$ for r

APPLICATIONS

113. *Safe Curve Speed*   (Refer to Example 8.) Find a safe speed limit x for an airport taxiway curve with the given radius R by using $R = \frac{1}{2}x^2$.
  (a) $R = 450$ feet   **(b)** $R = 800$ feet

114. *Braking Distance*   The braking distance y in feet that it takes for a car to stop on wet, level pavement can be estimated by $y = \frac{1}{9}x^2$, where x is the speed of the car in miles per hour. Find the speed associated with each braking distance. (*Source:* L. Haefner, *Introduction to Transportation Systems.*)
  (a) 25 feet   **(b)** 361 feet   **(c)** 784 feet

115. *Falling Object*   (Refer to Example 3.) How long does it take for a toy to hit the ground if it is dropped out of a window 60 feet above the ground? Does it take twice as long as it takes to fall from a window 30 feet above the ground?

116. *Falling Object*   If a metal ball is thrown *downward* with an initial velocity of 22 feet per second (15 mph)

from a 100-foot water tower, its height h in feet above the ground after t seconds is modeled by

$$h(t) = -16t^2 - 22t + 100.$$

(a) Determine symbolically when the height of the ball is 62 feet.

(b) Support your result in part (a) either graphically or numerically.

(c) If the ball is thrown *upward* at 22 feet per second, then its height is given by

$$h(t) = -16t^2 + 22t + 100.$$

Determine when the height of the ball is 80 feet.

*Exercises 117 and 118: **Television Size** (Refer to the introduction of this chapter.) The size S of the television screen recommended for a person who sits x feet from the screen ($6 \le x \le 15$) is given by*

$$S(x) = -0.227x^2 + 8.155x - 8.8.$$

If a person buys a television set with a size S screen, how far from the screen should the person sit?

117. $S = 42$ inches   **118.** $S = 50$ inches

119. *Distance*   Two athletes start jogging at the same time. One jogs north at 6 miles per hour while the second jogs east at 8 miles per hour. After how long are the two athletes 20 miles apart?

120. *Geometry* A triangle has an area of 35 square inches, and its base is 3 inches more than its height. Find the base and height of the triangle.

121. *Construction* A rectangular plot of land has an area of 520 square feet and is 6 feet longer than it is wide.
 (a) Write a quadratic equation in the form $ax^2 + bx + c = 0$, whose solution gives the width of the plot of land.
 (b) Solve the equation.

122. *Modeling Motion* The height y in feet of a tennis ball after x seconds is shown in the graph. Estimate when the ball was 25 feet above the ground.

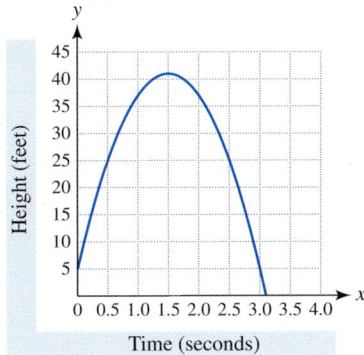

Time (seconds)

123. *Seedling Growth* (Refer to Exercise 87, Section 11.1.) The heights of melon seedlings grown at different temperatures are shown in the following graph. At what temperatures were the heights of the seedlings about 22 centimeters?   (*Source:* R. Pearl, "The growth of *Cucumis melo* seedlings at different temperatures.")

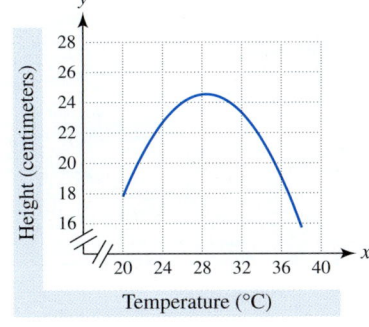

Temperature (°C)

124. *Internet Users* (Refer to Example 9.) Estimate either graphically or numerically when the number of Internet users in Western Europe reached 150 million.

125. *U.S. Population* The table shows the population of the United States in millions from 1800 through 2000 at 20-year intervals.

Year	1800	1820	1840	1860
Population	5	10	17	31

Year	1880	1900	1920	1940
Population	50	76	106	132

Year	1960	1980	2000
Population	179	226	269

Source: U.S. Census Bureau.

 (a) Without plotting the data, how do you know that the data are nonlinear?
 (b) These data are modeled by

$$f(x) = 0.0066(x - 1800)^2 + 5.$$

 Find the vertex of the graph of f and interpret it.
 (c) Estimate when the U.S. population reached 85 million.

126. *Trade Deficit* The U.S. trade deficit in billions of dollars for the years 1997, 1998, and 1999 can be computed by

$$f(x) = 16x^2 - 63,861x + 63,722,378,$$

where x is the year. In which year was the trade deficit \$164 billion?   (*Source:* Department of Commerce.)

WRITING ABOUT MATHEMATICS

127. Suppose that you are asked to solve

$$ax^2 + bx + c = 0.$$

Explain how the graph of $y = ax^2 + bx + c$ can be used to find any real solutions to the equation.

128. Explain why a quadratic equation could not have more than two solutions. (*Hint:* Consider the graph of $y = ax^2 + bx + c$.)

GROUP ACTIVITY
WORKING WITH REAL DATA

Directions: Form a group of 2 to 4 people. Select someone to record the group's responses for this activity. All members of the group should work cooperatively to answer the questions. If your instructor asks for the results, each member of the group should be prepared to respond.

Minimum Wage   The table shows the minimum wage for three different years.

Year	1940	1968	1997
Wage ($)	0.25	1.60	5.15

Source: Bureau of Labor Statistics.

 (a) Make a scatterplot of the data in the viewing rectangle [1930, 2010, 10] by [0, 6, 1].

(b) Find a quadratic function given by
$$f(x) = a(x - h)^2 + k$$
that models the data.

(c) Estimate the minimum wage in 1976 and compare it to the actual value of $2.30.

(d) Estimate when the minimum wage was $1.00.

(e) If current trends continue, predict the minimum wage in 2009. Compare it to the projected value of $7.25.

11.4   THE QUADRATIC FORMULA

Solving Quadratic Equations ▪ The Discriminant ▪ Quadratic Equations Having Complex Solutions

A LOOK INTO MATH ▷   To model the stopping distance of a car, highway engineers compute two quantities. The first quantity is the *reaction distance*, which is the distance a car travels from the time a driver first recognizes a hazard until the brakes are applied. The second quantity is *braking distance*, which is the distance a car travels after a driver applies the brakes. *Stopping distance* equals the sum of the reaction distance and the braking distance. If a car is traveling x miles per hour, highway engineers estimate the reaction distance in feet as $\frac{11}{3}x$ and the braking distance in feet as $\frac{1}{9}x^2$. See Figure 11.41. (**Source:** L. Haefner, *Introduction to Transportation Systems.*)

Figure 11.41 Stopping Distance

To estimate the total stopping distance d in feet, add the two expressions to obtain

$$d(x) = \frac{1}{9}x^2 + \frac{11}{3}x.$$

If a car's headlights don't illuminate the road beyond **500** feet, a safe nighttime speed limit x for the car can be determined by solving the quadratic equation

$$\frac{1}{9}x^2 \quad + \quad \frac{11}{3}x \quad = \quad 500.$$

Braking Distance + Reaction Distance = Stopping Distance

In this section we learn how to solve this equation with the quadratic formula.

Solving Quadratic Equations

Recall that any quadratic equation can be written in the form

$$ax^2 + bx + c = 0.$$

If we solve this equation for x in terms of a, b, and c by completing the square, we obtain the **quadratic formula**. We assume that $a > 0$ and derive it as follows.

$ax^2 + bx + c = 0$	Quadratic equation
$ax^2 + bx = -c$	Subtract c.
$x^2 + \dfrac{b}{a}x = -\dfrac{c}{a}$	Divide by a.
$x^2 + \dfrac{b}{a}x + \dfrac{b^2}{4a^2} = -\dfrac{c}{a} + \dfrac{b^2}{4a^2}$	Add $\left(\frac{b/a}{2}\right)^2 = \frac{b^2}{4a^2}$.
$\left(x + \dfrac{b}{2a}\right)^2 = -\dfrac{c}{a} + \dfrac{b^2}{4a^2}$	Perfect square trinomial
$\left(x + \dfrac{b}{2a}\right)^2 = -\dfrac{c \cdot 4a}{a \cdot 4a} + \dfrac{b^2}{4a^2}$	Multiply $-\frac{c}{a}$ by $\frac{4a}{4a}$.
$\left(x + \dfrac{b}{2a}\right)^2 = -\dfrac{4ac}{4a^2} + \dfrac{b^2}{4a^2}$	Simplify.
$\left(x + \dfrac{b}{2a}\right)^2 = \dfrac{-4ac + b^2}{4a^2}$	Add fractions.
$\left(x + \dfrac{b}{2a}\right)^2 = \dfrac{b^2 - 4ac}{4a^2}$	Rewrite.
$x + \dfrac{b}{2a} = \pm\sqrt{\dfrac{b^2 - 4ac}{4a^2}}$	Square root property
$x = -\dfrac{b}{2a} \pm \sqrt{\dfrac{b^2 - 4ac}{4a^2}}$	Add $-\frac{b}{2a}$.
$x = -\dfrac{b}{2a} \pm \dfrac{\sqrt{b^2 - 4ac}}{2a}$	Property of square roots
$x = \dfrac{-b \pm \sqrt{b^2 - 4ac}}{2a}$	Combine fractions.

QUADRATIC FORMULA

The solutions to $ax^2 + bx + c = 0$ with $a \neq 0$ are given by

$$x = \frac{-b \pm \sqrt{b^2 - 4ac}}{2a}.$$

NOTE: The quadratic formula can be used to solve *any* quadratic equation. It always "works."

MAKING CONNECTIONS

Completing the Square and the Quadratic Formula

The quadratic formula results from completing the square for the equation $ax^2 + bx + c = 0$. When you use the quadratic formula, the work of completing the square has already been done for you. However, you can always complete the square rather than use the quadratic formula. See Example 10.

The next three examples show symbolic and graphical solutions to quadratic equations.

EXAMPLE 1   Solving a quadratic equation having two solutions

Solve the equation $2x^2 - 3x - 1 = 0$. Support your results graphically.

Solution
Symbolic Solution   Let $a = \mathbf{2}, b = \mathbf{-3},$ and $c = \mathbf{-1}$.

$$x = \frac{-b \pm \sqrt{b^2 - 4ac}}{2a}$$ Quadratic formula

$$x = \frac{-(\mathbf{-3}) \pm \sqrt{(\mathbf{-3})^2 - 4(\mathbf{2})(\mathbf{-1})}}{2(\mathbf{2})}$$ Substitute for a, b, and c.

$$x = \frac{3 \pm \sqrt{17}}{4}$$ Simplify.

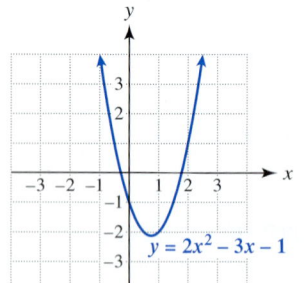

Figure 11.42 Two *x*-intercepts

The solutions are $\frac{3 + \sqrt{17}}{4} \approx 1.78$ and $\frac{3 - \sqrt{17}}{4} \approx -0.28$.

Graphical Solution   The graph of $y = 2x^2 - 3x - 1$ is shown in Figure 11.42. Note that the two *x*-intercepts correspond to the two solutions to $2x^2 - 3x - 1 = 0$. Estimating from this graph, we see that the solutions are approximately -0.25 and 1.75, which supports our symbolic solution. (You could also use a graphing calculator to find the *x*-intercepts.)

Now Try Exercise 9

CRITICAL THINKING

Use the results of Example 1 to evaluate each expression mentally.

$$2\left(\frac{3 + \sqrt{17}}{4}\right)^2 - 3\left(\frac{3 + \sqrt{17}}{4}\right) - 1 \quad \text{and} \quad 2\left(\frac{3 - \sqrt{17}}{4}\right)^2 - 3\left(\frac{3 - \sqrt{17}}{4}\right) - 1$$

EXAMPLE 2   Solving a quadratic equation having one solution

Solve the equation $25x^2 + 20x + 4 = 0$. Support your result graphically.

Solution

Symbolic Solution   Let $a = \mathbf{25}$, $b = \mathbf{20}$, and $c = \mathbf{4}$.

$$x = \frac{-b \pm \sqrt{b^2 - 4ac}}{2a} \qquad \text{Quadratic formula}$$

$$= \frac{-\mathbf{20} \pm \sqrt{\mathbf{20}^2 - 4(\mathbf{25})(\mathbf{4})}}{2(\mathbf{25})} \qquad \text{Substitute for } a, b, \text{ and } c.$$

$$= \frac{-20 \pm \sqrt{0}}{50} \qquad \text{Simplify.}$$

$$= \frac{-20}{50} = -0.4 \qquad \sqrt{0} = 0$$

There is one solution, -0.4.

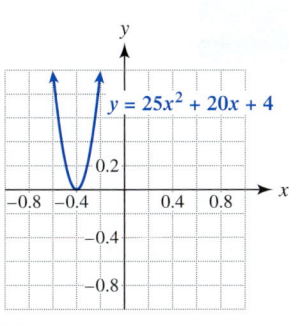

Figure 11.43 One x-intercept

Graphical Solution   The graph of $y = 25x^2 + 20x + 4$ is shown in Figure 11.43. Note that the one x-intercept, -0.4, corresponds to the solution to $25x^2 + 20x + 4 = 0$.

Now Try Exercise 11

EXAMPLE 3   Recognizing a quadratic equation having no real solutions

Solve the equation $5x^2 - x + 3 = 0$. Support your result graphically.

Solution

Symbolic Solution   Let $a = \mathbf{5}$, $b = \mathbf{-1}$, and $c = \mathbf{3}$.

$$x = \frac{-b \pm \sqrt{b^2 - 4ac}}{2a} \qquad \text{Quadratic formula}$$

$$= \frac{-(\mathbf{-1}) \pm \sqrt{(\mathbf{-1})^2 - 4(\mathbf{5})(\mathbf{3})}}{2(\mathbf{5})} \qquad \text{Substitute for } a, b, \text{ and } c.$$

$$= \frac{1 \pm \sqrt{-59}}{10} \qquad \text{Simplify.}$$

There are no real solutions to this equation because $\sqrt{-59}$ *is not a real number.* (Later in this section we discuss how to find complex solutions to quadratic equations like this one.)

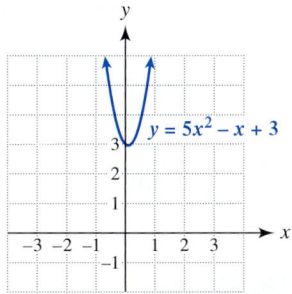

Figure 11.44 No x-intercepts

Graphical Solution   The graph of $y = 5x^2 - x + 3$ is shown in Figure 11.44. There are no x-intercepts, indicating that the equation $5x^2 - x + 3 = 0$ has no real solutions.

Now Try Exercise 12

▶ **REAL-WORLD CONNECTION**   Earlier in this section we discussed how engineers estimate safe stopping distances for automobiles. In the next example we solve the equation presented in the introduction.

EXAMPLE **Modeling stopping distance**

If a car's headlights do not illuminate the road beyond 500 feet, estimate a safe nighttime speed limit x for the car by solving $\frac{1}{9}x^2 + \frac{11}{3}x = 500$.

Solution

Begin by subtracting 500 from each side of the given equation.

$$\frac{1}{9}x^2 + \frac{11}{3}x - 500 = 0. \qquad \text{Subtract 500.}$$

To eliminate fractions, multiply each side by the LCD, which is 9. (This step is not necessary, but it makes the problem easier to work.)

$$x^2 + 33x - 4500 = 0. \qquad \text{Multiply by 9.}$$

Now let $a = 1$, $b = 33$, and $c = -4500$ in the quadratic formula.

$$x = \frac{-b \pm \sqrt{b^2 - 4ac}}{2a} \qquad \text{Quadratic formula}$$

$$= \frac{-33 \pm \sqrt{33^2 - 4(1)(-4500)}}{2(1)} \qquad \text{Substitute for } a, b, \text{ and } c.$$

$$= \frac{-33 \pm \sqrt{19,089}}{2} \qquad \text{Simplify.}$$

The solutions are

$$\frac{-33 + \sqrt{19,089}}{2} \approx 52.6 \quad \text{and} \quad \frac{-33 - \sqrt{19,089}}{2} \approx -85.6.$$

The negative solution has no physical meaning because negative speeds are not possible. The other solution is 52.6, so an appropriate speed limit might be 50 miles per hour.

Now Try Exercise 101

The Discriminant

The expression $b^2 - 4ac$ in the quadratic formula is called the **discriminant**. It provides information about the number of solutions to a quadratic equation.

THE DISCRIMINANT AND QUADRATIC EQUATIONS

To determine the number of solutions to the quadratic equation $ax^2 + bx + c = 0$, evaluate the discriminant $b^2 - 4ac$.

1. If $b^2 - 4ac > 0$, there are two real solutions.
2. If $b^2 - 4ac = 0$, there is one real solution.
3. If $b^2 - 4ac < 0$, there are no real solutions; there are two complex solutions.

EXAMPLE 5   Using the discriminant

Use the discriminant to determine the number of solutions to $4x^2 + 25 = 20x$. Then solve the equation, using the quadratic formula.

Solution

Write the equation as $4x^2 - 20x + 25 = 0$ so that $a = 4$, $b = -20$, and $c = 25$. The discriminant evaluates to

$$b^2 - 4ac = (-20)^2 - 4(4)(25) = 0.$$

Thus there is one real solution.

$$x = \frac{-b \pm \sqrt{b^2 - 4ac}}{2a} \qquad \text{Quadratic formula}$$

$$= \frac{-(-20) \pm \sqrt{0}}{2(4)} \qquad \text{Substitute.}$$

$$= \frac{20}{8} = 2.5 \qquad \text{Simplify.}$$

The only solution is 2.5.   Now Try Exercise 37(a), (b)

We also need to be able to analyze graphs of quadratic functions, which we demonstrate in the next example.

EXAMPLE 6   Analyzing graphs of quadratic functions

A graph of $f(x) = ax^2 + bx + c$ is shown in Figure 11.45.
(a) State whether $a > 0$ or $a < 0$.
(b) Solve the equation $ax^2 + bx + c = 0$.
(c) Determine whether the discriminant is positive, negative, or zero.

Figure 11.45

Solution

(a) The parabola opens downward, so $a < 0$.
(b) The graph of $f(x) = ax^2 + bx + c$ intersects the x-axis at -3 and 2. Therefore $f(-3) = 0$ and $f(2) = 0$. The solutions to $ax^2 + bx + c = 0$ are -3 and 2.
(c) There are two solutions, so the discriminant is positive.   Now Try Exercise 31

Quadratic Equations Having Complex Solutions

The quadratic equation written as $ax^2 + bx + c = 0$ has no real solutions if the discriminant, $b^2 - 4ac$, is negative. For example, the quadratic equation $x^2 + 4 = 0$ has $a = 1$, $b = 0$, and $c = 4$. Its discriminant is

$$b^2 - 4ac = 0^2 - 4(1)(4) = -16 < 0,$$

so this equation has no real solutions. However, if we use complex numbers, we can solve this equation as follows.

$$x^2 + 4 = 0 \qquad \text{Given equation}$$

$$x^2 = -4 \qquad \text{Subtract 4.}$$

$$x = \pm\sqrt{-4} \qquad \text{Square root property}$$

$$x = \sqrt{-4} \quad \text{or} \quad x = -\sqrt{-4} \qquad \text{Meaning of } \pm$$

$$x = 2i \quad \text{or} \quad x = -2i \qquad \text{The expression } \sqrt{-a}$$

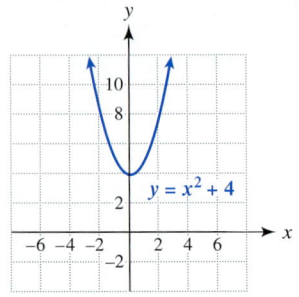

Figure 11.46 No *x*-intercepts

The solutions are $\pm 2i$. We check each solution to $x^2 + 4 = 0$ as follows.

$$(2i)^2 + 4 = (2)^2 i^2 + 4 = 4(-1) + 4 = 0 \qquad \text{It checks.}$$
$$(-2i)^2 + 4 = (-2)^2 i^2 + 4 = 4(-1) + 4 = 0 \qquad \text{It checks.}$$

The fact that the equation $x^2 + 4 = 0$ has only imaginary solutions can be seen visually from the graph of $y = x^2 + 4$, shown in Figure 11.46. This parabola does not intersect the *x*-axis, and so the equation $x^2 + 4 = 0$ has no real solutions.

These results can be generalized as follows.

THE EQUATION $x^2 + k = 0$

If $k > 0$, the solutions to $x^2 + k = 0$ are given by $x = \pm i\sqrt{k}$.

NOTE: This result is a form of the *square root property* that includes complex solutions.

EXAMPLE 7 Solving a quadratic equation having complex solutions

Solve $x^2 + 5 = 0$.

Solution
The solutions are $\pm i\sqrt{5}$. That is, $x = i\sqrt{5}$ or $x = -i\sqrt{5}$.   Now Try Exercise 55

When $b \neq 0$, the preceding method cannot be used. Consider the quadratic equation $2x^2 + x + 3 = 0$, which has $a = 2$, $b = 1$, and $c = 3$. Its discriminant is

$$b^2 - 4ac = 1^2 - 4(2)(3) = -23 < 0.$$

This equation has two complex solutions, as demonstrated in the next example.

EXAMPLE 8 Solving a quadratic equation having complex solutions

Solve $2x^2 + x + 3 = 0$. Write your answer in standard form: $a + bi$.

CALCULATOR HELP

To set your calculator in $a + bi$ mode or to access the imaginary unit *i*, see Appendix A (page AP-11).

Solution
Let $a = 2$, $b = 1$, and $c = 3$.

$$x = \frac{-b \pm \sqrt{b^2 - 4ac}}{2a} \qquad \text{Quadratic formula}$$

$$= \frac{-1 \pm \sqrt{1^2 - 4(2)(3)}}{2(2)} \qquad \text{Substitute for } a, b, \text{ and } c.$$

$$= \frac{-1 \pm \sqrt{-23}}{4} \qquad \text{Simplify.}$$

$$= \frac{-1 \pm i\sqrt{23}}{4} \qquad \sqrt{-23} = i\sqrt{23}$$

$$= -\frac{1}{4} \pm i\frac{\sqrt{23}}{4} \qquad \text{Divide each term by 4.}$$

The solutions are $-\frac{1}{4} + i\frac{\sqrt{23}}{4}$ and $-\frac{1}{4} - i\frac{\sqrt{23}}{4}$.

Now Try Exercise 71

CRITICAL THINKING

Use the results of Example 8 to evaluate each expression mentally.

$$2\left(-\frac{1}{4} + i\frac{\sqrt{23}}{4}\right)^2 + \left(-\frac{1}{4} + i\frac{\sqrt{23}}{4}\right) + 3 \quad \text{and} \quad 2\left(-\frac{1}{4} - i\frac{\sqrt{23}}{4}\right)^2 + \left(-\frac{1}{4} - i\frac{\sqrt{23}}{4}\right) + 3$$

Sometimes we can use properties of radicals to simplify a solution to a quadratic equation, as demonstrated in the next example.

EXAMPLE 9 Solving a quadratic equation having complex solutions

Solve $\frac{3}{4}x^2 + 1 = x$. Write your answer in standard form: $a + bi$.

Solution

Begin by subtracting x from each side of the equation and then multiply by 4 to clear fractions. The resulting equation is $3x^2 - 4x + 4 = 0$. Substitute $a = 3$, $b = -4$, and $c = 4$ in the quadratic formula.

$$x = \frac{-b \pm \sqrt{b^2 - 4ac}}{2a} \qquad \text{Quadratic formula}$$

$$= \frac{-(-4) \pm \sqrt{(-4)^2 - 4(3)(4)}}{2(3)} \qquad \text{Substitute.}$$

$$= \frac{4 \pm \sqrt{-32}}{6} \qquad \text{Simplify.}$$

$$= \frac{4 \pm 4i\sqrt{2}}{6} \qquad \sqrt{-32} = i\sqrt{32} = i\sqrt{16}\sqrt{2} = 4i\sqrt{2}$$

$$= \frac{2}{3} \pm \frac{2}{3}i\sqrt{2} \qquad \text{Divide 6 into each term and simplify.}$$

Now Try Exercise 77

In the next example we use completing the square to obtain complex solutions to a quadratic equation.

EXAMPLE 10   Completing the square to find complex solutions

Solve $x(x + 2) = -2$ by completing the square.

Solution
After applying the distributive property, the equation becomes $x^2 + 2x = -2$. Because $b = \mathbf{2}$, add $\left(\frac{b}{2}\right)^2 = \left(\frac{\mathbf{2}}{2}\right)^2 = \mathbf{1}$ to each side of the equation.

$$
\begin{array}{ll}
x^2 + \mathbf{2}x = -2 & \text{Equation to be solved} \\
x^2 + \mathbf{2}x + \mathbf{1} = -2 + \mathbf{1} & \text{Add 1 to each side.} \\
(x + 1)^2 = -1 & \text{Perfect square trinomial; add.} \\
x + 1 = \pm\sqrt{-1} & \text{Square root property} \\
x + 1 = \pm i & \sqrt{-1} = i, \text{ the imaginary unit} \\
x = -1 \pm i & \text{Add } -1 \text{ to each side.}
\end{array}
$$

The solutions are $-1 + i$ and $-1 - i$.

Now Try Exercise **75**

11.4 PUTTING IT ALL TOGETHER

Quadratic equations can be solved symbolically by using factoring, the square root property, completing the square, and the quadratic formula. Graphical and numerical methods can also be used to solve quadratic equations. In this section we discussed the quadratic formula and its discriminant, which we summarize in the following table.

Concept	Explanation	Examples
Quadratic Formula	The quadratic formula can be used to solve *any* quadratic equation written as $ax^2 + bx + c = 0$. The solutions are given by $$x = \frac{-b \pm \sqrt{b^2 - 4ac}}{2a}.$$	For the equation $$2x^2 - 3x + 1 = 0$$ with $a = 2$, $b = -3$, and $c = 1$, the solutions are $$\frac{-(-3) \pm \sqrt{(-3)^2 - 4(2)(1)}}{2(2)} = \frac{3 \pm \sqrt{1}}{4} = 1, \frac{1}{2}.$$
The Discriminant	The expression $b^2 - 4ac$ is called the discriminant. 1. $b^2 - 4ac > 0$ indicates two real solutions. 2. $b^2 - 4ac = 0$ indicates one real solution. 3. $b^2 - 4ac < 0$ indicates no real solutions; rather, there are two complex solutions.	For the equation $$x^2 + 4x - 1 = 0$$ with $a = 1$, $b = 4$, and $c = -1$, the discriminant is $$b^2 - 4ac = 4^2 - 4(1)(-1) = 20 > 0,$$ indicating two real solutions.

Concept	Explanation	Examples
Quadratic Formula and Complex Solutions	If the discriminant is negative $(b^2 - 4ac < 0)$, the solutions to a quadratic equation are complex numbers. If $k > 0$, the solutions to $x^2 + k = 0$ are given by $x = \pm i\sqrt{k}$.	$2x^2 - x + 3 = 0$ $$x = \frac{1 \pm \sqrt{(-1)^2 - 4(2)(3)}}{2(2)}$$ $$= \frac{1 \pm \sqrt{-23}}{4} = \frac{1}{4} \pm i\frac{\sqrt{23}}{4}$$ $x^2 + 7 = 0$ is equivalent to $x = \pm i\sqrt{7}$.

Like other equations that we have studied, quadratic equations can be solved symbolically, numerically, and graphically. The equation $x(x - 1) = 2$ is solved by these three methods.

NOTE: Symbolic solutions are *exact*. Numerical and graphical solutions are often *approximate*, particularly when solutions contain fractions or square roots.

Symbolic Solution

$$x(x - 1) = 2$$
$$x^2 - x = 2$$
$$x^2 - x - 2 = 0$$
$$(x + 1)(x - 2) = 0$$
$$x = -1 \text{ or } x = 2$$

Solutions are **−1** and **2**.

Numerical Solution

$x(x - 1) = 2$ when
$x = -1$ or **2**.

Graphical Solution

Points of intersectios are $(-1, 2)$ and $(2, 2)$, where $y_1 = x(x - 1)$ and $y_2 = 2$.

11.4 Exercises

MyMathLab   Math XL PRACTICE   WATCH   DOWNLOAD   READ   REVIEW

CONCEPTS

1. What is the quadratic formula used for?

2. What basic algebraic technique is used to derive the quadratic formula?

3. Write the discriminant.

4. If the discriminant evaluates to 0, what does that indicate about the quadratic equation?

5. Name four symbolic techniques for solving a quadratic equation.

6. Does every quadratic equation have at least one real solution? Explain.

7. Solve $x^2 - 4 = 0$.   8. Solve $x^2 + 4 = 0$.

THE QUADRATIC FORMULA

Exercises 9–12: Use the quadratic formula to solve the equation. Support your result graphically or numerically. If there are no real solutions, say so.

9. $2x^2 + 11x - 6 = 0$ 10. $x^2 + 2x - 24 = 0$

11. $-x^2 + 2x - 1 = 0$ 12. $3x^2 - x + 1 = 0$

Exercises 13–30: Solve by using the quadratic formula. If there are no real solutions, say so.

13. $x^2 - 6x - 16 = 0$ 14. $2x^2 - 9x + 7 = 0$

15. $4x^2 - x - 1 = 0$ 16. $-x^2 + 2x + 1 = 0$

17. $-3x^2 + 2x - 1 = 0$ 18. $x^2 + x + 3 = 0$

19. $36x^2 - 36x + 9 = 0$

20. $4x^2 - 5.6x + 1.96 = 0$

21. $2x(x - 3) = 2$ **22.** $x(x + 1) + x = 5$

23. $(x - 1)(x + 1) + 2 = 4x$

24. $\frac{1}{2}(x - 6) = x^2 + 1$ **25.** $\frac{1}{2}x(x + 1) = 2x^2 - \frac{3}{2}$

26. $\frac{1}{2}x^2 - \frac{1}{4}x + \frac{1}{2} = x$ **27.** $2x(x - 1) = 7$

28. $3x(x - 4) = 4$ **29.** $-3x^2 + 10x - 5 = 0$

30. $-2x^2 + 4x - 1 = 0$

THE DISCRIMINANT

Exercises 31–36: A graph of $y = ax^2 + bx + c$ is shown.
   (a) State whether $a > 0$ or $a < 0$.
   (b) Solve $ax^2 + bx + c = 0$, if possible.
   (c) Determine whether the discriminant is positive, negative, or zero.

31.

32.

33.

34.

35.

36.
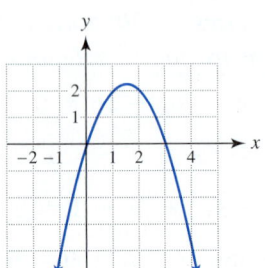

Exercises 37–44: Do the following for the given equation.
   (a) Evaluate the discriminant.
   (b) How many real solutions are there?
   (c) Support your answer for part (b) by using graphing.

37. $3x^2 + x - 2 = 0$ **38.** $5x^2 - 13x + 6 = 0$

39. $x^2 - 4x + 4 = 0$ **40.** $\frac{1}{4}x^2 + 4 = 2x$

41. $\frac{1}{2}x^2 + \frac{3}{2}x + 2 = 0$ **42.** $x - 3 = 2x^2$

43. $x(x + 3) = 3$

44. $(4x - 1)(x - 3) = -25$

Exercises 45–54: Use the quadratic formula to find any x-intercepts on the graph of the equation.

45. $y = x^2 - 2x - 1$ **46.** $y = x^2 + 3x + 1$

47. $y = -2x^2 - x + 3$ **48.** $y = -3x^2 - x + 4$

49. $y = x^2 + x + 5$ **50.** $y = 3x^2 - 2x + 5$

51. $y = x^2 + 9$ **52.** $y = x^2 + 11$

53. $y = 3x^2 + 4x - 2$ **54.** $y = 4x^2 - 2x - 3$

COMPLEX SOLUTIONS

Exercises 55–86: Solve the equation. Write complex solutions in standard form.

55. $x^2 + 9 = 0$ **56.** $x^2 + 16 = 0$

57. $x^2 + 80 = 0$ **58.** $x^2 + 20 = 0$

59. $x^2 + \frac{1}{4} = 0$ **60.** $x^2 + \frac{9}{4} = 0$

61. $16x^2 + 9 = 0$ **62.** $25x^2 + 36 = 0$

63. $x^2 = -6$ **64.** $x^2 = -75$

65. $x^2 - 3 = 0$ **66.** $x^2 - 8 = 0$

67. $x^2 + 2 = 0$ **68.** $x^2 + 4 = 0$

69. $x^2 - x + 2 = 0$ **70.** $x^2 + 2x + 3 = 0$

71. $2x^2 + 3x + 4 = 0$ **72.** $3x^2 - x = 1$

73. $x^2 + 1 = 4x$ **74.** $3x^2 + 2 = x$

75. $x^2 + x = -2$ **76.** $x(x - 4) = -8$

77. $5x^2 + 2x + 4 = 0$ **78.** $7x^2 - 2x + 4 = 0$

79. $\frac{1}{2}x^2 + \frac{3}{4}x + 1 = 0$ **80.** $-\frac{1}{3}x^2 + x - 2 = 0$

81. $x(x + 2) = x - 4$ **82.** $x - 5 = 2x(2x + 1)$

83. $x(2x - 1) = 1 + x$ **84.** $2x = x(3 - 4x)$

85. $x^2 = x(1 - x) - 2$ **86.** $2x^2 = 2x(5 - x) - 8$

YOU DECIDE THE METHOD

Exercises 87–100: Find exact solutions to the quadratic equation, using a method of your choice. Explain why you chose the method you did. Answers may vary.

87. $x^2 - 3x + 2 = 0$ **88.** $x^2 + 2x + 1 = 0$

89. $0.5x^2 - 1.75x - 1 = 0$ **90.** $\frac{3}{5}x^2 + \frac{9}{10}x - \frac{3}{5} = 0$

91. $x^2 - 5x + 2 = 0$ **92.** $2x^2 - x - 4 = 0$

93. $2x^2 + x = -8$ **94.** $4x^2 = 2x - 3$

95. $4x^2 - 1 = 0$ **96.** $3x^2 = 9$

97. $3x^2 + 6 = 0$ **98.** $4x^2 + 7 = 0$

99. $9x^2 + 1 = 6x$ **100.** $10x^2 + 15x = 25$

APPLICATIONS

*Exercises 101–104: **Modeling Stopping Distance** (Refer to Example 4.) Use $d = \frac{1}{9}x^2 + \frac{11}{3}x$ to find a safe speed x for the following stopping distances d.*

101. 42 feet   **102.** 152 feet

103. 390 feet   **104.** 726 feet

105. *U.S. AIDS Deaths* The cumulative numbers in thousands of AIDS deaths from 1984 through 1994 may be modeled by

$$f(x) = 2.39x^2 + 5.04x + 5.1,$$

where $x = 0$ corresponds to 1984, $x = 1$ to 1985, and so on until $x = 10$ corresponds to 1994. See the accompanying graph. Use the formula for $f(x)$ to estimate the year when the total number of AIDS

deaths reached 200 thousand. Compare your result with that shown in the graph.

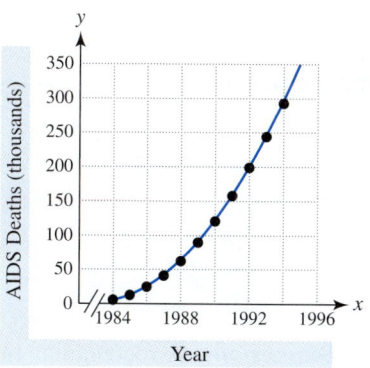

106. *Canoeing* A camper paddles a canoe 2 miles downstream in a river that has a 2-mile-per-hour current. To return to camp, the canoeist travels upstream on a different branch of the river. It is 4 miles long and has a 1-mile-per-hour current. The total trip (both ways) takes 3 hours. Find the average speed of the canoe in still water. (*Hint:* Time equals distance divided by rate.)

107. *Distance* Two cars leave an intersection, one traveling south and one traveling east, as shown in the figure. After 1 hour the two cars are 50 miles apart and the car traveling east has traveled 10 miles farther than the car traveling south. How far did each car travel?

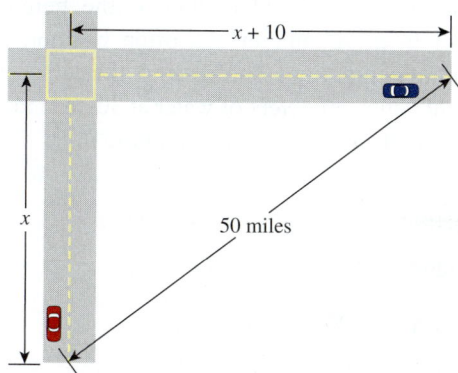

108. *Airplane Speed* A pilot flies 500 miles against a 20-mile-per-hour wind. On the next day, the pilot flies back home with a 10-mile-per-hour tail wind. The total trip (both ways) takes 4 hours. Find the speed of the airplane without a wind.

109. *Screen Dimensions*   The width of a rectangular computer screen is 3 inches more than its height. If the area of the screen is 154 square inches, find its dimensions

  (a) graphically,
  (b) numerically, and
  (c) symbolically.

110. *Sidewalk Dimension*   A rectangular flower garden in a park is 30 feet wide and 40 feet long. A sidewalk around the perimeter of the garden is being planned, as shown in the figure. The gardener has enough money to pour 624 square feet of cement sidewalk. Find the width of the sidewalk.

111. *Modeling Water Flow*   When water runs out of a hole in a cylindrical container, the height of the water in the container can often be modeled by a quadratic function. The data in the table show the height y in centimeters of water at 30-second intervals in a metal can that has a small hole in it.

Time	0	30	60	90
Height	16	11.9	8.4	5.3

Time	120	150	180
Height	3.1	1.4	0.5

These data are modeled by

$$f(x) = 0.0004x^2 - 0.15x + 16.$$

  (a) Explain why a linear function would not be appropriate for modeling these data.

(b) Use the table to estimate the time at which the height was 7 centimeters.

(c) Use $f(x)$ and the quadratic formula to solve part (b).

112. *Hospitals*   The general trend in the number of hospitals in the United States from 1945 through 2000 is modeled by

$$f(x) = -1.38x^2 + 84x + 5865,$$

where $x = 5$ corresponds to 1945, $x = 10$ to 1950, and so on until $x = 60$ represents 2000. See the scatterplot and accompanying graph.

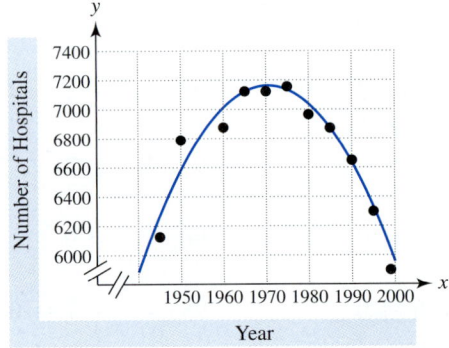

(a) Describe any trends in the numbers of hospitals from 1945 to 2000.

(b) What information does the vertex give?

(c) Use the formula for $f(x)$ to estimate the number of hospitals in 1970. Compare your result with that shown in the graph.

(d) Use the formula for $f(x)$ to estimate the year (or years) when there were 6300 hospitals. Compare your result with that shown in the graph.

WRITING ABOUT MATHEMATICS

113. Explain how the discriminant $b^2 - 4ac$ can be used to determine the number of solutions to a quadratic equation.

114. Let $f(x) = ax^2 + bx + c$. If you know the value of $b^2 - 4ac$, what information does this give you about the graph of f? Explain your answer.

1. Solve the quadratic equation $2x^2 - 7x + 3 = 0$ symbolically and graphically.

2. Use the square root property to solve $x^2 = 5$.

3. Complete the square to solve $x^2 - 4x + 1 = 0$.

4. Solve the equation $x^2 + y^2 = 1$ for y.

5. Use the quadratic formula to solve each equation.
   (a) $2x^2 = 3x + 1$

   (b) $9x^2 - 24x + 16 = 0$
   (c) $x^2 + x + 2 = 0$

6. Calculate the discriminant for each equation and give the number of *real* solutions.
   (a) $x^2 - 5x + 5 = 0$
   (b) $2x^2 - 5x + 4 = 0$
   (c) $49x^2 - 56x + 16 = 0$

11.5 QUADRATIC INEQUALITIES

Basic Concepts ■ Graphical and Numerical Solutions ■ Symbolic Solutions

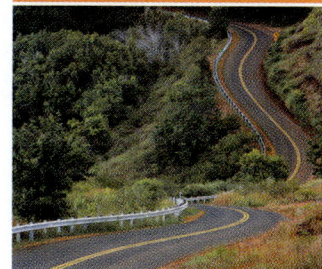

A LOOK INTO MATH ▷ Parabolas are frequently used in highway design. Sometimes it is necessary for engineers to determine where a highway is below a particular elevation. To do this, they may need to solve a *quadratic inequality*. Quadratic inequalities are nonlinear inequalities. These inequalities are often simple enough to be solved by hand. In this section we discuss how to solve them graphically, numerically, and symbolically.

Basic Concepts

If the equals sign in a quadratic equation is replaced with $>, \geq, <,$ or $\leq$, a **quadratic inequality** results. Examples of quadratic inequalities include

$$x^2 + 4x - 3 < 0, \quad 5x^2 \geq 5, \quad \text{and} \quad 1 - z \leq z^2.$$

Any quadratic equation can be written as

$$ax^2 + bx + c = 0, \qquad a \neq 0,$$

so any quadratic inequality can be written as

$$ax^2 + bx + c > 0, \qquad a \neq 0,$$

where $>$ may be replaced with $\geq, <,$ or $\leq$.

EXAMPLE **Identifying a quadratic inequality**

Determine whether the inequality is quadratic.
(a) $5x + x^2 - x^3 \leq 0$ (b) $4 + 5x^2 > 4x^2 + x$

Solution

(a) The inequality $5x + x^2 - x^3 \leq 0$ is not quadratic because it has an x^3-term.

(b) Write the inequality as follows.

$$4 + 5x^2 > 4x^2 + x \qquad \text{Given inequality}$$

$$4 + 5x^2 - 4x^2 - x > 0 \qquad \text{Subtract } 4x^2 \text{ and } x.$$

$$4 + x^2 - x > 0 \qquad \text{Combine like terms.}$$

$$x^2 - x + 4 > 0 \qquad \text{Rewrite.}$$

Because the inequality can be written in the form $ax^2 + bx + c > 0$ with $a = 1$, $b = -1$, and $c = 4$, it is a quadratic inequality. **Now Try Exercises** 7, 11

Graphical and Numerical Solutions

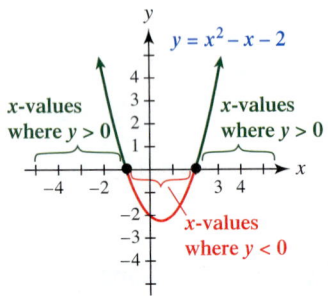

Figure 11.47

Equality often is the boundary between *greater than* and *less than*, so a first step in solving an inequality is to determine the x-values where equality occurs. We begin by using this concept with graphical and numerical techniques.

A graph of $y = x^2 - x - 2$ with x-intercepts -1 and 2 is shown in Figure 11.47. The solutions to $x^2 - x - 2 = 0$ are given by $x = -1$ or $x = 2$. Between the x-intercepts the graph dips below the x-axis and the y-values are negative. Thus the solutions to $x^2 - x - 2 < 0$ satisfy $-1 < x < 2$. To check this result we select a **test value**. For example, 0 lies between -1 and 2. If we substitute $x = 0$ in $x^2 - x - 2 < 0$, it results in a true statement.

$$0^2 - 0 - 2 < 0 \qquad \text{True}$$

When $x < -1$ or $x > 2$, the graph lies above the x-axis and the y-values are positive. Thus the solutions to $x^2 - x - 2 > 0$ satisfy $x < -1$ or $x > 2$. For example, 3 is greater than 2 and -3 is less than -1. Therefore both 3 and -3 are solutions. We can verify this result by substituting **3** and **-3** as test values in $x^2 - x - 2 > 0$.

$$3^2 - 3 - 2 > 0 \qquad \text{True}$$

$$(-3)^2 - (-3) - 2 > 0 \qquad \text{True}$$

In the next three examples we use these concepts to solve quadratic inequalities.

EXAMPLE 2 Solving a quadratic inequality

Make a table of values for $y = x^2 - 3x - 4$ and then sketch the graph. Use the table and graph to solve $x^2 - 3x - 4 \leq 0$. Write your answer in interval notation.

Solution

The points calculated for Table 11.14 are plotted in Figure 11.48 and connected with a smooth ∪-shaped graph.

Numerical Solution Table 11.14 shows that $x^2 - 3x - 4$ equals 0 when $x = -1$ or $x = 4$. Between these values, $x^2 - 3x - 4$ is negative so the solution set to $x^2 - 3x - 4 \leq 0$ is given by $-1 \leq x \leq 4$ or in interval notation, $[-1, 4]$.

Graphical Solution In Figure 11.48 the graph of $y = x^2 - 3x - 4$ shows that the x-intercepts are -1 and 4. Between these values, the graph dips *below* the x-axis. Thus the solution set is $[-1, 4]$.

TABLE 11.14

x	$y = x^2 - 3x - 4$
-2	6
-1	0
0	-4
1	-6
2	-6
3	-4
4	0
5	6

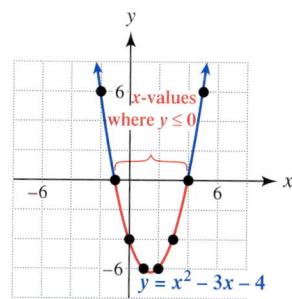

Figure 11.48

Now Try Exercises 19, 25

EXAMPLE 3 **Solving a quadratic inequality**

Solve $x^2 > 1$. Write your answer in interval notation.

Solution

First, rewrite $x^2 > 1$ as $x^2 - 1 > 0$. The graph $y = x^2 - 1$ is shown in Figure 11.49 with x-intercepts -1 and 1. The graph lies *above* the x-axis and is shaded green to the left of $x = -1$ and to the right of $x = 1$. Thus the solution set is given by $x < -1$ or $x > 1$, which can be written in interval notation as $(-\infty, -1) \cup (1, \infty)$.

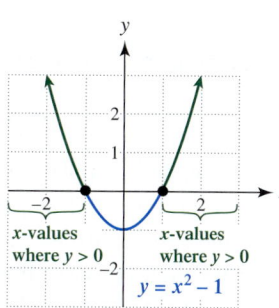

Figure 11.49

Now Try Exercise 33

EXAMPLE 4 **Solving some special cases**

Solve each of the inequalities graphically.
(a) $x^2 + 1 > 0$ **(b)** $x^2 + 1 < 0$ **(c)** $(x - 1)^2 \le 0$

Solution

(a) Because the graph of $y = x^2 + 1$, shown in Figure 11.50, is always above the x-axis, $x^2 + 1$ is always greater than 0. The solution set includes all real numbers, or $(-\infty, \infty)$.

(b) Because the graph of $y = x^2 + 1$, shown in Figure 11.50, never goes below the x-axis, $x^2 + 1$ is never less than 0. Thus there are no real solutions.

(c) Because the graph of $y = (x - 1)^2$, shown in Figure 11.51 on the next page, never goes below the x-axis, $(x - 1)^2$ is never less than 0. When $x = 1$, $y = 0$, so 1 is the only solution to the inequality $(x - 1)^2 \le 0$.

Figure 11.50

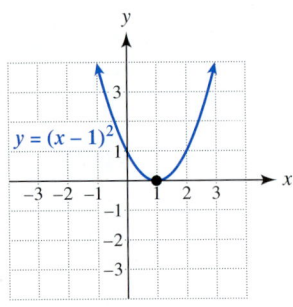

Figure 11.51

Now Try Exercises 35, 39, 41

▶ **REAL-WORLD CONNECTION** In the next example we show how quadratic inequalities are used in highway design.

EXAMPLE 5 Determining elevations on a sag curve

Parabolas are frequently used in highway design to model hills and sags (valleys) along a proposed route. Suppose that the elevation E in feet of a sag, or *sag curve*, is given by

$$E(x) = 0.00004x^2 - 0.4x + 2000,$$

where x is the horizontal distance in feet along the sag curve and $0 \le x \le 10{,}000$. See Figure 11.52. Estimate graphically the x-values where the elevation is 1500 feet or less.

(***Source:*** F. Mannering and W. Kilareski, *Principles of Highway Engineering and Traffic Analysis.*)

Figure 11.52

Solution

Graphical Solution We must solve the quadratic inequality

$$0.00004x^2 - 0.4x + 2000 \le 1500.$$

To do so, we let $Y_1 = .00004X^2 - .4X + 2000$ represent the sag or valley in the road and $Y_2 = 1500$ represent a horizontal line with an elevation of 1500 feet. Their graphs intersect at $x \approx 1464$ and $x \approx 8536$, as shown in Figure 11.53. The elevation of the proposed route is less than 1500 feet between these x-values. Therefore the elevation of the road is 1500 feet or less when $1464 \le x \le 8536$ (approximately).

CALCULATOR HELP

To find a point of intersection, see Appendix A (page AP-7).

[0, 10000, 1000] by [0, 2500, 500]   [0, 10000, 1000] by [0, 2500, 500]

Intersection
X=1464.4661 Y=1500

Intersection
X=8535.5339 Y=1500

(a)   (b)

Figure 11.53

Now Try Exercise 63

Symbolic Solutions

To solve a quadratic inequality symbolically, we first solve the corresponding equality. We can then write the solution to the inequality, using the following method.

SOLUTIONS TO QUADRATIC INEQUALITIES

Let $ax^2 + bx + c = 0$, $a > 0$, have two real solutions p and q, where $p < q$.

$ax^2 + bx + c < 0$ is equivalent to $p < x < q$ (see left-hand figure).

$ax^2 + bx + c > 0$ is equivalent to $x < p$ or $x > q$ (see right-hand figure).

Quadratic inequalities involving $\leq$ or $\geq$ can be solved similarly.

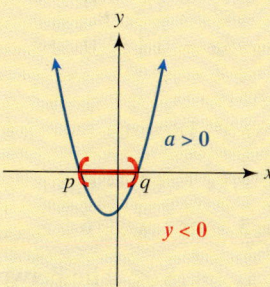

Solutions lie between p and q.

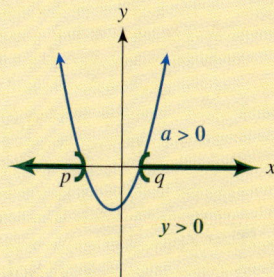

Solutions lie "outside" p and q.

One way to handle the situation where $a < 0$, is to multiply each side of the inequality by -1, in which case we must be sure to *reverse* the inequality symbol. For example, the inequality $-2x^2 + 8 \leq 0$ has $a = -2$, which is negative. If we multiply each side of this inequality by -1, we obtain $2x^2 - 8 \geq 0$ and now $a = 2$, which is positive. Another way to solve this inequality is to graph $y = -2x^2 + 8$, which is a parabola that opens downward. The following Critical Thinking addresses this situation. However, when $a < 0$ we will usually rewrite the quadratic inequality so that $a > 0$.

CRITICAL THINKING

The graph of $y = -x^2 + x + 12$ is a parabola opening downward with x-intercepts -3 and 4. Solve each inequality.

i. $-x^2 + x + 12 > 0$ **ii.** $-x^2 + x + 12 < 0$

EXAMPLE 6 **Solving quadratic inequalities**

Solve each inequality symbolically. Write your answer in interval notation.
(a) $6x^2 - 7x - 5 \geq 0$ **(b)** $x(3 - x) > -18$

Solution
(a) Begin by solving $6x^2 - 7x - 5 = 0$.

$$6x^2 - 7x - 5 = 0 \qquad \text{Quadratic equation}$$
$$(2x + 1)(3x - 5) = 0 \qquad \text{Factor.}$$
$$2x + 1 = 0 \quad \text{or} \quad 3x - 5 = 0 \qquad \text{Zero-product property}$$
$$x = -\frac{1}{2} \quad \text{or} \quad x = \frac{5}{3} \qquad \text{Solve.}$$

Therefore the solutions to $6x^2 - 7x - 5 \geq 0$ lie "outside" these two values and satisfy $x \leq -\frac{1}{2}$ or $x \geq \frac{5}{3}$. In interval notation the solution set is $\left(-\infty, -\frac{1}{2}\right] \cup \left[\frac{5}{3}, \infty\right)$.

(b) First rewrite the inequality as follows.

$$x(3 - x) > -18 \qquad \text{Given inequality}$$
$$3x - x^2 > -18 \qquad \text{Distributive property}$$
$$3x - x^2 + 18 > 0 \qquad \text{Add 18.}$$
$$-x^2 + 3x + 18 > 0 \qquad \text{Rewrite.}$$
$$x^2 - 3x - 18 < 0 \qquad \text{Multiply by } -1 \text{ because } a < 0;$$
$$\text{reverse the inequality symbol.}$$

Next, solve $x^2 - 3x - 18 = 0$.

$$(x + 3)(x - 6) = 0 \qquad \text{Factor.}$$
$$x = -3 \quad \text{or} \quad x = 6 \qquad \text{Solve.}$$

Solutions to $x^2 - 3x - 18 < 0$ lie between these two values and satisfy $-3 < x < 6$. In interval notation the solution set is $(-3, 6)$.   **Now Try Exercises** 29, 31

EXAMPLE 7   Finding the dimensions of a building

A rectangular building needs to be 7 feet longer than it is wide, as illustrated in Figure 11.54. The area of the building must be at least 450 square feet. What widths x are possible for this building? Support your results with a table of values.

Figure 11.54

Solution

Symbolic Solution   If x is the width of the building, $x + 7$ is the length of the building and its area is $x(x + 7)$. The area must be at least 450 square feet, so the inequality $x(x + 7) \geq 450$ must be satisfied. First solve the following quadratic equation.

$$x(x + 7) = 450 \qquad \text{Quadratic equation}$$
$$x^2 + 7x = 450 \qquad \text{Distributive property}$$
$$x^2 + 7x - 450 = 0 \qquad \text{Subtract 450.}$$
$$x = \frac{-7 \pm \sqrt{7^2 - 4(1)(-450)}}{2(1)} \qquad \begin{array}{l}\text{Quadratic formula; } a = 1, \\ b = 7, \text{ and } c = -450\end{array}$$
$$= \frac{-7 \pm \sqrt{1849}}{2} \qquad \text{Simplify.}$$
$$= \frac{-7 \pm 43}{2} \qquad \sqrt{1849} = 43$$
$$= 18, -25 \qquad \text{Evaluate.}$$

Thus the solutions to $x(x + 7) \geq 450$ are $x \leq -25$ or $x \geq 18$. The width is positive, so the building width must be 18 feet or more.

X	Y1
15	330
16	368
17	408
18	450
19	494
20	540
21	588

Y1■X(X+7)

Figure 11.55

Numerical Solution   A table of values is shown in Figure 11.55, where $Y_1 = X(X + 7)$ equals 450 when $x = 18$. For $x \geq 18$ the area is *at least* 450 square feet.

Now Try Exercise 67

11.5   PUTTING IT ALL TOGETHER

The following table summarizes solutions of quadratic inequalities containing the symbols $<$ or $>$. Cases involving $\leq$ or $\geq$ are solved similarly.

Method	Explanation
Solving a Quadratic Inequality Symbolically	Let $ax^2 + bx + c = 0$, $a > 0$, have two real solutions p and q, where $p < q$. $ax^2 + bx + c < 0$ is equivalent to $p < x < q$. $ax^2 + bx + c > 0$ is equivalent to $x < p$ or $x > q$. *Examples:* The solutions to $x^2 - 3x + 2 = 0$ are given by $x = 1$ or $x = 2$. The solutions to $x^2 - 3x + 2 < 0$ are given by $1 < x < 2$. The solutions to $x^2 - 3x + 2 > 0$ are given by $x < 1$ or $x > 2$.
Solving a Quadratic Inequality Graphically	Given $ax^2 + bx + c < 0$ with $a > 0$, graph $y = ax^2 + bx + c$ and locate any x-intercepts. If there are two x-intercepts, then solutions correspond to x-values between the x-intercepts. Solutions to $ax^2 + bx + c > 0$ correspond to x-values "outside" the x-intercepts. See the box on page 783.
Solving a Quadratic Inequality Numerically	If a quadratic inequality is expressed as $ax^2 + bx + c < 0$ with $a > 0$, then we can solve $y = ax^2 + bx + c = 0$ with a table. If there are two solutions, then the solutions to the given inequality lie between these values. Solutions to $ax^2 + bx + c > 0$ lie "outside" these values.

11.5   Exercises

CONCEPTS

1. How is a quadratic inequality different from a quadratic equation?

2. Do quadratic inequalities typically have two solutions? Explain.

3. Is 3 a solution to $x^2 < 7$?

4. Is 5 a solution to $x^2 \geq 25$?

5. The solutions to $x^2 - 2x - 8 = 0$ are -2 and 4. What are the solutions to $x^2 - 2x - 8 < 0$? Write your answer as an inequality.

6. The solutions to $x^2 + 2x - 3 = 0$ are -3 and 1. What are the solutions to $x^2 + 2x - 3 > 0$? Write your answer as an inequality.

QUADRATIC INEQUALITIES

Exercises 7–12: Is the inequality quadratic?

7. $x^2 + 4x + 5 < 0$

8. $x > x^3 - 5$

9. $x^2 > 19$

10. $x(x - 1) - 2 \geq 0$

11. $4x > 1 - x$

12. $2x(x^2 + 3) < 0$

Exercises 13–18: Is the given value of x a solution?

13. $2x^2 + x - 1 > 0$ $x = 3$

14. $x^2 - 3x + 2 \leq 0$ $x = 2$

15. $x^2 + 2 \leq 0$ $x = 0$

16. $2x(x - 3) \geq 0$ $x = 1$

17. $x^2 - 3x \leq 1$ $x = -3$

18. $4x^2 - 5x + 1 > 30$ $x = -2$

SOLVING QUADRATIC INEQUALITIES

Exercises 19–24: The graph of $y = ax^2 + bx + c$ *is given. Solve each equation or inequality.*

    (a) $ax^2 + bx + c = 0$

    (b) $ax^2 + bx + c < 0$

    (c) $ax^2 + bx + c > 0$

19.

20.

21.

22.

23.

24.

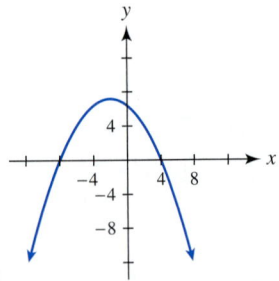

Exercises 25–28: A table for $y = ax^2 + bx + c$ *is given. Solve each equation or inequality.*

    (a) $ax^2 + bx + c = 0$

    (b) $ax^2 + bx + c < 0$

    (c) $ax^2 + bx + c > 0$

25. $y = x^2 - 4$

x	-3	-2	-1	0	1	2	3
y	5	0	-3	-4	-3	0	5

26. $y = x^2 - x - 2$

x	-3	-2	-1	0	1	2	3
y	10	4	0	-2	-2	0	4

27. $y = x^2 + 4x$

x	-5	-4	-3	-2	-1	0	1
y	5	0	-3	-4	-3	0	5

28. $y = -2x^2 - 2x + 1.5$

x	-2	-1.5	-1	-0.5	0	0.5	1
y	-2.5	0	1.5	2	1.5	0	-2.5

Exercises 29–46: Solve the inequality by any method. Write your answer in interval notation when appropriate.

29. $x^2 + 4x + 3 < 0$ **30.** $x^2 + x - 2 \leq 0$

31. $2x^2 - x - 15 \geq 0$ **32.** $3x^2 - 3x - 6 > 0$

33. $2x^2 \leq 8$ **34.** $x^2 < 9$

35. $x^2 > -5$ **36.** $-x^2 \geq 1$

37. $-x^2 + 3x > 0$ **38.** $-8x^2 - 2x + 1 \leq 0$

39. $x^2 + 2 \leq 0$ **40.** $x^2 + 3 \geq -5$

41. $(x - 2)^2 \leq 0$ **42.** $(x + 2)^2 \leq 0$

43. $(x + 1)^2 > 0$ **44.** $(x - 3)^2 > 0$

45. $x(1 - x) \geq -2$ **46.** $x(x - 2) < 3$

Exercises 47–50: Solve the equation in part (a). Use the results to solve the inequalities in parts (b) and (c).

47. (a) $x^2 - 4 = 0$ **48.** (a) $x^2 - 5 = 0$

    (b) $x^2 - 4 < 0$ (b) $x^2 - 5 \leq 0$

    (c) $x^2 - 4 > 0$ (c) $x^2 - 5 \geq 0$

49. (a) $x^2 + x - 1 = 0$

    (b) $x^2 + x - 1 < 0$

    (c) $x^2 + x - 1 > 0$

50. (a) $x^2 + 4x - 5 = 0$

    (b) $x^2 + 4x - 5 \leq 0$

    (c) $x^2 + 4x - 5 \geq 0$

Exercises 51–60: Solve the quadratic inequality symbolically. Write your answer in interval notation.

51. $x^2 + 10x + 21 \leq 0$ **52.** $x^2 - 7x - 18 < 0$

53. $3x^2 - 9x + 6 > 0$ **54.** $7x^2 + 34x - 5 \geq 0$

55. $x^2 < 10$ **56.** $x^2 \geq 64$

57. $x(x - 6) > 0$ **58.** $1 - x^2 \leq 0$

59. $x(4 - x) \leq 2$ **60.** $2x(1 - x) \geq 2$

 Exercises 61 and 62: Solve the inequality graphically to the nearest thousandth.

61. $\pi x^2 - \sqrt{3}x \leq \frac{3}{11}$ **62.** $\sqrt{5}x^2 - \pi^2 x \geq 10.3$

APPLICATIONS

 63. *Highway Design*   (Refer to Example 5.) The elevation E of a sag curve in feet is given by

$$E(x) = 0.0000375x^2 - 0.175x + 1000,$$

where $0 \leq x \leq 4000$.
   (a) Estimate graphically the x-values for which the elevation is 850 feet or less. (*Hint:* Use [0, 4000, 1000] by [500, 1200, 100] as a viewing rectangle.)
   (b) For what x-values is the elevation 850 feet or more?

64. *Early Cellular Phone Use*   Our society is in transition from an industrial to an informational society. Cellular communication has played an increasingly large role in this transition. The number of cellular subscribers in the United States in thousands from 1985 to 1991 can be modeled by

$$f(x) = 163x^2 - 146x + 205,$$

where x is the year and $x = 0$ corresponds to 1985, $x = 1$ to 1986, and so on.   (*Source:* M. Paetsch, *Mobile Communication in the U.S. and Europe.*)
   (a) Write a quadratic inequality whose solution set represents the years when there were 2 million subscribers or more.
   (b) Solve this inequality.

65. *Heart Disease Death Rates*   From 1960 to 2000, age-adjusted heart disease rates decreased dramatically. The number of deaths per 100,000 people can be modeled by

$$f(x) = -0.05107x^2 + 194.74x - 184{,}949,$$

where x is the year, as illustrated in the accompanying figure.   (*Source:* Department of Health and Human Services.)

   (a) Evaluate $f(1985)$, using both the formula and the graph. How do your results compare?
   (b) Use the graph to estimate the years when this death rate was 500 or less.
   (c) Solve part (b) by using the quadratic formula.

66. *Accidental Deaths*   From 1910 to 2000 the number of accidental deaths per 100,000 people generally decreased and can be modeled by

$$f(x) = -0.001918x^2 + 6.93x - 6156,$$

where x is the year, as shown in the accompanying figure.   (*Source:* Department of Health and Human Services.)

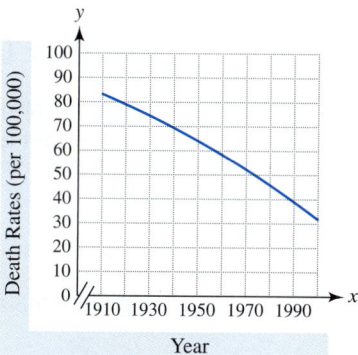

   (a) Evaluate $f(1955)$, using both the formula and the graph. How do your results compare?
   (b) Use the graph to estimate when this death rate was 60 or more.
   (c) Solve part (b) by using the quadratic formula.

67. *Dimensions of a Pen*   A rectangular pen for a pet is 5 feet longer than it is wide. Give possible values for the width w of the pen if its area must be between 176 and 500 square feet, inclusively.

68. *Dimensions of a Cylinder*   The volume of a cylindrical can is given by $V = \pi r^2 h$, where r is its radius and h is its height. See the accompanying figure. If $h = 6$ inches and the volume of the can must be 50 cubic inches or more, estimate to the nearest tenth of an inch possible values for r.

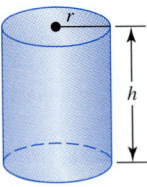

WRITING ABOUT MATHEMATICS

69. Consider the inequality $x^2 < 0$. Discuss the solutions to this inequality and explain your reasoning.

70. Explain how the graph of $y = ax^2 + bx + c$ can be used to solve the inequality

$$ax^2 + bx + c > 0$$

when $a < 0$. Assume that the x-intercepts of the graph are p and q with $p < q$.

11.6   EQUATIONS IN QUADRATIC FORM

Higher Degree Polynomial Equations ▪ Equations Having Rational Exponents ▪ Equations Having Complex Solutions

A LOOK INTO MATH ▷   Although many equations are *not* quadratic equations, they can sometimes be written in quadratic form. These equations are *reducible to quadratic form*. To express such an equation in quadratic form, we often use substitution. In this section we discuss this process.

Higher Degree Polynomial Equations

Sometimes a fourth degree polynomial can be factored like a quadratic trinomial, provided it does not have an x-term or an x^3-term. Let's consider the equation $x^4 - 5x^2 + 4 = 0$.

$$x^4 - 5x^2 + 4 = 0 \qquad \text{Given equation}$$
$$(x^2)^2 - 5(x^2) + 4 = 0 \qquad \text{Properties of exponents}$$

We use the substitution $u = x^2$.

$$u^2 - 5u + 4 = 0 \qquad \text{Let } u = x^2.$$
$$(u - 4)(u - 1) = 0 \qquad \text{Factor.}$$
$$u - 4 = 0 \quad \text{or} \quad u - 1 = 0 \qquad \text{Zero-product property}$$
$$u = 4 \quad \text{or} \quad u = 1 \qquad \text{Solve each equation.}$$

Because the given equation uses the variable x, we must give the solutions in terms of x. We substitute x^2 for u and then solve to obtain the following four solutions.

$$x^2 = 4 \quad \text{or} \quad x^2 = 1 \qquad \text{Substitute } x^2 \text{ for } u.$$
$$x = \pm 2 \quad \text{or} \quad x = \pm 1 \qquad \text{Square root property}$$

The solutions are -2, -1, 1, and 2.

In the next example we solve a sixth degree polynomial equation.

EXAMPLE 1   Solving equations by substitution

Solve $2x^6 + x^3 = 1$.

Solution

Start by subtracting 1 from each side.

$$2x^6 + x^3 - 1 = 0 \qquad \text{Subtract 1.}$$
$$2(x^3)^2 + (x^3) - 1 = 0 \qquad \text{Properties of exponents}$$
$$2u^2 + u - 1 = 0 \qquad \text{Let } u = x^3.$$
$$(2u - 1)(u + 1) = 0 \qquad \text{Factor.}$$
$$2u - 1 = 0 \quad \text{or} \quad u + 1 = 0 \qquad \text{Zero-product property}$$
$$u = \tfrac{1}{2} \quad \text{or} \quad u = -1 \qquad \text{Solve.}$$

Now substitute x^3 for u, and solve for x to obtain the following two solutions.

$$x^3 = \tfrac{1}{2} \quad \text{or} \quad x^3 = -1 \qquad \text{Substitute } x^3 \text{ for } u.$$
$$x = \sqrt[3]{\tfrac{1}{2}} \quad \text{or} \quad x = -1 \qquad \text{Take cube root of each side.}$$

Now Try Exercise **3**

Equations Having Rational Exponents

Equations that have negative exponents are sometimes reducible to quadratic form. Consider the following example, in which two methods are presented.

EXAMPLE 2   Solving an equation having negative exponents

Solve $-6m^{-2} + 13m^{-1} + 5 = 0$.

Solution

Method I   Use the substitution $u = m^{-1} = \frac{1}{m}$ and $u^2 = m^{-2} = \frac{1}{m^2}$.

$$-6m^{-2} + 13m^{-1} + 5 = 0 \qquad \text{Given equation}$$
$$-6u^2 + 13u + 5 = 0 \qquad \text{Let } u = m^{-1} \text{ and } u^2 = m^{-2}.$$
$$6u^2 - 13u - 5 = 0 \qquad \text{Multiply by } -1.$$
$$(2u - 5)(3u + 1) = 0 \qquad \text{Factor.}$$
$$2u - 5 = 0 \quad \text{or} \quad 3u + 1 = 0 \qquad \text{Zero-product property}$$
$$u = \tfrac{5}{2} \quad \text{or} \quad u = -\tfrac{1}{3} \qquad \text{Solve for } u.$$

Because $u = \frac{1}{m}$, $m = \frac{1}{u}$. Thus the solutions are given by $m = \frac{2}{5}$ or $m = -3$.

Method II   Another way to solve this equation is to multiply each side by the LCD, m^2.

$$-6m^{-2} + 13m^{-1} + 5 = 0 \qquad \text{Given equation}$$
$$m^2(-6m^{-2} + 13m^{-1} + 5) = m^2 \cdot 0 \qquad \text{Multiply by } m^2.$$
$$-6m^2 m^{-2} + 13m^2 m^{-1} + 5m^2 = 0 \qquad \text{Distributive property}$$
$$-6 + 13m + 5m^2 = 0 \qquad \text{Add exponents.}$$
$$5m^2 + 13m - 6 = 0 \qquad \text{Rewrite the equation.}$$
$$(5m - 2)(m + 3) = 0 \qquad \text{Factor.}$$
$$5m - 2 = 0 \quad \text{or} \quad m + 3 = 0 \qquad \text{Zero-product property}$$
$$m = \tfrac{2}{5} \quad \text{or} \quad m = -3 \qquad \text{Solve.}$$

Now Try Exercise **5**

In the next example we solve an equation having fractional exponents.

EXAMPLE 3 **Solving an equation having fractional exponents**

Solve $x^{2/3} - 2x^{1/3} - 8 = 0$.

Solution

Use the substitution $u = x^{1/3}$.

$$x^{2/3} - 2x^{1/3} - 8 = 0 \qquad \text{Given equation}$$
$$(x^{1/3})^2 - 2(x^{1/3}) - 8 = 0 \qquad \text{Properties of exponents}$$
$$u^2 - 2u - 8 = 0 \qquad \text{Let } u = x^{1/3}.$$
$$(u - 4)(u + 2) = 0 \qquad \text{Factor.}$$
$$u - 4 = 0 \quad \text{or} \quad u + 2 = 0 \qquad \text{Zero-product property}$$
$$u = 4 \quad \text{or} \quad u = -2 \qquad \text{Solve.}$$

Because $u = x^{1/3}$, $u^3 = (x^{1/3})^3 = x$. Thus $x = 4^3 = 64$ or $x = (-2)^3 = -8$. The solutions are -8 and 64.   Now Try Exercise **13**

Equations Having Complex Solutions

Sometimes an equation that is reducible to quadratic form also has complex solutions. This situation is discussed in the next two examples.

EXAMPLE 4 **Solving the fourth degree equation**

Find all complex solutions to $x^4 - 1 = 0$.

Solution

$$x^4 - 1 = 0 \qquad \text{Given equation}$$
$$(x^2)^2 - 1 = 0 \qquad \text{Properties of exponents}$$
$$u^2 - 1 = 0 \qquad \text{Let } u = x^2.$$
$$(u - 1)(u + 1) = 0 \qquad \text{Factor difference of squares.}$$
$$u - 1 = 0 \quad \text{or} \quad u + 1 = 0 \qquad \text{Zero-product property}$$
$$u = 1 \quad \text{or} \quad u = -1 \qquad \text{Solve for } u.$$

Now substitute x^2 for u, and solve for x.

$$x^2 = 1 \quad \text{or} \quad x^2 = -1 \qquad \text{Let } x^2 = u.$$
$$x = \pm 1 \quad \text{or} \quad x = \pm i \qquad \text{Square root property}$$

There are four complex solutions: -1, 1, $-i$, and i.   Now Try Exercise **25**

EXAMPLE **5**   **Solving a rational equation**

Find all complex solutions to $\frac{1}{x} + \frac{1}{x^2} = -1$.

Solution

This equation is a rational equation. However, if we multiply through by the LCD, x^2, we clear fractions and obtain a quadratic equation with complex solutions.

$$\frac{1}{x} + \frac{1}{x^2} = -1 \qquad \text{Given equation}$$

$$\frac{x^2}{x} + \frac{x^2}{x^2} = -1x^2 \qquad \text{Multiply each term by } x^2.$$

$$x + 1 = -x^2 \qquad \text{Simplify.}$$

$$x^2 + x + 1 = 0 \qquad \text{Add } x^2.$$

$$x = \frac{-1 \pm \sqrt{1^2 - 4(1)(1)}}{2(1)} \qquad \text{Quadratic formula}$$

$$x = \frac{-1 \pm i\sqrt{3}}{2} \qquad \sqrt{-3} = i\sqrt{3}$$

$$x = -\frac{1}{2} \pm \frac{i\sqrt{3}}{2} \qquad \frac{a \pm b}{c} = \frac{a}{c} \pm \frac{b}{c}$$

Now Try Exercise **31**

11.6 PUTTING IT ALL TOGETHER

The following table demonstrates how to reduce some types of equations to quadratic form.

Equation	Substitution	Examples
Higher Degree Polynomial	Let $u = x^n$ for some integer n.	To solve $x^4 - 3x^2 - 4 = 0$, let $u = x^2$. This equation becomes $$u^2 - 3u - 4 = 0.$$
Rational Exponents	Pick a substitution that reduces the equation to quadratic form.	To solve $n^{-2} + 6n^{-1} + 9 = 0$, let $u = n^{-1}$. This equation becomes $$u^2 + 6u + 9 = 0.$$ To solve $6x^{2/5} - 5x^{1/5} - 4 = 0$, let $u = x^{1/5}$. This equation becomes $$6u^2 - 5u - 4 = 0.$$
Equations Having Complex Solutions	Both polynomial and rational equations can have complex solutions. Use the fact that if $a > 0$, then $\sqrt{-a} = i\sqrt{a}$.	$$1 + \frac{1}{x^2} = 0$$ $$x^2 \cdot \left(1 + \frac{1}{x^2}\right) = 0 \cdot x^2$$ $$x^2 + 1 = 0$$ $$x^2 = -1$$ $$x = \pm i$$

11.6 Exercises

EQUATIONS REDUCIBLE TO QUADRATIC FORM

Exercises 1–6: Use the given substitution to solve the equation.

1. $x^4 - 7x^2 + 6 = 0$ $u = x^2$

2. $2k^4 - 7k^2 + 6 = 0$ $u = k^2$

3. $3z^6 + z^3 - 10 = 0$ $u = z^3$

4. $2x^6 + 17x^3 + 8 = 0$ $u = x^3$

5. $4n^{-2} + 17n^{-1} + 15 = 0$ $u = n^{-1}$

6. $m^{-2} + 24 = 10m^{-1}$ $u = m^{-1}$

Exercises 7–24: Solve. Find all real solutions.

7. $x^4 = 8x^2 + 9$ **8.** $3x^4 = 10x^2 + 8$

9. $3x^6 - 5x^3 - 2 = 0$ **10.** $6x^6 + 11x^3 + 4 = 0$

11. $2z^{-2} + 11z^{-1} = 40$ **12.** $z^{-2} - 10z^{-1} + 25 = 0$

13. $x^{2/3} - 2x^{1/3} + 1 = 0$ **14.** $3x^{2/3} + 18x^{1/3} = 48$

15. $x^{2/5} - 33x^{1/5} + 32 = 0$

16. $x^{2/5} - 80x^{1/5} - 81 = 0$

17. $x - 13\sqrt{x} + 36 = 0$

18. $x - 17\sqrt{x} + 16 = 0$

19. $z^{1/2} - 2z^{1/4} + 1 = 0$

20. $z^{1/2} - 4z^{1/4} + 4 = 0$

21. $(x + 1)^2 - 5(x + 1) - 14 = 0$

22. $2(x - 5)^2 + 5(x - 5) + 3 = 0$

23. $(x^2 - 1)^2 - 4 = 0$

24. $(x^2 - 9)^2 - 8(x^2 - 9) + 16 = 0$

EQUATIONS HAVING COMPLEX SOLUTIONS

Exercises 25–34: Find all complex solutions.

25. $x^4 - 16 = 0$ **26.** $\frac{1}{3}x^4 - 27 = 0$

27. $x^3 + x = 0$ **28.** $4x^3 + x = 0$

29. $x^4 - 2 = x^2$ **30.** $x^4 - 3 = 2x^2$

31. $\dfrac{1}{x} + \dfrac{1}{x^2} = -\dfrac{1}{2}$ **32.** $\dfrac{2}{x - 1} - \dfrac{1}{x} = -1$

33. $\dfrac{2}{x - 2} - \dfrac{1}{x} = -\dfrac{1}{2}$ **34.** $\dfrac{1}{x} - \dfrac{1}{x^2} = \dfrac{1}{2}$

WRITING ABOUT MATHEMATICS

35. Explain how to solve $ax^4 - bx^2 + c = 0$. Assume that the left side of the equation factors.

36. Explain what it means for an equation to be reducible to quadratic form.

CHECKING BASIC CONCEPTS
SECTIONS 11.5 AND 11.6

1. Solve the inequality $x^2 - x - 6 > 0$. Write your answer in interval notation.

2. Solve the inequality $3x^2 + 5x + 2 \leq 0$. Write your answer in interval notation.

3. Solve $x^6 + 6x^3 - 16 = 0$.

4. Solve $x^{2/3} - 7x^{1/3} - 8 = 0$.

5. Find all complex solutions to
$$x^4 + 2x^2 + 1 = 0.$$

CHAPTER 11 | SUMMARY

SECTION 11.1 ■ QUADRATIC FUNCTIONS AND THEIR GRAPHS

Quadratic Function   Any quadratic function f can be written as

$$f(x) = ax^2 + bx + c \quad (a \neq 0).$$

Graph of a Quadratic Function   Its graph is a parabola that is wider than the graph of $y = x^2$, if $0 < |a| < 1$ and narrower, if $|a| > 1$. The y-intercept is c.

 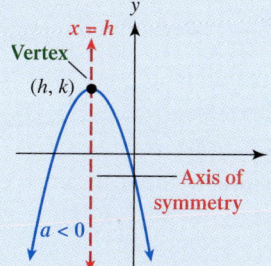

Axis of Symmetry   The parabola is symmetric with respect to this vertical line. The axis of symmetry passes through the vertex.

Vertex Formula   The x-coordinate of the vertex is $-\frac{b}{2a}$.

Example: Let $y = x^2 - 4x + 1$.

$$x = -\frac{-4}{2(1)} = 2 \quad \text{and} \quad y = 2^2 - 4(2) + 1 = -3. \text{ The vertex is } (2, -3).$$

SECTION 11.2 ■ PARABOLAS AND MODELING

Vertical and Horizontal Translations   Let h and k be positive numbers.

To graph	*shift the graph of $y = x^2$ by k units*
$y = x^2 + k$	upward.
$y = x^2 - k$	downward.

To graph	*shift the graph of $y = x^2$ by h units*
$y = (x - h)^2$	right.
$y = (x + h)^2$	left.

Example: Compared to $y = x^2$, the graph of $y = (x - 1)^2 + 2$ is translated right 1 unit and upward 2 units.

Vertex Form   Any quadratic function can be expressed as $f(x) = a(x - h)^2 + k$. In this form the point (h, k) is the vertex. A quadratic function can be put in this form by completing the square.

Example:
$$\begin{aligned} y &= x^2 + 10x - 4 && \text{Given equation} \\ &= (x^2 + 10x + 25) - 25 - 4 && \left(\tfrac{b}{2}\right)^2 = \left(\tfrac{10}{2}\right)^2 = 25 \\ &= (x + 5)^2 - 29 && \text{Perfect square trinomial; add.} \end{aligned}$$

The vertex is $(-5, -29)$.

SECTION 11.3 ■ QUADRATIC EQUATIONS

Quadratic Equations   Any quadratic equation can be written as $ax^2 + bx + c = 0$ and can have no real solutions, one real solution, or two real solutions. These solutions correspond to the x-intercepts on the graph of $y = ax^2 + bx + c$.

Example: $x^2 + x - 2 = 0$

$(x + 2)(x - 1) = 0$

$x = -2 \quad \text{or} \quad x = 1$

The x-intercepts for $y = x^2 + x - 2$ are -2 and 1.

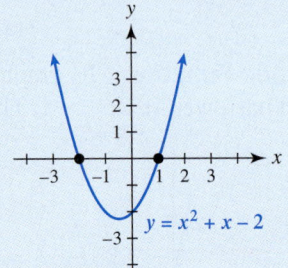

Completing the Square   Write the equation in the form $x^2 + bx = d$. Complete the square by adding $\left(\frac{b}{2}\right)^2$ to each side of the equation.

SECTION 11.4 ■ THE QUADRATIC FORMULA

The Quadratic Formula   The solutions to $ax^2 + bx + c = 0 \ (a \neq 0)$ are given by

$$x = \frac{-b \pm \sqrt{b^2 - 4ac}}{2a}.$$

Example: Solve $2x^2 + 3x - 1 = 0$ by letting $a = 2$, $b = 3$, and $c = -1$.

$$x = \frac{-3 \pm \sqrt{3^2 - 4(2)(-1)}}{2(2)} = \frac{-3 \pm \sqrt{17}}{4} \approx 0.28, \ -1.78$$

The Discriminant   The expression $b^2 - 4ac$ is called the discriminant. If $b^2 - 4ac > 0$, there are two real solutions; if $b^2 - 4ac = 0$, there is one real solution; and if $b^2 - 4ac < 0$, there are no real solutions, rather there are two complex solutions.

Example: For $2x^2 + 3x - 1 = 0$, the discriminant is

$$b^2 - 4ac = 3^2 - 4(2)(-1) = 17 > 0.$$

There are two real solutions to this quadratic equation.

SECTION 11.5 ■ QUADRATIC INEQUALITIES

Quadratic Inequalities   When the equals sign in a quadratic equation is replaced with $<, >, \leq$, or $\geq$, a quadratic inequality results. For example,

$$3x^2 - x + 1 = 0$$

is a quadratic equation and

$$3x^2 - x + 1 > 0$$

is a quadratic inequality. Like quadratic equations, quadratic inequalities can be solved symbolically, graphically, and numerically. An important first step in solving a quadratic inequality is to solve the corresponding quadratic equation.

Examples: The solutions to $x^2 - 5x - 6 = 0$ are given by $x = -1, 6$.

The solutions to $x^2 - 5x - 6 < 0$ satisfy $-1 < x < 6$.

The solutions to $x^2 - 5x - 6 > 0$ satisfy $x < -1$ or $x > 6$.

SECTION 11.6 ■ EQUATIONS IN QUADRATIC FORM

Equations Reducible to Quadratic Form An equation that is not quadratic, but can be put into quadratic form by using a substitution is reducible to quadratic form.

Example: To solve $x^{2/3} - 2x^{1/3} - 15 = 0$ let $u = x^{1/3}$. This equation becomes

$$u^2 - 2u - 15 = 0.$$

Factoring results in $(u + 3)(u - 5) = 0$, or $u = -3$ or 5. Because $u = x^{1/3}$, $x = u^3$ and $x = (-3)^3 = -27$ or $x = (5)^3 = 125$.

CHAPTER 11 REVIEW EXERCISES

SECTION 11.1

Exercises 1 and 2: Identify the vertex, axis of symmetry, and whether the parabola opens upward or downward. State where the graph is increasing and where it is decreasing.

1.

2.

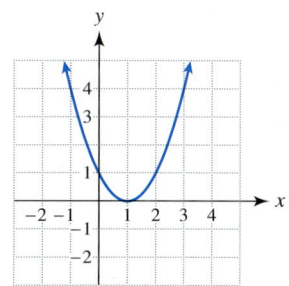

Exercises 3–6: Do the following.
    (a) Graph f.
    (b) Identify the vertex and axis of symmetry.
    (c) Evaluate f(x) at the given value of x.

3. $f(x) = x^2 - 2$, $x = -1$

4. $f(x) = -x^2 + 4x - 3$, $x = 3$

5. $f(x) = -\frac{1}{2}x^2 + x + \frac{3}{2}$, $x = -2$

6. $f(x) = 2x^2 + 8x + 5$, $x = -3$

7. Find the minimum y-value located on the graph of $y = 2x^2 - 6x + 1$.

8. Find the maximum y-value located on the graph of $y = -3x^2 + 2x - 5$.

Exercises 9–12: Find the vertex of the parabola.

9. $f(x) = x^2 - 4x - 2$ **10.** $f(x) = 5 - x^2$

11. $f(x) = -\frac{1}{4}x^2 + x + 1$ **12.** $f(x) = 2 + 2x + x^2$

SECTION 11.2

Exercises 13–18: Do the following.
    (a) Graph f.
    (b) Compare the graph of f with the graph of $y = x^2$.

13. $f(x) = x^2 + 2$ **14.** $f(x) = 3x^2$

15. $f(x) = (x - 2)^2$ **16.** $f(x) = (x + 1)^2 - 3$

17. $f(x) = \frac{1}{2}(x + 1)^2 + 2$

18. $f(x) = 2(x - 1)^2 - 3$

19. Write the vertex form of a parabola with $a = -4$ and vertex $(2, -5)$.

20. Write the vertex form of a parabola that opens downward with vertex $(-4, 6)$. Assume that $a = \pm 1$.

Exercises 21–24: Write the equation in vertex form. Identify the vertex.

21. $y = x^2 + 4x - 7$

22. $y = x^2 - 7x + 1$

23. $y = 2x^2 - 3x - 8$

24. $y = 3x^2 + 6x - 2$

Exercises 25 and 26: Find a value for the constant a so that $f(x) = ax^2 - 1$ models the data.

25.

x	1	2	3
$f(x)$	2	11	26

26.

x	-1	0	1
$f(x)$	$-\frac{3}{4}$	-1	$-\frac{3}{4}$

Exercises 27 and 28: Write $f(x)$ in the form given by $f(x) = ax^2 + bx + c$. Identify the y-intercept on the graph of f.

27. $f(x) = -5(x - 3)^2 + 4$

28. $f(x) = 3(x + 2)^2 - 4$

SECTION 11.3

Exercises 29–32: Use the graph of $y = ax^2 + bx + c$ to solve $ax^2 + bx + c = 0$.

29.

30.

31.

32.

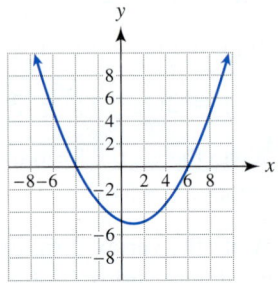

Exercises 33 and 34: A table of $y = ax^2 + bx + c$ is given. Solve $ax^2 + bx + c = 0$.

33.

34.

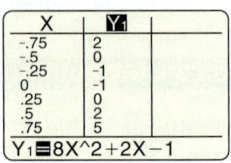

Exercises 35–38: Solve the quadratic equation (a) graphically and (b) numerically.

35. $x^2 - 5x - 50 = 0$

36. $\frac{1}{2}x^2 + x - \frac{3}{2} = 0$

37. $\frac{1}{4}x^2 + \frac{1}{2}x = 2$

38. $\frac{1}{2}x + \frac{3}{4} = \frac{1}{4}x^2$

Exercises 39–42: Solve by factoring.

39. $x^2 + x - 20 = 0$

40. $x^2 + 11x + 24 = 0$

41. $15x^2 - 4x - 4 = 0$

42. $7x^2 - 25x + 12 = 0$

Exercises 43–46: Use the square root property to solve.

43. $x^2 = 100$

44. $3x^2 = \frac{1}{3}$

45. $4x^2 - 6 = 0$

46. $5x^2 = x^2 - 4$

Exercises 47–50: Solve by completing the square.

47. $x^2 + 6x = -2$

48. $x^2 - 4x = 6$

49. $x^2 - 2x - 5 = 0$

50. $2x^2 + 6x - 1 = 0$

Exercises 51 and 52: Solve for the specified variable.

51. $F = \dfrac{k}{(R + r)^2}$ for R

52. $2x^2 + 3y^2 = 12$ for y

SECTION 11.4

Exercises 53–58: Use the quadratic formula to solve.

53. $x^2 - 9x + 18 = 0$

54. $x^2 - 24x + 143 = 0$

55. $6x^2 + x = 1$

56. $5x^2 + 1 = 5x$

57. $x(x - 8) = 5$

58. $2x(2 - x) = 3 - 2x$

Exercises 59–64: Solve by any method.

59. $x^2 - 4 = 0$

60. $4x^2 - 1 = 0$

61. $2x^2 + 15 = 11x$

62. $2x^2 + 15 = 13x$

63. $x(5 - x) = 2x + 1$

64. $-2x(x - 1) = x - \frac{1}{2}$

Exercises 65–68: A graph of $y = ax^2 + bx + c$ is shown.
  (a) State whether $a > 0$ or $a < 0$.
  (b) Solve $ax^2 + bx + c = 0$.
  (c) Determine whether the discriminant is positive, negative, or zero.

65.

66.

67.

68.
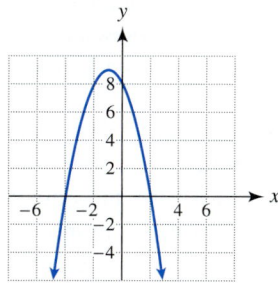

Exercises 69–72: Do the following for the given equation.
  (a) Evaluate the discriminant.
  (b) How many real solutions are there?
  (c) Support your answer for part (b) graphically.

69. $2x^2 - 3x + 1 = 0$ **70.** $7x^2 + 2x - 5 = 0$

71. $3x^2 + x + 2 = 0$

72. $4.41x^2 - 12.6x + 9 = 0$

Exercises 73–76: Solve. Write any complex solutions in standard form.

73. $x^2 + x + 5 = 0$ **74.** $2x^2 + 8 = 0$

75. $2x^2 = x - 1$ **76.** $7x^2 = 2x - 5$

SECTION 11.5

Exercises 77 and 78: The graph of $y = ax^2 + bx + c$ is shown. Solve each equation or inequality.

  (a) $ax^2 + bx + c = 0$
  (b) $ax^2 + bx + c < 0$
  (c) $ax^2 + bx + c > 0$

77.

78.
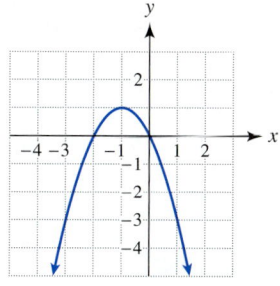

Exercises 79 and 80: A table of $y = ax^2 + bx + c$ is shown. Solve each equation or inequality.

  (a) $ax^2 + bx + c = 0$
  (b) $ax^2 + bx + c < 0$
  (c) $ax^2 + bx + c > 0$

79. $y = x^2 - 16$

x	-6	-4	-2	0	2	4	6
y	20	0	-12	-16	-12	0	20

80. $y = x^2 + x - 2$

x	-3	-2	-1	0	1	2	3
y	4	0	-2	-2	0	4	10

Exercises 81 and 82: Solve the quadratic equation in part (a). Use the results to solve the inequalities in parts (b) and (c).

81. (a) $x^2 - 2x - 3 = 0$
   (b) $x^2 - 2x - 3 < 0$
   (c) $x^2 - 2x - 3 > 0$

82. (a) $2x^2 - 7x - 15 = 0$
   (b) $2x^2 - 7x - 15 \le 0$
   (c) $2x^2 - 7x - 15 \ge 0$

Exercises 83–88: Solve the quadratic inequality. Write your answer in interval notation.

83. $x^2 + 4x + 3 \le 0$ **84.** $5x^2 - 16x + 3 < 0$

85. $6x^2 - 13x + 2 > 0$ **86.** $x^2 \ge 5$

87. $(x - 1)^2 \ge 0$ **88.** $x^2 + 3 < 2$

SECTION 11.6

Exercises 89–92: Solve the equation.

89. $x^4 - 14x^2 + 45 = 0$

90. $2z^{-2} + z^{-1} - 28 = 0$

91. $x^{2/3} - 9x^{1/3} + 8 = 0$

92. $(x - 1)^2 + 2(x - 1) + 1 = 0$

Exercises 93 and 94: Find all complex solutions.

93. $4x^4 + 4x^2 + 1 = 0$ **94.** $\dfrac{1}{x - 2} - \dfrac{3}{x} = -1$

APPLICATIONS

95. *Construction*   A rain gutter is being fabricated from a flat sheet of metal so that the cross section of the gutter is a rectangle, as shown in the accompanying figure. The width of the metal sheet is 12 inches.

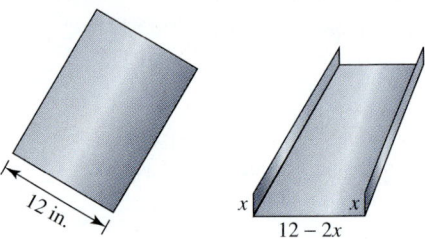

(a) Write a formula $f(x)$ that gives the area of the cross section.

(b) To hold the greatest amount of rainwater, the cross section should have maximum area. Find the dimensions that result in this maximum.

96. *Height of a Stone*   Suppose that a stone is thrown upward with an initial velocity of 44 feet per second (30 miles per hour) and is released 4 feet above the ground. Its height h in feet after t seconds is given by

$$h(t) = -16t^2 + 44t + 4.$$

(a) When does the stone reach a height of 32 feet?

(b) After how many seconds does the stone reach maximum height? Estimate this height.

97. *Maximizing Revenue*   Hotel rooms normally cost $90 per night. However, for a group rate the management is considering reducing the cost of a room by $3 for every room rented.

(a) Write a formula $f(x)$ that gives the revenue from renting x rooms at the group rate.

(b) Graph f in [0, 30, 5] by [0, 800, 100].

(c) How many rooms should be rented to receive revenue of $600?

(d) How many rooms should be rented to maximize revenue?

98. *Airline Complaints*   In 2000, major U.S. airlines promised better customer service. From 1997 through 1999, the number of complaints per 100,000 passengers can be modeled by

$$f(x) = 0.4(x - 1997)^2 + 0.8,$$

where x represents the year.   (**Source:** Department of Transportation.)

(a) Evaluate $f(1999)$. Interpret the result.

(b) Graph f in [1997, 1999, 1] by [0.5, 3, 0.5]. Discuss how complaints changed over this period.

99. *Braking Distance*   On dry pavement a safe braking distance d in feet for a car traveling x miles per hour is $d = \frac{x^2}{12}$. For each distance d, find x.   (**Source:** F. Mannering, *Principles of Highway Engineering and Traffic Control.*)

(a) $d = 144$ feet   **(b)** $d = 300$ feet

100. *Numbers*   The product of two numbers is 143. One number is 2 more than the other.

(a) Write an equation whose solution gives the smaller number x.

(b) Solve the equation.

101. *Educational Attainment*   From 1940 through 1991, the percentage of people with a high school diploma increased dramatically, as shown in the table.

Year	1940	1950	1960
H.S. Diploma (%)	25	34	44

Year	1970	1980	1991
H.S. Diploma (%)	55	69	78

Source: U.S. Census Bureau.

(a) Plot the data.

(b) Would it be reasonable to model these data with a linear function rather than a quadratic function? Explain your reasoning.

(c) Find a function that models the data.

(d) Use your answer from part (c) to estimate the percentage of people with a high school diploma in 1995 and 2005. Compare your answer to the actual values of 82% and 85%, respectively.

102. *U.S. Energy Consumption* From 1950 to 1970 per capita consumption of energy in millions of Btu can be modeled by $f(x) = \frac{1}{4}(x - 1950)^2 + 220$, where x is the year. (**Source:** Department of Energy.)

(a) Find and interpret the vertex.

(b) Graph f in [1950, 1970, 5] by [200, 350, 25]. What happened to energy consumption during this time period?

(c) Use f to predict the consumption in 1996. Actual consumption was 354 million Btu. Did f provide a good model for 1996? Explain.

103. *Screens* A square computer screen has an area of 123 square inches. Approximate its dimensions to the nearest tenth of an inch.

104. *Flying a Kite* A kite is being flown, as illustrated in the accompanying figure. If 130 feet of string have been let out, find the value of x.

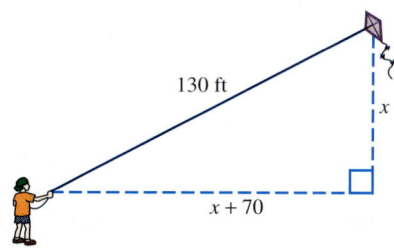

105. *Area* A uniform strip of grass is to be planted around a rectangular swimming pool, as illustrated in the accompanying figure. The swimming pool is 30 feet wide and 50 feet long. If there is only enough grass seed to cover 250 square feet, estimate the width x that the strip of grass should be.

106. *Dimensions of a Cone* The volume V of a cone is given by $V = \frac{1}{3}\pi r^2 h$, where r is its base radius and h is its height. See the accompanying figure. If $h = 20$ inches and the volume of the cone must be between 750 and 1700 cubic inches, inclusively, estimate to the nearest tenth of an inch possible values for r.

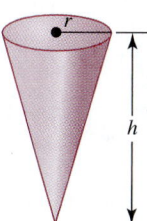

CHAPTER 11 TEST

Pass the Test · Video solutions to all test exercises

1. Find the vertex and axis of symmetry for the graph of $f(x) = -\frac{1}{2}x^2 + x + 1$. Evaluate $f(-2)$.

2. Find the minimum y-value located on the graph of $y = x^2 + 3x - 5$.

3. Find the exact value for the constant a so that $f(x) = ax^2 + 2$ models the data in the table.

x	-2	0	2	4
$f(x)$	0	2	0	-6

4. Graph $f(x) = \frac{1}{2}(x - 3)^2 + 2$ and compare the graph of f to the graph of $y = x^2$.

5. Write $y = x^2 - 6x + 2$ in vertex form. Identify the vertex and axis of symmetry.

6. Use the graph of $f(x) = ax^2 + bx + c$ to solve $ax^2 + bx + c = 0$. Then evaluate $f(1)$.

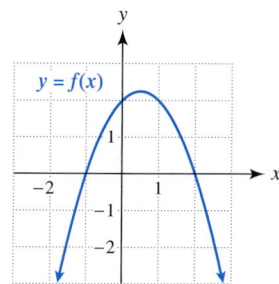

Exercises 7 and 8: Solve the quadratic equation.

7. $3x^2 + 11x - 4 = 0$ **8.** $2x^2 = 2 - 6x^2$

9. Solve $x^2 - 8x = 1$ by completing the square.

10. Solve $x(-2x + 3) = -1$ by using the quadratic formula.

11. Solve $9x^2 - 16 = 0$.

12. Solve $F = \dfrac{Gm^2}{r^2}$ for m.

13. A graph of $y = ax^2 + bx + c$ is shown.
   (a) State whether $a > 0$ or $a < 0$.
   (b) Solve $ax^2 + bx + c = 0$.
   (c) Determine whether the discriminant is positive, negative, or zero.

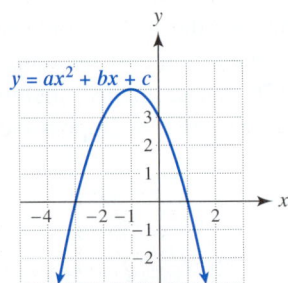

14. Complete the following for $-3x^2 + 4x - 5 = 0$.
   (a) Evaluate the discriminant.
   (b) How many real solutions are there?
   (c) Support your answer for part (b) graphically.

Exercises 15 and 16: The graph of $y = ax^2 + bx + c$ is shown. Solve each equation or inequality.
   (a) $ax^2 + bx + c = 0$
   (b) $ax^2 + bx + c < 0$
   (c) $ax^2 + bx + c > 0$

15.

16.

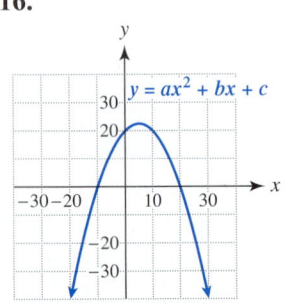

17. Solve the quadratic equation in part (a). Use the result to solve the inequalities in parts (b) and (c) and write your answer in interval notation.
   (a) $8x^2 - 2x - 3 = 0$
   (b) $8x^2 - 2x - 3 \le 0$
   (c) $8x^2 - 2x - 3 \ge 0$

18. Solve $x^2 + 2x \le 0$.

19. Solve $x^6 - 3x^3 + 2 = 0$. Find all real solutions.

20. Solve $2x^2 + 4x + 3 = 0$. Find all complex solutions.

21. Solve $\sqrt{2} - \pi x^2 = 2.12x - 0.5\pi$ graphically. Round your answers to the nearest hundredth.

22. *Braking Distance*   On wet pavement a safe braking distance d in feet for a car traveling x miles per hour is $d = \frac{x^2}{9}$. What speed corresponds to a braking distance of 250 feet?   (***Source:*** F. Mannering, *Principles of Highway Engineering and Traffic Control.*)

23. *Construction*   A fence is being constructed along a 20-foot building, as shown in the accompanying figure. No fencing is used along the building.
   (a) If 200 feet of fence are available, find a formula $f(x)$ (in terms of x) that gives the area enclosed.
   (b) What value of x gives the greatest area?

24. *Height of a Stone*   Suppose that a stone is thrown upward with an initial velocity of 88 feet per second (60 miles per hour) and is released 8 feet above the ground. Its height h in feet after t seconds is given by

$$h(t) = -16t^2 + 88t + 8.$$

   (a) Graph h in [0, 6, 1] by [0, 150, 50].
   (b) When does the stone strike the ground?
   (c) After how many seconds does the stone reach maximum height? Estimate this height.

CHAPTER 11 EXTENDED AND DISCOVERY EXERCISES

MODELING DATA WITH A QUADRATIC FUNCTION

 1. *Survival Rate of Birds* The survival rate of sparrowhawks varies according to their age. The following table summarizes the results of one study by listing the age in years and the percentage of birds that survived the previous year. For example, 52% of sparrowhawks that reached age 6 lived to be 7 years old. (***Source:*** D. Brown and P. Rothery, *Models in Biology.*)

Age	1	2	3	4	5
Percent (%)	45	60	71	67	67

Age	6	7	8	9
Percent (%)	61	52	30	25

(a) Try to explain the relationship between age and the likelihood of surviving the next year.

(b) Make a scatterplot of the data. What type of function might model the data? Explain.

(c) Graph each function. Which of the following functions models the data better?

$$f_1(x) = -3.57x + 71.1$$
$$f_2(x) = -2.07x^2 + 17.1x + 33$$

(d) Use one of these functions to estimate the likelihood of a 5.5-year-old sparrowhawk surviving for 1 more year.

 2. *Photosynthesis and Temperature* Photosynthesis is the process by which plants turn sunlight into energy. At very cold temperatures photosynthesis may halt even though the sun is shining. In one study the efficiency of photosynthesis for an Antarctic species of grass was investigated. The following table lists results for various temperatures. The temperature x is in degrees Celsius, and the efficiency y is given as a percent. The purpose of the research was to determine the temperature at which photosynthesis is most efficient. (***Source:*** D. Brown.)

x (°C)	−1.5	0	2.5	5	7	10	12
y (%)	33	46	55	80	87	93	95

x (°C)	15	17	20	22	25	27	30
y (%)	91	89	77	72	54	46	34

(a) Plot the data.

(b) What type of function might model these data? Explain your reasoning.

(c) Find a function f that models the data.

(d) Use f to estimate the temperature at which photosynthesis is most efficient in this type of grass.

TRANSLATIONS OF PARABOLAS IN COMPUTER GRAPHICS

Exercises 3 and 4: In older video games with two-dimensional graphics, the background is often translated to give the illusion that a character in the game is moving. The simple scene on the left shows a mountain and an airplane. To make it appear that the airplane is flying, the mountain can be translated to the left, as shown in the figure on the right. (***Reference:*** C. Pokorny and C. Gerald, *Computer Graphics.*)

 3. *Video Games* Suppose that the mountain in the figure on the left is modeled by $f(x) = -0.4x^2 + 4$ and that the airplane is located at the point $(1, 5)$.

(a) Graph f in $[-4, 4, 1]$ by $[0, 6, 1]$, where the units are kilometers. Plot the point $(1, 5)$ to show the location of the airplane.

(b) Assume that the airplane is moving horizontally to the right at 0.2 kilometer per second. To give a video game player the illusion that the airplane is moving, graph the image of the mountain and the position of the airplane after 10 seconds.

4. *Video Games* (Refer to Exercise 3.) Discuss how you could create the illusion of the airplane moving to the left and gaining altitude as it passes over the mountain. Try to perform a translation of this type. Explain your reasoning.

Exercises 5–8: ***Factoring and the Discriminant*** *If the discriminant of the trinomial $ax^2 + bx + c$ with integer coefficients is a perfect square, then it can be factored. For example, on the one hand, the discriminant of $6x^2 + x - 2$ is*

$$1^2 - 4(6)(-2) = 49,$$

which is a perfect square ($7^2 = 49$), so we can factor the trinomial as

$$6x^2 + x - 2 = (2x - 1)(3x + 2).$$

On the other hand, the discriminant for $x^2 + x - 1$ is

$$1^2 - 4(1)(-1) = 5,$$

which is not a perfect square, so we cannot factor this trinomial by using integers as coefficients. Similarly, if the discriminant is negative, the trinomial cannot be factored by using integer coefficients. Use the discriminant to predict whether the trinomial can be factored. Then test your prediction.

5. $10x^2 - x - 3$ **6.** $4x^2 - 3x - 6$

7. $3x^2 + 2x - 2$ **8.** $2x^2 + x + 3$

Exercises 9–14: ***Polynomial Inequalities*** *The solution set for a polynomial inequality can be found by first determining the boundary numbers. For example, to solve $f(x) = x^3 - 4x > 0$, begin by solving $x^3 - 4x = 0$. The solutions (boundary numbers) are -2, 0, and 2. The function $f(x) = x^3 - 4x$ is either only positive or only negative on intervals between consecutive zeros. To determine the solution set, we can evaluate test values for each interval as shown on the left below.*

Interval	Test Value	$f(x) = x^3 - 4x$
$(-\infty, -2)$	$x = -3$	$f(-3) = -15 < 0$
$(-2, 0)$	$x = -1$	$f(-1) = 3 > 0$
$(0, 2)$	$x = 1$	$f(1) = -3 < 0$
$(2, \infty)$	$x = 3$	$f(3) = 15 > 0$

$[-6, 6, 1]$ by $[-4, 4, 1]$

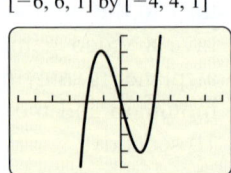

We can see that $f(x) > 0$ for $(-2, 0) \cup (2, \infty)$. These results also are supported graphically in the figure on the right above, where the graph of f is above the x-axis when $-2 < x < 0$ or when $x > 2$.

Use these concepts to solve the polynomial inequality.

9. $x^3 - x^2 - 6x > 0$

10. $x^3 - 3x^2 + 2x < 0$

11. $x^3 - 7x^2 + 14x \leq 8$

12. $9x - x^3 \geq 0$

13. $x^4 - 5x^2 + 4 > 0$

14. $1 < x^4$

Exercises 15–20: ***Rational Inequalities*** *Rational inequalities can be solved using many of the same techniques that are used to solve other types of inequalities. However, there is one important difference. For a rational inequality, the boundary between greater than and less than can either be an x-value where equality occurs or an x-value where a rational expression is undefined. For example, consider the inequality $f(x) = \frac{2 - x}{2x} > 0$. The solution to the equation $\frac{2 - x}{2x} = 0$ is 2. The rational expression $\frac{2 - x}{2x}$ is undefined when $x = 0$. Therefore we select test values on the intervals $(-\infty, 0)$, $(0, 2)$, and $(2, \infty)$. The table reveals that $f(x) > 0$ for $(0, 2)$.*

Interval	Test Value	$f(x) = \dfrac{2 - x}{2x}$
$(-\infty, 0)$	$x = -0.5$	$f(-0.5) = -2.5 < 0$
$(0, 2)$	$x = 1$	$f(1) = 0.5 > 0$
$(2, \infty)$	$x = 2.5$	$f(2.5) = -0.1 < 0$

$[-4.7, 4.7, 1]$ by $[-3.1, 3.1, 1]$

Note that $f(x)$ changes from negative to positive at $x = 0$, where $f(x)$ is undefined. These results are supported graphically in the figure on the right above.

Solve the rational inequality.

15. $\frac{3 - x}{3x} \geq 0$ **16.** $\frac{x - 2}{x + 2} > 0$

17. $\frac{3 - 2x}{1 + x} < 3$ **18.** $\frac{x + 1}{4 - 2x} \geq 1$

19. $\frac{5}{x^2 - 4} < 0$ **20.** $\frac{x}{x^2 - 1} \geq 0$

CHAPTERS 1–11 CUMULATIVE REVIEW EXERCISES

1. Evaluate $F = \dfrac{5}{z^2 + 1}$ when $z = -2$.

2. Classify each number as one or more of the following: natural number, whole number, integer, rational number, or irrational number: $0.\overline{4}$, $\sqrt{7}$, 0, -5, $\sqrt[3]{8}$, $-\frac{4}{3}$.

3. Simplify each expression. Write the result using positive exponents.
   (a) $\left(\dfrac{x^2 y^6}{x^{-3}}\right)^2$
   (b) $\dfrac{(xy^{-3})^2}{x(y^{-2})^{-1}}$
   (c) $(a^2 b)^2 (ab^3)^{-4}$

4. Write 9,290,000 in scientific notation.

5. Find $f(-2)$, if $f(x) = \sqrt{2 - x}$. What is the domain of f?

6. If $f(2) = 5$, then what point lies on the graph of f?

Exercises 7 and 8: Graph f by hand.

7. $f(x) = x^2 + 2x$

8. $f(x) = |2x - 4|$

9. Find the slope–intercept form of the line that passes through $(4, -1)$ and is parallel to the line passing through $(0, 1)$ and $(-2, 4)$.

10. Find the equation of a vertical line that passes through $(-3, 4)$.

11. Solve $2x - 3(x + 2) = 6$.

Exercises 12–14: Solve the inequality. Write your answer in interval notation.

12. $7 - x > 3x$

13. $|3x - 2| \le 1$

14. $-4 \le 1 - x < 2$

15. Solve the system.
$$-x - 4y = -3$$
$$5x + y = -4$$

16. Shade the solution set in the xy-plane.
$$3x + y \le 3$$
$$x - 3y \le 3$$

17. Solve the system.
$$x + y - z = 3$$
$$x - y + z = 1$$
$$2x - y - z = 1$$

Exercises 18–20: Multiply the expression.

18. $(3x - 2)(2x + 7)$

19. $3xy(x^2 + y^2)$

20. $(\sqrt{x} + 3)(\sqrt{x} - 3)$

Exercises 21 and 22: Factor the expression.

21. $x^3 - x^2 - 2x$

22. $4x^2 - 25$

Exercises 23 and 24: Solve each equation.

23. $x^2 - 3 = 0$

24. $x^2 + 1 = 2x$

Exercises 25 and 26: Simplify the expression.

25. $\dfrac{(x + 3)^2}{x + 2} \cdot \dfrac{x + 2}{2x + 6}$

26. $\dfrac{1}{x + 2} - \dfrac{1}{x}$

Exercises 27–30: Simplify the expression. Assume all variables are positive.

27. $\sqrt{16x^6}$

28. $16^{-3/2}$

29. $\dfrac{\sqrt[3]{81x}}{\sqrt[3]{3x}}$

30. $\sqrt{8x} + \sqrt{2x}$

31. Graph $f(x) = \sqrt{4x}$.

32. Find the distance between $(-1, 2)$ and $(4, 3)$.

33. Write $\dfrac{3 - i}{2 + i}$ in standard form.

34. Solve $3\sqrt{x + 1} = 2x$.

35. Solve the equation $2x = \sqrt{2.1 - x} + \sqrt[3]{0.1x}$ to the nearest hundredth.

36. Sketch a graph of $f(x) = x^2 - 2x + 3$.
    (a) Find the vertex.
    (b) Evaluate $f(-1)$.
    (c) What is the axis of symmetry?
    (d) Where is f increasing?

37. Write $f(x) = 2x^2 - 4x - 1$ in vertex form.

38. Compare the graph of $f(x) = 4(x + 1)^2 - 2$ to the graph of $y = x^2$.

39. Solve $x^2 + 6x = 2$ by completing the square.

40. Solve $2x^2 - 3x = 1$ by using the quadratic formula.

41. Solve $x(4 - x) = 3$.

42. The graph of $y = ax^2 + bx + c$ is shown. Solve each equation or inequality.
    (a) $ax^2 + bx + c = 0$ **(b)** $ax^2 + bx + c \leq 0$

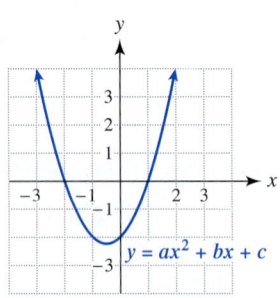

43. Solve $x^2 - 3x + 2 > 0$.

44. Solve $x^4 - 256 = 0$. Find all complex solutions.

Exercises 45–52: **Thinking Generally** *Match the graph (a.–h.) with its equation. Assume that a, b, and c are positive constants.*

45. $y = ax - b$ **46.** $y = b$

47. $y = -ax^2 + c$ **48.** $y = \dfrac{a}{x}$

49. $y = ax^3$ **50.** $y = |ax + b|$

51. $y = a\sqrt{x}$ **52.** $y = a\sqrt[3]{x}$

a.

b.

c.

d.

e.

f.

g.

h.
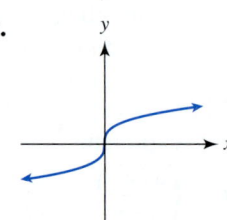

APPLICATIONS

53. *Calculating Water Flow* Water is being pumped out of a tank. The gallons G of water in the tank after t minutes is shown in the figure, where $y = G(t)$.
    (a) Evaluate $G(0)$. Interpret your answer.
    (b) What is the t-intercept? Interpret your answer.
    (c) What is the slope of the graph of G? Interpret your answer.
    (d) Find a formula for $G(t)$.

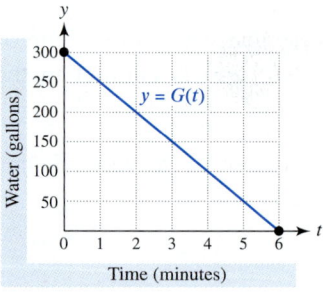

54. *Investment* Suppose $4000 are deposited in three accounts paying 4%, 5%, and 6% annual interest. The amount invested at 6% is $1000 more than the amount invested at 5%. The interest after 1 year is $216. How much is invested at each rate?

55. *Maximizing Area* There are 490 feet of fence available to surround the perimeter of a rectangular garden. On one side, there is a 10-foot gate that requires no fencing. What dimensions for the garden give the largest area?

56. *Height of a Tree* A 6-foot person casts a shadow that is 10 feet while a nearby tree casts a 55-foot shadow. Find the height of the tree.

CHAPTER 12

Exponential and Logarithmic Functions

The Swedish scientist Svante Arrhenius first predicted in 1900 a greenhouse effect resulting from emissions of carbon dioxide by industrialized countries. His classic calculation made use of logarithms and predicted that a doubling of the carbon dioxide concentration in the atmosphere would raise the average global temperature by 7°F to 11°F. An increase in world population has resulted in higher emissions of greenhouse gases, such as carbon dioxide, and these emissions have the potential to alter Earth's climate and destroy portions of the ozone layer.

Today there is renewed interest in global warming. Since 1958 the amount of carbon dioxide in the atmosphere has increased by 21%. According to the United Nation's Intergovernmental Panel on Climate Change, the world faces an average temperature rise of 3°C, or 5.4°F, this century, if greenhouse gas emissions continue to rise at their current pace.

In this chapter we use exponential and logarithmic functions to model a wide variety of phenomena, such as greenhouse gases, acid rain, the decline of the bluefin tuna, the demand for liver transplants, diversity of bird species, hurricanes, and earthquakes. Mathematics plays a key role in understanding, controlling, and predicting natural phenomena and people's effect on them.

> *Human history becomes more and more a race between education and catastrophe.*
>
> —H. G. WELLS

Source: M. Kraljic, *The Greenhouse Effect.*

12.1 COMPOSITE AND INVERSE FUNCTIONS

Composition of Functions ▪ One-to-One Functions ▪ Inverse Functions ▪ Tables and Graphs of Inverse Functions

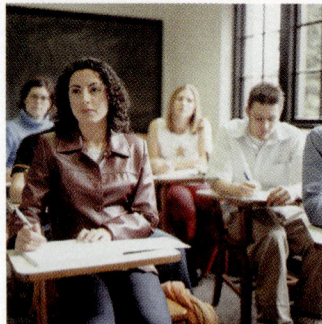

A LOOK INTO MATH ▷ Suppose that you walk into a classroom, turn on the lights, and sit down at your desk. How could you undo or reverse these actions? You might stand up from the desk, turn off the lights, and walk out of the classroom. Note that you must not only perform the "inverse" of each action, but you also must do them in the *reverse order*. In mathematics, we undo an arithmetic operation by performing its inverse operation. For example, the inverse operation of addition is subtraction, and the inverse operation of multiplication is division. In this section we explore these concepts further by discussing inverse functions.

Composition of Functions

▶ **REAL-WORLD CONNECTION**   Many tasks in life are performed in *sequence*, such as putting on your socks and then your shoes. These types of situations also occur in mathematics. For example, suppose that we want to calculate the number of ounces in 3 tons. Because there are 2000 pounds in one ton, we might first multiply **3** by 2000 to obtain **6000** pounds. There are 16 ounces in a pound, so we could multiply 6000 by 16 to obtain **96,000** ounces. This particular calculation involves a *sequence* of calculations that can be represented by the diagram shown in Figure 12.1.

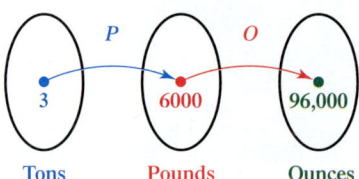

Tons   Pounds   Ounces

Figure 12.1

The results shown in Figure 12.1 can be calculated by using functions. Suppose that we let $P(x) = 2000x$ convert x tons to P pounds and also let $O(x) = 16x$ convert x pounds to O ounces. We can calculate the number of ounces in 3 tons by performing the *composition of O and P*. This method can be expressed symbolically as

$$(O \circ P)(3) = O\big(P(3)\big) \qquad \text{First compute } P(3).$$
$$= O(2000 \cdot 3) \qquad \text{Let } x = 3 \text{ in } P(x) = 2000x.$$
$$= O(6000) \qquad \text{Simplify.}$$
$$= 16(6000) \qquad \text{Let } x = 6000 \text{ in } O(x) = 16x.$$
$$= 96,000. \qquad \text{Multiply.}$$

Note that the output for function P—namely, $P(3)$—becomes the input for function O. That is, we evaluate $O\big(P(3)\big)$ by calculating $P(3)$ first and then substituting the result of 6000 in function O.

COMPOSITION OF FUNCTIONS

If f and g are functions, then the **composite function** $g \circ f$, or **composition** of g and f, is defined by

$$(g \circ f)(x) = g\big(f(x)\big).$$

NOTE: We read $g\big(f(x)\big)$ as "g of f of x."

The compositions $g \circ f$ and $f \circ g$ represent evaluating functions f and g in two different ways. When evaluating $g \circ f$, function f is performed first followed by function g, whereas for $f \circ g$ function g is performed first followed by function f. Note that in general $(g \circ f)(x) \ne (f \circ g)(x)$. That is, *the order in which functions are applied makes a difference,* the same way that putting on your socks and then your shoes is quite different from putting on your shoes and then your socks.

EXAMPLE 1 **Finding composite functions**

Evaluate $(g \circ f)(2)$ and then find a formula for $(g \circ f)(x)$.

(a) $f(x) = x^3, g(x) = 3x - 2$ **(b)** $f(x) = 5x, g(x) = x^2 - 3x + 1$

(c) $f(x) = \sqrt{2x}, g(x) = \dfrac{1}{x - 1}$

Solution

(a) $(g \circ f)(2) = g\big(f(2)\big)$　　　　　Composition of functions

$\qquad\qquad = g(8)$　　　　　　　　$f(2) = 2^3 = 8$

$\qquad\qquad = 22$　　　　　　　　$g(8) = 3(8) - 2 = 22$

Note that the output from f—namely, $f(2)$—becomes the input for g.

$(g \circ f)(x) = g\big(f(x)\big)$　　　　　Composition of functions

$\qquad\qquad = g(x^3)$　　　　　　　　$f(x) = x^3$

$\qquad\qquad = 3x^3 - 2$　　　　　　$g(x) = 3x - 2$

(b) $(g \circ f)(2) = g\big(f(2)\big)$　　　　　Composition of functions

$\qquad\qquad = g(10)$　　　　　　　　$f(2) = 5(2) = 10$

$\qquad\qquad = 71$　　　　　　　　$g(10) = 10^2 - 3(10) + 1 = 71$

$(g \circ f)(x) = g\big(f(x)\big)$　　　　　Composition of functions

$\qquad\qquad = g(5x)$　　　　　　　　$f(x) = 5x$

$\qquad\qquad = (5x)^2 - 3(5x) + 1$　　$g(x) = x^2 - 3x + 1$

$\qquad\qquad = 25x^2 - 15x + 1$　　　Simplify.

(c) $(g \circ f)(2) = g\big(f(2)\big)$　　　　　Composition of functions

$\qquad\qquad = g(2)$　　　　　　　　$f(2) = \sqrt{2(2)} = 2$

$\qquad\qquad = 1$　　　　　　　　$g(2) = \dfrac{1}{2 - 1} = 1$

$(g \circ f)(x) = g\big(f(x)\big)$　　　　　Composition of functions

$\qquad\qquad = g(\sqrt{2x})$　　　　　　$f(x) = \sqrt{2x}$

$\qquad\qquad = \dfrac{1}{\sqrt{2x} - 1}$　　　　　$g(x) = \dfrac{1}{x - 1}$

Now Try Exercises 13, 15, 19

Composite functions can also be evaluated numerically and graphically, as demonstrated in the next two examples.

EXAMPLE 2   **Evaluating composite functions with tables**

Use Tables 12.1 and 12.2 to evaluate each expression.

(a) $(f \circ g)(2)$ **(b)** $(g \circ f)(3)$ **(c)** $(f \circ f)(0)$

TABLE 12.1

x	0	1	2	3
$f(x)$	3	2	0	1

TABLE 12.2

x	0	1	2	3
$g(x)$	1	3	2	0

Solution

(a) $(f \circ g)(2) = f\big(g(2)\big)$ Composition of functions

$\qquad\qquad\quad = f(2)$ $g(2) = 2$

$\qquad\qquad\quad = 0$ $f(2) = 0$

(b) $(g \circ f)(3) = g\big(f(3)\big)$ Composition of functions

$\qquad\qquad\quad = g(1)$ $f(3) = 1$

$\qquad\qquad\quad = 3$ $g(1) = 3$

(c) $(f \circ f)(0) = f\big(f(0)\big)$ Composition of functions

$\qquad\qquad\quad = f(3)$ $f(0) = 3$

$\qquad\qquad\quad = 1$ $f(3) = 1$

Now Try Exercises 23, 25

EXAMPLE 3   **Evaluating composite functions graphically**

Use Figure 12.2 to evaluate $(g \circ f)(2)$.

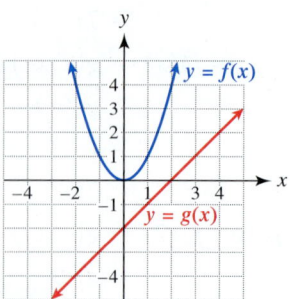

Figure 12.2

Solution

Because $(g \circ f)(2) = g\big(f(2)\big)$, start by using Figure 12.2 to evaluate $f(2)$. Figure 12.3(a) shows that $f(2) = 4$, which becomes the input for g. Figure 12.3(b) reveals that $g(4) = 2$.

$$(g \circ f)(2) = g\big(f(2)\big) = g(4) = 2$$

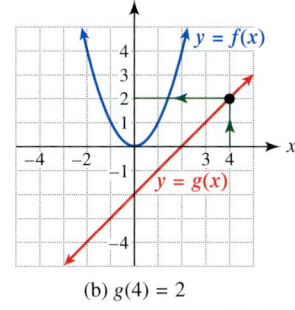

(a) $f(2) = 4$ (b) $g(4) = 2$

Figure 12.3

Now Try Exercise 29

One-to-One Functions

If we change the input for a function, does the output also change? Do *different inputs* always result in *different outputs* for every function? The answer is no. For example, if $f(x) = x^2 + 1$, then the inputs -2 and 2 result in the *same* output, 5. That is, $f(-2) = 5$ and $f(2) = 5$. However, for $g(x) = 2x$, *different inputs* always result in *different outputs*. Thus we say that g is a *one-to-one function*, whereas f is not.

> ### ONE-TO-ONE FUNCTION
>
> A function f is **one-to-one** if, for any c and d in the domain of f,
>
> $$c \neq d \quad \text{implies that} \quad f(c) \neq f(d).$$
>
> That is, different inputs always result in different outputs.

One way to determine whether a function f is one-to-one is to look at its graph. Suppose that a function has two different inputs that result in the same output. Then there must be two points on its graph that have the same y-value but different x-values. For example, if $f(x) = 2x^2$, then $f(-1) = 2$ and $f(1) = 2$. Thus the points $(-1, 2)$ and $(1, 2)$ both lie on the graph of f, as shown in Figure 12.4(a). Two points with different x-values and the same y-value determine a horizontal line, as shown in Figure 12.4(b). This horizontal line intersects the graph of f more than once, indicating that different inputs do *not* always have different outputs. Thus $f(x) = 2x^2$ is *not* one-to-one.

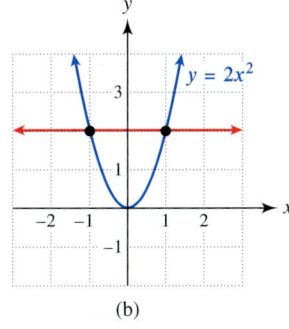

(a)   (b)

Figure 12.4

This discussion motivates the **horizontal line test**.

HORIZONTAL LINE TEST

If every horizontal line intersects the graph of a function f at most once, then f is a one-to-one function.

We apply the horizontal line test in the next example.

EXAMPLE 4   **Using the horizontal line test**

Determine whether each graph in Figure 12.5 represents a one-to-one function.

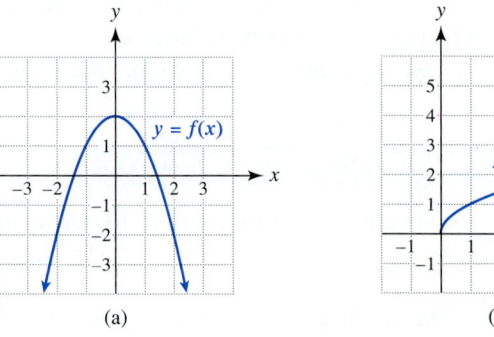

Figure 12.5

Solution

Figure 12.6(a) shows one of many horizontal lines that intersect the graph of $y = f(x)$ twice. Therefore function f is *not* one-to-one.

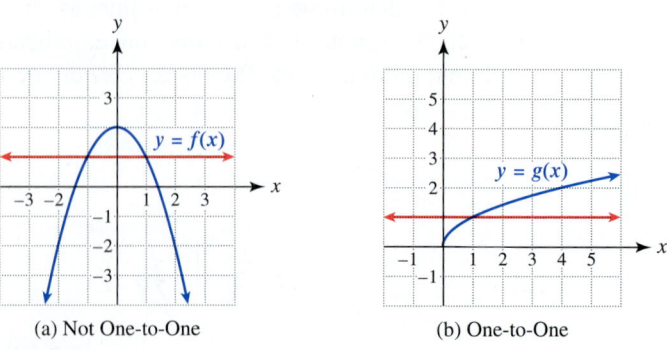

Figure 12.6

Figure 12.6(b) suggests that every horizontal line will intersect the graph of $y = g(x)$ *at most* once. Therefore function g is one-to-one.

Now Try Exercises 37, 39

MAKING CONNECTIONS

Vertical and Horizontal Line Tests

The *vertical line test* is used to identify functions, whereas the *horizontal line test* is used to identify one-to-one functions. For example, consider the graph of $f(x) = x^2$. A vertical line never intersects the graph more than once, so f is a function. A horizontal line can intersect the graph twice, so f *is not* a one-to-one function.

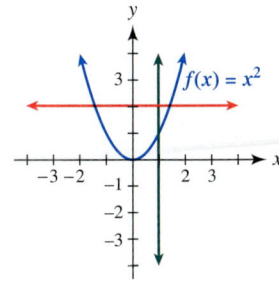

Inverse Functions

▶ **REAL-WORLD CONNECTION**   Turning on a light and turning off a light are inverse operations from ordinary life. Inverse operations undo each other. In mathematics, adding **5** to x, and subtracting **5** from x are inverse operations because

$$x + 5 - 5 = x.$$

Similarly, multiplying x by **5** and dividing x by **5** are inverse operations because

$$\frac{5x}{5} = x.$$

In general, addition and subtraction are inverse operations and multiplication and division are inverse operations.

EXAMPLE 5   **Finding inverse operations**

State the inverse operations for each statement. Then write a function f for the given statement and a function g for its inverse operations.
(a) Divide x by 3.
(b) Cube x and then add 1 to the result.

Solution
(a) The inverse of dividing x by 3 is to *multiply x by 3*. Thus

$$f(x) = \frac{x}{3} \quad \text{and} \quad g(x) = 3x.$$

(b) The inverse of cubing a number is taking a cube root, and the inverse of adding 1 is subtracting 1. The inverse operations of "cubing a number and then adding 1" are "subtracting 1 and then taking a cube root." For example, 2 cubed plus 1 is $2^3 + 1 = 9$. For the inverse operations, we *first* subtract 1 from 9 and then take the cube root to obtain 2. That is, $\sqrt[3]{9 - 1} = 2$. When there is more than one operation, we must perform the inverse operations in *reverse order*. Thus

$$f(x) = x^3 + 1 \quad \text{and} \quad g(x) = \sqrt[3]{x - 1}.$$

Now Try Exercises **43, 49**

Functions f and g in each part of Example 5 are examples of *inverse functions*. Note that in part (a), if $f(x) = \frac{x}{3}$ and $g(x) = 3x$, then

$$f(15) = 5 \quad \text{and} \quad g(5) = 15.$$

In general, *if f and g are inverse functions, then $f(a) = b$ implies $g(b) = a$.* Thus

$$(g \circ f)(a) = g\big(f(a)\big) = g(b) = a$$

for any a in the domain of f, whenever g and f are inverse functions. The composition of a function with its inverse leaves the input unchanged.

INVERSE FUNCTIONS

Let f be a one-to-one function. Then f^{-1} is the **inverse function** of f, if

$$(f^{-1} \circ f)(x) = f^{-1}\big(f(x)\big) = x, \qquad \text{for every } x \text{ in the domain of } f, \qquad \text{and}$$
$$(f \circ f^{-1})(x) = f\big(f^{-1}(x)\big) = x, \qquad \text{for every } x \text{ in the domain of } f^{-1}.$$

NOTE: In the expression $f^{-1}(x)$, the -1 is *not* an exponent. That is, $f^{-1}(x) \neq \frac{1}{f(x)}$. Rather, if $f(x) = \frac{x}{3}$, then $f^{-1}(x) = 3x$ and, if $f(x) = x^3 + 1$, then $f^{-1}(x) = \sqrt[3]{x - 1}$.

EXAMPLE 6   Verifying inverses

Verify that $f^{-1}(x) = 3x$ if $f(x) = \frac{x}{3}$.

Solution
We must show that $(f^{-1} \circ f)(x) = x$ and that $(f \circ f^{-1})(x) = x$.

$$
\begin{aligned}
(f^{-1} \circ f)(x) &= f^{-1}\big(f(x)\big) && \text{Composition of functions} \\
&= f^{-1}\left(\frac{x}{3}\right) && f(x) = \frac{x}{3} \\
&= 3\left(\frac{x}{3}\right) && f^{-1}(x) = 3x \\
&= x && \text{Simplify.} \\
(f \circ f^{-1})(x) &= f\big(f^{-1}(x)\big) && \text{Composition of functions} \\
&= f(3x) && f^{-1}(x) = 3x \\
&= \frac{3x}{3} && f(x) = \frac{x}{3} \\
&= x && \text{Simplify.} \qquad \text{Now Try Exercise } 51
\end{aligned}
$$

The definition of inverse functions states that f must be a one-to-one function. To understand why, consider Figure 12.7. In Figure 12.7(a) a one-to-one function f is represented by a diagram. To find f^{-1} the arrows are reversed. For example, $f(1) = 3$ implies that $f^{-1}(3) = 1$, so the arrow from 1 to 3 for f must be redrawn from 3 to 1 for f^{-1}.

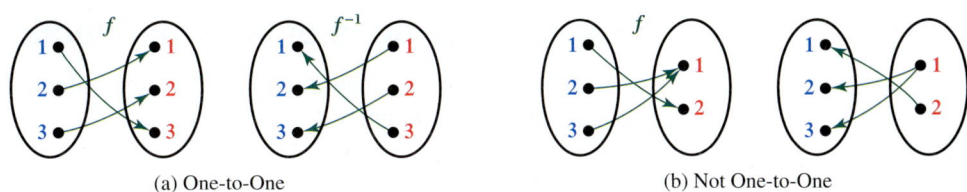

(a) One-to-One   (b) Not One-to-One

Figure 12.7

To be a *function* each input must correspond to exactly one output, which is the case in Figure 12.7(a). However, a different function f that is *not* a one-to-one function because inputs 2 and 3 both result in output 1 is shown in Figure 12.7(b). If the arrows for f are reversed to represent its inverse, then input 1 has two outputs, 2 and 3. Because no inverse *function* can satisfy both $f^{-1}(1) = 2$ and $f^{-1}(1) = 3$ at once, f^{-1} does not exist here.

The following steps can be used to find the inverse of a function symbolically.

FINDING AN INVERSE FUNCTION

To find f^{-1} for a one-to-one function f perform the following steps.

STEP 1: Let $y = f(x)$.

STEP 2: Interchange x and y.

STEP 3: Solve the formula for y. The resulting formula is $y = f^{-1}(x)$.

We apply these steps in the next example.

EXAMPLE 7   **Finding an inverse function**

Find the inverse of each one-to-one function.
(a) $f(x) = 3x - 7$ **(b)** $g(x) = (x + 2)^3$

Solution
(a) STEP 1: Let $y = 3x - 7$.

STEP 2: Write the formula as $x = 3y - 7$.

STEP 3: To solve for y, start by adding 7 to each side.

$$x + 7 = 3y \qquad \text{Add 7 to each side.}$$

$$\frac{x + 7}{3} = y \qquad \text{Divide each side by 3.}$$

Thus $f^{-1}(x) = \frac{x + 7}{3}$ or $f^{-1}(x) = \frac{1}{3}x + \frac{7}{3}$.

(b) STEP 1: Let $y = (x + 2)^3$.

STEP 2: Write the formula as $x = (y + 2)^3$.

STEP 3: To solve for y, start by taking the cube root of each side.

$$\sqrt[3]{x} = y + 2 \qquad \text{Take cube root of each side.}$$

$$\sqrt[3]{x} - 2 = y \qquad \text{Subtract 2 from each side.}$$

Thus $g^{-1}(x) = \sqrt[3]{x} - 2$.

Now Try Exercises **63, 71**

Tables and Graphs of Inverse Functions

TABLES   Inverse functions can be represented with tables and graphs. Table 12.3 shows a table of values for a function f.

TABLE 12.3

x	1	2	3	4	5
$f(x)$	3	6	9	12	15

Because $f(1) = 3$, $f^{-1}(3) = 1$. Similarly, $f(2) = 6$ implies that $f^{-1}(6) = 2$ and so on. Table 12.4 lists values for $f^{-1}(x)$.

TABLE 12.4

x	3	6	9	12	15
$f^{-1}(x)$	1	2	3	4	5

Note that the domain of f is $\{1, 2, 3, 4, 5\}$ and that the range of f is $\{3, 6, 9, 12, 15\}$, whereas the domain of f^{-1} is $\{3, 6, 9, 12, 15\}$ and the range of f^{-1} is $\{1, 2, 3, 4, 5\}$. *The domain of f is the range of f^{-1}, and the range of f is the domain of f^{-1}.* This statement is true in general for a function and its inverse.

GRAPHS   If $f(a) = b$, then the point (a, b) lies on the graph of f. This statement also means that $f^{-1}(b) = a$ and that the point (b, a) lies on the graph of f^{-1}. These points are shown in Figure 12.8(a) with a blue line segment connecting them. The line $y = x$ is a perpendicular bisector of this line segment. As a result, the graph of f^{-1} can be obtained from the graph of f by reflecting the graph of f across the line $y = x$. For example, the graphs of a function f and its inverse are shown in Figure 12.8(b).

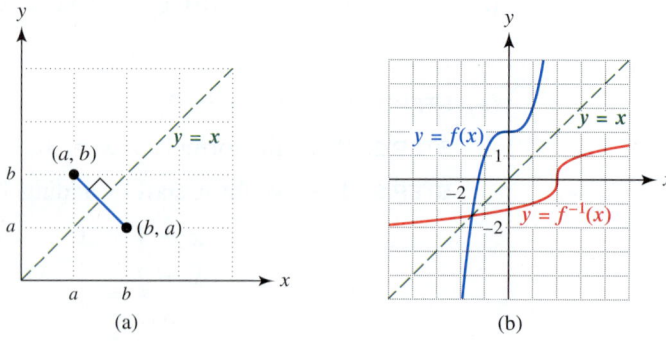

(a)   (b)

Figure 12.8

The relationship between the graph of a function and the graph of its inverse is summarized as follows.

GRAPHS OF FUNCTIONS AND THEIR INVERSES

The graph of f^{-1} is a reflection of the graph of f across the line $y = x$.

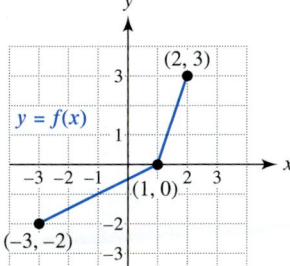

Figure 12.9

EXAMPLE 8 Graphing an inverse function

The line graph shown in Figure 12.9 represents a function f.
(a) Is f a one-to-one function?
(b) Sketch a graph of $y = f^{-1}(x)$.

Solution
(a) Every horizontal line intersects the graph of f at most once. By the horizontal line test, the graph represents a one-to-one function.
(b) The points $(-3, -2)$, $(1, 0)$, and $(2, 3)$ lie on the graph of f. It follows that the points $(-2, -3)$, $(0, 1)$, and $(3, 2)$ lie on the graph of f^{-1}. Plot these three points and then connect them with line segments, as shown in Figure 12.10(a). Note that the graph of $y = f^{-1}(x)$ is a reflection of the graph of $y = f(x)$ across the line $y = x$, as shown in Figure 12.10(b).

(a)

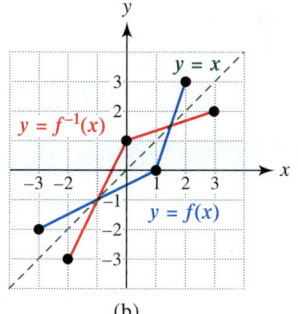

(b)

Figure 12.10

Now Try Exercise **81**

CRITICAL THINKING

The graph of a linear function f passes through the points (1, 2) and (2, 1). What two points does the graph of f^{-1} pass through? Find $f(x)$ and $f^{-1}(x)$.

EXAMPLE 9 Graphing an inverse function

The graph of $y = f(x)$ is shown in Figure 12.11. Sketch a graph of $y = f^{-1}(x)$.

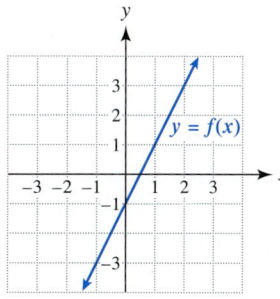

Figure 12.11

Solution

The graph of $y = f^{-1}(x)$ is the reflection of the graph of $y = f(x)$ across the line $y = x$ and is shown in Figure 12.12.

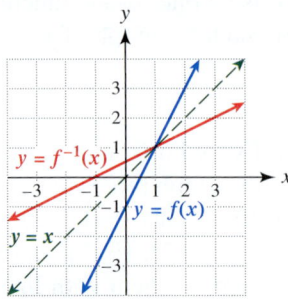

Figure 12.12

Now Try Exercise **83**

12.1 PUTTING IT ALL TOGETHER

The following table summarizes some concepts about composite and inverse functions.

Concept	Explanation	Examples
Composite Functions	The composite of g and f is given by $$(g \circ f)(x) = g(f(x)),$$ and represents a *new* function whose name is $g \circ f$.	If $f(x) = 1 - 4x$ and $g(x) = x^3$, then $$(g \circ f)(x) = g(f(x))$$ $$= g(1 - 4x)$$ $$= (1 - 4x)^3.$$
One-to-One Functions	Function f is one-to-one if different inputs always give different outputs.	$f(x) = x^2$ is not one-to-one because $f(-4) = f(4) = 16$, whereas $g(x) = x + 1$ is one-to-one because, if two inputs differ, then adding 1 to each does not change this difference.
Horizontal Line Test	This test is used to determine whether a function is one-to-one from its graph.	$f(x) = x^2$ is not one-to-one because a horizontal line can intersect its graph more than once.
Inverse Functions	f^{-1} will undo the operations performed by f. That is, $$(f^{-1} \circ f)(x) = x \quad \text{and}$$ $$(f \circ f^{-1})(x) = x.$$	If $f(x) = x^3$, then $f^{-1}(x) = \sqrt[3]{x}$ because cubing a number x and then taking its cube root results in the number x.

12.1 Exercises

CONCEPTS

1. $(g \circ f)(7) =$ _____

2. $(f \circ g)(x) =$ _____

3. Does $(f \circ g)(x)$ always equal $(g \circ f)(x)$?

4. If a function f is one-to-one, then different _____ always result in different _____ .

5. If $f(3) = 5$ and $f(7) = 5$, could f be one-to-one?

6. If every horizontal line intersects the graph of f at most once, then f is _____ .

7. The inverse operation of subtracting 10 is _____ .

8. $(f^{-1} \circ f)(7) =$ _____

9. If $f(6) = 8$, then $f^{-1}(\underline{\quad}) =$ _____ .

10. If $f^{-1}(y) = x$, then $f(\underline{\quad}) =$ _____ .

11. For f to have an inverse function, f must be _____ .

12. The graph of f^{-1} is a _____ of the graph of f across the line _____ .

COMPOSITE FUNCTIONS

Exercises 13–22: For the given $f(x)$ and $g(x)$, find each of the following.

    (a) $(g \circ f)(-2)$ *(b)* $(f \circ g)(4)$
    (c) $(g \circ f)(x)$ *(d)* $(f \circ g)(x)$

13. $f(x) = x^2$ $g(x) = x + 3$

14. $f(x) = 4x^2$ $g(x) = 5x$

15. $f(x) = 2x$ $g(x) = x^3 - 1$

16. $f(x) = 3x + 1$ $g(x) = x^2 + 4x$

17. $f(x) = \frac{1}{2}x$ $g(x) = |x - 2|$

18. $f(x) = 6x$ $g(x) = \dfrac{2}{x - 5}$

19. $f(x) = \dfrac{1}{x}$ $g(x) = 3 - 5x$

20. $f(x) = \sqrt{x + 3}$ $g(x) = x^3 - 3$

21. $f(x) = 2x$ $g(x) = 4x^2 - 2x + 5$

22. $f(x) = 9x - \dfrac{1}{3x}$ $g(x) = \dfrac{x}{3}$

Exercises 23–28: Evaluate each expression numerically.

x	-2	-1	0	1	2
$f(x)$	2	1	0	-1	-2

x	-2	-1	0	1	2
$g(x)$	0	1	-1	2	-2

23. (a) $(f \circ g)(0)$ **(b)** $(g \circ f)(-1)$

24. (a) $(f \circ g)(1)$ **(b)** $(g \circ f)(-2)$

25. (a) $(f \circ f)(-1)$ **(b)** $(g \circ g)(0)$

26. (a) $(g \circ g)(2)$ **(b)** $(f \circ f)(1)$

27. (a) $(f^{-1} \circ g)(-2)$ **(b)** $(g^{-1} \circ f)(2)$

28. (a) $(f \circ g^{-1})(1)$ **(b)** $(g \circ f^{-1})(-2)$

Exercises 29 and 30: Evaluate each expression graphically.

29. (a) $(f \circ g)(0)$
    (b) $(g \circ f)(1)$
    (c) $(f \circ f)(-1)$

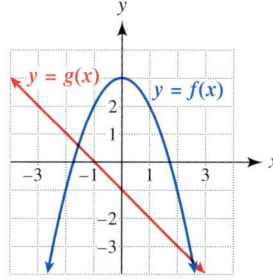

30. (a) $(f \circ g)(1)$
    (b) $(g \circ f)(-2)$
    (c) $(g \circ g)(-2)$

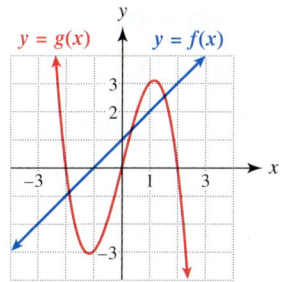

Exercises 31–36: Show that f is not a one-to-one function by finding two inputs that result in the same output. Answers may vary.

31. $f(x) = 5x^2$

32. $f(x) = 4 - x^2$

33. $f(x) = x^4 + 100$

34. $f(x) = \dfrac{x^2}{x^2 + 1}$

35. $f(x) = x^4 - 3x^2$

36. $f(x) = \sqrt{x^2 - 1}$

Exercises 37–42: Use the horizontal line test to determine whether the graph represents a one-to-one function.

37.

38.

39.

40.

41.

42.

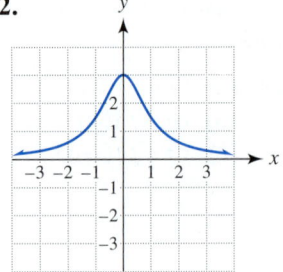

Exercises 43–50: (Refer to Example 5.) Give the inverse operation for the statement. Then write a function f for the given statement and a function g for its inverse.

43. Multiply x by 7.

44. Subtract 10 from x.

45. Add 5 to x and then divide the result by 2.

46. Multiply x by 6 and then add 8 to the result.

47. Multiply x by $\frac{1}{2}$ and then subtract 3 from the result.

48. Divide x by 10 and then add 20 to the result.

49. Cube the sum of x and 5.

50. Take the cube root of x and then subtract 2.

Exercises 51–58: (Refer to Example 6.) Verify that f(x) and $f^{-1}(x)$ are indeed inverse functions.

51. $f(x) = 4x$ $\qquad$ $f^{-1}(x) = \dfrac{x}{4}$

52. $f(x) = \dfrac{2x}{3}$ $\qquad$ $f^{-1}(x) = \dfrac{3x}{2}$

53. $f(x) = 3x + 5$ $\qquad$ $f^{-1}(x) = \dfrac{x - 5}{3}$

54. $f(x) = x + 7$ $\qquad$ $f^{-1}(x) = x - 7$

55. $f(x) = x^3$ $\qquad$ $f^{-1}(x) = \sqrt[3]{x}$

56. $f(x) = \sqrt[3]{x} - 4$ $\qquad$ $f^{-1}(x) = x^3 + 4$

57. $f(x) = \dfrac{1}{x}$ $\qquad$ $f^{-1}(x) = \dfrac{1}{x}$

58. $f(x) = \dfrac{x + 7}{7}$ $\qquad$ $f^{-1}(x) = 7x - 7$

Exercises 59–74: (Refer to Example 7.) Find $f^{-1}(x)$.

59. $f(x) = 12x$ $\qquad$ **60.** $f(x) = \frac{3}{4}x$

61. $f(x) = x + 8$ $\qquad$ **62.** $f(x) = x - 3$

63. $f(x) = 5x - 2$ $\qquad$ **64.** $f(x) = 3x + 4$

65. $f(x) = -\frac{1}{2}x + 1$ $\qquad$ **66.** $f(x) = \frac{3}{4}x - \frac{1}{4}$

67. $f(x) = 8 - x$ $\qquad$ **68.** $f(x) = 5 - x$

69. $f(x) = \dfrac{x + 1}{2}$ $\qquad$ **70.** $f(x) = \dfrac{3 - x}{5}$

71. $f(x) = \sqrt[3]{2x}$ $\qquad$ **72.** $f(x) = \sqrt[3]{x + 4}$

73. $f(x) = x^3 - 8$ $\qquad$ **74.** $f(x) = (x - 5)^3$

Exercises 75–78: Use the table to make a table of values for $f^{-1}(x)$. State the domain and range for f and for f^{-1}.

75.

x	0	1	2	3	4
$f(x)$	0	5	10	15	20

76.

x	−4	−2	0	2	4
f(x)	1	2	3	4	5

77.

x	−5	0	5	10	15
f(x)	4	2	0	−2	−4

78.

x	0	2	4	6	8
f(x)	8	6	4	2	0

Exercises 79–82: (Refer to Example 8.) Use the graph of $y = f(x)$ to sketch a graph of $y = f^{-1}(x)$.

79.

80.

81.

82.

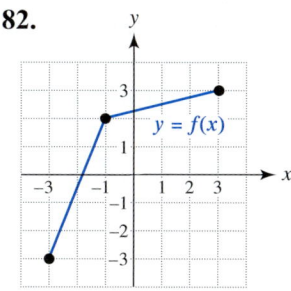

Exercises 83–86: (Refer to Example 9.) Use the graph of $y = f(x)$ to sketch a graph of $f^{-1}(x)$. Include the graph of f and the line $y = x$ in your graph.

83.

84.

85.

86.

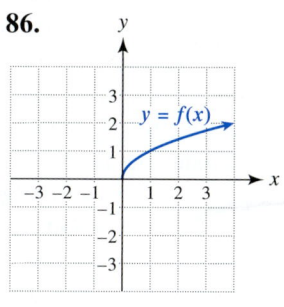

OPERATIONS ON FUNCTIONS

Exercises 87–90: Find each of the following for the given $f(x)$ and $g(x)$.

    (a) $(fg)(2)$ *(b)* $(f - g)(x)$ *(c)* $(f \circ g)(x)$

87. $f(x) = x^2 - 2, g(x) = x^2 + 2$

88. $f(x) = 2x^2, g(x) = 2x - 1$

89. $f(x) = \dfrac{1}{x}, g(x) = \dfrac{2}{x}$

90. $f(x) = x^3, g(x) = \sqrt[3]{x}$

APPLICATIONS

91. *Circular Wave* A stone is dropped in a lake, creating a circular wave. The radius r of the wave in feet after t seconds is $r(t) = 2t$.
  (a) Its circumference C is given by $C(r) = 2\pi r$. Evaluate $(C \circ r)(5)$ and interpret your result.
  (b) Find $(C \circ r)(t)$.

92. *Volume of a Balloon* The volume V of a spherical balloon with radius r is given by $V(r) = \frac{4}{3}\pi r^3$. Suppose that the balloon is being inflated so that the radius in inches after t seconds is $r(t) = \sqrt[3]{t}$.
  (a) Evaluate $(V \circ r)(3)$ and interpret your result.
  (b) Find $(V \circ r)(t)$.

93. *College Degree* The table lists the percentage P of people 25 or older who have completed 4 or more years of college during year x.

x	1960	1980	2000
P(x)	8	16	27

Source: U.S. Census Bureau.

  (a) Evaluate $P(1980)$ and interpret the results.
  (b) Make a table for $P^{-1}(x)$.
  (c) Evaluate $P^{-1}(16)$.

94. *Skin Cancer and Ozone* Ozone in the stratosphere filters out most of the harmful ultraviolet (UV) rays from the sun. However, depletion of the ozone layer is affecting this protection. The formula $U(x) = 1.5x$ calculates the percent increase in UV radiation for an x percent decrease in the thickness of the ozone layer. The formula $C(x) = 3.5x$ calculates the percent increase in skin cancer cases when the UV radiation increases by x percent. (*Source:* R. Turner, D. Pierce, and I. Bateman, *Environmental Economics.*)

  (a) Evaluate $U(2)$ and $C(3)$ and interpret each result.
  (b) Find $(C \circ U)(2)$ and interpret the result.
  (c) Find $(C \circ U)(x)$. What does it calculate?

95. *Temperature and Mosquitoes* Temperature can affect the number of mosquitoes observed on a summer night. Graphs of two functions, T and M, are shown. Function T calculates the temperature on a summer evening h hours past midnight, and M calculates the number of mosquitoes observed per 100 square feet when the outside temperature is T.

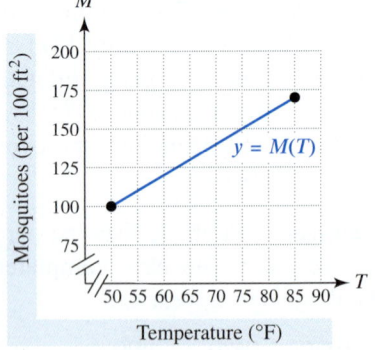

  (a) Find $T(1)$ and $M(75)$.
  (b) Evaluate $(M \circ T)(1)$ and interpret your result.
  (c) What does $(M \circ T)(h)$ calculate?
  (d) Find equations for the lines in each graph.
  (e) Use your answers from part (d) to write a formula for $(M \circ T)(h)$.

96. *High School Grades* The table lists the percentage of college freshmen with a high school grade average of A or A– during year x.

x	1970	1980	1990	2000
$P(x)$	20	26	29	43

Source: Department of Education.

  (a) Evaluate $P(1970)$ and interpret the results.
  (b) Make a table for $P^{-1}(x)$.
  (c) Evaluate $P^{-1}(43)$.

97. *Temperature* The function given by $f(x) = \frac{9}{5}x + 32$ converts x degrees Celsius to an equivalent temperature in degrees Fahrenheit.

  (a) Is f a one-to-one function? Why or why not?
  (b) Find $f^{-1}(x)$ and interpret what it calculates.

98. *Feet and Yards* The function given by $f(x) = 3x$ converts x yards to feet.

  (a) Is f a one-to-one function? Why or why not?
  (b) Find $f^{-1}(x)$ and interpret what it calculates.

99. *Quarts and Gallons* Write a function f that converts x gallons to quarts. Then find $f^{-1}(x)$ and interpret what it computes.

100. *One-to-One Function* The table lists monthly average wind speeds at Hilo, Hawaii, in miles per hour from July through December, where x is the month.

x	July	Aug	Sept	Oct	Nov	Dec
$f(x)$	7	7	7	7	7	7

  (a) Is function f one-to-one? Explain.
  (b) Does f^{-1} exist?
  (c) What happens if you try to make a table for f^{-1}?
  (d) Could f be one-to-one if it were computed at a different location? What would have to be true about the monthly average wind speeds?

WRITING ABOUT MATHEMATICS

101. Explain the difference between $(g \circ f)(2)$ and $(f \circ g)(2)$. Are they always equal? If f and g are inverse functions, evaluate $(g \circ f)(2)$ and $(f \circ g)(2)$.

102. Explain what it means for a function to be one-to-one.

12.2 EXPONENTIAL FUNCTIONS

Basic Concepts ■ Graphs of Exponential Functions ■
Models Involving Exponential Functions ■ The Natural Exponential Function

A LOOK INTO MATH ▷

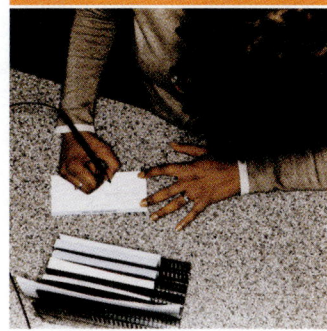

Many times the growth of a quantity depends on the amount or number present. The more money deposited in an account, the more interest the account earns; that is, the interest earned is proportional to the amount of money in an account. For example, if a person has $100 in an account and receives 10% annual interest, the interest accrued the first year will be $10, and the balance in the account will be $110 at the end of 1 year. Similarly, if a person begins with $1000 in a similar account, the balance in the account will be $1100 at the end of 1 year. This type of growth is called *exponential growth* and can be modeled by an *exponential function*.

Basic Concepts

▶ **REAL-WORLD CONNECTION**   Suppose that an insect population doubles each week. Table 12.5 shows the populations after x weeks. Note that, as the population of insects becomes larger, the *increase* in population each week becomes greater. The population is increasing by 100%, or doubling numerically, each week. When a quantity increases by a constant percentage (or constant factor) at regular intervals, its growth is exponential.

TABLE 12.5

Week	0	1	2	3	4	5
Population	100	200	400	800	1600	3200

We can model the data in Table 12.5 by using the exponential function

$$f(x) = 100(2)^x.$$

For example,

$$f(0) = 100(2)^0 = 100 \cdot 1 = 100,$$
$$f(1) = 100(2)^1 = 100 \cdot 2 = 200,$$
$$f(2) = 100(2)^2 = 100 \cdot 4 = 400,$$

and so on. Note that the exponential function f has a *variable as an exponent*.

EXPONENTIAL FUNCTION

A function represented by

$$f(x) = Ca^x, \quad a > 0 \quad \text{and} \quad a \neq 1,$$

is an **exponential function with base a and coefficient C**. (Unless stated otherwise, we assume that $C > 0$.)

In the formula $f(x) = Ca^x$, a is called the **growth factor** when $a > 1$ and the **decay factor** when $0 < a < 1$. For an exponential function, each time x increases by 1 unit $f(x)$

increases by a *factor* of a when $a > 1$ and decreases by a factor of a when $0 < a < 1$. Moreover, as

$$f(\mathbf{0}) = Ca^{\mathbf{0}} = C(1) = C,$$

the value of C equals the value of $f(x)$ when $x = 0$. If x represents time, C represents the initial value of f when time equals 0. Figure 12.13 illustrates **exponential growth** and **exponential decay** for $x > 0$.

(a) Exponential Growth   (b) Exponential Decay

Figure 12.13

The set of valid inputs (domain) for an exponential function includes all real numbers. The set of corresponding outputs (range) includes all positive real numbers.

In the next example we evaluate some exponential functions. When evaluating an exponential function, we evaluate a^x before multiplying by C. This standard order of precedence is much like doing multiplication before addition. For example, $2(3)^2$ should be evaluated as

$$2(9) = 18 \quad not \text{ as} \quad (6)^2 = 36.$$

EXAMPLE 1   **Evaluating exponential functions**

Evaluate $f(x)$ for the given value of x.

(a) $f(x) = 10(3)^x \qquad x = 2$ **(b)** $f(x) = 5\left(\tfrac{1}{2}\right)^x \qquad x = 3$

(c) $f(x) = \tfrac{1}{3}(2)^x \qquad x = -1$

Solution

(a) $f(\mathbf{2}) = 10(3)^{\mathbf{2}} = 10 \cdot 9 = 90$

(b) $f(\mathbf{3}) = 5\left(\tfrac{1}{2}\right)^{\mathbf{3}} = 5 \cdot \tfrac{1}{8} = \tfrac{5}{8}$

(c) $f(\mathbf{-1}) = \tfrac{1}{3}(2)^{\mathbf{-1}} = \tfrac{1}{3} \cdot \tfrac{1}{2} = \tfrac{1}{6}$ Now Try Exercises 9, 11, 13

MAKING CONNECTIONS

The Expressions a^{-x} and $\left(\tfrac{1}{a}\right)^x$

Using properties of exponents, we can write 2^{-x} as

$$2^{-x} = \frac{1}{2^x} = \left(\frac{1}{2}\right)^x.$$

In general, the expressions a^{-x} and $\left(\tfrac{1}{a}\right)^x$ are equal for positive a.

In the next example, we determine whether a function is linear or exponential.

EXAMPLE 2   **Finding linear and exponential functions**

For each table, determine whether f is a linear function or an exponential function. Find a formula for f.

(a)

x	0	1	2	3	4
$f(x)$	16	8	4	2	1

(b)

x	0	1	2	3	4
$f(x)$	5	7	9	11	13

(c)

x	0	1	2	3	4
$f(x)$	1	3	9	27	81

Solution

(a) Each time x increases by 1 unit, $f(x)$ decreases by a factor of $\frac{1}{2}$. Therefore f is an exponential function with a decay factor of $\frac{1}{2}$. Because $f(0) = 16$, $C = 16$ and so $f(x) = 16\left(\frac{1}{2}\right)^x$. This formula can also be written as $f(x) = 16(2)^{-x}$.

(b) Each time x increases by 1 unit, $f(x)$ increases by 2 units. Therefore f is a linear function, and the slope of its graph equals 2. The y-intercept is 5, so $f(x) = 2x + 5$.

(c) Each time x increases by 1 unit, $f(x)$ increases by a factor of 3. Therefore f is an exponential function with a growth factor of 3. Because $f(0) = 1$, $C = 1$ and so $f(x) = 1(3)^x$, or $f(x) = 3^x$.

Now Try Exercises 23, 25, 27

MAKING CONNECTIONS

Linear and Exponential Functions

For a *linear function*, given by $f(x) = ax + b$, each time x increases by 1 unit y increases (or decreases) by a units, where a equals the slope of the graph of f.

For an *exponential function*, given by $f(x) = Ca^x$, each time x increases by 1 unit y increases *by a factor of a* when $a > 1$ and decreases by a factor of a when $0 < a < 1$. The constant a equals either the growth factor or the decay factor.

▶ **REAL-WORLD CONNECTION**   If $100 are deposited in a savings account paying 10% annual interest, the interest earned after 1 year equals $100 \times 0.10 = \$10$. The total amount of money in the account after 1 year is $100(1 + 0.10) = \$110$. Each year the money in the account increases by a factor of 1.10, so after x years there will be $100(1.10)^x$ dollars in the account. Thus compound interest is an example of exponential growth.

COMPOUND INTEREST

If C dollars are deposited in an account and if interest is paid at the end of each year with an annual rate of interest r, expressed in decimal form, then after x years the account will contain A dollars, where

$$A = C(1 + r)^x.$$

The growth factor is $(1 + r)$.

NOTE: The compound interest formula takes the form of an exponential function with

$$a = 1 + r.$$

EXAMPLE 3   Calculating compound interest

A 20-year-old worker deposits $2000 in a retirement account that pays 13% annual interest at the end of each year. How much money will be in the account when the worker is 65 years old? What is the growth factor?

Solution

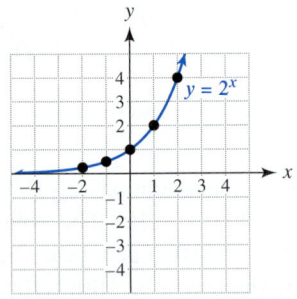

Figure 12.14

Here, $C = \mathbf{2000}$, $r = \mathbf{0.13}$, and $x = \mathbf{45}$. The amount in the account after 45 years is

$$A = \mathbf{2000}(1 + \mathbf{0.13})^{45} \approx \$489{,}282.80,$$

which is supported by Figure 12.14. In this dramatic example of exponential growth, $2000 grows to nearly half a million dollars in 45 years. Each year the amount of money on deposit is multiplied by a factor of $(1 + 0.13)$, so the growth factor is 1.13.

Now Try Exercise 59

Graphs of Exponential Functions

We can graph $f(x) = 2^x$ by first plotting some points, as in Table 12.6.

TABLE 12.6

x	-2	-1	0	1	2
2^x	$\frac{1}{4}$	$\frac{1}{2}$	1	2	4

If we plot these points and sketch the graph, we obtain Figure 12.15. Note that, for negative values of x, $0 < 2^x < 1$ and that, for positive values of x, $2^x > 1$. The graph of $y = 2^x$ passes through the point $(0, 1)$, never intersects the x-axis, and always lies above the x-axis.

We can investigate the graphs of exponential functions further by graphing $y = \mathbf{1.3}^x$, $y = \mathbf{1.7}^x$, and $y = \mathbf{2.5}^x$ (see Figure 12.16). For $a > 1$ the graph of $y = a^x$ *increases* at a faster rate for larger values of a. We now graph $y = \mathbf{0.7}^x$, $y = \mathbf{0.5}^x$, and $y = \mathbf{0.15}^x$ (see Figure 12.17). Note that, if $0 < a < 1$, the graph of $y = a^x$ *decreases* more rapidly for smaller values of a. The graph of $y = a^x$ is *increasing* when $a > 1$ and *decreasing* when $0 < a < 1$ (from left to right).

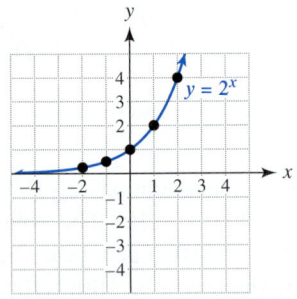

Figure 12.15

CRITICAL THINKING

Every graph of $y = a^x$ passes through what point? Why?

Figure 12.16

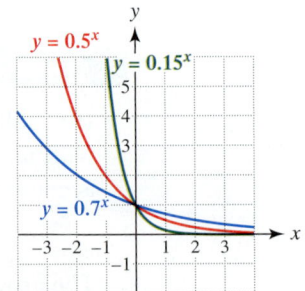

Figure 12.17

In the next example we show the dramatic difference between the outputs of linear and exponential functions.

EXAMPLE 4   Comparing exponential and linear functions

Compare $f(x) = 3^x$ and $g(x) = 3x$ graphically and numerically for $x \geq 0$.

Solution

Graphical Comparison   The graphs of $Y_1 = 3\char`\^X$ and $Y_2 = 3X$ are shown in Figure 12.18. The graph of the exponential function y_1 increases much faster than the graph of the linear function y_2.

Numerical Comparison   The tables of $Y_1 = 3\char`\^X$ and $Y_2 = 3X$ are shown in Figure 12.19. The values for y_1 increase much faster than the values for y_2.

[0, 5, 1] by [0, 120, 20]

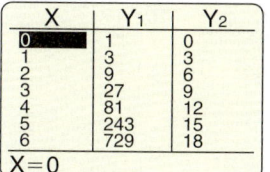

Figure 12.18   Figure 12.19   Now Try Exercise **37**

The results of Example 4 are true in general: For large enough inputs, exponential functions with $a > 1$ grow far faster than any linear function.

MAKING CONNECTIONS

Exponential and Polynomial Functions

The function $f(x) = 2^x$ is an exponential function. The base 2 is a constant and the exponent x is a variable, so $f(3) = 2^3 = 8$.

The function $g(x) = x^2$ is a polynomial function. The base x is a variable and the exponent 2 is a constant, so $g(3) = 3^2 = 9$.

The table clearly shows that the exponential function grows much faster than the polynomial function for larger values of x.

x	0	2	4	6	8	10	12
2^x	1	4	16	64	256	1024	4096
x^2	0	4	16	36	64	100	144

Models Involving Exponential Functions

▶ **REAL-WORLD CONNECTION**   Traffic flow on highways can be modeled by exponential functions whenever traffic patterns occur randomly. In the next example we model traffic at an intersection by using an exponential function.

EXAMPLE 5   ## Modeling traffic flow

On average, a particular intersection has 360 vehicles arriving randomly each hour. High-way engineers use $f(x) = (0.905)^x$ to estimate the likelihood, or probability, that *no* vehicle will enter the intersection within an interval of x seconds.   (*Source:* F. Mannering and W. Kilareski, *Principles of Highway Engineering and Traffic Analysis.*)

(a) Compute $f(5)$ and interpret the results.

(b) A graph of $y = f(x)$ is shown in Figure 12.20. Discuss this graph.

(c) Is this function an example of exponential growth or decay?

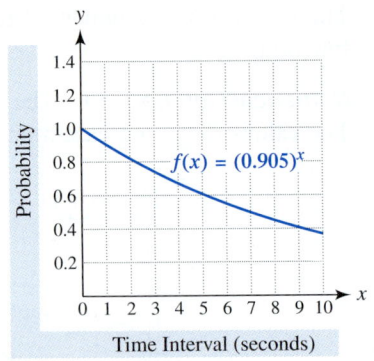

Figure 12.20

Solution

(a) The result $f(5) = (0.905)^5 \approx 0.61$ indicates that there is a 61% chance that no vehicle will enter the intersection during any particular 5-second interval.

(b) The graph decreases, which means that as the interval of time increases there is less chance (probability) that a car will not enter the intersection.

(c) Because the graph is decreasing and $a = 0.905 < 1$, this function is an example of exponential decay.   Now Try Exercise **87**

In the next example we use an exponential function to model how trees grow in a forest.

EXAMPLE 6   ## Modeling tree density in a forest

[0, 20, 5] by [0, 1, 0.1]

Ecologists studied the spacing of individual trees in a British Columbia forest. This pine forest was 40 to 50 years old and contained approximately 1600 randomly spaced trees per acre. The probability or likelihood that *no* tree is located within a circle of radius x feet can be estimated by $P(x) = (0.892)^x$. For example, $P(4) \approx 0.63$ means that, if a person picks a point at random in the forest, there is a 63% chance that no tree will be located within 4 feet of the person.   (*Source:* E. Pielou, *Populations and Community Ecology.*)

(a) Evaluate $P(8)$ and interpret the result.

(b) Graph P in [0, 20, 5] by [0, 1, 0.1] and discuss the graph.

Solution

(a) The probability $P(8) = (0.892)^8 \approx 0.40$ means that there is a 40% chance that no tree is growing within any particular circle of radius 8 feet.

(b) The graph of $Y_1 = 0.892\verb|^|X$, as shown in Figure 12.21, indicates that the larger the circle, the less the likelihood is of no tree being inside the circle.   Now Try Exercise **88**

Figure 12.21

The Natural Exponential Function

▶ **REAL-WORLD CONNECTION**   A special type of exponential function is called the *natural exponential function*, expressed as $f(x) = e^x$. The base e is a special number in mathematics similar to π. The number π is approximately 3.14, whereas the number e is approximately 2.72. The number e is named for the great Swiss mathematician, Leonhard Euler (1707–1783). Most calculators have a special key that can be used to compute the natural exponential function.

NATURAL EXPONENTIAL FUNCTION

The function represented by

$$f(x) = e^x$$

is the **natural exponential function**, where $e \approx 2.71828$.

▶ **REAL-WORLD CONNECTION**   The natural exponential function is frequently used to model **continuous growth**. For example, the fact that births and deaths occur throughout the year, not just at one time during the year, must be recognized when population growth is being modeled. If a population P is growing continuously at r percent per year, expressed as a decimal, we can model this population after x years by

$$P = Ce^{rx},$$

where C is the initial population. To evaluate natural exponential functions, we use a calculator, as in the next example.

EXAMPLE 7   **Modeling population**

CALCULATOR HELP

To evaluate the natural exponential function, see Appendix A (page AP-1).

In 2000 Florida's population was 16 million people and was growing at a continuous rate of 2% per year. This population in millions x years after 2000 can be modeled by

$$f(x) = 16e^{0.02x}.$$

Estimate the population in 2010.

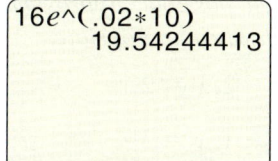

Figure 12.22

Solution
As 2010 is 10 years after 2000, we evaluate $f(10)$ to obtain

$$f(\mathbf{10}) = 16e^{0.02(\mathbf{10})} \approx 19.5,$$

which is supported by Figure 12.22. (Be sure to include parentheses around the exponent of e.) This model estimates the population of Florida to be about 19.5 million in 2010.

Now Try Exercise 77

CRITICAL THINKING

Sketch a graph of $y = 2^x$ and $y = 3^x$ in the same xy-plane. Then use these two graphs to sketch a graph of $y = e^x$. How do these graphs compare?

12.2 PUTTING IT ALL TOGETHER

The following table summarizes some important concepts of exponential functions and compound interest.

Topic	Explanation	Example
Exponential Function	An exponential function can be written as $f(x) = Ca^x$, where $a > 0$ and $a \neq 1$. If $a > 1$, the function models exponential growth, and if $0 < a < 1$, the function models exponential decay. The natural exponential function has $C = 1$ and $a = e \approx 2.71828$; that is, $f(x) = e^x$.	$f(x) = 3(2)^x$ models exponential growth and $g(x) = 2\left(\frac{1}{3}\right)^x$ models exponential decay.
Compound Interest	If C dollars are deposited in an account and if interest is paid at the end of each year with an annual rate of interest r, expressed as a decimal, then after x years the account will contain A dollars, where $$A = C(1 + r)^x.$$ The growth factor is $(1 + r)$.	If \$1000 are deposited in an account paying 5% annual interest, then after 6 years the amount A in the account is $$A = 1000(1 + 0.05)^6 \approx \$1340.10.$$

12.2 Exercises

CONCEPTS

1. Give a general formula for an exponential function f.

2. Sketch a graph of an exponential function that illustrates exponential decay.

3. Does the graph of $f(x) = a^x$, $a > 1$, illustrate exponential growth or decay?

4. Evaluate the expressions 2^x and x^2 for $x = 5$.

5. Approximate e to the nearest thousandth.

6. Evaluate e^2 and π^2 using your calculator.

7. If a quantity y grows exponentially, then for each unit increase in x, y increases by a constant _____.

8. If $f(x) = 1.5^x$ what is the growth factor?

EVALUATING AND GRAPHING EXPONENTIAL FUNCTIONS

Exercises 9–20: Evaluate the exponential function for the given values of x by hand when possible. Approximate answers to the nearest hundredth when appropriate.

9. $f(x) = 3^x$ $x = -2, x = 2$

10. $f(x) = 5^x$ $x = -1, x = 3$

11. $f(x) = 5(2^x)$ $x = 0, x = 5$

12. $f(x) = 3(7^x)$ $x = -2, x = 0$

13. $f(x) = \left(\frac{1}{2}\right)^x$ $x = -2, x = 3$

14. $f(x) = \left(\frac{1}{4}\right)^x$ $x = 0, x = 2$

15. $f(x) = 5(3)^{-x}$ $x = -1, x = 2$

16. $f(x) = 4\left(\frac{3}{7}\right)^x$ $\qquad x = 1, x = 4$

17. $f(x) = 1.8^x$ $\qquad x = -3, x = 1.5$

18. $f(x) = 0.91^x$ $\qquad x = 5.1, x = 10$

19. $f(x) = 3(0.6)^x$ $\qquad x = -1, x = 2$

20. $f(x) = 5(4.5)^{-x}$ $\qquad x = -2.1, x = 5.9$

Exercises 21 and 22: **Thinking Generally** *For the given exponential function, evaluate $f(0)$ and $f(-1)$.*

21. $f(x) = a^x$

22. $f(x) = (1 + r)^{2x}$

Exercises 23–28: (Refer to Example 2.) A table for a function f is given.
   (a) *Determine whether function f represents exponential growth, exponential decay, or linear growth.*
   (b) *Find a formula for f.*

23.

x	0	1	2	3	4
$f(x)$	64	16	4	1	$\frac{1}{4}$

24.

x	0	1	2	3	4
$f(x)$	$\frac{1}{2}$	1	2	4	8

25.

x	0	1	2	3	4
$f(x)$	8	11	14	17	20

26.

x	-2	-1	0	1	2
$f(x)$	4	2	1	$\frac{1}{2}$	$\frac{1}{4}$

27.

x	-2	-1	0	1	2
$f(x)$	2.56	3.2	4	5	6.25

28.

x	-2	-1	0	1	2
$f(x)$	-6	-2	2	6	10

Exercises 29–32: Use the graph of $y = Ca^x$ to determine the constants C and a.

29.

30.

31.

32.

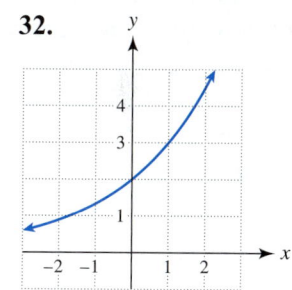

Exercises 33–36: Match the formula with its graph (a.–d.). Do not use a calculator.

33. $f(x) = 1.5^x$

34. $f(x) = \frac{1}{4}(2^x)$

35. $f(x) = 4\left(\frac{1}{2}\right)^x$

36. $f(x) = \left(\frac{1}{3}\right)^x$

a.

b.

c.

d.

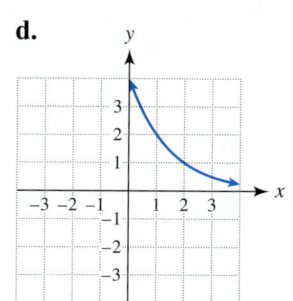

Exercises 37–48: Graph $y = f(x)$. *State whether the graph depicts exponential growth or exponential decay.*

37. $f(x) = 2^x$

38. $f(x) = 3^x$

39. $f(x) = \left(\frac{1}{4}\right)^x$

40. $f(x) = \left(\frac{1}{2}\right)^x$

41. $f(x) = 2^{-x}$

42. $f(x) = 3^{-x}$

43. $f(x) = 3^x - 1$

44. $f(x) = 2^x + 1$

45. $f(x) = 2^{x-1}$

46. $f(x) = 2^{x+1}$

47. $f(x) = 4\left(\frac{1}{3}\right)^x$

48. $f(x) = 3\left(\frac{1}{2}\right)^x$

Exercises 49–58: **Exponents** *Use properties of exponents to simplify the expression.*

49. $e^2 e^5$

50. $a^2 \cdot a^{-4}$

51. $\dfrac{a^4 a^{-2}}{a^3}$

52. $\dfrac{e^2 e^{-1}}{e}$

53. $e^x e^{y-x}$

54. $a^{1-x} a^{x-1}$

55. $\dfrac{2^{x+3}}{2^x}$

56. $\dfrac{e^{2x}}{e^x}$

57. $5^{-y} \cdot 5^{3y}$

58. $4^{-x} \cdot 4^{2x} \cdot 4^x$

COMPOUND INTEREST

Exercises 59–64: (Refer to Example 3.) If C dollars are deposited in an account paying r percent annual interest, approximate the amount in the account after x years.

59. $C = \$1500$ $r = 9\%$ $x = 10$ years

60. $C = \$1500$ $r = 15\%$ $x = 10$ years

61. $C = \$200$ $r = 20\%$ $x = 50$ years

62. $C = \$5000$ $r = 8.4\%$ $x = 7$ years

63. $C = \$560$ $r = 1.4\%$ $x = 25$ years

64. $C = \$750$ $r = 10\%$ $x = 13$ years

65. *Interest* Suppose that $1000 are deposited in an account paying 8% annual interest for 10 years. If $2000 had been deposited instead of $1000, would there be twice the money in the account after 10 years? Explain.

66. *Interest* Suppose that $500 are deposited in an account paying 5% annual interest for 10 years. If the interest rate had been 10% instead of 5%, would the total interest earned after 10 years be twice as much? Explain.

67. *Federal Debt* In 2003 the federal budget deficit was about $300 billion. At the same time, 30-year treasury bonds were paying 4.95% interest. Suppose that U.S. citizens loaned $300 billion to the federal government at 4.95%. If the federal government waited 30 years to pay the entire amount back, including the interest, how much would it be? (*Source:* U.S. Treasury Department.)

68. *Federal Debt* Repeat Exercise 67 but suppose that the interest rate is 2% higher. How much would the federal government owe after 30 years? Is the national debt sensitive to interest rates?

THE NATURAL EXPONENTIAL FUNCTION

Exercises 69–72: Evaluate f(x) for the given value of x. Approximate answers to the nearest hundredth.

69. $f(x) = e^x$ $x = 1.2$

70. $f(x) = 2e^x$ $x = 2$

71. $f(x) = 1 - e^x$ $x = -2$

72. $f(x) = 4e^{-x}$ $x = 1.5$

Exercises 73–76: Graph f(x) in $[-4, 4, 1]$ *by* $[0, 8, 1]$. *State whether the graph illustrates exponential growth or exponential decay.*

73. $f(x) = e^{0.5x}$

74. $f(x) = e^x + 1$

75. $f(x) = 1.5e^{-0.32x}$

76. $f(x) = 2e^{-x} + 1$

APPLICATIONS

Exercises 77–80: **Population Growth** *(Refer to Example 7.) The population P in 2000 for a state is given along with r, its annual percentage rate of continuous growth.*
  (a) *Write the formula* $f(x) = Pe^{rx}$, *where r is in decimal notation, that models the population in millions x years after 2000.*
  (b) *Estimate the population in 2010.*

77. California: $P = 34$ million, $r = 1.3\%$

78. Georgia: $P = 8$ million, $r = 2.3\%$

79. North Carolina: $P = 8$ million, $r = 1.9\%$

80. Texas: $P = 21$ million, $r = 2.1\%$

81. *Modeling Population* (Refer to Example 7.) In 2000 the population of Arizona was 5 million and growing continuously at a rate of 3.1% per year.
(a) Write a function f that models Arizona's population in millions x years after 2000.
(b) Graph f in [0, 10, 1] by [4, 7, 1].
(c) Estimate the population of Arizona in 2010.

82. *Dating Artifacts* Radioactive carbon-14 is found in all living things and is used to date objects containing organic material. Suppose that an object initially contains C grams of carbon-14. After x years it will contain A grams, where

$$A = C(0.99988)^x.$$

(a) Let $C = 10$ and graph A over a 20,000-year period. Is this function an example of exponential growth or decay?
(b) How many grams are left after 5700 years? What fraction of the carbon-14 is left?

83. *E. coli Bacteria* A strain of bacteria that inhabits the intestines of animals is named *Escherichia coli* (*E. coli*). These bacteria are capable of rapid growth and can be dangerous to humans—particularly children. The table shows the results of one study of the growth of *E. coli* bacteria, where concentrations are listed in thousands of bacteria per milliliter.

t (minutes)	0	50	100
Concentration	500	1000	2000

t (minutes)	150	200
Concentration	4000	8000

Source: G. S. Stent, *Molecular Biology of Bacterial Viruses.*

(a) Find C and a so that $f(t) = Ca^{t/50}$ models these data.
(b) Use $f(t)$ to estimate the concentration of bacteria after 170 minutes.
(c) Discuss the growth of this strain of bacteria over a 200-minute time period.

84. *Internet Use* Internet use in Western Europe has grown rapidly. The table at the top of the next column shows the number of Internet users y in millions, where $x = 0$ corresponds to 2000, $x = 1$ to 2001, and $x = 2$ to 2002.

x (year)	0	1	2
y (millions)	52	67	85

Source: Nortel Networks.

(a) Approximate C and a so that $f(x) = Ca^x$ models the data. (*Hint:* To find a, estimate the factor by which y increases each year.)
(b) Use $f(x)$ to estimate the number of users in 2004.
(c) How long is this type of growth likely to continue?

85. *Cellular Phone Use* In 1985, there were about 203,000 cellular phone subscribers in the United States. This number increased to about 84 million users in 2000, as illustrated in the figure. The rapid growth in cellular phone subscribers in millions can be modeled by $f(x) = 0.0272(1.495)^{x-1980}$, where x is the year. (*Source:* Cellular Telecommunications Industry Association.)
(a) Evaluate $f(1995)$ and interpret the result.
(b) What is the growth factor for $f(x)$? Explain what the growth factor indicates about cellular phone subscribers from 1985 to 2000.

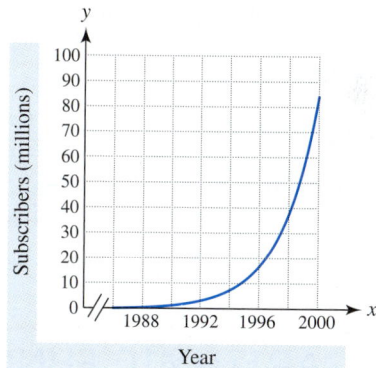

86. *Swimming Pool Maintenance* Chlorine is frequently used to disinfect swimming pools. The chlorine concentration should remain between 1.5 and 2.5 parts per million (ppm). After a warm, sunny day only 80% of the chlorine may remain in the water, with the other 20% dissipating into the air or combining with other chemicals in the water. (*Source:* D. Thomas, *Swimming Pool Operator's Handbook.*)
(a) Let $f(x) = 3(0.8)^x$ model the concentration of chlorine in parts per million after x days. What is the initial concentration of chlorine in the pool?
(b) If no more chlorine is added, estimate when the chlorine level drops below 1.6 parts per million.

87. *Modeling Traffic Flow*   (Refer to Example 5.) Construct a table of $f(x) = (0.905)^x$, starting at $x = 0$ and incrementing by 10, until $x = 50$.
  (a) Evaluate $f(0)$ and interpret the result.
  (b) For a time interval of what length is there only a 5% chance that no cars will enter the intersection?

88. *Modeling Tree Density*   (Refer to Example 6.)
  (a) Evaluate $P(10)$, $P(20)$, and $P(30)$. Interpret the results.
  (b) What happens to $P(x)$ as x becomes large? Explain how this probability relates to the spacing of trees in a forest.

WRITING ABOUT MATHEMATICS

89. A student evaluates $f(x) = 4(2)^x$ at $x = 3$ and obtains 512. Did the student evaluate the function correctly? What was the student's error?

90. For a set of data, how can you distinguish between linear growth and exponential growth? Give an example of each type of data.

CHECKING BASIC CONCEPTS
SECTIONS 12.1 AND 12.2

1. If $f(x) = 2x^2 + 5x - 1$ and $g(x) = x + 1$, find each expression.
  (a) $(g \circ f)(1)$ **(b)** $(f \circ g)(x)$

2. Sketch a graph of $f(x) = x^2 - 1$.
  (a) Is f a one-to-one function? Explain.
  (b) Does f have an inverse function?

3. If $f(x) = 4x - 3$, find $f^{-1}(x)$.

4. Evaluate $f(-2)$ if $f(x) = 3(2^x)$.

5. Sketch a graph of $f(x) = \left(\frac{1}{3}\right)^x$.

6. Use the graph of $y = Ca^x$ to determine the constants C and a.

12.3   LOGARITHMIC FUNCTIONS

The Common Logarithmic Function ▪ The Inverse of the Common Logarithmic Function ▪ Logarithms with Other Bases

A LOOK INTO MATH ▷

Logarithmic functions are used in many applications. For example, if one airplane weighs twice as much as another, does the heavier airplane typically need a runway that is twice as long? Using a logarithmic function, we can answer this question. (See Example 9.) Logarithmic functions are also used to measure the intensity of sound. In this section we discuss logarithmic functions and several of their applications.

The Common Logarithmic Function

In applications, measurements can vary greatly in size. Table 12.7 lists some examples of objects, with the approximate distances in meters across each.

TABLE 12.7

Object	Distance (meters)
Atom	10^{-9}
Protozoan	10^{-4}
Small Asteroid	10^2
Earth	10^7
Universe	10^{26}

Source: C. Ronan, *The Natural History of the Universe.*

Each distance is listed in the form 10^k for some k. The value of k distinguishes one measurement from another. The *common logarithmic function* or *base-10 logarithmic function*, denoted *log* or log_{10}, outputs k if the input x can be expressed as 10^k for some real number k. For example, $\log 10^{-9} = -9$, $\log 10^2 = 2$, and $\log 10^{1.43} = 1.43$. For any real number k, $\log 10^k = k$. Some values for $f(x) = \log x$ are given in Table 12.8.

TABLE 12.8

x	10^{-4}	10^{-3}	10^{-2}	10^{-1}	10^0	10^1	10^2	10^3	10^4
$\log x$	-4	-3	-2	-1	0	1	2	3	4

NOTE: A common logarithm is an **exponent** having base 10.

We use this information to define the common logarithm.

COMMON LOGARITHM

The **common logarithm of a positive number x**, denoted $\log x$, is calculated as follows. If x is written as $x = 10^k$, then

$$\log x = k,$$

where k is a real number. That is, $\log 10^k = k$.
The function given by

$$f(x) = \log x$$

is called the **common logarithmic function**.

The common logarithmic function outputs an exponent k, which may be positive, negative, or zero. However, a valid input must be positive because 10^k is always positive. *The expression log x equals the exponent k on base 10 that gives the number x.* For example, $\log 1000 = 3$ because $1000 = 10^3$.

NOTE: Previously we have always used one letter, such as f or g, to name a function. The common logarithm is the *first* function for which we use *three* letters, *log*, to name it. Thus $f(x)$, $g(x)$, and $\log(x)$ all represent functions. Generally, $\log(x)$ is written without parentheses as $\log x$. We can also define a function f to be the common logarithmic function by writing $f(x) = \log x$.

EXAMPLE Evaluating common logarithms

Simplify each common logarithm.

(a) $\log 100$ **(b)** $\log \frac{1}{10}$ **(c)** $\log \sqrt{1000}$ **(d)** $\log 45$

Solution

(a) $100 = 10^2$, so $\log 100 = \log 10^2 = \mathbf{2}$

(b) $\log \frac{1}{10} = \log 10^{-1} = \mathbf{-1}$

(c) $\log \sqrt{1000} = \log (1000)^{1/2} = \log (10^3)^{1/2} = \log 10^{3/2} = \frac{3}{2}$

(d) How to write 45 as a power of 10 is not obvious. However, we can use a calculator to determine that $\log 45 \approx 1.6532$. Thus $10\text{^}(1.6532) \approx 45$. Figure 12.23 supports these answers.

CALCULATOR HELP

To evaluate the common logarithmic function, see Appendix A (page AP-1).

(a)

Figure 12.23

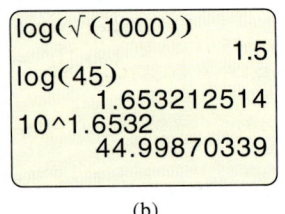

(b)

Now Try Exercises 13, 19, 25, 63

The points $(10^{-1}, -1)$, $(10^0, 0)$, $(10^{0.5}, 0.5)$, and $(10^1, 1)$ are on the graph of $y = \log x$. Plotting these points, as shown in Figure 12.24(a), and sketching the graph of $y = \log x$ results in Figure 12.24(b). Note some important features of this graph.

- The graph of the common logarithm increases very slowly for large values of x. For example, x must be 100 for $\log x$ to reach 2 and x must be 1000 for $\log x$ to reach 3.
- The graph passes through the point $(1, 0)$. Thus $\log 1 = 0$.
- The graph does not exist for negative values of x. The domain of $\log x$ includes only positive numbers. The range of $\log x$ includes all real numbers.
- When $0 < x < 1$, $\log x$ outputs negative values. The y-axis is a vertical asymptote, so as x approaches 0, $\log x$ approaches $-\infty$.

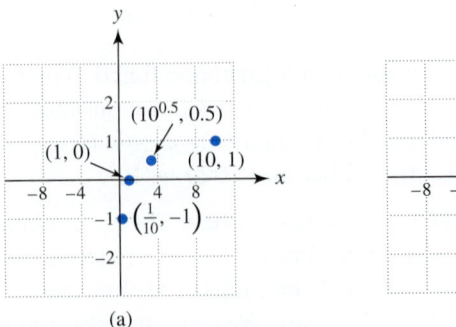

Figure 12.24

The Common Logarithmic Function and the Square Root Function

Much like the square root function, the common logarithmic function does not have an easy-to-evaluate formula. For example, we can calculate $\sqrt{4} = 2$ and $\sqrt{100} = 10$ mentally, but for $\sqrt{2}$ we usually rely on a calculator. Similarly, we can mentally calculate $\log 1000 = \log 10^3 = 3$, whereas we can use a calculator to approximate $\log 45$. The notation $\log x$ is implied to be $\log_{10} x$ much as $\sqrt{x}$ equals $\sqrt[2]{x}$. Another similarity between the square root function and the common logarithmic function is that their domains do not include negative numbers. If only real numbers are allowed as outputs, both $\sqrt{-3}$ and $\log(-3)$ are undefined expressions.

The Inverse of the Common Logarithmic Function

The graph of $y = \log x$ shown in Figure 12.24(b) is a one-to-one function because it passes the horizontal line test. Different inputs always result in different outputs. Thus the common logarithmic function has an inverse function. To determine this inverse function for $\log x$, consider Tables 12.9 and 12.10.

TABLE 12.9

x	-2	-1	0	1	2
10^x	10^{-2}	10^{-1}	10^0	10^1	10^2

TABLE 12.10

x	10^{-2}	10^{-1}	10^0	10^1	10^2
$\log x$	-2	-1	0	1	2

If we start with the number 2, compute 10^2, and then calculate $\log 10^2$, the result is 2.

$$\log(10^2) = \mathbf{2}$$

In general, $\log 10^x = x$ for any real number x. Now suppose that we perform the calculations in reverse order by taking the common logarithm and then computing a power of 10. For example, suppose that we start with the number 100. The result is

$$10^{\log 100} = 10^2 = \mathbf{100}.$$

In general, $10^{\log x} = x$ for any positive number x.

The *inverse function* of $f(x) = \log x$ is $f^{-1}(x) = 10^x$. That is, if $\log x = y$, then $10^y = x$. Note that composition of these two functions satisfies the definition of an inverse function.

$$
\begin{aligned}
(f \circ f^{-1})(x) &= f\big(\mathbf{f^{-1}(x)}\big) \\
&= f(\mathbf{10^x}) \\
&= \log 10^x \\
&= \mathbf{x}
\end{aligned}
\qquad \text{and} \qquad
\begin{aligned}
(f^{-1} \circ f)(x) &= f^{-1}\big(\mathbf{f(x)}\big) \\
&= f^{-1}(\mathbf{\log x}) \\
&= 10^{\log x} \\
&= \mathbf{x}
\end{aligned}
$$

In general, the graph of $y = f^{-1}(x)$ is a reflection of the graph of $y = f(x)$ across the line $y = x$. The graphs of $y = \log x$ and $y = 10^x$ are shown in Figure 12.25. Note that the graph of $y = 10^x$ is a reflection of the graph of $y = \log x$ across the line $y = x$.

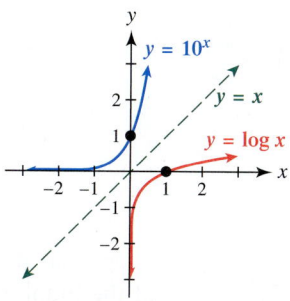

Figure 12.25

These inverse properties of log x and 10^x are summarized as follows.

INVERSE PROPERTIES OF THE COMMON LOGARITHM

The following properties hold for common logarithms.

$$\log 10^x = x, \qquad \text{for any real number } x$$

$$10^{\log x} = x, \qquad \text{for any positive real number } x$$

EXAMPLE 2   Applying inverse properties

Use inverse properties to simplify each expression.

(a) $\log 10^\pi$ **(b)** $\log 10^{x^2+1}$ **(c)** $10^{\log 7}$ **(d)** $10^{\log 3x}, x > 0$

Solution
(a) Because $\log 10^x = x$ for any real number x, $\log 10^\pi = \pi$.
(b) $\log 10^{x^2+1} = x^2 + 1$
(c) Because $10^{\log x} = x$ for any positive real number x, $10^{\log 7} = 7$.
(d) $10^{\log 3x} = 3x$, provided x is a positive number.   Now Try Exercises **21, 47, 49, 59**

We can also graph logarithmic functions, as demonstrated in the next example.

EXAMPLE 3   Graphing logarithmic functions

Graph each function f and compare its graph to $y = \log x$.
(a) $f(x) = \log(x - 2)$ **(b)** $f(x) = \log(x) + 1$

Solution
(a) We can use our knowledge of translations to sketch the graph of $y = \log(x - 2)$. For example, the graph of $y = (x - 2)^2$ is similar to the graph of $y = x^2$, except that it is translated 2 units to the *right*. Thus the graph of $y = \log(x - 2)$ is similar to the graph of $y = \log x$ (see Figure 12.25) except that it is translated 2 units to the right, as shown in Figure 12.26. The graph of $y = \log x$ passes through $(1, 0)$, so the graph of $y = \log(x - 2)$ passes through $(3, 0)$. Also, instead of the y-axis being a vertical asymptote, the line $x = 2$ is the vertical asymptote.

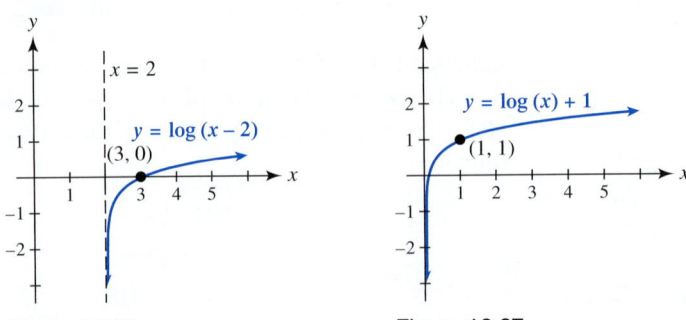

Figure 12.26   Figure 12.27

(b) The graph of $y = \log(x) + 1$ is similar to the graph of $y = \log x$, except that it is translated 1 unit *upward*. This graph is shown in Figure 12.27. Note that the graph of $y = \log(x) + 1$ passes through the point $(1, 1)$.   Now Try Exercises **83, 85**

NOTE: The graph of $y = \log x$ has a vertical asympotote when $x = 0$ and is undefined when $x < 0$. Thus the graph of $\log(x - 2)$ has a vertical asymptote when $x - 2 = 0$, or $x = 2$. It is undefined when $x - 2 < 0$ or $x < 2$.

▶ **REAL-WORLD CONNECTION** Logarithms are used to model quantities that vary greatly in intensity. For example, the human ear is extremely sensitive and able to detect intensities on the eardrum ranging from 10^{-16} watts per square centimeter (w/cm^2) to 10^{-4} w/cm^2, which usually is painful. The next example illustrates modeling sound with logarithms.

EXAMPLE **4** Modeling sound levels

Sound levels in decibels (dB) can be computed by $f(x) = 160 + 10 \log x$, where x is the intensity of the sound in watts per square centimeter. Ordinary conversation has an intensity of 10^{-10} w/cm^2. What decibel level is this? (**Source:** R. Weidner and R. Sells, *Elementary Classical Physics, Vol. 2.*)

Solution

To find the decibel level for ordinary conversation, evaluate $f(10^{-10})$.

$$f(10^{-10}) = 160 + 10\,\textbf{log}\,(\mathbf{10^{-10}}) \qquad \text{Substitute } x = 10^{-10}.$$
$$= 160 + 10(\mathbf{-10}) \qquad \text{Evaluate } \log(10^{-10}).$$
$$= 60 \qquad \text{Simplify.}$$

Ordinary conversation corresponds to 60 dB.

Now Try Exercise **93**

CRITICAL THINKING

If the sound level increases by 10 dB, by what factor does the intensity x increase?

Logarithms with Other Bases

Common logarithms are base-10 logarithms, but we can define logarithms having other bases. For example, base-2 logarithms are frequently used in computer science. Some values for the base-2 logarithmic function, denoted $f(x) = \log_2 x$, are shown in Table 12.11. If x can be expressed as $x = 2^k$ for some real number k, then $\log_2 x = \log_2 2^k = k$.

TABLE 12.11

x	2^{-3}	2^{-2}	2^{-1}	2^0	2^1	2^2	2^3
$\log_2 x$	-3	-2	-1	0	1	2	3

NOTE: A base-2 logarithm is an *exponent* having base 2.

Logarithms with other bases are evaluated in the next three examples.

EXAMPLE **5** Evaluating base-2 logarithms

Simplify each logarithm.
(a) $\log_2 8$ **(b)** $\log_2 \frac{1}{4}$

Solution
(a) The logarithmic expression $\log_2 8$ represents the exponent on base 2 that gives 8. Because $8 = 2^3$, $\log_2 8 = \log_2 \mathbf{2^3} = \mathbf{3}$.
(b) Because $\frac{1}{4} = \frac{1}{2^2} = 2^{-2}$, $\log_2 \frac{1}{4} = \log_2 \mathbf{2^{-2}} = \mathbf{-2}$.

Now Try Exercises **35, 37**

Some values of base-e logarithms are shown in Table 12.12. A base-e logarithm is referred to as a **natural logarithm** and denoted either $\log_e x$ or $\ln x$. Natural logarithms are used in mathematics, science, economics, electronics, and communications.

CALCULATOR HELP

To evaluate the natural logarithmic function, see Appendix A (page AP-1).

TABLE 12.12

x	e^{-3}	e^{-2}	e^{-1}	e^0	e^1	e^2	e^3
$\ln x$	-3	-2	-1	0	1	2	3

NOTE: A natural logarithm is an **exponent** having base e.

To evaluate natural logarithms, we usually use a calculator.

EXAMPLE 6 Evaluating natural logarithms

Approximate to the nearest hundredth.
(a) $\ln 10$ **(b)** $\ln \frac{1}{2}$

Solution
(a) Figure 12.28 shows that $\ln 10 \approx 2.30$.
(b) Figure 12.28 shows that $\ln \frac{1}{2} \approx -0.69$.

Now Try Exercises 67, 69

We now define base-a logarithms.

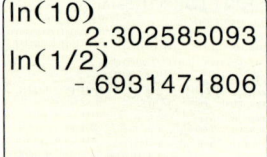

Figure 12.28

BASE-a LOGARITHMS

The **logarithm with base a of a positive number x**, denoted $\log_a x$, is calculated as follows. If x is written as $x = a^k$, then

$$\log_a x = k,$$

where $a > 0$, $a \neq 1$, and k is a real number. That is, $\log_a a^k = k$.
The function given by

$$f(x) = \log_a x$$

is called the **logarithmic function with base a**.

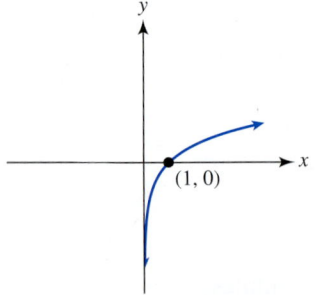

Figure 12.29 $y = \log_a x$, $a > 1$

Remember that *a logarithm is an exponent*. The expression $\log_a x$ equals the exponent k such that $a^k = x$. The graph of $y = \log_a x$ with $a > 1$ is shown in Figure 12.29. Note that the graph passes through the point $(1, 0)$. Thus $\log_a 1 = 0$.

CRITICAL THINKING

Explain why $\log_a 1 = 0$ for any positive base a, $a \neq 1$.

NOTE: The natural logarithm, $\ln x$, is a base-a logarithm with $a = e$. That is, $\ln x = \log_e x$.

EXAMPLE 7   **Evaluating base-*a* logarithms**

Simplify each logarithm.

(a) $\log_5 25$ **(b)** $\log_4 \frac{1}{64}$ **(c)** $\log_7 1$ **(d)** $\log_3 9^{-1}$

Solution

(a) $25 = 5^2$, so $\log_5 25 = \log_5 5^2 = 2$.

(b) $\frac{1}{64} = \frac{1}{4^3} = 4^{-3}$, so $\log_4 \frac{1}{64} = \log_4 4^{-3} = -3$.

(c) $1 = 7^0$, so $\log_7 1 = \log_7 7^0 = 0$. (The logarithm of 1 is 0, regardless of the base.)

(d) $9^{-1} = (3^2)^{-1} = 3^{-2}$, so $\log_3 9^{-1} = \log_3 3^{-2} = -2$.   Now Try Exercises 39, 40, 41, 45

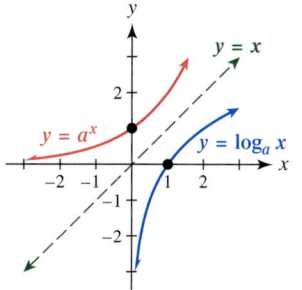

Figure 12.30 $a > 1$

The graph of $y = \log_a x$ in Figure 12.29 passes the horizontal line test, so it is a one-to-one function and it has an inverse. If we let $f(x) = \log_a x$, then $f^{-1}(x) = a^x$. This statement is a generalization of the fact that $\log x$ and 10^x represent inverse functions. The graphs of $y = \log_a x$ and $y = a^x$ with $a > 1$ are shown in Figure 12.30. Note that the graph of $y = a^x$ is a reflection of the graph of $y = \log_a x$ across the line $y = x$.

These inverse properties for logarithmic and exponential functions are summarized by the following.

INVERSE PROPERTIES OF BASE-*a* LOGARITHMS

The following properties hold for logarithms with base a.

$$\log_a a^x = x, \qquad \text{for any real number } x$$
$$a^{\log_a x} = x, \qquad \text{for any positive real number } x$$

EXAMPLE 8   **Applying inverse properties**

Simplify each expression.

(a) $\ln e^{0.5x}$ **(b)** $e^{\ln 4}$ **(c)** $2^{\log_2 7x}$ **(d)** $10^{\log(9x-3)}$

Solution

(a) $\ln e^{0.5x} = 0.5x$ because $\ln e^k = k$ for all k.

(b) $e^{\ln 4} = 4$ because $e^{\ln k} = k$ for all positive k.

(c) $2^{\log_2 7x} = 7x$ for $x > 0$ because $a^{\log_a k} = k$ for all positive k.

(d) $10^{\log(9x-3)} = 9x - 3$ for $x > \frac{1}{3}$ because $10^{\log k} = k$ for all positive k.

Now Try Exercises 51, 53, 55, 57

▶ **REAL-WORLD CONNECTION**   Logarithms occur in many applications. One application is runway length for airplanes, which we discuss in the next example.

EXAMPLE 9   **Calculating runway length**

There is a mathematical relationship between an airplane's weight x and the runway length required at takeoff. For certain types of airplanes, the minimum runway length L in thousands of feet may be modeled by $L(x) = 1.3 \ln x$, where x is in thousands of pounds. (*Source:* L. Haefner, *Introduction to Transportation Systems.*)

(a) Estimate the runway length needed for an airplane weighing 10,000 pounds.

(b) Does a 20,000-pound airplane need twice the runway length that a 10,000-pound airplane needs? Explain.

Solution

(a) Because $L(x) = 1.3 \ln x$, it follows that $L(10) = 1.3 \ln(10) \approx 3$. An airplane weighing 10,000 pounds requires a runway (at least) 3000 feet long.

(b) Because $L(20) = 1.3 \ln(20) \approx 3.9$, a 20,000-pound airplane does not need twice the runway length needed by a 10,000-pound airplane. Rather the heavier airplane needs roughly 3900 feet of runway, or only an extra 900 feet.   **Now Try Exercise 94**

12.3 PUTTING IT ALL TOGETHER

Common logarithms are base-10 logarithms. If a positive number x is written as $x = 10^k$, then $\log x = k$. The value of $\log x$ represents the exponent on the base 10 that gives x. We can define logarithms that have other bases. For example, the natural logarithm is a base-e logarithm that is usually evaluated using a calculator. The following table summarizes some important concepts related to base-a logarithms.

Concept	Description	Examples
Base-a Logarithms	*Definition:* $\log_a x = k$ means $x = a^k$, where $a > 0$ and $a \neq 1$. *Domain:* all *positive* real numbers *Range:* all real numbers *Graph:* $a > 1$ (shown to the right) Passes through $(1, 0)$; vertical asymptote: y-axis *Common Logarithm:* Base-10 logarithm and denoted $\log x$ *Natural Logarithm:* Base-e logarithm, where $e \approx 2.718$, and denoted $\ln x$	$\log 1000 = \log 10^3 = 3$, $\log_2 16 = \log_2 2^4 = 4$, and $\log_3 \dfrac{1}{81} = \log_3 3^{-4} = -4$ $y = \log_a x$, $a > 1$, passes through $(1,0)$
Inverse Properties	The following properties hold for base-a logarithms. $\log_a a^x = x$, for any real number x $a^{\log_a x} = x$, for any positive number x	$\log 10^{7.48} = 7.48$ and $2^{\log_2 63} = 63$

12.3 Exercises

CONCEPTS

1. What is the base of the common logarithm?

2. What is the base of the natural logarithm?

3. What are the domain and range of $\log x$?

4. What are the domain and range of $\log_a x$?

5. $\log 10^k =$ _____

6. $\ln e^k =$ _____

7. If $\log x = k$, then $10^k =$ _____.

8. If $x > 0$, then $10^{\log x} =$ _____.

9. What does k equal if $10^k = 5$?

10. What does k equal if $2^k = 5$?

11. $\log_a 1 =$ _____

12. $\log_a (-1)$ is _____ .

EVALUATING AND GRAPHING LOGARITHMIC FUNCTIONS

Exercises 13–62: Simplify the expression, if possible.

13. $\log 10^5$

14. $\log 10$

15. $\log 10^{-4}$

16. $\log 10^{-1}$

17. $\log 1$

18. $\log \sqrt[3]{100}$

19. $\log \frac{1}{100}$

20. $\log \frac{1}{10}$

21. $\log 10^{4.7}$

22. $\log 10^{2x+4}$

23. $\log_5 5^{6x}$

24. $\log_3 3^3$

25. $\log \sqrt{\frac{1}{1000}}$

26. $\log_2 2^6$

27. $\log_2 2^8$

28. $\log_2 2^{-5}$

29. $\log_2 \sqrt{8}$

30. $\log_2 \sqrt{32}$

31. $\log_2 \sqrt[3]{\frac{1}{4}}$

32. $\log_2 \frac{1}{64}$

33. $\log_2 -8$

34. $\log_2 -7$

35. $\log_2 4$

36. $\log_2 \frac{1}{32}$

37. $\log_2 \frac{1}{16}$

38. $\log_3 27$

39. $\log_3 \frac{1}{9}$

40. $\log_4 16$

41. $\ln 1$

42. $\ln e^2$

43. $\log 0.001$

44. $\log 0.0001$

45. $\log_5 \frac{1}{25}$

46. $\log_8 64$

47. $10^{\log 2}$

48. $10^{\log 7.5}$

49. $10^{\log x^2}$

50. $10^{\log |x|}$

51. $5^{\log_5 17}$

52. $9^{\log_9 73}$

53. $4^{\log_4 (2x)^2}$

54. $b^{\log_b (x-1)}$

55. $10^{\log 5}$

56. $\ln e^{3/4}$

57. $\ln e^{-5x}$

58. $e^{\ln 2x}$

59. $\log 10^{(2x-7)}$

60. $\log 10^{(8-4x)}$

61. $5^{\log_5 0.6z}$

62. $7^{\log_7 (x-9)}$

Exercises 63–70: Evaluate the logarithm, using a calculator. Round values to the nearest thousandth.

63. $\log 25$

64. $\log 0.501$

65. $\log 1.45$

66. $\log \frac{1}{35}$

67. $\ln 7$

68. $\ln 126$

69. $\ln \frac{4}{7}$

70. $\ln 0.67$

Exercises 71–74: Graph $y = f(x)$.

71. $f(x) = \log_2 x$

72. $\log_2 (x - 2)$

73. $f(x) = 2 + \log_2 x$

74. $\log_2 (x + 2)$

Exercises 75–78: Graph f in $[-4, 4, 1]$ by $[-4, 4, 1]$. Compare this graph to the graph of $y = \ln x$. Identify the domain of f.

75. $f(x) = \ln |x|$

76. $f(x) = \ln (x) - 2$

77. $f(x) = \ln (x + 2)$

78. $f(x) = 2 \ln x$

Exercises 79–82: Without using a calculator match $f(x)$ with its graph (a.–d.).

79. $f(x) = \log x$

80. $f(x) = \log_3 x$

81. $f(x) = \log_3 (x) + 2$

82. $f(x) = \log (x + 1)$

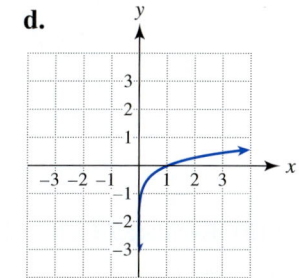

Exercises 83–90: Graph $y = f(x)$. Compare the graph to the graph of $y = \log x$.

83. $f(x) = \log(x) - 1$ **84.** $f(x) = \log(x) + 2$

85. $f(x) = \log(x + 1)$ **86.** $f(x) = \log(x + 2)$

87. $f(x) = \log(x - 1)$ **88.** $f(x) = \log(x - 3)$

89. $f(x) = 2 \log x$ **90.** $f(x) = -\log x$

Exercises 91 and 92: Complete the table.

91.

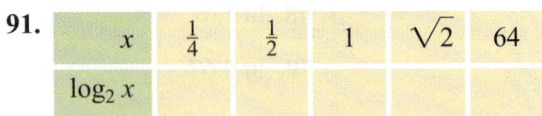

x	$\frac{1}{4}$	$\frac{1}{2}$	1	$\sqrt{2}$	64
$\log_2 x$					

92.

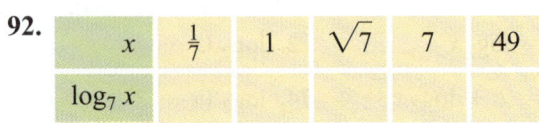

x	$\frac{1}{7}$	1	$\sqrt{7}$	7	49
$\log_7 x$					

APPLICATIONS

93. *Modeling Sound*   (Refer to Example 4.) At professional football games in domed stadiums the decibel level may reach 110. The eardrum usually experiences pain when the intensity of the sound reaches 10^{-4} watts per square centimeter. How many decibels does this quantity represent? Is the noise at a football game likely to hurt some people's eardrums?

94. *Runway Length*   (Refer to Example 9.)
  (a) A graph of $L(x) = 1.3 \ln x$ is shown in the figure, where $y = L(x)$. As the weight of the plane increases, what can be said about the length of the runway required?

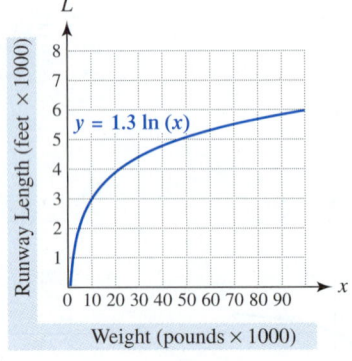

  (b) Evaluate $L(50)$ and interpret the result.

95. *Hurricanes*   The barometric air pressure P in inches of mercury at a distance of d miles from the eye of a severe hurricane can sometimes be modeled by the formula $P(d) = 0.48 \ln(d + 1) + 27$. Average air pressure is about 30 inches of mercury.   (*Source:* A. Miller and R. Anthes, *Meteorology.*)

  (a) Evaluate $P(0)$ and $P(50)$. Interpret the results.
  (b) A graph of $y = P(d)$ is shown in the figure. Describe how the air pressure changes as the distance from the eye of the hurricane increases.

  (c) Is the eye of the hurricane a low pressure area or a high pressure area?

96. *Predicting Wind Speed*   Wind speed typically varies in the first 20 meters above the ground. Close to the ground, wind speed is often less than it is at 20 meters above the ground. For this reason wind speeds are usually measured at heights from 5 to 10 meters by the U.S. Weather Service. For a particular day, let

$$W(h) = 2.76 \log(h + 1) + 2.3$$

compute the wind speed W in meters per second at a height h meters above the ground.   (*Source:* A. Miller.)
  (a) Find the wind speed at a height of 10 meters.
  (b) A graph of $y = W(h)$ is shown in the figure at the top of the next column. Interpret the graph.

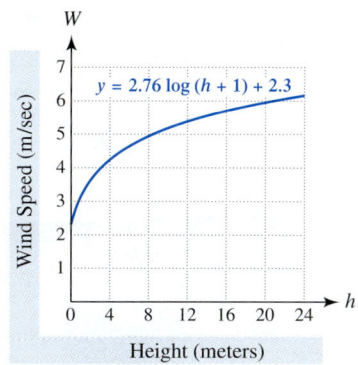

Height (meters)

97. *Growth in Salary*   Suppose that a person's salary is initially $40,000 and could be determined by either $f(x)$ or $g(x)$, where x represents the number of years of experience.
  i. $f(x) = 40{,}000(1.1)^x$
  ii. $g(x) = 40{,}000 \log(10 + x)$
Would most people prefer that their salaries increase exponentially or logarithmically? Explain.

98. *Calories Consumed and Land Ownership*   In developing countries there is a relationship between the amount of land a person owns and the average daily calories consumed. This relationship is modeled by

$$C(x) = 645 \log(x + 1) + 1925,$$

where x is the amount of land owned in acres and $0 \le x \le 4$.   (*Source:* D. Grigg, *The World Food Problem.*)
  (a) Find the average daily caloric intake for a person who owns 1 acre of land.
  (b) A graph of $y = C(x)$ is shown in the figure. How is the number of calories consumed each day affected by the amount of land owned?

Land Area Owned (acres)

  (c) Does a person who owns twice as much land as another person consume twice the calories? Explain your answer.

99. *Magnitude of a Star*   The first stellar brightness scale was developed 2000 years ago by two Greek astronomers, Hipparchus and Ptolemy. The brightest star in the sky was given a magnitude of 1, and the faintest star was given a magnitude of 6. In 1856 this scale was described mathematically by the formula

$$M = 6 - 2.5 \log \frac{I}{I_0},$$

where M is the magnitude of a star with an intensity of I and I_0 is the intensity of the faintest star seen in the sky.   (*Source:* M. Zeilik, *Introductory Astronomy and Astrophysics.*)

NOTE: The intensity of a star can be measured with instruments.

  (a) Find M if $I = 10$ and $I_0 = 1$.
  (b) What is the magnitude of a star that is 100 times more intense than the faintest star?
  (c) If the intensity of a star increases by a factor of 10, what happens to its magnitude?

100. *Population of Urban Regions*   Although less industrialized urban regions of the world are experiencing exponential population growth, industrialized urban regions are experiencing logarithmic population growth. Population in less industrialized urban regions can be modeled by

$$f(x) = 0.338(1.035)^x,$$

whereas the population in industrialized urban regions can be modeled by

$$g(x) = 0.36 + 0.15 \ln(x + 1).$$

In these formulas the output is in billions of people and x is in years, where $x = 0$ corresponds to 1950, $x = 10$ to 1960, and so on until $x = 80$ corresponds to 2030.   (*Source:* D. Meadows, *Beyond The Limits.*)
  (a) Evaluate $f(50)$ and $g(50)$. Interpret the results.
  (b) Graph f and g in [0, 80, 10] by [0, 5, 1]. Compare the two graphs.
  (c) If x increases from 20 to 40, by what factor does $f(x)$ increase? By what factor does $g(x)$ increase?

101. *Earthquakes*   The Richter scale is used to determine the intensity of earthquakes, which corresponds to the amount of energy released. If an earthquake has an intensity of x, its *magnitude*, as computed by the Richter scale, is given by $R(x) = \log \frac{x}{I_0}$, where I_0 is the intensity of a small, measurable earthquake.

   (a) On July 26, 1963, an earthquake in Yugoslavia had a magnitude of 6.0 on the Richter scale, and on August 19, 1977, an earthquake in Indonesia measured 8.0. Find the intensity x for each of these earthquakes if $I_0 = 1$.

   (b) How many times more intense was the Indonesian earthquake than the Yugoslavian earthquake?

102. *Path Loss for Cellular Phones*   For cellular phones to work throughout a country, large numbers of cellular towers are necessary. How well the signal is propagated throughout a region depends on the location of these towers. One quick way to estimate the strength of a signal at x kilometers is to use the formula

$$D(x) = -121 - 36 \log x.$$

This formula computes the decrease in the signal, using decibels, so it is always negative. For example,

$D(1) = -121$ means that at a distance of 1 kilometer the signal has decreased in strength by 121 decibels.
(**Source:** C. Smith, *Practical Cellular & PCS Design.*)

   (a) Evaluate $D(3)$ and interpret the result.
   (b) Graph D in $[1, 10, 1]$ by $[-160, -120, 10]$.
   (c) What happens to the signal as x increases?

WRITING ABOUT MATHEMATICS

103. Explain what $\log_a x$ means and give an example.

104. How would you explain to a student that $\log_a 1 \neq 1$?

GROUP ACTIVITY
WORKING WITH REAL DATA

Directions: Form a group of 2 to 4 people. Select someone to record the group's responses for this activity. All members of the group should work cooperatively to answer the questions. If your instructor asks for your results, each member of the group should be prepared to respond.

 Greenhouse Gases   Carbon dioxide (CO_2) is a greenhouse gas in the atmosphere that may raise average temperatures on Earth. The burning of fossil fuels could be responsible for the increased levels of carbon dioxide. If current trends continue, future concentrations of atmospheric carbon dioxide in parts per million (ppm) could reach the levels shown in the accompanying table. The CO_2 concentration in the year 2000 was greater than it had been at any time in the previous 160,000 years.

Year	2000	2050	2100	2150	2200
CO_2 (ppm)	364	467	600	769	987

Source: R. Turner, *Environmental Economics.*

   (a) Let x be in years, where $x = 0$ corresponds to 2000, $x = 1$ to 2001, and so on. Find values for C and a so that $f(x) = Ca^x$ models the data.
   (b) Graph f and the data in the same viewing rectangle.
   (c) Use $f(x)$ to estimate graphically the year when the carbon dioxide concentration will be double the preindustrial level of 280 ppm.

12.4 PROPERTIES OF LOGARITHMS

Basic Properties ▪ Change of Base Formula

A LOOK INTO MATH ▷

The discovery of logarithms by John Napier (1550–1617) played an important role in the history of science. Logarithms were instrumental in allowing Johannes Kepler (1571–1630) to calculate the positions of the planet Mars, which led to his discovery of the laws of planetary motion. Kepler's laws were used by Isaac Newton (1642–1727) to discover the universal laws of gravity. Although calculators and computers have made tables of logarithms obsolete, applications involving logarithms still play an important role in modern-day computation. One reason for their continued importance is that logarithms possess several important properties.

Basic Properties

In this subsection we discuss three important properties of logarithms. The first property is the product rule for logarithms.

> **PRODUCT RULE FOR LOGARITHMS**
>
> For positive numbers m, n, and $a \neq 1$,
> $$\log_a mn = \log_a m + \log_a n.$$

Figure 12.31

This property may be verified by using properties of exponents and the fact that $\log_a a^k = k$ for any real number k. Here we verify the product property for logarithms. Other properties presented can be verified in a similar manner.

If m and n are positive numbers, we can write $m = a^c$ and $n = a^d$ for some real numbers c and d.

$$\log_a mn = \log_a(a^c a^d) = \log_a(a^{c+d}) = c + d \quad \text{and}$$
$$\log_a m + \log_a n = \log_a a^c + \log_a a^d = c + d$$

Thus $\log_a mn = \log_a m + \log_a n$.

This property is illustrated in Figure 12.31, which shows that
$$\log 10 = \log(2 \cdot 5) = \log 2 + \log 5.$$

In the next two examples we demonstrate various operations involving logarithms.

EXAMPLE **1**   **Writing logarithms as sums**

Write each expression as a sum of logarithms. Assume that x is positive.
(a) $\log 21$ **(b)** $\ln 5x$ **(c)** $\log_2 x^3$

Solution
(a) $\log 21 = \log(3 \cdot 7) = \log 3 + \log 7$
(b) $\ln 5x = \ln(5 \cdot x) = \ln 5 + \ln x$
(c) $\log_2 x^3 = \log_2(x \cdot x \cdot x) = \log_2 x + \log_2 x + \log_2 x$ Now Try Exercises 13, 15, 16

EXAMPLE 2   **Combining logarithms**

Write each expression as one logarithm. Assume that x and y are positive.
(a) $\log 5 + \log 6$ **(b)** $\ln x + \ln xy$ **(c)** $\log_3 2x + \log_3 5x$

Solution
(a) $\log 5 + \log 6 = \log(5 \cdot 6) = \log 30$
(b) $\ln x + \ln xy = \ln(x \cdot xy) = \ln x^2 y$
(c) $\log_3 2x + \log_3 5x = \log_3(2x \cdot 5x) = \log_3 10x^2$ Now Try Exercises 25, 27, 29

The second property is the quotient rule for logarithms.

> ### QUOTIENT RULE FOR LOGARITHMS
>
> For positive numbers m, n, and $a \neq 1$,
> $$\log_a \frac{m}{n} = \log_a m - \log_a n.$$

```
log(10)
              1
log(20)-log(2)
              1
```

Figure 12.32

This property is illustrated in Figure 12.32, which shows that
$$\log 10 = \log \frac{20}{2} = \log 20 - \log 2.$$

EXAMPLE 3   **Writing logarithms as differences**

Write each expression as a difference of two logarithms. Assume that variables are positive.
(a) $\log \dfrac{3}{2}$ **(b)** $\ln \dfrac{3x}{y}$ **(c)** $\log_5 \dfrac{x}{z^4}$

Solution
(a) $\log \dfrac{3}{2} = \log 3 - \log 2$ **(b)** $\ln \dfrac{3x}{y} = \ln 3x - \ln y$

(c) $\log_5 \dfrac{x}{z^4} = \log_5 x - \log_5 z^4$ Now Try Exercises 19, 21, 23

NOTE: $\log_a(m + n) \neq \log_a m + \log_a n;$ $\log_a(m - n) \neq \log_a m - \log_a n;$

$$\log_a(mn) \neq \log_a m \cdot \log_a n;   \log_a\left(\frac{m}{n}\right) \neq \frac{\log_a m}{\log_a n}$$

EXAMPLE 4   **Combining logarithms**

Write each expression as one term. Assume that x is positive.
(a) $\log 50 - \log 25$ **(b)** $\ln x^3 - \ln x$ **(c)** $\log_4 15x - \log_4 5x$

Solution
(a) $\log 50 - \log 25 = \log \dfrac{50}{25} = \log 2$

(b) $\ln x^3 - \ln x = \ln \dfrac{x^3}{x} = \ln x^2$

(c) $\log_4 15x - \log_4 5x = \log_4 \dfrac{15x}{5x} = \log_4 3$ Now Try Exercises 33, 35, 37

The third property is the power rule for logarithms. To illustrate this rule we use

$$\log x^3 = \log(x \cdot x \cdot x) = \log x + \log x + \log x = 3 \log x.$$

Thus $\log x^3 = \mathbf{3} \log x$. This example is generalized in the following rule.

> ### POWER RULE FOR LOGARITHMS
>
> For positive numbers m and $a \neq 1$ and any real number r,
>
> $$\log_a(m^r) = r \log_a m.$$

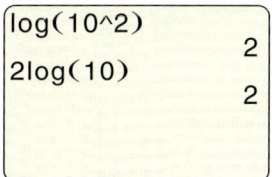

Figure 12.33

We use a calculator and the equation

$$\log 10^2 = \mathbf{2} \log 10$$

to illustrate this property. Figure 12.33 shows the result. We apply the power rule in the next example.

EXAMPLE 5 Applying the power rule

Rewrite each expression, using the power rule.
(a) $\log 5^6$ **(b)** $\ln(0.55)^{x-1}$ **(c)** $\log_5 8^{kx}$

Solution
(a) $\log 5^6 = \mathbf{6} \log 5$
(b) $\ln(0.55)^{x-1} = (\mathbf{x} - \mathbf{1}) \ln(0.55)$
(c) $\log_5 8^{kx} = \mathbf{kx} \log_5 8$

Now Try Exercises 39, 41, 47

Sometimes we use more than one property to simplify an expression. We assume that all variables are positive in the next two examples.

EXAMPLE 6 Combining logarithms

Write each expression as the logarithm of a single expression.
(a) $3 \log x + \log x^2$ **(b)** $2 \ln x - \ln \sqrt{x}$

Solution

(a) $3 \log x + \log x^2 = \log x^3 + \log x^2$ Power rule
$\qquad\qquad\qquad\quad = \log(x^3 \cdot x^2)$ Product rule
$\qquad\qquad\qquad\quad = \log x^5$ Properties of exponents

(b) $2 \ln x - \ln \sqrt{x} = 2 \ln x - \ln x^{1/2}$ $\sqrt{x} = x^{1/2}$
$\qquad\qquad\qquad = \ln x^2 - \ln x^{1/2}$ Power rule
$\qquad\qquad\qquad = \ln \dfrac{x^2}{x^{1/2}}$ Quotient rule
$\qquad\qquad\qquad = \ln x^{3/2}$ Properties of exponents

Now Try Exercises 49, 57

EXAMPLE 7   **Expanding logarithms**

Write each expression in terms of logarithms of x, y, and z.

(a) $\log \dfrac{x^2 y^3}{\sqrt{z}}$ **(b)** $\ln \sqrt[3]{\dfrac{xy}{z}}$

Solution

(a) $\log \dfrac{x^2 y^3}{\sqrt{z}} = \log x^2 y^3 - \log \sqrt{z}$ Quotient rule

$\qquad\qquad = \log x^2 + \log y^3 - \log z^{1/2}$ Product rule; $\sqrt{z} = z^{1/2}$

$\qquad\qquad = 2 \log x + 3 \log y - \dfrac{1}{2} \log z$ Power rule

(b) $\ln \sqrt[3]{\dfrac{xy}{z}} = \ln \left(\dfrac{xy}{z} \right)^{1/3}$ $\sqrt[3]{m} = m^{1/3}$

$\qquad\qquad = \dfrac{1}{3} \ln \dfrac{xy}{z}$ Power rule

$\qquad\qquad = \dfrac{1}{3} (\ln xy - \ln z)$ Quotient rule

$\qquad\qquad = \dfrac{1}{3} (\ln x + \ln y - \ln z)$ Product rule

$\qquad\qquad = \dfrac{1}{3} \ln x + \dfrac{1}{3} \ln y - \dfrac{1}{3} \ln z$ Distributive property

Now Try Exercises 63, 65

EXAMPLE 8   **Applying properties of logarithms**

Using only properties of logarithms and the approximations $\ln 2 \approx 0.7$, $\ln 3 \approx 1.1$, and $\ln 5 \approx 1.6$, find an approximation for each expression.

(a) $\ln 8$ **(b)** $\ln 15$ **(c)** $\ln \dfrac{10}{3}$

Solution

(a) $\ln 8 = \ln 2^3 = 3 \ln 2 \approx 3(0.7) = 2.1$

(b) $\ln 15 = \ln(3 \cdot 5) = \ln 3 + \ln 5 \approx 1.1 + 1.6 = 2.7$

(c) $\ln \dfrac{10}{3} = \ln \left(\dfrac{2 \cdot 5}{3} \right) = \ln 2 + \ln 5 - \ln 3 \approx 0.7 + 1.6 - 1.1 = 1.2$

Now Try Exercises 77, 79, 83

Change of Base Formula

▶ **REAL-WORLD CONNECTION**   Most calculators only have keys to evaluate common and natural logarithms. Occasionally, it is necessary to evaluate a logarithmic function with a base other than 10 or e. In these situations we use the following change of base formula, which we illustrate in the next example.

CHANGE OF BASE FORMULA

Let x and $a \neq 1$ be positive real numbers. Then

$$\log_a x = \frac{\log x}{\log a} \quad \text{or} \quad \log_a x = \frac{\ln x}{\ln a}.$$

EXAMPLE 9   Change of base formula

Approximate $\log_2 14$ to the nearest thousandth.

```
log(14)/log(2)
      3.807354922
ln(14)/ln(2)
      3.807354922
```

Figure 12.34

Solution

Using the change of base formula,

$$\log_2 14 = \frac{\log 14}{\log 2} \approx 3.807 \quad \text{or} \quad \log_2 14 = \frac{\ln 14}{\ln 2} \approx 3.807.$$

Figure 12.34 supports these results.

Now Try Exercise **89**

12.4 PUTTING IT ALL TOGETHER

The following table summarizes some important properties for base-a logarithms. Common and natural logarithms satisfy the same properties.

Concepts	Description	Examples
Properties of Logarithms	The following properties hold for positive numbers m, n, and $a \neq 1$ and for any real number r.	
1. Product Rule	**1.** $\log_a mn = \log_a m + \log_a n$	**1.** $\log 20 = \log 10 + \log 2$
2. Quotient Rule	**2.** $\log_a \frac{m}{n} = \log_a m - \log_a n$	**2.** $\log \frac{45}{6} = \log 45 - \log 6$
3. Power Rule	**3.** $\log_a (m^r) = r \log_a m$	**3.** $\ln x^6 = 6 \ln x$
Change of Base Formula	Let x and $a \neq 1$ be positive numbers. Then $\log_a x = \dfrac{\log x}{\log a}$ and $\log_a x = \dfrac{\ln x}{\ln a}$.	The expression $\log_3 6$ is equivalent to either $\dfrac{\log 6}{\log 3}$ or $\dfrac{\ln 6}{\ln 3}$.

12.4 Exercises

NOTE: Assume that variables are positive and that expressions are defined in this exercise set.

CONCEPTS

1. $\log 12 = \log 3 + \log (\underline{\quad})$

2. $\ln 5 = \ln 20 - \ln (\underline{\quad})$

3. $\log 8 = (\underline{\quad}) \log 2$

4. $\log mn = \underline{\quad}$

5. $\log \dfrac{m}{n} = \underline{\quad}$

6. $\log (m^r) = \underline{\quad}$

7. Does $\log x + \log y$ equal $\log (x + y)$?

8. Does $\log x - \log y$ equal $\log \left(\frac{x}{y}\right)$?

9. Does $\log (xy)$ equal $(\log x)(\log y)$?

10. Does $\log \left(\frac{x}{y}\right)$ equal $\frac{\log x}{\log y}$?

11. Give the change of base formula.

12. $\log_a 1 = \underline{\quad}$ and $\log_a a = \underline{\quad}$.

BASIC PROPERTIES OF LOGARITHMS

Exercises 13–18: Write the expression as a sum of two or more logarithms.

13. $\ln(3 \cdot 5)$ **14.** $\log(7 \cdot 11)$

15. $\log_3 xy$ **16.** $\log_5 y^2$

17. $\ln 10z$ **18.** $\log x^2 y$

Exercises 19–24: Write the expression as a difference of two logarithms.

19. $\log \frac{7}{3}$ **20.** $\ln \frac{11}{13}$

21. $\ln \frac{x}{y}$ **22.** $\log \frac{2x}{z}$

23. $\log_2 \frac{45}{x}$ **24.** $\log_7 \frac{5x}{4z}$

Exercises 25–32: Write the expression as one logarithm.

25. $\log 45 + \log 5$ **26.** $\log 30 - \log 10$

27. $\ln x + \ln y$ **28.** $\ln m + \ln n - \ln n$

29. $\ln 7x^2 + \ln 2x$ **30.** $\ln x + \ln y - \ln z$

31. $\ln x + \ln y^2 - \ln y$

32. $\ln \sqrt{z} - \ln z^3 + \ln y^3$

Exercises 33–38: Write the expression as one term.

33. $\log 20 - \log 4$ **34.** $\log 900 - \log 9$

35. $\ln x^4 - \ln x^2$ **36.** $\ln 9x^2 - \ln 3x$

37. $\log_2 12x - \log_2 3x$ **38.** $\log_5 18x^2 - \log_5 2x^2$

Exercises 39–48: Rewrite using the power rule.

39. $\log 3^6$ **40.** $\log x^7$

41. $\ln 2^x$ **42.** $\ln(0.77)^{x+1}$

43. $\log_2 5^{1/4}$ **44.** $\log_3 \sqrt{x}$

45. $\log_4 \sqrt[3]{z}$ **46.** $\log_7 3^\pi$

47. $\log x^{y-1}$ **48.** $\ln a^{2b}$

Exercises 49–60: Use properties of logarithms to write the expression as the logarithm of a single expression.

49. $4\log z - \log z^3$ **50.** $2\log_5 y + \log_5 x$

51. $\log x + 2\log x + 2\log y$

52. $\log x^2 + 3\log z - 5\log y$

53. $\log x - 2\log \sqrt{x}$ **54.** $\ln y^2 - 6\ln \sqrt[3]{y}$

55. $\ln 2^{x+1} - \ln 2$ **56.** $\ln 8^{1/2} + \ln 2^{1/2}$

57. $2\log_3 \sqrt{x} - 3\log_3 x$ **58.** $\ln \sqrt[3]{x} + \ln \sqrt{x}$

59. $2\log_a(x+1) - \log_a(x^2-1)$

60. $\log_b(x^2-9) - \log_b(x-3)$

Exercises 61–72: Use properties of logarithms to write the expression in terms of logarithms of x, y, and z.

61. $\log xy^2$ **62.** $\log \frac{x^2}{y^3}$

63. $\ln \frac{x^4 y}{z}$ **64.** $\ln \frac{\sqrt{x}}{y}$

65. $\log_4 \frac{\sqrt[3]{z}}{\sqrt{y}}$ **66.** $\log_2 \sqrt{\frac{x}{y}}$

67. $\log(x^4 y^3)$ **68.** $\log(x^2 y^4 z^3)$

69. $\ln \frac{1}{y} - \ln \frac{1}{x}$ **70.** $\ln \frac{1}{xy}$

71. $\log_4 \sqrt{\frac{x^3 y}{z^2}}$ **72.** $\log_3 \left(\frac{x^2 \sqrt{z}}{y^3}\right)$

Exercises 73–76: Graph f and g in the window $[-6, 6, 1]$ by $[-4, 4, 1]$. If the two graphs appear to be identical, prove that they are, using properties of logarithms.

73. $f(x) = \log x^3, g(x) = 3\log x$

74. $f(x) = \ln x + \ln 3, g(x) = \ln 3x$

75. $f(x) = \ln(x + 5), g(x) = \ln x + \ln 5$

76. $f(x) = \log(x - 2), g(x) = \log x - \log 2$

Exercises 77–86: (Refer to Example 8.) Using only properties of logarithms and the approximations $\log 2 \approx 0.3$, $\log 5 \approx 0.7$, and $\log 13 \approx 1.1$, find an approximation for the expression.

77. $\log 16$ **78.** $\log 125$

79. $\log 65$ **80.** $\log 26$

81. $\log 130$ **82.** $\log 100$

83. $\log \frac{5}{2}$ **84.** $\log \frac{26}{5}$

85. $\log \frac{1}{13}$ **86.** $\log \frac{1}{65}$

CHANGE OF BASE FORMULA

Exercises 87–92: Use the change of base formula to approximate each expression to the nearest hundredth.

87. $\log_3 5$

88. $\log_5 12$

89. $\log_2 25$

90. $\log_7 8$

91. $\log_9 102$

92. $\log_6 293$

APPLICATIONS

93. *Modeling Sound* (See Example 4, Section 12.3.) The formula $f(x) = 10 \log(10^{16} x)$ can be used to calculate the decibel level of a sound with an intensity x. Use properties of logarithms to simplify this formula to $f(x) = 160 + 10 \log x$.

94. *Cellular Phone Technology* A formula used to calculate the strength of a signal for a cellular phone is

$$L = 110.7 - 19.1 \log h + 55 \log d,$$

where h is the height of the cellular phone tower and d is the distance the phone is from the tower. Use properties of logarithms to write an expression for L that contains only one logarithm. (*Source:* C. Smith, *Practical Cellular & PCS Design.*)

WRITING ABOUT MATHEMATICS

95. State the three basic properties of logarithms and give an example of each.

96. A student insists that $\log(x - y)$ is equal to the expression $\log x - \log y$. How could you convince the student otherwise?

CHECKING BASIC CONCEPTS

SECTIONS 12.3 AND 12.4

1. Simplify each expression by hand.
   (a) $\log 10^4$ **(b)** $\ln e^x$
   (c) $\log_2 \frac{1}{8}$ **(d)** $\log_5 \sqrt{5}$

2. Sketch a graph of $f(x) = \log x$.
   (a) What are the domain and range of f?
   (b) Evaluate $f(1)$.
   (c) Can the common logarithm of a positive number be negative? Explain.
   (d) Can the common logarithm of a negative number be positive? Explain.

3. Write in terms of logarithms of x, y, and z. Assume that variables are positive.
   (a) $\log xy$ **(b)** $\ln \dfrac{x}{yz}$
   (c) $\ln x^2$ **(d)** $\log \dfrac{x^2 y^3}{\sqrt{z}}$

4. Write as the logarithm of a single expression.
   (a) $\log x + \log y$
   (b) $\ln 2x - 3 \ln y$
   (c) $2 \log_2 x + 3 \log_2 y - \log_2 z$

12.5 EXPONENTIAL AND LOGARITHMIC EQUATIONS

Exponential Equations and Models ▪ Logarithmic Equations and Models

A LOOK INTO MATH ▷

Although we have solved many equations throughout this course, one equation that we have not solved *symbolically* is $a^x = k$. This exponential equation occurs frequently in applications and is used to model either exponential growth or decay. For example, one bluefin tuna can be worth over $30,000. As a result, the numbers of bluefin tuna declined exponentially from 1974 to 1991. (See Exercise 99.) Logarithmic equations contain logarithms and are also used in modeling real-world data. To solve exponential equations, we use logarithms; to solve logarithmic equations, we use exponential expressions.

Exponential Equations and Models

To solve the equation $10 + x = 100$, we subtract 10 from each side because addition and subtraction are inverse operations.

$$10 + x - \mathbf{10} = 100 - \mathbf{10}$$
$$x = 90$$

To solve the equation $10x = 100$, we divide each side by 10 because multiplication and division are inverse operations.

$$\frac{10x}{\mathbf{10}} = \frac{100}{\mathbf{10}}$$
$$x = 10$$

Now suppose that we want to solve the exponential equation

$$10^x = 100.$$

What is new about this type of equation is that the variable x is an *exponent*. The inverse operation of 10^x is log x. Rather than subtracting 10 from each side or dividing each side by 10, we take the base-10 logarithm of each side. Doing so results in

$$\mathbf{log}\ 10^x = \mathbf{log}\ 100.$$

Because $\log 10^x = x$ for all real numbers x, the equation becomes

$$x = \log 100, \quad \text{or equivalently,} \quad x = 2.$$

These concepts are applied in the next example.

$\boxed{\text{EXAMPLE}\ \ 1}$ **Solving exponential equations**

Solve.
(a) $10^x = 150$ **(b)** $e^x = 40$ **(c)** $2^x = 50$ **(d)** $0.9^x = 0.5$

Solution
(a) $10^x = 150$ Given equation

$\mathbf{log}\ 10^x = \mathbf{log}\ 150$ Take the common logarithm of each side.

$x = \log 150 \approx 2.18$ Inverse property: $\log 10^k = k$ for all k

(b) The inverse operation of e^x is ln x, so we take the natural logarithm of each side.

$e^x = 40$ Given equation

$\mathbf{ln}\ e^x = \mathbf{ln}\ 40$ Take the natural logarithm of each side.

$x = \ln 40 \approx 3.69$ Inverse property: $\ln e^k = k$ for all k

(c) The inverse operation of 2^x is $\log_2 x$. Calculators do not usually have a base-2 logarithm key, so we take the common logarithm of each side and then apply the power rule.

$2^x = 50$ Given equation

$\mathbf{log}\ 2^x = \mathbf{log}\ 50$ Take the common logarithm of each side.

$x \log 2 = \log 50$ Power rule: $\log (m^r) = r \log m$

$x = \dfrac{\log 50}{\log 2} \approx 5.64$ Divide by log 2 and approximate.

(d) This time we begin by taking the natural logarithm of each side.

$$0.9^x = 0.5 \qquad \text{Given equation}$$

$$\ln 0.9^x = \ln 0.5 \qquad \text{Take the natural logarithm of each side.}$$

$$x \ln 0.9 = \ln 0.5 \qquad \text{Power rule: } \ln (m^r) = r \ln m$$

$$x = \frac{\ln 0.5}{\ln 0.9} \approx 6.58 \qquad \text{Divide by } \ln 0.9 \text{ and approximate.}$$

Now Try Exercises 13, 15, 25, 27

MAKING CONNECTIONS

Logarithms of Quotients and Quotients of Logarithms

The solution in Example 1(c) is $\frac{\log 50}{\log 2}$. Note that

$$\frac{\log 50}{\log 2} \neq \log 50 - \log 2.$$

```
log(50)/log(2)
          5.64385619
log(50)-log(2)
          1.397940009
log(25)
          1.397940009
```

However, $\log 50 - \log 2 = \log \frac{50}{2} = \log 25$ by the quotient rule for logarithms, as shown in the figure.

The next two examples illustrate methods for solving exponential equations.

EXAMPLE 2   **Solving exponential equations**

Solve each equation.
(a) $2e^x - 1 = 5$ **(b)** $3^{x-5} = 15$ **(c)** $e^{2x} = e^{x+5}$ **(d)** $3^{2x} = 2^{x+3}$

Solution
(a) Begin by solving for e^x.

$$2e^x - 1 = 5 \qquad \text{Given equation}$$

$$2e^x = 6 \qquad \text{Add 1 to each side.}$$

$$e^x = 3 \qquad \text{Divide each side by 2.}$$

$$\ln e^x = \ln 3 \qquad \text{Take the natural logarithm.}$$

$$x = \ln 3 \approx 1.10 \qquad \text{Inverse property: } \ln e^k = k$$

(b) Start by taking the common logarithm of each side. (We could also take the natural logarithm of each side.)

$$3^{x-5} = 15 \qquad \text{Given equation}$$

$$\log 3^{x-5} = \log 15 \qquad \text{Take the common logarithm of each side.}$$

$$(x - 5) \log 3 = \log 15 \qquad \text{Power rule for logarithms}$$

$$x - 5 = \frac{\log 15}{\log 3} \qquad \text{Divide by } \log 3.$$

$$x = \frac{\log 15}{\log 3} + 5 \approx 7.46 \qquad \text{Add 5 to each side and approximate.}$$

(c) In $e^{2x} = e^{x+5}$, the bases are equal, so the exponents must also be equal. To verify this assertion, take the natural logarithm of each side.

$$e^{2x} = e^{x+5} \qquad \text{Given equation}$$
$$\ln e^{2x} = \ln e^{x+5} \qquad \text{Take the natural logarithm.}$$
$$2x = x + 5 \qquad \text{Inverse property: } \ln e^{k} = k$$
$$x = 5 \qquad \text{Subtract } x.$$

(d) In $3^{2x} = 2^{x+3}$, the bases are not equal. However, we can still solve the equation by taking the common logarithm of each side. A logarithm of any base could be used.

$$3^{2x} = 2^{x+3} \qquad \text{Given equation}$$
$$\log 3^{2x} = \log 2^{x+3} \qquad \text{Take the common logarithm.}$$
$$2x \log 3 = (x + 3) \log 2 \qquad \text{Power rule for logarithms}$$
$$2x \log 3 = x \log 2 + 3 \log 2 \qquad \text{Distributive property}$$
$$2x \log 3 - x \log 2 = 3 \log 2 \qquad \text{Subtract } x \log 2.$$
$$x(2 \log 3 - \log 2) = 3 \log 2 \qquad \text{Factor out } x.$$
$$x = \frac{3 \log 2}{2 \log 3 - \log 2} \qquad \text{Divide by } 2 \log 3 - \log 2.$$
$$x \approx 1.38 \qquad \text{Approximate.}$$

Now Try Exercises 29, 31, 35, 43

EXAMPLE **3**   **Solving an exponential equation**

Graphs for $f(x) = 0.2e^x$ and $g(x) = 4$ are shown in Figure 12.35.
(a) Use the graphs to estimate the solution to the equation $f(x) = g(x)$.
(b) Check your estimate by solving the equation symbolically.

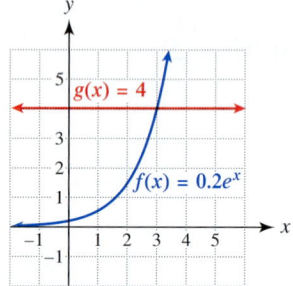

Figure 12.35

Solution
(a) The graphs intersect near the point $(3, 4)$. Therefore the solution is given by $x \approx 3$.
(b) We must solve the equation $0.2e^x = 4$.

$$0.2e^x = 4 \qquad \text{Given equation}$$
$$e^x = 20 \qquad \text{Divide each side by 0.2.}$$
$$\ln e^x = \ln 20 \qquad \text{Take the natural logarithm of each side.}$$
$$x = \ln 20 \qquad \text{Inverse property: } \ln e^k = k$$
$$x \approx 2.996 \qquad \text{Approximate.}$$

NOTE: The graphical estimate did not give the *exact* solution of $\ln 20$.

Now Try Exercise 45

▶ **REAL-WORLD CONNECTION**   In Section 12.2 we showed that if $1000 are deposited in a savings account paying 10% annual interest at the end of each year, the amount A in the account after x years is given by

$$A(x) = 1000(1.1)^x.$$

After 10 years there will be

$$A(\mathbf{10}) = 1000(1.1)^{\mathbf{10}} \approx \$2593.74$$

in the account. To calculate how long it will take for $4000 to accrue in the account, we need to solve the exponential equation

$$1000(1.1)^x = 4000.$$

We do so in the next example.

EXAMPLE 4 Solving exponential equations

Solve $1000(1.1)^x = 4000$ symbolically. Give graphical support for your answer.

Solution
Symbolic Solution Begin by dividing each side of the equation by 1000.

$1000(1.1)^x = 4000$	Given equation
$1.1^x = 4$	Divide by 1000.
$\log 1.1^x = \log 4$	Take the common logarithm of each side.
$x \log 1.1 = \log 4$	Power rule for logarithms
$x = \dfrac{\log 4}{\log 1.1} \approx 14.5$	Divide by log 1.1 and approximate.

[0, 20, 5] by [0, 6000, 1000]

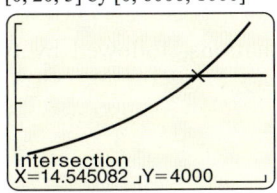

Intersection
X=14.545082 Y=4000

Figure 12.36

Interest is paid at the end of the year, so it will take 15 years for $1000 earning 10% annual interest to grow to (at least) $4000.

Graphical Solution Graphical support is shown in Figure 12.36, where the graphs of $Y_1 = 1000*1.1\text{^}X$ and $Y_2 = 4000$ intersect when $x \approx 14.5$. **Now Try Exercise 93**

▶ **REAL-WORLD CONNECTION** In the next example we model the life span of a robin with an exponential function.

EXAMPLE 5 Modeling the life spans of robins

The life spans of 129 robins were monitored over a 4-year period in one study. The formula $f(x) = 10^{-0.42x}$ can be used to calculate the percentage of robins remaining after x years. For example, $f(1) \approx 0.38$ means that after 1 year 38% of the robins were still alive. (*Source:* D. Lack, *The Life Span of a Robin.*)

(a) Evaluate $f(2)$ and interpret the result.
(b) Determine when 5% of the robins remained.

Solution
(a) $f(2) = 10^{-0.42(2)} \approx 0.145$. After 2 years about 14.5% of the robins were still alive.
(b) Use $5\% = 0.05$ and solve the following equation.

$10^{-0.42x} = 0.05$	Equation to solve
$\log 10^{-0.42x} = \log 0.05$	Take the common logarithm of each side.
$-0.42x = \log 0.05$	Inverse property: $\log 10^k = k$
$x = \dfrac{\log 0.05}{-0.42} \approx 3.1$	Divide by -0.42.

After about 3 years only 5% of the robins were still alive. **Now Try Exercise 96**

Logarithmic Equations and Models

To solve an exponential equation we use logarithms. To solve a logarithmic equation we *exponentiate* each side of the equation. To do so we use the fact that if $x = y$, then $a^x = a^y$ for any positive base a. For example, to solve

$$\log x = 3$$

we exponentiate each side of the equation, using base 10.

$$10^{\log x} = 10^3$$

Because $10^{\log x} = x$ for all positive x,

$$x = 10^3 = 1000.$$

To solve logarithmic equations, we frequently use the inverse property

$$a^{\log_a x} = x.$$

Examples of this inverse property include

$$e^{\ln 2k} = 2k, \quad 2^{\log_2 x} = x, \quad \text{and} \quad 10^{\log (x+5)} = x + 5.$$

The next two examples show how to solve logarithmic equations, followed by two applications of these methods.

EXAMPLE 6   Solving logarithmic equations

Solve and approximate solutions to the nearest hundredth when appropriate.
(a) $2 \log x = 4$ **(b)** $\ln 3x = 5.5$ **(c)** $\log_2 (x + 4) = 7$

Solution

(a)

$2 \log x = 4$	Given equation
$\log x = 2$	Divide each side by 2.
$10^{\log x} = 10^2$	Exponentiate each side, using base 10.
$x = 100$	Inverse property: $10^{\log k} = k$

(b)

$\ln 3x = 5.5$	Given equation
$e^{\ln 3x} = e^{5.5}$	Exponentiate each side, using base e.
$3x = e^{5.5}$	Inverse property: $e^{\ln k} = k$
$x = \dfrac{e^{5.5}}{3} \approx 81.56$	Divide each side by 3 and approximate.

(c)

$\log_2 (x + 4) = 7$	Given equation
$2^{\log_2 (x+4)} = 2^7$	Exponentiate each side, using base 2.
$x + 4 = 2^7$	Inverse property: $2^{\log_2 k} = k$
$x = 2^7 - 4$	Subtract 4 from each side.
$x = 124$	Simplify.

Now Try Exercises 61, 63, 69

Because the domain of any logarithmic function includes only positive numbers, it is important to check answers, as emphasized in the next example.

EXAMPLE 7   Solving a logarithmic equation

Solve $\log(x + 2) + \log(x - 2) = \log 5$. Check any answers.

Solution
Start by applying the product rule for logarithms.

$$\log(x + 2) + \log(x - 2) = \log 5 \qquad \text{Given equation}$$
$$\log((x + 2)(x - 2)) = \log 5 \qquad \text{Product rule}$$
$$\log(x^2 - 4) = \log 5 \qquad \text{Multiply.}$$
$$10^{\log(x^2-4)} = 10^{\log 5} \qquad \text{Exponentiate using base 10.}$$
$$x^2 - 4 = 5 \qquad \text{Inverse properties}$$
$$x^2 = 9 \qquad \text{Add 4.}$$
$$x = \pm 3 \qquad \text{Square root property}$$

Check each answer.

$$\log(\mathbf{3} + 2) + \log(\mathbf{3} - 2) \stackrel{?}{=} \log 5 \qquad \log(\mathbf{-3} + 2) + \log(\mathbf{-3} - 2) \stackrel{?}{=} \log 5$$
$$\log 5 + \log 1 \stackrel{?}{=} \log 5 \qquad\qquad \log(-1) + \log(-5) \neq \log 5$$
$$\log 5 + 0 \stackrel{?}{=} \log 5 \qquad\qquad\qquad\qquad \text{Undefined}$$
$$\log 5 = \log 5$$

Although 3 is a solution -3 is not, because both $\log(-1)$ and $\log(-5)$ are undefined expressions. Be sure to check your answers.   **Now Try Exercise 77**

EXAMPLE 8   Modeling runway length

For some types of airplanes with weight x, the minimum runway length L required at takeoff is modeled by

$$L(x) = 3 \log x.$$

In this equation L is measured in thousands of feet and x is measured in thousands of pounds. Estimate the weight of the heaviest airplane that can take off from a runway 5100 feet long.   (**Source:** L. Haefner, *Introduction to Transportation Systems.*)

Solution
Runway length is measured in thousands of feet, so we must solve the equation $L(x) = 5.1$.

$$3 \log x = 5.1 \qquad L(x) = 5.1$$
$$\log x = 1.7 \qquad \text{Divide each side by 3.}$$
$$10^{\log x} = 10^{1.7} \qquad \text{Exponentiate each side, using base 10.}$$
$$x = 10^{1.7} \qquad \text{Inverse property: } 10^{\log k} = k$$
$$x \approx 50.1 \qquad \text{Approximate.}$$

The largest airplane that can take off from this runway weighs about 50,000 pounds.

Now Try Exercise 97

CRITICAL THINKING

In Example 9, Section 12.3, we used the formula $L(x) = 1.3 \ln x$ to model runway length. Are $L(x) = 1.3 \ln x$ and $L(x) = 3 \log x$ equivalent formulas? Explain.

EXAMPLE   **Modeling bird populations**

Near New Guinea there is a relationship between the number of species of birds and the size of an island. Larger islands tend to have a greater variety of birds. Table 12.13 lists the number of species of birds y found on islands with an area of x square kilometers.

TABLE 12.13

x (km^2)	0.1	1	10	100	1000
y (species)	10	15	20	25	30

Source: B. Freedman, *Environmental Ecology.*

(a) Find values for the constants a and b so that $y = a + b \log x$ models the data.
(b) Predict the number of bird species on an island of 4000 square kilometers.

Solution
(a) Because $\log 1 = 0$, substitute $x = 1$ and $y = 15$ in the equation to find a.

$$15 = a + b \,\mathbf{log\ 1}$$
$$15 = a + b \cdot \mathbf{0}$$
$$15 = a$$

Thus $y = 15 + b \log x$. To find b, substitute $x = 10$ and $y = 20$.

$$20 = 15 + b \,\mathbf{log\ 10}$$
$$20 = 15 + b \cdot \mathbf{1}$$
$$5 = b$$

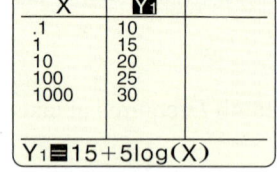

Figure 12.37

The data in Table 12.13 are modeled by $y = 15 + 5 \log x$. This result is supported by Figure 12.37.

(b) To predict the number of species on an island of 4000 square kilometers, let $x = \mathbf{4000}$ and find y.

$$y = 15 + 5 \log \mathbf{4000} \approx 33$$

The model estimates about 33 species of birds on this island.

Now Try Exercise 101

12.5 PUTTING IT ALL TOGETHER

Basic steps for solving exponential and logarithmic equations are summarized in the following table.

Type of Equation	Procedure	Example	
Exponential	Begin by solving for the exponential expression a^x. Then take a logarithm of each side.	$4e^x + 1 = 9$	Given equation
		$e^x = 2$	Solve for e^x.
		$\ln e^x = \ln 2$	Take the natural logarithm.
		$x = \ln 2$	Inverse property: $\ln e^k = k$

Type of Equation	Procedure	Example	
Logarithmic	Begin by solving for the logarithm in the equation. Then exponentiate each side of the equation, using the same base as the logarithm. That is, if $x = y$, then $a^x = a^y$ for any positive base a.	$\frac{1}{3}\log 2x = 1$	Given equation
		$\log 2x = 3$	Multiply by 3.
		$10^{\log 2x} = 10^3$	Exponentiate using base 10.
		$2x = 1000$	Inverse property: $10^{\log k} = k$
		$x = 500$	Divide by 2.

12.5 Exercises

 MyMathLab Math XL PRACTICE WATCH DOWNLOAD READ REVIEW

CONCEPTS

1. To solve $x - 5 = 50$, what should be done?

2. To solve $5x = 50$, what should be done?

3. To solve $10^x = 50$, what should be done?

4. To solve $\log x = 5$, what should be done?

5. $\log 10^x = $ _____

6. $10^{\log x} = $ _____

7. $\ln e^{2x} = $ _____

8. $e^{\ln (x+7)} = $ _____

9. Does $\frac{\log 5}{\log 4}$ equal $\log \frac{5}{4}$? Explain.

10. Does $\frac{\log 5}{\log 4}$ equal $\log 5 - \log 4$? Explain.

11. How many solutions are there to $\log x = k$, where k is any real number?

12. How many solutions are there to $10^x = k$, where k is a positive number?

EXPONENTIAL EQUATIONS

Exercises 13–44: Solve the exponential equation. Approximate answers to the nearest hundredth when appropriate.

13. $10^x = 1000$

14. $10^x = 0.01$

15. $2^x = 64$

16. $3^x = 27$

17. $2^{x-3} = 8$

18. $3^{2x} = 81$

19. $4^x + 3 = 259$

20. $3(5^{2x}) = 300$

21. $10^{0.4x} = 124$

22. $0.75^x = 0.25$

23. $e^{-x} = 1$

24. $0.5^{-5x} = 5$

25. $e^x = 25$

26. $e^x = 0.4$

27. $0.4^x = 2$

28. $0.7^x = 0.3$

29. $e^x - 1 = 6$

30. $2e^{4x} = 15$

31. $2(10)^{x+2} = 35$

32. $10^{3x} + 10 = 1500$

33. $3.1^{2x} - 4 = 16$

34. $5.4^{x-1} = 85$

35. $e^{3x} = e^{2x-1}$

36. $e^{x^2} = e^{3x-2}$

37. $5^{4x} = 5^{x^2-5}$

38. $2^{4x} = 2^{x+3}$

39. $e^{2x} \cdot e^x = 10$

40. $10^{x-2} \cdot 10^x = 1000$

41. $e^x = 2^{x+2}$

42. $2^{2x} = 3^{x-1}$

43. $4^{0.5x} = 5^{x+2}$

44. $3^{2x} = 7^{x+1}$

Exercises 45–48: (Refer to Example 3.) The symbolic and graphical representations of f and g are given.

(a) *Use the graph to solve* $f(x) = g(x)$.

(b) *Solve* $f(x) = g(x)$ *symbolically.*

45. $f(x) = 0.2(10^x)$,
    $g(x) = 2$

46. $f(x) = e^x$,
    $g(x) = 7.4$

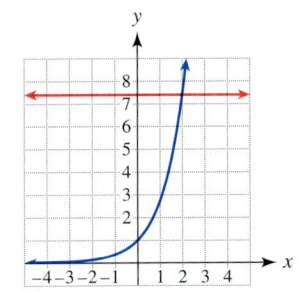

47. $f(x) = 2^{-x}$,
    $g(x) = 4$

48. $f(x) = 0.1(3^x)$,
    $g(x) = 0.9$

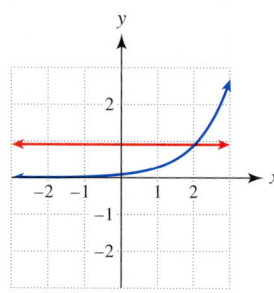

77. $\ln x + \ln(x + 1) = \ln 30$

78. $\log(x - 1) + \log(2x + 1) = \log 14$

79. $\log_3 3x - \log_3(x + 2) = \log_3 2$

80. $\log_4(x^2 - 1) - \log_4(x - 1) = \log_4 6$

81. $\log_2(x - 1) + \log_2(x + 1) = 3$

82. $\log_4(x^2 + 2x + 1) - \log_4(x + 1) = 2$

Exercises 49–56: Solve the equation symbolically. Give graphical or numerical support. Approximate answers to the nearest hundredth when appropriate.

49. $10^x = 0.1$

50. $2(10^x) = 2000$

51. $4e^x + 5 = 9$

52. $e^x + 6 = 36$

53. $4^x = 1024$

54. $3^x = 729$

55. $(0.55)^x + 0.55 = 2$

56. $5(0.9)^x = 3$

 Exercises 57–60: The given equation cannot be solved symbolically. Find any solutions either graphically or numerically to the nearest hundredth.

57. $e^x - x = 2$

58. $x \log x = 1$

59. $\ln x = e^{-x}$

60. $10^x - 2 = \log(x + 2)$

LOGARITHMIC EQUATIONS

Exercises 61–82: Solve the equation. Approximate answers to the nearest hundredth when appropriate.

61. $\log x = 2$

62. $\log x = 0.01$

63. $\ln x = 5$

64. $2 \ln x = 4$

65. $\log 2x = 7$

66. $6 \ln 4x = 12$

67. $\log_2 x = 4$

68. $\log_2 x = 32$

69. $\log_2 5x = 2.3$

70. $2 \log_3 4x = 10$

71. $2 \log x + 5 = 7.8$

72. $\ln(x - 1) = 3.3$

73. $5 \ln(2x + 1) = 55$

74. $5 - \log(x + 3) = 2.6$

75. $\log x^2 = \log x$

76. $\ln x^2 = \ln(3x - 2)$

Exercises 83–86: Two functions, f and g, are given.
    (a) Use the graph to solve $f(x) = g(x)$.
    (b) Solve $f(x) = g(x)$ symbolically.

83. $f(x) = \ln x$,
    $g(x) = 0.7$

84. $f(x) = \log_2 x$,
    $g(x) = 1.6$

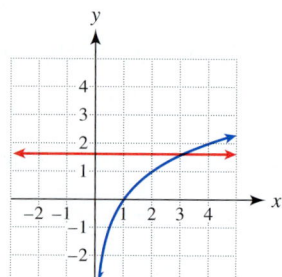

85. $f(x) = 5 \log 2x$,
    $g(x) = 3$

86. $f(x) = 2 \ln(x) - 3$,
    $g(x) = 0.9$

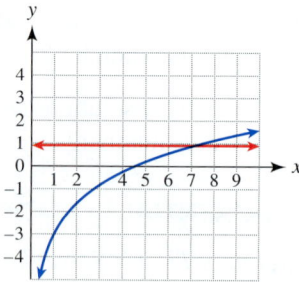

Exercises 87–92: Solve the equation symbolically. Give graphical or numerical support. Approximate answers to the nearest hundredth.

87. $\log x = 1.6$

88. $\ln x = 2$

89. $\ln(x + 1) = 1$

90. $2 \log(2x + 3) = 8$

91. $17 - 6 \log_3 x = 5$

92. $4 \log_2 x + 7 = 12$

APPLICATIONS

93. *Growth of a Mutual Fund* (Refer to Example 4.) An investor deposits $2000 in a mutual fund that returns 15% at the end of 1 year. Determine the length of time required for the investment to triple its value if the annual rate of return remains the same.

94. *Savings Account* (Refer to Example 4.) If a savings account pays 6% annual interest at the end of each year, how many years will it take for the account to double in value?

95. *Liver Transplants* In the United States the gap between available organs for liver transplants and people who need them has widened. The number of individuals waiting for liver transplants can be modeled by

$$f(x) = 2339(1.24)^{x-1988},$$

where x is the year. (**Source:** United Network for Organ Sharing.)
(a) Evaluate $f(1994)$ and interpret the result.
(b) Determine when the number of individuals waiting for liver transplants was 20,000.

96. *Life Span of a Robin* (Refer to Example 5.) Determine when 50% of the robins in the study were still alive.

97. *Runway Length* (Refer to Example 8.) Determine the weight of the heaviest airplane that can take off from a runway having a length of $\frac{3}{4}$ mile. (*Hint:* 1 mile = 5280 feet.)

98. *Runway Length* (Refer to Example 8.)
(a) Suppose that an airplane is 10 times heavier than a second airplane. How much longer should the runway be for the heavier airplane than for the lighter airplane? (*Hint:* Let the heavier airplane have weight 10x.)
(b) If the runway length is increased by 3000 feet, by what factor can the weight of an airplane that uses the runway be increased?

99. *The Decline of Bluefin Tuna* Bluefin tuna are large fish that can weigh 1500 pounds and swim at a speed of 55 miles per hour. They are used for sushi, and a prime fish can be worth more than $30,000. As a result, the number of western Atlantic bluefin tuna has declined dramatically. Their numbers in thousands between 1974 and 1991 can be modeled by $f(x) = 230(10^{-0.055x})$, where x is the number of years after 1974. See the accompanying graph. (**Source:** B. Freedman.)

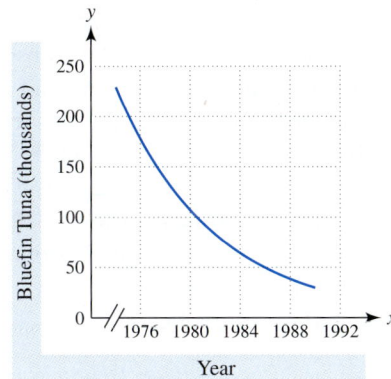

(a) Evaluate $f(1)$ and interpret the result.
(b) Use the graph to estimate the year in which bluefin tuna numbered 115 thousand.
(c) Solve part (b) symbolically.

100. *Insect Populations* (Refer to Example 9.) The table lists numbers of species of insects y found on islands having areas of x square miles.

x (square miles)	1	2	4
y (species)	1000	1500	2000

x (square miles)	8	16
y (species)	2500	3000

(a) Find values for constants a and b so that $y = a + b \log_2 x$ models the data.
(b) Construct a table for y and verify that your equation models the data.
(c) Estimate the number of species of insects on an island having an area of 12 square miles.

Exercises 101 and 102: (Refer to Example 9.) Find values for a and b so that $y = a + b \log x$ models the data.

101.

x	0.1	1	10	100
y	22	25	28	31

102.

x	0.01	1	100	1000
y	−10	−2	6	10

103. *Calories Consumed and Land Ownership* In developing countries there is a relationship between the amount of land a person owns and the average number of calories consumed daily. This relationship is modeled by $f(x) = 645 \log(x + 1) + 1925$, where x is the amount of land owned in acres and $0 \le x \le 4$. (***Source:*** D. Grigg, *The World Food Problem.*)

(a) Estimate graphically the number of acres owned by a person consuming 2200 calories per day.
(b) Solve part (a) symbolically.

104. *Population of Industrialized Urban Regions* The number of people living in industrialized urban regions throughout the world has not grown exponentially. Instead it has grown logarithmically and is modeled by

$$f(x) = 0.36 + 0.15 \ln(x - 1949).$$

In this formula the output is billions of people and the input x is the year, where $1950 \le x \le 2030$. (***Source:*** D. Meadows, *Beyond The Limits.*)

(a) Determine either graphically or numerically when this population may reach 1 billion.
(b) Solve part (a) symbolically.

105. *Fertilizer Use* Between 1950 and 1980, the use of chemical fertilizers increased worldwide. The table lists worldwide average use y in kilograms per acre of cropland during year x. (***Source:*** D. Grigg, *The World Food Problem.*)

x	1950	1963	1972	1979
y	5.0	11.3	22.0	31.2

(a) Are the data linear or nonlinear? Explain.
(b) The equation $y = 5(1.06)^{(x-1950)}$ may be used to model the data. The growth factor is 1.06. What does this growth factor indicate about fertilizer use during this time period?
(c) Estimate the year when fertilizer use was 15 kilograms per acre of cropland.

106. *Greenhouse Gases* If current trends continue, concentrations of atmospheric carbon dioxide (CO_2) in parts per million (ppm) are expected to increase. This increase in concentration of CO_2 has been accelerated by burning fossil fuels and deforestation. The exponential equation $y = 364(1.005)^x$ may be used to model CO_2 in parts per million, where $x = 0$ corresponds to 2000, $x = 1$ to 2001,

and so on. Estimate the year when the CO_2 concentration could be double the preindustrial level of 280 parts per million. (***Source:*** R. Turner, *Environmental Economics.*)

107. *Modeling Sound* The formula

$$f(x) = 160 + 10 \log x$$

is used to calculate the decibel level of a sound with intensity x measured in watts per square centimeter. The noise level at a basketball game can reach 100 decibels. Find the intensity x of this sound.

108. *Loudness of a Sound* (Refer to Exercise 107.)
(a) Show that, if the intensity of a sound increases by a factor of 10 from x to $10x$, the decibel level increases by 10 decibels. *Hint:* Show that

$$160 + 10 \log 10x = 170 + 10 \log x.$$

(b) Find the increase in decibels if the intensity x increases by a factor of 1000.
(c) Find the increase in the intensity x if the decibel level increases by 20.

109. *Hurricanes* (Refer to Exercise 95, Section 12.3.) The barometric air pressure in inches of mercury at a distance of x miles from the eye of a severe hurricane is given by $f(x) = 0.48 \ln(x + 1) + 27$. (***Source:*** A. Miller and R. Anthes, *Meteorology.*) How far from the eye is the pressure 28 inches of mercury?

110. *Earthquakes* The Richter scale is used to determine the intensity of earthquakes, which corresponds to the amount of energy released. If an earthquake has an intensity of x, its magnitude, as computed by the Richter scale, is given by $R(x) = \log \frac{x}{I_0}$, where I_0 is the intensity of a small, measurable earthquake.
(a) If x is 1000 times greater than I_0, how large is this increase on the Richter scale?
(b) If the Richter scale increases from 5 to 8, by what factor does the intensity x increase?

WRITING ABOUT MATHEMATICS

111. Explain the basic steps for solving the equation $a(10^x) - b = c$. Then write the solution.

112. Explain the basic steps for solving the equation $a \log 3x = b$. Then write the solution.

CHECKING BASIC CONCEPTS
SECTION 12.5

1. Solve the equation. Approximate answers to the nearest hundredth when appropriate.
   (a) $2(10^x) = 40$ (b) $2^{3x} + 3 = 150$
   (c) $\ln x = 4.1$ (d) $4 \log 2x = 12$

2. Solve $\log (x + 4) + \log (x - 4) = \log 48$. Check the answers.

3. If $500 are deposited in a savings account that pays 3% annual interest at the end of each year, the amount of money A in the account after x years is given by $A = 500(1.03)^x$. Estimate the number of years required for this amount to reach $900.

CHAPTER 12 SUMMARY
SECTION 12.1 ■ COMPOSITE AND INVERSE FUNCTIONS

Composition of Functions If f and g are functions, then the composite function $g \circ f$, or composition of g and f, is defined by $(g \circ f)(x) = g(f(x))$.

Example: If $f(x) = x - 5$ and $g(x) = 2x^2 + 4x - 6$, then $(g \circ f)(x)$ is

$$g(f(x)) = g(x - 5)$$
$$= 2(x - 5)^2 + 4(x - 5) - 6.$$

One-to-One Function A function f is one-to-one if, for any c and d in the domain of f,

$$c \neq d \quad \text{implies that} \quad f(c) \neq f(d).$$

That is, different inputs always result in different outputs.

Example: $f(x) = x^2 + 4$ is *not* one-to-one; $-3 \neq 3$, but $f(-3) = f(3) = 13$.

Horizontal Line Test If every horizontal line intersects the graph of a function f at most once, then f is a one-to-one function.

Examples:

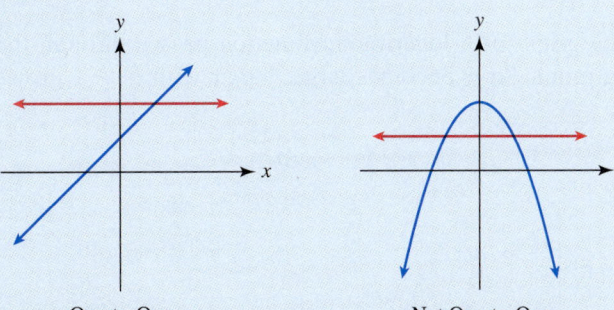

One-to-One   Not One-to-One

Inverse Functions If f is one-to-one, then f has an inverse function, denoted f^{-1}, that satisfies $(f^{-1} \circ f)(x) = x$ and $(f \circ f^{-1})(x) = x$.

Example: $f(x) = 7x$ and $f^{-1}(x) = \frac{x}{7}$ are inverse functions.

SECTION 12.2 ■ EXPONENTIAL FUNCTIONS

Exponential Function   An exponential function is defined by $f(x) = Ca^x$, where $a > 0$, $C > 0$, and $a \neq 1$. Its domain (set of valid inputs) is all real numbers and its range (outputs) is all positive real numbers.

Example: $f(x) = e^x$ is the natural exponential function and $e \approx 2.71828$.

Exponential Growth and Decay   When $a > 1$, the graph of $f(x) = Ca^x$ models exponential growth, and when $0 < a < 1$, it models exponential decay. The base a either represents the growth factor or the decay factor. The constant C equals $f(0)$.

Example: $f(x) = 1.5(2)^x$ is an exponential function with $a = 2$ and $C = 1.5$. It models exponential growth because $a > 1$. The growth factor is 2 because for each unit increase in x, the output from $f(x)$ increases by a *factor* of 2.

SECTION 12.3 ■ LOGARITHMIC FUNCTIONS

Base-a Logarithms   The logarithm with base a of a positive number x is denoted $\log_a x$. If $\log_a x = b$, then $x = a^b$. That is, $\log_a x$ represents the exponent on base a that results in x.

Example: $\log_2 16 = 4$ because $16 = 2^4$.

Domain and Range of Logarithmic Functions   The domain (set of valid inputs) of a logarithmic function is the set of all positive real numbers and the range (outputs) is the set of real numbers.

Graph of a Logarithmic Function   The graph of a logarithmic function passes through the point $(1, 0)$, as illustrated in the following graph. As x becomes large, $\log_a x$ with $a > 1$ grows very slowly.

SECTION 12.4 ■ PROPERTIES OF LOGARITHMS

Basic Properties   Logarithms have several important properties. For positive numbers m, n, and $a \neq 1$ and any real number r,

1. $\log_a mn = \log_a m + \log_a n$.

2. $\log_a \frac{m}{n} = \log_a m - \log_a n$.

3. $\log_a (m^r) = r \log_a m$.

Examples: 1. $\log 5 + \log 20 = \log (5 \cdot 20) = \log 100 = 2$

2. $\log 100 - \log 5 = \log \frac{100}{5} = \log 20 \approx 1.301$

3. $\ln 2^6 = 6 \ln 2 \approx 4.159$

NOTE: $\log_a 1 = 0$ for any valid base a. Thus $\log 1 = 0$ and $\log_2 1 = 0$.

Inverse Properties   The following inverse properties are important for solving exponential and logarithmic equations.

1. $\log_a a^x = x$, for any real number x

2. $a^{\log_a x} = x$,   for any positive number x

Examples: 1. $\log_2 2^\pi = \pi$

2. $10^{\log 2.5} = 2.5$

SECTION 12.5 ■ EXPONENTIAL AND LOGARITHMIC EQUATIONS

Solving Equations   The calculations a^x and $\log_a x$ are inverse operations, much like addition and subtraction or multiplication and division. When solving an exponential equation, we usually take a logarithm of each side. When solving a logarithmic equation, we usually exponentiate each side.

Examples:

$2(5)^x = 22$	Exponential equation
$5^x = 11$	Divide by 2.
$\log 5^x = \log 11$	Take the common logarithm.
$x \log 5 = \log 11$	Power rule
$x = \dfrac{\log 11}{\log 5}$	Divide by $\log 5$.
$\log 2x = 2$	Logarithmic equation
$10^{\log 2x} = 10^2$	Exponentiate each side.
$2x = 100$	Inverse properties
$x = 50$	Divide by 2.

CHAPTER 12   REVIEW EXERCISES

SECTION 12.1

Exercises 1 and 2: Find the following.

   (a) $(g \circ f)(-2)$ *(b)* $(f \circ g)(x)$

1. $f(x) = 2x^2 - 4x,$ $g(x) = 5x + 1$

2. $f(x) = \sqrt[3]{x - 6},$ $g(x) = 4x^3$

3. Use the tables to evaluate each expression.
   (a) $(f \circ g)(2)$ **(b)** $(g \circ f)(1)$

x	0	1	2	3
$f(x)$	3	2	1	0

x	0	1	2	3
$g(x)$	1	2	3	0

4. Use the graph to evaluate each expression.
   (a) $(f \circ g)(-1)$
   (b) $(g \circ f)(2)$
   (c) $(f \circ f)(1)$

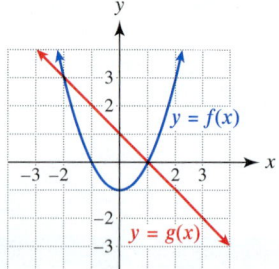

Exercises 5 and 6: Show that f is not one-to-one by finding two inputs that result in the same output. Answers may vary.

5. $f(x) = \dfrac{4}{1 + x^2}$ 6. $f(x) = x^2 - 2x + 1$

Exercises 7 and 8: Use the horizontal line test to determine whether the graph represents a one-to-one function.

7.

8.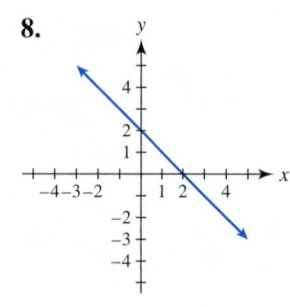

Exercises 9 and 10: Verify that f(x) and $f^{-1}(x)$ are indeed inverse functions.

9. $f(x) = 2x - 9$ $f^{-1}(x) = \dfrac{x + 9}{2}$

10. $f(x) = x^3 + 1$ $f^{-1}(x) = \sqrt[3]{x - 1}$

Exercises 11–14: Find $f^{-1}(x)$.

11. $f(x) = 5x$ 12. $f(x) = x - 11$

13. $f(x) = 2x + 7$ 14. $f(x) = \dfrac{4}{x}$

15. Use the table to make a table of values for $f^{-1}(x)$. What are the domain and range for f^{-1}?

x	0	1	2	3
$f(x)$	10	8	7	3

16. Use the graph of $y = f(x)$ to sketch a graph of $y = f^{-1}(x)$. Include the graph of f and the line $y = x$.

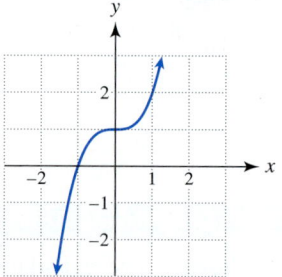

SECTIONS 12.2 AND 12.3

Exercises 17–20: Evaluate the exponential function for the given values of x.

17. $f(x) = 6^x$ $x = -1,$ $x = 2$

18. $f(x) = 5(2^{-x})$ $x = 0,$ $x = 3$

19. $f(x) = \left(\frac{1}{3}\right)^x$ $x = -1,$ $x = 4$

20. $f(x) = 3\left(\frac{1}{6}\right)^x$ $x = 0,$ $x = 1$

Exercises 21–24: Graph f. State whether the graph illustrates exponential growth, exponential decay, or logarithmic growth.

21. $f(x) = 2^x$ 22. $f(x) = \left(\frac{1}{2}\right)^x$

23. $f(x) = \ln(x + 1)$ 24. $f(x) = 3^{-x}$

Exercises 25 and 26: A table for a function f is given.
   (a) *Determine whether f represents linear or exponential growth.*
   (b) *Find a formula for f.*

25.

x	0	1	2	3	4
$f(x)$	5	10	20	40	80

26.

x	0	1	2	3	4
$f(x)$	5	10	15	20	25

27. Use the graph of $y = Ca^x$ to find C and a.

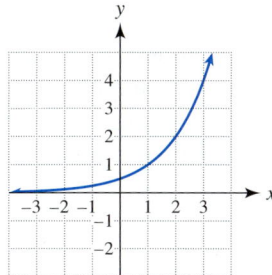

28. Use the graph of $y = k \log_2 x$ to find k.

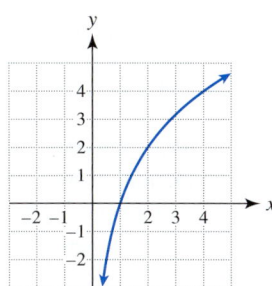

Exercises 29 and 30: If C dollars are deposited in an account that pays r percent annual interest at the end of each year, approximate the amount in the account after x years.

29. $C = \$1200$ $r = 10\%$ $x = 9$ years

30. $C = \$900$ $r = 18\%$ $x = 40$ years

Exercises 31–34: Evaluate f(x) for the given value of x. Approximate answers to the nearest hundredth.

31. $f(x) = 2e^x - 1$ $x = 5.3$

32. $f(x) = 0.85^x$ $x = 2.1$

33. $f(x) = 2 \log x$ $x = 55$

34. $f(x) = \ln(2x + 3)$ $x = 23$

Exercises 35–38: Evaluate the logarithm by hand.

35. $\log 0.001$ **36.** $\log \sqrt{10,000}$

37. $\ln e^{-4}$ **38.** $\log_4 16$

Exercises 39–42: Approximate to the nearest thousandth.

39. $\log 65$ **40.** $\ln 0.85$

41. $\ln 120$ **42.** $\log_2 \frac{2}{5}$

Exercises 43–46: Simplify, using inverse properties.

43. $10^{\log 7}$ **44.** $\log_2 2^{5/9}$

45. $\ln e^{6-x}$ **46.** $e^{2 \ln x}$

SECTION 12.4

Exercises 47–52: Write the expression by using sums and differences of logarithms of x, y, and z.

47. $\ln xy$ **48.** $\log \dfrac{x}{y}$

49. $\ln(x^2 y^3)$ **50.** $\log \dfrac{\sqrt{x}}{z^3}$

51. $\log_2 \dfrac{x^2 y}{z}$ **52.** $\log_3 \sqrt[3]{\dfrac{x}{y}}$

Exercises 53–56: Write as the logarithm of one expression.

53. $\log 45 + \log 5 - \log 3$

54. $\log_4 2x + \log_4 5x$

55. $2 \ln x - 3 \ln y$

56. $\log x^4 - \log x^3 + \log y$

Exercises 57–60: Rewrite, using the power rule.

57. $\log 6^3$ **58.** $\ln x^2$

59. $\log_2 5^{2x}$ **60.** $\log_4 (0.6)^{x+1}$

SECTION 12.5

Exercises 61–70: Solve the equation. Approximate answers to the nearest hundredth when appropriate.

61. $10^x = 100$ **62.** $2^{2x} = 256$

63. $3e^x + 1 = 28$ **64.** $0.85^x = 0.2$

65. $5 \ln x = 4$ **66.** $\ln 2x = 5$

67. $2 \log x = 80$ **68.** $3 \log x - 5 = 1$

69. $2^{x+4} = 3^x$

70. $\ln (2x + 1) + \ln (x - 5) = \ln 13$

Exercises 71 and 72: Do the following.
   (a) Solve $f(x) = g(x)$ graphically.
   (b) Solve $f(x) = g(x)$ symbolically.

71. $f(x) = \frac{1}{2}(2^x)$,   **72.** $f(x) = \log_2 2x$,
   $g(x) = 4$ $g(x) = 3$

     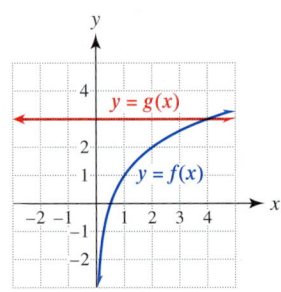

APPLICATIONS

73. Surface Area of a Balloon   The surface area S of a spherical balloon with radius r is given by the formula $S(r) = 4\pi r^2$. Suppose that the balloon is being inflated so that its radius in inches after t seconds is $r(t) = \sqrt{2t}$.
   (a) Evaluate $(S \circ r)(8)$ and interpret your result.
   (b) Find $(S \circ r)(t)$.

74. Sales Tax   Suppose that $f(x) = 0.08x$ calculates the sales tax in dollars on an item that costs x dollars.
   (a) Is f a one-to-one function? Why?
   (b) Find a formula for f^{-1} and interpret what it calculates.

75. Growth of a Mutual Fund   An investor deposits $1500 in a mutual fund that returns 11% annually. Determine the time required for the investment to double in value.

76. Modeling Data   Find values for the constants a and b so that $y = a + b \log x$ models these data.

x	0.1	1	10	100	1000
y	50	100	150	200	250

77. Modeling Data   Find values for the constants C and a so that $y = Ca^x$ models these data.

x	0	1	2	3	4
y	3	6	12	24	48

78. Earthquakes   The Richter scale, used to determine the magnitude of earthquakes, is based on the formula $R(x) = \log \frac{x}{I_0}$, where x is the measured intensity. Let $I_0 = 1$. Find the intensity x for an earthquake with $R = 7$.

79. Modeling Population   In 2000 the population of Nevada was 2 million and growing continuously at an annual rate of 5.1%. The population of Nevada in millions x years after 2000 can be modeled by

$$f(x) = 2e^{0.051x}.$$

   (a) Graph f in $[0, 10, 2]$ by $[0, 4, 1]$. Does this function represent exponential growth or decay?
   (b) Predict the population of Nevada in 2010.
   (c) Estimate the year when the population might reach 3 million.

80. Modeling Bacteria   A colony of bacteria can be modeled by $N(t) = 1000e^{0.0014t}$, where N is measured in bacteria per milliliter and t is in minutes.
   (a) Evaluate $N(0)$ and interpret the result.
   (b) Estimate how long it takes for N to double.

81. Modeling Wind Speed   Wind speeds are usually measured at heights from 5 to 10 meters above the ground. For a particular day, $f(x) = 1.2 \ln (x) + 5$ computes the wind speed in meters per second x meters above the ground, where $x \geq 1$. (**Source:** A. Miller and R. Anthes, *Meteorology*.)
   (a) Find the wind speed at a height of 5 meters.
   (b) Estimate the height at which the wind speed is 8 meters per second.

CHAPTER 12   TEST   🎓Pass the Test   Video solutions to all test exercises

1. If $f(x) = 4x^3 - 5x$ and $g(x) = x + 7$, evaluate $(g \circ f)(1)$ and $(f \circ g)(x)$.

2. Use the graph to evaluate each expression.
   (a) $(f \circ g)(-1)$ **(b)** $(g \circ f)(1)$

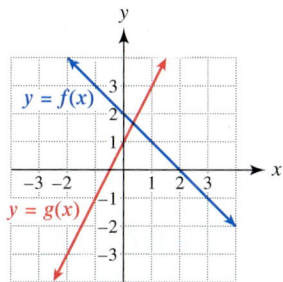

3. Explain why $f(x) = x^2 - 25$ is not a one-to-one function.

4. If $f(x) = 5 - 2x$, find $f^{-1}(x)$.

5. Use the graph of $y = f(x)$ to sketch a graph of $y = f^{-1}(x)$. Include the graph of f and the line $y = x$ in your graph.

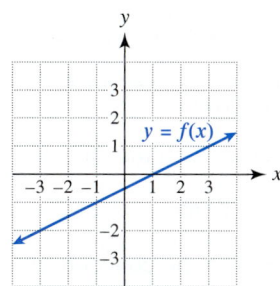

6. Use the table to write a table of values for $f^{-1}(x)$. What are the domain and range of f^{-1}?

x	1	2	3	4
$f(x)$	8	6	4	2

7. Evaluate $f(x) = 3\left(\frac{1}{4}\right)^x$ at $x = 2$.

8. Graph $f(x) = 1.5^{-x}$. State whether the graph of f illustrates exponential growth, exponential decay, or logarithmic growth.

Exercises 9 and 10: A table for a function f is given.
   (a) *Determine whether f represents linear or exponential growth.*
   (b) *Find a formula for f(x).*

9.

x	-2	-1	0	1	2
$f(x)$	0.75	1.5	3	6	12

10.

x	-2	-1	0	1	2
$f(x)$	-4	-2.5	-1	0.5	2

11. Use the graph of $y = Ca^x$ to find C and a.

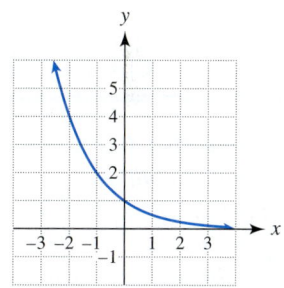

12. If \$750 are deposited in an account paying 7% annual interest at the end of each year, approximate the amount in the account after 5 years.

13. Let $f(x) = 1.5 \ln(x - 5)$. Approximate $f(21)$ to the nearest hundredth.

14. Evaluate $\log \sqrt{10}$ by hand.

15. Approximate $\log_2 43$ to the nearest thousandth.

16. Graph $f(x) = \log(x - 2)$. Compare this graph to the graph of $y = \log x$.

17. Write $\log \frac{x^3 y^2}{\sqrt{z}}$, using sums and differences of logarithms of x, y, and z.

18. Write $4 \ln x - 5 \ln y + \ln z$ as one logarithm.

19. Rewrite $\log 7^{2x}$, using the power rule.

20. Simplify $\ln e^{1-3x}$, using inverse properties.

Exercises 21–24: Solve the given equation. Approximate answers to the nearest hundredth when appropriate.

21. $2e^x = 50$

22. $3(10)^x - 7 = 143$

23. $5 \log x = 9$

24. $3 \ln 5x = 27$

25. *Modeling Data*   Find values for constants a and b so that $y = a + b \log x$ models the data.

x	0.01	0.1	1	10	100
y	-1	2	5	8	11

26. *Modeling Bacteria Growth*   A sample of bacteria is growing continuously at a rate of 9% per hour and can be modeled by $f(x) = 4e^{0.09x}$, where the input x represents elapsed time in hours and the output $f(x)$ is in millions of bacteria.

(a) What was the initial number of bacteria?

(b) Evaluate $f(5)$ and interpret the result.

(c) Does this function represent exponential growth or exponential decay?

(d) Determine the elapsed time when there were 6 million bacteria.

CHAPTER 12   EXTENDED AND DISCOVERY EXERCISES

*Exercises 1–4: **Radioactive Carbon Dating***   *While an animal is alive, it breathes both carbon dioxide and oxygen. Because a small portion of normal atmospheric carbon dioxide is made up of radioactive carbon-14, a fixed percentage of the animal's body is composed of carbon-14. When the animal dies, it quits breathing and the carbon-14 disintegrates without being replaced. One method used to determine when an animal died is to estimate the percentage of carbon-14 remaining in its bones. The **half-life** of carbon-14 is 5730 years. That is, half the original amount of carbon-14 in bones of a fossil will remain after 5730 years. The percentage P, in decimal form, of carbon-14 remaining after x years is modeled by $P(x) = a^x$.*

1. Find the value of a. (*Hint*: $P(5730) = 0.5$.)

2. Calculate the percentage of carbon-14 that remains after 10,000 years.

3. Estimate the age of a fossil with $P = 0.9$.

4. Estimate the age of a fossil with $P = 0.01$.

*Exercises 5–8: **Modeling Blood Flow in Animals***   *For medical reasons, dyes are injected into the bloodstream to determine the health of internal organs. In one study that involved animals, the dye BSP was injected to assess blood flow in the liver. The results are listed in the accompanying table, where x represents the elapsed time in minutes and y is the concentration of the dye in the bloodstream in milligrams per milliliter (mg/mL). Scientists modeled the data with $f(x) = 0.133(0.878(0.73^x) + 0.122(0.92^x))$.*

x (minutes)	1	2	3	4
y (mg/mL)	0.102	0.077	0.057	0.045

x (minutes)	5	7	9	13
y (mg/mL)	0.036	0.023	0.015	0.008

x (minutes)	16	19	22
y (mg/mL)	0.005	0.004	0.003

Source: F. Harrison, "The measurement of liver blood flow in conscious calves."

5. Graph f together with the data. Comment on the fit.

6. Determine the y-intercept and interpret the result.

7. What happens to the concentration of the dye after a long period of time? Explain.

8. Estimate graphically the time at which the concentration of the dye reached 40% of its initial amount. Would you want to solve this problem symbolically? Explain.

*Exercises 9 and 10: **Acid Rain***   *Air pollutants frequently cause acid rain. A measure of acidity is pH, which measures the concentration of the hydrogen ions in a solution, and ranges from 1 to 14. Pure water is neutral and has a pH of 7, acid solutions have a pH less than 7, and alkaline solutions have a pH greater than 7. The pH of a substance can be computed by $f(x) = -\log x$, where x represents the hydrogen ion concentration in moles per liter. Pure water exposed to normal carbon dioxide in the atmosphere has a pH of 5.6.*

If the pH of a lake drops below this level, it is indicative of an acid lake.   (**Source:** G. Howells, *Acid Rain and Acid Water.*)

9. In rural areas of Europe, rainwater typically has a hydrogen ion concentration of $x = 10^{-4.7}$. Find its pH. What effect might this rain have on a lake with a pH of 5.6?

10. Seawater has a pH of 8.2. Compared to seawater, how many times greater is the hydrogen ion concentration in rainwater from rural Europe?

Exercises 11 and 12: **Investment Account**   *If x dollars are deposited every 2 weeks (26 times per year) in an account paying an annual interest rate r, expressed in decimal form, the amount A in the account after n years can be approximated by the formula*

$$A = x\left[\frac{(1 + r/26)^{26n} - 1}{(r/26)}\right].$$

11. If $100 are deposited every 2 weeks in an account paying 9% interest, approximate the amount in the account after 10 years.

12. Suppose that your retirement account pays 12% annual interest. Determine how much you should deposit in this account every 2 weeks, in order to have one million dollars at age 65.

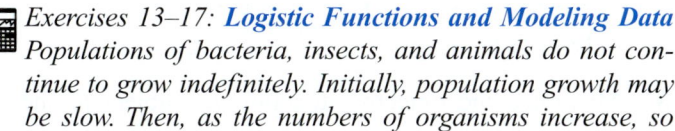 *Exercises 13–17:* **Logistic Functions and Modeling Data** *Populations of bacteria, insects, and animals do not continue to grow indefinitely. Initially, population growth may be slow. Then, as the numbers of organisms increase, so*

does the rate of growth. After a region has become heavily populated or saturated, the growth in population usually levels off because of limited resources. This type of growth may be modeled by a **logistic function.** *One of the earliest studies of population growth was done with yeast plants in 1913. A small amount of yeast was placed in a container with a fixed amount of nourishment. The units of yeast were recorded every 2 hours, giving the data shown in the table.*

Time (hours)	0	2	4	6
Yeast (units)	9.6	29.0	71.1	174.6

Time (hours)	8	10	12
Yeast (units)	350.7	513.3	594.8

Time (hours)	14	16	18
Yeast (units)	640.8	655.9	661.8

Source: D. Brown, *Models in Biology.*

13. Make a scatterplot of the data.

14. Use the graph to describe the growth of the yeast.

15. The data are modeled by the logistic function

$$Y(t) = \frac{663}{1 + 71.6(0.579)^t}.$$

Graph *Y* and the data in the same viewing rectangle.

16. Determine graphically when the amount of yeast equals 400 units.

17. Solve Exercise 16 symbolically.

CHAPTERS 1–12   CUMULATIVE REVIEW EXERCISES

1. Write the number 0.000429 in scientific notation.

2. Classify each real number as one or more of the following: natural number, whole number, integer, rational number, or irrational number.

$$-\frac{11}{7}, -3, 0, \sqrt{6}, \pi, 5.\overline{18}$$

3. Select the formula that models the data best.

x	−2	−1	0	1	2
y	−7	−5	−3	−1	1

   (i) $y = 3x + 1$ (ii) $y = x - 3$ (iii) $y = 2x - 3$

4. State whether the equation illustrates an identity, commutative, associative, or distributive property.

$$(5 - y) + 9 = 9 + (5 - y)$$

Exercises 5–8: Simplify the expression. Write the result with positive exponents.

5. $\left(\dfrac{1}{d^2}\right)^{-2}$

6. $\left(\dfrac{8a^2}{2b^3}\right)^{-3}$

7. $\dfrac{(2x^{-2}y^3)^2}{xy^{-2}}$

8. $\dfrac{x^{-3}y}{4x^2y^{-3}}$

9. Use the graph to express the equation of the line in slope–intercept form.

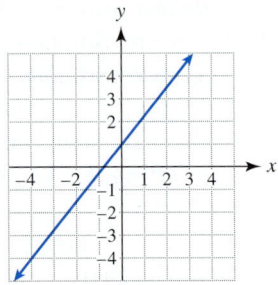

10. Find the domain of $f(x) = \dfrac{10}{x + 3}$.

11. Use the table to write the formula for $f(x) = ax + b$.

x	-2	-1	0	1	2
$f(x)$	-11	-7	-3	1	5

12. Write the equation of the vertical line passing through the point $(4, 7)$.

13. Calculate the slope of the line passing through the points $(4, -1)$ and $(2, -3)$.

14. Sketch the graph of a line passing through the point $(-1, -2)$ with slope $m = 3$.

Exercises 15 and 16: Write the slope–intercept form for a line satisfying the given conditions.

15. Perpendicular to $y = -\frac{1}{7}x - 8$, passing through $(1, 1)$

16. Parallel to $y = 3x - 1$, passing through $(0, 5)$

17. Use the graph to solve the equation $y_1 = y_2$.

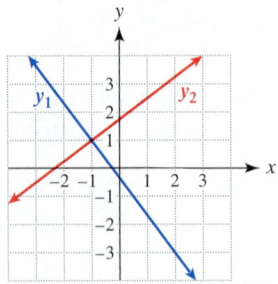

18. Use the table to solve $y < -4$, where y represents a linear function. Write the solution set as an inequality.

x	-2	-1	0	1	2
y	-24	-14	-4	6	16

Exercises 19–24: Solve the equation or inequality. Write the solutions to the inequalities in interval notation.

19. $\frac{2}{3}(x - 3) + 8 = -6$

20. $\frac{1}{3}z + 6 < \frac{1}{4}z - (5z - 6)$

21. $\left(\dfrac{t + 2}{3}\right) - 10 = \frac{1}{3}t - (5t + 8)$

22. $-10 \le -\frac{3}{5}x - 4 < -1$

23. $-2|t - 4| \ge -12$ **24.** $\left|\frac{1}{2}x - 5\right| = 3$

25. Shade the solution set in the xy-plane.

$$x + y > 3$$
$$2x - y \ge 3$$

26. Evaluate $\det A$ if $A = \begin{bmatrix} -1 & -2 \\ 3 & 4 \end{bmatrix}$.

Exercises 27–30: Solve the system of equations, if possible. Write the solution as an ordered pair or ordered triple where appropriate.

27. $4x - 3y = 1$
$5x + 2y = 7$

28. $2x - 3y = -2$
$-6x + 9y = 5$

29. $2x - y + 3z = -2$
$x + 5y - 2z = -8$
$-3x - y - 3z = 6$

30. $x + y - z = -1$
$-x - y - z = -1$
$x - 2y + z = 1$

31. Maximize the objective function R, subject to the given constraints.

$$R = 2x + 5y$$
$$3x + y \le 12$$
$$x + 3y \le 12$$
$$x \ge 0, y \ge 0$$

32. Find the area of the triangle by using a determinant. Assume that units are inches.

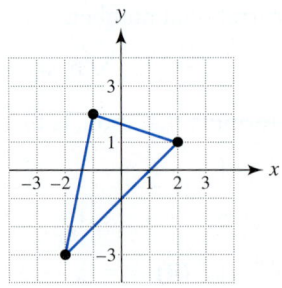

Exercises 33–36: Factor completely.

33. $2x^3 - 4x^2 + 2x$ **34.** $4a^2 - 25b^2$

35. $8t^3 - 27$

36. $4a^3 - 2a^2 + 10a - 5$

Exercises 37–40: Solve the equation.

37. $6x^2 - 7x - 10 = 0$ **38.** $9x^2 = 4$

39. $x^4 - 2x^3 = 15x^2$ **40.** $5x - 10x^2 = 0$

Exercises 41 and 42: Simplify the expression.

41. $\dfrac{x^2 + 5x + 6}{x^2 - 9} \cdot \dfrac{x - 3}{x + 2}$

42. $\dfrac{x^2 - 2x - 8}{x^2 + x - 12} \div \dfrac{(x - 4)^2}{x^2 - 16}$

Exercises 43 and 44: Solve the rational equation. Check your result.

43. $\dfrac{2}{x + 2} - \dfrac{1}{x - 2} = \dfrac{-3}{x^2 - 4}$

44. $\dfrac{3y}{y^2 + y - 2} = \dfrac{1}{y - 1} - 2$

45. Solve the equation for J.
$$P = \frac{J + 2z}{J}$$

46. Simplify the complex fraction.
$$\frac{\dfrac{3}{x^2} + x}{x - \dfrac{3}{x^2}}$$

47. Suppose that y varies directly with x. If $y = 15$ when $x = 3$, find y when x is 8.

48. Divide $(3x^3 - 2x - 15) \div (x - 2)$.

Exercises 49–54: Simplify the expression. Assume that all variables are positive.

49. $\left(\dfrac{x^6}{y^9}\right)^{2/3}$ **50.** $\sqrt[3]{-x^4} \cdot \sqrt[3]{-x^5}$

51. $\sqrt{5ab} \cdot \sqrt{20ab}$ **52.** $2\sqrt{24} - \sqrt{54}$

53. $\sqrt[3]{a^5b^4} + 3\sqrt[3]{a^5b}$

54. $(5 + \sqrt{5})(5 - \sqrt{5})$

55. Rationalize the denominator.
$$\frac{2}{5 - \sqrt{3}}$$

56. Find the domain of f. Write your answer in interval notation.
$$f(x) = \frac{3}{\sqrt{x - 4}}$$

Exercises 57 and 58: Solve. Check your answer.

57. $2(x + 1)^2 = 50$ **58.** $\sqrt{x + 6} = x$

Exercises 59 and 60: Write in standard form.

59. $(-2 + 3i) - (-5 - 2i)$

60. $\dfrac{3 - i}{1 + 3i}$

61. Find the vertex on the graph of the equation $f(x) = 3x^2 - 12x + 13$.

62. Find the maximum y-value on the parabola determined by $y = -2x^2 + 6x - 1$.

63. Compare the graph of $f(x) = (x - 3)^2 + 2$ to the graph of $y = x^2$.

64. Write the equation $y = x^2 + 6x - 2$ in vertex form and identify the vertex.

Exercises 65–68: Solve the quadratic equation by using the method of your choice.

65. $x^2 - 13x + 40 = 0$ **66.** $2d^2 - 5 = d$

67. $z^2 - 4z = -2$ **68.** $x^4 - 10x^2 + 24 = 0$

69. A graph of $y = ax^2 + bx + c$ is shown.
  (a) Solve $ax^2 + bx + c = 0$.
  (b) State whether $a > 0$ or $a < 0$.
  (c) Determine whether the discriminant is positive, negative, or zero.

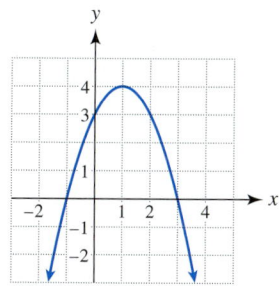

70. Solve $x^2 + 5x - 14 \geq 0$. Write your answer in interval notation.

71. For $f(x) = x^2 - 2$ and $g(x) = 2x + 1$, find the following.
(a) $(f \circ g)(1)$ (b) $(g \circ f)(x)$

72. Show that f is not one-to-one by finding two inputs that result in the same output. Answers may vary.
$$f(x) = x^2 + x - 6$$

73. Find $f^{-1}(x)$ for $f(x) = \frac{3}{x}$.

74. If $800 are deposited in an account that pays 7.5% annual interest at the end of each year, approximate the amount in the account after 15 years.

Exercises 75 and 76: Evaluate without a calculator.

75. $\log_3 81$ **76.** $e^{\ln 2x}$

77. Write the expression by using sums and differences of logarithms of x and y. Assume x and y are positive.
$$\log \frac{\sqrt{x}}{y^2}$$

78. Write $2 \ln x + \ln 5x$ as the logarithm of a single expression.

Exercises 79 and 80: Solve the equation. Approximate answers to the nearest hundredth.

79. $6 \log x - 2 = 9$

80. $2^{3x} = 17$

APPLICATIONS

81. *Population Growth*   The population P of a community with an annual percentage growth rate r (expressed as a decimal) after t years is given by $P = P_0(1 + r)^t$, where P_0 represents the initial population of the community. If a community having an initial population of $P_0 = 12,000$ grew to a population of $P = 14,600$ in $t = 5$ years, find the annual percentage growth rate for this community.

82. *Wing Span of a Bird*   The wing span L of a bird with weight W can sometimes be modeled by $L = 27.4 \sqrt[3]{W}$, where L is in inches and W is in pounds. Estimate the weight of a bird with a wing span of 36 inches.   (*Source:* C. Pennycuick, *Newton Rules Biology.*)

83. *U.S. Energy Consumption*   From 1950 to 1970, per capita consumption of energy in millions of Btu can be modeled by $f(x) = 0.25x^2 - 975x + 950,845$, where x is the year.   (*Source:* Department of Energy.)
(a) During what year was per capita energy consumption at its lowest?
(b) Find this minimum value.

84. *Braking Distance*   On dry, level pavement a safe braking distance d in feet for a car traveling x miles per hour is $d = \frac{x^2}{12}$. What speed corresponds to a braking distance of 350 feet?   (*Source:* F. Mannering, *Principles of Highway Engineering and Traffic Control.*)

85. *Investing for Retirement*   A college student invests $8000 in an account that pays interest annually. If the student would like this investment to be worth $1,000,000 in 45 years, what annual interest rate would the account need to pay?

86. *Modeling Wind Speed*   Wind speeds are usually measured at heights from 5 to 10 meters above the ground. For a particular day, $f(x) = 1.4 \ln(x) + 7$ computes the wind speed in meters per second x meters above the ground, where $x \geq 1$.   (*Source:* A. Miller and R. Anthes, *Meteorology.*)
(a) Find the wind speed at a height of 8 meters.
(b) Estimate the height at which the wind speed is 10 meters per second.

CHAPTER 13

Conic Sections

Chapter Contents

13.1 Parabolas and Circles

13.2 Ellipses and Hyperbolas

13.3 Nonlinear Systems of Equations and Inequalities

Throughout history people have been fascinated with the universe around them and compelled to understand its mysteries. Conic sections, which include parabolas, circles, ellipses, and hyperbolas, have played an important role in gaining this understanding. Although conic sections were described and named by the Greek astronomer Apollonius in 200 B.C., not until much later were they used to model motion in the universe. In the sixteenth century Tycho Brahe, the greatest observational astronomer of the age, recorded precise data on planetary movement in the sky. Using Brahe's data, in 1619 Johannes Kepler determined that planets move in elliptical orbits around the sun. In 1686 Newton used Kepler's work to show that elliptical orbits are the result of his famous theory of gravitation. We now know that all celestial objects—including planets, comets, asteroids, and satellites—travel in paths described by conic sections.

Today scientists search the sky for information about the universe with enormous radio telescopes in the shape of parabolic dishes. The Hubble telescope also makes use of a parabolic mirror. As a result, our understanding of the universe has changed dramatically in recent years.

Conic sections have had a profound influence on people's understanding of their world and the cosmos. In this chapter we introduce you to these age-old curves.

> *The art of asking the right questions in mathematics is more important than the art of solving them.*
>
> —GEORG CANTOR

Source: Historical Topics for the Mathematics Classroom, Thirty-first Yearbook, NCTM.

13.1   PARABOLAS AND CIRCLES

Types of Conic Sections ■ Parabolas with Horizontal Axes of Symmetry ■
Equations of Circles

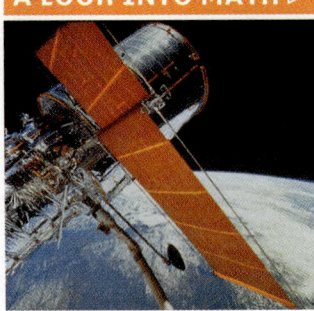

In this section we discuss two types of conic section: parabolas and circles. Recall that we discussed parabolas with vertical axes of symmetry in Chapter 11. In this section we discuss parabolas with horizontal axes of symmetry, but first we introduce the three basic types of conic section. The Hubble telescope travels in a path described by a conic section called an ellipse.

Types of Conic Sections

Conic sections are named after the different ways that a plane can intersect a cone. The three basic curves are parabolas, ellipses, and hyperbolas. A circle is a special case of an ellipse. Figure 13.1 shows the three types of conic section along with an example of the graph associated with each.

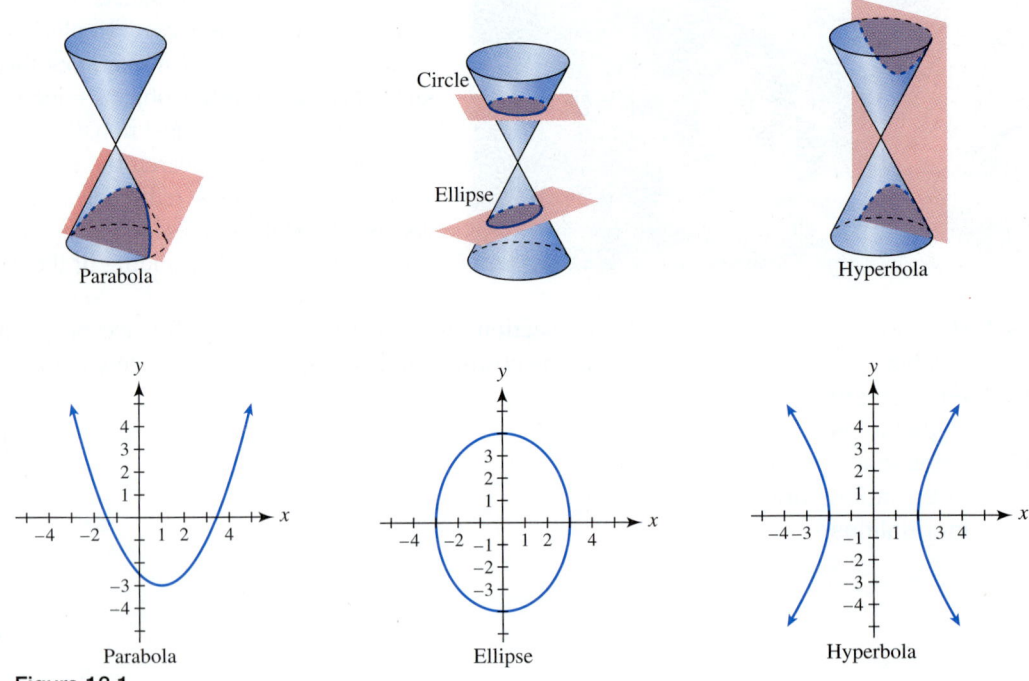

Figure 13.1

Parabolas with Horizontal Axes of Symmetry

Recall that the *vertex form of a parabola* with a vertical axis of symmetry is

$$y = a(x - h)^2 + k,$$

where (h, k) is the vertex. If $a > 0$, the parabola opens upward; if $a < 0$, the parabola opens downward, as shown in Figure 13.2. The preceding equation can also be expressed in the form

$$y = ax^2 + bx + c.$$

In this form the x-coordinate of the vertex is $x = -\frac{b}{2a}$.

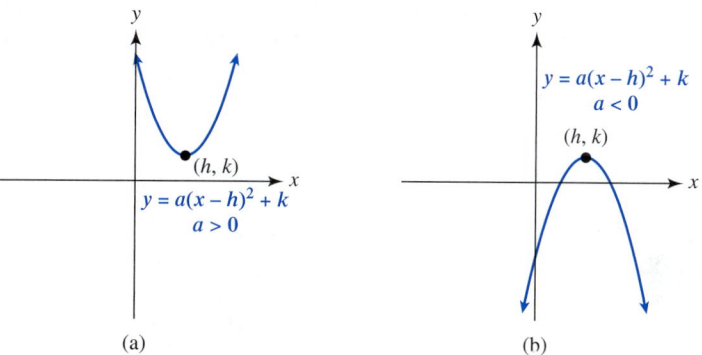

(a) (b)

Figure 13.2 Vertical Axes of Symmetry

Interchanging the roles of x and y (and also h and k) gives equations for parabolas that open to the right or the left. In this case, their axes of symmetry are horizontal.

PARABOLAS WITH HORIZONTAL AXES OF SYMMETRY

The graph of $x = a(y - k)^2 + h$ is a parabola that opens to the right if $a > 0$ and to the left if $a < 0$. The vertex of the parabola is located at (h, k).

The graph of $x = ay^2 + by + c$ is a parabola opening to the right if $a > 0$ and to the left if $a < 0$. The y-coordinate of its vertex is given by $y = -\frac{b}{2a}$.

These parabolas are illustrated in Figure 13.3.

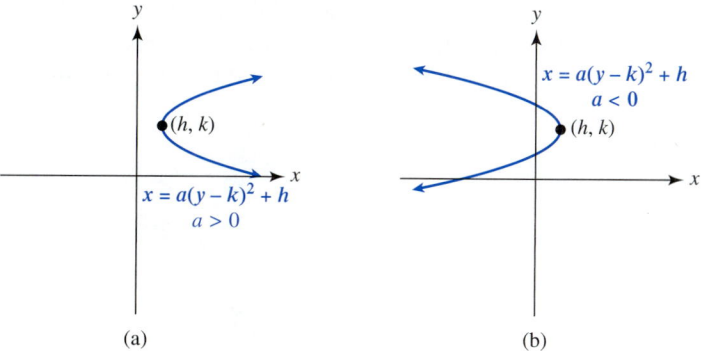

(a) (b)

Figure 13.3 Horizontal Axes of Symmetry

EXAMPLE 1 Graphing a parabola

Graph $x = -\frac{1}{2}y^2$. Find its vertex and axis of symmetry.

Solution
The equation can be written in vertex form because $x = -\frac{1}{2}(y - \mathbf{0})^2 + \mathbf{0}$. The vertex is $(\mathbf{0}, \mathbf{0})$, and because $a = -\frac{1}{2} < 0$, the parabola opens to the left. We can make a table of values, as shown in Table 13.1, and plot a few points to help determine the location and shape of the graph. To obtain Table 13.1, we first choose a y-value and then calculate an x-value. The resulting graph is shown in Figure 13.4. Its axis of symmetry is the x-axis, or $y = 0$.

TABLE 13.1

y	x
-2	-2
-1	$-\frac{1}{2}$
0	0
1	$-\frac{1}{2}$
2	-2

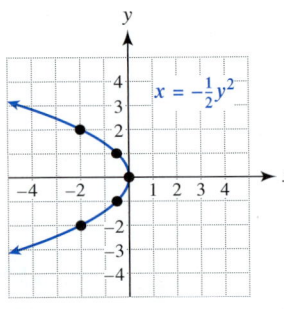

Figure 13.4

Now Try Exercise 11

EXAMPLE 2 Graphing a parabola

Graph $x = (y - 3)^2 + 2$. Find its vertex and axis of symmetry.

Solution
As $h = 2$ and $k = 3$ in the equation $x = a(y - k)^2 + h$, the vertex is $(2, 3)$, and because $a = 1 > 0$, the parabola opens to the right. This parabola has the same shape as $y = x^2$, except that it opens to the right rather than upward. To graph this parabola we can make a table of values and plot a few points. Table 13.2 can be obtained by first choosing y-values and then calculating corresponding x-values using $x = (y - 3)^2 + 2$.

Sometimes, finding the x- and y-intercepts of the parabola is helpful when you are graphing. To find the x-intercept let $y = 0$ in $x = (y - 3)^2 + 2$. The x-intercept is $x = (\mathbf{0} - 3)^2 + 2 = 11$. To find any y-intercepts let $x = 0$ in $x = (y - 3)^2 + 2$. Here $\mathbf{0} = (y - 3)^2 + 2$ means that $(y - 3)^2 = -2$, which has no real solutions, and that this parabola has no y-intercepts.

Both the graph of the parabola and the points from Table 13.2 are shown in Figure 13.5. Note that there are no y-intercepts and that the x-intercept is 11. The axis of symmetry is $y = 3$ because, if we fold the graph on the horizontal line $y = 3$, the two sides match.

TABLE 13.2

y	x
1	6
2	3
3	2
4	3
5	6

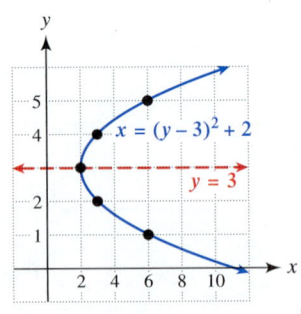

Figure 13.5

Now Try Exercise 17

EXAMPLE 3   Graphing a parabola and finding its vertex

Identify the vertex and then graph each parabola.
(a) $x = -y^2 + 1$ **(b)** $x = y^2 - 2y - 1$

Solution

(a) If we rewrite $x = -y^2 + 1$ as $x = -(y - 0)^2 + 1$, then $h = 1$ and $k = 0$, so the vertex is $(1, 0)$. By letting $y = 0$ in $x = -y^2 + 1$, we find that the x-intercept is $x = -0^2 + 1 = 1$. Similarly, we let $x = 0$ in $x = -y^2 + 1$ to find the y-intercepts. The equation $0 = -y^2 + 1$ has solutions -1 and 1.

The parabola opens to the left because $a = -1 < 0$. Additional points given in Table 13.3 will help in graphing the parabola shown in Figure 13.6.

TABLE 13.3

y	x
-2	-3
-1	0
0	1
1	0
2	-3

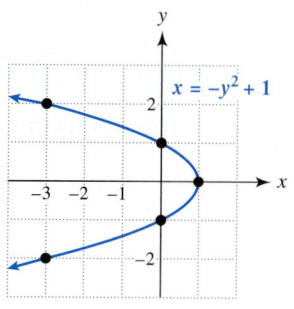

Figure 13.6

(b) The y-coordinate of the vertex for the graph of $x = y^2 - 2y - 1$ is given by

$$y = -\frac{b}{2a} = -\frac{-2}{2(1)} = 1.$$

To find the x-coordinate of the vertex, substitute $y = 1$ into the given equation.

$$x = (1)^2 - 2(1) - 1 = -2$$

The vertex is $(-2, 1)$. The parabola opens to the right because $a = 1 > 0$. The additional points given in Table 13.4 help in graphing the parabola shown in Figure 13.7. Note that the y-intercepts do not have integer values and that the quadratic formula could be used to find approximations for these values. The x-intercept is -1.

TABLE 13.4

y	x
-1	2
0	-1
1	-2
2	-1
3	2

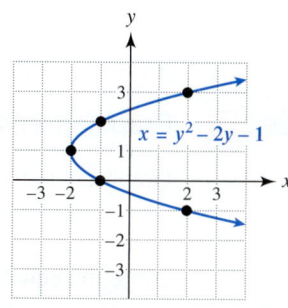

Figure 13.7   Now Try Exercises 13, 31

Equations of Circles

A **circle** consists of the set of points in a plane that are the same distance from a fixed point. The fixed distance is called the **radius**, and the fixed point is called the **center**. In Figure 13.8 all points lying on the circle are a distance of 2 units from the center (2, 1). Therefore the radius of the circle equals 2.

We can find the equation of the circle shown in Figure 13.8 by using the distance formula. If a point (x, y) lies on the graph of a circle, its distance from the center $(\mathbf{2}, \mathbf{1})$ is $\mathbf{2}$ and

$$\sqrt{(x - \mathbf{2})^2 + (y - \mathbf{1})^2} = \mathbf{2}.$$

Squaring each side gives

$$(x - \mathbf{2})^2 + (y - \mathbf{1})^2 = \mathbf{2}^2.$$

This equation represents the standard equation for a circle with center (2, 1) and radius 2.

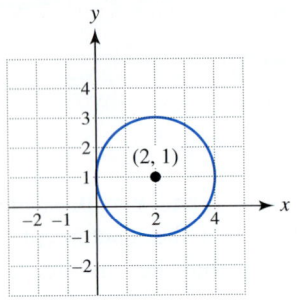

Figure 13.8

> ### STANDARD EQUATION OF A CIRCLE
>
> The **standard equation of a circle** with center (h, k) and radius r is
> $$(x - h)^2 + (y - k)^2 = r^2.$$

EXAMPLE 4 Graphing a circle

Graph $x^2 + y^2 = 9$. Find the radius and center.

Solution
The equation $x^2 + y^2 = 9$ can be written in standard form as

$$(x - \mathbf{0})^2 + (y - \mathbf{0})^2 = \mathbf{3}^2.$$

Therefore the center is $(\mathbf{0}, \mathbf{0})$ and the radius is $\mathbf{3}$. Its graph is shown in Figure 13.9.

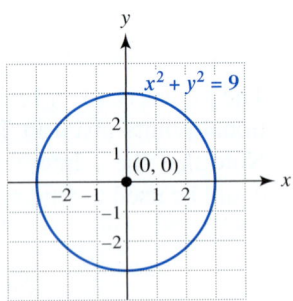

Figure 13.9

Now Try Exercise **58**

EXAMPLE 5 Graphing a circle

Graph $(x + 1)^2 + (y - 3)^2 = 4$. Find the radius and center.

Solution
Write the equation as

$$(x - (\mathbf{-1}))^2 + (y - \mathbf{3})^2 = \mathbf{2}^2.$$

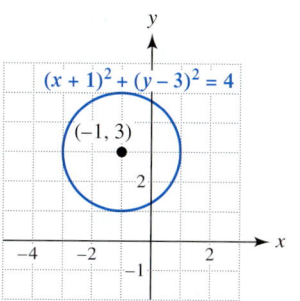

Figure 13.10

The center is $(-1, 3)$, and the radius is 2. Its graph is shown in Figure 13.10.

Now Try Exercise **61**

TECHNOLOGY NOTE: Graphing Circles

The graph of a circle does not represent a function. One way to graph a circle with a graphing calculator is to solve the equation for y and obtain two equations. One equation gives the upper half of the circle, and the other equation gives the lower half.

For example, to graph $x^2 + y^2 = 4$, begin by solving for y.

$$y^2 = 4 - x^2 \qquad \text{Subtract } x^2.$$

$$y = \pm \sqrt{4 - x^2} \qquad \text{Square root property}$$

Then graph $Y_1 = \sqrt{(4 - X^2)}$ and $Y_2 = -\sqrt{(4 - X^2)}$. The graph of y_1 is the upper half of the circle, and the graph of y_2 is the lower half of the circle, as shown in Figure 13.11.

$[-4.7, 4.7, 1]$ by $[-3.1, 3.1, 1]$ $[-4.7, 4.7, 1]$ by $[-3.1, 3.1, 1]$ $[-4.7, 4.7, 1]$ by $[-3.1, 3.1, 1]$

        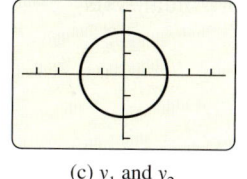

(a) y_1 (b) y_2 (c) y_1 and y_2

Figure 13.11

In the next example we use the *method of completing the square* to find the center and radius of a circle. (To review completing the square, refer to Sections 11.2 and 11.3.)

EXAMPLE 6 Finding the center of a circle

Find the center and radius of the circle given by $x^2 + 4x + y^2 - 6y = 5$.

Solution

Begin by writing the equation as

$$(x^2 + 4x + \underline{\quad}) + (y^2 - 6y + \underline{\quad}) = 5.$$

To complete the square, add $\left(\frac{4}{2}\right)^2 = 4$ and $\left(\frac{-6}{2}\right)^2 = 9$ to each side of the equation.

$$(x^2 + 4x + 4) + (y^2 - 6y + 9) = 5 + 4 + 9$$

Factoring each perfect square trinomial yields

$$(x + 2)^2 + (y - 3)^2 = 18.$$

The center is $(-2, 3)$, and because $18 = \left(\sqrt{18}\right)^2$, the radius is $\sqrt{18}$, or $3\sqrt{2}$.

Now Try Exercise **63**

CRITICAL THINKING

Does the following equation represent a circle? If so, give its center and radius.

$$x^2 + y^2 + 10y = -32$$

[−4, 4, 1] by [−5, 5, 1]

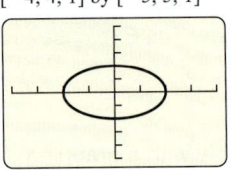

Figure 13.12

NOTE: If a circle is not graphed in a *square viewing rectangle*, it will appear to be an oval rather than a circle. In a square viewing rectangle a circle will appear circular. Figure 13.12 shows the circle graphed in a viewing rectangle that is not square.

13.1 PUTTING IT ALL TOGETHER

The following table summarizes some basic concepts about parabolas and circles.

Concept	Explanation	Example
Parabola with Horizontal Axis	Vertex form: $x = a(y - k)^2 + h$. If $a > 0$, it opens to the right; if $a < 0$, it opens to the left. The vertex is (h, k). See Figure 13.3. These parabolas may also be expressed as $x = ay^2 + by + c$, where the y-coordinate of the vertex is $y = -\frac{b}{2a}$.	$x = 2(y - 1)^2 + 4$ opens to the right and its vertex is $(4, 1)$.
Standard Equation of a Circle	Standard equation: $(x - h)^2 + (y - k)^2 = r^2$. The radius is r and the center is (h, k).	$(x + 2)^2 + (y - 1)^2 = 16$ has center $(-2, 1)$ and radius 4.

13.1 Exercises

CONCEPTS

1. Name the three general types of conic sections.

2. What is the difference between the graphs of $y = ax^2 + bx + c$ and $x = ay^2 + by + c$?

3. If a parabola has a horizontal axis of symmetry, does it represent a function?

4. Sketch a graph of a parabola with a horizontal axis of symmetry.

5. If a parabola has two y-intercepts, does it represent a function? Why or why not?

6. If $x = a(y - k)^2 + h$, what is the vertex?

7. The graph of $x = -y^2$ opens to the _____.

8. The graph of $x = 2y^2 + y - 1$ opens to the _____.

9. The graph of $(x - h)^2 + (y - k)^2 = r^2$ is a _____ with center _____.

10. The graph of $x^2 + y^2 = r^2$ is a circle with center _____ and radius _____.

PARABOLAS

Exercises 11–32: Graph the parabola. Find the vertex and axis of symmetry.

11. $x = y^2$

12. $x = -y^2$

13. $x = y^2 + 1$

14. $x = y^2 - 1$

15. $x = 2y^2$

16. $x = \frac{1}{4}y^2$

17. $x = (y - 1)^2 + 2$

18. $x = (y - 2)^2 + 1$

19. $y = (x + 2)^2 + 1$

20. $y = (x - 4)^2 + 5$

21. $x = \frac{1}{2}(y + 1)^2 - 3$

22. $x = -2(y + 3)^2 + 1$

23. $x = -3(y - 1)^2$

24. $x = \frac{1}{4}(y + 2)^2 - 3$

25. $y = 2x^2 - x + 1$

26. $y = -x^2 + 2x + 2$

27. $x = -2y^2 + 3y + 2$

28. $x = \frac{1}{2}y^2 + y - 1$

29. $x = 3y^2 + y$

30. $x = -\frac{3}{2}y^2 - 2y + 1$

31. $x = y^2 + 2y + 1$

32. $x = y^2 - 3y - 4$

Exercises 33–36: Use the graph to determine the equation of the parabola. (Hint: Either $a = 1$ or $a = -1$.)

33.

34.

35.

36.

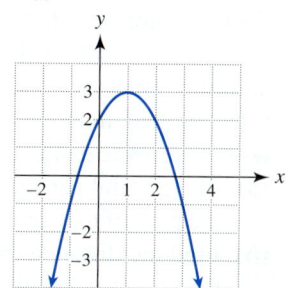

Exercises 37–40: Determine the direction that the parabola opens if it satisfies the given conditions.

37. Passing through $(2, 0)$, $(-2, 0)$, and $(0, -2)$

38. Passing through $(0, -3)$, $(0, 2)$, and $(1, 1)$

39. Vertex $(1, 2)$ passing through the point $(-1, -2)$ with a vertical axis

40. Vertex $(-1, 3)$ passing through the point $(0, 0)$ with a horizontal axis

41. What x-values are possible for the graph of the equation $x = 2y^2$?

42. What y-values are possible for the graph of the equation $x = 2y^2$?

43. **Thinking Generally** How many y-intercepts does a parabola

$$x = a(y - k)^2 + h$$

have if $a > 0$ and $h < 0$?

44. **Thinking Generally** Does the graph of the equation $x = ay^2 + by + c$ always have a y-intercept? Explain.

45. What is the x-intercept for the graph of the equation given by

$$x = 3y^2 - y + 1?$$

46. What are the y-intercepts for the graph of the equation given by

$$x = y^2 - 3y + 2?$$

CIRCLES

Exercises 47–52: Write the standard equation of the circle with the given radius r and center C.

47. $r = 1$ $C = (0, 0)$

48. $r = 4$ $C = (2, 3)$

49. $r = 3$ $C = (-1, 5)$

50. $r = 5$ $C = (5, -3)$

51. $r = \sqrt{2}$ $C = (-4, -6)$

52. $r = \sqrt{6}$ $C = (0, 4)$

Exercises 53–56: Use the graph to find the standard equation of the circle.

53.

54.

55.

56.

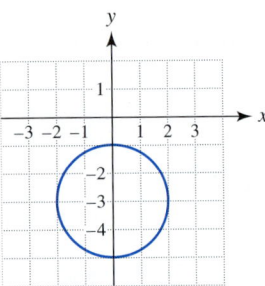

Exercises 57–66: Find the radius and center of the circle. Then graph the circle.

57. $x^2 + y^2 = 9$

58. $x^2 + y^2 = 1$

59. $(x - 1)^2 + (y - 3)^2 = 9$

60. $(x + 2)^2 + (y + 1)^2 = 4$

61. $(x + 5)^2 + (y - 5)^2 = 25$

62. $(x - 4)^2 + (y + 3)^2 = 16$

63. $x^2 + 6x + y^2 - 2y = -1$

64. $x^2 + y^2 + 12y + 32 = 0$

65. $x^2 + 6x + y^2 - 2y + 3 = 0$

66. $x^2 - 4x + y^2 + 4y = -3$

APPLICATIONS

67. *Radio Telescopes* The Parks radio telescope has the shape of a parabolic dish, as depicted in the figure at the top of the next column. A cross section of this telescope can be modeled by $x = \frac{32}{11,025}y^2$, where $-105 \le y \le 105$; the units are feet. (*Source:* J. Mar, *Structure Technology for Large Radar and Telescope Systems.*)

(a) Graph the cross-sectional shape of the dish in $[-40, 40, 10]$ by $[-120, 120, 20]$.

(b) Find the depth d of the dish.

68. *Train Tracks* To make a curve safer for trains, parabolic curves are sometimes used instead of circular curves. See the accompanying figures. (*Source:* F. Mannering and W. Kilareski, *Principles of Highway Engineering and Traffic Analysis.*)

(a) Suppose that a curve must pass through the points $(-1, 0)$, $(0, 2)$, and $(0, -2)$, where the units are kilometers. Find an equation for the train tracks in the form

$$x = a(y - h)^2 + k.$$

(b) Find another point that lies on the train tracks.

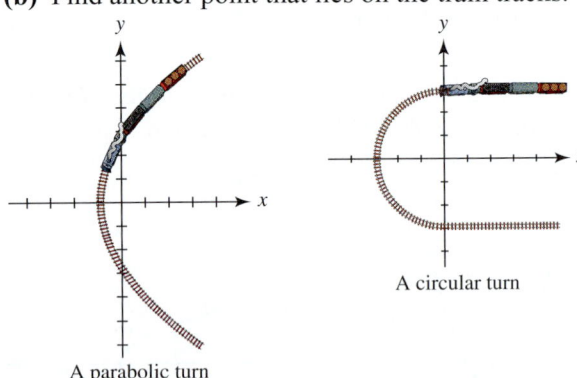

A parabolic turn

A circular turn

69. *Trajectories of Comets* Under certain circumstances, a comet can pass by the sun once and never return. In this situation the comet may travel in a parabolic path, as illustrated in the accompanying figure. Suppose that a comet's path is given by $x = -2.5y^2$, where the sun is located at $(-0.1, 0)$ and the units are astronomical units (A.U.). One astronomical unit equals 93 million miles. (*Source:* W. Thomson, *Introduction to Space Dynamics.*)

(a) Plot a point for the sun's location and then graph the path of the comet.

(b) Find the distance from the sun to the comet when the comet is located at $(-2.5, 1)$.

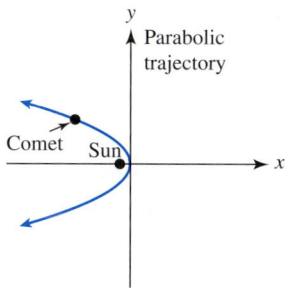

y
Parabolic trajectory
Comet   Sun
x

70. *Speed of a Comet*   (Continuation of Exercise 69.) The velocity V in meters per second of a comet traveling in a parabolic trajectory about the sun is given by $V = \frac{k}{\sqrt{D}}$, where D is the distance from the sun in meters and $k = 1.15 \times 10^{10}$.

(a) How does the velocity of the comet change as its distance from the sun changes?

(b) Calculate the velocity of the comet when it is closest to the sun. (*Hint:* 1 mile $\approx$ 1609 meters.)

WRITING ABOUT MATHEMATICS

71. Suppose that you are given the equation
$$x = a(y - k)^2 + h.$$

(a) Explain how you can determine the direction that the parabola opens.

(b) Explain how to find the axis of symmetry and the vertex.

(c) If the points $(0, 4)$ and $(0, -2)$ lie on the graph of x, what is the axis of symmetry?

(d) Generalize part (c) if $(0, y_1)$ and $(0, y_2)$ lie on the graph of x.

72. Suppose that you are given the vertex of a parabola. Can you determine the axis of symmetry? Explain.

GROUP ACTIVITY
WORKING WITH REAL DATA

Directions: Form a group of 2 to 4 people. Select someone to record the group's responses for this activity. All members of the group should work cooperatively to answer the questions. If your instructor asks for your results, each member of the group should be prepared to respond.

Radio Telescope   The U.S. Naval Research Laboratory designed a giant radio telescope weighing 3450 tons. Its parabolic dish has a diameter of 300 feet and a depth of 44 feet, as shown in the accompanying figure.   (**Source:** J. Mar, *Structure Technology for Large Radio and Radar Telescope Systems.*)

(a) Determine an equation of the form $x = ay^2$, $a > 0$, that models a cross section of the dish.

 (b) Graph your equation in an appropriate viewing rectangle.

13.2   ELLIPSES AND HYPERBOLAS

Equations of Ellipses ■ Equations of Hyperbolas

A LOOK INTO MATH ▷

Celestial objects travel in paths or trajectories determined by conic sections. For this reason, conic sections have been studied for centuries. In modern times physicists have learned that subatomic particles can also travel in trajectories determined by conic sections. Recall that the three main types of conic sections are parabolas, ellipses, and hyperbolas and that circles are a special type of ellipse. In Section 11.2 and Section 13.1, we discussed parabolas and circles. In this section we focus on ellipses and hyperbolas and some of their applications.

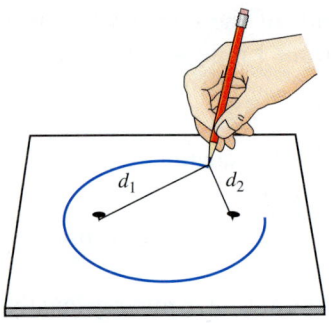

Equations of Ellipses

One method used to sketch an ellipse is to tie the ends of a string to two nails driven into a flat board. If a pencil is placed against the string anywhere between the nails, as shown in Figure 13.13, and is used to draw a curve, the resulting curve is an ellipse. The sum of the distances d_1 and d_2 between the pencil and each of the nails is always fixed by the length of the string. The location of the nails corresponds to the foci of the ellipse. An **ellipse** is the set of points in a plane, the sum of whose distances from two fixed points is constant. Each fixed point is called a **focus** (plural *foci*) of the ellipse.

Figure 13.13

CRITICAL THINKING

What happens to the shape of the ellipse shown in Figure 13.13 as the nails are moved farther apart? What happens to its shape as the nails are moved closer together? When would a circle be formed?

In Figure 13.14 the **major axis** and the **minor axis** are labeled for each ellipse. The major axis is the longer of the two axes. Figure 13.14(a) shows an ellipse with a *horizontal* major axis, and Figure 13.14(b) shows an ellipse with a *vertical* major axis. The **vertices**, V_1 and V_2, of each ellipse are located at the endpoints of the major axis, and the **center** of the ellipse is the midpoint of the major axis (or the intersection of the major and minor axes).

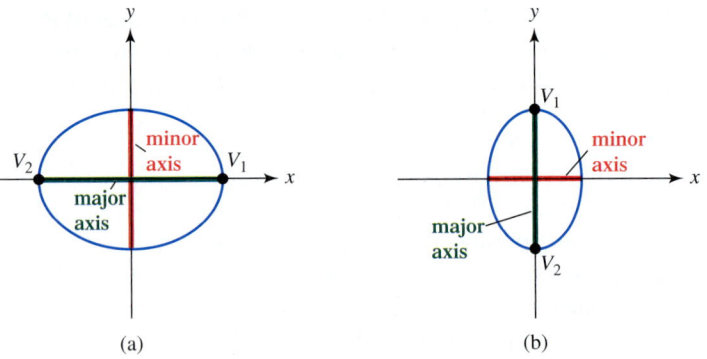

(a)   (b)

Figure 13.14

A vertical line can intersect the graph of an ellipse twice, so an ellipse cannot be represented by a function. However, some ellipses can be represented by the following equations.

STANDARD EQUATIONS FOR ELLIPSES CENTERED AT (0, 0)

The ellipse with center at the origin, *horizontal* major axis, and equation

$$\frac{x^2}{a^2} + \frac{y^2}{b^2} = 1, \qquad a > b > 0,$$

has vertices $(\pm a, 0)$ and endpoints of the minor axis $(0, \pm b)$.

The ellipse with center at the origin, *vertical* major axis, and equation

$$\frac{x^2}{b^2} + \frac{y^2}{a^2} = 1, \qquad a > b > 0,$$

has vertices $(0, \pm a)$ and endpoints of the minor axis $(\pm b, 0)$.

Figure 13.15(a) shows an ellipse having a horizontal major axis; Figure 13.15(b) shows one having a vertical major axis. The coordinates of the vertices V_1 and V_2 and endpoints of the minor axis U_1 and U_2 are labeled.

CRITICAL THINKING

Suppose that $a = b$ for an ellipse centered at (0, 0). What can be said about the ellipse? Explain.

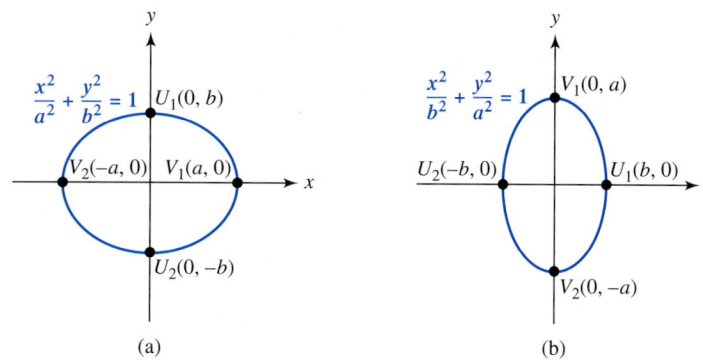

(a)   (b)

Figure 13.15

In the next example we show how to sketch graphs of ellipses.

EXAMPLE 1   **Sketching ellipses**

Sketch a graph of each ellipse. Label the vertices and endpoints of the minor axes.

(a) $\dfrac{x^2}{25} + \dfrac{y^2}{4} = 1$ **(b)** $9x^2 + 4y^2 = 36$

Solution

(a) The equation $\dfrac{x^2}{25} + \dfrac{y^2}{4} = 1$ describes an ellipse with $a^2 = 25$ and $b^2 = 4$. (When you are deciding whether 25 or 4 represents a^2, let a^2 be the larger of the two numbers.) Thus $a = 5$ and $b = 2$, so the ellipse has a horizontal major axis with vertices $(\pm 5, 0)$ and the endpoints of the minor axis are $(0, \pm 2)$. Plot these four points and then sketch the ellipse, as shown in Figure 13.16(a).

(b) To put $9x^2 + 4y^2 = 36$ in standard form, divide each term by 36.

$$9x^2 + 4y^2 = 36 \qquad \text{Given equation}$$

$$\frac{9x^2}{36} + \frac{4y^2}{36} = \frac{36}{36} \qquad \text{Divide each side by 36.}$$

$$\frac{x^2}{4} + \frac{y^2}{9} = 1 \qquad \text{Simplify.}$$

This ellipse has a vertical major axis with $a = 3$ and $b = 2$. The vertices are $(0, \pm 3)$, and the endpoints of the minor axis are $(\pm 2, 0)$, as shown in Figure 13.16(b).

(a)

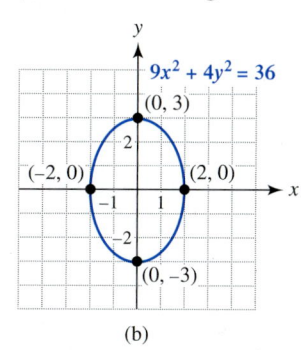

(b)

Figure 13.16

Now Try Exercises **11, 19**

EXAMPLE 2   **Finding the equation of an ellipse**

Use the graph in Figure 13.17 to determine an equation of the ellipse.

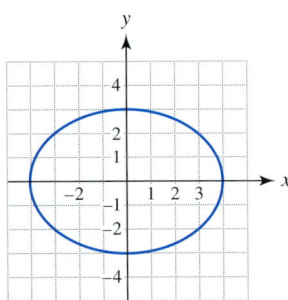

Figure 13.17

Solution

The ellipse is centered at $(0, 0)$ with a horizontal major axis. The length of the major axis is 8, so $a = 4$. The length of the minor axis is 6, so $b = 3$. Thus $a^2 = 16$ and $b^2 = 9$, and the equation of the ellipse is

$$\frac{x^2}{16} + \frac{y^2}{9} = 1.$$

Now Try Exercise 23

▶ **REAL-WORLD CONNECTION**   Planets travel around the sun in elliptical orbits. Astronomers have measured the values of a and b for each planet. Using this information, we can find the equation of a planet's orbit, as illustrated in the next example.

EXAMPLE 3   **Modeling the orbit of Mercury**

The planet Mercury has the least circular orbit of the eight major planets. For Mercury $a = 0.387$ and $b = 0.379$. The units are astronomical units (A.U.), where 1 A.U. equals 93 million miles—the distance between Earth and the sun. Graph $\frac{x^2}{a^2} + \frac{y^2}{b^2} = 1$ to model the orbit of Mercury in $[-0.6, 0.6, 0.1]$ by $[-0.4, 0.4, 0.1]$. Then plot the sun at the point $(0.08, 0)$.   (**Source:** M. Zeilik, *Introductory Astronomy and Astrophysics.*)

Solution
The orbit of Mercury is given by

$$\frac{x^2}{0.387^2} + \frac{y^2}{0.379^2} = 1.$$

To graph an ellipse with some graphing calculators, we must solve the equation for y. Doing so results in two equations.

$$\frac{x^2}{0.387^2} + \frac{y^2}{0.379^2} = 1$$

$$\frac{y^2}{0.379^2} = 1 - \frac{x^2}{0.387^2}$$

$$\frac{y}{0.379} = \pm\sqrt{1 - \frac{x^2}{0.387^2}}$$

$$y = \pm 0.379\sqrt{1 - \frac{x^2}{0.387^2}}$$

The orbit of Mercury results from graphing these two equations. See Figures 13.18(a) and (b). The point $(0.08, 0)$ represents the position of the sun in Figure 13.18(b).

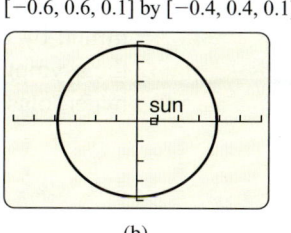

$[-0.6, 0.6, 0.1]$ by $[-0.4, 0.4, 0.1]$

(a)   (b)

Figure 13.18

Now Try Exercise **45**

CRITICAL THINKING

Use Figure 13.18 and the information in Example 3 to estimate the minimum and maximum distances that Mercury is from the sun.

Equations of Hyperbolas

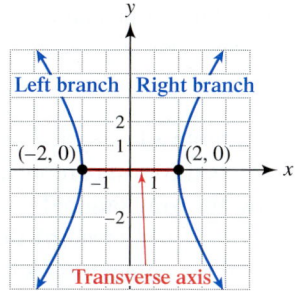

Figure 13.19

The third type of conic section is the **hyperbola**, which is the set of points in a plane, the difference of whose distances from two fixed points is constant. Each fixed point is called a **focus** of the hyperbola. Figure 13.19 shows a hyperbola whose equation is

$$\frac{x^2}{4} - \frac{y^2}{9} = 1.$$

This hyperbola is centered at the origin and has two **branches**, a *left branch* and a *right branch*. The **vertices** are $(-2, 0)$ and $(2, 0)$, and the line segment connecting the vertices is called the **transverse axis**. (The transverse axis is not part of the hyperbola.)

By the vertical line test, a hyperbola cannot be represented by a function, but many hyperbolas can be described by the following equations.

STANDARD EQUATIONS FOR HYPERBOLAS CENTERED AT $(0, 0)$

The hyperbola with center at the origin, *horizontal* transverse axis, and equation

$$\frac{x^2}{a^2} - \frac{y^2}{b^2} = 1$$

has vertices $(\pm a, 0)$.

The hyperbola with center at the origin, *vertical* transverse axis, and equation

$$\frac{y^2}{a^2} - \frac{x^2}{b^2} = 1$$

has vertices $(0, \pm a)$.

Hyperbolas, along with the coordinates of their vertices, are shown in Figure 13.20. The two parts of the hyperbola in Figure 13.20(a) are the *left branch* and *right branch*, whereas in Figure 13.20(b) the hyperbola has an *upper branch* and a *lower branch*. The dashed rectangle in each figure is called the **fundamental rectangle**, and its four vertices are determined by either $(\pm a, \pm b)$ or $(\pm b, \pm a)$. If its diagonals are extended, they correspond to the asymptotes of the hyperbola. The lines $y = \pm \frac{b}{a}x$ and $y = \pm \frac{a}{b}x$ are **asymptotes** for the hyperbolas, respectively, and may be used as an aid to graphing them.

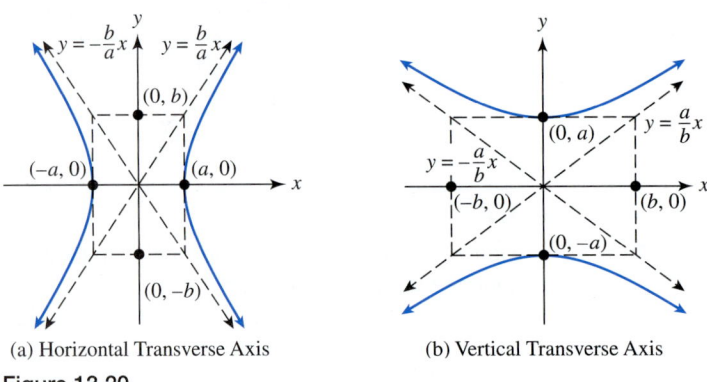

(a) Horizontal Transverse Axis   (b) Vertical Transverse Axis

Figure 13.20

NOTE: A hyperbola consists of two solid curves, or branches. The dashed lines and rectangles are not part of the actual graph but are used as an aid for sketching the hyperbola.

▶ **REAL-WORLD CONNECTION**   One interpretation of an asymptote of a hyperbola can be based on trajectories of comets as they approach the sun. Comets travel in parabolic, elliptic, or hyperbolic trajectories. If the speed of a comet is too slow, the gravitational pull of the sun captures the comet in an elliptic orbit (see Figure 13.21(a)). If the speed of the comet is too fast, the sun's gravity is too weak to capture the comet and the comet passes by it in a hyperbolic trajectory. Near the sun the gravitational pull is stronger, and the comet's trajectory is curved. Farther from the sun, the gravitational pull becomes weaker, and the comet eventually returns to a straight-line trajectory determined by the *asymptote* of the hyperbola (see Figure 13.21(b)). Finally, if the speed is neither too slow nor too fast, the comet will travel in a parabolic path (see Figure 13.21(c)).

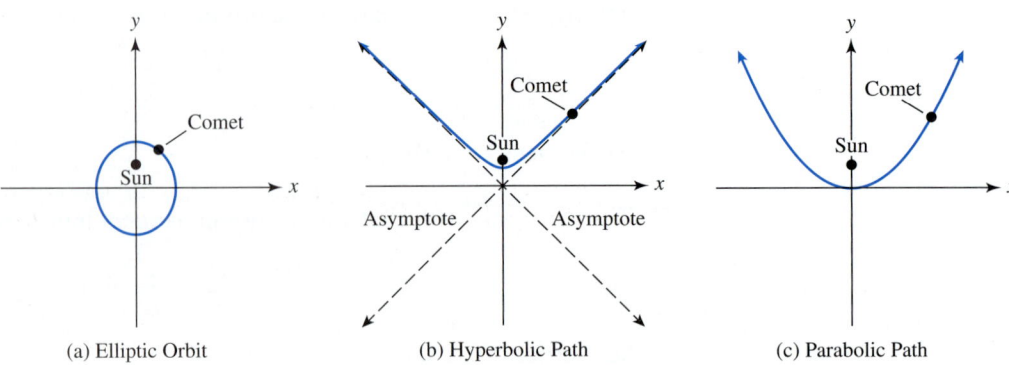

(a) Elliptic Orbit   (b) Hyperbolic Path   (c) Parabolic Path

Figure 13.21

EXAMPLE **4**   **Sketching a hyperbola**

Sketch a graph of $\frac{y^2}{4} - \frac{x^2}{9} = 1$. Label the vertices and show the asymptotes.

Solution

The equation is in standard form with $a^2 = 4$ and $b^2 = 9$, so $a = 2$ and $b = 3$. It has a vertical transverse axis with vertices $(0, -2)$ and $(0, 2)$. The vertices of the fundamental rectangle are $(\pm 3, \pm 2)$, that is, $(3, 2), (3, -2), (-3, 2),$ and $(-3, -2)$. The asymptotes are the diagonals of this rectangle and are given by $y = \pm \frac{a}{b}x$, or $y = \pm \frac{2}{3}x$. Figure 13.22 shows the hyperbola and these features.

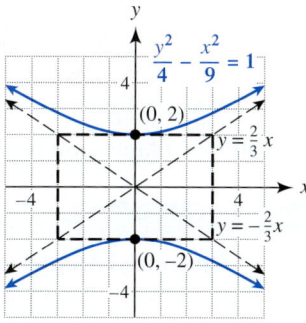

Figure 13.22

Now Try Exercise **27**

TECHNOLOGY NOTE: Graphing a Hyperbola

The graph of a hyperbola does not represent a function. One way to graph a hyperbola with a graphing calculator is to solve the equation for *y* and obtain two equations. One equation gives the upper (or right) half of the hyperbola and the other equation gives the lower (or left) half. For example, to graph $\frac{y^2}{4} - \frac{x^2}{8} = 1$, begin by solving for *y*.

$$\frac{y^2}{4} = 1 + \frac{x^2}{8} \qquad \text{Add } \tfrac{x^2}{8}.$$

$$y^2 = 4\left(1 + \frac{x^2}{8}\right) \qquad \text{Multiply by 4.}$$

$$y = \pm 2\sqrt{1 + \frac{x^2}{8}} \qquad \text{Square root property}$$

Graph $Y_1 = 2\sqrt{(1 + X^2/8)}$ and $Y_2 = -2\sqrt{(1 + X^2/8)}$. See Figure 13.23.

$[-6, 6, 1]$ by $[-4, 4, 1]$

(a)

(b)

Figure 13.23

EXAMPLE   **Determining the equation of a hyperbola from its graph**

Use the graph shown in Figure 13.24 to determine an equation of the hyperbola.

Solution

The hyperbola has a horizontal transverse axis, so the x^2-term must come first in the equation. The vertices of the hyperbola are $(\pm 3, 0)$, which indicates that $a = 3$ and so $a^2 = $ **9**. The value of b can be found by noting that one of the asymptotes passes through the point $(3, 4)$. This asymptote has the equation $y = \frac{b}{a}x$ or $y = \frac{4}{3}x$, so let $b = 4$ and $b^2 = $ **16**. The equation of the hyperbola is $\frac{x^2}{9} - \frac{y^2}{16} = 1$.

Now Try Exercise **39**

Figure 13.24

13.2 PUTTING IT ALL TOGETHER

The following table summarizes some basic concepts about ellipses and hyperbolas.

Concept	Description	
Ellipses Centered at $(0, 0)$ with $a > b > 0$	***Horizontal Major Axis*** Vertices: $(a, 0)$ and $(-a, 0)$ Endpoints of minor axis: $(0, b)$ and $(0, -b)$ $\dfrac{x^2}{a^2} + \dfrac{y^2}{b^2} = 1$	***Vertical Major Axis*** Vertices: $(0, a)$ and $(0, -a)$ Endpoints of minor axis: $(-b, 0)$ and $(b, 0)$ $\dfrac{x^2}{b^2} + \dfrac{y^2}{a^2} = 1$
Hyperbolas Centered at $(0, 0)$ with $a > 0$ and $b > 0$	***Horizontal Transverse Axis*** Vertices: $(a, 0)$ and $(-a, 0)$ Asymptotes: $y = \pm\dfrac{b}{a}x$ $\dfrac{x^2}{a^2} - \dfrac{y^2}{b^2} = 1$	***Vertical Transverse Axis*** Vertices: $(0, a)$ and $(0, -a)$ Asymptotes: $y = \pm\dfrac{a}{b}x$ $\dfrac{y^2}{a^2} - \dfrac{x^2}{b^2} = 1$

13.2 Exercises

CONCEPTS

1. Sketch an ellipse with a horizontal major axis.

2. Sketch a hyperbola with a vertical transverse axis.

3. The ellipse whose standard equation is $\frac{x^2}{a^2} + \frac{y^2}{b^2} = 1$, $a > b > 0$, has a _____ major axis.

4. The ellipse whose standard equation is $\frac{x^2}{b^2} + \frac{y^2}{a^2} = 1$, $a > b > 0$, has a _____ major axis.

5. What is the maximum number of times that a line can intersect an ellipse?

6. What is the maximum number of times that a parabola can intersect an ellipse?

7. The hyperbola whose equation is $\frac{x^2}{a^2} - \frac{y^2}{b^2} = 1$ has _____ and _____ branches.

8. The hyperbola whose equation is $\frac{y^2}{a^2} - \frac{x^2}{b^2} = 1$ has _____ and _____ branches.

9. How are the asymptotes of a hyperbola related to the fundamental rectangle?

10. Could an ellipse be centered at the origin and have vertices $(4, 0)$ and $(0, -5)$?

ELLIPSES

Exercises 11–22: Graph the ellipse. Label the vertices and endpoints of the minor axis.

11. $\dfrac{x^2}{9} + \dfrac{y^2}{25} = 1$

12. $\dfrac{y^2}{9} + \dfrac{x^2}{25} = 1$

13. $\dfrac{x^2}{9} + \dfrac{y^2}{4} = 1$

14. $\dfrac{x^2}{3} + \dfrac{y^2}{9} = 1$

15. $x^2 + \dfrac{y^2}{4} = 1$

16. $\dfrac{x^2}{9} + y^2 = 1$

17. $\dfrac{y^2}{5} + \dfrac{x^2}{7} = 1$

18. $\dfrac{y^2}{11} + \dfrac{x^2}{6} = 1$

19. $36x^2 + 4y^2 = 144$

20. $25x^2 + 16y^2 = 400$

21. $6y^2 + 7x^2 = 42$

22. $9x^2 + 5y^2 = 45$

Exercises 23–26: Use the graph to determine the equation of the ellipse.

23.

24.

25.

26.

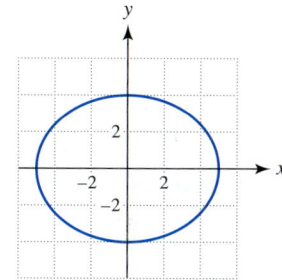

HYPERBOLAS

Exercises 27–38: Graph the hyperbola. Show the asymptotes and vertices.

27. $\dfrac{x^2}{4} - \dfrac{y^2}{9} = 1$

28. $\dfrac{y^2}{4} - \dfrac{x^2}{9} = 1$

29. $\dfrac{x^2}{25} - \dfrac{y^2}{16} = 1$

30. $\dfrac{y^2}{25} - \dfrac{x^2}{16} = 1$

31. $x^2 - y^2 = 1$

32. $y^2 - x^2 = 1$

33. $\dfrac{x^2}{3} - \dfrac{y^2}{4} = 1$

34. $\dfrac{y^2}{5} - \dfrac{x^2}{8} = 1$

35. $9y^2 - 4x^2 = 36$

36. $36x^2 - 25y^2 = 900$

37. $16x^2 - 4y^2 = 64$

38. $y^2 - 9x^2 = 9$

Exercises 39–42: Use the graph to determine an equation of the hyperbola.

39.

40.

41.

42.

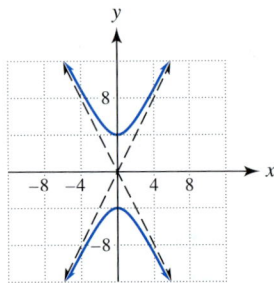

APPLICATIONS

43. *Geometry of an Ellipse* The area inside an ellipse is given by $A = \pi ab$, and its perimeter can be approximated by

$$P = 2\pi\sqrt{\frac{a^2 + b^2}{2}}.$$

Approximate A and P to the nearest hundredth for each ellipse.

(a) $\dfrac{x^2}{16} + \dfrac{y^2}{25} = 1$ **(b)** $\dfrac{x^2}{7} + \dfrac{y^2}{2} = 1$

44. *Geometry of an Ellipse* (Refer to Exercise 43.) If $a = b$ in the equation for an ellipse, the ellipse is a circle. Let $a = b$ in the formulas for the area and perimeter of an ellipse. Do the equations simplify to the area and perimeter for a circle? Explain.

45. *Planet Orbit* (Refer to Example 3.) Pluto's orbit is less circular than that of any of the eight planets. For Pluto $a = 39.44$ A.U. and $b = 38.20$. A.U.

(a) Graph the elliptic orbit of Pluto in the window $[-60, 60, 10]$ by $[-40, 40, 10]$. Plot the point $(9.81, 0)$ to show the position of the sun. Assume that the major axis is horizontal.

(b) Use the information in Exercise 43 to determine how far Pluto travels in one orbit around the sun and approximate the area inside its orbit.

46. *Halley's Comet* (Refer to Example 3.) The famous Halley's comet travels in an elliptical orbit with $a = 17.95$ and $b = 4.44$ and passes by Earth roughly every 76 years. The most recent pass by Earth was in February 1986.   (*Source:* M. Zeilik.)

(a) Graph the orbit of Halley's comet in $[-21, 21, 5]$ by $[-14, 14, 5]$. Assume that the major axis is horizontal and that all units are in astronomical units. Plot a point at $(17.39, 0)$ to represent the position of the sun.

(b) Use the formula in Exercise 43 to estimate how many miles Halley's comet travels in one orbit around the sun.

(c) Estimate the average speed of Halley's comet in miles per hour.

47. *Satellite Orbit* The orbit of Explorer VII and the outline of Earth's surface are shown in the accompanying figure. This orbit is described by

$$\frac{x^2}{4464^2} + \frac{y^2}{4462^2} = 1,$$

and the surface of Earth is described by

$$\frac{(x - 164)^2}{3960^2} + \frac{y^2}{3960^2} = 1.$$

Find the maximum and minimum heights of the satellite above Earth's surface if all units are miles.
(*Source:* W. Thomson, *Introduction to Space Dynamics.*)

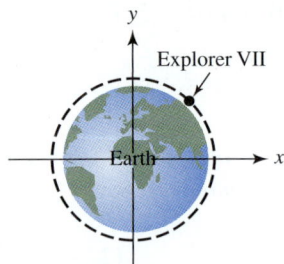

48. *Weight Machines* Elliptic shapes are used rather than circular shapes in modern weight machines. Suppose that the ellipse shown in the accompanying figure is represented by the equation

$$\frac{x^2}{16} + \frac{y^2}{100} = 1,$$

where the units are inches. Find r_1 and r_2.

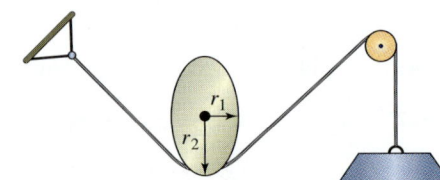

49. *Arch Bridge*   The arch under a bridge is designed as the upper half of an ellipse, as illustrated in the accompanying figure. Its equation is modeled by

$$400x^2 + 10,000y^2 = 4,000,000,$$

where the units are feet. Find the height and width of the arch.

50. *Population Growth*   Suppose that the population y of a country can be modeled by the upper right branch of the hyperbola

$$\frac{x^2}{a^2} - \frac{y^2}{b^2} = 1,$$

where x represents time in years. What happens to the population after a long period of time?

WRITING ABOUT MATHEMATICS

51. Explain how the values of a and b affect the graph of $\frac{x^2}{a^2} + \frac{y^2}{b^2} = 1$. Assume that $a > b > 0$.

52. Explain how the values of a and b affect the graph of $\frac{x^2}{a^2} - \frac{y^2}{b^2} = 1$. Assume that a and b are positive.

CHECKING BASIC CONCEPTS

SECTIONS 13.1 AND 13.2

1. Graph the parabola $x = (y - 2)^2 + 1$.   Find the vertex and axis of symmetry.

2. Find the equation of the circle with center $(1, -2)$ and radius 2. Graph the circle.

3. Find the x- and y-intercepts on the graph of
$$\frac{x^2}{4} + \frac{y^2}{9} = 1.$$

4. Graph the following. Label any vertices and state the type of conic section that it represents.

(a) $x = y^2$ (b) $\dfrac{x^2}{16} + \dfrac{y^2}{25} = 1$

(c) $\dfrac{x^2}{4} - \dfrac{y^2}{9} = 1$

(d) $(x - 1)^2 + (y + 2)^2 = 9$

13.3   NONLINEAR SYSTEMS OF EQUATIONS AND INEQUALITIES

Basic Concepts ▪ Solving Nonlinear Systems of Equations ▪ Solving Nonlinear Systems of Inequalities

A LOOK INTO MATH ▷

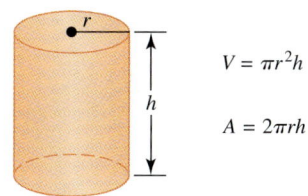

$V = \pi r^2 h$

$A = 2\pi r h$

Figure 13.25 Cylindrical Container

To describe characteristics of curved objects, we often need *nonlinear equations*. The equations of the conic sections discussed in this chapter are but a few examples of nonlinear equations. For instance, cylinders have a curved shape, as illustrated in Figure 13.25. If the radius of a cylinder is denoted r and its height h, then its volume V is given by the nonlinear equation $V = \pi r^2 h$ and its side area A is given by the nonlinear equation $A = 2\pi r h$.

If we want to manufacture a cylindrical container that holds 35 cubic inches and whose side area is 50 square inches, we need to solve the following **nonlinear system of equations**. (This system is solved in Example 4.)

$$\pi r^2 h = 35$$
$$2\pi r h = 50$$

In this section we solve nonlinear systems of equations and inequalities.

Basic Concepts

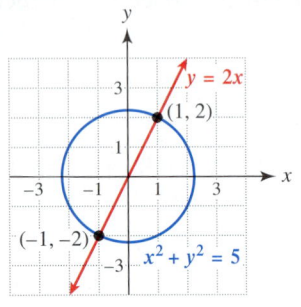

Figure 13.26

One way to locate the points at which the line $y = 2x$ intersects the circle $x^2 + y^2 = 5$, is to graph both equations (see Figure 13.26).

The equation describing the circle is nonlinear. Another way to locate the points of intersection is symbolically, by solving the nonlinear system of equations.

$$
\begin{aligned}
y &= 2x \\
x^2 + y^2 &= 5
\end{aligned}
$$

Linear systems of equations can have no solutions, one solution, or infinitely many solutions. It is possible for a nonlinear system of equations to have *any number* of solutions. Figure 13.26 shows that this nonlinear system of equations has two solutions: $(-1, -2)$ and $(1, 2)$.

Solving Nonlinear Systems of Equations

Nonlinear systems of equations in two variables can sometimes be solved graphically, numerically, and symbolically. One symbolic technique is the **method of substitution**, which we demonstrate in the next example.

EXAMPLE **1**   Solving a nonlinear system of equations symbolically

Solve the system of equations symbolically. Check any solutions.

$$
\begin{aligned}
y &= 2x \\
x^2 + y^2 &= 5
\end{aligned}
$$

Solution
Substitute $2x$ for y in the second equation and solve for x.

$$
\begin{array}{ll}
x^2 + (2x)^2 = 5 & \text{Let } y = 2x \text{ in the second equation.} \\
x^2 + 4x^2 = 5 & \text{Properties of exponents} \\
5x^2 = 5 & \text{Combine like terms.} \\
x^2 = 1 & \text{Divide by 5.} \\
x = \pm 1 & \text{Square root property}
\end{array}
$$

To determine corresponding y-values, substitute $x = \pm 1$ in $y = 2x$; the solutions are $(1, 2)$ and $(-1, -2)$. To check $(1, 2)$, substitute $x = 1$ and $y = 2$ in the given equations.

$$
\begin{array}{ll}
2 \overset{?}{=} 2(1) & \text{True} \\
(1)^2 + (2)^2 \overset{?}{=} 5 & \text{True}
\end{array}
$$

To check $(-1, -2)$, substitute $x = -1$ and $y = -2$ in the given equations.

$$
\begin{array}{ll}
-2 \overset{?}{=} 2(-1) & \text{True} \\
(-1)^2 + (-2)^2 \overset{?}{=} 5 & \text{True}
\end{array}
$$

The solutions check.

Now Try Exercise **13**

In the next example we solve a nonlinear system of equations graphically and symbolically. The symbolic solution gives the exact solutions.

EXAMPLE 2   Solving a nonlinear system of equations

Solve the nonlinear system of equations graphically and symbolically.

$$x^2 - y = 2$$
$$x^2 + y = 4$$

Solution

Graphical Solution   Begin by solving each equation for y.

$$y = x^2 - 2$$
$$y = 4 - x^2$$

$[-6, 6, 1]$ by $[-4, 4, 1]$

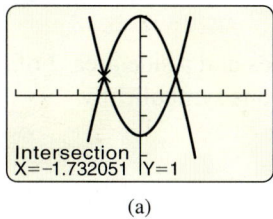

(a)

$[-6, 6, 1]$ by $[-4, 4, 1]$

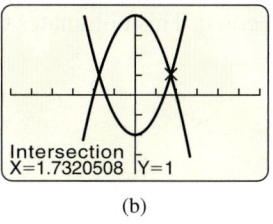

(b)

Figure 13.27

Graph $Y_1 = X^2 - 2$ and $Y_2 = 4 - X^2$. The solutions are approximately $(-1.73, 1)$ and $(1.73, 1)$, as shown in Figure 13.27. The graphs consist of two parabolas intersecting at two points.

Symbolic Solution   Solving the first equation for y gives $y = x^2 - 2$. Substitute this expression for y in the second equation and solve for x.

$x^2 + \boldsymbol{y} = 4$	Second equation
$x^2 + (\boldsymbol{x^2 - 2}) = 4$	Substitute $y = x^2 - 2$.
$2x^2 = 6$	Combine like terms; add 2.
$x^2 = 3$	Divide by 2.
$x = \pm\sqrt{3}$	Square root property

To determine y, substitute $x = \pm\sqrt{3}$ in $y = x^2 - 2$.

$$y = (\sqrt{3})^2 - 2 = 3 - 2 = \boldsymbol{1}$$
$$y = (-\sqrt{3})^2 - 2 = 3 - 2 = \boldsymbol{1}$$

The *exact* solutions are $(\sqrt{3}, \boldsymbol{1})$ and $(-\sqrt{3}, \boldsymbol{1})$.   Now Try Exercise 29(a), (b)

EXAMPLE 3   Solving a nonlinear system of equations

Solve the nonlinear system of equations symbolically and graphically.

$$x^2 - y^2 = 3$$
$$x^2 + y^2 = 5$$

Solution

Symbolic Solution   Instead of using substitution on this nonlinear system of equations, we use elimination. Note that if we add the two equations, the y-variable will be eliminated.

$$
\begin{array}{rl}
x^2 - y^2 &= 3 \\
\underline{x^2 + y^2} &= 5 \\
2x^2 &= 8
\end{array}
$$
Add equations.

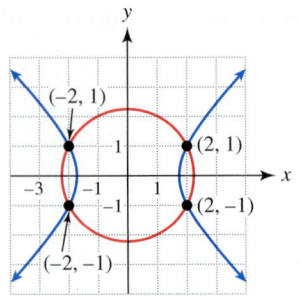

Figure 13.28

Solving gives $x^2 = 4$, or $x = \pm 2$. To determine y, substitute **4** for x^2 in $x^2 + y^2 = 5$.

$$4 + y^2 = 5 \quad \text{or} \quad y^2 = 1$$

Because $y^2 = 1$, $y = \pm 1$. There are four solutions: $(2, 1)$, $(2, -1)$, $(-2, 1)$, and $(-2, -1)$.

Graphical Solution The graph of the first equation is a hyperbola, and the graph of the second is a circle with radius $\sqrt{5}$. The four points of intersection (solutions) are $(\pm 2, \pm 1)$, as shown in Figure 13.28.

Now Try Exercise 23

In the next example we solve the system of equations presented in the introduction.

EXAMPLE 4 **Modeling the dimensions of a can**

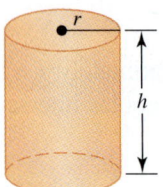

Find the dimensions of a can having a volume V of 35 cubic inches and a side area A of 50 square inches by solving the following nonlinear system of equations symbolically.

$$\pi r^2 h = 35$$
$$2\pi r h = 50$$

Solution
We can find r by solving each equation for h and then setting them equal. This eliminates the variable h.

$$\frac{50}{2\pi r} = \frac{35}{\pi r^2} \qquad h = \tfrac{50}{2\pi r} \text{ and } h = \tfrac{35}{\pi r^2}$$

$$50\pi r^2 = 70\pi r \qquad \text{Clear fractions (cross multiply).}$$

$$50\pi r^2 - 70\pi r = 0 \qquad \text{Subtract } 70\pi r.$$

$$10\pi r(5r - 7) = 0 \qquad \text{Factor out } 10\pi r.$$

$$10\pi r = 0 \quad \text{or} \quad 5r - 7 = 0 \qquad \text{Zero-product property}$$

$$r = 0 \quad \text{or} \quad r = \tfrac{7}{5} = 1.4 \qquad \text{Solve.}$$

Because $h = \frac{50}{2\pi r}$, $r = 0$ is not possible, but we can find h by substituting **1.4** for r in the formula.

$$h = \frac{50}{2\pi(\mathbf{1.4})} \approx 5.68$$

A can having a volume of 35 cubic inches and a side area of 50 square inches has a radius of 1.4 inches and a height of about 5.68 inches.

Now Try Exercise 51

Solving Nonlinear Systems of Inequalities

In Section 4.4 we solved systems of linear inequalities. A **nonlinear system of inequalities** in two variables can be solved similarly by using graphical techniques. For example, consider the nonlinear system of inequalities

$$y \geq x^2 - 2$$
$$y \leq 4 - x^2.$$

The graph of $y = x^2 - 2$ is a parabola opening upward. The solution set to $y \geq x^2 - 2$ includes all points lying on or above this parabola. See Figure 13.29(a). Similarly, the graph of $y = 4 - x^2$ is a parabola opening downward. The solution set to $y \leq 4 - x^2$ includes all points lying on or below this parabola. See Figure 13.29(b).

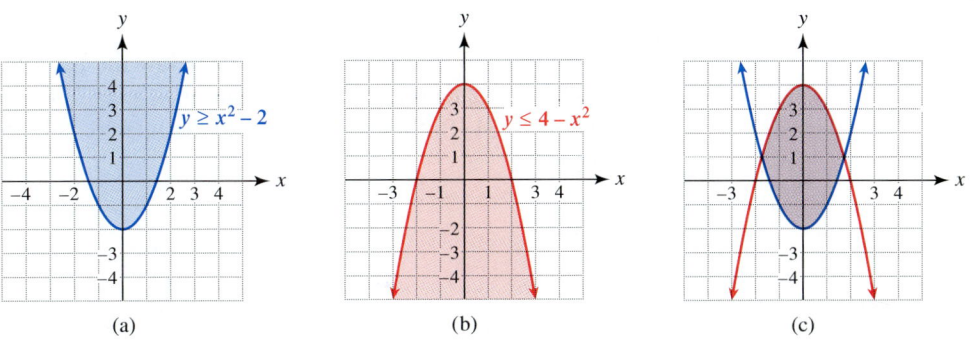

(a)   (b)   (c)

Figure 13.29

The solution set for this nonlinear *system* of inequalities includes all points (x, y) in *both* shaded regions. The *intersection* of the shaded regions is shown in Figure 13.29(c).

EXAMPLE 5 Solving a nonlinear system of inequalities graphically

Shade the solution set for the system of inequalities.

$$\frac{x^2}{4} + \frac{y^2}{9} < 1$$

$$y > 1$$

Solution

The solutions to $\frac{x^2}{4} + \frac{y^2}{9} < 1$ lie *inside* the ellipse $\frac{x^2}{4} + \frac{y^2}{9} = 1$. See Figure 13.30(a). Solutions to $y > 1$ lie above the line $y = 1$, as shown in Figure 13.30(b). The intersection of these two regions is shown in Figure 13.30(c). Any point in this region is a solution. For example, the point $(0, 2)$ lies in the shaded region and is a solution to the system. Note that a dashed curve and line are used when equality is not included.

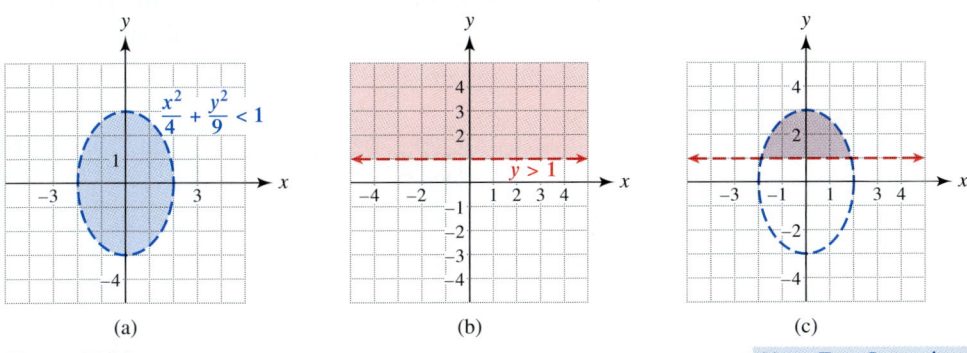

(a)   (b)   (c)

Figure 13.30

Now Try Exercise **39**

EXAMPLE 6   **Solving a nonlinear system of inequalities graphically**

Shade the solution set for the following system of inequalities.

$$x^2 + y \leq 4$$
$$-x \; + y \geq 2$$

Solution

The solutions to $x^2 + y \leq 4$ lie on or below the parabola $y = -x^2 + 4$, and the solutions to $-x + y \geq 2$ lie on or above the line $y = x + 2$. The appropriate shaded region is shown in Figure 13.31. Both the parabola and the line are solid because equality is included in both inequalities.

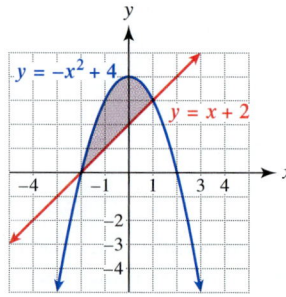

Figure 13.31

Now Try Exercise 37

In the next example we use a graphing calculator to shade a region that lies above both graphs, using the "$Y_1 =$" menu. This feature allows us to shade either above or below the graph of a function.

EXAMPLE 7   **Solving a system of inequalities with a graphing calculator**

Shade the solution set for the following system of inequalities.

$$y \geq x^2 - 2$$
$$y \geq -1 - x$$

Solution

Enter $Y_1 = X^2 - 2$ and $Y_2 = -1 - X$, as shown in Figure 13.32(a). Note that the option to shade above the graphs of Y_1 and Y_2 was selected to the left of Y_1 and Y_2. Then the two inequalities were graphed in Figure 13.32(b). The solution set corresponds to the region where there are both vertical and horizontal lines.

CALCULATOR HELP

To shade the solution set to a system of inequalities, see Appendix A (page AP-7).

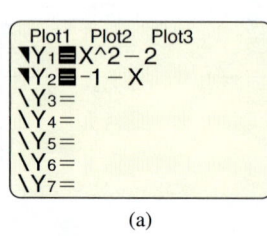

(a)

$[-4.7, 4.7, 1]$ by $[-3.1, 3.1, 1]$

(b)

Figure 13.32

Now Try Exercise 49

13.3 PUTTING IT ALL TOGETHER

In this section we discussed nonlinear systems of equations in two variables. Unlike a linear system of equations, a nonlinear system of equations can have *any number of solutions*. Nonlinear systems of equations can be solved symbolically and graphically. Nonlinear systems of inequalities involving two variables usually have infinitely many solutions, which can be represented by a shaded region in the xy-plane. The following table summarizes these concepts.

Concept	Explanation
Nonlinear Systems of Equations in Two Variables	To solve the following system of equations symbolically, using *substitution*, solve the first equation for y. $$x + y = 5 \quad \text{or} \quad y = 5 - x$$ $$x^2 - y = 1$$ Substitute $5 - x$ for y in the second equation and solve the resulting quadratic equation. (*Elimination* can also be used on this system.) $$x^2 - (5 - x) = 1 \quad \text{or} \quad x^2 + x - 6 = 0$$ implies that $$x = -3 \quad \text{or} \quad x = 2.$$ Then $y = 5 - (-3) = 8$ or $y = 5 - 2 = 3$. The solutions are $(-3, 8)$ and $(2, 3)$. Graphical support is shown in the accompanying figure. 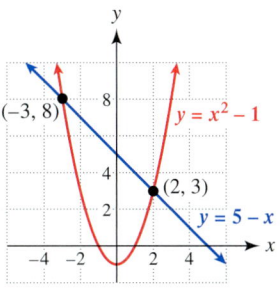
Nonlinear Systems of Inequalities in Two Variables	To solve the following system of inequalities graphically, solve each inequality for y. $$x + y \leq 5 \quad \text{or} \quad y \leq 5 - x$$ $$x^2 - y \leq 1 \quad \text{or} \quad y \geq x^2 - 1$$ The solutions lie on or above the parabola and on or below the line, as shown in the figure.

13.3 Exercises

CONCEPTS

1. How many solutions can a nonlinear system of equations have?

2. If a nonlinear system of equations has two equations, how many equations does a solution have to satisfy?

3. Determine visually the number of solutions to the following system of equations. Explain your reasoning.

$$y = x$$
$$x^2 + y^2 = 4$$

4. Describe the solution set to $x^2 + y^2 \leq 1$.

5. Does $(-2, -1)$ satisfy $5x^2 - 2y^2 > 18$?

6. Does $(3, 4)$ satisfy $x^2 - 2y \geq 4$?

7. Sketch a parabola and ellipse with four points of intersection.

8. Sketch a line and a hyperbola with two points of intersection.

NONLINEAR SYSTEMS OF EQUATIONS

Exercises 9–12: Use the graph to estimate all solutions to the system of equations. Check each solution.

9. $x^2 + y^2 = 10$
    $y = 3x$

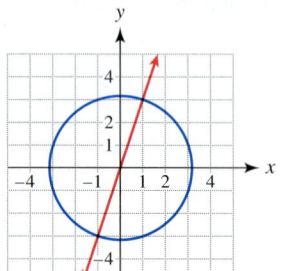

10. $x^2 + 3y^2 = 16$
    $y = -x$

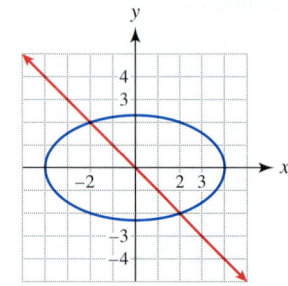

11. $y^2 - x^2 = 1$
    $x^2 + 3y^2 = 3$

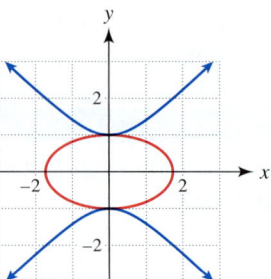

12. $y = 1 - x^2$
    $x^2 + y^2 = 1$

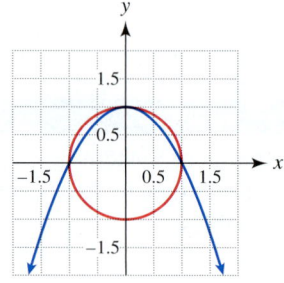

Exercises 13–24: Solve the system of equations symbolically. Check your solutions.

13. $y = 2x$
    $x^2 + y^2 = 45$

14. $y = x$
    $y^2 = 3 - 2x^2$

15. $x + y = 1$
    $x^2 - y^2 = 3$

16. $y - x = -1$
    $y = 2x^2$

17. $y - x^2 = 0$
    $x^2 + y^2 = 6$

18. $x^2 - y^2 = 4$
    $x^2 + y^2 = 4$

19. $3x^2 + 2y^2 = 5$
    $x - y = -2$

20. $x^2 + 2y^2 = 9$
    $2x^2 - y^2 = -2$

21. $x^2 + y^2 = 4$
    $x^2 - 9y^2 = 9$

22. $x^2 + y^2 = 15$
    $9x^2 + 4y^2 = 36$

23. $x^2 + y^2 = 10$
    $2x^2 - y^2 = 17$

24. $2x^2 + 3y^2 = 5$
    $3x^2 - 4y^2 = -1$

Exercises 25–28: Solve the system of equations graphically. Check your solutions.

25. $y = x^2 - 3$
    $2x^2 - y = 1 - 3x$

26. $x + y = 2$
    $x - y^2 = 3$

27. $y - x = -4$
    $x - y^2 = -2$

28. $xy = 1$
    $y = x$

Exercises 29–32: Solve the system of equations
   (a) *symbolically,*
   (b) *graphically, and*
   (c) *numerically.*

29. $\quad y = -2x$ **30.** $4x - y = 0$

$\quad\quad x^2 + y = 3$ $\quad x^3 - y = 0$

31. $\quad xy = 1$ **32.** $x^2 + y^2 = 4$

$\quad\quad x - y = 0$ $\quad y - x = 2$

NONLINEAR SYSTEMS OF INEQUALITIES

Exercises 33–36: Shade the solution set in the xy-plane.

33. $y \geq x^2$ **34.** $y \leq x^2 - 1$

35. $\dfrac{x^2}{4} + \dfrac{y^2}{9} > 1$ **36.** $x^2 + y^2 \leq 1$

Exercises 37–44: Shade the solution set in the xy-plane. Then use the graph to select one solution.

37. $y > x^2 + 1$ **38.** $y > x^2$

$\quad\quad y < 3$ $\quad y < x + 2$

39. $x^2 + y^2 \leq 1$ **40.** $y > x^2 - 2$

$\quad\quad\quad y < x$ $\quad y \leq 2 - x^2$

41. $\quad\quad x^2 + y^2 \leq 1$ **42.** $\quad\quad x^2 - y \geq 2$

$\quad\quad (x - 2)^2 + y^2 \leq 1$ $\quad (x + 1)^2 + y^2 \leq 4$

43. $x^2 - y^2 \leq 4$ **44.** $3x + 2y < 6$

$\quad\quad x^2 + y^2 \leq 9$ $\quad x^2 + y^2 \leq 16$

Exercises 45 and 46: Match the inequality or system of inequalities with its graph (a. or b.).

45. $y \leq \dfrac{1}{2}x^2$ **46.** $y \geq x^2 + 1$

$\quad\quad\quad\quad\quad\quad\quad\quad\quad y \leq 5$

a.

b.

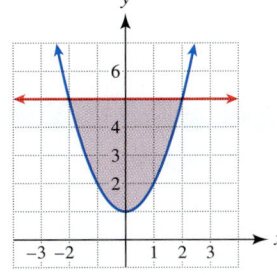

Exercises 47 and 48: Use the graph to write the inequality or system of inequalities.

47.   **48.**

   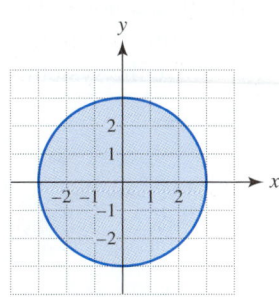

Exercises 49 and 50: Use a graphing calculator to shade the solution set to the system of inequalities.

49. $y \geq x^2 - 1$ **50.** $y \leq 4 - x^2$

$\quad\quad y \geq 2 - x$ $\quad y \leq x^2 - 4$

APPLICATIONS

51. *Dimensions of a Can* (Refer to Example 4.) Find the dimensions of a cylindrical container with a volume of 40 cubic inches and a side area of 50 square inches **(a)** graphically and **(b)** symbolically.

52. *Dimensions of a Can* (Refer to Example 4.) Is it possible to design an aluminum can with volume of 60 cubic inches and side area of 60 square inches? If so, find the dimensions of the can.

53. *Dimensions of a Cone* The volume V of a cone is given by $V = \frac{1}{3}\pi r^2 h$, and the surface area S of its side is given by $S = \pi r \sqrt{r^2 + h^2}$, where h is the height and r is the radius of the base (see the accompanying figure).

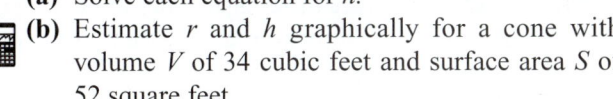

   (a) Solve each equation for h.

   (b) Estimate r and h graphically for a cone with volume V of 34 cubic feet and surface area S of 52 square feet.

54. *Area and Perimeter*   The area of a room is 143 feet, and its perimeter is 48 feet. Let x be the width and y be the length of the room. See the accompanying figure.
   (a) Write a nonlinear system of equations that models this situation.
   (b) Solve the system.

WRITING ABOUT MATHEMATICS

55. A student *incorrectly* changes the system of inequalities

$$\begin{array}{c} x^2 - y \geq 6 \\ 2x - y \leq -3 \end{array} \quad \text{to} \quad \begin{array}{c} y \geq x^2 - 6 \\ y \leq 2x + 3. \end{array}$$

The student discovers that $(1, 2)$ satisfies the second system but not the first. Explain the student's error.

56. Explain graphically how nonlinear systems of equations can have any number of solutions. Sketch graphs of different systems with zero, one, two, and three solutions.

CHECKING BASIC CONCEPTS
SECTION 13.3

1. Solve the following system of equations symbolically and graphically.

$$x^2 - y = 2x$$
$$2x - y = 3$$

2. Determine visually the number of solutions to the following system of equations.

$$y = x^2 - 4$$
$$y = x$$

3. The solution set for a system of inequalities is shown in the accompanying figure.
   (a) Find one ordered pair (x, y) that is a solution and one that is not.

(b) Write the system of inequalities represented by the graph.

4. Shade the solution set for the following system of inequalities.

$$x^2 + y^2 \leq 4$$
$$y < 1$$

CHAPTER 13   SUMMARY
SECTION 13.1 ■ PARABOLAS AND CIRCLES

Parabolas   There are three basic types of conic sections: parabolas, ellipses, and hyperbolas. A parabola can have a vertical or a horizontal axis. Two forms of an equation for a parabola with a *vertical axis* are

$$y = ax^2 + bx + c \quad \text{and} \quad y = a(x - h)^2 + k.$$

If $a > 0$ the parabola opens upward, and if $a < 0$ it opens downward (see the figure on the left). The vertex is located at (h, k). Two forms of an equation for a parabola with a *horizontal axis* are

$$x = ay^2 + by + c \quad \text{and} \quad x = a(y - k)^2 + h.$$

If $a > 0$ the parabola opens to the right, and if $a < 0$ it opens to the left (see the figure on the right). The vertex is located at (h, k).

Vertical Axis   Horizontal Axis

Circles   The standard equation for a circle with center (h, k) and radius r is

$$(x - h)^2 + (y - k)^2 = r^2.$$

SECTION 13.2 ■ ELLIPSES AND HYPERBOLAS

Ellipses   The standard equation for an ellipse centered at the origin with a *horizontal major axis* is $\frac{x^2}{a^2} + \frac{y^2}{b^2} = 1$, $a > b > 0$, and the vertices are $(\pm a, 0)$, as shown in the figure on the left. The standard equation for an ellipse centered at the origin with a *vertical major axis* is $\frac{x^2}{b^2} + \frac{y^2}{a^2} = 1$, $a > b > 0$, and the vertices are $(0, \pm a)$, as shown in the figure on the right. Circles are a special type of ellipse, with the major and minor axes having equal lengths.

Horizontal Major Axis   Vertical Major Axis

Hyperbolas   The standard equation for a hyperbola centered at the origin with a *horizontal transverse axis* is $\frac{x^2}{a^2} - \frac{y^2}{b^2} = 1$, the asymptotes are given by $y = \pm \frac{b}{a}x$, and the vertices are $(\pm a, 0)$, as shown in the figure on the left on the next page. The standard equation for a hyperbola centered at the origin with a *vertical transverse axis* is $\frac{y^2}{a^2} - \frac{x^2}{b^2} = 1$, the asymptotes are $y = \pm\frac{a}{b}x$, and the vertices are $(0, \pm a)$, as shown in the figure on the right.

Horizontal Transverse Axis

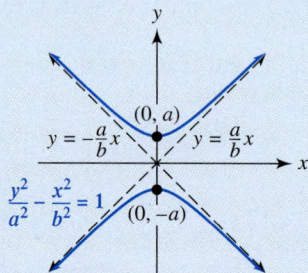

Vertical Transverse Axis

SECTION 13.3 ▪ NONLINEAR SYSTEMS OF EQUATIONS AND INEQUALITIES

Nonlinear Systems   Nonlinear systems of equations can have any number of solutions. The methods of substitution or elimination can often be used to solve a nonlinear system of equations symbolically. Nonlinear systems can also be solved graphically. The solution set for a nonlinear system of two inequalities in two variables is typically a region in the xy-plane. A solution is an ordered pair (x, y) that satisfies both inequalities.

Example: Solve $y \geq x^2 - 2$
$$y \leq 4 - \tfrac{1}{2}x^2.$$

CHAPTER 13   REVIEW EXERCISES

SECTION 13.1

Exercises 1–6: Graph the parabola. Find the vertex and axis of symmetry.

1. $x = 2y^2$

2. $x = -(y + 1)^2$

3. $x = -2(y - 2)^2$

4. $x = (y + 2)^2 - 1$

5. $x = -3y^2 + 1$

6. $x = \tfrac{1}{2}y^2 + y - 3$

7. Use the graph to determine the equation of the parabola.

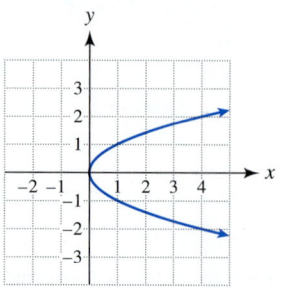

8. Use the graph to find the equation of the circle.

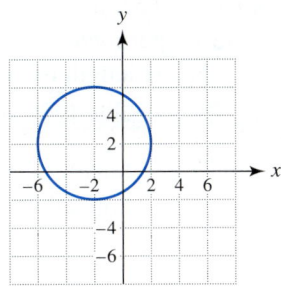

9. Write the equation of the circle with radius 1 and center $(0, 0)$.

10. Write the equation of the circle with radius 4 and center $(2, -3)$.

Exercises 11–14: Find the radius and center of the circle. Then graph the circle.

11. $x^2 + y^2 = 25$

12. $(x - 2)^2 + y^2 = 9$

13. $(x + 3)^2 + (y - 1)^2 = 5$

14. $x^2 - 2x + y^2 + 2y = 7$

SECTION 13.2

Exercises 15–18: Graph the ellipse. Label the vertices and endpoints of the minor axis.

15. $\dfrac{x^2}{4} + \dfrac{y^2}{25} = 1$ **16.** $x^2 + \dfrac{y^2}{4} = 1$

17. $25x^2 + 20y^2 = 500$ **18.** $4x^2 + 9y^2 = 36$

19. Use the graph to determine the equation of the ellipse.

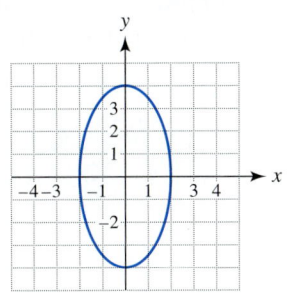

20. Use the graph to find the equation of the hyperbola.

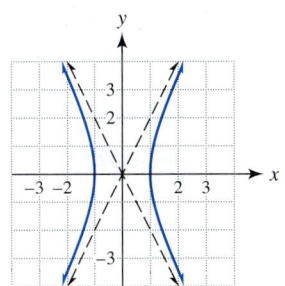

Exercises 21–24: Graph the hyperbola. Include the asymptotes in your graph.

21. $\dfrac{x^2}{9} - \dfrac{y^2}{4} = 1$ **22.** $\dfrac{y^2}{25} - \dfrac{x^2}{16} = 1$

23. $y^2 - x^2 = 1$ **24.** $25x^2 - 16y^2 = 400$

SECTION 13.3

Exercises 25–28: Use the graph to estimate all solutions to the system of equations. Check each solution.

25. $x^2 + y^2 = 9$ **26.** $xy = 2$
     $x + y = 3$ $y = 2x$

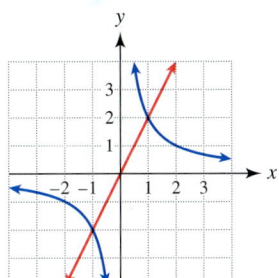

27. $x^2 - y = x$ **28.** $x^2 + y^2 = 5$
     $y = x$ $x^2 - y^2 = 3$

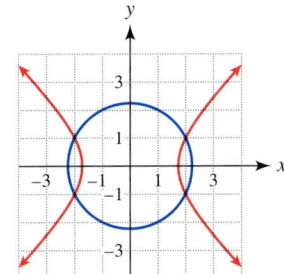

Exercises 29–32: Solve the system of equations. Check your solutions.

29. $y = x$
$x^2 + y^2 = 32$

30. $x - y = 4$
$x^2 + y^2 = 16$

31. $y = x^2$
$2x^2 + y = 3$

32. $y = x^2 + 1$
$2x^2 - y = 3x - 3$

Exercises 33 and 34: Solve the system graphically.

33. $2x - y = 4$
$x^2 + y = 4$

34. $x^2 + y = 0$
$x^2 + y^2 = 2$

Exercises 35 and 36: Solve the system of equations
   (a) *symbolically,*
   (b) *graphically, and*
   (c) *numerically.*

35. $y = x$
$x^2 + 2y = 8$

36. $y = x^3$
$x^2 - y = 0$

Exercises 37–44: Shade the solution set in the xy-plane.

37. $y \geq 2x^2$

38. $y < 2x - 3$

39. $y < -x^2$

40. $\dfrac{x^2}{9} + \dfrac{y^2}{16} \leq 1$

41. $y - x^2 \geq 1$
$y \leq 2$

42. $x^2 + y \leq 4$
$3x + 2y \geq 6$

43. $y > x^2$
$y < 4 - x^2$

44. $\dfrac{x^2}{4} + \dfrac{y^2}{9} > 1$
$x^2 + y^2 < 16$

Exercises 45 and 46: Use the graph to write the system of inequalities.

45.

46.

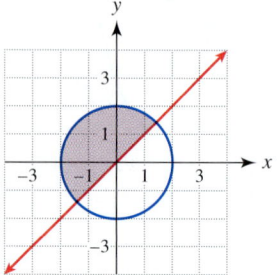

APPLICATIONS

47. *Area and Perimeter*   The area of a desktop is 1000 square inches, and its perimeter is 130 inches. Let x be the width and y be the length of the desktop.
   (a) Write a system of equations that models this situation.
   (b) Solve the system graphically.
   (c) Solve the system symbolically.

48. *Numbers*   The product of two positive numbers is 60, and their difference is 7. Let x be the smaller number and y be the larger number.
   (a) Write a system of equations whose solution gives the two numbers.
   (b) Solve the system graphically.
   (c) Solve the system symbolically.

49. *Dimensions of a Container*   The volume of a cylindrical container is $V = \pi r^2 h$, and its surface area, *excluding* the top and bottom, is $A = 2\pi r h$. Find the dimensions of a container with $A = 100$ square feet and $V = 50$ cubic feet. Is your answer unique?

50. *Dimensions of a Container*   The volume of a cylindrical container is $V = \pi r^2 h$, and its surface area, *including* the top and bottom, is $A = 2\pi r h + 2\pi r^2$. Graphically find the dimensions of a can with $A = 80$ square inches and $V = 35$ cubic inches. Is your answer unique?

51. *Geometry of an Ellipse*   The area inside an ellipse is given by $A = \pi ab$, and its perimeter P can be approximated by

$$P = 2\pi\sqrt{\frac{a^2 + b^2}{2}}.$$

   (a) Graph $\dfrac{x^2}{5} + \dfrac{y^2}{12} = 1.$
   (b) Estimate its area and perimeter.

52. *Orbit of Mars*   Mars has an elliptical orbit that is nearly circular, with $a = 1.524$ and $b = 1.517$, where the units are astronomical units (1 A.U. equals 93 million miles).   (*Source:* M. Zeilik.)

(a) Graph the orbit of Mars in $[-3, 3, 1]$ by $[-2, 2, 1]$. Plot the point $(0.15, 0)$ to show the position of the sun. Assume that the major axis is horizontal.

(b) Use the information in Exercise 51 to estimate how far Mars travels in one orbit around the sun. Approximate the area inside its orbit.

CHAPTER 13   TEST   Pass the Test   Video solutions to all test exercises

1. Graph the parabola $x = (y - 4)^2 - 2$. Find the vertex and axis of symmetry.

2. Use the graph to find the equation of the parabola.

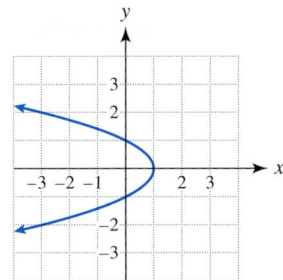

3. Use the graph to find the equation of the circle.

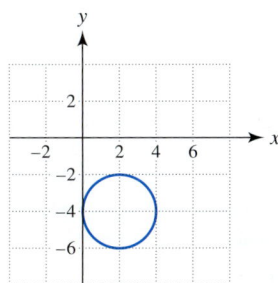

4. Write the equation of the circle with radius 10 and center $(-5, 2)$.

5. Find the radius and center of the circle given by

$$x^2 + 4x + y^2 - 6y = 3.$$

Then graph the circle.

6. Graph the ellipse $\frac{x^2}{16} + \frac{y^2}{49} = 1$. Label the vertices and endpoints of the minor axis.

7. Use the graph to determine the equation of the ellipse.

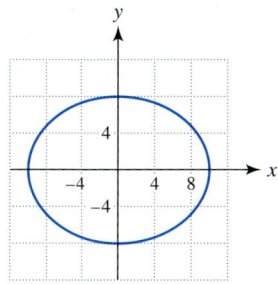

8. Graph the hyperbola $4x^2 - 9y^2 = 36$. Include the asymptotes in your graph.

9. Use the graph to estimate all solutions to the system of equations. Check each solution by substitution in the system of equations.

$$x^2 + y^2 = 16 \quad \text{and} \quad x - y = 4$$

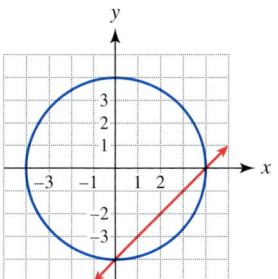

10. Solve the system of equations symbolically.

$$\begin{aligned} x - y &= 3 \\ x^2 + y^2 &= 17 \end{aligned}$$

11. Solve the system of equations graphically.

$$\begin{aligned} 2x^2 - y &= 4 \\ x^2 + y &= 8 \end{aligned}$$

12. Shade the solution set in the xy-plane.
$$3x + y > 6$$
$$x^2 + y^2 < 25$$

13. Use the graph to write the system of inequalities.

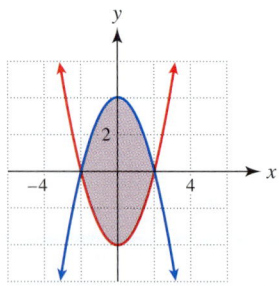

14. *Area and Perimeter*   The area of a rectangular swimming pool is 5000 square feet, and its perimeter is 300 feet. Let x be the width and y be the length of the pool.
  (a) Write a nonlinear system of equations that models this situation.
  (b) Solve the system.

 15. *Dimensions of a Box*   The volume of a rectangular box with a square bottom and open top is $V = x^2 y$, and its surface area is $A = x^2 + 4xy$, where x represents its width and length and y represents its height. Estimate graphically the dimensions of a box with $V = 1183$ cubic inches and $A = 702$ square inches. Is your answer unique? (*Hint:* Substitute appropriate values for V and A, and then solve each equation for y.)

16. *Orbit of Uranus*   The planet Uranus has an elliptical orbit that is nearly circular, with $a = 19.18$ and $b = 19.16$, where the units are astronomical units (1 A.U. equals 93 million miles).   (**Source:** M. Zeilik.)
  (a) Graph the orbit of Uranus in $[-30, 30, 10]$ by $[-20, 20, 10]$. Plot the point $(0.9, 0)$ to show the position of the sun. Assume that the major axis is horizontal.
  (b) Find the minimum distance between Uranus and the sun.

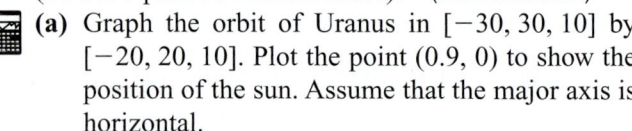

CHAPTER 13   EXTENDED AND DISCOVERY EXERCISES

Exercises 1–2: **Foci of Parabolas**   *The focus of a parabola is a point that has special significance. When a parabola is rotated about its axis, it sweeps out a shape called a **paraboloid**, as illustrated in the top figure on the following page. Paraboloids have an important reflective property. When incoming rays of light from the sun or distant stars strike the surface of a paraboloid, each ray is reflected toward the focus, as shown in the figure at the top of the next page labeled "Reflective Property." If the rays are sunlight, intense heat is produced, which can be used to generate solar heat. Radio signals from distant space also concentrate at the focus, and scientists can measure these signals by placing a receiver there.*

The same reflective property of a paraboloid can be used in reverse. If a light source is placed at the focus, the light is reflected straight ahead, as depicted in the figure labeled "Headlight." Searchlights, flashlights, and car headlights make use of this reflective property.

The focus is always located inside a parabola, on its axis of symmetry. If the distance between the vertex and the focus is $|p|$, the following equations can be used to locate the focus. Note that the value of p may be either positive or negative.

EQUATION OF A PARABOLA WITH VERTEX (0, 0)

Vertical Axis

The parabola with a focus at $(0, p)$ has the equation
$$x^2 = 4py.$$

Horizontal Axis

The parabola with a focus at $(p, 0)$ has the equation
$$y^2 = 4px.$$

Paraboloid

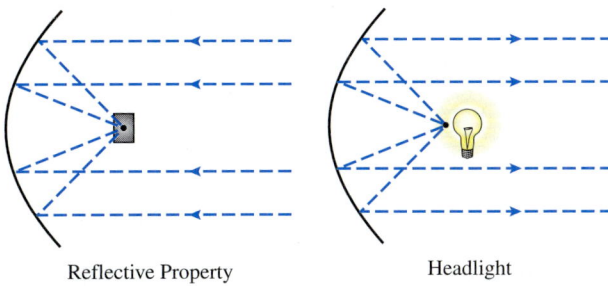

Reflective Property

Headlight

1. Graph each parabola. Label the vertex and focus.
   (a) $x^2 = 4y$ **(b)** $y^2 = -8x$ **(c)** $x = 2y^2$

2. The reflective property of paraboloids is used in satellite dishes and radio telescopes. The U.S. Naval Research Laboratory designed a giant radio telescope weighing 3450 tons. Its parabolic dish has a diameter of 300 feet and a depth of 44 feet, as shown in the accompanying figures.   (**Source:** J. Mar, *Structure Technology for Large Radio and Radar Telescope Systems.*)
   (a) Determine an equation in the form $y = ax^2$ that describes a cross section of this dish.
   (b) If the receiver is located at the focus, how far should it be from the vertex?

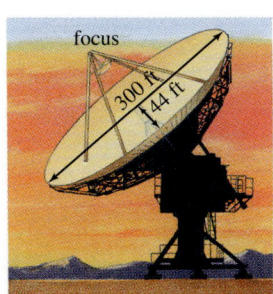

focus

300 ft
44 ft

Radio Telescope

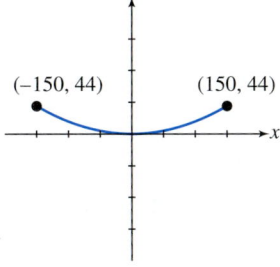

$(-150, 44)$ $(150, 44)$

Exercises 3 and 4:   ***Translations of Ellipses and Hyperbolas*** *Ellipses and hyperbolas can be translated so that they are centered at a point* (h, k), *rather than at the origin. These techniques are the same as those used for parabolas and circles. To translate a conic section so that it is centered at* (h, k) *rather than* $(0, 0)$, *replace x with* $(x - h)$ *and replace y with* $(y - k)$. *For example, to center* $\frac{x^2}{9} + \frac{y^2}{4} = 1$ *at* $(-1, 2)$, *change its equation to* $\frac{(x + 1)^2}{9} + \frac{(y - 2)^2}{4} = 1$. *See the accompanying figures.*

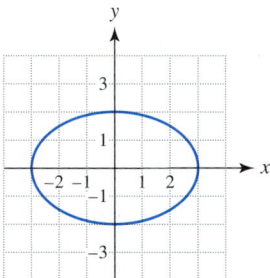

Ellipse Centered at $(0, 0)$

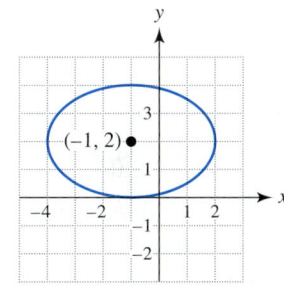

$(-1, 2)$ •

Ellipse Centered at $(-1, 2)$

3. Graph each conic section. For hyperbolas give the equations of the asymptotes.
   (a) $\dfrac{(x - 3)^2}{25} + \dfrac{(y - 1)^2}{9} = 1$
   (b) $\dfrac{(x + 1)^2}{4} + \dfrac{(y + 2)^2}{16} = 1$
   (c) $\dfrac{(x + 1)^2}{4} - \dfrac{(y - 3)^2}{9} = 1$
   (d) $\dfrac{(y - 4)^2}{16} - \dfrac{(x + 1)^2}{4} = 1$

4. Write the equation of an ellipse having the following properties.
   (a) Horizontal major axis of length 8, minor axis of length 4, and centered at $(-3, 5)$.
   (b) Vertical major axis of length 10, minor axis of length 6, and centered at $(2, -3)$.

5. Determine the center of the conic section.
   (a) $9x^2 - 18x + 4y^2 + 24y + 9 = 0$
   (b) $25x^2 + 150x - 16y^2 + 32y - 191 = 0$

CHAPTERS 1–13   CUMULATIVE REVIEW EXERCISES

1. Evaluate $K = x^2 + y^2$ when $x = 4$ and $y = -3$.

2. Simplify $\dfrac{(a^{-2}b)^2}{a^{-1}(b^3)^{-2}}$.

3. Write 7.345×10^{-3} in standard (decimal) notation.

4. Find $f(-4)$, if $f(x) = \frac{x}{x-4}$. What is the domain of f?

5. If the point $(3, 4)$ is on the graph of $y = f(x)$, then $f(\_\_) = \_\_$.

Exercises 6 and 7: Graph f by hand.

6. $f(x) = x^2 - x$ 7. $f(x) = 1 - 2x$

8. Find the slope–intercept form of the line that passes through the point $(2, -2)$ and is perpendicular to the line given by $y = -\frac{2}{3}x + 1$.

9. Solve $2(1 - x) - 4x = x$.

Exercises 10–12: Solve the inequality. Write your answer in interval notation.

10. $-5 \le 1 - 2x < 3$ 11. $x^2 - 4 \le 0$

12. $|1 - x| \ge 2$

13. Solve the system.
$$-2x + y = 1$$
$$5x - y = 2$$

14. Shade the solution set in the xy-plane.
$$x + y \le 4$$
$$x - y \ge 2$$

Exercises 15 and 16: Multiply the expression.

15. $(2x - 1)(x + 5)$ 16. $xy(2x - 3y^2 + 1)$

Exercises 17 and 18: Factor the expression.

17. $6x^2 - 13x - 5$ 18. $x^3 - 4x$

Exercises 19 and 20: Solve the equation.

19. $x^2 + 3x + 2 = 0$ 20. $x^2 + 1 = -3x$

Exercises 21 and 22: Simplify the expression.

21. $\dfrac{x - 2}{x + 2} \div \dfrac{2x - 4}{3x + 6}$ 22. $\dfrac{1}{x + 1} + \dfrac{1}{x - 1}$

Exercises 23–26: Simplify the expression. Assume all variables are positive.

23. $\sqrt{8x^2}$ 24. $8^{2/3}$

25. $\sqrt[3]{2x} \cdot \sqrt[3]{32x^2}$ 26. $3\sqrt{3x} + \sqrt{12x}$

27. Graph $f(x) = -\sqrt{x}$.

28. Find the distance between $(2, -3)$ and $(-2, 0)$.

29. Write $(2 + 3i)(2 - 3i)$ in standard form.

30. Solve $\sqrt{x + 2} = x$.

31. Find the vertex of the graph of $y = x^2 - 6x + 3$.

32. Write $f(x) = x^2 - 2x + 3$ in vertex form.

33. Compare the graph of $f(x) = \sqrt{x} - 4$ to the graph of $y = \sqrt{x}$.

34. Solve $x(3 - x) = 2$.

35. Solve $x^3 + x = 0$. Find all complex solutions.

36. Simplify by hand.
   (a) $\log 10{,}000$ (b) $\log_2 8$
   (c) $\log_3 3^x$ (d) $e^{\ln 6}$
   (e) $\log 2 + \log 50$ (f) $\log_2 24 - \log_2 3$

37. If $f(x) = x^2 + 1$ and $g(x) = 2x$, find the following.
   (a) $(f \circ g)(2)$ (b) $(g \circ f)(x)$

38. If $f(x) = 2 - 3x$, find $f^{-1}(x)$.

39. If \$1000 are deposited in an account that pays 5% annual interest, approximate the amount in the account after 6 years.

40. Write $\log \dfrac{x^2 \sqrt{y}}{z^3}$ in terms of logarithms of x, y, and z.

41. Solve $2e^x - 1 = 17$.

42. Solve $3 + \log 4x = 5$.

43. Graph each conic section.
   (a) $x = (y - 1)^2$ (b) $(x - 1)^2 + (y + 1)^2 = 4$
   (c) $\dfrac{x^2}{4} + \dfrac{y^2}{25} = 1$ (d) $4x^2 - 9y^2 = 36$

44. Solve the nonlinear system of equations.
$$x^2 + y^2 = 1$$
$$x^2 + 9y^2 = 9$$

45. Shade the solution set in the xy-plane.

$$x^2 + y^2 \le 4$$
$$x^2 - y \le 2$$

APPLICATIONS

46. *Calculating Distance* The distance D in miles that a driver of a car is from home after x hours is given by $D(x) = 400 - 50x$.

  (a) Evaluate $D(0)$. Interpret your answer.

  (b) What is the x-intercept on the graph of D? Interpret your answer.

  (c) What is the slope of the graph of D? Interpret your answer.

47. *Investment* Suppose $2000 are deposited in three accounts paying 5%, 6%, and 7% annual interest. The amount invested at 6% is $500 more than the amount invested at 5%. The interest after 1 year is $120. How much is invested at each rate?

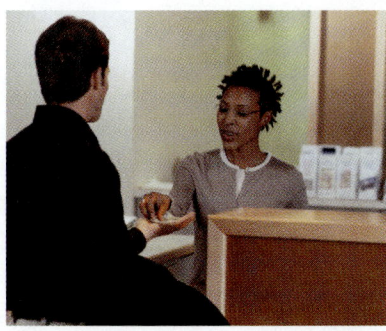

48. *Area* There are 1200 feet of fence available to surround the perimeter of a rectangular garden. What dimensions for the garden give the largest area?

49. *Population* The population of a city in millions after x years is given by $P = 2e^{0.02x}$. How long will it take for the population to double?

50. *Dimensions of a Can* Find the radius of a cylindrical can with a volume V of 60 cubic inches and a side area S of 50 square inches. (*Hint:* $V = \pi r^2 h$ and $S = 2\pi rh$.)

CHAPTER 14

Sequences and Series

I n this final chapter we present sequences and series, which are essential topics because they are used to model and approximate important quantities. Complicated population growth can be modeled with sequences, and accurate approximations for numbers such as π and e are made with series. For example, in about 2000 B.C. the Babylonians thought that π equaled 3.125, whereas at the same time the Chinese thought π equaled 3. By 1700 mathematicians were able to calculate π to 100 decimal places due to the invention of series. Series also are essential to the solution of many modern applied mathematics problems.

Although you may not always recognize the impact of mathematics on everyday life, its influence is nonetheless profound. Mathematics is the *language of technology*—it allows experiences to be quantified. In the preceding chapters we showed numerous examples of mathematics being used to model the real world. Computers, DVD players, cars, highway design, weather, hurricanes, electricity, government data, cellular phones, medicine, ecology, business, sports, and psychology represent only some of the applications of mathematics. In fact, if a subject is studied in enough detail, mathematics usually appears in one form or another. Although predicting what the future may bring is difficult, one thing *is* certain—mathematics will continue to play an important role in both theoretical research and new technology.

The essence of mathematics is not to make simple things complicated, but to make complicated things simple.

—STANLEY GUDDER

Source: Mathforum.org, Drexel University 1994–2007.

14.1 SEQUENCES

Basic Concepts ▪ Representations of Sequences ▪ Models and Applications

A LOOK INTO MATH ▷ Sequences are *ordered lists*. For example, names listed alphabetically represent a sequence. Figure 14.1 shows an insect population in thousands per acre over a 6-year period. Listing populations by year is another example of a sequence. In mathematics a sequence is a function, for which valid inputs must be natural numbers. For example, we can use a function f to define this sequence by letting $f(1)$ represent the insect population after 1 year, $f(2)$ represent the insect population after 2 years, and in general let $f(n)$ represent the population after n years. In this section we discuss sequences.

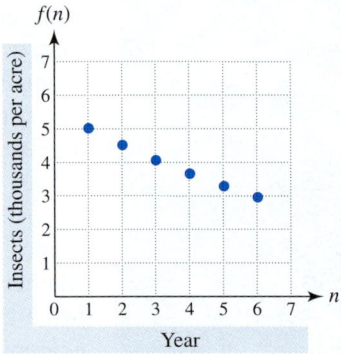

Figure 14.1 An Insect Population

Basic Concepts

▶ **REAL-WORLD CONNECTION** Suppose that an individual's starting salary is $40,000 per year and that the person's salary is increased by 10% each year. This situation is modeled by the formula

$$f(n) = 40,000(1.10)^n.$$

We do not allow the input n to be any real number, but rather limit n to a *natural number* because the individual's salary is constant throughout a particular year. The first 5 terms of the sequence are

$$f(1), f(2), f(3), f(4), f(5).$$

They can be computed as follows.

$$f(\mathbf{1}) = 40,000(1.10)^1 = 44,000$$
$$f(\mathbf{2}) = 40,000(1.10)^2 = 48,400$$
$$f(\mathbf{3}) = 40,000(1.10)^3 = 53,240$$
$$f(\mathbf{4}) = 40,000(1.10)^4 = 58,564$$
$$f(\mathbf{5}) = 40,000(1.10)^5 \approx 64,420$$

This sequence is represented *numerically* in Table 14.1.

TABLE 14.1   Numerical Representation

n	1	2	3	4	5
$f(n)$	44,000	48,400	53,240	58,564	64,420

A *graphical* representation results when each data point in Table 14.1 is plotted, as illustrated in Figure 14.2.

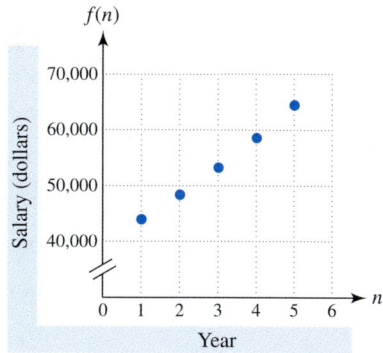

Figure 14.2 Graphical Representation

NOTE: Graphs of sequences are scatterplots.

The preceding sequence is an example of a *finite sequence* of numbers. The even natural numbers,

$$2, 4, 6, 8, 10, 12, 14, \ldots$$

are an example of an *infinite sequence* represented by $f(n) = 2n$, where n is a natural number. The three dots, or periods (called an *ellipsis*), indicate that the pattern continues indefinitely.

SEQUENCES

A **finite sequence** is a function whose domain is $D = \{1, 2, 3, \ldots, n\}$ for some fixed natural number n.

An **infinite sequence** is a function whose domain is the set of natural numbers.

Because sequences are functions, many of the concepts discussed in previous chapters apply to sequences. Instead of letting y represent the output, however, the convention is to write $a_n = f(n)$, where n is a natural number in the domain of the sequence. The *terms* of a sequence are

$$a_1, a_2, a_3, \ldots, a_n, \ldots.$$

The first term is $a_1 = f(1)$, the second term is $a_2 = f(2)$, and so on. The **nth term**, or **general term**, of a sequence is $a_n = f(n)$.

EXAMPLE 1   Computing terms of a sequence

Write the first four terms of each sequence for $n = 1, 2, 3,$ and 4.

(a) $f(n) = 2n - 1$ **(b)** $f(n) = 3(-2)^n$ **(c)** $f(n) = \dfrac{n}{n+1}$

Solution

(a) For $a_n = f(n) = 2n - 1$, we write

$$a_1 = f(1) = 2(1) - 1 = 1;$$
$$a_2 = f(2) = 2(2) - 1 = 3;$$
$$a_3 = f(3) = 2(3) - 1 = 5;$$
$$a_4 = f(4) = 2(4) - 1 = 7.$$

The first four terms are $1, 3, 5,$ and 7.

(b) For $a_n = f(n) = 3(-2)^n$, we write

$$a_1 = f(1) = 3(-2)^1 = -6;$$
$$a_2 = f(2) = 3(-2)^2 = 12;$$
$$a_3 = f(3) = 3(-2)^3 = -24;$$
$$a_4 = f(4) = 3(-2)^4 = 48.$$

The first four terms are $-6, 12, -24,$ and 48.

(c) For $a_n = f(n) = \dfrac{n}{n+1}$, we write

$$a_1 = f(1) = \frac{1}{1+1} = \frac{1}{2};$$
$$a_2 = f(2) = \frac{2}{2+1} = \frac{2}{3};$$
$$a_3 = f(3) = \frac{3}{3+1} = \frac{3}{4};$$
$$a_4 = f(4) = \frac{4}{4+1} = \frac{4}{5}.$$

The first four terms are $\frac{1}{2}, \frac{2}{3}, \frac{3}{4},$ and $\frac{4}{5}$. Note that, although the input to a sequence is a natural number, the output need not be a natural number.   Now Try Exercises 9, 11, 13

TECHNOLOGY NOTE: Generating Sequences

Many graphing calculators can generate sequences if you change the MODE from function (Func) to sequence (Seq). In Figures 14.3 and 14.4 the sequences from Example 1 are generated. On some calculators the sequence utility is found in the LIST OPS menus. The expression

$$\text{seq}(2n - 1, n, 1, 4)$$

represents terms 1 through 4 of the sequence $a_n = 2n - 1$ with the variable n.

```
seq(2n-1,n,1,4)
        {1 3 5 7}
seq(3(-2)^n,n,1,
4)
      {-6 12 -24 48}
```

```
seq(n/(n+1),n,1,
4)▶Frac
{1/2 2/3 3/4 4/...
```

Figure 14.3   Figure 14.4

Representations of Sequences

Because sequences are functions, they can be represented symbolically, graphically, and numerically. The next two examples illustrate such representations.

EXAMPLE 2 **Using a graphical representation**

Use Figure 14.5 to write the terms of the sequence.

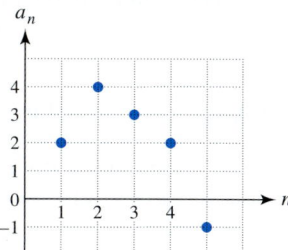

Figure 14.5 Finite Sequence

Solution
The points $(1, 2)$, $(2, 4)$, $(3, 3)$, $(4, 2)$, and $(5, -1)$ are shown in the graph. The terms of the sequence are $2, 4, 3, 2,$ and -1. Now Try Exercise **29**

EXAMPLE 3 **Representing a sequence**

In 2002 the average person in the United States used 100 gallons of water each day. Give symbolic, numerical, and graphical representations for a sequence that models the total amount of water used over a 7-day period.

Solution
Symbolic Representation Let $a_n = 100n$ for $n = 1, 2, 3, \ldots, 7$.

Numerical Representation Table 14.2 contains the sequence.

Graphical Representation Plot the points $(1, 100)$, $(2, 200)$, $(3, 300)$, $(4, 400)$, $(5, 500)$, $(6, 600)$, and $(7, 700)$, as shown in Figure 14.6.

TABLE 14.2

n	a_n
1	100
2	200
3	300
4	400
5	500
6	600
7	700

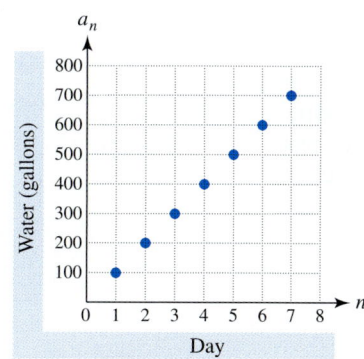

Figure 14.6 Water Usage   Now Try Exercise **39**

Models and Applications

▶ **REAL-WORLD CONNECTION**   A population model for a species of insect with a life span of 1 year can be described with a sequence. Suppose that each adult female insect produces, on average, r female offspring that survive to reproduce the following year. Let a_n represent the female insect population at the beginning of year n. Then the number of female insects is given by

$$a_n = Cr^{n-1},$$

where C is the initial population of female insects.   (**Source:** D. Brown and P. Rothery, *Models in Biology*.)

EXAMPLE 4   Modeling numbers of insects

Suppose that the initial population of adult female insects is 500 per acre and that $r = 1.04$. Then the average number of female insects per acre at the beginning of year n is described by

$$a_n = 500(1.04)^{n-1}.$$

Represent the female insect population numerically and graphically for 7 years. Discuss the results. By what percent is the population increasing each year?

Solution

Numerical Representation   Table 14.3 contains *approximations* for the first 7 terms of the sequence. The insect population increases from 500 to about 633 insects per acre during this time period.

TABLE 14.3

n	1	2	3	4	5	6	7
a_n	500	520	540.8	562.43	584.93	608.33	632.66

Graphical Representation   Plot the points (1, 500), (2, 520), (3, 540.8), (4, 562.43), (5, 584.93), (6, 608.33), and (7, 632.66), as shown in Figure 14.7. These results indicate that the insect population gradually increases. Because the growth factor is 1.04, the population is increasing by 4% each year.

CRITICAL THINKING

Explain how the value of r in Example 4 affects the population of female insects over time. Assume that $r > 0$.

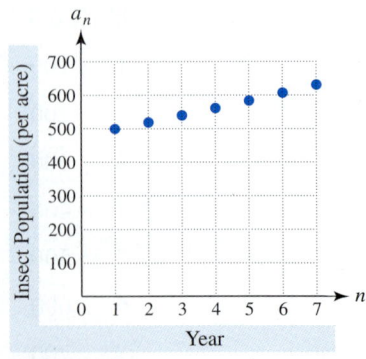

Figure 14.7

Now Try Exercise **45**

TECHNOLOGY NOTE: Graphs and Tables of Sequences

In the sequence mode, many graphing calculators are capable of representing sequences graphically and numerically. Figure 14.8(a) shows how to enter the sequence from Example 4 to produce the table of values shown in Figure 14.8(b).

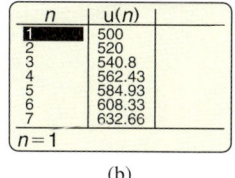

(a)   (b)

Figure 14.8

Figures 14.9(a) and (b) show the set-up for graphing the sequence from Example 4 to produce the graph shown in Figure 14.9(c).

[0, 8, 1] by [0, 800, 100]

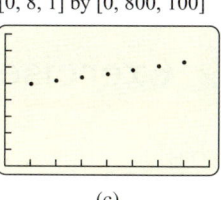

(a)   (b)   (c)

Figure 14.9

14.1 PUTTING IT ALL TOGETHER

An infinite sequence is a function whose domain is the set of natural numbers. A finite sequence has the domain $D = \{1, 2, 3, \ldots, n\}$ for some fixed natural number n. Graphs of sequences are scatterplots, *not* continuous lines and curves.

Sequences are functions that may be represented symbolically, numerically, and graphically. Examples of representations of sequences are shown in the following table.

Representation	Example
Symbolic	$a_n = n - 3$ represents a sequence. The first four terms of the sequence are $-2, -1, 0$, and 1: $$a_1 = 1 - 3 = -2, \quad a_2 = 2 - 3 = -1,$$ $$a_3 = 3 - 3 = 0, \quad \text{and} \quad a_4 = 4 - 3 = 1.$$
Numerical	A numerical representation for $a_n = n - 3$ with $n = 1, 2, 3$, and 4 is shown in the table. <table><tr><td>n</td><td>1</td><td>2</td><td>3</td><td>4</td></tr><tr><td>a_n</td><td>-2</td><td>-1</td><td>0</td><td>1</td></tr></table>

continued on next page

continued from previous page

Representation	Example
Graphical	For a graphical representation of the first four terms of $a_n = n - 3$, the points $(1, -2)$, $(2, -1)$, $(3, 0)$, and $(4, 1)$ are plotted.

14.1 Exercises

CONCEPTS

1. Give an example of a finite sequence.

2. Give an example of an infinite sequence.

3. An infinite sequence is a _____ whose domain is the set of _____.

4. An ordered list is a _____.

5. The third term in the sequence $4, -5, 6, -7, 8$ is _____.

6. The graph of a sequence is not a continuous graph but rather a _____.

7. If $f(n)$ represents a sequence, the second term of the sequence is given by _____.

8. If a_n represents a sequence, the fourth term of the sequence is given by _____.

EVALUATING AND REPRESENTING SEQUENCES

Exercises 9–16: Write the first four terms of the sequence for $n = 1, 2, 3,$ and 4.

9. $f(n) = n^2$

10. $f(n) = 3n + 4$

11. $f(n) = \dfrac{1}{n + 5}$

12. $f(n) = 3^n$

13. $f(n) = 5\left(\dfrac{1}{2}\right)^n$

14. $f(n) = n^2 + 2n$

15. $f(n) = 9$

16. $f(n) = (-1)^n$

Exercises 17–24: Write the first three terms of the sequence for $n = 1, 2,$ and 3.

17. $a_n = n^3$

18. $a_n = 5 - n$

19. $a_n = \dfrac{4n}{3 + n}$

20. $a_n = 3^{-n}$

21. $a_n = 2n^2 + n - 1$

22. $a_n = n^4 - 1$

23. $a_n = -2$

24. $a_n = n^n$

Exercises 25 and 26: **Thinking Generally** *Let b and c be fixed numbers (constants). Find a_1 and a_2.*

25. $a_n = bn + c$

26. $a_n = \dfrac{n + b}{n - c}, c \neq 1$ and $c \neq 2$

Exercises 27 and 28: Use the numerical representation to evaluate $\frac{1}{2}(a_1 + a_4)$.

27.

n	1	2	3	4	5
a_n	10	8	6	4	2

28.

n	1	2	3	4	5
a_n	-5	0	10	30	60

Exercises 29–32: Write the terms of the sequence.

29.

30.

31.

32.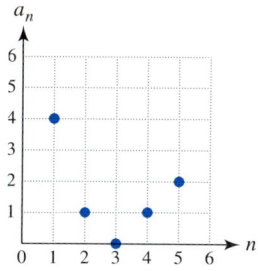

Exercises 33–38: Represent the first seven terms of the sequence numerically and graphically.

33. $a_n = n + 1$

34. $a_n = \frac{1}{2}n - \frac{1}{2}$

35. $a_n = n^2 - n$

36. $a_n = \frac{1}{2}n^2$

37. $a_n = 2^n$

38. $a_n = 2(0.5)^n$

APPLICATIONS

39. *Solid Waste*   On average, each U.S. resident in 2000 generated about 30 pounds of solid waste per week. Give symbolic, numerical, and graphical representations for a sequence that models the total amount of waste produced over a 7-week period.   (**Source:** Environmental Protection Agency.)

40. *Carbon Dioxide Emitters*   Because people burn fossil fuels, the United States emits more carbon dioxide than any other country in the world, or about 5.8 billion metric tons per year. (A metric ton is about 2200 pounds.) Give symbolic, numerical, and graphical representations for a sequence that models the total amount of carbon dioxide emitted in the United States in a 5-year period.   (**Source:** Energy Information Administration.)

41. *Geometry*   The lengths of the sides of a sequence of squares are given by 1, 2, 3, and 4, as shown in the following figure. Write sequences that give
(a) the areas of the squares, and

(b) the perimeters of the squares.

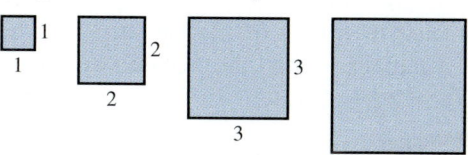

42. *Salaries*   An individual's starting salary is $50,000, and the individual receives an increase of 8% per year. Give symbolic, numerical, and graphical representations for this person's salary over 5 years.

43. *Depreciation*   Automobiles usually depreciate in value over time. Often, a newer automobile may be worth only 80% of its previous year's value. Suppose that a car is worth $25,000 new.
(a) How much is it worth after 1 year? After 2 years?
(b) Write a formula for a sequence that gives the car's value after n years.
(c) Make a table that shows how much the car was worth each year during the first 7 years.

44. *Falling Object*   The distance d that an object falls during *consecutive* seconds is shown in the table. For example, during the third second (from $n = 2$ to $n = 3$) an object falls a distance of 80 feet.

n (seconds)	1	2	3	4	5
d (feet)	16	48	80	112	144

(a) Find values for c and b so that $d = cn + b$ models these data.
(b) How far does an object fall during the sixth second?

45. *Modeling Insect Populations*   (Refer to Example 4.) Suppose that the initial population of insects is 2048 per acre and that $r = 0.5$. Use a sequence to represent the insect population over a 7-year period
(a) symbolically,
(b) numerically, and
(c) graphically.

46. *Auditorium Seating*   An auditorium has 50 seats in the first row, 55 seats in the second row, 60 seats in the third row, and so on.
(a) Make a table that shows the number of seats in the first seven rows.
(b) Write a formula that gives the number of seats in row n.
(c) How many seats are there in row 23?
(d) Graph the number of seats in each row for $n = 1, 2, 3, \dots, 10$.

47. Compare the graph of $f(x) = 2x + 1$, where x is a real number, with the graph of the sequence $f(n) = 2n + 1$, where n is a natural number.

48. Explain what a sequence is. Describe the difference between a finite and an infinite sequence.

14.2 ARITHMETIC AND GEOMETRIC SEQUENCES

Representations of Arithmetic Sequences ∎ Representations of Geometric Sequences ∎ Applications and Models

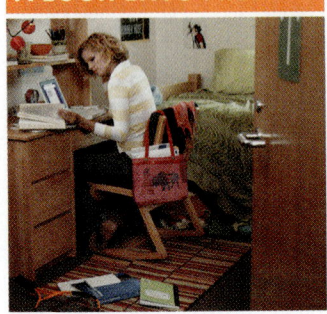

A LOOK INTO MATH ▷ Indoor air pollution has become more hazardous as people spend 80% to 90% of their time in tightly sealed, energy-efficient buildings, which often lack proper ventilation. Many contaminants such as tobacco smoke, formaldehyde, radon, lead, and carbon monoxide are often allowed to increase to unsafe levels. One way to alleviate this problem is to use efficient ventilation systems. Mathematics plays an important role in determining the proper amount of ventilation. In this section we use sequences to model ventilation in classrooms. Before implementing this model, however, we discuss the basic concepts relating to two special types of sequences.

Representations of Arithmetic Sequences

If a sequence is defined by a linear function, it is an *arithmetic sequence*. For example,

$$f(n) = 2n - 3$$

represents an arithmetic sequence because $f(x) = 2x - 3$ defines a linear function. The first five terms of this sequence are shown in Table 14.4.

Each time n increases by 1, the next term is 2 more than the previous term. We say that the *common difference* of this arithmetic sequence is $d = 2$. That is, the difference between successive terms equals 2. When the points associated with these terms are graphed, they lie on the line $y = 2x - 3$, as illustrated in Figure 14.10. Arithmetic sequences are represented by linear functions and so their graphical representations consist of collinear points (points that lie on a line). The slope m of the line equals the common difference d.

TABLE 14.4

n	$f(n)$
1	−1
2	1
3	3
4	5
5	7

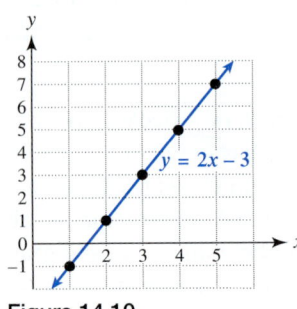

Figure 14.10

ARITHMETIC SEQUENCE

An **arithmetic sequence** is a linear function given by $a_n = dn + c$ whose domain is the set of natural numbers. The value of d is called the **common difference**.

NOTE: If each term after the first term is obtained by adding a fixed number to the previous term, then the sequence is an arithmetic sequence. The fixed number is the *common difference*. For example, 1, 6, 11, 16, 21, . . . is an arithmetic sequence because each term (after the first) is found by adding the common difference of 5 to the previous term.

EXAMPLE 1 Recognizing arithmetic sequences

Determine whether f is an arithmetic sequence. If it is, identify the common difference d.
(a) $f(n) = 2 - 3n$

(b)

n	$f(n)$
1	10
2	5
3	0
4	-5
5	-10

(c) A graph of f is shown in Figure 14.11.

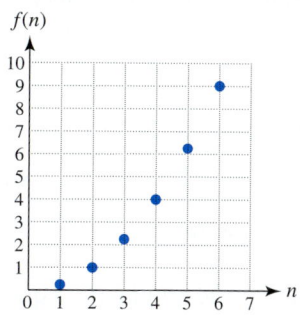

Figure 14.11

Solution
(a) This sequence is arithmetic because $f(x) = -3x + 2$ defines a linear function. The common difference is $d = -3$.
(b) The table reveals that each term is found by adding -5 to the previous term. This represents an arithmetic sequence with common difference $d = -5$.
(c) The sequence shown in Figure 14.11 is not an arithmetic sequence because the points are not collinear. That is, there is no common difference.   Now Try Exercises 11, 17, 25

MAKING CONNECTIONS

Common Difference and Slope

The common difference d of an arithmetic sequence equals the slope of the line passing through the collinear points. For example, if $a_n = -2n + 4$, the common difference is -2, and the slope of the line passing through the points on the graph of a_n is also -2 (see Figure 14.12).

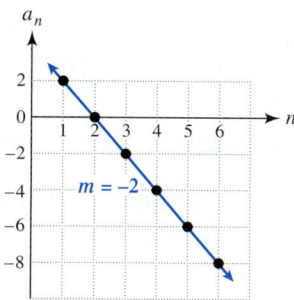

Figure 14.12

EXAMPLE 2   Finding symbolic representations

Find the general term a_n for each arithmetic sequence.
(a) $a_1 = 3$ and $d = 4$ **(b)** $a_1 = 3$ and $a_4 = 12$

Solution
(a) Let $a_n = dn + c$. For $d = 4$, we write $a_n = 4n + c$, and to find c we use $a_1 = 3$.

$$a_1 = 4(1) + c = 3 \quad \text{or} \quad c = -1$$

Thus $a_n = 4n - 1$.
(b) Because $a_1 = 3$ and $a_4 = 12$, the common difference d equals the slope of the line passing through the points $(1, 3)$ and $(4, 12)$, or

$$d = \frac{12 - 3}{4 - 1} = 3.$$

Therefore $a_n = 3n + c$. To find c we use $a_1 = 3$ and obtain

$$a_1 = 3(1) + c = 3 \quad \text{or} \quad c = 0.$$

Thus $a_n = 3n$.   Now Try Exercises **29, 31**

Consider the arithmetic sequence

$$1, 5, 9, 13, 17, 21, 25, 29, \ldots.$$

The common difference is $d = 4$, and the first term is $a_1 = 1$. To find the second term we add d to the first term. To find the third term we add $2d$ to the first term, and to find the fourth term we add $3d$ to the first term a_1. That is,

$$a_1 = 1,$$
$$a_2 = a_1 + 1d = 1 + 1 \cdot 4 = 5,$$
$$a_3 = a_1 + 2d = 1 + 2 \cdot 4 = 9,$$
$$a_4 = a_1 + 3d = 1 + 3 \cdot 4 = 13,$$

and, in general, a_n is determined by

$$a_n = a_1 + (n - 1)d = 1 + (n - 1)4.$$

This result suggests the following formula.

GENERAL TERM OF AN ARITHMETIC SEQUENCE

The nth term a_n of an arithmetic sequence is given by

$$a_n = a_1 + (n - 1)d,$$

where a_1 is the first term and d is the common difference.

EXAMPLE 3   Finding terms of an arithmetic sequence

If $a_1 = 5$ and $d = 3$, find a_{54}.

Solution
To find a_{54}, apply the formula $a_n = a_1 + (n - 1)d$.

$$a_{54} = 5 + (54 - 1)3 = 164$$ Now Try Exercise **35**

Representations of Geometric Sequences

If a sequence is defined by an exponential function, it is a *geometric sequence*. For example,

$$f(n) = 3(2)^{n-1}$$

represents a geometric sequence because $f(x) = 3(2)^{x-1}$ defines an exponential function. The first five terms of this sequence are shown in Table 14.5.

TABLE 14.5

TABLE 14.5

n	1	2	3	4	5
$f(n)$	3	6	12	24	48

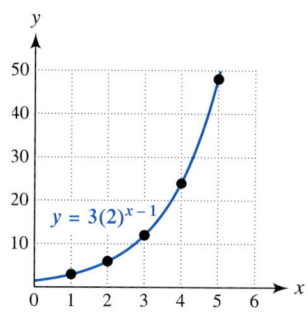

Figure 14.13

Successive terms are found by multiplying the previous term by 2. We say that the *common ratio* of this geometric sequence equals 2. Note that the ratios of successive terms are $\frac{6}{3}, \frac{12}{6}, \frac{24}{12}$, and $\frac{48}{24}$ and that they all equal the common ratio 2. When the points associated with the terms in Table 14.5 are graphed, they do *not* lie on a line. Rather they lie on the exponential curve $y = 3(2)^{x-1}$, as shown in Figure 14.13. A geometric sequence with a positive common ratio is an exponential function whose domain is the set of natural numbers. Its terms reflect either *exponential growth* or *exponential decay*.

GEOMETRIC SEQUENCE

A **geometric sequence** is given by $a_n = a_1(r)^{n-1}$, where n is a natural number and $r \neq 0$ or 1. The value of r is called the **common ratio**, and a_1 is the first term of the sequence.

NOTE: If each term after the first term is obtained by multiplying the previous term by a fixed number, the sequence is a geometric sequence. The fixed number is the *common ratio* and cannot be 0 or 1. For example, 3, 6, 12, 24, 48, . . . is a geometric sequence. Each term (after the first) is found by multiplying the previous term by the common ratio of 2.

EXAMPLE 4 Recognizing geometric sequences

Determine whether f is a geometric sequence. If it is, identify the common ratio.
(a) $f(n) = 2(0.9)^{n-1}$

(b)

n	1	2	3	4	5
$f(n)$	8	4	2	1	$\frac{1}{2}$

(c) A graph of f is shown in Figure 14.14.

Figure 14.14

Solution
(a) This sequence is geometric because $f(x) = 2(0.9)^{x-1}$ defines an exponential function. The common ratio is $r = 0.9$.
(b) The table shows that each successive term is half the previous term. This sequence represents a geometric sequence with a common ratio of $r = \frac{1}{2}$.
(c) The sequence shown in Figure 14.14 is not a geometric sequence because the points are collinear. There is no common ratio.   Now Try Exercises 41, 45, 53

MAKING CONNECTIONS

Common Ratios and Growth or Decay Factors

If the common ratio r of a geometric sequence is positive, then r equals either the growth factor or the decay factor for an exponential function.

EXAMPLE 5   **Finding symbolic representations**

Find a general term a_n for each geometric sequence.
(a) $a_1 = \frac{1}{2}$ and $r = 5$ **(b)** $a_1 = 2$, $a_3 = 18$, and $r < 0$.

CRITICAL THINKING

If we are given a_1 and a_5, can we determine the common ratio of a geometric series? Explain.

Solution
(a) Let $a_n = a_1(r)^{n-1}$. Because $a_1 = \frac{1}{2}$ and $r = 5$, we can write $a_n = \frac{1}{2}(5)^{n-1}$.
(b) $a_1 = 2$ and $a_3 = 18$, so

$$a_3 = a_1(r)^{3-1}$$
$$18 = 2(r)^2.$$

This equation simplifies to

$$r^2 = 9 \quad \text{or} \quad r = \pm 3.$$

It is specified that $r < 0$, so $r = -3$ and $a_n = 2(-3)^{n-1}$.   Now Try Exercises 55, 59

EXAMPLE 6   **Finding a term of a geometric sequence**

If $a_1 = 5$ and $r = 3$, find a_{10}.

Solution
To find a_{10}, apply the formula $a_n = a_1(r)^{n-1}$ with $a_1 = 5$, $r = 3$, and $n = 10$.

$$a_{10} = 5(3)^{10-1} = 5(3)^9 = 98,415$$ Now Try Exercise 61

Applications and Models

▶ **REAL-WORLD CONNECTION**   Sequences are frequently used to describe a variety of situations. In the next example we use a sequence to model classroom ventilation.

EXAMPLE 7   **Modeling classroom ventilation**

Ventilation is an effective means for removing indoor air pollutants. According to the American Society of Heating, Refrigerating, and Air-Conditioning Engineers (ASHRAE), a classroom should have a ventilation rate of 900 cubic feet per hour per person.
(a) Write a sequence that gives the hourly ventilation necessary for 1, 2, 3, 4, and 5 people in a classroom. Is this sequence arithmetic, geometric, or neither?

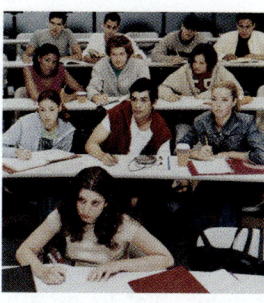

(b) Write the general term for this sequence. Why is it reasonable to limit the domain to natural numbers?

(c) Find a_{30} and interpret the result.

Solution

(a) One person requires 900 cubic feet of air circulated per hour, two people require 1800, three people 2700, and so on. The first five terms of this sequence are

$$900, 1800, 2700, 3600, 4500.$$

This sequence is arithmetic, with a common difference of 900.

(b) The *n*th term equals $900n$, so we let $a_n = 900n$. Because we cannot have a fraction of a person, limiting the domain to the natural numbers is reasonable.

(c) The result $a_{30} = 900(\mathbf{30}) = \mathbf{27{,}000}$ indicates that a classroom with **30** people should have a ventilation rate of **27,000** cubic feet per hour.   Now Try Exercise **65**

▶ **REAL-WORLD CONNECTION** Chlorine is frequently added to the water to disinfect swimming pools. The chlorine concentration should remain between 1.5 and 2.5 parts per million (ppm). On a warm, sunny day 30% of the chlorine may dissipate from the water. In the next example we use a sequence to model the amount of chlorine in a pool at the beginning of each day.   (***Source:*** D. Thomas, *Swimming Pool Operator's Handbook.*)

EXAMPLE 8   Modeling chlorine in a swimming pool

A swimming pool on a warm, sunny day begins with a high chlorine content of 4 parts per million. (Assume that each day 30% of the chlorine dissipates.)

(a) Write a sequence that models the amount of chlorine in the pool at the beginning of the first 3 days, assuming that no additional chlorine is added and that the days are warm and sunny. Is this sequence arithmetic, geometric, or neither?

(b) Write the general term for this sequence.

(c) At the beginning of what day does the chlorine first drop below 1.5 parts per million?

Solution

(a) Because 30% of the chlorine dissipates, 70% remains in the water at the beginning of the next day. If the concentration at the beginning of the first day is 4 parts per million, then at the beginning of the second day it is

$$4 \cdot 0.70 = 2.8 \text{ parts per million,}$$

and at the start of the third day it is

$$2.8 \cdot 0.70 = 1.96 \text{ parts per million.}$$

The first three terms are 4, 2.8, 1.96. Successive terms are found by multiplying the previous term by 0.7. Thus the sequence is geometric, with common ratio 0.7.

(b) The initial amount is $a_1 = \mathbf{4}$ and the common ratio is $r = \mathbf{0.7}$, so the sequence can be represented by $a_n = \mathbf{4}(\mathbf{0.7})^{n-1}$.

(c) The table shown in Figure 14.15 reveals that $a_4 = 4(0.7)^{4-1} \approx 1.372 < 1.5$. Thus, at the beginning of the fourth day, the chlorine level in the swimming pool drops below the recommended minimum of 1.5 parts per million.   Now Try Exercise **67**

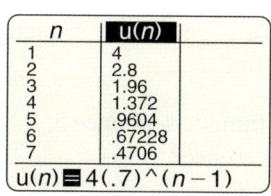

n	$u(n)$
1	4
2	2.8
3	1.96
4	1.372
5	.9604
6	.67228
7	.4706

$u(n) \equiv 4(.7)^{\wedge}(n-1)$

Figure 14.15

14.2   PUTTING IT ALL TOGETHER

In this section we discussed two types of sequences: arithmetic and geometric. Arithmetic sequences are linear functions, and geometric sequences with positive r are exponential functions. The inputs for both are limited to the natural numbers. Each successive term in an arithmetic sequence is found by adding the common difference d to the previous term. For a geometric sequence each successive term is found by multiplying the previous term by the common ratio r. The graph of an arithmetic sequence consists of points that lie on a line, whereas the graph of a geometric sequence (with a positive r) consists of points that lie on an exponential curve. Examples are shown in the following table.

Sequence	Formula	Example
Arithmetic	$a_n = dn + c$ or $a_n = a_1 + (n - 1)d$, where d is the common difference and a_1 is the first term.	If $a_n = 5n + 2$, then the common difference is $d = 5$ and the terms of the sequence are $$7, 12, 17, 22, 27, 32, 37, \ldots.$$ Each term after the first is found by adding 5 to the previous term. The general term can be written as $$a_n = 7 + 5(n - 1).$$
Geometric	$a_n = a_1(r)^{n-1}$, where r is the common ratio ($r \neq 0, r \neq 1$) and a_1 is the first term.	If $a_n = 4(-2)^{n-1}$, then the common ratio is $r = -2$ and the first term is $a_1 = 4$. The terms of the sequence are $$4, -8, 16, -32, 64, -128, 256, \ldots.$$ Each term after the first is found by multiplying the previous term by -2.

14.2   Exercises

MyMathLab

PRACTICE   WATCH   DOWNLOAD   READ   REVIEW

CONCEPTS

1. An arithmetic sequence is a(n) _(linear/exponential)_ function.

2. A geometric sequence with $r > 0$ is a(n) _(linear/exponential)_ function.

3. Give an example of an arithmetic sequence. State the common difference.

4. Give an example of a geometric sequence. State the common ratio.

5. To find successive terms in an arithmetic sequence, _____ the common difference to the _____ term.

6. To find successive terms in a geometric sequence, _____ the previous term by the _____.

7. Find the next term in the arithmetic sequence 3, 7, 11, 15. What is the common difference?

8. Find the next term in the geometric sequence 2, -4, 8, -16. What is the common ratio?

9. Write the general term a_n for a geometric sequence, using a_1 and r.

10. Write the general term a_n for an arithmetic sequence using a_1 and d.

ARITHMETIC SEQUENCES

Exercises 11–28: (Refer to Example 1.) Determine whether f is an arithmetic sequence. Identify the common difference when possible.

11. $f(n) = 10n - 5$ 12. $f(n) = -3n - 5$

13. $f(n) = 6 - n$ 14. $f(n) = 6 + \frac{1}{2}n$

15. $f(n) = n^3 + 1$ 16. $f(n) = 5\left(\frac{1}{3}\right)^{n-1}$

17.

n	1	2	3	4
$f(n)$	3	6	9	12

18.

n	1	2	3	4
$f(n)$	-7	-5	-3	-1

19.

n	1	2	3	4
$f(n)$	10	7	4	1

20.

n	1	2	3	4
$f(n)$	1	2	4	8

21.

n	1	2	3	4
$f(n)$	-4	0	8	12

22.

n	1	2	3	4
$f(n)$	1	2.5	4	5.5

23.

24.

25.

26.

27.

28.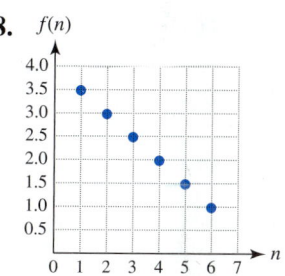

Exercises 29–34: (Refer to Example 2.) Find the general term a_n for the arithmetic sequence.

29. $a_1 = 7$ and $d = -2$ 30. $a_1 = 5$ and $a_2 = 9$

31. $a_1 = -2$ and $a_3 = 6$

32. $a_2 = 7$ and $a_3 = 10$

33. $a_8 = 16$ and $a_{12} = 8$

34. $a_3 = 7$ and $d = -5$

Exercises 35–38: (Refer to Example 3.)

35. If $a_1 = -3$ and $d = 2$, find a_{32}.

36. If $a_1 = 2$ and $d = -3$, find a_{19}.

37. If $a_1 = -3$ and $a_2 = 0$, find a_9.

38. If $a_3 = -3$ and $d = 4$, find a_{62}.

GEOMETRIC SEQUENCES

Exercises 39–54: (Refer to Example 4.) Determine whether f is a geometric sequence. Identify the common ratio when possible.

39. $f(n) = 3^n$ 40. $f(n) = 2(4)^n$

41. $f(n) = \frac{2}{3}(0.8)^{n-1}$ 42. $f(n) = 7 - 3n$

43. $f(n) = 2(n - 1)^2$ 44. $f(n) = 2\left(-\frac{3}{4}\right)^{n-1}$

45.

n	1	2	3	4
$f(n)$	2	4	8	16

46.

n	1	2	3	4
$f(n)$	-6	3	-1.5	0.75

47.

n	1	2	3	4
$f(n)$	1	4	9	16

48.

n	1	2	3	4
$f(n)$	7	4	-1	-8

49.

n	1	2	3	4
$f(n)$	2	8	32	128

50.

n	1	2	3	4
$f(n)$	1	$\frac{1}{2}$	$\frac{1}{4}$	$\frac{1}{8}$

51. $f(n)$

52. $f(n)$

53. $f(n)$

54. $f(n)$

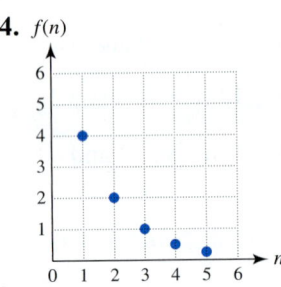

Exercises 55–60: (Refer to Example 5.) Find the general term a_n for the geometric sequence.

55. $a_1 = 1.5$ and $r = 4$ **56.** $a_1 = 3$ and $r = \frac{1}{4}$

57. $a_1 = -3$ and $a_2 = 6$ **58.** $a_1 = 2$ and $a_4 = 54$

59. $a_1 = 1$, $a_3 = 16$, and $r > 0$

60. $a_2 = 3$, $a_4 = 12$, and $r < 0$

Exercises 61–64: (Refer to Example 6.)

61. If $a_1 = 2$ and $r = 3$, find a_8.

62. If $a_1 = 4$ and $a_2 = 2$, find a_9.

63. If $a_1 = -1$ and $a_2 = 3$, find a_6.

64. If $a_3 = 5$ and $r = -3$, find a_7.

APPLICATIONS

65. *Room Ventilation*   (Refer to Example 7.) In areas such as bars and lounges that allow smoking, the ventilation rate should be 3000 cubic feet per hour per person.   (*Source:* ASHRAE.)
  (a) Write a sequence that gives the hourly ventilation necessary for 1, 2, 3, 4, and 5 people in a barroom. Is this sequence arithmetic, geometric, or neither?
  (b) Write the general term for this sequence.
  (c) Find a_{20} and interpret the result.
  (d) Give a graphical representation for this sequence, using $n = 1, 2, 3, \ldots, 8$. Are the points collinear?

66. *Salary*   Suppose that an employee receives a $2000 raise each year and that the sequence a_n models the employee's salary after n years. Is this sequence arithmetic, geometric, or neither? Explain.

67. *Chlorine in Swimming Pools*   (Refer to Example 8.) Suppose that the water in a swimming pool initially has a chlorine content of 3 parts per million and that 20% of the chlorine dissipates each day.
  (a) If no additional chlorine is added, write the general term for a sequence that gives the chlorine concentration at the beginning of each day.
  (b) Give a graphical representation for this sequence, using $n = 1, 2, 3, \ldots, 8$. Are the points collinear? Is this sequence arithmetic, geometric, or neither?

68. *Salary*   Suppose that an employee receives a 7% increase in salary each year and that the sequence a_n models the employee's salary after n years. Is this sequence arithmetic, geometric, or neither? Explain.

69. *Bouncing Ball*   A tennis ball bounces back to 85% of the height from which it was dropped and then to 85% of the height of each successive bounce.
  (a) Write the general term for a sequence a_n that gives the maximum height of the ball on the nth bounce. Let $a_1 = 5$ feet.
  (b) Is the sequence arithmetic or geometric? Explain.
  (c) Find a_8 and interpret the result.

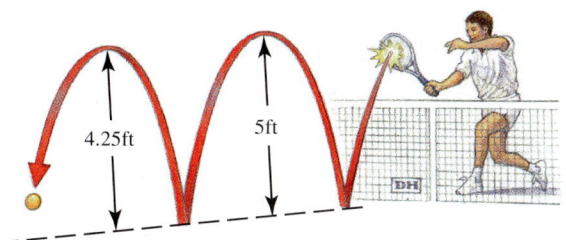

4.25ft   5ft

DH

70. *Falling Object*   The total distance D_n that an object falls in n seconds is shown in the table. Is the sequence arithmetic, geometric, or neither? Explain your reasoning.

n (seconds)	1	2	3	4	5
D_n (feet)	16	64	144	256	400

71. *Theater Seating*   A theater has 40 seats in the first row, 42 seats in the second row, 44 seats in the third row, and so on.

(a) Can the number of seats in each row be modeled by an arithmetic or geometric sequence? Explain.
(b) Write the general term for a sequence a_n that gives the number of seats in row n.
(c) How many seats are there in row 20?

72. *Appreciation of Lake Property*   A certain type of lake property in northern Minnesota is increasing in value by 15% per year. Let the sequence a_n give the value of this type of lake property at the beginning of year n.
  (a) Is a_n arithmetic, geometric, or neither? Explain your reasoning.
  (b) Write the general term for this sequence if $a_1 = \$100,000$.
  (c) Find a_7 and interpret the result.
  (d) Give a graph for a_n, where $n = 1, 2, 3, \ldots, 10$.

WRITING ABOUT MATHEMATICS

73. If you have a table of values for a sequence, how can you determine whether it is geometric? Give an example.

74. If you have a graph of a sequence, how can you determine whether it is arithmetic? Give an example.

CHECKING BASIC CONCEPTS
SECTIONS 14.1 AND 14.2

1. Write the first four terms of the sequence defined by $a_n = \dfrac{n}{n+4}$

2. Represent the sequence $a_n = n + 1$ numerically and graphically for $n = 1, 2, 3, 4, 5$.

3. Use the table to determine whether the sequence is arithmetic or geometric. Write the general term for the sequence.
  (a)

n	1	2	3	4	5
a_n	−2	1	4	7	10

  (b)

n	1	2	3	4	5
a_n	3	−6	12	−24	48

4. Find the general term a_n for an arithmetic sequence with $a_1 = 5$ and $d = 2$.

5. Find the general term a_n for a geometric sequence with $a_1 = 5$ and $r = 2$.

14.3 SERIES

Basic Concepts ▪ Arithmetic Series ▪ Geometric Series ▪ Summation Notation

Although the terms *sequence* and *series* are sometimes used interchangeably in everyday life, they represent different mathematical concepts. In mathematics a sequence is a function whose domain is the set of natural numbers, whereas a series is a summation of the terms in a sequence. Series have played a central role in the development of modern mathematics. Today series are often used to approximate functions that are too complicated to have formulas. Series are also used to calculate accurate approximations for π and e.

Basic Concepts

Suppose that a person has a starting salary of \$30,000 per year and receives a \$2000 raise each year. Then the *sequence*

$$30{,}000,\ 32{,}000,\ 34{,}000,\ 36{,}000,\ 38{,}000$$

lists these salaries over a 5-year period. The total amount earned is given by the *series*

$$30{,}000\ +\ 32{,}000\ +\ 34{,}000\ +\ 36{,}000\ +\ 38{,}000,$$

whose sum is \$170,000. We now define the concept of a series.

FINITE SERIES

A **finite series** is an expression of the form

$$a_1 + a_2 + a_3 + \cdots + a_n.$$

EXAMPLE 1   Computing total reported AIDS cases

Table 14.6 presents a sequence a_n that computes the number of AIDS cases diagnosed each year from 1997 through 2003, where $n = 1$ corresponds to 1997.

TABLE 14.6   Annual U.S. AIDS Cases

n	1	2	3	4	5	6	7
a_n	50,167	43,225	41,314	41,239	41,227	42,136	43,171

Source: Department of Health and Human Services.

(a) Write a series whose sum represents the total number of AIDS cases diagnosed from 1997 to 2003. Find its sum.

(b) Interpret the series $a_1 + a_2 + a_3 + \cdots + a_{10}$.

Solution

(a) The required series and sum are given by

$$50{,}167 + 43{,}225 + 41{,}314 + 41{,}239 + 41{,}227 + 42{,}136 + 43{,}171 = 302{,}479.$$

(b) The series $a_1 + a_2 + a_3 + \cdots + a_{10}$ represents the total number of AIDS cases diagnosed over 10 years from 1997 through 2006.

Now Try Exercise 41

Sequences and Series

A *sequence* is an *ordered list*; a *series* is the *sum of the terms of a sequence*. For example, the even integers from 2 to 20 are represented by the sequence

$$2, 4, 6, 8, 10, 12, 14, 16, 18, 20. \qquad \text{Sequence}$$

The corresponding series is

$$2 + 4 + 6 + 8 + 10 + 12 + 14 + 16 + 18 + 20, \qquad \text{Series}$$

which sums to 110.

Arithmetic Series

Summing the terms of an arithmetic sequence results in an **arithmetic series**. For example, $a_n = 2n - 1$ for $n = 1, 2, 3, \ldots, 7$ defines the arithmetic sequence

$$1, 3, 5, 7, 9, 11, 13.$$

The corresponding arithmetic *series* is

$$1 + 3 + 5 + 7 + 9 + 11 + 13,$$

whose sum is 49. The following formula gives the sum of the first n terms of an arithmetic sequence.

SUM OF THE FIRST n TERMS OF AN ARITHMETIC SEQUENCE

The **sum of the first n terms of an arithmetic sequence**, denoted S_n, is found by averaging the first and nth terms and then multiplying by n. That is,

$$S_n = a_1 + a_2 + a_3 + \cdots + a_n = n\left(\frac{a_1 + a_n}{2}\right).$$

The series $1 + 3 + 5 + 7 + 9 + 11 + 13$ consists of **7** terms, where the first term is **1** and the last term is **13**. Substituting in the formula gives

$$S_7 = 7\left(\frac{1 + 13}{2}\right) = 49,$$

which agrees with the sum obtained by adding the 7 terms.

Because $a_n = a_1 + (n - 1)d$ for an arithmetic sequence, S_n can also be written

$$S_n = n\left(\frac{a_1 + a_n}{2}\right)$$

$$= \frac{n}{2}(a_1 + a_n)$$

$$= \frac{n}{2}(a_1 + a_1 + (n - 1)d)$$

$$= \frac{n}{2}(2a_1 + (n - 1)d).$$

EXAMPLE 2   Finding the sum of a finite arithmetic series

Suppose that a person has a starting annual salary of $30,000 and receives a $1500 raise each year. Calculate the total amount earned after 10 years.

Solution
The sequence that gives the salary during year n is given by

$$a_n = 30{,}000 + 1500(n - 1).$$

One way to calculate the sum of the first 10 terms, denoted S_{10}, is to find a_1 and a_{10}, or

$$a_1 = 30{,}000 + 1500(\mathbf{1} - 1) = \mathbf{30{,}000}$$
$$a_{10} = 30{,}000 + 1500(\mathbf{10} - 1) = \mathbf{43{,}500}.$$

Thus the total amount earned during this 10-year period is

$$S_{10} = 10\left(\frac{a_1 + a_{10}}{2}\right)$$
$$= 10\left(\frac{\mathbf{30{,}000} + \mathbf{43{,}500}}{2}\right)$$
$$= \$367{,}500.$$

This sum can also be found with the second formula by letting $d = \mathbf{1500}$.

$$S_n = \frac{n}{2}(2a_1 + (n - 1)d)$$
$$= \frac{10}{2}(2 \cdot \mathbf{30{,}000} + (\mathbf{10} - 1)\mathbf{1500})$$
$$= 5(60{,}000 + 9 \cdot 1500)$$
$$= \$367{,}500.$$

Now Try Exercise **49**

EXAMPLE 3   Finding the sum of an arithmetic series

Find the sum of the series $2 + 4 + 6 + \cdots + 100$.

Solution
The first term of this series is $a_1 = \mathbf{2}$, and the common difference is $d = 2$. This series represents the even numbers from $\mathbf{2}$ to $\mathbf{100}$, so the number of terms is $n = \mathbf{50}.$ Using the formula

$$S_n = n\left(\frac{a_1 + a_n}{2}\right)$$

for the sum of an arithmetic series, we obtain

$$S_{50} = 50\left(\frac{\mathbf{2} + \mathbf{100}}{2}\right) = 2550.$$

Now Try Exercise **13**

TECHNOLOGY NOTE: Sum of a Series

The "seq(" utility, found on some calculators under the LIST OPS menus, generates a sequence. The "sum(" utility found on some calculators under the LIST MATH menus, calculates the sum of the sequence inside the parentheses. To verify the result in Example 2, let $a_n = 30{,}000 + 1500(n - 1)$. The value 367,500 for S_{10} is shown in Figure 14.16(a). The result found in Example 3 is shown in Figure 14.16(b).

    (a)   (b)

Figure 14.16

Geometric Series

A **geometric series** is the sum of the terms of a geometric sequence. For example,

$$1, 2, 4, 8, 16, 32$$

is a geometric sequence with $a_1 = 1$ and $r = 2$. Then

$$1 + 2 + 4 + 8 + 16 + 32$$

is a geometric series. We can use the following formula to sum a finite geometric series.

SUM OF THE FIRST n TERMS OF A GEOMETRIC SEQUENCE

If its first term is a_1 and its common ratio is r, then the **sum of the first n terms of a geometric sequence** is given by

$$S_n = a_1\left(\frac{1 - r^n}{1 - r}\right),$$

provided $r \neq 1$.

EXAMPLE 4 **Finding the sum of finite geometric series**

Find the sum of each series.
(a) $1 + 2 + 4 + 8 + 16 + 32 + 64 + 128 + 256$
(b) $\frac{1}{2} - \frac{1}{4} + \frac{1}{8} - \frac{1}{16} + \frac{1}{32}$

Solution
(a) This series is geometric, with $n = 9$, $a_1 = 1$, and $r = 2$, so

$$S_9 = 1\left(\frac{1 - 2^9}{1 - 2}\right) = 511.$$

(b) This series is geometric, with $n = 5$, $a_1 = \frac{1}{2} = 0.5$, and $r = -\frac{1}{2} = -0.5$, so

$$S_5 = 0.5\left(\frac{1 - (-0.5)^5}{1 - (-0.5)}\right) = \frac{11}{32} = 0.34375.$$

Now Try Exercises 17, 19

▶ **REAL-WORLD CONNECTION**   A sum of money from which regular payments are made is called an **annuity**. An annuity may be purchased with a lump sum deposit or by deposits made at various intervals. Suppose that $1000 are deposited at the end of each year in an annuity account that pays an annual interest rate I expressed as a decimal. At the end of the first year the account contains $1000. At the end of the second year $1000 are deposited again. In addition, the first deposit of $1000 would have received interest during the second year. Therefore the value of the annuity after 2 years is

$$1000 + 1000(1 + I).$$

After 3 years the balance is

$$1000 + 1000(1 + I) + 1000(1 + I)^2,$$

and after n years this amount is given by

$$1000 + 1000(1 + I) + 1000(1 + I)^2 + \cdots + 1000(1 + I)^{n-1}.$$

This series is a geometric series with its first term $a_1 = 1000$ and the common ratio $r = (1 + I)$. The sum of the first n terms is given by

$$S_n = a_1\left(\frac{1 - (1 + I)^n}{1 - (1 + I)}\right) = a_1\left(\frac{(1 + I)^n - 1}{I}\right).$$

EXAMPLE 5   **Finding the future value of an annuity**

Suppose that a 20-year-old worker deposits $1000 into an annuity account at the end of each year. If the interest rate is 12%, find the future value of the annuity when the worker is 65 years old.

Solution
Let $a_1 = 1000$, $I = 0.12$, and $n = 45$. The future value of the annuity is

$$S_n = a_1\left(\frac{(1 + I)^n - 1}{I}\right)$$

$$= 1000\left(\frac{(1 + 0.12)^{45} - 1}{0.12}\right)$$

$$\approx \$1,358,230. \qquad\qquad \text{Now Try Exercise } \boxed{23}$$

Summation Notation

Summation notation is used to write series efficiently. The symbol Σ, the uppercase Greek letter *sigma*, is used to indicate a sum.

> **SUMMATION NOTATION**
>
> $$\sum_{k=1}^{n} a_k = a_1 + a_2 + a_3 + \cdots + a_n$$

The letter k is called the **index of summation**. The numbers 1 and n represent the subscripts of the first and last terms in the series. They are called the **lower limit** and **upper limit** of the summation, respectively.

EXAMPLE 6 **Using summation notation**

Evaluate each series.

(a) $\displaystyle\sum_{k=1}^{5} k^2$ (b) $\displaystyle\sum_{k=1}^{4} 5$ (c) $\displaystyle\sum_{k=3}^{6} (2k - 5)$

Solution

(a) $\displaystyle\sum_{k=1}^{5} k^2 = 1^2 + 2^2 + 3^2 + 4^2 + 5^2 = 55$

(b) $\displaystyle\sum_{k=1}^{4} 5 = 5 + 5 + 5 + 5 = 20$

(c) $\displaystyle\sum_{k=3}^{6} (2k - 5) = \underset{k=3}{(2(3) - 5)} + \underset{k=4}{(2(4) - 5)} + \underset{k=5}{(2(5) - 5)} + \underset{k=6}{(2(6) - 5)}$

$= 1 + 3 + 5 + 7 = 16$ Now Try Exercises 27, 29, 31

Summation notation is used frequently in statistics. The next example demonstrates how averages can be expressed in summation notation.

EXAMPLE 7 **Applying summation notation**

Express the average of the n numbers $x_1, x_2, x_3, \ldots, x_n$ in summation notation.

Solution

The average of n numbers can be written as

$$\frac{x_1 + x_2 + x_3 + \cdots + x_n}{n}.$$

This expression is equivalent to $\frac{1}{n}\left(\sum_{k=1}^{n} x_k\right)$.   Now Try Exercise 35

NOTE: $\sum_{k=1}^{n} x_k$ is equivalent to $\displaystyle\sum_{k=1}^{n} x_k$.

▶ **REAL-WORLD CONNECTION**   Series play an essential role in various applications.

EXAMPLE 8 **Modeling air filtration**

Suppose that an air filter removes 90% of the impurities entering it.
(a) Find a series that represents the amount of impurities removed by a sequence of n air filters. Express this answer in summation notation.
(b) How many air filters would be necessary to remove 99.99% of the impurities?

Solution

(a) The first filter removes 90% of the impurities, so 10%, or 0.1, passes through it. Of the 0.1 that passes through the first filter, 90% is removed by the second filter, while 10% of 10%, or 0.01, passes through. Then, 10% of 0.01, or 0.001, passes through the third

Figure 14.17 Impurities
Passing Through Air Filters

filter. Figure 14.17 depicts these results, from which we can establish a pattern. If we let 100%, or 1, represent the amount of impurities entering the first air filter, the amount removed by n filters equals

$$(0.9)(1) + (0.9)(0.1) + (0.9)(0.01) + (0.9)(0.001) + \cdots + (0.9)(0.1)^{n-1}.$$

In summation notation we write this series as $\sum_{k=1}^{n} 0.9(0.1)^{k-1}$.

(b) To remove 99.99%, or 0.9999, of the impurities requires 4 air filters, because

$$\sum_{k=1}^{4} 0.9(0.1)^{k-1} = (0.9)(1) + (0.9)(0.1) + (0.9)(0.01) + (0.9)(0.001)$$

$$= 0.9 + 0.09 + 0.009 + 0.0009$$

$$= 0.9999.$$

Now Try Exercise **43**

14.3 PUTTING IT ALL TOGETHER

A finite sequence is an ordered list such as

$$a_1, a_2, a_3, a_4, a_5, \ldots, a_n.$$

A finite series is the summation of the terms of a sequence and can be expressed as

$$a_1 + a_2 + a_3 + a_4 + a_5 + \cdots + a_n.$$

The following table summarizes concepts related to arithmetic and geometric series.

Series	Description	Example
Finite Arithmetic	$a_1 + a_2 + a_3 + \cdots + a_n$, where $a_n = dn + c$ or $a_n = a_1 + (n-1)d$. The sum of the first n terms is $$S_n = n\left(\frac{a_1 + a_n}{2}\right) \quad \text{or}$$ $$S_n = \frac{n}{2}(2a_1 + (n-1)d),$$ where a_1 is the first term and d is the common difference.	The series $4 + 7 + 10 + 13 + 16 + 19 + 22$ is obtained from the sequence $a_n = 3n + 1 \quad \text{or} \quad a_n = 4 + 3(n-1)$. Its sum is $$S_7 = 7\left(\frac{4 + 22}{2}\right) = 91 \quad \text{or}$$ $$S_7 = \frac{7}{2}(2 \cdot 4 + (7-1)3) = 91.$$
Finite Geometric	$a_1 + a_2 + a_3 + \cdots + a_n$, where $a_n = a_1(r)^{n-1}$ for nonzero constants a_1 and r. The sum of the first n terms is $$S_n = a_1\left(\frac{1 - r^n}{1 - r}\right),$$ where a_1 is the first term and r is the common ratio ($r \neq 1$).	The series $3 + 6 + 12 + 24 + 48 + 96$ has $n = 6$, $a_1 = 3$, and $r = 2$. Its sum is $$S_6 = 3\left(\frac{1 - 2^6}{1 - 2}\right) = 189.$$

14.3 Exercises

CONCEPTS

1. The summation of the terms of a sequence is called a(n) _____.

2. Find the sum of the series $1 + 2 + 3 + 4$.

3. The series $1 + 3 + 5 + 7 + 9$ is an example of a(n) _____ series.

4. The series $1 + 3 + 9 + 27 + 81$ is an example of a(n) _____ series.

5. If $a_1 + a_2 + a_3 + \cdots + a_n$ is an arithmetic series, its sum is $S_n = $ _____.

6. If $a_1 + a_2 + a_3 + \cdots + a_n$ is a geometric series with the common ratio $r \neq 1$, its sum is $S_n = $ _____.

7. The symbol $\sum$ is used to indicate a _____.

8. Write $\sum_{k=1}^{4} a_k$ as a sum.

9. $\sum_{n=1}^{5} a_1 + (n - 1)d$ is an example of a(n)_____ series.

10. $\sum_{n=1}^{4} a_1 r^{n-1}$ is an example of a(n) _____ series.

SUMS OF SERIES

Exercises 11–16: Find the sum of the arithmetic series by using a formula.

11. $3 + 5 + 7 + 9 + 11 + 13$

12. $7.5 + 6 + 4.5 + 3 + 1.5 + 0 + (-1.5)$

13. $1 + 2 + 3 + 4 + \cdots + 40$

14. $1 + 3 + 5 + 7 + \cdots + 99$

15. $-7 + (-4) + (-1) + 2 + 5$

16. $89 + 84 + 79 + 74 + 69 + 64 + 59 + 54$

Exercises 17–22: Find the sum of the geometric series by using a formula.

17. $3 + 9 + 27 + 81 + 243 + 729 + 2187$

18. $2 - 1 + \frac{1}{2} - \frac{1}{4} + \frac{1}{8} - \frac{1}{16} + \frac{1}{32}$

19. $1 - 2 + 4 - 8 + 16 - 32 + 64 - 128$

20. $2 + \frac{1}{2} + \frac{1}{8} + \frac{1}{32} + \frac{1}{128} + \frac{1}{512}$

21. $0.5 + 1.5 + 4.5 + 13.5 + 40.5 + 121.5$

22. $0.6 + 0.3 + 0.15 + 0.075 + 0.0375$

*Exercises 23–26: **Annuities** (Refer to Example 5.) Find the future value of the annuity.*

23. $a_1 = \$2000$ $I = 0.08$ $n = 20$

24. $a_1 = \$500$ $I = 0.15$ $n = 10$

25. $a_1 = \$10{,}000$ $I = 0.11$ $n = 5$

26. $a_1 = \$3000$ $I = 0.19$ $n = 45$

SUMMATION NOTATION

Exercises 27–34: Write the terms of the series and find their sum.

27. $\sum_{k=1}^{4} 2k$

28. $\sum_{k=1}^{6} (k - 1)$

29. $\sum_{k=1}^{8} 4$

30. $\sum_{k=2}^{6} (5 - 2k)$

31. $\sum_{k=1}^{7} k^2$

32. $\sum_{k=1}^{4} 5(2)^{k-1}$

33. $\sum_{k=4}^{5} (k^2 - k)$

34. $\sum_{k=1}^{4} \log k$

Exercises 35–38: Write in summation notation.

35. $1^4 + 2^4 + 3^4 + 4^4 + 5^4 + 6^4$

36. $1 + \frac{1}{5^1} + \frac{1}{5^2} + \frac{1}{5^3} + \frac{1}{5^4}$

37. $1 + \frac{1}{2^2} + \frac{1}{3^2} + \frac{1}{4^2} + \frac{1}{5^2}$

38. $1 + \frac{1}{10} + \frac{1}{100} + \frac{1}{1000} + \frac{1}{10{,}000}$

39. Verify that $\sum_{k=1}^{n} k = \frac{n(n+1)}{2}$ by using a formula for the sum of the first n terms of an arithmetic series.

40. Use Exercise 39 to find the sum of the series $\sum_{k=1}^{200} k$.

APPLICATIONS

41. *Prison Escapees*   The table lists the number of escapees from state prisons each year.

Year	1990	1991	1992
Escapees	8518	9921	10,706

Year	1993	1994	1995
Escapees	14,035	14,307	12,249

Source: Bureau of Justice Statistics.

(a) Write a series whose sum is the total number of escapees from 1990 to 1995.
(b) Find its sum.

42. *Captured Prison Escapees*   (Refer to Exercise 41.) The table lists the number of escapees from state prisons who were captured, including inmates who may have escaped during a previous year.

Year	1990	1991	1992
Captured	9324	9586	10,031

Year	1993	1994	1995
Captured	12,872	13,346	12,166

Source: Bureau of Justice Statistics.

(a) Write a series whose sum is the total number of escapees captured from 1990 to 1995.
(b) Find its sum.
(c) Compare the number of escapees to the number captured during this time period.

43. *Air Filtration*   (Refer to Example 8.) Suppose that an air filter removes 80% of the impurities entering it.
(a) Find a series that represents the amount of impurities removed by a sequence of n air filters. Express the answer in summation notation.
(b) How many filters would be necessary to remove 96% of the impurities?

44. *Air Filtration*   Suppose that an air filter removes 70% of the impurities entering it.
(a) Find a series that represents the amount of impurities removed by a sequence of n air filters. Express the answer in summation notation.
(b) How many filters would be necessary to remove 97.3% of the impurities?

45. *Area*   A sequence of smaller squares is formed by connecting the midpoints of the sides of a larger square, as shown in the figure.

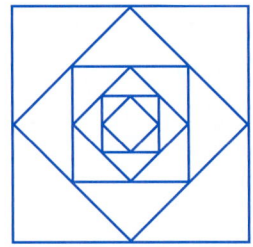

(a) If the area of the largest square is 1 square unit, determine the first five terms of a sequence that describes the area of each successive square.
(b) Use a formula to sum the areas of the first 10 squares.

46. *Perimeter*   (Refer to Exercise 45.) Use a formula to find the sum of the perimeters of the first 10 squares.

47. *Stacking Logs*   A stack of logs is made in layers, with one log less in each layer, as shown in the accompanying figure. If the top layer has 6 logs and the bottom row has 14 logs, what is the total number of logs in the pile? Use a formula to find this sum.

48. *Stacking Logs*   (Refer to Exercise 47.) Suppose that a stack of logs has 15 logs in the top layer and a total of 10 layers. How many logs are in the stack?

49. *Salaries*   (Refer to Example 2.) Suppose that an individual's starting salary is $35,000 per year and that the individual receives a $2000 raise each year. Find the total amount earned over 20 years.

50. *Salaries*   Suppose that an individual's starting salary is $35,000 per year and that the individual receives a 10% raise each year. Find the total amount earned over 20 years.

51. *Bouncing Ball*   A tennis ball first bounces to 75% of the height from which it was dropped and then to 75% of the height of each successive bounce. If it is dropped from a height of 10 feet, find the distance it *falls* between the fourth and fifth bounce.

52. *Bouncing Ball*   A tennis ball first bounces to 75% of the height from which it was dropped and then to 75% of the height of each successive bounce. If it is dropped from a height of 10 feet, find the *total* distance it travels before it reaches its fifth bounce. (*Hint:* Make a sketch.)

WRITING ABOUT MATHEMATICS

53. Discuss the difference between a sequence and a series. Give an example of each.

54. Suppose that an arithmetic series has $a_1 = 1$ and a common difference of $d = 2$, whereas a geometric series has $a_1 = 1$ and a common ratio of $r = 2$. Discuss how their sums compare as the number of terms n becomes large. (*Hint:* Calculate each sum for $n = 10, 20,$ and 30.)

GROUP ACTIVITY
WORKING WITH REAL DATA

Directions: Form a group of 2 to 4 people. Select someone to record the group's responses for this activity. All members of the group should work cooperatively to answer the questions. If your instructor asks for your results, each member of the group should be prepared to respond.

Depreciation   For tax purposes, businesses frequently depreciate equipment. Two methods of depreciation are *straight-line depreciation* and *sum-of-the-years'-digits*. Suppose that a college student buys a $3000 computer to start a business that provides Internet services. This student estimates the life of the computer at 4 years, after which its value will be $200. The difference between $3000 and $200, or $2800, may be deducted from the student's taxable income over a 4-year period.

In straight-line depreciation, equal portions of $2800 are deducted each year over the 4 years. The sum-of-the-years'-digits method calculates depreciation differently. For a computer having a useful life of 4 years, the sum of the years is computed by

$$1 + 2 + 3 + 4 = 10.$$

With this method, $\frac{4}{10}$ of $2800 is deducted the first year, $\frac{3}{10}$ the second year, and so on, until $\frac{1}{10}$ is deducted the fourth year. Both depreciation methods yield a total deduction of $2800 over the 4 years. (*Source:* Sharp Electronics Corporation, *Conquering the Sciences*.)

(a) Find an arithmetic sequence that gives the amount depreciated each year by each method.
(b) Write a series whose sum is the amount depreciated over 4 years by each method.

14.4 THE BINOMIAL THEOREM

Pascal's Triangle ■ **Factorial Notation and Binomial Coefficients** ■ **Using the Binomial Theorem**

A LOOK INTO MATH ▷

In this section we demonstrate how to expand expressions of the form $(a + b)^n$, where n is a natural number. These expressions occur in statistics, finite mathematics, computer science, and calculus. The two methods that we discuss are Pascal's triangle and the binomial theorem.

Pascal's Triangle

Expanding $(a + b)^n$ for increasing values of n gives the following results.

$$(a + b)^0 = \qquad\qquad 1$$
$$(a + b)^1 = \qquad\qquad 1a + 1b$$
$$(a + b)^2 = \qquad\qquad 1a^2 + 2ab + 1b^2$$
$$(a + b)^3 = \qquad\qquad 1a^3 + 3a^2b + 3ab^2 + 1b^3$$
$$(a + b)^4 = \qquad\qquad 1a^4 + 4a^3b + 6a^2b^2 + 4ab^3 + 1b^4$$
$$(a + b)^5 = \qquad 1a^5 + 5a^4b + 10a^3b^2 + 10a^2b^3 + 5ab^4 + 1b^5$$

```
            1
          1   1
        1   2   1
      1   3   3   1
    1   4   6   4   1
  1   5  10  10   5   1
```
Figure 14.18 Pascal's Triangle

Note that $(a + b)^1$ has two terms, starting with a and ending with b; $(a + b)^2$ has three terms, starting with a^2 and ending with b^2; and in general, $(a + b)^n$ has $n + 1$ terms, starting with a^n and ending with b^n. The exponent on a decreases by 1 each successive term, and the exponent on b increases by 1 each successive term.

The triangle formed by the highlighted numbers is called **Pascal's triangle**. This triangle consists of 1s along the sides, and each element inside the triangle is the sum of the two numbers above it, as shown in Figure 14.18. Pascal's triangle is usually written without variables and can be extended to include as many rows as needed.

We can use this triangle to expand $(a + b)^n$, where n is a natural number. For example, the expression $(m + n)^4$ consists of five terms written as

$$(m + n)^4 = \_m^4 + \_m^3n^1 + \_m^2n^2 + \_m^1n^3 + \_n^4.$$

Because there are five terms, the coefficients can be found in the fifth row of Pascal's triangle, which is

$$1 \quad 4 \quad 6 \quad 4 \quad 1.$$

Thus

$$(m + n)^4 = \underline{1}\, m^4 + \underline{4}\, m^3n^1 + \underline{6}\, m^2n^2 + \underline{4}\, m^1n^3 + \underline{1}\, n^4$$
$$= m^4 + 4m^3n + 6m^2n^2 + 4mn^3 + n^4.$$

EXAMPLE **Expanding a binomial**

Expand each binomial, using Pascal's triangle.
(a) $(x + 2)^5$ **(b)** $(2m - n)^3$

Solution

(a) To find the coefficients, use the sixth row in Pascal's triangle.

$$(x + 2)^5 = \underline{1}x^5 + \underline{5}x^4 \cdot 2^1 + \underline{10}x^3 \cdot 2^2 + \underline{10}x^2 \cdot 2^3 + \underline{5}x^1 \cdot 2^4 + \underline{1}(2^5)$$
$$= x^5 + 10x^4 + 40x^3 + 80x^2 + 80x + 32$$

(b) To find the coefficients, use the fourth row in Pascal's triangle.

$$(2m - n)^3 = \underline{1}\,(2m)^3 + \underline{3}\,(2m)^2(-n)^1 + \underline{3}\,(2m)^1(-n)^2 + \underline{1}\,(-n)^3$$
$$= 8m^3 - 12m^2n + 6mn^2 - n^3$$

Now Try Exercises `11, 15`

Factorial Notation and Binomial Coefficients

An alternative to Pascal's triangle is the binomial theorem, which uses **factorial notation**.

(a)

(b)

Figure 14.19

---

n FACTORIAL (*n*!)

For any positive integer n,

$$n! = 1 \cdot 2 \cdot 3 \cdot \cdots \cdot n.$$

We also define $0! = 1$.

NOTE: Because multiplication is commutative, n factorial can also be defined as

$$n! = n \cdot (n - 1) \cdot (n - 2) \cdot \cdots \cdot 2 \cdot 1.$$

Examples include the following.

$$0! = 1$$
$$1! = 1$$
$$2! = 1 \cdot 2 = 2$$
$$3! = 1 \cdot 2 \cdot 3 = 6$$
$$4! = 1 \cdot 2 \cdot 3 \cdot 4 = 24$$
$$5! = 1 \cdot 2 \cdot 3 \cdot 4 \cdot 5 = 120$$

Figure 14.19 supports these results. On some calculators, factorial (!) can be accessed in the MATH PRB menus.

EXAMPLE 2 Evaluating factorial expressions

Simplify the expression.

(a) $\dfrac{5!}{3!2!}$ **(b)** $\dfrac{4!}{4!0!}$

Solution

(a) $\dfrac{5!}{3!2!} = \dfrac{1 \cdot 2 \cdot 3 \cdot 4 \cdot 5}{(1 \cdot 2 \cdot 3)(1 \cdot 2)} = \dfrac{120}{6 \cdot 2} = 10$

(b) $0! = 1$, so $\dfrac{4!}{4!0!} = \dfrac{4!}{4!(1)} = \dfrac{4!}{4!} = 1$

Now Try Exercises `23, 25`

The expression $_nC_r$ represents a *binomial coefficient* that can be used to calculate the numbers in Pascal's triangle.

BINOMIAL COEFFICIENT $_nC_r$

For n and r nonnegative integers, $n \geq r$,

$$_nC_r = \frac{n!}{(n-r)!\,r!}$$

is a **binomial coefficient**.

Values of $_nC_r$ for $r = 0, 1, 2, \ldots, n$ correspond to the $n + 1$ numbers in row $n + 1$ of Pascal's triangle.

EXAMPLE 3   Calculating $_nC_r$

Calculate $_3C_r$ for $r = 0, 1, 2, 3$ by hand. Check your results on a calculator. Compare these numbers with the fourth row in Pascal's triangle.

Solution

$$_3C_0 = \frac{3!}{(3-0)!\,0!} = \frac{6}{6 \cdot 1} = 1 \qquad _3C_1 = \frac{3!}{(3-1)!\,1!} = \frac{6}{2 \cdot 1} = 3$$

$$_3C_2 = \frac{3!}{(3-2)!\,2!} = \frac{6}{1 \cdot 2} = 3 \qquad _3C_3 = \frac{3!}{(3-3)!\,3!} = \frac{6}{1 \cdot 6} = 1$$

These results are supported in Figure 14.20. The fourth row of Pascal's triangle is

$$1 \quad 3 \quad 3 \quad 1,$$

which agrees with the calculated values for $_3C_r$. On some calculators, the MATH PRB menus are used to calculate $_nC_r$.

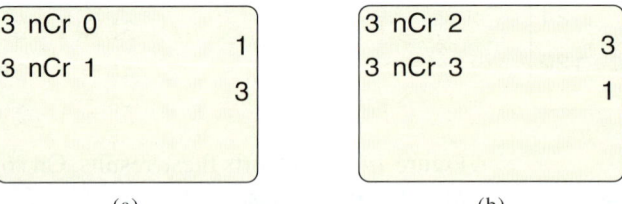

(a)   (b)

Figure 14.20   Now Try Exercises 27, 31

Using the Binomial Theorem

The binomial coefficients can be used to expand expressions of the form $(a + b)^n$. To do so, we use the **binomial theorem**.

BINOMIAL THEOREM

For any positive integer n and any numbers a and b,

$$(a + b)^n = {_nC_0}a^n + {_nC_1}a^{n-1}b^1 + \cdots + {_nC_{n-1}}a^1b^{n-1} + {_nC_n}b^n.$$

Using the results of Example 3, we write

$$(a + b)^3 = {}_3C_0a^3 + {}_3C_1a^2b^1 + {}_3C_2a^1b^2 + {}_3C_3b^3$$
$$= 1a^3 + 3a^2b + 3ab^2 + 1b^3$$
$$= a^3 + 3a^2b + 3ab^2 + b^3.$$

EXAMPLE 4   **Expanding a binomial**

Use the binomial theorem to expand each expression.
(a) $(x + y)^5$ **(b)** $(3 - 2x)^4$

Solution
(a) The coefficients are calculated as follows.

$${}_5C_0 = \frac{5!}{(5 - 0)!\,0!} = 1, \qquad {}_5C_1 = \frac{5!}{(5 - 1)!\,1!} = 5, \qquad {}_5C_2 = \frac{5!}{(5 - 2)!\,2!} = 10$$

$${}_5C_3 = \frac{5!}{(5 - 3)!\,3!} = 10, \qquad {}_5C_4 = \frac{5!}{(5 - 4)!\,4!} = 5, \qquad {}_5C_5 = \frac{5!}{(5 - 5)!\,5!} = 1$$

Using the binomial theorem, we arrive at the following result.

$$(x + y)^5 = {}_5C_0x^5 + {}_5C_1x^4y^1 + {}_5C_2x^3y^2 + {}_5C_3x^2y^3 + {}_5C_4x^1y^4 + {}_5C_5y^5$$
$$= 1x^5 + 5x^4y + 10x^3y^2 + 10x^2y^3 + 5xy^4 + 1y^5$$
$$= x^5 + 5x^4y + 10x^3y^2 + 10x^2y^3 + 5xy^4 + y^5$$

(b) The coefficients are calculated as follows.

$${}_4C_0 = \frac{4!}{(4 - 0)!\,0!} = 1, \qquad {}_4C_1 = \frac{4!}{(4 - 1)!\,1!} = 4, \qquad {}_4C_2 = \frac{4!}{(4 - 2)!\,2!} = 6,$$

$${}_4C_3 = \frac{4!}{(4 - 3)!\,3!} = 4, \qquad {}_4C_4 = \frac{4!}{(4 - 4)!\,4!} = 1$$

Using the binomial theorem with $a = 3$ and $b = (-2x)$, we arrive at the following result.

$$(3 - 2x)^4 = {}_4C_0(3)^4 + {}_4C_1(3)^3(-2x) + {}_4C_2(3)^2(-2x)^2$$
$$+ {}_4C_3(3)(-2x)^3 + {}_4C_4(-2x)^4$$
$$= 1(81) + 4(27)(-2x) + 6(9)(4x^2) + 4(3)(-8x^3) + 1(16x^4)$$
$$= 81 - 216x + 216x^2 - 96x^3 + 16x^4 \qquad \text{Now Try Exercises 39, 49}$$

The binomial theorem gives *all* of the terms of $(a + b)^n$. However, we can find any individual term by noting that the $(r + 1)$st term in the binomial expansion for $(a + b)^n$ is given by the formula ${}_nC_ra^{n-r}b^r$, for $0 \le r \le n$. The next example shows how to use this formula to find the $(r + 1)$st term of $(a + b)^n$.

EXAMPLE 5   **Finding the kth term in a binomial expansion**

Find the third term of $(x - y)^5$.

Solution
In this example the $(r + 1)$st term is the *third* term in the expansion of $(x - y)^5$. That is, $r + 1 = 3$, or $r = 2$. Also, the exponent in the expression is $n = 5$. To get this binomial

into the form $(a + b)^n$, we note that the first term in the binomial is $a = x$ and that the second term in the binomial is $b = -y$. Substituting the values for r, n, a, and b in the formula $_nC_r a^{n-r}b^r$ for the $(r + 1)$st term yields

$$_5C_2(x)^{5-2}(-y)^2 = 10x^3y^2.$$

The third term in the binomial expansion of $(x - y)^5$ is $10x^3y^2$.

Now Try Exercise 53

14.4 PUTTING IT ALL TOGETHER

In this section we showed how to expand the expression $(a + b)^n$ by using Pascal's triangle and the binomial theorem. The following table outlines important topics from this section.

Topic	Explanation	Example
Pascal's Triangle	1 1 1 1 2 1 **1 3 3 1** 1 4 6 4 1 1 5 10 10 5 1	$(a + b)^3 = \mathbf{1}a^3 + \mathbf{3}a^2b + \mathbf{3}ab^2 + \mathbf{1}b^3$ (Row 4) To expand $(a + b)^n$, use row $n + 1$ in the triangle.
Factorial Notation	The expression $n!$ equals $$1 \cdot 2 \cdot 3 \cdot \cdots \cdot n.$$	$5! = 1 \cdot 2 \cdot 3 \cdot 4 \cdot 5 = 120$
Binomial Coefficient $_nC_r$	$$_nC_r = \frac{n!}{(n - r)!r!}$$	$$_6C_4 = \frac{6!}{(6 - 4)!4!} = \frac{6!}{2!4!} = \frac{720}{2 \cdot 24} = 15$$
Binomial Theorem	$(a + b)^n = {_nC_0}a^n + {_nC_1}a^{n-1}b^1 + \cdots$ $\quad + {_nC_{n-1}}a^1b^{n-1} + {_nC_n}b^n$	$(a + b)^4 = {_4C_0}a^4 + {_4C_1}a^3b + {_4C_2}a^2b^2$ $\quad + {_4C_3}ab^3 + {_4C_4}b^4$ $\quad = 1a^4 + 4a^3b + 6a^2b^2 + 4ab^3 + 1b^4$ $\quad = a^4 + 4a^3b + 6a^2b^2 + 4ab^3 + b^4$

14.4 Exercises

PRACTICE   WATCH   DOWNLOAD   READ   REVIEW

CONCEPTS

1. How many terms result from expanding $(a + b)^4$?

2. How many terms result from expanding $(a + b)^n$?

3. To find the coefficients for the expansion of $(a + b)^3$, what row of Pascal's triangle do you use?

4. Write down the first 5 rows of Pascal's triangle.

5. $4! = $ _____

6. $1 \cdot 2 \cdot 3 \cdot 4 \cdot 5 \cdot 6 = $ _____

7. $_nC_r = $ _____   8. $(a + b)^2 = $ _____

9. **Thinking Generally** $\dfrac{n!}{(n - 1)!} = $ _____

10. **Thinking Generally** $_nC_n = $ _____

USING PASCAL'S TRIANGLE

Exercises 11–18: Use Pascal's triangle to expand the given expression.

11. $(x + y)^3$ **12.** $(x + y)^4$

13. $(2x + 1)^4$ **14.** $(2x - 1)^4$

15. $(a - b)^5$ **16.** $(3x + 2y)^3$

17. $(x^2 + 1)^3$ **18.** $\left(\frac{1}{2} - x^2\right)^5$

FACTORIALS AND BINOMIAL COEFFICIENTS

Exercises 19–32: Evaluate the expression.

19. $3!$ **20.** $6!$

21. $\frac{4!}{3!}$ **22.** $\frac{6!}{3!}$

23. $\frac{2!}{0!}$ **24.** $\frac{5!}{1!}$

25. $\frac{5!}{2!3!}$ **26.** $\frac{6!}{4!2!}$

27. $_5C_4$ **28.** $_3C_1$

29. $_6C_5$ **30.** $_2C_2$

31. $_4C_0$ **32.** $_4C_3$

Exercises 33–38: Evaluate the binomial coefficient with a calculator.

33. $_{12}C_7$ **34.** $_{13}C_8$

35. $_9C_5$ **36.** $_{25}C_{14}$

37. $_{19}C_{11}$ **38.** $_{10}C_6$

THE BINOMIAL THEOREM

Exercises 39–50: Use the binomial theorem to expand the expression.

39. $(m + n)^3$ **40.** $(m + n)^5$

41. $(x - y)^4$ **42.** $(1 - 3x)^4$

43. $(2a + 1)^3$ **44.** $(x^2 - 1)^3$

45. $(x + 2)^5$ **46.** $(a - 3)^5$

47. $(3 + 2m)^4$ **48.** $(m - 3n)^3$

49. $(2x - y)^3$ **50.** $(2a + 3b)^4$

Exercises 51–56: The $(r + 1)$st term of the expression $(a + b)^n$, $0 \le r \le n$, is given by $_nC_r a^{n-r}b^r$. Find the specified term. Refer to Example 5.

51. The first term of $(a + b)^8$

52. The second term of $(a - b)^{10}$

53. The fourth term of $(x + y)^7$

54. The sixth term of $(a + b)^9$

55. The first term of $(2m + n)^9$

56. The eighth term of $(2a - b)^8$

WRITING ABOUT MATHEMATICS

57. Explain how to find the numbers in Pascal's triangle.

58. Compare the expansion of $(a + b)^n$ to the expansion of $(a - b)^n$. Give an example.

CHECKING BASIC CONCEPTS
SECTIONS 14.3 AND 14.4

1. Determine whether the series is arithmetic or geometric.
  (a) $\frac{1}{2} + \frac{1}{4} + \frac{1}{8} + \cdots + \frac{1}{256}$
  (b) $\frac{1}{2} + \frac{5}{2} + \frac{9}{2} + \frac{13}{2} + \frac{17}{2}$

2. Use a formula to find the sum of the arithmetic series
$$4 + 8 + 12 + \cdots + 48.$$

3. Use a formula to find the sum of the geometric series
$$1 - 2 + 4 - 8 + 16 - 32 + 64 - 128 + 256 - 512.$$

4. Use Pascal's triangle to expand $(x - y)^4$.

5. Use the binomial theorem to expand $(x + 2)^3$.

CHAPTER 14   SUMMARY

SECTION 14.1 ■ SEQUENCES

Sequences   An *infinite sequence* is a function whose domain is the natural numbers. A *finite sequence* is a function whose domain is $D = \{1, 2, 3, \ldots, n\}$ for some natural number n.

Example: $a_n = 2n$ is a symbolic representation of the even natural numbers. The first six terms of this sequence are represented numerically and graphically in the table and figure.

n	a_n
1	2
2	4
3	6
4	8
5	10
6	12

SECTION 14.2 ■ ARITHMETIC AND GEOMETRIC SEQUENCES

Two common types of sequences are arithmetic and geometric.

Arithmetic Sequence   An arithmetic sequence is determined by a linear function of the form $f(n) = dn + c$ or $f(n) = a_1 + (n - 1)d$. Successive terms in an arithmetic sequence are found by adding the common difference d to the previous term. The sequence $1, 3, 5, 7, 9, 11, \ldots$ is an arithmetic sequence with its first term $a_1 = 1$, common difference $d = 2$, and general term $a_n = 2n - 1$.

Geometric Sequence   The general term for a geometric sequence is given by $f(n) = a_1 r^{n-1}$. Successive terms in a geometric sequence are found by multiplying the previous term by the common ratio r. The sequence $3, 6, 12, 24, 48, \ldots$ is a geometric sequence with its first term $a_1 = 3$, common ratio $r = 2$, and general term $a_n = 3(2)^{n-1}$.

SECTION 14.3 ■ SERIES

Series   A series results when the terms of a sequence are summed. The series associated with the sequence $2, 4, 6, 8, 10$ is

$$2 + 4 + 6 + 8 + 10,$$

and its sum equals 30. An arithmetic series results when the terms of an arithmetic sequence are summed, and a geometric series results when the terms of a geometric sequence are summed. In this chapter, we discussed formulas for finding sums of arithmetic and geometric series. See Putting It All Together for Section 14.3.

Summation Notation   Summation notation can be used to write series efficiently.

Example: $1^2 + 2^2 + 3^2 + 4^2 + 5^2 = \displaystyle\sum_{k=1}^{5} k^2$.

SECTION 14.4 ■ THE BINOMIAL THEOREM

Pascal's triangle may be used to find the coefficients for the expansion of $(a + b)^n$, where n is a natural number.

$$
\begin{array}{ccccccccccc}
 & & & & & 1 & & & & & \\
 & & & & 1 & & 1 & & & & \\
 & & & 1 & & 2 & & 1 & & & \\
 & & 1 & & 3 & & 3 & & 1 & & \\
 & 1 & & 4 & & 6 & & 4 & & 1 & \\
1 & & 5 & & 10 & & 10 & & 5 & & 1 \\
\end{array}
$$

Example: To expand $(x + y)^4$, use the fifth row of Pascal's triangle.

$$
\begin{aligned}
(x + y)^4 &= 1x^4 + 4x^3y + 6x^2y^2 + 4xy^3 + 1y^4 \\
&= x^4 + 4x^3y + 6x^2y^2 + 4xy^3 + y^4
\end{aligned}
$$

The binomial theorem can also be used to expand powers of binomials.

CHAPTER 14   REVIEW EXERCISES

SECTION 14.1

Exercises 1–4: Write the first four terms of the sequence for n = 1, 2, 3, and 4.

1. $f(n) = n^3$

2. $f(n) = 5 - 2n$

3. $f(n) = \dfrac{2n}{n^2 + 1}$

4. $f(n) = (-2)^n$

Exercises 5–6: Write the terms of the sequence.

5.

6.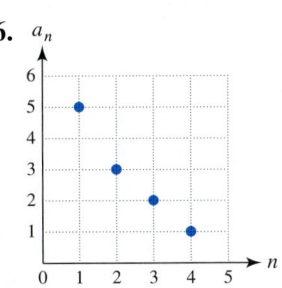

Exercises 7–10: Represent the first seven terms of the sequence numerically and graphically.

7. $a_n = 2n$

8. $a_n = n^2 - 4$

9. $a_n = 4\left(\frac{1}{2}\right)^n$

10. $a_n = \sqrt{n}$

SECTION 14.2

Exercises 11–18: Determine whether f is an arithmetic sequence. Identify the common difference when possible.

11. $f(n) = 5n - 1$

12. $f(n) = 4 - n^2$

13. $f(n) = 2^n$

14. $f(n) = 4 - \frac{1}{3}n$

15.

n	1	2	3	4
$f(n)$	20	17	14	11

16.

n	1	2	3	4
$f(n)$	-3	0	6	12

17.

18.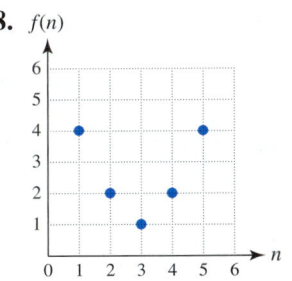

Exercises 19 and 20: Find the general term a_n for the arithmetic sequence.

19. $a_1 = -3$ and $d = 4$ **20.** $a_1 = 2$ and $a_2 = -3$

Exercises 21–28: Determine whether f is a geometric sequence. Identify the common ratio when possible.

21. $f(n) = 2(4)^n$ **22.** $f(n) = 2n^4$

23. $f(n) = 1 - 2n$ **24.** $f(n) = 5(0.7)^n$

25.

n	1	2	3	4
$f(n)$	5	4	3	1

26.

n	1	2	3	4
$f(n)$	27	-9	3	-1

27.   **28.**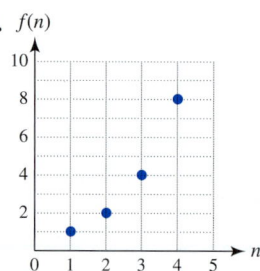

Exercises 29 and 30: Find the general term a_n for the geometric sequence.

29. $a_1 = 5$ and $r = 0.9$ **30.** $a_1 = 2$ and $a_2 = 8$

SECTION 14.3

Exercises 31–34: Find the sum, using a formula.

31. $4 + 9 + 14 + 19 + 24 + 29 + 34 + 39 + 44$

32. $4.5 + 3.0 + 1.5 + 0 - 1.5$

33. $1 - 4 + 16 - 64 + \cdots + 4096$

34. $1 + \frac{1}{2} + \frac{1}{4} + \frac{1}{8} + \frac{1}{16} + \cdots + \frac{1}{256}$

Exercises 35–38: Write the terms of the series.

35. $\sum_{k=1}^{5} 2k + 1$ **36.** $\sum_{k=1}^{4} \frac{1}{k+1}$

37. $\sum_{k=1}^{4} k^3$ **38.** $\sum_{k=2}^{7} (1-k)$

Exercises 39–42: Write the series in summation notation.

39. $1 + 2 + 3 + \cdots + 20$

40. $1 + \frac{1}{2} + \frac{1}{3} + \cdots + \frac{1}{20}$

41. $\frac{1}{2} + \frac{2}{3} + \frac{3}{4} + \cdots + \frac{9}{10}$

42. $1^2 + 2^2 + 3^2 + 4^2 + 5^2 + 6^2 + 7^2$

SECTION 14.4

Exercises 43–46: Use Pascal's triangle to expand the given expression.

43. $(x + 4)^3$ **44.** $(2x + 1)^4$

45. $(x - y)^5$

46. $(a - 1)^6$

Exercises 47–50: Evaluate the expression.

47. $3!$ **48.** $\dfrac{5!}{3!2!}$

49. $_6C_3$ **50.** $_4C_3$

Exercises 51–54: Use the binomial theorem to expand the given expression.

51. $(m + 2)^4$ **52.** $(a + b)^5$

53. $(x - 3y)^4$ **54.** $(3x - 2)^3$

APPLICATIONS

55. *Salaries* An individual's starting salary is $45,000, and the individual receives a 10% raise each year. Give symbolic, numerical, and graphical representations for this person's salary over 7 years. What type of sequence is it?

56. *Salaries* An individual's starting salary is $45,000, and the individual receives an increase of $5000 each year. Give symbolic, numerical, and graphical representations for this person's salary over 7 years. What type of sequence is it?

57. *Rain Forests* Rain forests are defined as forests that grow in regions that receive more than 70 inches of rain each year. The world is losing an estimated 49 million acres of rain forests annually. Give symbolic, numerical, and graphical representations for a sequence that models the total number of acres (in millions) lost over a 7-year period. (*Source:* New York Times Almanac, 1999.)

58. *Home Mortgage Payments*   The average home mortgage payment in 1996 was $1087 per month. Since then, mortgage payments have risen, on average, by 2.5% per year.
    (a) Write a sequence a_n that models the average mortgage payment, where $n = 1$ corresponds to 1996, $n = 2$ to 1997, and so on.

    (b) Is a_n arithmetic, geometric, or neither? Explain your reasoning.
    (c) Find a_5 and interpret the result.
    (d) Give a graphical representation for a_n, where $n = 1, 2, 3, \ldots, 10$.

CHAPTER 14   TEST   Pass the Test   Video solutions to all test exercises

1. Write the first four terms of the sequence for $n = 1, 2, 3,$ and 4.

$$f(n) = \frac{n^2}{n + 1}.$$

2. Use the graph to write the terms of the sequence.

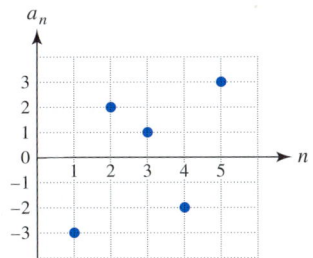

3. List the first seven terms of $a_n = n^2 - n$ in a table. Let $n = 1, 2, \ldots, 7$.

4. Expand the expression $(2x - 1)^4$.

Exercises 5 and 6: Determine whether the sequence is arithmetic or geometric. Identify either the common difference or the common ratio.

5. $f(n) = 7 - 3n$

6.

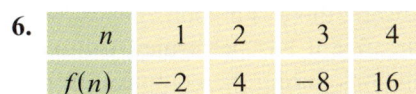

n	1	2	3	4
$f(n)$	-2	4	-8	16

7. Find the general term a_n for the arithmetic sequence if $a_1 = 2$ and $d = -3$.

8. Find the general term a_n for the geometric sequence if $a_1 = 2$ and $a_3 = 4.5$.

Exercises 9 and 10: Determine whether $a_n = f(n)$ is a geometric sequence. Identify the common ratio when possible.

9. $f(n) = -3(2.5)^n$ **10.** a_n

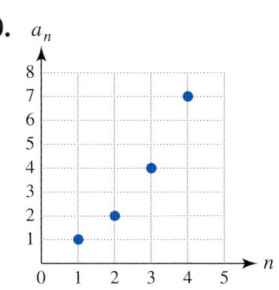

Exercises 11–12: Find the sum, using a formula.

11. $-1 + 2 + 5 + 8 + 11 + 14 + 17 + 20 + 23$

12. $1 - \frac{2}{3} + \frac{4}{9} - \frac{8}{27} + \frac{16}{81} - \frac{32}{243} + \frac{64}{729}$

13. Write the terms of the series $\sum_{k=2}^{7} 3k$.

14. Write the series $1^3 + 2^3 + 3^3 + \cdots + 60^3$ in summation notation.

15. Evaluate $\frac{7!}{4!\,3!}$.   **16.** Evaluate $_5C_3$.

17. *Auditorium Seating*   An auditorium has 50 seats in the first row, 57 seats in the second row, 64 seats in the third row, and so on. Use a formula to find the total number of seats in the first 45 rows.

18. *Median Home Price*   In 1997 the median price of a single-family home was $159,700 and was increasing at a rate of 4% per year. Give symbolic, numerical, and graphical representations for the median home price over a 7-year period, starting in 1997. What type of sequence is it?

19. *Tent Worms*   Large numbers of tent worms can defoliate trees and ruin crops. After they mature, they spin a cocoon and develop into moths that lay eggs. Suppose that an initial population of 2000 tent worms doubles every 5 days.

   (a) Write a formula for a_n that models the number of tent worms after $n - 1$ five-day time periods. (*Hint:* $a_1 = 2000$, $a_2 = 4000$, and $a_3 = 8000$.)

   (b) Is a_n arithmetic, geometric, or neither? Explain your reasoning.

   (c) Find a_6 and interpret the result.

   (d) Give a graph for a_n, where $n = 1, 2, 3, 4, 5, 6$.

CHAPTER 14   EXTENDED AND DISCOVERY EXERCISES

SEQUENCES AND SERIES

*Exercises 1 and 2: Recursive Sequences   Some sequences are not defined by a formula for a_n. Instead they are defined recursively. With a **recursive formula** you must find terms a_1 through a_{n-1} before you can find a_n. For example, let*

$$a_1 = 2$$
$$a_n = a_{n-1} + 3, \qquad \text{for } n \geq 2.$$

To find a_2, a_3, and a_4, we let $n = 2, 3, 4$.

$$a_2 = a_1 + 3 = 2 + 3 = 5$$
$$a_3 = a_2 + 3 = 5 + 3 = 8$$
$$a_4 = a_3 + 3 = 8 + 3 = 11$$

The first four terms of the sequence are 2, 5, 8, 11.

1. *Fibonacci Sequence*   The Fibonacci sequence dates back to 1202 and is one of the most famous sequences in mathematics. It can be defined recursively as follows.

$$a_1 = 1, \qquad a_2 = 1$$
$$a_n = a_{n-1} + a_{n-2}, \qquad \text{for } n \geq 3$$

Find the first 12 terms of this sequence.

2. *Insect Populations*   Frequently the population of a particular insect does not continue to grow indefinitely. Instead, its population grows rapidly at first and then levels off because of competition for limited resources. In one study, the behavior of the winter moth was modeled with a sequence similar to the following, where a_n gives the population density in thousands per acre during year n.

$$a_1 = 1$$
$$a_n = 2.85a_{n-1} - 0.19a_{n-1}^2, \qquad n \geq 2$$

(*Source:* G. Varley and G. Gradwell, "Population models for the winter moth.")

   (a) Make a table for $n = 1, 2, 3, \ldots, 7$. Describe what happens to the population density of the winter moth.

   NOTE: Many graphing calculators are capable of generating tables and graphs for a recursive sequence.

   (b) Graph the sequence for $n = 1, 2, 3, \ldots, 20$. Discuss the graph.

3. *Calculating π*   The quest for an accurate estimation for π is a fascinating story covering thousands of years. Because π is an irrational number, it cannot be represented exactly by a fraction. Its decimal expansion neither repeats nor has a pattern. The ability to compute π was essential to the development of societies because π appears in formulas used in construction, surveying, and geometry. In early historical records, π was given the value of 3. Later the Egyptians used a value of

$$\frac{256}{81} \approx 3.1605.$$

Not until the discovery of series was an exceedingly accurate decimal approximation of π possible. In 1989, π was computed to 1,073,740,000 digits, which required 100 hours of supercomputer time. Why would anyone want to compute π to so many decimal places? One practical reason is to test electrical circuits in new computers. If a computer has a small defect in its hardware, there is a good chance that an error will appear after it has performed trillions of arithmetic calculations during the computation of π. (*Source:* P. Beckmann, *A History of Pi.*) The series given by

$$\frac{\pi^4}{90} \approx \frac{1}{1^4} + \frac{1}{2^4} + \frac{1}{3^4} + \frac{1}{4^4} + \frac{1}{5^4} + \cdots + \frac{1}{n^4}$$

can be used to estimate π, where larger values of n give better approximations.

(a) Approximate π by finding the sum of the first four terms.

 (b) Use a calculator to approximate π by summing the first 50 terms. Compare the result to the actual value of π.

4. *Infinite Series*   The sum S of an infinite geometric series can be found if its common ratio r satisfies $|r| < 1$. It is given by

$$S = \frac{a_1}{1 - r}.$$

If $|r| \geq 1$, this sum does not exist. For example, the infinite geometric series

$$1 + \frac{1}{2} + \frac{1}{4} + \frac{1}{8} + \frac{1}{16} + \cdots$$

has $a_1 = 1$ and $r = \frac{1}{2}$. Therefore its sum S equals

$$S = \frac{1}{1 - \frac{1}{2}} = 2.$$

You might want to add terms of this series to see how increasing the number of terms results in a sum closer to 2. Find the sum of each infinite geometric series.

(a) $2 - 1 + \frac{1}{2} - \frac{1}{4} + \frac{1}{8} - \frac{1}{16} + \cdots$

(b) $1 + \frac{1}{3} + \frac{1}{9} + \frac{1}{27} + \frac{1}{81} + \cdots$

(c) $0.1 + 0.01 + 0.001 + 0.0001 + \cdots$

(d) $0.12 + 0.0012 + 0.000012$
$\qquad\qquad + 0.00000012 + \cdots$

CHAPTERS 1–14   CUMULATIVE REVIEW EXERCISES

1. State whether the equation illustrates an identity, commutative, associative, or distributive property.

$$29(102) = 29(100) + 29(2)$$

2. Identify the domain and range of the relation given by $S = \{(-6, 5), (-2, 1), (0, 3), (2, 0)\}$.

Exercises 3–6: Simplify the expression. Write the result using positive exponents.

3. $\dfrac{x^{-2}y^3}{(3xy^{-2})^3}$

4. $\left(\dfrac{3b}{6a^2}\right)^{-4}$

5. $\left(\dfrac{1}{z^2}\right)^{-5}$

6. $\dfrac{8x^{-3}y^2}{4x^3y^{-1}}$

7. Find the domain of $f(x) = \dfrac{-5}{x - 8}$.

8. Use the table to write the formula for $f(x) = ax + b$.

x	-2	-1	0	1	2
$f(x)$	5	3	1	-1	-3

9. Write the equation of the horizontal line that passes through the point $(2, 3)$.

10. Find the slope and the y-intercept of the graph of $f(x) = -3x + 5$.

Exercises 11 and 12: Write the slope–intercept form for a line satisfying the given conditions.

11. Perpendicular to $y = -\frac{2}{3}x - 4$, passing through $(1, 4)$

12. Parallel to $y = 2x - 7$, passing through $(5, 2)$

Exercises 13–18: Solve the equation or inequality. Write the solutions to inequalities in interval notation.

13. $\frac{2}{5}(x - 4) = -12$

14. $\frac{2}{5}z + \frac{1}{4}z > 2 - (z - 1)$

15. $-3|t - 5| \leq -18$

16. $\left|4 + \frac{2}{3}x\right| = 6$

17. $\dfrac{1}{4}t - (2t + 5) + 6 = \dfrac{t + 3}{4}$

18. $-3 \leq \frac{2}{3}x + 5 < 11$

19. Determine which of the following is a solution to the given system of equations.

$$(3, -2), \quad (-1, 3)$$
$$3x + y = 7$$
$$-2x - 3y = 0$$

20. Shade the solution set in the xy-plane.

$$x - y < 4$$
$$x + 2y \geq 7$$

Exercises 21 and 22: Solve the system of equations. Write the solution as an ordered pair or ordered triple where appropriate.

21. $x - 2y = 1$
$-2x + 7y = 4$

22. $x + y + z = 5$
$-2x - y + z = -10$
$x + 2y + 8z = 1$

23. Maximize the objective function R subject to the given constraints.

$$R = 3x + 8y$$

$$x + 4y \leq 10$$
$$4x + y \leq 10$$
$$x \geq 0, y \geq 0$$

24. Evaluate det A.

$$A = \begin{bmatrix} 4 & -3 \\ 3 & 2 \end{bmatrix}$$

Exercises 25 and 26: Multiply the expressions.

25. $2x^3(4x^4 - 3x^3 + 5)$ **26.** $(2z - 7)(3z + 4)$

Exercises 27 and 28: Factor completely.

27. $4x^2 - 9y^2$ **28.** $2a^3 - a^2 + 8a - 4$

Exercises 29 and 30: Solve the equation.

29. $4x^2 - x - 3 = 0$ **30.** $x^4 - 10x^3 = -24x^2$

Exercises 31 and 32: Simplify the expression.

31. $\dfrac{x^2 - 7x + 10}{x^2 - 25} \cdot \dfrac{x + 5}{x + 1}$

32. $\dfrac{x^2 + 7x + 12}{x^2 - 9} \div \dfrac{x^2 - 5x + 6}{(x - 3)^2}$

Exercises 33 and 34: Solve the rational equation. Check your result.

33. $\dfrac{2}{x + 5} = \dfrac{-3}{x^2 - 25} + \dfrac{1}{x - 5}$

34. $\dfrac{2y}{y^2 - 3y + 2} = \dfrac{1}{y - 2} + 2$

35. Solve the equation for W.

$$R = \frac{3C - 2W}{5}$$

36. Simplify the complex fraction.

$$\frac{\dfrac{1}{x^2} + \dfrac{2}{x}}{\dfrac{1}{x^2} - \dfrac{4}{x}}$$

Exercises 37 and 38: Simplify the expression. Assume that all variables are positive.

37. $\sqrt[3]{x^4 y^4} - 2\sqrt[3]{xy}$ **38.** $(4 + \sqrt{2})(4 - \sqrt{2})$

Exercises 39 and 40: Solve. Check your answer.

39. $8(x - 3)^2 = 200$ **40.** $3\sqrt{2x + 6} = 6x$

Exercises 41 and 42: Write the complex expression in standard form.

41. $(-3 + i)(-4 - 2i)$ **42.** $\dfrac{2 - 6i}{1 + 2i}$

43. Find the minimum y-value located on the graph of $y = 3x^2 + 8x + 5$.

44. Write the equation $y = 2x^2 + 8x + 17$ in vertex form and identify the vertex.

Exercises 45 and 46: Solve by using the method of your choice. Write any complex solutions in standard form.

45. $x^2 - 4x + 13 = 0$ **46.** $z^2 - 4z = 32$

47. A graph of $y = ax^2 + bx + c$ is shown.
  (a) Solve $ax^2 + bx + c = 0$.
  (b) State whether $a > 0$ or $a < 0$.
  (c) Determine whether the discriminant is positive, negative, or zero.

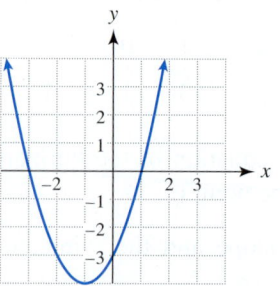

48. Solve the quadratic inequality. Write your answer in interval notation.
$$x^2 + 2x + 3 > 0$$

49. For $f(x) = x^2 + 1$ and $g(x) = 3x - 2$, find each of the following.
(a) $(f \circ g)(-2)$ **(b)** $(g \circ f)(x)$

50. Find $f^{-1}(x)$ for the one-to-one function
$$f(x) = \frac{3x + 1}{2}.$$

51. Write $\ln(x^3 \sqrt{y})$ by using sums and differences of logarithms of x and y.

52. Write $2 \log x - \log 4xy$ as one logarithm. Assume x and y are positive.

Exercises 53 and 54: Solve the equation. Approximate answers to the nearest hundredth.

53. $8 \log x + 3 = 17$ **54.** $4^{2x} = 5$

55. Graph the parabola $x = (y - 3)^2 + 1$. Find the vertex and the axis of symmetry.

56. Find the center and the radius of the circle whose equation is $x^2 - 6x + y^2 + 2y = -6$.

Exercises 57 and 58: Graph the ellipse or hyperbola. Label the vertices and the endpoints of the minor axis on the ellipse. Show the asymptotes on the hyperbola.

57. $\dfrac{x^2}{4} + \dfrac{y^2}{9} = 1$ **58.** $\dfrac{x^2}{16} - \dfrac{y^2}{4} = 1$

Exercises 59 and 60: Use the graph to determine the equation of the ellipse or hyperbola.

59. **60.**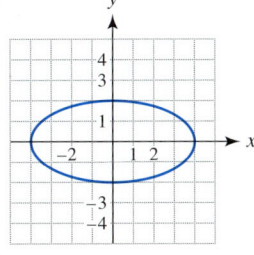

61. Solve the system of equations. Check your solutions.
$$y = x^2 + 1$$
$$x^2 + 2y = 5$$

62. Shade the solution set in the xy-plane.
$$y \geq x^2 - 2$$
$$y \leq -x$$

Exercises 63–66: Determine whether f is an arithmetic or a geometric sequence. If it is arithmetic, find the common difference. If it is geometric, find the common ratio.

63. $f(n) = 5 - 2n$ **64.** $f(n) = 3(0.2)^n$

65. $f(n) = 7(4)^n$ **66.** $f(n) = 6n + 1$

67. Find the general term a_n for the arithmetic sequence where $a_1 = 2$ and $a_2 = 5$.

68. Find the general term a_n for the geometric sequence where $a_1 = 4$ and $a_2 = 12$.

Exercises 69 and 70: Find the sum using a formula.

69. $3 + 7 + 11 + 15 + 19 + \cdots + 35$

70. $1 - 2 + 4 - 8 + 16 - \cdots + 1024$

Exercises 71 and 72: Expand the binomial expression.

71. $(2x + 3)^4$ **72.** $(2a - 5b)^3$

APPLICATIONS

73. *Radius of a Circle* If a circle has an area of A square units, its radius r is given by $r = \sqrt{\frac{A}{\pi}}$. Approximate the radius of a circle with an area of 14 square inches to the nearest hundredth of an inch.

74. *Distance from Work* Starting at a warehouse, a delivery truck driver travels down a straight highway for 3 hours at 40 miles per hour, stops and unloads the truck for 1 hour, and then returns to the warehouse at 60 miles per hour. Sketch a graph that shows the distance between the truck and the warehouse during this period of time.

75. *Exercise and Fluid Consumption* When a person exercises, the total amount of fluid he or she will need that day increases depending on the person's weight and the duration of the exercise. To determine the number of ounces of fluid needed, divide the person's weight by 2 and then add 0.4 ounces for every minute of exercise. (*Source:* Runner's World.)
(a) Write a function that gives the fluid requirements for a person weighing 170 pounds who exercises for x minutes a day.
(b) If an athlete who exercises for 90 minutes requires 130 ounces of fluid, find the athlete's weight.

76. *Airplane Speed*   An airplane travels 1080 miles into the wind in 3 hours. The return trip with the wind takes 2.7 hours. Find the average speed of the airplane and the average wind speed.

77. *Size of a Tent*   The length of a rectangular tent floor is 6 feet shorter than twice the width. If the area of the tent floor is 108 square feet, what are the dimensions of the tent?

78. *Working Together*   Suppose that one person can weed a garden in 60 minutes and a second person can weed the same garden in 90 minutes. How long would it take these two people to weed the garden if they worked together?

79. *Numbers*   The product of two positive numbers is 96. If the larger number is subtracted from 3 times the smaller number, the result is 12. Let x be the smaller number and let y be the larger number.
(a) Write a system of equations for this situation.
(b) What are the two numbers?

80. *Marching Band*   A band is marching in a triangular formation so that 1 person is in the first row, 3 people are in the second row, 5 people are in the third row, and so on. Use a formula to find the total number of musicians in the marching band if the last row contains 23 people.

Appendix A   Using the Graphing Calculator

Overview of the Appendix

This appendix provides instruction for the TI-83, TI-83 Plus, and TI-84 Plus graphing calculators that may be used in conjunction with this textbook. It includes specific keystrokes needed to work several examples from the text. Students are advised to consult the *Graphing Calculator Guidebook* provided by the manufacturer.

Entering Mathematical Expressions

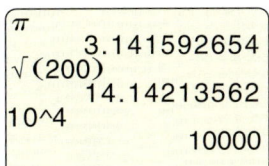

Figure A.1

EVALUATING π   To evaluate π, use the following keystrokes, as shown in the first and second lines of Figure A.1. (Do *not* use 3.14 or $\frac{22}{7}$ for π.)

$$\boxed{\text{2nd}}\ \boxed{\wedge [\pi]}\ \boxed{\text{ENTER}}$$

EVALUATING A SQUARE ROOT   To evaluate a square root, such as $\sqrt{200}$, use the following keystrokes, as shown in the third and fourth lines of Figure A.1.

$$\boxed{\text{2nd}}\ \boxed{x^2 [\sqrt{\ }]}\ \boxed{2}\ \boxed{0}\ \boxed{0}\ \boxed{)}\ \boxed{\text{ENTER}}$$

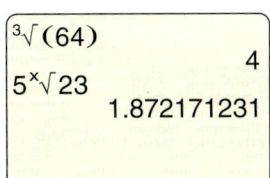

Figure A.2

EVALUATING AN EXPONENTIAL EXPRESSION   To evaluate an exponential expression, such as 10^4, use the following keystrokes, as shown in the last two lines of Figure A.1.

$$\boxed{1}\ \boxed{0}\ \boxed{\wedge}\ \boxed{4}\ \boxed{\text{ENTER}}$$

EVALUATING A CUBE ROOT   To evaluate a cube root, such as $\sqrt[3]{64}$, use the following keystrokes, as shown in the first and second lines of Figure A.2.

$$\boxed{\text{MATH}}\ \boxed{4}\ \boxed{6}\ \boxed{4}\ \boxed{)}\ \boxed{\text{ENTER}}$$

EVALUATING OTHER ROOTS   To evaluate a fifth root, such as $\sqrt[5]{23}$, use the following keystrokes, as shown in the third and fourth lines of Figure A.2.

$$\boxed{5}\ \boxed{\text{MATH}}\ \boxed{5}\ \boxed{2}\ \boxed{3}\ \boxed{\text{ENTER}}$$

Figure A.3

EVALUATING THE NATURAL EXPONENTIAL FUNCTION   To evaluate $14e^{0.0153(8)}$, use the following keystrokes, as shown in the first and second lines of Figure A.3.

$$\boxed{1}\ \boxed{4}\ \boxed{\text{2nd}}\ \boxed{\text{LN} [e^x]}\ \boxed{.}\ \boxed{0}\ \boxed{1}\ \boxed{5}\ \boxed{3}\ \boxed{\times}\ \boxed{8}\ \boxed{)}\ \boxed{\text{ENTER}}$$

EVALUATING THE COMMON LOGARITHMIC FUNCTION   To evaluate $\log(100)$, use the following keystrokes, as shown in the third and fourth lines of Figure A.3.

$$\boxed{\text{LOG}}\ \boxed{1}\ \boxed{0}\ \boxed{0}\ \boxed{)}\ \boxed{\text{ENTER}}$$

EVALUATING THE NATURAL LOGARITHMIC FUNCTION   To evaluate $\ln(10)$, use the following keystrokes, as shown in the last two lines of Figure A.3.

$$\boxed{\text{LN}}\ \boxed{1}\ \boxed{0}\ \boxed{)}\ \boxed{\text{ENTER}}$$

SUMMARY: ENTERING MATHEMATICAL EXPRESSIONS

To access the *number π*, use $\boxed{\text{2nd}}\,\boxed{\wedge [\pi]}$.

To evaluate a *square root*, use $\boxed{\text{2nd}}\,\boxed{x^2 [\sqrt{}\ ]}$.

To evaluate an *exponential expression*, use the $\boxed{\wedge}$ key. To square a number, the $\boxed{x^2}$ key can also be used.

To evaluate a *cube root*, use $\boxed{\text{MATH}}\,\boxed{4}$.

To evaluate a *kth root*, use $\boxed{k}\,\boxed{\text{MATH}}\,\boxed{5}$.

To access the *natural exponential function*, use $\boxed{\text{2nd}}\,\boxed{\text{LN }[e^x]}$.

To access the *common logarithmic function*, use $\boxed{\text{LOG}}$.

To access the *natural logarithmic function*, use $\boxed{\text{LN}}$.

Expressing Answers as Fractions

To evaluate $\frac{1}{3} + \frac{2}{5} - \frac{4}{9}$ in fraction form, use the following keystrokes, as shown in the last three lines of Figure 1.26(a) on page 56.

SUMMARY: EXPRESSING ANSWERS AS FRACTIONS

Enter the arithmetic expression. To access the "Frac" feature, use the keystrokes $\boxed{\text{MATH}}\,\boxed{1}$. Then press $\boxed{\text{ENTER}}$.

Displaying Numbers in Scientific Notation

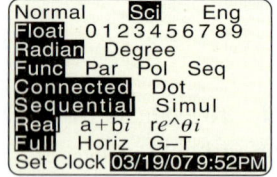

Figure A.4

To display numbers in scientific notation, set the graphing calculator in scientific mode (Sci), by using the following keystrokes. See Figure A.4. (These keystrokes assume that the calculator is in normal mode.)

$$\boxed{\text{MODE}}\,\boxed{\triangleright}\,\boxed{\text{ENTER}}\,\boxed{\text{2nd}}\,\boxed{\text{MODE [QUIT]}}$$

In scientific mode we can display the numbers 5432 and 0.00001234 in scientific notation, as shown in Figure A.5.

Figure A.5

SUMMARY: SETTING SCIENTIFIC MODE

If your calculator is in normal mode, it can be set in scientific mode by pressing

$$\boxed{\text{MODE}}\,\boxed{\triangleright}\,\boxed{\text{ENTER}}\,\boxed{\text{2nd}}\,\boxed{\text{MODE [QUIT]}}.$$

These keystrokes return the graphing calculator to the home screen.

Entering Numbers in Scientific Notation

Figure A.6

Numbers can be entered in scientific notation. For example, to enter 4.2×10^{-3} in scientific notation, use the following keystrokes. (Be sure to use the negation key $(-)$ rather than the subtraction key.)

$$\boxed{4}\ \boxed{.}\ \boxed{2}\ \boxed{\text{2nd}}\ \boxed{,[\text{EE}]}\ \boxed{(-)}\ \boxed{3}$$

This number can also be entered using the following keystrokes. See Figure A.6.

$$\boxed{4}\ \boxed{.}\ \boxed{2}\ \boxed{\times}\ \boxed{1}\ \boxed{0}\ \boxed{\wedge}\ \boxed{(}\ \boxed{(-)}\ \boxed{3}\ \boxed{)}$$

SUMMARY: ENTERING NUMBERS IN SCIENTIFIC NOTATION

One way to enter a number in scientific notation is to use the keystrokes

$$\boxed{\text{2nd}}\ \boxed{,[\text{EE}]}$$

to access an exponent (EE) of 10.

Making a Table

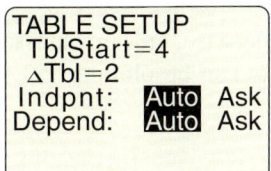

Figure A.7

To make a table of values for $y = 3x + 1$ starting at $x = 4$ and incrementing by 2, begin by pressing $\boxed{Y=}$ and then entering the formula $Y_1 = 3X + 1$, as shown in Figure A.7. To set the table parameters, press the following keys. See Figure A.8.

$$\boxed{\text{2nd}}\ \boxed{\text{WINDOW [TBLSET]}}\ \boxed{4}\ \boxed{\text{ENTER}}\ \boxed{2}$$

These keystrokes specify a table that starts at $x = 4$ and increments the x-values by 2. Therefore, the values of Y_1 at $x = 4, 6, 8, \ldots$ appear in the table. To create this table, press the following keys.

$$\boxed{\text{2nd}}\ \boxed{\text{GRAPH [TABLE]}}$$

We can scroll through x- and y-values by using the arrow keys. See Figure A.9. Note that there is no first or last x-value in the table.

```
TABLE SETUP
 TblStart=4
 △Tbl=2
Indpnt:  Auto  Ask
Depend:  Auto  Ask
```

Figure A.8

X	Y₁
4	13
6	19
8	25
10	31
12	37
14	43
16	49

Y₁ ▓ 3X + 1

Figure A.9

SUMMARY: MAKING A TABLE

1. Enter the formula for the equation using $\boxed{Y=}$.
2. Press $\boxed{\text{2nd}}\ \boxed{\text{WINDOW [TBLSET]}}$ to set the starting x-value and the increment between x-values appearing in the table.
3. Create the table by pressing $\boxed{\text{2nd}}\ \boxed{\text{GRAPH [TABLE]}}$.

Figure A.10

Figure A.11

Setting the Viewing Rectangle (Window)

There are at least two ways to set the standard viewing rectangle of $[-10, 10, 1]$ by $[-10, 10, 1]$. The first involves pressing $\boxed{\text{ZOOM}}$ followed by $\boxed{6}$. See Figure A.10. The second method for setting the standard viewing rectangle is to press $\boxed{\text{WINDOW}}$ and enter the following keystrokes. See Figure A.11.

$$\boxed{(-)}\ \boxed{1}\ \boxed{0}\ \boxed{\text{ENTER}}\ \boxed{1}\ \boxed{0}\ \boxed{\text{ENTER}}\ \boxed{1}\ \boxed{\text{ENTER}}$$
$$\boxed{(-)}\ \boxed{1}\ \boxed{0}\ \boxed{\text{ENTER}}\ \boxed{1}\ \boxed{0}\ \boxed{\text{ENTER}}\ \boxed{1}\ \boxed{\text{ENTER}}$$

(Be sure to use the negation key $(-)$ rather than the subtraction key.) Other viewing rectangles can be set in a similar manner by pressing $\boxed{\text{WINDOW}}$ and entering the appropriate values. To see the viewing rectangle, press $\boxed{\text{GRAPH}}$. An example is shown in Figure 8.18 on page 531.

SUMMARY: SETTING THE VIEWING RECTANGLE

To set the standard viewing rectangle, press $\boxed{\text{ZOOM}}\boxed{6}$. To set any viewing rectangle, press $\boxed{\text{WINDOW}}$ and enter the necessary values. To see the viewing rectangle, press $\boxed{\text{GRAPH}}$.

NOTE: You do not need to change "Xres" from 1.

Making a Scatterplot or a Line Graph

To make a scatterplot with the points $(-5, -5)$, $(-2, 3)$, $(1, -7)$, and $(4, 8)$, begin by following these steps.

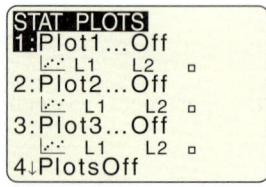

Figure A.12

1. Press $\boxed{\text{STAT}}$ followed by $\boxed{1}$.
2. If list L1 is not empty, use the arrow keys to place the cursor on L1, as shown in Figure A.12. Then press $\boxed{\text{CLEAR}}$ followed by $\boxed{\text{ENTER}}$. This deletes all elements in the list. Similarly, if L2 is not empty, clear the list.
3. Input each x-value into list L1 followed by $\boxed{\text{ENTER}}$. Input each y-value into list L2 followed by $\boxed{\text{ENTER}}$. See Figure A.13.

It is essential that both lists have the same number of values—otherwise, an error message appears when a scatterplot is attempted. Before these four points can be plotted, "STAT-PLOT" must be turned on. It is accessed by pressing

$$\boxed{\text{2nd}}\ \boxed{\text{Y = [STAT PLOT]}},$$

as shown in Figure A.14.

There are three possible "STATPLOTS," numbered 1, 2, and 3. Any one of the three can be selected. The first plot can be selected by pressing $\boxed{1}$. Next, place the cursor over "On" and press $\boxed{\text{ENTER}}$ to turn "Plot1" on. There are six types of plots that can be selected. The first type is a *scatterplot* and the second type is a *line graph*, so place the cursor over the first type of plot and press $\boxed{\text{ENTER}}$ to select a scatterplot. (To make the line graph, place the cursor over the second type of plot and press $\boxed{\text{ENTER}}$.) The x-values are stored in list L1, so select L1 for "Xlist" by pressing $\boxed{\text{2nd}}\boxed{1}$. Similarly, press $\boxed{\text{2nd}}\boxed{2}$ for the "Ylist," since the y-values are stored in list L2. Finally, there are three styles of marks that can be used to show data points in the graph. We will usually use the first because it is largest and shows up the best.

Figure A.13

Figure A.14

Figure A.15

$[-10, 10, 1]$ by $[-10, 10, 1]$

Figure A.16

Make the screen appear as in Figure A.15. Before plotting the four data points, be sure to set an appropriate viewing rectangle. Then press (GRAPH). The data points appear as in Figure A.16.

REMARK 1:   A fast way to set the viewing rectangle for any scatterplot is to select the "ZOOMSTAT" feature by pressing (ZOOM)(9). This feature automatically scales the viewing rectangle so that all data points are shown.

REMARK 2:   If an equation has been entered into the (Y =) menu and selected, it will be graphed with the data. This feature is used frequently to model data.

SUMMARY: MAKING A SCATTERPLOT OR A LINE GRAPH

The following are basic steps necessary to make either a scatterplot or a line graph.

1. Use (STAT)(1) to access lists L1 and L2.
2. If list L1 is not empty, place the cursor on L1 and press (CLEAR)(ENTER). Repeat for list L2, if it is not empty.
3. Enter the x-values into list L1 and the y-values into list L2.
4. Use (2nd)(Y = [STAT PLOT]) to select appropriate parameters for the scatterplot or line graph.
5. Set an appropriate viewing rectangle. Press (GRAPH). Otherwise, press (ZOOM)(9). This feature automatically sets the viewing rectangle and plots the data.

NOTE: (ZOOM)(9) *cannot* be used to set a viewing rectangle for the graph of a function.

Figure A.17

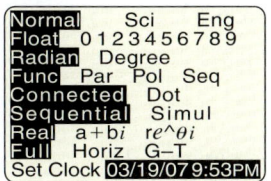

Figure A.18

Entering a Formula

To enter a formula, press (Y =). For example, use the following keystrokes after "$Y_1 = $" to enter $y = x^2 - 4$. See Figure A.17.

$$(Y =)\,(CLEAR)\,(X, T, \theta, n)\,(\wedge)\,(2)\,(-)\,(4)$$

Note that there is a built-in key to enter the variable X. If "$Y_1 = $" does not appear after pressing (Y =), press (MODE) and make sure the calculator is set in *function mode,* denoted "Func". See Figure A.18.

SUMMARY: ENTERING A FORMULA

To enter a formula, press (Y =). To delete formula, press (CLEAR).

$[-10, 10, 1]$ by $[-10, 10, 1]$

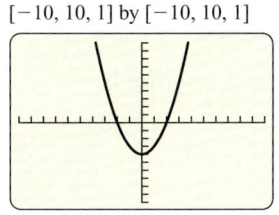

Figure A.19

Graphing a Function

To graph a function, such as $y = x^2 - 4$, start by pressing (Y =) and enter $Y_1 = X^2 - 4$. If there is an equation already entered, remove it by pressing (CLEAR). The equals signs in "$Y_1 = $" should be in reverse video (a dark rectangle surrounding a white equals sign), which indicates that the equation will be graphed. If the equals sign is not in reverse video, place the cursor over it and press (ENTER). Set an appropriate viewing rectangle and then press (GRAPH). The graph of f will appear in the specified viewing rectangle. See Figures A.17 and A.19.

1. Use the $\boxed{Y=}$ menu to enter the formula for the function.
2. Use the $\boxed{\text{WINDOW}}$ menu to set an appropriate viewing rectangle.
3. Press $\boxed{\text{GRAPH}}$.

Graphing a Vertical Line

Figure A.20

Set an appropriate window (or viewing rectangle). Then return to the home screen by pressing

$$\boxed{\text{2nd}}\ \boxed{\text{MODE [QUIT]}}.$$

To graph a vertical line, such as $x = -4$, press

$$\boxed{\text{2nd}}\ \boxed{\text{PRGM [DRAW]}}\ \boxed{4}\ \boxed{(-)}\ \boxed{4}.$$

See Figure A.20. Pressing $\boxed{\text{ENTER}}$ will make the vertical line appear, as shown in Figure A.21.

$[-6, 6, 1]$ by $[-6, 6, 1]$

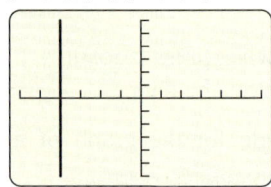

Figure A.21

SUMMARY: GRAPHING THE VERTICAL LINE $x = h$

1. Set an appropriate window by pressing $\boxed{\text{WINDOW}}$.
2. Return to the home screen by pressing $\boxed{\text{2nd}}\ \boxed{\text{MODE [QUIT]}}$.
3. Draw a vertical line by pressing $\boxed{\text{2nd}}\ \boxed{\text{PRGM [DRAW]}}\ \boxed{4}\ \boxed{h}\ \boxed{\text{ENTER}}$.

Squaring a Viewing Rectangle

In a square viewing rectangle the graph of $y = x$ is a line that makes a 45° angle with the positive x-axis, a circle appears circular, and all sides of a square have the same length. An approximate square viewing rectangle can be set if the distance along the x-axis is 1.5 times the distance along the y-axis. Examples of viewing rectangles that are (approximately) square include

$$[-6, 6, 1] \text{ by } [-4, 4, 1] \quad \text{and} \quad [-9, 9, 1] \text{ by } [-6, 6, 1].$$

Square viewing rectangles can be set automatically by pressing either

$$\boxed{\text{ZOOM}}\ \boxed{4} \quad \text{or} \quad \boxed{\text{ZOOM}}\ \boxed{5}.$$

ZOOM 4 provides a *decimal window*, which is discussed later. See Figure A.22.

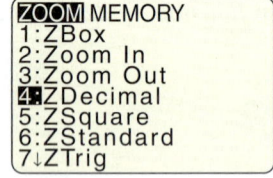

Figure A.22

SUMMARY: SQUARING A VIEWING RECTANGLE

Either $\boxed{\text{ZOOM}}\ \boxed{4}$ or $\boxed{\text{ZOOM}}\ \boxed{5}$ may be used to produce a square viewing rectangle. An (approximately) square viewing rectangle has the form

$$[-1.5k, 1.5k, 1] \text{ by } [-k, k, 1],$$

where k is a positive number.

Locating a Point of Intersection

Figure A.23

To find the point of intersection for the graphs of

$$y_1 = 3(1 - x) \quad \text{and} \quad y_2 = 2,$$

start by entering Y_1 and Y_2, as shown in Figure A.23. Set the window, and graph both equations. Then press the following keys to find the intersection point.

$$\boxed{\text{2nd}}\ \boxed{\text{TRACE [CALC]}}\ \boxed{5}$$

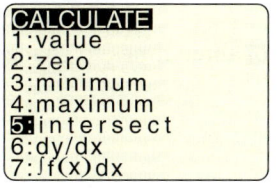

Figure A.24

See Figure A.24, where the "intersect" utility is being selected. The calculator prompts for the first curve, as shown in Figure A.25. Use the arrow keys to locate the cursor near the point of intersection and press $\boxed{\text{ENTER}}$. Repeat these steps for the second curve. Finally we are prompted for a guess. For each of the three prompts, place the free-moving cursor near the point of intersection and press $\boxed{\text{ENTER}}$. The approximate coordinates of the point of intersection will be shown.

$[-6, 6, 1]$ by $[-4, 4, 1]$

Figure A.25

SUMMARY: FINDING A POINT OF INTERSECTION

1. Graph the two equations in an appropriate viewing rectangle.
2. Press $\boxed{\text{2nd}}\ \boxed{\text{TRACE [CALC]}}\ \boxed{5}$.
3. Use the arrow keys to select an approximate location for the point of intersection. Press $\boxed{\text{ENTER}}$ to make the three selections for "First curve?", "Second curve?", and "Guess?". (Note that if the cursor is near the point of intersection, you usually do not need to move the cursor for each selection. Just press $\boxed{\text{ENTER}}$ three times.)

Accessing the Absolute Value

Figure A.26

To graph $y_1 = |x - 50|$, begin by entering $Y_1 = \text{abs}(X - 50)$. The absolute value (abs) is accessed by pressing

$$\boxed{\text{MATH}}\ \boxed{\triangleright}\ \boxed{1}.$$

See Figure A.26.

SUMMARY: ACCESSING THE ABSOLUTE VALUE

1. Press $\boxed{\text{MATH}}$.
2. Position the cursor over "NUM".
3. Press $\boxed{1}$ to select the absolute value.

Shading a System of Inequalities

Figure A.27

To shade the solution set for the system of linear inequalities $2x + y \leq 5$, $-2x + y \geq 1$, begin by solving each inequality for y to obtain $y \leq 5 - 2x$ and $y \geq 2x + 1$. Then let $Y_1 = 5 - 2X$ and $Y_2 = 2X + 1$, as shown in Figure A.27. Position the cursor to the left of

$[-15, 15, 5]$ by $[-10, 10, 5]$

Figure A.28

Y_1 and press $\boxed{\text{ENTER}}$ three times. The triangle that appears indicates that the calculator will shade the region below the graph of Y_1. Next locate the cursor to the left of Y_2 and press $\boxed{\text{ENTER}}$ twice. This triangle indicates that the calculator will shade the region above the graph of Y_2. After setting the viewing rectangle to $[-15, 15, 5]$ by $[-10, 10, 5]$, press $\boxed{\text{GRAPH}}$. The result is shown in Figure A.28.

SUMMARY: SHADING A SYSTEM OF INEQUALITIES

1. Solve each inequality for y.
2. Enter each formula as Y_1 and Y_2 in the $\boxed{Y =}$ menu.
3. Locate the cursor to the left of Y_1 and press $\boxed{\text{ENTER}}$ two or three times to shade either above or below the graph of Y_1. Repeat for Y_2.
4. Set an appropriate viewing rectangle.
5. Press $\boxed{\text{GRAPH}}$.

NOTE: The "Shade" utility in the DRAW menu can also be used to shade the region *between* two graphs.

Entering the Elements of a Matrix

In Example 6(b), Section 9.2 on page 627, the elements of a matrix are entered. The augmented matrix A is given by

$$A = \begin{bmatrix} 1 & 1 & 2 & \bigm| & 1 \\ -1 & 0 & 1 & \bigm| & -2 \\ 2 & 1 & 5 & \bigm| & -1 \end{bmatrix}.$$

Use the following keystrokes on the TI-83 Plus or TI-84 Plus to define a matrix A with dimension 3×4. (*Note:* On the TI-83 the matrix menu is found by pressing $\boxed{\text{MATRX}}$.)

$$\boxed{\text{2nd}}\ \boxed{x^{-1}\,[\text{MATRIX}]}\ \boxed{\triangleright}\ \boxed{\triangleright}\ \boxed{1}\ \boxed{3}\ \boxed{\text{ENTER}}\ \boxed{4}\ \boxed{\text{ENTER}}$$

See Figure 9.4(a).

Then input the 12 elements of the matrix A, row by row. Finish each entry by pressing $\boxed{\text{ENTER}}$. After these elements have been entered, press

$$\boxed{\text{2nd}}\ \boxed{\text{MODE [QUIT]}}$$

to return to the home screen. To display the matrix A, press

$$\boxed{\text{2nd}}\ \boxed{x^{-1}\,[\text{MATRIX}]}\ \boxed{1}\ \boxed{\text{ENTER}}.$$

See Figure A.29.

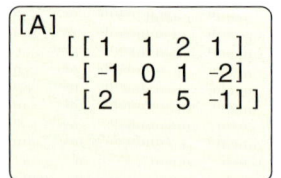

Figure A.29

SUMMARY: ENTERING THE ELEMENTS OF A MATRIX A

1. Begin by accessing the matrix A by pressing $\boxed{\text{2nd}}\ \boxed{x^{-1}\,[\text{MATRIX}]}\ \boxed{\triangleright}\ \boxed{\triangleright}\ \boxed{1}$.
2. Enter the dimension of A by pressing $\boxed{m}\ \boxed{\text{ENTER}}\ \boxed{n}\ \boxed{\text{ENTER}}$, where the dimension of the matrix is $m \times n$.
3. Input each element of the matrix, row by row. Finish each entry by pressing $\boxed{\text{ENTER}}$. Use $\boxed{\text{2nd}}\ \boxed{\text{MODE [QUIT]}}$ to return to the home screen.

NOTE: On the TI-83, replace the keystrokes $\boxed{\text{2nd}}\ \boxed{x^{-1}\,[\text{MATRIX}]}$ with $\boxed{\text{MATRX}}$.

Reduced Row–Echelon Form

In Example 6(b), Section 9.2 on page 627, the reduced row–echelon form of a matrix is found. To find this reduced row–echelon form, use the following keystrokes from the home screen on the TI-83 Plus or TI-84 Plus.

(2nd) (x^{-1} [MATRIX]) (▷) (ALPHA) (APPS [B]) (2nd) (x^{-1} [MATRIX]) (1) ( )) (ENTER)

The resulting matrix is shown in Figure 9.4(b). On the TI-83 graphing calculator, use the following keystrokes to find the reduced row–echelon form.

(MATRX) (▷) (ALPHA) (MATRX [B]) (MATRX) (1) ( )) (ENTER)

SUMMARY: FINDING REDUCED ROW–ECHELON FORM OF A MATRIX

1.   To make rref([A]) appear on the home screen, use the following keystrokes for the TI-83 Plus or TI-84 Plus graphing calculator.

(2nd) (x^{-1} [MATRIX]) (▷) (ALPHA) (APPS [B]) (2nd) (x^{-1} [MATRIX]) (1) ( ))

2.   Press (ENTER) to calculate the reduced row–echelon form.
3.   Use arrow keys to access elements that do not appear on the screen.

NOTE: On the TI-83, replace the keystrokes (2nd) (x^{-1} [MATRIX]) with (MATRX) and (APPS [B]) with (MATRX [B]).

Evaluating a Determinant

In Example 3(a), Section 9.3 on page 634, a graphing calculator is used to evaluate a determinant of a matrix. Start by entering the 9 elements of the 3×3 matrix A, as shown in Figure 9.6(a). To compute det A, perform the following keystrokes from the home screen.

(2nd) (x^{-1} [MATRIX]) (▷) (1) (2nd) (x^{-1} [MATRIX]) (1) ( )) (ENTER)

The results are shown in the last two lines of Figure 9.6(b).

SUMMARY: EVALUATING A DETERMINANT OF A MATRIX

1.   Enter the dimension and elements of the matrix A.
2.   Return to the home screen by pressing

(2nd) (MODE [QUIT]).

3.   On the TI-83 Plus or TI-84 Plus, perform the following keystrokes.

(2nd) (x^{-1} [MATRIX]) (▷) (1) (2nd) (x^{-1} [MATRIX]) (1) ( )) (ENTER)

NOTE: On the TI-83, replace the keystrokes (2nd) (x^{-1} [MATRIX]) with (MATRX).

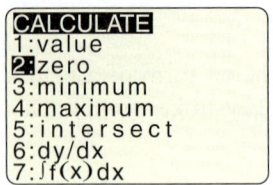

Figure A.30

Locating an *x*-Intercept or Zero

To locate an *x*-intercept or *zero* of $f(x) = x^2 - 4$, start by entering $Y_1 = X^2 - 4$ into the $\boxed{Y=}$ menu. Set the viewing rectangle to $[-9, 9, 1]$ by $[-6, 6, 1]$ and graph Y_1. Afterwards, press the following keys to invoke the zero finder. See Figure A.30.

$$\boxed{2nd}\ \boxed{TRACE\ [CALC]}\ \boxed{2}$$

The graphing calculator prompts for a left bound. Use the arrow keys to set the cursor to the left of the *x*-intercept and press $\boxed{ENTER}$. The graphing calculator then prompts for a right bound. Set the cursor to the right of the *x*-intercept and press $\boxed{ENTER}$. Finally, the graphing calculator prompts for a guess. Set the cursor roughly at the *x*-intercept and press $\boxed{ENTER}$. See Figures A.31–A.33. The calculator then approximates the *x*-intercept or zero automatically, as shown in Figure A.34. The zero of -2 can be found similarly.

Figure A.31

Figure A.32

Figure A.33

Figure A.34

SUMMARY: LOCATING AN *x*-INTERCEPT OR ZERO

1. Graph the function in an appropriate viewing rectangle.
2. Press $\boxed{2nd}\ \boxed{TRACE\ [CALC]}\ \boxed{2}$.
3. Select the left and right bounds, followed by a guess. Press $\boxed{ENTER}$ after each selection. The calculator then approximates the *x*-intercept or zero.

Setting Connected and Dot Mode

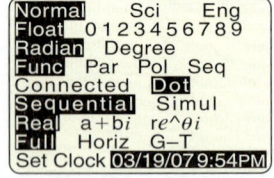

Figure A.35

To set your graphing calculator in dot mode, press $\boxed{MODE}$, position the cursor over "Dot," and press $\boxed{ENTER}$. See Figure A.35. Graphs will now appear in dot mode rather than connected mode.

SUMMARY: SETTING CONNECTED OR DOT MODE

1. Press $\boxed{MODE}$.
2. Position the cursor over "Connected" or "Dot". Press $\boxed{ENTER}$.

Setting a Decimal Window

With a decimal window, the cursor stops on convenient *x*-values. In the decimal window $[-9.4, 9.4, 1]$ by $[-6.2, 6.2, 1]$ the cursor stops on *x*-values that are multiples of 0.2. If we reduce the viewing rectangle to $[-4.7, 4.7, 1]$ by $[-3.1, 3.1, 1]$, the cursor stops on *x*-values

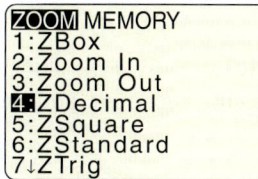

Figure A.36

that are multiples of 0.1. To set this smaller window automatically, press ZOOM 4 . See Figure A.36. Decimal windows are also useful when graphing rational functions with asymptotes in connected mode.

SUMMARY: SETTING A DECIMAL WINDOW

1. Press ZOOM 4 to set the viewing rectangle $[-4.7, 4.7, 1]$ by $[-3.1, 3.1, 1]$.
2. A larger decimal window is $[-9.4, 9.4, 1]$ by $[-6.2, 6.2, 1]$.

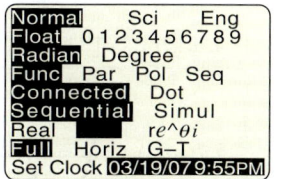

Figure A.37

Setting $a + bi$ Mode

To evaluate expressions containing square roots of negative numbers, such as $\sqrt{-25}$, set your calculator in $a + bi$ mode by using the following keystrokes.

MODE ▽ ▽ ▽ ▽ ▽ ▽ ▷ ENTER 2nd MODE [QUIT]

See Figures A.37 and A.38.

Figure A.38

SUMMARY: SETTING $a + bi$ MODE

1. Press MODE .
2. Move the cursor to the seventh line, highlight $a + bi$, and press ENTER .
3. Press 2nd MODE [QUIT] and return to the home screen.

Evaluating Complex Arithmetic

Complex arithmetic can be performed much like other arithmetic expressions. This is done by entering

2nd . [i]

Figure A.39

to obtain the imaginary unit i from the home screen. For example, to find the sum $(-2 + 3i) + (4 - 6i)$, perform the following keystrokes on the home screen.

((−) 2 + 3 2nd . [i]) + (4 − 6 2nd . [i]) ENTER

The result is shown in Figure A.39. Other complex arithmetic operations are done similarly.

SUMMARY: EVALUATING COMPLEX ARITHMETIC

Enter a complex expression in the same way as you would any arithmetic expression. To obtain the complex number i, use 2nd . [i] .

Finding Maximum and Minimum Values

```
CALCULATE
1:value
2:zero
3:minimum
4:maximum
5:intersect
6:dy/dx
7:∫f(x)dx
```

Figure A.40

To find a minimum y-value (or vertex) on the graph of $f(x) = 1.5x^2 - 6x + 4$, start by entering $Y_1 = 1.5X\text{^}2 - 6X + 4$ from the (Y =) menu. Set the viewing rectangle and then perform the following keystrokes to find the minimum y-value.

(2nd) (TRACE [CALC]) (3)

See Figure A.40.

The calculator prompts for a left bound. Use the arrow keys to position the cursor left of the vertex and press (ENTER). Similarly, position the cursor to the right of the vertex for the right bound and press (ENTER). Finally, the graphing calculator asks for a guess between the left and right bounds. Place the cursor near the vertex and press (ENTER). See Figures A.41–A.43. The minimum value is shown in Figure A.44.

$[-4.7, 4.7, 1]$ by $[-3.1, 3.1, 1]$

Figure A.41

$[-4.7, 4.7, 1]$ by $[-3.1, 3.1, 1]$

Figure A.42

$[-4.7, 4.7, 1]$ by $[-3.1, 3.1, 1]$

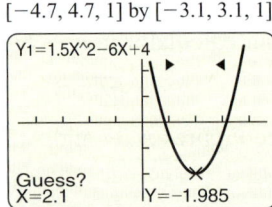

Figure A.43

$[-4.7, 4.7, 1]$ by $[-3.1, 3.1, 1]$

Figure A.44

A maximum of the function f on an interval can be found in a similar manner, except enter

(2nd) (TRACE [CALC]) (4).

The calculator prompts for left and right bounds, followed by a guess. Press (ENTER) after the cursor has been located appropriately for each prompt. The graphing calculator will display the maximum y-value. For example, see the Technology Note in Section 11.1 on page 736.

SUMMARY: FINDING MAXIMUM AND MINIMUM VALUES

1. Graph the function in an appropriate viewing rectangle.
2. Press (2nd) (TRACE [CALC]) (3) to find a minimum y-value.
3. Press (2nd) (TRACE [CALC]) (4) to find a maximum y-value.
4. Use the arrow keys to locate the left and right x-bounds, followed by a guess. Press (ENTER) to select each position of the cursor.

Basic Terminology

A **set** is a collection of things, and the members of a set are called **elements**. A set can be described by listing its elements between braces. For example, the set W containing the *weekdays* is

$$W = \{\text{Monday, Tuesday, Wednesday, Thursday, Friday}\}.$$

This set has 5 elements. For example, Monday *is an element of* W, which is denoted

$$\text{Monday} \in W.$$

However, Sunday *is not an element of* W, which is denoted

$$\text{Sunday} \notin W.$$

If a set contains no elements, then it is called the **empty set** or **null set**. The empty set is denoted $\varnothing$ or $\{\ \}$. For example, the set Z that contains the names of U.S. states starting with the letter Z is the empty set. That is, $Z = \varnothing$ or $Z = \{\ \}$.

NOTE: Do *not* write the empty set as $\{\varnothing\}$.

EXAMPLE 1   Listing the elements of a set

Use set notation to list the elements of each set S described.
(a) The natural numbers from 1 to 12 that are odd
(b) The days of the week that start with the letter T
(c) The last names of U.S. presidents in office during the 1990s

Solution
(a) The list of the natural numbers from 1 to 12 is

$$1, 2, 3, 4, 5, 6, 7, 8, 9, 10, 11, \text{ and } 12.$$

The set of odd natural numbers from this list is

$$S = \{1, 3, 5, 7, 9, 11\}.$$

(b) $S = \{\text{Tuesday, Thursday}\}$
(c) $S = \{\text{Bush, Clinton}\}$

Now Try Exercises **1**, **3**

EXAMPLE 2   Determining the elements of sets

Use $\in$ or $\notin$ to make each statement true.
(a) $5 \underline{\hspace{1cm}} \{1, 2, 3, 4, 5, 6\}$
(b) $-2 \underline{\hspace{1cm}} \{-4, 0, 2, 4, 6\}$
(c) $\frac{1}{2} \underline{\hspace{1cm}} \{0, 0.5, 1.0, 1.5, 2.0\}$

Solution
(a) Because 5 is an element of $\{1, 2, 3, 4, \mathbf{5}, 6\}$, we write

$$5 \in \{1, 2, 3, 4, 5, 6\}.$$

(b) Because -2 is not an element of $\{-4, 0, 2, 4, 6\}$, we write

$$-2 \notin \{-4, 0, 2, 4, 6\}.$$

(c) Because $\frac{1}{2} = 0.5$, we write

$$\frac{1}{2} \in \{0, 0.5, 1.0, 1.5, 2.0\}.$$

Now Try Exercises 11, 13, 15

Universal Set

When discussing sets, we assume that there is a *universal set*. The **universal set** contains all elements under consideration. For example, if the universal set U is all days of the week, then the set S containing the days that start with the letter S is

$$S = \{\text{Sunday, Saturday}\}.$$

However, if the universal set U is only the weekdays, then the set S containing the days starting with the letter S is the *empty set*, or $S = \{\ \}$.

EXAMPLE 3   **Using different universal sets**

Determine the set O of odd integers that belong to each universal set U.
(a) $U = \{1, 2, 3, 4, 5, 6, 7, 8, 9, 10\}$
(b) $U = \{1, 6, 11, 16, 21, 26\}$
(c) $U = \{1, 2, 3, 4, \ldots\}$

Solution
(a) The odd integers in U are 1, 3, 5, 7, and 9, so

$$O = \{1, 3, 5, 7, 9\}.$$

(b) The odd integers in $U = \{\mathbf{1}, 6, \mathbf{11}, 16, \mathbf{21}, 26\}$ are $\mathbf{1}$, $\mathbf{11}$, and $\mathbf{21}$. Thus

$$O = \{1, 11, 21\}.$$

(c) The three dots in $\{1, 2, 3, 4, \ldots\}$ indicate that U contains all natural numbers. Thus the set O contains all odd natural numbers, or

$$O = \{1, 3, 5, 7, 9, \ldots\}.$$

Now Try Exercises 21, 23

Subsets

If every element in a set B is contained in a set A, then we say that B is a **subset** of A, denoted $B \subseteq A$. For example, if $A = \{1, 2, 3, 4\}$ and $B = \{2, 4\}$, then $B \subseteq A$ because every element in B belongs to A. However, A is not a subset of B, denoted $A \nsubseteq B$. because the elements 1 and 3 are in A but *not* in B. The symbol $\nsubseteq$ is read "is not a subset of."

If every element in set A is in set B and every element in set B is in set A, then A and B are **equal sets**, denoted $A = B$. Note that if $A \subseteq B$ and $B \subseteq A$, then $A = B$. Why?

EXAMPLE 4   **Determining subsets**

Let $A = \{a, b, c, d, e\}$, $B = \{b, c, d\}$, $C = \{b, e\}$, and $D = \{e, b\}$. Determine whether each statement is true or false.
(a) $A \subseteq B$ **(b)** $B \subseteq A$ **(c)** $C = D$ **(d)** $C \nsubseteq A$ **(e)** $\varnothing \subseteq B$ **(f)** $A \subseteq A$

Solution

(a) False; the elements a and e in A are not in B, so A is *not* a subset of B.

(b) True; every element in B is in A, so B is a subset of A.

(c) True; although the elements in C and D are listed in a different order, they contain exactly the same elements, so C and D are equal.

(d) False; every element in C is in A, so C *is* a subset of A.

(e) True; the empty set, or null set, is a subset of *every* set.

(f) True; every element in A is in A, so A is a subset of itself.

NOTE: Every set is a subset of itself.   Now Try Exercises **29, 31, 33, 35**

Venn Diagrams

Venn diagrams are often used to depict relationships among sets. A large rectangle typically represents the universal set, and subsets of the universal set are represented by regions within the universal set. In Figure B.1 the universal set U is represented by everything inside the large rectangle. The set A is represented by the red circular region within this rectangle because A is a subset of U.

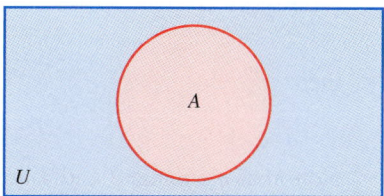

Figure B.1 $A \subseteq U$

The **complement** of a set A, denoted A', is the set containing all elements in the universal set that are *not* in A. That is, if $a \notin A$, then $a \in A'$. For example, if

$$U = \{1, 2, 3, 4, 5, 6\} \quad \text{and} \quad A = \{1, 2, 3\},$$

then

$$A' = \{4, 5, 6\}$$

because the elements 4, 5, and 6 are found in U but not in A. This situation is illustrated by the Venn diagram in Figure B.2. The red region is A, and the blue region is A'. Together, the red and blue regions comprise the universal set U.

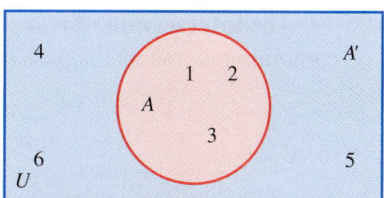

Figure B.2 A and A'

NOTE: Every element in U must be in either A or A', but not both.

EXAMPLE　5　**Determining complements**

Let the universal set U be $U = \{$red, blue, yellow, green, black, white$\}$, and let two subsets of U be $A = \{$red, blue, yellow$\}$ and $B = \{$black, white$\}$. Find each of the following.
(a) A'　　**(b)** B'　　**(c)** U'

Solution
(a) The elements in U that are not in A are in $A' = \{$green, black, white$\}$.
(b) $B' = \{$red, blue, yellow, green$\}$
(c) Because the universal set U contains every element under consideration, the complement of U, or U', is empty. That is, $U' = \{\ \}$.
Now Try Exercises 73, 75

Union and Intersection

Although we do not perform arithmetic operations, such as multiplication or division, on sets, we can find the *union* or *intersection* of two or more sets. The **union** of two sets A and B, denoted $A \cup B$ and read "A union B," is the set containing any element that can be found in *either* set A or set B. If an element is in both A and B, then this element is listed only once in $A \cup B$. For example, if

$$A = \{1, 2, 3, 4\} \quad \text{and} \quad B = \{3, 4, 5, 6\},$$

then

$$A \cup B = \{1, 2, 3, 4, 5, 6\}.$$

Note that elements 3 and 4 are in both A and B but are listed only once in $A \cup B$. A Venn diagram of this situation is shown in Figure B.3. The region that represents the union is shaded blue.

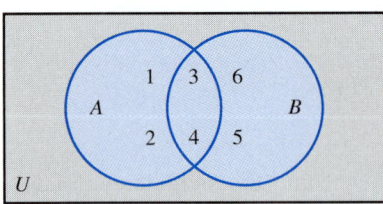

Figure B.3

The **intersection** of two sets A and B, denoted $A \cap B$ and read "A intersect B," is the set containing elements that can be found in *both* set A *and* set B. For example, if

$$A = \{1, 2, 3, 4\} \quad \text{and} \quad B = \{3, 4, 5, 6\},$$

then

$$A \cap B = \{3, 4\}.$$

Note that elements 3 and 4 belong to *both A and B*. This situation is illustrated by the Venn diagram in Figure B.4. The purple region containing elements 3 and 4 represents the intersection of A and B.

Figure B.4

EXAMPLE **6** Finding unions and intersections of sets

Let $A = \{a, b, c, x, z\}$, $B = \{a, x, y\}$, and $C = \{c, x, y, z\}$. Find each set.
(a) $A \cup B$ **(b)** $B \cap C$ **(c)** $A \cup C$ **(d)** $A \cap B \cap C$

Solution

(a) The union contains the elements belonging to either A *or* B or both. Thus

$$A \cup B = \{a, b, c, x, y, z\}.$$

(b) The intersection contains the elements belonging to both B *and* C. Thus

$$B \cap C = \{x, y\}.$$

(c) $A \cup C = \{a, b, c, x, y, z\}$

(d) For an element to be in the intersection of three sets, the element must belong to each set. The only element belonging to A *and* B *and* C is x. Thus $A \cap B \cap C = \{x\}$.

Now Try Exercises 63, 65, 79

EXAMPLE **7** Using set operations

Let $U = \{1, 2, 3, 4, 5, 6, 7, 8\}$, $A = \{2, 4, 6, 8\}$, $B = \{1, 3, 5, 6\}$, and $C = \{5, 6, 7, 8\}$. Find each set.
(a) $A \cap B$ **(b)** $B \cap B'$ **(c)** $B \cup C$ **(d)** $A \cup C'$

Solution

(a) $A \cap B = \{6\}$

(b) Because $B = \{1, 3, 5, 6\}$ and $B' = \{2, 4, 7, 8\}$, $B \cap B' = \{\ \}$.

NOTE: The intersection of a set and its complement is always the empty set.

(c) $B \cup C = \{1, 3, 5, 6, 7, 8\}$

(d) Because $A = \{2, 4, 6, 8\}$ and $C' = \{1, 2, 3, 4\}$, $A \cup C' = \{1, 2, 3, 4, 6, 8\}$.

Now Try Exercises 64, 71, 77

Finite and Infinite Sets

Sets can have either a finite number of elements or infinitely many elements. The elements of a **finite set** can be listed explicitly, whereas the elements of an **infinite set** cannot be listed because there are infinitely many elements. For example, the set of integers S from -3 to 3 is a finite set with 7 elements because

$$S = \{-3, -2, -1, 0, 1, 2, 3\}.$$

In contrast, the set of natural numbers,

$$N = \{1, 2, 3, 4, \ldots\},$$

is an infinite set because the list continues without end.

EXAMPLE **8** **Recognizing finite and infinite sets**

Identify each set as finite or infinite.
(a) The set of integers
(b) The set of natural numbers between 1 and 5, inclusive
(c) The set of rational numbers between 1 and 5, inclusive

Solution
(a) There are infinitely many integers, so the set of integers is infinite.
(b) The set of natural numbers between 1 and 5, inclusive, is $\{1, 2, 3, 4, 5\}$. Because we can list each element of this set, it is a finite set.
(c) Because there are infinitely many rational numbers (fractions) between 1 and 5, this set is an infinite set.   Now Try Exercises **81, 83**

B Exercises

BASIC CONCEPTS

Exercises 1–10: Use set notation to list the elements of the set described.

1. The natural numbers less than 8

2. The natural numbers greater than 4 and less than or equal to 10

3. The days of the week starting with the letter S

4. U.S. states whose names start with the letter A

5. The letters of the alphabet from A to G

6. The letters of the alphabet from D to M

7. Dogs that can fly under their own power

8. People having an income of $10 trillion in 2008

9. The even integers greater than −2 and less than 11

10. The odd integers less than 21 and greater than 12

Exercises 11–20: Insert $\in$ or $\notin$ to make the statement true.

11. 10 _____ $\{5, 10, 15, 20\}$

12. 7 _____ $\{1, 3, 5, 7, 9, 11\}$

13. 5 _____ $\{2, 4, 6, 8, 10\}$

14. −1 _____ $\{0, 1, 2, 3, 4, 5\}$

15. $\frac{1}{4}$ _____ $\{0, 0.25, 0.5, 0.75, 1\}$

16. $\frac{2}{5}$ _____ $\{0, 0.2, 0.4, 0.6, 0.8, 1\}$

17. $\frac{1}{3}$ _____ $\{0, 0.33, 0.67, 1\}$

18. $\frac{4}{2}$ _____ $\{1, 3, 4, 5, 7\}$

19. red _____ $\{$blue, green, red, yellow$\}$

20. M _____ $\{$M, T, W, R, F$\}$

Exercises 21–24: Determine the set E of even integers that belong to the given universal set U.

21. $U = \{1, 2, 3, 4, 5, 6, 7, 8, 9, 10\}$

22. $U = \{-3, -2, -1, 0, 1, 2, 3\}$

23. $U = \{1, 2, 3, 4, \ldots\}$

24. $U = \{1, 3, 5, 7, 9, 11, \ldots\}$

Exercises 25–28: Determine the set A of words starting with the letter A that belong to the given universal set U.

25. $U = \{$Apple, Orange, Pear, Apricot$\}$

26. $U = \{$Apple, Orange$\}$

27. $U = \{$Calculus, Algebra, Geometry$\}$

28. $U = \{$Calculus, Algebra, Geometry, Arithmetic$\}$

SUBSETS

Exercises 29–36: Let $A = \{1, 2, 3, 4\}$, $B = \{3, 4, 5, 6\}$, and $C = \{3, 6\}$. Determine whether the statement is true or false.

29. $A \subseteq B$

30. $B \subseteq A$

31. $C \subseteq B$

32. $\varnothing \subseteq C$

33. $C \subseteq C$

34. $A \nsubseteq B$

35. $B \nsubseteq C$

36. $C \subseteq A$

Exercises 37–44: Let $A = \{a, b, c, d, e\}$, $B = \{b, d\}$, and $C = \{a, c, e\}$. Determine whether the statement is true or false.

37. $B \subseteq A$

38. $C \subseteq A$

39. $B \nsubseteq C$

40. $C = B$

41. $C \subseteq \varnothing$

42. $C \nsubseteq B$

43. $A \subseteq A$

44. $C \nsubseteq C$

VENN DIAGRAMS

Exercises 45–52: Let A and B be two sets and U be the universal set. Sketch a Venn diagram and shade the region that illustrates the given set.

45. $A \cup B$

46. $A \cap B$

47. A'

48. B'

49. $A \cap U$

50. $B \cup U$

51. $(A \cap B)'$

52. $(A \cup B)'$

UNIONS, INTERSECTIONS, AND COMPLEMENTS

Exercises 53–62: Write the expression in terms of one set.

53. $\{1, 2, 3\} \cup \{3, 4\}$

54. $\{1, 2, 3\} \cup \{6\}$

55. $\{a, b, c\} \cap \{a, b, d\}$

56. $\{x, y, z\} \cap \{a, b, c\}$

57. $\{a, b, c\} \cup \{\ \}$

58. $\{a, b, c\} \cap \varnothing$

59. $\{a, b\} \cup \{c, b\} \cup \{d, a\}$

60. $\{1, 3, 5\} \cup \{4, 5\} \cup \{4, 5, 6\}$

61. $\{4, 5, 8, 9\} \cap \{3, 5, 8, 9\} \cap \{4, 5, 8\}$

62. $\{1, 2, 5\} \cap \{2, 1\} \cap \{1, 2, 3\}$

Exercises 63–80: Let

$$U = \{1, 2, 3, 4, 5, 6, 7, 8, 9, 10\},$$
$$A = \{1, 3, 5, 7, 9\}, B = \{2, 4, 6, 8, 10\},$$
$$C = \{3, 4, 5, 6, 7\}, \text{ and } D = \{4, 5, 6, 9\}.$$

Use set notation to list the elements in the given set.

63. $A \cup B$

64. $A \cap B$

65. $C \cap D$

66. $A \cup D$

67. $U \cap \varnothing$

68. $\varnothing \cup U$

69. $A \cup \varnothing$

70. $\varnothing \cap B$

71. $D \cap D'$

72. $A \cup A$

73. U'

74. A'

75. B'

76. D'

77. $A \cup D'$

78. $A' \cap B'$

79. $A \cap B \cap C$

80. $B \cup C \cup D$

FINITE AND INFINITE SETS

Exercises 81–86: Determine whether the given set is finite or infinite.

81. The set of whole numbers

82. The set of real numbers

83. The set of natural numbers less than 1000

84. The set of natural numbers greater than 4 and less than 20

85. The days of the week

86. The names of the students in your class

Appendix C   Linear Programming

Basic Concepts

▶ **REAL-WORLD CONNECTION**   Suppose that a small business sells candy for $3 per pound and fresh ground coffee for $5 per pound. All inventory is sold by the end of the day. The revenue R collected in dollars is given by

$$R = 3x + 5y,$$

where x is the pounds of candy sold and y is the pounds of coffee sold. For example, if the business sells **80** pounds of candy and **40** pounds of coffee during a day, then its revenue is

$$R = 3(\mathbf{80}) + 5(\mathbf{40}) = \$440.$$

The function $R = 3x + 5y$ is called an **objective function**.

Suppose also that the company cannot package more than 150 pounds of candy and coffee per day. Then the inequality

$$x + y \le 150$$

represents a **constraint** on the objective function, which limits the company's revenue for any one day. A goal of this business might be to maximize

$$R = 3x + 5y,$$

subject to the constraints

$$x + y \le 150$$
$$x \ge 0, \ y \ge 0.$$

Note that the constraints $x \ge 0$ and $y \ge 0$ are included because the number of pounds of candy or coffee cannot be negative. The problem that we have described is called a *linear programming problem*. Before learning how to solve a linear programming problem, we need to discuss the set of *feasible solutions*.

Region of Feasible Solutions

The constraints for a linear programming problem consist of linear inequalities. These inequalities are satisfied by some points in the xy-plane but not by others. The set of solutions to these constraints is called the **feasible solutions**. For example, the region of feasible solutions to the constraints for the business just described is shaded in Figure C.1.

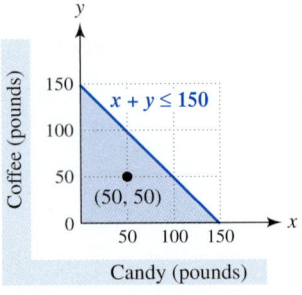

Figure C.1 Constraints on Sales

The point (50, 50) lies in the shaded region and represents the business selling 50 pounds of candy and 50 pounds of coffee. In the next example, we shade the region of feasible solutions to a set of constraints.

 Finding the region of feasible solutions

Shade the region of feasible solutions to the following constraints.

$$x + 2y \le 30$$
$$2x + y \le 30$$
$$x \ge 0, \; y \ge 0$$

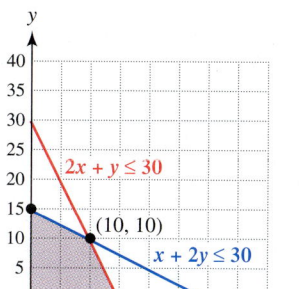

Figure C.2

Solution

The feasible solutions are the ordered pairs (x, y) that *satisfy all four* inequalities. They lie below the lines $x + 2y = 30$ and $2x + y = 30$, and above the line $y = 0$ and to the right of $x = 0$, as shown in Figure C.2. Note that the inequalities $x \ge 0$ and $y \ge 0$ restrict the feasible solutions to quadrant I.

Now Try Exercise **15**

Solving Linear Programming Problems

A **linear programming problem** consists of an *objective function* and a system of linear inequalities called *constraints*. The solution set for the system of linear inequalities is called the *region of feasible solutions*. The objective function describes a quantity that is to be optimized. The **optimal value** to a linear programming problem often results in maximum revenue or minimum cost.

When the system of constraints has only two variables, the boundary of the region of feasible solutions often consists of line segments intersecting at points called *vertices* (plural of **vertex**.) To solve a linear programming problem, we use the *fundamental theorem of linear programming*.

FUNDAMENTAL THEOREM OF LINEAR PROGRAMMING

If the optimal value for a linear programming problem exists, then it occurs at a vertex of the region of feasible solutions.

The fundamental theorem of linear programming is used to solve the following linear programming problem.

EXAMPLE 2 **Maximizing an objective function**

Maximize the objective function $R = 2x + 3y$ subject to

$$x + 2y \le 30$$
$$2x + y \le 30$$
$$x \ge 0, \; y \ge 0.$$

Solution

The region of feasible solutions is shaded in Figure C.2 on the previous page. Note that the vertices on the boundary of feasible solutions are (0, 0), (15, 0), (10, 10), and (0, 15). To find the maximum value of R, substitute each vertex in the formula for R, as shown in Table C.1. The maximum value of R is 50 when $x = 10$ and $y = 10$.

TABLE C.1

Vertex	$R = 2x + 3y$
(0, 0)	$2(0) + 3(0) = 0$
(15, 0)	$2(15) + 3(0) = 30$
(10, 10)	$2(10) + 3(10) = 50$ ← **Maximum R**
(0, 15)	$2(0) + 3(15) = 45$

Now Try Exercise **33**

The following steps are helpful in solving linear programming word problems.

STEPS FOR SOLVING A LINEAR PROGRAMMING WORD PROBLEM

STEP 1: Read the problem carefully. Consider making a table.

STEP 2: Write the objective function and all the constraints.

STEP 3: Sketch a graph of the region of feasible solutions. Identify all vertices.

STEP 4: Evaluate the objective function at each vertex. A maximum (or a minimum) occurs at a vertex.

NOTE: If the region is unbounded, a maximum (or minimum) may not exist.

EXAMPLE **3**   Minimizing the cost of vitamins

A breeder is mixing two different vitamins, Brand X and Brand Y, into pet food. Each serving of pet food should contain at least 60 units of vitamin A and 30 units of vitamin C. Brand X costs 80 cents per ounce and Brand Y costs 50 cents per ounce. Each ounce of Brand X contains 15 units of vitamin A and 10 units of vitamin C, whereas each ounce of Brand Y contains 20 units of vitamin A and 5 units of vitamin C. Determine how much of each brand of vitamin should be mixed to produce a minimum cost per serving.

Solution

STEP 1: Begin by listing the information, as illustrated in Table C.2.

TABLE C.2

Brand	Amount	Vitamin A	Vitamin C	Cost
X	x	15	10	80 cents
Y	y	20	5	50 cents
Minimum		60	30	

STEP 2: If x ounces of Brand X are purchased at 80 cents per ounce and if y ounces of Brand Y are purchased at 50 cents per ounce, then the total cost C is given by $C = 80x + 50y$. Because each ounce of Brand X contains 15 units of vitamin A and each ounce of Brand Y contains 20 units of vitamin A, the total number of units of vitamin A is $15x + 20y$. If each serving of pet food must contain at least 60 units of vitamin A, the constraint is $15x + 20y \geq 60$. Similarly, because each serving requires at least 30 units of vitamin C, $10x + 5y \geq 30$. The linear programming problem then becomes the following.

Minimize:	$C = 80x + 50y$	Cost (in cents)
Subject to:	$15x + 20y \geq 60$	Vitamin A
	$10x + 5y \geq 30$	Vitamin C
	$x \geq 0, \ y \geq 0$	

STEP 3: The region containing the feasible solutions is shown in Figure C.3.

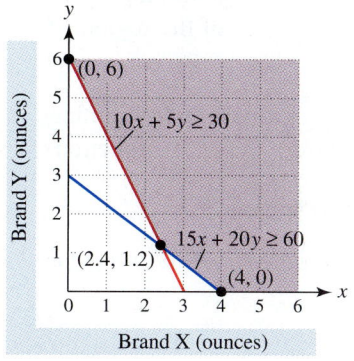

Figure C.3

NOTE: To determine the vertex (2.4, 1.2) solve the system of equations

$$15x + 20y = 60$$
$$10x + 5y = 30$$

by using elimination.

STEP 4: The vertices for this region are (0, 6), (2.4, 1.2), and (4, 0). Evaluate the objective function C at each vertex, as shown in Table C.3.

TABLE C.3

Vertex	$C = 80x + 50y$	
(0, 6)	$80(0) + 50(6) = 300$	
(2.4, 1.2)	$80(2.4) + 50(1.2) = 252$	← **Minimum cost (cents)**
(4, 0)	$80(4) + 50(0) = 320$	

The minimum cost occurs when 2.4 ounces of Brand X and 1.2 ounces of Brand Y are mixed, at a cost of $2.52 per serving.

Now Try Exercise 45

C Exercises

CONCEPTS

1. A procedure used in business to optimize quantities such as cost and profit is called _____.

2. In linear programming, the function to be optimized is called the _____ function.

3. The region in the xy-plane that satisfies the constraints is called the region of _____.

4. In linear programming, the constraints typically consist of a system of _____.

5. If the optimal value for a linear programming problem exists, then it occurs at a _____ of the region of feasible solutions.

6. To find the optimal value in a linear programming problem, substitute each vertex in the _____ function.

REGIONS OF FEASIBLE SOLUTIONS

Exercises 7–20: Shade the region of feasible solutions for the following constraints.

7. $x + y \le 3$
   $x \ge 0, y \ge 0$

8. $2x + y \le 4$
   $x \ge 0, y \ge 0$

9. $4x + 3y \le 12$
   $x \ge 0, y \ge 0$

10. $5x + 3y \le 15$
    $x \ge 0, y \ge 0$

11. $x \le 5$
    $y \le 2$
    $x \ge 0, y \ge 0$

12. $x \le 3$
    $y \le 4$
    $x \ge 1, y \ge 1$

13. $x + y \le 5$
    $x + y \ge 2$
    $x \ge 0, y \ge 0$

14. $2x + y \le 6$
    $x + y \ge 3$
    $x \ge 0, y \ge 0$

15. $3x + 2y \le 6$
    $2x + 3y \le 6$
    $x \ge 0, y \ge 0$

16. $5x + 3y \le 30$
    $3x + 5y \le 30$
    $x \ge 0, y \ge 0$

17. $x + y \le 3$
    $x + 3y \ge 3$
    $x \ge 0, y \ge 0$

18. $3x + y \ge 6$
    $x + 2y \ge 6$
    $x \ge 0, y \ge 0$

19. $x + 2y \ge 4$
    $3x + 2y \ge 6$
    $x \ge 0, y \ge 0$

20. $4x + 3y \ge 12$
    $3x + 4y \ge 12$
    $x \ge 0, y \ge 0$

LINEAR PROGRAMMING

Exercises 21–24: Find the maximum of R on the region of feasible solutions shown in the figure.

21. $R = 4x + 5y$

22. $R = 2x + 3y$

23. $R = x + 3y$

24. $R = 12x + 9y$

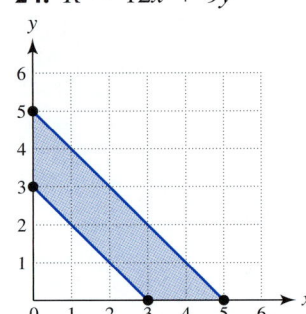

Exercises 25–28: Use the figures in Exercises 21–24 to complete Exercises 25–28, respectively, to minimize C.

25. $C = 2x + 3y$

26. $C = 3x + y$

27. $C = 5x + y$

28. $C = 2x + 7y$

Exercises 29–36: Maximize the objective function R, subject to the given constraints.

29. $R = 3x + 5y$
    $x + y \le 150$
    $x \ge 0, y \ge 0$

30. $R = 6x + 5y$
    $x + y \le 8$
    $x \ge 0, y \ge 0$

31. $R = 3x + 2y$
    $2x + y \le 6$
    $x \ge 0, y \ge 0$

32. $R = x + 3y$
    $x + 2y \le 4$
    $x \ge 0, y \ge 0$

33. $R = 12x + 9y$
$3x + y \le 6$
$x + 3y \le 6$
$x \ge 0, y \ge 0$

34. $R = 10x + 30y$
$x + 3y \le 12$
$3x + y \le 12$
$x \ge 0, y \ge 0$

35. $R = 4x + 5y$
$x + y \ge 2$
$x + 2y \le 4$
$x \ge 0, y \ge 0$

36. $R = 3x + 7y$
$x + y \ge 1$
$3x + y \le 3$
$x \ge 0, y \ge 0$

Exercises 37–42: Minimize the objective function C, subject to the given constraints.

37. $C = x + 2y$
$x \le 3, y \le 2$
$x \ge 0, y \ge 0$

38. $C = 3x + y$
$x \le 5, y \le 3$
$x \ge 1, y \ge 1$

39. $C = 8x + 15y$
$x + y \ge 4$
$x \ge 0, y \ge 0$

40. $C = x + 2y$
$3x + 4y \ge 12$
$x \ge 0, y \ge 0$

41. $C = 30x + 40y$
$2x + y \le 6$
$x + y \ge 2$
$x \ge 0, y \ge 0$

42. $C = 50x + 70y$
$2x + 3y \le 6$
$x + y \ge 1$
$x \ge 0, y \ge 0$

APPLICATIONS

Exercises 43–48: Solve the linear programming problem.

43. *Maximizing Revenue*   A small business sells candy for $4 per pound and coffee for $6 per pound. The business can package and sell at most a total of 100 pounds of candy and coffee per day, but at least 20 pounds of candy must be sold each day. Determine how many pounds of candy and coffee need to be sold each day to maximize revenue.

44. *Minimizing Cost*   It costs a business $20 to make one compact disc player and $10 to make one radio. Each week the company must make a combined total of at least 50 compact disc players and radios. At least as many compact disc players as radios must be manufactured. Determine how many compact disc players and radios should be made to minimize weekly costs.

45. *Vitamin Cost*   A pet owner is mixing two different vitamins, Brand X and Brand Y, into pet food. Brand X costs 90 cents per ounce and Brand Y costs 60 cents per ounce. Each serving is a mixture of the two brands and should contain at least 40 units of vitamin A and 30 units of vitamin C. Each ounce of Brand X contains 20 units of vitamin A and 10 units of vitamin C, whereas each ounce of Brand Y contains 10 units of vitamin A and 10 units of vitamin C. Determine how much of each brand of vitamin should be mixed to produce a minimum cost per serving.

46. *Pet Food Cost*   A pet owner is buying two brands of food, X and Y, for his animals. Each serving of the mixture of the two foods should contain at least 60 grams of protein and 40 grams of fat. Brand X costs 75 cents per unit and Brand Y costs 50 cents per unit. Each unit of Brand X contains 20 grams of protein and 10 grams of fat, whereas each unit of Brand Y contains 10 grams of protein and 10 grams of fat. Determine how much of each brand should be bought to obtain a minimum cost per serving.

47. *Raising Animals*   A breeder can raise no more than 50 hamsters and mice but no more than 20 hamsters. If she sells the hamsters for $15 each and the mice for $10 each, find the maximum revenue.

48. *Maximizing Profit*   A business manufactures two parts, X and Y. Machines A and B are needed to make each part. To make part X, machine A is needed 3 hours and machine B is needed 1 hour. To make part Y, machine A is needed 1 hour and machine B is needed 2 hours. Machine A is available 60 hours per week and machine B is available 50 hours per week. The profit from part X is $300 and the profit from part Y is $250. How many parts of each type should be made to maximize weekly profit?

Appendix D Synthetic Division

Basic Concepts

A shortcut called **synthetic division** can be used to divide $x - k$, where k is a number, into a polynomial. For example, to divide $x - 2$ into $3x^3 - 8x^2 + 7x - 6$, we do the following (with the equivalent long division shown at the right).

Synthetic Division

$$
\begin{array}{r|rrrr}
2 & 3 & -8 & 7 & -6 \\
  &   & 6 & -4 & 6 \\
\hline
  & 3 & -2 & 3 & 0
\end{array}
$$

Long Division of Polynomials

$$
\begin{array}{r}
3x^2 - 2x + 3 \\
x - 2 \overline{)3x^3 - 8x^2 + 7x - 6} \\
\underline{3x^3 - 6x^2} \\
-2x^2 + 7x \\
\underline{-2x^2 + 4x} \\
3x - 6 \\
\underline{3x - 6} \\
0
\end{array}
$$

Note that the highlighted numbers in the expression for long division correspond to the third row in synthetic division. The remainder is 0, which is the last number in the third row. The quotient is $3x^2 - 2x + 3$. Its coefficients are 3, -2, and 3 and are located in the third row. To divide $x - 2$ into $3x^3 - 8x^2 + 7x - 6$ with synthetic division, use the following steps.

STEP 1. In the top row write 2 (the value of k) on the left and then write the coefficients of the dividend $3x^3 - 8x^2 + 7x - 6$.

STEP 2. **(a)** Copy the leading coefficient 3 of $3x^3 - 8x^2 + 7x - 6$ in the third row and multiply it by 2 (the value of k). Write the result 6 in the second row below -8. Add -8 and 6 in the second column to obtain the -2 in the third row.

(b) Repeat the process by multiplying -2 by 2 and place the result -4 below 7. Then add 7 and -4 to obtain 3.

(c) Multiply 3 by 2 and place the result 6 below the -6. Adding 6 and -6 gives 0.

STEP 3. The last number in the third row is 0, which is the remainder. The other numbers in the third row are the coefficients of the quotient, which is $3x^2 - 2x + 3$.

EXAMPLE 1 Performing synthetic division

Use synthetic division to divide $x^4 - 5x^3 + 9x^2 - 10x + 3$ by $x - 3$.

Solution
Because the divisor is $x - 3$, the value of k is 3.

$$
\begin{array}{r|rrrrr}
3 & 1 & -5 & 9 & -10 & 3 \\
  &   & 3 & -6 & 9 & -3 \\
\hline
  & 1 & -2 & 3 & -1 & 0
\end{array}
$$

The quotient is $x^3 - 2x^2 + 3x - 1$ and the remainder is 0. This result is also expressed by

$$\frac{x^4 - 5x^3 + 9x^2 - 10x + 3}{x - 3} = x^3 - 2x^2 + 3x - 1 + \frac{0}{x - 3}, \quad \text{or}$$

$$\frac{x^4 - 5x^3 + 9x^2 - 10x + 3}{x - 3} = x^3 - 2x^2 + 3x - 1.$$

Now Try Exercise **3**

EXAMPLE **2** **Performing synthetic division**

Use synthetic division to divide $2x^3 - x + 5$ by $x + 1$.

Solution

Write $2x^3 - x + 5$ as $2x^3 + 0x^2 - x + 5$. The divisor $x + 1$ can be written as

$$x + 1 = x - (-1),$$

so we let $k = -1$.

$$
\begin{array}{r|rrrr}
-1 & 2 & 0 & -1 & 5 \\
   &   & -2 & 2 & -1 \\
\hline
   & 2 & -2 & 1 & 4
\end{array}
$$

The remainder is 4, and the quotient is $2x^2 - 2x + 1$. This result can also be expressed as

$$\frac{2x^3 - x + 5}{x + 1} = 2x^2 - 2x + 1 + \frac{4}{x + 1}.$$

Now Try Exercise **13**

D Exercises

Exercises 1–16: Use synthetic division to divide.

1. $\dfrac{x^2 + 3x - 1}{x - 1}$

2. $\dfrac{2x^2 + x - 1}{x - 3}$

3. $(3x^2 - 22x + 7) \div (x - 7)$

4. $(5x^2 + 29x - 6) \div (x + 6)$

5. $\dfrac{x^3 + 7x^2 + 14x + 8}{x + 4}$

6. $\dfrac{2x^3 + 3x^2 + 2x + 4}{x + 1}$

7. $\dfrac{2x^3 + x^2 - 1}{x - 2}$

8. $\dfrac{x^3 + x - 2}{x + 3}$

9. $(x^3 - 2x^2 - 2x + 4) \div (x - 4)$

10. $(2x^4 + 3x^2 - 4) \div (x + 2)$

11. $(x^3 - 3x^2 - 8x - 10) \div (x - 5)$

12. $(x^4 - 1) \div (x + 1)$

13. $(2x^4 - x) \div (x + 2)$

14. $(2x^3 + x - 1) \div (x - 3)$

15. $(b^4 - 1) \div (b - 1)$

16. $(a^2 + a) \div (a + 2.5)$

Answers to Selected Exercises

SECTION 1.1 (pp. 8–10)

1. counting **2.** zero **3.** 1 **4.** composite **5.** prime
6. composite **7.** factors **8.** formula **9.** variable
10. equals sign **11.** sum **12.** product **13.** quotient
14. difference **15.** Composite; $4 = 2 \times 2$ **17.** Neither
19. Prime **21.** Composite; $92 = 2 \times 2 \times 23$
23. Composite; $225 = 3 \times 3 \times 5 \times 5$ **25.** Prime
27. $6 = 2 \times 3$ **29.** $12 = 2 \times 2 \times 3$
31. $32 = 2 \times 2 \times 2 \times 2 \times 2$ **33.** $39 = 3 \times 13$
35. $294 = 2 \times 3 \times 7 \times 7$
37. $300 = 2 \times 2 \times 3 \times 5 \times 5$ **39.** Yes **41.** No
43. Yes **45.** Yes **47.** 10 **49.** 7 **51.** 4 **53.** 18
55. 4 **57.** 9 **59.** 8 **61.** 14 **63.** 1 **65.** 28 **67.** 7
69. 5 **71.** 18 **73.** 3 **75.** $3s$; s is the cost of soda.
77. $x + 5$; x is the number. **79.** $n + 5$; n is the number.
81. $3n$; n is the number. **83.** $p - 200$; p is the population.
85. $\frac{z}{6}$; z is the number. **87.** st; s is the speed and t is the time.
89. $\frac{x + 7}{y}$; x is one number, y is the other number.
91.

Yards (y)	1	2	3	4	5	6	7
Feet (F)	3	6	9	12	15	18	21

$F = 3y$
93. $P = 100D$ **95.** $M = 3x$; 108 mi **97.** $B = 6D$
99. $C = 12x$ **101.** 198 square feet

SECTION 1.2 (pp. 23–26)

1. $\frac{3}{4}; \frac{1}{4}$ **2.** 11; 21 **3.** 0 **4.** 1 **5.** numerators
6. denominators **7.** $\frac{1}{4}$ **8.** $\frac{a}{b}$ **9.** multiply **10.** $\frac{1}{4}$
11. $\frac{1}{a}$ **12.** $\frac{1}{5}$ **13.** $\frac{ac}{bd}$ **14.** $\frac{ad}{bc}$ **15.** $\frac{a + c}{b}$ **16.** $\frac{a - c}{b}$
17. 2 **18.** 12 **19.** 9 **20.** $\frac{6}{6}$ **21.** 4 **23.** 25 **25.** 10
27. $\frac{3}{5}$ **29.** $\frac{3}{5}$ **31.** $\frac{1}{2}$ **33.** $\frac{1}{3}$ **35.** $\frac{2}{5}$ **37.** $\frac{1}{3}$ **39.** $\frac{2}{5}$
41. $\frac{1}{4}$ **43.** $\frac{1}{6}$ **45.** $\frac{3}{20}$ **47.** 1 **49.** $\frac{3}{5}$ **51.** $\frac{12}{5}$ **53.** $\frac{3}{4}$
55. 1 **57.** $\frac{3a}{2b}$ **59.** $\frac{3}{16}$ **61.** 4 **63.** $\frac{1}{3}$
65. (a) $\frac{1}{5}$ (b) $\frac{1}{7}$ (c) $\frac{7}{4}$ (d) $\frac{8}{9}$
67. (a) 2 (b) 9 (c) $\frac{101}{12}$ (d) $\frac{17}{31}$
69. $\frac{3}{2}$ **71.** $\frac{3}{2}$ **73.** 8 **75.** $\frac{4}{3}$ **77.** $\frac{63}{8}$ **79.** 12 **81.** $\frac{3}{10}$
83. $\frac{a}{2}$ **85.** 1 **87.** (a) 1 (b) $\frac{1}{3}$ **89.** (a) 2 (b) 1
91. (a) $\frac{7}{33}$ (b) $\frac{1}{11}$ **93.** 10 **95.** 45 **97.** 15 **99.** 24
101. 12 **103.** 24 **105.** $\frac{3}{6}, \frac{4}{6}$ **107.** $\frac{28}{36}, \frac{15}{36}$ **109.** $\frac{3}{48}, \frac{28}{48}$
111. $\frac{4}{12}, \frac{9}{12}, \frac{10}{12}$ **113.** $\frac{5}{6}$ **115.** $\frac{13}{16}$ **117.** $\frac{1}{4}$ **119.** $\frac{1}{6}$
121. $\frac{59}{70}$ **123.** $\frac{13}{36}$ **125.** $\frac{17}{600}$ **127.** $\frac{9}{8}$ **129.** $4\frac{3}{4}$ ft

131. $\frac{961}{1260}$ **133.** $32\frac{5}{16}$ inches **135.** $\frac{5}{8}$ yd^2
137. $8\frac{1}{4}$ miles **139.** $\frac{6}{125}$

CHECKING BASIC CONCEPTS 1.1 & 1.2 (p. 26)

1. (a) Prime (b) Composite; $28 = 2 \times 2 \times 7$
(c) Neither (d) Composite; $180 = 2 \times 2 \times 3 \times 3 \times 5$
2. 2 **3.** 30 **4.** $x + 5$ **5.** $I = 12F$
6. (a) 3 (b) 8 **7.** (a) $\frac{5}{7}$ (b) $\frac{2}{3}$ **8.** $\frac{3}{4}$
9. (a) $\frac{1}{2}$ (b) $\frac{1}{4}$ (c) $\frac{2}{5}$ (d) $\frac{7}{12}$ **10.** $3\frac{1}{3}$ cups

SECTION 1.3 (pp. 32–33)

1. add **2.** multiply **3.** Six **4.** 2 **5.** a^6
6. base; exponent **7.** 6^2 **8.** 8^3 **9.** 4^5 **10.** x^3
11. 17; multiplication; addition
12. 2; exponents; subtraction **13.** 4; left; right
14. 2; left; right **15.** No; $2^3 = 8$, but $3^2 = 9$.
16. $5 \cdot 5$ **17.** 2^5 **19.** 3^4 **21.** $\left(\frac{1}{2}\right)^4$ **23.** a^5
25. $(x + 3)^2$ **27.** (a) 16 (b) 16 **29.** (a) 6 (b) 1
31. (a) 32 (b) 1000 **33.** (a) $\frac{4}{9}$ (b) $\frac{1}{32}$
35. (a) $\frac{8}{125}$ (b) $\frac{81}{49}$ **37.** 2^3 **39.** 5^2 **41.** 7^2 **43.** 10^3
45. $\left(\frac{1}{2}\right)^4$ **47.** $\left(\frac{2}{3}\right)^5$ **49.** 29 **51.** 4 **53.** 90 **55.** 3
57. 2 **59.** 19 **61.** 32 **63.** 125 **65.** 3 **67.** 4
69. 80 **71.** $\frac{49}{16}$ **73.** $2^3 - 8$; 0 **75.** $30 - 4 \cdot 3$; 18
77. $\frac{4^2}{2^3}$; 2 **79.** $\frac{40}{10} + 2$; 6 **81.** $100(2 + 3)$; 500
83. About 536,870,912 bytes **85.** (a) $k = 7$ (b) 32
87. (a) 8 yr (b) \$80,000

SECTION 1.4 (pp. 42–44)

1. $-b$ **2.** 7 **3.** b **4.** -9 **5.** natural **6.** rational
7. 4 **8.** principal **9.** real **10.** irrational
11. rational **12.** irrational **13.** $0.\overline{27}$ **14.** 1; 4
15. $\sqrt{2}$ (answers may vary) **16.** not equal
17. approximately equal **18.** 3 **19.** 0 **20.** left
21. origin **22.** b **23.** $>$ **25.** $=$
27. (a) -9 (b) 9 **29.** (a) $-\frac{2}{3}$ (b) $\frac{2}{3}$
31. (a) -8 (b) 8 **33.** (a) $-a$ (b) a
35. 6 **37.** $\frac{1}{2}$ **39.** 0.25 **41.** 0.875 **43.** 1.5
45. 0.05 **47.** $0.\overline{6}$ **49.** $0.\overline{7}$
51. Natural, whole, integer, and rational
53. Natural, whole, integer, and rational
55. Whole, integer, and rational **57.** Rational
59. Rational **61.** Irrational
63. Natural, integer, and rational
65. Natural, integer, and rational **67.** Rational

69.

71.

73.

75.

77.

79. 5.23   **81.** 7   **83.** 4   **85.** $\pi - 3$ **87.** $-b$
89. $<$ **91.** $>$ **93.** $>$ **95.** $<$ **97.** $>$
99. $<$ **101.** $-9, -2^3, -3, 0, 1$ **103.** $-2, -\frac{3}{2}, \frac{1}{3}, \sqrt{5}, \pi$
105. $-4^2, -\frac{17}{28}, -\frac{4}{7}, \sqrt{2}, \sqrt{7}$ **107. (a)** 16.5
(b) Answers may vary.   **(c)** 17.125; answers may vary.

CHECKING BASIC CONCEPTS 1.3 & 1.4 (p. 44)

1. (a) 5^4 **(b)** 7^5 **2. (a)** 8   **(b)** 10,000   **(c)** $\frac{8}{27}$
(d) -81 **3. (a)** 4^3 **(b)** 2^6 **4. (a)** 26   **(b)** 9   **(c)** 2
(d) $\frac{1}{2}$ **(e)** 4   **(f)** 0   **5.** $5^3 \div 3$, or $\frac{5^3}{3}$ **6. (a)** 17   **(b)** $-a$
7. (a) 0.15   **(b)** 0.625   **8. (a)** Natural, integer, and rational
(b) Integer and rational   **(c)** Irrational   **(d)** Rational
9.

10. (a) 12   **(b)** a **11. (a)** $<$ **(b)** $<$ **(c)** $<$
12. $-7, -1.6, 0, \frac{1}{3}, \sqrt{3}, 3^2$

SECTION 1.5 (pp. 49–51)

1. two   **2.** addends   **3.** sum   **4.** zero   **5.** positive
6. negative   **7.** absolute value   **8.** difference
9. addition   **10.** opposite; $(-b)$ **11.** addition
12. subtraction   **13.** $-25; 0$ **15.** $\sqrt{21}; 0$
17. $-5.63; 0$ **19.** 4   **21.** 2   **23.** -3 **25.** 4   **27.** 2
29. -1 **31.** 10   **33.** -150 **35.** 1   **37.** -7 **39.** $\frac{1}{4}$
41. $-\frac{9}{14}$ **43.** $-\frac{5}{4}$ **45.** -1.1 **47.** 34   **49.** -1
51. 0   **53.** -3 **55.** 7   **57.** 1   **59.** $-\frac{1}{14}$ **61.** $\frac{1}{2}$
63. 2.9   **65.** -164 **67.** -11 **69.** -9 **71.** 42
73. -5 **75.** 50   **77.** 8.8   **79.** $\frac{1}{2}$ **81.** -1
83. $2 + (-5); -3$ **85.** $-5 + 7; 2$ **87.** $-(2^3); -8$
89. $-6 - 7; -13$ **91.** $6 + (-10) - 5; -9$ **93.** \$230
95. 25 yards   **97.** 64,868 feet

SECTION 1.6 (pp. 59–60)

1. factors   **2.** product   **3.** negative   **4.** positive
5. quotient   **6.** dividend; divisor   **7.** $\frac{1}{a}$ **8.** reciprocal
9. $-\frac{4}{3}$ **10.** reciprocal, or multiplicative inverse   **11.** $-a$
12. 25   **13.** -25 **14.** $\frac{1}{b}$ **15.** positive   **16.** negative
17. subtraction   **18.** 5; 8   **19.** -12 **21.** -18 **23.** 0

25. 60   **27.** $\frac{1}{4}$ **29.** -1 **31.** 200   **33.** -5000 **35.** 120
37. $-\frac{3}{2}$ **39.** 1   **41.** Negative   **43.** -1 **45.** -16 **47.** 8
49. -40 **51.** -2 **53.** 10   **55.** -4 **57.** -3 **59.** -32
61. 0   **63.** 0   **65.** $-\frac{1}{22}$ **67.** $\frac{4}{15}$ **69.** $-\frac{15}{16}$ **71.** Undefined
73. -1 **75.** $-\frac{4}{3}$ **77.** 0.5   **79.** 0.1875   **81.** 3.5   **83.** $5.\overline{6}$
85. 1.4375   **87.** 0.875   **89.** $\frac{1}{4}$ **91.** $\frac{4}{25}$ **93.** $\frac{5}{8}$ **95.** $\frac{11}{16}$
97. $2.\overline{3}; \frac{7}{3}$ **99.** $2.1; \frac{21}{10}$ **101.** $-1.8\overline{3}; -\frac{11}{6}$ **103.** $0.8; \frac{4}{5}$
105. About 131 million   **107.** 0.156

CHECKING BASIC CONCEPTS 1.5 & 1.6 (pp. 60–61)

1. (a) 0   **(b)** -19 **2. (a)** $\frac{8}{9}$ **(b)** -3.2
3. (a) $-1 + 5; 4$ **(b)** $4 - (-3); 7$ **4.** $144°F$
5. (a) 35   **(b)** $\frac{4}{15}$ **6. (a)** -9 **(b)** -32 **(c)** 25
7. (a) $-\frac{15}{2}$ **(b)** $\frac{15}{32}$ **8.** $-\frac{6}{7}$ **9. (a)** -5 **(b)** -5
(c) -5 **(d)** 5   **10. (a)** 0.6   **(b)** 3.875

SECTION 1.7 (pp. 69–71)

1. commutative; addition   **2.** commutative; multiplication
3. associative; addition   **4.** associative; multiplication
5. subtraction; division   **6.** subtraction; division
7. distributive   **8.** distributive   **9.** b **10.** a
11. identity; addition   **12.** identity; multiplication
13. $-a$ **14.** $\frac{1}{a}$ **15.** $10 + (-6)$ **17.** $6 \cdot (-5)$
19. $10 + a$ **21.** $7b$ **23.** $1 + (2 + 3)$
25. $(2 \cdot 3) \cdot 4$ **27.** $a + (5 + c)$ **29.** $x \cdot (3 \cdot 4)$
31. $a + b + c = (a + b) + c$
$\qquad\qquad = c + (a + b)$
$\qquad\qquad = c + (b + a)$
$\qquad\qquad = c + b + a$
33. 20   **35.** $ab - 8a$ **37.** $-t - z$
39. $-5 + a$ **41.** $3a + 15$ **43.** $3z - 18$
45. $a \cdot (b + c + d) = a \cdot \big((b + c) + d\big)$
$\qquad\qquad = a \cdot (b + c) + ad$
$\qquad\qquad = ab + ac + ad$
47. $11x$ **49.** $-b$ **51.** $2a$ **53.** $-14w$
55. Commutative (multiplication)   **57.** Associative (addition)
59. Distributive   **61.** Distributive, commutative (multiplication)
63. Distributive   **65.** Associative (multiplication)
67. Distributive   **69.** Identity (addition)
71. Identity (multiplication)   **73.** Identity (multiplication)
75. Inverse (multiplication)   **77.** Inverse (addition)
79. 30   **81.** 100   **83.** 178   **85.** 477   **87.** 79   **89.** 90
91. 816   **93.** 1   **95.** $\frac{1}{6}$ **97. (a)** 410   **(b)** 9970
(c) -6300 **(d)** $-140,000$ **99. (a)** 19,000   **(b)** $-45,100$
(c) 60,000   **(d)** $-7,900,000$ **101. (a)** 1.256   **(b)** 0.96
(c) 0.0987   **(d)** -0.0056 **(e)** 120   **(f)** 457.8
103. Commutative (addition)   **105.** 19.8 miles
107. (a) $13 \times (5 \times 2)$; multiplying by 10 is easy.
(b) Associative (multiplication)

SECTION 1.8 (pp. 77–79)

1. term   **2.** is not; is   **3.** coefficient   **4.** -1 **5.** 6
6. factors; terms   **7.** like   **8.** like; unlike   **9.** like
10. distributive   **11.** Yes; 91   **13.** Yes; -6 **15.** No
17. Yes; 1   **19.** No   **21.** Yes; -9 **23.** Like   **25.** Like
27. Unlike   **29.** Unlike   **31.** Like   **33.** Unlike
35. Unlike   **37.** Like   **39.** $8x$ **41.** $14y$ **43.** $41a$
45. 0   **47.** Not possible   **49.** Not possible   **51.** $3x^2$
53. $-xy$ **55.** 0   **57.** $3x + 2$ **59.** $-2z + \frac{1}{2}$ **61.** $11y$
63. $4z - 1$ **65.** $12y - 7z$ **67.** $-3x - 4$ **69.** $-6x - 1$
71. $-\frac{1}{3}x + \frac{2}{3}$ **73.** $\frac{2}{5}x + \frac{3}{5}y + \frac{1}{5}$ **75.** $0.4x^2$ **77.** $7x^2 - 7x$
79. $2b$ **81.** $7x^3 + 2$ **83.** x **85.** 3   **87.** z **89.** $3x - 2$
91. $2z + 3$ **93.** $5x + 6x$; $11x$ **95.** $x^2 + 2x^2$; $3x^2$
97. $6x - 4x$; $2x$ **99. (a)** $1570w$ **(b)** 65,940 square feet
101. (a) $50x$ **(b)** 2400 cubic feet   **(c)** 24 minutes

CHECKING BASIC CONCEPTS 1.7 & 1.8 (p. 79)

1. (a) $18y$ **(b)** $x + 10$ **2.** $20y$ **3. (a)** $5 - x$
(b) $5x - 35$ **4.** Distributive   **5.** 0   **6. (a)** 60   **(b)** 7
(c) 368   **7. (a)** Unlike   **(b)** Like   **8. (a)** $14z$
(b) $-3y + 3$ **9. (a)** $-3y - 3$ **(b)** $-14y$
(c) x **(d)** 35   **10.** $3x + 5x$; $8x$

CHAPTER 1 REVIEW (pp. 84–87)

1. Prime   **2.** Composite; $27 = 3 \times 3 \times 3$
3. Composite; $108 = 2 \times 2 \times 3 \times 3 \times 3$
4. Composite; $91 = 7 \times 13$ **5.** Neither   **6.** Neither
7. 3   **8.** 5   **9.** 12   **10.** 6   **11.** 7   **12.** 7   **13.** 8
14. 6   **15.** $5c$, where c is the cost of the CD
16. $x - 5$, where x is the number   **17.** $3^2 + 5$
18. $2^3 \div (3 + 1)$ **19.** $3x$, where x is the number
20. $x - 4$, where x is the number   **21. (a)** $\frac{5}{8}$ **(b)** $\frac{3}{4}$
22. (a) $\frac{3}{4}$ **(b)** $\frac{3}{5}$ **23.** $\frac{5}{8}$ **24.** $\frac{2}{9}$ **25.** $\frac{3}{11}$ **26.** $\frac{12}{23}$ **27.** 2
28. $\frac{3x}{2y}$ **29.** $\frac{3}{35}$ **30. (a)** $\frac{1}{8}$ **(b)** 1   **(c)** $\frac{19}{5}$ **(d)** $\frac{2}{3}$ **31.** 9
32. $\frac{9}{14}$ **33.** 12   **34.** $\frac{1}{8}$ **35.** 24   **36.** 42   **37.** $\frac{1}{3}$ **38.** $\frac{1}{2}$
39. $\frac{19}{24}$ **40.** $\frac{9}{22}$ **41.** $\frac{5}{12}$ **42.** $\frac{13}{18}$ **43.** 5^6 **44.** $\left(\frac{7}{6}\right)^3$
45. 3^4 **46.** x^5 **47.** $(x + 1)^2$ **48.** $(a - 5)^3$
49. (a) 64   **(b)** 49   **(c)** 8   **50.** 5   **51.** 25   **52.** 7
53. 3   **54.** 25   **55.** 1   **56.** 2   **57.** 0   **58.** 0   **59.** 10
60. 1   **61.** 5   **62.** 19   **63. (a)** 8   **(b)** 3
64. (a) $-\frac{3}{7}$ **(b)** $-\frac{2}{5}$ **65. (a)** 0.8   **(b)** 0.15
66. (a) $0.\overline{5}$ **(b)** $0.\overline{63}$ **67.** Whole, integer, and rational
68. Rational   **69.** Integer and rational   **70.** Irrational
71. Irrational   **72.** Rational
73.

74. (a) 5   **(b)** π **(c)** $\sqrt{2} - 1$
75. (a) $<$ **(b)** $>$ **(c)** $<$ **(d)** $>$

76. $-3, -\frac{2}{3}, \sqrt{3}, \pi - 1, 3$ **77.** 4   **78.** -3 **79.** 1
80. -5 **81.** 1   **82.** -2 **83.** -44 **84.** 40   **85.** $-\frac{11}{4}$
86. -7 **87.** $-\frac{2}{9}$ **88.** $-\frac{5}{4}$ **89.** $\frac{2}{7}$ **90.** $-\frac{4}{35}$ **91.** 4
92. $-\frac{3}{4}$ **93.** $3 + (-5); -2$ **94.** $2 - (-4); 6$ **95.** $0.\overline{7}$
96. 2.2   **97.** $\frac{3}{5}$ **98.** $\frac{3}{8}$ **99.** Commutative (multiplication)
100. Associative (addition)   **101.** Distributive
102. Commutative (addition)   **103.** Identity (multiplication)
104. Associative (multiplication)   **105.** Distributive
106. Identity (addition)   **107.** Inverse (addition)
108. Inverse (multiplication)   **109.** 40   **110.** 301
111. 2475   **112.** 6580   **113.** 549.8   **114.** 43.56
115. Yes; 55   **116.** Yes; -1 **117.** No   **118.** No
119. $-6x$ **120.** $15z$ **121.** $4x^2$ **122.** $3x + 1$
123. $\frac{1}{2}z + 2$ **124.** $x - 18$ **125.** $9x^2 - 6$ **126.** 0
127. 5   **128.** c **129.** $3y + 2$ **130.** $2x - 5$
131. (a) $7x$ **(b)** 420 square feet   **(c)** 24 minutes
132. 16 square feet
133.

Gallons (G)	1	2	3	4	5	6
Pints (P)	8	16	24	32	40	48

$P = 8G$

134. $C = 0.25x$ **135.** $\frac{3}{20}$ **136.** \$200,000   **137.** $1\frac{3}{20}$ feet
138. $15\frac{3}{8}$ miles   **139.** \$1101   **140.** $124°$F
141. About 129 million

CHAPTER 1 TEST (pp. 87–88)

1. (a) Prime   **(b)** Composite; $56 = 2 \times 2 \times 2 \times 7$
2. $\frac{15}{7}$ **3.** $4^2 - 3$; 13   **4.** $\frac{3}{4}$ **5. (a)** $\frac{3}{4}$ **(b)** $\frac{16}{45}$ **(c)** $\frac{2}{7}$
(d) $\frac{15}{4}$ **(e)** $\frac{31}{36}$ **(f)** $\frac{2}{13}$ **6.** y^4 **7. (a)** 8   **(b)** 71
(c) -40 **(d)** 9   **8. (a)** Integer and rational   **(b)** Irrational
9.

10. (a) $<$ **(b)** $>$ **11. (a)** -6 **(b)** 21
12. (a) Distributive   **(b)** Associative (multiplication)
(c) Commutative (addition)   **13.** 1734   **14. (a)** $12 - 4z$
(b) $15x - 6$ **(c)** $x - 19$ **15. (a)** $\frac{19}{12}x$ **(b)** $12\frac{2}{3}$ acres
16. $2\frac{3}{5}$ feet   **17. (a)** $C = 13x$ **(b)** \$221   **18.** \$640

CHAPTER 2: LINEAR EQUATIONS AND INEQUALITIES

SECTION 2.1 (pp. 97–99)

1. true   **2.** false   **3.** solution   **4.** solution set
5. solutions   **6.** Equivalent   **7.** $b + c$ **8.** subtraction
9. bc **10.** division   **11.** equivalent   **12.** given
13. 22   **15.** 3   **17.** -5 **19.** 8   **21.** -34 **23.** 9
25. 17   **27.** 2   **29.** 5   **31.** -15 **33.** 1963   **35.** a
37. $\frac{1}{5}$ **39.** 6   **41.** 3   **43.** 0   **45.** 7   **47.** -6

49. 3   **51.** $\frac{5}{4}$ **53.** 5   **55.** 7   **57.** $\frac{3}{2}$ **59.** a

61. (a)

Hours (x)	0	1	2	3	4	5	6
Rainfall (R)	3	3.5	4	4.5	5	5.5	6

(b) $R = 0.5x + 3$ **(c)** 4.5 inches; yes   **(d)** 4.125 inches
63. (a) $L = 300x$ **(b)** $870 = 300x$ **(c)** 2.9
65. 17.5 years   **67.** $25,000

SECTION 2.2 (pp. 108–110)

1. $ax + b = 0$ **2.** 3; 2   **3.** Exactly one   **4.** is not
5. Addition, multiplication   **6.** Distributive   **7.** LCD
8. 100   **9.** None   **10.** Infinitely many   **11.** Infinitely
many   **12.** None   **13.** Yes; $a = 3, b = -7$ **15.** Yes;
$a = \frac{1}{2}, b = 0$ **17.** No   **19.** No   **21.** Yes;
$a = 1.1, b = -0.9$ **23.** Yes; $a = 2, b = -6$ **25.** No
27.

x	0	1	2	3	4
$-3x + 7$	7	4	1	-2	-5

2

29.

x	-2	-1	0	1	2
$4 - 2x$	8	6	4	2	0

-1

31. $\frac{3}{11}$ **33.** 23   **35.** 7   **37.** $-\frac{11}{5}$ **39.** $-\frac{7}{2}$ **41.** 3
43. $\frac{9}{4}$ **45.** $-\frac{5}{2}$ **47.** $\frac{4}{7}$ **49.** $\frac{9}{8}$ **51.** 1   **53.** -0.055
55. 8   **57.** $\frac{1}{7}$ **59.** 1   **61.** $-\frac{b}{a}$ **63.** No solutions
65. One solution   **67.** Infinitely many solutions
69. No solutions   **71.** Infinitely many solutions
73. (a)

Hours (x)	0	1	2	3	4
Distance (D)	4	12	20	28	36

(b) $D = 8x + 4$ **(c)** 28 miles; yes   **(d)** 2.25 hours; the
bicyclist is 22 miles from home after 2 hours and 15 minutes.
75. 2001   **77.** 2004

CHECKING BASIC CONCEPTS 2.1 & 2.2 (p. 110)

1. (a) No   **(b)** Yes
2.

x	3	3.5	4	4.5	5
$4x - 3$	9	11	13	15	17

4

3. (a) 18   **(b)** $\frac{1}{6}$ **(c)** $2.\overline{6}$ or $\frac{8}{3}$ **(d)** 3
4. (a) One solution   **(b)** Infinitely many solutions
(c) No solutions   **5. (a)** $D = 300 - 75x$
(b) $0 = 300 - 75x$ **(c)** 4 hours

SECTION 2.3 (pp. 120–123)

1. Check your solution.   **2.** $n + 1$ and $n + 2$
3. $\frac{x}{100}$ **4.** 0.01   **5.** left   **6.** right   **7.** $\frac{P_2 - P_1}{P_1} \times 100$
8. increase; decrease   **9.** rt **10.** distance; time
11. $2 + x = 12$; 10   **13.** $\frac{x}{5} = x - 24$; 30
15. $\frac{x + 5}{2} = 7$; 9   **17.** $\frac{x}{2} = 17$; 34   **19.** 31, 32, 33
21. 34   **23.** 8   **25.** 21   **27.** 11   **29.** 7   **31.** 13
33. 96 billion kilowatt-hours   **35.** $\frac{37}{100}$; 0.37

37. $\frac{37}{25}$; 1.48   **39.** $\frac{69}{1000}$; 0.069   **41.** $\frac{1}{2000}$; 0.0005   **43.** 45%
45. 180%   **47.** 0.6%   **49.** 40%   **51.** 75%   **53.** $83.\overline{3}$%
55. About 41.4%   **57.** $988 per credit
59. About 130.4 million   **61.** About 52,223
63. 25%; -20%   **65.** $d = 8$ miles
67. $r = 20$ feet per second   **69.** $t = 5$ hours   **71.** 60 mph
73. $\frac{3}{8}$ hour   **75.** 0.6 hr at 5 mph; 0.5 hr at 8 mph
77. 3.75 hours   **79.** 30 ounces
81. $1500 at 6%; $2500 at 5%   **83.** 6 gallons   **85.** 10 g

SECTION 2.4 (pp. 133–135)

1. formula   **2.** LW **3.** $\frac{D}{G}$ **4.** $\frac{1}{2}bh$ **5.** $\frac{1}{360}$ **6.** 360
7. 180   **8.** LWH **9.** $2LW + 2WH + 2LH$ **10.** $2\pi r$
11. πr^2 **12.** $\pi r^2 h$ **13.** $\frac{x}{5}$ **14.** $\frac{1}{2}(a + b)h$ **15.** 18 ft^2
17. 9 in^2 **19.** $16\pi \approx 50.3$ ft^2 **21.** 11 ft^2 **23.** 91 in^2
25. 36 in^2 **27.** $8\pi \approx 25.1$ in.   **29.** 4602 ft^2 **31.** 65°
33. 81°   **35.** 30°   **37.** 36°, 36°, 108°
39. $C = 12\pi \approx 37.7$ in.; $A = 36\pi \approx 113.1$ in^2
41. $r = 1$ in.; $A = \pi \approx 3.14$ in^2
43. $V = 2640$ in^3; $S = 1208$ in^2
45. $V = 2$ ft^3; $S = \frac{32}{3}$ ft^2 **47.** 20π in^3 **49.** 600π in^3
51. $\frac{147}{64}\pi \approx 7.2$ ft^3 **53.** $W = \frac{A}{L}$ **55.** $h = \frac{V}{\pi r^2}$
57. $a = \frac{2A}{h} - b$ **59.** $W = \frac{V}{LH}$ **61.** $b = 2s - a - c$
63. $b = a - c$ **65.** $a = \frac{cd}{b - d}$ **67.** 15 in.   **69.** 2.98
71. 2.24   **73.** 77°F   **75.** -40°F   **77.** -5°C
79. -20°C   **81.** 11 mpg   **83.** 2.4 mi

CHECKING BASIC CONCEPTS 2.3 & 2.4 (pp. 135–136)

1. (a) $3x = 36$; 12   **(b)** $35 - x = 43$; -8
2. $-32, -31, -30$ **3.** 0.095   **4.** 125%
5. About 3536   **6.** 6.5 hr   **7.** $5000 at 6%; $3000 at 7%
8. 18 gal   **9.** 12 in.
10. $A = 9\pi \approx 28.3$ ft^2; $C = 6\pi \approx 18.8$ ft   **11.** 30°
12. $l = \frac{A - \pi r^2}{\pi r}$

SECTION 2.5 (pp. 146–150)

1. equals sign; $<, \leq, >, \geq$ **2.** greater than; less than
3. solution   **4.** equivalent   **5.** one   **6.** infinitely many
7. number line   **8.** is not   **9.** $>$ **10.** $<$ **11.** $>$
12. No; $-4x < 8$ is equivalent to $x > -2$
13.

15.

17.

19.

21. $x < 0$ **23.** $x \le 3$ **25.** $x \ge 10$ **27.** $[6, \infty)$
29. $(-2, \infty)$ **31.** $(-\infty, 7]$ **33.** Yes   **35.** Yes   **37.** Yes
39. Yes   **41.** No   **43.** $x > -2$ **45.** $x < 1$
47. $x \ge 3$

x	1	2	3	4	5
$-2x + 6$	4	2	0	-2	-4

49. $x < -1$

x	-3	-2	-1	0	1
$5 - x$	8	7	6	5	4
$x + 7$	4	5	6	7	8

51. $x > 3$;

53. $y \ge -2$;

55. $z > 8$;

57. $t \le -5$;

59. $x < 5$;

61. $t \le -2$;

63. $y > -\frac{3}{20}$;

65. $z \ge -\frac{14}{3}$;

67. $\{x \mid x > 1\}$ **69.** $\{x \mid x \ge -7\}$ **71.** $\{x \mid x < 6\}$
73. $x < 7$ **75.** $x \le -\frac{4}{3}$ **77.** $x > -\frac{39}{2}$ **79.** $x \le \frac{3}{2}$
81. $x > \frac{1}{2}$ **83.** $x \le \frac{5}{4}$ **85.** $x > -\frac{8}{3}$ **87.** $x \le -\frac{3}{10}$
89. $x \le -\frac{1}{3}$ **91.** $x < 5$ **93.** $x \le -\frac{3}{2}$ **95.** $x > -\frac{9}{8}$
97. $x \le \frac{4}{7}$ **99.** $x < \frac{21}{11}$ **101.** $x > 60$ **103.** $x \ge 21$
105. $x > 40{,}000$ **107.** $x \le 70$ **109.** Less than 10 feet
111. It is less than 20 inches.   **113.** 86 or more
115. 4.5 hours   **117.** 4 days   **119. (a)** $C = 1.5x + 2000$
(b) $R = 12x$ **(c)** $P = 10.5x - 2000$ **(d)** 191 or more
compact discs   **121. (a)** After 48 minutes   **(b)** After more
than 48 minutes   **123.** Altitudes more than 4.5 miles
125. 2000 and later

CHECKING BASIC CONCEPTS 2.5 (p. 150)

1.

2. $x < 1$ **3.** Yes
4.   ; $x \ge -1$

x	-2	-1	0	1	2
$5 - 2x$	9	7	5	3	1

5. (a) $x > 3$ **(b)** $x \ge -35$ **(c)** $x \le -1$
6. $x \le 12$ **7.** More than 13 inches

CHAPTER 2 REVIEW (pp. 154–157)

1. -6 **2.** 2   **3.** $\frac{9}{4}$ **4.** $-\frac{1}{2}$ **5.** 3   **6.** $-\frac{7}{3}$ **7.** $-\frac{5}{2}$
8. $-\frac{7}{2}$ **9.** Yes; $a = 5, b = -3$ **10.** Yes; $a = -4, b = 1$
11. No   **12.** No   **13.** 2   **14.** 22   **15.** $\frac{27}{5}$ **16.** 7   **17.** -7
18. $-\frac{2}{3}$ **19.** $\frac{9}{4}$ **20.** 0   **21.** $\frac{7}{8}$ **22.** $\frac{11}{4}$ **23.** No solutions
24. Infinitely many solutions   **25.** Infinitely many solutions
26. One solution
27.   1.5

x	0.5	1.0	1.5	2.0	2.5
$-2x + 3$	2	1	0	-1	-2

28.   0

x	-2	-1	0	1	2
$-(x + 1) + 3$	4	3	2	1	0

29. $6x = 72$; 12   **30.** $x + 18 = -23$; -41
31. $2x - 5 = x + 4$; 9   **32.** $x + 4 = 3x$; 2
33. 16, 17, 18, 19   **34.** $-52, -51, -50$ **35.** $\frac{17}{20}$; 0.85
36. $\frac{7}{125}$; 0.056   **37.** $\frac{3}{10{,}000}$; 0.0003   **38.** $\frac{171}{50}$; 3.42
39. 89%   **40.** 0.5%   **41.** 230%   **42.** 100%
43. $d = 24$ mi   **44.** $d = 3850$ ft
45. $r = 25$ yards per second   **46.** $t = \frac{25}{3}$ hr
47. 7.5 m^2 **48.** $36\pi \approx 113.1$ ft^2 **49.** 864 in^2, or 6 ft^2
50. 45.5 in^2 **51.** $18\pi \approx 56.5$ ft   **52.** $25\pi \approx 78.5$ in^2
53. 50°   **54.** 22.5°   **55.** $625\pi \approx 1963.5$ in^3
56. 1620 in^2, or 11.25 ft^2 **57.** 174 in^2
58. About 60.6 ft^2 **59.** $x = a - y$ **60.** $x = \frac{P - 2y}{2}$
61. $y = \frac{z}{2x}$ **62.** $b = 3S - a - c$ **63.** $b = \frac{12T - 4a}{3}$
64. $c = \frac{ab}{d - b}$ **65.** 2.82   **66.** 3.12   **67.** 59°F   **68.** 45°F
69.

70.

71.

72.

73. $x < 3$ **74.** $x \ge -1$ **75.** Yes   **76.** Yes   **77.** Yes
78. No
79. $x < 2$

x	0	1	2	3	4
$5 - x$	5	4	3	2	1

80. $x \le 2.5$

x	1	1.5	2	2.5	3
$2x - 5$	-3	-2	-1	0	1

81. $x > 3$ **82.** $x \ge -5$ **83.** $x \le -1$ **84.** $x < \frac{23}{3}$
85. $x \le \frac{1}{9}$ **86.** $x \ge \frac{9}{4}$ **87.** $x < 50$ **88.** $x \le 45{,}000$
89. $x \ge 16$ **90.** $x < 1995$
91. (a)

Time	12:00	1:00	2:00	3:00	4:00	5:00
Rainfall (R)	2	2.75	3.5	4.25	5	5.75

(b) $R = \frac{3}{4}x + 2$ **(c)** $\frac{23}{4} = 5\frac{3}{4}$ in.; yes   **(d)** $\frac{77}{16} = 4\frac{13}{16}$ in.
92. $2125

93. (a)

Hours (x)	1	2	3	4	5
Distance (D)	40	30	20	10	0

(b) $D = 50 - 10x$ **(c)** 20 miles; yes   **(d)** 3 hours or less or from noon to 3:00 p.m.   **94.** 1998   **95.** 4
96. About 82%   **97.** $\frac{1}{6}$ hr, or 10 min   **98.** 28 gal
99. 50 mL   **100.** $500 at 3%; $800 at 5%   **101.** 33 in. by 23 in.   **102.** 1.8 mi   **103.** 25 inches or less   **104.** 74 or more
105. 6 hr   **106. (a)** $C = 85x + 150{,}000$ **(b)** $R = 225x$
(c) $P = 140x - 150{,}000$ **(d)** 1071 or fewer

CHAPTER 2 TEST (pp. 157–158)

1. -6 **2.** $\frac{5}{2}$ **3.** $-\frac{16}{11}$ **4.** $\frac{25}{6}$ **5.** No solutions
6.

x	0	1	2	3	4
6 − 2x	6	4	2	0	−2

3

7. $x + (-7) = 6; 13$ **8.** $2x + 6 = x - 7; -13$
9. 111, 112, 113   **10.** $\frac{4}{125}$; 0.032   **11.** 34.5%
12. $37.50   **13.** 1056 ft/sec   **14.** 7.5 in^2
15. $C = 30\pi \approx 94.2$ in.; $A = 225\pi \approx 706.9$ in^2
16. 30°   **17.** $x = \frac{y - z}{3y}$ **18.** $x = \frac{20R - 4y}{5}$
19. $x \le 6$ **20.** $x > -\frac{1}{7}$ **21. (a)** $S = 2x + 5$ **(b)** 21 in.
(c) 17.5 in.   **22.** 2000 mL   **23.** 300%

CHAPTERS 1 AND 2 CUMULATIVE REVIEW (p. 159)

1. Composite; $3 \times 3 \times 5$ **2.** Prime   **3.** $\frac{1}{2}$ **4.** $\frac{1}{9}$
5. $\frac{13}{24}$ **6.** $\frac{13}{15}$ **7.** Integer and rational   **8.** Irrational
9. 3   **10.** 10   **11.** 15   **12.** 9   **13.** -7 **14.** $\frac{3}{7}$
15. $4x^3$ **16.** $5x + 3$ **17.** 14   **18.** -4 **19.** $\frac{1}{2}$ **20.** 1
21. Infinitely many   **22.** No solutions   **23.** 29, 30, 31
24. 0.047   **25.** 17%   **26.** 65 mph   **27.** 20
28. $25\pi \approx 78.5$ in^2 **29.** $x = \frac{a + 4}{3y}$ **30.** $x = 3A - y - z$
31. $x < 1$

32. $x \le 2$

33. $I = 36Y$ **34.** $618   **35.** 300 mL
36. $1500 at 4%; $1750 at 6%

CHAPTER 3:   GRAPHING EQUATIONS

SECTION 3.1 (pp. 168–171)

1. xy-plane   **2.** origin   **3.** $(0, 0)$ **4.** 4   **5.** II   **6.** III
7. axes   **8.** $x; y$ **9.** scatterplot   **10.** line
11. $(-2, -2), (-2, 2), (0, 0), (2, 2)$
13. $(-1, 0), (0, -3), (0, 2), (2, 0)$
15. (a) I   **(b)** III   **17. (a)** None   **(b)** I
19. (a) II   **(b)** IV   **21.** I and III

23. **25.**

27. **29.**

31. **33.**

35. **37.**

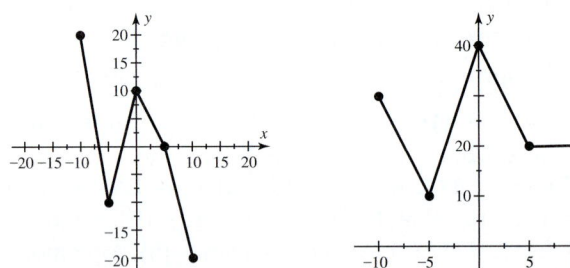

39. (1940, 182), (1960, 484), (1980, 632), (2000, 430); in 1940 there were 182 billion cigarettes consumed in the U.S. (answers may vary slightly).

41. (a)

(b) Federal income tax receipts increased.

43. (a)

45. (a)

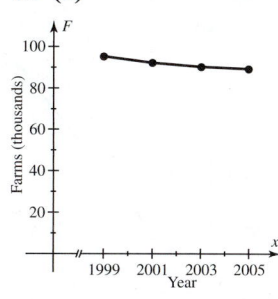

39.

x	−2	0	2
y	3	2	1

Table values may vary.

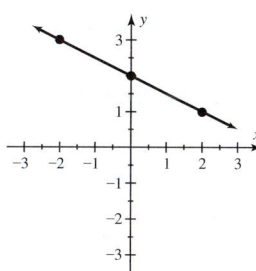

(b) The number of welfare beneficiaries increased and then decreased.

(b) The number of farms decreased.

47. (a) The rate decreased.   **(b)** 9   **(c)** About −73.1%

SECTION 3.2 (pp. 178–180)

1. two   **2.** ordered pair   **3.** solution   **4.** linear   **5.** is
6. solutions   **7.** graph   **8.** line   **9.** Yes   **11.** No
13. No   **15.** Yes   **17.** No

19.

x	−2	−1	0	1	2
y	−8	−4	0	4	8

21.

x	−8	−4	0	4	8
y	−4	0	4	8	12

23.

x	6	3	0	−3	−9
y	−2	0	2	4	8

25.

x	−3	0	3	6
y	−9	0	9	18

27.

x	−8	−4	0	4
y	−2	0	2	4

29.

x	8	6	4	2
y	−2	0	2	4

31.

x	$-\frac{1}{2}$	$-\frac{1}{4}$	0	$\frac{1}{4}$
y	−2	−1	0	1

33. They must be multiples of 5.

35.

x	−1	0	1
y	2	0	−2

Table values may vary.

37.

x	0	1	2
y	3	2	1

Table values may vary.

41.

43.

45.

47.

49.

51.

53.

55.

57.

59.

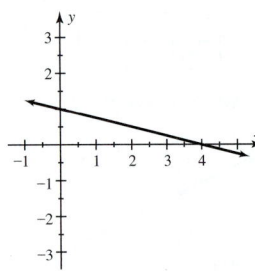

73. (a) 2.5; 91

(b)

(c) 1998

61.

63.

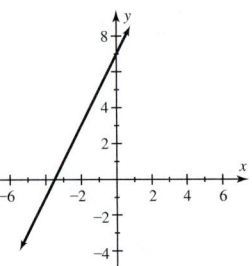

CHECKING BASIC CONCEPTS 3.1 & 3.2 (pp. 180–181)

1. $(-2, 2)$, II; $(-1, -2)$, III; $(1, 3)$, I; $(3, 0)$, none

2.   **3.**

The percentage increased.

65.

67.

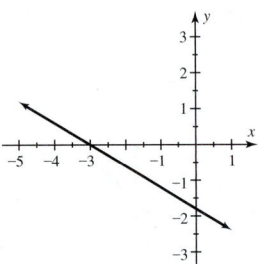

4. Yes

5.

x	-2	-1	0	1	2
y	5	3	1	-1	-3

6. (a)   **(b)**

69. (a)

t	2010	2020	2030	2040	2050
P	13.2	15.0	16.7	18.5	20.3

(b) 2030

71. (a)

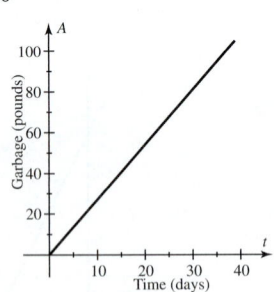

(b) About 37 days

7. (a) In 1995, receipts were $1.425 trillion; in 2001, receipts were $2.115 trillion.

(b)

(c) 1996

SECTION 3.3 (pp. 189–192)

1. Two   **2.** one; one   **3.** x-intercept   **4.** 0
5. y-intercept   **6.** 0   **7.** 3   **8.** horizontal   **9.** $y = b$
10. 3   **11.** vertical   **12.** $x = k$ **13.** 3; -2 **15.** 0; 0
17. -2 and 2; 4   **19.** 1; 1

21.

x	-2	-1	0	1	2
y	0	1	2	3	4

-2; 2

23.

x	-4	-2	0	2	4
y	-6	-4	-2	0	2

2; -2

25. x-int: 3; y-int: -2

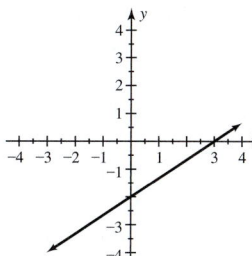

27. x-int: 5; y-int: -3

29. x-int: 6; y-int: -2

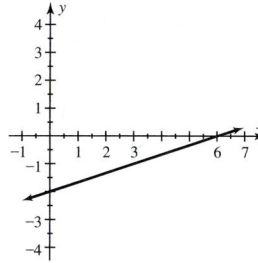

31. x-int: -1; y-int: 6

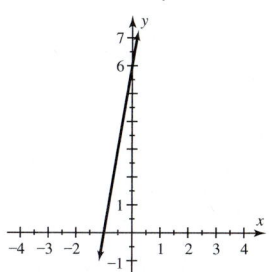

33. x-int: 7; y-int: 3

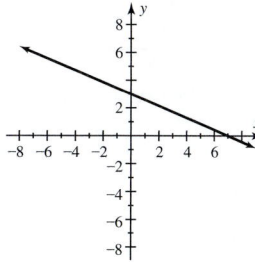

35. x-int: 4; y-int: -3

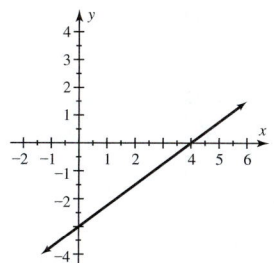

37. x-int: 4; y-int: -2

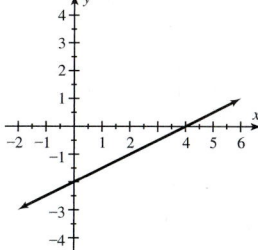

39. x-int: -4; y-int: 3

41. x-int: 3; y-int: 2

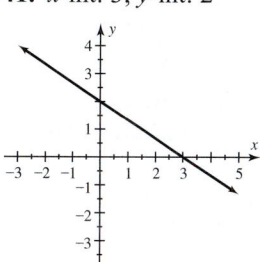

43. x-int: -2; y-int: 5

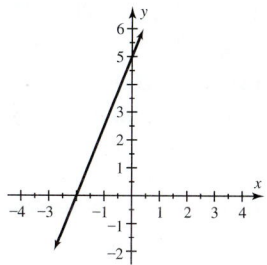

45. x-int: $\frac{C}{A}$; y-int: $\frac{C}{B}$

47. (a)   (b)

49. (a)   (b)

51. (a)   (b)

53. (a)   (b)

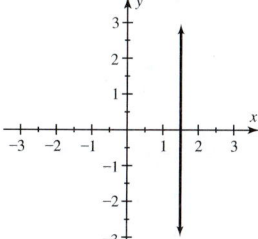

55. $y = 4$ **57.** $x = -1$ **59.** $y = -6$ **61.** $x = 5$

63. $y = 1$ **65.** $x = -6$ **67.** $y = 2; x = 1$
69. $y = -45; x = 20$ **71.** $y = 5; x = 0$ **73.** $x = -1$
75. $y = -\frac{5}{6}$ **77.** $x = 4$ **79.** $x = -\frac{2}{3}$ **81.** $y = 0$
83. (a) y-int: 200; x-int: 4   **(b)** The driver was initially
200 miles from home; the driver arrived home after 4 hours.
85. (a) v-int: 128; t-int: 4

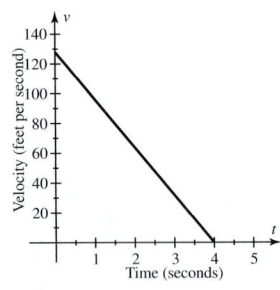

(b) The initial velocity was 128 ft/sec; the velocity after
4 seconds was 0.
87. (a) y-int: 2000; x-int: 4   **(b)** The pool initially contained
2000 gallons; the pool was empty after 4 hours.

SECTION 3.4 (pp. 202–207)

1. horizontal   **2.** vertical   **3.** rise; run   **4.** horizontal
5. vertical   **6.** $\frac{y_2 - y_1}{x_2 - x_1}$ **7.** positive   **8.** negative
9. Positive   **11.** Zero   **13.** Negative
15. 0; the rise always equals 0.
17. 1; the graph rises 1 unit for each unit of run.
19. 2; the graph rises 2 units for each unit of run.
21. Undefined; the run always equals 0.
23. $\frac{1}{2}$; the graph rises 1 unit for each 2 units of run.
25. -1; the graph falls 1 unit for each unit of run.
27. 2;   **29.** Undefined;

31. $-\frac{2}{3}$;   **33.** 0;

    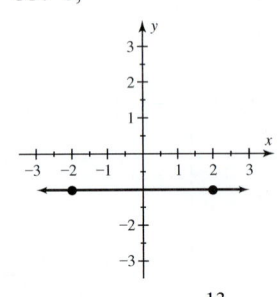

35. $\frac{5}{7}$ **37.** $-\frac{6}{7}$ **39.** 0   **41.** Undefined   **43.** $\frac{13}{20}$
45. $\frac{9}{100}$ **47.** $-\frac{5}{7}$ **49.** 49

51.

53.

55.

57.

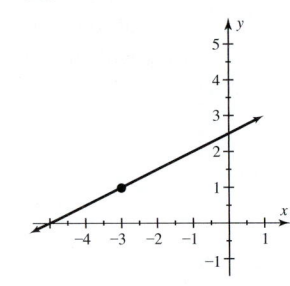

59. 2; 1; -2 **61.** -9; -1; -9
63.

x	0	1	2	3
y	-4	-2	0	2

65.

x	1	2	3	4
y	4	1	-2	-5

67.

x	-4	-2	0	2
y	0	3	6	9

69. (a)   **71. (a)**

    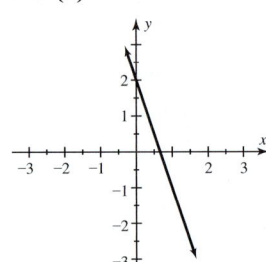

(b) 2   **(b)** -3
73. (a)   **75. (a)**

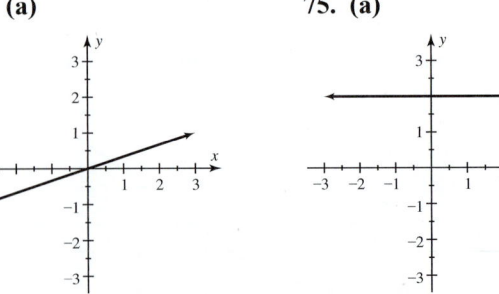

(b) $\frac{1}{3}$ **(b)** 0
77. c.   **79.** b.
81. (a) $m_1 = 1000$; $m_2 = -1000$ **(b)** $m_1 = 1000$: Water
is being added to the pool at a rate of 1000 gallons per hour.

$m_2 = -1000$: Water is being removed from the pool at a rate of 1000 gallons per hour.   **(c)** Initially the pool contained 2000 gallons of water. Over the first 3 hours, water was pumped into the pool at a rate of 1000 gallons per hour. For the next 2 hours, water was pumped out of the pool at a rate of 1000 gallons per hour.

83. (a) $m_1 = 50$; $m_2 = 0$; $m_3 = -50$ **(b)** $m_1 = 50$: The car is moving away from home at a rate of 50 mph. $m_2 = 0$: The car is not moving. $m_3 = -50$: The car is moving toward home at a rate of 50 mph.   **(c)** Initially the car is at home. Over the first 2 hours, the car travels away from home at a rate of 50 mph. Then the car is parked for 1 hour. Finally, the car travels toward home at a rate of 50 mph.

85. m^3/min

87.

89.

91. (a) 10   **(b)** The birth rate increased on average by 10,000 children per year.

93. (a) 2.5   **(b)** The revenue is $2.50 per screwdriver.

95. (a) 1; 2   **(b)** $\frac{1}{50}$ **(c)** Oil should be added at a rate of 1 gallon of oil per 50 gallons of gasoline.

97. (a)

(b) $m_1 = 20$; $m_2 = 30$; $m_3 = 40$ **(c)** $m_1 = 20$: Each mile between 0 and 5 miles is worth $20 per mile. $m_2 = 30$: Each mile between 5 and 10 miles is worth $30 per mile. $m_3 = 40$: Each mile between 10 and 15 miles is worth $40 per mile.

99. (a) 1200   **(b)** Median family income increased on average by $1200 per year over this time period.   **(c)** $48,000

101. (a) Toward   **(b)** 10 ft   **(c)** 2 ft/min   **(d)** 5

CHECKING BASIC CONCEPTS 3.3 & 3.4 (pp. 207–208)

1. -2; 3

2.

x	-2	-1	0	1	2
y	-6	-4	-2	0	2

1; -2

3. (a) x-int: 6; y-int: -3 **(b)** y-int: 2

(c) x-int: -1

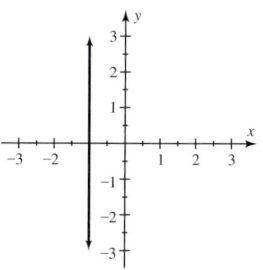

4. $y = 4$; $x = -2$ **5. (a)** $\frac{3}{4}$ **(b)** 0   **(c)** Undefined

6. $-\frac{1}{2}$

7.

8. (a) $m_1 = 0$; $m_2 = 2$; $m_3 = -\frac{2}{3}$ **(b)** $m_1 = 0$: The depth is not changing. $m_2 = 2$: The depth increased at a rate of 2 feet per hour. $m_3 = -\frac{2}{3}$: The depth decreased at a rate of $\frac{2}{3}$ foot per hour.   **(c)** Initially the pond had a depth of 5 feet. For the first hour, there was no change in the depth of the pond. For the next hour, the depth of the pond increased at a rate of 2 feet per hour to a depth of 7 feet. Finally, the depth of the pond decreased for 3 hours at a rate of $\frac{2}{3}$ foot per hour until it was 5 feet deep.

9. (a) 28   **(b)** For counties between 450 and 600 square miles, the population increases at an average rate of 28 people per square mile.   **(c)** No. We do not know if this trend continues.

SECTION 3.5 (pp. 214–217)

1. $y = mx + b$ **2.** slope   **3.** y-intercept   **4.** horizontal

5. origin   **6.** $y = 2x + 5$ **7.** parallel   **8.** slope

9. perpendicular   **10.** reciprocals   **11.** f.

13. a.   **15.** e.   **17.** $y = x - 1$ **19.** $y = -2x + 1$

21. $y = \frac{1}{2}x - 2$ **23.** $y = -2x$ **25.** $y = \frac{3}{4}x + 2$
27. $y = x + 2$ **29.** $y = 2x - 1$

    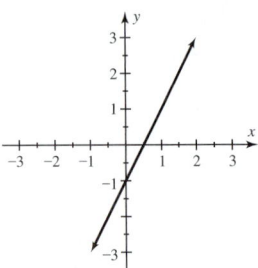

31. $y = -\frac{1}{2}x - 2$ **33.** $y = \frac{1}{3}x$

    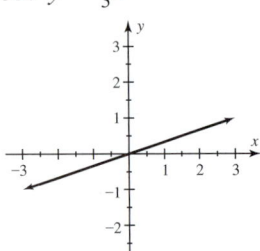

35. $y = -2x + 1$

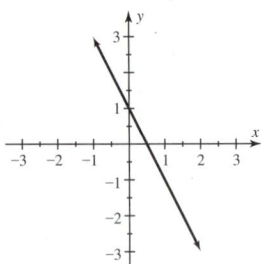

37. (a) $y = -x + 4$ **(b)** $-1; 4$
39. (a) $y = -2x + 4$ **(b)** $-2; 4$
41. (a) $y = \frac{1}{2}x + 2$ **(b)** $\frac{1}{2}; 2$
43. (a) $y = \frac{2}{3}x - 2$ **(b)** $\frac{2}{3}; -2$
45. (a) $y = \frac{1}{4}x + \frac{3}{2}$ **(b)** $\frac{1}{4}; \frac{3}{2}$
47. (a) $y = -\frac{1}{3}x + \frac{2}{3}$ **(b)** $-\frac{1}{3}; \frac{2}{3}$
49.   **51.**

53.   **55.**

57.   **59.**

    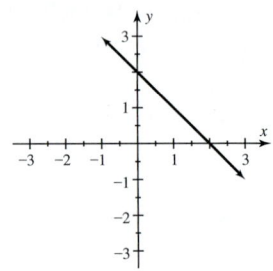

61. $y = 2x + 2$ **63.** $y = x - 2$ **65.** $y = \frac{4}{7}x + 3$
67. $y = 3x$ **69.** $y = -\frac{1}{2}x + \frac{5}{2}$ **71.** $y = 2x$
73. $y = \frac{1}{3}x + \frac{1}{3}$ **75. (a)** \$25   **(b)** 25 cents   **(c)** 25;
the fixed cost of renting the car   **(d)** 0.25; the cost per mile
of driving the car   **77. (a)** \$7.45   **(b)** $C = 0.07x + 3.95$
(c) 67 min   **79. (a)** $m = 0.35; b = 164.3$
(b) The fixed cost of owning the car for one month

SECTION 3.6 (pp. 224–227)

1. One   **2.** One   **3.** $y = mx + b$
4. $y - y_1 = m(x - x_1)$ or $y = m(x - x_1) + y_1$
5. $1; 3$ **6.** distributive   **7.** Yes; every nonvertical line
has exactly one slope and one y-intercept.
8. No; it depends on the point used.
9. Yes   **11.** No   **13.** Yes   **15.** $y - 2 = \frac{3}{4}(x - 1)$
17. $y + 1 = -\frac{1}{2}(x - 3)$ **19.** $y - 1 = 4(x + 3)$
21. $y + 3 = \frac{1}{2}(x + 5)$ **23.** $y - 30 = 1.5(x - 2000)$
25. $y - 4 = \frac{7}{3}(x - 2)$ **27.** $y = \frac{3}{5}(x - 5)$
29. $y - 15 = 5(x - 1990)$ **31.** $y = 3x - 2$
33. $y = \frac{1}{3}x$ **35.** $y = \frac{2}{3}x + \frac{1}{12}$ **37.** $y = -2x + 9$
39. $y = \frac{3}{5}x - 2$ **41.** $y = -16x - 19$ **43.** $-mx_1 + y_1$
45. $y = -2x + 5$ **47.** $y = -x + 1$ **49.** $y = -\frac{1}{9}x + \frac{1}{3}$
51. $y = 2x - 7$ **53.** $y = 2x - 15$ **55.** $y = -2x - 1$
57. $y = \frac{1}{2}x - \frac{5}{2}$ **59. (a)** Toward   **(b)** After 1 hour the person is 250 miles from home. After 4 hours the person is 100
miles from home.   **(c)** 50 mph   **(d)** $y = -50x + 300$; the
car is traveling toward home at 50 mph.

61. (a) Entering; 300 gallons   **(b)** 100; initially the tank contains 100 gallons.   **(c)** $y = 50x + 100$; the amount of water is increasing at a rate of 50 gallons per minute.
(d) 8 minutes
63. (a) $-5°$F per hour   **(b)** $T = -5x + 45$; the temperature is decreasing at a rate of $5°$F per hour.
(c) 9; at 9 A.M. the temperature was $0°$F.
(d)

(e) At 4 A.M. the temperature was $25°$F.
65. (a) 1,130,000; 1,450,000   **(b)** 0.032; the number of inmates increased on average by 32,000 per year.
67. (a) In 2002, tuition and fees were \$4081; in 2004, tuition and fees were \$5133.   **(b)** $y - 4081 = 526(x - 2002)$ or $y - 5133 = 526(x - 2004)$; tuition and fees increased on average by \$526/yr.   **(c)** About \$5700; \$5659
69. (a) $y = 4.9x - 9780.6$ **(b)** 68.4 million

CHECKING BASIC CONCEPTS 3.5 & 3.6 (pp. 227–228)

1. $y = -3x + 1$ **2.** $y = \frac{4}{5}x - 4; \frac{4}{5}; -4$
3.

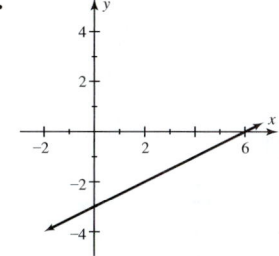

4. (a) $y = 3x - 2$ **(b)** $y = -\frac{3}{2}x$ **(c)** $y = -\frac{7}{3}x - \frac{5}{3}$
5. $y - 3 = -2(x + 1)$ **6.** $y = -2x + 1$
7. $y = 2x + 3$ **8. (a)** $y = -12x + 48$ **(b)** 12 mph
(c) 4 P.M.   **(d)** 48 miles   **9. (a)** 5 inches   **(b)** 2 inches per hour   **(c)** 5; total inches of snow that fell before noon
(d) 2; the rate of snowfall was 2 inches per hour.

SECTION 3.7 (pp. 233–235)

1. abstraction   **2.** predict   **3.** linear   **4.** exact
5. approximate   **6.** constant   **7.** constant rate of change
8. initial amount   **9.** f.   **11.** a.   **13.** e.   **15.** Yes
17. No   **19.** No   **21.** Exact; $y = 2x - 2$
23. Approximate; $y = 2x + 2$ **25.** Approximate; $y = 2$
27. $y = 5x + 40$ **29.** $y = -20x - 5$ **31.** $y = 8$

33. (a) Yes   **(b)**

 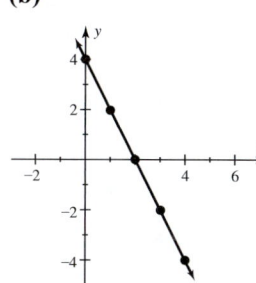

(c) $y = -2x + 4$
35. (a) No   **(b)**

 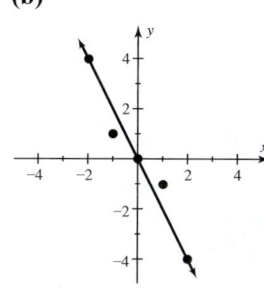

(c) $y = -2x$
37. (a) Yes   **(b)**

 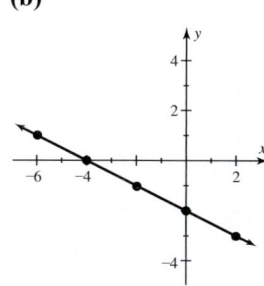

(c) $y = -\frac{1}{2}x - 2$
39. (a) No   **(b)**

 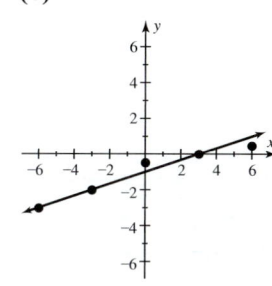

(c) $y = \frac{1}{3}x - 1$
41. $g = 5t + 200$, where g represents gallons of water and t represents time in minutes.
43. $d = 6t + 5$, where d represents distance in miles and t represents time in hours.
45. $p = 8t + 200$, where p represents total pay in dollars and t represents time in hours.
47. $r = t + 5$, where r represents total number of roofs shingled and t represents time in days.

49. (a) $-\frac{2}{45}$ acre per year   **(b)** $A = -\frac{2}{45}t + 5$

51. (a) $y = 8x - 15,864$ **(b)** About 208,000

53. (a)   **(b)**

(c) 20; the mileage is 20 miles per gallon.

(d) $y = 20x$ **(e)** 140 miles

CHECKING BASIC CONCEPTS 3.7 (p. 236)

1. Yes   **2.** Approximate; $y = x - 1$

3. (a) $y = 10x + 50$ **(b)** $y = -2x + 200$

4. (a) Yes   **(b)**

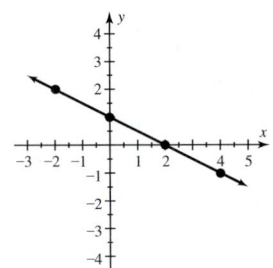

(c) $y = -\frac{1}{2}x + 1$

5. (a) $T = 0.075x$ **(b)** $4.5°F$

CHAPTER 3 REVIEW (pp. 241–246)

1. $(-2, 0)$: none; $(-1, 2)$: II; $(0, 0)$: none; $(1, -2)$: IV; $(1, 3)$: I

2.   **3.**

4.

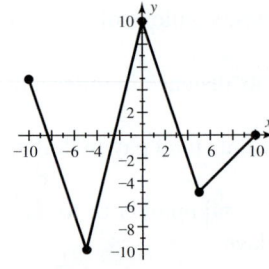

5. Yes   **6.** No   **7.** No   **8.** Yes

9.

x	-2	-1	0	1	2
y	6	3	0	-3	-6

10.

x	4	3	2.5	2	1
y	-3	-1	0	1	3

11.   **12.**

x	-2	0	2	4
y	-4	2	8	14

x	1	2	3	4
y	6	5	4	3

13.   **14.**

x	-0.5	0	0.5	1
y	-1	0	1	2

x	-1	-3	-5	-7
y	1	2	3	4

15.   **16.**

17.   **18.**

19.   **20.**

21.   **22.**

23. 3; −2   **24.** −2 and 2; −4

25.

x	−2	−1	0	1	2
y	4	3	2	1	0

2; 2

26.

x	−4	−2	0	2	4
y	−4	−3	−2	−1	0

4; −2

27. x-int: 3; y-int: −2   **28.** x-int: 1; y-int: −5

29. x-int: 4; y-int: −2   **30.** x-int: 2; y-int: 3

31. (a)   **(b)**

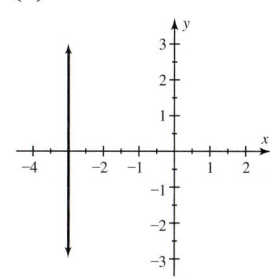

32. $x = -1$; $y = 1$ **33.** $y = 1$ **34.** $x = 3$
35. $y = 3$; $x = -2$ **36. (a)** y-int: 90; x-int: 3   **(b)** The driver is initially 90 miles from home; the driver arrives home after 3 hours.   **37.** 2   **38.** $-\frac{2}{5}$ **39.** 0   **40.** Undefined
41. 3   **42.** $-\frac{1}{2}$

43. (a).   **44. (a)**

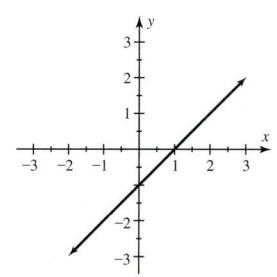

(b) −2; 0   **(b)** 1; −1

45. (a)   **46. (a)**

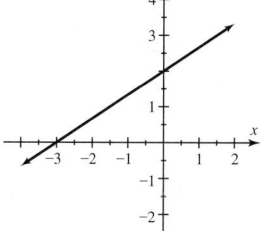

(b) $-\frac{1}{2}$; 2   **(b)** $\frac{2}{3}$; 2
47.   **48.**

49.   **50.**

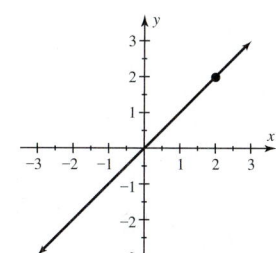

51. slope: 2; x-int: 1; y-int: −2

52.

x	0	1	2	3
y	1	$\frac{3}{2}$	2	$\frac{5}{2}$

53. $y = x + 1$ **54.** $y = -2x + 2$
55. $y = 2x - 2$ **56.** $y = -\frac{3}{4}x + 3$

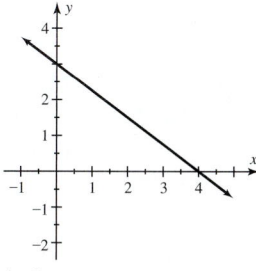

57. (a) $y = -x + 3$ **(b)** −1; 3
58. (a) $y = \frac{3}{2}x - 3$ **(b)** $\frac{3}{2}$; −3
59. (a) $y = 2x - 20$ **(b)** 2; −20
60. (a) $y = \frac{5}{6}x - 5$ **(b)** $\frac{5}{6}$; −5

61.

62.

63.

64.

65.

66.

67.

68.

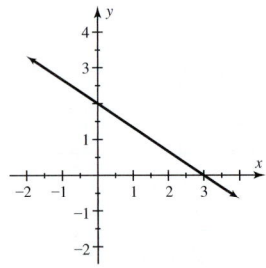

69. $y = 5x - 5$ **70.** $y = -2x$ **71.** $y = -\frac{5}{6}x + 2$

72. $y = -2x - 3$ **73.** $y = \frac{2}{3}x - 2$ **74.** $y = -\frac{1}{5}x - 2$

75. Yes   **76.** No   **77.** $y = 5x - 3$ **78.** $y = 20x - 65$

79. $y = -\frac{2}{3}x - \frac{1}{3}$ **80.** $y = 3x - 90$ **81.** $y = \frac{4}{3}x - 4$

82. $y = 2x - 1$ **83.** $y = 2x - 3$ **84.** $y = -\frac{2}{3}x - \frac{2}{3}$

85. $y = 3x + 5$ **86.** $y = \frac{1}{3}x + 7$

87. $y = 2x + 11$ **88.** $y = -\frac{1}{4}x + 3$ **89.** No

90. Approximate; $y = -x + 5$ **91.** $y = -2x + 40$

92. $y = 20x + 200$ **93.** $y = 50$ **94.** $y = 5x - 20$

95. (a) Yes

(b)

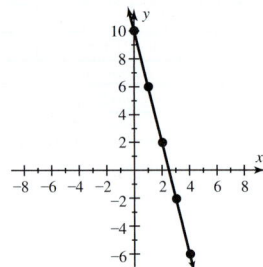

(c) $y = -4x + 10$

96. (a) No

(b)

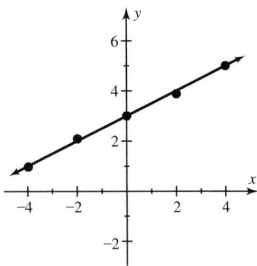

(c) $y = \frac{1}{2}x + 3$

97. (a)

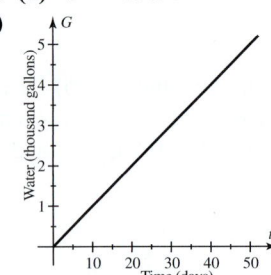

(b) The number of divorces increased significantly between 1960 and 1980, and then remained unchanged from 1980 to 2000.

98. (a) $G = 100t$

(b)

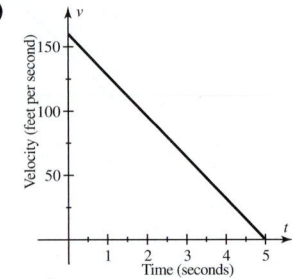

(c) 50 days

99. (a)

(b) v-int: 160; t-int: 5; the initial velocity was 160 ft/sec, and the velocity after 5 seconds was 0.

100. (a) $m_1 = 0$; $m_2 = -1500$; $m_3 = 500$ **(b)** $m_1 = 0$: The population remained unchanged. $m_2 = -1500$: The population decreased at a rate of 1500 insects per week. $m_3 = 500$: The population increased at a rate of 500 insects per week.   **(c)** For the first week the population did not change from its initial value of 4000. Over the next two weeks the population decreased at a rate of 1500 insects per week until it reached 1000. Finally, the population increased at a rate of 500 per week for two weeks, reaching 2000.

101.

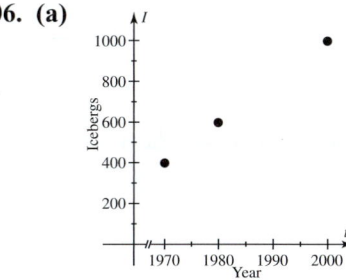

102. (a) -105 **(b)** The number of nursing homes decreased at an average rate of 105 per year.

103. (a) 0.6   **(b)** Enrollment increased at an average rate of 0.6 million students per year.   **(c)** 86 million students

104. (a) \$35   **(b)** 20¢   **(c)** 35; the fixed cost of renting the car   **(d)** 0.2; the cost for each mile driven

105. (a) Toward; the slope is negative.   **(b)** After 1 hour the car is 200 miles from home; after 3 hours the car is 100 miles from home.   **(c)** $y = -50x + 250$; the car is moving toward home at 50 mph.   **(d)** 150 miles; 150 miles

106. (a)

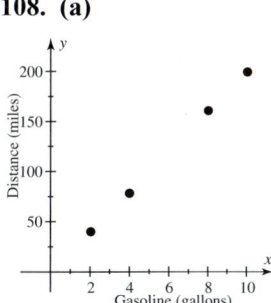

(b) $I = 20t - 39{,}000$; the number of icebergs increased at an average rate of 20 per year.   **(c)** Yes   **(d)** 1100 icebergs

107. (a) 90,100; 65,900   **(b)** -6.05; deaths from pneumonia decreased at an average rate of 6050 per year.

108. (a)   **(b)**

(c) 20; the mileage is 20 miles per gallon.
(d) $y = 20x$; no   **(e)** About 180 miles

CHAPTER 3 TEST (pp. 246–247)

1. $(-2, -2)$: III; $(-2, 1)$: II; $(0, -2)$: none; $(1, 0)$: none; $(2, 3)$: I; $(3, -1)$: IV; $(3, 1)$: I

2.

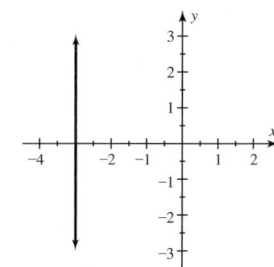

3.

x	-2	-1	0	1	2
y	-8	-6	-4	-2	0

$2; -4$ **4.** Yes

5.

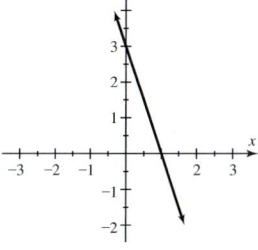

6. x-int: 3; y-int: -5

7.   **8.**

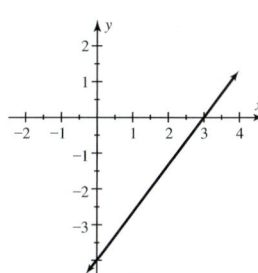

9.   **10.**

11. $y = -x + 2$ **12.** $y = 2x + \frac{1}{2}$; 2; $\frac{1}{2}$

13. $y = -5$; $x = 1$ **14.** $-\frac{2}{9}$ **15.** $y = -\frac{4}{3}x - 5$

16. $y = 3x - 11$ **17.** $y = -3x + 5$ **18.** $y = -\frac{1}{2}x$

19. $y = \frac{1}{2}x + 5$ **20.** Approximate; $y = x - 1$

21.

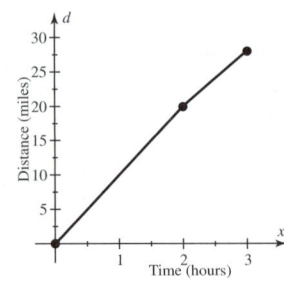

22. (a) $m_1 = 2; m_2 = -9; m_3 = 2; m_4 = 5$
(b) $m_1 = 2$: The population increased at a rate of 2000 fish per year. $m_2 = -9$: The population decreased at a rate of 9000 fish per year. $m_3 = 2$: The population increased at a rate of 2000 fish per year. $m_4 = 5$: The population increased at a rate of 5000 fish per year.   **(c)** For the first year the population increased from an initial value of 8000 to 10,000 at a rate of 2000 fish per year. During the second year the population dropped dramatically to 1000 at a rate of 9000 fish per year. Over the third year the population grew to 3000 at a rate of 2000 fish per year. Finally, over the fourth year, the population grew at a rate of 5000 fish per year to reach 8000.
23. $N = 100x + 2000$

CHAPTERS 1–3 CUMULATIVE REVIEW (pp. 248–250)

1. Composite; $40 = 2 \times 2 \times 2 \times 5$ **2.** Prime
3. $n + 10$ **4.** $n^2 - 2$ **5.** $\frac{7}{6}$ **6.** $\frac{2}{3}$ **7.** 2   **8.** $\frac{17}{30}$ **9.** 14
10. 7   **11.** $\frac{5}{3}$ **12.** -9 **13.** Rational   **14.** Irrational
15.

			(b)	(a)		(c)				
-5	-4	-3	-2	-1	0	1	2	3	4	5

16. 2   **17.** 18   **18.** -4 **19.** $7x + 1$ **20.** $x - 4$
21. -3 **22.** 21   **23.** 2   **24.** 0
25.

x	-2	-1	0	1	2
y	10	8	6	4	2

1

26. $2n + 2 = n - 5; -7$ **27.** $-26, -25, -24, -23$
28. 8 hours   **29.** 720 ft^2 **30.** 20 ft^2 **31.** $b = \frac{2A}{h}$
32. $L = \frac{P - 2W}{2}$ **33.** $x < 2$
34.

x	0	1	2	3	4
$2x - 3$	-3	-1	1	3	5

$x \le 2$

35. $\{x \mid x > 0\}$ **36.** $\{x \mid x \le \frac{1}{2}\}$
37. $(-2, -3)$: III; $(0, 3)$: none; $(2, -2)$: IV; $(2, 1)$: I
38.

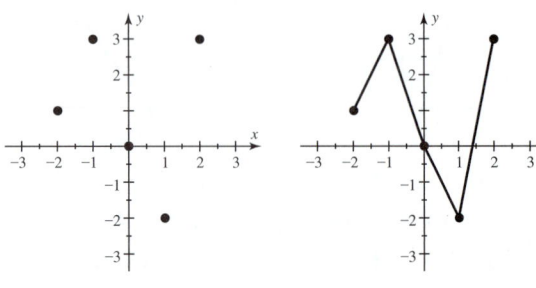

39. Yes
40.

x	-2	-1	0	1	2
y	3	2.5	2	1.5	1

41.

42.

43.

44.

45.

46.

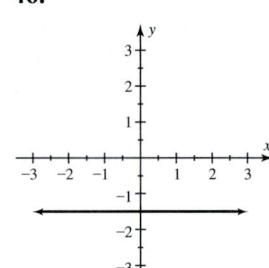

47. $\frac{3}{2}; -3; y = 2x - 3$ **48.** $2; 1; y = -\frac{1}{2}x + 1$
49. x-int: -10; y-int: 8   **50.** $x = \frac{3}{2}; y = -2$
51.

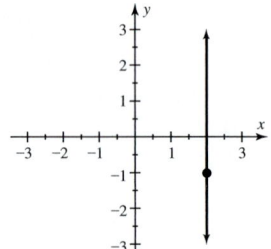

52.

53. $y = 3x - 3$

54. $y = \frac{3}{5}x - 3$;

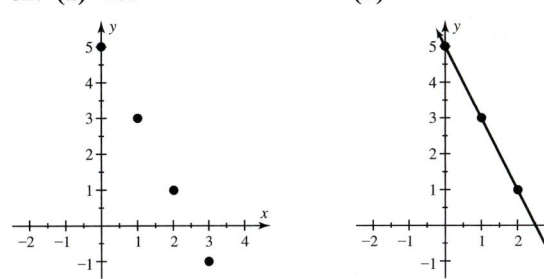

55. $y = -\frac{1}{3}x + 7$ **56.** $y = -\frac{2}{3}x - 3$ **57.** $y = -2x + 1$
58. $y = \frac{1}{4}x + \frac{1}{2}$ **59.** Approximate; $y = \frac{1}{2}x + 1$
60. $y = 5x + 100$ **61.** $I = 5000x + 20{,}000$
62. (a) Yes   **(b)**

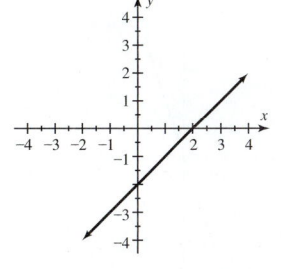

(c) $y = -2x + 5$
63. $C = 8x$ **64.** $\frac{99}{200}$ **65. (a)** \$807.40   **(b)** 1990
66. About 3.1%   **67.** \$1000 at 3%; \$2000 at 4%
68. (a) 0.16   **(b)** Participation increased at an average
rate of 0.16 million students per year.   **(c)** 28 million
69. (a) \$85   **(b)** \$25   **(c)** 30¢

CHAPTER 4: **SYSTEMS OF LINEAR EQUATIONS IN TWO VARIABLES**

SECTION 4.1 (pp. 259–261)

1. ordered   **2.** intersection-of-graphs   **3.** one   **4.** $(11, 9)$
5. No   **6.** table   **7.** the same   **8.** intercepts   **9.** 1   **11.** 2
13. -2 **15.** 2   **17.** $(1, 1)$ **19.** $(2, -3)$ **21.** $(2, 0)$
23. $(2, 1)$ **25.** $(3, 2)$ **27.** $(-1, 1)$ **29.** $(2, 4)$ **31.** $(3, 1)$
33. $(1, 3)$;

x	0	1	2	3
$y = x + 2$	2	3	4	5
$y = 4 - x$	4	3	2	1

35. (a) $(-1, 1)$ **(b)** $(-1, 1)$ **37. (a)** $(3, 1)$ **(b)** $(3, 1)$
39. (a) $(1, 3)$ **(b)** $(1, 3)$ **41.** $(-1, -2)$ **43.** $(1, -1)$
45. $(1, 2)$ **47.** $(3, 2)$ **49.** $(3, 1)$ **51.** $(1, 2)$
53. (a) $x + y = 4, x - y = 0$ **(b)** 2, 2
55. (a) $2x + y = 7, x - y = 2$ **(b)** 3, 1
57. (a) $x = 3y, x - y = 4$ **(b)** 6, 2
59. (a) $C = 0.5x + 50$ **(b)** 60 mi   **(c)** 60 mi
61. (a) $x + y = 36; x = 2y$ **(b)** $(24, 12)$
63. (a) $x - y = 4; 2x + 2y = 28$ where x is length, y is
width   **(b)** $(9, 5)$; the rectangle is 9 in. $\times$ 5 in.

SECTION 4.2 (pp. 268–271)

1. substitution   **2.** no solutions; one solution; infinitely many
3. exact   **4.** parentheses   **5.** It has infinitely many solutions.
6. It has no solutions.   **7.** consistent   **8.** inconsistent
9. independent   **10.** dependent   **11.** $(3, 6)$ **13.** $(2, 1)$
15. $(-1, 0)$ **17.** $(3, 0)$ **19.** $\left(\frac{1}{2}, 0\right)$ **21.** $(0, 4)$ **23.** $(1, 2)$
25. $(6, -4)$ **27.** $(4, -3)$ **29.** $\left(1, -\frac{1}{2}\right)$ **31.** $(1, 3)$
33. $(0, -2)$ **35.** $(2, 2)$ **37.** $(-3, 5)$ **39.** One;
consistent; independent   **41.** Infinitely many; consistent;
dependent   **43.** None; inconsistent   **45.** No solutions;
inconsistent   **47.** $(2, 1)$; consistent; independent
49. Infinitely many; consistent; dependent   **51.** No solutions;
inconsistent   **53.** No solutions   **55.** Infinitely many
57. $(3, 1)$ **59.** Infinitely many   **61.** No solutions
63. $\left(-\frac{4}{3}, \frac{11}{9}\right)$ **65.** $(0, 0)$ **67.** No solutions
69. No solutions   **71.** They are a single line.
73. (a) $L - W = 10, 2L + 2W = 72$ **(b)** $(23, 13)$
75. (a) $x = \frac{1}{2}y; x + y = 90$ **(b)** $(30, 60)$ **(c)** $(30, 60)$
77. (a) $x - y = 1.68; y = 0.98x$ **(b)** $(84, 82.32)$
79. 94 ft $\times$ 50 ft   **81.** 17, 53   **83.** 10 mph; 2 mph
85. $3.\overline{3}$ L of 20% solution, $6.\overline{6}$ L of 50% solution
87. Superior: 32,000 mi^2; Michigan: 22,000 mi^2

CHECKING BASIC CONCEPTS 4.1 & 4.2 (p. 272)

1. (a) -2 **(b)** 6   **2.** $(4, 2)$ **3.** $(2, 1)$ **4. (a)** No solutions
(b) $(1, -1)$; one   **(c)** Infinitely many   **5. (a)** $x + y = 300$,
$150x + 200y = 55{,}000$ **(b)** $(100, 200)$

SECTION 4.3 (pp. 280–283)

1. Substitution; elimination   **2.** addition   **3.** $=$ **4.** $=$
5. Add the equations.   **6.** Multiply the first equation by -3.
7. It has infinitely many solutions.   **8.** It has no solutions.
9. $(1, 1)$ **11.** $(-2, 1)$ **13.** No solutions   **15.** Infinitely
many   **17.** $(6, 1)$ **19.** $(-1, 4)$ **21.** $(2, 4)$ **23.** $(2, 1)$
25. $(-4, 1)$ **27.** $(5, 8)$ **29.** $(-2, -5)$ **31.** $(1, 0)$
33. $(2, 2)$ **35.** $(4, -1)$ **37.** $\left(\frac{4}{5}, \frac{9}{5}\right)$ **39.** $(3, 2)$ **41.** $(0, 1)$
43. $(2, 1)$ **45.** $(4, -3)$ **47.** $(1, 0)$
49. Infinitely many;   **51.** One;

53. No solutions;

55. One;

57. Infinitely many;

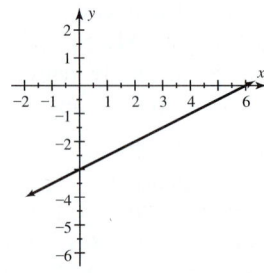

59. Men: 38,000; women: 28,000   **61.** Bicycle:18 min; stair climber: 12 min   **63.** Current: 2 mph; boat: 6 mph
65. $2000 at 3%; $3000 at 5%   **67.** $-43, 26$
69. (a) 20 in. $\times$ 40 in.   **(b)** 20 in. $\times$ 40 in.

SECTION 4.4 (pp. 292–295)

1. All points below and including the line $y = k$
2. All points to the right of the line $x = k$.
3. All points above and including the line $y = x$ **4.** test
5. dashed   **6.** solid   **7.** $Ax + By = C$ **8.** both inequalities   **9.** intersect   **10.** No   **11.** Yes   **13.** No
15. No   **17.** Yes   **19.** Yes   **21.** No   **23.** $x > 1$
25. $y \geq 2$ **27.** $y < x$ **29.** $-x + y \leq 1$
31.

33.

35.

37.

39.

41.

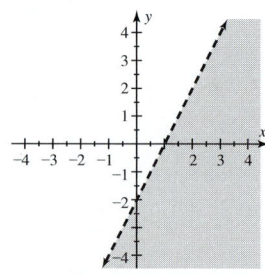

43. Yes   **45.** No   **47.** Yes   **49.** The region containing $(1, 2)$
51. The region containing $(1, 0)$
53.

55.

57.

59.

61.

63.

65.

67.

69.

71.

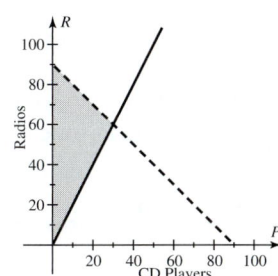

73. (a) 200 bpm; 150 bpm

(b)

(c) Possible heart rates for ages 20 to 70   **75.** 150 to 200 lb

CHECKING BASIC CONCEPTS 4.3 & 4.4 (p. 295)

1. $(1, 1)$ **2. (a)** $(0, -1)$; one   **(b)** No solutions
(c) Infinitely many   **3.** Infinitely many
4. (a)

(b)

5.

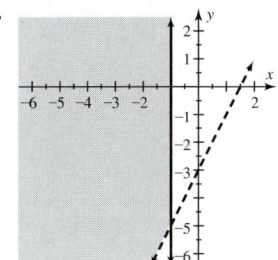

6. (a) $x + y = 11, x - y = 5$ **(b)** $(8, 3)$

CHAPTER 4 REVIEW (pp. 298–301)

1. 3   **2.** 2   **3.** $(1, 2)$ **4.** $(5, 2)$ **5.** $(4, 3)$ **6.** $(2, -4)$
7. $(2, 2)$ **8.** $(1, 2)$ **9.** $(2, 6)$ **10.** $(1, 1)$ **11.** $(4, -3)$
12. $(1, 2)$ **13.** $(1, 1)$ **14.** $(1, 2)$ **15.** $(1, 1)$
16. $(-2, -1)$ **17.** $(2, 6)$ **18.** $(2, -10)$ **19.** $(1, 3)$
20. $(5, 10)$ **21.** $(-2, 1)$ **22.** $(-4, -4)$ **23. (a)** None

(b) Inconsistent   **24. (a)** One   **(b)** Consistent; independent
25. (a) Infinitely many   **(b)** Consistent; dependent
26. (a) None   **(b)** Inconsistent   **27.** No solutions
28. No solutions   **29.** Infinitely many   **30.** $(1, 1)$
31. $(2, 1)$ **32.** $(-1, 2)$ **33.** $(11, -1)$ **34.** $(1, 0)$
35. $\left(\frac{9}{4}, \frac{7}{4}\right)$ **36.** $\left(\frac{5}{2}, 1\right)$ **37.** $(5, -7)$ **38.** $(8, 2)$
39. $(-2, 3)$ **40.** $(1, 0)$ **41.** $(1, 3)$ **42.** $(2, -1)$
43. Infinitely many   **44.** Infinitely many   **45.** No solutions
46. One   **47.** Yes   **48.** No   **49.** No   **50.** Yes
51. $y > 1$ **52.** $y \leq 2x + 1$

53.

54.

55.

56.

57.

58.

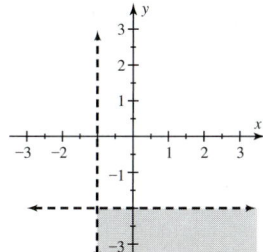

59. Yes   **60.** No   **61.** The region containing $(2, -2)$
62. The region containing $(1, 3)$
63.

64.

65.

66.

67.

68.

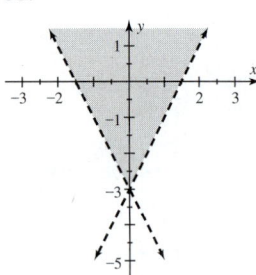

69. 3100 deaths in 1912; 42,625 deaths in 2003
70. Men: 102,500 cases; women: 82,500 cases
71. (a) $C = 0.2x + 40$ **(b)** 250 mi   **(c)** 250 mi
72. 75°, 105°   **73. (a)** $2x + y = 180, 2x - y = 40$
(b) (55, 70)   **(c)** (55, 70)   **74.** 20 ft × 24 ft
75. (a) $x + y = 10, 80x + 120y = 920$; x is $80 rooms,
y is $120 rooms.   **(b)** (7, 3)   **76.** 7 lb of $2 candy; 11 lb of
$3 candy   **77.** Bicycle: 35 min; stair climber: 25 min
78. 2 mph   **79. (a)** 16 ft × 24 ft (answers may vary slightly)
(b) 16 ft × 24 ft
80.

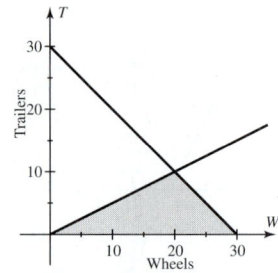

81. (a) 136 bpm; 108 bpm
(b)

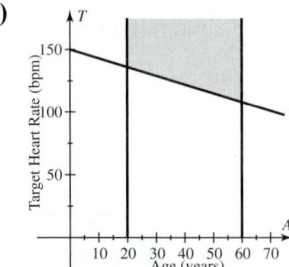

(c) Target heart rates above 70% of the maximum heart rate
for ages 20 to 60

CHAPTER 4 TEST (pp. 302–303)

1. (1, 2)   **2.** (−2, −1)   **3.** (−1, −2)   **4.** (−2, 3)
5. (1, 3)   **6. (a)** (2, 1); one; consistent   **(b)** No solutions;
zero; inconsistent   **7. (a)** Infinitely many   **(b)** Consistent;
dependent   **8. (a)** None   **(b)** Inconsistent   **9.** (−3, 4)
10. No solutions   **11.** Infinitely many solutions
12. (−1, 3)   **13.** $y \leq -\frac{1}{2}x$ **14.** Yes
15.

16.

17.

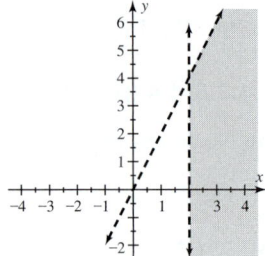

18.

19. $1.9 trillion in 2003; $2.0 trillion in 2004
20. $\frac{2}{3}$ hr at 6 mph; $\frac{1}{3}$ hr at 9 mph

CHAPTERS 1–4 CUMULATIVE REVIEW (pp. 304–305)

1. $2 \times 2 \times 2 \times 3 \times 5$ **2. (a)** 4   **(b)** 8   **3. (a)** Rational
(b) Irrational   **4. (a)** <   **(b)** >   **5.** Associative
(multiplication)   **6.** 3060   **7. (a)** $4x^2$ **(b)** $5x - 2$
8. $\frac{2}{9}$ **9.** $\frac{1}{2}$ **10.** No solutions   **11.** 11, 12, 13, 14
12. 432 in² or 3 ft²   **13.** $x = \frac{W + 7y}{3}$ **14.** $x \geq 1$
15.

16. (a)   **(b)**

          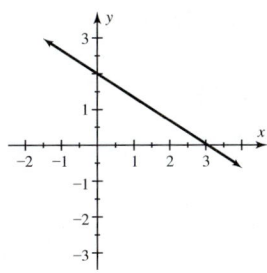

17. $2; -8$

18. (a)   **(b)**

          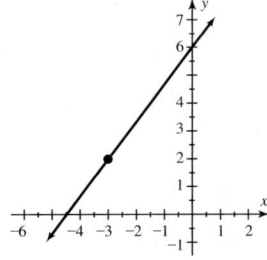

19. (a) $y = -\frac{1}{2}x - 1$ **(b)** $y = 4x - 3$

20. $y = -\frac{1}{4}x + 3$ **21.** $y = -3x + 4$ **22.** $(4, 4)$

23. $(1, -3)$ **24.** No solutions   **25.** Infinitely many solutions   **26.** $(-1, 2)$

27.   **28.**

          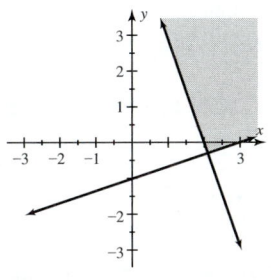

29. $F = 16.5R$ **30.** $94°C$ **31.** 245 **32.** 158.05
33. $G = -3x + 30$ **34.** 1200 at 5%; $1200 at 6%

CHAPTER 5: POLYNOMIALS AND EXPONENTS

SECTION 5.1 (pp. 315–317)

1. base; exponent   **2.** 1   **3.** $\left(\frac{1}{2}\right)^3$ **4.** x^4 **5.** -1 **6.** 1
7. a^{m+n} **8.** a^{mn} **9.** $a^n b^n$ **10.** $\frac{a^n}{b^n}$ **11.** 64   **13.** -8
15. -8 **17.** 1   **19.** 32   **21.** 26   **23.** 8   **25.** $\frac{1}{2}$
27. 3^3 or 27   **29.** 4^8 or 65,536   **31.** 2^5 or 32   **33.** x^9
35. z^4 **37.** x^6 **39.** $20x^7$ **41.** $-3x^3 y^4$ **43.** 2^6 or 64
45. n^{12} **47.** x^7 **49.** $49b^2$ **51.** $a^3 b^3$ **53.** $4x^4$
55. $-64b^6$ **57.** $x^{14}y^{21}$ **59.** $x^{12}y^9$ **61.** $a^{10}b^8$ **63.** $\frac{1}{27}$

65. $\frac{a^5}{b^5}$ **67.** $\frac{(x-y)^3}{27}$ **69.** $\frac{25}{(a+b)^2}$ **71.** $\frac{8x^3}{125}$ **73.** $\frac{27x^6}{125y^{12}}$
75. $(x + y)^4$ **77.** $(a + b)^5$ **79.** 6   **81.** $a^3 + 2ab^2$
83. $12a^5 + 6a^3 b$ **85.** $r^2 t + rt^2$
87. $a = 3, b = 1, m = 1, n = 0$ (answers may vary)
89. $10x^4$ **91.** $8x^3$ **93.** $9\pi x^4$ **95.** $1157.63
97.

					4
	4				4
	4	×	4	=	4

99.

										2		3
										2		
	2									2		3
	2	×	3	×	3	×	2	=		2	×	3
	2		3				2			2		3

SECTION 5.2 (pp. 325–327)

1. monomial   **2.** -3 **3.** polynomial   **4.** degree
5. binomial   **6.** trinomial   **7.** 2; 3   **8.** like   **9.** opposite
10. vertically   **11.** 2; 3   **13.** 2; -1 **15.** 2; -5
17. 0; -6 **19.** Yes; 1; 1; 1   **21.** Yes; 3; 1; 2   **23.** No
25. Yes; 2; 2; 4   **27.** No   **29.** Yes; 1; 3; 6   **31.** Yes; x
33. Yes; $-5x^3$ **35.** No   **37.** Yes; $2ab$ **39.** Yes; $4xy^2$
41. $-x + 9$ **43.** $4x^2 + 8x - 5$ **45.** $5a^3 + 1$
47. $4y^3 + 3y^2 + 4y - 9$ **49.** $4xy + 1$ **51.** $2a^2 b^3$
53. $9x^2 + x - 6$ **55.** $x^2 - 7x - 1$ **57.** $-5x^2$
59. $-3a^2 + a - 4$ **61.** $2t^2 + 3t - 4$ **63.** $4x - 2$
65. $-3x^2 + 5x + 10$ **67.** $-3a^2 - 5a$
69. $z^3 - 6z^2 - 6z - 1$ **71.** $3xy + 2x^2 y^2$ **73.** $-a^3 b$
75. $-x^2 - 5x - 4$ **77.** $-2x^3 - 6x - 2$
79. (a) 200 bpm   **(b)** 100 bpm   **(c)** It decreases quickly at first, then more slowly.   **81.** $2z^2$; 200 in^2
83. $2x^2 + x^2$ or $3x^2$; 108 ft^2 **85.** $\pi x^2 + \pi y^2$; 13π ft^2
87. (a) $m_1 \approx 0.077$; $m_2 \approx 0.083$; $m_3 \approx 0.077$. A line is reasonable but not exact.   **(b)** For the given years, its estimates are reasonable.

CHECKING BASIC CONCEPTS 5.1 & 5.2 (p. 327)

1. (a) -25 **(b)** 1   **2. (a)** 10^8 **(b)** $-12x^7$ **(c)** $a^6 b^2$
(d) $\frac{x^4}{z^{12}}$ **3.** 3; 2; 4   **4. (a)** $2W^2 H$ **(b)** 2880 in^3
5. (a) $3a^2 + 6$ **(b)** $2z^3 + 7z - 8$ **(c)** $4xy$ **6.** 34

SECTION 5.3 (pp. 333–335)

1. The product rule   **2.** Distributive   **3.** one; two; three
4. term; term   **5.** No   **6.** vertically   **7.** x^7 **9.** $-12a^2$
11. $20x^5$ **13.** $4x^2 y^3$ **15.** $-12x^3 y^3$ **17.** $3x + 12$
19. $-45x - 5$ **21.** $4z - z^2$ **23.** $-5y - 3y^2$
25. $15x^3 - 12x$ **27.** $6x^3 - 6x^2$ **29.** $-32t^2 - 8t - 8$

31. $-5n^4 + n^3 - 2n^2$ **33.** $x^2y + xy^2$ **35.** $x^4y - x^3y^2$
37. $-a^4b + 2ab^4$ **39.** $x^2 + 3x$ **41.** $x^2 + 4x + 4$
43. $x^2 + 9x + 18$ **45.** $x^2 + 8x + 15$
47. $x^2 - 17x + 72$ **49.** $6z^2 - 19z + 10$ **51.** $64b^2 - 1$
53. $10y^2 - 3y - 7$ **55.** $5 - 13a + 6a^2$ **57.** $1 - 9x^2$
59. $x^3 - x^2 + x - 1$ **61.** $4x^3 - 3x^2 + 16x - 12$
63. $2n^3 + n^2 + 6n + 3$ **65.** $m^3 + 4m^2 + 4m + 1$
67. $6x^3 - 7x^2 + 14x - 8$ **69.** $x^3 + 1$
71. $4b^4 + 3b^3 + 19b^2 + 9b + 21$
73. $x^3 - x^2 - 5x + 2$ **75.** $a^3 - 8$
77. $6x^4 - 2x^3 + 5x^2 - x + 1$ **79.** $m \cdot n$
81. (a) $h^3 - 2h^2 - 8h$ (b) 14,175 in^3
83. (a) $50x - x^2$ (b) 625   **85.** $6x^2 + 12x + 6$
87. (a) $64t - 16t^2$ (b) 64; 64   (c) Yes; yes

SECTION 5.4 (pp. 340–342)

1. $a^2 - b^2$ **2.** Yes; the product is a sum and a difference.
3. $a^2 + 2ab + b^2$ **4.** $a^2 - 2ab + b^2$
5. No; $(x + y)^2 = x^2 + 2xy + y^2$
6. No; $(r - t)^2 = r^2 - 2rt + t^2$
7. No; $(z + 5)^3 = z^3 + 15z^2 + 75z + 125$ **8.** $(a + b)^2$
9. $x^2 - 9$ **11.** $x^2 - 1$ **13.** $16x^2 - 1$ **15.** $1 - 4a^2$
17. $4x^2 - 9y^2$ **19.** $a^2b^2 - 25$ **21.** $a^2b^2 - 16$
23. $a^4 - b^4$ **25.** $x^6 - y^6$ **27.** 9999   **29.** 391
31. 9900   **33.** $x^2 + 2x + 1$ **35.** $a^2 - 4a + 4$
37. $4x^2 + 12x + 9$ **39.** $9b^2 + 30b + 25$
41. $\frac{9}{16}a^2 - 6a + 16$ **43.** $1 - 2b + b^2$
45. $25 + 10y^3 + y^6$ **47.** $a^4 + 2a^2b + b^2$
49. $a^3 + 3a^2 + 3a + 1$ **51.** $x^3 - 6x^2 + 12x - 8$
53. $8x^3 + 12x^2 + 6x + 1$
55. $216u^3 - 108u^2 + 18u - 1$
57. $t^4 + 6t^3 + 12t^2 + 8t$ **59.** $20x + 36$
61. $x^2 + 2x - 35$ **63.** $9x^2 - 30x + 25$
65. $25x^2 + 35x + 12$ **67.** $16b^2 - 25$
69. $-20x^3 + 35x^2 - 10x$ **71.** $64 - 48a + 12a^2 - a^3$
73. $x^3 + 6x^2 + 9x$ **75.** $x^4 - 5x^2 + 4$ **77.** $a^{2n} - b^{2n}$
79. (a) $x^2 + 4x + 4$ (b) $x^2 + 4x + 4$
81. (a) $4x^2 + 12x + 9$ (b) $4x^2 + 12x + 9$
83. (a) $6x^2 + 60x + 150$ (b) $x^3 + 15x^2 + 75x + 125$
85. (a) $1 + 2x + x^2$ (b) 1.21; the money increases by
1.21 times in 2 years if the interest rate is 10%.
87. (a) $1 - 2x + x^2$ (b) 0.25; if the chance of rain on each
day is 50%, then there is a 25% chance that it will not rain on
either day.   **89.** (a) $32z + 256$ (b) 2176; the area of an
8-foot-wide sidewalk around a 60×60 foot pool is 2176 ft^2.

CHECKING BASIC CONCEPTS 5.3 & 5.4 (p. 343)

1. (a) $-15x^3y^5$ (b) $-6x + 4x^2$
(c) $3a^3b - 6a^2b^2 + 3ab^3$
2. (a) $4x^2 + 9x - 9$ (b) $2x^4 - 2$ (c) $x^3 + y^3$

3. (a) $25x^2 - 4$ (b) $x^2 + 6x + 9$
(c) $4 - 28x + 49x^2$ (d) $t^3 + 6t^2 + 12t + 8$
4. (a) $4\pi a^2 + 56\pi a + 196\pi$ (b) 900π
5. (a) $m^2 + 10m + 25$ (b) $m^2 + 10m + 25$

SECTION 5.5 (pp. 351–354)

1. $\frac{1}{a^n}$ **2.** $\frac{1}{8}$ **3.** a^n **4.** 8   **5.** a^{m-n} **6.** 8   **7.** $\frac{b^m}{a}$ **8.** $\frac{25}{8}$
9. $\left(\frac{b}{a}\right)^n$ **10.** $\frac{25}{9}$ **11.** $1 \le |b| < 10$ **12.** No   **13.** (a) $\frac{1}{4}$
(b) 9   **15.** (a) 2   (b) 100   **17.** (a) $\frac{1}{81}$ (b) $\frac{1}{8}$
19. (a) $\frac{1}{576}$ (b) 64   **21.** (a) 1   (b) 64   **23.** (a) 25
(b) $\frac{49}{4}$ **25.** (a) $\frac{1}{x}$ (b) $\frac{1}{a^4}$ **27.** (a) $\frac{1}{x^2}$ (b) $\frac{1}{a^8}$
29. (a) $\frac{y^3}{x^3}$ (b) $\frac{1}{x^3y^3}$ **31.** (a) $\frac{1}{16t^4}$ (b) $\frac{1}{(x + 1)^7}$
33. (a) a^8 (b) $\frac{1}{r^2t^6}$ **35.** (a) $\frac{b^2}{a^4}$ (b) x^2 **37.** (a) a^{13}
(b) $\frac{2}{z^3}$ **39.** (a) $-\frac{2y^3}{3x^2}$ (b) $\frac{1}{x^3}$ **41.** (a) $2b$ (b) $\frac{a^3}{b^3}$
43. (a) y^5 (b) $2t^3$ **45.** (a) x^2y^2 (b) a^6b^3
47. (a) $\frac{b^2}{a^2}$ (b) $\frac{4v}{u}$ **49.** $\frac{a^n}{a^m} = a^{n-m} = a^{-(m-n)} = \frac{1}{a^{m-n}}$
51. Thousand   **53.** Billion   **55.** Hundredth   **57.** 2000
59. 45,000   **61.** 0.008   **63.** 0.000456   **65.** 39,000,000
67. $-500,000$ **69.** 2×10^3 **71.** 5.67×10^2
73. 1.2×10^7 **75.** 4×10^{-3} **77.** 8.95×10^{-4}
79. -5×10^{-2} **81.** 1,500,000   **83.** -1.5
85. 2000   **87.** 0.002   **89.** (a) About 5.859×10^{12} mi
(b) About 2.5×10^{13} mi   **91.** About 9.2×10^9 mi
93. (a) 1.246×10^{13} (b) About $41,812

SECTION 5.6 (pp. 359–360)

1. $\frac{a}{d} + \frac{b}{d}$ **2.** $\frac{a}{d} + \frac{b}{d} - \frac{c}{d}$ **3.** term   **4.** No   **5.** No
6. 9; 4; 1   **7.** $x + 1$; $2x^2 - 2x + 1$; 4   **8.** $0x^2$ **9.** $2x$
11. $z^3 + z^2$ **13.** $\frac{a^2}{2} - 3$ **15.** $\frac{4}{x} - 7x^2$ **17.** $3x^3 - 1 + \frac{2}{x}$
19. $4y^2 - 1 + \frac{2}{y}$ **21.** $3m^2 - 2m + 4$
23. $2x + 1 + \frac{3}{x - 2}$ **25.** $x + 1$ **27.** $x^2 + 1 + \frac{-1}{x - 1}$
29. $x^2 - x + 2 + \frac{1}{4x + 1}$ **31.** $x^2 + 2x + 3 + \frac{8}{x - 2}$
33. $3x^2 + 3x + 3 + \frac{5}{x - 1}$ **35.** $x + 3 + \frac{-x - 2}{x^2 + 1}$
37. $x + 1$ **39.** $x^2 - 2x + 4$ **41.** They are the same.
43. $4x$
45. $x + 2$;

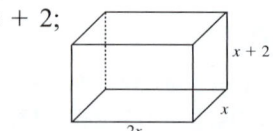

CHECKING BASIC CONCEPTS 5.5 & 5.6 (p. 361)

1. (a) $\frac{1}{81}$ (b) $\frac{1}{2x^7}$ (c) $\frac{b^8}{16a^2}$ **2.** (a) z^5 (b) $\frac{y^6}{x^3}$ (c) $\frac{x^6}{27}$
3. (a) 4.5×10^4 (b) 2.34×10^{-4} (c) 1×10^{-2}

4. (a) 47,100 **(b)** 0.006 **5.** $5a - 3$ **6.** $3x + 2$; R: -2
7. (a) 9.3×10^7 **(b)** 500 sec (8 min 20 sec)

CHAPTER 5 REVIEW (pp. 365–367)

1. 125 **2.** -81 **3.** 4 **4.** 11 **5.** -5 **6.** 1 **7.** 6^5
8. 10^{12} **9.** z^9 **10.** y^6 **11.** $30x^9$ **12.** a^4b^4 **13.** 2^{10}
14. m^{20} **15.** a^3b^3 **16.** x^8y^{12} **17.** x^7y^{11} **18.** 1
19. $(r - t)^9$ **20.** $(a + b)^6$ **21.** $\frac{9}{(x-y)^2}$ **22.** $\frac{(x+y)^3}{8}$
23. 7; 6 **24.** 5; -1 **25.** Yes; 1; 1; 1 **26.** Yes; 4; 1; 3
27. Yes; 3; 2; 2 **28.** No **29.** $5x^2 - x + 3$
30. $-6x^2 + 3x + 7$ **31.** $3x + 4$ **32.** $-2x^2 - 13$
33. $-2x^2 + 9x + 5$ **34.** $2a^3 - a^2 + 7a$
35. $5y^2 - 3xy$ **36.** $4xy$ **37.** $-x^5$ **38.** $-r^3t^4$
39. $-6t + 15$ **40.** $2y - 12y^2$ **41.** $18x^5 + 30x^4$
42. $-x^3 + 2x^2 - 9x$ **43.** $-a^3b + 2a^2b^2 - ab^3$
44. $a^2 + 3a - 10$ **45.** $8x^2 + 13x - 6$
46. $-2x^2 + 3x - 1$ **47.** $2y^3 + y^2 + 2y + 1$
48. $2y^4 - y^2 - 1$ **49.** $z^3 + 1$
50. $4z^3 - 15z^2 + 13z - 3$ **51.** $z^2 + z$ **52.** $2x^2 + 4x$
53. $z^2 - 4$ **54.** $25z^2 - 81$ **55.** $1 - 9y^2$
56. $25x^2 - 16y^2$ **57.** $r^2t^2 - 1$ **58.** $4m^4 - n^4$
59. $x^2 + 2x + 1$ **60.** $16x^2 + 24x + 9$
61. $y^2 - 6y + 9$ **62.** $4y^2 - 20y + 25$
63. $16 + 8a + a^2$ **64.** $16 - 8a + a^2$
65. $x^4 + 2x^2y^2 + y^4$ **66.** $x^2y^2 - 4xy + 4$
67. $z^3 + 15z^2 + 75z + 125$
68. $8z^3 - 12z^2 + 6z - 1$ **69.** 3599 **70.** 396 **71.** $\frac{1}{9}$
72. $\frac{1}{9}$ **73.** 4 **74.** $\frac{1}{1000}$ **75.** 36 **76.** $\frac{1}{25}$ **77.** $\frac{1}{z^2}$
78. $\frac{1}{y^4}$ **79.** $\frac{1}{a^2}$ **80.** $\frac{1}{x^2}$ **81.** $\frac{1}{4t^2}$ **82.** $\frac{1}{a^3b^6}$ **83.** $\frac{1}{y^3}$
84. x^4 **85.** $\frac{2}{x^3}$ **86.** $\frac{2x^4}{3y^3}$ **87.** $\frac{a^5}{b^5}$ **88.** $4t^4$ **89.** $\frac{27}{x^3}$
90. $2ab$ **91.** $\frac{y^2}{x^2}$ **92.** $\frac{2v}{3u}$ **93.** 600 **94.** 52,400
95. 0.0037 **96.** 0.06234 **97.** 1×10^4
98. 5.61×10^7 **99.** 5.4×10^{-5} **100.** 1×10^{-3}
101. 24,000,000 **102.** 0.2 **103.** $\frac{5}{3}x + 1$
104. $3b^2 - 2 + \frac{1}{b^2}$ **105.** $3x + 2 + \frac{4}{x-1}$
106. $3x - 4 + \frac{6}{3x+2}$ **107.** $x^2 - 3x - 1$
108. $x^2 + \frac{-1}{2x-1}$ **109.** $x - 1 + \frac{-2x+2}{x^2+1}$
110. $x^2 + 2x + 5$ **111. (a)** 60 bpm **(b)** 160 bpm
(c) It increases. **112.** $6xy$; 72 ft^2
113. $10z^2 + 25z$; 510 in^2 **114.** x^4y^2 **115.** $833.71
116. $\frac{4}{3}\pi x^3 + 8\pi x^2 + 16\pi x + \frac{32}{3}\pi$ **117. (a)** $96t - 16t^2$
(b) 128; after 2 sec the ball is 128 ft high.
118. (a) $600L - L^2$ **(b)** 27,500; a rectangular building
with a perimeter of 1200 ft and a side of length 50 ft has an
area of 27,500 ft^2. **119. (a)** $x^2 + 10x + 25$
(b) $x^2 + 10x + 25$ **120. (a)** $16x$ **(b)** 1600

121. About $8795 per person **122.** About 519,590,000 gal
or 5.1959×10^8 gal

CHAPTER 5 TEST (pp. 367–368)

1. Yes; 3; 2; 3 **2.** $x^3 - 4x + 8$ **3.** $4x + 6$
4. $-3y^3 - 2y + 1$ **5.** $x^2 + x - 7$ **6.** $4a^3 + 2ab$
7. (a) -6 **(b)** $\frac{1}{64}$ **(c)** 8 **(d)** -3 **8.** $24y^{11}$
9. a^5b^8 **10.** x^4 **11.** $\frac{a^3}{b^6}$ **12.** $a^3b - ab^3$ **13.** $\frac{4}{9a^4b^6}$
14. $\frac{2y^3}{x}$ **15.** $\frac{16}{(a+b)^4}$ **16.** $12x^5 - 18x^3 + 3x^2$
17. $2z^2 - 2z - 12$ **18.** $49y^4 - 9$ **19.** $9x^2 - 12x + 4$
20. $m^3 + 9m^2 + 27m + 27$ **21.** $y^3 - y + 6$
22. 6396 **23.** 0.0061 **24.** 5.41×10^3 **25.** $3x - 2 + \frac{1}{x}$
26. $x^2 - x + 1 + \frac{-1}{x+2}$ **27. (a)** $20t$ **(b)** $2t + 2000$
(c) $18t - 2000$; profit from selling t tickets
28. $12x^2$; 1200 ft^2 **29.** $3x^3 + 27x^2 + 54x$
30. (a) $88t - 16t^2$ **(b)** 120; after 3 sec the ball is
120 ft high.

CHAPTERS 1–5 CUMULATIVE REVIEW (pp. 369–370)

1. (a) 8 **(b)** 8 **2. (a)** 29 **(b)** $\frac{3}{2}$ **3. (a)** 7
(b) No solutions **4. (a)** Infinitely many solutions **(b)** -2
5. 68 mph **6.** $-39, -38, -37$ **7. (a)** $\frac{21}{50}$ **(b)** $\frac{19}{250}$
8. $x = \frac{Y}{W-4}$
9.

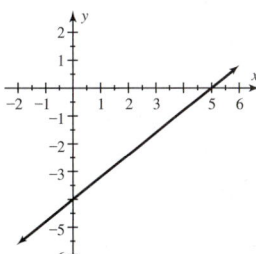

10.

11. $y = -2x + 1$ **12.** 2; -3 **13.** $y = \frac{1}{2}x - 4$
14. $y = 3x + 1$ **15.** No solutions **16.** $(2, -1)$
17. Infinitely many solutions **18.** $(0, -6)$
19.

20.

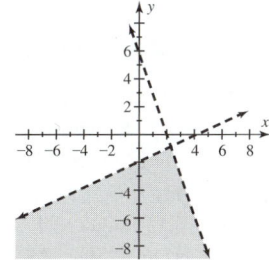

21. (a) $15x^5$ **(b)** $x^{10}y^7$ **22. (a)** $2x^2 - x + 3$
(b) $12a^3 + a - 5$ **23. (a)** $2x^2 - 11x - 21$
(b) $y^3 - 10y - 3$ **(c)** $16x^2 - 49$ **(d)** $25a^2 + 30a + 9$
24. (a) $\frac{1}{x}$ **(b)** $\frac{x^9}{8}$ **(c)** $\frac{x^4}{2y^2}$ **(d)** $\frac{x^7}{y^8}$ **25.** 2.4×10^{10}
26. 0.000000471 **27. (a)** $4x^2 - 1$ **(b)** $2x - 5 + \frac{1}{x + 3}$
28. -25% **29.** $200\ \text{mL}$ **30. (a)** $2x^2 + 10x$
(b) $x^2 + 7x + 10$ **(c)** $2x^2 + 4x$ **(d)** $10x^2 + 42x + 20$

CHAPTER 6: FACTORING POLYNOMIALS AND SOLVING EQUATIONS

SECTION 6.1 (pp. 378–380)

1. factoring   **2.** a **3.** multiplying
4. prime; multiplication   **5.** greatest common factor (GCF)
6. grouping   **7.** $1, 2, x, 2x$ **8.** $2x$
9. $2(x + 2)$;

	2x	4
2		

$x \quad 2$

11. $z(z + 4)$;

	z^2	$4z$
z		

$z \quad 4$

13. $3x(x + 3)$ **15.** $2y^2(2y - 1)$ **17.** $2z(z^2 + 4z - 2)$
19. $3xy(2x - y)$ **21.** $6x; 6x(1 - 3x)$
23. $4y^2; 4y^2(2y - 3)$ **25.** $3z; 3z(2z^2 + z + 3)$
27. $x^2; x^2(x^2 - 5x - 4)$
29. $5y^2; 5y^2(y^3 + 2y^2 - 3y + 2)$ **31.** $x; x(y + z)$
33. $ab; ab(b - a)$ **35.** $5x^2y^3; 5x^2y^3(y + 2x)$
37. $ab; ab(a + b + 1)$ **39.** $(x - 2)(x + 1)$
41. $(z + 4)(z + 5)$ **43.** $2x(2x^2 - 1)(x - 5)$
45. $(x^2 + 3)(x + 2)$ **47.** $(y^2 + 1)(2y + 1)$
49. $(2z^2 + 5)(z - 3)$ **51.** $(4t^2 + 3)(t - 5)$
53. $(3r^2 - 2)(3r + 2)$ **55.** $(7x^2 - 2)(x + 3)$
57. $(y^2 - 2)(2y - 7)$ **59.** $(z^2 - 7)(z - 4)$
61. $(x^3 + 2)(2x - 3)$ **63.** $(x + y)(a + b)$
65. $3(x^2 + 1)(x + 2)$ **67.** $2y(3y^2 - 1)(y - 4)$
69. $x^2(x^2 - 3)(x + 2)$ **71.** $2x^2(x^2 - 3)(2x + 1)$
73. $xy(x + y)(x - 2)$ **75.** $a(x^2 + \frac{b}{a}x + \frac{c}{a})$
77. (a) $16t$ **(b)** $16t(5 - t)$
79. (a) $168\ \text{in}^3$ **(b)** $4x(x^2 - 15x + 50)$

SECTION 6.2 (pp. 385–387)

1. 1 **2.** $x^2 + (m + n)x + mn; m + n; mn$
3. $mn; m + n$ **4.** prime   **5.** $1, 12; 2, 6; 3, 4$
6. $-1, -30; -2, -15; -3, -10; -5, -6$ **7.** $4, 7$
9. $3, -10$ **11.** $-5, 10$ **13.** $-7, -4$ **15.** $(x + 1)(x + 2)$
17. $(y + 2)(y + 2)$ **19.** Prime   **21.** $(x + 3)(x + 5)$
23. $(m + 4)(m + 9)$ **25.** $(n + 10)(n + 10)$

27. $(x - 1)(x - 5)$ **29.** $(y - 3)(y - 4)$
31. $(z - 5)(z - 8)$ **33.** $(a - 7)(a - 9)$ **35.** Prime
37. $(b - 5)(b - 25)$ **39.** $(x - 5)(x + 18)$
41. $(m - 5)(m + 9)$ **43.** Prime   **45.** $(n - 10)(n + 20)$
47. $(x - 1)(x + 23)$ **49.** $(a - 4)(a + 8)$
51. $(b + 4)(b - 5)$ **53.** Prime   **55.** $(x + 8)(x - 9)$
57. $(y + 2)(y - 17)$ **59.** $(z + 6)(z - 11)$
61. $5(x - 4)(x + 2)$ **63.** $y(y - 2)(y - 5)$
65. $3a(a + 1)(a + 6)$ **67.** $2x(x^2 - 3x + 4)$
69. $2m^2(m - 7)(m + 2)$ **71.** $(x + 1)(x + 5)$
73. $(x - 1)(x - 3)$ **75.** $(6 - x)(2 + x)$
77. $(8 + x)(4 - x)$ **79.** $(x + 1)(x + k)$
81. $x + 1$;

	x^2	x
x		
1	x	1

$x \quad 1$

83. $x + 1, x + 2$;

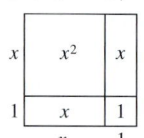

	x^2	$2x$
x		
1	x	2

$x \quad 2$

85. $x + 1$ **87.** $(x + 2)(x + 6)$

CHECKING BASIC CONCEPTS 6.1 & 6.2 (p. 387)

1. $4x$ **2.** $6z^2(2z - 3)$ **3. (a)** $(6y + 5)(y - 2)$
(b) $(x^2 + 5)(2x + 1)$ **(c)** $4(z^2 + 1)(z - 3)$
4. (a) $(x + 2)(x + 4)$ **(b)** $(x - 7)(x + 6)$ **(c)** Prime
(d) $4a(a + 2)(a + 3)$ **5.** $(x + 5)(x + 5)$

SECTION 6.3 (pp. 393–395)

1. $ac; b$ **2.** $ax^2; c$ **3.** $+; +$ **4.** $+; -$ **5.** $-; -$
6. $-; +$ **7.** $3; x$ **8.** $2; 4x$ **9.** $1; 2x$ **10.** $1; 2x$
11. $(x + 3)(2x + 1)$ **13.** Prime   **15.** $(x + 1)(3x + 1)$
17. $(2x + 3)(3x + 1)$ **19.** $(x - 2)(5x - 1)$
21. $(y - 1)(2y - 5)$ **23.** Prime   **25.** $(z - 5)(7z - 2)$
27. $(t - 3)(3t + 2)$ **29.** $(3r + 2)(5r - 3)$
31. $(3m - 4)(8m + 3)$ **33.** $(5x - 1)(5x + 2)$
35. $(x + 2)(6x - 1)$ **37.** Prime   **39.** $(3n + 1)(7n - 1)$
41. $(2y + 3)(7y + 1)$ **43.** $(4z - 3)(7z - 1)$
45. $(3x - 2)(10x - 3)$ **47.** Prime   **49.** $(2t + 3)(9t - 2)$
51. $3(2a - 1)(2a + 3)$ **53.** $y(3y - 2)(4y - 1)$
55. $3x(4x - 3)(2x - 1)$ **57.** $2x^2(4x^2 - 3x + 1)$
59. $7x^2(2x + 1)(2x + 3)$ **61.** $(3x + 1)(x + k)$
63. $(7x + 1)(x + 2)$ **65.** $(2x - 1)(x - 2)$
67. $-(4x + 3)(2x - 1)$ **69.** $-(x + 5)(2x - 3)$
71. $-(x - 3)(5x + 1)$
73. $3x + 2$ by $2x + 1$;

	$6x^2$	$4x$
2x		
1	$3x$	2

$3x \quad 2$

75. $(2x + 1)(x + 3)$

SECTION 6.4 (pp. 401–402)

1. $(a - b)(a + b)$ **2.** cannot   **3.** $6x$; $7y$ **4.** $(a + b)^2$
5. $(a - b)^2$ **6.** $6x$ **7.** $20rt$ **8.** $(a + b)(a^2 - ab + b^2)$
9. $(a - b)(a^2 + ab + b^2)$ **10.** $2x$; $3y$ **11.** $-$; $+$
12. $+$; $-$ **13.** $(x - 1)(x + 1)$ **15.** $(z - 10)(z + 10)$
17. $(2y - 1)(2y + 1)$ **19.** $(6z - 5)(6z + 5)$
21. $(3 - x)(3 + x)$ **23.** $(1 - 3y)(1 + 3y)$
25. $(2a - 3b)(2a + 3b)$ **27.** $(6m - 5n)(6m + 5n)$
29. $(9r - 7t)(9r + 7t)$ **31.** $(x + 4)^2$
33. Not possible   **35.** $(x - 3)^2$ **37.** $(3y + 1)^2$
39. $(2z - 1)^2$ **41.** Not possible   **43.** $(3x + 5)^2$
45. $(2a - 9)^2$ **47.** $(x + y)^2$ **49.** $(r - 5t)^2$
51. Not possible   **53.** $(z + 1)(z^2 - z + 1)$
55. $(x + 4)(x^2 - 4x + 16)$ **57.** $(y - 2)(y^2 + 2y + 4)$
59. $(n - 1)(n^2 + n + 1)$ **61.** $(2x + 1)(4x^2 - 2x + 1)$
63. $(m - 4n)(m^2 + 4mn + 16n^2)$
65. $(2x + 5y)(4x^2 - 10xy + 25y^2)$
67. $4(x - 2)(x + 2)$ **69.** $2(y - 7)^2$
71. $5(z + 2)(z^2 - 2z + 4)$ **73.** $xy(x - y)(x + y)$
75. $2m(m^2 - 5m + 9)$ **77.** $7x^2(10x - 3y)(10x + 3y)$
79. $2(2a + b)(4a^2 - 2ab + b^2)$ **81.** $4b^2(b + 3)^2$
83. $4(5r - 2t)(25r^2 + 10rt + 4t^2)$
85. $2x + 3$;

$2x$	$4x^2$	$6x$
3	$6x$	9
	$2x$	3

CHECKING BASIC CONCEPTS 6.3 & 6.4 (p. 402)

1. **(a)** $(x - 4)(2x + 3)$ **(b)** $(2x + 7)(3x - 2)$
2. **(a)** Prime   **(b)** $2y(3y + 1)(y - 2)$
3. $(3x + 2)(x + 3)$ **4.** **(a)** $(z - 8)(z + 8)$
(b) $(3r - 2t)(3r + 2t)$ **5.** **(a)** $(x + 6)^2$ **(b)** $(3a - 2b)^2$
6. **(a)** $(m - 3)(m^2 + 3m + 9)$
(b) $(5n + 3)(25n^2 - 15n + 9)$
7. **(a)** $4(2x - 1)(2x + 1)$ **(b)** $3y(y + 2)(y^2 - 2y + 4)$

SECTION 6.5 (pp. 407–408)

1. Greatest common factor   **2.** GCF   **3.** Grouping
4. No; a sum of squares cannot be factored.
5. Yes; a sum of cubes can be factored.   **6.** completely
7. $2(2x - 1)$ **9.** $2(y^2 - 2y + 2)$ **11.** $(z - 2)(z + 2)$
13. $(a + 2)(a^2 - 2a + 4)$ **15.** $(2b - 3)^2$
17. Not possible   **19.** $(x^2 + 5)(x - 1)$
21. $(y - 4)(y - 1)$ **23.** $(x - 3)(x + 3)(x + 4)$
25. $8(a - 2)(a^2 + 2a + 4)$ **27.** $2x(3x^2 + 1)(2x - 3)$
29. $2t(3t + 2)(9t^2 - 6t + 4)$

31. $2(r - 1)(r + 1)(r + 3)$ **33.** $3z^2(2z + 3)(z - 5)$
35. $2b^2(3b - 1)(2b - 1)$ **37.** $6z(y - 2z)(y + 2z)$
39. $3y(x - 5)^2$ **41.** $(3m - 2n)(9m^2 + 6mn + 4n^2)$
43. $3(x - 2)(x + 2)(x - 1)(x^2 + x + 1)$
45. $(5a + 3)(a - 6)$ **47.** $3r(t + 5)(t + 6)$
49. $(3b^2 + 4)(3b + 2)$ **51.** $2n(3n^2 + n - 5)$
53. $4(x - 3y)(x + 3y)$ **55.** $2a(a - 4)^2$
57. $4x(2y + 1)(4y^2 - 2y + 1)$
59. $2b(2b - 1)(2b + 1)(b + 3)$ **61.** $3x + 1$

SECTION 6.6 (pp. 414–416)

1. 0; 0 **2.** No; one side of the equation must be zero.
3. $2x = 0$; $x + 6 = 0$ **4.** Subtract $4x$ from each side.
5. Apply the zero-product property.   **6.** solving
7. $ax^2 + bx + c = 0$ with $a \neq 0$ **8.** $x^2 - 6x + 1 = 0$
9. zero   **10.** 2   **11.** $x = 0$ or $y = 0$ **13.** $-8, 0$
15. $1, 2$ **17.** $\frac{1}{2}, \frac{3}{4}$ **19.** $\frac{1}{3}, \frac{3}{7}$ **21.** $0, 5, 8$ **23.** $0, 1$
25. $0, 5$ **27.** $-\frac{3}{2}, 0$ **29.** $-1, 1$ **31.** $-\frac{1}{2}, \frac{1}{2}$
33. $-2, -1$ **35.** $5, 7$ **37.** $-2, \frac{1}{2}$ **39.** $-\frac{5}{2}, -\frac{2}{3}$
41. $-5, 5$ **43.** $0, 5$ **45.** $-3, 0$ **47.** $-1, 6$ **49.** $-\frac{5}{3}, \frac{3}{4}$
51. $-2, 1$ **53.** $-3, \frac{1}{2}$ **55.** $-\frac{1}{2}$ **57.** $-1, -\frac{2}{3}$ **59.** 12 ft
61. 4   **63.** 4
65. **(a)** 6 sec
(b)

Time (t)	0	1	2	3	4	5	6
Height (h)	0	80	128	144	128	80	0

; 3 sec

67. **(a)** 81.8 ft; 327.3 ft; when the speed doubles, the braking distance quadruples.   **(b)** About 19 mph   **(c)** About 19 mph; yes
69. **(a)** About 11 million; about 66 million   **(b)** About 1981
71. 40 by 50 pixels

CHECKING BASIC CONCEPTS 6.5 & 6.6 (p. 416)

1. **(a)** $9(a^2 - 2a + 3)$ **(b)** $7x(y^2 + 4)$
2. **(a)** $2z^2(3z - 2)(z - 4)$ **(b)** $2r^2(t - 3)(t + 3)$
3. **(a)** $4x(3x - 2)^2$ **(b)** $3(2b - 3)(4b^2 + 6b + 9)$
4. **(a)** $0, \frac{3}{2}$ **(b)** $-1, \frac{3}{5}$ **5.** $-3, 1$ **6.** After $\frac{11}{2}$ sec

SECTION 6.7 (pp. 421–422)

1. GCF   **2.** zero-product   **3.** factors
4. $(x^2 - y^2)^2$ **5.** $(z^2 + 1)(z^2 + 2)$
6. No; $x^4 - 1 = (x - 1)(x + 1)(x^2 + 1)$
7. Subtract x from each side.   **8.** Factor out x^2.
9. $5(x - 3)(x + 2)$ **11.** $-4(y + 2)(y + 6)$
13. $-10(z + 5)(2z + 1)$ **15.** $-4(t + 3)(7t - 5)$
17. $r(r - 1)(r + 1)$ **19.** $3x(x - 2)(x + 3)$
21. $12z(2z - 1)(3z + 2)$ **23.** $x^2(x - 2)(x + 2)$

25. $t^2(t-1)(t+2)$ **27.** $(x^2-3)(x^2-2)$
29. $(x^2+3)(2x^2+1)$ **31.** $(y^2+3)^2$
33. $(x^2-3)(x^2+3)$ **35.** $(x-3)(x+3)(x^2+9)$
37. $z^3(z+1)^2$ **39.** $(x+y)(2x-y)$
41. $(a+b)^2(a-b)^2$ **43.** $x(x+y)(x-y)$
45. $x(2x+y)^2$ **47.** (a) $x(x-2)(x+2)$ (b) $-2,0,2$
49. (a) $2y(y-6)(y+3)$ (b) $-3,0,6$ **51.** $-8,-3$
53. $-1,3$ **55.** $-1,0,4$ **57.** $-6,0,4$ **59.** $-6,0,6$
61. $-7,0,1$ **63.** $-3,-2,2,3$ **65.** $-1,1$ **67.** $-3,3$
69. $-1,1,2$ **71.** 5   **73.** (a) $x<7.5$ in. because the
width is 15 in.   (b) $300-4x^2$ (c) 2.5 in.
75. 18.6 trillion ft^3 **77.** Factoring is very difficult
(answers may vary).

CHECKING BASIC CONCEPTS 6.7 (p. 423)

1. (a) $3(x-4)(x+2)$ (b) $-5(2y+1)(y-1)$
2. (a) $(z^2-5)(z^2+5)$ (b) $7(t-1)(t+1)(t^2+1)$
3. (a) $(x-2)^2(x+2)^2$ (b) $y(y+10)(2y-3)$
4. $-4,0,3$ **5.** 3

CHAPTER 6 REVIEW (pp. 426–428)

1. $4z^2; 4z^2(2z-1)$ **2.** $3x^2; 3x^2(2x^2+x-4)$
3. $3y; 3y(3x+5z^2)$ **4.** $a^2b^2; a^2b^2(b+a)$
5. $(x-3)(x+2)$ **6.** $y(y+3)(x-5)$
7. $(z^2+5)(z-2)$ **8.** $(t^2+8)(t+1)$
9. $(x^2+6)(x-3)$ **10.** $(x-y)(a+b)$
11. $x^2(x^2-2)(x+3)$ **12.** $2y(y^2+1)(y+3)$
13. $4,5$ **14.** $-3,7$ **15.** $-9,-4$ **16.** $-25,4$
17. $(x-4)(x+3)$ **18.** $(x+4)(x+6)$
19. $(x-2)(x+8)$ **20.** $(x-7)(x+6)$
21. $(x-1)(x+3)$ **22.** $(x+10)(x+12)$
23. $2x(x-2)(x+5)$ **24.** $x^2(x+4)(x-7)$
25. $(2-x)(5-x)$ **26.** $(6-x)(4+x)$
27. $(3x-1)(3x+2)$ **28.** $(x-1)(2x+5)$
29. $(x+3)(3x+5)$ **30.** $(5x-1)(7x+1)$
31. $(3x+1)(8x-5)$ **32.** $(x+9)(4x-3)$
33. $3x(2x+7)(2x+1)$ **34.** $2x^2(x+3)(4x-5)$
35. $(3-2x)(4+x)$ **36.** $(1-2x)(1+5x)$
37. $(z-2)(z+2)$ **38.** $(3z-8)(3z+8)$
39. $(6-y)(6+y)$ **40.** $(10a-9b)(10a+9b)$
41. $(x+7)^2$ **42.** $(x-5)^2$ **43.** $(2x-3)^2$
44. $(3x+8)^2$ **45.** $(2t-1)(4t^2+2t+1)$
46. $(3r+2t)(9r^2-6rt+4t^2)$
47. $2x(x-5)(x+5)$ **48.** $3(2x+3)(4x^2-6x+9)$
49. $2x(x+7)^2$ **50.** $2x(x-4)(x^2+4x+16)$

51. $3(3y^2-2y+2)$ **52.** $y(z-3)(z+3)$
53. $x(x-2)(x+2)(x+7)$ **54.** $3x(2x+3)^2$
55. $3a(b-2)(b^2+2b+4)$ **56.** $5x(x^2+4)$
57. $6x(2x-y)(2x+y)$ **58.** $y(x+3)(x^2-3x+9)$
59. $m=0$ or $n=0$ **60.** 0   **61.** $-9,\frac{3}{4}$ **62.** $-\frac{6}{5},\frac{1}{4}$
63. $0,1,2$ **64.** $0,7$ **65.** $-8,8$ **66.** $-7,-2$
67. $-2,3$ **68.** $-\frac{3}{2},\frac{2}{5}$ **69.** $-4,18$ **70.** $-2,\frac{5}{2}$
71. $5(x-5)(x+2)$ **72.** $-3(x-3)(x+5)$
73. $y(y-2)(y+2)$ **74.** $3y(y-1)(y+3)$
75. $2z^2(z+2)(z+5)$ **76.** $8z^2(z-2)(z+2)$
77. $(x^2-3)^2$ **78.** $(x-3)(x+3)(2x^2+3)$
79. $(a+5b)^2$ **80.** $x(x+y)(x-y)$ **81.** $-\frac{1}{2},5$
82. $0,\frac{5}{2},3$ **83.** $-5,0,5$ **84.** $0,3,4$ **85.** $-2,2$
86. $-4,4$ **87.** -4 **88.** $-1,1$
89. $3x+7$;   **90.** $x+1$ by $x+5$;

	$3x$	7
$3x$	$9x^2$	$21x$
7	$21x$	49

	x	5
x	x^2	$5x$
1	x	5

91. $x+1$ **92.** $(x+1)(x+3)$ **93.** $(2x+3)(x+6)$
94. 2   **95.** 8 ft by 15 ft   **96.** After 2 sec and 3 sec
97. (a) 390 ft   (b) 15 mph   (c) 15 mph; yes
98. (a) $10,000   (b) $50 or $150   (c) $50 or $150; yes
99. (a) 372 million   (b) 1975   **100.** 50 by 80 pixels
101. (a) $2000-4x^2$ (b) 5 in.

CHAPTER 6 TEST (p. 429)

1. $4xy; 4xy(x-5y+3)$ **2.** $3a^2b^2; 3a^2b^2(3a+1)$
3. $(y+z)(a+b)$ **4.** $(x^2-5)(3x+1)$
5. $(y-2)(y+6)$ **6.** $(2x+5)^2$ **7.** $(z-4)(4z-3)$
8. $(t-7)(2t-3)$ **9.** $3x(x+1)(2x-1)$
10. $2(z-3)(z+3)(z^2+3)$ **11.** $4y(3y-5)(3y+5)$
12. $7x(x+2)(x^2-2x+4)$ **13.** $a^2(4a+3)^2$
14. $2(b-2)(b+2)(b^2+4)$ **15.** $-4,4$ **16.** $-4,5$
17. $\frac{4}{3}$ **18.** $-6,11$ **19.** $-3,0,3$ **20.** $-2,-1,1,2$
21. $3x+5$ **22.** $(x+2)(x+3)$ **23.** (a) 275 ft
(b) 33 mph   **24.** 1 sec and 2 sec

CHAPTERS 1–6 CUMULATIVE REVIEW (pp. 430–432)

1. $2\times2\times2\times2\times3\times3$ **2.** 16   **3.** $\frac{3}{7}$ **4.** $\frac{7}{10}$
5. 17   **6.** $-\frac{7}{2}$ **7.** 8
8. 1;

x	-2	-1	0	1	2
$2x+3$	-1	1	3	5	7

9. $3n-5=n-7; -1$ **10.** $\frac{57}{1000}; 0.057$ **11.** 12.3%
12. $W=\frac{P-2L}{2}$ **13.** $z>2$

14.

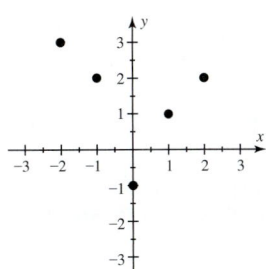

15. x-int: $\frac{2}{3}$; y-int: -2

16. y-int: -2

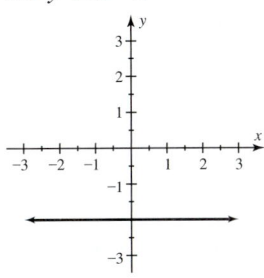

17. -1; -2; $y = -2x - 2$ **18.** $y = -\frac{3}{2}x + \frac{7}{2}$

19. $y = \frac{4}{3}x + \frac{11}{3}$ **20.** $(1, 2)$ **21.** $(1, -1)$ **22.** $(-2, 5)$

23. (a) One   **(b)** Consistent; independent

24. (a) None   **(b)** Inconsistent

25.

26.

27.

28.

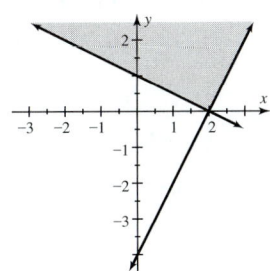

29. -16 **30.** 1 **31.** $\frac{x^{10}}{y^4}$ **32.** $-14x^5 + 21x^4$

33. $21x^2 + 29x - 10$ **34.** $x^2 - 14x + 49$ **35.** $\frac{1}{a^2}$

36. $\frac{1}{4t^6}$ **37.** $\frac{1}{x^2y^5}$ **38.** $32x^5y^{10}$ **39.** 0.00623

40. 5.43×10^5 **41.** $2x^2 + 4x$ **42.** $3x + \frac{-4x + 1}{x^2 + 1}$

43. $(2y^2 - 5)(x + 2)$ **44.** $(t^2 + 1)(t + 6)$

45. $(x - 4)(x + 7)$ **46.** $(2y + 3)(3y - 4)$

47. $(5x - 2y)(5x + 2y)$ **48.** $(8x - 1)^2$

49. $(3t - 2)(9t^2 + 6t + 4)$ **50.** $-4(x - 3)(x + 2)$

51. $(x - 3)(x + 3)(x^2 - 3)$ **52.** $x^2y(x - y)$

53. $-5, 0, 5$ **54.** $-2, 1$ **55.** $-\frac{7}{2}, 0, \frac{7}{2}$ **56.** $-3, 3$

57. (a) $10x + 8x$ or $18x$ **(b)** 50 min   **58.** $6\frac{11}{12}$ mi

59. \$21,000   **60.** 25 min skiing; 35 min running

61. (a) $C = 0.25x + 20$ **(b)** 320 mi

62. (a) $2x + y = 180$; $2x - y = 20$ **(b)** $(50, 80)$, the angles are 50°, 50°, and 80°.

63.

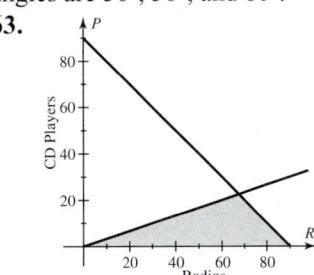

64. $18xy$; 108 yd^2 **65.** $8x^3y^6$ **66. (a)** $x^2 + 4x + 4$

(b) $x^2 + 2x + 2x + 4 = x^2 + 4x + 4$

67. $x + 6$ **68. (a)** 4 sec   **(b)** After 1 sec and 3 sec

CHAPTER 7:   RATIONAL EXPRESSIONS

SECTION 7.1 (pp. 440–443)

1. $\frac{P}{Q}$; polynomials   **2.** Yes; both x and $2x^2 + 1$ are polynomials.   **3.** the denominator equals zero   **4.** a **5.** $\frac{a}{b}$

6. rational   **7.** $\frac{3}{10}$ **8.** $\frac{3}{2x}$ **9.** -1 **10.** $1 - x$ **11.** $-\frac{3}{7}$

13. $-\frac{4}{9}$ **15.** $-\frac{1}{4}$ **17.** Undefined   **19.** Undefined

21. -1 **23.** 1

25.

x	-2	-1	0	1	2
$\frac{x}{x+1}$	2	—	0	$\frac{1}{2}$	$\frac{2}{3}$

27.

x	-2	-1	0	1	2
$\frac{3x}{2x^2+1}$	$-\frac{2}{3}$	-1	0	1	$\frac{2}{3}$

29. 0 **31.** 3 **33.** $-\frac{4}{5}$ **35.** None   **37.** $-5, 5$

39. $-3, -2$ **41.** $1, \frac{5}{2}$ **43.** $\frac{2}{3}$ **45.** $\frac{1}{2}$ **47.** $-\frac{2}{5}$

49. $-\frac{1}{3}$ **51. (a)** $\frac{1}{2}$ **(b)** $\frac{1}{2}$ **53. (a)** -1 **(b)** -1

55. $\frac{1}{2x^2}$ **57.** $\frac{4y}{3x}$ **59.** $\frac{1}{2}$ **61.** $\frac{3}{5}$ **63.** $\frac{x+1}{x+6}$ **65.** $\frac{5y+3}{y+2}$

67. -1 **69.** -1 **71.** $-\frac{1}{3}$ **73.** $-\frac{1}{2}$ **75.** 1

77. $-\frac{3x+5}{3x-5}$ **79.** $\frac{n-1}{n-5}$ **81.** $\frac{x}{6}$ **83.** $\frac{z-2}{z-3}$ **85.** $\frac{x+4}{3x+2}$

87. $\frac{1}{3x-2}$ **89.** 1 **91.** $-\frac{1}{2}$ **93. (a)** Equation   **(b)** 6

95. (a) Expression   **(b)** $\frac{1}{x+1}$ **97. (a)** Expression

(b) $x - 2$ **99. (a)** Equation   **(b)** 8

101. (a) $\frac{1}{2}$; when traffic arrives at a rate of 3 vehicles per minute, the average wait is $\frac{1}{2}$ minute.

(b)

x	2	4	4.5	4.9	4.99
T	$\frac{1}{3}$	1	2	10	100

As x nears 5 vehicles per minute, a small increase in x increases the wait dramatically.

103. $\frac{1}{2}$ **105.** (a) $\frac{3}{n}$ (b) $\frac{n-3}{n}$; $\frac{97}{100}$; there is a 97% chance that a winning ball will not be drawn.   **107.** (a) 6 hr   (b) $\frac{M}{60}$

109. (a) $x = 5$ (b) As the average rate nears 5 cars per minute, a small increase in x increases the wait dramatically.

SECTION 7.2 (pp. 448–450)

1. $\frac{10}{21}$ **2.** $\frac{14}{15}$ **3.** $\frac{AC}{BD}$ **4.** $\frac{AD}{BC}$ **5.** $\frac{x+7}{x+1}$ **6.** $\frac{4}{B}$ **7.** No; it is equal to $1 + \frac{2}{x}$.   **8.** $\frac{1}{x}; \frac{z}{y}$ **9.** $\frac{2}{5}$ **11.** $\frac{12}{7}$ **13.** $\frac{2}{3}$

15. $\frac{2}{11}$ **17.** 4   **19.** $\frac{8}{15}$ **21.** 10   **23.** $\frac{12}{5}$ **25.** 1

27. $\frac{z+1}{z+4}$ **29.** $\frac{2}{3}$ **31.** $\frac{x+2}{x-2}$ **33.** $\frac{8(x+1)}{x^2}$ **35.** $\frac{x-3}{x}$

37. $\frac{z+3}{z-7}$ **39.** 1   **41.** $(t+1)(t+2)$ **43.** $\frac{x(x+4)}{x^2+4}$

45. $\frac{z-1}{z+2}$ **47.** $\frac{y}{y-1}$ **49.** $\frac{x+1}{x-3}$ **51.** $\frac{x-3}{x-1}$ **53.** $\frac{2}{2x+3}$

55. -2 **57.** $\frac{z-1}{z+1}$ **59.** $\frac{y+2}{2y+1}$ **61.** $\frac{4(t-1)}{t^2+1}$

63. $\frac{y-3}{y-5}$ **65.** $2x$ **67.** $\frac{2z+1}{z+5}$ **69.** $\frac{1}{t+6}$ **71.** $\frac{2a+3b}{a+b}$

73. $x - y$ **75.** 1   **77.** (a) $D = \frac{30}{x}$ (b) 300 ft; 75 ft; one-fourth as far   **79.** (a) $\frac{1}{n+1}$ (b) $\frac{1}{100}$

CHECKING BASIC CONCEPTS 7.1 & 7.2 (p. 450)

1. Undefined; $\frac{3}{8}$ **2.** (a) $\frac{2x}{5y}$ (b) 5   (c) $\frac{x+2}{x+4}$

3. (a) $\frac{4}{9}$ (b) $\frac{2}{x-1}$ **4.** (a) $\frac{5z}{6}$ (b) $x + 1$

5. (a)

x	0.5	1.0	1.5	1.9
T	$\frac{2}{3}$	1	2	10

(b) As x nears 2 persons per minute, a small increase in x increases the wait dramatically.

SECTION 7.3 (pp. 457–458)

1. add; numerators; denominators   **2.** subtract; numerators; denominators   **3.** $\frac{3}{5}$ **4.** $\frac{4}{7}$ **5.** $\frac{A+B}{C}$ **6.** $\frac{A-B}{C}$

7. $4 - x$ **8.** $\frac{4-x}{x-1}$ **9.** $x + 5$ **10.** $\frac{x+5}{7}$ **11.** 1

13. $\frac{6}{5}$ **15.** 1   **17.** $\frac{3}{7}$ **19.** $\frac{1}{2}$ **21.** $\frac{1}{2}$ **23.** $\frac{2}{3}$ **25.** $\frac{3}{x}$

27. $\frac{1}{2}$ **29.** 3   **31.** 1   **33.** $\frac{4z}{4z+3}$ **35.** 2   **37.** 1

39. $\frac{1}{x-1}$ **41.** $z - 1$ **43.** 2   **45.** $y - 1$ **47.** $\frac{x}{2}$

49. $z - 2$ **51.** $x - 3$ **53.** $\frac{7n}{2n^2-n+5}$ **55.** $\frac{6}{x+3}$

57. $\frac{9}{ab}$ **59.** $\frac{1}{x+y}$ **61.** $\frac{16}{a-b}$ **63.** $\frac{b}{2a}$ **65.** 1

67. $a - b$ **69.** $2x - 3y$ **71.** It equals 7.

73. (a) $\frac{14}{n+1}$ (b) $\frac{7}{50}$; there are 7 chances out of 50 chances that a defective battery is chosen.

SECTION 7.4 (pp. 465–467)

1. Examples include 36 and 54 (answers may vary).   **2.** xy **3.** $\frac{3}{3}$ **4.** $\frac{x+1}{x+1}$ **5.** 12   **6.** $6x^2$ **7.** 12   **9.** 6   **11.** 30   **13.** 72   **15.** $12x$ **17.** $10x^2$ **19.** $x(x+1)$

21. $(2x+1)(x+3)$ **23.** $36x^3$ **25.** $x(x-1)(x+1)$ **27.** $(x+1)^2$ **29.** $(2x-1)^3(x+3)$

31. $(2x-1)(2x+1)$ **33.** $(x-1)(x+1)$ **35.** $4x(x^2+4)$ **37.** $(2x+3)(x+2)(x+3)$ **39.** $\frac{3}{9}$

41. $\frac{15}{21}$ **43.** $\frac{2x^2}{8x^3}$ **45.** $\frac{x-2}{x^2-4}$ **47.** $\frac{x}{x^2+x}$

49. $\frac{2x^2+2x}{x^2+2x+1}$ **51.** $\frac{13}{10}$ **53.** $\frac{2}{9}$ **55.** $\frac{14}{25}$ **57.** $\frac{9}{20}$ **59.** $\frac{13}{12x}$

61. $\frac{5z-7}{z^3}$ **63.** $\frac{y-x}{xy}$ **65.** $\frac{a^2+b^2}{ab}$ **67.** $\frac{7}{2(x+2)}$

69. $\frac{t+2}{t(t-2)}$ **71.** $\frac{n^2+4n+5}{(n-1)(n+1)}$ **73.** $-\frac{3}{x-3}$ **75.** 0

77. $\frac{6x-4}{(x-1)^2}$ **79.** $\frac{3}{2y-1}$ **81.** $\frac{x-1}{x(x+2)}$

83. $\frac{3x+5}{(x-2)(x+2)}$ **85.** $\frac{x+9}{x(x-3)(x+3)}$

87. $\frac{x^2+4x-4}{x(x-2)(x+2)}$ **89.** $\frac{2x+2}{(x+2)^2}$

91. $\frac{x^2+2x-1}{(x+1)(x+2)(x+3)}$ **93.** $-\frac{2b}{(a+b)(a-b)}$ **95.** 0

97. $\frac{13}{12a}$ **99.** 0   **101.** $\frac{3}{x}$ **103.** $\frac{1}{75}; \frac{1}{75}$; yes   **105.** $\frac{D-F}{FD}$

CHECKING BASIC CONCEPTS 7.3 & 7.4 (p. 467)

1. (a) 1   (b) $\frac{2-x}{3x}$ (c) z **2.** (a) $15x$ (b) $4x(x+1)$

(c) $(x+1)(x-1)$ **3.** (a) $\frac{6x+5}{x(x+1)}$ (b) $\frac{4}{x-3}$

(c) $-\frac{x+4}{2x+1}$ **4.** $\frac{a^2+b^2}{(a-b)(a+b)}$

SECTION 7.5 (pp. 475–477)

1. $\frac{1}{2} \cdot \frac{4}{3} = \frac{2}{3}$ **2.** $\frac{a}{b} \cdot \frac{d}{c} = \frac{ad}{bc}$ **3.** fractions   **4.** $\frac{x+\frac{1}{2}}{x-\frac{1}{2}}$

5. Division   **6.** $\frac{\frac{x}{2}}{\frac{1}{x-1}}$ **7.** $\frac{\frac{a}{b}}{\frac{c}{d}}$ **8.** $x(x+2)$ **9.** 30

11. $(x-1)(x+1)$ **13.** $x(2x-1)(2x+1)$

15. $\frac{4}{5}$ **17.** $\frac{10}{7}$ **19.** $\frac{9}{14}$ **21.** $\frac{1}{2}$ **23.** $\frac{3y}{x}$ **25.** 3   **27.** $\frac{p+1}{p+2}$

29. $\frac{5}{z}$ **31.** $\frac{y}{y-3}$ **33.** $\frac{x^2-1}{x^2+1}$ **35.** $\frac{x^2}{3}$ **37.** 1   **39.** $\frac{n+m}{n-m}$

41. $\frac{2x+y}{2x-y}$ **43.** $\frac{1+b}{1-a}$ **45.** $\frac{q+2}{q}$ **47.** $2x + 3$

49. $\frac{2x}{(x+1)(2x-1)(2x+1)}$ **51.** $\frac{1}{ab}$ **53.** $\frac{ab}{a+b}$

55. $\frac{P(1+\frac{r}{26})^{52}-P}{\frac{r}{26}}$ **57.** $R = \frac{ST}{S+T}$

SECTION 7.6 (pp. 487–491)

1. rational   **2.** $\frac{2x+5}{3x}; \frac{2x+5}{3x} = 9$ (answers may vary)

3. $ad = bc; b; d$ **4.** Yes   **5.** $12x$ **6.** $V; T$ **7.** $\frac{3}{2}$ **9.** $\frac{5}{2}$

11. $\frac{7}{6}$ **13.** 0   **15.** $\frac{1}{4}$ **17.** $\frac{10}{11}$ **19.** 3   **21.** 25   **23.** 5

25. $\frac{5}{16}$ **27.** 1   **29.** No solutions $\left(\text{extraneous: } -\frac{1}{2}\right)$ **31.** $\frac{1}{2}$

33. -1 **35.** $-\frac{3}{2}, -1$ **37.** 4 **39.** 4 **41.** -6 **43.** 6
45. -3 (extraneous: 1) **47.** 1 (extraneous: -2) **49.** $\frac{7}{12}$
51. $\frac{1}{6}, 5$ **53.** 3 **55.** No solutions (extraneous: 1)
57. 3 **59.** No solutions (extraneous: -1) **61.** No solutions
63. -2 **65.** Expression; 1; 1 **67.** Equation; 2
69. Equation; 4 **71.** Expression; $x + 2$; 4 **73.** $-\frac{1}{2}; \frac{1}{2}$
75. $-1; \frac{1}{2}$ **77. (a)** $-3, 1$ **(b)** $-3, 1$ **79. (a)** -1
(b) -1 **81. (a)** 2 **(b)** 2 **83. (a)** $-2, 2$ **(b)** $-2, 2$
85. 0.300, 1.100 **87.** 1.084 **89.** $a = \frac{F}{m}$
91. $r = \frac{V}{I} - R$ **93.** $b = \frac{2A}{h}$ **95.** $z = \frac{15}{k - 3}$
97. $b = \frac{aT}{a - T}$ **99.** $x = \frac{ky}{3y + 2k}$ **101.** 9 cars per minute
103. $\frac{12}{7} \approx 1.7$ hours **105.** $\frac{8}{3} \approx 2.7$ days **107.** 18 mph;
20 mph **109. (a)** 120; the braking distance is 120 feet when
the slope of the road is -0.05. **(b)** -0.1 **111.** 15 mph
113. 200 mph **115.** 10 mph running; 5 mph walking
117. 3 **119.** 32

CHECKING BASIC CONCEPTS 7.5 & 7.6 (p. 491)

1. (a) $\frac{5}{6}$ **(b)** $\frac{1}{9x^2}$ **(c)** $\frac{b - a}{b + a}$ **(d)** $\frac{r + t}{2rt}$ **2. (a)** $\frac{1}{5}$
(b) -4 **3. (a)** 2 **(b)** $-2, \frac{1}{2}$ **4.** -3 (extraneous: 1)
5. (a) $x = \frac{2(b + 3y)}{a}$ **(b)** $m = \frac{k}{2k - 1}$ **6. (a)** 300; when
the slope of the hill is 0.1, the braking distance is 300 feet.
(b) 0.3; the braking distance is 200 feet when the slope of
the road is 0.3.

SECTION 7.7 (pp. 501–506)

1. A statement that two ratios are equal **2.** $\frac{5}{6} = \frac{x}{7}$
3. It doubles. **4.** It is halved. **5.** constant **6.** constant
7. Directly; if the number being fed doubles, the bill will
double. **8.** Inversely; doubling the number of painters will
halve the time. **9.** 15 **11.** 21 **13.** 48 **15.** $\frac{21}{8}$
17. $-4, 4$ **19.** $-\frac{7}{2}, \frac{7}{2}$ **21.** $b = \frac{ad}{c}$ **23. (a)** $\frac{5}{8} = \frac{9}{x}$
(b) $\frac{72}{5}$ **25. (a)** $\frac{4}{8} = \frac{10}{x}$ **(b)** 20 **27. (a)** $\frac{98}{7} = \frac{x}{11}$
(b) \$154 **29. (a)** $\frac{3}{180} = \frac{7}{x}$ **(b)** 420 min **31. (a)** 2
(b) 12 **33. (a)** $\frac{3}{2}$ **(b)** 9 **35. (a)** $-\frac{15}{2}$ **(b)** -45
37. (a) 24 **(b)** 3 **39. (a)** 40 **(b)** 5 **41. (a)** 400
(b) 50 **43. (a)** $k = 0.25$ **(b)** $z = 8.75$ **45. (a)** $k = 11$
(b) $z = 385$ **47. (a)** $k = 10$ **(b)** $y = 350$
49. (a) Direct **(b)** $y = \frac{3}{2}x$ **51. (a)** Inverse **(b)** $y = \frac{36}{x}$
(c)   **(c)**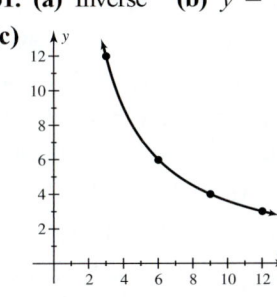

53. (a) Neither **(b)** NA **(c)** NA **55.** Direct; 2
57. Neither **59.** Inverse; 8 **61.** 47.6 minutes
63. 1.625 inches **65.** About 2123 lb **67.** $9\frac{1}{3}$ c
69. (a) Direct; the ratios **71. (a)** Direct; the ratios
$\frac{R}{W}$ always equal 0.012. $\frac{G}{A}$ always equal 27.
(b) $R = 0.012W$ **(b)** $G = 27A$

(c) 38.4 pounds **(c)** For each square-inch
increase in the cross-sectional
area of the hose, the flow
increases by 27 gallons
per minute.

73. (a) $F = \frac{1200}{L}$ **(b)** 80 pounds
75. (a) Direct **(b)** $y = -19x$ **(c)** Negative; for each
1-mile increase in altitude the temperature decreases by 19°F.
(d) 47.5°F decrease **77.** 1.8 ohms **79.** $z = 0.5104x^2y^3$
81. About 133 pounds **83.** About 35.2 pounds **85.** 750

CHECKING BASIC CONCEPTS 7.7 (p. 506)

1. (a) $\frac{18}{5}$ **(b)** $\frac{15}{4}$ **2. (a)** $\frac{4}{6} = \frac{8}{x}$; 12 **(b)** $\frac{2}{148} = \frac{5}{x}$;
370 minutes **3.** 60; 6 **4. (a)** Direct; the ratios $\frac{y}{x}$ always
equal $\frac{3}{2}; \frac{3}{2}$. **(b)** Inverse; the products xy always equal 24; 24.
5. \$160

CHAPTER 7 REVIEW (pp. 510–513)

1. $-\frac{3}{5}$ **2.** -3 **3.** Undefined **4.** Undefined
5.

x	-2	-1	0	1	2
$\frac{3x}{x-1}$	2	$\frac{3}{2}$	0	—	6

6. $-2, 2$ **7.** $\frac{5y^3}{3x^2}$ **8.** $x - 6$ **9.** -1 **10.** $\frac{x - 5}{5}$
11. $\frac{x + 3}{x + 1}$ **12.** $\frac{x + 4}{x + 1}$ **13.** 2 **14.** $\frac{1}{x + 5}$ **15.** $\frac{1}{z + 3}$
16. $\frac{x}{x - 2}$ **17.** $\frac{5}{6}$ **18.** $\frac{8}{x(x + 1)}$ **19.** $\frac{1}{2}$ **20.** 1
21. $x + y$ **22.** $2(a^2 + ab + b^2)$ **23.** $\frac{10}{x + 10}$
24. $\frac{1}{x - 1}$ **25.** 1 **26.** 1 **27.** $\frac{1}{x + 1}$ **28.** $\frac{2}{x - 5}$
29. $\frac{2}{xy}$ **30.** $\frac{x}{y}$ **31.** $15x$ **32.** $10x^2$ **33.** $x(x - 5)$
34. $10x^2(x - 1)$ **35.** $(x - 1)(x + 1)^2$
36. $x(x - 4)(x + 4)$ **37.** $\frac{9}{24}$ **38.** $\frac{16}{12x}$ **39.** $\frac{3x^2 + 6x}{x^2 - 4}$
40. $\frac{2x}{x^2 + x}$ **41.** $\frac{3x - 3}{5x^2 - 5x}$ **42.** $\frac{2x^2 + 4x}{2x^2 + x - 6}$ **43.** $\frac{19}{24}$
44. $\frac{7}{4x}$ **45.** $-\frac{1}{9x}$ **46.** $\frac{4x + 3}{x(x - 1)}$ **47.** $\frac{2x}{(x - 1)(x + 1)}$

48. $\frac{8 - 9x}{6x^2}$ **49.** $\frac{2x - 7}{6x}$ **50.** $-\frac{1}{(x - 1)(x + 1)}$

51. $\frac{5y - x}{(x - y)(x + y)}$ **52.** $\frac{13}{6x}$ **53.** $\frac{3x + y}{2xy}$ **54.** -1

55. $\frac{33}{28}$ **56.** $\frac{7}{10}$ **57.** $\frac{n}{2}$ **58.** $\frac{3(p + 1)}{p - 1}$ **59.** $\frac{3(m + 1)}{2(m - 1)^2}$

60. $\frac{2n - 1}{4(2n + 1)}$ **61.** $\frac{1}{3}$ **62.** $\frac{2 - x}{2 + x}$ **63.** $\frac{1}{x^2}$ **64.** $x + 3$

65. $\frac{20}{7}$ **66.** $\frac{8}{3}$ **67.** $\frac{1}{5}$ **68.** -5 **69.** -4 **70.** $-3, -1$

71. 4 **72.** -2 **73.** 5 **74.** 7 **75.** $-2, 5$

76. $-3, 3$ **77.** 8 **78.** No solutions   **79.** No solutions
(extraneous: -1)   **80.** $-\frac{3}{2}$ (extraneous: 0)   **81.** -3

82. -12 **83.** $\frac{4}{5}$ **84.** -3 **85.** $b = \frac{2ac}{3a - c}$

86. $x = \frac{y}{y - 1}$ **87.** 2 **88.** $\frac{15}{7}$ **89.** (a) $\frac{6}{x} = \frac{13}{20}$ (b) $\frac{120}{13}$

90. (a) $\frac{341}{11} = \frac{x}{8}$ (b) \$248   **91.** (a) 4 (b) 20

92. (a) 3 (b) 15 **93.** (a) 10 (b) 2 **94.** (a) 21 (b) $\frac{21}{5}$

95. $k = 3$ **96.** $z = 720$

97. (a) Inverse   (b) $y = \frac{60}{x}$ **98.** (a) Direct   (b) $y = 3x$
(c)   (c)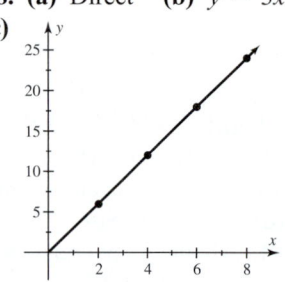

99. Direct; $\frac{1}{2}$ **100.** Inverse; 12   **101.** (a) $\frac{1}{5} = 0.2$; when
the average rate of arrival is 10 cars per minute, the average
wait is 0.2 minute, or 12 seconds.

(b)

x	5	10	13	14	14.9
T	$\frac{1}{10}$	$\frac{1}{5}$	$\frac{1}{2}$	1	10

(c) It increases dramatically.   **102.** 60 mph
103. $\frac{800}{13} \approx 61.5$ hours   **104.** 10 mph and 12 mph
105. 8 mph   **106.** About 33.3 feet   **107.** 200 vehicles
108. \$468   **109.** 36 pounds   **110.** 1.6 inches
111. 750 to 2250 seconds, or 12.5 to 37.5 minutes
112. About 771 lb   **113.** 6493.8 watts   **114.** 702 pounds

CHAPTER 7 TEST (p. 514)

1. $\frac{9}{5}$ **2.** -2 **3.** $x + 5$ **4.** $x - 5$ **5.** 3 **6.** 2

7. $\frac{x - 1}{10x}$ **8.** $\frac{6}{x(x + 3)}$ **9.** $\frac{4x + 1}{x + 4}$ **10.** $\frac{t + 7}{2t - 3}$

11. $-\frac{1}{y + 1}$ **12.** $\frac{x^2y - x + y}{xy^2}$ **13.** $\frac{b}{15}$ **14.** $\frac{p}{p - 2}$

15. $\frac{35}{2}$ **16.** 3 **17.** 1 **18.** $-1, 2$ **19.** $\frac{2}{7}$ **20.** -12

21. No solutions $\left(\text{extraneous: } \frac{1}{2}\right)$ **22.** 0 (extraneous: 5)

23. $x = \frac{2 + 5y}{3y}$ **24.** $b = \frac{a}{a - 1}$ **25.** (a) $\frac{7}{2}$ (b) 21

26. Inversely; 32   **27.** 24 hours   **28.** 67.5 feet

29. $\frac{16}{5} = 3.2$; when the arrival rate is 24 people per hour,
there are about 3 people in line.

CHAPTERS 1–7 CUMULATIVE REVIEW (pp. 515–517)

1. $24\pi \approx 75.4$ **2.** $2x - 2$ **3.** $-\frac{5}{8}$ **4.** $\frac{2}{5}$ **5.** $\frac{3}{4}$

6. $\frac{2}{9}$ **7.** $2x + 2$ **8.** $y - 11$ **9.** -1 **10.** 4.5%

11. $W = \frac{V}{6L}$ **12.** $x \leq \frac{1}{4}$

13.   **14.**

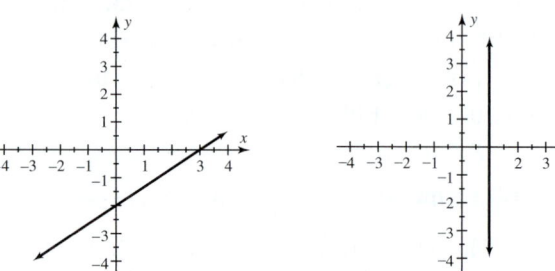

15. $y = 3x + 5$

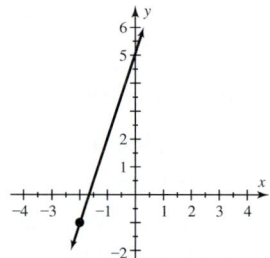

16. $y = 2x - 1$ **17.** $y = -\frac{2}{3}x + \frac{1}{3}$ **18.** $y = \frac{2}{3}x + \frac{8}{3}$
19. $N = 200x + 2000$ **20.** $(1, -2)$ **21.** $(1, 2)$
22. $(2, -8)$ **23.** Infinitely many solutions
24. No solutions   **25.** $(1, 2)$
26. $(0, 0)$ (answers may vary)

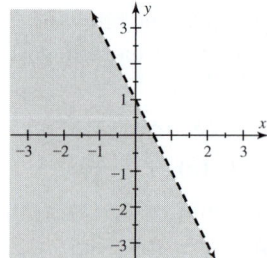

27. $4x^2 + 1$ **28.** $15z^8$ **29.** a^3b^3 **30.** $10y^2 - 11y - 6$
31. $x^4 - 2x^2y^2 + y^4$ **32.** $4t^2 + 20t + 25$ **33.** 2
34. 9 **35.** $\frac{1}{27x^6}$ **36.** $\frac{2}{x^2}$ **37.** 1.23×10^{-3}
38. $2x + 1 + \frac{4}{x - 1}$ **39.** $5z^2; 5z^2(4z - 3)$
40. $3xy; 3xy(4x + 5y)$ **41.** $(3 - x)(2 + 5x)$
42. $(3z - 2)(3z + 2)$ **43.** $(t + 8)^2$
44. $x(x - 4)(x + 4)$ **45.** $x(x - 9)(x + 11)$
46. $(a + 3b)^2$ **47.** $-5, \frac{7}{2}$ **48.** $0, 2$ **49.** $-7, 2$
50. $-2, 0, 2$ **51.** $1; 2$ **52.** $\frac{x + 1}{x - 1}$ **53.** 1 **54.** $\frac{x}{x^2 - 1}$
55. $\frac{1}{12x}$ **56.** $\frac{1}{2(x + 2)}$ **57.** $\frac{x + 2}{x - 2}$ **58.** $\frac{40}{7}$ **59.** $\frac{7}{12}$

60. $-1, 2$ **61.** $\frac{z + 2y}{3}$ **62.** 5.5   **63.** 80   **64.** Inverse; $y = \frac{20}{x}$ **65. (a)** $21x$ **(b)** 90 min   **66.** 35 min running, 25 min walking   **67. (a)** 24.1   **(b)** Head Start participation increased by 24,100 per year, on average.
(c) 1099 thousand, or 1.1 million
68. (a) $2x + y = 180$, $2x - y = 32$ **(b)** (53, 74) or 53°, 53°, 74°   **69. (a)** 21.6; in 1995 natural gas consumption was 21.6 trillion ft³.   **(b)** 2000   **70.** About 500 lb

CHAPTER 8:   INTRODUCTION TO FUNCTIONS

SECTION 8.1 (pp. 534–539)

1. function   **2.** y equals f of x **3.** symbolic   **4.** numerical
5. domain   **6.** range   **7.** one   **8.** To identify graphs of functions   **9.** Verbal, numerical, symbolic, and graphical or diagrammatic   **10.** (3, 4); 3; 6   **11.** (a, b) **12.** d
13. 1   **14.** equal   **15.** Yes   **17.** No   **19.** -6; -2
21. 0; $\frac{3}{2}$ **23.** 25; $\frac{9}{4}$ **25.** 3; 3   **27.** 13; -22 **29.** $-\frac{1}{2}$; $\frac{2}{5}$
31. (a) $I(x) = 36x$ **(b)** $I(10) = 360$
33. (a) $A(r) = \pi r^2$ **(b)** $A(10) = 100\pi \approx 314.2$
35. (a) $A(x) = 43,560x$ **(b)** $A(10) = 435,600$
37. $f = \{(1, 3), (2, -4), (3, 0);$
$D = \{1, 2, 3\}; R = \{-4, 0, 3\}$
39. $f = \{(a, b), (c, d), (e, a), (d, b)\};$
$D = \{a, c, d, e\}; R = \{a, b, d\}$
41.   **43.**

45.   **47.**

49.

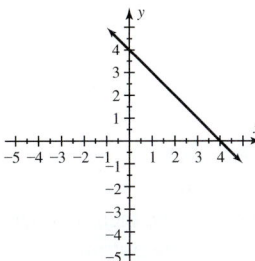

51. 3; -1 **53.** 0; 2   **55.** -4; -3 **57.** 5.5; 3.7
59. 26.9; in 1990 average fuel efficiency was 26.9 mpg.
61. Numerical

x	-3	-2	-1	0	1	2	3
$y = f(x)$	2	3	4	5	6	7	8

Graphical

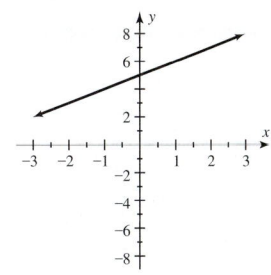

Symbolic: $y = x + 5$
63. Numerical

x	-3	-2	-1	0	1	2	3
$y = f(x)$	-17	-12	-7	-2	3	8	13

Graphical

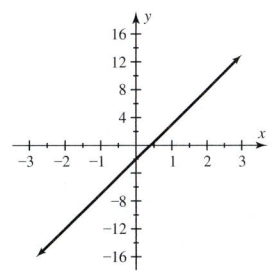

Symbolic: $y = 5x - 2$
65. Numerical   Graphical

$[-10, 10, 1]$ by $[-10, 10, 1]$

X	Y1
-3	0
-2	1
-1	1.4142
0	1.7321
1	2
2	2.2361
3	2.4495

Y1 ≡ √(X+3)

67. Numerical   Graphical (Dot Mode)

$[-10, 10, 1]$ by $[-10, 10, 1]$

X	Y1
-3	4
-2	2.3333
-1	1.5
0	1.
1	.66667
2	.42857
3	.25

Y1 ≡ (5−X)/(5+X)

69. Subtract $\frac{1}{2}$ from the input x to obtain the output y.
71. Divide the input x by 3 to obtain the output y.
73. Subtract 1 from the input x and then take the square root to obtain the output y.

75. $f(x) = 0.50x$

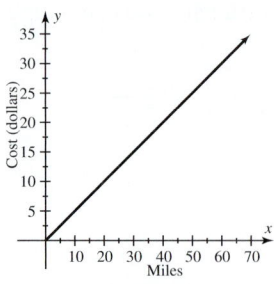

Miles	10	20	30	40	50	60	70
Cost	$5	$10	$15	$20	$25	$30	$35

77. 178.5; in 1995 the average price was $178,500.
79. $D: -2 \le x \le 2$; $R: 0 \le y \le 2$
81. $D: -2 \le x \le 4$; $R: -2 \le y \le 2$
83. D: All real numbers; $R: y \ge -1$
85. $D: -3 \le x \le 3$; $R: -3 \le y \le 2$
87. $D = \{1, 2, 3, 4\}$; $R = \{5, 6, 7\}$ **89.** All real numbers
91. All real numbers   **93.** $x \ne 5$ **95.** All real numbers
97. $x \ge 1$ **99.** All real numbers   **101.** $x \ne 0$
103. (a) 60.3   (b) $D = \{1910, 1930, 1950, 1970, 1990, 2000\}$;
$R = \{35.5, 36.9, 56.2, 60.3, 80.5, 84.4\}$ (c) Decreased
105. No   **107.** Yes   **109.** (a) 0.2   (b) Yes. Each month
has one average amount of precipitation.   (c) 2, 3, 7, 11
111. Yes. D: All real numbers; R: All real numbers   **113.** No
115. Yes. $D: -4 \le x \le 4$; $R: 0 \le y \le 4$ **117.** Yes. D: All
real numbers; $R: y = 3$ **119.** No   **121.** Yes   **123.** No
125. The person walks away from home, then turns around and
walks back a little slower.

127. 10; 10
[−10, 10, 1] by [−10, 10, 1]

129. 10; 5
[0, 100, 10] by [−50, 50, 10]

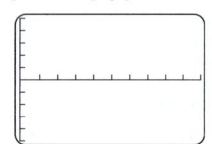

131. 16; 5
[1980, 1995, 1] by [12000, 16000, 1000]

133. [−6, 6, 1] by
[−6, 6, 1]

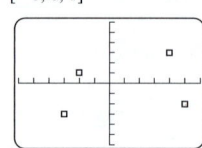

135. [−30, 30, 5] by
[−50, 50, 5]

137. [−200, 200, 50] by
[−250, 250, 50]

SECTION 8.2 (pp. 551–556)

1. $ax + b$ **2.** b **3.** line   **4.** horizontal   **5.** 7   **6.** 0
7. Carpet costs $2 per square foot. Ten square feet of carpet
costs $20.   **8.** The rate at which water is leaving the tank is
4 gallons per minute. After 5 minutes the tank contains
80 gallons of water.   **9.** Yes; $a = \frac{1}{2}$, $b = -6$ **11.** No
13. Yes; $a = 0$, $b = -9$ **15.** Yes; $a = -9$, $b = 0$
17. Yes   **19.** No   **21.** Yes; $f(x) = 3x - 6$
23. Yes; $f(x) = -\frac{3}{2}x + 3$ **25.** No
27. Yes; $f(x) = 2x - 2$ **29.** $-16; 20$ **31.** $\frac{17}{3}; 2$
33. $-22; -22$ **35.** $-2; 0$ **37.** $-1; -4$ **39.** 1; 1
41. $f(x) = 6x; 18$ **43.** $f(x) = \frac{x}{6} - \frac{1}{2}; 0$ **45.** d.   **47.** b.
49.

51.

53.

55.

57.

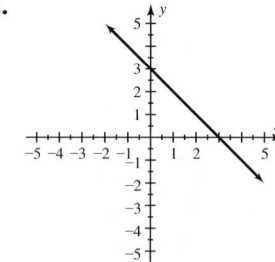

59. $f(x) = \frac{1}{16}x$ **61.** $f(t) = 65t$ **63.** $f(x) = 24$ **65.** a
67. (a) f multiplies the
input x by -2 and then adds 1
to obtain the output y.
(b)

x	-2	0	2
$y = f(x)$	5	1	-3

69. (a) f multiplies the
input x by $\frac{1}{2}$ and then subtracts
1 to obtain the output y.
(b)

x	-2	0	2
$y = f(x)$	-2	-1	0

(c)

(c)

71. (a)

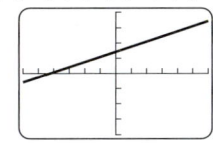

(b) $[-6, 6, 1]$ by $[-4, 4, 1]$

73. (a)

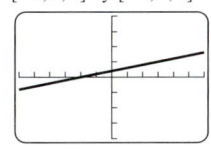

(b) $[-6, 6, 1]$ by $[-4, 4, 1]$

75. b.   **77.** c.   **79.** -1 **81.** $(-1, 0)$ **83.** $(-8, -1)$
85. $\left(-1, -\frac{1}{6}\right)$ **87.** $(0.2, 0.25)$ **89.** $(2005, 9)$
91. $(2a, 2b)$ **93. (a)** 1   **(b)** 1   **(c)** equal
95. 79.6 years   **97.** 227 million
99. (a) Symbolic: $f(x) = 70$
Graphical

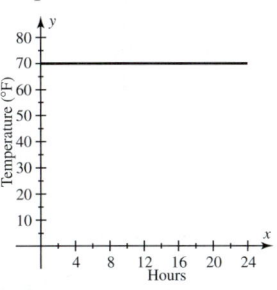

(b)

Hours	0	4	8	12	16	20	24
Temp. (°F)	70	70	70	70	70	70	70

(c) Constant
101. $D(t) = 60t + 50$ **103. (a)** $f(x) = 2.7x$; 162
(b) $g(x) = 4.3x$; 258
105. (a) $[1, 7, 1]$ by $[0, 300, 100]$

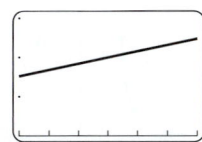

(b) Malware increased.   **(c)** 184; in March there were 184
malware viruses.   **(d)** The number of malware viruses
increased, on average, at a rate of 16 per month.

107. (a) $V(T) = 0.5T + 137$ **(b)** 162 cm³
109. $f(x) = 40x$; about 2.92 pounds
111. (a) 180 minutes   **(b)** Decreased by 4 minutes
(c) $f(x) = -4x + 180$ **(d)** 164 minutes

CHECKING BASIC CONCEPTS 8.1 & 8.2 (p. 556)

1. Symbolic: $f(x) = x^2 - 1$
Graphical

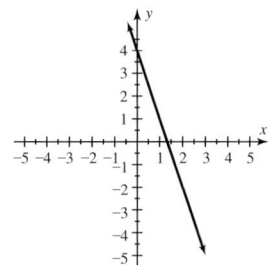

2. (a) $D: -3 \leq x \leq 3$; $R: -4 \leq y \leq 4$ **(b)** 0; 4
(c) No. The graph is not a line.   **3. (a)** Yes   **(b)** No
(c) Yes   **(d)** Yes
4. $f(-2) = 10$

5. $f(x) = \frac{1}{2}x - 1$ **6. (a)** 32.98; in 1990 the median age
was about 33 years.   **(b)** 0.264: the median age is increasing by 0.264 years each year. 27.7: in 1970 the median age
was 27.7 years.   **7.** $(1, -1)$

SECTION 8.3 (pp. 564–567)

1. $x > 1$ and $x \leq 7$ (answers may vary)
2. $x \leq 3$ or $x > 5$ (answers may vary)   **3.** No   **4.** Yes
5. Yes   **6.** Numerically, graphically, symbolically
7. Yes, no   **9.** No, yes   **11.** No, yes   **13.** $[2, 10]$
15. $(5, 8]$ **17.** $(-\infty, 4)$ **19.** $(-2, \infty)$ **21.** $[-2, 5)$
23. $(-8, 8]$ **25.** $(3, \infty)$ **27.** $(-\infty, -2] \cup [4, \infty)$
29. $(-\infty, 1) \cup [5, \infty)$ **31.** $(-3, 5]$ **33.** $(-\infty, -2)$
35. $(-\infty, 4)$ **37.** $(-\infty, 1) \cup (2, \infty)$

39. $\{x \mid -1 \le x \le 3\}$

41. $\{x \mid -2 < x < 2.5\}$

43. $\{x \mid x \le -1 \text{ or } x \ge 2\}$

45. All real numbers

47. $[-6, 7]$ **49.** $\left(\frac{13}{4}, \infty\right)$ **51.** $(-\infty, \infty)$
53. $\left(-\infty, -\frac{11}{5}\right) \cup (-1, \infty)$ **55.** No solutions
57. $[-6, 1)$ **59.** $\left[-\frac{1}{4}, \frac{11}{8}\right)$ **61.** $[-9, 3]$ **63.** $\left[-4, -\frac{1}{4}\right)$
65. $(-1, 1]$ **67.** $[-2, 1]$ **69.** $(0, 2)$ **71.** $(9, 21]$
73. $\left[\frac{12}{5}, \frac{22}{5}\right]$ **75.** $\left(-\frac{13}{3}, 1\right)$ **77.** $[0, 12]$ **79.** $[-1, 2]$
81. $(-1, 2)$ **83.** $[-3, 1]$ **85.** $(-\infty, -2) \cup (0, \infty)$
87. (a) Toward, because distance is decreasing
(b) 4 hours, 2 hours   **(c)** From 2 to 4 hours   **(d)** During
the first 2 hours   **89.** $[1, 4]$ **91.** $(-\infty, -2) \cup (0, \infty)$
93. $[2002, 2010]$ **95.** $[1, 3]$ **97.** $(-\infty, \infty)$
99. $(-\infty, 1) \cup [3, \infty)$ **101.** $(c - b, d - b]$
103. From 1995 to 2000   **105.** From 0.5 to 1.5 miles
107. From $5.\overline{6}$ to 9 feet   **109.** From 1.5 to 2.5 miles
111. $-67.\overline{7}°\text{C}$ to $36.\overline{6}°\text{C}$

SECTION 8.4 (pp. 578–581)

1. domain   **2.** range   **3.** $[1, \infty)$ **4.** $(-\infty, -1)$
5. $(-\infty, \infty)$ **6.** $(-\infty, 5) \cup (5, \infty)$ **7.** absolute value
8. exponent   **9.** 2   **10.** 1   **11.** rational   **12.** $-\frac{1}{2}$
13. $D: (-\infty, \infty); R: (-\infty, \infty)$
15. $D: (-\infty, \infty); R: (-\infty, \infty)$ **17.** $D: (-\infty, \infty); R: [2, \infty)$
19. $D: (-\infty, \infty); R: (-\infty, 0]$ **21.** $D: [-1, \infty); R: [0, \infty)$
23. $D: (-\infty, \infty); R: [0, \infty)$ **25.** $(-\infty, 1) \cup (1, \infty)$
27. $(-\infty, 2) \cup (2, \infty)$
29. $(-\infty, -2) \cup (-2, 2) \cup (2, \infty)$
31. $(-\infty, 0) \cup (0, 2) \cup (2, \infty)$ **33.** $(-\infty, 1) \cup (1, \infty)$
35. $(-\infty, -1) \cup (-1, 3) \cup (3, \infty)$
37. $D: (-\infty, \infty); R: (-\infty, \infty)$ **39.** $D: [-2, 2]; R: [-2, 2]$
41. $D: [-2, 3]; R: [-2, 2]$ **43.** Yes; 1; linear
45. Yes; 3; cubic   **47.** No   **49.** Yes; 2; quadratic
51. No   **53.** Yes; 4; fourth degree   **55.** 12; 0   **57.** 1; $\frac{11}{4}$
59. 0; 6   **61.** 14; -14 **63.** -6; 9   **65.** $\frac{1}{11}$; $-\frac{1}{7}$
67. $-\frac{5}{6}$; undefined   **69.** $\frac{5}{6}$; undefined   **71.** 1; -1
73. -2; -2 **75.** -4; 0   **77.** -1; undefined

79.

81.

83.

85.

87.

89.

91.

93.

95.

97.

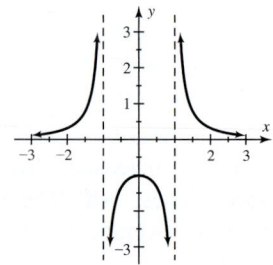

99. (a) 19   **(b)** -9 **(c)** 150   **(d)** 0
101. (a) 41   **(b)** -21 **(c)** 900   **(d)** Undefined
103. (a) $2x + 3$ **(b)** -1 **(c)** $x^2 + 3x + 2$ **(d)** $\frac{x+1}{x+2}$
105. (a) $x^2 - x + 1$ **(b)** $1 - x - x^2$ **(c)** $x^2 - x^3$
(d) $\frac{1-x}{x^2}$ **107. (a)** No; answers may vary.   **(b)** Yes
(c) No; $0 \le t \le 10$ **109. (a)** Yes; walking for 0 minutes
burns 0 calories.   **(b)** Linear; the data lie on a straight line.
(c) $C(t) = \frac{16}{3}t$ **111.** c   **113.** d
115. (a) $8.4\overline{6}$; the elevation of the outer rail should be
about 8.5 inches when the radius of the curve is 300 feet.
(b) $[0, 600, 100]$ by $[0, 50, 10]$ **(c)** It decreases.

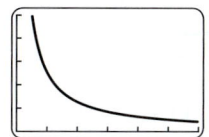

CHECKING BASIC CONCEPTS 8.3 & 8.4 (p. 582)

1. (a) Yes   **(b)** No   **2. (a)** $[-3, 1]$
(b) $(-\infty, -1] \cup [3, \infty)$ **(c)** $\left[-\frac{8}{3}, \frac{8}{3}\right]$ **3. (a)** $(-\infty, \infty)$
(b) $(-\infty, 1) \cup (1, \infty)$ **(c)** $[0, \infty)$
4. (a) D: $[-2, 1]$; R: $[-3, 1]$ **(b)** 1; -3
5.

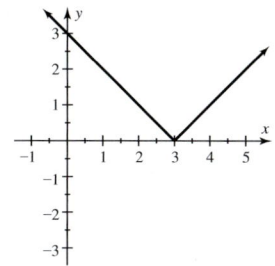

SECTION 8.5 (pp. 589–592)

1. $|3x + 2| = 6$ (answers may vary)
2. $|2x - 1| \le 17$ (answers may vary)   **3.** Yes   **4.** Yes
5. Yes   **6.** No, it is equivalent to $-3 < x < 3$.   **7.** 2
8. 0   **9.** No, yes   **11.** No, yes   **13.** Yes, yes   **15.** 0, 4
17. $-7, 7$ **19.** No solutions   **21.** $-\frac{9}{4}, \frac{9}{4}$ **23.** $-4, 4$
25. $-6, 5$ **27.** 1, 2   **29.** $-8, 12$ **31.** No solutions
33. $-15, 18$ **35.** $-1, \frac{1}{3}$ **37.** $-5, \frac{3}{5}$ **39.** -6
41. (a) $-4, 4$ **(b)** $\{x \mid -4 < x < 4\}$
(c) $\{x \mid x < -4 \text{ or } x > 4\}$ **43. (a)** $\frac{1}{2}, 2$
(b) $\{x \mid \frac{1}{2} \le x \le 2\}$ **(c)** $\{x \mid x \le \frac{1}{2} \text{ or } x \ge 2\}$
45. $[-3, 3]$ **47.** $(-\infty, -4) \cup (4, \infty)$ **49.** No solutions
51. $(-\infty, 0) \cup (0, \infty)$ **53.** $\left(-\infty, -\frac{7}{2}\right) \cup \left(\frac{7}{2}, \infty\right)$
55. $(-3, 5)$ **57.** $(-\infty, -9] \cup [-1, \infty)$ **59.** $\left[\frac{5}{6}, \frac{11}{6}\right]$
61. $[-10, 14]$ **63.** No solutions   **65.** $(-\infty, \infty)$

67. No solutions   **69.** $(-\infty, \infty)$ **71.** $(-\infty, -13] \cup [17, \infty)$
73. $[0.9, 1.1]$ **75.** $(-\infty, 9.5) \cup (10.5, \infty)$ **77. (a)** $-1, 3$
(b) $(-1, 3)$ **(c)** $(-\infty, -1) \cup (3, \infty)$ **79. (a)** $-1, 0$
(b) $[-1, 0]$ **(c)** $(-\infty, -1] \cup [0, \infty)$
81. $(-\infty, -1] \cup [1, \infty)$ **83.** $[-2, 4]$
85. $(-\infty, 1) \cup (3, \infty)$ **87.** $(2, 4.\overline{6})$ **89.** $(-\infty, \infty)$
91. $\{x \mid -3 \le x \le 3\}$ **93.** $\{x \mid x < 2 \text{ or } x > 3\}$
95. $|x| \le 4$ **97.** $|y| > 2$ **99.** $|2x + 1| \le 0.3$
101. $|\pi x| \ge 7$ **103.** two   **105. (a)** $\{T \mid 19 \le T \le 67\}$
(b) Monthly average temperatures vary from 19°F to 67°F.
107. (a) $\{T \mid -26 \le T \le 46\}$ **(b)** Monthly average
temperatures vary from -26°F to 46°F.   **109. (a)** About
19,058 feet   **(b)** Africa and Europe   **(c)** South America,
North America, Africa, Europe, and Antarctica
111. $\{d \mid 2.498 \le d \le 2.502\}$; the diameter can vary from
2.498 to 2.502 inches.   **113.** $|d - 3.8| \le 0.03$
115. Values between 19 and 21, exclusively

CHECKING BASIC CONCEPTS 8.5 (p. 593)

1. $-\frac{28}{3}, 12$ **2. (a)** $-\frac{2}{3}, \frac{14}{3}$ **(b)** $\left(-\frac{2}{3}, \frac{14}{3}\right)$
(c) $\left(-\infty, -\frac{2}{3}\right) \cup \left(\frac{14}{3}, \infty\right)$ **3.** $(0, 6)$; $(-\infty, 0] \cup [6, \infty)$
4. (a) 1, 3   **(b)** $[1, 3]$ **(c)** $(-\infty, 1] \cup [3, \infty)$

CHAPTER 8 REVIEW (pp. 597–601)

1. $-7; 0$ **2.** $-22; 2$ **3.** $-2; 1$ **4.** 5; 5
5. (a) $P(q) = 2q$ **(b)** $P(5) = 10$; there are 10 pints in
5 quarts.   **6. (a)** $f(x) = 4x - 3$ **(b)** $f(5) = 17$; three
less than four times 5 is 17.   **7.** $(3, -2)$ **8.** 4; -6
9.   **10.**

11.   **12.**

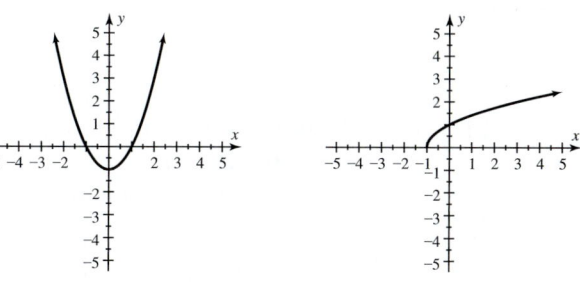

13. 1; 4   **14.** 1; -2 **15.** 7; -1

16. Numerical

x	-3	-2	-1	0	1	2	3
$y = f(x)$	-11	-8	-5	-2	1	4	7

Symbolic: $f(x) = 3x - 2$

Graphical

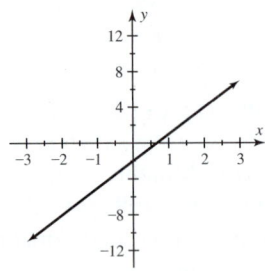

17. D: All real numbers; $R: y \le 4$ **18.** $D: -4 \le x \le 4$; $R: -4 \le y \le 0$ **19.** Yes   **20.** No

21. $D = \{-3, -1, 2, 4\}$; $R = \{-1, 3, 4\}$; yes

22. $D = \{-1, 0, 1, 2\}$; $R = \{-2, 2, 3, 4, 5\}$; no

23. All real numbers   **24.** $x \ge 0$ **25.** $x \ne 0$

26. All real numbers   **27.** $x \le 5$ **28.** $x \ne 2$

29. All real numbers   **30.** All real numbers

31. No   **32.** Yes   **33.** Yes; $a = -4, b = 5$ **34.** No

35. Yes; $f(x) = \frac{3}{2}x - 3$ **36.** No   **37.** 1

38. $f(-2) = -3; f(1) = 0$

39.

40.

41.

42.

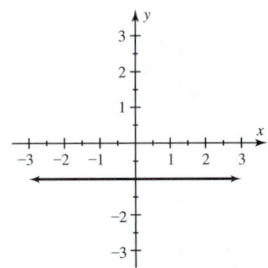

43. $H(x) = 24x$; $H(2) = 48$; there are 48 hours in 2 days.

44. (a)

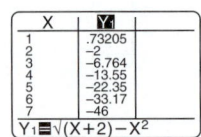

(b) Domain: $x \ge 2$

$[-10, 10, 1]$ by $[-10, 10, 1]$

45. $(0.5, -3)$ **46.** $\left(\frac{5}{12}, \frac{3}{8}\right)$

47. $[-2, 2]$

48. $(-\infty, -3]$

49. $\left(-\infty, \frac{4}{5}\right] \cup (2, \infty)$

50. $(-\infty, \infty)$

51. $[-2, 1]$ **52. (a)** -4 **(b)** 2 **(c)** $[-4, 2]$
(d) $(-\infty, 2)$ **53. (a)** 2 **(b)** $(2, \infty)$ **(c)** $(-\infty, 2)$

54. (a) 4 **(b)** 2 **(c)** $(2, 4)$ **55.** $\left[-3, \frac{2}{3}\right]$ **56.** $(-6, 45]$

57. $\left(-\infty, \frac{7}{2}\right)$ **58.** $[1.8, \infty)$ **59.** $(-3, 4)$

60. $(-\infty, 4) \cup (10, \infty)$ **61.** $(-5, 5)$ **62.** $[8, 28]$

63. $(-9, 21)$ **64.** $[8, 28]$ **65.** $\left(-\frac{11}{5}, \frac{7}{5}\right]$ **66.** $[88, 138)$

67. $D = (-\infty, \infty)$; $R = [0, \infty)$

68. $D = (-\infty, \infty)$; $R = [0, \infty)$

69. $(-\infty, 4) \cup (4, \infty)$ **70.** $D = [-3, 1]$; $R = [-3, 6]$

71. Yes; 2; quadratic   **72.** Yes; 1; linear

73. Yes; 3; cubic   **74.** No   **75.** 11; 2   **76.** $-\frac{4}{5}$; undefined

77.

78.

79.

80.

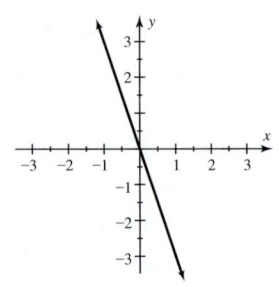

81. (a) 12   **(b)** 27   **82. (a)** $x^2 - x$ **(b)** $x + 1$

83. No, no   **84.** No, yes   **85.** No, yes   **86.** No, yes

87. (a) 0, 4   **(b)** $(0, 4)$ **(c)** $(-\infty, 0) \cup (4, \infty)$

88. (a) $-3, 1$ **(b)** $[-3, 1]$ **(c)** $(-\infty, -3] \cup [1, \infty)$

89. $-22, 22$ **90.** 1, 8   **91.** $-26, 42$ **92.** $-\frac{23}{3}, \frac{25}{3}$

93. $0, 2$ **94.** $-3, \frac{9}{5}$ **95. (a)** $-8, 6$ **(b)** $[-8, 6]$
(c) $(-\infty, -8] \cup [6, \infty)$ **96. (a)** $-\frac{5}{2}, \frac{7}{2}$ **(b)** $\left[-\frac{5}{2}, \frac{7}{2}\right]$
(c) $\left(-\infty, -\frac{5}{2}\right] \cup \left[\frac{7}{2}, \infty\right)$ **97.** $(-\infty, -3) \cup (3, \infty)$
98. $(-4, 4)$ **99.** $[-3, 4]$ **100.** $\left(-\infty, -\frac{5}{2}\right] \cup [5, \infty)$
101. $[4.4, 4.6]$ **102.** $\left[\frac{3}{13}, \frac{7}{13}\right]$ **103.** $(-\infty, \infty)$ **104.** $\frac{3}{2}$
105. $(-\infty, -1.5] \cup [1.5, \infty)$ **106.** $[-2, 6]$
107. $|x| \leq 0.05$ **108.** $|5x - 1| > 4$
109. (a) About 25.1
(b) Decreased
[1885, 1965, 10] by [22, 26, 1]

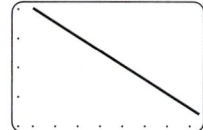

(c) -0.0492; the median age decreased by about 0.0492
year per year.
110. (a) 2.4 million   **(b)** The number of marriages each
year did not change.   **111. (a)** $f(x) = 8x$ **(b)** 8
(c) The total fat increases at the rate of 8 grams per cup.
112. (a)

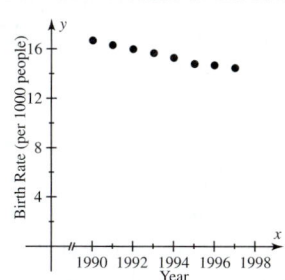

(b) $f(x) = -0.31x + 633.6$ **(c)** About 13.6 per 1000
people (answers may vary)
113. (a) 113; in 1995 there were 113 unhealthy days.
(b) $D = \{1995, 1997, 1999, 2000, 2003\}$
$R = \{56, 60, 87, 88, 113\}$
(c) It decreased and then increased.
114. (a) Linear

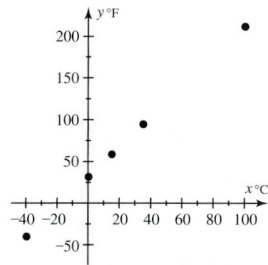

(b) $f(x) = \frac{9}{5}x + 32$; a 1°C change equals $\frac{9}{5}$°F change.
(c) 68°F

115. (a) 73.6°F   **(b)** July

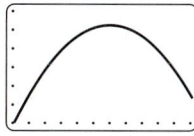

(c) Temperatures increase from January through July and
then decrease from July through December.
[1, 12, 1] by [30, 90, 10]

116. 13 feet by 31 feet   **117.** $-44.\overline{4}$°C to $41.\overline{6}$°C
118. $|L - 160| \leq 1$; $159 \leq L \leq 161$
119. (a) $|A - 3.9| \leq 1.7$ **(b)** $2.2 \leq A \leq 5.6$
120. Values between 32.2 and 37.8

CHAPTER 8 TEST (pp. 602–603)

1. $46, (4, 46)$ **2.** $C(x) = 4x$; $C(5) = 20$, 5 pounds of
candy costs $20.
3. (a)   **(b)**

(c)   **(d)**

        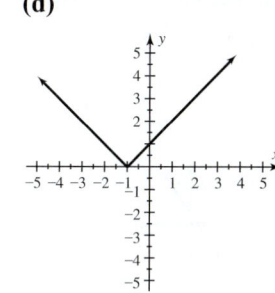

4. $0, -3$; $D: -3 \leq x \leq 3$, $R: -3 \leq y \leq 0$

5. Symbolic: $f(x) = x^2 - 5$

Numerical

x	-3	-2	-1	0	1	2	3
y = f(x)	4	-1	-4	-5	-4	-1	4

Graphical

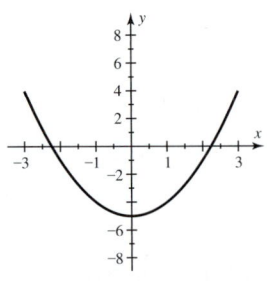

6. No, it fails the vertical line test.

7. (a) $D = \{-2, -1, 0, 5\}$ **(b)** All real numbers
(c) $x \geq -4$ **(d)** All real numbers **(e)** $x \neq 5$

8. It is; $f(x) = -8x + 6$

9.

10. $(-\infty, -2) \cup (1, \infty)$ **11. (a)** -5 **(b)** 5
(c) $[-5, 5]$ **(d)** $(-\infty, 5)$ **12.** $(-8, 0)$ **13.** $-12, 24$
14. (a) $[-5, 5]$ **(b)** $(-\infty, 0) \cup (0, \infty)$ **(c)** No solutions
(d) $(-\infty, -2) \cup (2, \infty)$ **(e)** $(-\infty, \infty)$ **(f)** $(-2, 8)$
15. (a) $-\frac{2}{5}, \frac{4}{5}$ **(b)** $\left[-\frac{2}{5}, \frac{4}{5}\right]$ **(c)** $\left(-\infty, -\frac{2}{5}\right] \cup \left[\frac{4}{5}, \infty\right)$
16. $D: (-\infty, \infty); R: [0, \infty)$ **17.** Yes; 3; cubic
18. $\frac{8}{7}; (-\infty, 5) \cup (5, \infty)$ **19. (a)** 9 **(b)** $2x^3 + 2x$
20. (a) $f(x) = 0.4x + 75$ **(b)** 35 minutes
21. (a) Yes; $P(0) = 100, P(20) = 134, P(30) = 150,$
$P(50) = 180$ **(b)** Probably not because the race is over
after 50 seconds; $[0, 50]$.

CHAPTERS 1–8 CUMULATIVE REVIEW (pp. 604–605)

1. $2^3 \cdot 3 \cdot 5$ **2.** $3n - 4 = n; 2$ **3.** $-\frac{1}{6}$ **4.** 6 **5.** 15
6. 7.5% **7.** $a = \frac{3A + b}{2}$ **8.** $\{t \mid t > 2\}$
9.   **10.**

11. $y = -\frac{1}{2}x + 2$ **12.** $(1, 1)$

13.

14.

15.

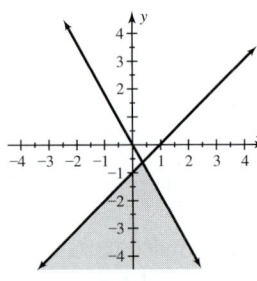

16. $-x^2 + 2x - 9$ **17.** $\frac{1}{x}$ **18.** $6b$ **19.** $24t^3$ **20.** $\frac{1}{4x^2y^2}$
21. $2x^5 - 8x^4 - 10x^2$ **22.** $10x^2 - 33x - 7$ **23.** $y^2 - 9$
24. $x^2 - 8xy + 16y^2$ **25.** 25,000 **26.** 2.8×10^{-2}
27. $3x^2 - 2x + 4$ **28.** $3x^2 + 5x + 5 + \frac{6}{x - 1}$
29. $5x^2y^2(2y - 3x)$ **30.** $(x - 1)(x + 1)(x + 3)$
31. $(z - 1)(2z + 3)$ **32.** $(4x - 5)(4x + 5)$
33. $(a - 2)(a^2 + 2a + 4)$ **34.** $(z^2 + 1)(z^2 + 6)$
35. $-5, 0$ **36.** 0 **37.** $-3, \frac{1}{2}$ **38.** $-1, 0, 1$
39. $x + 2$ **40.** $-\frac{5}{3};$ undefined **41.** $2x - 4$
42. $\frac{3x + 7}{(x + 1)(x + 3)}$ **43.** -8 **44.** 10 **45.** 12
46. $D:(-\infty, \infty); R:[-2, \infty)$

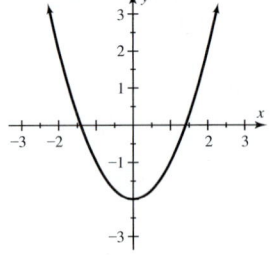

47. $(-\infty, 5) \cup (8, \infty)$ **48.** $-1, 5$ **49.** $[-1, 5]$
50. $(-\infty, 2) \cup (6, \infty)$ **51.** $d = 72t$
52. (a) $y = \frac{1}{4}x + 2$ **(b)** $\frac{1}{4}$ **(c)** Rain is falling at a rate
of $\frac{1}{4}$ inch per hour. **(d)** 3 inches **53.** 2100 at 5%, $2900
at 7% **54.** 104 feet **55. (a)** $20x$ **(b)** 50 min **56.** $1650

CHAPTER 9:   SYSTEMS OF LINEAR EQUATIONS

SECTION 9.1 (pp. 617–620)

1. No; three planes cannot intersect at exactly 2 points.
2. $x + y + z = 5, 2x - 3y + z = 7, x + 2y - 4z = 2$
(answers may vary)   **3.** Yes
4. No; a solution must be an ordered triple.   **5.** Two
6. Three   **7.** no   **8.** Infinitely many   **9.** $(1, 2, 3)$
11. $(-1, 1, 2)$ **13.** $(3, -1, 1)$ **15.** $\left(\frac{17}{2}, -\frac{3}{2}, 2\right)$
17. $(5, -2, -2)$ **19.** $(11, 2, 2)$ **21.** $(1, -1, 2)$
23. $(0, -3, 2)$ **25.** $(-1, 3, -2)$ **27.** $(-1, 2, 4)$
29. $(8, 5, 2)$ **31.** No solutions   **33.** $(4 - z, 2, z)$
35. $(3, -3, 0)$ **37.** $\left(-\frac{3}{2}, 5, -10\right)$ **39.** $\left(\frac{3}{2}, 1, -\frac{1}{2}\right)$
41. No solutions   **43.** $(4 - z, 1, z)$ **45.** No solutions
47. (a) $x + 2y + 4z = 10, x + 4y + 6z = 15,$
$3y + 2z = 6$ **(b)** $(2, 1, 1.5)$; a hamburger costs \$2,
fries \$1, and a soft drink \$1.50.   **49. (a)** $x + y + z = 180,$
$x - z = 55, x - y - z = -10$ **(b)** $x = 85°, y = 65°,$
and $z = 30°$ **(c)** These values check.   **51.** $110°, 50°, 20°$
53. (a) $a + 600b + 4c = 525, a + 400b + 2c = 365,$
$a + 900b + 5c = 805$ **(b)** $a = 5, b = 1, c = -20,$
$F = 5 + A - 20W$ **(c)** 445 fawns
55. (a) $N + P + K = 80, N + P - K = 8, 9P - K = 0$
(b) $(40, 4, 36)$; 40 pounds nitrogen, 4 pounds phosphorus,
36 pounds potassium   **57.** \$7500 at 8%, \$8500 at 10%, and
\$14,000 at 15%

SECTION 9.2 (pp. 629–632)

1. A rectangular array of numbers

2. $\begin{bmatrix} 2 & 1 & 3 \\ 0 & -4 & 2 \end{bmatrix}$ is 2×3 (answers may vary).

3. $\left[\begin{array}{cc|c} 1 & 3 & 10 \\ 2 & -6 & 4 \end{array}\right]$ is 2×3 (answers may vary).

4. 3×4 **5.** $\left[\begin{array}{cc|c} 1 & 0 & -3 \\ 0 & 1 & 5 \end{array}\right]$ (answers may vary)

6. $4, 2, -3$ **7.** 3×3 **9.** 3×2

11. $\left[\begin{array}{cc|c} 1 & -3 & 1 \\ -1 & 3 & -1 \end{array}\right]$ **13.** $\left[\begin{array}{ccc|c} 2 & -1 & 2 & -4 \\ 1 & -2 & 0 & 2 \\ -1 & 1 & -2 & -6 \end{array}\right]$

15. $x + 2y = -6, 5x - y = 4$ **17.** $x - y + 2z = 6,$
$2x + y - 2z = 1, -x + 2y - z = 3$
19. $x = 4, y = -2, z = 7$ **21.** $(1, 3)$ **23.** $\left(-\frac{3}{2}, 2\right)$
25. $\left(\frac{7}{2}, -\frac{3}{2}\right)$ **27.** $\left(\frac{1}{2}, \frac{3}{2}\right)$ **29.** $(1, 2, 3)$ **31.** $(3, -3, 3)$
33. $(-1, 1, 0)$ **35.** $(1, 1, 1)$ **37.** $(-3, 2, 2)$
39. $(-7, 5)$ **41.** $(1, 2, 3)$ **43.** $(1, 0.5, -3)$
45. $(0.5, 0.25, -1)$ **47.** $(0.5, -0.2, 1.7)$
49. Dependent   **51.** Inconsistent   **53.** Dependent

55. $\left(\frac{1}{a}, \frac{1}{b}, \frac{2}{ab}\right)$ **57.** 214 pounds
59. (a) $a + 2b + 1.4c = 3, a + 1.5b + 0.65c = 2,$
$a + 4b + 3.4c = 6$ **(b)** $a = 0.6, b = 0.5,$ and $c = 1$
(c) $4.1 \approx 4$ people   **61.** $\frac{1}{2}$ hour at 5 miles per hour, 1 hour
at 6 miles per hour, and $\frac{1}{2}$ hour at 8 miles per hour
63. \$500 at 5%, \$1000 at 8%, and \$1500 at 12%

CHECKING BASIC CONCEPTS 9.1 & 9.2 (p. 632)

1. $(1, 3, -1)$ **2.** $(1, 2, 3)$ **3.** $(-2, 2, -1)$

SECTION 9.3 (pp. 638–639)

1. square   **2.** number   **3.** system of linear equations
4. 0   **5.** -2 **7.** -53 **9.** -323 **11.** -5 **13.** -36
15. -42 **17.** -50 **19.** 0   **21.** -3555 **23.** -7466.5
25. abc **27.** 15 square feet   **29.** 52 square feet
31. 25.5 square feet   **33.** $(2, -2)$
35. $\left(-\frac{29}{11}, -\frac{34}{11}\right)$ **37.** $(-5, 7)$

CHECKING BASIC CONCEPTS 9.3 (p. 640)

1. (a) -1 **(b)** 17   **2.** $x = -4, y = 6$
3. 21 square units

CHAPTER 9 REVIEW (pp. 642–643)

1. Yes   **2.** $(1, -1, 2)$ **3.** $(-4, 3, 2)$
4. $(-1, 5, -3)$ **5.** $(-1, 3, 2)$ **6.** $(1, 1, 1)$
7. No solutions   **8.** $(2 - z, 2z - 1, z)$

9. $\left[\begin{array}{ccc|c} 1 & 1 & 1 & -6 \\ 1 & 2 & 1 & -8 \\ 0 & 1 & 1 & -5 \end{array}\right]$; $(-1, -2, -3)$

10. $\left[\begin{array}{ccc|c} 1 & 1 & 1 & -3 \\ -1 & 1 & 0 & 5 \\ 0 & 1 & 1 & -1 \end{array}\right]$; $(-2, 3, -4)$

11. $\left[\begin{array}{ccc|c} 1 & 2 & -1 & 1 \\ -1 & 1 & -2 & 5 \\ 0 & 2 & 1 & 10 \end{array}\right]$; $(-5, 4, 2)$

12. $\left[\begin{array}{ccc|c} 2 & 2 & -2 & -14 \\ -2 & -3 & 2 & 12 \\ 1 & 1 & -4 & -22 \end{array}\right]$; $(-4, 2, 5)$

13. $(-7, 4, 2)$ **14.** $(5.4, 2.1, 9.7)$ **15.** -8 **16.** 30
17. 89   **18.** 130   **19.** 181,845   **20.** 67.688
21. 46 square feet   **22.** 128 square feet   **23.** $(2, -1)$
24. $(-5, 3)$ **25.** $\left(\frac{3}{2}, \frac{1}{2}\right)$ **26.** $(7, -3)$ **27.** 5489 in 1994;
4641 in 2004   **28.** \$8 tickets: 285; \$12 tickets: 195
29. (a) $m + 3c + 5b = 14, m + 2c + 4b = 11,$
$c + 3b = 5$ **(b)** Malts: \$3; cones: \$2; bars: \$1

30. $100°, 65°$, and $15°$ **31.** 2 pounds of $1.50 candy, 4 pounds of $2 candy, 6 pounds of $2.50 candy
32. (a) $a + 202b + 63c = 40, a + 365b + 70c = 50,$ $a + 446b + 77c = 55$ **(b)** $a \approx 27.134; b \approx 0.061,$ $c \approx 0.009$ **(c)** About 46 inches

CHAPTER 9 TEST (pp. 643–644)

1. No. Three planes cannot intersect at exactly three points.
2. $(1, -2, -2)$ **3.** $(-4, 2, -5)$
4. No solutions   **5.** $(1, 2, 3)$

6. (a) $\begin{bmatrix} 2 & -4 & | & -10 \\ -3 & -2 & | & 7 \end{bmatrix}$ **(b)** $(-3, 1)$

7. (a) $\begin{bmatrix} 1 & 1 & 1 & | & 2 \\ 1 & -1 & -1 & | & 3 \\ 2 & 2 & 1 & | & 6 \end{bmatrix}$ **(b)** $\left(\frac{5}{2}, \frac{3}{2}, -2\right)$

8. 114   **9.** $\left(-\frac{47}{2}, -\frac{83}{2}\right)$

10. 6 mph: 30 min; 7 mph: 12 min; 9 mph: 18 min
11. $85°, 60°$, and $35°$
12. (a) $a + 20b + 25c = 168; a + 24b + 40c = 270;$ $a + 30b + 50c = 405$ **(b)** $a = -270, b = 20.1,$ $c = 1.44$ **(c)** About 316 lb

CHAPTERS 1–9 CUMULATIVE REVIEW (pp. 646–648)

1. $2^3 \cdot 3^2 \cdot 5$ **2.** $2n + 7 = n - 2; -9$
3. $\frac{23}{21}$ **4.** $-\frac{1}{6}$ **5.** 18   **6.** $\frac{1}{10}$ **7.** $\frac{6}{5}$
8. 2   **9.** $\frac{31}{25}; 1.24$ **10.** 6 miles
11. $b = \frac{2A}{h} - a$ **12.** $t < \frac{4}{7}$

13.
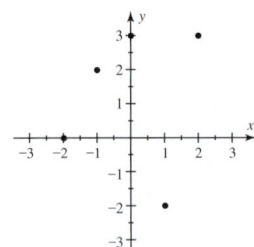

14. x-int: 4; y-int: 2

15. x-int: -4; y-int: 3
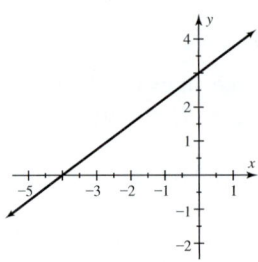

16. x-int: -2; y-int: none
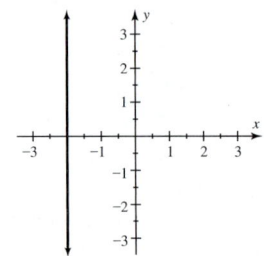

17. x-int: 2; y-int: 4; $y = -2x + 4$
18. $y = -2x + 1$ **19.** $y = \frac{3}{5}x + \frac{34}{5}$ **20.** $y = 2x + 3$
21. $P = 500x + 4000$ **22.** $(7.5, 10)$
23.   **24.**

25.   **26.**

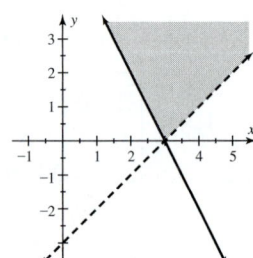

27. -76 **28.** $\frac{27}{4}$ **29.** 1   **30.** $24t^4$ **31.** $2r^4t^4$ **32.** $\frac{1}{4t^6}$
33. $2ab^3$ **34.** $\frac{a^6}{8b^6}$ **35.** $2a^4 - 4a^3 + 6a^2$
36. $5x^2 - 34x - 7$ **37.** $2y^2 - 3y - 9$ **38.** $a^2 - b^2$
39. $4x^2 + 12xy + 9y^2$ **40.** 2.13×10^6
41. $2x^2 - 4x + 3$ **42.** $x^3 - 4x^2 + 3x - 2 + \frac{1}{x - 5}$
43. $5ab^2(2 - 5a^2b^3)$ **44.** $(y - 3)(y^2 + 2)$
45. $(2z + 3)(3z - 1)$ **46.** $(2z - 3)(2z + 3)$
47. $(2y - 5)^2$ **48.** $(a - 3)(a^2 + 3a + 9)$

49. $(z^2 - 3)(4z^2 - 5)$ **50.** $ab(a + b)(2a - b)$
51. $-2, 1$ **52.** $0, 9$ **53.** $-\frac{1}{3}, \frac{3}{2}$ **54.** $-2, 0, 2$
55. $x - 4$ **56.** $\frac{x - 6}{3x - 4}$ **57.** $2(x - 1)$ **58.** 1
59. $\frac{2x + 3}{(x - 1)(x + 1)}$ **60.** $\frac{y - x}{y + x}$
61. 8 **62.** $\frac{17}{12}$ **63.** 1 **64.** 10 **65.** 13
66. $D = (-\infty, \infty);\ R = [-2, \infty)$

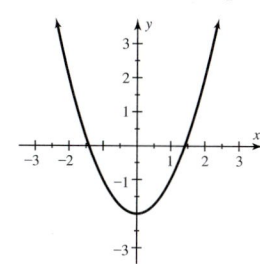

67. $2, -2$ **68.** $2, -4$ **69.** $(3, \infty)$ **70.** $(-\infty, 0]$
71. $(-1, 5]$

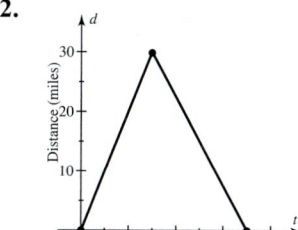

72. $\left(-\infty, -\frac{3}{2}\right) \cup [3, \infty)$

73. $[-2, 4]$ **74.** $[-16, 10]$
75. **(a)** $-3, 1$ **(b)** $[-3, 1]$ **(c)** $(-\infty, -3] \cup [1, \infty)$
76. $-6, 18$ **77.** $(-\infty, -6) \cup \left(\frac{8}{3}, \infty\right)$ **78.** $[4, 7]$
79. $(1, 1, 2)$ **80.** $(-1, 3, -2)$ **81.** 14
82.

83. 3 children and 5 adults   **84.** $\frac{6}{7}$ hour
85. **(a)** 112 feet   **(b)** 5.5 seconds
86. 61.25 feet
87. **(a)** $\frac{1}{5}$ minute; the average wait is $\frac{1}{5}$ minute or 12 seconds when vehicles arrive at 15/min.
    (b)

x	5	10	15	19	19.9
T	$\frac{1}{15}$	$\frac{1}{10}$	$\frac{1}{5}$	1	10

    (c) The waiting time increases dramatically.
88. $73.63 \le L \le 74.37$
89. **(a)** $4b + 3f + 4m = 23;\ b + 2f + m = 7;$
$3b + f + 2m = 13$ **(b)** Burgers: \$2; fries: \$1; malt: \$3

CHAPTER 10: RADICAL EXPRESSIONS AND FUNCTIONS

SECTION 10.1 (pp. 659–661)

1. ± 3 **2.** 3 **3.** 2 **4.** Yes   **5.** b **6.** $x^{1/2}$ **7.** $a^{4/3}$
8. $\sqrt[n]{a}$ **9.** $(\sqrt[n]{a})^m$ or $\sqrt[n]{a^m}$ **10.** Yes   **11.** 3 **13.** 0.6
15. $\frac{4}{5}$ **17.** x **19.** 3 **21.** -4 **23.** $\frac{2}{3}$ **25.** $-x^3$
27. $4x^2$ **29.** 3 **31.** -3 **33.** Not possible   **35.** -2.24
37. 1.71 **39.** -1.48 **41.** 1.74 **43.** 1.71 **45.** 3.74
47. 0.55 **49.** $\sqrt{9};\ 3$ **51.** $\sqrt[3]{8};\ 2$ **53.** $\sqrt{\frac{4}{9}};\ \frac{2}{3}$
55. $\sqrt[3]{(-8)^2}$ or $\left(\sqrt[3]{-8}\right)^2;\ 4$ **57.** $\sqrt[3]{8};\ 2$
59. $\frac{1}{\sqrt[4]{16^3}}$ or $\frac{1}{\left(\sqrt[4]{16}\right)^3};\ \frac{1}{8}$ **61.** $\frac{1}{\left(\sqrt{4}\right)^3};\ \frac{1}{8}$ **63.** $\sqrt[4]{z}$
65. $\frac{1}{\sqrt[5]{y^2}}$ **67.** $\sqrt[3]{3x}$ **69.** $y^{1/2}$ **71.** x **73.** $2x^{2/3}$
75. $\frac{7}{x^{1/6}}$ **77.** b^3 **79.** x^3 **81.** xy^2 **83.** $\frac{y^2}{x}$
85. $y^{13/6}$ **87.** $\frac{x^4}{9}$ **89.** $\frac{y^3}{x}$ **91.** $y^{1/4}$ **93.** $\frac{1}{a^{2/3}}$
95. $\frac{1}{k^2}$ **97.** $b^{3/4}$ **99.** $p^2 + p$ **101.** $x^{5/6} - x$
103. 4 **105.** $|y|$ **107.** $|x - 5|$ **109.** $|x - 1|$
111. $|y|$ **113.** $|x^3|$ **115.** x **117.** $R \approx 1.17$ miles
119. About 1 step/sec   **121.** $A = 6$

SECTION 10.2 (pp. 667–669)

1. Yes   **2.** No   **3.** $\sqrt[3]{ab}$ **4.** $\frac{a}{b}$ **5.** $\frac{a}{b}$ **6.** $\frac{1}{3}$ **7.** No
8. Yes   **9.** No, since $1^3 \ne 3$ **10.** Yes; $4^3 = 64$
11. 3 **13.** 10 **15.** 4 **17.** $\frac{3}{5}$ **19.** $\frac{1}{4}$ **21.** $\frac{2}{3}$ **23.** x^3
25. $\frac{\sqrt[3]{7}}{3}$ **27.** $\frac{\sqrt[4]{x}}{3}$ **29.** $\frac{3}{z}$ **31.** $\frac{x}{4}$ **33.** 3 **35.** 4 **37.** 3
39. -2 **41.** a **43.** 3 **45.** $2x^2$ **47.** $-a^2\sqrt[3]{5}$
49. $2x\sqrt[4]{y}$ **51.** $6x$ **53.** $2x^2yz^3$ **55.** $\frac{3}{2}$
57. $\sqrt[3]{12ab}$ **59.** $5\sqrt{z}$ **61.** $\frac{1}{a}$ **63.** $\sqrt{x^2 - 16}$
65. $\sqrt[3]{a^3 + 1}$ **67.** $\sqrt{x + 1}$ **69.** 10 **71.** 2
73. 3 **75.** $10\sqrt{2}$ **77.** $3\sqrt[3]{3}$ **79.** $2\sqrt{2}$
81. $-2\sqrt[5]{2}$ **83.** $b^2\sqrt{b}$ **85.** $2n\sqrt{2n}$
87. $2ab^2\sqrt{3b}$ **89.** $-5xy\sqrt[3]{xy^2}$ **91.** $5\sqrt[3]{5t^2}$
93. $\frac{3t}{r\sqrt[4]{5r^3}}$ **95.** $\frac{3\sqrt[3]{x^2}}{y}$ **97.** $\frac{7a\sqrt{a}}{9}$
99. $\left(\sqrt[mn]{a^m b^m}\right)^n = (a^m b^m)^{n/mn}$
$= (a^m b^m)^{1/m}$
$= a^{m/m} b^{m/m}$
$= ab$
101. $\sqrt[6]{3^5}$ **103.** $2\sqrt[12]{2^5}$ **105.** $3\sqrt[12]{3^{11}}$ **107.** $x\sqrt[12]{x}$
109. $\sqrt[12]{r^{11}t^7}$ **111.** **(a)** 400 in^2 **(b)** $A = 100W^{2/3}$

CHECKING BASIC CONCEPTS 10.1 & 10.2 (p. 669)

1. (a) ± 7 (b) 7 **2.** (a) -2 (b) -3

3. (a) $\sqrt{x^3}$ or $(\sqrt{x})^3$ (b) $\sqrt[3]{x^2}$ or $(\sqrt[3]{x})^2$

(c) $\frac{1}{\sqrt[5]{x^2}}$ or $\frac{1}{(\sqrt[5]{x})^2}$ **4.** $|x - 1|$ **5.** (a) $\frac{1}{8}$ (b) 10

(c) $-2x\sqrt[3]{xy}$ (d) $\frac{4b^2}{5}$ **6.** $\sqrt[6]{7^5}$

SECTION 10.3 (pp. 679–680)

1. $2\sqrt{a}$ **2.** $3\sqrt[3]{b}$ **3.** like

4. Yes; $4\sqrt{15} - 3\sqrt{15} = \sqrt{15}$ **5.** No **6.** $\frac{\sqrt{7}}{\sqrt{7}}$

7. $\sqrt{t + 5}$ **8.** $\frac{5 + \sqrt{2}}{5 + \sqrt{2}}$ **9.** Not possible

11. $\sqrt{7}, 2\sqrt{7}, 3\sqrt{7}$ **13.** $2\sqrt[3]{2}, -3\sqrt[3]{2}$

15. Not possible **17.** $2\sqrt[3]{xy}, xy\sqrt[3]{xy}$ **19.** $9\sqrt{3}$

21. $6\sqrt[3]{5}$ **23.** Not possible **25.** Not possible

27. Not possible **29.** $5\sqrt[3]{2}$ **31.** $8\sqrt{2}$ **33.** $6\sqrt{11}$

35. $2\sqrt{x} - \sqrt{y}$ **37.** $2\sqrt[3]{z}$ **39.** $-5\sqrt[3]{6}$ **41.** $y^2 - y$

43. $9\sqrt{7}$ **45.** $6\sqrt[4]{3}$ **47.** $7\sqrt{x}$ **49.** $8\sqrt{2k}$

51. $-2\sqrt{11}$ **53.** $5\sqrt[3]{2} - \sqrt{2}$ **55.** $-\sqrt[3]{xy}$

57. $3\sqrt{x + 2}$ **59.** $4x\sqrt{x}$ **61.** $\frac{\sqrt[3]{7x}}{6}$ **63.** $\frac{3\sqrt{3}}{2}$

65. $\frac{71\sqrt{2}}{10}$ **67.** $2\sqrt{2}$ **69.** $(5x - 1)\sqrt[4]{x}$

71. $(8x + 2)\sqrt{x}$ or $2\sqrt{x}(4x + 1)$ **73.** $(3ab - 1)\sqrt[4]{ab}$

75. $(n - 2)\sqrt[3]{n}$ **77.** $x - \sqrt{x} - 6$ **79.** 2 **81.** 119

83. $x - 64$ **85.** $ab - c$ **87.** $x + \sqrt{x} - 56$ **89.** $\frac{\sqrt{7}}{7}$

91. $\frac{4\sqrt{3}}{3}$ **93.** $\frac{\sqrt{5}}{3}$ **95.** $\frac{\sqrt{3b}}{6}$ **97.** $\frac{t\sqrt{r}}{2r}$ **99.** $\frac{3 + \sqrt{2}}{7}$

101. $\sqrt{10} - 2\sqrt{2}$ **103.** $\frac{11 - 4\sqrt{7}}{3}$ **105.** $\sqrt{7} + \sqrt{6}$

107. $\frac{z + 3\sqrt{z}}{z - 9}$ **109.** $\frac{a + 2\sqrt{ab} + b}{a - b}$ **111.** $\sqrt{x + 1} + \sqrt{x}$

113. $\frac{3\sqrt[3]{x^2}}{x}$ **115.** $\frac{\sqrt[3]{x}}{x}$ **117.** $12\sqrt{3} \approx 20.8$ cm

SECTION 10.4 (pp. 689–692)

1.

2.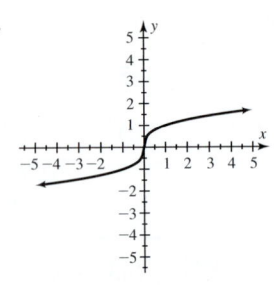

3. $\{x \mid x \geq 0\}$ **4.** All real numbers **5.** $f(x) = x^p$,

p is rational. **6.** $f(x) = \sqrt[n]{x}$, where n is an integer greater than 1. **7.** $\{x \mid x \geq 0\}$ **8.** All real numbers **9.** 3, Not possible **11.** $\sqrt{6}$, Not possible **13.** $2\sqrt{5}, \sqrt{6}$ **15.** 1, 2

17. $-2, 1$ **19.** $-1, \sqrt[3]{-6}$ or $-\sqrt[3]{6}$ **21.** 4 **23.** 5

25.

x	$\sqrt{x} + 1$
-1	—
0	1
1	2
4	3
9	4

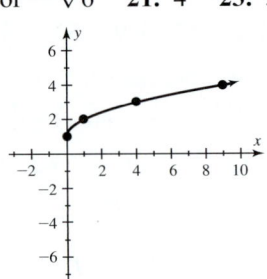

27.

x	$\sqrt{3x}$
-1	—
0	0
$\frac{1}{3}$	1
$\frac{4}{3}$	2
3	3

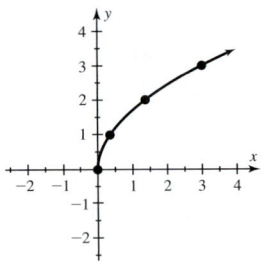

29.

x	$2\sqrt[3]{x}$
-8	-4
-1	-2
0	0
1	2
8	4

 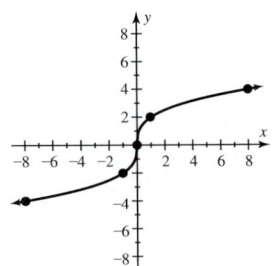

31.

x	$\sqrt[3]{x - 1}$
-7	-2
0	-1
1	0
2	1
9	2

 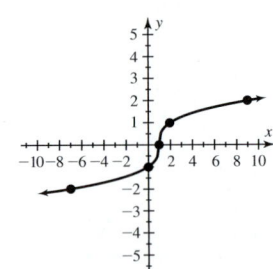

33. $[-2, \infty)$ **35.** $[2, \infty)$ **37.** $[2, \infty)$ **39.** $(-\infty, 1]$

41. $\left(-\infty, \frac{8}{5}\right]$ **43.** $(-\infty, \infty)$ **45.** $\left(-\frac{1}{2}, \infty\right)$

47.

49.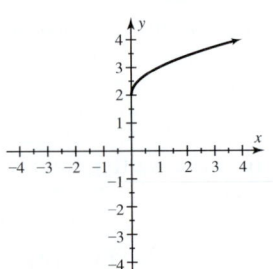

Shifted 2 units left   Shifted 2 units upward

51.

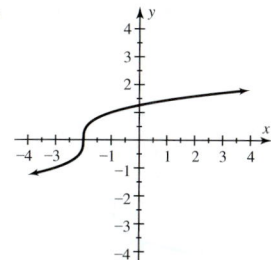

Shifted 2 units left

53. $f(x) = \sqrt{x}$ **55.** $f(x) = \sqrt[3]{x^2}$ **57.** $f(x) = \frac{1}{\sqrt[5]{x}}$

59. 32; 55.90   **61.** $-\frac{1}{128} \approx -0.01$; 0.04

63. 4; Not possible   **65.** 4; 4

67. $[0, 6, 1]$ by $[0, 6, 1]$ **69.** $[0, 6, 1]$ by $[0, 6, 1]$

   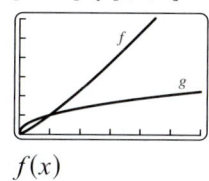

$g(x)$ $f(x)$

71. $x^p > x^q$ **73. (a)** 6   **(b)** $\sqrt{2x}$ **(c)** $4|x|$

(d) 2   **75.** b.   **77.** c.   **79.** 1 second

81. (a) 6 years   **(b)** The twin in the spaceship will be 4 years younger than the twin on Earth.

83. (a) $[0, 5, 1]$ by $[0, 0.5, 0.1]$ **(b)** $k \approx 0.16$

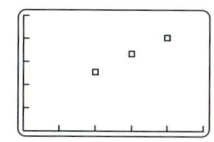

(c) $[0, 5, 1]$ by $[0, 0.5, 0.1]$; yes   **(d)** 0.295 m^2

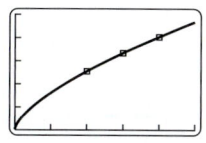

(answers may vary)

85. (a) $k \approx 0.91$

(b) $[0, 1.5, 0.1]$ by $[0, 1, 0.1]$

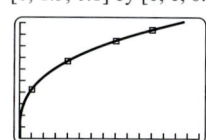

It increases.

(c) About 0.808 m   **(d)** About 0.788; a bird weighing 0.65 kg has a wing span of about 0.788 m.   **87.** 2877 in^2

CHECKING BASIC CONCEPTS 10.3 & 10.4 (p. 692)

1. (a) 6   **(b)** 3   **(c)** $6x^3$ **2. (a)** $7\sqrt{6} + \sqrt{7}$

(b) $5\sqrt[3]{x}$ **(c)** $\sqrt{x}$ **3. (a)** $(y - x)\sqrt[3]{xy}$ **(b)** 14

4. $\frac{\sqrt{6}}{2}$ **5.** $\frac{\sqrt{5} + 1}{2}$

6. (a)   **(b)**

          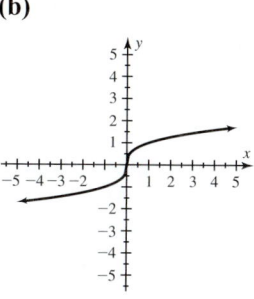

$f(-1)$ is undefined. $f(-1) = -1$

(c)

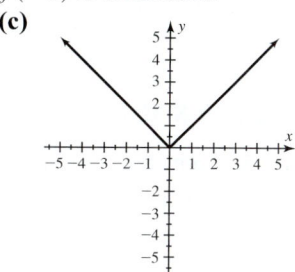

$f(-1) = 1$

7. 3.2   **8.** $[4, \infty)$

SECTION 10.5 (pp. 703–707)

1. Square each side.   **2.** Cube each side.   **3.** Yes

4. Check them.   **5.** Finding an unknown side of a right triangle (answers may vary)   **6.** 5

7. $d = \sqrt{(x_2 - x_1)^2 + (y_2 - y_1)^2}$ **8.** $x^{1/2} + x^{3/4} = 2$

9. 2   **11.** x **13.** $2x + 1$ **15.** $5x^2$ **17.** 64   **19.** 81

21. 6   **23.** 48   **25.** 6   **27.** 3   **29.** 27   **31.** -2

33. 15   **35.** $\frac{1}{2}$ **37.** 0, 1   **39.** 2   **41.** -4 **43.** 9

45. 9   **47.** ± 7 **49.** ± 10 **51.** $-5, 3$ **53.** $-3, 7$

55. 4   **57.** -4 **59.** 1   **61.** $\frac{7}{5}$ **63.** ± 2 **65.** $\sqrt[5]{12}$

67. $-5, 1$ **69.** 3   **71.** 1.88   **73.** $-1, 0.70$ **75.** 1.79

77. -0.47 **79.** 16   **81.** 11   **83.** $L = \frac{8T^2}{\pi^2}$ **85.** $A = \pi r^2$

87. Yes   **89.** Yes   **91.** Yes   **93.** No   **95.** $\sqrt{32} = 4\sqrt{2}$

97. 7   **99.** $c = 5$ **101.** $b = \sqrt{61}$ **103.** $a = 14$

105. $\sqrt{20} = 2\sqrt{5}$ **107.** $\sqrt{4000} = 20\sqrt{10}$ **109.** $\sqrt{89}$

111. 5   **113.** 4   **115.** 2, 122   **117.** $W = 8$ pounds

119. About 3 miles   **121.** About 269 feet

123. 19 inches   **125.** $h \approx 16.3$ inches, $d \approx 33.3$ inches

127. (a) 121 feet   **(b)** About 336 feet

129. (a) About 38 miles per hour; the vehicle involved in the accident was traveling about 38 miles per hour.

(b) About 453 feet   **131. (a)** It increases by a factor of 8.

(b) $v = \sqrt[3]{\frac{W}{3.8}}$ **(c)** 20 mph   **133.** $a\sqrt{2}$

SECTION 10.6 (pp. 714–715)

1. $2 + 3i$ (answers may vary)   **2.** No; any real number a can be written as $a + 0i$.   **3.** i **4.** -1 **5.** $i\sqrt{a}$
6. $10 - 7i$ **7.** $a + bi$ **8.** $1 + 2i$ **9.** 4 **10.** -5
11. 0 **12.** pure   **13.** $i\sqrt{5}$ **15.** $10i$ **17.** $12i$
19. $2i\sqrt{3}$ **21.** $3i\sqrt{2}$ **23.** 3 **25.** $-8 + 7i$
27. $1 - 9i$ **29.** $-10 + 7i$ **31.** $-13 + 13i$
33. $20 - 12i$ **35.** 41 **37.** 20 **39.** $1 + 5i$
41. $3 + 4i$ **43.** $-2 - 6i$ **45.** -2 **47.** $a^2 + 9b^2$
49. $-i$ **51.** i **53.** -1 **55.** 1 **57.** $3 - 4i$ **59.** $6i$
61. $5 + 4i$ **63.** -1 **65.** $1 - i$ **67.** $-\frac{6}{29} + \frac{15}{29}i$
69. $2 + i$ **71.** $-1 + 6i$ **73.** $-1 - 2i$ **75.** $-\frac{3}{2}i$
77. $-\frac{1}{2} + \frac{3}{2}i$ **79.** $\frac{290}{13} + \frac{20}{13}i$

CHECKING BASIC CONCEPTS 10.5 & 10.6 (p. 715)

1. (a) 4 **(b)** 28 **(c)** 3 **2.** 13 **3.** 9.6 inches   **4.** $-3, 1$
5. (a) $8i$ **(b)** $i\sqrt{17}$ **6. (a)** $3 - 4i$ **(b)** $-2 + 3i$
(c) $5 + i$ **(d)** $\frac{3}{4} + \frac{3}{4}i$

CHAPTER 10 REVIEW (pp. 719–722)

1. 2 **2.** 6 **3.** $3|x|$ **4.** $|x - 1|$ **5.** -4
6. -5 **7.** x^2 **8.** $3x$ **9.** 2 **10.** -1 **11.** x^2
12. $x + 1$ **13.** $\sqrt{14}$ **14.** $\sqrt[3]{-5}$ **15.** $\left(\sqrt{\frac{x}{y}}\right)^3$ or $\sqrt{\left(\frac{x}{y}\right)^3}$
16. $\frac{1}{\sqrt[3]{(xy)^2}}$ or $\frac{1}{(\sqrt[3]{xy})^2}$ **17.** 9 **18.** 2 **19.** 64 **20.** 27
21. z^2 **22.** xy^2 **23.** $\frac{x^3}{y^9}$ **24.** $\frac{y^2}{x}$ **25.** 8 **26.** -2
27. x^2 **28.** 2 **29.** $-\frac{\sqrt[3]{x}}{2}$ **30.** $\frac{1}{3}$ **31.** $4\sqrt{3}$ **32.** $3\sqrt{6}$
33. $\frac{3}{x}$ **34.** $4ab\sqrt{2a}$ **35.** $9xy$ **36.** $5z\sqrt[3]{z}$ **37.** $x + 1$
38. $\frac{2a\sqrt[4]{a}}{b}$ **39.** $2\sqrt[6]{x^5}$ **40.** $\sqrt[6]{r^5t^8}$ or $t\sqrt[6]{r^5t^2}$ **41.** $4\sqrt{3}$
42. $3\sqrt[3]{x}$ **43.** $-3\sqrt[3]{5}$ **44.** $-\sqrt[4]{y}$ **45.** $11\sqrt{3}$
46. $7\sqrt{2}$ **47.** $13\sqrt[3]{2}$ **48.** $3\sqrt{x + 1}$ **49.** $(2x - 1)\sqrt{x}$
50. $(b + 2a)\sqrt[3]{ab}$ **51.** $5 + 4\sqrt{2}$
52. $7 + 7\sqrt{3} - \sqrt{5} - \sqrt{15}$ **53.** 3 **54.** 95
55. $a - 2b$ **56.** $xy + \sqrt{xy} - 2$ **57.** $\frac{4\sqrt{5}}{5}$ **58.** $\frac{r\sqrt{t}}{2t}$
59. $\frac{3 - \sqrt{2}}{7}$ **60.** $\frac{5 + \sqrt{7}}{9}$ **61.** $\sqrt{8} + \sqrt{7}$
62. $\frac{a - 2\sqrt{ab} + b}{a - b}$
63.

64.

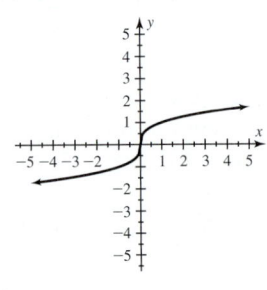

65. $f(x) = \sqrt{x}; 2$ **66.** $f(x) = \sqrt[7]{x^2}; \sqrt[7]{16}$
67.

68.

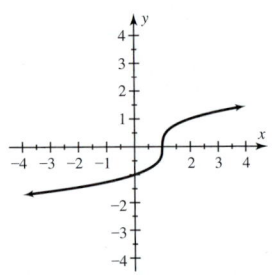

Shifted 2 units downward   Shifted 1 unit to the right
69. $[1, \infty)$ **70.** $(-\infty, 3]$ **71.** $(-\infty, \infty)$ **72.** $(-2, \infty)$
73. 2 **74.** 4 **75.** 9 **76.** 9 **77.** 8 **78.** $\frac{1}{4}$ **79.** 4.5
80. 1.62 **81.** $c = \sqrt{65}$ **82.** $b = \sqrt{39}$ **83.** $\sqrt{41}$
84. $\sqrt{52} = 2\sqrt{13}$ **85.** ± 11 **86.** ± 4 **87.** $-3, 5$
88. 4 **89.** 3 **90.** 2 **91.** ± 4 **92.** -1 **93.** 1
94. $-2, 0$ **95.** -2 **96.** $-2 + 4i$ **97.** $5 + i$
98. $1 + 2i$ **99.** $-\frac{14}{13} + \frac{5}{13}i$ **100.** $2 - 2i$
101. About 85 feet   **102.** $\sqrt{16,200} = 90\sqrt{2} \approx 127.3$ feet
103. About 0.79 second   **104. (a)** 5 square units
(b) $5\sqrt{5}$ cubic units   **(c)** $\sqrt{10}$ units   **(d)** $\sqrt{15}$ units
105. About 0.82 foot   **106.** About 0.13 foot; it is shorter.
107. $r = \sqrt[210]{\frac{281}{4}} - 1 \approx 0.02$; from 1790 through 2000 the average annual percentage growth rate was about 2%.
108. (a) About 43 miles per hour   **(b)** About 34 miles per hour; a steeper bank allows for a higher speed limit; yes
109. $\sqrt{7} \approx 2.65$ feet   **110.** About 2108 square inches
111. (a) $0.5 = \frac{1}{2}$ **(b)** About 0.09, or $\frac{9}{100}$

CHAPTER 10 TEST (pp. 722–723)

1. -3 **2.** $|z + 1|$ **3.** $5x^2$ **4.** $2z^2$ **5.** $2xy\sqrt[4]{y}$
6. 1 **7.** $\sqrt[5]{7^2}$ or $(\sqrt[5]{7})^2$ **8.** $\sqrt[3]{\left(\frac{y}{x}\right)^2}$ or $\left(\sqrt[3]{\frac{y}{x}}\right)^2$ **9.** 16
10. $\frac{1}{216}$ **11.** $x^{4/3}$ **12.** $x^{7/10}$ **13.** $(-\infty, 4]$
14.

(graph)

15. $8z^{3/2}$ **16.** $\frac{z}{y^{2/3}}$ **17.** 9 **18.** $\frac{y}{2}$ **19.** $4\sqrt{7} + \sqrt{5}$
20. $6\sqrt[3]{x}$ **21.** $14\sqrt{2}$ **22.** 2 **23. (a)** 27 **(b)** 7
(c) 3 **(d)** 17 **24. (a)** $\frac{2\sqrt{7}}{21}$ **(b)** $\frac{-1 + \sqrt{5}}{4}$ **25.** 2.63
26. $\sqrt{120} \approx 10.95$ **27.** $\sqrt{8} = 2\sqrt{2}$ **28.** $2 - 19i$

29. $-6 + 8i$ **30.** $\frac{5}{4}$ **31.** $\frac{4}{29} + \frac{10}{29}i$ **32. (a)** $r = \sqrt[3]{\frac{3V}{4\pi}}$
(b) About 2.29 inches   **33.** 1.31 pounds

CHAPTERS 1–10 CUMULATIVE REVIEW (pp. 724–725)

1. 36π **2.** $D = \{-1, 0, 1\}, R = (2, 4\}$
3. (a) $\frac{1}{a^3 b^9}$ **(b)** $x^4 x^9$ **(c)** $r^8 t^5$ **4.** 4.3×10^{-4}
5. $3; x \neq 2$ **6.** All real numbers
7.   **8.**

9.

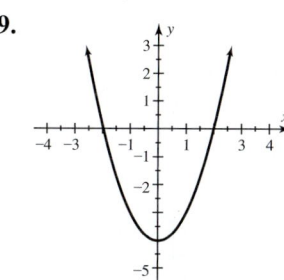

(a) D: all real numbers; R: $y \geq -4$ **(b)** 0   **(c)** $-2, 2$
(d) $-2, 2$ **10.** $y = \frac{1}{2}x + \frac{5}{2}$ **11.** $f(x) = 1 - 3x$
12.

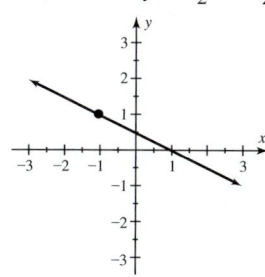

13. $\frac{6}{11}$ **14.** $(-\infty, 3]$ **15.** $[-1, 5]$ **16.** $\left[-\frac{5}{2}, 1\right]$
17. (a) $(4, 4)$ **(b)** No solutions
18.

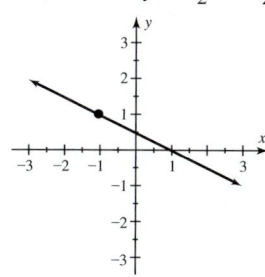

19. $(3, 2, 1)$ **20.** 4   **21.** $-4x^4 + 16x$ **22.** $x^2 - 16$
23. $5x^2 - 7x - 6$ **24.** $16x^2 + 72x + 81$

25. $(3x - 4)(3x + 4)$ **26.** $(x - 2)^2$ **27.** $3x^2(5x - 3)$
28. $(4x - 3)(3x + 1)$ **29.** $(r - 1)(r^2 + r + 1)$
30. $(x - 3)(x^2 + 5)$ **31.** 1, 2   **32.** $-2, 0, 2$
33. $2(x + 2)$ **34.** $\frac{7x - 5}{x(x - 1)}$ **35.** $6x$ **36.** 4   **37.** $\frac{1}{64}$
38. 5   **39.** $4x$ **40.** $x^{3/4}$ or $\sqrt[4]{x^3}$ **41.** $2x$ **42.** $6\sqrt{3x}$
43. $2x^2 - \sqrt{3x} - 3$ **44.** $(-\infty, 1]$
45.

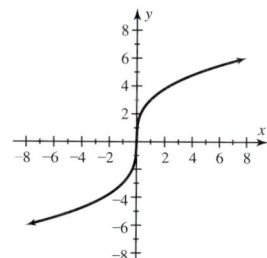

46. $\sqrt{10}$ **47.** $5 + i$ **48.** 13   **49.** 6   **50.** 28
51. $\frac{4}{9}, 4$ **52.** ± 3 **53.** 1.41   **54.** 4.06
55. Water is leaving the tank at 15 gal/min.
56.

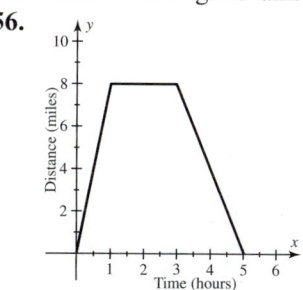

57. $1300 at 5%, $700 at 4%   **58.** 8 in. by 8 in. by 4 in.
59. 22.4 feet   **60.** $45°, 55°, 80°$

CHAPTER 11: QUADRATIC FUNCTIONS AND EQUATIONS

SECTION 11.1 (pp. 738–742)

1. parabola   **2.** The vertex
3. axis of symmetry   **4.** $(0, 0)$
5.

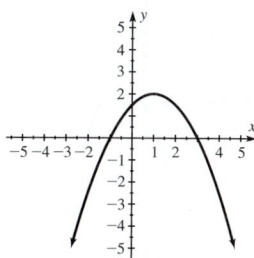

6. $-\frac{b}{2a}$ **7.** narrower   **8.** reflected   **9.** $ax^2 + bx + c$
with $a \neq 0$ **10.** vertex   **11.** $0, -4$ **13.** $-2, -2$
15. $(1, -2); x = 1$; upward; incr: $x \geq 1$; decr: $x \leq 1$
17. $(-2, 3); x = -2$; downward; incr: $x \leq -2$;
decr: $x \geq -2$

19. (a)

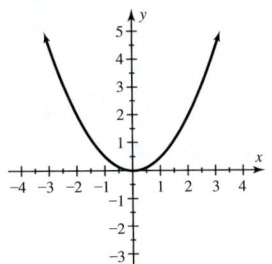

(b) $(0, 0); x = 0$
(c) $2; 4.5$

23. (a)

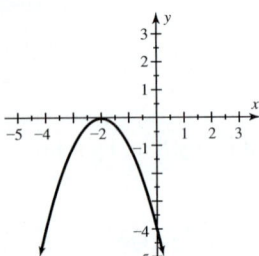

(b) $(0, 1); x = 0$
(c) $-11; -26$

27. (a)

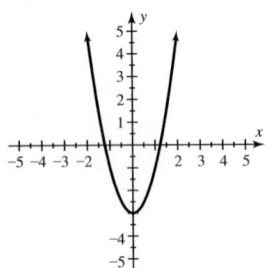

(b) $(-2, 0); x = -2$
(c) $0; -25$

31. (a)

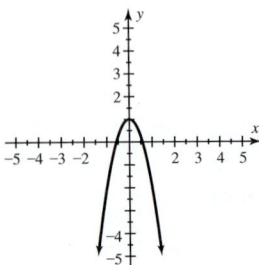

(b) $(0, -3); x = 0$
(c) $5; 15$

21. (a)

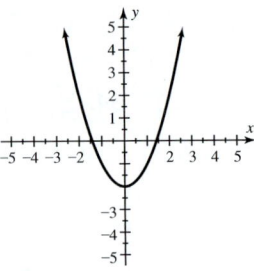

(b) $(0, -2); x = 0$
(c) $2; 7$

25. (a)

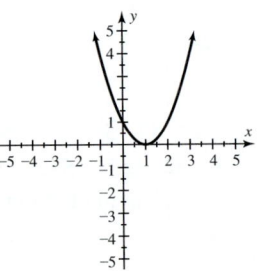

(b) $(1, 0); x = 1$
(c) $9; 4$

29. (a)

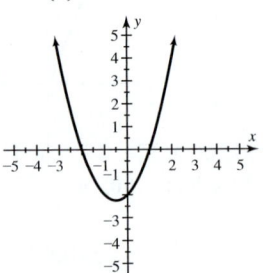

(b) $(-0.5, -2.25); x = -0.5$
(c) $0; 10$

33. (a)

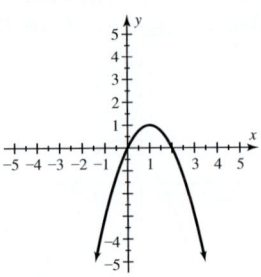

(b) $(1, 1); x = 1$
(c) $-8; -3$

35. (a)

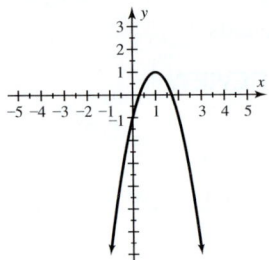

(b) $(1, 1); x = 1$
(c) $-17; -7$

39. $(2, -6)$ **41.** $(-3, 4)$

47.

Reflected across the x-axis

51.

Wider

37. (a)

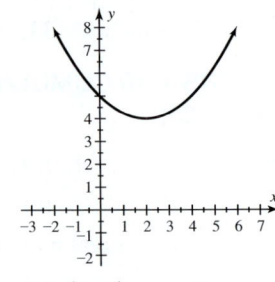

(b) $(2, 4); x = 2$
(c) $8; 4.25$

43. $(0, 3)$ **45.** $(1, 1.4)$

49.

Narrower

53.

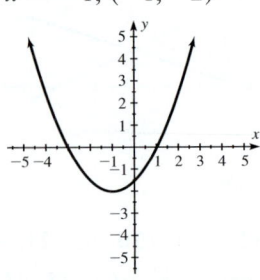

Reflected across the x-axis
and wider

55. -2; incr: $x \geq -1$; decr: $x \leq -1$

57. $-\frac{25}{4}$; incr: $x \geq \frac{5}{2}$; decr: $x \leq \frac{5}{2}$

59. $-\frac{7}{2}$; incr: $x \geq -\frac{1}{2}$; decr: $x \leq -\frac{1}{2}$

61. 6; incr: $x \leq 1$; decr: $x \geq 1$ **63.** 4; incr: $x \leq 2$;
decr: $x \geq 2$ **65.** $-\frac{39}{8}$; incr: $x \leq \frac{1}{4}$; decr: $x \geq \frac{1}{4}$

67. (a) Upward; wider   **(b)** $x = -1; (-1, -2)$
(c) y-int: $-\frac{3}{2}$; x-int: $-3, 1$ **(d)**

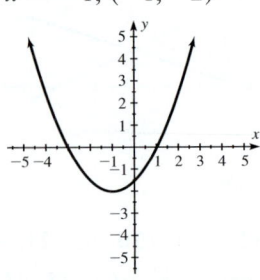

69. (a) Downward; the same   **(b)** $x = 1$; $(1, 1)$
(c) y-int: 0; x-int: 0, 2   **(d)**

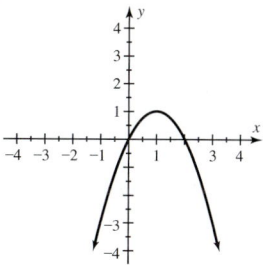

71. (a) Upward; narrower   **(b)** $x = -\frac{1}{2}$; $\left(-\frac{1}{2}, -\frac{9}{2}\right)$
(c) y-int: -4; x-int: $-2, 1$
(d)

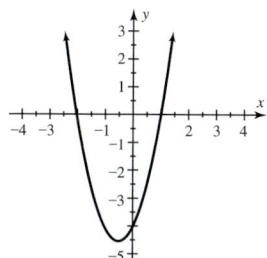

73. 10, 10   **75.** d.   **77.** a.   **79. (a)** 2 feet   **(b)** 2 seconds
(c) 66 feet   **81.** $\frac{66}{32} \approx 2$ seconds; about 74 feet
83. (a) The revenue increases when $x \leq 50$ and it
decreases when $x \geq 50$.   **(b)** \$2500; 50
(c) $f(x) = x(100 - x)$ **(d)** \$2500; 50
85. 300 feet by 600 feet
87. (a) [20, 40, 5] by [0, 30, 5]

(b) About 28.4°C   **(c)** About 28.4°C   **89.** 42 in.

SECTION 11.2 (pp. 750–752)

1. $x^2 + 2$ **2.** $(x - 2)^2$ **3.** $(1, 2)$ **4.** $(-1, -2)$
5. $f(x) = ax^2 + bx + c$; $f(x) = a(x - h)^2 + k$
6. $y = a(x - h)^2 + k$; (h, k) **7.** downward   **8.** $-\frac{b}{2a}$
9. (a)   **11. (a)**

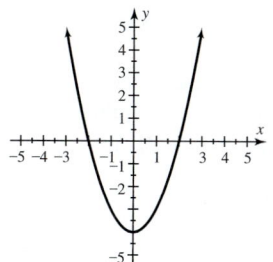

(b) $(0, -4)$ **(b)** $(0, 1)$
(c) Down 4 units   **(c)** Narrower and up 1 unit

13. (a)

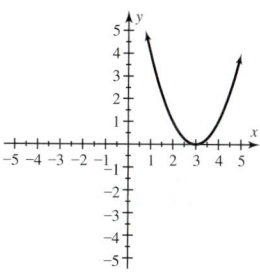

(b) $(3, 0)$
(c) Right 3 units
17. (a)

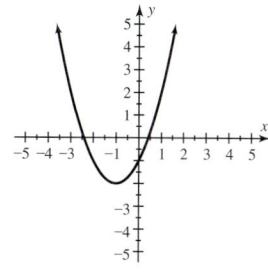

(b) $(0, 2)$
(c) Reflected across the
x-axis and up 2 units
21. (a)

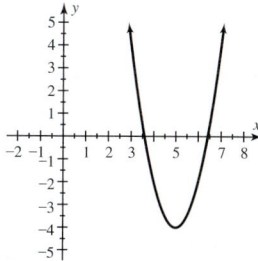

(b) $(-1, -2)$
(c) Left 1 unit and down
2 units
25. (a)

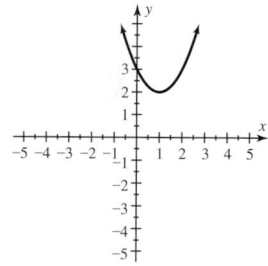

(b) $(5, -4)$
(c) Narrower, right 5 units
and down 4 units

15. (a)

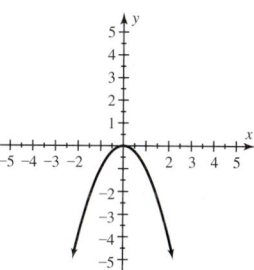

(b) $(0, 0)$
(c) Reflected across the x-axis
19. (a)

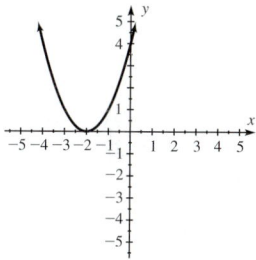

(b) $(-2, 0)$
(c) Left 2 units
23. (a)

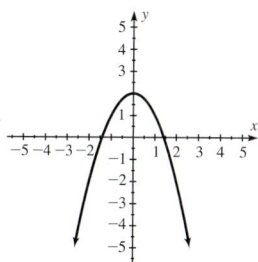

(b) $(1, 2)$
(c) Right 1 unit and up
2 units
27. (a)

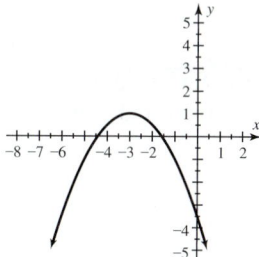

(b) $(-3, 1)$
(c) Wider, reflected across
the x-axis, left 3 units and
up 1 unit

29. Translated 1 unit right, 2 units downward, and is wider

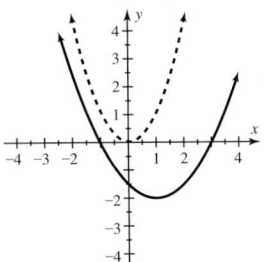

31. Translated 1 unit left, 3 units upward, opens downward, and is narrower

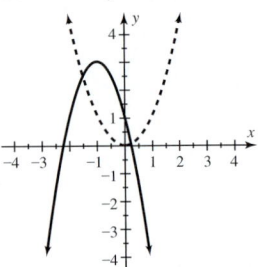

33. $[-20, 20, 2]$ by $[-20, 20, 2]$

35. $y = 3(x - 3)^2 + 4$; $y = 3x^2 - 18x + 31$

37. $y = -\frac{1}{2}(x - 5)^2 - 2$; $y = -\frac{1}{2}x^2 + 5x - \frac{29}{2}$

39. $y = (x - 1)^2 + 2$ **41.** $y = -(x - 0)^2 - 3$

43. $y = (x - 0)^2 - 3$ **45.** $y = -(x + 1)^2 + 2$

47. $y = (x + 1)^2 - 4$; $(-1, -4)$

49. $y = (x - 2)^2 + 1$; $(2, 1)$

51. $y = \left(x + \frac{3}{2}\right)^2 - \frac{17}{4}$; $\left(-\frac{3}{2}, -\frac{17}{4}\right)$

53. $y = \left(x - \frac{7}{2}\right)^2 - \frac{45}{4}$; $\left(\frac{7}{2}, -\frac{45}{4}\right)$

55. $y = 3(x + 1)^2 - 4$; $(-1, -4)$

57. $y = 2\left(x - \frac{3}{4}\right)^2 - \frac{9}{8}$; $\left(\frac{3}{4}, -\frac{9}{8}\right)$

59. $y = -2(x + 2)^2 + 13$; $(-2, 13)$ **61.** $a = 2$

63. $a = 0.3$ **65.** $y = 2(x - 1)^2 - 3$

67. $y = 0.5(x - 1980)^2 + 6$

69. (a)

(b) $D(x) = \frac{1}{12}x^2$

71. (a) $f(x) = 2(x - 1982)^2 + 1$ (answers may vary)
(b) 201 thousand (answers may vary)

CHECKING BASIC CONCEPTS 11.1 & 11.2 (p. 752)

1. (a)

(b)

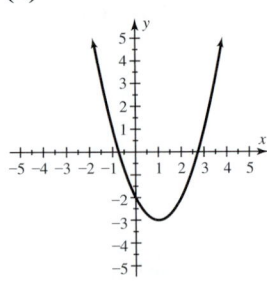

$(0, -2)$; $x = 0$ $(1, -3)$; $x = 1$

2. y_1 opens upward, whereas y_2 opens downward, y_1 is narrower than y_2.   **3.** 7; incr: $x \leq 2$; decr: $x \geq 2$

4. (a)

(b)

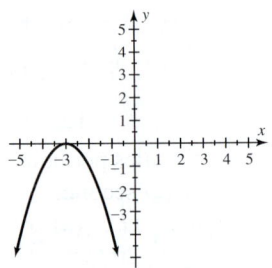

1 unit right, 2 units up   Reflected across the x-axis, 3 units left

5. (a) $y = (x + 7)^2 - 56$ **(b)** $y = 4(x + 1)^2 - 6$

SECTION 11.3 (pp. 762–765)

1. $x^2 + 3x - 2 = 0$ (answers may vary); it can have 0, 1, or 2 solutions   **2.** Nonlinear   **3.** Factoring, square root property, completing the square

4.

5.

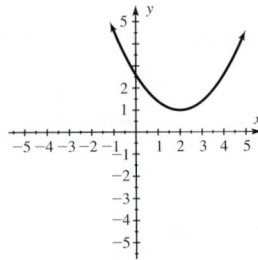

(answers may vary)   (answers may vary)

6. Two; the solutions are the x-intercepts.   **7.** ± 8; the square root property (answers may vary)   **8.** $\left(\frac{b}{2}\right)^2$ **9.** Yes

11. No **13.** Yes **15.** No **17. (a)** 3.65, −1.65
(b) 1.32, −5.32 **19. (a)** 1.32, −0.12 **(b)** −0.28, −0.83
21. −2, 1 **23.** No real solutions **25.** −2, 3 **27.** −0.5
29. −1, 5 **31.** −3, 1 **33.** −3, 3 **35.** $-\frac{1}{2}, \frac{3}{2}$ **37.** −1
39. No real solutions **41.** −7, 5 **43.** $-\frac{1}{3}, \frac{1}{2}$ **45.** $1, \frac{3}{2}$
47. −2, 7 **49.** $\frac{6}{5}, \frac{3}{2}$ **51.** ±12 **53.** $\pm\frac{8}{\sqrt{5}}$ or $\pm\frac{8\sqrt{5}}{5}$
55. −6, 4 **57.** −7, 9 **59.** $\frac{1\pm\sqrt{5}}{2}$ **61.** $5\pm\sqrt{5}$
63. 4 **65.** $\frac{25}{4}$ **67.** 16; $(x-4)^2$ **69.** $\frac{81}{4}; \left(x+\frac{9}{2}\right)^2$
71. −4, 6 **73.** $-3\pm\sqrt{11}$ **75.** $\frac{3\pm\sqrt{29}}{2}$ **77.** $\frac{5\pm\sqrt{21}}{2}$
79. $1\pm\sqrt{5}$ **81.** $\frac{3\pm\sqrt{41}}{4}$ **83.** $\frac{2\pm\sqrt{11}}{2}$
85. $\frac{-3\pm\sqrt{5}}{12}$ **87.** −6, 2 **89.** $-2\pm\sqrt{2}$ **91.** $\pm\sqrt{2}$
93. $-5, \frac{7}{3}$ **95.** $4\pm\sqrt{14}$ **97.** $\pm\sqrt{\frac{c}{a}}$ **99.** −3, 6
101. 3, 5 **103.** 5, 7 **105.** $y=\pm\sqrt{x+1}$
107. $v=\pm\sqrt{\frac{2K}{m}}$ **109.** $r=\pm\sqrt{\frac{k}{E}}$ **111.** $f=\pm\frac{1}{2\pi\sqrt{LC}}$
113. (a) 30 miles per hour **(b)** 40 miles per hour
115. About 1.9 seconds; no **117.** About 8 feet
119. 2 hours **121. (a)** $x^2+6x-520=0$
(b) −26 or 20; 20 feet **123.** About 23°C and 34°C
125. (a) The rate of increase for each 20-year period is not constant. **(b)** (1800, 5); in 1800 the population was 5 million. **(c)** 1910

SECTION 11.4 (pp. 775–778)

1. To solve quadratic equations that are written in the form $ax^2+bx+c=0$ **2.** Completing the square
3. b^2-4ac **4.** One solution **5.** Factoring, square root property, completing the square, and the quadratic formula
6. No; not when $b^2-4ac<0$ **7.** ±2 **8.** ±2i
9. $-6, \frac{1}{2}$ **11.** 1 **13.** −2, 8 **15.** $\frac{1\pm\sqrt{17}}{8}$
17. No real solutions **19.** $\frac{1}{2}$ **21.** $\frac{3\pm\sqrt{13}}{2}$
23. $2\pm\sqrt{3}$ **25.** $\frac{1\pm\sqrt{37}}{6}$ **27.** $\frac{1\pm\sqrt{15}}{2}$
29. $\frac{5\pm\sqrt{10}}{3}$ **31. (a)** $a>0$ **(b)** −1, 2 **(c)** Positive
33. (a) $a>0$ **(b)** No real solutions **(c)** Negative
35. (a) $a<0$ **(b)** 2 **(c)** Zero **37. (a)** 25 **(b)** 2
39. (a) 0 **(b)** 1 **41. (a)** $-\frac{7}{4}$ **(b)** 0 **43. (a)** 21
(b) 2 **45.** $1\pm\sqrt{2}$ **47.** $-\frac{3}{2}, 1$ **49.** None **51.** None
53. $\frac{-2\pm\sqrt{10}}{3}$ **55.** ±3i **57.** $\pm4i\sqrt{5}$ **59.** $\pm\frac{1}{2}i$
61. $\pm\frac{3}{4}i$ **63.** $\pm i\sqrt{6}$ **65.** $\pm\sqrt{3}$ **67.** $\pm i\sqrt{2}$
69. $\frac{1}{2}\pm i\frac{\sqrt{7}}{2}$ **71.** $-\frac{3}{4}\pm i\frac{\sqrt{23}}{4}$ **73.** $2\pm\sqrt{3}$
75. $-\frac{1}{2}\pm i\frac{\sqrt{7}}{2}$ **77.** $-\frac{1}{5}\pm i\frac{\sqrt{19}}{5}$ **79.** $-\frac{3}{4}\pm i\frac{\sqrt{23}}{4}$

81. $-\frac{1}{2}\pm i\frac{\sqrt{15}}{2}$ **83.** $\frac{1\pm\sqrt{3}}{2}$ **85.** $\frac{1}{4}\pm i\frac{\sqrt{15}}{4}$
87. 1, 2 **89.** $-\frac{1}{2}, 4$ **91.** $\frac{5\pm\sqrt{17}}{2}$ **93.** $-\frac{1}{4}\pm\frac{3}{4}i\sqrt{7}$
95. $\pm\frac{1}{2}$ **97.** $\pm i\sqrt{2}$ **99.** $\frac{1}{3}$ **101.** 9 miles per hour
103. 45 miles per hour **105.** $x\approx8.04$, or about 1992; this agrees with the graph. **107.** 30 mi, 40 mi **109. (a)–(c)** 11 in. by 14 in. **111. (a)** The rate of change is not constant.
(b) 75 seconds (answers may vary) **(c)** 75 seconds

CHECKING BASIC CONCEPTS 11.3 & 11.4 (p. 779)

1. $\frac{1}{2}, 3$ **2.** $\pm\sqrt{5}$ **3.** $2\pm\sqrt{3}$ **4.** $y=\pm\sqrt{1-x^2}$
5. (a) $\frac{3\pm\sqrt{17}}{4}$ **(b)** $\frac{4}{3}$ **(c)** $-\frac{1}{2}\pm i\frac{\sqrt{7}}{2}$
6. (a) 5; two real solutions **(b)** −7; no real solutions
(c) 0; one real solution

SECTION 11.5 (pp. 785–788)

1. It has an inequality symbol rather than an equals sign.
2. No, they often have infinitely many. **3.** No **4.** Yes
5. $-2<x<4$ **6.** $x<-3$ or $x>1$ **7.** Yes **9.** Yes
11. No **13.** Yes **15.** No **17.** No **19. (a)** −3, 2
(b) $-3<x<2$ **(c)** $x<-3$ or $x>2$ **21. (a)** −2, 2
(b) $-2<x<2$ **(c)** $x<-2$ or $x>2$ **23. (a)** −10, 5
(b) $x<-10$ or $x>5$ **(c)** $-10<x<5$
25. (a) −2, 2 **(b)** $-2<x<2$ **(c)** $x<-2$ or $x>2$
27. (a) −4, 0 **(b)** $-4<x<0$ **(c)** $x<-4$ or $x>0$
29. $(-3, -1)$ **31.** $(-\infty, -2.5]\cup[3, \infty)$ **33.** $[-2, 2]$
35. $(-\infty, \infty)$ **37.** $(0, 3)$ **39.** No real solutions **41.** 2
43. $(-\infty, -1)\cup(-1, \infty)$ **45.** $[-1, 2]$ **47. (a)** −2, 2
(b) $-2<x<2$ **(c)** $x<-2$ or $x>2$
49. (a) $\frac{-1\pm\sqrt{5}}{2}$ **(b)** $\frac{-1-\sqrt{5}}{2}<x<\frac{-1+\sqrt{5}}{2}$
(c) $x<\frac{-1-\sqrt{5}}{2}$ or $x>\frac{-1+\sqrt{5}}{2}$
51. $[-7, -3]$ **53.** $(-\infty, 1)\cup(2, \infty)$
55. $(-\sqrt{10}, \sqrt{10})$ **57.** $(-\infty, 0)\cup(6, \infty)$
59. $(-\infty, 2-\sqrt{2}]\cup[2+\sqrt{2}, \infty)$ **61.** $[-0.128, 0.679]$
63. (a) From 1131 feet to 3535 feet (approximately)
(b) Before 1131 feet or after 3535 feet (approximately)
65. (a) About 383; they agree (approx.). **(b)** About 1969 or after **(c)** About 1969 or after **67.** From 11 feet to 20 feet

SECTION 11.6 (p. 792)

1. $\pm1, \pm\sqrt{6}$ **3.** $-\sqrt[3]{2}, \sqrt[3]{\frac{5}{3}}$ **5.** $-\frac{4}{5}, -\frac{1}{3}$
7. −3, 3 **9.** $-\sqrt[3]{\frac{1}{3}}, \sqrt[3]{2}$ **11.** $-\frac{1}{8}, \frac{2}{5}$ **13.** 1
15. 1, $32^5=33,554,432$ **17.** 16, 81 **19.** 1
21. −3, 6 **23.** $-\sqrt{3}, \sqrt{3}$ **25.** $\pm2, \pm2i$ **27.** 0, ±i
29. $\pm\sqrt{2}, \pm i$ **31.** $-1\pm i$ **33.** ±2i

CHECKING BASIC CONCEPTS 11.5 & 11.6
(p. 792)

1. $(-\infty, -2) \cup (3, \infty)$ **2.** $\left[-1, -\frac{2}{3}\right]$ **3.** $-2, \sqrt[3]{2}$
4. $-1, 8^3 = 512$ **5.** $\pm i$

CHAPTER 11 REVIEW (pp. 795–799)

1. $(-3, 4); x = -3$; downward; incr: $x \le -3$;
decr: $x \ge -3$ **2.** $(1, 0); x = 1$; upward; incr: $x \ge 1$;
decr: $x \le 1$

3. (a)

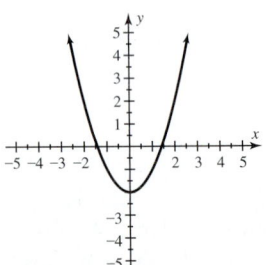

(b) $(0, -2); x = 0$
(c) -1

4. (a)

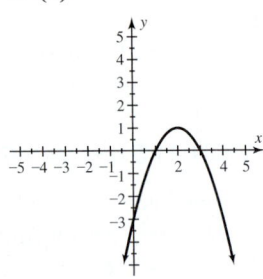

(b) $(2, 1); x = 2$
(c) 0

5. (a)

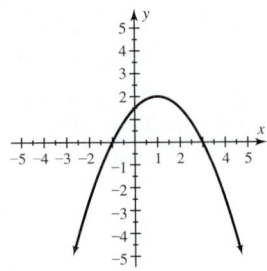

(b) $(1, 2); x = 1$
(c) -2.5

6. (a)

(b) $(-2, -3); x = -2$
(c) -1

7. $-\frac{7}{2}$ **8.** $-\frac{14}{3}$ **9.** $(2, -6)$ **10.** $(0, 5)$ **11.** $(2, 2)$
12. $(-1, 1)$

13. (a)

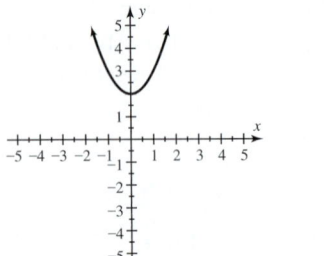

(b) Up 2 units

14. (a)

(b) Narrower

15. (a)

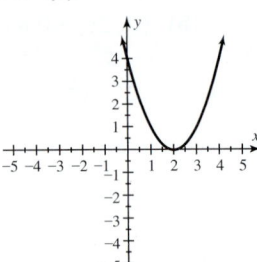

(b) Right 2 units
17. (a)

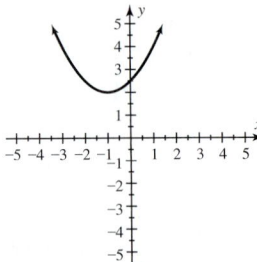

(b) Wider, left 1 unit,
up 2 units

16. (a)

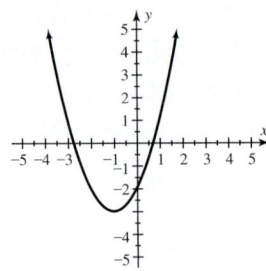

(b) Left 1 unit, down 3 units
18. (a)

(b) Narrower, right 1 unit,
down 3 units

19. $y = -4(x - 2)^2 - 5$ **20.** $y = -(x + 4)^2 + 6$
21. $y = (x + 2)^2 - 11; (-2, -11)$
22. $y = \left(x - \frac{7}{2}\right)^2 - \frac{45}{4}; \left(\frac{7}{2}, -\frac{45}{4}\right)$
23. $y = 2\left(x - \frac{3}{4}\right)^2 - \frac{73}{8}; \left(\frac{3}{4}, -\frac{73}{8}\right)$
24. $y = 3(x + 1)^2 - 5; (-1, -5)$ **25.** $a = 3$
26. $a = \frac{1}{4}$ **27.** $f(x) = -5x^2 + 30x - 41; -41$
28. $f(x) = 3x^2 + 12x + 8; 8$ **29.** $-2, 3$ **30.** -1
31. No real solutions   **32.** $-4, 6$ **33.** $-10, 5$
34. $-0.5, 0.25$ **35.** $-5, 10$ **36.** $-3, 1$ **37.** $-4, 2$
38. $-1, 3$ **39.** $-5, 4$ **40.** $-8, -3$ **41.** $-\frac{2}{5}, \frac{2}{3}$
42. $\frac{4}{7}, 3$ **43.** ± 10 **44.** $\pm \frac{1}{3}$ **45.** $\pm \frac{\sqrt{6}}{2}$ **46.** No real
solutions   **47.** $-3 \pm \sqrt{7}$ **48.** $2 \pm \sqrt{10}$ **49.** $1 \pm \sqrt{6}$
50. $\frac{-3 \pm \sqrt{11}}{2}$ **51.** $R = -r \pm \sqrt{\frac{k}{F}}$
52. $y = \pm\sqrt{\frac{12 - 2x^2}{3}}$ **53.** $3, 6$ **54.** $11, 13$ **55.** $-\frac{1}{2}, \frac{1}{3}$
56. $\frac{5 \pm \sqrt{5}}{10}$ **57.** $4 \pm \sqrt{21}$ **58.** $\frac{3 \pm \sqrt{3}}{2}$
59. ± 2 **60.** $\pm \frac{1}{2}$ **61.** $\frac{5}{2}, 3$ **62.** $\frac{3}{2}, 5$ **63.** $\frac{3 \pm \sqrt{5}}{2}$
64. $\frac{1 \pm \sqrt{5}}{4}$ **65. (a)** $a > 0$ **(b)** $-2, 3$ **(c)** Positive
66. (a) $a > 0$ **(b)** 2   **(c)** Zero   **67. (a)** $a < 0$
(b) No real solutions   **(c)** Negative   **68. (a)** $a < 0$
(b) $-4, 2$ **(c)** Positive   **69. (a)** 1   **(b)** 2   **70. (a)** 144
(b) 2   **71. (a)** -23 **(b)** 0   **72. (a)** 0   **(b)** 1

73. $-\frac{1}{2} \pm i\frac{\sqrt{19}}{2}$ **74.** $\pm 2i$ **75.** $\frac{1}{4} \pm i\frac{\sqrt{7}}{4}$ **76.** $\frac{1}{7} \pm i\frac{\sqrt{34}}{7}$
77. (a) $-2, 6$ **(b)** $-2 < x < 6$ **(c)** $x < -2$ or $x > 6$
78. (a) $-2, 0$ **(b)** $x < -2$ or $x > 0$ **(c)** $-2 < x < 0$
79. (a) $-4, 4$ **(b)** $-4 < x < 4$ **(c)** $x < -4$ or $x > 4$
80. (a) $-2, 1$ **(b)** $-2 < x < 1$ **(c)** $x < -2$ or $x > 1$
81. (a) $-1, 3$ **(b)** $-1 < x < 3$ **(c)** $x < -1$ or $x > 3$
82. (a) $-\frac{3}{2}, 5$ **(b)** $-\frac{3}{2} \le x \le 5$ **(c)** $x \le -\frac{3}{2}$ or $x \ge 5$
83. $[-3, -1]$ **84.** $\left(\frac{1}{5}, 3\right)$ **85.** $\left(-\infty, \frac{1}{6}\right) \cup (2, \infty)$
86. $(-\infty, -\sqrt{5}] \cup [\sqrt{5}, \infty)$ **87.** $(-\infty, \infty)$ **88.** No
solutions   **89.** $\pm\sqrt{5}, \pm 3$ **90.** $-\frac{1}{4}, \frac{2}{7}$ **91.** $1, 512$ **92.** 0
93. $\pm i\frac{\sqrt{2}}{2}$ **94.** $2 \pm i\sqrt{2}$ **95. (a)** $f(x) = x(12 - 2x)$
(b) 6 inches by 3 inches   **96. (a)** After 1 second and
1.75 seconds   **(b)** 1.375 seconds; 34.25 feet
97. (a) $f(x) = x(90 - 3x)$
(b) [0, 30, 5] by [0, 800, 100]

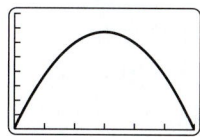

(c) 10 or 20 rooms   **(d)** 15 rooms
98. (a) 2.4; in 1999, there were 2.4 complaints per
100,000 passengers.
(b) [1997, 1999, 1] by [0.5, 3, 0.5]

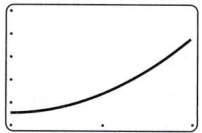

Complaints have increased.
99. (a) $\sqrt{1728} \approx 41.6$ miles per hour   **(b)** 60 miles per
hour   **100. (a)** $x(x + 2) = 143$ **(b)** $x = -13$ or
$x = 11$; the numbers are -13 and -11 or 11 and 13.
101. (a) [1935, 1995, 10] by [0, 100, 10]

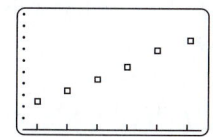

(b) Yes, the data are nearly linear.
(c) $f(x) = 1.04(x - 1940) + 25$ (answers may vary)
(d) 82%, 93%; answers may vary.
102. (a) (1950, 220); in 1950, the per capita consumption
was at a low of 220 million Btu.
(b) [1950, 1970, 5] by [200, 350, 25]

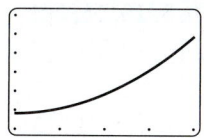

It increased.

(c) $f(1996) = 749$; no; the trend represented by this model
did not continue after 1970.
103. About 11.1 inches by 11.1 inches   **104.** 50 feet
105. About 1.5 feet   **106.** About 6.0 to 9.0 inches

CHAPTER 11 TEST (pp. 799–800)

1. $\left(1, \frac{3}{2}\right); x = 1; -3$ **2.** $-\frac{29}{4}$ **3.** $a = -\frac{1}{2}$
4.

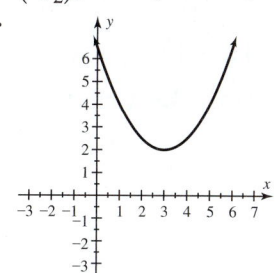

It is wider, shifted right 3 units and upward 2 units.
5. $y = (x - 3)^2 - 7; (3, -7); x = 3$ **6.** $-1, 2; 2$
7. $-4, \frac{1}{3}$ **8.** $-\frac{1}{2}, \frac{1}{2}$ **9.** $4 \pm \sqrt{17}$ **10.** $\frac{3 \pm \sqrt{17}}{4}$
11. $\pm\frac{4}{3}$ **12.** $m = \pm\sqrt{\frac{Fr^2}{G}}$ **13. (a)** $a < 0$ **(b)** $-3, 1$
(c) Positive   **14. (a)** -44 **(b)** No real solutions
(c) The graph of $y = -3x^2 + 4x - 5$ does not intersect
the x-axis.   **15. (a)** $-1, 1$ **(b)** $-1 < x < 1$
(c) $x < -1$ or $x > 1$ **16. (a)** $-10, 20$
(b) $x < -10$ or $x > 20$ **(c)** $-10 < x < 20$
17. (a) $-\frac{1}{2}, \frac{3}{4}$ **(b)** $\left[-\frac{1}{2}, \frac{3}{4}\right]$ **(c)** $\left(-\infty, -\frac{1}{2}\right] \cup \left[\frac{3}{4}, \infty\right)$
18. $[-2, 0]$ **19.** $\sqrt[3]{2}, 1$ **20.** $-1 \pm i\frac{\sqrt{2}}{2}$
21. $-1.37, 0.69$ **22.** $\sqrt{2250} \approx 47.4$ miles per hour
23. (a) $f(x) = (x + 20)(90 - x)$ **(b)** 35
24. (a) [0, 6, 1] by [0, 150, 50]

(b) After about 5.6 seconds   **(c)** 2.75 seconds; 129 feet

CHAPTERS 1–11 CUMULATIVE REVIEW (pp. 803–804)

1. 1   **2.** Natural: $\sqrt[3]{8}$; whole: 0, $\sqrt[3]{8}$; integer: 0, -5, $\sqrt[3]{8}$;
rational: $0.\overline{4}$, 0, -5, $\sqrt[3]{8}$, $-\frac{4}{3}$; irrational: $\sqrt{7}$ **3. (a)** $x^{10}y^{12}$
(b) $\frac{x}{y^8}$ **(c)** $\frac{1}{b^{10}}$ **4.** 9.29×10^6 **5.** $2; x \le 2$ **6.** $(2, 5)$

7.

8.

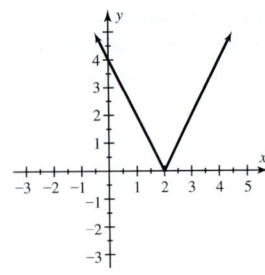

9. $y = -\frac{3}{2}x + 5$ **10.** $x = -3$ **11.** -12

12. $\left(-\infty, \frac{7}{4}\right)$ **13.** $\left[\frac{1}{3}, 1\right]$ **14.** $(-1, 5]$ **15.** $(-1, 1)$

16.

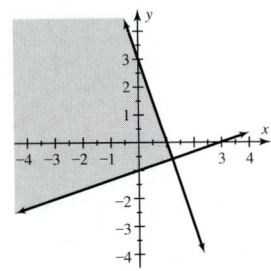

17. $(2, 2, 1)$ **18.** $6x^2 + 17x - 14$ **19.** $3x^3y + 3xy^3$
20. $x - 9$ **21.** $x(x + 1)(x - 2)$ **22.** $(2x + 5)(2x - 5)$
23. $\pm\sqrt{3}$ **24.** 1   **25.** $\frac{1}{2}(x + 3)$ **26.** $-\frac{2}{x(x + 2)}$
27. $4x^3$ **28.** $\frac{1}{64}$ **29.** 3   **30.** $3\sqrt{2x}$
31.

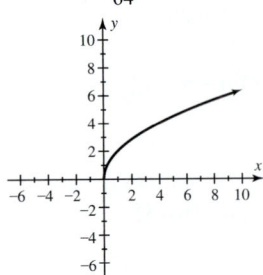

32. $\sqrt{26}$ **33.** $1 - i$ **34.** 3   **35.** 0.79
36.

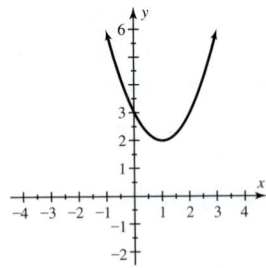

(a) $(1, 2)$ **(b)** 6   **(c)** $x = 1$ **(d)** $x \geq 1$
37. $f(x) = 2(x - 1)^2 - 3$ **38.** Shifted 1 unit left, 2 units

downward; narrower   **39.** $-3 \pm \sqrt{11}$ **40.** $\frac{3 \pm \sqrt{17}}{4}$
41. 1, 3   **42. (a)** $-2, 1$ **(b)** $-2 \leq x \leq 1$
43. $x < 1$ or $x > 2$ **44.** $\pm 4, \pm 4i$ **45.** c.   **46.** f.
47. g.   **48.** e.   **49.** d.   **50.** a.   **51.** b.   **52.** h.

53. (a) 300; initially, the tank holds 300 gal.
(b) 6; after 6 minutes the tank is empty.   **(c)** -50; water
is pumped out at 50 gal/min.   **(d)** $G(t) = 300 - 50t$
54. $2200 at 6%, $1200 at 5%, $600 at 4%
55. 125 ft by 125 ft   **56.** 33 feet

CHAPTER 12:   EXPONENTIAL AND LOGARITHMIC FUNCTIONS

SECTION 12.1 (pp. 817–820)

1. $g(f(7))$ **2.** $f(g(x))$ **3.** No   **4.** inputs; outputs
5. No   **6.** one-to-one   **7.** adding 10   **8.** 7   **9.** 8; 6
10. $x; y$ **11.** one-to-one   **12.** reflection; $y = x$
13. (a) 7   **(b)** 49   **(c)** $(g \circ f)(x) = x^2 + 3$
(d) $(f \circ g)(x) = (x + 3)^2$ **15. (a)** -65 **(b)** 126
(c) $(g \circ f)(x) = 8x^3 - 1$ **(d)** $(f \circ g)(x) = 2x^3 - 2$
17. (a) 3   **(b)** 1   **(c)** $(g \circ f)(x) = \left|\frac{1}{2}x - 2\right|$
(d) $(f \circ g)(x) = \frac{1}{2}|x - 2|$ **19. (a)** $\frac{11}{2}$ **(b)** $-\frac{1}{17}$
(c) $(g \circ f)(x) = 3 - \frac{5}{x}$ **(d)** $(f \circ g)(x) = \frac{1}{3 - 5x}$
21. (a) 77   **(b)** 122   **(c)** $(g \circ f)(x) = 16x^2 - 4x + 5$
(d) $(f \circ g)(x) = 8x^2 - 4x + 10$ **23. (a)** 1   **(b)** 2
25. (a) -1 **(b)** 1   **27. (a)** 0   **(b)** 2   **29. (a)** 2
(b) -3 **(c)** -1 **31.** $f(1) = f(-1) = 5$
33. $f(1) = f(-1) = 101$ **35.** $f(2) = f(-2) = 4$
37. Yes   **39.** No   **41.** Yes   **43.** Divide x by 7;
$f(x) = 7x; g(x) = \frac{x}{7}$ **45.** Multiply x by 2 then subtract 5;
$f(x) = \frac{x + 5}{2}; g(x) = 2x - 5$ **47.** Add 3 to x and
multiply the result by 2; $f(x) = \frac{1}{2}x - 3; g(x) = 2(x + 3)$
49. Take the cube root of x and subtract 5;
$f(x) = (x + 5)^3; g(x) = \sqrt[3]{x} - 5$
51. $(f \circ f^{-1})(x) = 4\left(\frac{x}{4}\right) = x; (f^{-1} \circ f)(x) = \frac{4x}{4} = x$
53.–57. Show $(f \circ f^{-1})(x) = (f^{-1} \circ f)(x) = x$. See the
answer to Exercise 51 above.   **59.** $f^{-1}(x) = \frac{x}{12}$
61. $f^{-1}(x) = x - 8$ **63.** $f^{-1}(x) = \frac{x + 2}{5}$
65. $f^{-1}(x) = -2(x - 1)$ **67.** $f^{-1}(x) = 8 - x$
69. $f^{-1}(x) = 2x - 1$ **71.** $f^{-1}(x) = \frac{x^3}{2}$
73. $f^{-1}(x) = \sqrt[3]{x} + 8$
75.

x	0	5	10	15	20
$f^{-1}(x)$	0	1	2	3	4

Domain of f = range of $f^{-1} = \{0, 1, 2, 3, 4\}$
Range of f = domain of $f^{-1} = \{0, 5, 10, 15, 20\}$
77.

x	4	2	0	-2	-4
$f^{-1}(x)$	-5	0	5	10	15

Domain of f = range of $f^{-1} = \{-5, 0, 5, 10, 15\}$
Range of f = domain of $f^{-1} = \{-4, -2, 0, 2, 4\}$

79.

81.

83.

85.

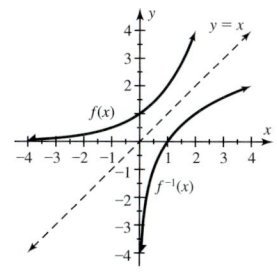

87. (a) 12   **(b)** -4 **(c)** $x^4 + 4x^2 + 2$ **89. (a)** $\frac{1}{2}$

(b) $-\frac{1}{x}$ **(c)** $\frac{x}{2}$ **91. (a)** 20π; after 5 seconds, the wave
has a circumference of $20\pi \approx 62.8$ feet.

(b) $(C \circ r)(t) = 4\pi t$ **93. (a)** 16; in 1980, 16% of people
25 or older completed four or more years of college.

(b)

x	8	16	27
$P^{-1}(x)$	1960	1980	2000

(c) 1980

95. (a) $75°$; 150   **(b)** 150; One hour after midnight there are
150 mosquitoes per 100 square feet.   **(c)** The number of
mosquitoes per 100 square feet, h hours after midnight
(d) $T(h) = -5h + 80$; $M(T) = 2T$
(e) $(M \circ T)(h) = -10h + 160$
97. (a) Yes, different inputs result in different outputs.
(b) $f^{-1}(x) = \frac{5}{9}(x - 32)$ converts x degrees Fahrenheit to
an equivalent temperature in degrees Celsius.
99. $f(x) = 4x$; $f^{-1}(x) = \frac{x}{4}$ converts x quarts to gallons.

SECTION 12.2 (pp. 828–832)

1. $f(x) = Ca^x$
2.

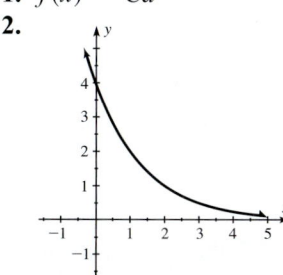

3. Growth   **4.** 32; 25   **5.** 2.718   **6.** 7.389; 9.870
(approximately)   **7.** factor   **8.** 1.5   **9.** $\frac{1}{9}$; 9   **11.** 5; 160

13. $4; \frac{1}{8}$ **15.** $15; \frac{5}{9}$ **17.** 0.17; 2.41   **19.** 5; 1.08   **21.** $1; \frac{1}{a}$
23. (a) Exponential decay   **(b)** $f(x) = 64\left(\frac{1}{4}\right)^x$
25. (a) Linear growth   **(b)** $f(x) = 3x + 8$
27. (a) Exponential growth   **(b)** $f(x) = 4(1.25)^x$
29. $C = 1, a = 2$ **31.** $C = 4, a = \frac{1}{4}$ **33.** c.   **35.** d.
37.

Growth

39.

Decay

41.

Decay

43.

Growth

45.

Growth

47.

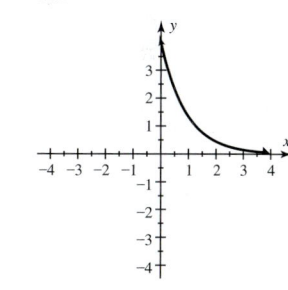

Decay

49. e^7 **51.** $\frac{1}{a}$ **53.** e^y **55.** 2^3 or 8   **57.** 5^{2y} or 25^y
59. $3551.05   **61.** $1,820,087.63   **63.** $792.75
65. Yes; this is equivalent to having two accounts, each containing $1000 initially.   **67.** About $1.28 trillion
69. 3.32   **71.** 0.86
73. $[-4, 4, 1]$ by $[0, 8, 1]$

Growth

75. $[-4, 4, 1]$ by $[0, 8, 1]$

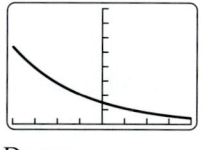

Decay

77. (a) $f(x) = 34e^{0.013x}$ **(b)** 38.7 million
79. (a) $f(x) = 8e^{0.019x}$ **(b)** 9.7 million

81. (a) $f(x) = 5e^{0.031x}$
(b) [0, 10, 1] by [4, 7, 1]

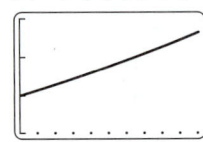

(c) About 6.82 million
83. (a) $C = 500$, $a = 2$ **(b)** About 5278 thousand per milliliter **(c)** The growth is exponential.
85. (a) About 11.3; in 1995 there were about 11.3 million cellular phone subscribers. **(b)** 1.495; each year from 1985 to 2000 the number of subscribers increased by a factor of 1.495, or by 49.5%.
87.

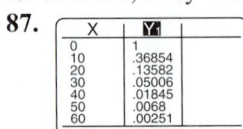

(a) 1; the probability that no vehicle will enter the intersection during a period of 0 seconds is 1 or 100%.
(b) About 30 seconds

CHECKING BASIC CONCEPTS 12.1 & 12.2 (p. 832)

1. (a) 7 **(b)** $(f \circ g)(x) = 2x^2 + 9x + 6$
2.

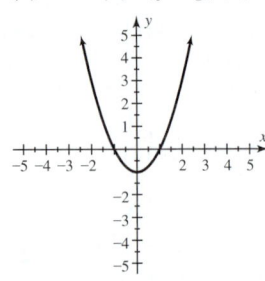

(a) No, it does not pass the horizontal line test. **(b)** No
3. $f^{-1}(x) = \frac{x+3}{4}$ **4.** $\frac{3}{4}$
5.

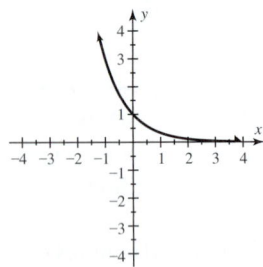

6. $C = 2$; $a = \frac{1}{2}$

SECTION 12.3 (pp. 840–844)

1. 10 **2.** e **3.** $D = \{x \mid x > 0\}$; R: all real numbers
4. $D = \{x \mid x > 0\}$; R: all real numbers **5.** k **6.** k
7. x **8.** x **9.** $\log 5$ **10.** $\log_2 5$ **11.** 0 **12.** undefined
13. 5 **15.** -4 **17.** 0 **19.** -2 **21.** 4.7 **23.** $6x$
25. $-\frac{3}{2}$ **27.** 8 **29.** $\frac{3}{2}$ **31.** $-\frac{2}{3}$ **33.** Undefined

35. 2 **37.** -4 **39.** -2 **41.** 0 **43.** -3 **45.** -2
47. 2 **49.** x^2 **51.** 17 **53.** $(2x)^2$ **55.** 5 **57.** $-5x$
59. $2x - 7$ **61.** $0.6z$ **63.** 1.398 **65.** 0.161
67. 1.946 **69.** -0.560
71. **73.**

 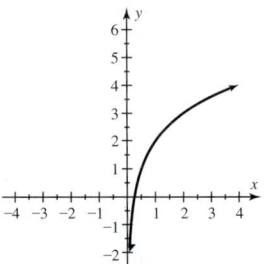

75. [-4, 4, 1] by [-4, 4, 1] **77.** [-4, 4, 1] by [-4, 4, 1]

Reflected across the y-axis 2 units to the left;
together with the graph of $D = \{x \mid x > -2\}$
$y = \ln x$; $D = \{x \mid x \neq 0\}$
79. d. **81.** a.
83. **85.**

 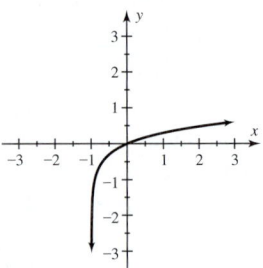

1 unit downward 1 unit to the left
87. **89.**

 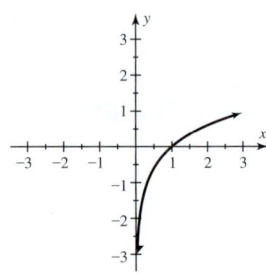

1 unit to the right Increases faster
91.

x	1/4	1/2	1	$\sqrt{2}$	64
$\log_2 x$	-2	-1	0	1/2	6

93. 120 dB; yes **95. (a)** $P(0) = 27$; the air pressure at the eye of the hurricane is 27 inches of mercury. $P(50) \approx 28.9$; the air pressure 50 miles from the eye is about 28.9 inches of mercury. **(b)** It increases rapidly at first and then more slowly. **(c)** Low

97. Exponentially, because it grows faster over time.
99. (a) 3.5   **(b)** 1   **(c)** It decreases by 2.5.
101. (a) 10^6; 10^8 **(b)** 100 times

SECTION 12.4 (pp. 849–851)

1. 4   **2.** 4   **3.** 3   **4.** $\log m + \log n$ **5.** $\log m - \log n$
6. $r \log m$ **7.** No   **8.** Yes   **9.** No   **10.** No
11. $\log_a x = \frac{\log x}{\log a}$ or $\log_a x = \frac{\ln x}{\ln a}$ **12.** 0; 1
13. $\ln 3 + \ln 5$ **15.** $\log_3 x + \log_3 y$
17. $\ln 2 + \ln 5 + \ln z$ **19.** $\log 7 - \log 3$
21. $\ln x - \ln y$ **23.** $\log_2 45 - \log_2 x$ **25.** $\log 225$
27. $\ln xy$ **29.** $\ln 14x^3$ **31.** $\ln xy$ **33.** $\log 5$ **35.** $\ln x^2$
37. 2   **39.** $6 \log 3$ **41.** $x \ln 2$ **43.** $\frac{1}{4}\log_2 5$ **45.** $\frac{1}{3}\log_4 z$
47. $(y-1)\log x$ **49.** $\log z$ **51.** $\log x^3 y^2$ **53.** 0
55. $\ln 2^x$ **57.** $\log_3 \frac{1}{x^2}$ **59.** $\log_a \frac{x+1}{x-1}$ **61.** $\log x + 2 \log y$
63. $4 \ln x + \ln y - \ln z$ **65.** $\frac{1}{3}\log_4 z - \frac{1}{2}\log_4 y$
67. $4 \log x + 3 \log y$ **69.** $\ln x - \ln y$
71. $\frac{3}{2}\log_4 x + \frac{1}{2}\log_4 y - \log_4 z$
73. $[-6, 6, 1]$ by $[-4, 4, 1]$ $[-6, 6, 1]$ by $[-4, 4, 1]$

By the power rule, $\log x^3 = 3 \log x$
75. $[-6, 6, 1]$ by $[-4, 4, 1]$ $[-6, 6, 1]$ by $[-4, 4, 1]$

Not the same
77. 1.2   **79.** 1.8   **81.** 2.1   **83.** 0.4   **85.** -1.1
87. 1.46   **89.** 4.64   **91.** 2.10
93. $10 \log (10^{16}x) = 10(\log 10^{16} + \log x) =$
$10(16 + \log x) = 160 + 10 \log x$

CHECKING BASIC CONCEPTS 12.3 & 12.4 (p. 851)

1. (a) 4   **(b)** x **(c)** -3 **(d)** $\frac{1}{2}$
2.

(a) $D = \{x \mid x > 0\}$; R: all real numbers   **(b)** 0
(c) Yes; for example, $\log \frac{1}{10} = -1$.   **(d)** No; negative numbers are not in the domain of $\log x$.

3. (a) $\log x + \log y$ **(b)** $\ln x - \ln y - \ln z$ **(c)** $2 \ln x$
(d) $2 \log x + 3 \log y - \frac{1}{2}\log z$ **4. (a)** $\log xy$
(b) $\ln \frac{2x}{y^3}$ **(c)** $\log_2 \frac{x^2 y^3}{z}$

SECTION 12.5 (pp. 859–862)

1. Add 5 to each side.   **2.** Divide each side by 5.
3. Take the common logarithm of each side.
4. Exponentiate each side using base 10.   **5.** x **6.** x
7. $2x$ **8.** $x + 7$ **9.** No; $\log \frac{5}{4} = \log 5 - \log 4$
10. No; $\log 5 - \log 4 = \log \frac{5}{4}$ **11.** 1   **12.** 1
13. 3   **15.** 6   **17.** 6   **19.** 4   **21.** $\frac{\log 124}{0.4} \approx 5.23$
23. 0   **25.** $\ln 25 \approx 3.22$ **27.** $\frac{\ln 2}{\ln 0.4} \approx -0.76$
29. $\ln 7 \approx 1.95$ **31.** $\log \frac{35}{2} - 2 \approx -0.76$
33. $\frac{\log 20}{2 \log 3.1} \approx 1.32$ **35.** -1 **37.** $-1, 5$ **39.** $\frac{\ln 10}{3} \approx 0.77$
41. $\frac{2 \ln 2}{1 - \ln 2} \approx 4.52$ **43.** $\frac{2 \log 5}{0.5 \log 4 - \log 5} \approx -3.51$
45. (a) 1   **(b)** 1   **47. (a)** -2 **(b)** -2 **49.** -1
51. 0   **53.** 5   **55.** $\frac{\log 1.45}{\log 0.55} \approx -0.62$ **57.** $-1.84, 1.15$
59. 1.31   **61.** 100   **63.** $e^5 \approx 148.41$ **65.** 5,000,000
67. 16   **69.** $\frac{2^{2.3}}{5} \approx 0.98$ **71.** $10^{1.4} \approx 25.12$
73. $\frac{e^{11} - 1}{2} \approx 29,936.57$ **75.** 1   **77.** 5   **79.** 4   **81.** 3
83. (a) 2   **(b)** $e^{0.7} \approx 2.01$ **85. (a)** 2
(b) $\frac{1}{2}(10^{0.6}) \approx 1.99$ **87.** $10^{1.6} \approx 39.81$
89. $e - 1 \approx 1.72$ **91.** 9   **93.** 8 years
95. (a) About 8503; in 1994 about 8503 people were waiting for liver transplants.   **(b)** In about 1998   **97.** About 20,893 pounds   **99. (a)** About 203; in 1975 there were about 203 thousand bluefin tuna.   **(b)** In 1979   **(c)** In 1979
101. $a = 25, b = 3$ **103. (a)** About 1.67 acres
(b) About 1.67 acres   **105. (a)** Nonlinear; they do not increase at a constant rate.   **(b)** Each year the amount of fertilizer increases by a factor of 1.06, or by 6%.   **(c)** In 1968
107. 10^{-6} watts/square centimeter   **109.** About 7 miles

CHECKING BASIC CONCEPTS 12.5 (p. 863)

1. (a) $\log 20 \approx 1.30$ **(b)** $\frac{\log 147}{3 \log 2} \approx 2.40$
(c) $e^{4.1} \approx 60.34$ **(d)** 500   **2.** $x = 8$ **3.** 20 years

CHAPTER 12 REVIEW (pp. 866–868)

1. (a) 81   **(b)** $(f \circ g)(x) = 50x^2 - 2$ **2. (a)** -32
(b) $(f \circ g)(x) = \sqrt[3]{4x^3 - 6}$ **3. (a)** 0   **(b)** 3
4. (a) 3   **(b)** -2 **(c)** -1 **5.** $f(1) = f(-1) = 2$
6. $f(0) = f(2) = 1$ **7.** No   **8.** Yes
9. $(f \circ f^{-1})(x) = 2\left(\frac{x+9}{2}\right) - 9 = x$
$(f^{-1} \circ f)(x) = \frac{(2x-9)+9}{2} = x$

10. $(f \circ f^{-1}) = (\sqrt[3]{x-1})^3 + 1 = x$
$(f^{-1} \circ f)(x) = \sqrt[3]{(x^3+1)} - 1 = x$

11. $f^{-1}(x) = \frac{x}{5}$ **12.** $f^{-1}(x) = x + 11$

13. $f^{-1}(x) = \frac{x-7}{2}$ **14.** $f^{-1}(x) = \frac{4}{x}$

15.

x	10	8	7	3
$f^{-1}(x)$	0	1	2	3

$D = \{3, 7, 8, 10\}; R = \{0, 1, 2, 3\}$

16.

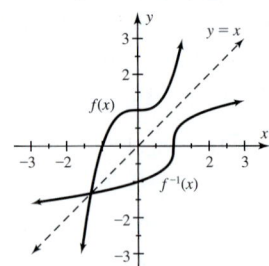

17. $\frac{1}{6}$; 36   **18.** 5; $\frac{5}{8}$ **19.** 3; $\frac{1}{81}$ **20.** 3; $\frac{1}{2}$

21.

Exponential growth

22.

Exponential decay

23.

Logarithmic growth

24.

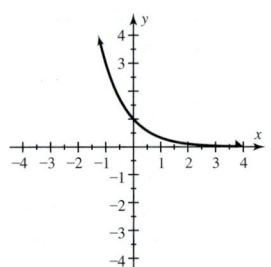

Exponential decay

25. (a) Exponential growth   **(b)** $f(x) = 5(2)^x$
26. (a) Linear growth   **(b)** $f(x) = 5x + 5$
27. $C = \frac{1}{2}, a = 2$ **28.** $k = 2$ **29.** \$2829.54
30. \$675,340.51   **31.** 399.67   **32.** 0.71   **33.** 3.48
34. 3.89   **35.** -3 **36.** 2   **37.** -4 **38.** 2   **39.** 1.813
40. -0.163 **41.** 4.787   **42.** -1.322 **43.** 7   **44.** $\frac{5}{9}$
45. $6 - x$ **46.** x^2 **47.** $\ln x + \ln y$ **48.** $\log x - \log y$
49. $2 \ln x + 3 \ln y$ **50.** $\frac{1}{2} \log x - 3 \log z$
51. $2 \log_2 x + \log_2 y - \log_2 z$ **52.** $\frac{1}{3} \log_3 x - \frac{1}{3} \log_3 y$
53. $\log 75$ **54.** $\log_4 (10x^2)$ **55.** $\ln \frac{x^2}{y^3}$ **56.** $\log xy$
57. $3 \log 6$ **58.** $2 \ln x$ **59.** $2x \log_2 5$

60. $(x+1) \log_4 0.6$ **61.** 2   **62.** 4   **63.** $\ln 9 \approx 2.20$

64. $\frac{\log 0.2}{\log 0.85} \approx 9.90$ **65.** $e^{0.8} \approx 2.23$ **66.** $\frac{1}{2}e^5 \approx 74.21$

67. 10^{40} **68.** 100   **69.** $\frac{4 \log 2}{\log 3 - \log 2} \approx 6.84$ **70.** 6

71. (a) 3   **(b)** 3   **72. (a)** 4   **(b)** 4
73. (a) 64π; after 8 seconds, the balloon has a surface area of $64\pi \approx 201$ in². **(b)** $(S \circ r)(t) = 8\pi t$ **74. (a)** Yes, different inputs result in different outputs.   **(b)** $f^{-1}(x) = \frac{x}{0.08}$ calculates the cost of an item whose sales tax is x dollars.
75. 7 years   **76.** $a = 100, b = 50$ **77.** $C = 3, a = 2$
78. 10^7
79. (a) [0, 10, 2] by [0, 4, 1]

Growth

(b) About 3.3 million   **(c)** 2008   **80. (a)** 1000; there were 1000 bacteria/mL initially.   **(b)** About 495.11 minutes
81. (a) About 6.93 meters/second   **(b)** 12.18 meters

CHAPTER 12 TEST (pp. 869–870)

1. 6; $(f \circ g)(x) = 4(x+7)^3 - 5(x+7)$
2. (a) 3   **(b)** 3   **3.** $-5 \neq 5$, but $f(-5) = f(5) = 0$
(answers may vary)   **4.** $f^{-1}(x) = \frac{5-x}{2}$
5.

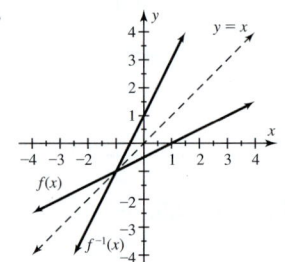

6.

x	8	6	4	2
$f^{-1}(x)$	1	2	3	4

$D = \{2, 4, 6, 8\}; R = \{1, 2, 3, 4\}$

7. $\frac{3}{16}$

8.

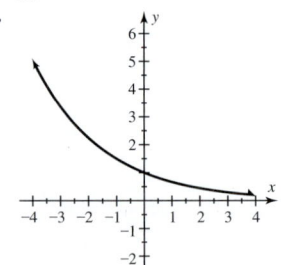

Exponential decay

9. (a) Exponential growth   **(b)** $f(x) = 3(2)^x$
10. (a) Linear growth   **(b)** $f(x) = 1.5x - 1$

11. $C = 1, a = \frac{1}{2}$ **12.** $1051.91 **13.** 4.16
14. $\frac{1}{2}$ **15.** 5.426
16.

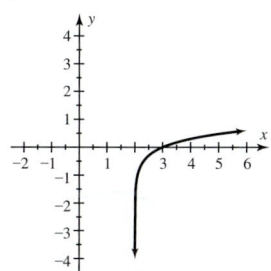

Shifted to the right 2 units
17. $3 \log x + 2 \log y - \frac{1}{2} \log z$ **18.** $\ln \frac{x^4 z}{y^5}$ **19.** $2x \log 7$
20. $1 - 3x$ **21.** $\ln 25 \approx 3.22$ **22.** $\log 50 \approx 1.70$
23. $10^{1.8} \approx 63.10$ **24.** $\frac{1}{5}e^9 \approx 1620.62$ **25.** $a = 5, b = 3$
26. (a) 4 million **(b)** $4e^{0.45} \approx 6.27$; after 5 hours there
were about 6.27 million bacteria. **(c)** Growth **(d)** After
4.51 hours

CHAPTERS 1–12 CUMULATIVE REVIEW
(pp. 871–874)

1. 4.29×10^{-4} **2.** Natural: none; whole: 0; integer:
$-3, 0$; rational: $-\frac{11}{7}, -3, 0, 5.\overline{18}$; irrational: $\sqrt{6}, \pi$
3. (iii) **4.** Commutative **5.** d^4 **6.** $\frac{b^9}{64a^6}$ **7.** $\frac{4y^8}{x^5}$
8. $\frac{y^4}{4x^5}$ **9.** $y = \frac{5}{4}x + 1$ **10.** $\{x \mid x \neq -3\}$
11. $f(x) = 4x - 3$ **12.** $x = 4$ **13.** 1
14.

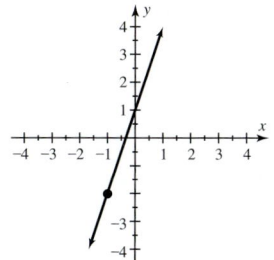

15. $y = 7x - 6$ **16.** $y = 3x + 5$ **17.** -1 **18.** $x < 0$
19. -18 **20.** $(-\infty, 0)$ **21.** $\frac{4}{15}$ **22.** $(-5, 10]$
23. $[-2, 10]$ **24.** $4, 16$
25.

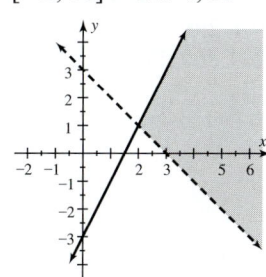

26. 2 **27.** $(1, 1)$ **28.** No solutions **29.** $(-4, 0, 2)$
30. $(0, 0, 1)$ **31.** $R = 21$ **32.** 8 square inches

33. $2x(x - 1)^2$ **34.** $(2a - 5b)(2a + 5b)$
35. $(2t - 3)(4t^2 + 6t + 9)$ **36.** $(2a^2 + 5)(2a - 1)$
37. $-\frac{5}{6}, 2$ **38.** $-\frac{2}{3}, \frac{2}{3}$ **39.** $-3, 0, 5$ **40.** $0, \frac{1}{2}$
41. 1 **42.** $\frac{x + 2}{x - 3}$ **43.** 3 **44.** -3 **45.** $J = \frac{2z}{P - 1}$
46. $\frac{x^3 + 3}{x^3 - 3}$ **47.** 40 **48.** $3x^2 + 6x + 10 + \frac{5}{x - 2}$
49. $\frac{x^4}{y^6}$ **50.** x^3 **51.** $10ab$ **52.** $\sqrt[6]{6}$ **53.** $(b + 3)a\sqrt[3]{a^2b}$
54. 20 **55.** $\frac{5 + \sqrt{3}}{11}$ **56.** $(4, \infty)$ **57.** $-6, 4$ **58.** 3
59. $3 + 5i$ **60.** $-i$ **61.** $(2, 1)$ **62.** $\frac{7}{2}$
63. Shifted right 3 units and up 2 units
64. $y = (x + 3)^2 - 11; (-3, -11)$ **65.** 5, 8
66. $\frac{1 \pm \sqrt{41}}{4}$ **67.** $2 \pm \sqrt{2}$ **68.** $\pm 2, \pm \sqrt{6}$
69. (a) $-1, 3$ **(b)** $a < 0$ **(c)** Positive
70. $(-\infty, -7] \cup [2, \infty)$ **71. (a)** 7
(b) $(g \circ f)(x) = 2x^2 - 3$ **72.** $f(-4) = f(3) = 6$
73. $f^{-1}(x) = \frac{3}{x}$ **74.** $2367.10 **75.** 4 **76.** $2x$
77. $\frac{1}{2}\log x - 2 \log y$ **78.** $\ln (5x^3)$ **79.** $10^{11/16} \approx 68.13$
80. $\frac{\log 17}{3 \log 2} \approx 1.36$ **81.** 4% **82.** 2.27 pounds
83. (a) 1950 **(b)** 220 million Btu **84.** $\sqrt{4200} \approx 64.8$ mph
85. About 11.3% **86. (a)** 9.91 meters per second
(b) 8.52 meters

CHAPTER 13: CONIC SECTIONS

SECTION 13.1 (pp. 882–885)

1. Parabola, ellipse, hyperbola **2.** The axes of symmetry
are vertical and horizontal, respectively. **3.** No
4.

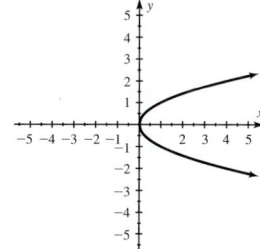

5. No; it does not pass the vertical line test. **6.** (h, k)
7. left **8.** right **9.** circle; (h, k) **10.** $(0, 0); r$
11.   **13.**

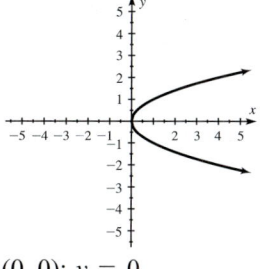

$(0, 0); y = 0$ $(1, 0); y = 0$

15.

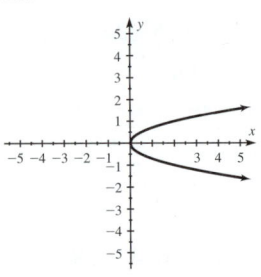

$(0, 0); y = 0$

17.

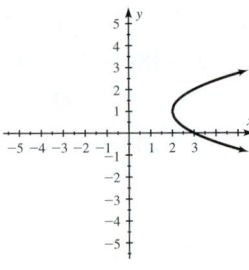

$(2, 1); y = 1$

19.

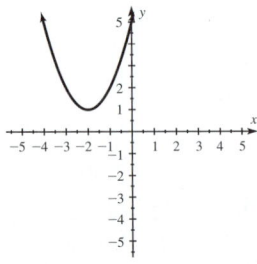

$(-2, 1); x = -2$

21.

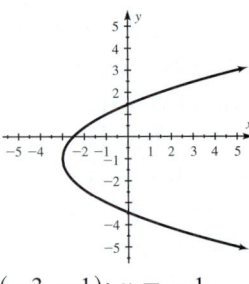

$(-3, -1); y = -1$

23.

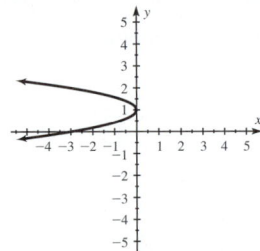

$(0, 1); y = 1$

25.

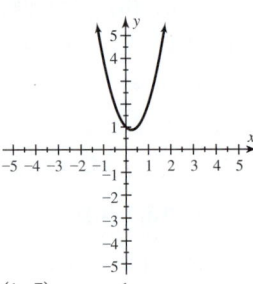

$\left(\frac{1}{4}, \frac{7}{8}\right); x = \frac{1}{4}$

27.

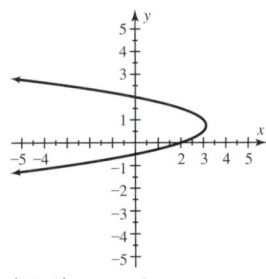

$\left(\frac{25}{8}, \frac{3}{4}\right); y = \frac{3}{4}$

29.

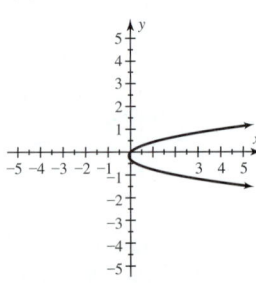

$\left(-\frac{1}{12}, -\frac{1}{6}\right); y = -\frac{1}{6}$

31.

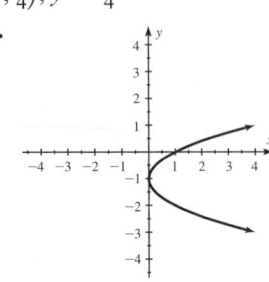

$(0, -1); y = -1$

33. $y = x^2$ **35.** $x = (y + 1)^2 - 2$ **37.** Upward
39. Downward   **41.** $x \geq 0$ **43.** Two   **45.** 1
47. $x^2 + y^2 = 1$ **49.** $(x + 1)^2 + (y - 5)^2 = 9$
51. $(x + 4)^2 + (y + 6)^2 = 2$ **53.** $x^2 + y^2 = 16$
55. $(x + 3)^2 + (y - 2)^2 = 1$
57. $3; (0, 0)$ **59.** $3; (1, 3)$

    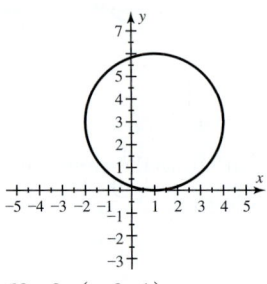

61. $5; (-5, 5)$ **63.** $3; (-3, 1)$

65. $\sqrt{7}; (-3, 1)$

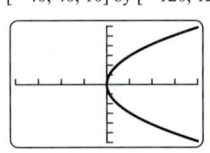

67. (a)
[−40, 40, 10] by [−120, 120, 20]

(b) 32 ft

69. (a)
[−1.5, 1.5, 0.5] by [−1, 1, 0.5]

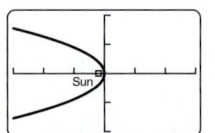

(b) 2.6 A.U., or
241,800,000 miles

SECTION 13.2 (pp. 893–895)

1.

2.

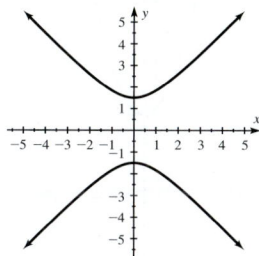

(answers may vary)   (answers may vary)

3. horizontal   **4.** vertical   **5.** 2   **6.** 4   **7.** left; right
8. lower; upper   **9.** They are the diagonals extended.   **10.** No

11.

13.

15.

17.

19.

21.

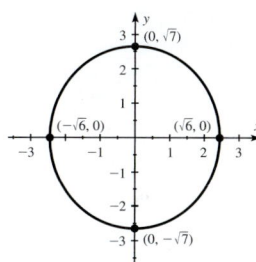

23. $\frac{x^2}{9} + \frac{y^2}{4} = 1$ **25.** $\frac{y^2}{25} + \frac{x^2}{16} = 1$
27.

29.

31.

33.

35.

37.

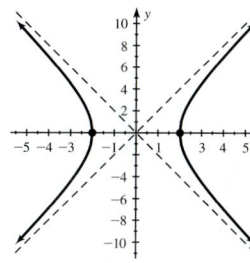

39. $x^2 - y^2 = 1$ **41.** $\frac{y^2}{4} - \frac{x^2}{9} = 1$
43. (a) $A \approx 62.83; P \approx 28.45$
(b) $A \approx 11.75; P \approx 13.33$
45. (a) $[-60, 60, 10]$ by $[-40, 40, 10]$

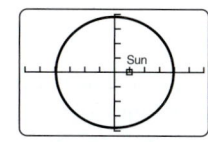

(b) $P \approx 243.9$ A.U., or about 2.27×10^{10} miles; $A \approx 4733$
square A.U., or about 4.09×10^{19} square miles
47. Maximum: 668 miles; minimum: 340 miles
49. Height: 20 feet; width: 200 feet

CHECKING BASIC CONCEPTS 13.1 & 13.2 (p. 895)

1.

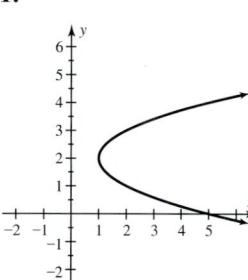

2. $(x - 1)^2 + (y + 2)^2 = 4$

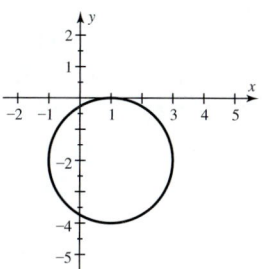

$(1, 2); y = 2$
3. x-intercepts: ±2; y-intercepts: ±3

4. (a)

Parabola

(b)

Ellipse

(c)

Hyperbola

(d)

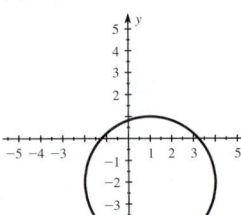

Circle (and ellipse)

SECTION 13.3 (pp. 902–904)

1. Any number   **2.** Two   **3.** Two; the line intersects the circle twice.   **4.** All points inside and including a circle of radius 1 centered at the origin   **5.** No   **6.** No

7.

(answers may vary)

8.

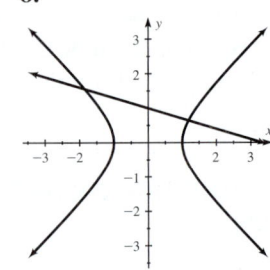

(answers may vary)

9. $(1, 3), (-1, -3)$ **11.** $(0, -1), (0, 1)$

13. $(3, 6), (-3, -6)$ **15.** $(2, -1)$

17. $(-\sqrt{2}, 2), (\sqrt{2}, 2)$ **19.** $\left(-\frac{3}{5}, \frac{7}{5}\right), (-1, 1)$

21. No solutions   **23.** $(\pm 3, \pm 1)$ **25.** $(-1, -2), (-2, 1)$

27. $(7, 3), (2, -2)$ **29.** $(-1, 2), (3, -6)$

31. $(-1, -1), (1, 1)$

33.

35.

37.

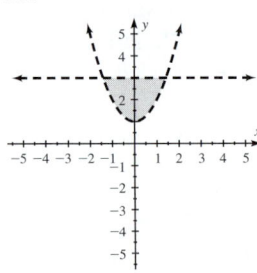

$(0, 2)$, (answers may vary)

39.

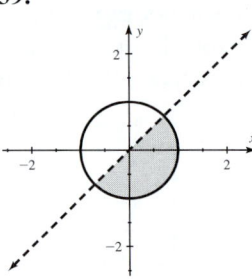

$\left(\frac{1}{2}, -\frac{1}{2}\right)$, (answers may vary)

41.

$(1, 0)$

43.

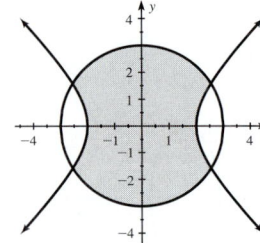

$(0, 0)$, (answers may vary)

45. a.   **47.** $y \geq x^2; y < 4 - x$

49. $[-10, 10, 1]$ by $[-10, 10, 1]$

51. $r = 1.6$ inches; $h \approx 4.97$ inches

53. (a) $h = \frac{3V}{\pi r^2}; h = \sqrt{\left(\frac{S}{\pi r}\right)^2 - r^2}$ **(b)** $r \approx 2.02$ feet, $h \approx 7.92$ feet; $r \approx 3.76$ feet, $h \approx 2.30$ feet

CHECKING BASIC CONCEPTS 13.3 (p. 904)

1. $(1, -1), (3, 3)$ **2.** Two   **3. (a)** $(0, 3), (4, 4)$ (answers may vary)   **(b)** $y \geq 2 - x$ and $y \leq 4 - x^2$

4.

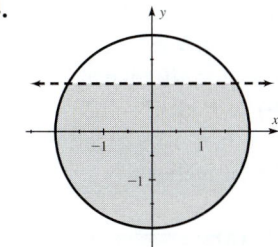

CHAPTER 13 REVIEW (pp. 906–909)

1.

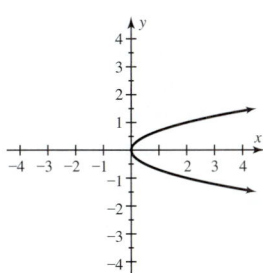

$(0, 0); y = 0$

2.

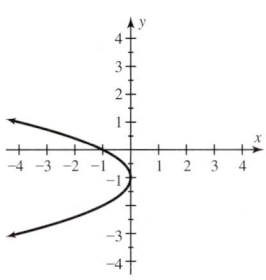

$(0, -1); y = -1$

3.

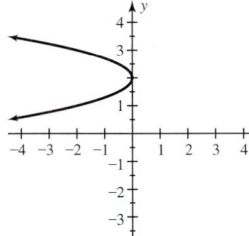

$(0, 2); y = 2$

4.

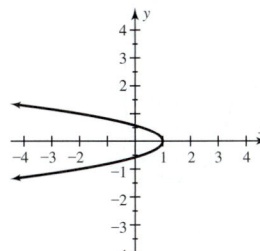

$(-1, -2); y = -2$

5.

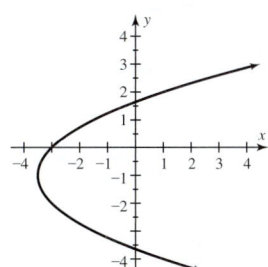

$(1, 0); y = 0$

6.

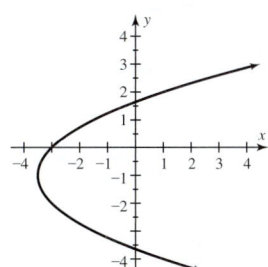

$\left(-\frac{7}{2}, -1\right); y = -1$

7. $x = y^2$ **8.** $(x + 2)^2 + (y - 2)^2 = 16$

9. $x^2 + y^2 = 1$ **10.** $(x - 2)^2 + (y + 3)^2 = 16$

11. $5; (0, 0)$

12. $3; (2, 0)$

13. $\sqrt{5}; (-3, 1)$

14. $3; (1, -1)$

15.

16.

17.

18.

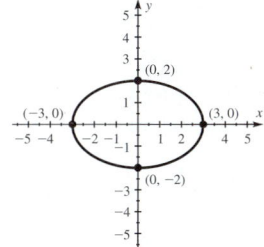

19. $\dfrac{y^2}{16} + \dfrac{x^2}{4} = 1$ **20.** $x^2 - \dfrac{y^2}{4} = 1$

21.

22.

23.

24.

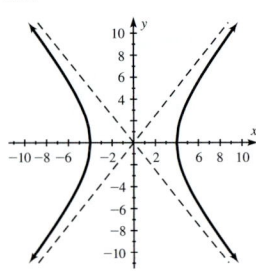

25. $(0, 3), (3, 0)$ **26.** $(-1, -2), (1, 2)$ **27.** $(0, 0), (2, 2)$
28. $(-2, -1), (-2, 1), (2, -1), (2, 1)$
29. $(-4, -4), (4, 4)$ **30.** $(0, -4), (4, 0)$
31. $(-1, 1), (1, 1)$ **32.** $(1, 2), (2, 5)$
33. $(-4, -12), (2, 0)$ **34.** $(-1, -1), (1, -1)$
35. $(2, 2), (-4, -4)$ **36.** $(1, 1), (0, 0)$
37.

38.

39.

40.

41.

42.

43.

44.

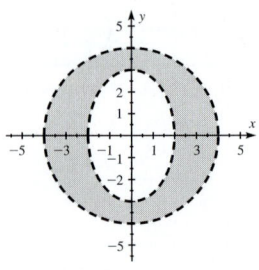

45. $y \geq x^2 - 2, y \leq 2 - x$ **46.** $y \geq x, x^2 + y^2 \leq 4$
47. (a) $xy = 1000, 2x + 2y = 130$ **(b)** $x = 25$ inches;
$y = 40$ inches   **(c)** $x = 25$ inches; $y = 40$ inches
48. (a) $xy = 60, y - x = 7$ **(b)** $x = 5; y = 12$
(c) $x = 5; y = 12$ **49.** $r = 1$ foot, $h \approx 15.92$ feet; yes
50. Either $r \approx 0.94$ inches, $h \approx 12.60$ inches or $r \approx 3.00$
inches, $h \approx 1.23$ inches; no
51. (a)
$[-7.5, 7.5, 1]$ by $[-5, 5, 1]$

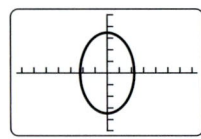

52. (a)
$[-3, 3, 1]$ by $[-2, 2, 1]$

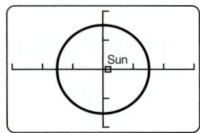

(b) $A \approx 24.33, P \approx 18.32$

(b) $P \approx 9.55$ A.U., or about
8.9×10^8 miles; $A \approx 7.26$
square A.U., or about
6.3×10^{16} square miles

CHAPTER 13 TEST (pp. 909–910)

1.

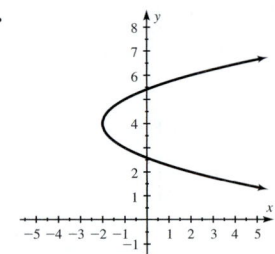

$(-2, 4); y = 4$
2. $x = -y^2 + 1$ **3.** $(x - 2)^2 + (y + 4)^2 = 4$
4. $(x + 5)^2 + (y - 2)^2 = 100$
5. $r = 4$, center $= (-2, 3)$ **6.**

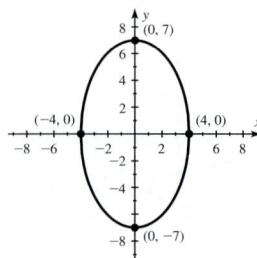

7. $\frac{x^2}{100} + \frac{y^2}{64} = 1$
8.

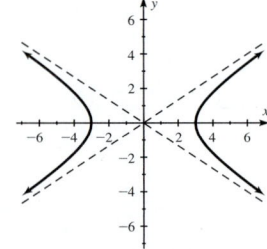

9. $(0, -4), (4, 0)$ **10.** $(-1, -4), (4, 1)$

11. $(-2, 4), (2, 4)$

12.

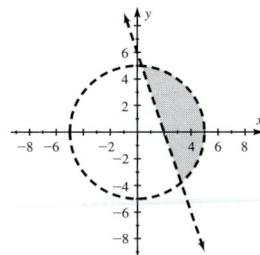

13. $y \le 4 - x^2, y \ge x^2 - 4$ **14. (a)** $xy = 5000$,

$2x + 2y = 300$ **(b)** 50 feet by 100 feet

15. Either $x \approx 22.08$ in., $y \approx 2.43$ in. or

$x \approx 7.29$ in., $y \approx 22.24$ in.; no

16. (a) $[-30, 30, 10]$ by $[-20, 20, 10]$

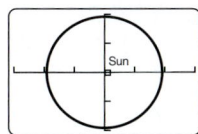

(b) 18.28 A.U., or about 1,700,040,000 miles

CHAPTERS 1–13 CUMULATIVE REVIEW
(pp. 912–913)

1. 25   **2.** $\dfrac{b^8}{a^3}$ **3.** 0.007345   **4.** $\dfrac{1}{2}; x \ne 4$ **5.** 3; 4

6.   **7.**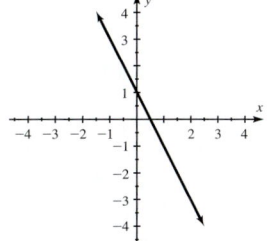

8. $y = \dfrac{3}{2}x - 5$ **9.** $\dfrac{2}{7}$ **10.** $(-1, 3]$ **11.** $[-2, 2]$

12. $(-\infty, -1] \cup [3, \infty)$ **13.** $(1, 3)$

14.

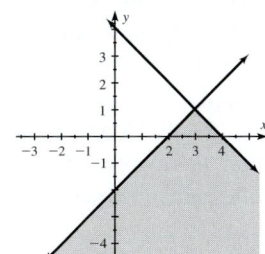

15. $2x^2 + 9x - 5$ **16.** $2x^2y - 3xy^3 + xy$

17. $(3x + 1)(2x - 5)$ **18.** $x(x + 2)(x - 2)$

19. $-2, -1$ **20.** $\dfrac{3 \pm \sqrt{5}}{2}$ **21.** $\dfrac{3}{2}$ **22.** $\dfrac{2x}{x^2 - 1}$

23. $2x\sqrt{2}$ **24.** 4   **25.** $4x$ **26.** $5\sqrt{3x}$

27.

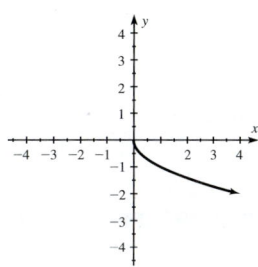

28. 5   **29.** 13   **30.** 2   **31.** $(3, -6)$

32. $f(x) = (x - 1)^2 + 2$ **33.** Shifted 4 units right

34. 1, 2   **35.** 0, $\pm i$ **36. (a)** 4   **(b)** 3   **(c)** x **(d)** 6

(e) 2   **(f)** 3   **37. (a)** 17   **(b)** $2x^2 + 2$ **38.** $\dfrac{2}{3} - \dfrac{1}{3}x$

39. \$1340.10   **40.** $2 \log x + \dfrac{1}{2} \log y - 3 \log z$ **41.** $\ln 9$

42. 25

43. (a)   **(b)**

(c)   **(d)**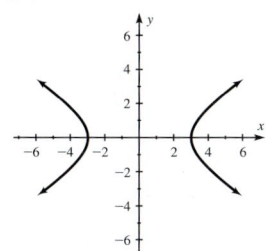

44. $(0, -1), (0, 1)$

45.

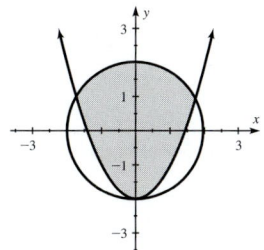

46. (a) 400; initially, the driver is 400 miles from home.

(b) 8; after 8 hours, the driver arrives at home.

(c) -50; the driver is traveling 50 mph toward home.

47. \$500 at 5%, \$1000 at 6%, \$500 at 7%   **48.** 300 ft by

300 ft   **49.** $50 \ln 2 \approx 34.7$ yr   **50.** 2.4 in.

CHAPTER 14: SEQUENCES AND SERIES

SECTION 14.1 (pp. 922–924)

1. 1, 2, 3, 4 (answers may vary)   **2.** 1, 3, 5, 7, . . . (answers may vary)   **3.** function; natural numbers
4. sequence   **5.** 6   **6.** scatterplot   **7.** $f(2)$ **8.** a_4
9. 1, 4, 9, 16   **11.** $\frac{1}{6}, \frac{1}{7}, \frac{1}{8}, \frac{1}{9}$ **13.** $\frac{5}{2}, \frac{5}{4}, \frac{5}{8}, \frac{5}{16}$ **15.** 9, 9, 9, 9
17. 1, 8, 27   **19.** $1, \frac{8}{5}, 2$ **21.** 2, 9, 20   **23.** $-2, -2, -2$
25. $b + c$; $2b + c$ **27.** 7   **29.** 3, 4, 5, 3, 1
31. 6, 5, 4, 3, 2, 1
33.

n	1	2	3	4	5	6	7
a_n	2	3	4	5	6	7	8

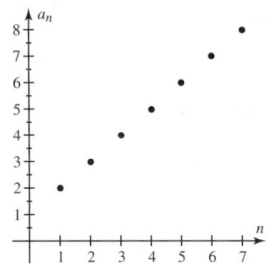

35.

n	1	2	3	4	5	6	7
a_n	0	2	6	12	20	30	42

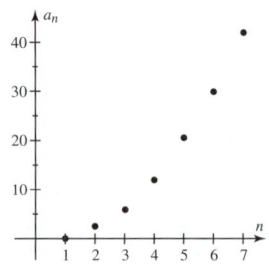

37.

n	1	2	3	4	5	6	7
a_n	2	4	8	16	32	64	128

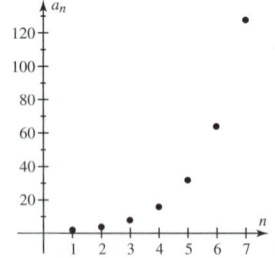

39. $a_n = 30n$ for $n = 1, 2, 3, \ldots, 7$

n	1	2	3	4	5	6	7
a_n	30	60	90	120	150	180	210

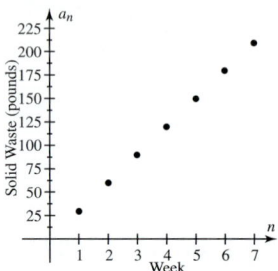

41. (a) 1, 4, 9, 16   **(b)** 4, 8, 12, 16
43. (a) \$20,000; \$16,000   **(b)** $a_n = 25{,}000(0.8)^n$
(c)

n	1	2	3	4	5	6	7
a_n	20,000	16,000	12,800	10,240	8192	6553.6	5242.9

45. (a) $a_n = 2048(0.5)^{n-1}$, for $n = 1, 2, 3, \ldots, 7$
(b)

n	1	2	3	4	5	6	7
a_n	2048	1024	512	256	128	64	32

(c)

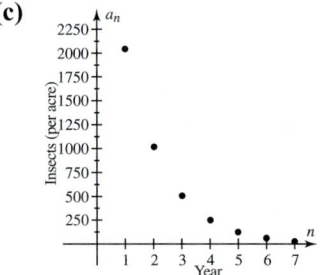

SECTION 14.2 (pp. 930–933)

1. linear   **2.** exponential   **3.** $a_n = 3n + 1$; 3 (answers may vary)   **4.** $a_n = 5(2)^{n-1}$; 2 (answers may vary)
5. add; previous   **6.** multiply; common ratio   **7.** 19; 4
8. 32; -2 **9.** $a_n = a_1(r)^{n-1}$ **10.** $a_n = a_1 + (n-1)d$
11. Yes; 10   **13.** Yes; -1 **15.** No   **17.** Yes; 3
19. Yes; -3 **21.** No   **23.** Yes; 1   **25.** No   **27.** Yes; 2
29. $a_n = -2n + 9$ **31.** $a_n = 4n - 6$
33. $a_n = -2n + 32$ **35.** 59   **37.** 21   **39.** Yes; 3
41. Yes; 0.8   **43.** No   **45.** Yes; 2   **47.** No   **49.** Yes; 4
51. Yes; 2   **53.** No   **55.** $a_n = 1.5(4)^{n-1}$
57. $a_n = -3(-2)^{n-1}$ **59.** $a_n = 1(4)^{n-1}$ **61.** 4374
63. 243   **65. (a)** 3000, 6000, 9000, 12,000, 15,000; arithmetic
(b) $a_n = 3000n$ **(c)** 60,000; when there are 20 people, the ventilation rate should be 60,000 cubic feet per hour.

(d)

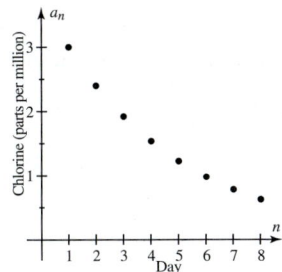

Yes

67. (a) $a_n = 3(0.8)^{n-1}$
(b) No; geometric

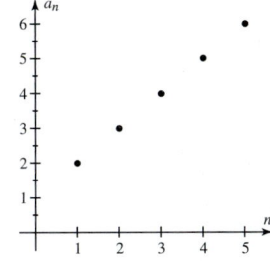

69. (a) $a_n = 5(0.85)^{n-1}$ **(b)** Geometric; the common ratio is 0.85.   **(c)** About 1.6; on the 8th bounce the ball reaches a maximum height of about 1.6 ft.
71. (a) Arithmetic; the common difference is 2.
(b) $a_n = 40 + 2(n-1)$ or $a_n = 38 + 2n$ **(c)** 78

CHECKING BASIC CONCEPTS 14.1 & 14.2 (p. 933)

1. $\frac{1}{5}, \frac{1}{3}, \frac{3}{7}, \frac{1}{2}$
2.

n	1	2	3	4	5
a_n	2	3	4	5	6

3. (a) Arithmetic; $a_n = 3n - 5$ **(b)** Geometric;
$a_n = 3(-2)^{n-1}$ **4.** $a_n = 2n + 3$ **5.** $a_n = 5(2)^{n-1}$

SECTION 14.3 (pp. 941–943)

1. series   **2.** 10   **3.** arithmetic   **4.** geometric
5. $n\left(\frac{a_1 + a_n}{2}\right)$ or $\frac{n}{2}(2a_1 + (n-1)d)$ **6.** $a_1\left(\frac{1 - r^n}{1 - r}\right)$
7. sum   **8.** $a_1 + a_2 + a_3 + a_4$ **9.** arithmetic

10. geometric   **11.** 48   **13.** 820   **15.** -5 **17.** 3279
19. -85 **21.** 182   **23.** \$91,523.93   **25.** \$62,278.01
27. $2 + 4 + 6 + 8$; 20
29. $4 + 4 + 4 + 4 + 4 + 4 + 4 + 4$; 32
31. $1 + 4 + 9 + 16 + 25 + 36 + 49$; 140
33. $12 + 20$; 32   **35.** $\sum_{k=1}^{6} k^4$ **37.** $\sum_{k=1}^{5} \frac{1}{k^2}$
39. $\sum_{k=1}^{n} k = n\left(\frac{a_1 + a_n}{2}\right) = n\left(\frac{1 + n}{2}\right) = \frac{n(n + 1)}{2}$
41. (a) $8518 + 9921 + 10{,}706 + 14{,}035 + 14{,}307 + 12{,}249$
(b) 69,736   **43. (a)** $\sum_{k=1}^{n} 0.8(0.2)^{k-1}$ **(b)** 2
45. (a) $1, \frac{1}{2}, \frac{1}{4}, \frac{1}{8}, \frac{1}{16}$ **(b)** $\frac{1023}{512}$ **47.** 90 logs   **49.** \$1,080,000
51. About 3.16 feet

SECTION 14.4 (pp. 948–949)

1. 5   **2.** $n + 1$ **3.** 4
4.
$$1$$
$$1 \quad 1$$
$$1 \quad 2 \quad 1$$
$$1 \quad 3 \quad 3 \quad 1$$
$$1 \quad 4 \quad 6 \quad 4 \quad 1$$
5. 24   **6.** $6! = 720$ **7.** $\frac{n!}{(n - r)!\, r!}$ **8.** $a^2 + 2ab + b^2$
9. n **10.** 1   **11.** $x^3 + 3x^2y + 3xy^2 + y^3$
13. $16x^4 + 32x^3 + 24x^2 + 8x + 1$
15. $a^5 - 5a^4b + 10a^3b^2 - 10a^2b^3 + 5ab^4 - b^5$
17. $x^6 + 3x^4 + 3x^2 + 1$ **19.** 6   **21.** 4   **23.** 2
25. 10   **27.** 5   **29.** 6   **31.** 1   **33.** 792   **35.** 126
37. 75,582   **39.** $m^3 + 3m^2n + 3mn^2 + n^3$
41. $x^4 - 4x^3y + 6x^2y^2 - 4xy^3 + y^4$
43. $8a^3 + 12a^2 + 6a + 1$
45. $x^5 + 10x^4 + 40x^3 + 80x^2 + 80x + 32$
47. $81 + 216m + 216m^2 + 96m^3 + 16m^4$
49. $8x^3 - 12x^2y + 6xy^2 - y^3$ **51.** a^8 **53.** $35x^4y^3$
55. $512m^9$

CHECKING BASIC CONCEPTS 14.3 & 14.4 (p. 949)

1. (a) Geometric   **(b)** Arithmetic   **2.** 312
3. -341 **4.** $x^4 - 4x^3y + 6x^2y^2 - 4xy^3 + y^4$
5. $x^3 + 6x^2 + 12x + 8$

CHAPTER 14 REVIEW (pp. 951–953)

1. $1, 8, 27, 64$ **2.** $3, 1, -1, -3$ **3.** $1, \frac{4}{5}, \frac{3}{5}, \frac{8}{17}$
4. $-2, 4, -8, 16$ **5.** $-2, 0, 4, 2$ **6.** $5, 3, 2, 1$

7.

n	1	2	3	4	5	6	7
a_n	2	4	6	8	10	12	14

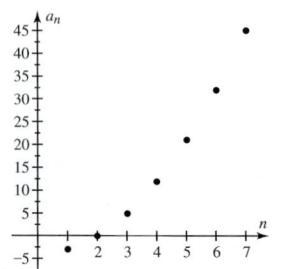

8.

n	1	2	3	4	5	6	7
a_n	−3	0	5	12	21	32	45

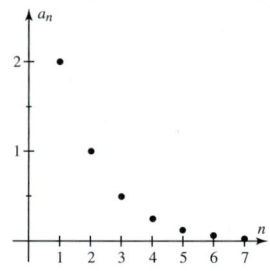

9.

n	1	2	3	4	5	6	7
a_n	2	1	0.5	0.25	0.125	0.0625	0.0313

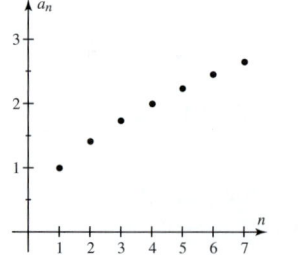

10.

n	1	2	3	4	5	6	7
a_n	1	1.4142	1.7321	2	2.2361	2.4495	2.6458

11. Yes; 5   **12.** No   **13.** No   **14.** Yes; $-\frac{1}{3}$
15. Yes; −3   **16.** No   **17.** Yes; −1   **18.** No
19. $a_n = 4n - 7$ **20.** $a_n = -5n + 7$ **21.** Yes; 4
22. No   **23.** No   **24.** Yes; 0.7   **25.** No   **26.** Yes; $-\frac{1}{3}$
27. No   **28.** Yes; 2   **29.** $a_n = 5(0.9)^{n-1}$
30. $a_n = 2(4)^{n-1}$ **31.** 216   **32.** 7.5   **33.** 3277
34. $\frac{511}{256}$ **35.** $3 + 5 + 7 + 9 + 11$ **36.** $\frac{1}{2} + \frac{1}{3} + \frac{1}{4} + \frac{1}{5}$

37. $1 + 8 + 27 + 64$
38. $-1 + (-2) + (-3) + (-4) + (-5) + (-6)$
39. $\sum\limits_{k=1}^{20} k$ **40.** $\sum\limits_{k=1}^{20} \frac{1}{k}$ **41.** $\sum\limits_{k=1}^{9} \frac{k}{k+1}$ **42.** $\sum\limits_{k=1}^{7} k^2$
43. $x^3 + 12x^2 + 48x + 64$
44. $16x^4 + 32x^3 + 24x^2 + 8x + 1$
45. $x^5 - 5x^4y + 10x^3y^2 - 10x^2y^3 + 5xy^4 - y^5$
46. $a^6 - 6a^5 + 15a^4 - 20a^3 + 15a^2 - 6a + 1$
47. 6   **48.** 10   **49.** 20   **50.** 4
51. $m^4 + 8m^3 + 24m^2 + 32m + 16$
52. $a^5 + 5a^4b + 10a^3b^2 + 10a^2b^3 + 5ab^4 + b^5$
53. $x^4 - 12x^3y + 54x^2y^2 - 108xy^3 + 81y^4$
54. $27x^3 - 54x^2 + 36x - 8$
55. $a_n = 45,000(1.1)^{n-1}$ for $n = 1, 2, 3, \ldots, 7$; geometric

n	1	2	3	4	5	6	7
a_n	45,000	49,500	54,450	59,895	65,885	72,473	79,720

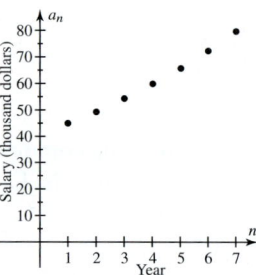

56. $a_n = 45,000 + 5000(n - 1)$ for $n = 1, 2, 3, \ldots, 7$; arithmetic

n	1	2	3	4	5	6	7
a_n	45,000	50,000	55,000	60,000	65,000	70,000	75,000

57. $a_n = 49n$ for $n = 1, 2, 3, \ldots, 7$

n	1	2	3	4	5	6	7
a_n	49	98	147	196	245	294	343

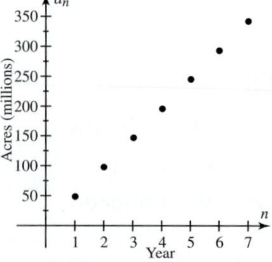

58. (a) $a_n = 1087(1.025)^{n-1}$ **(b)** Geometric; the common ratio is 1.025.   **(c)** About 1200; the average mortgage payment in 2000 was about $1200 per month.

(d)

CHAPTER 14 TEST (pp. 953–954)

1. $\frac{1}{2}, \frac{4}{3}, \frac{9}{4}, \frac{16}{5}$ **2.** $-3, 2, 1, -2, 3$

3.

n	1	2	3	4	5	6	7
a_n	0	2	6	12	20	30	42

4. $16x^4 - 32x^3 + 24x^2 - 8x + 1$

5. Arithmetic; -3 **6.** Geometric; -2

7. $a_n = 2 - 3(n - 1)$ or $a_n = 5 - 3n$

8. $a_n = 2(1.5)^{n-1}$ **9.** Yes; 2.5   **10.** No

11. 99   **12.** $\frac{463}{729}$ **13.** $6 + 9 + 12 + 15 + 18 + 21$

14. $\sum\limits_{k=1}^{60} k^3$ **15.** 35   **16.** 10   **17.** 9180

18. $a_n = 159{,}700(1.04)^{n-1}$ for $n = 1, 2, 3, \ldots, 7$; geometric

n	1	2	3	4	5	6	7
a_n	159,700	166,088	172,732	179,641	186,826	194,299	202,071

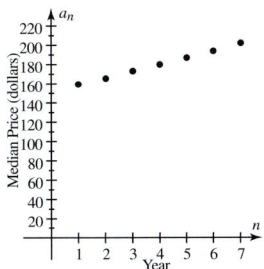

19. (a) $a_n = 2000(2)^{n-1}$ **(b)** Geometric; the common ratio is 2.   **(c)** 64,000; after 25 days there are 64,000 worms.

(d)

1. Distributive   **2.** $D = \{-6, -2, 0, 2\}$; $R = \{0, 1, 3, 5\}$

3. $\frac{y^9}{27x^5}$ **4.** $\frac{16a^8}{b^4}$ **5.** z^{10} **6.** $\frac{2y^3}{x^6}$ **7.** $D = \{x \mid x \neq 8\}$

8. $f(x) = -2x + 1$ **9.** $y = 3$ **10.** $-3; 5$

11. $y = \frac{3}{2}x + \frac{5}{2}$ **12.** $y = 2x - 8$ **13.** -26 **14.** $\left(\frac{20}{11}, \infty\right)$

15. $(-\infty, -1] \cup [11, \infty)$ **16.** $-15, 3$ **17.** $\frac{1}{8}$

18. $[-12, 9)$ **19.** $(3, -2)$

20.

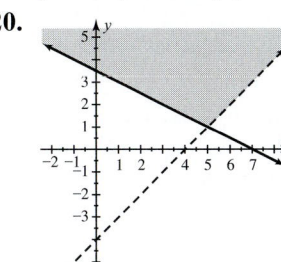

21. $(5, 2)$ **22.** $(3, 3, -1)$ **23.** $R = 22$ **24.** 17

25. $8x^7 - 6x^6 + 10x^3$ **26.** $6z^2 - 13z - 28$

27. $(2x - 3y)(2x + 3y)$ **28.** $(a^2 + 4)(2a - 1)$

29. $-\frac{3}{4}, 1$ **30.** $0, 4, 6$ **31.** $\frac{x - 2}{x + 1}$ **32.** $\frac{x + 4}{x - 2}$

33. 12   **34.** $\frac{1}{2}, 3$ **35.** $W = \frac{3C - 5R}{2}$ **36.** $\frac{1 + 2x}{1 - 4x}$

37. $(xy - 2)\sqrt[3]{xy}$ **38.** 14   **39.** $-2, 8$ **40.** $\frac{3}{2}$

41. $14 + 2i$ **42.** $-2 - 2i$ **43.** $-\frac{1}{3}$

44. $y = 2(x + 2)^2 + 9$; $(-2, 9)$ **45.** $2 \pm 3i$

46. $-4, 8$ **47. (a)** $-3, 1$ **(b)** $a > 0$ **(c)** Positive

48. $(-\infty, \infty)$ **49. (a)** 65   **(b)** $(g \circ f)(x) = 3x^2 + 1$

50. $f^{-1}(x) = \frac{2x - 1}{3}$ **51.** $3 \ln x + \frac{1}{2} \ln y$ **52.** $\log \frac{x}{4y}$

53. 56.23   **54.** 0.58

55.

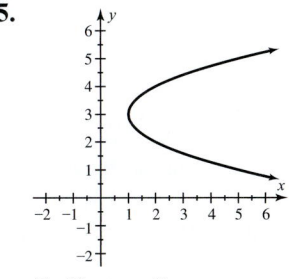

$(1, 3)$; $y = 3$

56. $(3, -1)$; 2

57.

58.

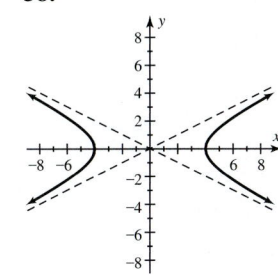

59. $\frac{y^2}{4} - \frac{x^2}{16} = 1$ **60.** $\frac{x^2}{16} + \frac{y^2}{4} = 1$ **61.** $(1, 2), (-1, 2)$

62.

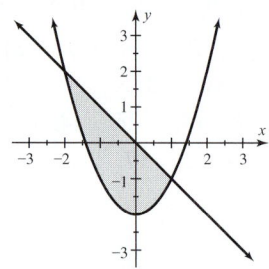

63. Arithmetic; -2 **64.** Geometric; 0.2
65. Geometric; 4   **66.** Arithmetic; 6   **67.** $a_n = 3n - 1$
68. $a_n = 4(3)^{n-1}$ **69.** 171   **70.** 683
71. $16x^4 + 96x^3 + 216x^2 + 216x + 81$
72. $8a^3 - 60a^2b + 150ab^2 - 125b^3$
73. $\sqrt{\frac{14}{\pi}} \approx 2.11$ inches

74.

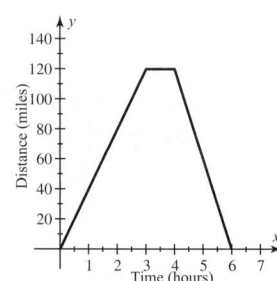

75. (a) $f(x) = 0.4x + 85$ (b) 188 pounds
76. Airplane: 380 mph, wind: 20 mph   **77.** 9 feet by 12 feet
78. 36 minutes   **79.** (a) $xy = 96, 3x - y = 12$
(b) 8 and 12   **80.** 144

APPENDIX B: SETS (pp. AP-18–AP-19)

1. $\{1, 2, 3, 4, 5, 6, 7\}$ **2.** $\{5, 6, 7, 8, 9, 10\}$ **3.** {Sunday, Saturday}   **4.** {Alabama, Alaska, Arizona, Arkansas}
5. {A, B, C, D, E, F, G}   **6.** {D, E, F, G, H, I, J, K, L, M}
7. $\varnothing$ **8.** $\varnothing$ **9.** $\{0, 2, 4, 6, 8, 10\}$ **10.** $\{13, 15, 17, 19\}$
11. $\in$ **13.** $\notin$ **15.** $\in$ **17.** $\notin$ **19.** $\in$
21. $E = \{2, 4, 6, 8, 10\}$ **23.** $E = \{2, 4, 6, 8, 10, \ldots\}$
25. $A = \{\text{Apple, Apricot}\}$ **27.** $A = \{\text{Algebra}\}$
29. False   **31.** True   **33.** True   **35.** True   **37.** True
39. True   **41.** False   **43.** True
45.

47.

49.

51.

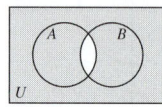

53. $\{1, 2, 3, 4\}$ **55.** $\{a, b\}$ **57.** $\{a, b, c\}$ **59.** $\{a, b, c, d\}$
61. $\{5, 8\}$ **63.** $\{1, 2, 3, 4, 5, 6, 7, 8, 9, 10\}$ **65.** $\{4, 5, 6\}$
67. $\varnothing$ **69.** $\{1, 3, 5, 7, 9\}$ **71.** $\varnothing$ **73.** $\varnothing$
75. $\{1, 3, 5, 7, 9\}$ **77.** $\{1, 2, 3, 5, 7, 8, 9, 10\}$ **79.** $\varnothing$
81. Infinite   **83.** Finite   **85.** Finite

APPENDIX C: LINEAR PROGRAMMING
(pp. AP-24–AP-25)

1. linear programming   **2.** objective   **3.** feasible solutions
4. linear inequalities   **5.** vertex   **6.** objective

7.   **9.**

11.   **13.**

15.   **17.**

19.

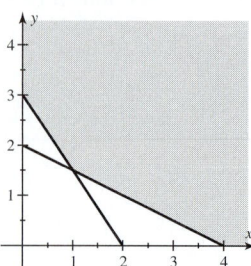

21. $R = 31$ **23.** $R = 15$

25. $C = 5$ **27.** $C = 2$

29. $R = 750$ **31.** $R = 12$

33. $R = 31.5$ **35.** $R = 16$

37. $C = 0$ **39.** $C = 32$

41. $C = 60$

43. 20 pounds of candy, 80 pounds of coffee

45. 1 ounce Brand X, 2 ounces Brand Y

47. $600

APPENDIX D: SYNTHETIC DIVISION
(p. AP-27)

1. $x + 4 + \frac{3}{x - 1}$ **3.** $3x - 1$ **5.** $x^2 + 3x + 2$

7. $2x^2 + 5x + 10 + \frac{19}{x - 2}$

9. $x^2 + 2x + 6 + \frac{28}{x - 4}$ **11.** $x^2 + 2x + 2$

13. $2x^3 - 4x^2 + 8x - 17 + \frac{34}{x + 2}$

15. $b^3 + b^2 + b + 1$

GLOSSARY

absolute value A real number a, written $|a|$, is equal to its distance from the origin on the number line.

absolute value equation An equation that contains an absolute value.

absolute value function The function defined by $f(x) = |x|$.

absolute value inequality An inequality that contains an absolute value.

addends In an addition problem, the two numbers that are added.

addition property of equality If a, b, and c are real numbers, then $a = b$ is equivalent to $a + c = b + c$.

additive identity The number 0.

additive inverse (opposite) The additive inverse or opposite of a number a is $-a$.

adjacency matrix A matrix used to represent a map showing distances between cities.

algebraic expression An expression consisting of numbers, variables, operation symbols, such as $+$, $-$, $\times$, and $\div$, and grouping symbols, such as parentheses.

annuity A sum of money from which regular payments are made.

approximately equal The symbol $\approx$ indicates that two quantities are nearly equal.

arithmetic sequence A linear function given by $a_n = dn + c$ whose domain is the set of natural numbers.

arithmetic series The sum of the terms of an arithmetic sequence.

associative property for addition For any real numbers a, b, and c, $(a + b) + c = a + (b + c)$.

associative property for multiplication For any real numbers a, b, and c, $(a \cdot b) \cdot c = a \cdot (b \cdot c)$.

asymptotes of a hyperbola The two lines determined by the diagonals of the hyperbola's fundamental rectangle.

augmented matrix A matrix used to represent a system of linear equations; a vertical line is positioned in the matrix where the equal signs occur in the system of equations.

average The result of adding up the numbers of a set and then dividing the sum by the number of elements in the set.

axis of symmetry of a parabola The line passing through the vertex of the parabola that divides the parabola into two symmetric parts.

base The value of b in the expression b^n.

basic principle of fractions When simplifying fractions, the principle that states $\frac{a \cdot c}{b \cdot c} = \frac{a}{b}$.

binary operation An operation that requires two numbers to calculate an answer.

binomial A polynomial with two terms.

binomial coefficient The expression $_nC_r = n!/((n - r)!r!)$, where n and r are nonnegative integers, $n \geq r$, that can be used to calculate the numbers in Pascal's triangle.

binomial theorem A theorem that provides a formula to expand expressions of the form $(a + b)^n$.

braces { }, used to enclose the elements of a set.

branches A hyperbola has two branches, a left branch and a right branch, or an upper branch and a lower branch.

byte A unit of computer memory, capable of storing one letter of the alphabet.

center of a circle The fixed point in the center of a circle.

center of an ellipse The midpoint of the major axis.

circle The set of points in a plane that are the same distance from a fixed point.

circumference The perimeter of a circle.

coefficient The numeric constant of a term.

coefficient of a monomial The number in a monomial.

common difference The value of d in an arithmetic sequence, $a_n = dn + c$.

common logarithmic function The function given by $f(x) = \log x$.

common logarithm of a positive number x Denoted $\log x$, it may be calculated as follows: If x is expressed as $x = 10^k$, then $\log x = k$, where k is a real number. That is, $\log 10^k = k$.

common ratio The value of r in a geometric sequence, $a_n = a_1(r)^{n-1}$.

commutative property for addition For any real numbers a and b, $a + b = b + a$.

commutative property for multiplication For any real numbers a and b, $a \cdot b = b \cdot a$.

complement The set containing all elements in the universal set that are not in A, denoted A'.

completing the square method An important technique in mathematics that involves adding a constant to a binomial so that a perfect square trinomial results.

complex conjugate The complex conjugate of $a + bi$ is $a - bi$.

complex fraction A rational expression that contains fractions in its numerator, denominator, or both.

complex number A complex number can be written in standard form as $a + bi$, where a and b are real numbers and i is the imaginary unit.

composite function If f and g are functions, then g of f, or composition of g and f, is defined by $(g \circ f)(x) = g(f(x))$ and is read "g of f of x."

composite number A natural number greater than 1 that is not a prime number.

compound inequality Two inequalities joined by the words *and* or *or*.

compound interest A type of interest paid at the end of each year using the formula $A = C(1 + r)^x$, where C is the amount deposited, r is the annual interest rate in decimal form, and x is the number of years the account will contain A dollars.

conic section The curve formed by the intersection of a plane and a cone.

conjugate The conjugate of $a + b$ is $a - b$.

consistent system A system of linear equations with at least one solution.

constant function A linear function with $a = 0$ and can be written as $f(x) = b$.

constant of proportionality (constant of variation) In the equation $y = kx$, the number k.

constraint In linear programming, an inequality which limits the objective function.

continuous growth Growth in a quantity that is directly proportional to the amount present.

Cramer's rule A method that uses determinants to solve linear systems of equations.

cube root The number b is a cube root of a if $b^3 = a$.

cube root function The function defined by $f(x) = \sqrt[3]{x}$.

cubic function A function f of degree 3 represented by $f(x) = ax^3 + bx^2 + cx + d$, where $a, b, c,$ and d are constants and $a \neq 0$.

decay factor The value of a in an exponential function, $f(x) = Ca^x$, when $0 < a < 1$.

degree A degree, (°), is 1/360 of a revolution.

degree of a monomial The sum of the exponents of the variables.

degree of a polynomial The degree of the term (or monomial) with highest degree.

dependent equations Equations in a linear system that have infinitely many solutions.

dependent variable The variable that represents the output of a function.

determinant A real number associated with a square matrix.

diagrammatic representation A function represented by a diagram.

difference The answer to a subtraction problem.

difference of two cubes Expression in the form $a^3 - b^3$, which can be factored as $(a - b)(a^2 + ab + b^2)$.

difference of two squares Expression in the form $a^2 - b^2$, which can be factored as $(a - b)(a + b)$.

dimension of a matrix The size expressed in number of rows and columns. For example, if a matrix has m rows and n columns, its dimension is $m \times n$ (m by n).

directly proportional A quantity y is directly proportional to x if there is a nonzero number k such that $y = kx$.

discriminant The expression $b^2 - 4ac$ in the quadratic formula.

distance formula The distance d between the points (x_1, y_1) and (x_2, y_2) in the xy-plane is $d = \sqrt{(x_2 - x_1)^2 + (y_2 - y_1)^2}$.

distributive properties For any real numbers $a, b,$ and c, $a(b + c) = ab + ac$ and $a(b - c) = ab - ac$.

dividend In a division problem, the number being divided.

divisor In a division problem, the number being divided *into* the dividend.

domain The set of all x-values of the ordered pairs in a function.

element of a matrix Each number in a matrix.

elements of a set The members of a set.

elimination method A symbolic method used to solve a system of equations that is based on the property that if "equals are added to equals the results are equal."

ellipse The set of points in a plane, the sum of whose distances from two fixed points is constant.

empty set (null set) A set that contains no elements.

equal sets If every element in a set A is in set B and every element in set B is in set A, then A and B are equal sets, denoted $A = B$.

equation A mathematical statement that two algebraic expressions are equal.

equivalent equations Equations that have the same solution set.

even root The nth root, $\sqrt[n]{a}$, where n is even.

expansion of a determinant by minors A method of finding a 3×3 determinant by using determinants of 2×2 matrices.

exponent The value of n in the expression b^n.

exponential decay When $0 < a < 1$, the graph of $f(x) = Ca^x$ models exponential decay.

exponential equation An equation that has a variable as an exponent.

exponential expression An expression that has an exponent.

exponential function with base a and coefficient C A function represented by $f(x) = Ca^x$, where $a > 0, C > 0,$ and $a \neq 1$.

exponential growth When $a > 1$, the graph of $f(x) = Ca^x$ models exponential growth.

extraneous solution A solution that does not satisfy the given equation.

factors In a multiplication problem, the two numbers multiplied.

factorial notation $n! = 1 \cdot 2 \cdot 3 \cdot \cdots \cdot n$ for any positive integer.

factoring The process of writing a polynomial as a *product* of lower degree polynomials.

factoring by grouping A technique that uses the associative and distributive properties by grouping four terms of a polynomial in such a way that the polynomial can be factored even though its greatest common factor is 1.

feasible solutions In linear programming, the set of solutions that satisfy the constraints.

finite sequence A function with domain $D = \{1, 2, 3, \ldots, n\}$ for some fixed natural number n.

finite series A series that contains a finite number of terms, and can be expressed in the form $a_1 + a_2 + a_3 + \cdots + a_n$ for some n.

finite set A set where the elements can be listed explicitly.

focus (plural foci) A fixed point used to determine the points that form a parabola, an ellipse, or a hyperbola.

FOIL A method for multiplying two binomials $(A + B)$ and $(C + D)$. Multiply First terms AC, Outside terms AD, Inside terms BC, and Last terms BD; then combine like terms.

formula A special type of equation used to calculate one quantity from given values of other quantities.

function A set of ordered pairs (x, y), where each x-value corresponds to exactly one y-value.

function notation The notation $y = f(x)$, where the input x produces output y.

fundamental rectangle The rectangle of a hyperbola whose four vertices are determined by either $(\pm a, \pm b)$ or $(\pm b, \pm a)$, where $\frac{x^2}{a^2} - \frac{y^2}{b^2} = 1$ or $\frac{y^2}{a^2} - \frac{x^2}{b^2} = 1$.

Gauss–Jordan elimination A method used to solve a linear system in which matrix row transformations are applied to an augmented matrix.

general term (nth term) of a sequence a_n, where n is a natural number in the domain of a sequence $a_n = f(n)$.

geometric sequence An exponential function given by $a_n = a_1(r)^{n-1}$, where n is a natural number and $r \neq 0$ or 1.

geometric series The sum of the terms of a geometric sequence.

graphical representation A graph of a function.

graphical solution A solution to an equation obtained by graphing.

greater than If a real number b is located to the right of a real number a on the number line, we say that b is greater than a, and write $b > a$.

greater than or equal to If a real number a is greater than or equal to b, denoted $a \geq b$, then either $a > b$ or $a = b$ is true.

greatest common factor (GCF) The term with the highest degree and greatest coefficient that is a factor of all terms in the polynomial.

growth factor The value of a in the exponential function, $f(x) = Ca^x$, when $a > 1$.

half-life The time it takes for a radioactive sample to decay to half its original amount.

horizontal line test If every horizontal line intersects the graph of a function f at most once, then f is a one-to-one function.

hyperbola The set of points in a plane, the difference of whose distances from two fixed points is constant.

identity property of 1 If any number a is multiplied by 1, the result is a, that is, $a \cdot 1 = 1 \cdot a = a$.

identity property of 0 If 0 is added to any real number, a, the result is a, that is, $a + 0 = 0 + a = a$.

imaginary number A complex number $a + bi$ with $b \neq 0$.

imaginary part The value of b in the complex number $a + bi$.

imaginary unit A number denoted i whose properties are $i = \sqrt{-1}$ and $i^2 = -1$.

improper fraction A fraction whose numerator is greater than its denominator.

inconsistent system A system of linear equations that has no solution.

independent equations Equations in a linear system that have one solution.

independent variable The variable that represents the input of a function.

index The value of n in the expression $\sqrt[n]{a}$.

index of summation The variable k in the expression $\sum_{k=1}^{n}$.

inequality When the equals sign in an equation is replaced with any one of the symbols $<$, $\leq$, $>$, or $\geq$.

infinite sequence A function whose domain is the set of natural numbers.

infinite set A set with infinitely many elements.

input An element of the domain of a function.

integers A set of numbers including natural numbers, their opposites, and 0, or $\ldots, -3, -2, -1, 0, 1, 2, 3, \ldots$.

intercept form A linear equation in the form $x/a + y/b = 1$.

intersection Denoted $A \cap B$ and read "A intersect B" is the set containing elements that belong to *both A and B*.

intersection-of-graphs method A graphical technique for solving two equations.

interval notation A notation for number line graphs that eliminates the need to draw the entire line.

inverse function If f is a one-to-one function, then f^{-1} is the inverse function of f, if $(f^{-1} \circ f)(x) = f^{-1}(f(x)) = x$, for every x in the domain of f, and $(f \circ f^{-1})(x) = f(f^{-1}(x)) = x$, for every x in the domain of f^{-1}.

inversely proportional A quantity y is inversely proportional to x if there is a nonzero number k such that $y = k/x$.

irrational numbers Real numbers that cannot be expressed as fractions, such as π or $\sqrt{2}$.

joint variation A quantity z varies jointly as x and y if there is a nonzero number k such that $z = kxy$.

leading coefficient In a polynomial of one variable, the coefficient of the monomial with highest degree.

least common denominator (LCD) The common denominator with the fewest factors.

least common multiple (LCM) The smallest number that two or more numbers will divide into evenly.

less than If a real number a is located to the left of a real number b on the number line, we say that a is less than b and write $a < b$.

less than or equal to If a real number a is less than or equal to b, denoted $a \leq b$, then either $a < b$ or $a = b$ is true.

like radicals Radicals that have the same index and the same radicand.

like terms Two terms, or monomials, that contain the same variables raised to the same powers.

linear equation An equation that can be written in the form $ax + b = 0$, where $a \neq 0$.

linear equation in two variables An equation that can be written in the form $Ax + By = C$, where A, B, and C are fixed numbers and A and B are not both equal to 0.

linear function A function f represented by $f(x) = ax + b$, where a and b are constants.

linear inequality A linear inequality results whenever the equals sign in a linear equation is replaced with any one of the symbols $<, \leq, >,$ or $\geq$.

linear inequality in two variables When the equals sign in a linear equation of two variables is replaced with $<, \leq, >,$ or $\geq$, a linear inequality in two variables results.

linear polynomial A polynomial of degree 1 that can be written as $ax + b$, where $a \neq 0$.

line graph The resulting graph when consecutive data points in a scatterplot are connected with straight line segments.

linear programming problem A problem consisting of an objective function and a system of linear inequalities called constraints.

logistic function A function used to model growth of a population.

logarithm with base a of a positive number x Denoted $\log_a x$, it may be calculated as follows: If x can be expressed as $x = a^k$, then $\log_a x = k$, where $a > 0$, $a \neq 1$, and k is a real number. That is, $\log_a a^k = k$.

logarithmic function with base a The function represented by $f(x) = \log_a x$.

lower limit In summation notation, the number representing the subscript of the first term of the series.

lowest terms A fraction is in lowest terms if its numerator and denominator have no factors in common.

main diagonal In an augmented matrix, the diagonal set of numbers from the upper left of the matrix to the lower right.

major axis The longest axis of an ellipse, which connects the vertices.

matrix A rectangular array of numbers.

method of substitution A symbolic method for solving a system of equations in which one equation is solved for one of the variables and then the result is substituted into the other equation.

minor axis The shortest axis of an ellipse.

minors The 2×2 matrices that are used to find a determinant of a 3×3 matrix.

monomial A number, a variable, or a product of numbers and variables raised to natural number powers.

multiplication property of equality If a, b, and c are real numbers with $c \neq 0$, then $a = b$ is equivalent to $ac = bc$.

multiplicative identity The number 1.

multiplicative inverse (reciprocal) The multiplicative inverse of a nonzero number a is $1/a$.

natural exponential function The function represented by $f(x) = e^x$, where $e \approx 2.71828$.

natural logarithm The base-e logarithm, denoted either $\log_e x$ or $\ln x$.

natural numbers The set of (counting) numbers expressed as $1, 2, 3, 4, 5, 6, \ldots$.

negative slope On a graph, the slope of a line that falls from left to right.

negative square root The negative square root is denoted $-\sqrt{a}$.

nonlinear data If data points do not lie on a (straight) line, the data are nonlinear.

nonlinear function A function that is *not* a linear function; its graph is not a line.

nonlinear system of equations Two or more equations, at least one of which is nonlinear.

nonlinear system of inequalities Two or more inequalities, at least one of which is nonlinear.

nth root The number b is an nth root of a if $b^n = a$, where n is a positive integer.

nth term (general term) of a sequence Denoted $a_n = f(n)$.

null set (empty set) A set that contains no elements.

numerical representation A table of values for a function.

numerical solution A solution often obtained by using a table of values.

objective function The given function in a linear programming problem.

odd root The nth root, $\sqrt[n]{a}$, where n is odd.

one-to-one function A function f in which for any c and d in the domain of f, $c \neq d$ implies that $f(c) \neq f(d)$. That is, different inputs always result in different outputs.

opposite (additive inverse) The opposite, or additive inverse, of a number a is $-a$.

opposite of a polynomial The polynomial obtained by negating each term in a given polynomial.

optimal value In linear programming, the value that often results in maximum revenue or minimum cost.

ordered pair A pair of numbers written in parentheses (x, y), in which the order of the numbers is important.

ordered triple Can be expressed as (x, y, z), where x, y, and z are numbers and represent a solution to a linear system in three variables.

origin On the number line, the point associated with the real number 0; in the xy-plane, the point where the axes intersect, $(0, 0)$.

output An element of the range of a function.

parabola The $\cup$-shaped graph of a quadratic function that either opens upward, downward, to the right, or to the left.

parallel lines Two or more lines in the same plane that never intersect; they have the same slope.

Pascal's triangle A triangle made up of numbers in which there are 1s along the sides and each element inside the triangle is the sum of the two numbers above it.

percent change If the quantity changes from x to y, then the percent change is $[(y - x)/x] \times 100$.

perfect cube An integer with an integer cube root.

perfect nth power The value of a if there exists an integer b such that $b^n = a$.

perfect square An integer with an integer square root.

perfect square trinomial A trinomial that can be factored as the square of a binomial, for example, $a^2 + 2ab + b^2 = (a + b)^2$.

perpendicular lines Two lines in a plane that intersect to form a right (90°) angle.

point–slope form The line with slope m passing through the point (x_1, y_1), given by the equation $y - y_1 = m(x - x_1)$, or equivalently, $y = m(x - x_1) + y_1$.

polynomial The sum of one or more monomials.

polynomial functions in one variable Functions that contain a polynomial in one variable.

polynomials of one variable Polynomials that contain one variable.

positive slope On a graph, the slope of a line that rises from left to right.

power function A function that can be represented by $f(x) = x^p$, where p is a rational number.

prime number A natural number greater than 1 that has *only* itself and 1 as natural number factors.

prime polynomial A polynomial with integer coefficients that cannot be factored by using integer coefficients.

principal square root The square root of a that is nonnegative, denoted $\sqrt{a}$.

probability A real number between 0 and 1. A probability of 0 indicates that an event is impossible, whereas a probability of 1 indicates that an event is certain.

product The answer to a multiplication problem.

proportion A statement that two ratios are equal.

Pythagorean theorem If a right triangle has legs a and b with hypotenuse c, then $a^2 + b^2 = c^2$.

quadrants The four regions determined by the xy-plane.

quadratic equation An equation that can be written in the form $ax^2 + bx + c = 0$, where a, b, and c are constants, with $a \neq 0$.

quadratic formula The solutions to the quadratic equation, $ax^2 + bx + c = 0, a \neq 0$, are $(-b \pm \sqrt{b^2 - 4ac})/(2a)$.

quadratic function A function f represented by the equation $f(x) = ax^2 + bx + c$, where a, b, and c are constants, with $a \neq 0$.

quadratic inequality If the equals sign in a quadratic equation is replaced with $>$, $\geq$, $<$, or $\leq$, a quadratic inequality results.

quadratic polynomial A polynomial of degree 2 that can be written as $ax^2 + bx + c$, with $a \neq 0$.

quotient The answer to a division problem.

radical expression An expression that contains a radical sign.

radical sign The symbol $\sqrt{}$ or $\sqrt[n]{}$ for some positive integer n.

radicand The expression under the radical sign.

radius The fixed distance between the center and any point on the circle.

range The set of all y-values of the ordered pairs in a function.

rate of change Slope can be interpreted as a rate of change. It indicates how fast the graph of a line is changing.

ratio A comparison of two quantities, expressed as a quotient.

rational equation An equation that contains one or more rational expressions.

rational expression A polynomial divided by a nonzero polynomial.

rational function A function defined by $f(x) = p(x)/q(x)$, where $p(x)$ and $q(x)$ are polynomials and the domain of f includes all x-values such that $q(x) \neq 0$.

rational number Any number that can be expressed as the ratio of two integers p/q, where $q \neq 0$; a fraction.

rationalizing the denominator The process of removing radicals from a denominator so that the denominator contains only rational numbers.

real numbers All rational and irrational numbers; any number that can be represented by decimal numbers.

real part The value of a in a complex number $a + bi$.

reciprocal (multiplicative inverse) The reciprocal of a nonzero number a is $1/a$.

rectangular coordinate system (xy-plane) The xy-plane used to plot points and graph data.

reduced row–echelon form A matrix form for representing a system of linear equations in which there are 1s on the main diagonal with 0s above and below each 1.

reflection If the point (x, y) is on the graph of a function, then $(x, -y)$ is on the graph of its reflection across the x-axis.

relation A set of ordered pairs.

rise The change in y between two points on a line, that is, $y_2 - y_1$.

root function In the power function, $f(x) = x^p$, if $p = 1/n$, where $n \geq 2$ is an integer, then f is also a root function, which is given by $f(x) = \sqrt[n]{x}$.

run The change in x between two points on a line, that is, $x_2 - x_1$.

scatterplot A graph of distinct points plotted in the xy-plane.

scientific notation A real number a written as $b \times 10^n$, where $1 \leq |b| < 10$ and n is an integer.

set A collection of things.

set-builder notation Notation to describe a set of numbers without having to list all of the elements. For example, $\{x \mid x > 5\}$ is read as "the set of all real numbers x such that x is greater than 5."

slope The ratio of the change in y (rise) to the change in x (run) along a line. The slope m of a line passing through the points (x_1, y_1) and (x_2, y_2) is $m = (y_2 - y_1)/(x_2 - x_1)$, where $x_1 \neq x_2$.

slope–intercept form The line with slope m and y-intercept b is given by $y = mx + b$.

solution Each value of the variable that makes the equation true.

solution set The set of all solutions to an equation.

solution to a system In a system of equations, an ordered pair, (x, y), that makes *both* equations true.

square matrix A matrix in which the number of rows and the number of columns are equal.

square root The number b is a square root of a number a if $b^2 = a$.

square root function The function given by $f(x) = \sqrt{x}$, where $x \geq 0$.

square root property If k is a nonnegative number, then the solutions to the equation $x^2 = k$ are given by $x = \pm \sqrt{k}$. If $k < 0$, then this equation has no real solutions.

standard equation of a circle The standard equation of a circle with center (h, k) and radius r is $(x - h)^2 + (y - k)^2 = r^2$.

standard form of a complex number $a + bi$, where a and b are real numbers.

standard form of an equation for a line The form $Ax + By = C$, where A, B, and C are constants, with A and B not both 0.

standard form of a quadratic equation The equation given by $ax^2 + bx + c = 0$, where $a \neq 0$.

standard viewing rectangle of a graphing calculator Xmin $= -10$, Xmax $= 10$, Xscl $= 1$, Ymin $= -10$, Ymax $= 10$, and Yscl $= 1$, denoted $[-10, 10, 1]$ by $[-10, 10, 1]$.

subscript The symbol x_1 has a subscript of 1 and is read "x sub one" or "x one".

subset If every element in a set B is contained in a set A, then we say that B is a subset of A, denoted $B \subseteq A$.

sum The answer to an addition problem.

summation notation Notation in which the uppercase Greek letter *sigma* represents the sum, for example,

$$\sum_{k=1}^{n} a_k = a_1 + a_2 + a_3 + \cdots + a_n.$$

sum of the first n terms of an arithmetic sequence Denoted S_n, is found by averaging the first and nth terms and then multiplying by n.

sum of the first n terms of a geometric sequence Given by $S_n = a_1(1 - r^n)/(1 - r)$, if its first term is a_1 and its common ratio is r, provided $r \neq 1$.

sum of two cubes Expression in the form $a^3 + b^3$, which can be factored as $(a + b)(a^2 - ab + b^2)$.

symbolic representation Representing a function with a formula; for example, $f(x) = x^2 - 2x$.

symbolic solution A solution to an equation obtained by using properties of equations; the resulting solution set is exact.

synthetic division A shortcut that can be used to divide $x - k$, where k is a number, into a polynomial.

system of linear equations in two variables A system of equations in which each equation can be written as $Ax + By = C$.

system of linear inequalities in two variables Two or more inequalities to be solved at the same time, the solution to which must satisfy both inequalities.

table of values An organized way to display the inputs and outputs of a function; a numerical representation.

term A number, a variable, or a product of numbers and variables raised to powers.

terms of a sequence $a_1, a_2, a_3, \ldots a_n, \ldots$ where the first term is $a_1 = f(1)$, the second term is $a_2 = f(2)$, and so on.

test value When graphing the solution set of an inequality, a point chosen to determine which region of the xy-plane to include in the solution set.

three-part inequality A compound inequality written in the form $a < x < b$, where $\leq$ may replace $<$.

translations The shifting of a graph upward, downward, to the right, or to the left in such a way that the shape of the graph stays the same.

transverse axis In a hyperbola, the line segment that connects the vertices.

trinomial A polynomial with three terms.

unary operation An operation that requires only one number.

union Denoted $A \cup B$ and read "A union B," it is the set containing any element that can be found in *either A or B*.

universal set A set that contains all elements under consideration.

upper limit In summation notation, the number representing the subscript of the last term of the series.

variable A symbol, such as x, y, or z, used to represent any unknown quantity.

varies directly A quantity y varies directly with x if there is a nonzero number k such that $y = kx$.

varies inversely A quantity y varies inversely with x if there is a nonzero number k such that $y = k/x$.

varies jointly A quantity z varies jointly as x and y if there is a nonzero number k such that $z = kxy$.

Venn diagrams Diagrams used to depict relationships between sets.

verbal representation A description of what a function computes in words.

vertex The lowest point on the graph of a parabola that opens upward or the highest point on the graph of a parabola that opens downward.

vertex form of a parabola The vertex form of a parabola with vertex (h, k) is $y = a(x - h)^2 + k$, where $a \neq 0$ is a constant.

vertical asymptote A vertical asymptote typically occurs in the graph of a rational function when the denominator of the rational expression is 0, but the numerator is not 0; it can be represented by a vertical line in the graph of a rational function.

vertical line test If every vertical line intersects a graph in no more than one point, then the graph represents a function.

vertices of an ellipse The endpoints of the major axis.

vertices of a hyperbola The endpoints of the transverse axis.

viewing rectangle (window) On a graphing calculator, the window that determines the x- and y-values shown in the graph.

whole numbers The set of numbers given $0, 1, 2, 3, 4, 5, \ldots$.

x-axis The horizontal axis in the xy-plane.

x-intercept The x-coordinate of a point where a graph intersects the x-axis.

Xmax Regarding the viewing rectangle of a graphing calculator, Xmax is the maximum x-value along the x-axis.

Xmin Regarding the viewing rectangle of a graphing calculator, Xmin is the minimum x-value along the x-axis.

Xscl The distance represented by consecutive tick marks on the x-axis.

y-axis The vertical axis in the xy-plane.

xy-plane (rectangular coordinate system) The system used to plot points and graph data.

y-intercept The y-coordinate of a point where a graph intersects the y-axis.

Ymax Regarding the viewing rectangle of a graphing calculator, Ymax is the maximum y-value along the y-axis.

Ymin Regarding the viewing rectangle of a graphing calculator, Ymin is the minimum y-value along the y-axis.

Yscl The distance represented by consecutive tick marks on the y-axis.

zero-product property If the product of two numbers is 0, then at least one of the numbers must be 0, that is, $ab = 0$ implies $a = 0$ or $b = 0$ (or both).

zero of a polynomial An x-value that results in an output of 0 when it is substituted into a polynomial; for example, the zeros of $f(x) = x^2 - 4$ are 2 and -2.

BIBLIOGRAPHY

Baase, S. *Computer Algorithms: Introduction to Design and Analysis.* 2nd ed. Reading, Mass.: Addison-Wesley Publishing Company, 1988.

Beckmann, P. *A History of Pi.* New York: Barnes and Noble, Inc., 1993.

Brown, D., and P. Rothery. *Models in Biology: Mathematics, Statistics and Computing.* West Sussex, England: John Wiley and Sons Ltd, 1993.

Burden, R., and J. Faires. *Numerical Analysis.* 5th ed. Boston: PWS-KENT Publishing Company, 1993.

Callas, D. *Snapshots of Applications in Mathematics.* Delhi, New York: State University College of Technology, 1994.

Conquering the Sciences. Sharp Electronics Corporation, 1986.

Eves, H. *An Introduction to the History of Mathematics.* 5th ed. Philadelphia: Saunders College Publishing, 1983.

Freedman, B. *Environmental Ecology: The Ecological Effects of Pollution, Disturbance, and Other Stresses.* 2nd ed. San Diego: Academic Press, 1995.

Garber, N., and L. Hoel. *Traffic and Highway Engineering.* Boston, Mass.: PWS Publishing Co., 1997.

Grigg, D. *The World Food Problem.* Oxford: Blackwell Publishers, 1993.

Haefner, L. *Introduction to Transportation Systems.* New York: Holt, Rinehart and Winston, 1986.

Harrison, F., F. Hills, J. Paterson, and R. Saunders. "The measurement of liver blood flow in conscious calves." *Quarterly Journal of Experimental Physiology* 71: 235–247.

Historical Topics for the Mathematics Classroom, Thirty-first Yearbook. National Council of Teachers of Mathematics, 1969.

Horn, D. *Basic Electronics Theory.* Blue Ridge Summit, Penn.: TAB Books, 1989.

Howells, G. *Acid Rain and Acid Waters.* 2nd ed. New York: Ellis Horwood, 1995.

Huffman, R. *Atmospheric Ultraviolet Remote Sensing.* San Diego: Academic Press, 1992.

Kincaid, D., and W. Cheney. *Numerical Analysis.* Pacific Grove, Calif.: Brooks/Cole Publishing Company, 1991.

Kraljic, M. *The Greenhouse Effect.* New York: The H. W. Wilson Company, 1992.

Lack, D. *The Life of a Robin.* London: Collins, 1965.

Lancaster, H. *Quantitative Methods in Biological and Medical Sciences: A Historical Essay.* New York: Springer-Verlag, 1994.

Mannering, F., and W. Kilareski. *Principles of Highway Engineering and Traffic Analysis.* New York: John Wiley and Sons, 1990.

Mar, J., and H. Liebowitz. *Structure Technology for Large Radio and Radar Telescope Systems.* Cambridge, Mass.: The MIT Press, 1969.

Mason, C. *Biology of Freshwater Pollution.* New York: Longman Scientific and Technical, John Wiley and Sons, 1991.

Meadows, D. *Beyond the Limits.* Post Mills, Vermont: Chelsea Green Publishing Co., 1992.

Miller, A., and R. Anthes. *Meteorology.* 5th ed. Columbus, Ohio: Charles E. Merrill Publishing Company, 1985.

Miller, A., and J. Thompson. *Elements of Meteorology.* 2nd ed. Columbus, Ohio: Charles E. Merrill Publishing Company, 1975.

Paetsch, M. *Mobile Communications in the U.S. and Europe: Regulation, Technology, and Markets.* Norwood, Mass.: Artech House, Inc., 1993.

Pearl, R., T. Edwards, and J. Miner. "The growth of *Cucumis melo* seedlings at different temperatures." *J. Gen. Physiol.* 17: 687–700.

Pennycuick, C. *Newton Rules Biology.* New York: Oxford University Press, 1992.

Pielou, E. *Population and Community Ecology: Principles and Methods.* New York: Gordon and Breach Science Publishers, 1974.

Pokorny, C., and C. Gerald. *Computer Graphics: The Principles behind the Art and Science.* Irvine, Calif.: Franklin, Beedle, and Associates, 1989.

Ronan, C. *The Natural History of the Universe.* New York: MacMillan Publishing Company, 1991.

Ryan, B., B. Joiner, and T. Ryan. *Minitab Handbook.* Boston: Duxbury Prerss, 1985.

Smith, C. *Practical Cellular and PCS Design.* New York: McGraw-Hill, 1998.

Stent, G. S. *Molecular Biology of Bacterial Viruses.* San Francisco: W. H. Freeman, 1963.

Taylor, J. *DVD Demystified.* New York: McGraw-Hill, 1998.

Taylor, W. *The Geometry of Computer Graphics.* Pacific Grove, Calif.: Wadsworth and Brooks/Cole, 1992.

Thomas, D. *Swimming Pool Operators Handbook.* National Swimming Pool Foundation of Washington, D.C., 1972.

Thomas, V. *Science and Sport.* London: Faber and Faber, 1970.

Thomson, W. *Introduction to Space Dynamics.* New York: John Wiley and Sons, 1961.

Toffler, A., and H. Toffler. *Creating a New Civilization: The Politics of the Third Wave.* Kansas City, Mo.: Turner Publications, 1995.

Triola, M. *Elementary Statistics.* 7th ed. Reading, Mass.: Addison-Wesley Publishing Company, 1998.

Tucker, A., A. Bernat, W. Bradley, R. Cupper, and G. Scragg. *Fundamentals of Computing I: Logic, Problem Solving, Programs, and Computers.* New York: McGraw-Hill, 1995.

Turner, R. K., D. Pierce, and I. Bateman. *Environmental Economics, An Elementary Approach.* Baltimore: The Johns Hopkins University Press, 1993.

Van Sickle, J. *GPS for Land Surveyors.* Chelsey, Mich.: Ann Arbor Press, 1996.

Varley, G., and G. Gradwell. "Population models for the winter moth." *Symposium of the Royal Entomological Society of London* 4: 132–142.

Wang, T. *ASHRAE Trans.* 81, Part 1 (1975): 32.

Weidner, R., and R. Sells. *Elementary Classical Physics,* Vol. 2. Boston: Allyn and Bacon, Inc., 1965.

Williams, J. *The Weather Almanac 1995.* New York: Vintage Books, 1994.

Wright, J. *The New York Times Almanac 1999.* New York: Penguin Group, 1998.

Zeilik, M., S. Gregory, and D. Smith. *Introductory Astronomy and Astrophysics.* 3rd ed. Philadelphia: Saunders College Publishers, 1992.

PHOTO CREDITS

INDEX OF APPLICATIONS

INDEX

Tab Your Way to Success

Use these tabs to mark important pages of your textbook for quick reference and review.

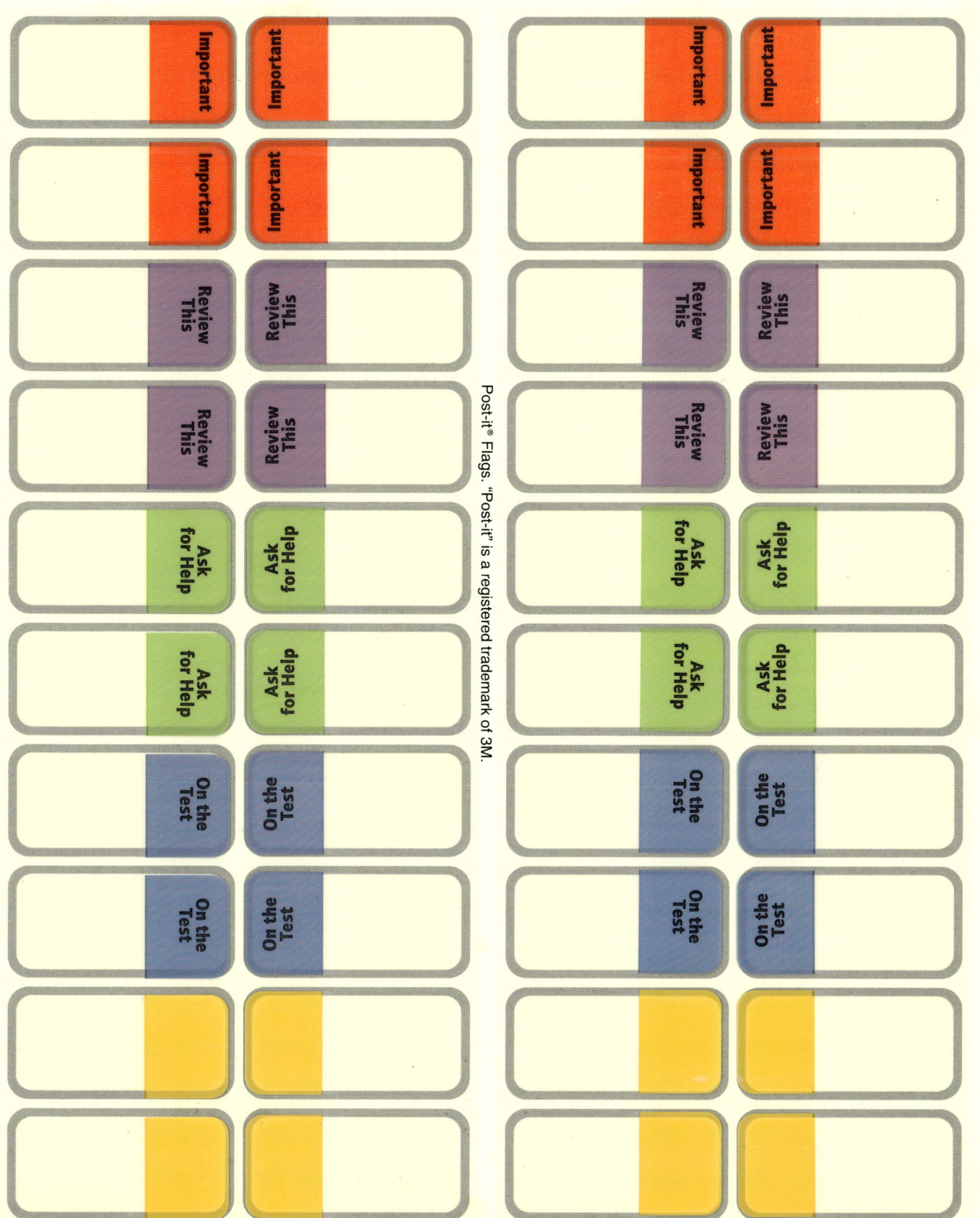

Post-it® Flags. "Post-it" is a registered trademark of 3M.

Using the Tabs

Customize Your Textbook and Make It Work for You!

These removable and reusable tabs offer you five ways to be successful in your math course by letting you bookmark pages with helpful reminders.

Use these tabs to flag anything your instructor indicates is important.

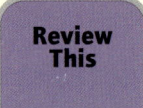

Mark important definitions, procedures, or key terms to review later.

Not sure of something? Need more instruction? Place these tabs in your textbook to address any questions with your instructor during your next class meeting or with your tutor during your next tutoring session.

If your instructor alerts you that something will be covered on a test, use these tabs to bookmark it.

Write your own notes or create more of the preceding tabs to help you succeed in your math course.

ISBN-13: 978-0-321-53636-5
ISBN-10:   0-321-53636-3

EAN

9 780321 536365

90000